Principles of Invertebrate Paleontology

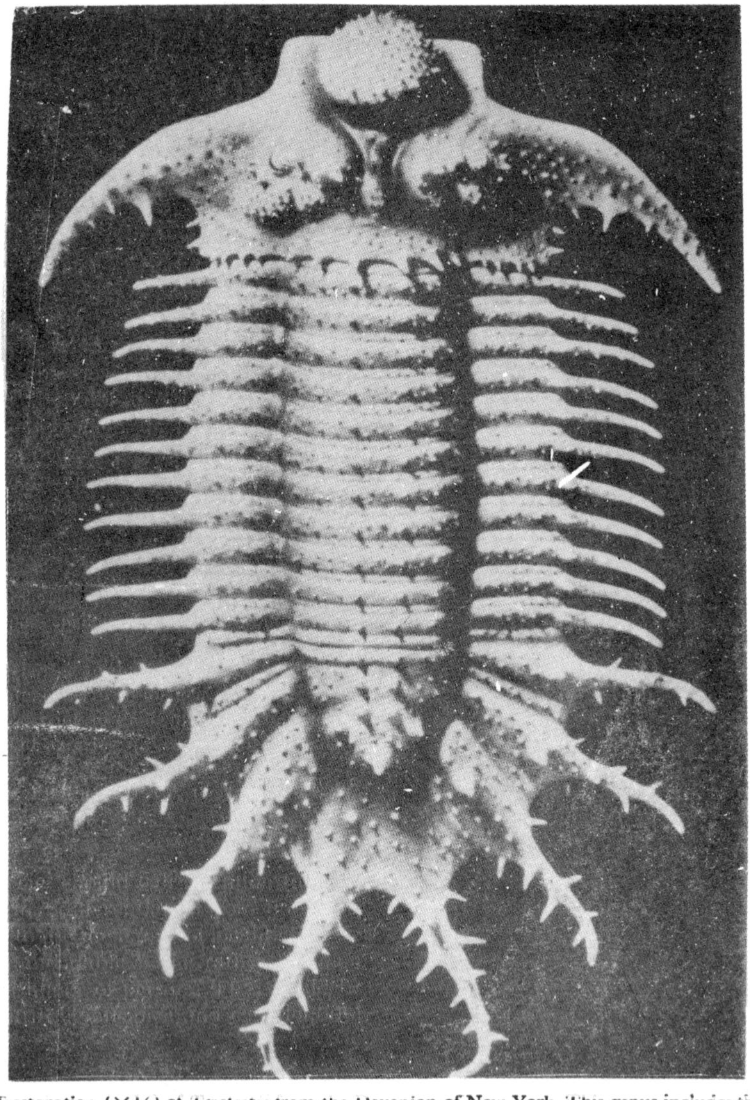

Restoration (×⅓) of *Terataspis* from the Devonian of New York. This genus includes the largest known trilobites. Representatives of one species attained a maximum length of 0.675 m. (27 in.). (*Photograph furnished by I. G. Reimann; also see Buffalo Mus. Sci., Hobbies, vol.* 23, *no.* 4, *p.* 77, 1943.)

Principles of Invertebrate Paleontology

A REVISED AND ENLARGED EDITION OF
Twenhofel and Shrock, *Invertebrate Paleontology*

Robert R. Shrock
Professor of Geology
Massachusetts Institute of Technology
and Associate in Invertebrate Palaeontology
Museum of Comparative Zoology
Harvard University

William H. Twenhofel

SECOND EDITION

CBS

CBS Publishers & Distributors Pvt. Ltd.

New Delhi • Bengaluru • Chennai • Kochi • Kolkata • Mumbai
Hyderabad • Nagpur • Patna • Pune • Vijayawada

ISBN: 81-239-1218-8

First Indian Edition: 1987
Reprint: 1999, 2002, 2005

This edition has been reprinted in India by arrangement with McGraw-Hill Book Company, Inc., USA

Published by:
Satish Kumar Jain for CBS Publishers & Distributors Pvt. Ltd.,
4819/XI Prahlad Street, 24 Ansari Road, Daryaganj, New Delhi - 110002
delhi@cbspd.com, cbspubs@airtelmail.in • www.cbspd.com
Ph.: 23289259, 23266861, 23266867 • Fax: 011-23243014

Corporate Office: 204 FIE, Industrial Area, Patparganj, Delhi - 110 092
Ph: 49344934 • Fax: 011-49344935
E-mail: publishing@cbspd.com • publicity@cbspd.com

Branches:
- ***Bengaluru:*** 2975, 17th Cross, K.R. Road, Bansankari 2nd Stage, Bengaluru - 70 • Ph: +91-80-26771678/79 • Fax: +91-80-26771680 E-mail: cbsbng@gmail.com, bangalore@cbspd.com
- ***Chennai:*** No. 7, Subbaraya Street, Shenoy Nagar, Chennai - 600030 Ph: +91-44-26681266, 26680620 • Fax: +91-44-42032115 E-mail: chennai@cbspd.com
- ***Kochi:*** Ashana House, 39/1904, A.M. Thomas Road, Valanjambalam, Ernakulum, Kochi • Ph: +91-484-4059061-65 Fax: +91-484-4059065 • E-mail: cochin@cbspd.com
- ***Kolkata:*** 6-B, Ground Floor, Rameshwar Shaw Road, Kolkata - 700014 Ph: +91-33-22891126/7/8 • E-mail: kolkata@cbspd.com
- ***Mumbai:*** 83-C, Dr. E. Moses Road, Worli, Mumbai - 400018 Ph: +91-9833017933, 022-24902340/41 • E-mail: mumbai@cbspd.com

Representatives:
- Hyderabad: 0-9885175004
- Patna: 0-9334159340
- Vijayawada: 0-9000660880
- Nagpur: 0-9021734563
- Pune: 0-9623451994

Printed at:
Neekunj Print Process, Delhi (India)

PREFACE TO SECOND EDITION

This work deals with the invertebrate life of the past, as that life is revealed in the fossil record, and is so organized that it can be used for either undergraduate or graduate instruction.

Ancient life, obviously, cannot be studied directly. Although the student of extinct life may in some cases have complete shells and skeletons upon which to base conclusions as to the nature of the living animal, more often than not he must depend upon fragments of hard parts, isolated structural components, vague impressions and outlines, and other evidence of the once living organism. Interpretations based on such incomplete materials may prove partly or wholly erroneous when complete specimens are later discovered, as new discoveries are likely to lead to a better understanding of systematic, genetic, evolutionary, and ecologic relations.

Great advancements in the understanding of certain invertebrate groups and in the interpretation of the relations of others, as well as the many paleontological discoveries since the appearance of our "Invertebrate Paleontology" (Twenhofel and Shrock) in 1935, made revision of that work not only desirable but imperative. As revision proceeded, it soon became obvious that a larger and more advanced work was in progress and that the final product would not be a revision but an entirely new book. Consequently, the present work bears a new title.

In this work twenty-two major divisions (phyla) of invertebrate animals are treated systematically, with major emphasis on the fossil record. Each phylum is discussed in proportion to its geological importance. Those phyla with a long and complex fossil record are considered in detail; those with little or no record, only briefly.

Chief emphasis is laid on the nature, occurrence, and relationships of the organic hard parts, since these are generally the only parts of an animal of which there is a definite fossil record, but considerable attention is also given to the soft parts of the organism because of their importance in interpreting the significance and relationships of the fossils. This attention to the morphology of soft parts is felt to be justified by the fact that many students come to the study of paleontology with little or no knowledge of systematic zoology.

The treatment of each of the more important phyla is basically systematic, *i.e.* taxonomic, as the authors feel that the students using the present work will wish to keep their newly acquired knowledge organized and classified, and there seems no better way of accomplishing this than by considering the phyla essentially in the order of increasing complexity (from Protozoa, the simplest, to Hemichordata, the most complex).

Each phylum is first considered as a natural group of organisms clearly different from all other phyla. A typical living representative is next described and illustrated in some detail to provide a general zoological background for the phylum as a whole. Then follows a detailed discussion of each major division and its subdivisions (generally class, order, and family, in descending taxonomic value). If justified by existing knowledge, the paleoecology is reviewed briefly. Likewise, the evolutionary history is outlined for those groups that have been intensively studied, but many divisions of ancient invertebrates have not yet been sufficiently investigated so that definite statements can be made about their origin, evolution, and descendants. Summaries of the geologic history and fossil record of the phylum complete the discussion.

A list of selected references is given at the end of each chapter, providing the interested reader with the chief sources on which the subject matter of the chapter is based.

We have drawn widely from our colleagues, and from some repeatedly, for many kinds of information. We are particularly appreciative of the assistance rendered by the following persons: Ruth Todd and E. H. Myers (Foraminifera); G. Colom (Tintinnids); V. Okulitch (Pleosponges and Corals); Wm. H. Easton, G. M. Ehlers, and R. E. Stumm (Corals); the Kenneth Casters (Scyphozoa); J. J. Galloway (Stromatoporoids); G. A. Cooper, P. E. Cloud, Jr., and W. C. Bell (Brachiopods); Wm. Clench and Ruth Turner (Mollusca); J. B. Knight (Gastropods); Rousseau Flower, A. K. Miller, and Wm. E. Schevill (Cephalopods); Elisabeth Deichmann (Holothurians); Christina Lochman Balk, Harold Brooks, Frank M. Carpenter, Robert V. Kesling, Alexander Petrunkevitch, P. E. Raymond, and Harry Whittington (Arthropods); O. M. B. Bulman (Graptolites); and W. H. Hass (Conodonts).

Illustrative material was taken from the works of many authors and from a wide variety of journals, reports, and other publications. Specific acknowledgment is made to each author in the legend accompanying the figure, and the statement of acknowledgment indicates the extent of indebtedness; *e.g.* (After Smith, 1940) means that the figure was taken without change from a work that Smith published somewhere in 1940. (If the work is not listed in the references at the end of the appropriate chapter, it can be found in one or the other of the two well-known sources—the *Bibliographies of North American Geology* or the *Zoological Record*.) When our figure represents a modification of or an adaptation from one of Smith's figures, that fact is indicated by an appropriate statement. All original figures, whether photographs or drawings, are duly credited.

It is a pleasant duty to acknowledge certain original figures prepared or requested specifically for the present work. Mrs. Carol Carriere, working under the direction of one of us (Shrock), prepared the figures numbered 7–2, 10–4, 10–15, 10–41, 10–42, 10–57, 10–82, 13–4E, 13–5, 13–6, 13–13, 14–41, and 14–53A. Miss Ruth Turner, Research Assistant, Department of Mollusks, Museum of Comparative Zoology (Harvard College), working under the direction of Wm. Clench, Curator of Mollusks in the same museum, prepared the numerous drawings of Mollusca credited to her in the legends. Dr. Elisabeth Deichmann, Curator of Marine Invertebrates, also of the Museum of Comparative Zoology, prepared the drawings for Fig. 4–14.

For permission to use certain illustrations, which are duly credited to appropriate authors in the legends, we are greatly indebted to the following publishers, organizations, journals, etc.: American Association of Petroleum Geologists (*Bulletin*); *American Journal of Science;* Appleton-Century-Crofts, Inc. ("Outlines of Zoology," Thomson, 1899); Edward Arnold & Co. ("Outlines of Palaeontology," Swinnerton, 1947); Australian and New Zealand Association for the Advancement of Science; A. & C. Black Ltd. ("A Treatise on Zoology," Lankester, 1900); Wm. Blackwood & Sons Ltd. ("Manual of Palaeontology," Nicholson, 1872; "Manual of Palaeontology," Nicholson and Lydekker, 1889); British Association for the Advancement of Science ("The Molluscs of South Australia," Cotton and Godfrey, 1938–1940); The British Museum (Natural History); Cambridge Philosophical Society (*Biological Reviews*); Cambridge University Press ("The Invertebrata," Borradaile *et al.*, 1945); Encyclopaedia Britannica, 14th ed. ("Echinoderma," Bather, 1929); Gebrüder Borntraeger ("Handbuch der Paläozoologie," Schindewolf *et al.*, 1939 *et seq.*); *Geological Magazine;* Geological Society of London (*Quarterly Journal*); *Journal of Paleontology;* McGraw-Hill Book Company, Inc. ("Paleontology," Berry, 1929; "Invertebrate Paleontology," Twenhofel and Shrock, 1935; "The Invertebrates: Protozoa through Ctenophora," Vol. 1, Hyman, 1940; "General Zoology," Storer, 1951); Macmillan & Co., Ltd. ("Cambridge Natural History," Harmer, Shipley *et al.*, 1906 *et seq.;* "Textbook of Zoology," Parker and Haswell, 1930); Marine Biological Association of the United Kingdom (*Journal*); Musei Zoologici Polonici (*Annales*); New Zealand Board of Science and Art (*Manual* 7, Thomson, 1927); *Palaeontologia Polonica;* Paleontological Research Institution (*Bulletins of American Paleontology; Palaeontographica Americana*); Queensland Museum (*Memoir*); Royal Society of London (*Philosophical Transactions*); Royal Society of New Zealand (*Transactions*, Percival, 1944); Schleicher Frères, eds. ("Traité de Zoologie concrète," Vol. 5, Délage et Herouard, 1901); The Linnean Society of London (*Proceedings*); University of Michigan Press (*Contributions Museum of Paleontology*); University of Toronto Press; University of Upsala (*Bulletin Geological Institution*); John Wiley & Sons, Inc. and The Technology Press ("Index Fossils of North America," Shimer and Shrock, 1944; "Principles of Micropalaeontology," Glaessner, 1947); *Zoologiska Bidrag från Uppsala.*

All other illustrations, particularly those from journals and official reports, have been used on the assumption that both authors and their publishers would approve such use of their work.

Pertinent literature is referred to in general by a combination of author's name, date of publication, and page number—*e.g.* (Smith, 1936, p. 361). Works that are not included in our list of references at the end of each chapter can be found in the appropriate annual volume of the *Bibliography of North American Geology*, or of the *Zoological Record*, or in the different numbers of *Fossilium Catalogus.*

The kindness of Dr. A. S. Romer, Director of the Museum of Comparative Zoology (Harvard College), in placing the facilities of that great museum at our disposal and the willing assistance of the museum's library staff merit our deep appreciation.

We have profited greatly from the many comments that our colleagues have sent us or made to us concerning our earlier book, "Invertebrate Paleontology" (McGraw-Hill Book Company, Inc., 1935) and we particularly appreciate the many helpful suggestions that have come to us from teachers and students of paleontology.

Efficient secretarial and editorial work by Mrs. Dorothy Morrow, Mrs. Marilyn Zeigler, and Miss Helen Binkley was of great help in preparing the manuscript, and Mrs. Donna Burt was most helpful with the proofs.

The first draft of the entire manuscript was prepared by one of us (Shrock) and then revised by both of us. All illustrations, other than those specifically credited, were prepared by the first author (Shrock), and most of them were examined and approved by the second author (Twenhofel). Jointly we assume complete responsibility for both text matter and illustrations.

ROBERT R. SHROCK
WILLIAM H. TWENHOFEL

LEXINGTON, MASS.
ORLANDO, FLA.
February, 1952

PREFACE TO FIRST EDITION

Paleontology deals with the life of the past. That life is now represented by fossils, which are mainly the preserved hard or skeletal parts of the original organisms. Any serious study of paleontology should be accompanied by careful study of the relations existing between the skeletal parts and soft tissues of living organisms so that restoration and interpretation of the fleshy parts of extinct organisms may be made from their skeletal remains. The present environmental organic relations, frequently extremely involved, should likewise be studied so that some understanding of the environmental relations among the fossil organisms themselves and between them and their physical environment may be acquired.

Approaching the study of fossils from these points of view, they are no longer mere lifeless stones but become entities of a onetime living world; they represent adaptations to and products of the environments in which they and their ancestors lived; they appear as living organisms in harmony with, but also in competition with, their plant and animal associates; and finally they evolve, in which evolution they take their proper place in the intricately interwoven fabric of organic development and become at the same time beacons along the road of geologic time, marking off successive stages in earth history.

With these various aspects in mind, and always with a studied effort to adapt the subject matter to the needs of the beginning student, the authors have attempted to treat each invertebrate phylum with about the same completeness, irrespective of the importance of the phylum at the present time, the importance and abundance of the fossil remains, or the extent to which either fossil or living representatives have been studied.

In a general way the discussion of successive phyla follows the same outline, but there are some exceptions where deviation seems to be warranted. The introductory part of the chapter sets forth the salient features of the phylum as a primary subdivision of the animal kingdom and attempts a general description of the phylum as a whole. Since, in any discussion of fossil organisms, it is of paramount importance to know the relations of preservable hard parts to the fleshy parts of the organism, typical representatives of each phylum are treated fully from this zoological point of view. This part of the chapter is followed by a detailed consideration of the composition, extent, architecture, and structure of the hard parts capable of preservation. A discussion of classification is next in order. Subdivisions down to or below orders in rank are described, illustrated, and traced in geologic history. Present environmental relations are next compared with those of the past, and finally a short discussion is devoted to the geologic history and stratigraphic importance of the several subdivisions of the phylum. After much consideration it seemed best to limit the bibliography to only a very few of the more recent and significant works, and to avoid as much as possible introducing references in the body of the text. Since the text is written for the American student of paleontology, it naturally follows that the majority of references, as well as illustrations, are of American origin.

No attempt has been made to limit the use of technical terms, but these are almost always defined the first time they appear. The index will indicate the page or pages on which the term is defined or described and, if illustrated, the page on which the illustration appears. The student of fossil invertebrates must ultimately acquire a technical vocabulary, for precise definitions and accurate descriptions are impossible without it, and little or nothing is gained by postponing its acquisition or avoiding its use.

The text was prepared jointly without any definite division of labor, hence we assume responsibility both jointly and severally. The illustrations were prepared by the junior author. The senior author assisted in and approved the selection of the materials to be illustrated and accepts responsibility for them. The illustrations are designed to show certain definite structures and features rather than to illustrate an entire specimen. In some instances it seemed advisable to sacrifice absolute accuracy for clarity and effectiveness; hence this fact should be considered in studying the numerous drawings.

Parts or all of Figs. 62, 72, 75, 114, 116–117, 126–128, 131–132, 137, 145, 159–161, 165 and 168–169 were prepared under the junior author's supervision by Miss Carol Haugh, whose assistance is hereby gratefully acknowledged. The authors are further deeply indebted to the following authors and publishers for kind permission to reproduce certain illustrations: Blackwood and Sons for numerous figures from "Manual of Paleontology" by Nicholson and Lydekker; Macmillan & Company, Ltd., for illustrations from "Textbook of Zoology" by Parker and Haswell; D. Appleton-Century Company for several figures from "Outlines of Zoology" by Thomson; A. & C. Black, Ltd., for certain illustrations from "A Treatise on Zoology" by Lankester and collaborators; and McGraw-Hill Book Company, Inc., for three figures from "Paleontology" by Berry. Many figures have been taken from American and English periodicals and geological reports for which acknowledgments to the authors are made in appropriate connections. Finally, parts of the manuscript were read by E. F. Bean, State Geologist of Wisconsin, whose criticism and suggestions are hereby acknowledged.

W. H. Twenhofel
R. R. Shrock

Madison, Wis.
October, 1935

CONTENTS

CHAPTER 6: **WORM PHYLA (Excluding ANNELIDA)** . . . 183

CHAPTER 7: **BRYOZOA (POLYZOA)** 195

CHAPTER 8: **PHORONIDA**. 255

CHAPTER 9: **BRACHIOPODA** 260

CHAPTER 1

INTRODUCTION

DEFINITION OF PALEONTOLOGY

Paleontology is commonly defined as the study of the life of past geologic ages. Inasmuch as it is based on the fossil remains of ancient plants and animals, it may also be defined as the study of fossils. Paleontologists, therefore, are students of fossils.[1] Invertebrate paleontology, the subject of the present work, is concerned with fossil invertebrate animals.

Biology, here defined as the study of living things through time, can be subdivided into the study of existing life (**neontology**) and ancient or extinct life (**paleontology**) (Table 1-1). **Botany,** the science of plants, and **zoology,** the science of animals, together make the subdivision of biology that has been designated neontology, whereas **paleobotany,** concerned with fossil plants, and **paleozoology,** treating of fossil animals, constitute the subdivision of biology defined above as paleontology. Logically the title of the present work should be "Principles of Invertebrate Paleozoology." This title has not been used, however, because it is general practice among American students of fossil animals to designate their science paleontology rather than paleozoology.

TABLE 1-1. RELATIONS OF PALEONTOLOGY TO CLOSELY ALLIED SUBJECTS

BIOLOGY The study of life through time	NEONTOLOGY The study of existing life	BOTANY The science of living plants	ZOOLOGY The science of living animals
	PALEONTOLOGY The study of the life of past geologic ages	PALEOBOTANY The science of fossil plants	PALEOZOOLOGY The science of fossil animals Invertebrate paleontology Vertebrate paleontology

The scope of paleontology was clearly outlined long ago as follows in a press notice by Major J. W. Powell, the second director of the U.S. Geological Survey:

[1] Several points of view as to what a paleontologist is or should be are discussed in the following papers: KNIGHT, J. B., 1947, Paleontologist or geologist, *Bull. Geol. Soc. Amer.*, vol. 58, pp. 281–286. WELLER, J. M., 1947, 1948, Relations of the invertebrate paleontologist to geology, *Jour. Paleontology*, vol. 21, pp. 570–575, 1947. *Ibid.*, vol. 22, pp. 268–269, 1948. NEWELL, N. D., and COLBERT, E. H., 1948, Paleontologist—biologist or geologist? *Jour. Paleontology*, vol. 22, pp. 264–267.

Paleontology, the science of fossils, is the geologist's clock, by which he determines the times in earth history when the beds containing the fossils were deposited. Geological time is divided into periods which are characterized by the existence of certain plants and animals. Without paleontology the geologic classification of formations, their correlation, and the determination of their mutual relations would be impossible. In fact, real and symmetrical progress in geology would be impossible without corresponding interrelated development and refinement in its handmaid, paleontology. The study of the economic geology of any region of complicated structure is blind and inconsequent unless the time relations are known. These relations are indicated by the fossils which the strata contain.

THE ORGANIC WORLD

The **organic world,** which comprises all living things, is divisible into the **animal kingdom** and the **plant kingdom.** The relation of these two kingdoms is shown diagrammatically in Fig. 1-1 in the form of the letter Y. At the upper ends of this Y are such complex animals as man, bird, elephant, snake, and fish and such advanced plants as the oak, the pine, and the rose. In intermediate positions are simpler animals such as the crayfish, mollusk, brachiopod, etc., and simpler plants such as the fern and horsetail. At the base of the two branches belong the simple and primitive animals, well represented by sponges and the protozoan *Amoeba*, and similar lowly plants, represented by sea lettuce and algae. To complete the diagram the primitive plants and animals should merge downward into some common ancestral organism from which both may be considered to have evolved. Some investigators think that this ancestral organism was an extremely simple plant, thus presupposing plants to have come into existence earlier than animals; others believe that the first organisms possessed characteristics common to both plants and animals. Whatever the case, some of these early one-celled organisms adopted sessile habits, acquired food from the immediate surroundings, and ultimately became plants. Others developed the ability to search for food, acquired the power of locomotion, and became animals. However the above changes took place and whatever the characteristics of the common ancestor of animals and plants, it remains reasonably certain that the first form of life to appear on earth was a single-celled organism. And, further, this hypothetical organism had the potentialities from which the processes of organic development, acting throughout past geologic ages, have created the dynasties of plants and animals that have moved across the stage of geologic time. The only record of the grand procession is found in fossils—*evidences of ancient life preserved in the sediments and rocks of the earth's crust.*

The great abundance and variety of life impress every student who extends his investigations beyond the larger and more easily observed organisms and below the surface of the continents and oceans. A census of the macroscopic organic remains in a woodland soil near Washington, D.C., over an area 2 ft. square and to a depth to which a bird can scratch easily (12 to 25 mm., or ½ to 1 in.), showed 112 animal items and 194 seeds and fruits. These totals indicate the presence of more than 1,000,000 animals and 2,000,000 seeds and fruits per acre. In a grassy meadow of the same latitude the soil yielded 1,254 animal items and 3,113 seeds, indicating the presence of more than

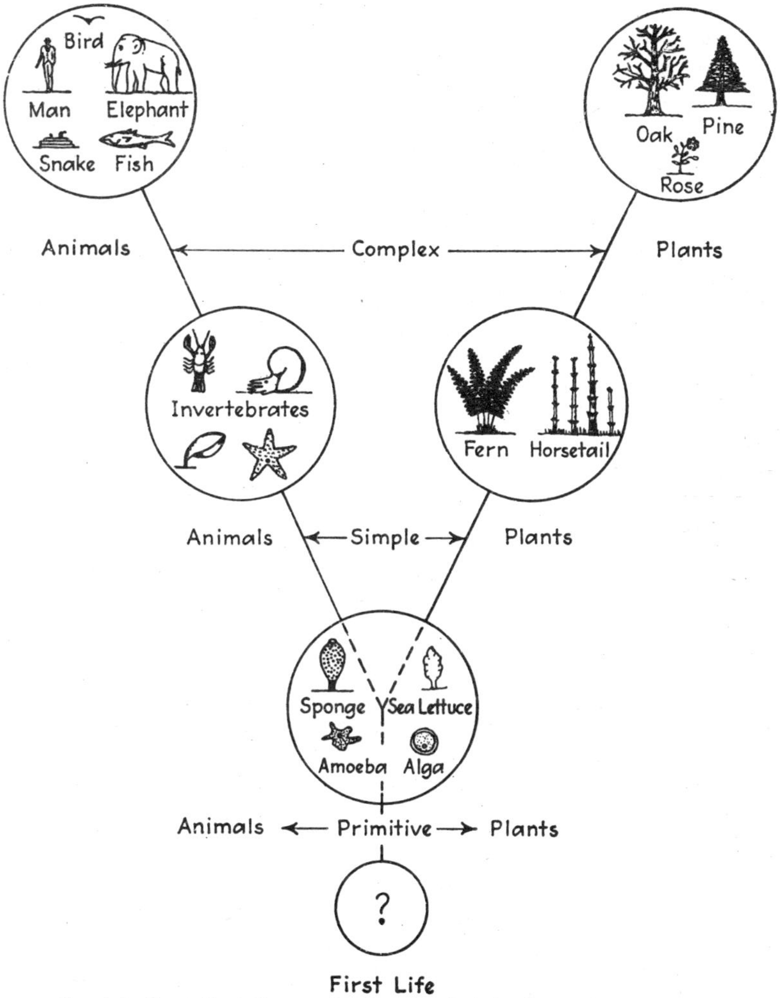

FIG. 1-1. Generalized diagram showing the relations of animals and plants.

13,000,000 animals and nearly 34,000,000 seeds per acre.[1] Large as these totals are, they would have been vastly greater had the investigation been carried deeper and had it included the microscopic organisms.

Similar investigations of life in fresh and salt waters, and in the bottom sediments of water bodies, show an equal abundance of plants and animals. For example, 974 macroscopic organisms—with a couple of hundred not

[1] McAtee, W. L., 1907, Census of four square feet, *Science*, vol. 26, pp. 447–449.

counted—were collected in a California estuary in muds taken from a circular excavation 1.125 m. (45 in.) in diameter and 45 cm. (18 in.) deep. Forty-four species were represented. These figures indicate more than 3,000,000 individuals per acre and do not include microscopic organisms.[1]

That a varied population of organisms can quickly exploit a favorable area is shown by the case of incrusted shells of *Atrina* collected in the shallow waters of the Gulf of Mexico. These fragile pelecypod shells, which were brought from their moderately deep habitat into shallow water during a 1935 hurricane, were covered from top to bottom in a period of five months with a medley of shallow-water marine animals. One shell picked at random showed 25 different kinds of animals, more than 100 individuals of eight different phyla, all living on an area of shell surface approximating 60 sq. in. As only macroscopic animals were counted, it follows that the extraneous population living on the drifted shell was far larger than the totals mentioned, by virtue of the microscopic organisms that undoubtedly were present.[2]

The ability of a single species to spread quickly and widely over a favorable environment is illustrated by the familiar marine gastropod *Littorina littorea*.[3] This species was accidently or intentionally brought to the Canadian shore about 1840 and in the intervening century has spread northward to Labrador and southward to Atlantic City, N.J. Its pelagic early stages allowed it to be carried southward with the plankton of the Labrador current. It prefers rocky bottoms of the lower intertidal zone but is also abundant on muddy bottoms.

A similar example is furnished by certain species of the foraminifer *Lepidocyclina* (*sensu lato*), which attained world-wide distribution in the early Cenozoic seas (Fig. 2-16).

The number of known living species of animals was estimated in 1912[4] to be 522,400 and in 1951[5] to be well over 900,000. It is certain that many thousands of animals are still undescribed or unknown. Probably a great number of these are marine creatures which frequent the oceanic depths and are as yet uncaptured or, at least, undescribed.

Beebe's descent in the sea to half a mile below the surface and Barton's 1949 descent to a depth of 4,500 ft.[6] showed an unknown population of almost unbelievably fantastic animals within those depths, and discoveries made during and since World War II leave no doubt but that the oceanic depths are extensively populated. As recently as 1948, during the cruise of the *Albatross*, at a station northeast of the Virgin Islands, hauls of fishes and invertebrate

[1] MacGinitie, G. E., 1945, Ecological aspects of a California marine estuary, *Amer. Midland Naturalist*, vol. 16, pp. 629–675.

[2] Perry, L. M., 1936, A marine tenement, *Science*, vol. 84, pp. 156–157.

[3] Bequaert, J. C., 1943, The genus *Littorina* in the Western Atlantic, *Johnsonia*, vol. 1, no. 7, pp. 1–27.

[4] Pratt, H. S., 1912, On the number of known species of animals, *Science*, vol. 35, pp. 467–468.

[5] Storer, T. I., 1951, "General Zoology," 2d ed, New York, McGraw-Hill Book Company, Inc., 832 pp

[6] Man's deepest dive, *Life*, vol. 27, pp. 21–23, 1949.

benthonic animals from a depth of more than 4,200 fathoms (8,000 – m., or 25,200 ft.) proved that even at these great depths organic life exists.[1]

Prior to 1800, fossil organisms had received little critical attention, and in 1820 only 127 species of fossil plants and 2,100 species of fossil animals were known. By 1847 the former had increased to 2,050 and the latter to 24,300. Since that time these numbers have multiplied tremendously.

It seems reasonable to assume that, if all living organisms are considered, not more than 1 out of 1,000,000 individuals will ever leave a fossil record. The ratio will vary with the species, habitat, and places of death and burial. Under some conditions almost every organism with hard parts may become fossil, but under other conditions the same organisms would have nothing preserved. Countless millions of American bison have roamed the Western prairies, yet only an occasional bone of one of these may now be found. It is probable that after the lapse of a thousand years not one in a million of these animals will have left a bone or any other indication of its former presence. Insects live in most lands in numbers reaching astronomic proportions, yet probably not more than one in many millions leaves any evidence of its existence.

It may be assumed that conditions favorable for the preservation of organisms have not always prevailed and that the fossil record gives an incomplete picture of the abundance and nature of ancient life. Furthermore, man certainly has seen only a small percentage of the countless billions of fossils present in the earth's crust and knows and has described only a part of the many species. Even when most of the fossil species have been found and described, and in spite of the great increase in knowledge of the life of past ages accruing from those discoveries, the record will still be fragmentary because millions of ancient organisms almost certainly left no fossil record of their existence. Nevertheless, the known fossils are believed to portray correctly in its broader outlines the life of ancient times, and it is thought that more extensive knowledge will do little more than add detail to the grander picture already outlined. Paleontologists have long recognized in this picture support for the doctrine of evolution—*descent with accumulative modifications*—and new knowledge, they feel, will only strengthen the foundation on which the doctrine rests.

THE ANIMAL KINGDOM

The animal kingdom, which comprises the left arm of the Y in Fig. 1-1, may be compared to a great tree, the **animal tree,** with many branches and branchlets (Fig. 1-2). In this work attention will be directed only to the invertebrate division of the tree.

The tips of the branches and branchlets of the animal tree represent modern life. By following the branchlets back to the larger branches and then along the latter to their place of junction with the trunk, it is seen that the nearer one approaches the trunk, and also the base of the tree, the simpler and less numerous become the groups of animals. The diagram clearly indicates that related forms of life evolved from common parent stocks and also that the

[1] PETTERSSON, H., 1949, The floor of the ocean, *Endeavour*, vol. 8, p. 186.

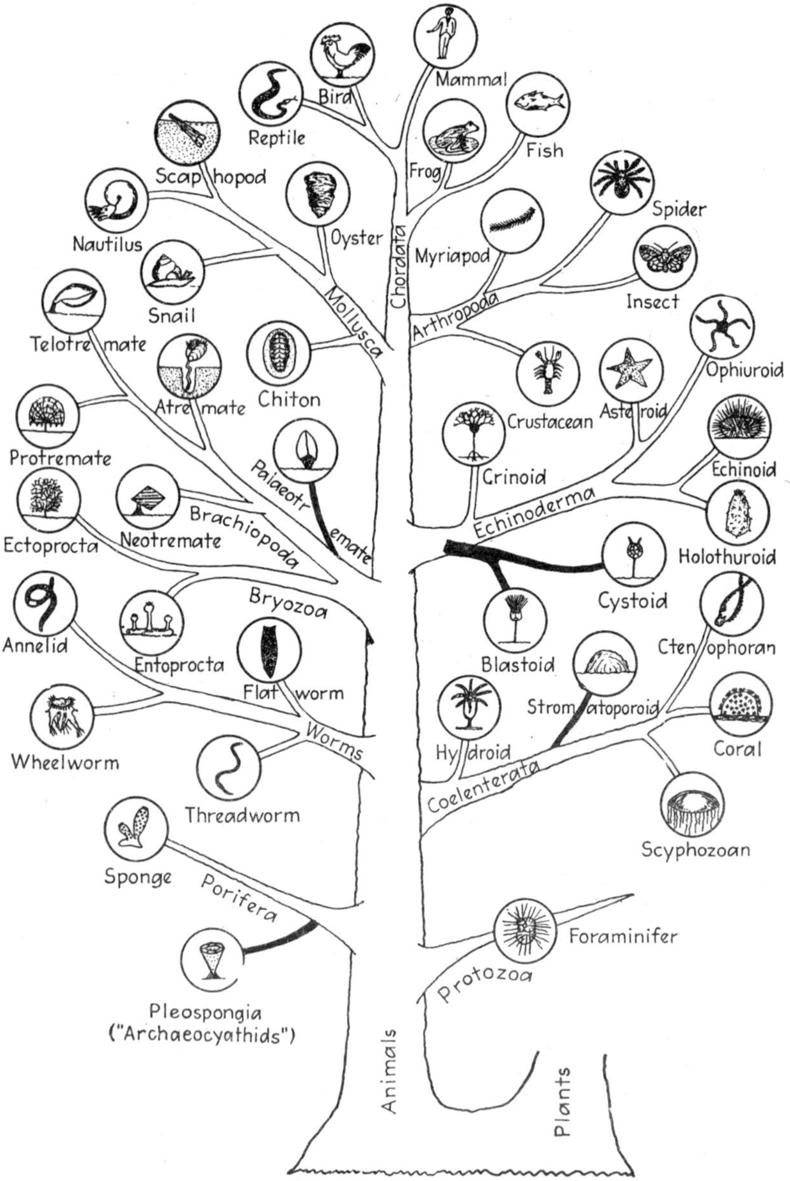

FIG. 1-2. Diagrammatic and generalized tree of animal life. The different phyla and their subdivisions are shown by distinct branches and branchlets, respectively. All branches are shown springing from the general animal stock, but their position on the tree does not necessarily indicate either the time when the phylum split away from the main trunk or the relative organic complexity of the members of the phylum. Extinct branchlets are shown in solid black.

latter branched from the main trunk of the animal tree at different times in geologic history. Going back still farther into geologic time, the trunk of the animal tree finally merges with that of the plant tree, indicating that both diverged from a common ancestral stock. The history and development of this stock fade into the impenetrable beginnings of the planet.

If one imagines that the tree just described has been covered with drifting sands, so that only the tips of the branchlets and branches are exposed, the relations between modern and ancient animals would not be apparent. The exposed tips represent modern life, but ancient life is concealed beneath the sands of the dune. It is the task of the paleontologist to exhume the buried tree, by working downward into the rocks of the earth's crust and backward into geologic time, and to demonstrate the relations that exist between the living species of animals and those that have become extinct.

CLASSIFICATION OF ANIMALS

Not only are organisms classified into two great kingdoms, but these in turn are subdivided into successively smaller and more restricted groups, ending finally with the individual, the only real biologic entity.

The smallest taxonomic division usually employed is the **species.** This term has been applied to a group of closely related organisms that possess in common one or more distinctive features, interbreed, and reproduce their characteristics in fertile young. Other definitions of species have been proposed, and estimates of the number of known species vary greatly depending upon the way in which the investigator defines the term. Fertile interbreeding seems to be necessary among individuals of a group if it is to constitute a species. In general, it appears that the fertility of interbreeding between animals of two groups depends upon the length of time that has elapsed since the two separated from the parent stock. A group of closely related species comprise a **genus,** a group of genera constitute a **family,** related families are grouped into **orders,** related orders into **classes,** and classes into **phyla**—the primary divisions of the animal and plant kingdoms.

The animal kingdom is divided into 21 principal **phyla** (singular, **phylum;** Gr. *phylon*, race), which are listed on the next page in approximate order of increasing complexity.[1]

This classification is based largely on the form and structure of the soft parts, complexity of organization, and nature of whatever hard parts the animal has. Inasmuch as most of the soft parts of organisms can be studied only in living species, it becomes extremely difficult to assign several important extinct groups of fossils to their correct taxonomic divisions, because little or nothing is known about their soft parts except as they may be revealed in certain structural features of skeletal or shell remains. Among these extinct groups are the coral-like **stromatoporoids;** the **graptolites,**

[1] We accept the 20 phyla recognized by Storer (1943) in the first edition of his "General Zoology," but change their order slightly, and add the Phylum Onychophora. In his recent second edition, Storer (1951) recognizes 23 phyla, which are in some cases different from those in his first edition.

which superficially resemble certain living hydrozoan coelenterates but have some features suggesting affinities with the Bryozoa on the one hand and the hemichordates on the other; and the **trilobites,** which seem to be related to both Crustacea and Arachnida.

PRINCIPAL ANIMAL PHYLA

(*Phyla with little or no fossil record are indicated by an asterisk.*)

I.	**Protozoa**	One-celled animals such as *Amoeba.*
II.	**Porifera**	Sponges.
III.	**Coelenterata**	Hydrozoans, sea anemones, and corals.
IV.	**Ctenophora**	Comb jellies.
V.	***Platyhelminthes**	Flatworms.
VI.	***Nemertea**	Ribbon worms.
VII.	***Nemathelminthes**	Roundworms or nematodes.
VIII.	***Gordiacea**	"Horsehair worms."
IX.	***Acanthocephala**	Spiny-headed worms.
X.	***Kinorhyncha**	Tiny marine worms.
XI.	***Trochelminthes**	Wheel animalcules.
XII.	***Chaetognatha**	Arrow worms.
XIII.	**Bryozoa**	Moss animals.
XIV.	**Phoronida**	Phoronids.
XV.	**Brachiopoda**	Lamp shells.
XVI.	**Mollusca**	Snails, clams, squids, and octopuses.
XVII.	**Annelida**	Segmented worms.
XVIII.	**Onychophora**	Caterpillar-like organisms most closely resembling certain arthropods.
XIX.	**Arthropoda**	Crustaceans, insects, spiders, centipedes, millepedes.
XX.	**Echinoderma**	Sea lilies, starfishes, and sea urchins.
XXI.	**Chordata**	Animals with notochord or backbone: hemichordates, tunicates, lancelets, fish, amphibians, reptiles, birds, mammals.

The chief characteristics of the principal animal phyla are shown in Table 1-2, and the geologic range and relative abundance in time of the same phyla are shown in Table 1-3. If these tables are compared with the list of phyla given in a preceding paragraph, it will be noted that most of the starred phyla—those with little or no fossil record—have been omitted.

These phyla have been omitted because so far as is known they do not seem to have left a fossil record of any consequence, even though there is good reason to suppose that certain of them have been represented geologically for a long time.

The width of each bar indicates relative abundance within a phylum, and the length shows the geologic range; both are based on the known or inferred fossil record. The bars are not comparable with reference to the abundance they indicate. Protozoa, Porifera, Coelenterata, and Brachiopoda have been reported from the Pre-Cambrian, but some paleontologists question either the reality, identification, or age of the so-called fossils.

The subdivision of a phylum is illustrated by the complete classification of the domestic dog and of man:

Kingdom	Animalia	Animalia.
Phylum	Chordata	Chordata.
Class	Mammalia	Mammalia.
Order	Carnivora	Primate.
Family	Canidae	Hominidae.
Genus	*Canis*	*Homo.*
Species	*familiaris*	*sapiens.*
Individual	Carlo	John Brown

Most fossils are designated by a double name, *e.g.*, *Homo sapiens.* The first name refers to the genus and the second to the species. In rare cases a third name is added to indicate a subspecies or variety.[1] The first or generic name is always capitalized. The second or specific name, which usually refers to some distinctive feature of the fossil, and the varietal name, which is also generally descriptive, are never capitalized. Both should always agree in gender with the generic name. The complete name of the fossil usually is followed by the name or names of the person or persons who first described it. The fossil brachiopod *Pentamerus oblongus cylindricus* Hall and Whitfield may be taken as an example to illustrate the significance of the different words in the designation of a fossil.

Pentamerus	Generic name, referring to the five compartments into which the hollow shell is divided by certain internal partitions.
oblongus	Specific name, referring to the oblong shape of the shell and agreeing in gender with the generic name.
cylindricus	Varietal name, referring to the cylindrical nature of the oblong shell and agreeing in gender with the generic name.
Hall and Whitfield	Hall (James) and Whitfield (R. P.), the first to describe the variety.

Although the names of genera and species should call attention to those features that were responsible for the original differentiation from closely related forms, it has also become customary to use the names of localities where the fossils were collected. Much less commendable is the rather common practice of naming genera and species in honor of persons. This should be discouraged because such names carry little significance and cast doubtful honor on the person concerned.

[1] For a discussion of how the terms subspecies and variety are used by paleontologists, the reader is referred to NEWELL, N. D., 1947, Infraspecific categories in invertebrate paleontology, *Evolution*, vol. 1, pp. 163–171.

TABLE 1-2. SOME CHARACTERISTICS OF THE PRINCIPAL ANIMAL PHYLA

Cells	Germ layers	Symmetry	Digestive tract	Excretory organs	Coelome	Circulatory system	Respiratory organs	Segmentation	Distinctive features (exceptions omitted)	PHYLUM
1	. .	v	. .	. .	. .	. .	. .	. .	Microscopic, one-celled or colonies of like cells	PROTOZOA
Cells many, arranged in layers or tissues (METAZOA)	2, diploblastic	Radial	O	O	O	O	O	O	Body perforated by pores and canals	PORIFERA
			Incomplete	O	O	O	O	O	Nematocysts; digestive tract sac-like	COELENTERATA
	3, triploblastic	Bilateral		+	O	O	O	O	Flat, soft; digestive tract branched or none	PLATYHELMINTHES
			Complete (with anus)	+	O	+	O	O	Slender, soft, ciliated; soft proboscis	NEMERTEA
				+	ps	O	O	O	Cylindrical, tough cuticle; no cilia	NEMATHELMINTHES
				+	ps	O	O	O	Microscopic, cilia on oral disc	TROCHELMINTHES
				+	+	C	O	O	Small; arrow-shaped, transparent; lateral fins	CHAETOGNATHA
				+	+	O	O	O	Grow as mosslike or incrusting colonies	BRYOZOA (POLYZOA)
				+	+	+	O	O	Limy shell composed of dorsal and ventral valves; a fleshy stalk	BRACHIOPODA
				+	+	+	+	O	External limy shell of 1, 2 or 8 parts, or none; body soft	MOLLUSCA
				+	+	+	+,O	+	Slender, of many like segments; fine setae as appendages	ANNELIDA
				+	+h	+	+	⊕	Caterpillar-like, with antennae; hemocoele; tracheae; numerous legs	ONYCHOPHORA
				+	+h	+	+	+	Segmented, with jointed appendages; exoskeleton of chitin	ARTHROPODA
		Radial		O	+	+	+	O	Symmetry 5-part radial; tube feet; spiny skeleton	ECHINODERMA
		Bilateral		+	+	+	+	+	Notochord, dorsal tubular nerve cord, gill slits; usually, fins or limbs. Hemichordata simpler	CHORDATA

+, present; O, absent; ⊕, poorly developed; +, O, present or absent; *ps*, pseudocoele, not lined by peritoneum; *h*, coelome reduced, body spaces a hemocoele; *v*, symmetry various, or none.

* Zoological data taken from Storer's "General Zoology," McGraw-Hill Book Company, Inc., 1951, and used with the kind permission of both author and publisher.

TABLE 1-3. GEOLOGIC RANGE OF THE PRINCIPAL ANIMAL PHYLA

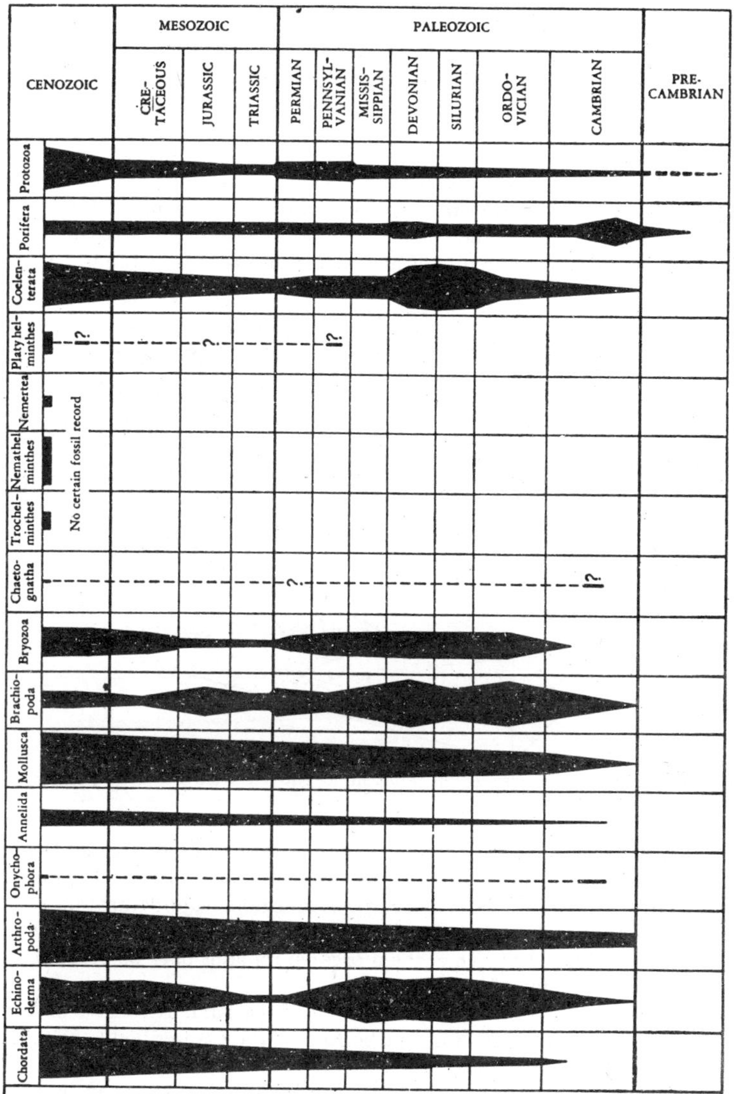

The width of the range band indicates roughly the abundance of representatives within the phylum; it is not comparable from phylum to phylum. Dashed lines indicate postulated record not verified by fossils.

HABITATS AND HABITS OF ANIMALS

All existing habitats and environments that can support life are occupied by populations of plants and animals adapted to prevailing conditions. Similar habitats and adaptations seem to have existed since the appearance of life upon the earth.

There are three fundamental life habitats—**air, land,** and **water.** The life of the extensive air habitat can have the remains of its inhabitants preserved only in the environments of the other two, for aerial organisms must fall to earth or into water when they die. The environments of the land are varied, including the treeless summits of mountains, moss-covered tundra, heavily

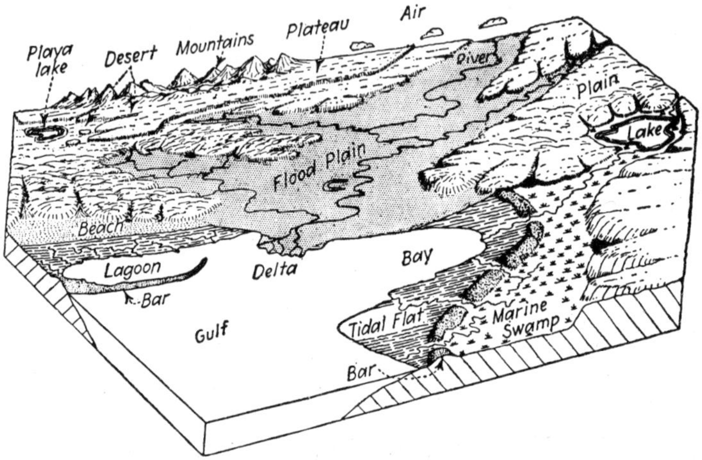

Fig. 1-3. Continental environments adjacent to and far from the ocean, and marine environments along the coast.

forested slopes of uplands and moist flatlands, treeless grass-covered prairies, barren deserts, dune areas of many coasts, lush forest growths of temperate and tropical rain-belt regions, and swamps in regions of immature topography and about bodies of water (Fig. 1-3). Each of these environments may in turn be subdivided. In some of them organic remains may be preserved; in others this is virtually impossible.

The aqueous habitat is as varied as that of the land. It consists of rivers, lakes, swamps, seas, and oceans. The bottoms of these water bodies, with their accumulating sediments, are probably the most favorable environments of all for preservation of organic remains provided the remains can be buried sufficiently rapidly to remove them from attack by scavengers and to prevent decay of delicate tissues and solution of microscopic structures.

Rivers frequently receive contributions of organic material from bordering lands and in times of flood themselves reach out to grasp some of the organisms of the flood plains through which they flow. The lake is somewhat like the river in that there are many opportunities for it to receive the remains of air and land organisms. The seas and oceans, however, are the great receptacles of indigenous organic accumulations and to these also come large contributions of organic materials from the air, land, and rivers. As a consequence, the larger proportion of fossils are found in marine formations.

The bottom of a sea or ocean may be divided into four great life zones: littoral, neritic, bathyal, and abyssal (Fig. 1-4). The **littoral** or **tidal zone** is the narrow strip of shore between highest and lowest tides. Under conditions of unusually high waters the landward margin of a beach becomes a part of the littoral zone, and with unusually low water the landward margin of the neritic zone becomes a part of the littoral. Living conditions in the littoral zone are extremely difficult because the diurnal ebb and flow of the tide causes alternate exposure and covering of the bottom materials and organ-

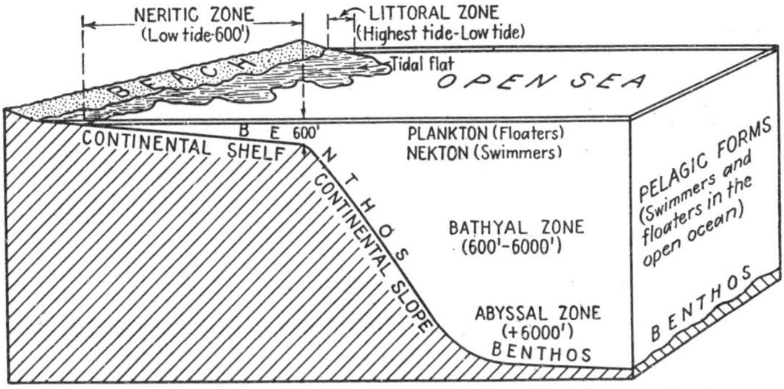

FIG. 1-4. Diagram showing the different types of life and the life zones in the ocean. It should be noted that the littoral, neritic, bathyal, and abyssal realms are actually depth zones.

isms. These frequently changing conditions are not favorable for preservation of organisms.

The **neritic life zone** is that of the continental shelf. It consists of the bottom and overlying waters between lowest tide and the 100-fathom (600-ft.) depth. This is the zone of lighted and agitated waters, of abundant plant life, and of abundant food for the prolific assemblage of animals that make their home on the bottom or in the water above. The bottom especially is usually densely populated, and conditions of preservation range from poor to excellent, depending largely upon the rate of deposition and the nature and abundance of scavengers. The majority of fossil invertebrate assemblages of the geologic past seem to have flourished upon bottoms of the neritic zone, and much of invertebrate evolution is thought to have taken place in the waters above the continental shelves of ancient seas.

The **bathyal life zone** includes the bottom and overlying water between the 100- and 1,000-fathom (600- and 6,000-ft.) depths. It merges on the landward side into lighted waters of the neritic zone and on the seaward side into the dark abyss. Only the very top of the zone has any light, hence any green plant life is limited to the surface waters. Life conditions are such, however, that there is a large and varied assemblage of animals in the zone. Deposition is slow, and the possibility of burial and preservation of organic remains does not seem to be as good as in the neritic zone.

The **abyssal life zone** of the oceans includes the bottom and overlying waters below the depth of 1,000 fathoms (6,000 ft.). Here the waters are dark and cold, the pressures are great, there is no green plant life, and there is little animal life. Deposition is slow, but how much organic material is preserved cannot be stated, though it seems likely that the quantity is relatively small.

Each of the aqueous environments just described has an assemblage of organisms that generally consists of three groups defined by their habits of life. Bottom-dwelling organisms constitute the **benthos;** those that wander over the bottom are designated **vagrant benthos,** whereas those fixed to the bottom are referred to as **sessile benthos.** Swimming organisms that live in the waters above the bottoms make up the **nekton.** Some nektonic organisms travel long distances during life and commonly die far from places of birth. Floating organisms constitute the **plankton.** These lack locomotory organs and structures or have them so little developed that they are essentially ineffective. Consequently they float about subject to the motions of the water in which they live. In some cases they are carried thousands of miles in this way. The term **pelagic** is commonly applied to the drifting, floating, and swimming organisms of the open ocean away from the coast.

DEFINITION OF FOSSIL

In spite of the fact that the term **fossil** (L. *fossilis*, something dug up, from *fodere*, to dig) has been defined in a great many ways by different authors, there is as yet little general agreement on a precise definition, though there are certain requirements common to most if not all of the definitions. Rather than attempt still another, it seems more advisable to point out the important characteristics of the objects that paleontologists have been calling fossils for several centuries.

Certainly a fossil must be *some evidence of the existence of an animal or plant that once lived.* Rocks composed of mineral substances precipitated as a result of organic activity are not fossils, though certain restricted parts of a stratum may be referred to as fossils, if they exhibit specific structures of organic origin such as scalloped laminations (*e.g.*, algal colonies), stromatolites,[1] etc. The innumerable products of human ingenuity, whether they be the crudest of eoliths, or more refined objects such as carvings, paintings, buildings, or manufactured articles, are not usually considered fossils because they give little or no indication of the physical nature of the organism that made them.

If, however, a footprint or handprint is inadvertently or purposely impressed in soft mud, wet concrete, or cooling lava, or if a clam shell is pressed into and buried with mud, all these impressions may become true fossils. Hence, the mold of the mother and child found in the cindery cover of Pompeii is potentially a true fossil. From the examples just given it becomes clear that not only must a fossil give evidence of an organism; *it must also furnish*

[1] Cloud, P. E., Jr., 1942, Notes on stromatolites, *Amer. Jour. Sci.*, vol. 240, pp. 363–379

some idea of the nature (size, shape, form, structure, ornamentation, etc.) of part or all of the organism.

The skeletons of many kinds of animals trapped long ago in the tar pits of California are regarded as fossils, yet one would hesitate to apply that term to a dead dog or butterfly were it to meet death in the same pit in the same manner tomorrow. Likewise, most anthropologists do not hesitate to apply the term fossil to the bones of Cro-Magnon man found buried in the caves of Europe, yet they would not think of applying the same term to the skeleton of John Smith buried a few years ago in the Crown Point cemetery. It becomes obvious, therefore, that *a fossil must have age,* but this particular requirement is so intangible and indeterminable that it cannot be defined. In the majority of cases the organisms represented by fossils lived prior to the present time unit. They need not be extinct, however.

Some authors insist that *fossils must have been preserved in the materials of the earth's crust by natural agencies and processes.* Immediately the question arises as to whether or not man, or for that matter any living organism, is a natural agent. The authors are of the opinion that any organism, regardless of the way in which it is buried, may become a fossil; likewise, that organic remains that were buried in a past epoch, whether by natural agencies and processes or by the intent of an intelligent and reasoning organism, are appropriately considered fossils.

NATURE OF FOSSIL RECORD

Every organism has the possibility of leaving two general types of fossil records. One consists of structures external to the body of the organism but caused by part or all of the body; the other includes the body itself, preserved in its entirety, in part, or in fragments that have been detached from the body. Such remains cannot be regarded as fossils, however, until they have been preserved in the materials of the earth's crust for posterity and have acquired that indefinite age quality that human sentiment demands.

Unaltered Remains. In rare cases ancient animals may be preserved *in toto* with little or no alteration from the living state. One of the most striking examples is that of the extinct mammoths of Siberia which have been preserved in cold storage in the frozen tundra for many thousands of years. Some of the animals are so perfectly preserved that the eyes, skin, blood, flesh, and even partly digested vegetation in the stomach remain much as they were when the animals died. In fact these animals were first discovered by dogs, which fed on their flesh and ultimately led men to the exposed bodies.[1]

Altered Remains. More common, but much less spectacular, than the preceding case are the many mammalian bones in the Cenozoic rocks of western United States, the countless molluscan shells from Tertiary strata in different parts of the world, and the common atremate brachiopod shells in the Upper Cambrian strata of the Upper Mississippi Valley. Many of these have suffered but little alteration since burial. The vast majority of fossils, however, have undergone more or less alteration since the death of the organism re-

[1] Tolmachoff, I. P., 1929, The carcasses of the mammoth and rhinoceros found in the frozen ground of Siberia, *Trans. Amer. Phil. Soc.*, n.s., vol. 23, pt. 1, art. 1, 74 pp.

sponsible for them. These altered remains have been divided into several groups, depending upon the degree to which original material, shape, size, and structure have been modified.

Certain fossils have been altered to a simpler chemical composition by **leaching,** as illustrated by many Cenozoic molluscan shells. In these the original outer chitinous covering (periostracum) has either completely disappeared or is represented by a carbonaceous film. The calcareous part, which generally constitutes most of the shell, also commonly shows pitting, roughening, dulling of the luster, and other evidences of solution.

In more ancient rocks the skeletal parts of graptolites and the exoskeletons of certain arthropods, supposedly composed in life of a substance like chitin, are now represented by films of carbonaceous material. Most of the nitrogen, oxygen, and hydrogen have escaped long since as a consequence of decay. This decomposition is known as **distillation.**

Shells and skeletal structures of a permeable nature are commonly altered by the addition of certain inorganic substances of which calcium carbonate and silica are perhaps the most common. Ground waters accomplish this alteration by invading the pores and depositing in them the substances that subsequently crystallize into minerals, without at the same time altering the original shell or skeletal matter. In this way the fossils gain weight, commonly swell to some extent, and nearly always become less susceptible to future destruction. Fossilized Cenozoic bones and molluscan shells illustrate this kind of alteration which has been called **permineralization.** Some woods and certain animal hard parts seem to have been soaked in silica, as they now have all pore spaces filled with that substance and, in some cases, are also surrounded by it. This type of preservation is appropriately designated **impregnation** or **embedding,** and organic remains thus preserved have been designated **embedded fossils.**[1]

Under certain conditions ground waters completely dissolve original shell or skeletal matter and deposit some other substance in its place. Calcite, dolomite, pyrite, and quartz are common replacing substances in this process of **mineralization** or **petrifaction.** In this kind of alteration no organic residue remains. The fossil is a **pseudomorph** of the original organic hard part. Rarely is the original microscopic structure of the hard part well preserved; more commonly it is destroyed altogether.

Impressions (Molds) and Casts. Any organic structure may leave an impression if it is pressed into or surrounded by a soft material capable of receiving or retaining the imprint (Figs. 1-5, 1-6). Shells, solid objects, and other organic structures that are at first preserved in rock and later removed by solution leave a cavity, on the wall of which is an impression of the exterior of the structure. This **external impression** commonly has been designated an **external mold** or **natural mold.** If some plastic substance like molding clay or gutta percha is pressed into such a cavity, the substance makes a filling known as an **artificial replica** of the original object. Percolating ground waters under certain conditions deposit mineral substances in external molds

[1] DARRAH, W. C., 1941, Changing views of petrifaction, *Pan-Amer. Geol.*, vol. 76, pp. 13–26.

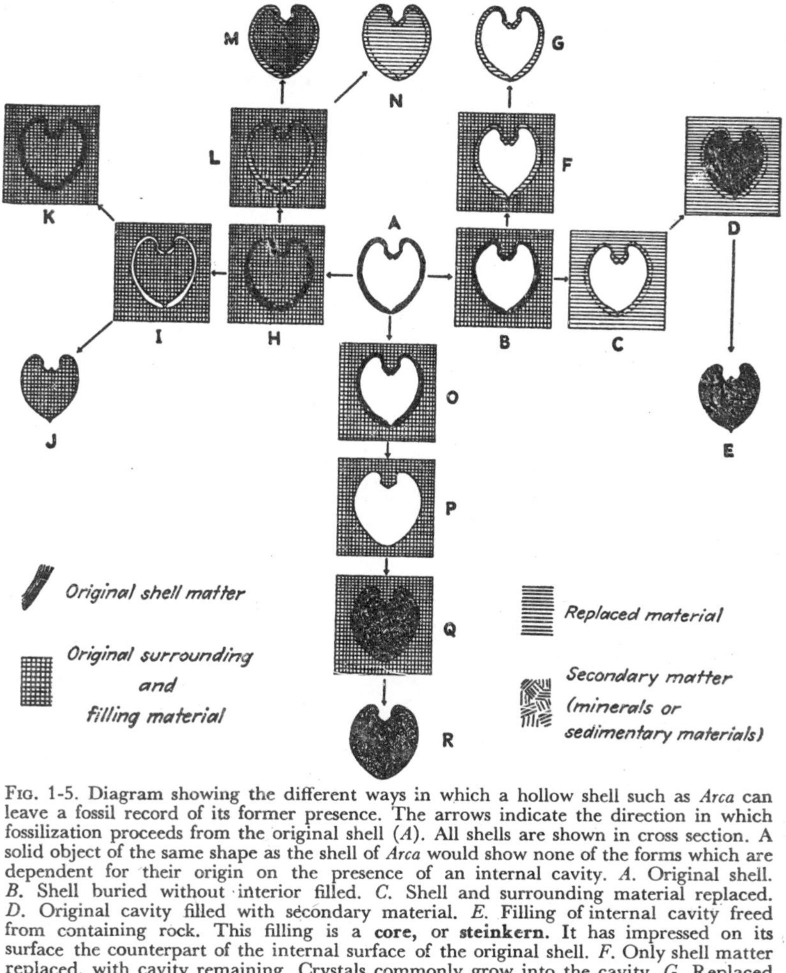

FIG. 1-5. Diagram showing the different ways in which a hollow shell such as *Arca* can leave a fossil record of its former presence. The arrows indicate the direction in which fossilization proceeds from the original shell (*A*). All shells are shown in cross section. A solid object of the same shape as the shell of *Arca* would show none of the forms which are dependent for their origin on the presence of an internal cavity. *A*. Original shell. *B*. Shell buried without interior filled. *C*. Shell and surrounding material replaced. *D*. Original cavity filled with secondary material. *E*. Filling of internal cavity freed from containing rock. This filling is a **core,** or **steinkern.** It has impressed on its surface the counterpart of the internal surface of the original shell. *F*. Only shell matter replaced, with cavity remaining. Crystals commonly grow into the cavity. *G*. Replaced shell released from containing rock. *H*. Shell filled and then buried. *I*. Shell matter removed by solution, with the remaining cavity representing the position of the original shell. *J*. Original filling released from rock. This is a core, or steinkern, exactly like *E* except in the manner of formation. *K*. Cavity left after solution of original shell filled with secondary material to form a **natural cast.** *L*. Shell alone replaced; filling and surrounding material unaffected. *M*. Replaced shell, with unaltered filling, freed from surrounding rock. *N*. Shell and filling replaced and then released from surrounding rock. *O*. Shell buried without interior filled. *P*. Shell matter dissolved leaving an **external mold.** The surface of the cavity bears an impression which is the counterpart of the exterior surface of the original shell. *Q*. Cavity left after solution of shell filled with secondary matter. *R*. Filling of cavity released from surrounding rock. This filling is a **replica** of the original surface of the shell.

to form **natural replicas** of the original exterior of the structures. The surface of such a replica is an exact duplicate of the original surface, and the size and shape are those of the original object. It should be emphasized, however, that a replica does not preserve any of the internal structures of an original object;

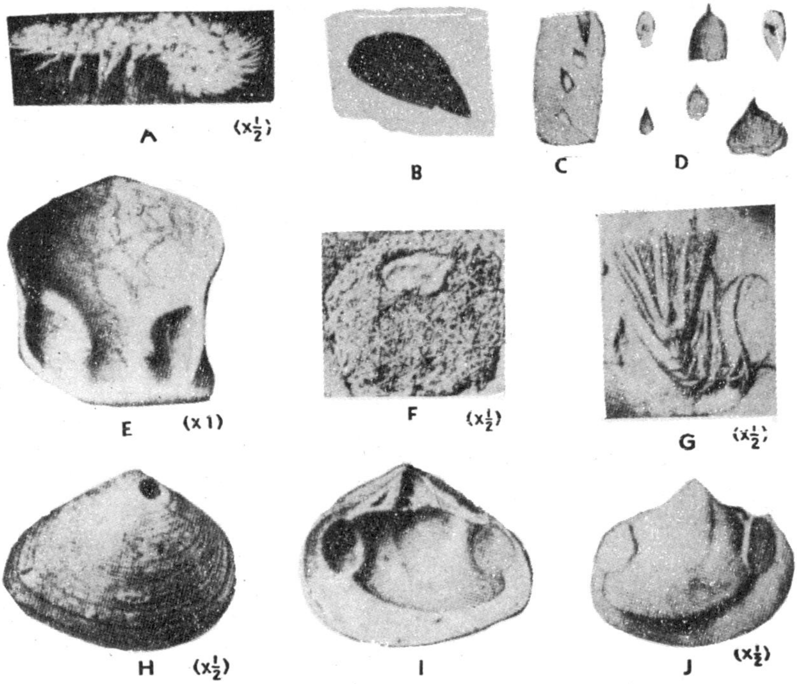

FIG. 1-6. Kinds of fossils and types of preservation. *A.* A scale- and spine-bearing worm (*Canadia*), from the Mid-Cambrian Burgess shale, which illustrates the exquisite preservation of many fossils in that black shale. *B. Beltina danai*, possibly a joint of a flattened arthropod appendage. This specimen came from the Algonkian Greyson shale of Montana, and if it is organic, it represents the oldest nonalgal Pre-Cambrian fossil known. *C–D.* Supposed egg cases or reproductive buds of Ordovician graptolites. These were horny capsules somewhat similar to the "cysts" that have been receiving attention recently. *E.* The internal surface of the glabella of a trilobite covered with the fine zoaria of a simple bryozoan (*Hederella*). Two shells of a tubicolar annelid worm (*Spirorbis*) are also cemented to the surface near the front. *F.* The spicular skeleton of a siliceous sponge. *G.* A crinoid with pinnulate arms and cirrate stem preserved in Devonian limestone. *H–J. Crassatellites*, an Upper Cretaceous pelecypod: *H*, exterior of a right valve showing growth lines and a prominent boring near the beak; *I*, interior of a right valve of a well-preserved shell; *J*, filling of the interior of the shell (internal filling, or steinkern). (*A after Walcott*, 1911; *B after Walcott*, 1899; *C–D after Nicholson*, 1872; *E after Clarke*, 1900; *F after Walcott*, 1920; *G after Talbot*, 1905; *H–J after Stephenson*, 1927.)

it is merely a solid occupying the space once bounded by the external surface of the original object. It should not, therefore, be confused with a cast or a core (see next two paragraphs).

Hollow organic structures present some interesting possibilities with respect to the several different kinds of fossils they can make. Suppose an ordinary clam shell with the two valves closed is buried in mud or sand and during burial is also filled by fine sediment entering the internal cavity through tiny openings along the irregular contact of the two valves. There now exist two kinds of potential fossils associated with the buried shell—an impression of the exterior and a filling of the internal cavity. Suppose the shell is totally removed by solution. There remains a cavity, the **natural mold,** bounded by the external impression and the surface of the internal filling. If this cavity is filled naturally with some mineral substance, the filling is a **natural cast** of the original object; if it is filled artificially, as with rubber or some similar substance, the filling is an **artificial cast.**

The filling of the internal cavity in the case mentioned in the preceding paragraph bears on its surface an impression of the internal surface of the two valves of the original shell. A replica of the original internal surface can be obtained by surrounding the filling with liquid rubber or molding clay of some kind, after which the filling can be removed for study. This filling has been designated a **core,** or **steinkern** (rock kernel); it should not be called an internal cast because such a designation is inaccurate and confusing.

Solid objects, such as calcareous sponge spicules and echinoderm spines, can leave an external mold and a cast, which in this case is also a replica, but they cannot leave a core or steinkern. Glauconitic cores are not uncommon among the Foraminifera; calcareous and dolomitic steinkerns of early Paleozoic gastropods and pelecypods are abundant and widespread; and clastic steinkerns of Cenozoic mollusks are familiar objects along many modern seacoasts where the rocks of this age are being eroded.

Trails, Tracks, and Footprints. Many animals, during their travels over mud and sand bottoms, leave impressions of certain parts of their bodies such as **trails, tracks,** and **footprints.** Although these are more often curiosities than significant fossils, they at least indicate something of the nature of the appendages and other locomotory structures possessed by the animal. Fossil structures of this sort have been ascribed to organisms as different as worms, mollusks, arthropods, and vertebrates. In each case the marking represents a single impression or a series of such impressions made in a fairly soft sediment by some part of an appendage or a continuous, variously formed groove or a series of such grooves left by the passage of some miniature plowlike structure of the animal. Such markings as these may or may not tell something concerning the animals that made them. Commonly they record tragedies of the past, such as that shown by markings found near the German city of Nierstein on the Rhine. Here in a sandstone, which was once a desert sand, are the small tracks of an insect. Death stalked the unwary insect, as shown by lizard tracks that converge upon the insect tracks. Soon the trails come together; beyond, the lizard walks alone!

Under unusual conditions soft-bodied animals, such as the jellyfish, which is more than 90 per cent water, may be compressed by the weight of overlying muds into films, which in some cases leave partial or complete impressions of the exterior of the body without furnishing much information concerning the shape of the original organism.

Burrows, Borings, and Tubes. Certain animals excavate burrows in the sand, bore holes into solid rock, gather granular bottom materials about their bodies, or secrete chitinous and calcareous tubes in which to live. If the shells of the organism are left in burrows, etc., as is sometimes the case with sea urchins and pelecypods, it is possible to determine the origin of the cavity. More often, however, the origin must be conjectured by comparison with similar features that are constructed or produced by living organisms, and in many cases any statement about the way in which some particular structure was formed is little more than a guess. Fossil burrows and borings are not uncommon, and tubes are present in many rocks.

Coprolites and Castings. The terms **coprolite** and **casting** have been applied to fossil excreta; hence the composing material has passed through the alimentary tract of some animal. In many cases the excreted matter has undergone physical and chemical changes and for these reasons stands out conspicuously on weathered surfaces. Coprolites have the form of straplike markings, discrete pellets, or an alternating series of markings and pellets. Certain fish coprolites found in the Devonian of New York contain ganoid scales, showing that some fish fed on others.

Miscellaneous Markings. Almost every serious collector of fossils has cached away in his cabinets peculiar markings or structures of one sort or another, which may have been made by some ancient animal. Although these are always intriguing because of their unknown origin, they are of little practical value to the paleontologist.

USES OF FOSSILS

The organisms now represented as fossils lived at definite times during the geologic past and hence have **chronological significance** and value. William Smith, the great English stratigrapher, long ago discovered that his stratigraphic units had distinct assemblages of fossil organisms, and one of the outstanding results of the study of the geologic column during the past century has been the demonstration that each important geologic time period is characterized by a distinctive group of animals and plants. The fossil remains of these assemblages, therefore, have value for determining **stratigraphic position** in the geologic column. Certain remains, because of their use in this way, are termed **index fossils.**[1]

The forms and structures of animals, to a greater or less degree, bear the impress of the environment in which the organisms lived, and the characteristics and arrangement of the fossil remains reflect conditions of burial. Hence fossils may be of considerable use in determining **ancient environments.** They may also indicate something of the nature of **ancient climates.**

It has often been possible to determine the relations of ancient lands and seas through the study of the distribution of certain fossils, and the same studies in some cases have shown the directions of currents and the routes of migration of ancient organisms. Fossils have always been of great importance, therefore, in the study of ancient geography, or **paleogeography.**

[1] See SHIMER, H. W., and SHROCK, R. R., "Index Fossils of North America," 1944. New York, John Wiley & Sons, Inc.; Cambridge, Mass., The Technology Press.

Living organisms exhibit a great variety of interrelations in the different environments—commensalism, parasitism, etc. Fossil organisms commonly furnish interesting insights into the relations of ancient organisms on coral reefs and shell banks, in lagoons, etc. It has even been possible in some cases to determine what the animal ate, how it was related to neighboring organisms, where and how it died, and what happened to it after death. Hence fossils may tell much concerning **paleoecology,** or the relations of ancient organisms.[1]

Fossils comprise one of the foundation stones of the doctrine of **organic evolution.** Inasmuch as each fossil represents a biologic entity, it soon became clear to the investigators who were building up the geologic column that the plants and animals of each stratigraphic division had evolved from those of the preceding time and were in turn ancestral to those that followed.

Important as they are, fossils must always be used with certain limitations. For example, one of the Tertiary formations of Kansas contains an assemblage of Pennsylvanian fossils derived by erosion from an earlier stratigraphic sequence. Obviously these fossils are of no value for correlation. In fact, if they were taken at face value they would lead to erroneous conclusions about the environment in which the strata bearing them were deposited. They would also cast doubt on the general concept of the development of organisms through geologic time. It should be emphasized, therefore, that a rock is no older than the youngest fossil in it.

Fossils are not always found in the places where the living organisms made their homes. Shells are frequently carried elsewhere before or after death and entombed in sediments of environments in which the living organisms could not have dwelt. The sediments containing fossils are also consequences of ancient environments, and the impress of the environment on these sediments may be so marked as to preclude the possibility of incorrect interpretation. On the other hand, the impress of the environment of life on the organisms represented by fossils may have been slight, and one may be led to consider the environment of deposition as that of life. It should be obvious, therefore, that great caution is essential. Agassiz found leaves of land plants, pieces of bamboo, stalks of sugar cane, and land shells in the Caribbean Sea at a depth of over 1,000 fathoms and at a distance of from 10 to 15 miles from land, in association with the shells of the bottom-dwelling organisms of relatively deep water. Shells of echinoids and mollusks and bones of fishes are present in the swamps of the lands bordering the Gulf of St. Lawrence. These examples illustrate the problems confronting the paleontologist. Floating forms of the extinct graptolites attained world-wide distribution, yet the remains of the group are found in greatest abundance and with finest preservation at occasional levels in black shales which do not contain a normal bottom-dwelling assemblage of other fossils. The indications are that the environment in which the black shales were deposited probably was not one to which the graptolites were normally adapted; hence it seems likely that the graptolite assemblage,

[1] See TWENHOFEL, W. H., 1936, Organisms and their environment, *Nat. Res. Council, Rept. Comm. Paleoecology*, 1935–1936, pp. 1–9. ALLAN, R. S., 1948, Geological correlation and paleoecology, *Bull. Geol. Soc. Amer.*, vol. 59, pp. 1–10.

and perhaps that of associated fossils as well, is one of death rather than of life. An assemblage thus brought together after death has been designated a **thanatocoenose.** Little or nothing of the life environment of the organisms of a thanatocoenose can be determined from the containing sediments. A natural assemblage of living organisms or fossils is a **biocoenose,** and in the case of the fossils both they and the containing sediments have a community of characteristics that reflect the nature of the environment of deposition.

Use of Fossils for Correlation. The value of any fossil, no matter what the objective of the study, depends upon its position in the geologic column and its geographic location. A fossil without this information is little more than a curiosity. It may tell of geography, but it is the geography of an unknown place or time. It represents some place in the geologic column, but that position is unknown. It represents a stage of evolution, but that stage is not determinate. Hence it is always of the utmost importance to know the location of a fossil both geographically and stratigraphically.

Students of sedimentary rocks since the days of William Smith in England and Cuvier, Lamarck, and Brongniart in France, about 150 years ago, have found that there is a definite relation between the fossil contents of rocks and the position of those rocks in the geologic column. It has been found that the more recently the rocks have been formed, the more complex and varied is the assemblage of the contained organisms; the earlier the formation, the simpler the organic content.

The relative ages of the units of the standard geologic column had first to be determined on the basis of **superposition.** The next step was to determine the fossil assemblage of each time-rock unit. Thereafter the age of any isolated sequence of strata could be determined by the fossils present in it, and once the fossils were identified the sequence could then be assigned to the proper unit in the standard section. This is **correlation** by fossils.

The oldest sedimentary rocks, those of the Pre-Cambrian, have few definitely recognizable fossils, and for that reason they cannot be identified or correlated by means of the organic content. Other criteria, therefore, must be sought, and all of these (except radioactivity) are useless for long-distance correlation. In all the major divisions of the geologic column since the Pre-Cambrian, however, the organic assemblages have been determined, at least in the broader outlines, so that it is now comparatively easy to recognize the equivalents of each such division in any part of the world.

Fossils as Indices of Ancient Geography. Paleogeography is the science that deals with ancient geography—with the distribution of former lands and seas, with ancient rivers and lakes, with former plains and mountains, and with the positions of ancient shore lines. The adaptations of organisms are characteristic of particular environments; hence those shown by fossil forms may indicate the existence and position of ancient flood plains, deltas, prairies, mountains, deserts, lakes, rivers, shore lines, and the positions of the deep and shallow parts of the seas. Environments may be indicated, therefore, by the associations exhibited by fossil organisms. In the delta, as an example, there is a mingling in the deposits of terrestrial, lacustrine, and marine organisms. The distribution of fossil marine organisms may indicate former connections between bodies of salt water, and, similarly, scattered occurrences of land animals may point to connections between lands now separated. Certain animals have always lived in the sea. Among these are the stony corals, the echinoderms, the brachiopods, and the cephalopods. Their presence in deposits, therefore, indicates the sea or proximity to the sea.

Fossils as Indices of Ancient Climates. One of the great factors of most environments is climate, and many organisms are narrowly adapted to climatic conditions. Among these are certain foraminifers, corals, mollusks, palms, and magnolias. The fossil palms from the Tertiary strata of Spitzbergen and the fossil magnolias from similar strata in Greenland; the great profusion of corals in the Ordovician and Silurian of Anticosti, in the Silurian of the Michigan Basin, in the Silurian of the island of Gotland in the Baltic Sea, and in the Devonian strata at Louisville, Ky.; and the fossil ferns and primitive sponges found on Antarctica all prove the existence in these localities of climates quite different from, and much warmer than, those of the present.

Fossils as Evidence of Evolution. No line of evidence more forcefully and clearly supports the fundamental postulate of evolution—*descent with accumulative modifications*—than that furnished by fossils. Through their study one sees dynasty after dynasty of organisms appear as simple, adaptable, and easily modified forms; one witnesses their deployment into the many natural environments where they succeed or fail; one sees the successful reach a zenith of abundance and complexity under optimum living conditions; and finally one observes the races passing into old age with loss of racial vitality and ease of adaptability. Ultimately, their last futile attempts to overcome an environment to which they are no longer adapted result in degenerate or overspecialized forms which still continue for a short time into a future of organic development in which they are not destined to play a part. At last they disappear from the scene, these helpless stragglers, never to return, and no new and vigorous races evolve from their degenerate or overspecialized bodies. Senile races never seem to take the backward trail to simplicity and renewed strength; rather, they march on to inevitable extinction, as no advancement is known to have come from overspecialized or degenerate forms. This is the panorama of organic development that the paleontologist visualizes from his study of the fossil record left in the rocks by former organisms.

The history of the earth has been divided into large time divisions known as **eras.** These major chapters in earth history have been postulated to begin and end with supposedly world-wide and profound crustal disturbances designated **orogenies** or **revolutions.** Their names—**Archeozoic** (ancient life), **Proterozoic** (earlier life), **Paleozoic** (old life), **Mesozoic** (medieval life), and **Cenozoic** (recent life)—suggest the stages of development of the life that their fossils record. There is no simple term for the rocks that were made during an era. It is customary simply to speak or write of the Paleozoic rocks, the Mesozoic rocks, etc. Eras have been subdivided, in descending rank, into **periods, epochs,** and **ages.** The rocks formed during the time

Table 1-4. Table of Geologic Time[1]

	Systems	Series North America		Series Europe			Organic characteristics
Cenozoic	Quaternary	Recent Pleistocene		Recent Pleistocene			Invertebrates of modern aspect; Foraminifera and Radiolaria abundant in all seas; corals largely limited to warmer tropical and semi-tropical seas; many worms but few with hard parts; Bryozoa much less abundant in modern seas than formerly; brachiopods few in number of individuals and species; mollusks, arthropods, and certain echinoderms numerous and varied. The Mollusca and Arthropoda are possibly at or near their climax.
Cenozoic	Tertiary	Pliocene Miocene Oligocene Eocene Paleocene		Pliocene Miocene Oligocene Eocene Paleocene			
Mesozoic	Cretaceous	Upper Cretaceous		European stages	Maestrichtian Campanian Santonian Coniacian Turonian Cenomanian		Extinction of ammonites; abundant planktonic Foraminifera; Mollusca abundantly represented; general beginning of modern forms of all kinds of invertebrates.
		Lower Cretaceous		European stages	Albian Aptian Barremian Hauterivian Valanginian Berriasian		
	Jurassic	(Represented by continental and marine deposits in United States. Stages not yet fully defined.)	Upper Jurassic	European stages: Malm	Purbeckian Portlandian Kimmeridgian Corallian Oxfordian Callovian		Great abundance of ammonites; culmination of belemnites; appearance of oyster-like pelecypods; few bryozoans or brachiopods; corals uncommon.
			Middle Jurassic	European stages: Dogger	Bathonian Bajocian		
			Lower Jurassic	European stages: Lias	Toarcian Pliensbachian Sinemurian Hettangian		
	Triassic	(Poorly represented in United States by marine deposits except on the Pacific coast. Stages not yet fully defined.)			Alps	Germany	Appearance of true ammonites and disappearance of ceratites; few Foraminifera, corals, bryozoans, and brachiopods.
			Upper Triassic	European stages	Rhaetian Norian Carnian	Keuper	
			Middle Triassic	European stages	Ladinian Anisian	Muschelkalk	
			Lower Triassic	European stages	Scythian	Bunter	

[1] Geologic time from the beginning of the Cambrian to the present has been estimated at 500,000,000 years. The time interval between the beginning of the Cambrian and the earliest earth history recorded in the rocks, not considering igneous activity, is placed at 1,500,000 years. (*See Ahrens, L. A., 1950.*)

The percentages and the actual numbers of years indicated for the different divisions are intended to express order of magnitude only and are to be regarded as good estimates, as their basis is radioactive determination.

European equivalents are at best only approximations and the stratigraphic position of several of the equivalents is still undecided.

TABLE 1-4. TABLE OF GEOLOGIC TIME[1] (*Continued*)

	Systems	Series North America		Series Europe	Organic characteristics
Paleozoic	Permian	Upper	Ochoan	Tartarian	Great extinction of many Paleozoic groups; more advanced fusulinid protozoans abundant but become extinct during period; unusually spinose brachiopods; last of the trilobites.
		Middle	Guadalupian Leonardian	Kazanian Kungurian Artinskian	
		Lower	Wolfcampian	Sakmarian	
	Pennsylvanian	Monongahelan = Virgilian		Stephanian	Climax of fusulinid protozoans; small cup corals abundant; many fenestrate bryozoans; dominance of productid brachiopods; early arthropods in Mazon Creek nodules; conodonts numerous. Coal-forming flora at its height.
		Conemaughian = Missourian			
		Alleghenian = Desmoinesian		Westphalian	
		Kanawhan = Lampasasian			
		Leean = Morrowan		U. Namurian	
	Mississippian	Chesterian		L. Namurian	First calcareous Foraminifera; fenestrate bryozoans; beginning of spiny productid brachiopods; goniatites; climax of crinoids and blastoids; extinction of the graptolites; conodonts numerous.
		Meramecian		Viséan	
		Osagean			
		Kinderhookian		Tournaisian	
	Devonian	Bradfordian Chautauquan		Famennian	Sponges abundant locally; corals abundant and varied; dominance of spiriferoid brachiopods and disappearance of pentameroid brachiopods; great decline of trilobites and graptolites.
		Senecan		Frasnian	
		Erian		Givetian Eifelian	
		Ulsterian		Coblenzian Gedinnian	
	Silurian	Cayugan		Ludlovian	Great extent of coral reefs; first abundance of spiriferoid brachiopods; culmination of pentameroid brachiopods; many orthoceratite cephalopods; trilobites on the decline; dendroid graptolites abundant.
		Niagaran		Wenlockian	
		Albian (Medinan)		Llandoverian	
	Ordovician	Cincinnatian			Rise of arenaceous Foraminifera; first true corals and coral reef; trepostomatous bryozoans abundant; dominance of orthoid and strophomenoid brachiopods; climax of ancient straight cephalopods; trilobites numerous and varied; graptolites at climax; conodonts numerous.
		Champlainian (Mohawkian)		Caradocian	
				Llandeillian	
		Canadian		Arenigian Tremadocian	
	Cambrian	Croixian			First representatives of most invertebrate phyla; archaeocyathids worldwide and abundant and becoming extinct in middle of period; most brachiopods of atrematous, neotrematous and palaeotrematous types; many types of soft-bodied worms; first appearance and climax of trilobites; a few echinoderms; rise of graptolites.
		Albertan		Menevian	
		Waucobian		Harlechian	
Algonkian (Proterozoic)					Evidence of life consists of extensive algal precipitations; siliceous sponge spicules; one impression supposed to be that of a jellyfish; trails of many kinds generally ascribed to crawling and floating animals or drifting plants; a simple brachiopod, if the inclosing rocks are correctly identified as to age; and numerous carbonaceous fragments supposed to be remains of ancient organisms. Most supposed fossils thus far reported from Algonkian rocks are questionable.
Archean (Archeozoic)					Direct evidence of life is lacking, except that the carbon in certain carbonaceous material is organic; deposits of iron (thought by some to have been precipitated by bacteria), marble, and carbonaceous rocks (graphite and carbonaceous slates) indirectly suggest that some kinds of life were present in Archeozoic seas.

represented by these divisions have been designated **systems, series,** and **stages** respectively; these have also been designated **time-rock units** because they refer to the rocks formed during a certain period of time.[1] The relations of time and time-rock units may be shown as follows:

Time Unit	*Time-rock Unit*
Era.	......
Period.	System.
Epoch.	Series.
Age.	Stage.

Names of eras end in the suffix **-zoic** (Gr. *zoon*, animal) and, as stated in a previous paragraph, refer to the life of the times, with the first part of the word (*e.g.*, archeo-, from Gr. *archaios*, first) indicating the nature of that life. Names of periods, epochs, and ages, and of systems, series, and stages as well, are generally of geographic origin and bear the suffix **-an** (*e.g.*, Mississippian), unless some other ending is desirable for purpose of euphony.

The eras and smaller time divisions together constitute the **geologic time scale.** The sequences of rocks formed during geologic time constitute the **geologic column.** Table 1-4 shows the time scale now in general use.[2] It is not possible to construct a simple diagram or descriptive table that will show the true nature of the geologic column, because it is highly complex and differs greatly in many respects from place to place.

[1] For discussions of the nomenclature of rock units the reader is referred to the following works: ASHLEY, G. H., *et al.*, 1933, 1939, Classification and nomenclature of rock units, *Bull. Geol. Soc. Amer.*, vol. 44, pp. 423–459; *Bull. Amer. Assoc. Petrol. Geol.*, vol. 17, pp. 843–868; vol. 23, pp. 1068–1088. SUTTON, A. H., 1940, Time and stratigraphic terminology, *Bull. Geol. Soc. Amer.*, vol. 51, pp. 1397–1412. TOMLINSON, C. W., 1940, Technique of stratigraphic nomenclature, *Bull. Amer. Assoc. Petrol. Geol.*, vol. 24, pp. 2038–2046. SCHENCK, H. G., and MULLER, S. W., 1941, Stratigraphic terminology, *Bull. Geol. Soc. Amer.*, vol. 52, pp. 1419–1426. SCHENCK, H. G., *et al.*, 1941, Stratigraphic nomenclature, *Bull. Amer. Assoc. Petrol. Geol.*, vol. 25, pp. 2195–2211. Stratigraphic commission notes (1947 to date; R. C. Moore, Chairman): Note 1, Organization and objectives of the Stratigraphic Commission, *Bull. Amer. Assoc. Petrol. Geol.*, vol. 31, pp. 513–518 (1947); Note 2, Nature and classes of stratigraphic units, *Ibid.*, vol. 31, pp. 519–528 (1947); Note 3, Rules of geological nomenclature of the Geological Survey of Canada, *Ibid.*, vol. 32, pp. 366–367 (1948); Note 4, Naming of subsurface stratigraphic units, *Ibid.*, vol. 32, pp. 367–371 (1948); Note 5, Definition and adoption of the terms stage and age, *Ibid.*, vol. 32, pp. 372–376 (1948); Note 6, Discussion of nature and classes of stratigraphic units, *Ibid.*, vol. 32, pp. 376–381 (1948).

[2] This time scale is a compilation from many sources. The more recent and important publications used are the following: DUNBAR, C. O., 1942, Correlation charts prepared by the Committee on Stratigraphy of the National Research Council, *Bull. Geol. Soc. Amer.*, vol. 53, pp. 429–434. SWARTZ, C. K., *et al.*, 1942 [Silurian] *Ibid.*, vol. 53, pp. 533–538. COOPER, G. A., *et al.*, 1942 [Devonian] *Ibid.*, vol. 53, pp. 1729–1794. MOORE, R. C., *et al.*, 1944 [Pennsylvanian] *Ibid.*, vol. 55, pp. 657–706. HOWELL, B. F., *et al.*, 1944 [Cambrian] *Ibid.*, vol. 55, pp. 993–1003. WELLER, J. M., *et al.*, 1948 [Mississippian] *Ibid.*, vol. 59, pp. 91–196. DUNBAR, C. O., 1949, "Historical Geology" (Appendix B: Correlation Tables), New York, John Wiley & Sons, Inc. MOORE, R. C., 1949, "Historical Geology," New York, McGraw-Hill Book Company, Inc.

REFERENCES[1]

Borradaile, L. A., Potts, F. A., Eastham, L. E. S., and Saunders, J. T. 1935. "The Invertebrata: A Manual for the Use of Students," 2d ed. New York, Cambridge University Press. 725 pp.

Buchsbaum, R. 1938. "Animals without Backbones: An Introduction to the Invertebrates." Chicago, University of Chicago Press. 371 pp. (rev. ed., 405 pp.)

Cambridge Natural History (see Harmer).

Edwards, W. N. 1931. "Guide to an Exhibition Illustrating the Early History of Palaeontology." London, British Museum (Natural History). 68 pp.

"Geology, 1888–1938." Fiftieth Anniversary Volume, Geological Society of America, 1941. (Oceanography, pp. 43–69, by H. C. Stetson; Invertebrate Paleontology, pp. 71–103, by P. E. Raymond; Stratigraphy, pp. 177–220, by R. C. Moore.)

Glaessner, M. F. 1945–1947. "Principles of Micropaleontology." New York, John Wiley & Sons, Inc. 296 pp. 1947. (First published in 1945 by Melbourne University Press, Carlton, Victoria, Australia.)

Goldring, W. F. 1929. "Handbook of Paleontology for Beginners and Amateurs." Pt. I, The Fossils. N.Y. State Museum Handbook 9. 356 pp.

———. 1931. *Ibid.*, Pt. II, The Formations. N.Y. State Museum Handbook 10. 488 pp.

Grabau, A. W. 1924. "Principles of Stratigraphy." New York, A. G. Seiler. 1185 pp.

Grassé, P. P., *et al.* 1948 to date. "Traité de zoologie: anatomie, systématique, biologie." Paris, Masson et Cie.

Harmer, S. F., Shipley, A. E., *et al.* 1895–1910. Cambridge Natural History. New York, Cambridge University Press. Vols. 1–10.

Kükenthal, W., and Krumbach, T. 1923 to date. "Handbuch der Zoologie." Berlin and Leipzig, Walter de Gruyter and Co.

Lankester, E. Ray, *et al.* 1900–1909. "A Treatise on Zoology." London. A. and C. Black, Ltd.

Lull, R. S. 1931. "Fossils. What They Tell Us of Plants and Animals of the Past." New York, The University Society. 114 pp.

Oakley, K. P., and Muir-Wood, H. M. 1949. "The Succession of Life through Geological Time." London, British Museum (Natural History). 92 pp.

Raymond, P. E. 1935. Pre-Cambrian life. *Bull. Geol. Soc. Amer.*, vol. 46, pp. 375–392.

———. 1939. "Prehistoric Life." Cambridge, Mass., Harvard University Press. 324 pp.

Shimer, H. W., and Shrock, R. R. 1944. "Index Fossils of North America." New York, John Wiley & Sons, Inc.: Cambridge, Mass., The Technology Press. 837 pp.

Storer, T. I. 1951. "General Zoology." 2d ed. New York, McGraw-Hill Book Company, Inc. 832 pp.

Swinnerton, H. H. 1947. "Outlines of Palaeontology," 3d ed. London, Edward Arnold & Co. 393 pp.

Twenhofel, W. H. 1932. "Treatise on Sedimentation." Baltimore, The Williams & Wilkins Company, 926 pp.

———. 1939. "Principles of Sedimentation." New York, McGraw-Hill Book Company. Inc., 610 pp.

Zittel, K. A. von. 1913. "Textbook of Paleontology." (Eastman and Broili editions.) Vol. 1. New York, The Macmillan Company; London, Macmillan & Co., Ltd., 839 pp.

[1] The interested student will find an enormous amount of printed material on paleontology and much pertinent matter available to him from the related fields of geology and biology. Accordingly, the following list is limited to only the most general works. The reference lists at the ends of the following chapters and more particularly from the bibliographies included in the works in the following lists will serve as more complete and specific guides.

CHAPTER 2

PHYLUM PROTOZOA

INTRODUCTION

The Phylum Protozoa[1] includes the simplest and most primitive of animals. The individual protozoan, which ranges in size from less than 1 micron ($1\ \mu = 0.001$ mm.) to several centimeters, consists fundamentally of a cell-like body of protoplasm that is invariably differentiated into **cytosome,** which is composed of **cytoplasm,** and a **nucleus** (Fig. 2-1*A*). Some Protozoa have a naked body and can undergo continuous change of form. The majority, however, have a constant and characteristic body form due to the development of a special elastic or rigid envelope, the **pellicle,** which is composed of an organic substance. In addition to the pellicle, which always envelops the protozoan body closely, other protective structures are produced by certain Protozoa. These are composed largely of inorganic materials and commonly encase the body rather loosely. They vary greatly in composition and architecture and are the only parts of protozoans likely to be preserved.

The individual protozoan body—which may be thought of as a single cell, although zoologists prefer to consider it as noncellular—acts as a complete and independent organism, performing all the functions necessary for life, such as acquisition (capture) and assimilation of food, rejection or excretion of waste products, secretion, reproduction, growth, and locomotion. Most protozoans are uninucleate and solitary. A few form loose aggregations and are referred to as **colonial Protozoa.** In such colonies the individuals are bound together by protoplasmic threads or are embedded in a common matrix of protoplasm. Inasmuch as the cells in these aggregations are essentially similar in structure and function, there is no differentiation of them into tissue or definite organs. However, the dividing line between the most complex of the colonial Protozoa and the simplest forms of **Metazoa**—multicelled animals with cells differentiated into tissue and organs—is not a sharp one in all cases.

Protozoa obtain nourishment in several ways. Many use other protozoans and microscopic plants, as well as general organic debris, as sources of food. They capture and ingest food, digest and assimilate part of it, and reject indigestible portions. In simpler forms, food is ingested on any part of the periphery, and undigested portions and waste products are ejected from any part of the periphery. More complex forms ingest their food through a special opening, the **cytostome,** and eject wastes through a definite opening, the **cytopyge.** Protozoa that obtain nourishment in either of the manners just described are said to have **holozoic** nutrition. Certain protozoans can syn-

[1] Protozoa—Gr. *protos,* first, + *zoon,* animal: referring to the fact that members of the phylum are first among the animals in simplicity.

thesize carbohydrates through the action of sunlight on chlorophyll, a process known as **photosynthesis.** In this process oxygen is liberated and the carbon combines with other elements derived from water and inorganic salts. These Protozoa have **holophytic** nutrition. A third type of nutrition, **saprozoic,** is used by many Protozoa, especially those that live within the bodies of other organisms. In this type of nutrition, the protozoan obtains nourishment by diffusion through the body surface. Parasitic forms can nourish themselves by absorbing the digested or decomposed substances of the host. Many Protozoa, nourishing themselves by more than one method at the same or different times, are said to have **mixotrophic** nutrition.

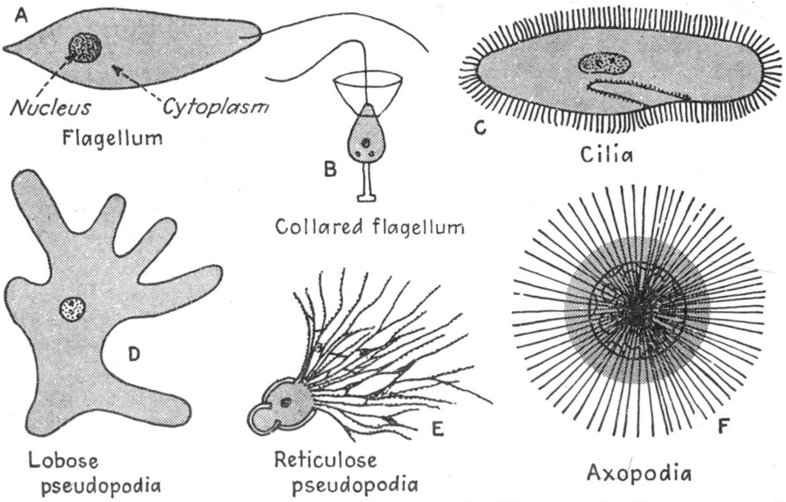

FIG. 2-1. PHYLUM PROTOZOA. Diagrams illustrating the different kinds of locomotory and food-capturing structures among the Protozoa. All diagrams much enlarged.

Locomotion i- possible for most protozoans and is accomplished by one of three different types of processes on the periphery of the body. Those protozoans without a definite pellicle move slowly by sending ıt fingerlike extensions (**pseudopodia**) from any part of the body surface. Inasmuch as the protoplasm is contractile and the body form plastic, the pseudopodia can change shape and size readily. They are lobose, filose, or reticulose (Figs. 2-1, 2-5, 2-7). A special kind, the **axopodia,** are composed of axial rods v··h a cytoplasmic envelope and are semipermanent structures (Fig. 2-1*F*). M..ny of the more complex protozoans h..ve part or all of the body surface covered with minute hairlike processes, the **cilia** (Fig. 2-1*C*), which drive the organism through the water by their rhythmic wavelike motion. Cilia also aid in the ingestion of food and serve a tactile function in some forms. The flagellate protozoans bear one or more whiplike **flagella,** generally on the anterior end of the body (Fig. 2-1*A*, 2-4). A front flagellum that is directed forward pulls the organism along by its characteristic movement. If there is a second flagellum at the anterior end, it may trail posteriorly and act as a steering device as well as for propulsion. A flagellum in the posterior part of the animal propels the body forward by its vibration. Parasitic Protozoa, and also a few others, are without locomotory processes in the adult stage.

Reproduction among most Protozoa is asexual, sexual, or by an alternation of sexual and asexual generations. Some highly specialized groups (*e.g.*, Sporozoa), however, reproduce by formation of spores. Asexual reproduction is accomplished by simple and direct, or complicated and indirect, nuclear and cytosomic division (binary fission; multiple division), or by budding. In the first method the protozoan separates into two or more parts; then each part, which contains some of the original nucleus and some of the original cytoplasm, ultimately develops into a distinct and complete organism. In budding, one or more smaller individuals form from the parent organism. Sexual reproduction takes place by complete (**syngamy**), or incomplete union (**conjugation**) of two special reproductive cells (**gametes**) from two individuals of one and the same species. In complete union the two cells unite, the nuclei ultimately become one, the cytoplasms fuse, and a single organism results. Some protozoans, notably the Foraminifera, reproduce by an orderly succession of asexual and sexual generations (Figs. 2-6, 2-7). Inasmuch as the individuals of the two generations build different shells, the species is said to be **dimorphic.** A more complete description of this type of reproduction is given on a later page under Foraminifera.

Most protozoans have in addition to the pellicle some kind of external protection, which may be an **envelope, lorica, test,** or **shell.** Envelopes and loricae are generally composed of a chitinous substance secreted by the organism. These are reinforced in some forms with sand grains or calcareous platelets (**coccoliths**) (Fig. 2-2*B–C*). Shells and tests are made of cellulose, chitin, pseudochitin (tectin), calcareous or siliceous material, or particles selected from the bottom sediments and cemented together by one of the substances just mentioned. Some shells have calcareous or siliceous platelets or disks as a part of the structure. Isolated spicules and platelets of silica, and complete internal skeletons of silica, are characteristic of the Heliozoa and Radiolaria. Calcareous tests are especially characteristic of the Foraminifera.

In addition to the hard parts described in the preceding paragraph, the bodies of certain protozoans contain crystals and granules of insoluble substances representing catabolic products. The crystals show considerable variety in geometric form, range from less than 1 μ to more than 30 μ long, and are composed most commonly of phosphates (calcium phosphate and calcium chlorophosphate), carbonates, oxalates, and urates. There is little likelihood that these crystals could ever be preserved. Protozoans with a test (*e.g.*, Foraminifera) generally have it completely surrounded by protoplasm, though part of the surface may at times be uncovered. It may be a single spheroidal chamber; a simple hollow tube open at one or at both ends; a series of tubular, globular, or ellipsoidal chambers connected by openings and arranged in a variety of ways; or a series of concentric, perforated shells separated from each other by delicate radial spines and other supporting structures (Figs. 2-3, 2-8, 2-12). The test wall is granular in agglutinated forms; solid (**imperforate**) or porous (**perforate**) in secreted shells. The protoplasm of agglutinate and imperforate tests flows out through the main aperture, or through many small apertures, and commonly forms a thin covering over the entire exterior (Figs. 2-5, 2-9). In perforate tests minute threads of protoplasm stream out through pores in the wall as well as through the apertures (Fig. 2-3*B*).

Protozoa were generally unknown as such until after the discovery of the microscope, although individuals larger than microscopic size now exist and are known to have lived during the geologic past. The first shells found were referred to several other animal groups before their true identity and biologic relations were determined.

Protozoa now live in almost every habitat or environment, and many live as parasites in other organisms. More than 15,000 living species have been described, and probably 20,000 fossil species are known. In the present oceans their abandoned microscopic tests and shells are settling to the bottoms in prodigious numbers to join there equally large numbers of shells of benthonic Protozoans. These accumulations form the extensive *Globigerina* and radiolarian oozes that cover great areas of sea bottom. Similar protozoan

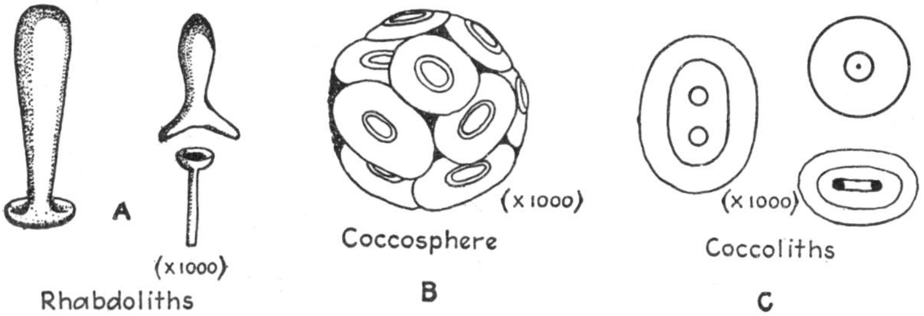

Fig. 2-2. Rhabdoliths and coccoliths, minute calcareous bodies recovered from deep-sea ooze, referred to protozoans by some investigators, to algae by others. (*After Murray.*)

shells accumulated on ancient sea bottoms and are today represented by *Fusulina* limestones of the Pennsylvanian and foraminiferal limestones of the Cretaceous and Tertiary, and radiolarites of the more recent rocks.

Although it is commonly stated that the first form of life on earth was an extremely simple, chlorophyll-bearing plant, it may be assumed that the first animals appeared soon afterward, and that the latter were probably simple protozoans. Radiolaria have been reported from the Pre-Cambrian, but grave doubt has been cast on the organic origin of the objects or the age of the rock containing them. Similar doubt has been expressed as to the organic origin of supposed Pre-Cambrian Foraminifera.

The earliest radiolarians thus far generally accepted are found in Lower Cambrian rocks, though there seems no good reason why they should not have existed in much earlier seas.

The major divisions of the Phylum Protozoa are based on the modes of locomotion (Fig. 2-1). The following classification, slightly modified from Kudo (1946), shows the larger and paleontologically more important divisions:

Phylum 1. Protozoa. Unicellular, aquatic, terrestrial, or parasitic animals which live singly or, in a few cases, as colonial aggregations. Since colonial Protozoa are merely cell aggregations without histological differentiation, they are readily distinguishable from the Metazoa, which have such differentiation. Certain protozoans build external tests and internal skeletons. The tests are composed of discrete inorganic particles, of calcareous and siliceous substances, or of organic materials. The internal skeletons are generally siliceous.

Class 1. Mastigophora. Free-living or parasitic protozoans with one to several flagella; dominantly with single nucleus, but a few multinucleated; body naked in some forms, encased in a more or less rigid test in others; habitat fresh or salt water; largely planktonic.

Order 1. Chrysomonadina. Fresh-water or marine, commonly planktonic, microscopic protozoans with one or two flagella and plastic body form. Representatives of some families have hard parts. Certain of the hard parts have been reported as fossils. (Coccolithidae and Silicoflagellidae.)

Order 2. Cryptomonadina. Fresh-water or marine, free-swimming or creeping protozoans with constant body form; one group has spherical cysts composed of cellulose. No fossils are known.

Order 3. Phytomonadina. Algal-like, free-living multiflagellate protozoans with definite body form and generally with a cellulose membrane. No fossils are known.

Order 4. Euglenoidina. Large ordinarily uniflagellate protozoans with the elongate body plastic or of definite form and without a preservable membrane.

Euglenoid fossils have been reported from flint nodules of the Baltic Cretaceous, from the Eocene Green River oil shale of western United States, and from a Pliocene shale of Madagascar.[1] The Green River specimen is remarkable in that it seems to represent the preservation of a naked protoplast (Bradley, 1931).

Order 5. Chloromonadina. Rare, uninucleate flagellates about which little is known. No fossils have been reported.

Order 6. Dinoflagellata. Biflagellate protozoans with body covered by a cellulose envelope consisting of a simple smooth piece, of two valves, or of numerous plates; dominantly marine and planktonic. Certain fossils from the Jurassic and Cretaceous of Europe have been referred to this order.[2]

Order 7. Rhizomastigina. Simple to complex protozoans representing borderline forms between the Mastigophora and Sarcodina; some forms have flagella, others have pseudopodia. No fossils are known.

Order 8. Protomonadina. Mainly parasitic flagellates with plastic or amoeboid bodies. No fossils are known.

Order 9. Polymastigina. Uninucleate or multinucleate, multiflagellate, dominantly parasitic protozoans without a preservable membrane or skeleton. No fossils are known.

Order 10. Hypermastigina. Uninucleate, multiflagellate protozoans inhabiting the alimentary tract of certain insects. No fossils are known.

Class 2. Sarcodina. Protozoans with changeable body form capable of sending out pseudopodia that function for both locomotion and food capturing; external tests

[1] DEFLANDRE, G., and LENOBLE, A., 1948, Sur le présence d'euglèniens fossiles du genre *Trachelomonas* Ehr. dans une schiste pliocène de Madagascar, *Compt. rend. acad. sci.*, vol. 226, pp. 509–511.

[2] A good summary of modern dinoflagellates is given in the following article: PAULSEN, O., 1949, Observations on dinoflagellates, *Kongelige Danske Videnskab. Selskab, Biol. Skrift.*, vol. 6, no. 4, 67 pp.

and internal skeletons of various forms and materials are developed in some groups; habit solitary or colonial; habitat fresh and salt water; reproduction by conjugation, by fission, or by a combination of both. Mastigophora and Sarcodina are thought by some to have descended from a common ancestor.

Order 1. Proteomyxa. Mainly parasitic protozoans with radiating threadlike pseudopodia that branch or anastomose with one another; most forms parasitic in plants in fresh or salt water; no preservable test. No fossils are known.

Order 2. Mycetozoa. Fungus-like organisms which seemingly occupy a borderline position between the Protozoa and Protophyta. No fossils are known.

Order 3. Amoebina. Protozoans with lobose pseudopodia inhabiting a wide range of both aquatic and semiterrestrial environments; without a test. The fossil record is uncertain. One genus has been reported from Cretaceous chert of Europe.

Order 4. Testacea. Amoeboid protozoans enveloped by a single-chambered test into which the entire body can be drawn; habitat aquatic and semiterrestrial; test composed of foreign particles and siliceous platelets or scales. The poor fossil record is probably due to the fact that most forms inhabit fresh water, hence have little chance of being preserved. The earliest known fossil is from the Eocene Green River shale of western United States.

Order 5. Foraminifera. Comparatively large, almost exclusively marine protozoans with reticulose or filose pseudopodia and with tests of chitinous, calcareous, or siliceous material or of agglutinated foreign particles. Most Foraminifera live on the ocean bottom, moving slowly over the mud and ooze by means of their pseudopodia; a few are attached to objects on the bottom; and some are pelagic. More than 300 genera of living and extinct Foraminifera are known, and several thousands of fossil species have been described from many parts of the world.

Order 6. Heliozoa. Spherical protozoans with numerous radiating axopodia (stiff pseudopodia which tend not to reticulate or fuse) and without a central capsule. The body may be naked, enclosed in a gelatinous mantle, or provided with a latticed test with or without spicules. Some of the skeletal elements are composed of silica, but there is no known fossil record. Most heliozoans live in fresh water; a few inhabit the sea.

Order 7. Radiolaria. Marine protozoans characterized by a central, perforated membranous capsule which varies in shape according to the form of the organism. Most radiolarians are spherical, but a few are hemispherical or tabular. Skeletons are composed of silica or strontium sulphate and show great range in structure. Enormous numbers of living Radiolaria are known but relatively few fossil species have been described, though the fossil record of the order goes back to the most ancient sedimentary beds in the Paleozoic, and possibly even to Pre-Cambrian rocks.

Class 3. Sporozoa. Exclusively parasitic, spore-bearing protozoans, usually incapable of locomotion, with fixed cell wall, but without preservable skeletal material.

Class 4. Ciliata. Ciliated protozoans of various habitats and body structures, generally free-living in both fresh and salt waters, but also parasitic in a few cases. There is rarely any preservable skeletal material and few known fossils, the tintinnids being the only ones of importance (Fig. 2-22).

Class 5. Suctoria. Protozoans that are ciliated only when young and free-moving. Cilia are lost when the organism develops tentacles and becomes attached. The body is spheroidal, ellipsoidal, or dendroidal, and generally without skeletal material. A few fresh-water forms, however, have a radiate test of sand grains and other bodies. No fossils are known

Class Mastigophora

The Mastigophora constitute a heterogeneous assemblage of simple organisms that have the common characteristics of one to several flagella. Certain forms bear chlorophyll and are commonly classified as plants by botanists. Most mastigophorans, however, possess typical animal characteristics. They are free-living or parasitic; the former live in all kinds of fresh and salt water where they are both planktonic and benthonic. Their supporting and skeletal structures are, in general, not composed of preservable material. However, living representatives of the several groups noted below

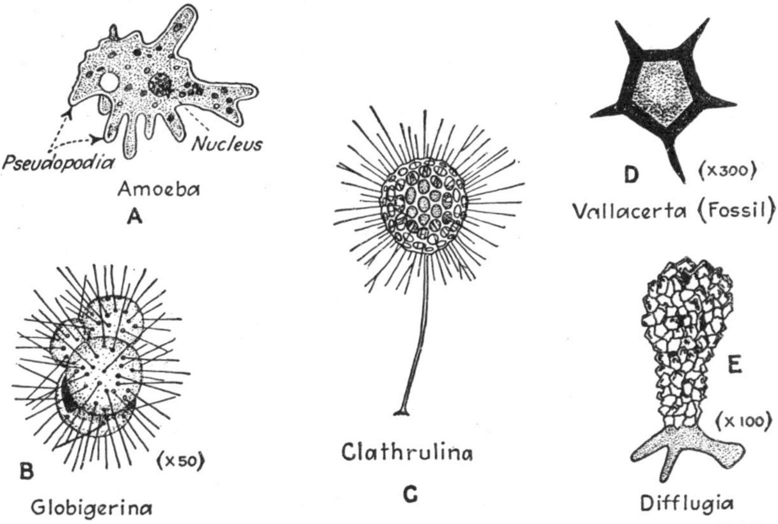

Fig. 2-3. Phylum Protozoa. *A*. *Amoeba*, a common, rather complex protozoan. *B*. *Globigerina*, a common foraminifer with multichambered calcareous test. Ooze composed of these tiny shells covers large areas of existing ocean bottoms. *C*. *Clathrulina*, a heliozoan with a perforated siliceous test, much magnified. *D*. *Vallacerta hortoni*, a fossil silicoflagellate from the Upper Cretaceous of California. *E*. *Difflugia*, a representative of the Testacea, with an agglutinated test. (*A*, *E after Parker and Haswell*, 1928; *B after Brady*, 1884; *C* from Kudo, 1931, after Leidy, 1879; *D after Hanna*, 1928.)

have calcareous and siliceous elements. Minute fossil bodies of similar nature have been referred to these groups.

The dominantly salt water Coccolithidae (Order Chrysomonadina), which are commonly assigned to the algae, include forms having both perforate (**tremalith**) and imperforate (**discolith**) calcareous disks known collectively as **coccoliths** (Fig. 2-2*C*). These minute bodies, which differ considerably in shape, are abundant in some marine oozes. Although they have been reported from sedimentary rocks as old as Upper Cambrian (Glaessner, 1945, page 17), they do not seem to be common as fossils until the Jurassic. They have been reported from fine-grained marine limestones of Upper Jurassic and Lower Cretaceous age in the Mediterranean region.

In North America they have been reported from the Cretaceous of Minnesota and Manitoba.

The exclusively marine planktonic Silicoflagellidae (Order Chrysomonadina) have a siliceous skeleton of unique structure. Fossil silicoflagellates have been found in the younger (Cretaceous to Recent) sedimentary rocks of western North America (Fig. 2-3*D*) and of Europe (Hanna, 1928).

Fossil representatives of one other mastigophoran order, the Dinoflagellata, have been reported from the European Cretaceous. It is possible that other orders have left fossils, but they have not been reported. The microscopic bodies found some years ago in Baltic Cretaceous flint nodules, so definitely

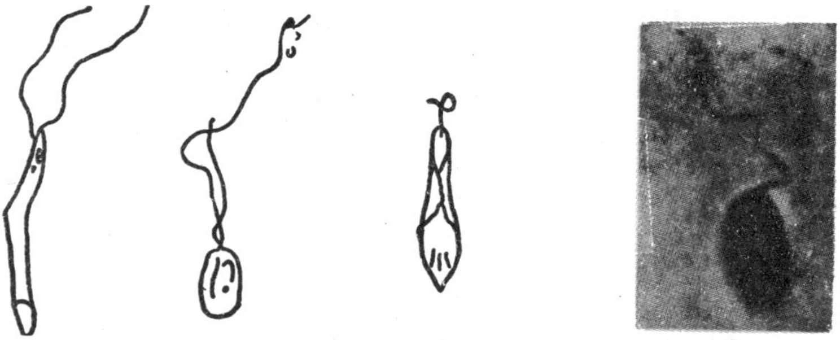

FIG. 2-4. Class MASTIGOPHORA. Supposed fossil flagellates (*Ophiobolus lapidaris*) from Cretaceous chert of the Baltic region. *A*. Elongate form with two equally long flagella, ×670. *B*. Oval form with a long and a shorter flagellum, ×320. *C*. Bag-shaped form with only one short flagellum, ×670. *D*. Photomicrograph of a supposed fossil flagellate with one whiplike flagellum, ×654. (*After Wetzel*, 1933.)

and distinctly preserved that the flagella are attached to the organism (Fig. 2-4), indicate new fields of paleontological exploration.[1]

CLASS SARCODINA

The protozoans of this class lack a definite pellicle, hence can change form by extending pseudopodia. The pseudopodia function for both locomotion and food capture. The Sarcodina are holozoic, feeding on other protozoans and small metazoans, and on simple plants. They live in both fresh and salt waters and at all depths from the surface to more than 5,000 m. (16,000 ft). They are planktonic, nektonic, and benthonic, and some species show considerable range in depth; certain supposed planktonic *Globigerina* have been dredged alive from a depth of 1,000 m. (3,000 ft.).

The cytoplasm is generally differentiated into a tough external **ectoplasm** and fluid internal **endoplasm** (Fig. 2-5), but this differentiation is not constant. In Radiolaria, a perforated membranous capsule marks the boundary between these two cytoplasmic regions (Figs. 2-18) Most Sarcodina are

[1] See review by M. I. GOLDMAN (*Jour. Paleontology*, vol. 8, pp. 482–483, 1934) of Wetzel's "Die in organischer substanz erhaltenen Mikrofossilien des Baltisches Kreidefeuersteins " *Palaeontographica*, vols. 77, 78A, 1933.

uninucleate, but numerous species of Mycetozoa and Foraminifera are multinucleate.

Reproduction is generally asexual (binary fission, multiple nuclear division, budding). Sexual reproduction is rare. A life cycle of alternating sexual and asexual generations has been worked out for some species.

The orders Proteomyxa, Mycetozoa, and Amoebina lack a shell or test. Testacea have a chitinous single-chambered test commonly armored with sand grains. Foraminifera have calcareous and siliceous tests of variable structure. Heliozoa and Radiolaria have complicated skeletons of siliceous material. The tests and shells of the Sarcodina vary greatly in size, shape, composition, and architecture. Almost all the known fossil protozoans, some 20,000 or more species in number, belong to the Foraminifera or Radiolaria. Although several other orders of Sarcodina include individuals having preservable skeletal elements, they do not seem to have left a fossil record of any magnitude.

Orders 1–3. Proteomyxa, Mycetozoa, Amoebina. No unquestioned fossil representatives of these orders are known, and from their nature and habits of life it seems doubtful that the organisms could leave any sort of record except under unusual conditions. Certain peculiar minute calcareous bodies resembling coccoliths have been considered by some investigators to have been made by organisms included in the Amoebina, but it seems more likely that these fossils should be ascribed to the Mastigophora (see page 34) or to the calcareous algae. Some species of *Amoeba* (Fig. 2-3*A*) contain microscopic crystals,[1] but it is doubtful if these could be preserved. Certain tiny bodies in siliceous nodules from the Baltic Cretaceous have been questionably referred to a living amoeboid genus, but the reference is open to doubt.

Order 4. Testacea. This order includes amoeboid protozoans that build around the body a single-chambered, agglutinated test into which they can completely withdraw. Usually there is only one aperture through which the pseudopodia are extruded. Testaceans live in aquatic and semiterrestrial habitats.

The test, which varies somewhat in form and structure, has, as a base, a chitinous or pseudochitinous (mucoid) membrane to which are commonly cemented foreign bodies acquired from the sea bottom, or siliceous (rarely calcareous) platelets secreted by the organism. The foreign bodies are sand grains (usually quartz), or diatom shells (Fig. 2-3*E*). The platelets, which may be denticulated, are in the form of round, oval, elliptical, or quadrangular scales or disks which are disposed in imbricating fashion in some forms and generally have a definite arrangement.

Fossil testaceans seem to be rare. The earliest known to the authors are three genera—*Difflugia* (Fig. 2-3*E*), *Quadrula*, and *Euglypha*, all with living representatives, from the Eocene Green River formation of western United States.

Order 5. Foraminifera. Foraminifera are comparatively large unicellular (noncellular) animals differing from all other Protozoa in the possession of a

[1] Luce, R. H., and Pohl, A. W., 1935, Nature of crystals found in *Amoeba*, *Science*, vol. 82, pp. 595–596.

network of branched or anastomosing threadlike pseudopodia and having a secreted or secreted-agglutinated test of variable composition and complexity. They are adapted to all aquatic habitats; most live in marine waters, but some can exist in salt lakes or brackish water, and the members of one primitive family (Allogromiidae) live in fresh water. Most are typically slow-moving bottom dwellers; some are pelagic. They live at all depths, commonly in great abundance, and in modern seas they are widely distributed in all latitudes. More than 35 per cent (48,000,000 sq. miles) of the present ocean bottom is covered with ooze largely composed of their tests. *Globigerina* (Fig. 2-3*B*) ooze is especially common. Such ooze is generally absent in the deepest parts of the oceans because the empty calcareous tests are dissolved before they reach the bottom.

Foraminiferal tests range in size from 0.01 to 190 mm. and in composition from calcareous, chitinous, or siliceous on the one hand to agglutinated tests of many kinds of foreign particles on the other. Structurally, the test ranges from a simple imperforate chamber to an aggregation of chambers arranged symmetrically with reference to each other (Fig. 2-8). Wall structure is simple in some forms, highly complex in others, and the external surface commonly bears ridges, spines, and nodes.

It has been stated that about 30,000 species of living and fossil Foraminifera have been described, and some investigators believe that many additional thousands of species are yet unknown.[1] Although Pre-Cambrian Foraminifera have been reported, they have been generally discredited. The earliest known species have been found in Cambrian strata, and the number and variety of tests increase upward in the geologic column.

Great impetus has been given the study of Foraminifera during the last 30 years. Micropaleontologists employed by petroleum companies have found the small tests, which commonly can be easily secured from well cuttings and cores, useful for correlating oil-bearing and associated strata. Biological oceanographers, interested in the life of the oceans, have studied living Foraminifera and the foraminiferal oozes of the sea bottom. Many other students have investigated both living and fossil Foraminifera, with the general result that the literature on these interesting protozoans is growing enormously and descriptions of new genera and species are constantly being published.

Nature of the Organism. The individual foraminifer is a single cell of **cytoplasm** with one or more nuclei. The cytoplasm is differentiated into the **endoplasm,** which contains the nuclei and occupies the test, and the **ectoplasm,** which constitutes the outer part, either filling the aperture or forming a thin coating on both the inner and outer surfaces of the test. The foraminiferal test is both external and internal, though most of the animal generally lies inside the test wall (Figs. 2-5, 2-9). In certain groups the cytoplasm inside and outside the test is connected through a simple or multiple aperture (Fig. 2-5) as well as through tiny perforations in the test wall. The ectoplasm streams out from the body, forming long threadlike pseudopodia that branch

[1] Hofker (*Jour. Paleontology*, vol. 22, p. 517, 1948) states that to these 30,000 so-called species more than 130,000 synonyms have been applied.

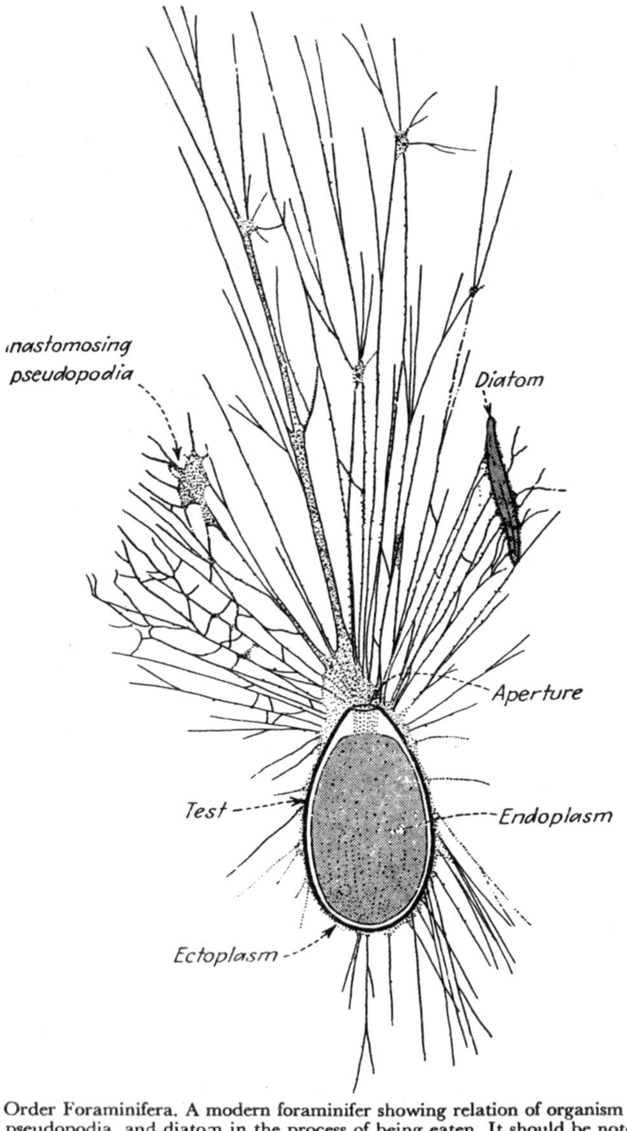

FIG. 2-5. Order Foraminifera. A modern foraminifer showing relation of organism to test, nature of pseudopodia, and diatom in the process of being eaten. It should be noted that the test is not only filled with cytoplasm but is also surrounded with it so that the test is actually an internal structure. (*Modified after Schultze*, 1854.)

and anastomose to present a tangled or characteristic appearance (Figs. 2-5, 2-9). The pseudopodia function as organs of locomotion, being able to change form and direction in a few minutes, and for capturing food and expelling undigested matter.

Growth and Reproduction. The first formed chamber of the foraminiferal test is the **proloculus.**[1] It is a tiny sphere with a small aperture in simple Foraminifera but becomes double and more complicated in advanced forms such as the orbitoids and fusulinids. It is one of the most important features of a foraminiferal shell because of the fact that each species has two forms of test—one with a large and one with a small proloculus. This dual nature of the test is known as **dimorphism** and is the result of a complicated reproductive cycle (Figs. 2-6, 2-7).

The wall of the foraminiferal test is either secreted wholly by the cytoplasm or partly secreted by the cytoplasm and later armored with granular material collected from the bottom by the pseudopodia. The formation of chambers goes on in a cyst built by the animal; hence observation of the process is difficult, with the result that relatively little is actually known about how the wall is made. If the test grows continuously from the proloculus, it is a simple sphere with a small neck or a hollow tube that is straight, curved, or coiled. Such a test is said to be **unilocular** or **monothalamous** (Fig. $2\text{-}8^{1-4}$). Periodic growth results in the formation of multichambered tests of great variety (Fig. 2-8). These are described as **multilocular** or **polythalamous.**

The life cycle of many Foraminifera consists of an orderly succession of sexual and asexual phases known as **alternation of generations.** This alternation results in the production of two different kinds of tests which illustrate dimorphism. A smaller shell, produced asexually, generally has a large proloculus, the **megalosphere,** hence is designated **megalospheric.** A larger shell, produced sexually, generally has a small proloculus, the **microsphere,** hence is designated **microspheric.** Five stages of the reproductive cycle for three foraminiferal species are shown diagrammatically in Fig. 2-6 and explained in detail in the caption for the figure. A brief description of the life cycle follows (see Fig. 2-7 for a simplified diagram illustrating similar reproduction in another genus).

The microspheric individual, the **agamont,** which has a relatively large test with a small proloculus, is multinucleate. In the adult state the cytoplasm subdivides into spherical masses of uniform size, each with one of the original nuclei. Either before or after leaving the parental test, each uninucleate mass secretes a calcareous proloculus much larger than the one in the parent test. This is the megalospheric proloculus of an individual of the new generation. By adding chambers to the proloculus the individual grows to an adult

[1] In most American literature on Foraminifera the initial chamber is called the **proloculum,** a term proposed by Cushman in 1905 (*Amer. Nat.*, vol. 39, p. 538). Recently, however, it has been pointed out (Schenck, H. G., 1944, Proloculus in Foraminifera, *Jour. Paleontology*, vol. 18, pp. 275–282) that since the Latin word *locus* is of masculine gender, its diminutive should be of the same gender, hence *loculus*. Therefore, several foraminiferal specialists have recommended that the form **proloculus** be adopted, and we follow this recommendation.

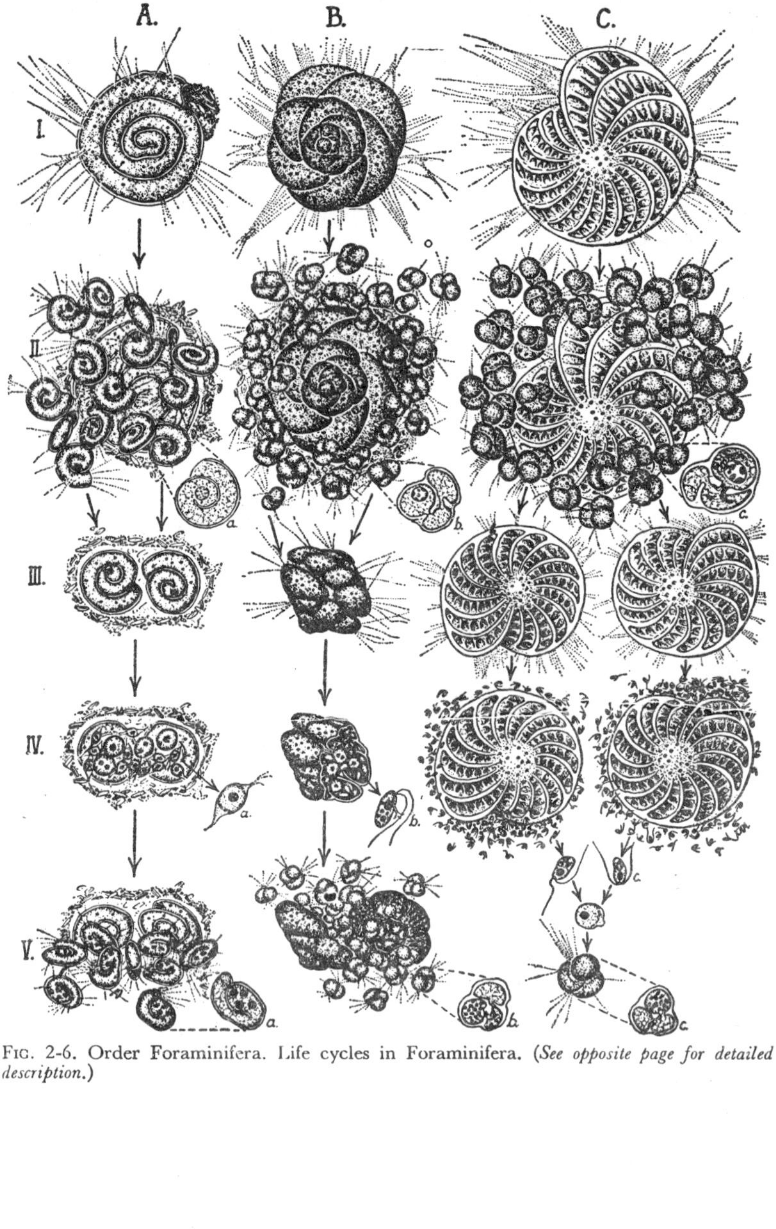

FIG. 2-6. Order Foraminifera. Life cycles in Foraminifera. (*See opposite page for detailed description.*)

gamont (megalospheric individual) building a relatively small test with a large proloculus. When the adult stage is reached in the gamont generation, the single nucleus breaks up into a multitude of tiny secondary nuclei, each of which is surrounded with cytoplasm. These then become flagellate **gametes,** or **zoospores,** leave the parental test, move about until they meet and conjugate with gametes from other gamonts of the same species, and then as pairs form embryos or **zygotes.** The zygote secretes a small initial shell, the microspheric proloculus, and develops into an adult agamont, becoming multinucleate and building during its growth a large, many chambered test. Thus is completed one alternation of generations.

Generally, megalospheric tests are much more abundant than microspheric, but both must be looked for in all species. Among simpler species of more advanced genera, such as *Polystomella*, the numerical ratio may be 30 to 1. The greater number of megalospheric tests in such genera is explained by the fact that the gametes, being free and pelagic, have less probability of

FIG. 2-6. The life cycle of many Foraminifera consists of an orderly succession of sexual and asexual phases. This alternation of generations results in **test dimorphism.** Five stages are diagrammatically shown for three species: (*A*) *Spirillina vivipara*, (*B*) *Discorbis patelliformis*, and (*C*) *Polystomella crispa.*

Asexual Generation (I and II)

Stage I. In *D. patelliformis* and *P. crispa*, as in many Foraminifera, the **initial chamber or** proloculus of the sexually produced **multinucleate agamont** is smaller than that of the asexually produced **mononucleate gamont,** and these two chambers are known as **microspheric** and **megalospheric** respectively, whereas in *S. vivipara*, a more primitive species, there is little difference in the diameter of the proloculus of the two generations. In all species studied, the agamont (microspheric) test is the larger.

Stage II. Following an orderly series of nuclear divisions, multiple fission results in as many mononucleate agamonts as there were nuclei produced. In *D. patelliformis* multiple fission and test secretion takes place within the parent test, while in *S. vivipara* and *P. crispa* these activities are preceded by the escape of the protoplasmic content from the test. Stage II (*a*, *b*, and *c*). Enlargement of sexual gamont, showing the nucleus.

Sexual Generation (III, IV, and V)

Stage III. In *S. vivipara*, sexual reproduction is preceded by the association of **gamonts** within a cyst composed of bottom debris and animal cement. A somewhat similar association takes place in *D. patelliformis* where no cyst is formed, whereas in *P. crispa* it is presumed that a close association between gamonts does not occur, the **gametes,** or **zoospores,** being free and pelagic.

Stage IV. The gametes of *S. vivipara* are amoeboid, whereas those of *D. patelliformis* and *P. crispa* are flagellated. **Gametogenesis** includes an orderly series of nuclear divisions of the equatorial type followed by a reduction division. Fertilization takes place between gametes derived from *different* gamonts. In *S. vivipara* the agamonts develop within a cyst; in *D. patelliformis*, within the space formed by the dissolution of septa between the chambers; whereas in *P. crispa* the gametes are pelagic, fertilization depending upon the chance meeting of gametes. Stage IV (*a*, *b*, and *c*). Enlargement of gametes.

Stage V. Juvenile (young) agamonts are multinucleate as a result of an orderly series of nuclear divisions immediately following fertilization and before the secretion of the test. Agamonts of *S. vivipara* contain but four nuclei, whereas those of *D. patelliformis* and *P. crispa* each have about 40 nuclei. The sexually produced young of *S. vivipara* develop within a cyst; those of *D. patelliformis*, within the space resulting from the dissolution of the septa between chambers and the ventral surface of associated tests; whereas those of *P. crispa* develop from the free **zygotic amoebula.** (In *P. crispa* the gametes from two different gamonts fuse to form a free-swimming **zygote** which ultimately develops into an asexual agamont.) Stage V (*a*, *b*, and *c*). Enlargement of asexual agamont showing nuclei.

(*The diagrams, and the accompanying explanatory text, both slightly modified, were taken by the kind permission of the author from Myers, E. H., 1938, The present state of our knowledge concerning the life cycle of the Foraminifera, Proc. Nat. Acad. Sci., vol.* 24, *pp.* 10–17.)

being fertilized, compared with those species in which the gametes are confined in cysts; hence fewer microspheric forms develop. As a general rule the megalosphere is larger than the microsphere, and the megalospheric test tends to have fewer chambers than the microspheric (Fig. 2-11). This differ-

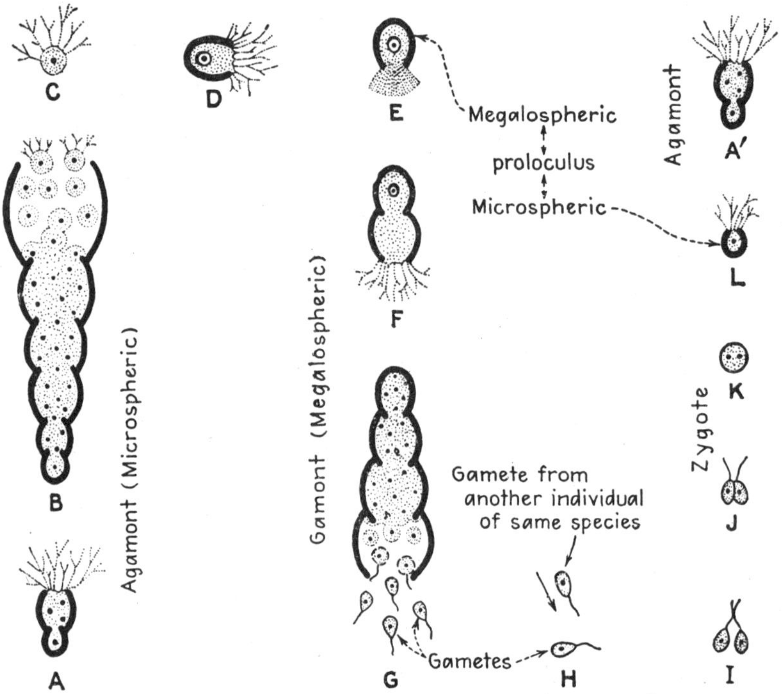

FIG. 2-7. Order Foraminifera. Diagram showing reproduction in Foraminifera by alternation of generations. *A*. A young agamont with proloculus and one chamber. *B*. An adult agamont in the process of asexual reproduction. Two flagellated offspring are leaving the parent test. *C*. A young gamont just after leaving test and before secreting a proloculus. *D–E*. Young gamonts with their megalospheric proloculus. *F*. Gamont with proloculus and one chamber. The individual has only one nucleus until it starts reproduction (*G*), when the single nucleus breaks up into many minute nuclei. *G*. Adult gamont giving off gametes, which will combine with gametes from other megalospheric individuals of the same species (*H–I*) to form zygotes (*J–K*), each of which develops into a young agamont (*L–A′*). *L*. Young agamont with microspheric proloculus. *A′*. Young agamont with proloculus and one chamber. This individual starts a new generation. (*Adapted from Kudo*, 1931, *and Kühn*, 1926.)

ence between the two kinds of tests increases with evolution, as is illustrated by the great difference of size of test in the fusulinids, lepidocyclinids, and camerinids. Youthful growth stages seem to be more completely developed in microspheric tests, an unfortunate situation because of the relative paucity of these tests among the more evolved species. A summary of the chief differences between the two types of tests follows:

Megalospheric	*Microspheric*
1. Asexually produced.	1. Sexually produced.
2. Proloculus large.	2. Proloculus small.
3. Chambers relatively few.	3. Chambers many.
4. Test generally smaller.	4. Test generally larger.
5. Ontogeny incomplete.	5. Ontogeny more complete.
6. Tests common and numerous.	6. Test uncommon or missing altogether; comparatively rare.

Dimorphism is of great importance to students of Foraminifera as it must be certain that both forms are known when a new species is erected. Dimorphic species have been reported among the fusulinids,[1] among certain Mesozoic[2] species, and in many Cenozoic and living species. In addition, some species show a series of megalospheric tests with a proloculus of varying size, followed by a microspheric test. This characteristic has been called **polymorphism,** and to certain types Hofker[3] has given the term **trimorphism,** because two or more different megalospheric forms intervene between successive microspheric tests (Fig. 2-10). As Cushman and Hofker point out, if all three forms can be proved to belong to a single species, then whatever names they have borne in the past must be suppressed in favor of that originally applied to the microspheric test, since it shows best the characteristics and ontogeny of the species. The problem of trimorphism and polymorphism is a most confusing and difficult one and will not be solved except with critical quantitative work on an extensive scale.

The Foraminiferal Test. The size, composition, shape, and architecture of foraminiferal tests vary to such an extent that each of these characteristics deserves separate consideration.

Size. The smaller tests of Foraminifera range from 0.01 mm. to twice that dimension in diameter, as in some of the *Lagynidae.* The common forms range from 0.1 to 5 mm. in diameter. Some forms attain unusually large dimensions, as *Parafusulina,* which is as much as 70 mm. long, and *Neusina,* which reaches the surprisingly large size of 190 mm. (almost 8 in.).

Composition. The tests of Foraminifera are composed of two strikingly different kinds of material. One kind is secreted by the organism; the other consists of foreign particles gathered from bottom sediments and cemented together to form agglutinated tests. Secreted tests are composed of gelatinous, pseudochitinous (mucoid), chitinous, and calcareous substances. Only the chitinous and calcareous tests are known fossil. A few of the siliceous tests may have been secreted by the organism, but it is likely that most of them

[1] DUNBAR, C. O., SKINNER, J. W., and KING, R. E., Dimorphism in Permian fusulines, *Geol. Soc. Amer. Proc.* 1934, p. 368, 1935; *Univ. Texas Bull.* 3501, pp. 173–190, 1936. Thompson (*Univ. Kans. Pub. Pal. Contr.*, Art. 1, p. 10, 1948) states, "It is not certain that the fusulinids display dimorphism."

[2] CUSHMAN, J. A., 1943, The megalospheric and microspheric forms of *Frondicularia sagittula* Van Den Broeck and their bearing on specific description, *Contr. Cushman Laboratory for Foraminiferal Research,* vol. 19, pp. 25–26.

[3] HOFKER, J., 1948, On *Asterigina gürichi* (Franke) and remarks on polymorphism and the stratigraphic use of Foraminifera, *Jour. Paleontology,* vol. 22, pp. 509–517.

were composed originally of some substance other than silica and that they have been subsequently silicified. Agglutinated tests are composed of mineral grains and bits of organic debris selected from the bottom sediments and ultimately added to the test wall. The following list indicates the different combinations of materials found in foraminiferal tests:

1. Organism with stiffened but flexible outer surface.
2. Test composed of gelatinous or chitinous matter.
3. Test of chitin with attached foreign particles.
4. Test of foreign particles cemented with chitin.
5. Test of foreign particles cemented with calcareous or siliceous material.
6. Test entirely of calcium carbonate.
7. Test partly or entirely of silica (doubtful).

At present there is some difference of opinion as to the order of appearance of the types of tests listed above. The evolution of the test seems to have followed a definite sequence. Most students of the Foraminifera believe that the earliest definite tests were chitinous, and that these were followed by agglutinated tests which developed when foreign particles were cemented to the exterior surface of the chitinous test.[1] The cement was probably at first chitinous, followed later by other substances, especially calcareous. As the cementing substances came to make up more and more of the test wall, less and less agglutinated material was used. The final result was a smooth imperforate or perforate calcareous test. All stages in this hypothetical sequence can be observed among living Foraminifera, and the change from a chitinous and agglutinated test to a purely calcareous one can actually be seen, according to Cushman (1948), in the development of a single individual. Galloway (1933, p. 19), on the other hand, argues that the agglutinated walls are found only in degenerate and specialized forms and that they can develop from all other kinds of walls.

Chitinous Tests. The most primitive foraminiferal test with a definite shape and aperture is composed of chitin, commonly thin and transparent. That chitin was one of the earliest substances used is suggested by the fact that many families of Foraminifera, regardless of the material of the adult test, have a thin chitinous inner layer in the earliest part of the test. Purely chitinous tests are not common among living Foraminifera. One of the earliest fossil species, *Chitinodendron franconianum*, from the Upper Cambrian of Wisconsin, is thought to have been a chitinous form (Ruedemann and Shrock, 1939).

Arenaceous Tests. These tests, commonly referred to as agglutinated, usually are found among the earlier and more primitive groups of Foraminifera. They are composed of quartz, mica, and other mineral grains or of bits of organic debris such as shell fragments and sponge spicules (Fig. 2-8). These foreign particles are selected from bottom sediments by the organism

[1] It is interesting to note here that the earliest known fossil Foraminifera, which have been reported from the Lower Cambrian of Labrador and Greenland (Howell and Dunn, 1942) and from the Upper Cambrian of Wisconsin (Ruedemann and Shrock, 1939), seem to have had chitinous tests armored with foreign particles.

and ultimately are loosely or firmly cemented together over the thin, chitinous inner layer, which is thought to represent the primitive chitinous test of still simpler groups.

The most primitive arenaceous Foraminifera, for example, *Astrorhiza*, simply cement the bottom materials about channels leading to the central chamber. Foreign particles of all sorts are indiscriminately included, and only the inner part of the test wall is firmly cemented.

Less primitive species show the power of selection in varying degrees. These characteristically select only certain constituents from the bottom sediments and discard all others. The species of *Rhabdammina* (Fig. 2-15*G*), for example, usually use sand grains, whereas *Marsipella* from the same bottom generally selects sponge spicules. *Psammosphaera fusca* selects nothing but sand grains, and these generally of a single color; *P. parva* not only uses sand grains of a fairly uniform size but also commonly adds one large acerose sponge spicule which is built into the test wall in such a way that each end of the spicule protrudes (Fig. 2-8[7]). Other forms select only mica flakes, even though these are uncommon in the bottom deposits; and still others show a preference for sponge spicules, utilizing the large unbroken ones for constructing a framework and the smaller broken fragments for filling in the polygonal spaces between the definitely arranged spicules in the main meshwork (Fig. 2-8[8]).

It has been suggested that the individuals having a high degree of selectivity ingest the particles they want and later move these to the surface of the body, where they are cemented to the thin, chitinous inner layer of the test. As Cushman (1948, page 13) has stated, "That this selection occurs in single-celled forms, which are but a speck of protoplasmic material, is the great wonder."

The cement of the test in the most primitive Foraminifera seems to be chitin like that constituting the inner wall of the test; the foreign particles are simply included in the outer part. In most arenaceous forms the cement is some shade of brown and is designated "ferruginous." It characterizes early Paleozoic tests and is commonly seen in living forms. The quantity is small in some tests, whereas in others it makes up so much of the wall that the agglutinated particles are inconspicuous. Cushman states that the "ferruginous" cement may, under certain conditions, be replaced by calcareous material, and that by one additional step an entirely calcareous test can be developed. Siliceous cement is present in a few groups, but is in no case common.

If the Lower Cambrian fossils from Greenland and Labrador reported by Howell and Dunn as fossil Foraminifera are correctly identified—and of this there is some doubt—then it seems probable that the earliest known Foraminifera had chitinous walls armored to some extent with arenaceous material. Although these problematical fossils are now calcified, it is believed that the original test wall was composed of chitin that has long since been replaced with calcite. It is worth noting that arenaceous material remained as a residue after the simple little globular tests were dissolved in a weak solution of hydrochloric acid. From this the authors concluded that the original test

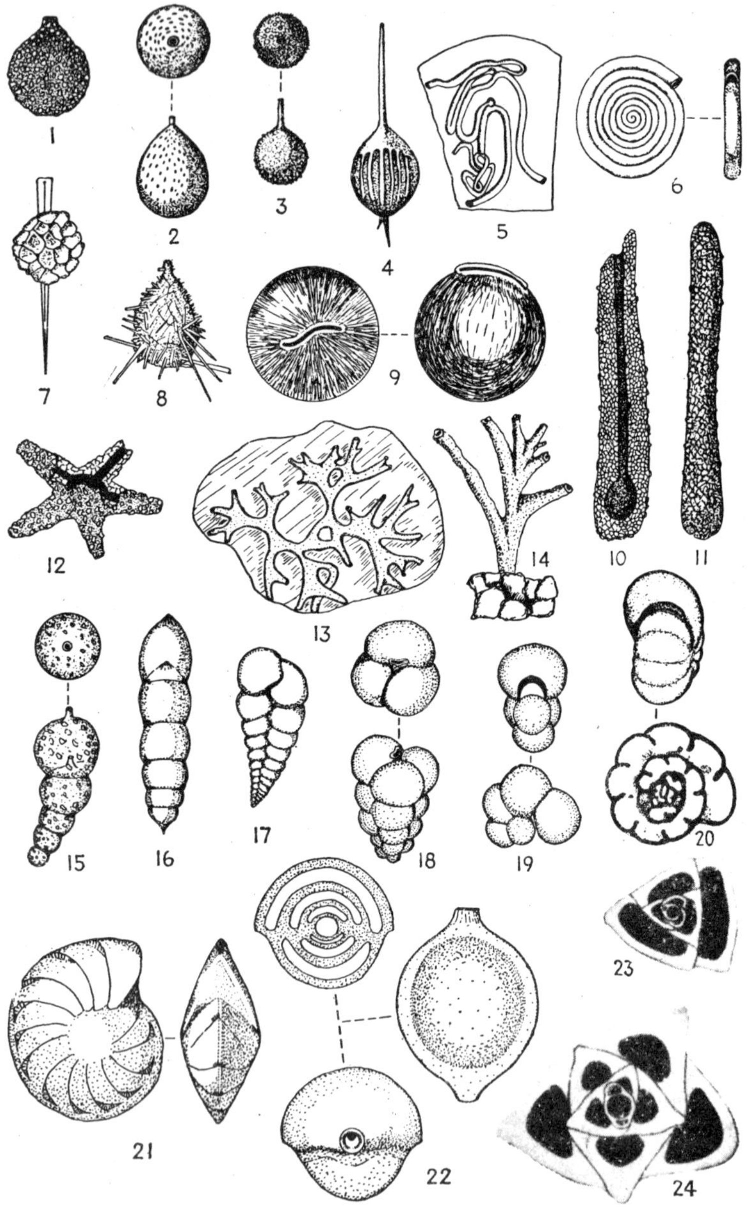

FIG. 2-8. Order Foraminifera. Variation in features of foraminiferal tests. (*See opposite page for detailed description.*)

was composed of chitin containing a small amount of arenaceous material.

Siliceous Tests. Siliceous walls are developed in only one important family of Foraminifera, the Silicinidae, which have their first known fossil representatives in Jurassic rocks. Siliceous tests have also been reported in certain other families, but it is questionable whether the walls of these were secreted by the animal or were secondarily silicified.

Calcareous Tests. Great numbers of living Foraminifera have calcareous tests, most post-Paleozoic species built their tests of calcareous material, and it is known that at least as early as the Mississippian some forms were using this substance for their tests (*e.g.*, *Endothyra*). By Pennsylvanian time one important group of Paleozoic Foraminifera, the fusulinids, were building calcareous tests in such numbers that they now constitute the larger part of some marine limestones. Similarly, the tests of species of *Globigerina* constitute vast quantities of sediment on the bottoms of existing oceans.

Tests with calcareous walls have several different kinds of microscopic structure to which specific terms have been applied. Great importance is attached to wall structure in assigning genera to the 50 or more families or to higher taxonomic categories.[1]

Cushman maintains that calcareous tests may develop directly from agglutinated forms by first having the calcium carbonate replace the "ferruginous" cement and then finally made to comprise the entire test, with complete exclusion of agglutinated foreign particles. He further holds that imperforate calcareous tests are more primitive than perforate ones. Galloway, on the other hand, considers that the agglutinated forms have never developed into any other type but represent rather the end products of several different lines of shell development. Calcareous tests are thought by him to have been formed directly from gelatinous forms. Only by chemical analyses of test substances and careful microscopic study of wall structure in both recent and fossil tests, together with all available morphological and stratigraphical information, can a decision on this argument be rendered.

Form and Architecture of Tests. The typical foraminiferal test consists fundamentally of a series of chambers arranged more or less symmetrically with respect to the initial globular chamber or proloculus. A few shells, such

[1] In an important recent article Alan Wood (1949) points out the taxonomic importance of the microstructure of the foraminiferal test.

FIG. 2-8. Foraminiferal tests showing shape, size, architecture, arrangement of chambers, nature of aperture, and character of test material. 1. *Saccammina* (agglutinated, ×7½). 2. *Lagena* (perforated, calcareous, ×37½). 3. *Lagena* (spinose, ×30). 4. *Lagena* (spinose, ×37½). 5. *Hyperammina* (×7½). 6. *Ammodiscus* (×5). 7. *Psammosphaera* (agglutinated, with single acerose sponge spicule, ×20). 8. *Reophax* (agglutinated, with sponge spicules, ×25). 9. *Pilulina* (agglutinated, fine felted sponge spicules, ×6). 10–11. *Hyperammina* (agglutinated, ×5). 12. *Rhadammina* (agglutinated, ×5). 13. *Sagenella* (agglutinated, ×5). 14. *Dendrophrya* (arborescent, agglutinated, ×15). 15. *Hormosina* (agglutinated, ×7½). 16. *Nodosaria* (smooth, calcareous, ×15). 17. *Textularia* (biserial, calcareous, ×12½). 18. *Verneuiliana* (triserial, calcareous, ×12½). 19. *Globigerina* (planispiral, calcareous, ×25). 20. *Endothyra* (calcareous, ×12–15). 21. *Lenticulina* (involute, calcareous, ×6). 22. *Pyrgo* (biloculine, calcareous, ×12½, ×20). 23. *Triloculina* (milioline, calcareous, ×40). 24. *Quinqueloculina* (quinqueloculine, calcareous, ×40). (20 *after Henbest*, 1944; 23, 24 *after Martinolli*, 1920; *all others after Brady*, 1884.)

as the flask-shaped *Lagena* (Fig. 2-8), have only one chamber, in which case proloculus and test are the same. Such a test is said to be monothalamous or unilocular. The majority of tests, however, are polythalamous or multilocular (Fig. 2-8).

Formation of Test. Probably the simplest of all tests is represented by *Lagena*, which is flask-shaped and unilocular, with the single opening or aperture at the end of the neck. Without the neck this test is exactly like the proloculus. Possibly more primitive, but not simpler, are the stellate tests of

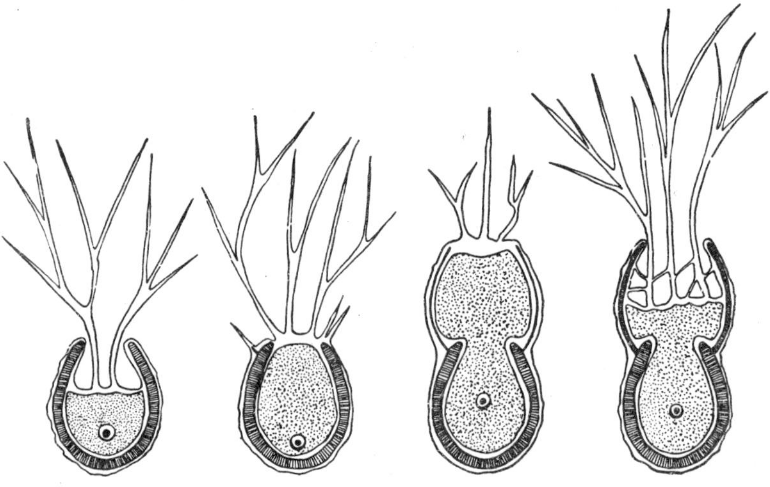

A B C D

FIG. 2-9. Order Foraminifera. Diagram showing the relation of a foraminifer (gamont) to its test at different stages of growth. In *A* the single chamber is the microspheric proloculus; in *B–D* another chamber is added. Note the outflowing pseudopodia, the thin layer of ectoplasm enclosing the test, and the position of the nucleus. If a third chamber were shown, the nucleus would lie in the middle chamber. (*From Thomson*, 1899, *after Dreyer, with modifications.*)

Rhabdammina (Fig. 2-8[12]), which are composed of a central body and radiating tubular arms made out of the material collected around the pseudopodia; or those of *Hyperammina*, in which the test is a tubular body (Fig. 2-8[10–11]). Slightly more complex than the single-chambered test is that which consists of an elongated, tubular chamber, usually coiled in some fashion about the proloculus (Fig. 2-8[6]).

After the first chamber of multilocular forms has been constructed, further growth of the organism causes the protoplasm to overflow through the aperture to the outside, where it collects and immediately begins to construct a new chamber, connected with the proloculus but not necessarily of the same shape or size (Fig. 2-9). If the protrusion of protoplasm is such that it surrounds the aperture in a symmetrical manner, the new chamber will surround the aperture of the proloculus in much the same manner as a collar. If the proloculus is entirely covered by the protruding protoplasm, then the new chamber will completely surround it. Collection of the protoplasm on only one side of the aperture will cause the new chamber to be formed in a corresponding position. It is obvious, therefore, that with growth many different arrangements of chambers result, depending upon the relations between the protruding protoplasm and the aperture of the proloculus. The overflowing of the protoplasm is not a random matter but follows a definite system for the individuals of each species, though the system is not always the same throughout the life of a given individual.

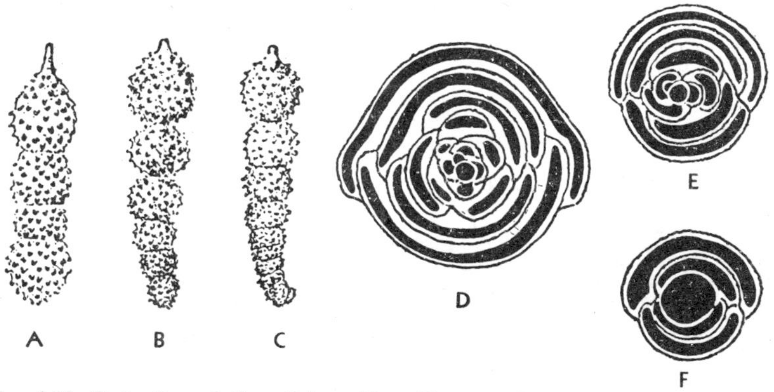

FIG. 2-10. Order Foraminifera. Trimorphism. The tests shown as *A*, *B*, and *C* occur together, and although they differ somewhat in form and have been given different generic assignments, they actually represent three forms of a single species and hence illustrate trimorphism. *A*. *Nodosaria aculeata* has a conspicuously large proloculus and but three additional chambers, all arranged in a straight line. *B*. *Dentalina floscula* has a smaller megalospheric proloculus and five additional chambers, all arranged along a slightly curved test axis. *C*. *Marginulina hirsuta* has a microspheric proloculus and nine or more chambers, with the first few somewhat compressed. Adult ornamentation is similar in all three. Only *C*, the microspheric form, has all the characteristics of the species, and hence it should be the one used in describing the species. Furthermore, it would seem best to drop the first two generic and specific terms and include all three under *Marginulina hirsuta*, as Cushman suggests.

The tests of *Idalina antiqua* (*D–F*) provide a second example of trimorphism in a somewhat more complex foraminifer. *D*. Section of a microspheric specimen with quinqueloculine early stage followed by triloculine and finally biloculine stage. *E*. Section of a microspheric specimen with triloculine early stage followed by a biloculine stage, the quinqueloculine stage being skipped. *F*. Section of a megalospheric specimen with the biloculine stage directly following the proloculus, the triloculine and quinqueloculine stages being skipped. For these three forms *Pyrgo* (*Biloculina*) is a good genus, inasmuch as it includes all three of the forms illustrated. (*After Cushman, with slight modifications*, 1948.)

In *Nodosaria* successive chambers are arranged in a linear series with the base of one chamber surrounding the aperture of the preceding (Figs. 2-8, 2-10*A*). Such a test is **uniserial.** In a **biserial** form the chambers are arranged in two rows or series (Figs. 2-8[17], 2-11*C*). This arrangement is illustrated in *Textularia* and has been designated **textularian.** **Triserial** forms have three chambers to a whorl, so that in such forms as *Verneuiliana* the chambers appear to be arranged in three series or rows (Fig. 2-8[18]). In *Robulus* the chambers are arranged about the proloculus in a planispiral fashion (Fig. 2-8), and the test is said to be **planispiral.** If the chambers are arranged along a spiral line which does not lie in a plane, the test is **trochoid.** A coiled test is **convolute (evolute),** if all coils are visible; or completely **involute,** if only the last chamber, or the chambers of the last coil or whorl, are visible. The chambers in a few shells are so arranged that they have a definite angular relation with the preceding chambers. In *Pyrgo* (Fig. 2-8) one chamber is 180° from the preceding chamber, in *Triloculina* the difference is 120°, and in *Quinqueloculina* the angular relation between adjacent chambers is 72°. Some foraminiferal tests closely resemble small bushes and are designated **arborescent.** Depending upon the shape, tests can also be described as **stellate** (star-shaped), **discoid** (disklike), **flabelliform** (fan-shaped), **pyriform** (pear-shaped), **fistulose** (having an abnormal opening into the test), etc. (see Fig. 2-8).

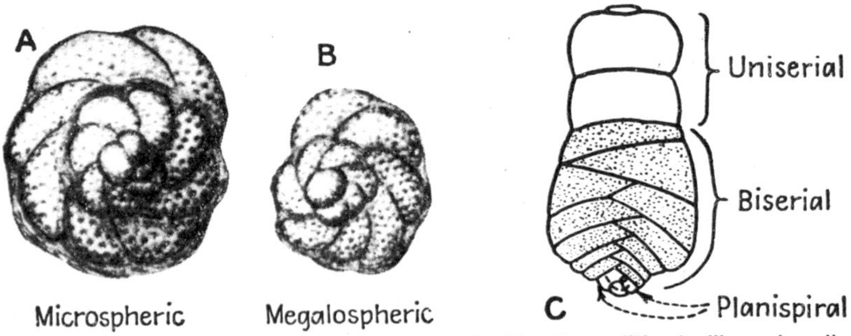

FIG. 2-11. Order Foraminifera. Test form. *A–B. Discorbis patelliformis*, illustrating dimorphism. *A* is the test of a microspheric agamont; *B* is the test of a megalospheric gamont. In this species the direction of rotation of microspheric tests is clockwise or right-handed (dextral); that of megalospheric tests, the reverse (sinistral). *C.* Diagrammatic sketch of *Vulvulina*, a test that shows the first four chambers planispirally arranged, as indicated by the dashed line, the next 12 (stippled) biserially arranged, and the last two uniserially arranged. (*A–B after Myers*, 1940; *C diagrammatic after Glaessner*, 1947.)

The arrangement of chambers commonly changes during the growth of an individual so that the adult test shows several types of architecture. *Vulvulina* (Fig. 2-11*C*), for example, adds the first 4 chambers planispirally, the next 12 biserially, and the last 2 uniserially. Development may be progressive or retrogressive. In certain forms, the chambers of the test are divided into smaller compartments (**chamberlets**) by a variety of dividing partitions, and the internal structure as a consequence becomes somewhat complicated (Fig. 2-13). Taking into account all the different architectural types of tests, as well as the combinations of the materials of which they are composed, it is almost unbelievable that single-celled organisms could ever have constructed such complex tests as are known.

The order in which the several types of tests have developed is not yet completely known. Within the life cycle of certain species, definite architectural types follow each other. The succession, however, is not always the same in different species; hence, it is not possible to consider such successions as evolutionary series. Until more data are available, this moot question must be left in its present unsatisfactory condition.

Apertures and Foramina. Every foraminiferal test has at least one conspicuous opening, or **aperture,** through which the pseudopodia can protrude and the reproductive bodies escape. The nature of the aperture is of great importance not only with regard to the formation of successive chambers, but also for taxonomic purposes (Fig. 2-12). Even though it changes in nature as the test develops, it is rather constant in the adult of a given species, and has proved to be an important feature in differentiating genera. Furthermore, the nature of the aperture at different growth stages of a single individual can be used in some cases to determine the evolutionary history of the species.

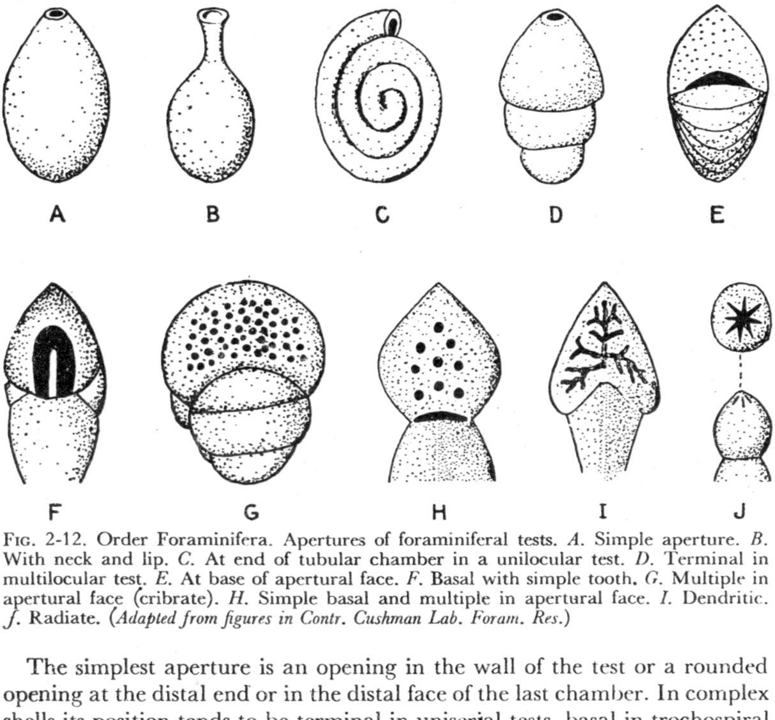

FIG. 2-12. Order Foraminifera. Apertures of foraminiferal tests. *A.* Simple aperture. *B.* With neck and lip. *C.* At end of tubular chamber in a unilocular test. *D.* Terminal in multilocular test. *E.* At base of apertural face. *F.* Basal with simple tooth. *G.* Multiple in apertural face (cribrate). *H.* Simple basal and multiple in apertural face. *I.* Dendritic. *J.* Radiate. (*Adapted from figures in Contr. Cushman Lab. Foram. Res.*)

The simplest aperture is an opening in the wall of the test or a rounded opening at the distal end or in the distal face of the last chamber. In complex shells its position tends to be terminal in uniserial tests, basal in trochospiral tests, and peripheral in planispiral forms. In shape the aperture ranges from a simple circular opening in primitive forms, through oval, crescentic, and star-shaped openings, to single slits or systems of slitlike openings. Commonly it is partly blocked by a simple or complex tooth, a plate, or a tube. In a few forms it is multiple, and in certain of these several distinct apertures differ from each other in shape and position (Fig. 2-12).

When a new chamber is added to the test, the apertural or distal face of the preceding chamber, with its single or multiple openings, becomes internal. The internal passages between successive chambers are designated **foramina** (singular, **foramen**), whence the name of the order, Foraminifera (foramina-bearing). Hence, the term aperture is properly applied only to the single or multiple opening in the distal wall of the last chamber.

Surface Ornamentation. The exteriors of many calcareous Foraminifera are highly ornamented with spines, nodes, knobs, costae, ridges, raised mesh-work, etc. These features are limited to certain chambers in some forms, whereas in others they cover the entire test. Commonly they are partly or entirely lost in the **gerontic** or old-age stage of growth.

Wall Structure. In addition to the ornamentation of the exterior, the wall of the foraminiferal test is thickened greatly in some species by the addition of shell matter to the exterior of the chambers, after they have been completed. This added material commonly gives the test a smoother surface.

Test walls that have tiny perforations are said to be **perforate,** whereas those that are solid are designated **imperforate.** The perforations are irregularly distributed in some tests but have a regular pattern in others. Both types of wall have been found in arenaceous as well as in calcareous tests, hence perforate and imperforate walls do not have the importance in classification that some authors have assumed (Wood, 1949).

The walls of foraminiferal tests exhibit a wide range of microscopic structure, and in most groups the nature of this structure is useful in classifying the different forms. On the basis of careful microscopical investigations, using polarized light, Wood (1949) was able to divide Foraminifera into the following main groups, of which only the Porcellanea and Fusulinidea are claimed to be natural:

A. Agglutinating Foraminifera. Test composed of foreign particles bound together with a cement secreted by the organism. Both perforate and imperforate walls have been noted.

B. Porcellanea. Original test wall composed of irregularly arranged crystals of calcite, equidimensional to somewhat elongate. Inasmuch as porcelaneous tests are rather susceptible to recrystallization, they are commonly altered on fossilization, undergoing several striking changes in the process. The characteristic brown color of the unaltered test completely vanishes, the wall becomes minutely granular, and the test has a characteristic porcelaneous appearance (like the surface of frosted glass) in reflected light. Although it is commonly stated that porcelaneous tests are imperforate, it is now known that both perforate and imperforate walls are developed.

C. Fusulinidea. Test wall composed of exceedingly fine, equidimensional calcite crystals that are subangular and closely fitted together. Inasmuch as this division of Foraminifera is extinct, it is impossible to say whether the tests (*e.g., Endothyra*) have recrystallized from an original minutely crystalline material secreted by the organism or were secreted in essentially their present state. The best evidence supports the former view, namely that the original tests were not agglutinated.

The wall structure of a typical fusulinid is basically a single primary wall, the **protheca,** to which is commonly added a veneer of dense and dark secondary deposition, the **epitheca,** on one or both surfaces. The spiral wall (**spirotheca**) of the test with **fusulinellid** structure has a protheca of thin diaphanous material, the **diaphanotheca,** which is finely perforate; a dark outer rindlike layer, the **tectum,** which is not always present; and an internal epithecal veneer, the **tectorium** (Fig. 2-13). Most Pennsylvanian and Permian fusulinids have a thick, rather coarsely porous protheca, the **keriotheca,** and a tectum.

The typical fusulinid test has a complicated internal structure consisting of axial and radial or spiral features, together with internal partitions and other features. A few of these are shown in Fig. 2-13.

D. Hyalina. Test wall composed of elongate calcite crystals with the *C* axis (*i.e.*, the longest axis) perpendicular to the surface of the chamber, so that a section of the test wall viewed under polarized light with Nicols crossed shows a black cross with concentric rings

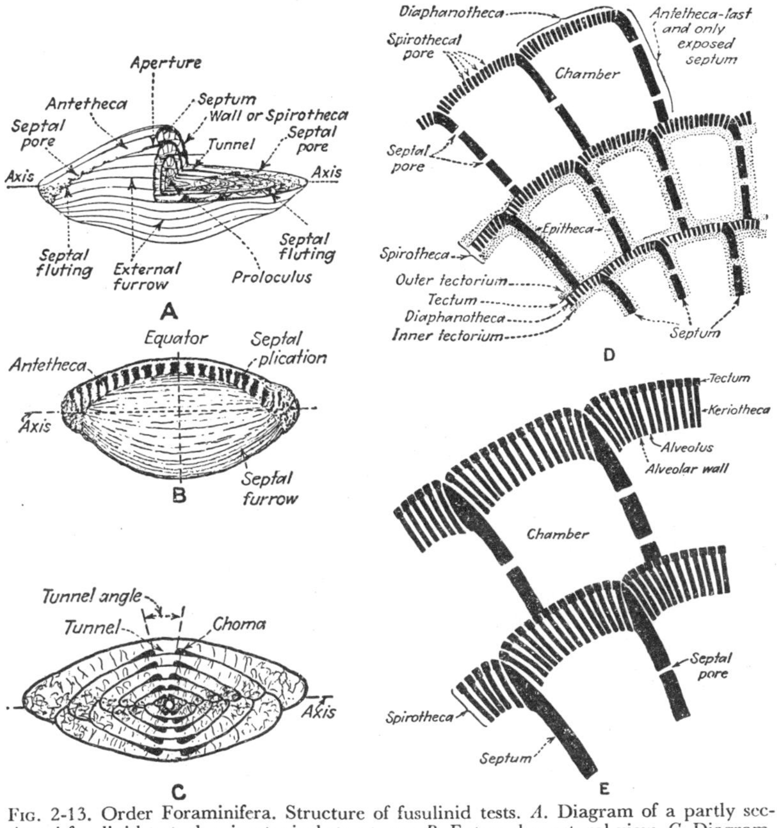

FIG. 2-13. Order Foraminifera. Structure of fusulinid tests. *A*. Diagram of a partly sectioned fusulinid test, showing typical structures. *B*. External apertural view. *C*. Diagrammatic axial section. *D*. Fusulinellid wall structure, much enlarged. *E*. Schwagerinid or keriothecal wall structure, much enlarged. (*A after Dunbar and Condra*, 1927; *B–E modified after Henbest*, 1944.)

of color. The test wall is generally transparent, in striking contrast to the typical porcelaneous test, but unfortunately some tests with hyaline structure lack transparency and may, without due caution, be confused with porcelaneous tests. Microscopic structure will immediately resolve any such confusion. The fact that hyaline tests are more generally perforated than those of the other groups may be connected in some way with the microstructure of the wall.

E. Abnormal Types. Two groups of Foraminifera cannot be assigned to the foregoing types because of their remarkably different wall structure.

Three closely related genera—*Spirillina*, *Patellina*, and *Patellinoides*—tend to have the test composed of a single crystal of calcite. In a few specimens the test has several large crystals with irregular boundaries, and it may even be irregularly fibrous in an occasional specimen of *Spirillina*.

Carteria spiculotesta is unique among animals in having its test composed of ellipsoidal spicules, each of which is a single crystal of calcite. Each crystal lies with its *C* axis parallel to the length of the spicule. It is thought that the spicules are directly secreted by the organism.

Classification of Foraminifera. The classification of any group of organisms should rest on certain well-defined and generally accepted biological and paleontological principles, among which are:

1. Every group of organisms has evolved through geologic time, undergoing accumulative modifications in the process, so that there is an actual succession of forms in time.
2. The order of development is as follows: simple, complex, specialized; or simple, complex, degenerate. Simple organisms do not seem to evolve from complex, specialized, or degenerate forms.

In addition to these general principles, certain features of the foraminiferal tests themselves are of prime importance in differentiating the several subdivisions of the order. Chief among these features, not necessarily in order of importance, are:

1. Nature of the test wall—composition of test material; microscopic structure; and general architecture of wall components.
2. Morphology of tests—shape, size, and arrangement of chambers; and internal architecture of entire test.
3. Nature and position of aperture.

The general tendency of leading students of the Foraminifera is to subdivide the order into families and lower taxonomic categories. Cushman's last classification, published in 1948, less than a year before his death, includes an even 50 families, to which might be added Henson's (1948) new family,[1] Meandropsinidae. Galloway (1933) proposed 35 families, many of which are the same as Cushman's, and Glaessner (1945) grouped his 37 families into seven superfamilies (as well as an additional superfamily without a fossil record). These are compared with Cushman's classification in Table 2-1. A representative genus of each of the more common of Cushman's 50 families is shown in Fig. 2-14.

Ecology and Paleoecology of Foraminifera. Most living Foraminifera are vagrant benthonic organisms which crawl about slowly over the bottom sediments, using their netlike pseudopodia to capture the microscopic organisms that serve as food and to select from the bottom the foreign particles that some ultimately build into arenaceous tests. Some forms, after an early free-moving

[1] In 1951 an additional family, Asterocyclinidae, was proposed (Bronnimann, P., 1951, A model of the internal structure of *Discocyclina* ss., *Jour. Paleontology*, vol. 25, pp. 208–211.), but it is not included in the present work. It is possible that still other recently proposed families have been overlooked because of the great volume of foraminiferal literature. Also see Thalmann (1952).

TABLE 2-1. APPROXIMATELY EQUIVALENT FAMILIES IN GLAESSNER'S (1947) AND CUSHMAN'S (1948) CLASSIFICATIONS

Glaessner's Classification Superfamily	*Glaessner's Classification* Family	*Cushman's Classification* Family
I. Astrorhizidea	1. Astrorhizidae	1. Allogromiidae 2. Astrorhizidae 3. Rhizamminidae 5. Hyperamminidae
	2. Saccamminidae	4. Saccamminidae
	3. Ammodiscidae	7. Ammodiscidae
II. Lituolidea	1. Reophacidae	6. Reophacidae
	2. Lituolidae	8. Lituolidae (part) 14. Loftusiidae 21. Placopsilinidae(?)
	3. Orbitolinidae	22. Orbitolinidae
	4. Textulariidae	9. Textulariidae
	5. Trochamminidae	20. Trochamminidae
	6. Verneuilinidae	10. Verneuilinidae 11. Valvulinidae
III. Endothyridea	1. Endothyridae	8. Lituolidae (part)
	2. Fusulinidae	12. Fusulinidae 13. Neoschwagerinidae
IV. Miliolidea	1. Miliolidae	17. Miliolidae
	2. Ophthalmidiidae	18. Ophthalmidiidae
	3. Peneroplidae	27. Peneroplidae
	4. Alveolinidae	28. Alveolinellidae
V. Lagenidea	1. Lagenidae	23. Lagenidae
	2. Polymorphinidae	24. Polymorphinidae
VI. Buliminidea	1. Buliminidae	30. Heterohelicidae (part) 31. Buliminidae
	2. Cassidulinidae	38. Cassidulinidae (part)
	3. Ellipsoidinidae	32. Ellipsoidinidae
	4. Chilostomellidae	39. Chilostomellidae
VII. Rotaliidea	1. Spirillinidae	33. Rotaliidae (part)
	2. Discorbidae	33. Rotaliidae (part) 43. Anomalinidae
	3. Globigerinidae	40. Globigerinidae 41. Hantkeninidae
	4. Globorotaliidae	42. Globorotaliidae
	5. Gümbelinidae	30. Heterohelicidae (part)
	6. Planorbulinidae	44. Planorbulinidae 45. Rupertiidae
	7. Cymbaloporidae	37. Cymbaloporidae
	8. Nonionidae	25. Nonionidae
	9. Ceratobuliminidae	38. Cassidulinidae (part)
	10. Amphisteginidae	35. Amphisteginidae
	11. Rotaliidae	33. Rotaliidae (part)
	12. Calcarinidae	36. Calcarinidae
	13. Miogypsinidae	50. Miogypsinidae
	14. Orbitoididae	48. Orbitoididae
	15. Discocyclinidae	49. Discocyclinidae
	16. Camerinidae	26. Camerinidae
	No categories	15. Neusinidae 16. Silicinidae 19. Fischerinidae 29. Keramosphaeridae 34. Pegidiidae 46. Victoriellidae 47. Homotremidae 51. Meandropsinidae

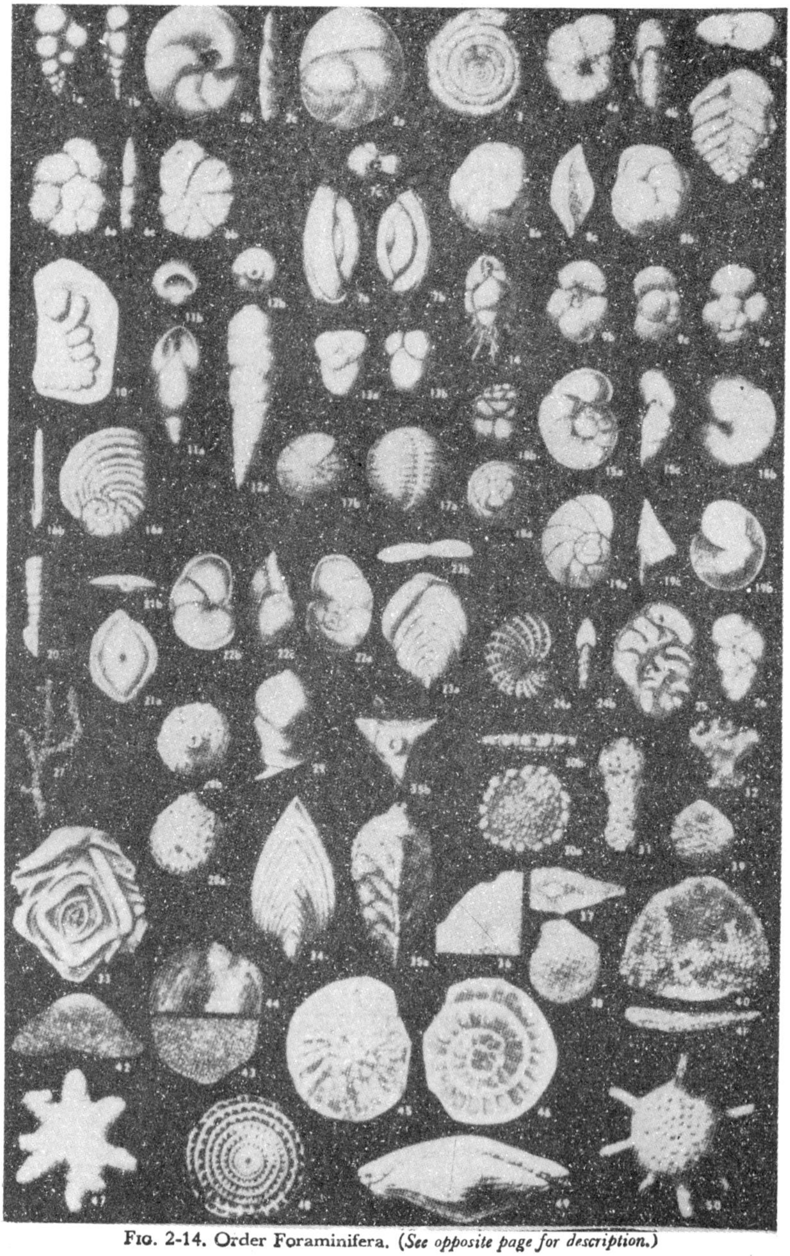

Fig. 2-14. Order Foraminifera. (*See opposite page for description.*)

stage, attach their tests to objects or other organisms on the bottom and remain fixed for part or all of the remainder of life.

Although only a few species, belonging to eight or nine genera, are planktonic and pelagic, these are represented in existing oceans by immense numbers of individuals. Tests of *Globigerina* (Fig. 2-3*B*) make up a large part of the bottom oozes over more than 48,000,000 square miles (35 per cent) of present ocean bottom.[1] Phleger found living specimens of *Globigerina* to be present in appreciable numbers at all depths from the surface down to 1,000 m. in the North Atlantic, and the bottom samples from depths of 3,000 to 5,000 m. in the same area contained great numbers of large and heavy-shelled tests of *Globigerina bulloides* and *Globigerinoides rubra*. A majority of the specimens found in most of the deep tows were empty tests which presumably were settling to the bottom from the overlying heavily inhabited layer.

There has been a growing interest in the ecology of living and fossil

[1] Crickmay *et al.* (Crickmay, G. W., Ladd, H. S., and Hoffmeister, J. E., Shallow-water *Globigerina* sediments, *Bull. Geol. Soc. Amer.*, vol. 52, pp. 79–106, 1941) reported (1941) a Neocene limestone from the Fiji Islands to consist of about 20 per cent by volume of the remains of Globigerinidae and to have accumulated in water probably not much deeper than 80 to 100 m. Further evidence that foraminiferal limestones can accumulate in shallow water has been presented by Myers (Biological evidence as to the rate at which tests of Foraminifera are contributed to marine sediments, *Jour. Paleontology*, vol. 16, pp. 397–398, 1942), who estimated (1942) that approximately 1,000 tests of *Elphidium crispum* (a calcareous shallow-water planktonic form) are contributed annually to every square foot of the bottom where they live.

Fig. 2-14. Representatives of 41 of the principal families of the Cushman classification. 1. *Gümbelina globulosa* (×60) (Heterohelicidae). 2. *Discorbis nitida* (×60) (Rotaliidae). 3. *Ammodiscus semiconstrictus regularis* (×45) (Ammodiscidae). 4. *Haplophragmoides canariensis* (×30) (Lituolidae). 5. *Textularia ripleyensis* (×30) (Textulariidae). 6. *Trochammina diagonis* (×25) (Trochamminidae). 7. *Quinqueloculina costata* (×25) (Miliolidae). 8. *Cassidulina laevigata* (×25) (Cassidulinidae). 9. *Globigerina bulloides* (×25) (Globigerinidae). 10. *Placopsilina redoakenesis* (×20) (Placopsilinidae). 11. *Pleurostomella clavata* (×20) (Ellipsoidinidae). 12. *Hippocrepina indivisa* (×20) (Hyperamminidae). 13. *Allomorphina trigona* (×20) (Chilostomellidae). 14. *Bulimina aculeata* (×20) (Buliminidae). 15. *Cibicides lobatulus* (×20) (Anomalinidae). 16. *Peneroplis planatus* (×20) (Peneroplidae). 17. *Borelis melo* (×20) (Alveolinellidae). 18. *Tretomphalus bulloides* (×20) (Cymbaloporidae). 19. *Asterigerina carinata* (×15) (Amphisteginidae). 20. *Clavulina parisiensis* (×15) (Valvulinidae). 21. *Rzehakina epigona lata* (×15) (Silicinidae). 22. *Globorotalia tumida* (×15) (Globorotaliidae). 23. *Polymorphina complanata* (×15) (Polymorphinidae). 24. *Elphidium lessonii* (×15) (Nonionidae). 25–26. *Eorupertia boninensis* (×4–6) (Victoriellidae). 27. *Rhizammina algaeformis* (×1.5–2.5) (Rhizamminidae). 28. *Saccammina spherica* (×4–6) (Saccamminidae). 29. *Carpenteria proteiformis* (×4–6) (Rupertiidae). 30. *Planorbulinella larvata* (×4–6) (Planorbulinidae). 31. *Reophax texana* (×8–10) (Reophacidae). 32. *Homotrema rubrum* (×1.5–2.5) (Homotremidae). 33. *Ophthalmidium inconstans* (×8–10) (Ophthalmidiidae). 34. *Frondicularia goldfussi* (×8–10) (Lagenidae). 35. *Tritaxia pyramidata* (×8–10) (Verneuilinidae). 36–37. *Discocyclina anconensis* (×8–10) (Discocyclinidae). 38. *Discocyclina anconensis* (×4–6) (Discocyclinidae). 39. *Miogypsina kotoi* (×4–6) (Miogypsinidae). 40–41. *Miogypsina kotoi* (×8–10) (Miogypsinidae). 42–43. *Orbitolinoides senni* (×4–6) (Orbitolinidae). 44. *Orbitolinoides senni* (×1.5–2.5) (Orbitolinidae). 45–46. "*Nummulites*" *senni* (×8–10) (Camerinidae). 47. *Astrorhiza arenaria* (×1.5–2.5) (Astrorhizidae). 48–49. *Triticites ventricosus* (×4–6) (Fusulinidae). 50. *Calcarina spengleri* (×8–10) (Calcarinidae). (*Prepared by Ruth Todd, Research Associate, Cushman Laboratory for Foraminiferal Research. All figures from Cushman's "Foraminifera,"* 1948, *after numerous authors, except the following:* 9, *after Cushman and Henbest,* 1940; 24, *after Cushman,* 1939; 36–38, *after Vaughan,* 1945; 39–41, *after Glaessner,* 1945; 42–44, *after Vaughan,* 1945; 45–46, *after Cizancourt,* 1948.)

Foraminifera in the last decade or so because of the possible implications fossil forms may have concerning depths, temperature, and other conditions in ancient seas.[1] It is much too early, however, to generalize on the relations of Foraminifera to such oceanographic factors as temperature, depth, salinity, bottom materials, currents, and productivity. Generalizations thus far made, with a few such exceptions as those of Norton, Natland, Martin, Myers, and Phleger, have been based largely on qualitative rather than quantitative data, with the result that they had better be evaluated critically (Myers, 1948).

It seems likely that only a small percentage of foraminiferal tests are preserved. They are generally fragile, hence easily broken. Calcareous tests dissolve as they sink in the deeper and colder waters and are not likely to attain depths exceeding 5,000 to 6,000 m. If the tests do reach the bottom there is still a good chance that they will be destroyed or at least broken by the action of scavengers and burrowing organisms. In the face of all these adverse conditions, however, vast quantities of tests do accumulate, as witness the extent of the *Globigerina* ooze over existing ocean bottoms and the common fusulinid and nummulitic limestones of the past, which show that similar accumulation of foraminiferal tests took place in the late Paleozoic and in the Tertiary.

Geologic History. Foraminifera are thought to have been present in Pre-Cambrian seas, but as yet no undoubted fossils of this order have been described from rocks older than the Cambrian. Reports of Pre-Cambrian Foraminifera have been discredited generally, because of doubt concerning the authenticity or identification of the alleged fossils or of uncertainty as to the true age of the strata (Raymond, 1935).

Cambrian Foraminifera have been reported from widely separated regions over the earth. Wood (1947) recently demonstrated that the supposed Upper Cambrian Foraminifera of the Malverns actually came from a fragment of Rhaetic or Lower Liassic limestone. This important discovery makes invalid many conclusions on the evolutionary sequence of foraminiferal types that had been based on the assumption that these "earliest known well-preserved foraminifera" were Cambrian.

Tiny globular objects believed to be fossil Foraminifera have been described from the Lower Cambrian of East Greenland by Poulsen, and by Howell and Dunn (1942) (Fig. 2-15*C*). They are questionably referred to the modern

[1] See the several articles on this subject by Cushman, Myers, and Phleger in the Reports of the Committee on Paleoecology, etc., of the National Research Council, 1937–1948.

The possibility of using certain ecological aspects of living Foraminifera to interpret paleoecological conditions of fossil forms has been demonstrated by Natland (The temperature—and depth—distribution of some recent and fossil Foraminifera in the Southern California region, *Bull. Scripps Inst. Oceanography, Univ. Calif., Tech. Ser.*, vol. 3, pp. 225–230, 1933); Myers (Ecologic relationships of some recent and fossil Foraminifera, *Nat. Res. Council, Rept. Comm. Marine Ecology as Related to Paleontology*, 1941–1942, pp. 31–36, 1942); and Hedberg (Some Recent and fossil brackish to fresh-water Foraminifera, *Jour. Paleontology*, vol. 8, pp. 469–476, 1934). Hedberg found that certain species of arenaceous and calcareous Foraminifera live in waters of low salinity in Venezuela and that these species also are present as fossils in the same region in Tertiary sediments. From these findings he concluded that the ancient sediments, previously suspected of brackish-water origin for other reasons, were probably deposited in water of low salinity.

genus *Psammosphaera*. Similar objects have been found in the Lower Cambrian Forteau formation of Labrador. Inasmuch as the calcified objects left an arenaceous residue when dissolved in weak hydrochloric acid, it was concluded that the "substance of which the original test of the 'foraminifera' was composed was evidently 'chitin' (tectine), containing a small amount of arenaceous material." If these are actually fossil Foraminifera, they seem to be the oldest yet known.

An additional Cambrian record is that of a peculiar chitinous and filamentous dendritic structure from the Upper Cambrian of Wisconsin that has been

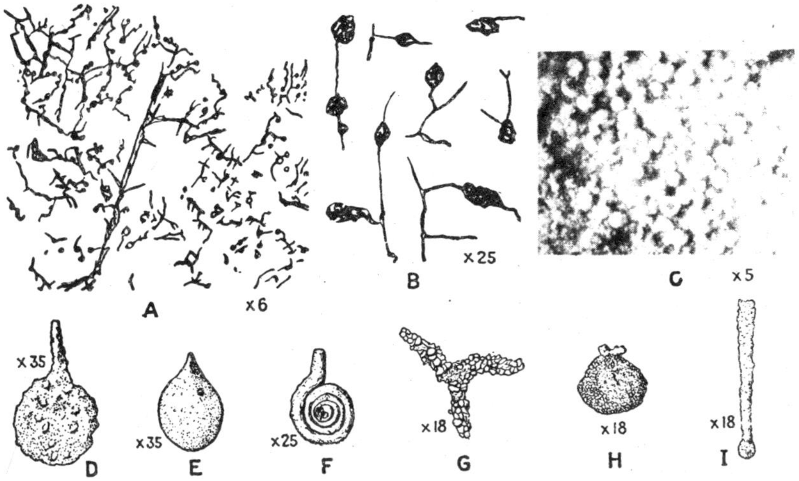

FIG. 2-15. Order Foraminifera. Earliest foraminifers. *A–B. Chitinodendron*, from the Upper Cambrian of Wisconsin (*A*) and the Baltic Silurian (*B*), a supposed chitinous foraminifer consisting of small bladders connected with tiny tubes. *C. Psammosphaera?*, a supposed globular foraminifer from the Lower Cambrian of Labrador. The tiny globular objects are composed of calcite but leave a fair amount of sandy residue when dissolved in weak hydrochloric acid. Similar objects from the Ordovician of Manchuria left thin round tests when treated in the same manner. *D–I.* Arenaceous foraminifera from the Ordovician and Silurian of Oklahoma: *D–E, Lagenammina; F, Ammodiscus; H, Saccammina;* and *I, Hyperammina*, all from the Silurian Chimney Hill limestone of Oklahoma; *G, Rhabdammina* from the Ordovician Viola limestone of Oklahoma. (*A after Ruedemann and Shrock*, 1939; *B from A, after Eisenack; C after Howell and Dunn*, 1942; *D–I after Moreman*, 1933.)

provisionally assigned to the living genus *Chitinodendron* (Fig. 2-15*A–B*) (Ruedemann and Shrock, 1939). Other Cambrian records similarly are open to some doubt for one reason or another.

The earliest undoubted Foraminifera now seem to be simple arenaceous tests described by Moreman (1930; 1933) from the Lower Ordovician Arbuckle, Upper Ordovician Viola, and Middle Silurian Chimney Hill limestones of Oklahoma (Fig. 2-15*D–I*). The variety and abundance of these earliest arenaceous forms indicate clearly that by Ordovician time the Foraminifera had evolved many genera and species that could make agglutinated tests. Since Moreman's discovery several dozen genera and more than a hundred species of arenaceous Foraminifera have been described from Ordovician, Silurian, and Devonian rocks of North America and other continents.

It would seem from the known fossil record that the first great evolutionary advance of the Foraminifera came in the earlier Paleozoic when the organisms, probably up to that time having only chitinous tests, started to agglutinate foreign particles to the primary tests, thus developing the simplest of the arenaceous forms (Astrorhizidea and Lituolidea of Glaessner) now known to be present in abundance in Ordovician, Silurian, and Devonian rocks. The Astrorhizidea seem to have reached a climax in evolution in Recent times, but this may be due to the fact that the fossil record, which is spotty, is incomplete. The Lituolidea, which include the multichambered septate arenaceous Foraminifera, seem to have reached their climax in the Cretaceous, declining thereafter through the Tertiary to Recent time.

In the later Paleozoic the simple arenaceous forms of the earlier periods were overshadowed by the second large and important group of Foraminifera, the calcareous fusulinids (Endothyridea of Glaessner), which became extinct by the end of the Paleozoic. The earliest of these, *Endothyra*[1] (Fig. 2-8[20]), first appeared and became abundant in the Mississippian. It is thought to have been the ancestor of the great line of fusulinids that characterize Pennsylvanian and Permian rocks the world over. At least eight major fusulinid faunal zones have been recognized in North America and two others in the Eastern Hemisphere (Thompson, 1948). Typical genera are *Millerella*, *Fusulinella*, *Triticites*, *Pseudoschwagerina*, and *Parafusulina*. Fusulinids are excellent index fossils.

The third great evolutionary surge in the history of the Foraminifera started in the early Mesozoic, after the once-abundant fusulinids had vanished from the scene at the close of the Permian, and reached flood tide in the Upper Cretaceous. By this time, all but a few of the existing families had been differentiated, though some were not to reach their climax until later in the Cenozoic. Outstanding among the Mesozoic Foraminifera were the lituolids mentioned previously, the lagenids (Lagenidea), which reached their climax in the Jurassic, but are still represented in modern seas, and the rotaliids (Rotaliidea), which had differentiated into 15 families by the close of the Mesozoic, though 10 of these did not attain maximum development until sometime in the Cenozoic.

Just as Paleozoic foraminiferal faunas differ from those of the Mesozoic, so those of the Mesozoic differ from Cenozoic faunas, though not so strikingly because many Mesozoic families and genera continue into the Cenozoic.

The latest step of great importance in the evolution of the Foraminifera was taken in the late Mesozoic when planktonic forms became abundant and widespread. These and many benthonic species attained world-wide distribution in the widespread seas of the Cenozoic, where their tests accumulated in vast numbers to make thick foraminiferal limestones. The blocks of the

[1] The Devonian species of *Endothyra*, *E. gallowayi* Thomas, has been made the genotype of a new genus, *Nanicella*, by Henbest (*Jour. Wash. Acad. Sci.*, vol. 25, pp. 34–35, 1935). Although this species seems to be related to *Endothyra* and the fusulinids, further details are needed before its true relationships can be established.

great pyramids of Gizeh, near Cairo, are composed of the calcareous tests of *Nummulites*, a large benthonic form that flourished in the Tethys Sea which separated Europe from Africa for much of the Cenozoic. *Lepidocyclina*, one of the largest of all calcareous Foraminifera, attained world-wide distribution between 45° N. lat. and 40° S. lat. in the Eocene and Miocene (Fig. 2-16). In recent geologic time the chief contributors to bottom deposits have been the several species of the pelagic genus, *Globigerina*, whose calcareous tests constitute a large percentage of the globigerina ooze.

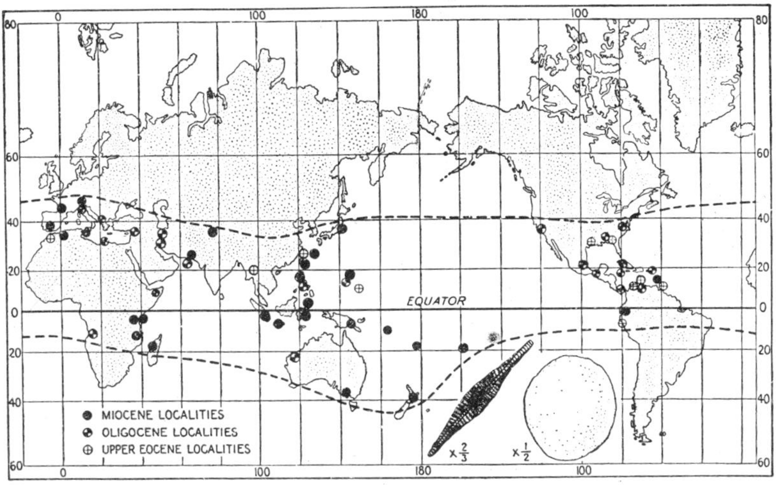

Fig. 2-16. Order Foraminifera. The geographic distribution of *Lepidocyclina* (*sensu lato*), a world-wide index fossil of the Eocene and Oligocene, with inset showing nature of test diagrammatically. (*Locality and other data from Schenck and Childs*, 1942.)

Uses of Foraminifera. Fossil Foraminifera have increased in importance so rapidly during the last 30 years that today they constitute one of the most vigorously studied, widely used, and fully classified orders in the animal kingdom. Many of the 18,000 to 20,000 so-called species are superb index fossils (Fig. 2-17), and not a few genera have proved to be world-wide in their distribution. Because of their value for correlation of strata locally, regionally, and from continent to continent, they have found wide use by petroleum micropaleontologists. With the recognition that ecological conditions favoring foraminiferal growth can be used to determine paleoecological conditions on ancient bottoms, a vigorous program of study is now in progress to exploit this interesting aspect of Foraminifera. Finally, since it has been shown that certain species of *Globigerina*, as well as of other genera, seem to be adjusted to definite temperature conditions in existing oceans, cores from the ocean bottom have been studied eagerly to see if the several glacial periods are reflected by changes in the nature of the foraminiferal content (Phleger, 1939). Certain colder and warmer faunas have been found; much additional information of similar nature is certain to be obtained as the coring devices penetrate deeper and deeper into the sediments on the ocean bottom.

Fortunately for American students of fossil Foraminifera, the great Cushman Laboratory for Foraminiferal Research, with its magnificant collection and library, brought together through a quarter of a century by its founder, the late Joseph Augustine Cushman, has been moved to the National Museum in Washington. There it is a part of one of the world's greatest collections of Foraminifera, and is available to qualified investigators (Henbest, 1947).

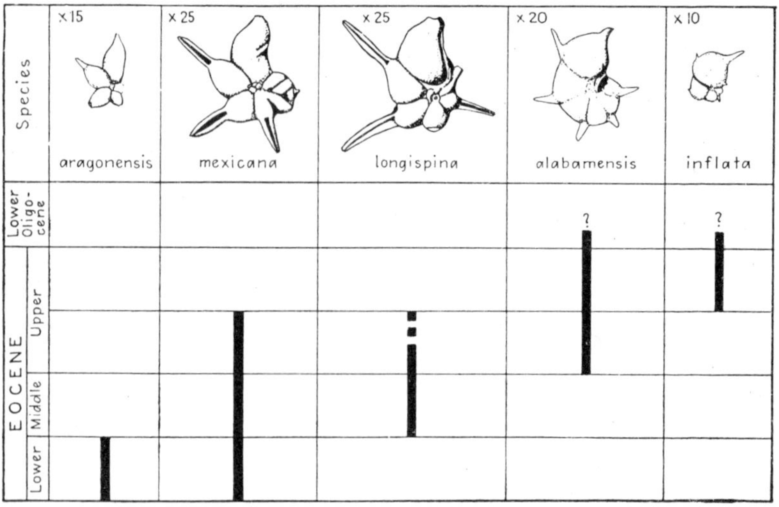

FIG. 2-17. Order Foraminifera. Stratigraphic distribution of distinctive species of *Hantkenina*. The geographically wide-ranging species of this genus are excellent index fossils for the Eocene. (*A after Nuttall*, 1930; *B–D after Cushman*, 1924; *E after Howe*, 1928; *also see Thalmann*, 1942, *and Glaessner*, 1945.)

Order 6. Heliozoa. Heliozoans are spheroidal protozoans with radiating axopodia. They resemble the radiolarians but differ from them in lacking the central capsule. The body is naked, covered with a gelatinous mantle, or provided with a spheroidal latticelike test that has spinelike spicules in some forms (Fig. 2-3*C*). Sand grains, diatom shells, and other foreign bodies are commonly attached to the exterior surface of the organism. Siliceous scales and spines are found on a few forms.

There seems to be no undoubted record of fossil heliozoans. Since most living forms that have preservable test material inhabit fresh-water environments, the probability of their leaving a fossil record is small.

Order 7. Radiolaria. Radiolarians are exclusively marine, planktonic organisms characterized by radiating filamentous pseudopodia (whence the name *Radiolaria*, radiating) and a central capsule (Fig. 2-18*A*). They live mostly in tropical waters at or near the surface and build an amazing variety of delicate microscopic tests of opaline silica (Figs. 2-18, 2-19, 2-20). When the siliceous tests are abandoned they sink slowly to the bottom, where they form important deposits if other substances are not present in sufficient quantity to mask them. These deposits are known as "radiolarian ooze" and cover some 2 to 3 per cent (2,000,000 sq. miles) of the present ocean bottoms. Radiolarian ooze can accumulate on any bottom, but usually it is masked by other sediments on bottoms of depths less than 3,750 m. (12,000 ft.). At greater depths foraminiferal and other soluble shells go into solution before reaching the bottom so that the tests of Radiolaria constitute much or most of the deposits. Radiolarian ooze resembles diatomaceous ooze formed by minute plants (**diatoms**), but it is exclusively marine, whereas the diatomaceous ooze may be formed in either fresh or salt water. The shells are always microscopic in size and almost always siliceous. Rocks composed dominantly of radiolarian tests are known as **radiolarite.**

The Organism. The body of the radiolarian, ordinarily spherical or spheroidal in shape, is composed of cytoplasm that can be differentiated into three parts: (1) a relatively thick **extracapsular layer,** (2) an **intracapsular** nucleated part, and (3) a **central capsule** that divides 1 from 2 (Fig. 2-18). The central capsule is a unique feature and distinguishes the Radiolaria from all other Protozoa.

The extracapsular cytoplasm is concerned with flotation, nutrition, the reception of stimuli, and secretion of the siliceous skeleton. It is divisible into an inner **assimilative layer** next to the central capsule; an intermediate layer, the **calymma,** that is flotational in function and also has the power to deposit silica; and a permeable enveloping membrane, the **sarcodictyum.** Radiating contractile pseudopodia arise in the extracapsular layer and extend through all three divisions.

The central capsule is generally a single, porous membrane of chitinoid or mucoid material. It is a more or less permanent structure and reflects the configuration of the test of the animal (Fig. 2-19). It is one of the important features in the classification of Radiolaria.

The intracapsular cytoplasm is a granular mass of several complex substances with one or more nuclei. It serves chiefly for food storage and for reproduction.

The Test. According to Haeckel, radiolarian tests show greater diversity of form than any other order of the animal kingdom. More than 5,000 living species, and perhaps as many as 1,000 fossil species, have been described. These are included in some 740 genera. It is interesting to note here that almost all fossil species, even those as old as Lower Cambrian, have been assigned to existing genera, indicating for the Radiolaria a remarkable persistence of characteristics through time.

As can be seen from Figs. 2-18 and 2-19 the radiolarian test is actually an internal skeleton for the most part. It generally exhibits marked symmetry about a point, a line, or a plane, but it may lack any apparent symmetry. The same basic shapes appear in tests of different materials.

The radiolarian test is generally composed of opaline silica (Spumellaria and Nassellaria), but in one group, the Phaeodaria, silica is combined with an organic compound and in another, the Acantharia, the skeletons are com-

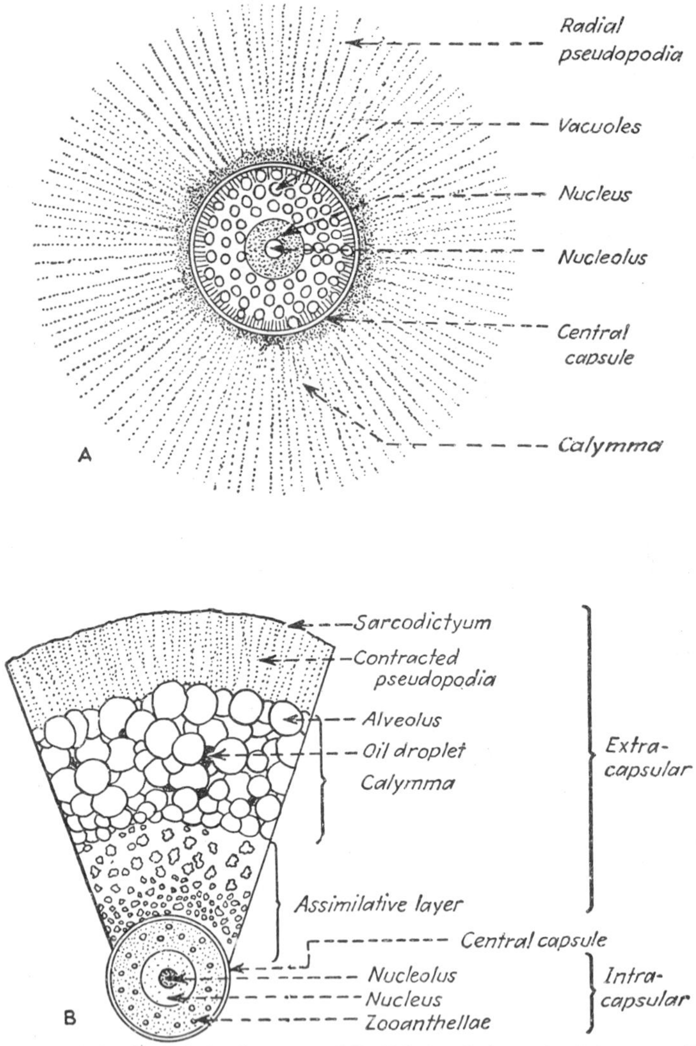

FIG. 2-18. Order Radiolaria. Structure of Radiolaria. *A*. An entire living spumellarian (*Actissa*), showing the spherical central capsule containing finely granulated cytoplasm, which is radially striated in the cortical zone. The radial pseudopodia pierce and extend beyond the calymma (shaded). *B*. Generalized diagram showing relationships of the several anatomical parts. (*A after Haeckel*, 1887; *B from Aberdeen*, 1940, *after Haeckel*, 1887.)

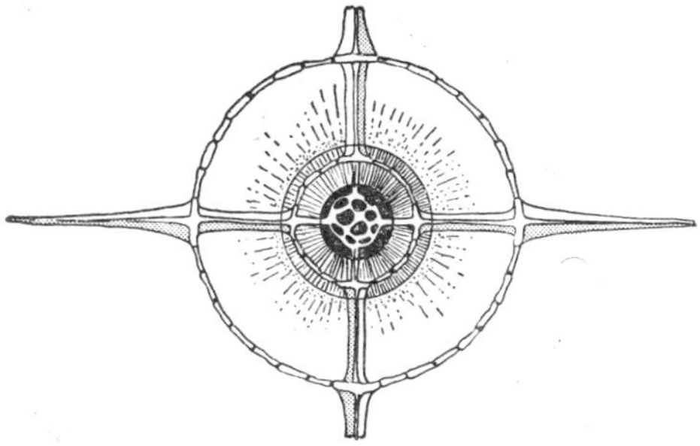

FIG. 2-19. Order Radiolaria. Diagram showing relation of a radiolarian to its test. The nucleus (in black) surrounds an inner shell; the cytoplasm of the central capsule is radially striped; and radial pseudopodia extend outward from the central capsule in life, but in the drawing they are shown contracted. (*After Hertwig*, 1879).

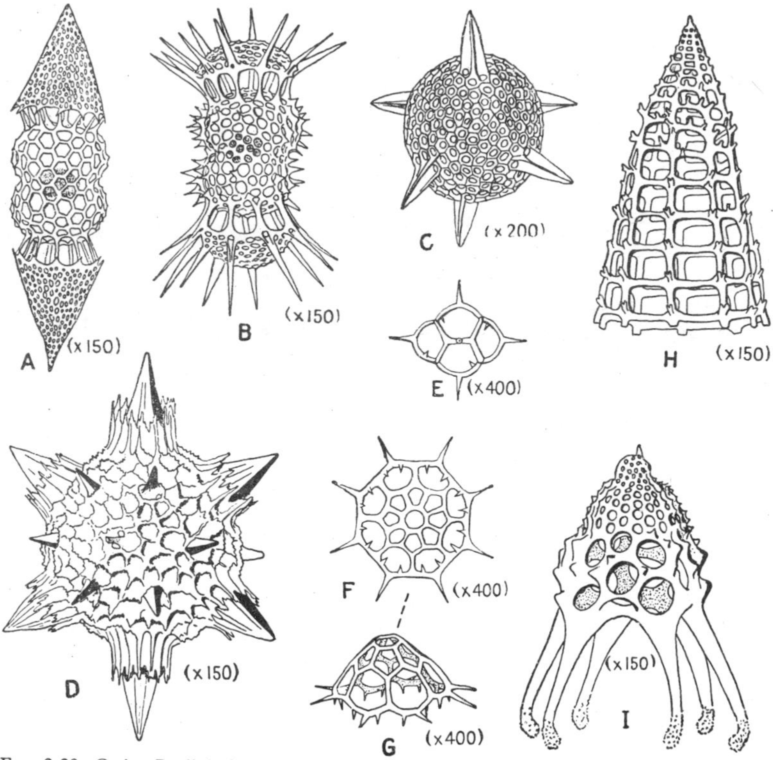

FIG. 2-20. Order Radiolaria. Tests of living genera. *A–C.* Spumellaria: *A*, *Panartus; B*, *Panicium; C*, *Hexastylus. D. Hexaconus*, a representative of the Acantharia. *E–G.* Phaeodaria: *E*, *Dictyocha*, a single piece of the spicular skeleton: *F–G. Cannopilus*, a complex spicule viewed from below and from the side. *H–I.* Nassellaria: *H*, *Bathropyramis; I*, *Alacorys.* (*Redrawn from Haeckel*, 1887.)

posed of strontium sulphate.[1] Representatives of these four groups of tests are shown in Figs. 2-19 and 2-20.

In every case the skeletal matter is secreted by the organism. It takes the form of (1) loosely woven rods, spicules, and spines either free or fused together (Fig. 2-20*E*); (2) a comparatively rigid meshwork of globular, conical, stellate, or discoidal shape, with the peripheral parts in some tests produced into simple or complex spines (Fig. 2-20*C*); and (3) a series of concentrically arranged perforated shells of different shapes, held together with minute supporting rods and spines (Fig. 2-20*B*).

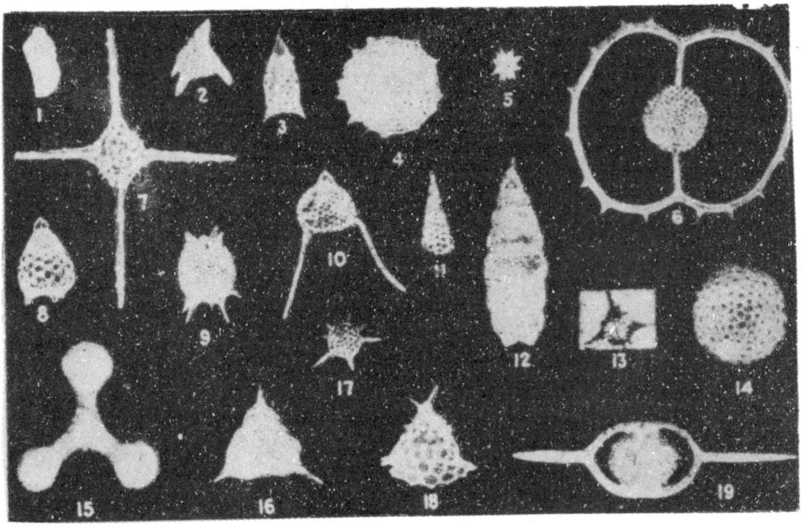

FIG. 2-21. Order Radiolaria. Fossil forms. 1–3, 8, 11, 12. Cretaceous Nassellaria. 4, 6–7, 9, 16–17, 19. Cretaceous Spumellaria. 5. Miocene Acantharia. 10, 18. Miocene Nassellaria. 14–15. Miocene Spumellaria. 13. Devonian Spumellaria. (1–12, 14–19 *after Campbell*, 1944, *and Campbell and Clark*, 1944; 13 *after Aberdeen*, 1940.)

Geologic History. The geologic record of the Radiolaria, although extending over a great length of time, is a rather meager one, especially when compared with that of the Foraminifera. The earliest fossil radiolarians were reported by Barrois (1892) and described and illustrated by Cayeux (1894) from supposed Pre-Cambrian Algonkian rocks of Saint-Lo, Brittany. Although the illustrations of the fossils leave little doubt about their authenticity as radiolarians, there is doubt as to whether or not the artist saw what he drew (Raymond, 1935, p. 379). Other aspects of the supposed fauna are such as to discredit the Radiolaria altogether. The supposed Pre-Cambrian Radiolaria reported from Australia turned out to be associated with *archaeocyathids*, hence Lower Cambrian (David, T. W. E., and Itowchin, W., 1896). There is no doubt, therefore, that Radiolaria were present in the earliest Cambrian seas, and possibly also in Pre-Cambrian seas. Beginning with the Cambrian,

[1] ODUM, H. T., 1951, Notes on the strontium content of sea water, celestite Radiolaria, and strontianite snail shells, *Science*, vol. 114, pp. 211–213.

records of fossil radiolarians are scattered through the Paleozoic, Mesozoic, and Cenozoic strata (Fig. 2-21) generally in rocks high in silica, such as siliceous shales and cherts. Only in radiolarites are skeletons abundant. Fossil Radiolaria, unlike the Foraminifera, have not been found useful for age determination or correlation, partly because of the fact that fossil species are much like existing ones, and partly because so few forms have been found.

It was once thought that radiolarites and other rocks rich in radiolarians were deep-sea deposits (*i.e.*, of the order of 3,700 + m., or 12,000 + ft.), but in recent years certain radiolarian faunas have been shown to have been deposited in only moderately deep or even relatively shallow waters. They have also been found in beach sands. It follows, therefore, that every aspect of a radiolarian fauna and its enclosing rock should be examined critically before conclusions are drawn concerning the depth of water in which the skeletons accumulated.

Class Sporozoa

The Sporozoa are exclusively parasitic and reproduce by means of spores. They possess neither cilia nor flagella and, except for a few forms which are able to move about by means of pseudopodia, lack the power of locomotion. Since the organism does not possess any preservable skeleton or test, a fossil record is not probable. Should the host of a sporozoan be perfectly preserved, however, as is the case with the hairy mammoths in the frozen tundra of Siberia, it is conceivable that the sporozoans, as inhabitants of the fossilized host, might also be preserved. Thus far, however, no undoubted fossil sporozoan seems to have been reported.

Class Ciliata

This class includes protozoans that bear cilia throughout life, live in all kinds of water, and build an organic test, the **lorica,** which has been reported fossil in a few cases. One group, the tintinnids of Order Oligotricha, construct an organic lorica which has the shape of a baseless hurricane globe (Fig. 2-22). The lorica wall, which is composed of a complex organic substance of unknown chemical composition, commonly includes or becomes incrusted with foreign particles which give it an agglutinated texture. Fossil tintinnids have been found in Pleistocene peat bogs and in fine-grained limestones of Upper Jurassic and Lower Cretaceous age in the Mediterranean region and The Himalaya (Colom, 1948). The loricas in the limestones have been completely replaced with calcite. Fossil tintinnids do not seem to have been reported from North America but should be sought for, particularly in fine-grained limestones and calcareous shales of marine origin, where they may be expected to be associated with radiolarians and coccoliths.

Class Suctoria

Suctorians are ciliated protozoans, commonly grouped with the Ciliata under the term Infusoria, which lose their cilia after an early free-moving stage, develop tentacles, and become attached. Marine forms are generally without test material (although one genus, *Coleps*, does have a test of platelets); a few fresh-water forms, however, agglutinate sand grains and other

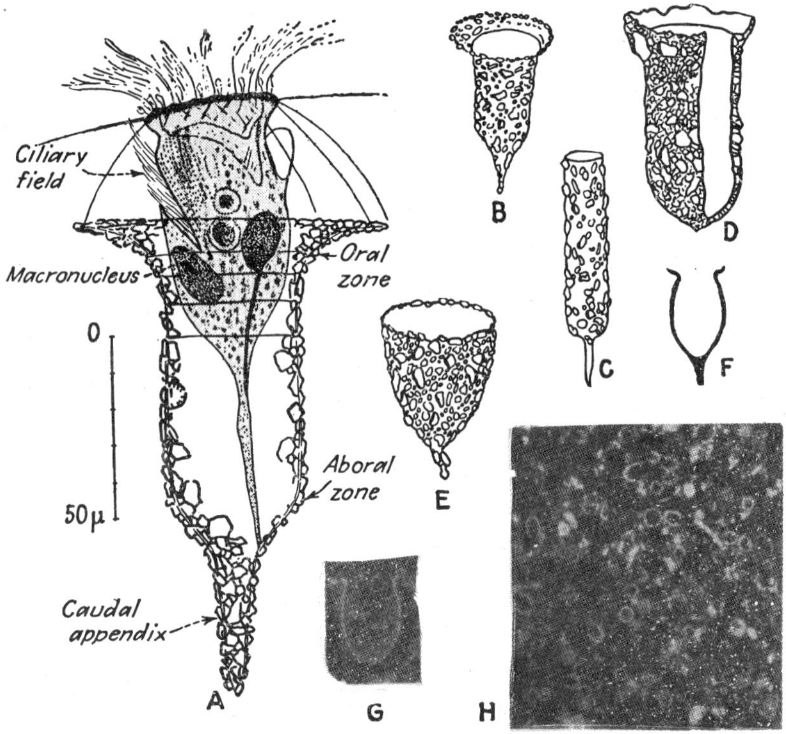

FIG. 2-22. Class CILIATA. Living and fossil tintinnids. *A.* Diagram showing relation of organism to agglutinated test in *Tintinnopsis campanula*, a living species. *B–E.* Tests of several species of *Tintinnopsis* showing range in shape and size (all figures ×175). *F–G.* Fossil loricas of *Tintinnopsella carpathica* from Tithonian (U. Jura.) and Neocomian (L. Cret.) of Majorca (*F* is ×250; *G*, a thin section, is ×335). *H.* Thin section of Tithonian limestone from Majorca showing tests of *Calpionella alpina* and *C. elliptica* (×135). (*A after Fauré-Fremiet*, 1924; *B–D after Colom*, 1948; *E from Colom*, 1948, *after Daday*, 1887; *F–H after Colom*, 1948.)

foreign particles to form radiate tests. No undoubted fossil suctorians seem to have been described, so nothing is known about the geologic history of the class.

REFERENCES[1]

General

BARROIS, C. 1892. Sur le présence de fossiles dans le terrain azoique de Bretagne. *Compt. rend. acad. sci.*, vol. 115, pp. 326–328.

BRADLEY, W. H. 1931. Origin and microfossils of the oil shale of the Green River formation of Colorado and Utah. *U.S. Geol. Surv., Prof. Paper* 168, p. 41.

[1] No attempt is made to list an extensive bibliography for this or following phyla. Only a few of the comprehensive works are included along with certain articles of special importance in the present discussion. Long reference lists will be found in several of the works cited; a few recent or specially important references are cited in the footnotes.

CAYEUX, M. L. 1894, Les preuves de l'existence d'organismes dans le terrain Précambrien. *Bull. soc. geol. France*, ser. 3, vol. 22, pp. 197–228.
GLAESSNER, M. F. 1945–1947. "Principles of Micropalaeontology." New York, John Wiley & Sons, Inc., 296 pp., 1947. (First published September, 1945, by Melbourne University Press, Carlton, Victoria, Australia.)
HANNA, G. D. 1928. Silicoflagellata from the Cretaceous of California. *Jour. Paleontology*, vol. 1, pp. 259–263.
HYMAN, L. H. 1940. "The Invertebrates." New York, McGraw-Hill Book Company, Inc. Vol. I, "Protozoa through Ctenophora." 720 pp. (Protozoa, pp. 44–232, Figs. 6–67.)
KUDO, R. R. 1946. "Protozoology," 3d ed. Springfield, Ill., Charles C Thomas. 778 pp.
MORET, L. 1940. "Manuel de paléontologie animale." Paris, Masson et Cie.
RAYMOND, P. E. 1935. Pre-Cambrian life. *Bull. Geol. Soc. Amer.*, vol. 46, pp. 379–380.
SHIMER, H. W., and SHROCK, R. R. 1944. "Index Fossils of North America." New York, John Wiley & Sons, Inc.; Cambridge, Mass., The Technology Press, 837 pp.
STORER, T. I. 1951. "General Zoology," 2d ed. New York, McGraw-Hill Book Company, Inc., 832 pp.
SWINNERTON, H. H. 1947. "Outlines of Palaeontology," 3d ed. London, Edward Arnold & Co., 393 pp.
WETZEL, O. 1933. Die in organischer Substanz erhaltenen Microfossilien des Baltischen Kreide-Feuersteins. *Palaeontographica*, vol. 77, abt. A, pp. 141–186.

Foraminifera

BRADY, B. H. 1884. Report on the Foraminifera dredged by *H.M.S. Challenger* during the years 1873–1876. *Zool.*, vol. 9, 814 pp.
CRICKMAY, G. W., LADD, H. S., and HOFFMEISTER, J. E. 1941. Shallow-water Globigerina sediments. *Bull. Geol. Soc. Amer.*, vol. 52, pp. 79–106.
CUSHMAN, J. A. 1937. Paleoecology of the Foraminifera. *Nat. Res. Council, Rept. Comm. Paleoecology*, 1936–1937, pp. 7–10.
——— 1948. "Foraminifera, Their Classification and Economic Use," 4th ed., revised and enlarged, with an illustrated key to the genera. Cambridge, Mass., Harvard University Press. 605 pp.
DUNBAR, C. O., and HENBEST, L. G. 1942. Pennsylvanian Fusulinidae of Illinois, etc. *Ill. Geol. Surv., Bull.* 67, 218 pp.
ELLIS, B. F., and MESSINA, A. R. 1942. "A Catalogue of the Foraminifera." New York, American Museum of Natural History, 30,000 pp.
GALLOWAY, J. J. 1933. "A Manual of Foraminifera." Bloomington, Ind., The Principia Press, Inc. 483 pp.
GLAESSNER, M. F. 1945–1947. "Principles of Micropalaeontology." New York, John Wiley & Sons, Inc. 296 pp., 1947. (First published September 1945, by Melbourne University Press, Carlton, Victoria, Australia.)
HENBEST, L. G. 1947. National collections of Foraminifera [U.S. National Museum, Washington, D.C.; Cushman Laboratory for Foraminiferal Research, Sharon, Mass.]. *Bull. Amer. Assoc. Petrol. Geol.*, vol. 31, pp. 2050–2054.
HENSON, F. R. S. 1948. "Larger Imperforate Foraminifera of South-Western Asia. Families Lituolidae, Orbitolinidae and Meandropsinidae." London, British Museum (Natural History). 127 pp. (Reviewed by J. A. Cushman in *Science*, vol. 109, p. 415, 1949.)
HOWELL, B. F., and DUNN, P. H. 1942. Early Cambrian "Foraminifera." *Jour. Paleontology*, vol. 16, pp. 638–639.
MOREMAN, W. L. 1930. Arenaceous Foraminifera from Ordovician and Silurian limestones of Oklahoma. *Jour. Paleontology*, vol. 4, pp. 42–59.
———. 1933. Arenaceous Foraminifera from the Lower Paleozoic rocks of Oklahoma. *Ibid.*, vol. 7, pp. 393–397.

Myers, E. H. 1942. Biological evidence as to the rate at which tests of Foraminifera are contributed to marine sediments. *Jour. Paleontology*, vol. 16, pp. 397–398.

———. 1942*a*. A quantitative study of the productivity of the Foraminifera in the sea. *Proc. Amer. Phil. Soc.*, vol. 85, pp. 325–342.

———. 1943. Biology, ecology, and morphogenesis of a pelagic foraminifer. *Stanford Univ. Pub., Biol. Ser.*, vol. 9, pp. 1–30.

———. 1948. Bibliography on the ecology of the Foraminifera. *Nat. Res. Council, Rept. Comm. Treatise Marine Ecology and Paleoecology*, 1946–1947, no. 7, pp. 28–37. (This is an excellent summary paper with a comprehensive list of references.)

Phleger, F. B. 1939, 1942. Foraminifera of submarine cores from the continental slope. *Bull. Geol. Soc. Amer.*, vol. 50, pp. 1395–1422; vol. 53, pp. 1073–1097.

Ruedemann, R. and Shrock, R. R. 1939. A new Wisconsin Upper Cambrian foraminifer. *Amer. Jour. Sci.*, vol. 237, pp. 66–71.

Thalmann, H. E. 1952. Twenty years of "Foraminiferal Statistics" (1931–1950) [abstract]. *Soc. Econ. Paleontologists and Mineralogists*, Program of 26th annual meeting, p. 37.

Thompson, M. L. 1948. Studies of American fusulinids. *Univ. Kan. Pub., Pal. Contr.*, art. 1, 184 pp.

Wood, A. 1947. The supposed Cambrian Foraminifera from the Malverns. *Quart. Jour. Geol. Soc. London*, vol. 102, pp. 447–460.

———. 1949. The structure of the wall of the test in the Foraminifera; its value in classification. *Ibid.*, vol. 04, pp. 229–255.

Radiolaria

Aberdeen, E. 1940. Radiolarian fauna of the Caballos formation, Marathon Basin, Texas. *Jour. Paleontology*, vol. 14, pp. 127–139.

Campbell, A. S. 1944. Miocene radiolarian faunas from Southern California. *Geol. Soc. Amer., Spec. Paper* 51, 76 pp.

——— and Clark, B. L. 1944*a*. Radiolaria from Upper Cretaceous of Middle California. *Ibid., Spec. Paper* 57, 61 pp.

Clark, B. L., and Campbell, A. S. 1942. Eocene radiolarian faunas from the Mt. Diablo area, California. *Ibid., Spec. Paper* 39, 112 pp.

——— and ———. 1945. Radiolaria from the Kreyenhagen formation near Los Banos, California. *Geol. Soc. Amer., Mem.* 10, 66 pp.

——— and ———. 1945*a*. Possible shallow-water origin of radiolarian shale of the Mount Diablo area, Middle California. *Nat. Res. Council, Rept. Comm. Marine Ecology as Related to Paleontology*, 1944–1945, no. 5, pp. 32–36.

David, T. W. E., and Itowchin, W. 1896. Note on the occurrence of casts of Radiolaria in Pre-Cambrian (?) rocks, South Australia. *Linn. Soc. N.S.W., Proc.* 21, pp. 571–583.

Haeckel, E. 1887. Die Radiolarien, Eine Monographie, I, II. Report on the Radiolaria collected by *H.M.S. Challenger* during the years 1873–1876, *Zool.*, vol. 18, 1803 pp.

Hertwig, R. 1879. "Der Organismus der Radiolarien." Jena, Gustav Fischer, 149 pp.

Rüst, D. 1885. Beiträge zur Kenntnis der fossilen Radiolarien aus Gesteinen der Jura. *Palaeontographica*, vol. 36, pp. 269–332.

———. 1891–1892. Beiträge zur Kenntnis der fossilen Radiolarien aus Gesteinen der Trias. *Ibid.*, vol. 38, pp. 107–200.

———. 1898–1899. Beitrage zur Kenntnis der fossilen Radiolarien und der paläeozoischen Schichten. *Ibid.*, vol. 45, pp. 1–68.

Ciliata

Colom, G. 1948. Fossil tintinnids: loricated Infusoria of the order of the Oligotricha. *Jour. Paleontology*, vol. 22, pp. 223–263.

Fauré-Fremiet, E. 1924. Contribution à la connaissance des Infusoires planktoniques. *Bull. biol. soc. France Belg.*, Supplement 6, pp. 1–171.

CHAPTER 3

PHYLUM PORIFERA

INTRODUCTION

The Phylum Porifera[1] includes a restricted group of simple animals, the most familiar, but by no means the most typical, of which are the several sponges of commerce. The Porifera constitute the simplest group of multicelled animals, *i.e.*, those composed of many cells that are interdependent and differentiated into definite tissues. They represent the first advance beyond the Protozoa. They are typically sessile, aquatic, colonial organisms and live mainly in marine environments. A few forms dwell in bodies of fresh water, where individuals may be present in great abundance. Lakes in which deposition is slow may have siliceous sponge spicules constituting a substantial portion of the sediments.

Because of close resemblance to plants, sponges were once referred to that kingdom, but the possession of a digestive system and other typical animal characteristics and the lack of either chlorophyll or cellulose stamp them as animals. They were also referred to the Coelenterata at one time.

NATURE OF THE ANIMAL

The body of a simple sponge may be compared to a vase, attached at the base, open at the top, and with the wall perforated by numerous **canals** opening externally as **ostia** (Fig. 3-1*A*). The canals open into a central cavity, the **spongocoel,** which itself opens to the exterior through the **osculum** at the top of the organism. Water enters through the canals, passes into the spongocoel, and leaves through the osculum. In some forms the spongocoel is large and deep; in others it is so shallow as to be little more than a depression or flat space on the upper surface. All known living sponges have some parts of the inner surface lined with collared flagellate cells. In simple forms these are confined to the spongocoel. In the more complex sponges the wall is thick, the structure is less simple, the canal system is considerably more complicated, the collared flagellate cells line parts of the canals, and oscula may be numerous (Fig. 3-1*B*–*D*).

Shape of Animal. The shape of a sponge varies greatly. It may be cylindrical, spherical, globose, pyriform, foliate or leaflike, explanate, or discoidal. Some species have long or short stems, by which they are attached to objects on the bottom; others are stemless and are attached directly by their undersurfaces; and still others are anchored by root tufts of spicules, as illustrated by the living *Euplectella* (Figs. 3-3*E*, 3-5*I*–*K*). Some forms are incrusting, and a few are more or less dendritic with the branches separated or united to form

[1] Porifera—L. *porus*, passage or pore, +*ferre*, to bear; referring to the fact that the wall of the sponge is perforated by simple pores or complicated passages.

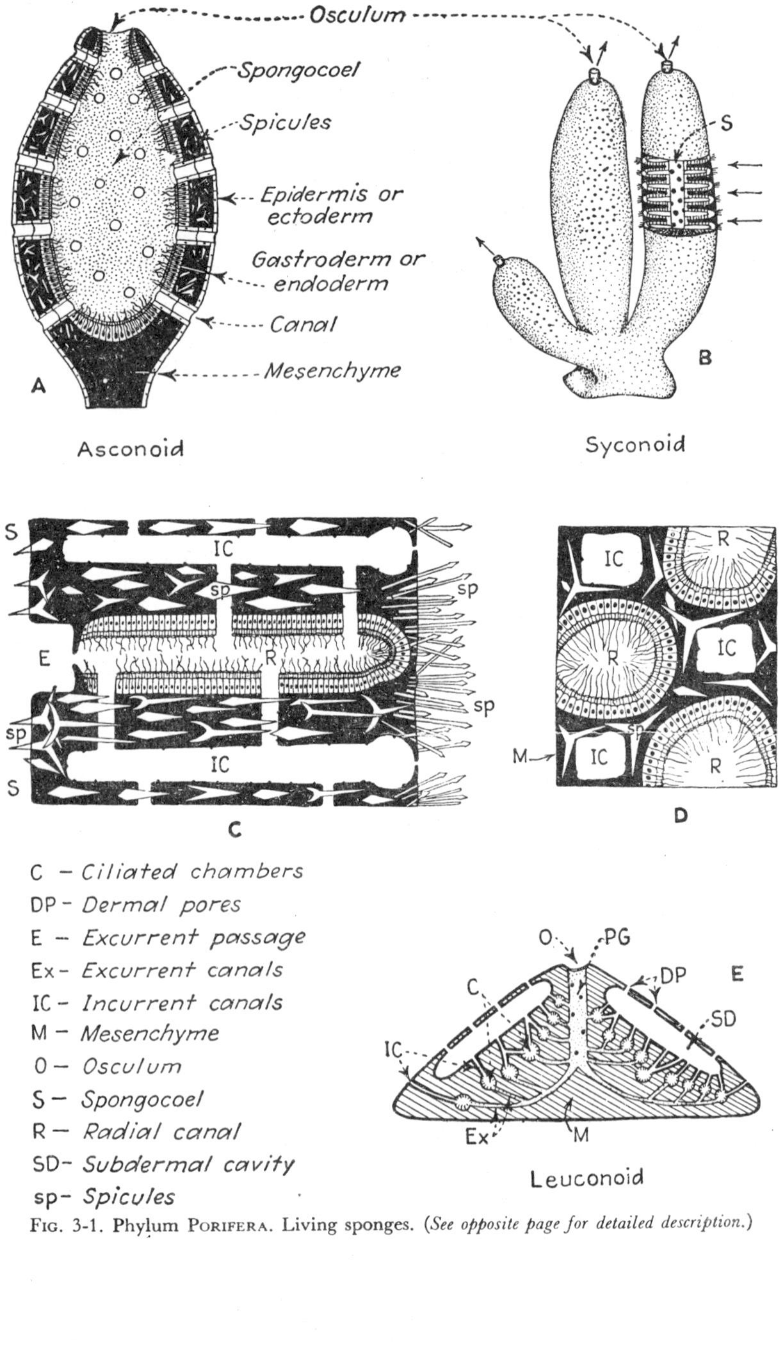

FIG. 3-1. Phylum PORIFERA. Living sponges. (*See opposite page for detailed description.*)

a more or less complex network. Body form is constant for some species, but it commonly varies with the environment regardless of the species, hence is of little use in determining taxonomic relationships. Sponges range from less than a millimeter to as much as 2 m. (6 ft.) in diameter or height.

The Body Wall. The wall of a sponge consists of two well-defined layers of cells separated by an irregular layer of protoplasm (Fig. 3-1). The outer layer, **epidermis** or **ectoderm,** is composed of a single layer of thin, flat cells. These cells extend into the canals that perforate the body wall. The ectoderm seems to function chiefly as a protective covering. The inner layer, or **endoderm,** lines the spongocoel and chambers and parts or the whole of the canals in the wall. It consists of flagellated columnar cells loosely packed together. The flagellate cells, designated **choanocytes,** have a collar surrounding the **flagellum** and closely resemble an individual of the *Choanoflagellata* of the Protozoa (Fig. 2-1) from which group some zoologists think the Porifera evolved. The motions of the flagella draw currents of water through the wall of the sponge into the spongocoel, whence it is expelled through the osculum or through oscula. The endodermal cells engulf food from passing water and yield waste products to it. Between the endoderm and ectoderm are numerous spaces filled with **mesenchyme** (**mesogloea**), a gelatinous substance composed of several kinds of free cells, the **amoebocytes,** and minute **spicules** of both inorganic (calcareous and siliceous) and organic composition. Some amoebocytes function in food transfer and excretion. Certain amoebocytes originate in the epidermis and later migrate to the mesenchyme where they secrete the spicular skeletal elements, which are slender rods (**monaxons**) or three- and four-rayed spicules (**triaxons** and **tetraxons** respectively).

The Canal System. Most of the body wall of a sponge is ramified by some system of canals through which move the currents of water produced by the choanocytes. These canals begin on the exterior surface in **dermal** or **inhalant pores** (**ostia**) and traverse the wall in various ways in different species to terminate finally in **exhalant pores** in the wall of the spongocoel or in numerous oscula on the exterior surface of the sponge.

In simple sponges of the **asconoid** type the wall of the organism is perforated by numerous canals which extend directly from the exterior to the central cavity. In these forms the spongocoel is always lined with choanocytes, and similar cells may line parts of the canals. Water is drawn in through the canals and expelled through the osculum at the top (Fig. 3-1*A*).

In somewhat more complex sponges, included in the **syconoid** type, the wall is thickened and is traversed by numerous straight radial passages which generally have a regular arrangement (Fig. 3-1*B*). Of these there are two kinds—the **incurrent canals,** which open to the exterior; and the **radial canals,** which communicate with the spongocoel through **excurrent passages.** The radial and incurrent canals are parallel and are separated by

---

FIG. 3-1. Diagrams showing the three types of canal systems, shape and general structure of the entire body, and relation of the skeletal elements to the body wall. *A*. Diagrammatic. *B–D*. *Sycon gelatinosum:* *B*, view of a colony, with a small part of one branch cut away to show the internal structure; *C*, longitudinal section of several of the canals in the wall; *D*, transverse section of several canals. *E*. *Spongilla*, a fresh-water sponge. The spicules lie in the cross-lined portion which is the mesenchyme, or mesogloea. (*B–E after Parker and Haswell*, 1930.)

a fairly thin layer of fleshy substance that is perforated by numerous small pores, which permit communication between the two passages. Each radial canal connects with an excurrent passage through a single pore, and the passage communicates directly with the spongocoel (Fig. 3-1*C*). The incurrent canal is lined with epidermal cells, the radial canal with choanocytes, and both the excurrent passages and the spongocoel with flattened nonflagellate endodermal cells. In this form of sponge the radial canals act as pumping stations. By the motion of the flagella, water is drawn into the chamber through the incurrent canals and expelled into the spongocoel through the excurrent passages. It then passes out of the spongocoel through the osculum. Lying between the epidermal and endodermal layers is the mesenchyme in which the spicules of the skeleton are embedded (Fig. 3-1*C–D*). The spicules are of four kinds: long straight rods (monaxons) about the osculum, short straight monaxons surrounding the ostia, T-shaped or triradiate spicules lining the spongocoel, and three-branched forms embedded in the body wall.

A third type of sponge, the **leuconoid,** is shown in Fig. 3-1*E*. It has a body of thick, dense mesenchyme ramified by a complex system of branching canals. Choanocytes are restricted to small spherical chambers that function as minute pumping stations, drawing in the water from the outside and forcing it toward the spongocoel. The water enters a large vestibular cavity through dermal pores; thence it moves through tiny canals into small spherical chambers lined with choanocytes; and from these it is forced into other canals that lead inward to the spongocoel, whence it flows out of the sponge through the osculum at the top. The skeletal elements lie embedded in the mesenchyme.

Reproduction. Two types of reproduction are exhibited by the Porifera—one sexual, the other asexual. In the sexual type the embryo is formed in the mesenchyme from the union of male and female cells developed there from certain amoebocytes. This embryo is released into the spongocoel, from which it escapes to the outside through the osculum. After a period of free life as a ciliated larva, it settles to the bottom, attaches itself to some object, and develops into a mature individual. The asexual method, designated **vegetative reproduction** and commonly referred to as **budding,** is by far the more common among the Porifera. In this method of reproduction the adult first gives off budlike extensions from the body wall. These then develop directly into new organisms which are internally connected with the parent (Fig. 3-1*B*). The buds may remain attached to the parent throughout life, or they may be set adrift to settle down ultimately as separate individuals. So-called "colonies" are built when the buds remain with the parents, though it is questionable if the new buds should any more be considered separate individuals than the twigs of a tree should be considered separate trees. If a piece of sponge is detached from the living individual, the fragment generally regenerates and develops into a complete individual.

THE SKELETON

The soft body of a sponge is generally supported by some sort of internal structure, commonly designated the **skeleton,** which is composed of organic fibers or crystalline spicules, or a combination of the two. The familiar bath sponge illustrates a skeleton composed of **spongin,** which is an insoluble and chemically inert, sulphur-containing protein (**scleroprotein**). Calcareous sponges, such as *Leucosolenia* and *Sycon*, have spicules of calcium carbonate, whereas the so-called **glass** sponges (*e.g.*, *Euplectella*) have skeletal elements and skeletons composed of silicic acid, $H_2Si_3O_7$ (opaline silica), or of some other opaline siliceous material. Spicules show considerable range in shape, size, composition, and structure (Figs. 3-2, 3-3, 3-5).

Crystalline spicules are secreted by special cells in the mesenchyme known as **scleroblasts,**[1] and spongin by others designated **spongioblasts.** Formation of the more complex spicules requires cooperation between several cells. As will be seen later, spicules are of importance in determining the nature of the continuous skeleton, if one is developed, and as a basis for classifying the different kinds of sponges.

The fossil record of sponges includes complete calcareous and siliceous skeletons or fragments and impressions of these, isolated spicules of similar material, and rare specimens preserving the outlines of complete bodies, which have been flattened and otherwise altered.

Form and Nature of Spicules and Skeletons. Almost all sponges have spicules of one kind or another, and many have a continuous flexible or relatively rigid skeleton. The nature and composition of spicules and skeletons constitute the basis for the following primary subdivisions of the phylum:

Class 1. Calcarea. Skeleton composed of separate simple calcareous spicules, which are one-, three-, or four-rayed (Fig. 3-2).

Class 2. Hexactinellida. Skeleton composed of siliceous six-rayed spicules, or some modification of this type, separate or united into networks (Figs. 3-2, 3-3).

Class 3. Demospongia. Skeleton composed of horny fibers or siliceous spicules or both; spicules of two different sizes and not six-rayed (Figs. 3-2, 3-3).

Class 4. Pleospongia. Skeleton a compact, single- or double-walled calcareous cup or cone. This class became extinct in the Middle Cambrian (Figs. 3-9, 3-11).

Sponge spicules, referred to by some zoologists as **sclerites,** are of two general kinds—**megascleres,** which are the larger skeletal spicules that make the supporting structure of the sponge, and **microscleres,** which are the smaller crystalline and flesh spicules that are scattered throughout the mesenchyme. The following discussion considers only the megascleres, since these are generally the only preservable spicules.

Sponge spicules have a wide range of shape. Many are beautiful in form and exhibit great variety of structural detail (Figs. 3-2, 3-3, etc.). They consist fundamentally of simple spines, or of spines radiating from a point. In the living organism they have an axis of organic material around which

[1] The calcium carbonate and silica used in the making of crystalline spicules are extracted from the water by the sponge. Inasmuch as sea water contains only a trace of silica, it follows that a sponge must circulate an enormous volume of sea water through its body in order to obtain enough silica for its spicules. Hyman ("The Invertebrates," McGraw-Hill Book Company, Inc., p. 300, 1940) states that when calcareous sponges are placed in calcium-free water they are unable to secrete spicules, forming only the soft organic axes, and the spicules already formed are dissolved. As a consequence the organism, robbed of the support furnished by the spicules, collapses and degenerates.

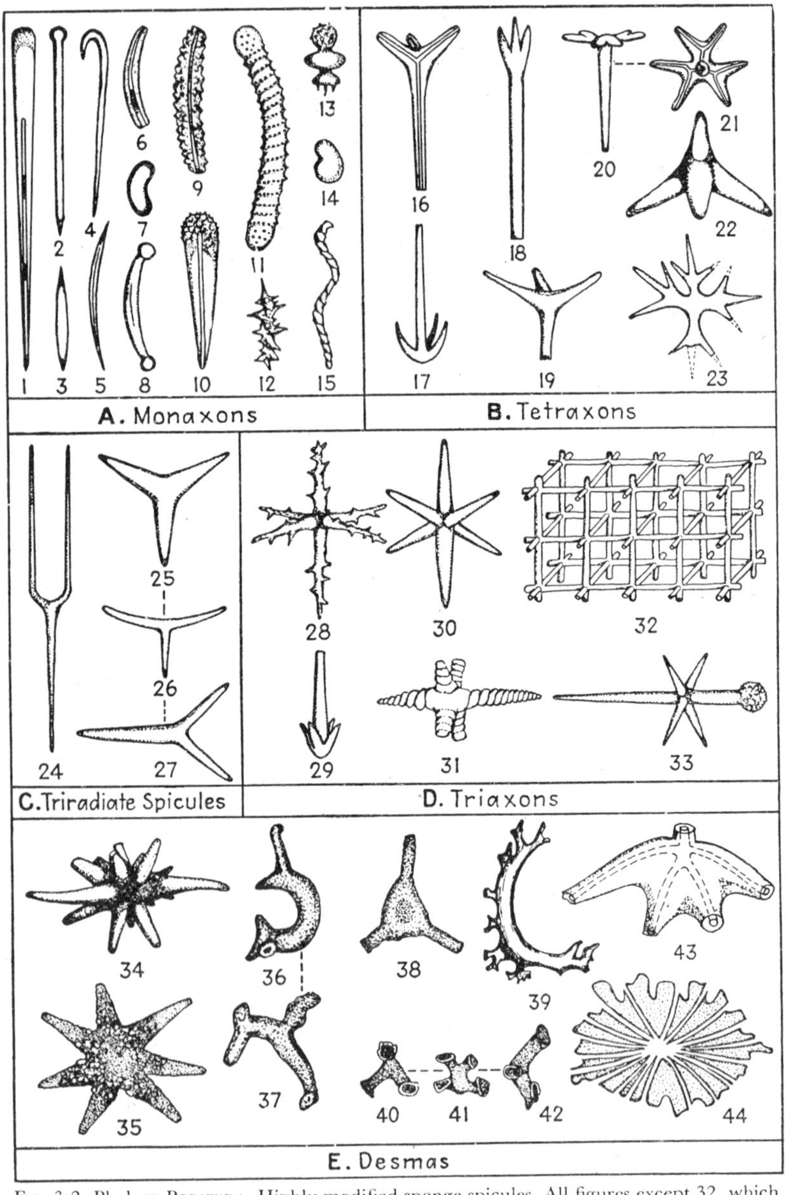

FIG. 3-2. Phylum PORIFERA. Highly modified sponge spicules. All figures except 32, which is diagrammatic, and 33, from bottom sediments of the Pacific, are fossil. (1, 4, 15, 17, 19–22, 39 *from Zittel's several papers;* 2–3, 5–14, 16, 18, 23–31, 34–38, 40–44 *after Hinde,* 1887–1893.)

either calcium carbonate or hydrated silica is deposited. Siliceous spicules, as a consequence, usually have a tiny axial tube; this feature, however, is nearly always obliterated in calcareous spicules because of recrystallization.

Isolated spicules from living species can rarely be classified more closely than families or suborders. This situation, which has long been recognized by students of living sponges, has resulted in the common practice of applying a form name to all spicules of similar architectural design (*e.g.*, a needle-shaped form is a **style,** a C-shaped one a **sigma,** etc.). The resulting nomenclature, consisting of more than 50 special terms, is fully discussed by Hyman (1940) for living sponges, and by Scott (1943) for fossil spicules. The latter convincingly points out why fossil spicules should not be assigned generic and specific names, and why some artificial classification based on spicule architecture is probably preferable at this stage of knowledge.

TABLE 3-1. TYPES OF SPONGE SKELETAL ELEMENTS (SCLERITES)
See Figs. 3-1, 3-2, 3-3, 3-5, etc.

Number of axes	Number of rays	Description of spicule
Monaxon	Monactinal or diactinal	Formed by growth in one or both directions along a single straight or curved axis
Tetraxon	Tetractinal	Consisting of four rays not in the same plane; loss of rays gives triactinal (triradiate), diactinal, and monactinal spicules
Triaxon	Hexactinal	Consisting of three axes crossing at right angles, with six rays extending at right angles from a central point
Polyaxon	Polyactinal	Consisting of several equal rays radiating from a central point; most common among flesh spicules and not to be confused with desmas (see below)
Desma		An ordinary monaxon, triradiate, or tetraxon spicule on which layers of silica have been irregularly deposited to form much-ornamented and somewhat massive spicules; desmas commonly unite to form a **lithistid** skeleton (Figs. 3-2*E*, 3-3*C*-*D*)
Spheres		Rounded bodies in which growth is concentric around a center (Fig. 3-3*C*)
Spongin fibers		Fibers of spongin arranged in a network or in a branching fashion; foreign particles may be associated with these fibers (Fig. 3-8)

Spicules are commonly classified according to the number of rays or of axes, as shown in Table 3-1. No attempt is made to list in the table the special terms mentioned in the previous paragraph, but somewhat more extended descriptions of the chief types of spicules follow.

Monaxons. The simplest spicules are tiny straight or slightly curved bodies produced by growth of skeletal matter in one or both directions along a single axis. These are designated monaxons (Gr. *mona*, one, + *aktis*, ray) and are described as **monactinal** or **diactinal** depending on whether growth was in one direction only or in both directions from a central point. Examples of both types are illustrated in Figs. 3-2, 3-3, and 3-5. Monaxons may be smooth, prickly, or knobby; symmetrical or asymmetrical; straight, curved, or twisted; and doubly or singly pointed. They are abundant in bottom sediments of present oceans, seas, lakes, and rivers, and fossil siliceous monaxons have been reported from rocks of all ages from the earliest Cambrian, and possibly even from Pre-Cambrian rocks.

Tetraxons. Spicules consisting of four rays, not in the same plane but radiating from a common point, are designated **tetraxons** and are described as **tetractinal** or **quadriradiate.** The four rays may be about equal, but more commonly one is much elongated with the remaining three branched, fused into a single plate, or otherwise modified (Figs. 3-2, 3-3).

By loss of rays a tetraxon becomes three-, two-, or one-rayed; the last two are commonly indistinguishable from diactinal and monactinal spicules. The triradiate or triactinal spicule, which is the most common form in calcareous spicules, is regarded as a modified tetraxon (Fig. 3-2).

Triaxons. The triaxon spicule consists fundamentally of three axes crossing perpendicularly, so that six rays radiate at right angles from a common point. This basic type of spicule is greatly modified by reduction or loss of rays, by curving and branching of rays, and by the development of ornamental surface features such as spines, nodes, etc. (Figs. 3-2, 3-3). Hexactinal spicules are found only in Class Hexactinellida (Fig. 3-5).

Polyaxons. Spicules having several rays that radiate from a central point are designated **polyaxons** and are described as **polyactinal.** These are common among microscleres, and certain forms are designated **asters** because of their starlike appearance (Fig. 3-2).

Desmas. Among the megascleres are spicules of the preceding types which have been greatly modified in form and ornamentation by the addition of silica irregularly deposited on the basic spicule. This silica at first follows closely the shape of the basic spicule, but later additions take the form of elaborate branches, tubercles, etc. Desmas are commonly united into flexible or relatively rigid skeletons (Figs. 3-2, 3-3). Such a continuous reticulated meshwork is described as **lithistid** (Fig. 3-3*C–D*), and is characteristic of Order Tetractinellida.

Spheres. Some sponges contain rounded spicular bodies in which growth is concentric around a center. These are known simply as spheres (Fig. 3-3*C*).

Spongin Fibers. Some sponges have a skeleton composed wholly or in part of fibers of spongin arranged either in some sort of network or in a branching manner (Fig. 3-8). Certain Monaxonida have siliceous spicules bound together by or incorporated in spongin material. The skeletons of Order Keratosa are composed entirely of spongin fibers which commonly contain foreign particles such as protozoan tests, sponge spicules, and sand grains (Fig. 3-8).

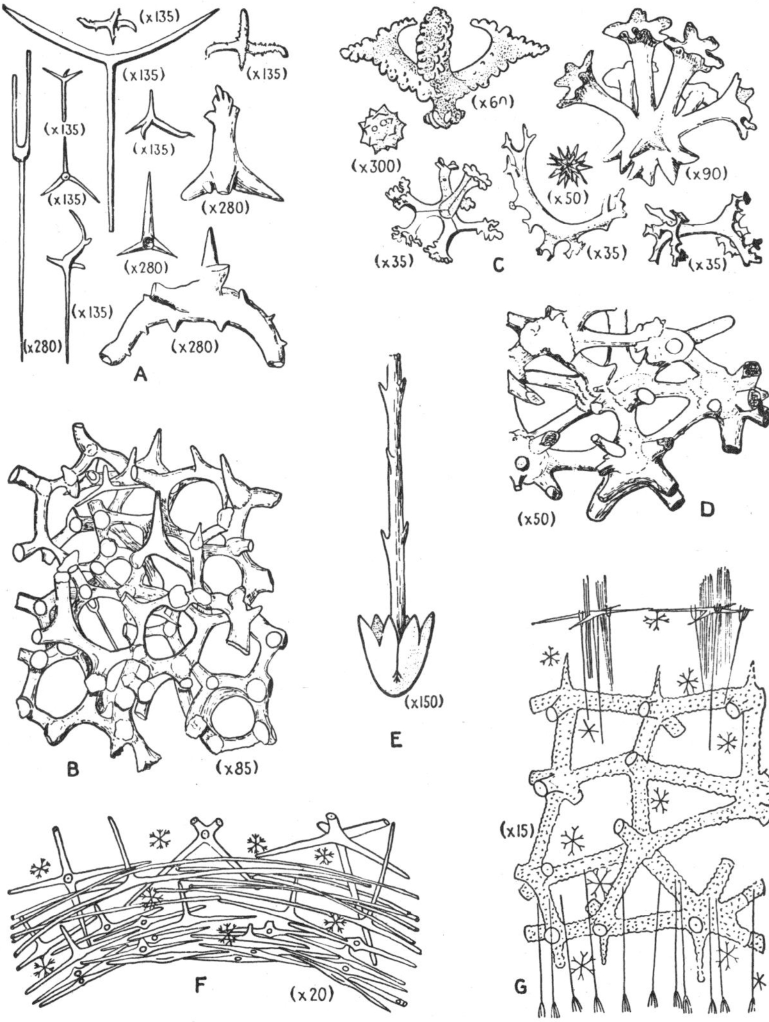

FIG. 3-3. Phylum PORIFERA. Types of sponge spicules and skeletons. *A–B*. Free and united spicules of a pharetrone sponge (Calcarea), *Petrosoma*. *C–D*. Free and united spicules from several lithistid sponges (Demospongia). *E–G*. Spicules and skeleton of several hexactinellid sponges (Hexactinellida): *E*, an anchor needle from root tuft of *Euplectella; F*, felted spicules of *Euplectella*, illustrating lyssacine skeletal structure; *G*, dictyonine skeleton and associated free spicules of *Hexactinella*, transverse section of wall without the soft parts. (*A–B after Döderlein*, 1897; *C–D after Rauff*, 1893–1894; *E–G after Schulze*, 1887.)

Continuous Skeletons. In certain sponges the spicules are matted and felted together or actually united to form a flexible or relatively strong and rigid skeleton. The living *Hyalonema* (Fig. 3-5) has long threads of silica interwoven in a ropelike mass, and *Euplectella*, the familiar glass sponge (Fig. 3-5), has a tubular skeleton not unlike the framework of a dirigible or giant plane fuselage. The glass sponges of the Cambrian and Devonian seem to have had essentially similar skeletons (Fig. 3-6).

Both living and fossil calcareous sponges are characterized by rather massive skeletons in which the identity of the spicules is commonly lost in the recrystallized fossil.

Fossilization of Skeletal Material. The environment created by decomposition of the flesh of a dead sponge is one in which the extremely delicate spicules can be dissolved rather easily.

Even if the spicules escape destruction in this environment and become a part of the bottom deposits, they are likely to be subjected to other modifying processes. Changes commonly take place during lithification of the deposits or after they have been elevated above sea level. Siliceous spicules may be dissolved by percolating ground waters, changed to a cryptocrystalline condition, or replaced in whole or in part by calcareous or other material. In the same way calcareous spicules may be dissolved or replaced by silica and other substances. It is clear, therefore, that the present condition of a fossil sponge is not necessarily its original one; hence, the classification of fossil sponges must rest on the form of the skeleton and spicules as well as on the composition of the material of which they are made.

CLASSIFICATION

Sponges are quite similar to certain colonial forms of Protozoa, except for the internal skeleton and the differentiation of the body wall into two definite layers. Some investigators have suggested that the Porifera evolved from some ancestral stock like the Mastigophora of the Protozoa, because of the fact that choanocytes constitute the endoderm of the body. In the typical sponge there are no localized nerve cells, no blood or circulatory system, and no organized muscular tissue, though certain cells in the mesenchyme can perform functions suggesting some of these. The differentiation of cells into epidermis, endoderm, and mesenchyme, in spite of the simplicity of the sponge body itself, marks that animal as a parazoan, though a primitive and perhaps aberrant one.

Classification of the Porifera is based on the nature (largely form) and composition of the skeletal elements. As these elements, in both living and fossil sponges, are least affected by the environment and mode of growth, they have been found to be the most reliable basis for subdividing the phylum into classes. In the smaller taxonomic categories, however, aspects of the sponge other than its spicules have to be used because of the fact that similar spicules are found in widely different genera.

Largely on the basis of the nature and composition of the skeletal elements or the complete skeleton, the Porifera, as previously stated, can be divided into four classes—Calcarea, Hexactinellida, Demospongia, and Pleospongia.

Class Calcarea (Calcispongia)

The sponges of this class form a sharply defined group and can be distinguished readily from all other sponges by their calcareous spicules and skeletons. The calcareous material is largely calcite, or in some cases arago-

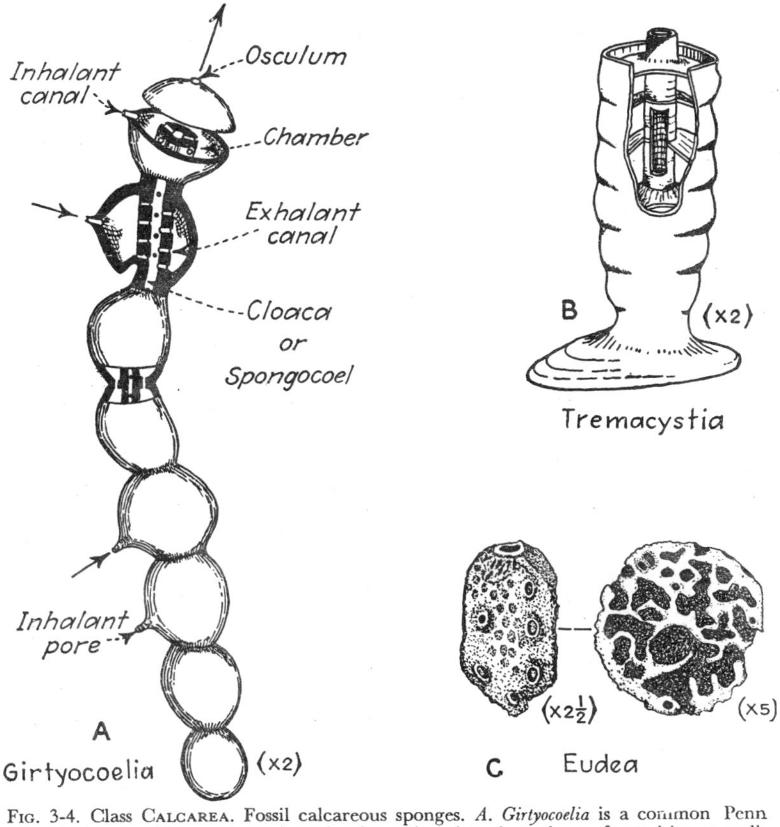

Fig. 3-4. Class Calcarea. Fossil calcareous sponges. *A. Girtyocoelia* is a common Pennsylvanian genus. The specimen is variously sectioned to show the perforated inner walls, the perforated partitions between adjacent chambers, and the internal structure. The walls as shown are probably somewhat thicker than they were in life. *B. Tremacystia* is a calcareous sycon sponge from the European Cretaceous. *C. Eudea* is a typical pharetrone sponge from the Triassic and Jurassic. (*B after Taylor*, 1910; *C after Hinde*, 1887–1893.)

nite, with small percentages of magnesium carbonate and several other compounds. The spicules[1] are monaxons, triradiates, and quadriradiates (Fig. 3-2). They are separate except in the **pharetrone** sponges, in which the spicules are enclosed by a calcareous cement or united into a network (Fig. 3-4*C*).

[1] Typical spicules are lacking in the living *Astrosclera willeyana*, which has calcareous spherules.

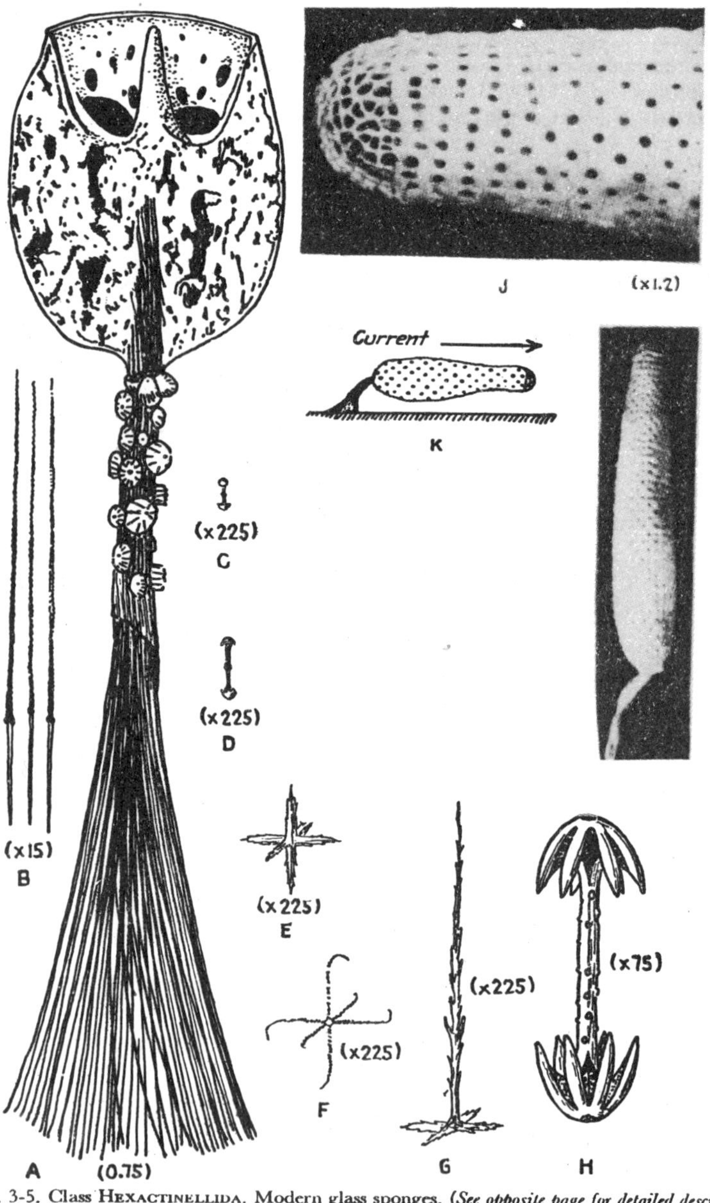

FIG. 3-5. Class HEXACTINELLIDA. Modern glass sponges. (*See opposite page for detailed description.*)

The sponges of this class exhibit all three types of structure—asconoid, syconoid, and leuconoid (Fig. 3-1). All asconoid sponges and all typical syconoid sponges belong to the Calcarea. They are typically small, seldom more than 15 cm. high, and generally consist of solitary or clustered vase-shaped bodies, each with a terminal osculum (Fig. 3-1). They commonly have a bristly appearance due to projecting spicules.

Living calcareous sponges are exclusively marine and are world-wide in distribution. They are confined almost entirely to the shallow waters of the neritic zone, and some live between tide levels.

Representatives of Class Calcarea are poorly preserved as fossils, probably because of the ease with which the calcareous spicules can be dissolved and the tendency for loss of identity upon replacement by other substances. Most fossil calcareous sponges belong to the pharetrones, which have reticulated skeletons composed of spicules that are united in some way (Fig. 3-3). The earliest known representative (*Camarocladia*) of the class appears in the Cambrian.

The Class Calcarea has been subdivided into the following two orders:

Order 1. Homocoela. Body wall thin, interior unfolded and lined continuously with choanocytes; structure asconoid; skeletal material not likely to be preserved; fossil record meager and fragmentary. Living example—*Leucosolenia.*

Order 2. Heterocoela. Body wall thick, interior folded; spongocoel lined with flattened endodermal cells, with choanocytes limited to canals or chambers; structure syconoid and leuconoid; skeleton well developed; substantial fossil record, limited mainly to the pharetrones. *Cambrian to Recent.*

The sycones are small, shallow water, marine Calcarea with thin walls perforated by radially disposed canals. The skeletal elements are regularly arranged in the fleshy wall and are mainly monaxons and triradiate spicules. A typical living genus is *Sycon* (Fig. 3-1); typical fossil representatives are *Girtyocoelia* from the Pennsylvanian and *Tremacystia* from the Cretaceous (Fig. 3-4).

The pharetrones are thick-walled calcareous sponges with an intricate canal system and anastomosing spicules that form a rigid skeleton of variable shape. *Petrosoma* (Fig. 3-3*A–B*) is a typical living pharetrone and *Eudea* from the Triassic and Jurassic is a fossil representative (Fig. 3-4*C*).

Class Hexactinellida (Hyalospongia)

The hexactinellids or glass sponges differ from all other sponges in having hexactinal siliceous spicules (whence the name—Gr. *hektos*, six, + *aktis*, a ray, + L. *-ella*, diminutive suffix). The beautiful glasslike skeleton consists of separate spicules, together with a network or meshwork of spicules united loosely or compactly to form a rather rigid structure (Figs. 3-2[32], 3-3*G*). The spicules are composed of silicic acid (opaline silica), and contain small

Fig. 3-5. *A–H. Hyalonema: A*, complete individual, showing vertical section of upper part and portion of basal tuft invested with a symbiotic zoanthid; *B–H*, spicules of different sizes and shapes that are felted together in the body wall. *I–K. Euplectella: I*, a complete skeleton with root tuft of silica threads; *J*, distal end of skeleton, showing nature and distribution of pores and of skeletal material, and the large osculum covered with a convex sieve plate; *K*, supposed living position on the bottom. It is also possible that the sponge may maintain a vertical position with the osculum up. (*A–H after Schulze*, 1887.)

amounts of an organic substance called **spiculin.** Although always retaining the fundamental symmetry of the typical hexaxon, the spicules show many modifications which produce a great variety of forms (Fig. 3-2). Skeletons with **lyssacine** structure are formed by the interlacing or juxtaposition of the elongated rays of hexaxons, giving a loose structure with meshes of irregular shape (Fig. 3-3*F*). More fully developed skeletons with **dictyonine** structure are three-dimensional networks in which the rays of regular hexaxons are

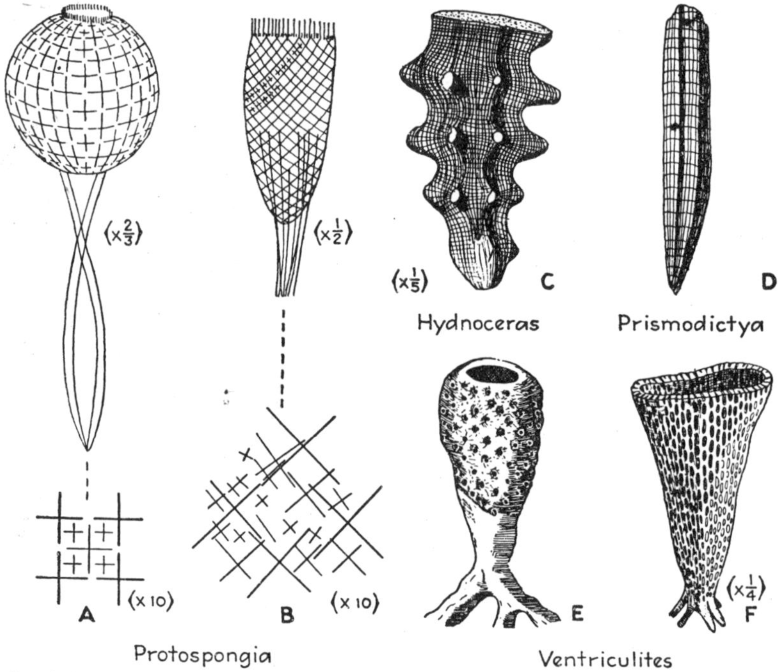

FIG. 3-6. Phylum PORIFERA. Fossil glass sponges (Hexactinellida). *A–B. Protospongia* from the Upper Cambrian. *C–D. Hydnoceras* (*C*) and *Prismodictya* (*D*) from the New York Devonian. *E–F. Ventriculites* from the European Cretaceous. (*A–B after Dawson*, 1868; *E–F after Nicholson and Lydekker*, 1889.)

fused at their tips in all six directions (Fig. 3-2[32]). These structures are relatively rigid and have meshes of regular shape (Figs. 3-2, 3-3, 3-6*A–B*).

The Hexactinellida are strongly individualized sponges with cylindrical, vaselike, or funnel-shaped bodies, which are attached either directly by the base or more commonly by a root tuft of spicules (Fig. 3-5). They are rarely more than a meter high, and most are less than half that size. They are exclusively marine and live at depths ranging from 100 to more than 5,000 m. (300 to 16,000 ft.). The class is well represented in present oceans and also as fossils in rocks as old as the Cambrian. The well-known glass sponges of the New York Devonian are typical of fossil hexactinellids. The Hexactinellida are generally subdivided into the following orders,[1] which are based largely on the way the megascleres are united to form the architecture of the skeleton:

Order 1. Lyssacina. Hexactinellid sponges with lyssacine skeleton. Typical living genera are *Hyalonema* and *Euplectella* (Fig. 3-5). Characteristic of fossil forms are *Protospongia* from the Cambrian and *Hydnoceras*, one of the common glass sponges of the New York Devonian (Fig. 3-6). *Cambrian to Recent.*

Order 2. Dictyonina. Hexactinellid sponges in which regular hexaxons are fused at their six tips to form a three-dimensional network with more or less symmetrically arranged interspaces. *Hexactinella* (Fig. 3-3*G*) is typical of living forms and *Ventriculites* (Fig. 3-6*E–F*) from the European Cretaceous is a characteristic fossil genus. *Triassic to Recent.*

Class Demospongia

The Demospongia include those sponges that have no spicules, those that have nonhexactinal siliceous spicules, those with spongin fibers, and finally those with a combination of siliceous spicules and spongin. The canal system in this diverse group is invariably leuconoid.

Many living sponges belong in this class, and there is a considerable fossil record which reaches back as far as the Cambrian. Zoologists recognize the following three orders (by some considered subclasses):

Order 1. Tetractinellida. Demospongia with skeletons of two- and four-rayed siliceous spicules (monaxial and tetractinal megascleres; whence the name—Gr. *tetra*, four, + *aktis*, ray, + L. *-ella*, diminutive suffix), which lie free in the mesenchyme or are united by spongin or silica. Skeletons formed by the fusion of the spicules are termed lithistid and are likely to be preserved. They have been included in a separate order, the *Lithistida*, by many paleontologists because of their importance as fossils. *Jerea* (Fig. 3-7*D*) and *Siphonia* (Fig. 3-7*F*) from the Cretaceous are typical of the Suborder Tetracladina; *Astylospongia* (Fig. 3-7*H–I*) from the Silurian represents the Eutaxicladina; *Cylindrophyma* (Fig. 3-7*E*) from the Jurassic typifies the Anomocladina; *Doryderma* from the Cretaceous illustrates the Megamorina; and *Jereica* (Fig. 3-7*G*) from the Cretaceous is representative of the Rhizomorina. The tetractinellids range from Cambrian to Recent, with greatest abundance and variety in Cretaceous deposits. *Cambrian to Recent.*

Order 2. Monaxonida. Monaxonid sponges have skeletons composed of one- and two-rayed siliceous monaxons which are felted or bundled together. The spicules are scattered through the mesenchyme, and when the animal dies they are freed and become a part of the bottom deposits. As most living marine siliceous sponges belong to this order, their spicules are a characteristic constituent of many bottom sediments. Fresh-water sponges (Spongillidae) are also included in this order, and their tiny spicules are common in the deposits of streams and lakes.

Monaxonids are the most common of all sponges. They live throughout the world in

[1] Zoologists subdivide the Hexactinellida on the basis of the microscleres into two subclasses or orders—Amphidiscophora, with microscleres composed of a shaft bearing a disk at each end, and Hexasterophora, with starlike microscleres. Because the microscleres of many fossil hexactinellids are not preserved, it is impossible to refer most fossil forms to the correct order. For this reason we prefer to subdivide the class on the method of union of the megascleres.

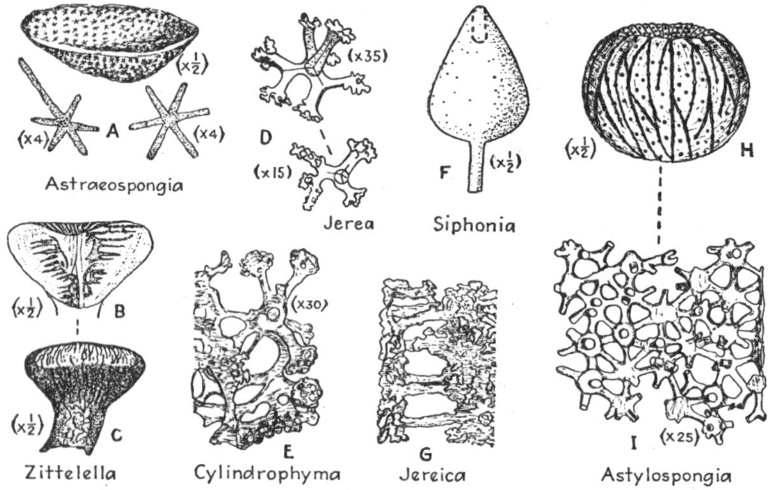

FIG. 3-7. Phylum PORIFERA. Fossil siliceous sponges. *A. Astraeospongia* from the Silurian of Tennessee. *B–C. Zittelella* from the Ordovician of Illinois. *D.* Spicules of *Jerea* from the European Cretaceous. *E. Cylindrophyma*, a magnified portion of the skeleton, from the Jurassic. *F. Siphonia*, a common sponge in the Upper Cretaceous of western Europe. *G. Jereica*, a magnified portion of the skeleton, *H–I. Astylospongia*, from the Silurian of Gotland: *H*, a complete specimen; *I*, a magnified portion of the skeleton. (*B–C after Ulrich and Everett*, 1890; *D after Rauff*, 1893–1894, *and Schrammen*, 1910–1912; *E, G, I after Zittel, several papers.*)

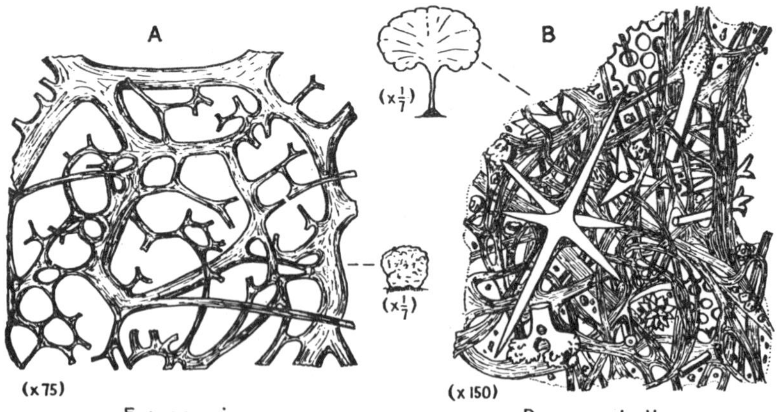

FIG. 3-8. Class DEMOSPONGIA. Living horny sponges. *A.* Small part of the skeleton, showing loose structure of spongin fibers; and diagrammatic outline of complete sponge. *B.* Part of sponge wall, showing network of spongin fibers in which radiolarian shells, sponge spicules, and sand grains are entangled; shape and size of complete sponge indicated by small diagram. (*A after Lendenfeld*, 1889; *B after Haeckel*, 1889.)

abundance in bodies of fresh water and in shallow marine waters along the shore, seldom below 50 m., where they attach themselves to objects on the bottom.

Typical representatives of the order are *Spongilla*, the common fresh-water form; *Cliona*, the well-known marine boring sponge; and *Halicliona*, a common marine form. The fossil record of this order is meager. No complete sponges have been reported, but monaxon megascleres are present in Middle Cambrian rocks. Evidence of boring sponges has been found in shells as old as Devonian. *Cambrian to Recent.*

Order 3. Keratosa.[1] The skeletons of this order consist exclusively of spongin fibers and are commonly described as horny. The skeleton, familiar to everyone in the common bath sponge *Euspongia*, is composed of a network of spongin fibers in which foreign particles and objects are commonly embedded (Fig. 3-8). Most horny sponges live on rocky, shallow bottoms in tropical and subtropical regions, though they are not limited to this environment.

There seems to be no certain fossil record of this large order, although impressions and crushed specimens (compressions) may be expected. *Recent.*

Class Pleospongia

The Pleospongia are an extinct group of calcareous, cup-shaped sponge-like organisms that appeared in the earliest Cambrian seas, quickly attained world-wide distribution (Fig. 3-12), and then became extinct in the Middle Cambrian (Fig. 3-10).

The organisms of this class have been given several names—Archaeos, Archaeocyatha, Archaeocyathacea, Archaeocyathinae, Cyathospongia, and Pleospongia. We adopt the last name because its definition most satisfactorily encompasses the many different genera of archaeocyathids now known.

Pleosponges have been assigned to organic groups of widely different nature—calcareous algae, foraminifers, calcareous sponges, siliceous sponges, receptaculitids, and corals. We prefer to follow most modern investigators of the group in considering it an extinct group related to the calcareous sponges (Raymond, 1931; Okulitch, 1935).

Nature of Skeleton

The typical pleosponge skeleton is a single- or double-walled calcareous cone. It ranges in shape from cylindrical or conical on the one hand to cup- or saucer-shaped on the other. The open central cavity enclosed by the skeletal wall is deep or shallow depending upon the shape of the skeleton as a whole. The simplest and presumably the most primitive, but not the most common, skeletons have a single perforate wall (*e.g.*, *Monocyathus*, Fig. 3-9*E*). The more advanced forms have an outer cup or cone and an inner cup fitting inside, so that the skeleton is double-walled (Fig. 3-9). The space between the inner and outer walls, the **intervallum,** is occupied by several different structures that support the inner cup and subdivide the space into compartments. In typical skeletons the intervallum contains numerous vertical (and radial) partitions designated **parieties.** Horizontal bars (**synapticula**) connect the parieties, and irregular thin laminae (**dissepiments**) extend from pariety to pariety in some skeletons; horizontal tabular

[1] Keratosa—Gr. *keras*, *keratos*, horn, referring to the horny fibers of the skeleton.

plates (**tabulae**) supplement or supplant them in others; and vesicular structure, composed of many curved plates, is commonly developed to a considerable degree in the intervallum (Fig. 3-11)

Both walls are perforate, the outer generally having smaller pores than the inner, and the inner cup is commonly open at the bottom. Most of the intervallum structures are also porous. The skeleton seems to lack spicular elements, being composed instead of calcareous spherules somewhat like those in the living calcareous sponge *Astrosclera*. Nothing is known of the

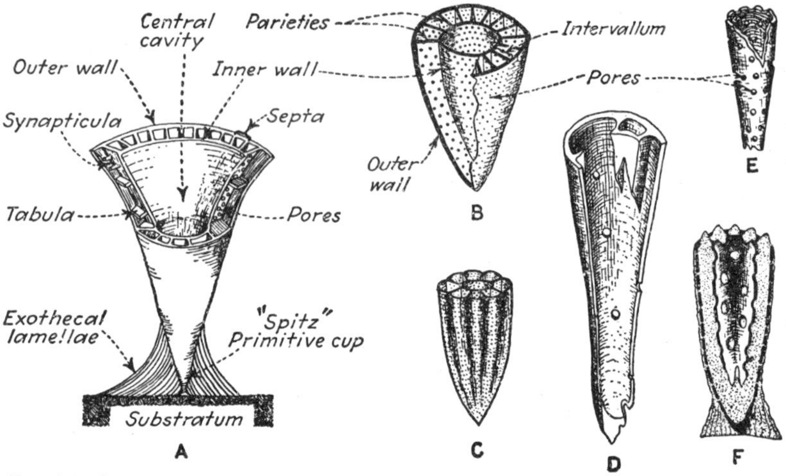

FIG. 3-9. Class PLEOSPONGIA. Structure of pleosponges. *A*. Generalized diagram of an archaeocyathid showing the more important structural features. *B*. A typical ajacicyathid, showing the major skeletal elements. *C*. *Archaeocyathellus*, which illustrates a vertically fluted skeleton. *D*. Reconstruction of a young pleosponge (*Ajacicyathus nevadensis*), which at this stage resembles the *Olynthus* stage of the calcareous sponges. *E*. *Monocyathus*, a pleosponge with a simple single perforated wall. *F*. Diagrammatic restoration of the living archaeocyathid, with the living tissue indicated by stippling. (*A after Taylor*, 1910; *B–E after Okulitch*, 1943; *F after Okulitch*, 1935.)

fleshy part of the animal, but it is assumed that the flesh completely surrounded the skeleton, without completely filling the large central cavity, so that in life there was a prominent depression on the upper side of the animal (Fig. 3-9*F*).

Ontogeny

Investigations of the life history of the pleosponges carried on during the past 15 years have led to the assumption that the free-floating pleospongian larva did not begin secreting a skeleton until it settled down on a suitable bottom and attached itself to the substratum or to some object on the bottom. Once attached, the larva then built a simple calcareous cone consisting of a single perforate wall. This cone was attached to the bottom by exothecal lamellae (Fig. 3-9*A*). At this stage the cup is thought to have been essentially identical, except for size, to a mature *Monocyathus* (Fig. 3-9*E*), which is a single-walled porous cup. Such a simple ancestral type seems to have given rise to *Monocyathus* by regularization of the pores, and to the double-walled genera by development of an inner wall and of rods, transverse partitions (parieties), and other intervallum structures. As the skeleton of the young pleosponge strongly suggests an animal similar to the *Olynthus* stage (Fig. 3-9*D*) that appears in the early development of all calcareous sponges, it would seem that the Pleospongia went through early growth stages similar to those of the calcareous sponges and are, therefore, related to them. They have been made a separate class of the Porifera, however, because, since they are extinct, the nature of their fleshy body, and hence their relations to living calcareous sponges, can never be known.[1]

Classification

The Pleospongia are divided into the following subclasses[2] with the supposed relationships shown in Fig. 3-10.

Subclass 1. *Monocyatha*

These are single-walled, conical pleosponges with a laminar and perforate wall. Some monocyathids closely resemble the *Olynthus* stage of the calcareous sponges. *Monocyathus* (Fig. 3-9*E*) is typical of the subclass.

Subclass 2. *Archaeocyatha*

This subclass includes all the typical pleosponges, *i.e.*, those which invariably have double-walled conical skeletons with intervallum structures of several different kinds. *Ajacicyathus* (Figs. 3-9*D*, 3-11*G–H*), *Archaeocyathellus* (Fig. 3-9*C*), and *Nevadacyathus* (Fig. 3-11*A*) are typical of Order Ajacicyathina; *Cambrocyathus* (Fig. 3-11*B–E*) and *Protopharetra* (Fig. 3-11*F*) are typical genera of Order Metacyathina.

[1] It should be pointed out that in the several areas where well-preserved pleosponges are present, there exists one of the most interesting and significant faunal associations of geologic time. Here, in calcareous rocks of earliest Cambrian time, is a host of widely diversified calcareous-secreting organisms that built a variety of structures different enough one from another to have characteristics of both sponge skeletons and coral exoskeletons. Did there exist in this population the beginnings of all three groups—pleosponges, calcareous corals, and calcareous sponges? Certainly there is a basis for suggesting that as early as the beginning of the Cambrian some organisms were already secreting complex skeletons like those of certain calcareous corals, whereas others were at the same time forming double-walled perforate skeletons like some calcareous sponges. Surely the tendency to build such complex structures was not developed in a day, and there is good reason, there fore, to support the contention of Okulitch (1943, p. 43), who states: "I believe that the Pleospongia with other sponges began in the Proterozoic. The common ancestor was a simple, skeletonless, gastrula-like organism essentially resembling the Olynthus stage of pleospongian and calcareous sponges."

[2] A fifth subclass, the Exocyatha, was erected to include independent or incrusting masses of calcareous material closely associated with pleosponges. These exothecal masses lack definite walls and consist of more or less irregular vesicular structure. Inasmuch as they now are considered to be extraneous growths around the cup of the pleosponge, hence excrescences of different regular forms, rather than independent organisms, the subclass will have to be dropped (Okulitch, personal correspondence, Mar. 20, 1947).

Subclass 3. Acanthocyatha

This subclass includes a group of pleosponges of uncertain affinities, which superficially resemble certain Anthozoa (corals). The inner cavity is filled with skeletal tissue; both walls and parieties lack pores; and the inner wall is not well developed. Members of this subclass may represent an aberrant

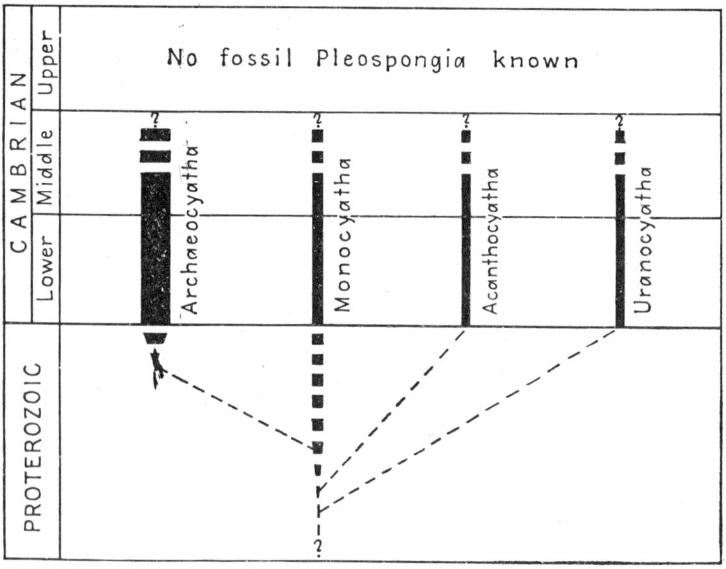

FIG. 3-10. Class PLEOSPONGIA. Supposed relationship of subclasses of Pleospongia. It has been suggested that the pleosponges arose and differentiated in Pre-Cambrian time but did not secrete calcareous skeletons until the beginning of the Cambrian, as no fossil Pre-Cambrian pleosponges are known. The exact time in the Middle Cambrian when the different subclasses became extinct is uncertain. (*Modified after Okulitch*, 1943.)

group of pleosponges, or a group transitional between pleosponges and corals.

Subclass 4. Uranocyatha

These are pleosponges of hollow, spheroidal, ovoidal, or gastrula-like form with a seemingly single wall composed of calcareous spicules. The members of this subclass ultimately may have to be assigned to other Pleospongia or possibly to Class Calcarea.

Geologic History

Well-preserved skeletons of pleosponges appear first in Lower Cambrian limestones and shales and are unknown after the Middle Cambrian. They have been found in the Lower Cambrian of Labrador, Quebec, New York, Virginia, Georgia, British Columbia, Washington, Nevada, California, Sonora (Mexico), Sardinia, Urals, southeastern Russia, Siberia, China, Australia, and Antarctica; and in the Middle Cambrian of New Brunswick, Spain, Moroccan Atlas, Sardinia, Urals, and Siberia (Fig. 3-12).

The Pleospongia undoubtedly had a long Pre-Cambrian history, although no fossil representatives have yet been found (Fig. 3-10). That they were an important part of the benthonic life of the early Cambrian seas is indicated by the fact that more than 450 species (assigned to 92 genera, 26 families, and 8 orders) evolved before the group became extinct. Some species grew in such profusion that they built up reefs (bioherms and biostromes) on the

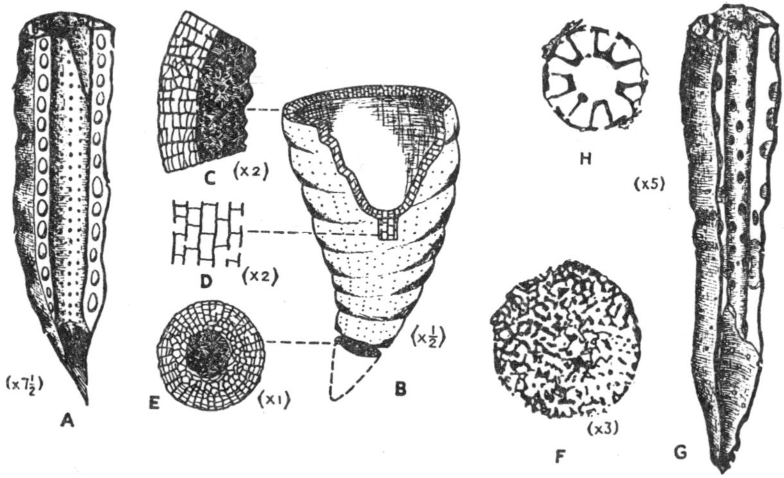

FIG. 3-11. Class PLEOSPONGIA. Typical archaeocyathid Pleospongia. *A. Nevadacyathus*, from the Lower Cambrian of Nevada, partly restored. *B–E. Cambrocyathus*, from the Lower Cambrian of Labrador: *B*, restoration based on actual specimen in which the central cavity is filled with stone; *C*, transverse section of a small part of the wall at the top of the cone showing the prominent perforated septa and numerous dissepiments; the minute pores in the outside wall are too small to be apparent; *D*, tangential section showing perforated septa and dissepiments; *E*, transverse section of cone near base, showing the outer septal zone and the inner vesiculose region; the central cavity is filled with stone. *F. Protopharetra raymondi*, from the Lower Cambrian of Nevada, showing narrow central cavity and vesicular tissue of intervallum. *G–H. Ajacicyathus*, common in the Lower Cambrian of Nevada, partly restored, in longitudinal and transverse sections. (*A*, *F–H after Okulitch*, 1943.)

bottom in Labrador, Virginia, California, and Australia. These are the earliest reefs known to have been built by animals.

Pleosponges have been found in association with brachiopods (*Rustella*), gastropods (*Helcionella* and *Scenella*), trilobites (*Olenellus*), and algae. The cause of their extinction is unknown. It has been suggested that they became extinct because (1) their small pores greatly handicapped them in the muddy waters of Middle Cambrian seas, or (2) they were overwhelmed and ultimately supplanted by calcareous algae. A third possibility is that with the appearance of new and more efficient competitors for the same ecological niche, they were unable to hold their own and so perished. All these suggestions deserve critical consideration. Another important feature of these ancient organisms is that some Labrador specimens of the Lower Cambrian *Cambrocyathus amourensis* appear to have preserved in them the outlines of cells that lined the vesicles of one part of the skeleton. If these are truly cell

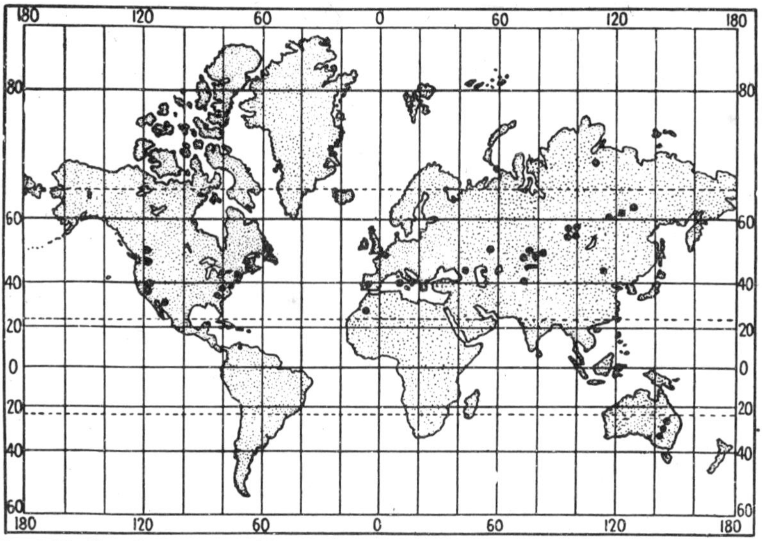

FIG. 3-12. Class PLEOSPONGIA. Map showing the world-wide distribution of the Lower and Middle Cambrian pleosponges. Localities where fossil pleosponges have been found are indicated by large dots.

outlines, they represent the earliest known fossil record of what was once soft living tissue!

SPONGELIKE ORGANISMS OF UNKNOWN AFFINITIES

Certain well-known fossils resemble sponges in several ways but cannot be included in the four classes described above. They are thought by most paleontologists to be sponges of some sort, nevertheless, and are included here in a separate category.

Receptaculites, the so-called "sunflower coral," and *Ischadites*, both characteristic of the lower and middle Paleozoic, seem to belong to the Porifera but cannot be placed in any of the divisions given above. The shape, internal structure, and arrangement and nature of the canals suggest affinity with the Porifera. Both genera seem to have had calcareous skeletons (Fig. 3-13).

Astraeospongia is a bowl-shaped, concavo-convex form without traces of attachment on its convex, supposedly under, surface. The skeleton is composed of six-rayed spicules with all six rays in one plane and with two additional rays at right angles to this plane and reduced to buttonlike prominences (Fig. 3-7*A*). The genus is characteristic of Silurian strata the world over but is also found in the Devonian.

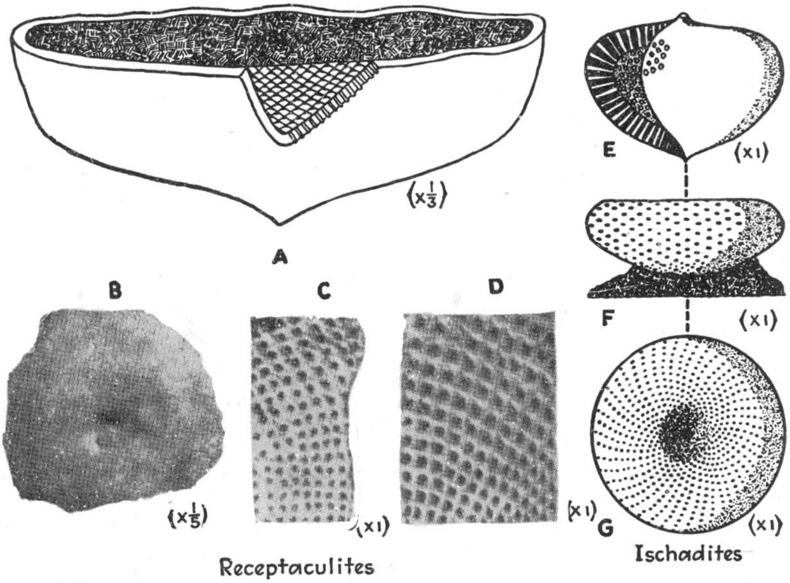

FIG. 3-13. Phylum PORIFERA. *A–D. Receptaculites oweni* from the Middle Ordovician at Freeport, Ill.: *A*, restoration of incomplete specimen, based on a mud filling of the hollow bowl (see Fig. 3-14 for restoration of a complete specimen); the surface of this filling bears an impression of the inner surface of the bowl; the wall is shown without the pores except in one small area; *B*, basal portion of a deformed specimen, showing the characteristic arrangement of the pores; *C–D*, external and internal views, respectively, of small portions of the wall. *E–G. Ischadites iowensis* from the Middle Ordovician (Prosser) at Elkader, Iowa: *E*, restoration of a complete skeleton, based on well-preserved but slightly deformed specimens; the radial spicules are indicated as white rods, and the central opening on the upper surface is shown; pores are indicated in one small area; in fossilized specimens the slightly pointed upper part has collapsed to form a central shallow depression; *F*, lateral view of a somewhat deformed specimen with the base still buried in the matrix; *G*, top view of same specimen, showing the characteristic arrangement of the radial spicules and pores; the shallow collapse depression around the opening is also shown.

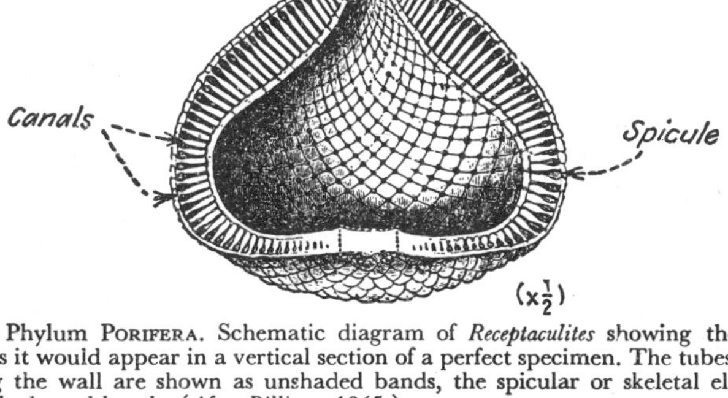

FIG. 3-14. Phylum PORIFERA. Schematic diagram of *Receptaculites* showing the internal structure as it would appear in a vertical section of a perfect specimen. The tubes or canals perforating the wall are shown as unshaded bands, the spicular or skeletal elements as shaded club-shaped bands. (*After Billings*, 1865.)

ECOLOGY AND PALEOECOLOGY[1]

The Porifera, being sessile benthos, exhibit characteristics resulting from adaptation to that environment. They live fastened to objects on the bottom and exhibit a wide variety of spheroidal, vaselike, and branching forms. All are marine except one family, the fresh-water Spongillidae, representatives of which are found in streams, lakes, and ponds throughout the world. The marine sponges are present in all seas and at all depths from the strand line to the deepest abysses of the ocean. The calcareous sponges (Calcarea) and the true horny sponges (Keratosa) are most abundant in relatively shallow waters (less than 450 m.), and it is assumed that ancient representatives of these two classes, as well as the pleosponges, lived in a similar environment. Modern siliceous sponges live at all depths but probably not beyond a few hundred miles from land, and from the fossil record it seems that the same was true of ancient siliceous sponges. The extensive colonies of certain glass sponges in modern seas can be duplicated by the well-known protospongin colonies of the Cambrian and those of the New York and Pennsylvania Devonian.

Living sponges in many cases are veritable aquariums. The interior of the organism literally teems with a great variety of life, consisting of crustaceans, worms, mollusks, and other invertebrates. In a recent investigation[2] a sponge about as large as a washtub was found to contain 17,128 individual animals—approximately two for every cubic inch of the sponge's bulk.

> Highest in zoological rank were five little inch-long fishes, very slenderly built so that they could get about in the sponge canals. There were many worms and a number of barnacles. Most numerous, however, were shrimp of a strange species with one claw much larger than the other, and in some specimens almost as large as its body. There were 16,352 of these shrimp.
>
> A different kind of exploitation of a growing sponge for protection is described by Dr. W. H. Longley, in charge of the laboratory at Tortugas. A crab that lives in the waters there tears off bits of living sponge and holds them to its shell until they take hold and continue their growth. Thereafter the crab has the benefit of concealment, enhanced by the inedibility of the sponge, which is full of disagreeable prickles and in addition has a most noxious odor.

The sulphur sponge *Cliona* bores into oyster shells and other bivalves for protection rather than for food, and at times is so abundant as to essentially wipe out an oyster bed.[3] Sponges commonly live in intimate association with other animals or with plants in such a way that both are mutually benefited by the double life.

The Porifera seem always to have been gregarious animals, and fossil remains, where found in any abundance, are usually of local distribution.

[1] See De Laubenfels, M. W., 1936, The oecology of Porifera and possibilities of deductions as to the paleoecology of sponges from their fossils, *Nat. Res. Council, Rept. Comm. Paleoecology*, 1935–1936, (App. J, *Ann. Rept. Div. Geol. Geog.*), pp. 44–54.

[2] Pearse, A. S. 1933. *Science News Letter*, p. 331.

[3] *Nat. Res. Council, Rept. Comm. Marine Ecology as Related to Paleontology*, 1942–1943, No. 3, pp. 2–3.

This statement does not apply, of course, to loose spicules, which may be buried far from the place where they once existed in the body of a living sponge.

GEOLOGIC DISTRIBUTION

The Porifera represent an ancient, aberrant branch of the animal kingdom, which probably developed from some sort of free-swimming colony of choanoflagellate cells far back in the Pre-Cambrian. So far as is known, they did not give rise to any other group of animals. They may well be said to have traveled a byroad of metazoan evolution, and for this reason some zoologists place them in a separate multicellular division, the Parazoa.

The phylum has shown a remarkable persistence in certain ways, especially as regards the nature of the skeletal elements. Spicules made by Cambrian sponges are indistinguishable from similar ones in living sponges. In all three classes of living sponges the spicules exhibit a strong tendency to fuse into some sort of continuous framework, and a similar tendency has prevailed from the earliest Paleozoic, as witness the fossil record. The Calcarea have spicules fused to form the pharetrones; the Hexactinellida have the dictyonine skeleton; and the Demospongia, the massive lithistid skeletons (Fig. 3-3*C–D*). There has also been a limited tendency toward complete loss of spicules.

The fossil record of the Porifera is not an abundant one except in a few scattered localities, but it is of long duration. The earliest record is of supposed siliceous spicules from the Algonkian Chuar formation of the Grand Canyon.[1] Supposed siliceous spicules from the Pre-Cambrian of Brittany, and *Atikokania lawsoni*, a supposed sponge from the ancient Seine River series at Steeprock Lake in Ontario, have now been rather generally discredited.[2]

Protospongia and *Cambrocyathus* from the Cambrian; *Receptaculites* and *Ischadites* from the Ordovician and later strata; *Astylospongia* and *Astraeospongia* from the Silurian; *Hydnoceras*, *Prismodictya*, and the numerous other forms from the famous "glass-sponge" assemblages in the Devonian of New York; and *Girtyocoelia* from the Pennsylvanian are important index fossils of the Paleozoic. Some have world-wide distribution, whereas others are limited to restricted localities.

Sponge remains are well represented in Mesozoic rocks and are especially abundant in Jurassic strata of Europe. In the Swabian Alps the lower part of a great reef or bioherm in the Upper Jurassic limestone is composed largely of sponge remains, some of which are exceptionally well preserved. Cretaceous sponges, among which lithistids, hexactinellids, and calcareous sponges are the common kinds, are well represented in the Lower Greensand and in the Chalk of England.

Tertiary sponges are represented chiefly by detached and isolated spicules.

[1] Although the statement has been made repeatedly in print that sponge spicules have been found in the Grand Canyon Pre-Cambrian, we have been unable to find any published statement or any illustration made by a person who actually saw the specimens. Consequently, this occurrence must be regarded with uncertainty until made more definite or invalidated.

[2] Raymond, P. E., 1935, Pre-Cambrian life, *Bull. Geol. Soc. Amer.*, vol. 46, pp. 375–392.

Few whole specimens have been found. Isolated sponge spicules have wide distribution in present marine and fresh-water deposits and are commonly encountered in residues from the acid digestion of ancient calcareous rocks. Certain types of flint and chert nodules contain great numbers of siliceous sponge spicules, and it has been suggested that some of these concretions owe their origin to the solution of spicules and the subsequent deposition of the dissolved silica.

REFERENCES

General

HYMAN, L. H. 1940. "The Invertebrates." New York, McGraw-Hill Book Company, Inc., Vol. I, "Protozoa through Ctenophora," pp. 284-364.

LAUBENFELS, M. W. DE. 1936. The oecology of Porifera and possibilities of deductions as to the paleoecology of sponges from their fossils. *Nat. Res. Council, Rept. Comm. Paleoecology*, 1935–1936, pp. 44–54.

Fossil Porifera (Excluding Pleospongia)

CASTER, K. E. 1939. Siliceous sponges from the Mississippian and Devonian strata of the Penn-York embayment. *Jour. Paleontology*, vol. 13, pp. 1–20. (Abstract) 1938. *Bull. Geol. Soc. Amer.*, vol. 49, pp. 1910–1911.

———. 1941. The Titusvilliidae; Paleozoic and Recent branching Hexactinellida. *Pal. Americana*, vol. 2, pp. 1–52.

CLARKE, J. M. 1920. The great glass-sponge colonies of the Devonian; their origin, rise and disappearance. *Jour. Geol.*, vol. 28, pp. 25–37.

HALL, J., and CLARKE, J. M. 1898. A memoir on the Paleozoic reticulate sponges constituting the Family Dictyospongidae. *N.Y. State Geol. Ann. Rept.* 15, pp. 741–984. 1899. *Ann. Rept.* 16, pp. 41–448. 1898. *N.Y. State Mus. Mem.* 2, 350 pp.

HINDE, G. J. 1887–1893. "Monograph of British Fossil Sponges." Palaeontographical Society.

KING, R. H. 1938. Pennsylvanian sponges of north-central Texas. *Jour. Paleontology*, vol. 12, pp. 498–504.

———. 1943. New Carboniferous and Permian sponges. *State Geol. Surv. Kan., Bull.* 46, pp. 1–36.

OAKLEY, K. P. 1937. Cretaceous sponges: some biological and geological considerations. *Proc. Geol. Assoc.*, vol. 48, pp. 330–347.

O'CONNELL, M. 1919. The Schrammen collection of Cretaceous Silicispongiae in the American Museum of Natural History. *Bull. Amer. Mus. Nat. Hist.*, vol. 41, pp. 1–261. (This article contains an excellent list of references.)

RAUFF, H. 1893, Palaeospongiologie. *Palaeontographica*, vol. 40, pp. 1–346. 1895. *Ibid.*, vol. 41, pp. 223–272.

RAYMOND, P. E., and OKULITCH, V. J. 1940. Some Chazyan sponges. *Bull. Mus. Comp. Zool.* (Harvard), vol. 86, pp. 195–214.

SCOTT, H. W. 1937. Classification of sponge spicules (abstract). *Geol. Soc. Amer. Proc.*, 1936, p. 359.

———. 1943. Siliceous sponge spicules from the Lower Pennsylvanian of Montana. *Amer. Midland Naturalist*, vol. 29, pp. 732–760.

SHIMER, H. W., and SHROCK, R. R. 1944. "Index Fossils of North America." New York, John Wiley & Sons, Inc.; Cambridge, Mass., The Technology Press, pp. 49–57.

ULRICH, E. O. 1890. American Paleozoic Sponges. *Ill. Geol. Surv.*, vol. 8, pp. 209–241.

WALCOTT, C. D. 1920. Cambrian geology and paleontology, vol. 4, no. 6. Middle Cambrian Spongiae. *Smith Misc. Coll.*, vol. 67, no. 6, pp. 261–364.

WELLER, J. M. 1930. Siliceous sponge spicules of Pennsylvanian age from Illinois and Indiana. *Jour. Paleontology*, vol. 4, pp. 223–251.

Pleospongia

BEDFORD, R., and BEDFORD, W. R. 1936. Further notes on Archaeocyathi (Cyathospongia) and other organisms from the Lower Cambrian of Beltana, South Australia. *Kyancutta Mus., Mem.* 2, pp. 9–20.

OKULITCH, V. J. 1935. Cyathospongia; a new class of Porifera to include the Archaeocyathinae. *Trans. Roy. Soc. Can.*, 3d ser., vol. 29, pp. 75–106. (*Abstract*) *Roy. Soc. Can. Proc.*, vol. 29, No. 3, p. xcix.

———. 1943. North American Pleospongia. *Geol. Soc. Amer., Spec. Paper* 48, 112 pp.

RAYMOND, P. E. 1931. The systematic position of the Archaeocyathinae. *Bull. Mus. Comp. Zool.* (Harvard), vol. 55, pp. 172–177.

SIMON, W. 1939. Archaeocyathacea. *Abhandl. Senckenberg. naturf. Gesel.*, vol. 448, pp. 1–87.

———. 1941. Archaeocyathaceea. III. Ergänzungen zur Taxonomie aus neureren Arbeiten. *Senckenbergiana*, vol. 23, pp. 1–19.

TAYLOR, T. G. 1910. Archaeocyathinae, from the Cambrian of South Australia, with an account of the morphology and affinities of the whole class. *Mems. Roy. Soc. South Australia*, vol. 2, No. 2, pp. 1–188.

TING, T. H. 1937. Revision der Archaeocyathinen. *Neues Jahrb. Min., Geol., Pal.*, vol. 78, pp. 327–379.

VOLOGDIN, A. G. 1937. Archaeocyatha and the results of their study in U.S.S.R. *Prob. Paleontology*, vols. II–III, pp. 481–500 (*Pub. Lab. Pal. Moscow Univ.*).

CHAPTER 4

PHYLUM COELENTERATA[1]

INTRODUCTION

The Coelenterata constitute a large and varied group of living and extinct organisms representing the lowest animals with definite tissues. The individuals are either solitary or colonial in habit, and they are of two distinctly different but essentially homologous types, **polyp** and **medusa** (Fig. 4-1). The polyp has a tubular body with one end closed and attached and with the other carrying a central mouth that is generally surrounded by one or more rings of soft tentacles. The medusa, which is free-swimming, has a body shaped like an umbrella, with tentacles along the margin and the mouth in a central position on the concave underside. Both polyps and medusae show considerable range in morphology, and both appear in the life cycle of many species. The polyp is generally considered to represent a persistent larval stage, whereas the medusa seems to be a completely evolved coelenterate. In species that reproduce by **alternation of generations,** polyps and medusae alternate with one another in definite order. Polyps may secrete (1) an exoskeleton of horny or calcareous matter, (2) an endoskeleton of loose spicular elements (**sclerites, sclerodermites**), or (3) an endoskeleton of sclerites that are fused and form a continuous structure, or (4) may lack hard parts altogether. Medusae have no skeleton of any sort (Fig. 4-1*B*). The fossil record, which reaches back to the earliest Cambrian, and possibly into the Pre-Cambrian, consists almost entirely of skeletal structures, but some remarkably preserved structures ascribed to medusae have been found in the most ancient rocks (Fig. 4-5).

The chief characteristics of coelenterates can be summarized as follows:

1. Symmetry is radial, or biradial about an oral-aboral axis that extends from mouth to base.
2. The body wall consists of two layers of cells, an outer **epidermis (ectoderm)** and an inner **gastrodermis (endoderm),** separated by little or much gelatinous **mesogloea (mesenchyme).** Small stinging cells (**nematocysts**), particularly characteristic of the Coelenterata, are present in either or both layers.
3. The mouth is surrounded by tentacles and opens into a gastrovascular cavity, the **enteron** (or **coelenteron,** whence the name of the phylum), which serves for both digestion and distribution of food. The enteron is a simple saclike cavity in lower coelenterates but is divided into compartments by radial partitions (**mesenteries**) in more advanced forms.
4. There is no special opening through which waste products and undigested food are expelled; these are carried out through the mouth.

[1] Coelenterata—Gr. *koilos*, hollow, + *enteron*, intestine; referring to the hollow internal cavity of the individual coelenterate.

Cnidaria is now being used for this phylum by some zoologists (*e.g.*, Hyman, 1940).

5. Coelenterata lack blood, respiratory and circulatory organs, and a central nervous system but have a crude system of nerve cells in the body wall. They also lack both a canal system and collared flagellate cells, in these respects differing strikingly from the Porifera.

6. Reproduction is variable. Some species reproduce sexually, others by alternation of generations, and still others by a modified sexual cycle.

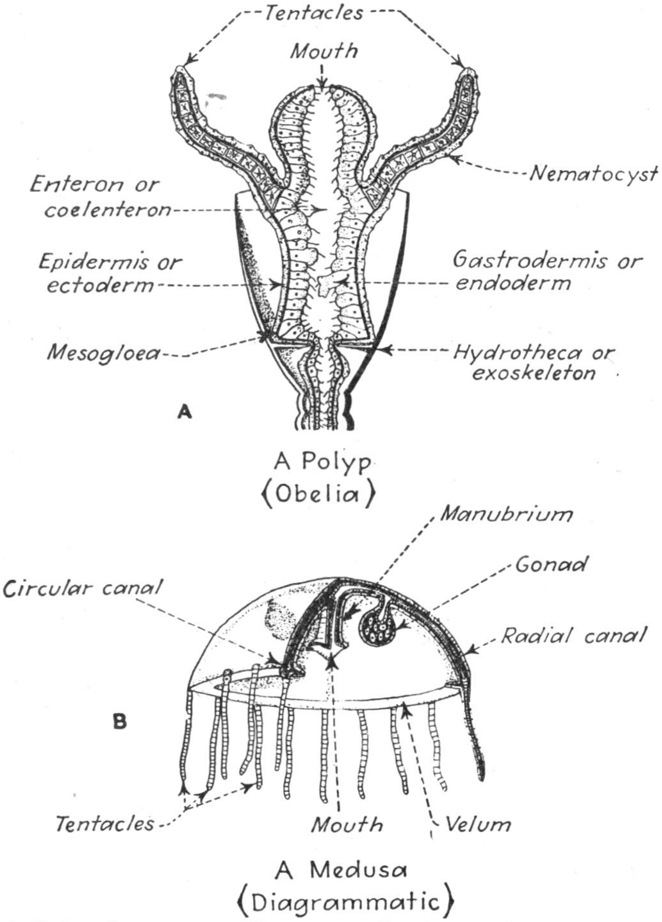

FIG. 4-1. Phylum COELENTERATA. The two types of coelenterate individuals. *A*. A **polyp** of the colonial *Obelia*, sectioned vertically to show the numerous structures of the body and the relation of organism to exoskeleton. *B*. A **medusa**, sectioned to show the more important internal and external structures. (*Modified after Parker and Haswell*, 1930.)

7. The skeleton (limited to polyps) is horny or calcareous, external or internal, and discontinuous or continuous. Medusae lack hard parts but have left a fossil record as impressions and fillings of the enteron (Figs. 4-2*E*-*F*, 4-5).

Throughout their long history coelenterates have lived mainly in salt water. A few species now live in fresh water, and some ancient forms may have been similarly adapted. All fossil stony corals, the extinct stromatoporoids, and the few known fossil hydrozoans and jellyfish seem to have been marine.

The phylum has an extensive fossil record. The excellence and completeness of this record, the abundance of fossils, and the great stratigraphic value of certain groups have stimulated extensive investigation of fossil coelenterates.

The Organism

The individual coelenterate is one of two definite morphological types: (1) a hollow, saclike polyp or (2) an umbrella-shaped medusa (Fig. 4-1). Polyps are almost invariably attached, commonly secrete a skeleton, and are typified by the marine colonial *Obelia* (Fig. 4-1*A*). Medusae are usually free-swimming (nektonic or nektonic-planktonic) and without skeletal hard parts. The common jellyfish and the familiar Portuguese man-of-war are typical examples of medusae.

The Polyp. The body of a typical polyp is a cylindrical tube closed at the bottom end and open at the top, with a circlet of tentacles surrounding the opening. In some coelenterates the basal part of the body is pointed and this end of the body or of the exoskeleton is simply stuck into the bottom mud; more commonly, however, attachment is by a **basal** or **pedal disk,** by skeletal secretion, or by rootlike **stolons.**

The free oral end of a polyp carries the oral opening, or mouth, on the summit of a vase-shaped or conical **hypostome,** which is encircled by hollow tentacles covered with stinging organs (nematocysts). In many corals the mouth is surrounded by an expanded **oral disk** from which rise many solid and hollow tentacles (Fig. 4-34*A*–*B*).

The mouth communicates directly with the enteron in hydrozoans, but in the more complex coelenterates the body wall is invaginated to form a **gullet** (Fig. 4-2*C*), which in the early growth stages is termed a **stomodeum** (Hyman, 1944).

Hydra, a solitary fresh-water hydrozoan, has 6 to 10 hollow threadlike tentacles; other hydrozoans have different numbers; and sea anemones are noted for the many tentacles that rise from the oral disk. When the animal is feeding the tentacles are extended, but when it is disturbed these are retracted and folded over the mouth.

The tentacles and body wall of a polyp are composed of two well-defined layers—an outer of epidermal cells, the **epidermis,** and an inner of gastrodermal cells, the **gastrodermis.** The two layers are separated by gelatinous mesogloea of variable thickness (Fig. 4-1*A*).

The epidermis (or **ectoderm**) of a polyp forms the outer layer of the organism and secretes the horny or calcareous exoskeleton. It contains muscular and nerve tissue, and commonly has minute stinging cells, the **nematocysts.** These cells, which are diagnostic of the Coelenterata (also present in a few protozoans), have not been demonstrated in extinct representatives of the phylum.

Each nematocyst is a capsule containing a coiled capillary tube. When the nematocyst is stimulated, the tube is discharged to the exterior, where it may paralyze prey by injecting a poisonous fluid or hold it by bristles. The tube can also serve for adhesion.

The inside of the enteron, and of those tentacles that are hollow, is lined with endodermal cells. These are not collared as in the Porifera but commonly have flagella and also nematocysts. Cells of the gastroderm acquire

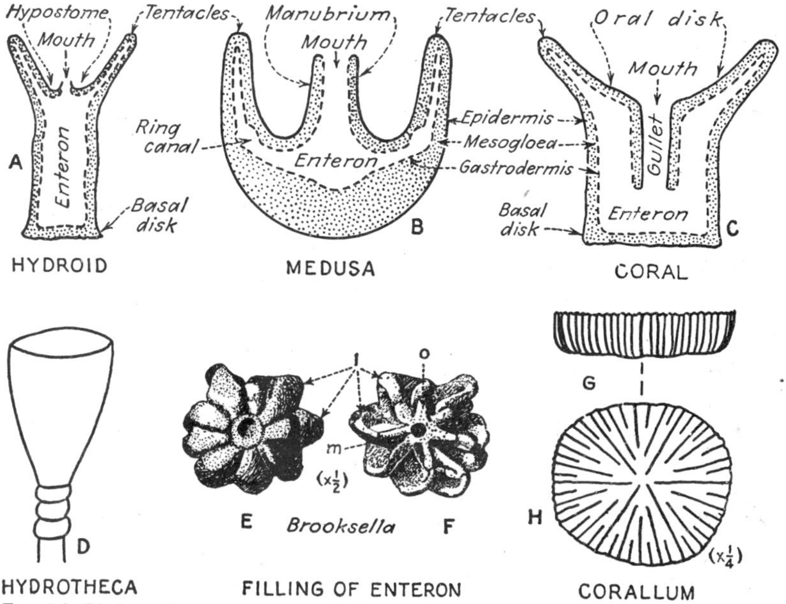

FIG. 4-2. Phylum COELENTERATA. *A–C.* Comparison of a hydroid polyp, a medusa (inverted), and an anthozoan or coral polyp. *D.* Chitinous hydrotheca made by a hydroid polyp. *E–F.* Supposed mud fillings of the enteron of a Cambrian jellyfish. *G–H.* Side and top views of a calcareous corallum of a solitary coralline polyp. (*A–C modified after Storer,* 1943.)

and assimilate food, discharge waste products into the enteron, and in some species produce the reproductive organs. Waste products and sex cells leave the polyp through the oral opening; food and foreign sex cells are brought into the enteron through the same opening. Water is drawn in and expelled by contraction and expansion of the body wall.

The mesogloea varies from a thin noncellular cement to a thick gelatinous mass or even a fibrous layer. Other than acting as a kind of connective tissue it seems to perform no important function except to secrete small calcareous sclerites which form a kind of endoskeleton in some hydroids and corals.

Polyps usually have well-developed radial and in the Anthozoa less obvious biradial symmetry. The one chief axis of radial symmetry passes through the mouth and base; soft and hard parts are arranged concentrically about this axis. The body wall is unwrinkled and the enteron is a simple tube in the hydroids, but in the Anthozoa the wall has many inward folds and the enteron is divided into compartments by partitions that have the symmetry of the organism. Certain exoskeletal features are closely related to these folds and partitions. (See later discussion of mesenteries and septa.) Most polyps are sessile and colonial, but a few live a solitary benthonic life. The Siphonophora differ in being completely pelagic. Skeleton formation is common among polypoid coelenterates, either as exoskeletal structures of chitinous and calcareous matter secreted by the endoderm on the outer surface or as endoskeletal particles in the form of calcareous sclerites and continuous structures made in the mesogloea.

The Medusa. The typical medusa resembles an umbrella with tentacles hanging down along the rim and with the mouth at the end of a short tube (**manubrium**) in a central position on the underside (Figs. 4-1*B*, 4-2*B*). A prominent inwardly projecting shelf, the **velum,** is characteristic of the Hydrozoa.

In addition to epidermis and gastrodermis, there is considerable mesogloea in many medusae, and the enteron occupies only part of the interior of the bell (Figs. 4-1*B*, 4-2*B*). Four (or some multiple of four) **radial canals** lead from the enteron to a **circular canal** lying in the bell margin. The four radial canals are 90° apart and, along with the marginal sense organs and tentacles, give the ordinary medusa a marked four-part (**tetramerous**) radial symmetry. This symmetry is conspicuous in certain fossil impressions that have been referred to medusae.

All medusae are more than 90 per cent water by weight, and some brackish-water species have been reported to contain as much as 98 per cent.[1] In spite of their soft nature, however, medusae are known to leave impressions in soft sediments and to have their gastral cavities filled with mud and sand (Fig. 4-2*E–F*). Certain fossil impressions and supposed gastral fillings seem to represent similar phenomena among ancient jellyfish (Fig. 4-5).

Reproduction. Reproduction among the Coelenterata is varied. Coelenterates having only the medusoid form reproduce sexually. Those having only the polyp stage reproduce either sexually or both sexually and asexually. Those coelenterates having alternation of the polypoid with the medusoid stage have asexual reproduction confined to the polyp and sexual reproduction to the medusa.

In **sexual reproduction** sperm cells are discharged into the enteron whence they find their way to the exterior through the mouth. They drift about until they enter the coelenteron of another individual of the same species, where they unite with ova produced in the ectoderm or gastroderm, depending on the group. A ciliated free-swimming larva, the **planula** (Fig. 4-20*A–D*), develops from this union and ultimately leaves the parent through the mouth. After a short free life the planula attaches itself by the anterior end, a new opening appears on the free end, and the planula ultimately

[1] HYMAN, L. H., 1938, The water content of medusae, *Science*, vol. 87, p. 166. This author states that the water content of medusae in sea water of typical salinity ranges from 94 to 96.5 per cent and in brackish water is as much as 98 per cent.

becomes an adult polyp. Coelenterates having only the medusoid stage release the sexual products into the water, and new medusae develop directly from these.

In **asexual reproduction** a new polyp buds from some part of the body wall or from the oral disk (calicinal) of the parent, from a rootlike extension (stolonal), or from some part of the common connective tissue. Reproduction of the asexual type may also take place by **fission,** in which one polyp divides to form two new ones; or by **rejuvenescence,** whereby a new polyp seems to form through the disintegration of its parent (Figs. 4-31, 4-32).

Budding takes place in the soft parts of the animal, and in some cases as many as three generations appear on a single organism. The buds in non-colonial forms leave the parent polyp and settle down by themselves. If they do not leave the parent immediately, however, skeletal material tends to be secreted around the base of the attached polyp. Reproduction by budding leads to the formation of colonies commensurate with the extent to which it proceeds, and the individuals of such colonies usually are connected by a common fleshy tissue known as **coenenchyme.** After a space has been fully occupied, further budding results in an immense mortality, because there is no room for the newly created individuals, and only those can survive that take the space of parents unable to maintain themselves in the intense competition for space.

In reproduction by alternation of generations some of the asexually produced buds of the colony develop the special function of producing small medusae, which ultimately become free-swimming. Upon attaining maturity these medusae release sexual products which unite with their opposites from other medusae of the same species. This union of male and female elements produces an embryo, which passes through a developmental period as a free-moving individual and then attaches itself and matures into a sessile polyp. The polyp then buds and thus starts a new colony.

The colonial habit is strongly developed in the Coelenterata, especially among the reef builders that make extensive structures composed to a greater or less degree of their stony exoskeletons.[1] Many living coelenterates lack the colonial habit, and many extinct corals likewise did not form colonies.

The Skeleton

Most sessile coelenterates secrete an internal or external skeleton of some sort.[2] The internal skeleton, commonly referred to as an endoskeleton, usually consists of minute skeletal elements (sclerites) secreted in the mesen-

[1] It should not be assumed, however, that organic reefs are made only by coelenterates. Many other organisms are also important builders (*e.g.*, calcareous algae).

[2] Bryan and Hill (1941) have shown that the skeletal elements of the scleractinians (hexacorals) grow by spherulitic crystallization of calcium carbonate, and they conclude that this is "due to the organic guidance of an inorganic process." Dr. H. C. Wang has advised us that he can recognize at least two, and possibly three, fundamental types of microscopic structure among the Paleozoic rugose corals (Tetracorallia): (1) a flaky type with lamellar structure; (2) a type characterized by fibers of calcite; and (3) a layered structure traversed by fibers. These structures are well enough preserved in many fossils to be used as a basis of ordinal and lower taxonomic divisions. (Also see Wang, 1951.)

chyme. These sclerites are generally scattered loosely through the mesenchyme, but in some species of corals they are joined together to form a continuous structure.

Exoskeletons are secreted by the ectodermal cells, hence reproduce the irregularities of the epiderm. If the epidermal surface is crenulated, for example, the wall of the exoskeleton will also have crenulations. The exoskeleton is typically a rigid framework of calcareous or horny material, and it can be the work of a solitary polyp or of a colony of individuals.

Development of the Exoskeleton. The embryonic skeleton of a typical coelenterate has the shape and appearance of a tiny, hollow cup and is known as the prototheca (earliest cup) (Fig. 4-20*G*). From this simple cup, by additions at its upper end, arise a great variety of skeletal types depending upon the way in which the polyps bud and grow.

The conical or tubular cup secreted by an adult polyp about and under itself is designated a **corallite.** This is more or less horn-shaped, if the polyp is young, but in the adults of many forms it tends to become cylindrical. Under conditions of excessive budding, necessitating utilization of all space, the corallites are crowded together and become prismatic. The upper part of the corallite, or that part occupied by the polyp, is the **calyx** (**calice**). The adult skeletal structure, whether composed of a single corallite or of a great number, is a **corallum** (plural, **coralla**). Hence in the case of the solitary corals, which remain conical or tubular throughout life, the skeleton is both a corallite and a corallum. In all other cases two or more corallites combine to form a corallum.

A mature corallum may be a single isolated cone partly filled with internal structures; a hemispherical mass of calcareous matter resembling a button or biscuit; a horny or calcareous framework resembling a bush or tree; or a massive irregularly shaped domal structure composed of many cylindrical or prismatic tubes more or less tightly packed together (Figs. 4-3, 4-4, 4-11, 4-33).

CLASSIFICATION

In the subdivision of a group of organisms as varied as those that are included in the Coelenterata, consideration has to be given to the following:

1. Phylogenetic relations where they can be determined.
2. Nature of the soft parts of the animal to the extent that they can be studied directly or inferred from fossil structures.
3. Nature of life cycle, including mode of reproduction.
4. Structural and architectural features of the exoskeletons.
5. Internal skeletal structures.

On the basis of these considerations the Phylum Coelenterata is divided into the following classes:

Class 1. Hydrozoa. Living and fossil, fresh-water and marine, individual or colonial coelenterates, some forms having both polyp and medusoid stages. The geologic range is from Lower Cambrian to Recent, and the fossil record is meager.

Class 2. Stromatoporoidea. A heterogeneous assemblage of extinct marine organisms bearing some resemblances to both sponges and hydrozoans. The established range is from the Ordovician to the end of the Devonian, but several genera have been described from the "Permo-Carboniferous" of Farther India.

Class 3. Scyphozoa. Living and fossil medusoid coelenterates without a polyp stage or with a very limited one. The medusoid stage is well developed. Fossils have been reported from the Lower and Middle Cambrian, Permian, and Jurassic.

Class 4. Anthozoa. Solitary and colonial, living and extinct, fleshy and stony corals with large exoskeletons of complicated structure and variable architecture. There is no medusoid stage. The group appears to begin in the early Ordovician[1] and extends throughout geologic time to the present. This is probably the most important group of the Phylum Coelenterata.

Class Hydrozoa[2]

The Organism

The Hydrozoa are typically bell-shaped or vaselike coelenterates with tentacles surrounding the mouth, which is situated in the center of the oral disk. The mouth opens directly into a simple, undivided tubular enteron. The corallites are simple cups and cones (Figs. 4-1*A*, 4-2*D*). Individuals are small, usually not exceeding 2 or 3 mm. in diameter and slightly more in height. They may be typical polyps, specialized polyps (generally with reproductive functions only), or small medusae. Hence they are sometimes described as **dimorphic** or **polymorphic.** The fully formed medusae have a velum.

Some groups of hydrozoans have an alternation of generations in which a polyp stage alternates with a medusoid stage. Medusae bud from specialized polyps of the colony, which differ in shape as well as in function from the normal food-getting polyps and are surrounded by a transparent bulblike, horny structure, the **gonotheca** (Fig. 4-3*C–D*). Normal polyps live in vaselike cups termed **hydrothecae** (Figs. 4-1*A*, 4-2*D*). The specialized polyps are nourished by normal polyps, with which they are connected by coenenchyme, and it is commonly difficult to differentiate between individual and colony. Any of the normal polyps can produce new polyps by budding, but the latter tend to remain attached. Medusoid forms lacking the polyp stage reproduce sexually. Polyp forms lacking the medusoid stage reproduce both sexually and asexually. In the cases just cited the reproductive organs are almost always of ectodermal origin.

Hydrozoa that reproduce asexually without detachment of the buds develop colonies that commonly expand into large masses of various shapes. The polyps in such colonies are connected in some manner with those adjacent by common fleshy coenenchyme, and ordinarily they do not tend to come into close contact The epidermis of the polyps and the underside of the connective coenenchyme secrete an exoskeletal framework of horny or cal-

[1] The genus *Mackenzia* from the Middle Cambrian of British Columbia was assigned to the holothurians by C. D. Walcott (*Smith, Misc. Coll.*, vol. 57, p. 54, 1911). A. L. Clark (*Science*, vol. 35, p. 277, 1912), however, suggested that the organism was an actiniarian, and if this is correct, the range of the Anthozoa extends to the Middle Cambrian.

[2] Hydrozoa—Gr. *hydor*, water, + *zoon*, animal; referring to the fact that the animal dwells in water.

careous material. That secreted by the coenenchyme connects and surrounds the cavities occupied by the polyps and consists of more or less structureless skeletal matter termed **sclerenchyme.** Hydrozoans having a horny exoskeleton of definite shape are illustrated by living *Obelia* (Fig. 4-1*A*) and *Tubularia*. Colonial associations of calcareous exoskeletons without definite shape are illustrated by the living *Millepora* (Fig. 4-4*A–E*). The extinct stromatoporoids have a skeletal structure somewhat analogous to that of *Millepora* (Fig. 4-7).

About 2,700 species of living hydrozoans are known. These are chiefly marine and include both solitary and colonial forms.

Habitat and Geologic History

The Hydrozoa live mainly in marine waters at the present time, and all species secreting skeletons of horny or calcareous substances are confined to that habitat. The few forms living in fresh water have no hard parts and hence have little chance of leaving any fossil evidence of their existence. Both polyps and medusae live in each of the aquatic habitats.

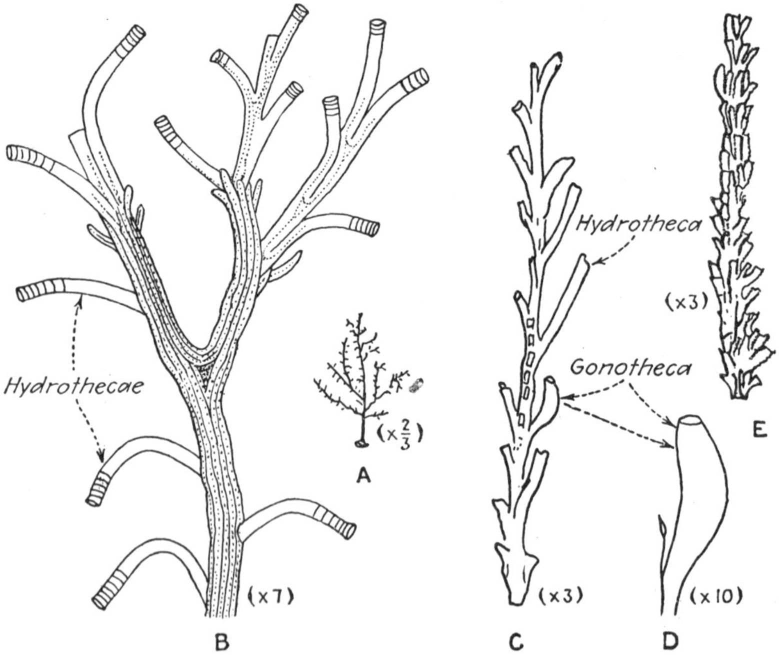

FIG. 4-3. Class HYDROZOA. Living and fossil Hydroida. *A–B*. *Cryptolaria*, a common marine hydroid: *A*, a colony, somewhat reduced; *B*, part of colony enlarged to show the hydrothecae. *C–E*. *Archaeocryptolaria*, an ancient fossil hydroid: *C*, *A. recta* from the Middle Cambrian of Australia; *D*, gonotheca of *C*; *E*, *A. compacta* from the Lower Ordovician Athens shale of Virginia. (*A–B after Allman*, 1888; *C–D after Chapman*, 1919; *E after Decker*, 1948.)

Fossil remains of the Hydrozoa are found in rocks ranging in age from the earliest Cambrian to the most recently formed deposits. Inclusion in the Hydrozoa of the extinct stromatoporoids, a common practice in textbooks of paleontology and zoology, greatly enlarges the group.

Classification

The Class Hydrozoa can be subdivided into the following orders:

Order 1. Hydroida.
Order 2. Hydrocorallina.
Order 3. Trachylina.
Order 4. Siphonophora.

Order 1. Hydroida. The Hydroida are hydrozoans with a well-developed polyp generation of solitary or colonial habit that usually buds off free medusae. A naked (athecate = Gymnoblastea) and an encased (thecate = Calyptoblastea) division are recognized. Nematocysts are limited to the epidermis. The exoskeleton is dendritic or flowerlike and is composed of horny or calcareous matter capable of preservation. The order, composed of some dozen or more families of living hydroids, includes the well-known *Hydra* (fresh-water, solitary, no medusae), which is not, however, typical of the order, and the marine *Hydractinia*, *Sertularia*, *Tubularia*, *Obelia* (Fig. 4-1*A*), *Cryptolaria* (Fig. 4-3*A–B*), and *Campanularia* (all sessile and colonial).

Fossil Hydroida are known from Lower and Middle Cambrian[1] rocks of North America and Australia (Fig. 4-3*C–E*) and have been reported from rocks of younger age in different parts of the world. Certain *Hydractinia*, for example, are said to be common in the Jurassic of the Mediterranean region. All fossil Hydroida thus far reported are of the colonial and polypoid type; no fossil medusae of this order are known.

Order 2. Hydrocorallina. The polyps of this order, subdivided by some authors into Milleporina and Stylasterina, are dimorphic, are connected with coenenchyme, and protrude from tiny pores in a massive calcareous skeleton termed the **coenosteum.** The skeleton, which is composed of calcareous fibers secreted by the coenenchyme, has tubular openings of two sizes corresponding to the two types of polyps. The larger tubes, **gastropores,** house the feeding polyps (**gastrozooids**), whereas the smaller **dactylopores** house protective polyps known as **dactylozooids.** Tabulae divide the older part of the tubes into compartments (Fig. 4–4).

Millepora (Fig. 4-4*A–E*), constituting the Milleporina, and *Stylaster* (Fig. 4-4*F–H*), a typical representative of the Stylasterina, are common in shallow

[1] RUEDEMANN, R., 1931, Some new Middle Cambrian fossils from British Columbia, *Proc. U.S. Nat. Mus.*, vol. 79, pp. 1–18; RUEDEMANN, R., *Camptostroma*, a Lower Cambrian floating hydrozoan, *ibid.*, vol. 82, pp. 1–8. CHAPMAN, F., 1919, Some hydroid remains of Lower Palaeozoic age from Monegetta, near Lancefield, *Proc. Roy. Soc. Victoria*, vol. 31, n.s., pp. 388–393. CHAPMAN, F., and THOMAS, D. E., 1936, The Cambrian Hydroida of the Heathcote and Monegetta districts, *ibid.*, vol. 48, n.s., pp. 193–210. See also DECKER, C. E., 1948, A new species of hydrozoan from the Athens shale of Virginia, *Jour. Paleontology*, vol. 22, pp. 528–529

tropical waters to a depth of 30 m. (about 100 ft). Their upright leaflike and branching calcareous skeletons, not uncommonly ½ m. in height, make up an important part of many coral reefs. The earliest known fossil representatives of the order appear in Triassic rocks.

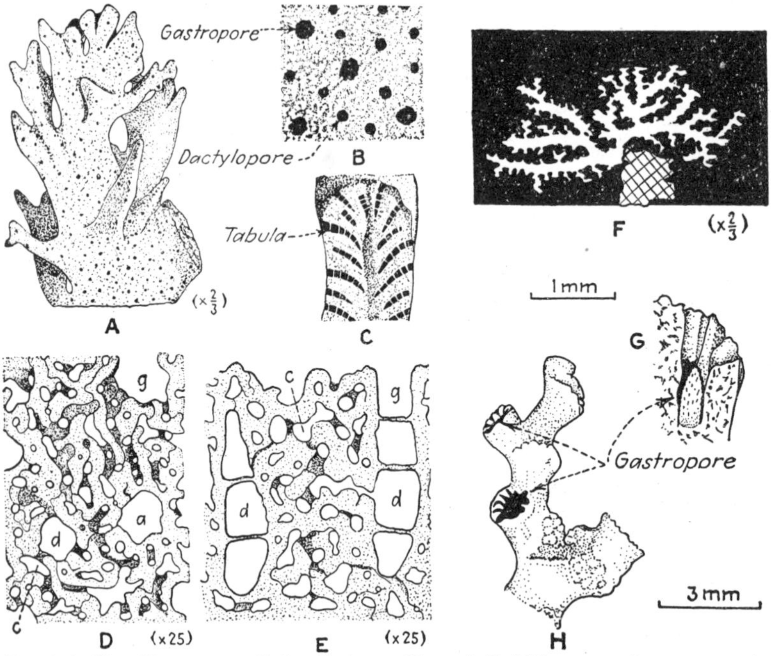

FIG. 4-4. Class HYDROZOA—Order Hydrocorallina. *A–E. Millepora*, an important rock-building hydrozoan, particularly on present coral reefs: *A*, complete colony; *B*, portion of surface, magnified, showing gastropores and dactylopores separated by sclerenchymal material; *C*, vertical section, magnified, showing tabulated zooidal tubes (gastropores and dactylopores); *D–E*. Tangential and longitudinal sections, much magnified, showing gastropores (*g*), dactylopores (*d*), and coenenchymal canals (*c*). *F–H. Stylaster* (*Eustylaster*) *crassior*, a typical example of the Stylasterina: *F*, a complete colony attached to a rock; *G*, part of a branch; *H*, longitudinal section of a gastropore, showing internal structure. (*A–E modified after Nicholson and Lydekker*, 1889; *F–H after Broch*, 1936.)

Order 3. Trachylina. The typical representative of this order, a small medusa with a velum, develops directly from an ovum (through a planula stage), or has the polyp generation much reduced. Trachyline medusae are considered to be the most primitive living coelenterates. Most of them are marine and pelagic, living in warmer waters from the surface to depths as great as 3,000 m. (10,000 ft.); some forms are confined to shallow marine waters; and a few species can live in brackish or fresh water.

The earliest supposed fossil representative of this order is the genus *Kirklandia* (Fig. 4-5*A*), recently reported from the Lower Cretaceous of

Texas[1] in the form of well-preserved imprints. As certain of the trachyline medusae are composed of tough and resistant jellylike substance, there seems to be a better possibility than some paleontologists seem willing to admit that the jellyfish of this order can leave a fossil record.

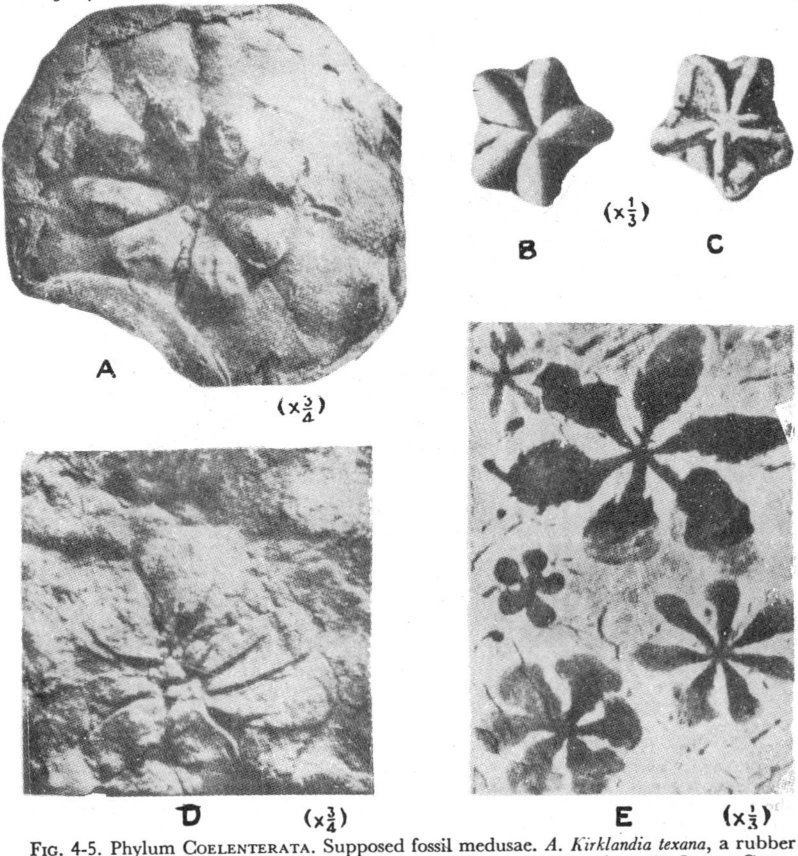

FIG. 4-5. Phylum COELENTERATA. Supposed fossil medusae. *A. Kirklandia texana*, a rubber cast of natural mold of a supposed trachyline hydrozoan medusa from the Lower Cretaceous of Texas. *B–C. Laotira cambria*, a well-preserved impression or gastric filling of supposed medusa from the Middle Cambrian of Alabama: *B* shows five principal exumbrella lobes and two small interradial lobes; *C* is a subumbrella view of *B*, showing strongly defined lobes. *D. Brooksella canyonensis*, a supposed jellyfish, from the Algonkian Grand Canyon series of Arizona. *E. Dactyloidites asteroides*, a slab with fine specimens of a supposed medusa, all showing extreme pressing out of the lobes of the umbrella, from the Middle Cambrian of Alabama. (*A after Caster*, 1945; *B–C, E after Walcott*, 1898; *D from original photograph by G. A. Cooper, also published in Proc. U.S. Nat. Mus., vol.* 89, 1941.)

Order 4. Siphonophora. The siphonophores are polymorphic, swimming or floating colonies of greatly modified polyps and medusae. The polyps are

[1] CASTER, K. E., 1945, A new jellyfish (*Kirklandia texana* Caster) from the Lower Cretaceous of Texas, *Pal. Americana*, vol. 3, no. 18, pp. 1–52.

of three types—feeding, protective, and reproductive. The medusoid forms include swimming bells, reproductive individuals, protective individuals, and a highly specialized **pneumatophore,** or float. The float has a chitinous internal lining and could be preserved as a fossil. *Physalia,* the Portuguese man-of-war, which has a large float (10 to 30 cm.) and tentacles hanging down for many meters, is probably the best known example of the order, although there are also many small and inconspicuous representatives.

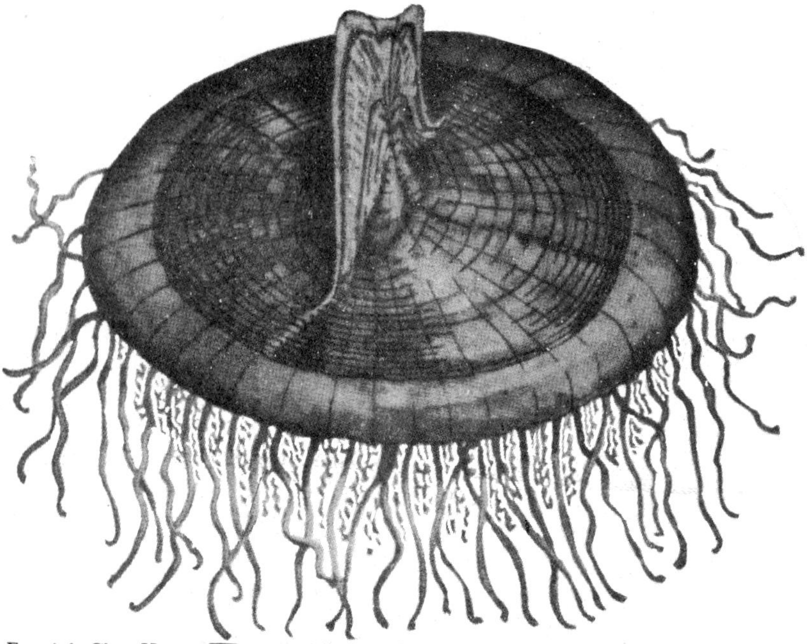

FIG. 4-6. Class HYDROZOA—Order Siphonophora. Restoration of *Plectodiscus cortlandensis,* a supposed "by-the-wind sailor," or disconectate siphonophore, from the New York Devonian (Ithaca). The drawing is based largely on the holotype pneumatophore, with the fleshy details inferred from preserved specimens of *Velella velella* (*mutica*), common "sailor" of the Gulf Stream. (*From original drawing by Anneliese S. Caster, which has also been published in Pal. Americana, vol.* 2, 1942.)

The oldest supposed fossil Siphonophora so far known seem to be closest to modern drifting siphonophores of Family Porpitidae. They have been reported from rocks of Middle Cambrian to Devonian age. Fossil representatives of the modern sailing siphonophores, the so-called "by-the-wind sailors" (Family Velellidae), have been reported recently from the Silurian[1] and Devonian[2] of New York (Fig. 4-6).

[1] FLOWER, R. H., and WAYLAND-SMITH, R., 1947, New fauna from the Vernon shale of New York, *Bull. Geol. Soc. Amer.,* vol. 58, p. 1180. (The authors report a new siphonophore from the Vernon shale.)

[2] CASTER, K. E., 1942, Two siphonophores from the Paleozoic, *Pal. Americana,* vol. 3, no. 14, pp. 1–34. (This important article reviews previous work and has an excellent bibliography.)

Class Stromatoporoidea

The Stromatoporoidea are a heterogeneous group of extinct marine organisms of uncertain taxonomic position[1] that built branching and massive calcareous skeletons on the shallow bottoms of Paleozoic and Mesozoic seas (Figs. 4-7, 4-8). These calcareous masses bear some resemblance to certain hydrocoralline exoskeletons, and for this reason the stromatoporoids have long been included in the Hydrozoa as an extinct order. As almost nothing is known about the soft parts of the organisms that built these structures, classification has had to be based on the features of the stony skeleton. Unfortunately, however, the skeletons exhibit a wide diversity of structure, so that students of stromatoporoids have differed as to how the group should be classified. It seems best, therefore, to set them apart as an extinct class but consider them most closely related to the Hydrozoa. From this heterogeneous group, which is admittedly a convenient one although almost certainly not a natural one, any genus can be removed as soon as its true biological relationships have been demonstrated.

Nature of the Skeleton

The calcareous skeleton of a stromatoporoid is termed a **coenosteum.** It has constant internal structure in the same species, but varies in dimension from tiny colonies less than an inch in diameter to large bodies several feet across and weighing hundreds of pounds. Shapes may be spheroidal, domal, columnar, dendroidal, laminar-incrusting, or some combination of these.

[1] Stromatoporoids (some or all) have been classified as calcareous algae, foraminifers of the *Gypsina plana* type, horny and calcareous sponges, hydrozoans, tabulate corals, bryozoans, and cephalopods.

The earliest serious study of the group seems to have been made by Baron von Rosen (see references at end of chapter) in 1867. He considered them horny sponges that became calcified. Nicholson, in his great monograph on British stromatoporoids, came to the conclusion that they are coelenterates, with affinities closest to certain Hydrozoa. He divided the Stromatoporoidea into a milleporoid group, characterized by so-called "zooidal tubes," and a hydractinoid division, which lacked such tubes. Parks, who investigated North American stromatoporoids more extensively than anyone else, accepted Nicholson's classification but doubted the existence of the zooidal tubes—the features on which Nicholson based his divisions! In 1914 Heinrich denied the existence of the zooidal tubes altogether. He designated the hydractinoids true stromatoporoids, but excluded the milleporoid group from the stromatoporoids altogether. Heinrich subdivided the true stromatoporoids into the Actinostromidae, characterized by massive fibers, and the Stromatoporidae, having porous or perforate fibers. Twitchell, a contemporary of Parks and Heinrich, concluded in 1929 that the true stromatoporoids of the latter were actually the ancestors of the modern *Demospongiae*, hence sponges, and not hydrozoans at all! Parks likewise turned away from the hydrozoan assignment for some so-called stromatoporoids, calling attention in 1933 (1934–1935) to the striking resemblance of such forms to certain Foraminifera. Still another stromatoporoid specialist, Kühn (1928, etc.), proposed an elaborate classification of the several hundred species of so-called stromatoporoids, assigning them to three extinct orders of the Hydrozoa—Stromatoporoidea, Labechioidea, and Sphaeractinoidea—which orders also appear under his authorship in Schindewolf's "Handbuch der Paläozoologie" Berlin, Gebrüder Borntraeger, 1939. This classification appears on a following page.

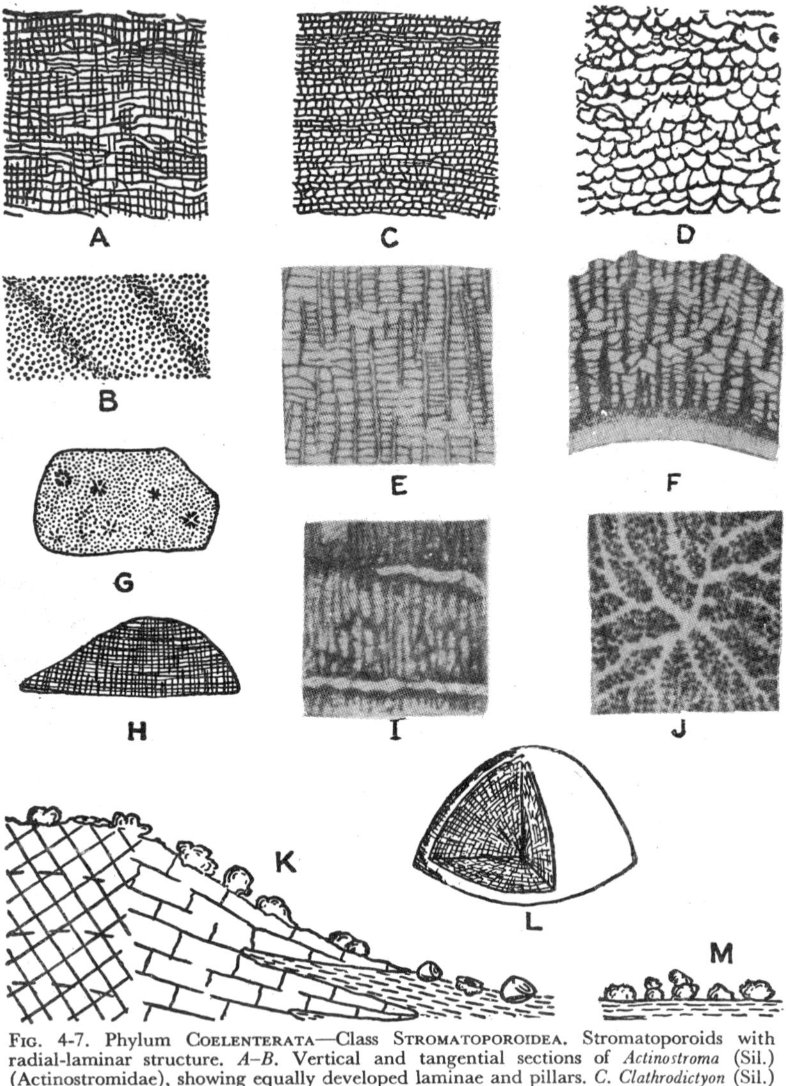

FIG. 4-7. Phylum COELENTERATA—Class STROMATOPOROIDEA. Stromatoporoids with radial-laminar structure. *A–B*. Vertical and tangential sections of *Actinostroma* (Sil.) (Actinostromidae), showing equally developed laminae and pillars. *C*. *Clathrodictyon* (Sil.) (Clathrodictyonidae), with laminae well developed and pillars confined to interlaminar spaces. *D*. *Clathrodictyon*, with cystose structure. *E*. *Stromatocerium* (Ord.) (Labechiidae), vertical section showing strongly developed pillars and short laminae. *F*. *Labechia* (Labechiidae), vertical section. *G–H*. Tangential and vertical sections (×2) of *Heptastylis* (Spongiomorphidae), a Triassic and Jurassic genus. *I*. *Stromatopora* (Ord.–Penn.) (Stromatoporidae), with thick, well-developed laminae. *J*. An astrorhiza of *Stromatopora*. *K*. Coenostea growing on a Silurian bioherm and being covered by sediment at the foot of the mound where they came to rest after rolling down the slope. *L*. A typical coenosteum sectioned to show internal structure. *M*. Coenostea forming a biostrome on a bottom of calcareous mud. *K–M* are diagrammatic; all others, ×7.5 unless otherwise indicated. (*A–F, I–J after Parks*, 1907, 1908, *and* 1936; *G–H after Nicholson*, 1886–1892.)

Clathrodictyon from the Upper Ordovician and Silurian exhibits all these shapes. The wide variation in shape is considered to be a response to variable bottom conditions.

The typical coenosteum is a framework of laminae, curved diaphragms, and connecting pillars, ramified to some extent by continuous or discontinuous galleries or interlaminar spaces. In a few forms the framework surrounds a prominent axis or tabulated axial cylinder.

It is supposed that the coenosteum was built mainly by ectodermal cells and is an exoskeleton. The nature of the skeletal structure, however, has led many investigators to conclude that although the organism probably sat on its coenosteum it also penetrated and to some extent completely enclosed the latest additions of stony matter.

The internal structure of coenostea is best studied on polished surfaces and in thin sections. Identifications based only on coenosteal shape and surficial characteristics are likely to be erroneous, because many genera with distinctly different internal structure have coenostea of similar size and shape.

Although the internal structure of the stromatoporoid coenosteum is reticulate or vesiculose in general plan, two somewhat distinct varieties can be recognized—**radial-laminar** and **beatricioid.**

The radial-laminar type of structure consists essentially of numerous closely spaced concentric laminae separated by outwardly directed porous or dense pillars and curved plates. The coenosteum of some species consists almost entirely of concentric laminae, so that it has a strongly layered appearance (*Stromatopora;* Fig. 4-7*I*). In a second group laminae and pillars are about equally developed, and a vertical section through a coenosteum has the appearance of a lattice (*Clathrodictyon, Actinostroma;* Fig. 4-7*A–D*). In a third group the radial pillars dominate and the concentric laminae are replaced by discontinuous regularly or irregularly disposed curved plates (*Stromatocerium, Labechia;* Fig. 4-7*E–F*). In still another group both laminar and radial elements are poorly developed and the entire framework is a labyrinthine mass of fibers and curved plates.[1]

It is thought that in life the entire surface of a coenosteum with radial-laminar structure was covered with living tissue in which distinct individuals were embedded or occupied zooidal tubes in some forms. These individuals seem to have been connected by minute tubes that ramified the porous outer part of the coenosteum. The way in which pillars and laminae were produced is unknown, but it may be that the individual organisms built pillars around themselves and at intervals added laminae for support as they grew outward. In some species the outer surface of a lamina bears conspicuous nodes and shallow pits with stellate or rosette systems of minute grooves or canals. These latter features are designated **astrorhizae** (Fig. 4-7*J*) and have been interpreted in several different ways.

Coenostea with structure of the beatricioid type are single or branching columns with a tabulated axial tube surrounded by one or more concentric zones of vesicular or reticulate structure of the radial-laminar type (Fig. 4-8).

[1] J. J. Galloway, on the basis of extensive unpublished research, does not consider these to be stromatoporoids (personal communication, April, 1949).

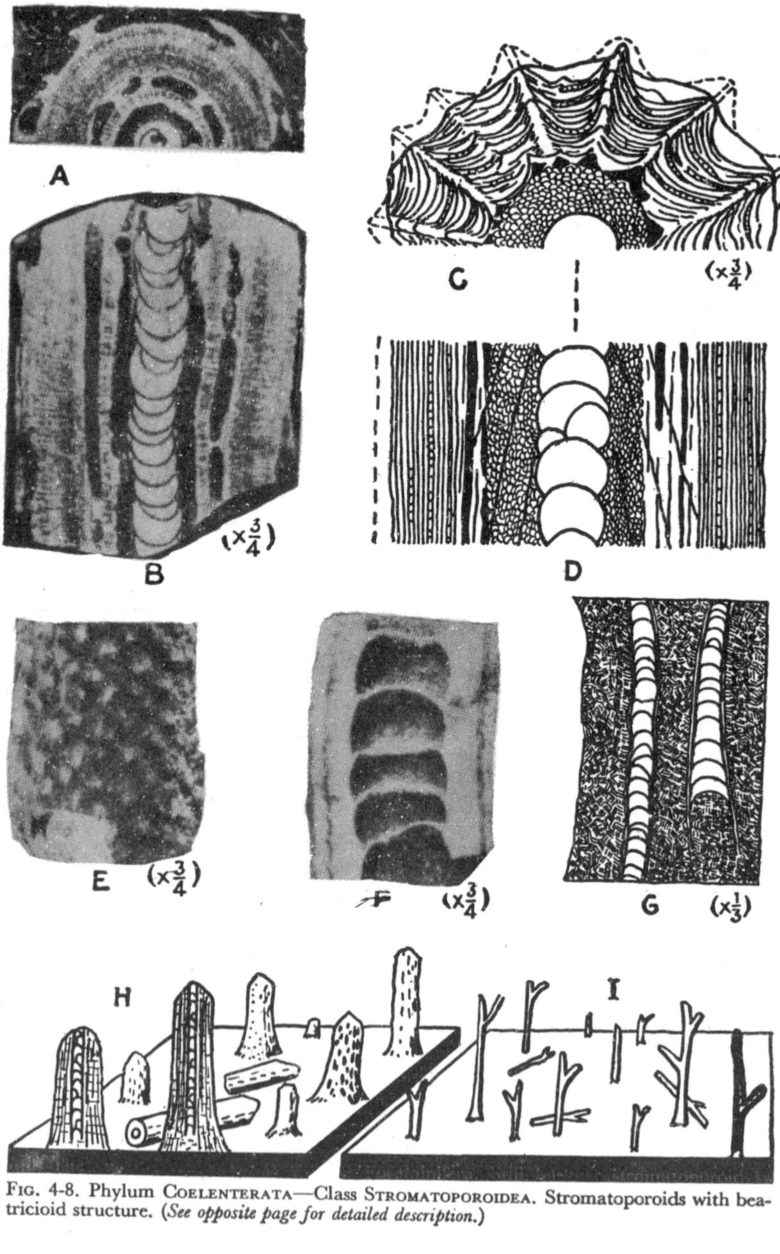

FIG. 4-8. Phylum COELENTERATA—Class STROMATOPOROIDEA. Stromatoporoids with beatricioid structure. (*See opposite page for detailed description.*)

The coenosteum of some species is a strong, well-integrated structure and consequently was preserved in its entirety (*Beatricea;* Fig. 4-8*E–F*). It is quite characteristic of certain forms, however, to have only the central tube well preserved (*Cryptophragmus;* Fig. 4-8*A–B*). The surrounding layers were probably destroyed in the fossilization process, though it is possible that they were only poorly developed or were never present at all in some forms.

Little can be determined about the nature of the organism that built coenostea of the beatricioid type. It seems reasonable to suppose that the vesicular radial-laminar structure of the concentric zones was built in the same way as outlined in a preceding paragraph. The purpose of the camerate axial tube and its relationship to the colonial organism as a whole are uncertain. It has been suggested that this tube was built by an alga and that the surrounding vesicular structure was made by some other organism that enveloped the algal structure with its stony secretions.[1] This interpretation does not seem likely and has not been generally accepted. We do not accept it here.

Classification

More than 65 genera and 320 species are assigned to the Stromatoporoidea. These range in age from Cambrian to Cretaceous and include several excellent index fossils of the middle Paleozoic. Taxonomic categories of higher than family rank generally have not been recognized because of the prevailing tendency to consider the Stromatoporoidea an extinct order of Hydrozoa. Kühn (1939), however, in discussing the Hydrozoa, has proposed a complete classification of the genera that we would include in the Stromatoporoidea, with that term having the broadest meaning. Kuhn's classification is shown in Table 4-1. Representatives of the more important of his families are illustrated in Figs. 4-7 and 4-8.

Geologic History of the Stromatoporoidea

The origin of the Stromatoporoidea is unknown. They appeared first in the Cambrian; they flourished in the mid-Paleozoic seas, where they contributed importantly to the construction of biostromes and bioherms (Fig. 4-7*K,M*) (Cumings, 1932; Cumings and Shrock, 1928; Fenton, M. A., 1931; Fenton, C. L., 1931; Raymond, 1924) and one large group persisted until the end of the Mesozoic. They do not seem to have been ancestral to any later group of organisms.

[1] SHIDELER, W. H., 1946, Beatricidae (abstract), *Bull. Geol. Soc. Amer.*, vol. 57, p. 1230.

FIG. 4-8. *A–B*. Transverse and longitudinal sections of *Cryptophragmus antiquatus* (Ord.), showing width of complete colony. *C–D*. Transverse and longitudinal sections of *Aulacera undulata* (Ord.) (Aulaceridae). *E–F*. Exterior surface and longitudinal section of *Beatricea nodulosa* (Ord.) (Aulaceridae), showing prominence of camerate tube. *G*. Two small specimens of *Beatricea gracilis*. *H–I*. Congregations of simple and forked stromatoporoid coenostea in growth position, and several broken coenostea that have fallen over (diagrammatic). It is assumed that the coenostea grew with the camerate tube vertical and the tabulae convex upward. (*A–B after Raymond*, 1914.)

TABLE 4-1. CLASSIFICATION AND GEOLOGIC RANGE OF THE FAMILIES OF STROMATOPOROIDEA (*sensu lato*)

Based largely on Kühn in Schindewolf's "Handbuch der Paläozoologie," Band 2A, 1939

Order	Family	Range									
		Camb.	Ord.	Sil.	Dev.	Miss.	Penn.	Perm.	Trias.	Jura.	Cret.
Stromatoporidea	Actinostromidae	x	x	x	x					x	x
	Clathrodictyonidae	x	x	x	x						
	Stromatoporidae		x	x	x	x	x	?		?	
	Disjectoporidae							x	x		
	Stromatoporinidae								x	x	x
Labechioidea	Labechiidae		x	x	x	x	x				
	Idiostromidae			x	x						
	Aulaceridae		x	x	x	x	x	x			
Sphaeractinoidea	Sphaeractinidae							x	x	x	x
	Heterastridiidae								x		x
	Spongiomorphidae								x	x	

CLASS SCYPHOZOA[1]

The Scyphozoa, or Scyphomedusae as they are commonly designated, are free-swimming umbrella-shaped medusae characterized by strongly developed tetramerous radial symmetry in both polypoid and medusoid generations. The gastrovascular cavity is divided by fleshy partitions into a central and four surrounding pouches of equal size and similar shape. Radial canals are present and their number is a multiple of four. Most scyphozoans have their parts symmetrically repeated around the oral-aboral axis to the number of four or some multiple of four; a few species, however, are built on a plan of six, and have hexamerous symmetry.

The umbrella, which lacks a velum, usually has a lobed margin from which long tentacles depend. Umbrellas as much as 2 m. (7 ft.) in diameter with tentacles 40 m. (130 ft.) long have been described. Living scyphozoans have many nematocysts and often become an annoyance to swimmers at seaside beaches. Most of a scyphozoan is water (as much as 96.5 per cent),[2] but the organic substance is commonly a tough gelatinous material of sufficient strength to leave impressions in soft sediments and to contain mud or sand that fills the gastrovascular cavity. Scyphozoa, therefore, can leave a fossil record and several supposed fossil medusae have been assigned to this class.

Objects and imprints believed to have been made by medusae have been found in rocks of many ages, the earliest being Lower Cambrian, or possibly

[1] Scyphozoa—Gr. *skyphos*, cup, + *zoon*, animal; referring to the resemblance of the scyphozoan to an inverted cup.

[2] HYMAN, *op. cit.*, pp. 166–167.

even Pre-Cambrian.[1] Although most have radial symmetry, not all have a tetramerous plan. The bell has from 4 to 12 lobes that continue to the center of the underside. There were radial canals, but traces of mouth and tentacles are generally missing, although certain markings in older Cambrian rocks have been ascribed to trailing of medusoid arms along a mud bottom.

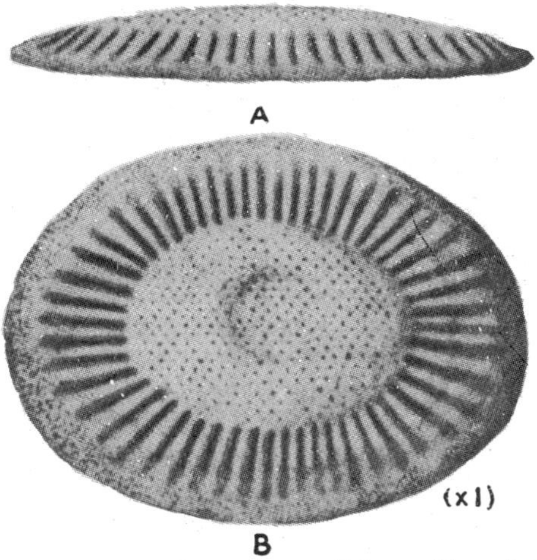

FIG. 4-9. Phylum COELENTERATA. *Camptostroma roddyi*, a problematical coelenterate fossil from the Lower Cambrian of Pennsylvania. It has been considered a floating tubularian hydrozoan and a coronate scyphozoan. It illustrates the difficulty paleontologists encounter in trying to assign fossils of extinct animals to living groups. The restoration shows the supposed top surface (lower figure) and a side view. (*After Ruedemann*, 1933.)

Zoologists and paleontologists now generally recognize the following four orders of Scyphozoa:

Order 1. Stauromedusae (Lucernariida). Goblet-shaped; sessile benthos, in colder marine coastal waters, living attached to seaweeds by an oral stalk. *No known fossils.*

Order 2. Cubomedusae (Carybdeida). Bells cubical, with four or multiples of four tentacles; widespread in warmer marine waters. The earliest known representative of this order is *Medusina quadrata* from the Jurassic limestone of Solenhofen.

Order 3. Coronatae. Deep-water scyphozoans with bell that is surrounded by a circular furrow above the scalloped margin; known fossil only in the Jurassic limestone of Bavaria, unless the Lower Cambrian *Camptostroma* (Fig. 4-9), which was considered by Ruedemann to be a tubularian hydrozoan, is actually a coronate scyphozoan as Kieslinger (1939) suggests. Some of the Solenhofen specimens are clearly recognizable as Coronatae, but others differ from any known living forms.

[1] A supposed fossil jellyfish has been reported from the Pre-Cambrian Nankoweap (Algonkian) sandstone of the Grand Canyon. Although the fossil does not resemble any known medusae, its lobate nature (Fig. 4-5*D*) has convinced many investigators of its medusoid affinities. If it is a true fossil jellyfish, then this type of coelenterate was present in seas far earlier than the Cambrian. For a description and illustrations of this important fossil, to which the binomial designation *Brooksella canyonensis* has been given, see Bassler (1941) and Van Gundy (1951).

Order 4. Discomedusae (Rhizostomeae and Semaeostomeae). These have the corners of the mouth prolonged as four grooved oral arms. Representatives of this order, which includes most jellyfishes, are widely distributed throughout existing oceans. *Aurelia* and *Rhizostoma* are typical living forms. Most supposed fossil medusae are assigned to this order.

Typical supposed fossil discomedusae include impressions of the upper and lower surfaces of the bell and mud fillings of the gastric pouches. Fossil specimens supposedly representing these types of preservation have been described from the Pre-Cambrian (Algonkian) of the Grand Canyon (*Brooksella canyonensis;* Fig. 4-5*D*), from the Lower Cambrian of New York, Sweden, Russia, and Bohemia; from the Middle Cambrian of British Columbia (*Peytoia*) and Alabama (*Brooksella* and *Laotira*) (Fig. 4-5); from the Upper Cambrian of Wyoming (*Laotira*); from the Permian of Saxony; and from the Jurassic of Solenhofen, Bavaria. The identification of some of these specimens has been questioned, and there seems to be doubt as to the exact nature of the animals that made many of them.

There now seems to be little doubt that Scyphozoa were present in the earliest Cambrian seas. Although no undoubted fossils referable to this class of coelenterates have been found in Pre-Cambrian rocks, unless the supposed jellyfish from the Grand Canyon is in truth a scyphozoan (see footnote on page 117), there seems no good reason to doubt that ancestors of the Cambrian Scyphozoa lived in the waters of Pre-Cambrian seas.

Group Conularida

The Conularida are a small group of extinct marine animals that had chitinous, pyramidal or flattened conical shells with marked quadrilateral radial symmetry (Fig. 4-10). The young were attached by the closed apical end, but with growth some became free-swimming. The apertural end of the shell is generally constricted by four incurved lobes, and in life tentacles are supposed to have extended outward from the apertural opening (Fig. 4-10*A*). Fossil conularid tests, which are commonly flattened in calcareous shales and argillaceous limestones, suggesting a somewhat flexible shell, have been found in rocks ranging in age from Lower Cambrian to Upper Triassic. Some were as long as 22 cm. (9 in.) and had an apertural diameter of 12 cm. (4¾ in.)

Fossil conularids were long considered to be extinct mollusks and generally were assigned to the Gastropoda. In 1937, however, Kiderlen,[1] supported by Knight[2] and later by Bouček,[3] presented evidence for referring the group, including *Serpulites* and several other genera, to the Scyphozoa. That assign-

[1] KIDERLEN, H., 1937, Die Conularien: ueber Bau und Leben der ersten Scyphozoa, *Neues Jahrb. Min., Geol., Pal.*, Abt. B, BB77, pp. 113–169.

[2] KNIGHT, J. B., 1937, *Conchopeltis* Walcott, an Ordovician genus of the Conularia, *Jour. Paleontology*, vol. 11, pp. 186–188.

[3] BOUČEK, B., 1939, Conularida *in* Schindewolf's "Handbuch der Paläozoologie," vol. 2A, pp. A111–131, Berlin, Gebrüder Borntraeger.

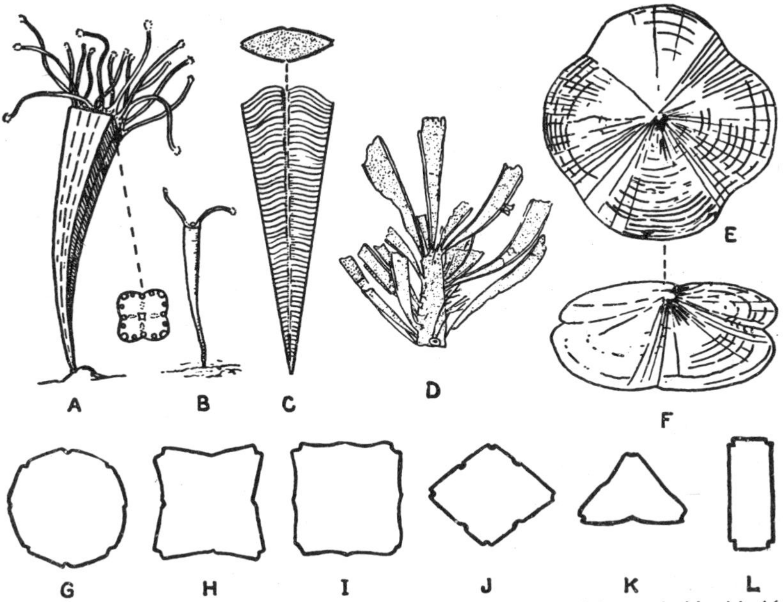

FIG. 4-10. Group CONULARIDA. *A*. Reconstruction of a living adult conularid with 16 tufted tentacles, and a cross section at the aperture showing oral opening and positions of the tentacle bases. *B*. Reconstruction of a young *Serpulites* with but two tentacles. *C*. Lateral view and transverse section of a flattened *Conularia* from the Wisconsin Devonian. This shell originally had a cross section similar to *J* below. *D*. A colony of *Serpulites* with young individuals attached to the parent stock. *E–F*. Top and oblique views of *Conchopeltis* showing its tetralobate flattened conical form. *G-K*. Outlines of cross sections (restored) of several subgenera of *Conularia*. *L*. Cross section of shell of *Conulariella*. (*A–B*, *G–L after Kiderlen*, 1937, *and Bouček*, 1939; *D after Ruedemann*, 1898; *E–F after Knight*, 1937.)

ment is followed here, but with reservation of opinion as to its correctness.[1]

The chief characteristic of the Conularida that supports their removal from the Mollusca and assignment to the Scyphozoa is the marked quadrilateral radial symmetry of the test—a symmetry that is in striking contrast to the bilateral symmetry of the gastropods. Furthermore, four symmetrically arranged pairs of markings resembling muscle scars have been found in one conularid genus, *Conchopeltis* (Fig. 4-10*E–F*).

The typical conularid had a thin, somewhat fragile chitinous shell of

[1] Knight suggested that the Hydrozoa deserved about equal consideration with the Scyphozoa as a receptacle for the conularids. He also pointed out that the radial symmetry of *Tentaculites*, and of certain other related animals generally assigned to the Mollusca, is more typical of the coelenterates than of the mollusks. These radially symmetrical fossils are not included in the Coelenterata in the present work but are discussed in a later chapter.

H. and G. Termier (1948) recently suggested that the conularids belong in a group including *Serpulites* and *Hyolithes* and descended from an ancient phoronid-like ancestor.

conical or pyramidal shape, open at the larger apertural end and closed at the smaller apical end, which seems to have served for attachment of young individuals. The shell has marked quadrilateral radial symmetry, particularly when viewed in transverse section (Fig. 4-10*G–L*). Each of the four lateral faces is generally ornamented with some combination of transverse and longitudinal ridges and furrows, and a prominent furrow typically occupies the edge along which adjacent faces meet.

Specialists (Bouček, 1939; Sinclair, 1952) have proposed four families to include the 20 conularid genera thus far described: Conulariidae, typified by the common genus *Conularia* (Fig. 4-10*C, G–K*); Conulariellidae, with the single genus *Conulariella* (Fig. 4-10*L*); Conulariopsidae, with the single genus *Conulariopsis;* and Serpulitidae, represented by *Serpulites* (Fig. 4-10*B,D*). *Conchopeltis* (Fig. 4-10*E–F*), a supposed conularid from the Ordovician, cannot be assigned to any of the four families mentioned, hence it is of uncertain position.

The Serpulitidae appear first, with *Serpulites*, and range from Lower Cambrian to Permian. *Conulariella* ranges from Middle Cambrian to Ordovician, and *Conchopeltis* is limited to the Ordovician. The Conulariidae include genera ranging from Lower Cambrian to Upper Triassic. *Conulariopsis* is limited to the Lower Triassic.

CLASS ANTHOZOA[1]

Introduction

The Organism. The Anthozoa are exclusively polypoid and marine coelenterates with biradial or bilateral symmetry based on a plan of six, eight, or more parts. The body, which is relatively short and stout compared to the typical hydrozoan, is attached to the substratum or to its own exoskeleton by its base and has the upper end expanded into an oral disk bearing hollow tentacles (Fig. 4-11). The mouth leads into a gullet that hangs down into the enteron. The gullet commonly bears one or more flagellated grooves, the **siphonoglyphs,** along which food-bearing currents of water are directed to the enteron.

Externally, typical anthozoans exhibit flowerlike radial symmetry; internally, however, the soft parts are arranged according to a bilateral plan which appears in the embryonic stage, hence is primary. Stony exoskeletons tend to have the same primary bilateral symmetry, but radial arrangement of hard parts is so perfect in some species that the bilateral plan is not readily apparent.

The enteron of the anthozoan is divided into compartments by longitudinal partitions, or **mesenteries,**[2] extending inward from the wall of the

[1] Anthozoa—Gr. *anthos*, flower, + *zoon*, animal; referring to the flowerlike appearance of the polyp when it has its tentacles extended.

[2] Many zoologists call these fleshy partitions **septa,** but we shall avoid using the term in this sense because paleontologists have long applied that designation to certain prominent radially directed vertical plates in the corallites of stony anthozoan exoskeletons (Fig. 4-11*A*). We shall use instead the term **mesentery,** even though this word has a definite meaning for zoologists with reference to coelomate animals (Hyman, 1940, p. 539).

enteron (Fig. 4-11*A*). Below the lower end of the gullet the inner margins of the mesenteries are free and bear nematocysts; above, they unite with the gullet to make a complete radial partition.

The body wall consists of epidermis, mesenchyme (mesogloea), and gastrodermis. An internal spicular skeleton secreted by the mesogloea is

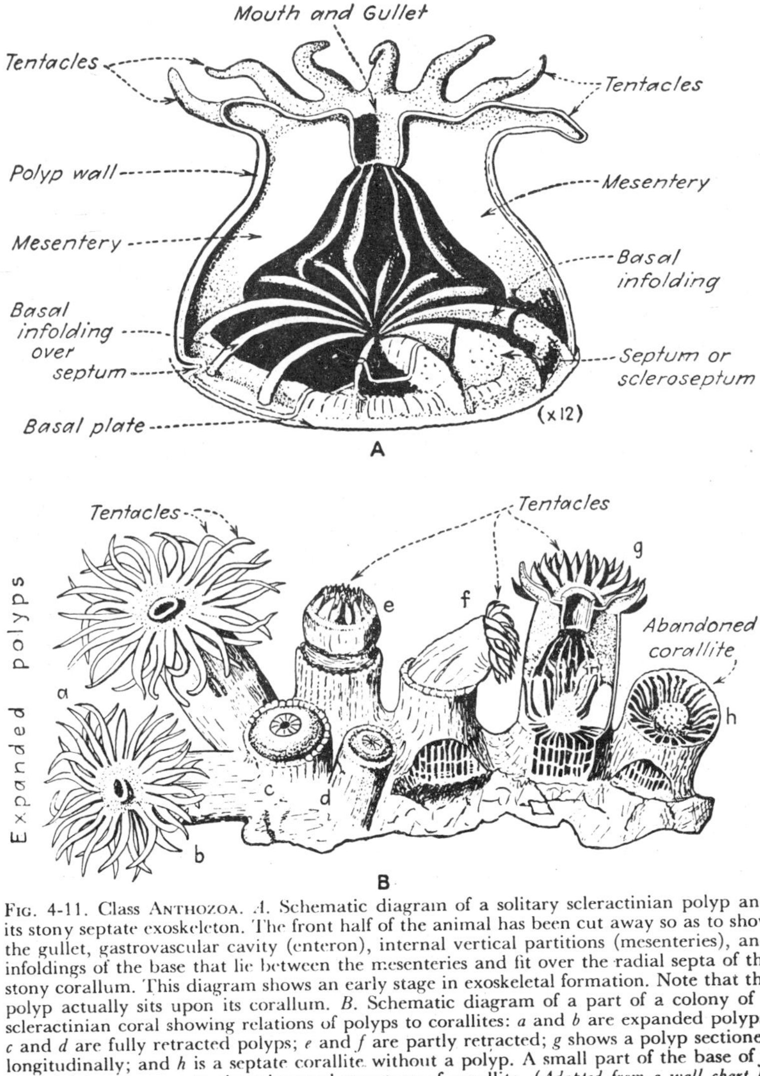

FIG. 4-11. Class ANTHOZOA. *A*. Schematic diagram of a solitary scleractinian polyp and its stony septate exoskeleton. The front half of the animal has been cut away so as to show the gullet, gastrovascular cavity (enteron), internal vertical partitions (mesenteries), and infoldings of the base that lie between the mesenteries and fit over the radial septa of the stony corallum. This diagram shows an early stage in exoskeletal formation. Note that the polyp actually sits upon its corallum. *B*. Schematic diagram of a part of a colony of a scleractinian coral showing relations of polyps to corallites: *a* and *b* are expanded polyps; *c* and *d* are fully retracted polyps; *e* and *f* are partly retracted; *g* shows a polyp sectioned longitudinally; and *h* is a septate corallite without a polyp. A small part of the base of *f*, *g*, and *h* is cut away to show internal structure of corallite. (*Adapted from a wall chart by Pfurtscheller.*)

present in some forms; many anthozoans have a massive exoskeleton secreted by the ectodermal cells; and those of one large group, the sea anemones, lack a skeleton altogether.

Relation of Animal to Skeleton. Those anthozoans that have a stony or chitinous skeleton generally secrete the skeletal substance around the epidermis so that the structure, which is a true exoskeleton, conforms to the external surface of the lower part of the polyp (Fig. 4-11). Each polyp thus builds around and under itself a cup or disk of some sort. This structure is known as a **theca** or **corallite.** Colonial anthozoans bind adjacent corallites together into a variety of structures by secreting additional skeletal material between individual polyps. A complete skeleton, whether the single theca of a solitary anthozoan or the united thecae of a colony, is designated a **corallum** (plural, **coralla**).

If the column and base of the polyp are smooth, the corallite has a smooth inner surface because it conforms closely to the epidermal surface of the polyp. This is not, however, a common condition among anthozoans.

More commonly the body wall in the lower part of the polyp is much wrinkled by radially arranged folds, the more prominent of which alternate in position with the mesenteries (Fig. 4-11*A*). From the base and wall of the theca in these anthozoans radially arranged vertical plates of stony substance project inward, following the inward folds of the body wall, and the major or longer plates thus alternate with the mesenteries (Fig. 4-11*A*). These plates are the **septa,** or better the **sclerosepta,** of the theca. They are of great importance in the study of coral evolution and in the classification of both extinct and living stony corals.

The shape, wall structure, and internal architecture of corallites, and in some cases of coralla also, are of great importance to the paleontologist studying fossil corals. Consequently, these aspects of the theca or corallite are described in detail as the anthozoan orders are discussed.

Some anthozoans secrete skeletal elements of horny or calcareous material in the mesogloea. These **sclerodermites,** as they are commonly called, may lie free in either mesogloea or epidermis, or they may be packed together to form an internal axis known as a **sclerobasis.** Anthozoans with this kind of internal skeleton also have an exoskeleton with septate corallites.

Reproduction and Ontogeny. Reproduction takes place both asexually and sexually. New individuals formed by budding ordinarily remain attached to the parent and thereby aid in building the colony. In sexual reproduction a free-swimming larva is formed. This ultimately attaches itself to some object on the bottom and develops into a new individual. Upon settling down, the larva assumes the shape of a thin biscuit, modifying that general shape to conform to any irregularities on the object to which it has become attached. It first secretes a **basal plate** (Fig. 4-20*G*). Soon afterward the basal part of the animal develops four or six folds, in which are quickly formed a corresponding number of bladelike radial upgrowths of the basal plate. These are the primary septa of the skeleton, and ultimately in almost all cases they become united at the outer edges by the theca. The septa of the first cycle are the **protosepta;** all later ones are **metasepta** (Figs. 4-20,

4-37[17–19]). This complete structure, roughly conical or cup-shaped, constitutes the embryonic exoskeleton, or **prototheca.** Growth ends soon after this stage in a few cases, and a buttonlike, septate skeleton results, as illustrated by *Hadrophyllum* (Fig. 4-39*B*) and *Fungia* (Fig. 4-26*H*). Usually, however, the animal continues to grow and develops many cycles of septa in the skeleton, until the septa may number more than a hundred. The number tends to be, but seldom is, a multiple of the number of primary septa.

In order to place itself in a favorable position with respect to other organisms on the bottom, the polyp more or less continuously builds its exoskeleton upward, necessitating progressive abandonment of the earlier parts of the corallite. As the polyp thus moves upward in its exoskeleton, it builds supporting structures of several different kinds beneath itself. Incomplete horizontal plates built in this way are termed **dissepiments,** and similar plates extending across the corallite are called **tabulae** (Figs. 4-34, 4-36).

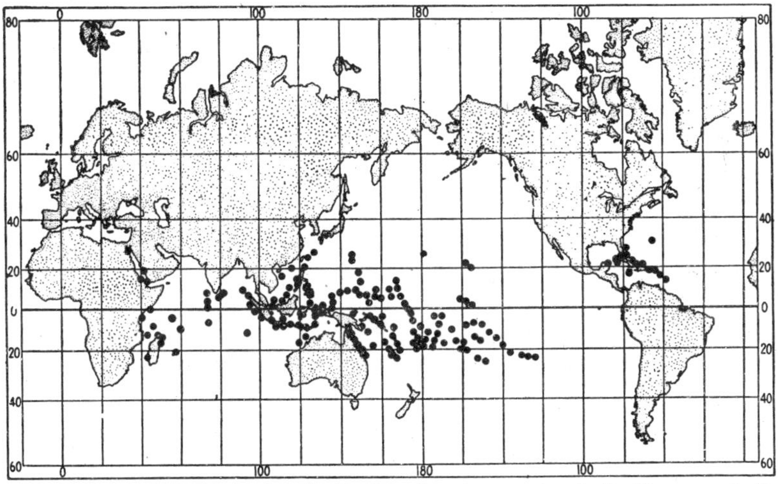

Fig. 4-12. Map of world showing distribution of modern coral reefs. Reef areas are indicated by large dots. No important coral reefs lie outside the tropical belt bounded by the 28°-latitude lines, but delicate, deep-sea reef-building corals live in the North Sea and along the Norwegian coast as far north as 70° N. lat.

Habitat. All the more than 6,000 known species of living Anthozoa, including both solitary and colonial forms, are marine, and most of these live in warm shallow waters less than 180 m. (600 ft.) deep. Some species, however, live in cold waters and others range to great depths.[1]

The reef-building corals do not grow in water that is deeper than 50 m. (150+ ft.), and they flourish best above 30 m. (100 ft.). However, relatives

[1] The Danish research vessel *Galathea* is reported to have trawled 17 sea anemones from a depth of more than 34,000 ft. in the Mindanao Deep northeast of the island of Mindanao in the Philippines (*New York Times*, July 25, 1951).

of these corals live in somewhat deeper water (50 to 75 m., or 150 to 250 ft.), but soon give way to deep-sea species that go down to several thousand meters and live in water with temperatures between 1 and 15°C. An interesting exception are reefs in the North Sea and along the Norwegian coast[1] built at depths of 200 to 600 m. (600+ to 2,000 ft.) by the branching coral *Lophohelia*. In general, corals flourish best in warm waters with temperatures about 22°C. (72°F.), and common reef-building corals cannot endure temperatures below 18°C. (64°F.). As a consequence, most modern coral reefs are limited to continental and island shores in tropical and subtropical waters between 28° N. and S. lat. (Fig. 4-12). In this belt are many coral islands and barrier reefs composed of the broken and macerated remains of coral colonies mixed with stony skeletons and shells of associated organisms. On many reefs the nullipores or coralline algae, the hydrozoan millepores (*Millepora*), and the calcareous alcyonarian corals are large contributors and not uncommonly have a dominant role in the reef building. Vaughan (1919, page 216) describes a typical reef as follows: "Coral reefs are ridges or mounds of limestone, the upper surfaces of which lie, or lay at the time of their formation, near the level of the sea, and are predominantly composed of calcium carbonate secreted by organisms, of which the most important are corals."

Ancient reefs, or **bioherms**[2] as they have been designated, not unlike their modern counterparts in essential respects, have been discovered in many parts of the world in rocks of all ages from Ordovician to Pleistocene.

It is not known that ancient reef-building corals had the same limitations of depth and temperature as living forms.[3] However, it is generally assumed that the 8,000 to 10,000 species of ancient and extinct anthozoans, including the reef-building corals long since extinct, lived under environmental conditions essentially similar to those now prevailing in modern seas where reef-building corals flourish. This assumption is based on analogy and may not be correct. The associated sediments usually show that the waters were shallow.

Classification

The Anthozoa are subdivided into five subclasses, of which the last three are extinct—Alcyonaria (Octocorallia), Zoantharia (Hexacorallia), Tetracorallia (Rugosa), Schizocorallia, and Tabulata. The subdivision is made on the basis of the following characteristics:

[1] C. Dons (1943) reports about 65 recent reefs and about 35 subfossil ones, 8 above sea level, along the Norwegian coast as far north as 70° 57′ N. lat. The scleractinian *Lophohelia pertusa* forms the main part of every reef.

[2] Cumings, E. R., and Shrock, R. R., 1928, Niagaran coral reefs of Indiana and adjacent states and their stratigraphic relations, *Bull. Geol. Soc. Amer.*, vol. 39, p. 599. Cumings, E. R., 1932, Reefs or bioherms? *ibid.*, vol. 43, pp. 331–352. Ladd, H. S., 1944, Reefs and other bioherms, *Nat. Res. Council, Rept. Comm. Marine Ecology as Related to Paleontology*, 1943–1944, pp. 26–29.

[3] It is doubtful if it is valid to compare modern corals, with their relative complexity and close association with certain symbiotic algae (Zooxanthellae), with ancient and extinct corals whose soft parts and biotic associations are essentially unknown.

1. Number and arrangement of tentacles and mesenteries.
2. Number, arrangement, and nature of septa (sclerosepta).
3. Size, shape, and relationship of corallites.
4. Nature of internal structures in corallites.
5. Nature of coralla.

The geologic history and supposed relationships of the five subclasses of Anthozoa are indicated on Fig. 4-30.

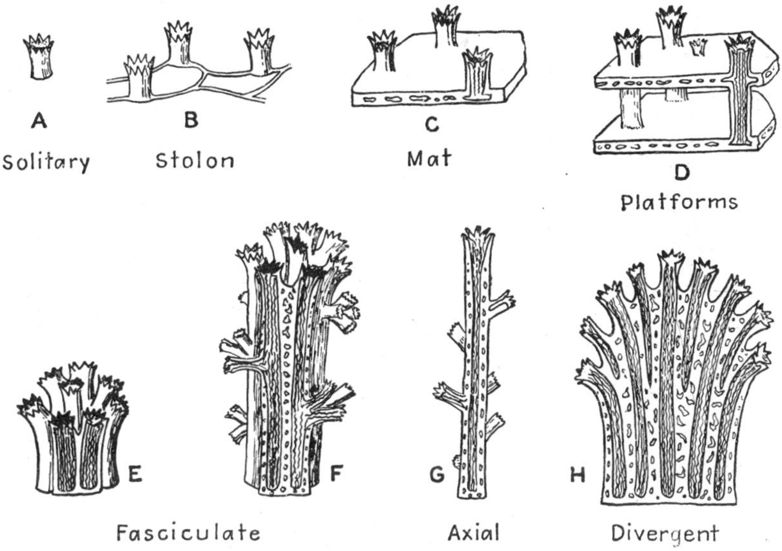

FIG. 4-13. Subclass ALCYONARIA. Sketches showing nature of colonial aggregations of polyps. *A*. Solitary individual (*Haimea*). *B*. Polyps rising from a system of stolons, or solenia (*Cornularia*). *C*. Polyps rising from a basal mat (*Clavularia*). *D*. Polyps connected by platforms (*Tubipora*). *E*. Small fasciculate colony (*Organidus*). *F*. Colony with long main polyps and bunches of lateral individuals (*Vaeringia*). *G*. Colony of one large axial polyp and numerous lateral individuals (*Telesto*). *H*. Colony of long slender polyps without lateral individuals (*Alcyonium*). (*All figures modified after Delage and Hérouard*, 1901.)

Subclass Alcyonaria (*Octocorallia*)

The alcyonarian polyp differs from all other coelenterate polyps in possessing eight pinnate tentacles, which make a single marginal circle on the flat oral disk, and eight mesenteries that are attached to the prominent gullet. The eight symmetrically arranged tentacles and mesenteries give the polyp what seems to be octamerous radial symmetry, whence the name Octocorallia previously used for this subclass. Actually, however, the fundamental symmetry of the alcyonarian is bilateral, as shown by the single siphonoglyph, the elongation of the mouth and gullet, the arrangement of the eight pairs of muscles, and the placement of other soft parts.

All Alcyonaria are colonial, with lobed or branching coralla. The polyps are not directly connected but communicate with each other by means of

solenia (singular, **solenium**), which are tubular continuations of the gastrovascular wall. New polyps sprout directly from the solenia and remain attached. The network of solenia is embedded in the common flesh, or coenenchyme, and the polyps of the colony rise from this network with their bases embedded in the coenenchyme (Fig. 4-13).

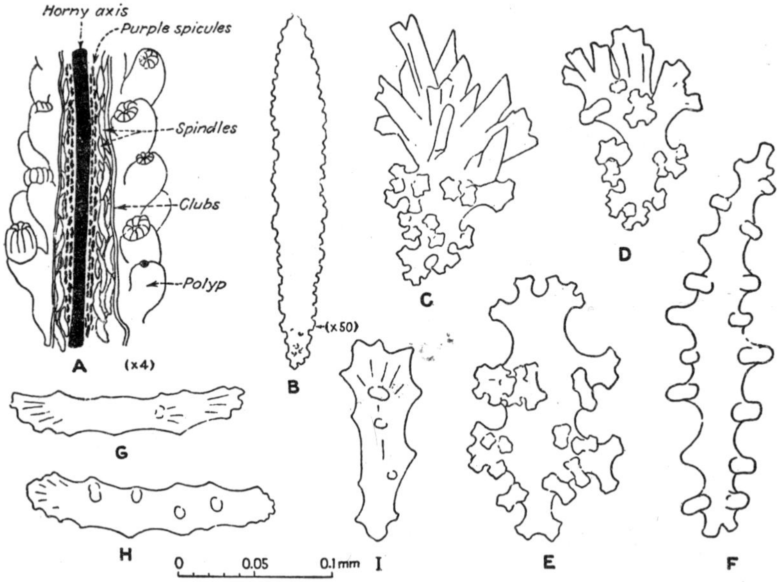

FIG. 4-14. Subclass ALCYONARIA. Spicules. *A*. Schematic diagram of part of a branch of a Plexaurid, partly cut open to show the horny axial core, the smaller purplish spicules of the inner layer, and the larger spindles in the outer layer, beneath the closely packed outermost layer of clubs indicated as a smooth layer. The individual polyps with tentacles infolded are shown surrounding the axial part of the colony. *B*. A large spindle from outer layer. *C–D*. Clubs from outermost layer. *E–F*. Purple spicules from inner layer. *G–I*. Flat rods from tentacles of polyps. (*Original drawings by Elisabeth Deichmann*, 1949.)

Alcyonarian colonies are supported by some kind of endoskeleton that is composed of horny fibers and calcareous spicules (Fig. 4-14) secreted in the mesogloea by certain ectodermal cells. The tiny spicules (sclerites) may lie free in the mesogloea or epidermis, but more commonly they are united by calcareous cement or horny fibers to make a solid or hollow axial structure, the sclerobasis, around which mesenchyme, solenia, and polyps are arranged (Fig. 4-14*A*). In some alcyonarians the skeletal elements are so combined as to produce regularly tabulated tubes (Helioporidae) (Fig. 4-15[6–8]). Coralla range in shape from simple rods (sea pens) and fan-shaped meshworks (sea fans) to bushy and massive structures composed of tubes and connecting stony matter (Figs. 4-15[6–8], 4-16, 4-17).

The oldest undoubted fossil Alcyonaria are found in Triassic rocks, but several much earlier corals have been assigned provisionally to the subclass.[1] Since the beginning of the Mesozoic, alcyonarian corals have become more and more numerous and varied, but their fossil record is relatively unimportant.

The Alcyonaria are generally subdivided into the following six orders:

Order 1. Stolonifera.
Order 2. Telestacea.
Order 3. Alcyonacea.
Order 4. Coenothecalia.
Order 5. Gorgonacea.
Order 6. Pennatulacea.

Order 1. Stolonifera. The polyps arise from one or more solenial tubes that are enclosed in flat basal stolons or in a mat attached to the bottom (Fig. 4-13*B–C*). They live in shallow tropical and temperate waters.

The skeleton when present consists of separate calcareous spicules, similar to those of the Alcyonacea, or of closely packed tubes and interconnecting platforms. *Tubipora*, the familiar red "organ-pipe coral" that is common on coral reefs, consists of closely packed erect tubes ioined together at regular levels by transverse platforms (Figs. 4-15$^{6-8}$, 4-42^{2}). The living colony consists of long tubular polyps that spring from a basal plate and occupy the long corallites. The polyps are united at definite levels by solenia in transverse stolons from which new polyps may spring. The spicules secreted in these stolons fuse to form the interconnecting stony platforms of the skeleton.

The geologic history of the Stolonifera is unknown, as no fossil representatives seem to have been discovered. However, there is no obvious reason why the substantial skeleton and even the microscopic spicules (Fig. 4-15$^{6-11}$) should not have been preserved on past sea bottoms under favorable conditions.

Order 2. Telestacea. Colonies of this order consist of simple or branched stems arising from a slender base (Fig. 4-15^{1}). Each stem is one elongated polyp to which are attached many lateral polyps. The spicules may be free or

[1] Okulitch (1939, p. 79) states that Vologdin (1932) described a coral, *Bija sibirica*, from the Cambrian of Siberia and provisionally assigned it to the Alcyonaria. However, little is known about this fossil, hence its taxonomic relations are uncertain.

Eisenack (1942) recently reported chitinous black spicular elements from certain Silurian rocks and assigned them to the Alcyonaria. If these are actually alcyonarian spicules, then the fossil record of the subclass must be greatly extended to include the Silurian, and if *Bija*, mentioned previously, is also an alcyonarian, the record must be further extended to include the Cambrian.

An extinct family of Paleozoic corals, the Heliolitidae (Fig. 4-45), are commonly assigned to the Alcyonaria. They appeared in the Ordovician and became extinct in the Devonian, without leaving known descendants. Inasmuch as they have 12 pseudosepta, rather than eight or some multiple of eight, and because there is such a long gap of time between their disappearance in the Devonian and the appearance in the Cretaceous of the Helioporidae, which they most closely resemble, we prefer to follow Okulitch and assign this family to the extinct Schizocorallia.

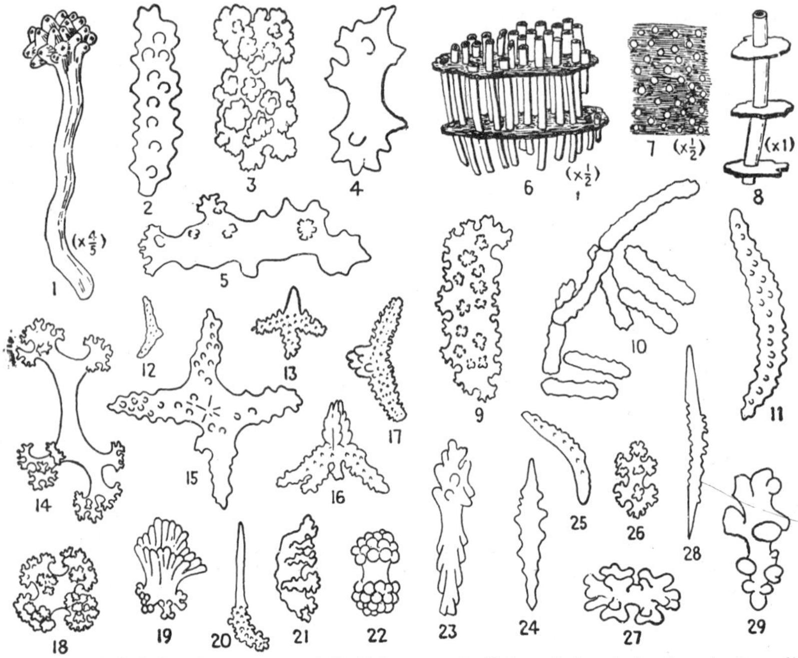

FIG. 4-15. Subclass ALCYONARIA. 1–5. Telestacea: 1, *Telesto;* 2–5, spicules from body wall of *Telesto* (×150). 6–11. Stolonifera: 6–8, side view, top view, and a single corallite of *Tubipora;* 9–11, spicules from *Cyathopodium*, (×150). 12–22, 25, 28. Gorgonacean spicules of several kinds (12–13, 16–17, 20, 25, and 28, ×35; 14–15, 18–19, 21–22, ×150). 23–24, 26–27, 29. Alcyonacean spicules, (×150). (*All figures except* 6–8 *after Deichmann*, 1936.)

united to some extent by horny and calcareous matter. *Telesto* (Fig. 4-15[1–5]) is a living representative; *no fossils are known.*

Order 3. Alcyonacea. These are the "soft corals," which have polyps with the lower part embedded in mesenchyme and only the oral part protruding. The skeletal elements are free and are oval disks, warty rods and spindles, and modified forms of the latter (Fig. 4-15). They are present in large numbers in some species. Fossil alcyonacean spicules have been reported from the Cretaceous of Bohemia and are probably more common than the record indicates.

Order 4. Coenothecalia. *Heliopora*, the single genus in the order and the common blue coral of the Indo-Pacific reefs, builds a massive, calcareous skeleton of aragonite fibers (not spicules) that are fused into layers. Two kinds of tubular corallites are developed. The larger, termed **autopores,** are occupied by the bases of the polyps; the smaller, called **siphonopores,** are occupied by erect solenial tubes (Fig. 4-42[1]). As the skeleton grows, tabulae are built across both types of corallites, so that only the outer surface of the skeleton is occupied by living tissue.

Fossil *Heliopora* are known from the Cretaceous to the Recent.

Order 5. Gorgonacea. These are the "horny corals" which are represented in modern oceans by more than 1,000 species. "*Gorgonia*" (*Rhipidogorgia*, Fig. 4-16), the sea fan, and *Corallium*, the red coral used for jewelry, typify the order.

The corallum has the shape of a fan or a bush, and is generally attached by a basal expansion or by stolons. Each branch or stem has an axial rodlike skeleton typically consisting of **gorgin,** a tough horny protein containing certain other organic substances. This axis is entirely horny in one large group, but contains calcareous spicules in another, and may be entirely calcareous as in the red coral, *Corallium*. It is surrounded by coenenchyme that contains the short polyps radiating from the axis, solenial tubes, and loose calcareous spicules in the form of disks, scales, and different types of warty rods and spindles (Figs. 4-15, 4-16).

Horny skeletons are known fossil from the Tertiary, and calcareous ones from Cretaceous to Recent.

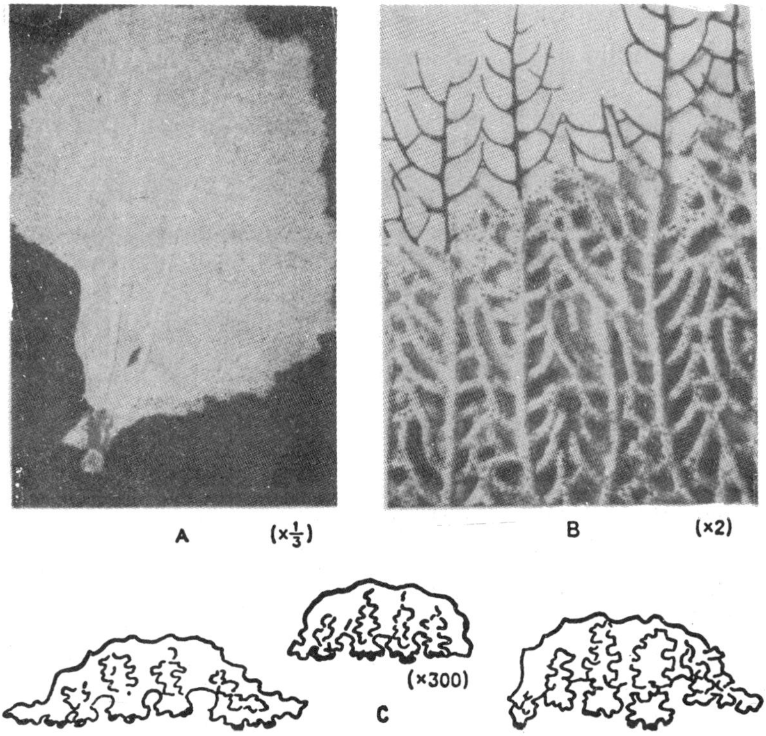

FIG. 4-16. Subclass ALCYONARIA—Order Gorgonacea. *A*. A complete colony of the fanlike fenestrate *Rhipidogorgia*. *B*. A small part of the periphery of the colony, showing the chitinous axes of several of the branches and the numerous pits on the branches where zooids lived. *C*. Three typical calcareous spicules from the coenenchyme of a rhipidogorgian polyp. Spicules characteristic of the flesh of other gorgonacean polyps are illustrated in Fig. 4-15. (*C after Deichmann*, 1936.)

Order 6. Pennatulacea. The colony in this order consists of large elongated **primary axial polyps** with numerous laterally arranged dimorphic **secondary polyps** (Fig. 4-17*A*). It has a slender frondlike or leaflike form. The skeleton consists of a horny or calcareous axis with a basal peduncle that resembles a slightly tapering pencil or pen, the smaller end of which is commonly embedded in sand or mud, and a variety of smooth calcareous spicules scattered loosely through the coenenchyme (Fig. 4-17*B–F*).

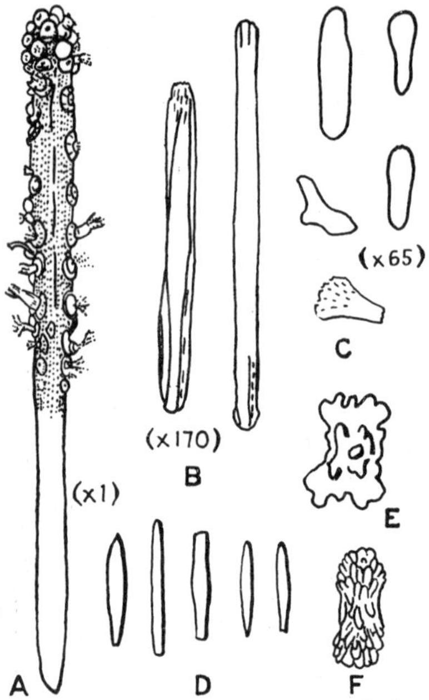

Fig. 4-17. Subclass Alcyonaria—Order Pennatulacea. *A. Kophobelemnon*, a complete colony with a prominent stalk. *B–C.* Calcareous spicules from same genus as *A. D–F.* Calcareous needles and irregular spicules, greatly enlarged, from three different pennatulacean genera. It should be noted that the spicules in this order are characteristically smooth and much less varied in architecture than those in the other orders of Alcyonaria (cf. Figs. 4-14 and 4-15). (*After Köllicker*, 1870–1872.)

Fossil pennatulids (calcified axes) have been reported from rocks as old as Triassic. *Pennatula*, the common sea pen, is a typical living representative.

Under favorable conditions alcyonarians make substantial contributions to calcareous sea-bottom deposits. In some places off the Florida coast, as much as 10 per cent of the ooze is composed of alcyonarian spicules.

Subclass Zoantharia (Hexacorallia)

The Zoantharia include all living anthozoans with mesenteries in cycles of 6, 12, or multiples of 6. Not uncommonly, and always secondarily, the mesenteries are arranged in a plan of 5 (pentameral), 7 (septameral), or 8 (octameral), but in the last case there are never just 8 single mesenteries as in the Alcyonaria.

The typical hexacoral polyp is a cylinder, radially symmetrical in external features, with the oral end expanded into an **oral disk** bearing many hollow tentacles (as many as hundreds in some forms) that are arranged in a number of different ways. The aboral end in some solitary forms is expanded into a **pedal disk,** whereas in others it is rounded or pointed and inserted into the substratum. The oval or slitlike mouth opens into a flattened **gullet,** which may bear one or two ciliated grooves, **siphonoglyphs,** that lie in the plane of the primary bilateral symmetry. Six single primary mesenteries or six pairs of them extend from the gullet to the wall of the enteron and divide the gastrovascular cavity into symmetrical compartments. Below the gullet the inner ends of the mesenteries are free. Shorter secondary mesenteries are intercalated between the major 6 or 12. All mesenteries are paired, *i.e.*, they lie symmetrically on each side of the plane of bilateral symmetry that bisects the corners of the mouth. The muscular system is commonly highly developed, and retractor muscles present in the mesenteries have symmetrical arrangements, which are of taxonomic importance in classifying living zoantharians.

One large group of hexacorals, the fleshy sea anemones, lack a skeleton of any sort. Another small group contains members that construct a horny skeleton. Of greatest interest to the paleontologist, however, is a third group, the scleractinians or madreporarians, which build solitary and colonial coralla of calcareous material. These are the stony corals that are found the world over on living reefs as well as on other marine bottoms. The skeleton is secreted by the ectodermal cells and is a massive external structure.

The Zoantharia are generally divided into five orders:

Order 1. Actiniaria (sea anemones).
Order 2. Scleractinia or Madreporaria (stony corals)
Order 3. Zoanthidea.
Order 4. Antipatharia (black corals).
Order 5. Ceriantharia

Order 1. Actiniaria. The sea anemones are fleshy corals without a skeleton. The polyp is fairly large as corals go, is columnar in shape, and commonly has a pedal disk, a gullet with siphonoglyphs, and paired mesenteries in multiples of six. Sea anemones are sessile on rocky and sandy bottoms but are not necessarily fixed. *Edwardsia*, for example, lives in burrows, whereas *Actinia* habitually grows on the shells of hermit crabs.

Since actiniarians have no hard parts, it follows that they have left no fossil record dependent on a skeleton. Interestingly enough, however, the oldest known true coral, *Mackenzia*[1] (Fig. 4-18*B*), is supposedly an actiniarian. Like so many of its associates in the Middle Cambrian Burgess shale of British Columbia, it is exquisitely pre-

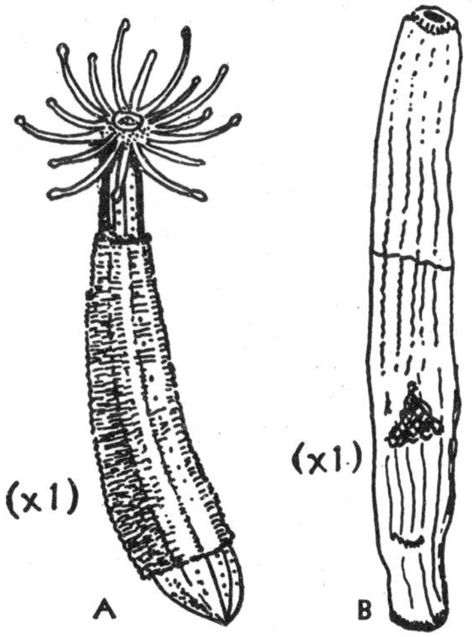

FIG. 4-18. Subclass ZOANTHARIA—Order Actiniaria. *A. Edwardsia*, a living burrow-dwelling actiniarian. *B. Mackenzia*, a supposed fossil actiniarian from the Middle Cambrian Burgess shale of British Columbia. This is the oldest known fossil coral. It shows six mesenteries and twelve tentacles that seem to be retracted around the mouth. (*A after Andres*, 1884; *B after Walcott*, 1911.)

[1] *Mackenzia* was first described by Walcott (*Smith. Misc. Coll.*, vol. 57, no. 3, 1911) as a holothurian, but A. H. Clark (Cambrian holothurians, *Amer. Nat.*, vol. 47, pp. 488–507, 1913) demonstrated that it could not be an echinoderm. Its similarity to the modern burrowing actiniarian, *Edwardsia* (Fig. 4-18*A*), indicates that it probably should be considered a Middle Cambrian coral.

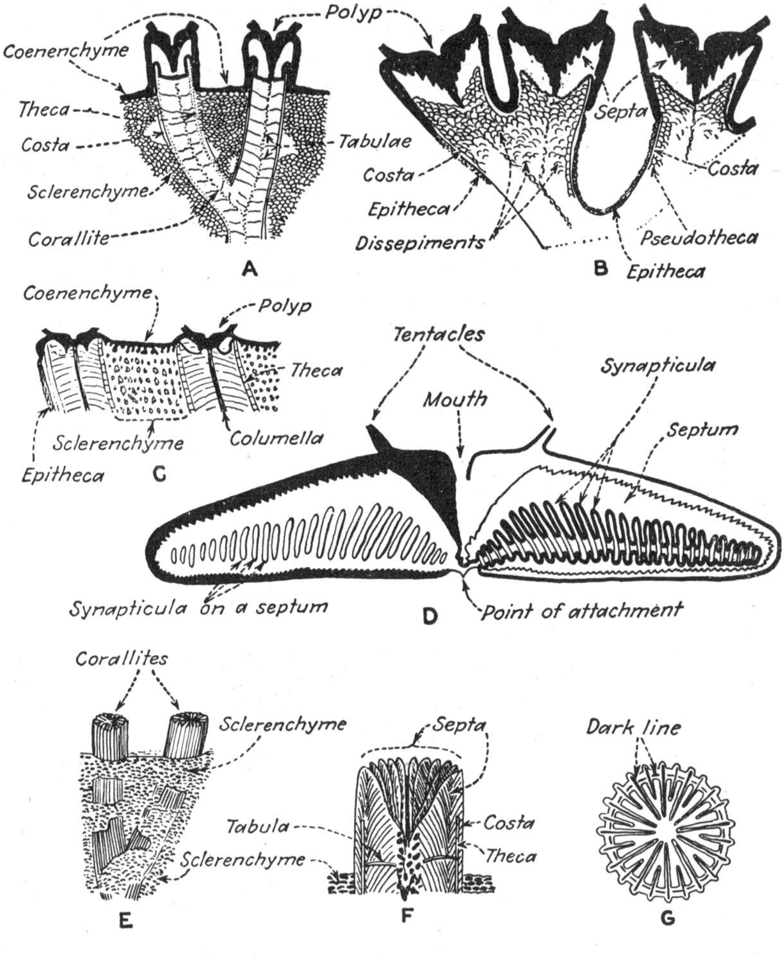

FIG. 4-19. Phylum COELENTERATA—Order Scleractinia. Structures of corallites and coralla and their relations to the polyps that made them. *A. Galaxea*, a form with a sclerenchymatous corallum. Two corallites surmounted by polyps are separated by sclerenchyme secreted by the coenenchyme connecting the polyps. *B. Mussa*, with a nonsclerenchymatous corallum. The structure is dominantly dissepimental. *C. Turbinaria*, a genus with well-developed sclerenchymatous structure between the tabulated corallites. *D. Fungia*, a solitary form with conspicuously developed septa. The body wall is shown in solid black in the left half of the diagram and in outline on the right. *Fungia* has a strongly septate nonsclerenchymatous corallum. *E–G. Galaxea: E*, a branching corallite with surrounding sclerenchyme and with the polyps removed; *F*, enlarged drawing of half the calyx of a corallite, showing the different internal structures; *G*, diagrammatic transverse section of a corallite showing the position of the "dark line" in the radial and tangential skeletal elements. (*Modified after Ogilvie*, 1895.)

served. Soft though it was, it left an imprint so delicate that the mesenteries and 16 tentacles retracted about the mouth are clearly indicated, seemingly proving its actiniarian nature (Fig. 4-18*B*). *Mackenzia* closely resembles the modern *Edwardsia* (Fig. 4-18*A*), which has 16 mesenteries and tentacles, and many hexacorals pass through a so-called *Edwardsia* stage in their early development. Here seems to be one of the longest and most baffling gaps in the evolution of any animal group—from Middle Cambrian *Mackenzia* to Recent *Edwardsia;* only the accidental burial of the former, under conditions so unusual that they have seldom been repeated in all geologic time, reveals the great span of the skeletonless anemones.

Order 2. Scleractinia (Madreporaria) This order of hexacorals, by far the most numerous and varied of all the living Anthozoa, includes the stony corals of modern seas and several thousand fossil species that go as far back as Middle Triassic. Both solitary and colonial forms build strongly septate calcareous exoskeletons secreted by the epidermis. The corallum consists essentially of well-developed septa and a variety of transverse and peripheral supporting and binding structures (basal disk, epitheca, dissepiments, synapticula, etc.), not all of which are present in any one individual. The corallum is a single discoidal or conoidal structure in solitary forms and a massive spheroidal, hemispheroidal, or dendritic structure in colonial species. Some of the latter reach dimensions of several meters.

This order of Anthozoa is so important and so well studied that there is justification for a somewhat detailed discussion of the structural features of the exoskeleton.

Polyp and Exoskeleton. The scleractinian individual is a typical zoantharian polyp with a basal disk, cylindrical column, oral disk, and one or more rows of tentacles around the centrally located mouth (Figs. 4-11, 4-20). The hollow enteron is subdivided by two series of mesenteries, and there is a prominent gullet. The base and lower part of the body wall of the polyp are convoluted or wrinkled, with involutions that fit over the stony sclerosepta of the exoskeleton. The polyp, therefore, *sits* on its exoskeleton, particularly if it is a solitary species. In the colonial coralla, the polyps occupy a tiny septate pit, the **calyx,** at the top of the corallite, and the corallites are commonly bound together by vesicular stony substance, **sclerenchyme,**[1] secreted by the coenenchyme (Fig. 4-19).

The forms of polyp and exoskeleton (theca, corallum) are closely related. In conoidal and cylindrical forms the polyp is as high as it is across, whereas in discoidal coralla it may be many times as wide as high (Figs. 4-11, 4-19).

Development of Exoskeleton. Shortly after the scleractinian larva (Fig. 4-20*A–G*) fixes itself to the substratum or to some object on the bottom, it

[1] Zoologists commonly use the terms **mesenchyme** and **coenenchyme** for fleshy connective tissue in the Coelenterata, whereas paleontologists employ the first term for the stony skeletal structures between corallites secreted by the common fleshy connective tissue. To avoid confusion that might arise from the double meaning of the term, we prefer to use the term **sclerenchyme** for all the vesicular skeletal structure between corallites in colonial coralla.

secretes a thin **basal plate** which later becomes much thickened by additions of calcareous material. As the polyp grows upward from the basal plate, the margin of the plate also grows upward and becomes the periphery of a tiny cup, the **prototheca.** As growth proceeds, the body wall of the lower part of the polyp becomes wrinkled, and vertical plates, the septa or sclerosepta,

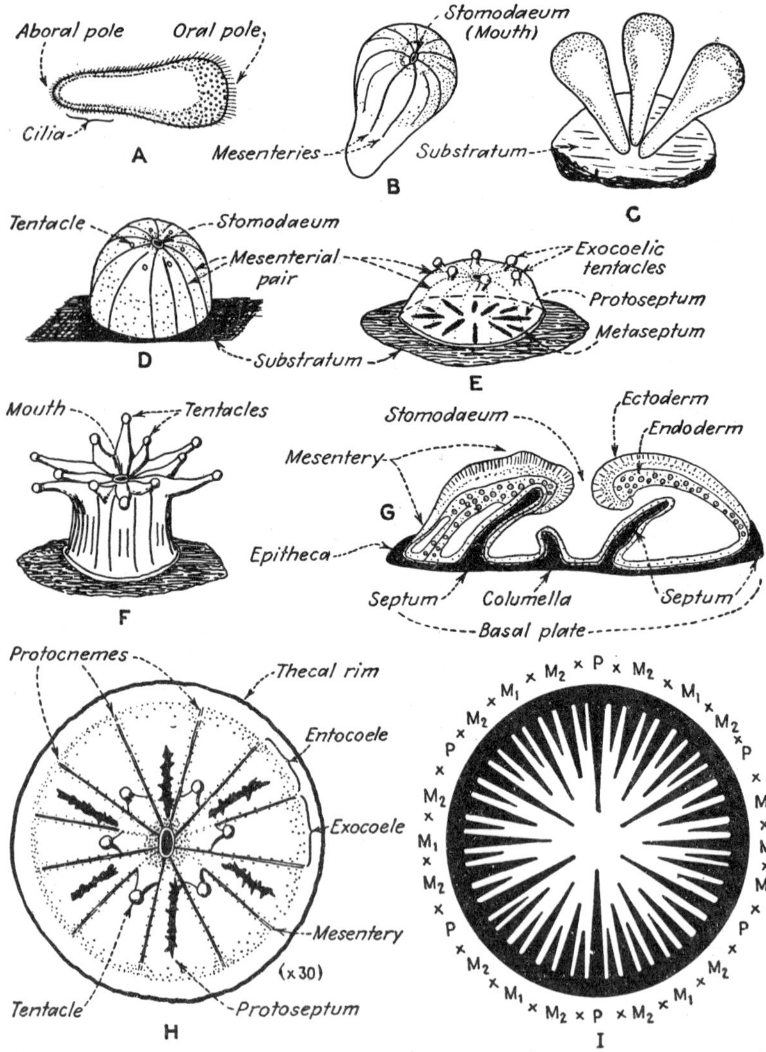

FIG. 4-20. Order Scleractinia. Development of the scleractinian *Siderastrea radians* and its highly septate corallum. (*See opposite page for detailed description.*)

rise from the basal plate in intermesenterial positions.[1] The upward inflection of the basal plate constitutes the **epitheca,** which marks the outer limit of the corallite or corallum. It may not be developed; in colonial forms it commonly surrounds each corallite, and it may cover the common outer wall of the entire corallum (Figs. 4-19, 4-24, 4-34).

Durham (1949) found from a study of 14 genera and 33 species of Scleractinia that the genera fall into two groups on the basis of septal developments. In the **monocyclic** forms the adult corallum develops by direct conical enlargement of the first formed prototheca, whereas in the **polycyclic** types it is developed by enlargement of a later thecal stage formed outside the prototheca, with one or more distinct stages of growth intervening between the prototheca and the stage from which the adult corallum is developed.

The distal parts of the septa may be united by the epitheca, by a wall (the **theca**) separate from it, or by both. Continuations of the septa beyond the peripheral epitheca are called **costae** (Figs. 4-19, 4-34).

Septa are the first skeletal structures to appear after deposition of the basal plate and formation of the prototheca. They may appear as the first skeletal elements if no basal plate is developed (Fig. 4-20*E*). They show great diversification in structure and development and constitute the most

[1] Much attention has been paid to the development of mesenteries and septa because of their importance in the classification and evolution of the corals in general. Mesenteries appear early in the development of a scleractinian and follow a definite order, as indicated in Figs. 4-20*F*, *H*, 4-21. When the first cycle is completed they number 6 and then 12 and are arranged in six pairs with hexamerous symmetry. Each mesentery of this cycle has a prominent retractor muscle on one side or the other. Later, several additional cycles of mesenteries usually are added. Along with mesenterial development, but only after the 12 primary mesenteries have been formed, there is infolding of the body wall to form inwardly and upwardly directed invaginations between the mesenteries. The calcareous septa are produced in these invaginations; hence they have an intermesenterial position, as shown in Fig. 4-11*A*. The concomitant development of secondary and later mesenteries and septa has been worked out for several species of living corals (Figs. 4-20, 4-21), and the septal development alone is known for a few extinct fossil species (Fig. 4-35).

FIG. 4-20. *A*. Larva (**planula**) immediately on extrusion. The narrower aboral pole is anterior in swimming. The mouth is not yet functional. *B*. A second-day larva just before settling. Six pairs of mesenteries are present, three pairs of which reach the stomodeum. *C*. Three larvae settling close together by the narrow aboral poles. *D*. A third-day larva which has flattened out upon attachment. Four pairs of mesenteries reach the stomodeum and the rudiments of six exocoelic tentacles have appeared. *E*. A living polyp fully expanded and viewed from the side as though transparent. Six entocoelic protosepta and six exocoelic metasepta are developed. A basal plate lies under the polyp but is not shown in the diagram. *F*. Expanded polyp viewed from the side to show manner of appearance of the second-cycle mesenteries (**metacnemes**) on the column wall. *G*. Radial vertical section through a young polyp showing its relationship to the early corallum. The exoskeletal parts are shown in solid black. This stage is designated the **prototheca.** *H*. A living polyp two or three days after settling viewed as though transparent. The basal plate is not visible but is present at this stage and binds the septa and theca together to form the prototheca. *I*. Diagram showing the order of development of the first three cycles of septa. The first cycle consists of six protosepta (*P*); the second of six metasepta (M_1); and the third of 12 metasepta (M_2). Additional metasepta (*X*), the shortest shown, are added in later cycles. (*Adapted from Duerden*, 1904.)

important skeletal feature on which classification can be based (Figs. 4-23, 4-37).

The septum is a thin plate made up of fibrous aragonitic structural units called **sclerodermites** (Figs. 4-22, 4-24*E*). These are variously arranged in

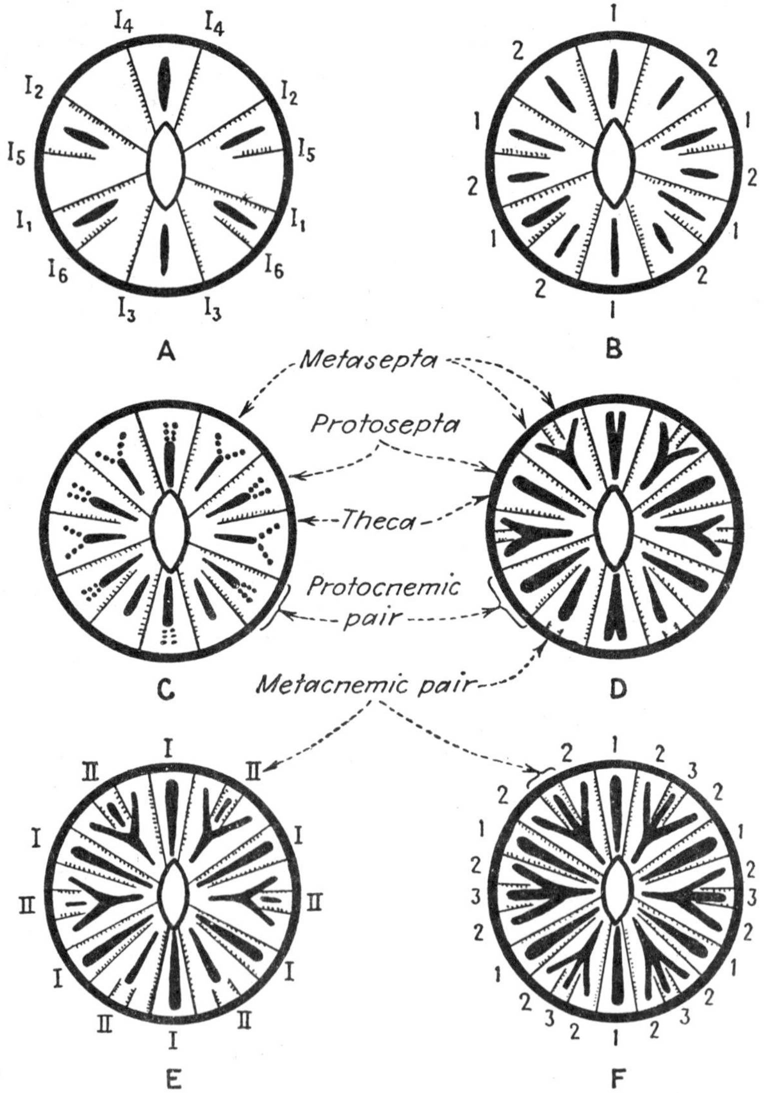

FIG. 4-21. Order Scleractinia. (*See opposite page for detailed description.*)

vertical series to make **trabeculae,** (Figs. 4-24*E–F*, 4-25*A–D*), which form solid **laminar septa** (Fig. 4-25*A*), if closely united, and **fenestrate** or **perforate septa** (Fig. 4-25*M–N*), if loosely connected. Several secondary structures are likely to be developed in the corallite along with the formation of septa (Figs. 4-24, 4-25). A few of these deserve brief notice here.

Dissepiments are more or less horizontal vesicular or tabular structures occupying the spaces between septa or costae. They support the lower surface of the soft parts and fill the void left by upward growth. Dissepiments are endothecal or exothecal depending on whether they lie inside or outside the thecal wall (Fig. 4-25).

Pali are vertical lamellae or pil-

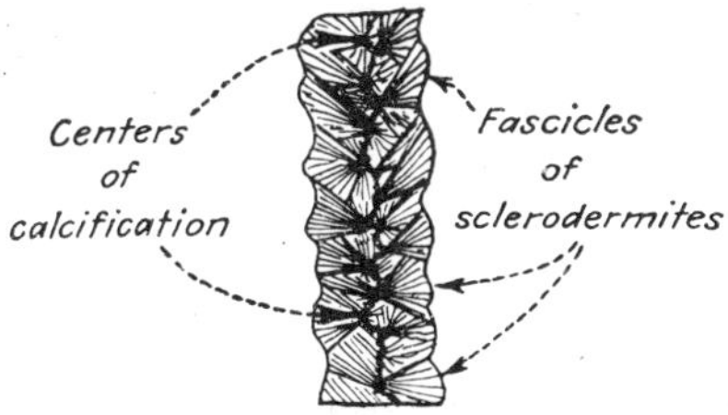

FIG. 4-22. Order Scleractinia. Structure of small part of a septum, showing centers of calcification, dark line joining them in zigzag pattern, and bundles, or fascicles, of calcareous fibers, or sclerodermites, radiating from them. The bundles are expressed on the surface as granulations. (*Modified after Ogilvie*, 1895.)

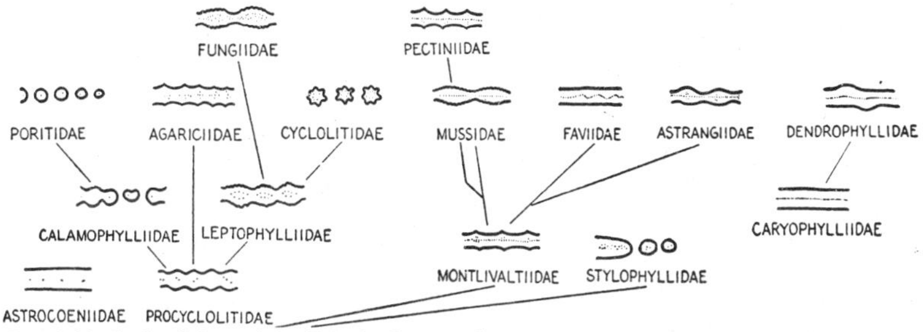

FIG. 4-23. Order Scleractinia. Principal types of septal structure and probable relationships of important groups. The individual figures, which are highly diagrammatic and not to scale, show the structure of septa as seen in horizontal section, with the dots indicating trabecular centers. (*Modified after Vaughan and Wells*, 1943.)

FIG. 4-21. Series of diagrammatic figures illustrating the relationships of mesenteries (I, II) and septa (1, 2, 3) in the establishment of the first two cycles of mesenteries and three cycles of septa in *Siderastrea radians*, a modern scleractinian coral. The mesenteries are shown by light lines with muscles indicated by much shorter and closely spaced perpendicular lines; septa are shown in solid black; and the theca is indicated by the heavy black outer circle. The upper border is regarded as dorsal; the lower, ventral. *A*. Arrangement of mesenteries, subnumbered in the order of development (the six pairs of mesenteries constituting the first cycle are designated **protocnemes**), and of septa on the appearance of the six primary or directive septa (**protosepta**). *B*. Appearance after establishment of the second septal cycle; there are now six protosepta (1) and six metasepta (2). The former continue to enlarge but remain simple, whereas the latter become bifurcate and then trifurcate as successive cycles of metasepta are added. It is to be noted that the protosepta lie in the spaces bounded by the two mesenteries of the same pair (*i.e.*, they lie in the **endocoeles**), whereas the first six metasepta all lie in spaces between the mesenteries of adjacent protocnemic pairs (*i.e.*, in **exocoeles**), and the third cycle of septa (*E*) are again endocoelic. *C–D*. Septal enlargement and increase by development of nodules that ultimately fuse and by bifurcation to produce forked septa. *E–F*. Endocoelic metasepta, constituting the third septal cycle, appear in the metacnemic pairs (**II**), and ultimately fuse with the forked septa of the second cycle. Compare with Fig. 4-20. (*Modified after Duerden*, 1904.)

lars lying near the axis of the corallite in front of the inner edges of certain septa. They are secondary features closely related to septal development and are especially characteristic of the Caryophylliidae (Figs. 4-27*Q–R*; 4-34*H*).

Columella is a term applied to several types of structure commonly developed in the axial part of a corallite. The columella may be a continuous, solid or porous rod, or a general cylindrical or conical zone of various kinds of structure (Figs. 4-34*G–L*, 4-37[6–12]).

Synapticula are rods or bars that connect the opposing faces of adjacent septa, perforating the intervening fleshy mesenteries. They are especially characteristic of fenestrate septa (Figs. 4-19*D*, 4-34*F*).

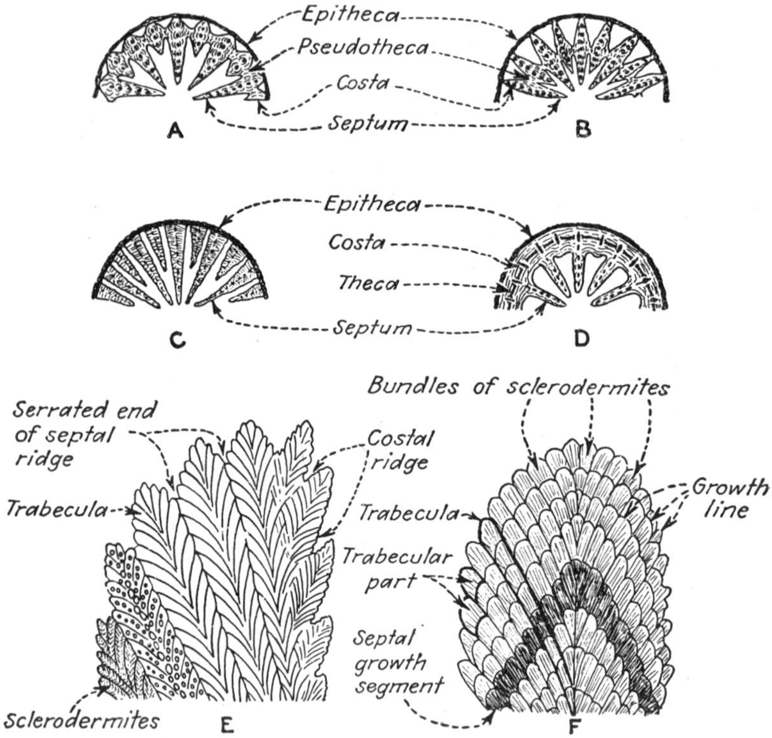

FIG. 4-24. Order Scleractinia. Septal structure. *A–B*. Transverse sections, showing the structural relations between septa, costae, pseudotheca, and epitheca. The pseudotheca is shown near the center of the calyx in *B*. *C–D*. Transverse sections, showing the relations between septa and associated structures. *E*. Septum with broad granulose septal ridges composed of trabeculae radiating from the axes of the ridges. The trabeculae are composed of bundles of sclerodermites (cf. Fig. 4-22). *F*. Radial structure in a septum. The septum grows by periodic increments, or growth segments. Each segment is composed of scalelike parts, the trabecular parts, and these in turn are composed of a bundle of sclerodermites (cf. Fig. 4–22). A trabecula consists of a series of successively younger parts, whereas a growth segment is composed of parts of the same age. The trabecula indicated has 8 parts; the growth segment, 19 parts. (*Modified after Ogilvie*, 1895.)

Peritheca is applied to a variety of skeletal structures uniting the corallites and secreted by the common fleshy coenenchyme of the colony. It may be composed of sclerodermites, or it may lack these and be a solid or porous deposit to which the terms **sclerenchyme** and **coenosteum** have been applied.

Stereome is a layer of calcareous material of variable thickness laid down secondarily on septa and similar structures.

Corallum is the entire skeleton deposited by a single polyp or by a colony of polyps. In a colonial corallum each tube occupied by a complete indi-

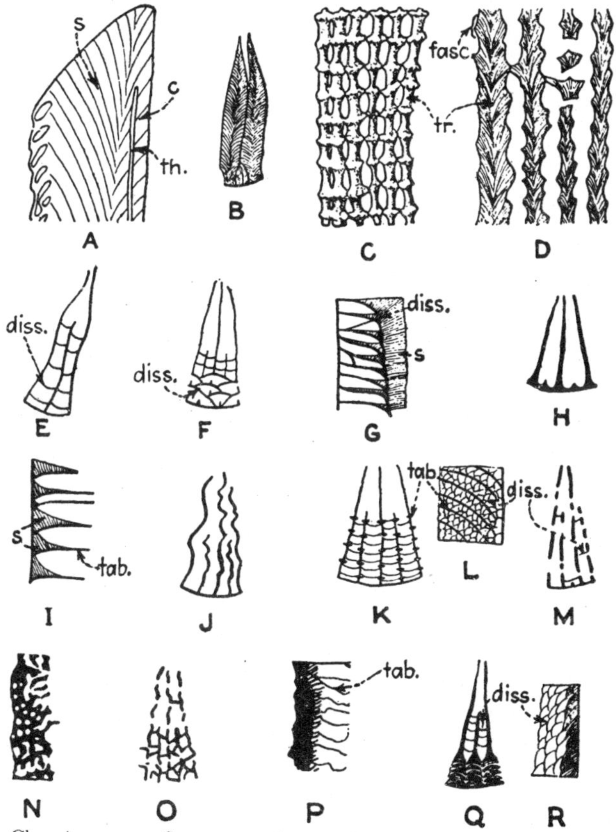

FIG. 4-25. Class ANTHOZOA. Structure and types of septa. *A*. Diagram showing relation of septum (*s*) and costa (*c*) to the theca (*th.*) in *Galaxea*. *B*. Dilated septum. *C–D*. Exterior and median vertical section of a septum of *Cyclolites* (*tr.* = trabeculae; *fasc.* = fascicle or bundle of fibers). *E–R*. Types of septa (*E–F*, *H*, *J–K*, *M*, *O*, and *Q* in transverse section; *G*, *I*, *L*, *N*, *P*, and *R* in median vertical section): *E*, attenuated; *F–G*, lonsdaleoid; *H*, long major septa, minor septa short and appearing as septal ridges; *I*, amplexoid; *J*, zigzag; *K–L*, carinate with yardarm carinae; *M–N*, perforate; *O*, retiform; *P*, acanthine; *Q–R*, naic. (*car.* = carina; *diss.* = dissepiment; *s.* = septum; *tab.* = tabula). (*From Hill*, 1935, *after several authors.*)

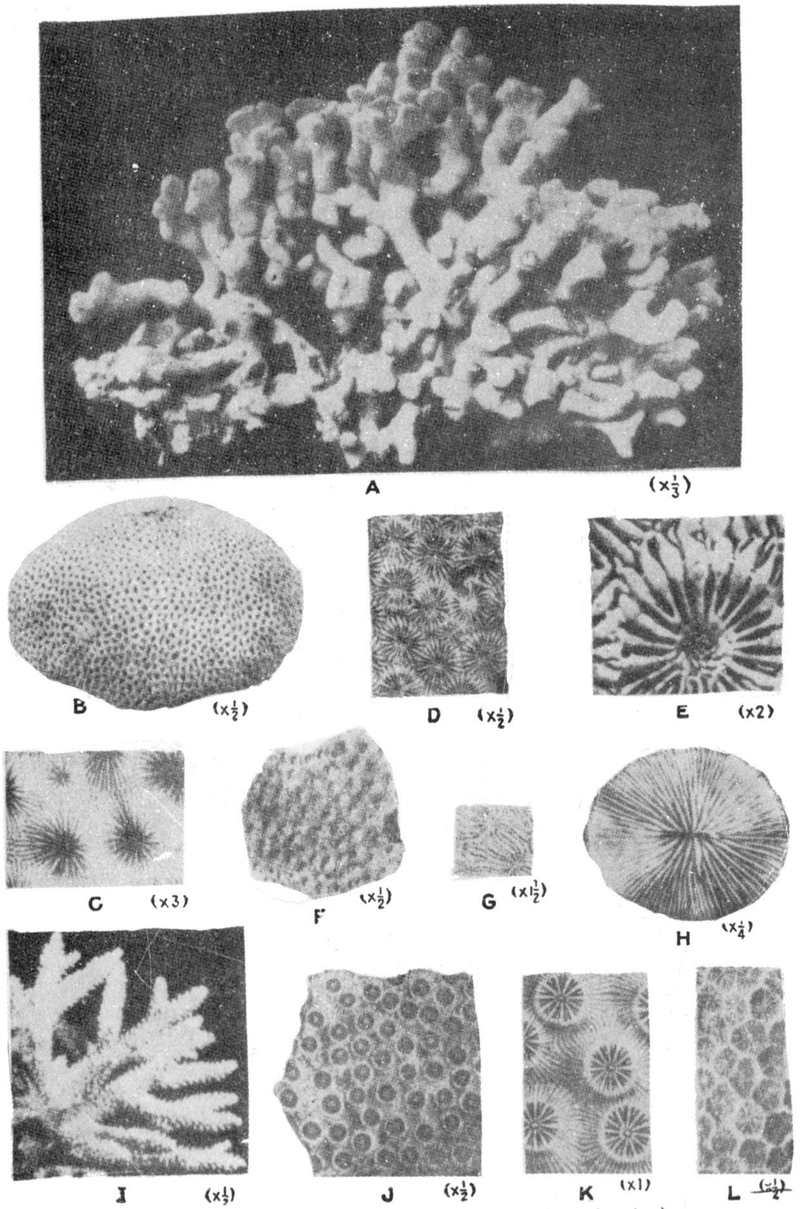

FIG. 4-26. Order Scleractinia. (*See opposite page for explanation.*)

vidual is a **corallite,** and corallites are bound together by sclerenchyme secreted by fleshy coenenchyme. The upper concavity of the corallite is the **calyx,** which commonly has a prominent troughlike depression, the **fossa** or **fossula** (Figs. 4-34*R*, 4-37$^{11-12}$).

Skeletal substance is continually secreted by the polyp so that the corallum grows both laterally and vertically. By the formation of tabulae and dissepiments in the corallites and sclerenchymal structure between, the fleshy part of a colony always remains on the surface of the stony corallum (Figs. 4-11, 4-34*A–B*).

The colony grows in numbers of individuals by methods of reproduction that are not too well understood. Buds form in a variety of positions and develop directly into mature individuals of the colony (Figs. 4-31, 4-32). Under favorable conditions coralla may grow to be more than 2 m. (7 ft.) across.

Classification. The 500 recognized genera of Scleractinia are assigned to 30 existing families and subfamilies and 18 extinct families, all of which are grouped into five suborders, largely on the basis of the diversity of septal structure (Fig. 4-23). The geological ranges and supposed relations of the several taxonomic divisions of the order are shown in Table 4-2.

The five suborders can be characterized briefly as follows:

Suborder 1. Astrocoeniida. Septa laminar or rudimentary, consisting generally of a few simple trabeculae strongly inclined from the axis of divergence. *Middle Triassic to Recent.*

Astrocoenia, *Acropora*, and *Stylina* are representative genera (Fig. 4-26).

Suborder 2. Fungiida. Septa fenestrate, but commonly appearing laminar in later stages; more or less porous, with margins beaded or dentate; one or more fan systems of trabeculae; synapticula present. *Middle Triassic to Recent.*

Thamnasteria, *Siderastrea*, *Porites*, *Diploastrea*, and *Fungia* (Fig. 4-19*D*) are typical genera (Fig. 4-26).

Suborder 3. Faviida. Septa laminar, nonporous, with margins dentate; one or more fan systems of trabeculae; synapticula lacking. *Middle Triassic to Recent.*

Thecosmilia, *Galaxea* (Fig. 4-19*A*, *E–G*), *Trochosmilia*, *Meandrina*, *Montastrea*, and *Solenastrea* are representative genera (Fig. 4-27).

Suborder 4. Caryophylliida. Septa consisting of one fan system of mostly simple trabeculae; always laminar and nonporous with smooth margins; synapticula lacking. *Jurassic to Recent.*

Caryophyllia, *Turbinolia*, and *Flabellum* are typical genera (Fig. 4-27).

Suborder 5. Dendrophyllidae. Septa secondarily thickened and irregularly porous; margins smooth or beaded; synapticula present. *Tertiary to Recent.*

Endopachys, *Balanophyllia*, and *Dendrophyllia* are representative genera (Fig. 4-27).

FIG. 4-26. Order Scleractinia. *A. Porites porites*, an almost complete corallum of one of the commonest modern reef builders from Puerto Rican waters. *B–C. Siderastrea: B, S. radians*, typical corallum from Florida; *C, S. siderea* from the Bahamas. *D–E. Diploastrea heliopora* from French Somaliland. *F–G. Thamnasteria rectilamellosa* (U. Trias.); part of a corallum and several calices somewhat enlarged. *H. Fungia*, a solitary corallum, from the Fiji Islands. *I.* Small part of a corallum of *Acropora cervicornis* from Florida. *J–K. Stylina girodi* from the Jurassic. *L. Astrocoenia guadalupae* from Texas Lower Cretaceous. (*A after Vaughan*, 1901; *B–G, I–L from Vaughan and Wells*, 1943, *after several authors.*)

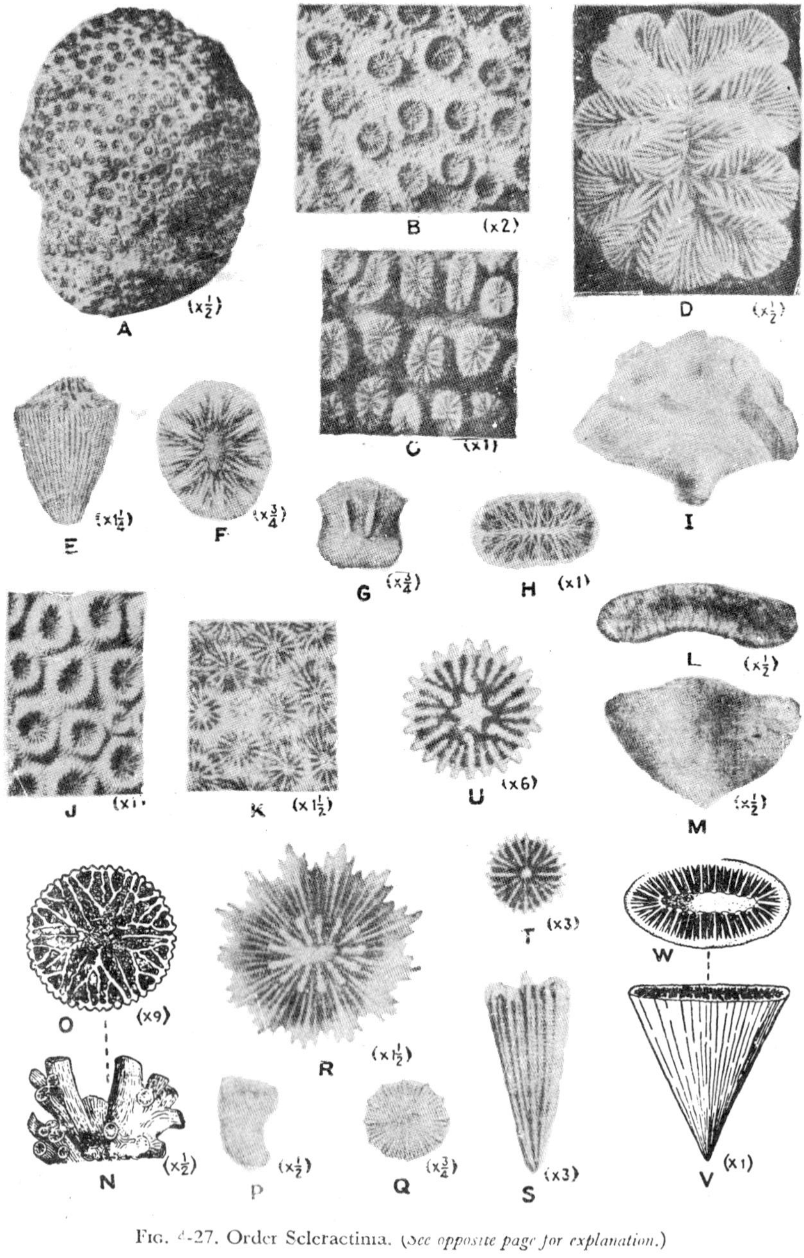

FIG. 4-27. Order Scleractinia. (*See opposite page for explanation.*)

Summary of Order Scleractinia. The exoskeletons of Scleractinia can be distinguished from those of the Tetracorallia by the marked hexameral arrangement of septa and the absence of distinct bilateral symmetry in adult specimens; from those of the Tabulata by the strongly septate corallites and coralla and general lack of tabulae entirely across the corallite; and from the Alcyonaria by the hexameral arrangement of septa.

Order 3. Zoanthidea. The order consists of a small group of solitary and colonial anemone-like corals without a pedal disk or skeleton of any sort. They have left no known fossil record, but are of interest to the paleontologist because in the development and arrangement of mesenteries they differ from all other living Anthozoa but somewhat resemble the extinct Tetracorallia. The mesenteries are paired and coupled and constitute a single cycle in which long mesenteries alternate with much shorter ones (Fig. 4-28).

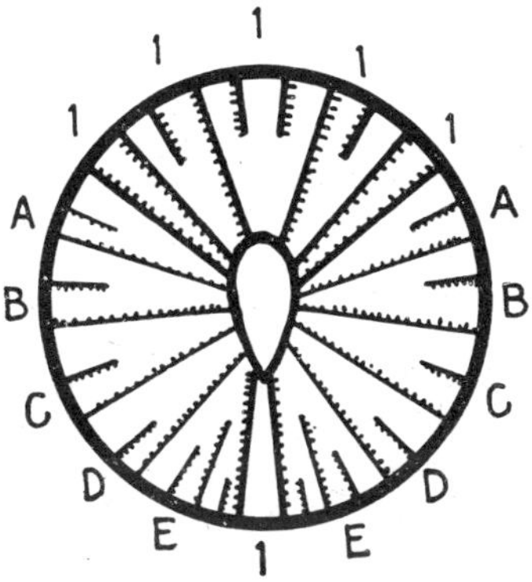

FIG. 4-28. Subclass ZOANTHARIA—Order Zoanthidea. Diagram of the mesenteries in a zoanthid. Five pairs of primary mesenteries (1), indicated by thicker lines, lie in the dorsal part of the polyp and a single pair (1) has a ventral position. New mesenteries, indicated by the fainter lines (*A–E*), are added in succession exclusively within the exocoelic chamber on each side of the ventral directives. It is this characteristic that distinguishes the zoanthids from all other living anthozoans and gives them resemblance to the extinct tetracorals, which added new septa in the same regions. The arrangement of mesenteries exhibits a strongly bilateral symmetry. (*After Duerden,* 1902.)

Order 4. Antipatharia. This order, commonly referred to as the "black corals," includes slender plantlike colonies which, like the gorgonians, consist of a horny and thorny skeletal axis covered by a thin coenenchyme bearing polyps (Fig. 4-29). The polyps protrude from the coenenchyme at right angles to the axis and bear six tentacles. Black corals are typical of the deeper and warmer waters, living chiefly below 100 m. (300 ft.). Although it would

FIG. 4-27. Order Scleratinia. *A–B. Solenastrea hyades* from Florida Pliocene. *C.* A few corallites of *Galaxea fascicularis* from Murray Islands. *D. Meandrina braziliensis*, a complete corallum. *E–F. Balanophyllia: E, B. irrorata* from Texas Eocene; *F, B.* (*Eupsammia*) *trochiformis*, transverse section of an Eocene corallum from Paris Basin. *G–H. Endopachys maclurii*, side and calicular views of an Eocene specimen. *I. Thecosmilia annularis* from the Coral Rag of England. *J–K. Montastrea;* portions of a Puerto Rican and a Brazilian species. *L–M. Trochosmilia didymophila* from the Upper Cretaceous; calicular and side views. *N–O. Dendrophyllia dendrophylloides* from the London Clay, England. *P–R. Caryophyllia: P–Q, C. bukowskii* from the Pliocene; *R*, enlarged calice of *C. alcocki* from the Hawaiian Islands. *S–U. Turbinolia* (Middle Eoc.): *S–T, T. sulcata; U, T. pharetra. V–W. Flabellum* from the Lower Cretaceous of England. (*A–B after Vaughan,* 1917; *C–H, J–M, P–W from Vaughan and Wells,* 1943, *after numerous authors; I after Nicholson and Lydekker,* 1889; *N–O after Edwards and Haime,* 1850.)

TABLE 4-2. GEOLOGICAL RANGES AND PROBABLE RELATIONSHIPS OF SUBORDERS, FAMILIES, AND SUBFAMILIES OF SCLERACTINIA (*After Vaughan and Wells*, 1943, *p.* 91)

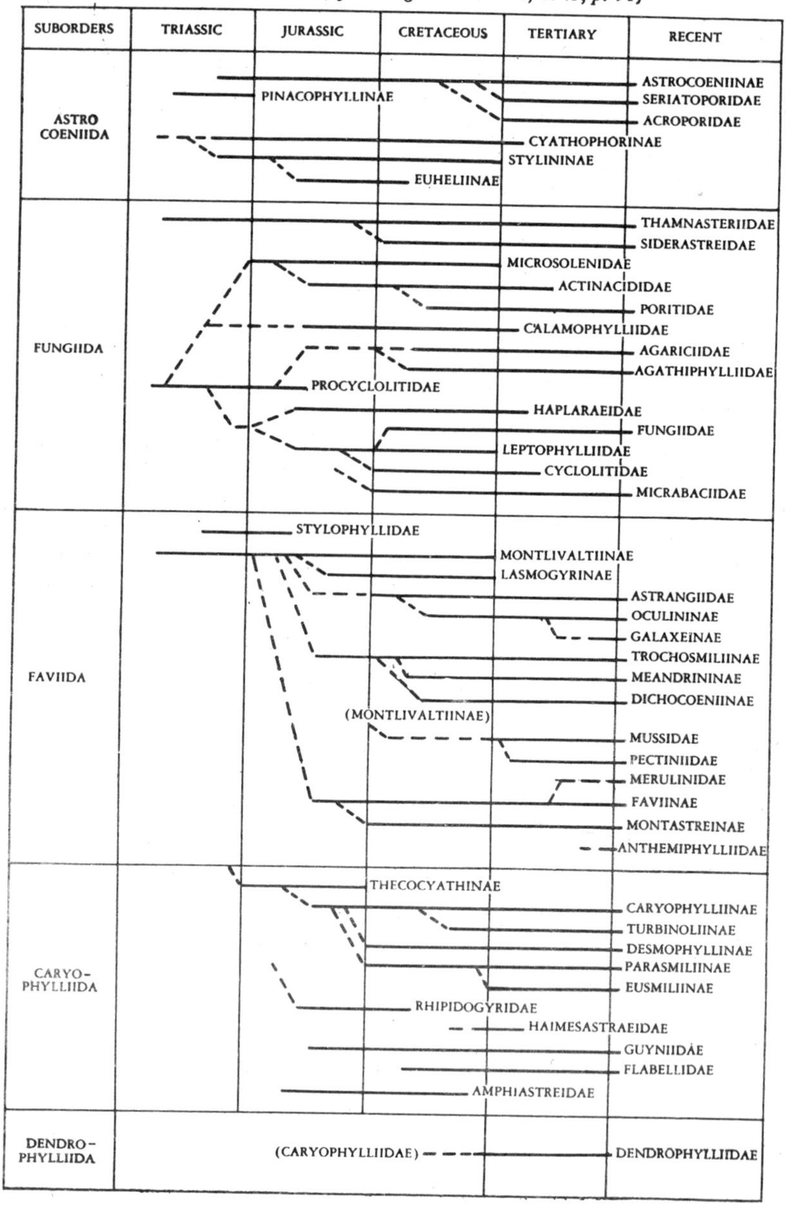

seem possible for the horny axis to be preserved, no fossil representatives of this order are known.

Order 5. Ceriantharia. These are slender cylindrical anemone-like polyps that lead a solitary life in slime-covered vertical tubes in the bottom sediments. Only the oral disk with its two sets of tentacles is exposed. The aboral end, which has a tiny pore, cannot be drawn upward, but the oral end, if disturbed, can be drawn downward into the tube. The mesenterial arrangement differs from that of all other anthozoans, consisting of one cycle of complete single coupled mesenteries of indefinite number. There is no skeleton.

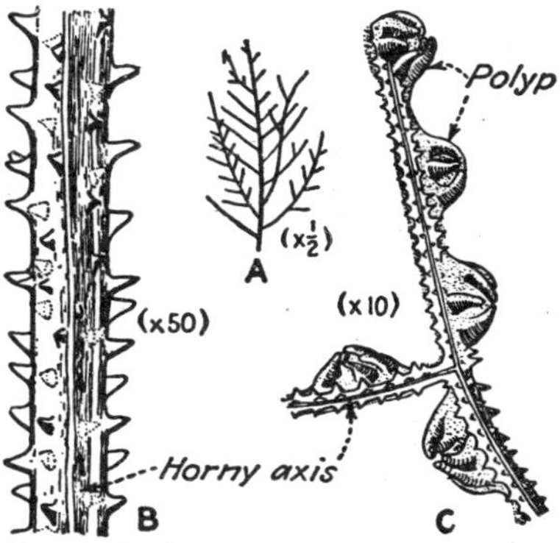

FIG. 4-29. Subclass ZOANTHARIA—Order Antipatharia. *A*. Diagrammatic sketch of an antipatharian colony. *B–C*. *Antipathes ternatensis*, a typical black coral: *B*, a small fragment of the horny axis, showing its spiny nature; *C*, a portion of two branches, showing five polyps and their relation to the horny axis beset with thornlike projections. (*After Schultze*, 1896.)

Cerianthid corals are widely distributed in present seas, living mainly in warmer waters but at all depths. Some individuals are as long as 35 cm. (18 in.), and tubes reach a length of a meter. No fossil cerianthids have been reported—which is not unexpected, since the polyp has no skeletal hard parts—but certain vertical tubular cavities in sandy shales and sandstones may have been formed by ancient representatives of this order.

Geologic History of Zoantharia. The stony Zoantharia, or Hexacorallia, are now generally assumed to have left no fossil record older than the Mesozoic (Middle Triassic). Whether or not hexacorals were present in Paleozoic seas is a moot question. In any case all supposed Paleozoic hexacorals have been assigned to some other subdivision of the Coelenterata.[1]

Two hypotheses have been proposed to explain the origin of the Zoantharia. The first postulates that the living Zoantharia, the Tabulata extinct since the end of the Paleozoic, and the exclusively Paleozoic Tetracorallia had a common ancestor in some as yet undetermined early Paleozoic anthozoan stem. It is assumed, but not proved, that the hexacorals existed in Paleozoic seas but lacked the ability to secrete a stony skeleton, hence left no fossil record of their existence (unless *Mackenzia* is a hexacoral), as did the members of the other classes. By Middle Triassic, however, they had developed the skeleton-forming habit, and they soon exploited the niches left by the then extinct tetracorals and tabulates and the rapidly disappearing

[1] For a discussion of this question see ROBINSON, W. J., 1917, The relationship of the Tetracoralla to the Hexacoralla, *Trans. Conn. Acad. Arts Sci.*, vol. 21, pp. 145–200. ROBINSON, W. J., 1923, The ancestry of the Hexacorallia, *Amer. Jour. Sci.*, 5th ser., vol. 6, pp. 424–426. RAYMOND, P. E., 1921, The history of corals and the "limeless" oceans, *Amer. Jour. Sci.*, 5th ser., vol. 2, pp. 343–347.

schizocorals. This hypothesis seems to have gained wide acceptance and is diagrammed in Fig. 4-30.

The second hypothesis holds that the Zoantharia descended directly from the Tetracorallia. The transition is supposed to have taken place late in the Permian and early in the Triassic, inasmuch as no true hexacorals have as yet been found in rocks older than Middle Triassic. Some facts support this hypothesis, but others do not.[1] It does not seem to have gained much support.

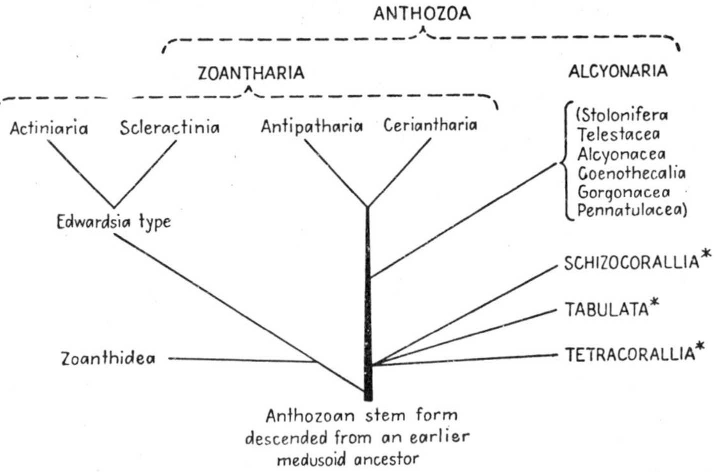

FIG. 4-30. Class ANTHOZOA. Diagram of a highly speculative evolutionary tree of the class. All extinct components are indicated by an asterisk. (*Based in part on Hyman*, 1940.)

The chief area of hexacoral evolution and distribution during Mesozoic and Cenozoic time was in the warm waters of the central and western Tethys. The chief evolutionary trends were:

1. Development of colonial habit from simple coralla by several different modes of colony formation. This tendency caused greater compactness of the corallum.
2. Reduction in size of polyps and increase in their numbers.
3. Increase in porosity of septa and related skeletal parts with consequent increase in growth rate.
4. Development of a marginal edge zone, and from it the coenenchyme, implying a relative increase in size and efficiency of the polyps with respect to the skeleton.

The earliest hexacorals, which are found in Mid-Triassic rocks, were of the reef-building type and are thought to have required the same environmental conditions as do modern corals. Non-reef-building types were derived from the simple reef builders, appearing first in the Jurassic and expanding rapidly throughout the remainder of the Mesozoic and Cenozoic. Today they are well adapted to almost every kind of clear marine water.[2]

[1] CLOUD, P. E., JR., 1948, Some problems and patterns of evolution exemplified by fossil invertebrates, *Evolution*, vol. 2, pp. 334–341.

[2] VAUGHAN, T. W., and WELLS, J. W., 1937, Revision of the suborders, families, and genera of the Madreporarian hexacorals (abstract), *Proc. Geol. Soc. Amer.* 1936, p. 359.

Subclass Tetracorallia (Rugosa)

The Tetracorallia are a large and important group of extinct, exclusively Paleozoic corals that are characterized by insertion of major septa at four points on the corallite periphery and arrangement of the well-developed septa in four areas commonly designated quadrants (whence the name of the subclass Gr. *tetra*, four, + *korallion*, coral). The solitary forms built straight or curved conical or conicocylindrical coralla and are commonly designated "horn corals" (*e.g.*, *Zaphrentis*). Colonial species built dendritic or massive

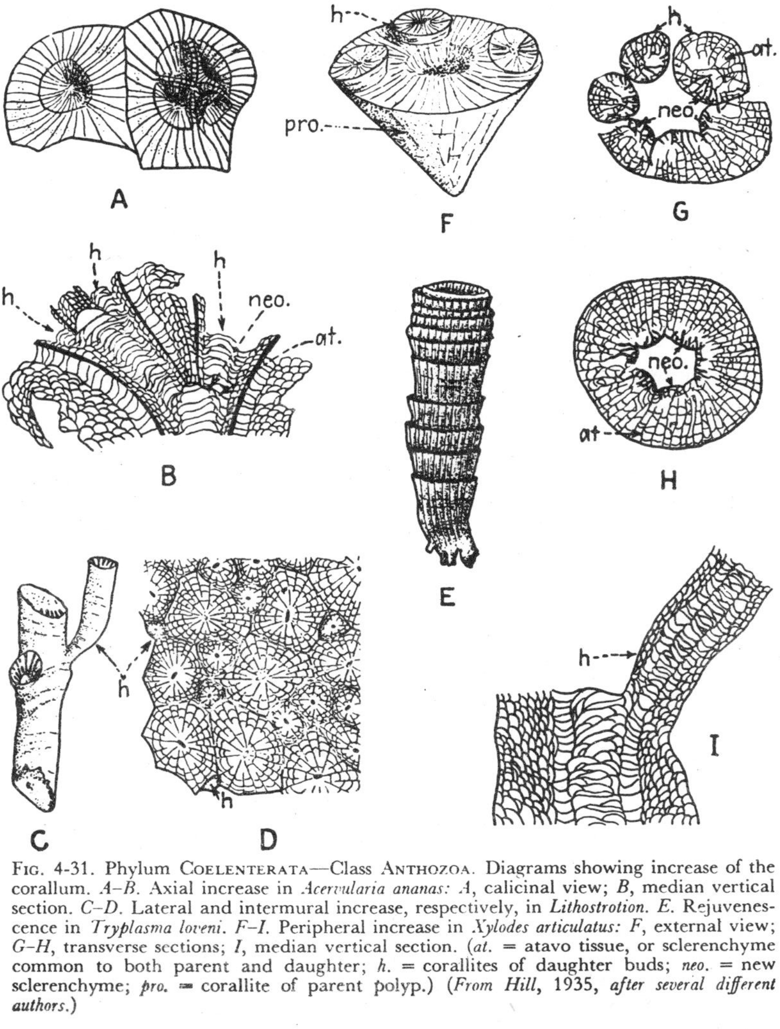

FIG. 4-31. Phylum COELENTERATA—Class ANTHOZOA. Diagrams showing increase of the corallum. *A–B*. Axial increase in *Acervularia ananas: A*, calicinal view; *B*, median vertical section. *C–D*. Lateral and intermural increase, respectively, in *Lithostrotion*. *E*. Rejuvenescence in *Tryplasma loveni*. *F–I*. Peripheral increase in *Xylodes articulatus: F*, external view; *G–H*, transverse sections; *I*, median vertical section. (*at.* = atavo tissue, or sclerenchyme common to both parent and daughter; *h.* = corallites of daughter buds; *neo.* = new sclerenchyme; *pro.* = corallite of parent polyp.) (*From Hill*, 1935, *after several different authors.*)

compound coralla of many closely packed septate corallites of both cylindrical and prismatic nature. The name *Rugosa* is commonly applied to this subclass, calling attention to the rough exterior of many of the corallites.

The shape and nature of the polyp are unknown except as they can be inferred from features of the exoskeleton. It is assumed that the polyp had the same relationship to the corallite, and to structures within and on it, as can be observed in living Zoantharia (Fig. 4-11). It would follow, therefore, that the tetracoral polyp had mesenteries and that these alternated in position with the major septa. This remains to be shown.

It is supposed that reproduction was both sexual and asexual, and that most solitary polyps and the first individual of a colony were produced sexually, whereas the colony developed by budding of new individuals from older polyps. The buds thus formed had lateral, peripheral, or intermural positions (Figs. 4-31, 4-32).

As in the Zoantharia, with which the Tetracorallia are supposed to be closely related, septal development has been extensively studied for the light it throws on the evolution and anthozoan affinities of the subclass. This problem is discussed in a following section.

Nature of Tetracoral Exoskeleton. The typical corallite of a tetracoral where not modified by crowding is a cone or cylinder, divided into compartments or spaces of many shapes by vertical plates both radially and concentrically arranged and by incomplete horizontal, inclined, and curved plates that are given a variety of designations. Other additions of skeletal matter are made here and there to produce spines, nodes, tubes, crossbars, ridges, solid and porous rods, and other delicate features. It is obvious, therefore, that the actual structure of many tetracorals is highly complex, and, as would be expected, an extensive terminology has been developed by specialists who find it necessary to describe specifically the many delicate exoskeletal structures, some of which are illustrated in Figs. 4-34, 4-36, and 4-37. For our purpose here it is necessary to select for definition only a few of the long list of terms.[1] Some structures have been given several different names (Jeffords, 1947).

The exoskeleton is composed of calcareous material (sclerenchyme) that generally has fibrous structure and by analogy with living corals is assumed to have been deposited as an exoskeleton by the ectodermal cells of the epidermis (Figs. 4-19, 4-34*A*–*B*). A solitary coral built a **simple** corallum (Fig. 4-33), discoid, conical, or tubular; a colony of polyps built a **compound** corallum invariably composed of many corallites loosely aggregated or closely

[1] The reader will find excellent discussions of the terminology of tetracoral exoskeletons in the following: GRABAU, A. W., 1922, Paleozoic corals of China, Pt. 1, Tetraseptata, *Paleontologica Sinica*, Ser. B, vol. 2, fasc. 1, pp. 1–70 (3–26). GROVE, B. H., 1934, Studies in Paleozoic corals, *Amer. Midland Naturalist*, vol. 15, pp. 97–137. HILL, D., 1935, British terminology for rugose corals, *Geol. Mag.*, vol. 72, pp. 481–519. SANFORD, W. G., 1939, A review of the families of tetracorals, *Amer. Jour. Sci.*, vol. 237, pp. 295–424; EASTON, W. H., 1944, Corals from the Chouteau and related formations of the Mississippi valley region, *Ill. Geol. Surv.*, *Rept. Inv.* 97, pp. 1–93 (especially 15–21). SMITH, S., 1945, Upper Devonian corals of the Mackenzie River region, Canada, *Geol. Soc. Amer.*, *Spec. Paper* 59, pp. 1–126 (4–9). SMITH, S., and JEFFORDS, R. M., 1947, Pennsylvanian lophophyllidid corals, *Univ. Kan. Pub.*, *Pal. Contr.* 1, pp. 1–84 (12–13).

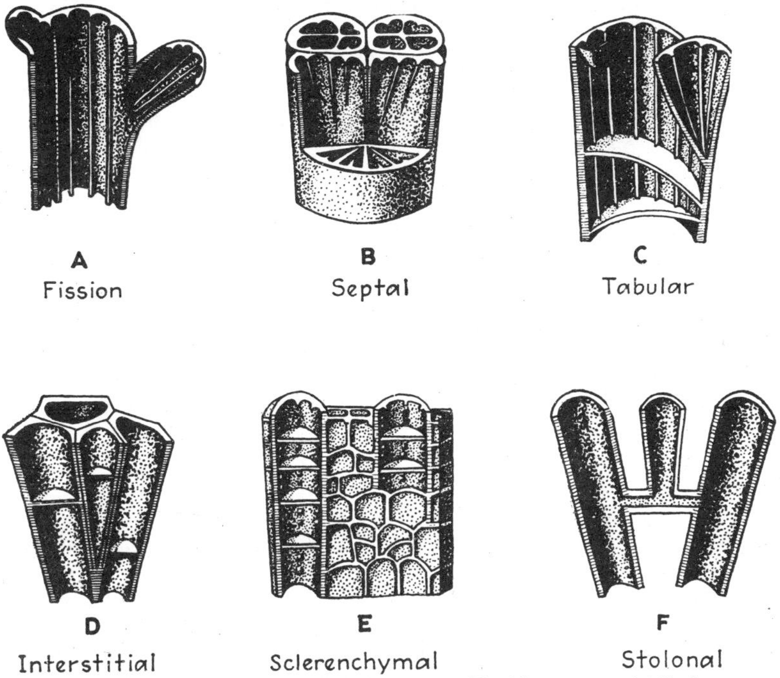

FIG. 4-32. Phylum COELENTERATA. Different modes of budding among the Coelenterata. (*From Grove*, 1934, *after Koch*, 1883, *with modifications*.)

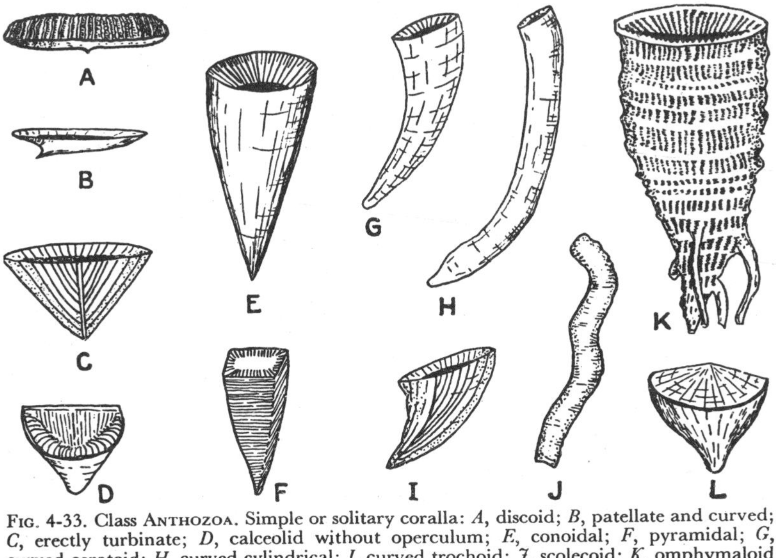

FIG. 4-33. Class ANTHOZOA. Simple or solitary coralla: *A*, discoid; *B*, patellate and curved; *C*, erectly turbinate; *D*, calceolid without operculum; *E*, conoidal; *F*, pyramidal; *G*, curved ceratoid; *H*, curved cylindrical; *I*, curved trochoid; *J*, scolecoid; *K*, omphymaloid with roots; *L*, calceolid with operculum closed. (*After Hill*, 1935.)

packed into massive structures (Figs. 4-39, 4-40). Vertical skeletal elements, of which the septa are the most important, lie in planes parallel with the upward growth of the polyp, and horizontal skeletal elements, typified by tabulae and dissepiments, extend across the corallite. Vertical and horizontal elements combine to make a framework that is enclosed within a calcareous

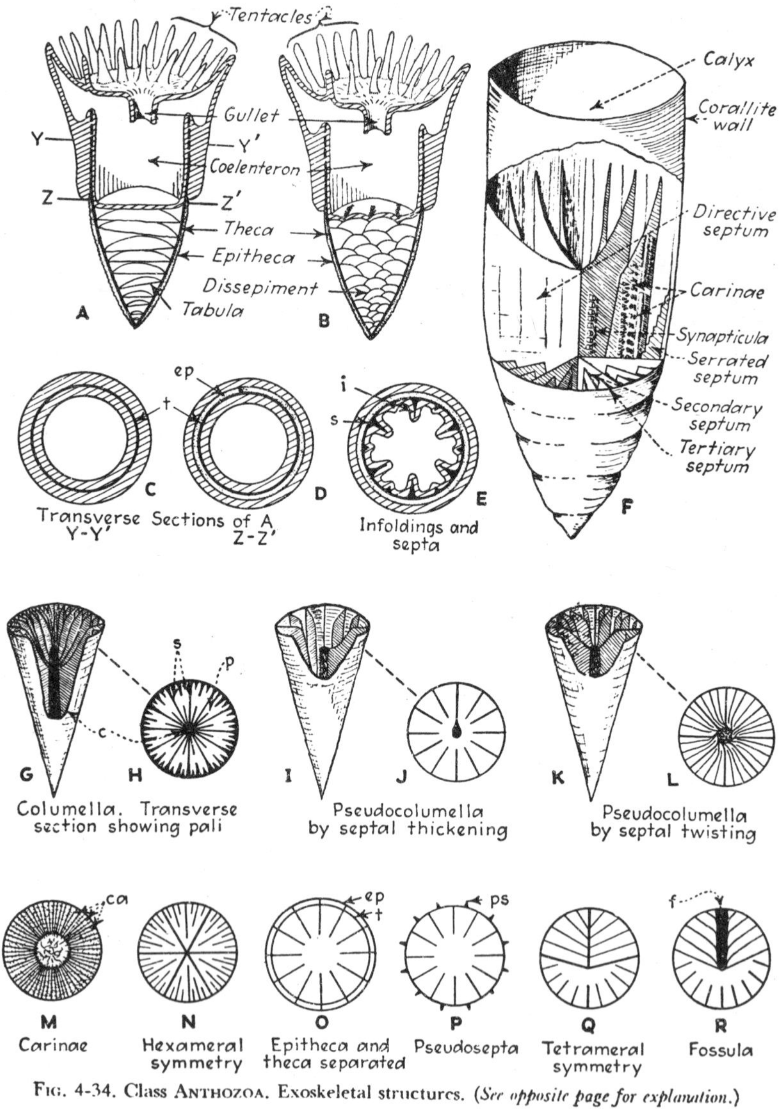

FIG. 4-34. Class ANTHOZOA. Exoskeletal structures. (*See opposite page for explanation.*)

sheath, the theca or epitheca, depending upon the time and mode of formation (Fig. 4-34).

Colonies arise from a simple, presumably sexually produced polyp by asexual reproducton. Baby polyps bud from the parent and build corallites of their own. On the basis of position in the corallite, there are four types of **increase** (here used to refer to the locations where new polyps bud off and start their corallite) of the corallum: *axial*, *peripheral*, *lateral*, and *intermural* (Fig. 4-31). Another series of types, based largely on structural features of the corallite, is illustrated in Fig. 4-32.

The calyx, or calice, is the depression at the end of the corallite last occupied by the polyp. It has a lid, or **operculum,** in a few species (Figs. 4-33*L*, 4-39*K*). Below the calyx in the remainder of the exoskeleton is the framework of vertical and horizontal elements invested by the epitheca. An entire corallum is commonly enclosed in a calcareous sheath called the **holotheca.** Both epitheca and holotheca commonly show external wrinklings, ridges, grooves, and other irregularities (whence the term **rugose** corals). These are generally thought to have been formed by ectoderm overhanging the margin of the corallite.

Septa are the chief vertical skeletal elements in most tetracorals. They extend vertically from apex to calyx floor and radially from corallite periphery toward or entirely to the axis. The septa may join in the center and even become twisted to form some sort of axial column (Fig. 4-34*K*–*L*). Septa are composed of fibers of calcium carbonate grouped together to form spines, or **trabeculae,** as in the Zoantharia (Fig. 4-24). In many corallites they bear sharp flanges, **carinae,** and tiny tubular or solid projections resembling the **synapticula** of the Zoantharia. Not uncommonly the free inner margin is serrated. (Figs. 4-34*F*, 4-36).

The septa are in two series, the longer or **major,** which alternate with the shorter, or **minor.** In the early growth stages a single **axial septum** appears in the plane of bilateral symmetry (Fig. 4-35). Later this septum becomes two diametrically opposed septa—one that is more than half the diameter of the corallite and designated the **cardinal septum,** the other much shorter and called the **counter septum.** Shortly after the axial septum has developed, a small vertical plate appears on each side of the axial septum at its cardinal end. These two plates remain attached to the wall of the corallite and to the axial septum, but gradually spread outward and eventually become the **alar septa.** In a third stage a pair of vertical plates appear at the counter end of

FIG. 4-34. Class ANTHOZOA. Exoskeletal structures. *A*. Polyp and tabulate corallum. *B*. Polyp and dissepimental corallum. *C*. Transverse section in upper part of *A*, showing only theca (*t*). *D*. Transverse section at base of calyx in *A*, showing epitheca (*ep*) and theca (*t*). *E*. Transverse section of a septate corallite occupied by a polyp, showing relation of septa (*s*) to infoldings (*i*) of the wall (mesenteries are omitted). *F*. Composite diagrammatic corallite, showing numerous internal structures. *G*. Corallum with columella. *H*. Transverse section of *G*, showing columella (*c*), pali (*p*), and septa (*s*). *I*. Corallum with pseudocolumella produced by thickening of the free edge of a septum. *J*. Transverse section of *I*. *K*. Corallum with pseudocolumella produced by twisting of inner edges of septa. *L*. Transverse section of *K*. *M*–*R*. Transverse sections of corallites showing septa, carinae (*ca*), epitheca (*ep*), theca (*t*), pseudosepta (*ps*), and fossula (*f*): *M*, *O*, and *P* show radial symmetry; *N* shows hexameral and radial symmetry; *Q* and *R* show tetrameral symmetry characteristic of the tetracorals.

the axial septum and spread outward, like the earlier two but to a more limited extent, eventually to become the **counterlateral septa.** There is now a pause in the development of the tetracoral; the six primary septa thus far developed—cardinal, counter, two alar, and two counterlateral—are called the **protosepta.** All major septa formed after the protosepta are **metasepta** (cf. Figs. 4-20, 4-21, 4-37), and these appear at definite places in the four unequal quadrants marked off by the cardinal, counter, and alar septa. Thus it is seen that although the tetracoral first passes through a stage with six sclerosepta, all subsequent major septa are added in pairs in the counter and cardinal quadrants. In the cardinal quadrants metasepta appear on each side of the cardinal septum; in the counter quadrants, they develop on the counter side of the alar septa (Fig. 4-35). As the counterlateral septa have little relation to the insertion and development of subsequent septa, some coral specialists exclude them from the protosepta and consider them as metasepta. After the protosepta are developed, metasepta and minor septa are added in several different ways and in variable numbers so that the number of septa in an adult corallite can range from a dozen to more than a hundred. With increase in corallite size and in number of septa the tetrameral symmetry may be more or less completely obliterated by a secondary radial symmetry, but the major and minor septa continue to have their respective lengths.

Many tetracorals have one or more prominent longitudinal depressions in the floor of the calyx in the position of one or more of the protosepta. This type of depression, designated a **fossula** (plural, **fossulae**), is formed by partial or complete abortion of a protoseptum. A **cardinal fossula** (Figs. 4-34*R*, 4-37[14]) is the most common, but **alar** and **counter fossulae** are present in some species. Fossulae are commonly marked by depression of the tabulae. In a few tetracorals a **pseudofossula** (Fig. 4-37[13]) is present. This is a depression or space between septa on the floor of the calyx, not formed (as are true fossulae) by partial or complete suppression of a septum. Pseudofossulae on the counter side of the alar septa, hence called **alar pseudofossulae** (Fig. 4-37[13]), are especially characteristic of the lophophyllid tetracorals.

In many parts of the exoskeleton, sclerenchyme is added to skeletal parts, giving them a dilated or enlarged form. Zones of such dilation are **stereozones** and are most commonly developed in the peripheral part of the corallite. The secondary sclerenchyme so added is referred to as **stereoplasm.**

Many tetracorals have some sort of compact or vesicular structure along the axis of the corallite, and commonly a distinct columnar zone or rod is present. Both **columella** (with several different subtypes, as shown in Figs. 4-34 and 4-37) and **pseudocolumella** (Fig. 4-34*I–L*) have generally been applied to this axial structure, depending on its nature and mode of formation. The noncommittal term **column** is used in a general sense for any sort of axial structure.

On the exterior of some tetracorals longitudinal depressions, the **septal grooves,** mark the inbending of the outer wall along lines at the edges of the septa. These grooves commonly indicate the positions of the first four protosepta (Fig. 4-38). By grinding away the outer millimeter or so of well-preserved horn corals such as *Streptelasma*, it is possible to determine the positions

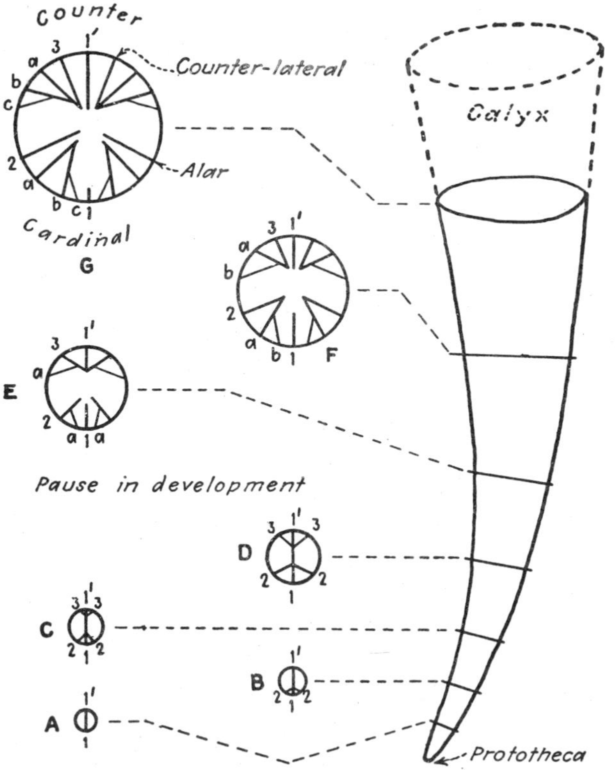

FIG. 4-35. Class ANTHOZOA—Subclass TETRACORALLIA. Diagrams showing order of development of septa in *Zaphrentis*, a Paleozoic tetracoral. The embryonic stage is represented by the prototheca which constitutes the apex of the conical corallum. Shortly after this tiny cup has been formed its cavity becomes divided by an axial septum (*A*), which later breaks near the center to become the cardinal (1) and counter (1′) septa. In stage *B* two new septa (2) appear at the cardinal end of the axial septum, one on each side. These gradually shift to a radial position and become the alar septa (*E–G*). In stage *C* a second pair of new septa appear (3), this time near the counter end of the axial septum, and these ultimately shift into a radial position and become the counterlateral septa. After stage *D* has been reached there is a pause in septal development. At this stage six primary septa, the protosepta, are present—the cardinal (1) and the counter (1′), which are still united as the axial septum, two alar (2) and two counterlaterals (3). These now exhibit marked biradial symmetry and are analogous to the six protosepta that appear simultaneously early in the development of a modern coral (*e.g.*, *Siderastrea*, Fig. 4–20). Following the pause in development, successive pairs of new septa (*a*, *b*, *c*), the metasepta, appear on the cardinal side of both alar and counterlateral septa, and these latter protosepta shift gradually into a radial position (*E–G*). In the adult corallum the number of septa in the counter quadrants (one on each side of the counter septum) is generally greater than that in the cardinal quadrants, and the septa themselves show marked bilateral symmetry in their arrangement. Generally, in the development of *Zaphrentis*, the cardinal septum is arrested and a gap, or fossula, marks its position in the septal cycle. Less conspicuous fossulae may develop on the counter side of the alar septa and on either side of the counter septum. The sections are taken from the prototheca to the calyx at approximately the intervals shown. (*Adapted from Carruthers*, 1906; *also see Swinnerton*, 1947.)

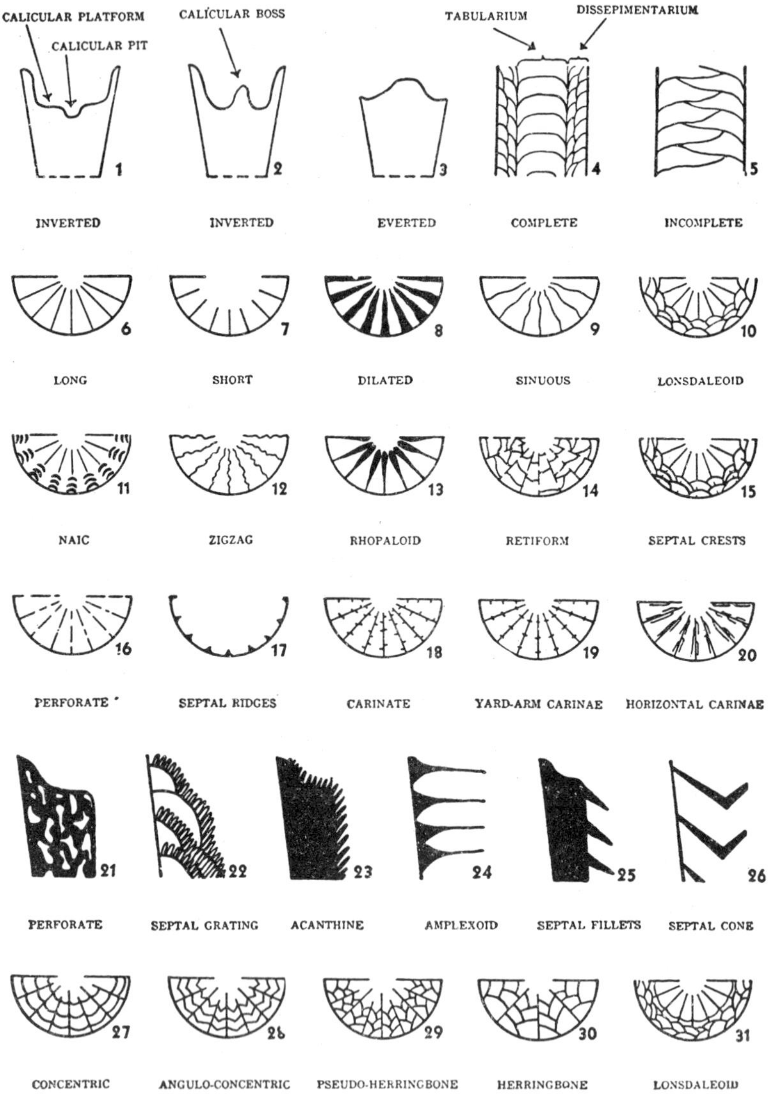

FIG. 4-36. Class ANTHOZOA. Structures of coral exoskeletons. 1–3. Longitudinal sections of distal portions of corallites. 4–5. Longitudinal sections showing types of tabulae. 6–20. Transverse sections showing types of septa. 21–26. Longitudinal sections of left half of corallites showing types of septa. 27–31. Transverse sections showing types of dissepiments. (*After Easton*, 1944.)

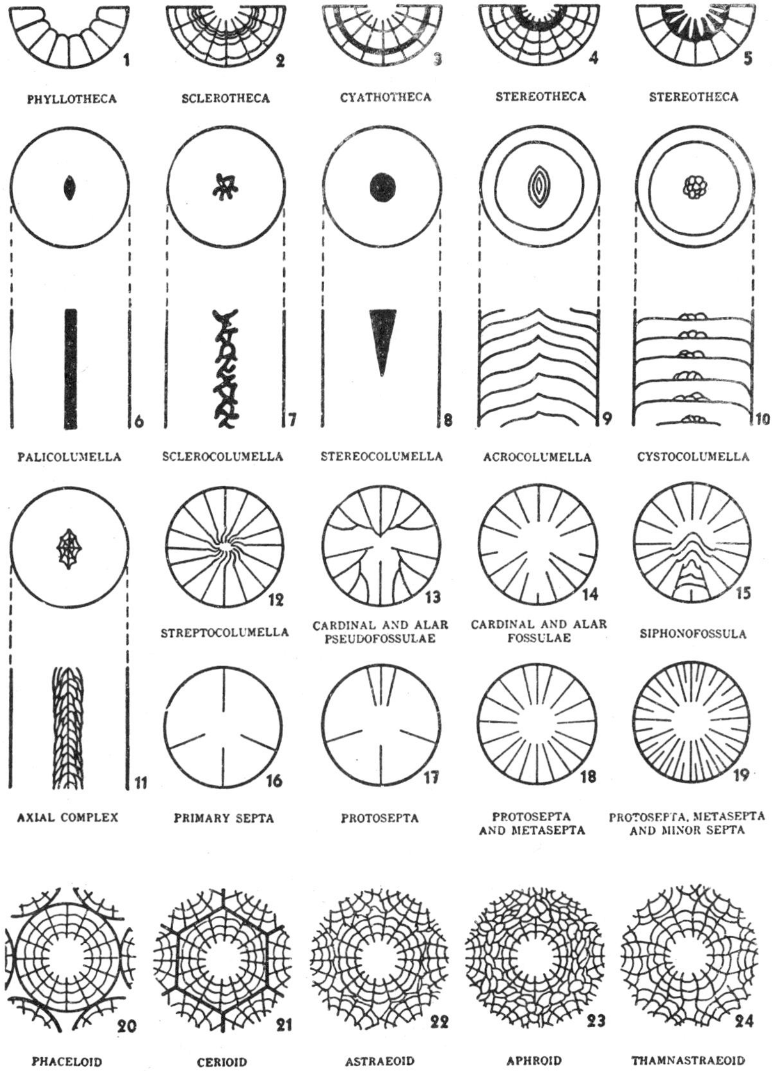

FIG. 4-37. Class ANTHOZOA. Structures of coral exoskeletons. 1–5. Transverse sections showing types of inner walls (four is a dissepimental stereotheca; five is a septal stereotheca). 6–12. Transverse (above) and longitudinal (below) sections showing types of axial columns. 13–15. Transverse sections showing types of fossulae. 16–19. Transverse sections showing appearances of various septal groups. 20. Transverse section of one corallite and portions of surrounding corallites in a phaceloid corallum. 21–24. Transverse sections of one corallite and portions of surrounding corallites in plocoid coralla. (*After Easton*, 1944.)

of the cardinal, counter, and alar septa by observing the arrangement of metasepta (Fig. 4-38).

Transverse skeletal elements, designated **tabulae** (sing. **tabula**) and in the form of subhorizontal, convex or concave platforms, extend part or all the way across the corallite and are not limited by the septa. If they reach entirely across the axial area without intersecting any axial structure they are **complete** (Fig. 4-36[4]); otherwise they are **incomplete** (Fig. 4-36[5]) or **anastomosing.** The strongly tabulate axial zone has been designated the **tabularium** (Fig. 4-36[4]).

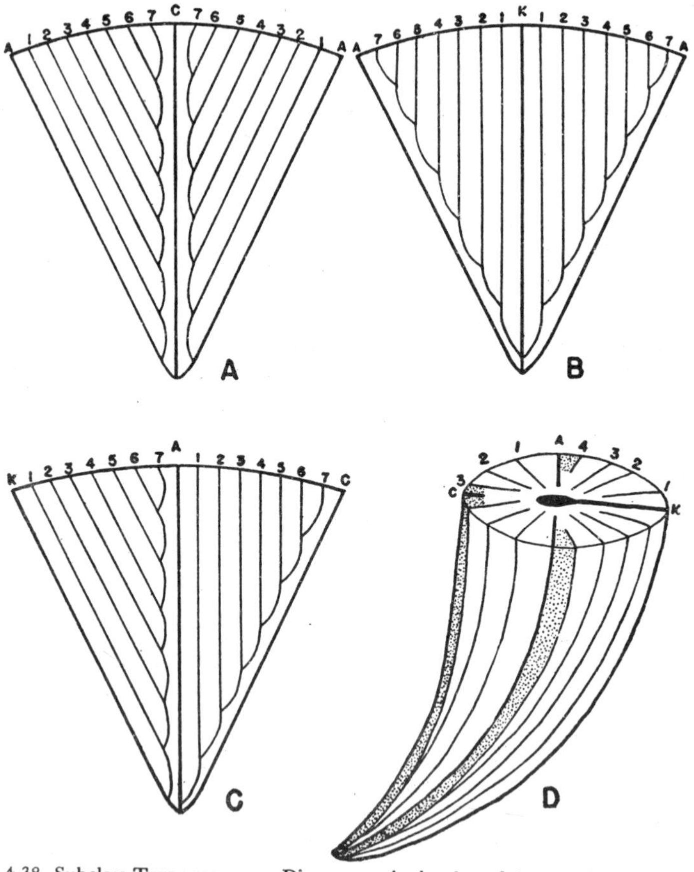

FIG. 4-38. Subclass TETRACORALLIA. Diagrammatic sketches of the septal groove pattern of tetracorals. *A*. Pattern in the cardinal quadrants. *B*. Pattern in the counter quadrants. *C*. Pattern of the cardinal and counter quadrants seen from the side. *D*. Sketch showing relationships of septal grooves to septa, fossulae, and pseudofossula (stippled). (*A* = alar septum; *C* = cardinal septum; *K* = counter septum. Numerals indicate order of insertion of septa.) (*After Jeffords*, 1942.)

Tetracorals in which minor septa are developed are characterized by small curved plates (dissepiments) built one on another so as to form vesicles. These plates are generally convex upward and toward the axis of the corallite (Fig. 4-34*B*). The peripheral zone, in which dissepiments are usually best developed, is the **dissepimentarium** (Fig. 4-36[4]).

A few species have corallites that seem in external view to be made of a series of superimposed cones. This **rejuvenation,** or **rejuvenescence** (Fig. 4-31*E*), is the result of a renewal of immature structural characteristics associated with constriction of the corallite after maturity has been attained.

Classification. The Subclass Tetracorallia contains more than 500 genera and several times that many species. These were formerly included in half a dozen "catchall" families, but in recent years, with intensive study of many of the old genera, it has been found necessary to erect new families with more precise limitations. At the present writing tetracoral specialists have differentiated more than 40 families. These, together with their geologic range, are shown in Table 4-3. Representative genera of a few of the more important families are illustrated in Figs. 4-39 and 4-40.

Evolution and Geologic History of the Tetracorallia. Opinions differ as to how the Tetracorallia are related to other extinct coral groups (Tabulata and Schizocorallia) and to existing Anthozoa. As pointed out on page 145, some investigators consider them the ancestors of the Scleractinia (Madreporaria), and possibly of the Zoantharia (Hexacorallia) as a whole, using as support for this contention the fact that the young tetracoral first passes through a stage with six sclerosepta. By other investigators, however, this septal sequence is interpreted to mean that the tetracorals are related to the antipatharian-cerianthid line from which they diverged early in coral evolution. Further relationship with the zoanthids is strongly indicated by the fact that after the six protosepta have been formed, subsequent septa are inserted in four locations, two of which are the same as the ones to which septal formation is limited in the zoanthids. Hyman (1940) concludes that this similarity suggests an affinity of the zoanthids to the tetracorals and probably justifies placing the zoanthids at the bottom of the zoanthid-anemone-coral line rather than as an offshoot of the anemones (see Fig. 4-30). This second view, therefore, holds that the Tetracorallia, the Zoanthidea-Scleractinia-Actiniaria line, and the Antipatharia-Ceriantharia line all branched from a common anthozoan stem form quite early in their history, as indicated in Fig. 4-30.[1]

The oldest known tetracorals are found in rocks of Lower Ordovician (Chazy) age. The subclass attained its maximum in both abundance of individuals and number of species in Mid-Paleozoic (Silurian and Devonian) seas, after which it declined to extinction at the end of the Permian. Fossil tetracorals are most abundant in calcareous rocks—limestones, argillaceous limestones, and calcareous shales—in which they commonly make extensive biostromes but rarely form reef-mounds or bioherms. Many are reliable index fossils.

[1] See footnote on p. 145 for additional discussion of this problem.

TABLE 4-3. KEY TO THE FAMILIES OF THE TETRACORALLIA
Erwin C. Stumm, 1952

Key					Major distinguishing character	Family name	Author and date	Geologic range
Typically simple forms, discoid, patelloid, ceratoid, trochoid, cylindroid	Tab. abs.	Axial structure absent	Fossula(e) absent	Dissepiments absent	Short, simple septa	Petraiidae	DeKoninck, 1872	Ord.-Dev.
Typically simple forms, discoid, patelloid, ceratoid, trochoid, cylindroid	Tabulae present	Axial structure absent	Fossula(e) absent	Dissepiments absent	Radial septa; smooth epithecae	Streptelasmidae	Grabau, 1922	Ord.-Sil.
Typically simple forms, discoid, patelloid, ceratoid, trochoid, cylindroid	Tabulae present	Axial structure absent	Fossula(e) absent	Dissepiments absent	Peripheral septal ridges; flat tabulae	Pycnostylidae	Stumm, 1952*	Sil.
Typically simple forms, discoid, patelloid, ceratoid, trochoid, cylindroid	Tabulae present	Axial structure absent	Fossula(e) absent	Dissepiments absent	Perforate septa	Calostylidae	Roemer, 1883	Sil.
Typically simple forms, discoid, patelloid, ceratoid, trochoid, cylindroid	Tabulae present	Axial structure absent	Fossula(e) absent	Dissepiments absent	Acanthine septa; discoid forms without tabulae	Rhabdocyclidae	Hill, 1940	Sil.
Typically simple forms, discoid, patelloid, ceratoid, trochoid, cylindroid	Tabulae present	Axial structure absent	Fossula(e) absent	Dissepiments absent	Vermiform shape; fluted exterior	Heterophyllidae	Hill, 1940	Miss.-Penn.
Typically simple forms, discoid, patelloid, ceratoid, trochoid, cylindroid	Tabulae present	Axial structure absent	Fossula(e) absent	Dissepiments absent	Short septa becoming longer over upper side of tabula	Amplexidae	Chapman, 1893	Miss.-Perm.
Typically simple forms, discoid, patelloid, ceratoid, trochoid, cylindroid	Tabulae present	Axial structure absent	Fossula(e) absent	Dissepiments absent	Radial septa; long protosepta producing four quadrants	Calophyllidae	Stumm, 1952*	Perm.
Typically simple forms, discoid, patelloid, ceratoid, trochoid, cylindroid	Tabulae present	Axial structure absent	Fossula(e) absent	Dissepiments absent	Long protosepta and counterlateral septa	Plerophyllidae	Koker, 1924	Perm.
Typically simple forms, discoid, patelloid, ceratoid, trochoid, cylindroid	Tabulae present	Axial structure absent	Fossula(e) present	Dissepiments absent	Discoid or globose; tabulae suppressed in most forms	Hadrophyllidae	Stumm, 1949	Dev.-Penn.
Typically simple forms, discoid, patelloid, ceratoid, trochoid, cylindroid	Tabulae present	Axial structure absent	Fossula(e) present	Dissepiments absent	Fossula on convex side of corallum	Zaphrentidae	Edwards and Haime, 1851	Sil.-Miss.
Typically simple forms, discoid, patelloid, ceratoid, trochoid, cylindroid	Tabulae present	Axial structure absent	Fossula(e) present	Dissepiments absent	Fossula on concave side of corallum	Hapsiphyllidae	Grabau, 1928	Miss.-Perm.
Typically simple forms, discoid, patelloid, ceratoid, trochoid, cylindroid	Tabulae present	Axial structure present	Fossula(e) present	Dissepiments absent	Axial aulos formed by septal dilation	Syringaxonidae	Hill, 1939	Sil.-Perm.
Typically simple forms, discoid, patelloid, ceratoid, trochoid, cylindroid	Tabulae present	Axial structure present	Fossula(e) present	Dissepiments absent	Axial stereocolumella	Metriophyllidae	Hill, 1939	Sil.-Dev.
Typically simple forms, discoid, patelloid, ceratoid, trochoid, cylindroid	Tabulae present	Axial structure present	Fossula(e) present	Dissepiments absent	Axial palicolumella	Cyathaxonidae	Edwards and Haime, 1849	Miss.-Perm.
Typically simple forms, discoid, patelloid, ceratoid, trochoid, cylindroid	Tabulae present	Axial structure present	Fossula(e) present	Dissepiments absent	Columella formed by rhopaloid counter septum	Lophophyllidiidae	Moore and Jeffords, 1945	Penn.-Perm.
Typically simple forms, discoid, patelloid, ceratoid, trochoid, cylindroid	Tabulae present	Axial structure absent	Fossula(e) present	Dissepiments present	Axial stereozone	Pycnactidae	Hill, 1940	Sil.-?Dev.
Typically simple forms, discoid, patelloid, ceratoid, trochoid, cylindroid	Tabulae present	Axial structure absent	Fossula(e) present	Dissepiments present	Domed tabulae	Acrophyllidae	Stumm, 1949	Sil.-Dev.
Typically simple forms, discoid, patelloid, ceratoid, trochoid, cylindroid	Tabulae present	Axial structure absent	Fossula(e) present	Dissepiments present	Septa in quadrants	Halliidae	Chapman, 1893	Sil.-Dev.
Typically simple forms, discoid, patelloid, ceratoid, trochoid, cylindroid	Tabulae present	Axial structure absent	Fossula(e) present	Dissepiments present	Thin septa; cardinal fossula on convex side	Bethanyphyllidae	Stumm, 1949	Sil.-Dev.
Typically simple forms, discoid, patelloid, ceratoid, trochoid, cylindroid	Tabulae present	Axial structure absent	Fossula(e) present	Dissepiments present	Keyhole fossula	Palaeosmiliidae	Hill, 1940	Miss.
Typically simple forms, discoid, patelloid, ceratoid, trochoid, cylindroid	Tabulae present	Axial structure absent	Fossula(e) present	Dissepiments present	Open tabular fossula on convex side	Caniniidae	Hill, 1940	Miss.-Perm.

TABLE 4-3. KEY TO THE FAMILIES OF THE TETRACORALLIA. (*Continued*)
Erwin C. Stumm, 1952

Key					Major distinguishing character	Family name	Author and date	Geologic range
Typically simple forms, discoid, patelloid, ceratoid, trochoid, cylindroid	Tabulae present	Axial structure absent	Fossula(e) absent	Dissepiments present	Discontinuous septa; obscure fossula in some forms	Omphymatidae	Wedekind, 1927	Sil.
					Thin radially arranged septa; erect calyxes	Leptoinophyllidae	Stumm, 1949	Sil.-Dev.
					Dilated or carinate septa; reflexed calyxes	Acanthophyllidae	Hill, 1939	Sil.-Dev.
					Wide septa in contact	Chonophyllidae	Holmes, 1889	Sil.-Dev.
				Degenerate septa	Axial stereozone	Digonophyllidae	Wedekind, 1924	?Sil.-Dev.
					Acanthine septal spines	Cystiphyllidae	Edwards and Haime, 1850	Sil.-?Dev.
					One operculum	Calceolidae	Lindström, 1883	Sil.-Dev.
					Four opercula	Goniophyllidae	Dybowski, 1873	Sil.
					Septal cones	Cystiphylloidae	Stumm, 1949	Dev.
		Ax. str. pres.			Complex axial structure	Clisiophyllidae	Nicholson and Thomson, 1883	Miss.-Perm.
Compound forms, phaceloid, cerioid, or astraeoid		Axial structure absent			Phaceloid forms with thin septa	Entelophyllidae	Hill, 1940	Sil.-Dev.
					Cerioid forms with thin, interrupted septa	Arachnophyllidae	Dybowski, 1873	Sil.
					Phaceloid and cerioid forms with inner walls	Acervulariidae	Stumm, 1952*	Sil.
					Phaceloid or cerioid; small corallites; no dissepiments in some forms	Columnariidae	Rominger, 1876	Ord.-Dev.
					Phaceloid, cerioid, or astraeoid. Carinate septa in most forms	Disphyllidae	Hill, 1939	Sil.-Dev.
					Phaceloid or cerioid with peripheral dissepimentaria between septa and walls; elongate dissepiments	Spongophyllidae	Hill, 1939	Sil.-Dev.
		Axial structure present			Axial aulos; carinate septa; phaceloid or cerioid forms	Eridophyllidae	Stumm, 1949	Dev.-Miss.
					Phaceloid or cerioid forms with fusiform columella	Lithostrotiontidae	Grabau, 1931	Miss.-Perm.
					Phaceloid, cerioid, or astraeoid forms with complex axial structures	Lonsdaleidae	Chapman, 1893	Miss.-Perm.

* These family names are to be credited to Erwin C. Stumm, who prepared Table 4-3 specifically for the present work.

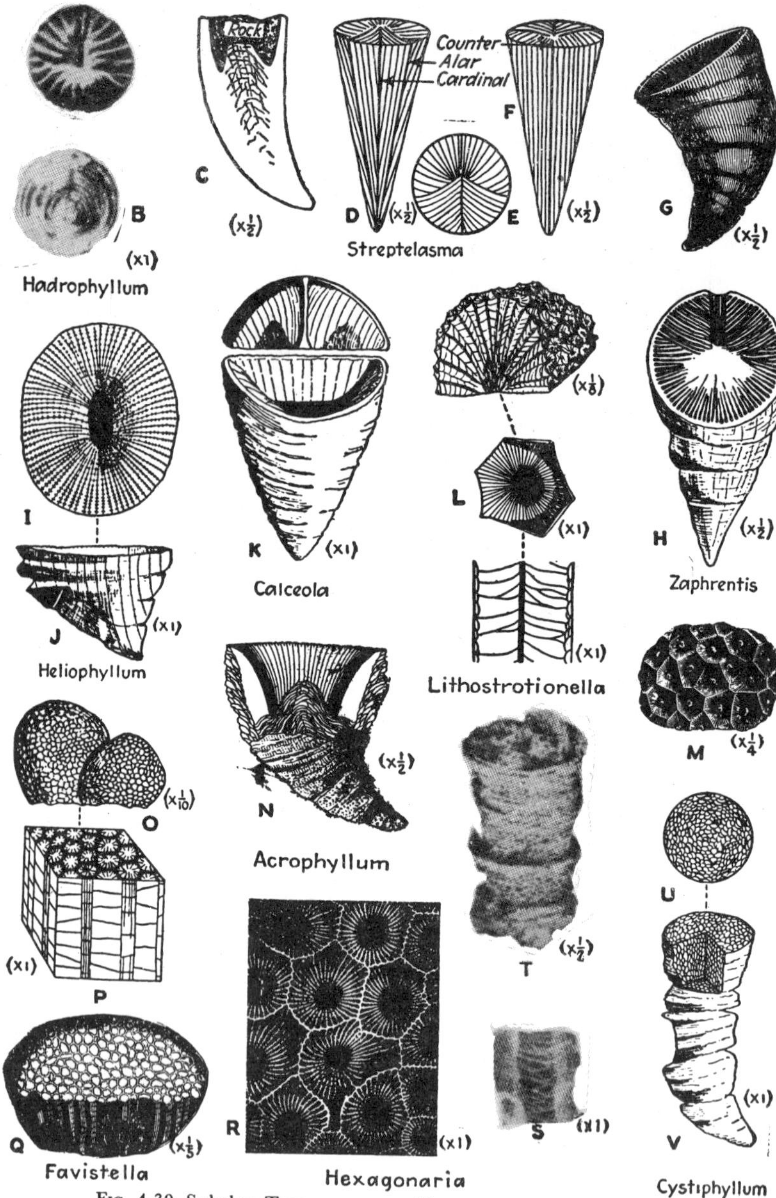

Fig. 4-39. Subclass Tetracorallia. (*See opposite page for explanation.*)

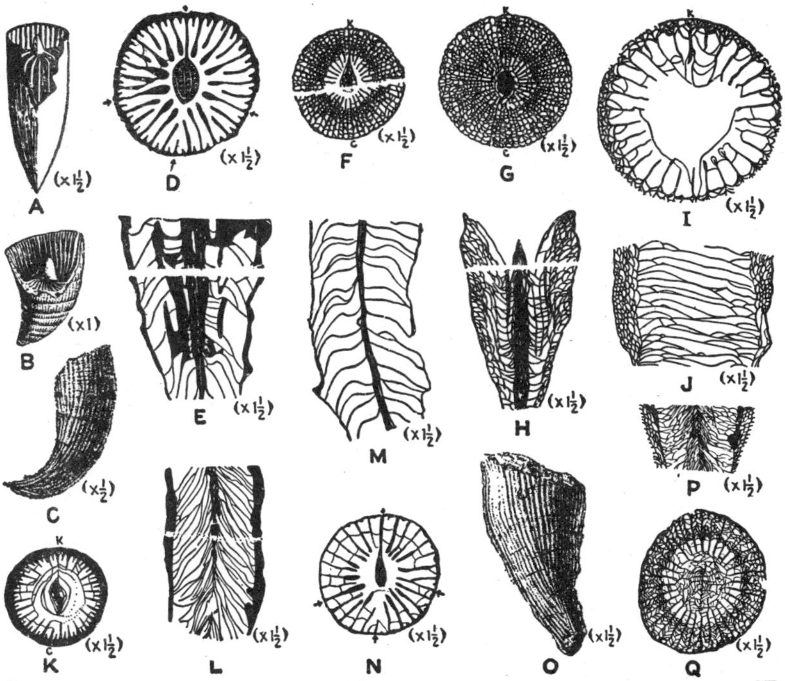

FIG. 4-40. Subclass TETRACORALLIA. *A–B*. Two species of the Carboniferous *Cyathaxonia* (Cyathaxonidae), showing prominent axial palicolumella. *C–E*. Exterior, transverse, and longitudinal sections of *Lophophyllidium* (Lophophyllidiidae), a common Pennsylvanian and Permian tetracoral. *F–H*. Two transverse sections and a longitudinal section of *Heritschia* (Clisiophyllidae) from Kansas Permian. *I–J*. Two sections of *Pseudozaphrentoides* (Caniniidae) from Kansas. *K–L*. *Leonardophyllum* (Clisiophyllidae) from Kansas Permian, characterized by prominent columella and strongly arched tabulae. *M–O*. Longitudinal and transverse sections and exterior of *Stereostylus* (Lophophyllidiidae) from the Kansas Pennsylvanian. *P–Q*. Two sections of *Dibunophyllum* (Clisiophyllidae) from Kansas. (*A modified after Zittel*, 1927; *B after Nicholson*, 1872; *C–Q after Jeffords or Moore and Jeffords*, 1941–1948.)

FIG. 4-39. Subclass TETRACORALLIA. *A–B*. Top and basal views of *Hadrophyllum* (Hadrophyllidae), from Kentucky Mississippian. *C–F*. *Streptelasma* (Streptelasmidae), a common Upper Ordovician genus from Kentucky: *C*, longitudinal section of *S. corniculum*, showing only partial preservation of tabulate and dissepimental structure in axial region; *D,F*, different views of *S. corniculum*, with theca ground away so that septal ends are visible; *E*, transverse section a little below base of calyx, showing tetrameral arrangement of septa. *G–H*. Side and calicinal views of different species of a Devonian *Zaphrentis* (Zaphrentidae): *H* shows the typical cardinal fossula. *I–J*. Two views of *Heliophyllum* (Acanthophyllidae) from Michigan Devonian, showing prominent carinae. *K*. *Calceola* (Calceolidae), an operculate Devonian tetracoral. *L*. *Lithostrotionella* (Lithostrotionellidae), a world-wide index fossil of the Mississippian: portion of a corallum, calicinal view of a single corallite, and longitudinal section of one corallite. *M*. Incomplete corallum of *Arachnophyllum* (Arachnophyllidae), an index tetracoral of the Silurian. *N*. *Acrophyllum* (Acrophyllidae), from the Devonian, showing characteristic domed axial tabulae. *O–Q*. *Favistella* (Columnariidae), an index tetracoral of the Upper Ordovician. *R–S*. Transverse and longitudinal sections of corallites of *Hexagonaria* (Disphyllidae), from the Ohio Devonian. *T–V*. A partial corallum and two views of a second corallum of *Cystiphyllum* (Cystiphyllidae), a common Silurian and Devonian tetracoral characterized by great development of dissepiments and vesiculose structure. (*A–B after Bassler*, 1937; *G, M, Q after Nicholson*, 1872; *K after Stebbing; N after Lambe*, 1899; *R–S after Stumm*, 1948; *T after Stewart*, 1938.)

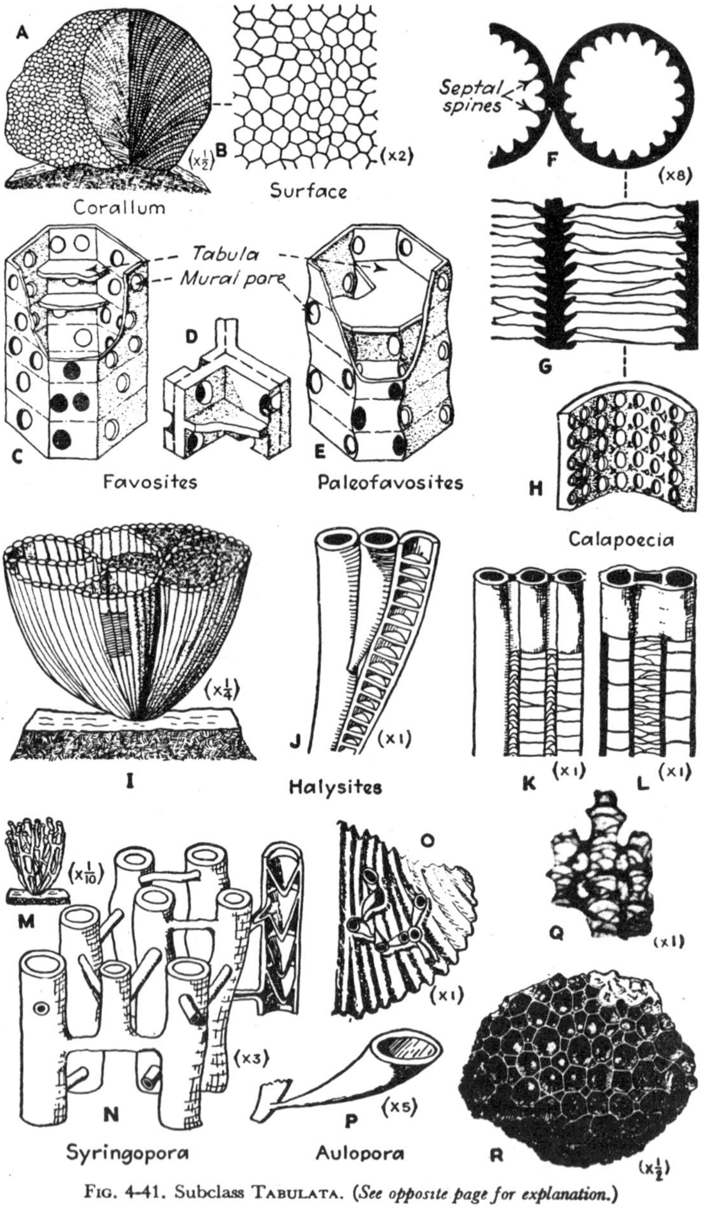

FIG. 4-41. Subclass TABULATA. (*See opposite page for explanation.*)

Subclass Tabulata

The Tabulata are an extinct, exclusively Paleozoic[1] group of compound corals characterized by strongly tabulate (whence the name Tabulata), generally nonseptate cylindrical or prismatic corallites. The coralla vary considerably in shape, from dendritic to spheroidal or laminar, depending on the species and to some extent on bottom conditions where they lived.

The corallite walls are generally thick and independent, so that individual corallites separate readily from the corallum. They are solid in most genera, but in the cleistoporids they are perforate and in the favositids they are perforated by **mural pores,** which show considerable variation in arrangement and by some have been interpreted as sites where buds failed to develop to maturity (Fig. 4-41*C–E*).

Septa are absent or rudimentary and when present commonly consist of rows of spines, nodes, or simple ridges on the inner surface of the corallite wall. In one genus (*Fossopora*) they number 6, in a few others 12, but generally the number seems variable—a situation which may be due to the difficulty encountered in cutting a transverse section in the exact plane of a complete cycle of septa. The typical condition, however, is for the corallites to have only indistinct septa or to lack them altogether.

Dissepiments, synapticula, columellae, and other intracorallite structures are not present in the Tabulata.

The taxonomic relationships of the Tabulata with other corals are uncertain, to a large extent because essentially nothing is known about the nature of the tabulate polyp or the way the septa were added, and probably also because the group is almost certainly not a natural one. Most students of

[1] The common statement that Tabulata range through the Mesozoic is based on the inclusion of the Chaetetida, which we assign to the Schizocorallia (see p. 169).

FIG. 4-41. Subclass TABULATA. *A–D. Favosites hemisphericus* from the Devonian at Louisville, Ky.: *A*, a small corallum in position of growth and sectioned to show internal intensely tabulate character; *B*, portion of the surface magnified to show polygonal outline of the corallites; *C*, a single corallite much magnified to show the relation between mural pores and tabulae; *D*, diagram to show the relation of mural pores to the double wall made by adjacent corallites. *E. Paleofavosites prolificus*, from the Upper Ordovician of Anticosti Island; diagram of a single corallite showing the relation of the mural pores to the tabulae and corners or angles of the corallite. *F–H. Calapoecia canadensis* from the Ordovician of Manitoba: *F*, transverse section of two corallites showing prominent septal spines; *G*, longitudinal section showing septal spines and tabulae; *H*, diagram of a portion of the corallite wall (minus the tabulae) showing the reticulated character caused by the large number of mural pores. *I–L. Halysites: I*, a complete corallum in position of growth, with part of the exoskeleton weathered out of the matrix and part still embedded in it; *J*, diagram of three corallites, showing the method of budding and presence of tabulae; *K*, *H. catenularia*, a species consisting of alternating large and small tabulated corallites, from the Silurian of Quebec; *L*, *H. catenularia amplitubulata*, showing different-sized and differently tabulated corallites, from the Lower Helderberg of Quebec. *M–N. Syringopora: M*, a complete corallum in position of growth; *N*, diagram to show relations of adjacent corallites and internal structure (infundibuliform tabulae). *O–P. Aulopora serpens* from the Devonian of Wisconsin: *O*, the corallum in position of growth attached to the surface of a brachiopod shell; *P*, a single corallite somewhat enlarged showing the method of budding. *Q. Chonostegites clappi*, from New York Devonian, showing arched tabulae. *R. Pleurodictyum connexum*, from New York Devonian, showing a large corallum with deep calices. (*F–H, K–L after Lambe*, 1899; *Q–R after Hall*, 1876.)

corals have assigned the Tabulata to the Alcyonaria,[1] usually as an appendix, on the basis of the supposed similarities of polyp growth and skeletal formation shown in Fig. 4-42. Some think there is good evidence, especially in the presence and number of septa, to support the view that they are closely related to the Zoantharia.[2] Because of this diversity in opinion, and also because there is some convenience to the systematist in having a separate category for groups of uncertain affinities, we prefer to consider the Tabulata a

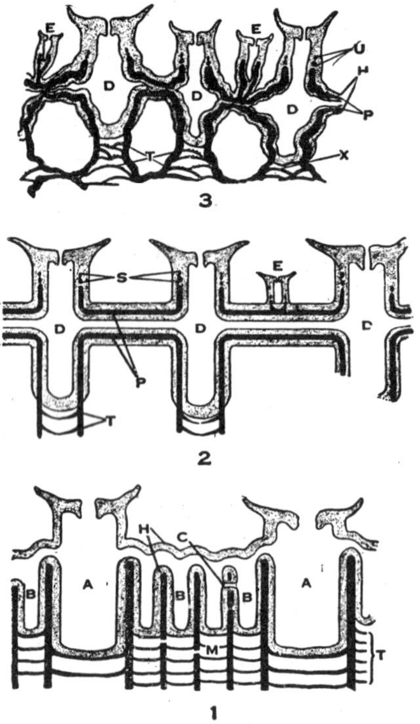

FIG. 4-42. Phylum COELENTERATA—Class ANTHOZOA. Anatomical similarities between living alcyonarians and fossil "tabulates." 1. Diagrammatic representation of soft and hard parts in *Heliopora*, sufficing also as a restoration of *Favosites canadensis* of the Onondagan. 2. Diagrammatic representation of anatomy and budding in the modern *Tubipora*. 3. Restoration of *Chonostegites* as a tubiporiform alcyonarian. In 1, the skeleton is external, secreted by the ectoderm. In 2, it is internal, secreted by ectodermal cells wandering into the mesenchyme. In 3, the skeleton is assumed to be primarily internal, but secondarily external. In all, skeletal parts are black; soft parts stippled. A = sexual polyp or anthozooid; B = extension of stolon sheet or siphonozooid; C = connecting canal; D = established polyp; E = polyp budding from stolon sheet; H = stolon sheet; M = mural pore; P = connecting plate; S = calcareous spicules; T = tabulae; U = hypothetical spicules; X = spines on tabulae.) (*After Fenton and Fenton*, 1936.)

[1] See FENTON, C. L., and FENTON, M. A., 1936 (The "Tabulate" corals of Hall's "Illustrations of Devonian Fossils," *Ann. Carnegie Mus.*, vol. 25, pp. 17–58) for a discussion and bibliography bearing on this subject.

[2] SWINNERTON, A. C., 1947. "Outlines of Palaeontology," 3d ed., London, Edward Arnold & Co., p. 49.

separate subclass of Anthozoa, though perhaps not equal in rank to the Alcyonaria and Zoantharia. We follow this course without accepting or rejecting the specialist's conclusions that the Tabulata are extinct relatives of (1) the living alcyonarians or (2) the zoantharians.

The Tabulata range from Lower Ordovician (*Lichenaria*, Fig. 4-43*B–C*) to Permian and include many index genera that are widespread geographically

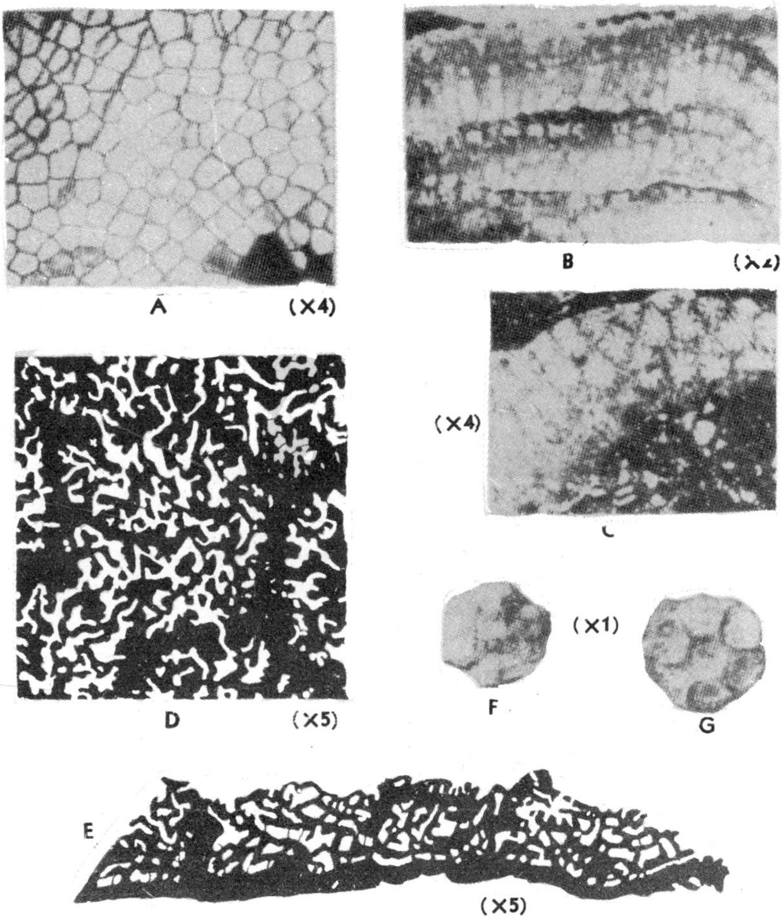

FIG. 4-43. Subclass TABULATA (Favositidae, Lichenaridae, and Cleistoporidae). *A*. Transverse section of *Lamottia heroënsis*, showing mural pores connecting adjacent corallites. *B–C*. Vertical and transverse sections of *Lichenaria prima*. *D–G*. *Cleistopora typa*: *D*, transverse section near surface, showing vague hexagonal outline of corallite; *E*, longitudinal section, showing reticulate structure in heavy horizontal rows, with open spaces containing thin diaphragms; the corallite walls are perforate and not well differentiated; corallites are flat-floored; *F–G*, lower and upper views of two coralla. (*A–C after Okulitch*, 1936; *D–G after Easton*, 1944.)

and long-ranging stratigraphically (*e.g.*, *Favosites*, *Halysites*, and *Syringopora*). Together with stromatoporoids and calcareous algae, they were important contributors to Paleozoic coral reefs (bioherms), especially in the Silurian of the Michigan Basin, Anticosti Island, and Gotland and in the Devonian of many parts of the world. As a matter of fact the oldest known true coral reef was built in the early Ordovician seas of Vermont by a team consisting of a favositoid coral, *Lamottia* (Fig. 4-43*A*), and a massive stromatoporoid, *Stromatocerium* (Raymond, 1924).

The Tabulata may be divided into the following six families:

Family 1. Favositidae.[1] Variously shaped coralla with prismatic corallites so arranged as to resemble the cells of a honeycomb, whence the designation "honeycomb corals" commonly applied to favositids. Mural pores are well developed, either along the edges of the corallites (Fig. 4-41*E*) or in the walls between tabulae (Fig. 4-41*C–D*). Septa are lacking or rudimentary and variable in number. Budding seemingly took place on all sides of the organism. *Lower Ordovician to Permian.*

Lamottia, *Paleofavosites*, *Favosites*, and *Calapoecia* show the important structural features characteristic of members of the family (Fig. 4-41).

Family 2. Cleistoporidae. Coralla of loosely packed cylindrical or prismatic, nonseptate corallites with perforate walls that may not be well differentiated and with tabulae that merge into vesicular structure within the corallite. The family is closely related to the Favositidae by reason of their compound nature, tabulate corallites, and perforate walls.[2] *Cleistopora* (Fig. 4-43*D–G*) from the Lower Devonian to Mississippian, the range of the family, is a representative genus.

Family 3. Halysitidae. Coralla composed of long, commonly nonseptate cylindrical corallites joined directly or by smaller parallel tubes into palisadelike walls with labyrinthine arrangement. In transverse section the palisade resembles a chain (whence the common name "chain coral" and the name of the family—Gr. *halysis*, chain). If septa are present they take the form of 6 or 12 rows of spines. No mural pores are developed. Budding took place on opposite sides of the long diameter of the organism. The single genus, *Halysites* (Fig. 4-41*I–L*), on which the family is based, ranges from Upper Ordovician to Lower Devonian.

Family 4. Syringoporidae.[3] Coralla composed of separate cylindrical, generally nonseptate corallites which are united at intervals by lateral tubes to form a bushlike or fasciculate corallum, commonly called an "organ-pipe coral," a designation also given the alcyonarian *Tubipora* (Figs. 4-15$^{6-8}$, 4-42^{2}). The tabulae are well developed and are characteristically inverted conical as shown in Fig. 4-41*N*. Mural pores are not present. New corallites developed in whorls along the corallite margin or sprouted singly from a connecting process (Fig. 4-32*F*). *Syringopora* (Fig. 4-41*M–N*), which has the same range as the family, Ordovician to Pennsylvanian, is a representative genus.

Family 5. Auloporidae.[4] Characterized by a creeping, branching, or reticulated corallum of small cylindrical or cornucopia-shaped corallites that lack septa and generally have few tabulae. A typical corallum resembles a string of small cones, each of which

[1] The reader will find an excellent analysis of this important Mid-Paleozoic family by Swann, D. H., 1947, The *Favosites alpenensis* lineage in the Middle Devonian Traverse group of Michigan, *Univ. Mich. Contr. Mus. Pal.*, vol. 6, art. 9, pp. 1–317.

[2] Easton, 1944, *op. cit.*

[3] Easton, W. H., 1947, Genera of syringoporoid corals and their relationships (abstract), *Bull. Geol. Soc. Amer.*, vol. 58, p. 1261.

[4] Fenton, M. A., and Fenton, C. L., 1937, *Aulopora:* a form-genus of tabulate corals and bryozoans, *Amer. Midland Naturalist*, vol. 18 pp. 109–128.

springs from an earlier corallite and in turn buds at least one new individual. *Aulopora* (Fig. 4-41*Q–P*), which is considered by some investigators to be the young or immature stage of individuals of the Syringoporidae, is typical and has the same range as the family, Ordovician to Permian.

Family 6. Lichenaridae. Primitive corals with small massive or encursting coralla of prismatic corallites that are tabulate but without septa or mural pores. *Lichenaria* (Fig. 43*B–C*), on which the family is based, is limited to the Ordovician (Chazy-Trenton). It is supposed to be the most primitive stony coral known. Okulitch (1939, page 78) would also place *Tetradium? simplex* from the Maryland Beekmantown here rather than in the Tetradidae.

Subclass Schizocorallia

The Schizocorallia are a small extinct group of Paleozoic and early Mesozoic corals that built simple, branching and massive, colonial coralla characterized by corallites that lack true septa and increase by fission. The

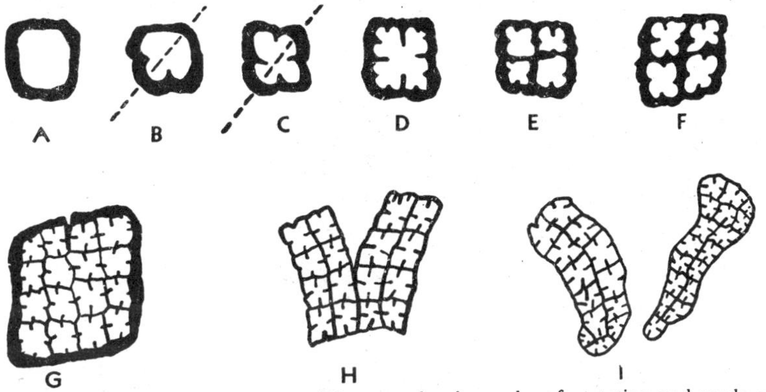

FIG. 4-44. Subclass SCHIZOCORALLIA. Diagrams showing order of septation and mode of increase of corallum in *Tetradium cellulosum*. All somewhat magnified, but *G* and *H* not to same scale as others. (*Modified after Okulitch*, 1935.)

calcareous coralla are composed of many long, slender corallites with imperforate walls and numerous complete, horizontal or slightly arched tabulae. The walls of most corallites are completely amalgamated. In some species the corallites have lamellar vertical plates projecting inward from the circumference; these resemble the septa present in zoantharian corallites, but they developed in a different way and seem to have been closely related to the asexual mode of reproduction by which the individuals of a colony multiplied. It is in the mode of reproduction that the Schizocorallia differ from all other anthozoans; hence a separate subclass has been erected and the name[1] of the subclass calls attention to the fact that increase was by simple fission[2]

[1] Schizocorallia—Gr. *schizein*, *schizo-*, to split, + *korallion*, coral.

[2] In actuality the mode of reproduction is not simple. Not only may the tabulated septate larger corallites arise from a number of the smaller tubes by disappearance of the walls of the latter, but the larger tubes may be replaced by the growth of dividing walls. Jones and Hill (1940, p. 188) call this type of reproduction increase *commutation* (to interchange). The smaller tubes multiply by the growth of new divisional walls—a fission similar to that in *Tetradium* and *Chaetetes*.

(Figs. 4-32*A–B*, 4-44). The geologic range is from Ordovician to Jurassic.

Four orders have been erected and these include corals that are similar in having a fission (or septal—see Figs. 4-32*B*, 4-44) type of increase.

Order 1. Tetradida. The order includes slender fascicles of prismatic corallites, square or rectangular in cross section, that have four long septa and reproduce by quadripartite fission, as shown in Fig. 4-44. *Ordovician.*

Tetradium (Fig. 4-44) is the best known genus in the order, and appears first in the Beekmantown (*T.? simplex* Bassler).

Order 2. Heliolitida. These are compound corals characterized by large cylindrical corallites with 12 equal septa and many smaller nonseptate tubes, in some genera

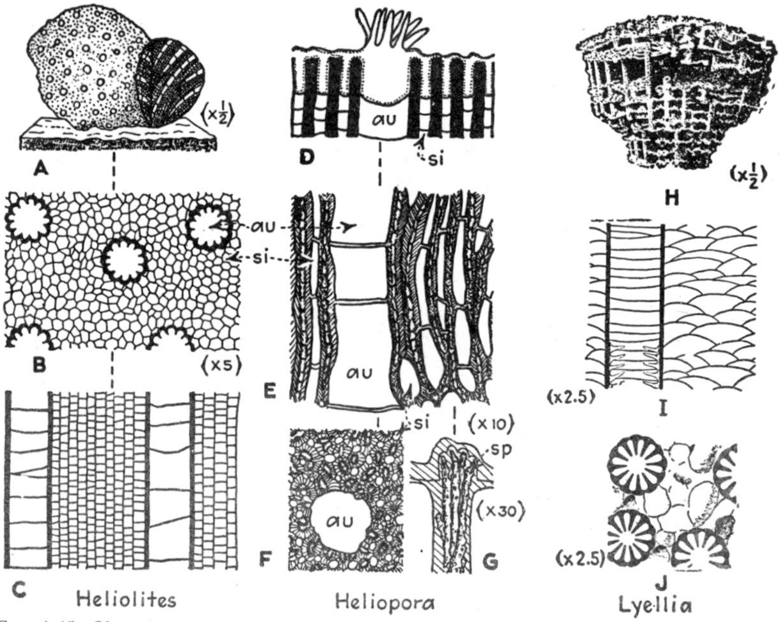

Fig. 4-45. Class Anthozoa. *Heliopora* and Schizocorallia compared. *A–C. Heliolites interstinctus*, from the Silurian: *A*, a small corallum in position of growth, sectioned to show the general internal structure; *B*, tangential section, enlarged to show the large septate cylindrical tubes and the numerous smaller nonseptate prismatic tubes; *C*, longitudinal section, enlarged to show difference between tabulae in the two types of tubes. The large and small tubes have been compared with the autopores and siphonopores, respectively, in *Heliopora*. *D–F. Heliopora coerula*, a living alcyonarian: *D*, diagram showing relationship of living organism to exoskeleton; the autopore (*au*) is occupied by a complete polyp, whereas the smaller siphonopores (*si*) are filled with siphonozooids; exoskeletal material is shown in black; *E*, longitudinal section of a small part of an exoskeleton, showing the microscopic structure of the walls of an autopore and of several siphonopores; *F*, transverse section of an autopore and several siphonopores, showing radial structure of the exoskeletal elements. *G. Sarcophyton*, longitudinal section of the wall between two siphonopores showing the calcareous spicules (*sp*) as they lie in the living organism. *H–J. Lyellia americana*, a supposed Silurian schizocoral: *H*, part of a weathered corallum; *I–J*, diagrammatic longitudinal and transverse sections, respectively, showing nature of internal structure. Note that the large tubes are strongly septate and tabulate, whereas the sclerenchyme between the tubes is vesicular. (*D, G adapted from Moseley; E–F after Nicholson and Lydekker*, 1889, *H after Hall*, 1881; *I–J after Lambe*, 1889.)

replaced by vesicular structure (*e.g.*, *Lyellia*—Fig. 4-45*H–J*), that increase by fission into two or four. The corallite walls are imperforate and fully amalgamated. *Ordovician to Devonian.*

The Heliolitidae, on which this order is based, have been subjected to much taxonomic shifting within the Coelenterata, having been assigned to the Alcyonaria, Tabulata,[1] Tetracorallia, Zoantharia (Hexacorallia), and as a separate section (Tubocoralla, Heliolitida[2]) of the so-called Madreporaria.

The Heliolitidae are here retained in the Schizocorallia, but we are well aware that strong arguments have been presented for assigning them to the Tabulata, as suggested by Weissermel (1939), or for recognizing them as a separate category of Zoantharia, as argued by Jones and Hill (1940).

Heliolites and *Lyellia* are typical genera (Fig. 4-45).

Order 3. Chaetetida. This order includes cerioid and meandroid coralla that are composed of simple slender prismatic corallites with fully amalgamated imperforate walls. The corallites increased by bipartite or, rarely, quadripartite fission. True septa are lacking but vertical plates that seem to have served in fission are present. Tabulae are usually well developed and numerous. *Ordovician to Jurassic, especially characteristic of the Pennsylvanian.*

Chaetetes (Fig. 4-46*A–C*), on which the order is largely based, is a perplexing fossil. There seems little doubt that it is an anthozoan, though it was once considered a bryozoan, but its position among the living and extinct Anthozoa is admittedly uncertain. It is included in the Schizocorallia because the method by which the corallites increase resembles the fission in the Tetradida and Heliolitida. The coralla of *Chaetetes* commonly form small reefs or bioherms in the Pennsylvanian rocks of Oklahoma.

Order 4. Multisolenida. This order was recently erected to include the single genus *Multisolenia* (Fig. 4-46*D–F*) from the Silurian of Canada.[3] This unique coral has a

[1] Weissermel, 1939, (Neue Bertrage zur Kenntnis der Geologie, Palaeontologie und Petrographie der Umgegend von Konstantinopel. 3. Obersilurische und devonische Korallen, Stromatoporiden und Trepostome von der Prinseninsel Antirovitha und aus Bithynien. *Abhandl. preuss. geol. Landesanst* [*NF*], Heft 190, pp. 1–131) has criticized Okulitch's assignment of the Heliolitidae to the Schizocorallia, and assigns the family to the Tabulata, which he considers a valid group of corals.

[2] Jones and Hill (1940), after fully discussing the nature and possible affinities of the Heliolitidae, created a new zoantharian division, Order Heliolitida, arguing as follows:

1. The Heliolitidae cannot be Alcyonaria because they have 12 (not 8) septa and the trabeculae of the skeleton have a different arrangement. (They also conclude that the heliolitid polyp was *not* dimorphic.)

2. They cannot be Tabulata because they have a fixed number of septa (12; the number of septa in the few tabulates having them is variable), lack mural pores, and have vesicular coenosteum (**reticulum**) between corallites.

3. They differ from the Tetracorallia in seemingly lacking two orders of septa, and in the uncertainty that the vesicular coenosteum between corallites is a true dissepimentarium.

4. They may be Zoantharia (Hexacorallia) by having a fixed number of 12 equal septa, although there is no proof that these appeared in cycles of six. Consequently, a new group of Zoantharia, the Heliolitida, is erected to include the Heliolitidae.

The authors do not comment on Okulitch's assignment of the heliolitids to the Schizocorallia.

[3] Fritz, M. A., 1937, *Multisolenia*, a new genus of Paleozoic corals, *Jour. Paleontology*, vol. 11, pp. 231–234. Fritz, M. A., 1938, Resemblance of the coral *Multisolenia* to *Desmipora*, *ibid.*, vol. 12, p. 299.

combination of characteristics previously unknown. It resembles the Chaetetida in having a similar corallum, thin-walled prismatic corallites that have tabulae, and spiny pseudoseptal plates which participate in the process of simple fission. It differs, nowever, in having the corallites connected by many solenia—a feature that suggests affinity with the Alcyonaria. The new order has been referred to the Schizocorallia because, as Fritz states, "While the solenia are outstanding features, nevertheless, the mode of fission is so like that shown by the Schizocoralla as a whole that inclusion in that subclass seems warranted."

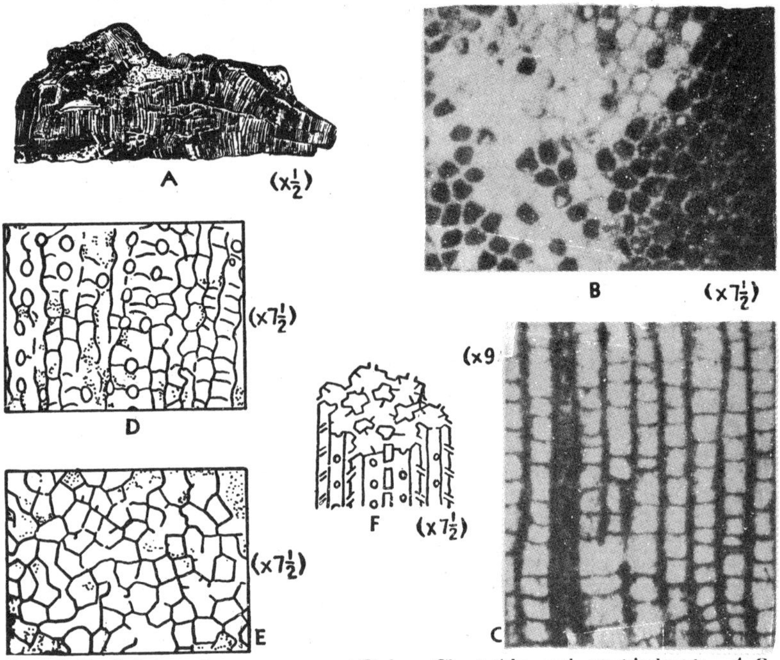

FIG. 4-46. Subclass SCHIZOCORALLIA—Orders Chaetetida and Multisolenida. *A–C*, *Chaetetes milleporaceous: A*, a typical corallum; *B–C*, transverse and longitudinal sections. *D–F. Multisolenia tortuosa: D–E.* longitudinal and transverse sections; note regularly spaced solenia, tortuous walls, arched tabulae, and spiny septa; *F*, Diagram showing prismatic corallites, numerous solenia, and septa in different stages of development. (*A after Keyes*, 1894; *D–F after Fritz*, 1937.)

The Schizocorallia are almost certainly not a natural group, and the genera now included, which range from Ordovician to Jurassic, are grouped in the order as much for the convenience of the systematist as for the reason that they all have the common characteristic of corallite increase by fission. Additional studies need to be made to determine the validity of the subclass and of the orders assigned to it.

COELENTERATA AS ROCK BUILDERS

Certain groups of coelenterates were outstanding rock builders in past geologic history, and their relatives and descendants are playing a similar role in existing oceans, as indicated by the wide distribution of coral reefs in the warmer tropical waters of today (Fig. 4-12).

The oldest known true coral reef or bioherm is preserved in a Lower Ordovician (Chazy) limestone of Vermont (Raymond, 1924). Here a favositoid coral (*Lamottia*) and a hydrocoralline (*Stromatocerium*), together with pelmatozoans and bryozoans, comprise a biostrome or bioherm nearly half a mile long. Although reef building was not important until the Silurian and later, widespread biostromes and more rarely bioherms of *Columnaria*, *Paleofavosites*, and *Halysites* (and also the stromatoporoids) were made in late Ordovician seas in the Baltic region of Europe and the Gulf of St. Lawrence region of North America. In the Silurian the reef builders came into their own in many parts of the world. One of the best known of these reef areas is the belt that almost encircles the Michigan Basin.

This broad belt of isolated bioherms stretches, with interruptions, from western Ontario into the northern peninsula of Michigan, thence south across eastern Wisconsin by way of Milwaukee and northeastern Illinois by way of Chicago, and thence eastward across northern Indiana and Ohio. In some places in northern Indiana the belt is 50 miles wide. The bioherms were built by tabulates, tetracorals, stromatoporoids, and lime-secreting algae, with the assistance of such usual reef denizens as echinoderms, worms, brachiopods, mollusks, and trilobites. Extensive coralline deposits are present in the Silurian of Anticosti Island. In Europe the famous "ball stones" of England, the reefs of the island of Gotland, and those of Estonia all belong to the Silurian.

Reef building continued into the Devonian, so that bioherms and biostromes of corals and stromatoporoids are abundant, though usually not on such a grand scale as during the preceding period. The well-known "Reef" at Louisville, Ky., which is really a coral biostrome, has long been known as a rich collecting ground. In Europe the great Devonian coral bioherms of the Eifel district and the Devonian and Carboniferous biohermal masses of Belgium have long been known.[1] The closing epochs of the Paleozoic do not seem to have been so widely favorable for reef-building corals, but typical bioherms were built in Mississippian seas in New Mexico[2] and elsewhere, and small bioherms of *Chaetetes* in the Pennsylvanian seas of Oklahoma.

By the close of the Permian the extinction of two great groups of Paleozoic corals, the Tetracorallia and Tabulata, had been essentially completed. Only a few stragglers of the Schizocorallia, the Chaetetidae, persisted into the Mesozoic, but they were never important and disappeared before the close of the era.

The early part of the Triassic had almost no fossil record of corals, but by the middle of the period the first of the modern stony corals had developed

[1] GRABAU, A. W., 1903, Paleozoic coral reefs, *Bull. Geol. Soc. Amer.*, vol. 14, pp. 337–352. DUPONT, E., 1881, Sur l'origine des calcaires devoniens de la Belgique, *Bull. acad. roy. Belgique*, ser. 3, vol. 2, pp. 264–280; DUPONT, E., 1883, Sur les origines du calcaire carbonifere de la Belgique, *ibid.*, ser. 3, vol. 5, pp. 211–229.

[2] LAUDON, L. R., and BOWSHER, A. L., 1941, Mississippian formations of Sacramento Mountains, *Bull. Amer. Assoc. Petrol. Geol.*, vol. 25, pp. 2107–2160.

the ability to secrete an exoskeleton, and before the end of the period they were building reefs in California, British Columbia, and elsewhere. Reef building has continued from then to the present time, and today reefs are widespread in tropical waters between 28° N. and S. lat. (Fig. 4-12). By far the best known of these is the Great Barrier Reef along the northeastern coast of Australia.[1] Many of the Pacific reefs were intensively studied previous to and during World War II, and several important descriptions of certain of them have already appeared in print.[2] Modern reefs and ancient bioherms are now being investigated vigorously on a world-wide scale—this interest is probably due in part to the recent discovery of an oil pool in a Silurian reef in Illinois[3] and the spectacular Leduc field in the Devonian of Alberta[4]—and the next decade or two are certain to add greatly to our knowledge of them.[5]

Although the stony corals are always prominently mentioned in any discussion of the building of coral reefs, it must be strongly emphasized that fully as important on many ancient and modern reefs, and much more important on others, have been the millepores (Hydrozoa), stromatoporoids (extinct hydrocorallinoid organisms), and lime-secreting algae (nullipores). Commonly, the coralla of corals make the crude massive framework of a reef structure that serves as a foundation for other organisms. As the stony debris from the reef organisms sifts downward through the openings of the framework, it fills up the voids, and so the reef structure gradually becomes more massive in its lower part as the surface portion continues to expand or to hold its own against the sea.

GEOLOGIC HISTORY AND EVOLUTION OF THE COELENTERATA

Fossil Coelenterata range from Lower Cambrian to Recent, and some divisions of the phylum have been more or less common in the marine waters of every geologic period since the beginning of the Paleozoic. At certain times corals were so abundant that the ancient seas, like certain oceanic areas of today, are referred to as "coral seas." And as today, ancient corals built banks and reefs that now lie as great structureless masses in the midst of stratified limestones and shales. In recent years these ancient reefs, which are also known as bioherms, have attracted considerable interest as possible reservoirs

[1] Saville-Kent, W., 1893, "The Great Barrier Reef of Australia, Its Products and Potentialities," London, W. H. Allen and Co. Yonge, C. M., 1930, "A Year on the Great Barrier Reef," New York, G. P. Putnam's Sons. Fairbridge, R. W., 1950, Recent and Pleistocene coral reefs of Australia, *Jour. Geol.*, vol. 58, pp. 330–401.

[2] Emery, K. O., 1948, Submarine geology of Bikini Atoll, *Bull. Geol. Soc. Amer.*, vol. 59, pp. 855–860. Emery, K. O., Tracey, J. I., Ladd, H. S., and Hoffmeister, J. E., 1948, Reefs of Bikini, Marshall Islands, *ibid.*, pp. 861–878.

[3] Lowenstam, H. A., 1948, Marine Pool, Madison County, Illinois Silurian-reef producer, *Ill. State Geol. Surv., Rept. Inv.*, no. 131, pp. 153–188.

[4] Layer, D. B., *et al.*, 1949, Leduc oil field, Alberta, a Devonian coral-reef discovery, *Bull. Amer. Assoc. Petrol. Geol.*, vol. 33, pp. 572–602.

[5] The interested reader will find a series of excellent symposial papers in vol. 34 of *Bull. Amer. Assoc. Petrol. Geol.*, 1950, and a comprehensive list of references to bioherms and biostromes in the timely "Bibliography of Organic Reefs, Bioherms, and Biostromes" published in 1950 by the Seismograph Service Corporation of Tulsa, Okla.

in which petroleum could accumulate, and an excellent annotated bibliography on them appeared recently (see footnote on page 172).

The Coelenterata almost certainly originated in the seas of the later Pre-Cambrian, because a representative of each of the three classes has been found in Lower and Middle Cambrian rocks (Fig. 4-47). The earliest fossil

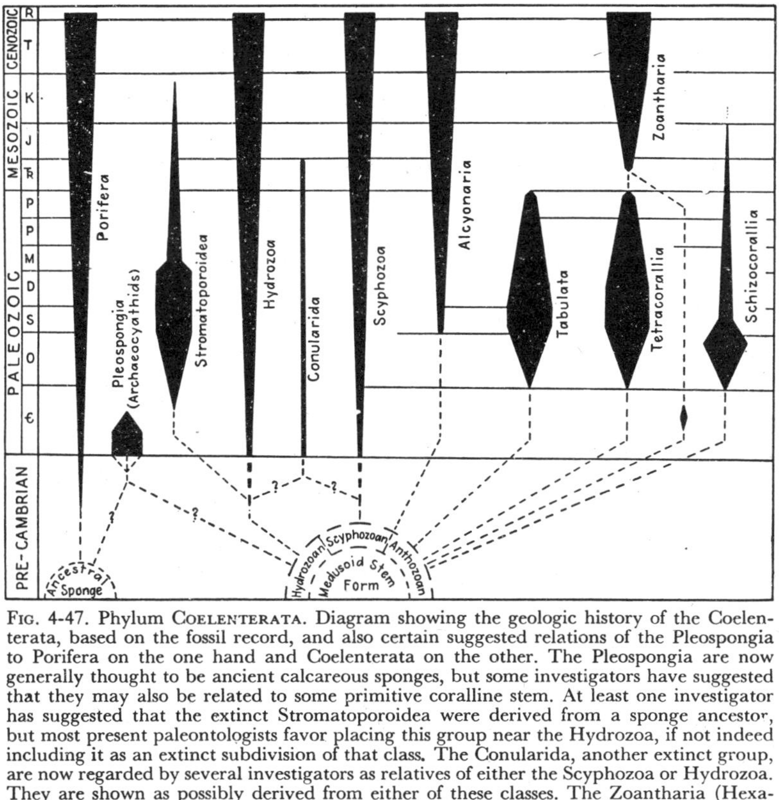

FIG. 4-47. Phylum COELENTERATA. Diagram showing the geologic history of the Coelenterata, based on the fossil record, and also certain suggested relations of the Pleospongia to Porifera on the one hand and Coelenterata on the other. The Pleospongia are now generally thought to be ancient calcareous sponges, but some investigators have suggested that they may also be related to some primitive coralline stem. At least one investigator has suggested that the extinct Stromatoporoidea were derived from a sponge ancestor, but most present paleontologists favor placing this group near the Hydrozoa, if not indeed including it as an extinct subdivision of that class. The Conularida, another extinct group, are now regarded by several investigators as relatives of either the Scyphozoa or Hydrozoa. They are shown as possibly derived from either of these classes. The Zoantharia (Hexacorallia) are thought by some to have descended from the Paleozoic Tetracorallia, though the supposed transition during the early Triassic is not represented by fossils. Another, and possibly more acceptable, suggestion is that the subclass existed throughout the Paleozoic but failed to leave a record because of lacking hard parts. *Mackenzia*, from the Middle Cambrian Burgess shale, is supposedly the earliest fossil representative of the Paleozoic zoantharians and also the earliest known true coral. The dashed lines in the diagram connect each of the large subdivisions with the supposed ancestral stock, but are not intended to indicate the times when the subdivisions separated from the stock. Question marks indicate uncertainty as to relationships. The width of the bar merely indicates when the group reached its maximum or minimum, hence the several bars are not to be compared as to importance of fossil record or number of genera and species. The Hydrozoa and Scyphozoa are questionably extended downward in the Pre-Cambrian on the supposition that *Brooksella canyonensis* is a fossil representative of one or the other of them. The Porifera are extended in the same manner on the basis of reported but unfigured sponge spicules that Walcott is supposed to have found in certain Pre-Cambrian formations of the Grand Canyon region.

representative of the Hydrozoa is a colonial form (*Archaeocryptolaria*, Fig. 4-3*C–E*) supposed to belong to Order Hydroida. By Middle Cambrian time several different orders of Hydrozoa had evolved, and from then until the present time hydrozoans have been common in marine, and possibly also in fresh, waters. The Hydrozoa are thought to have evolved from some simple medusa of the same primitive nature as members of the living order Trachylina (Fig. 4-48). The Siphonophora, an early supposed offshoot from the same primitive medusoid stem, seem to have been present as early as Middle Cambrian.

The Scyphozoa seem to have been abundant by Lower and Middle Cam-

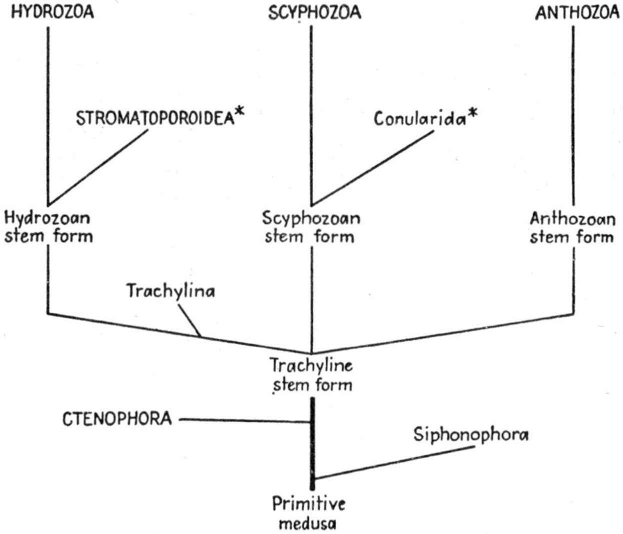

FIG. 4-48. Phylum COELENTERATA. A conjectural diagram of coelenterata relationships which is to be regarded as highly speculative. Extinct components are indicated by an asterisk. (*Based in part on suggestions by Hyman*, 1940.)

brian time (*i.e.*, assuming that *Brooksella*, etc., are gastric fillings), and if the object named *Brooksella canyonensis* from the Pre-Cambrian of the Grand Canyon is truly a fossil jellyfish, the class had representatives in the seas of the later Pre-Cambrian. Inasmuch as the soft body of a jellyfish is not likely to be preserved or to leave any sort of record except under unusual conditions, it is not surprising that so few fossil scyphozoans have been found.

The earliest fossil anthozoan thus far reported is *Mackenzia*, a Middle Cambrian anemone somewhat similar to the living *Edwardsia*. The first stony corals appear in Lower Ordovician rocks, and from their extensive fossil record it is obvious that they quickly spread over the bottoms of Ordovician and later seas. During the Paleozoic the stony corals belonged to the now extinct Tetracorallia, Schizocorallia, and Tabulata. From Middle Triassic to Recent the scleractinians have carried on, and they are well represented in modern coral seas (Fig. 4-12).

The evolution of the Coelenterata has long been a much-discussed subject among both zoologists and paleontologists. One of the latest so-called "trees," admittedly speculative, is shown in Fig. 4-48.

It is generally supposed that the phylum evolved from a primitive medusa in the seas of later Pre-Cambrian time. Certain of the larvae of this hypothetical medusa settled to the bottom, attached themselves with the mouth and tentacles directed upward, and developed into polyps. Polyps later developed the colonial habit and still later the ability to secrete a horny or calcareous skeleton. The first polyps are supposed to have had a simple mouth and cylindrical column, somewhat like the more primitive of the hydrozoans. With evolution the mouth became elongated and modified by a siphonoglyph, internal partitions were developed on a hexamerous plan, and the polyp body came to have prominent bilateral hexamerous symmetry. This symmetry, together with the hexamerous arrangement of mesenteries (as well as of the later developed protosepta), not only seems to have characterized many extinct coelenterates, but also persists in many modern anthozoans.[1]

Three main branches and one side branch (Siphonophora) evolved from the postulated primitive medusa. The Scyphozoa persisted as medusae from their beginning, evolving and becoming more complex only within the limits of that body form. The Hydrozoa developed by way of a supposed trachyline stem form with general characteristics probably similar to those of the more primitive living hydrozoans. Many developed the ability to secrete horny or calcareous skeletal elements and complete skeletons. A primitive anthozoan is postulated to have evolved from the primitive medusa and to have developed the general anthozoan stem from which the several classes of Anthozoa ultimately sprang. The extinct Tabulata, Schizocorallia, and Tetracorallia evolved from this stem form, the first two probably directly, and the tetracorals possibly by way of a line that ultimately produced the Alcyonaria, Ceriantharia, and Antipatharia. The reasons for this conclusion are discussed at some length on page 145. A second important line leading to the modern anemones and stony corals is supposed to have diverged from the Antipatharian line quite early, as indicated by the Middle Cambrian *Mackenzia*,

[1] Hyman (1944, pp. 640–641) ascribes the strong bilateral symmetry of many corals to the elongation of the mouth that was required in order to provide an ingoing respiratory current when the size of the enteron increased. The earliest anthozoan is assumed to have had one siphonoglyph, (hence bilateral symmetry was already established at this early stage) and probably three couples of mesenteries—one at each end of the elongated mouth, and a transverse couple. This six-mesenterial stage is not only present in the extinct Tetracorallia (being indicated by the six protosepta), but also appears in larval cerianthids, and probably in larval zoanthids, and is retained in many Antipatharia. As evolution progressed a fourth couple was added, thus producing the eight-mesenterial stage retained in the Alcyonaria and passed through, as the *Edwardsia* stage (see Fig. 4-30), by all anemones (Actiniaria) and stony corals (Scleractinia or Madreporaria). As pairing of mesenteries was established, biradial symmetry came to be imposed upon the ancient and fundamental bilateral symmetry and in many forms completely obliterated it.

but to have remained skeletonless until Middle Triassic, when the modern scleractinians first appeared.

The ancestors of the Tabulata are uncertain, and the group does not seem to have left any descendants. Okulitch considers the tabulate *Lichenaria* from the Lower Ordovician (Chazy) as the most primitive stony coral known. The Schizocorallia are almost certainly not a natural group and are as yet too little known to be placed certainly in coral evolution. The Tetracorallia may have been directly ancestral to the Scleractinia, but as Hyman (1944) concludes, it seems more probable that the scleractinians evolved from the actiniarians (anemones) and that skeleton development originated independently in the Tetracorallia and Scleractinia. It seems more likely that the Tetracorallia arose from an Alcyonarian—Ceriantharian—Antipatharian line as indicated in Fig. 4-30.

REFERENCES

Coelenterata (General)

HYMAN, L. H. 1940. "The Invertebrates." New York, McGraw-Hill Book Company, Inc. Vol. I, "Protozoa through Ctenophora," 720 pp.

RAYMOND, P. E. 1941. Invertebrate Paleontology *in* Fiftieth Anniversary Volume, Geological Society of America, pp. 79–80.

SHIMER, H. W., and SHROCK, R. R. 1944. "Index Fossils of North America." New York, John Wiley & Sons, Inc.; Cambridge, Mass., The Technology Press. Chap. IV, pp. 58–122.

Hydrozoa—Stromatoporoidea

FENTON, C. L. 1931. Niagaran stromatoporoid reefs of the Chicago region. *Amer. Midland Naturalist*, vol. 12, pp. 203–212.

FENTON, M. A. 1931. A Devonian stromatoporoid reef (Petoskey, Michigan). *Amer. Midland Naturalist*, vol. 12, pp. 195–202.

HEINRICH, M. 1914. On the structure and classification of the Stromatoporoidea. *Neues Jahrb. Min., Geol., Pal.* 1914, no. 23, pp. 732–736. (Translated by C. M. LeVene, *Jour. Geol.*, vol. 24, pp. 57–60, 1916.)

HICKSON, S. J. 1934. On *Gypsina plana* and on the systematic position of the stromatoporoids. *Quart. Jour. Micros. Sci.*, vol. 76, pt. 3, pp. 433–480.

KÜHN, O. 1927. Zur Systematik und Nomenklatur der Stromatoporen. *Centralb. Min. Stuttgart*, vol. 12, pp. 546–551.

———. 1928. Hydrozoa. Fossilium Catalogus I. Pars 36, pp. 1–144. Berlin.

———. 1939. Hydrozoa *in* Schindewolf's "Handbuch der Paläozoologie," vol. 2A, pp. A1–A68. Berlin, Gebrüder Borntraeger. (This article contains an excellent list of references.)

NICHOLSON, H. A. 1886–1892. "A Monograph of British Stromatoporoids." *Palaeontographical Society*, vols. 39, 42, 44, and 46.

PARKS, W. A. 1907. The stromatoporoids of the Guelph formation in Ontario. *Univ. Toronto Studies*, Geol. Ser., no. 4.

———. 1908. Niagaran stromatoporoids. *Ibid.*, no. 5.

———. 1909. Silurian stromatoporoids of America. *Ibid.*, no. 6.

———. 1910. Ordovician stromatoporoids. *Ibid.*, no. 7.

———. 1935. Systematic position of Stromatoporoidea. *Jour. Paleontology*, vol. 9, pp. 18–29.

———. 1936. Devonian stromatoporoids of North America. *Univ. Toronto Studies*, Geol. Ser., no. 39.

RAYMOND, P. E. 1914. A *Beatricea*-like organism from the Middle Ordovician. *Canada Dept. Min., Mus. Bull.* 5, Geol. Ser. 21, pp. 1–19.

———. 1931. Notes on invertebrate fossils, with descriptions of new species. 3. Further notes on *Beatricea*-like organisms. *Bull. Mus. Comp. Zool.* (Harvard), vol. 55, pp. 165–213.

ROSEN, F. VON. 1869. Ueber die Natur der Stromatoporen, und ueber die Erhaltung der Hornfasser der Spongien in fossilen Zustande. *Verh. K. russ. mineral. Ges.*, St. Petersburg, 2d ser., vol. 4, pp. 1–98.

TWITCHELL, G. B. 1928–1929. The structure and relationships of true stromatoporoids. *Amer. Midland Naturalist*, vol. 11, pp. 270–302.

YABE, H., and TOSHIO, S. 1935. Jurassic Stromatoporoids from Japan. *Sci. Rept. Tokoku Imp. Univ., Geol.* 14, pp. 135–192.

Scyphozoa

TERMIER, H., and TERMIER, G. 1948. Position systématique et biologie des Conulaires. *Rev. scientifique*, fasc. 12., 86th year, pp. 711–722.

WALCOTT, C. D. 1898. Fossil medusae. *U.S. Geol. Surv., Mon.* 30.

Anthozoa

BASSLER, R. S. 1937. The Paleozoic rugose coral family Paleocyclidae. *Jour. Paleontology*, vol. 11, pp. 189–201.

———. 1941. A supposed jellyfish from the Pre-Cambrian of the Grand Canyon. *Proc. U.S. Nat. Mus.*, vol. 89, pp. 519–522.

BROWN, T. C. 1907. Developmental stages in *Streptelasma rectum*, Hall. *Amer. Jour. Sci.*, vol. 23, pp. 277–284.

———. 1909. Studies on the morphology and development of certain rugose corals. *Ann. N.Y. Acad. Sci.*, vol. 19, pp. 45–97.

———. 1915. The development of the mesenteries in the zooids of Anthozoa and its bearing upon the systematic position of the Rugosa. *Amer. Jour. Sci.*, vol. 39, pp. 535–542.

BRYAN, W. H., and HILL, D. 1941. Spherulitic crystallization as a mechanism of skeletal growth in the hexacorals. *Proc. Roy. Soc. Queensland*, vol. 52, pp. 78–91.

CLARK, A. H. 1913. Cambrian holothurians. *Amer. Nat.*, vol. 47, pp. 488–507.

CUMINGS, E. R. 1932. Reefs or bioherms? *Bull. Geol. Soc. Amer.*, vol. 43, pp. 331–352.

DANA, J. D. 1890. "Corals and Coral Islands," 3d ed. New York, Dodd, Mead, & Company, Inc.

DARWIN, C. 1889 (1896). "The Structure and Distribution of Coral Reefs." New York, Appleton. 344 pp.

DAVIS, W. M. 1928. "The Coral Reef Problem." New York, American Geographical Society. 596 pp.

DONS, C. 1934. Ueber die nordlichsten Korallenriffe der Welt. *K. norske Videnskap. Selsk., Trondhjem*, vol. 6, pp. 206–209.

DUERDEN, J. E. 1904. The coral *Sidastrea radians* and its postlarval development. *Carnegie Inst. Wash., Pub.* 20, pp. 1–130.

———. 1904*a*. The morphology of the Madreporaria. V. Septal sequence. *Biol. Bull.*, vol. 7, pp. 79–104.

———. 1905. The fossula in rugose corals. *Ibid.*, vol. 9, pp. 27–52.

DURHAM, J. W. 1947. Corals from the Gulf of California and the North Pacific coast of America. *Geol. Soc. Amer., Mem.* 20, pp. 1–68.

———. 1949. Ontogenetic stages of some simple corals. *Univ. Cal. Pub., Bull. Dept. Geol. Sci.*, vol. 28, pp. 137–172.

EASTON, W. H. 1944. Corals from the Chouteau and related formations of the Mississippi Valley region. *Ill. Geol. Surv., Rept. Inv.* 97, pp. 1–93.

———. 1945. Amplexoid corals from the Chester of Illinois and Arkansas. *Jour. Paleontology*, vol. 19, pp. 625–532. (Reprinted in *Ill. Geol. Surv., Rept. Inv.* 113.)

EISENACK, A. 1942. Die Melanoskleritoiden eine neue Grupp silurischer Mikrofossilien aus dem Unterstamm der Nesseltiere. *Pal. Zeitschr.*, vol. 23, pp. 157–180.

FAIRBRIDGE, R. W. 1950. Recent and Pleistocene coral reefs of Australia. *Jour. Geol.*, vol. 50, pp. 330–401.

FENTON, C. L., and FENTON, M. A. 1936. The tabulate corals of Hall's "Illustration of Devonian Fossils." *Ann. Carnegie Mus.*, vol. 25, pp. 17–58.

FLOWER, R. H., and WAYLAND-SMITH, R. 1947. New fauna from the Vernon shale of New York (abstract). *Bull. Geol. Soc. Amer.*, vol. 58, p. 1180.

GARDINER, S. 1931. "Coral Reefs and Atolls." London, Macmillan & Co., Ltd. 181 pp.

HALL, J. 1843–1887. Illustrations of Devonian Fossils *in* "Paleontology of New York," vols. 1, 2, 3, and 6.

HICKSON, S. J. 1924. An introduction to the study of recent corals. *Pub. Univ. Manchester*, Biol. Ser. 4, 257 pp.

HILL, D. 1935. British terminology for rugose corals. *Geol. Mag.*, vol. 72, pp. 481–519. (This article contains an excellent list of references.)

———. 1936. The British Silurian rugose corals with acanthine septa. *Roy. Soc. London*, ser. B, no. 534, vol. 226.

———. 1939. The Devonian rugose corals of Lilydale and Loyola, Victoria. *Proc. Roy. Soc. Victoria*, vol. 51, pp. 219–256.

———. 1940. The Silurian Rugosa of the Yass-Browning district, N.S.W. *Jour. Proc. Roy. Soc. N.S.W.*, vol. 76, pp. 142–164.

———. 1942. The Middle Devonian rugose corals of Queensland, III. Burdekin Downs, Fanning R., and Reid Gap, North Queensland, *Proc. Roy. Soc. Queensland*, 14th ser., vol. 53, pp. 229–268.

HINDS, N. E. A. 1938. An Algonkian jellyfish from the Grand Canyon of the Colorado. *Science* (n.s.), vol. 88, pp. 186–187.

HOWELL, B. F. 1947. Eocene Alcyonaria in New Jersey (abstract). *Bull. Geol. Soc. Amer.*, vol. 58, p. 1195.

JEFFORDS, R. M. 1942. Lophophyllid corals from Lower Pennsylvanian rocks of Kansas and Oklahoma. *State Geol. Surv. Kan., Bull.* 41, pp. 185–260.

JOHNSON, J. H. 1946. Corals as builders of limestone. *Mines Mag.*, vol. 36, pp. 605–610.

JONES, O. A., and HILL, D. 1940. The Heliolitidae of Australia, with a discussion of the morphology and systematic position of the family. *Proc. Roy. Soc. Queensland*, vol. 51, pp. 150–168, 183–215.

KOWALEWSKY, A., and MARION, A. F. 1883. Documents pour l'histoire embryogénique des Alcyonaires. *Ann. mus. hist. nat. Marseille, Zool.* 1, pp. 1–43.

LEWIS, H. P. 1935. The Lower Carboniferous corals of Nova Scotia. *Ann. Mag. Nat. Hist.*, 10th ser., no. 91, vol. 16, pp. 118–142.

MILNE-EDWARDS, H., and HAIME, J. 1851. Monographie des polypiers fossiles des terraines paléozoiques. *Arch. du Muséum*, Paris, vol. V.

NICHOLSON, H. A. 1879. "On the Structure and Affinities of the 'Tabulate Corals' of the Palaeozoic Period." Edinburgh. 337 pp.

NYHOLM, K. G. 1944. Zur Entwicklung und Entwicklungsbiologie der Ceriantharien und Aktinien. *Zool. Bidrag.* Uppsala, pp. 87–249.

OGILVIE, M. M. 1897. Microscopic and systematic study of Madreporarian types of corals. *Phil. Trans. Roy. Soc. London*, vol. 187, pp. 83–345.

OKULITCH, V. J. 1935. Tetradidae—a revision of the genus *Tetradium*. *Trans. Roy. Soc. Can.*, 3d ser., sec. IV, vol., 29, pp. 49–74.

———. 1936. On the genera *Heliolites*, *Tetradium* and *Chaetetes*. *Amer. Jour. Sci.*, (5), vol. 32, pp. 361–379.

———. 1936*a*. Some Chazyan corals. *Trans. Roy. Soc. Can.*, 3d ser., sec. IV, vol. 30, pp. 59–73.

———. 1938. Some Black River Corals. *Ibid.*, vol. 32, pp. 87–111.

———. 1939. Evolutionary trends of some Ordovician corals. *Ibid.*, vol. 33, pp. 67–80.

Perkins, B. F. 1951. "An Annotated Bibliography of North American Upper Cretaceous Corals, 1795–1950." Southern Methodist University Press, Fondren Science Series, no. 3, 45 pp.

Pugh, W. E. 1950. "Bibliography of Organic Reefs, Bioherms. and Biostromes." Seismograph Service Corp., Tulsa, Okla. 139 pp.

Raymond, P. E. 1924. The oldest coral reef. *Vt. State Geologist*, 14th Rept., 1923–1924, pp. 72–76.

Sanford, W. G. 1939. A review of the families of Tetracorals. *Amer. Jour. Sci.*, vol. 237, pp. 295–323, 401–423.

Saville-Kent, W. 1893. "The Great Barrier Reef of Australia." London, W. H. Allen and Co.

Schultze, L. 1896. Beitrag zur Systematik der Antipatharien. *Abhandl. Senckenberg. natur. Gesel.*, vol. 23, pp. 1–39.

Shimer, H. W., and Shrock, R. R. 1944. "Index Fossils of North America." New York, John Wiley & Sons, Inc.; Cambridge, Mass., The Technology Press. Class Anthozoa, pp. 78–122. (This work has a comprehensive list of North American references.)

Sinclair, G. W. 1952. A classification of the Conularida. *Fieldiana: Geology* (Chicago Nat. Hist. Mus.), vol. 10, pp. 135–145.

Smith, S. 1945. Upper Devonian corals of the Mackenzie region of Canada. *Geol. Soc. Amer., Spec. Paper* 59, 126 pp.

Stewart, G. 1938. Middle Devonian corals of Ohio. *Geol. Soc. Amer., Spec. Paper* 8, 120 pp.

Stumm, E. C. 1937. The lower Middle Devonian tetracorals of the Nevada Limestone. *Jour. Paleontology*, vol. 11, pp. 423–443.

———. 1938. Upper Middle Devonian rugose corals of the Nevada Limestone. *Ibid.*, vol. 12, pp. 478–485.

———. 1940. Upper Devonian rugose corals of the Nevada limestone. *Ibid.*, vol. 14, pp. 57–67.

———. 1945. Revision of the families and genera of the Devonian Tetracoralla (abstract). *Bull. Geol. Soc. Amer.*, vol. 56, p. 1203.

Van Gundy, G. E. 1937. Jellyfish from Grand Canyon Algonkian. *Science* (n.s.), vol. 85, p. 314.

———. 1951. Nankoweap group of the Grand Canyon Algonkian of Arizona. *Bull. Geol. Soc. Amer.*, vol. 62, pp. 953–959.

Vaughan, T. W. 1917. The reef-coral (L. Pliocene) fauna of Carrizo Creek, Imperial County, California, and its significance. *U.S. Geol. Surv., Prof. Paper* 98T, pp. 353–395.

———. 1919. Corals and the formation of coral reefs. *Smith. Inst., Ann. Rept.* 1917, pp. 189–276.

———. 1944. Recent studies of the ecology of corals. *Nat. Research Council, Rept. Subcomm. Ecol. Marine Organisms*, pp. 46–51.

——— and Wells, J. W. 1943. Revision of the suborders, families, and genera of the Scleractinia. *Geol. Soc. Amer., Spec. Paper* 44, 363 pp.

Wang, H. C. 1948. The Middle Devonian rugose corals of Eastern Yunnan. *Nat. Univ. Peiping, Contr. Geol. Inst.*, no. 33, pp. 1–42.

———. 1951. A revision of the Zoantharia Rugosa in the light of their minute skeletal structures. *Phil. Trans. Roy. Soc. London*, vol. 234, pp. 175–246.

Wells, J. W. 1937. Individual variation in the rugose coral species. *Pal. Americana*, vol. 11, no. 6, pp. 1–20.

Yonge, C. M. 1930. "A Year on the Great Barrier Reef." New York, G. P. Putnam's Sons. 246 pp.

CHAPTER 5

PHYLUM CTENOPHORA

The Ctenophora, commonly referred to as comb jellies or sea walnuts, constitute a group of some 80 species of free-swimming exclusively marine animals with transparent gelatinous bodies characterized by biradial symmetry (Fig. 5-1). The outstanding external feature is the presence on the surface of the spheroidal body of eight meridional rows of ciliary plates, or comb plates (whence the common name comb jellies, and the name of the phylum—Gr. *kteis*, comb, + *phoros*, bearing).[1] Some ctenophores also have two retractile tentacles.

Previous to 1889 the ctenophores were included in the Coelenterata because of their biradial symmetry and general resemblance to jellyfishes, but they are now commonly placed in a separate phylum because of their distinctive features of structure and biology. Ctenophores resemble coelenterates in having the following:

1. Biradial symmetry.
2. Parts arranged on an oral-aboral axis.
3. A gastrovascular cavity with branches.
4. Gelatinous mesogloea.
5. No internal spaces except the digestive system.
6. No other systems of organs.

They differ from the coelenterates in having the following:

1. Eight meridional rows of comb plates.
2. Mesodermal muscles.
3. A more complicated digestive system.
4. An aboral sensory region.
5. No nematocysts.

By possessing the characteristics just listed, the Ctenophora are on a somewhat higher structural grade than the Coelenterata. They are considered to be an early offshoot from the trachyline stem form (Fig. 4-48) that differentiated considerably without leading to any higher forms (Hyman). Their geologic history is unknown.

There is no skeleton system in the ctenophores, and there is no known fossil record. The only hard parts in the organism are tiny spherules, or **lithocysts,** composed of calcium phosphate that are clustered together in the sensory region at the aboral pole to constitute the **statolith.** The statolith and certain associated structures seem to act as an organ of equilibrium.

This small and relatively unimportant phylum has been included for

[1] In life the comb plates beat with wavelike motion. Zoologists have studied this beating of the combs intensively because it is both an example of ciliary motion and a possible case of nervous control of cilia (Hyman).

several reasons. The ctenophores are among the most characteristic planktonic organisms of the ocean, inhabiting all seas from the surface to depths as great as 3,000 m. (9,000 + ft.). In colder waters, as off Newport, R.I., and off the Maine coast, the surface of the sea is sometimes covered for thousands of square meters by great submerged rafts composed of the luminescent

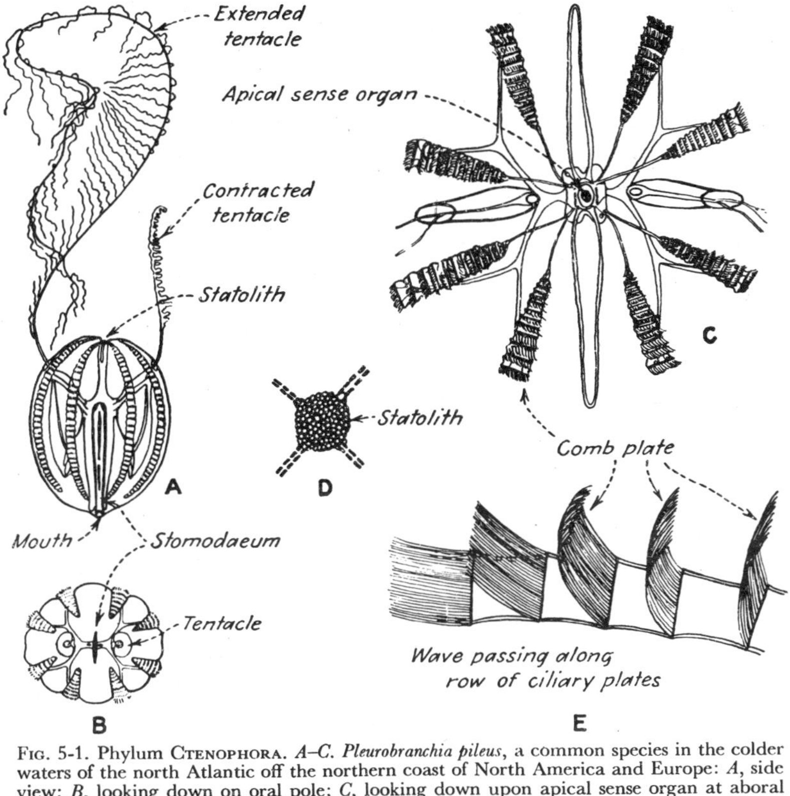

FIG. 5-1. Phylum CTENOPHORA. *A–C. Pleurobranchia pileus*, a common species in the colder waters of the north Atlantic off the northern coast of North America and Europe: *A*, side view; *B*, looking down on oral pole; *C*, looking down upon apical sense organ at aboral pole. *D*. Greatly magnified apical sense organ with extensions to the eight rows of ciliated plates. *E*. Diagram of a small part of one row, showing cilated plates in different positions as wave passes along. (*A–C modified after Mayer*, 1912; *D–E modified after Chun*, 1880.)

bodies of one or a few species that have been concentrated by tides and currents. These immense swarms wreak havoc with the small pelagic animals. including certain fish, on which they feed.

The fluid and solid content of ctenophores is similar to that of most medusae, but ctenophores are much more fragile than jellyfish, the gentle stroke of an oar being all that is required to rip them into shreds. Only under most unusual conditions, therefore, would ctenophores have any chance whatsoever of being preserved, yet soft and skeletonless bodies fully as fragile have been preserved; hence there may be fossil ctenophores that have been overlooked or are as yet undiscovered. The foregoing description, which is admittedly brief, is felt to be justified on the ground that the student of fossils needs to have a knowledge of all common invertebrates, whether they have hard parts or not.

REFERENCES

CHUN, C. 1880. Die Ctenophoren. Fauna Flora Golfes Neapel 1. 1898. Die Ctenophoren der Plankton-Expedition. *Ergeb. Plank. Exped.* 2Ka.

HYMAN, L. H. 1940. "The Invertebrates." New York, McGraw-Hill Book Company Inc. Vol. I, "Protozoa through Ctenophora." (Ctenophora, pp. 662–696.)

MAYER, A. G. 1912. Ctenophores of the Atlantic coast of North America. *Carnegie Inst. Wash., Pub.* 162, 58 pp.

STORER, T. I. 1951. "General Zoology," 2d ed. New York, McGraw-Hill Book Company, Inc. 832 pp. (Ctenophores, pp. 342–344.)

CHAPTER 6

WORM PHYLA (EXCLUDING ANNELIDA)

INTRODUCTION

For many years invertebrate paleontologists have used the indefinite term "worms" for a great variety of fossils of uncertain origin such as trails, burrows, castings, and carbonaceous films. Almost any vague fossil with a long round body or any marking or burrow that could be made by an organism of such shape was, and still is, likely to be referred to this heterogeneous assemblage. As different groups of these elongate bodies came to be studied carefully, some were shown to represent developmental stages in the life histories of higher animals, whereas others were found to possess structures allying them with other phyla. There still remains, however, a large and varied group of fossils supposedly referable to ancient wormlike creatures. Those fossils that have been referred to the nonsegmented worms are discussed in this chapter; those included in the segmented worms (Phylum Annelida) are considered in Chap. 11.

In earlier classifications of animals, organisms with elongate bodies but without conspicuous appendages were grouped together under the term Vermes (L. *vermes*, worms) or simply Worms. These animals differ from protozoans and from the radial sponges, coelenterates, and ctenophores in having a head or **anterior end** with sense organs, a **posterior end,** a **ventral surface** next to the substratum over which they move, a **dorsal surface** or back that is uppermost, and **bilateral symmetry** in both external and internal features. Although the unsegmented worms as a group share all these characteristics with most of the other higher animals, they differ so widely among themselves in morphological features that they are now separated into the following phyla, most of which seem to lack any appreciable fossil record.

PHYLUM PLATYHELMINTHES (FLATWORMS)

The Platyhelminthes (Gr. *platys*, flat, + *helminthos*, worm) constitute a phylum of unsegmented worms having a thin soft body that is dorsoventrally flattened. Although these flatworms, as they are commonly called, are the lowest of the worms, they are greatly advanced over the Porifera, Coelenterata, and Ctenophora in having (1) conspicuous bilateral symmetry, with clearly differentiated anterior and posterior ends and ventral and dorsal surfaces; (2) a well-developed nervous system, with enlarged anterior ganglia and nerve cords extending posteriorly along the body; (3) a third layer, the mesoderm, which lies between the ectoderm and endoderm and produces muscles and other organs; (4) well-developed muscles; and (5) specialized reproductory organs. They differ from most higher animals in having (1) no distinct body cavity; (2) the alimentary tract or gut branched to various parts of the body; and (3) the sexes united. The following three classes are generally recognized.

Class Turbellaria

The turbellarians are free-living flatworms, chiefly limited to cool and clear fresh waters where they commonly are found on the undersurfaces of submerged objects. A few forms inhabit marine waters, and some are endoparasites. There is no skeleton, but most flatworms have an elastic basement

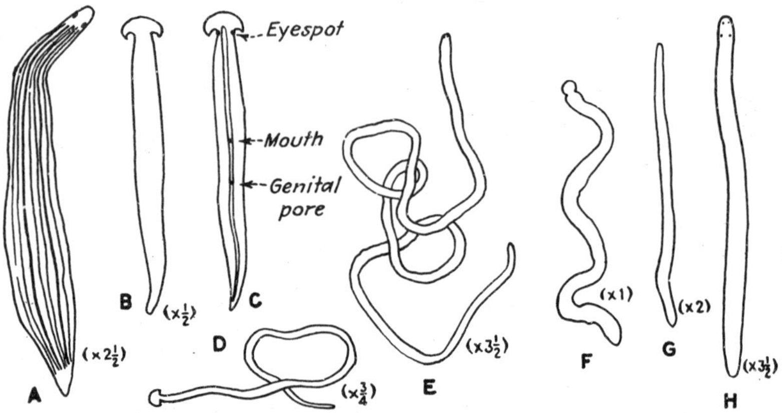

Fig. 6-1. *A–D.* Phylum Platyhelminthes—Class Turbellaria: *A*, *Platydemus; B–C.* dorsal and ventral views, respectively, of *Bipalium; D*, *Placocephalus*. *E–H.* Phylum Nemertea: Several different nemertean worms, showing range in size and shape. (*A–D generalized from Graff*, 1899; *E–H outlined from Buerger*, 1895.)

membrane. The common fresh-water planarians of North America are typical of the class (Fig. 6-1*A–D*). No fossil Turbellaria are known.

Class Trematoda

The trematodes, or flukes, are parasitic flatworms that live mostly in vertebrates and have an anterior sucking disk and a well-developed digestive system. The individuals of one group (Monogenea) live in only one host and are large **ectoparasites** (*i.e.*, they live on the outer surface of the host). The other group (Digenea) consists exclusively of **endoparasites** (*i.e.*, internal parasites), and these require two hosts for completion of their life cycle, which is perhaps as complex as that of any known animal. More than 3,000 species of living trematodes have been described, of which the common liver fluke of sheep may be the best known. No fossil trematodes have been certainly identified, but parasitic worms found in some Carboniferous and Tertiary insects have been referred to the Platyhelminthes and may represent the Trematoda.

CLASS CESTOIDEA

The cestodes, which include the notorious tapeworm of humans (Fig. 6-2), are endoparasitic flatworms without a digestive cavity. Externally they show little resemblance to other Platyhelminthes, but internally their anatomy indicates close relationship to the Turbellaria and Nematoda. Nothing is known of the geologic history of this class of flatworms, which includes more than 1,500 existing species, as no fossil cestodes have been certainly identified.

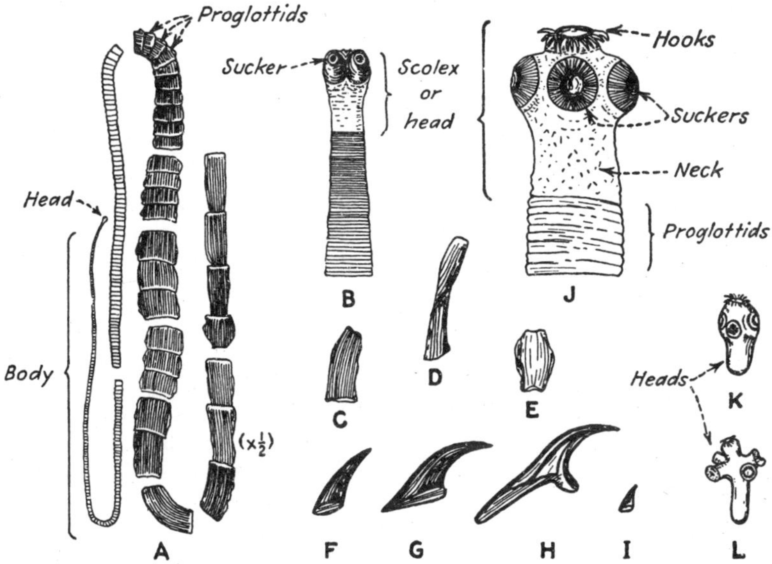

FIG. 6-2. Phylum PLATYHELMINTHES—Class CESTOIDEA. *Taenia*, the common tapeworm. *A*. Parts of a typical individual, about one-half natural size. *B*. Head of *A*, much enlarged. *C–E*. Typical body "joints" or proglottids of *A*. *F–I*. Hooks (greatly magnified) that are present on terminus of scolex and used by the parasite to attach itself to flesh. These horny hooks are hard and somewhat calcareous. *J*. Scolex or head, much enlarged to show typical features. *K–L*. Scolices just before beginning to develop the characteristic proglottids at the base of the neck. (*Modified after Leuckhart*, 1879–1886.)

PHYLUM NEMERTEA (RIBBON WORMS)

The nemerteans, or ribbon worms, are free-living slender worms having soft, flat bodies that are unsegmented and capable of great change in length through contraction or elongation (Fig. 6-1*E–H*). Some are as short as 5 mm. whereas others may stretch to 25 m. (80 ft.). Most of the 500 existing species are marine, living near the bottom among algae, on the bottom under stones, or in burrows between the tides.

Although the Nemertea are commonly included in the Platyhelminthes, which they resemble in several important respects, they differ from the flatworms sufficiently to warrant their inclusion in a separate phylum. Furthermore, their nervous system shows considerable resemblance to that of the hemichordates and chordates; hence there is good reason to recognize them as a distinct phylum.

The geologic history of the Nemertea is unknown, as no fossil representatives of the phylum have been certainly identified. In fact, unquestioned identification may never be possible. However, inasmuch as some nemerteans burrow and others move freely over the bottom, it seems reasonable to assume that ancient representatives of the family behaved in a similar manner. From this assumption it follows that some of the supposed fossil worm burrows and trails may have been made by Nemertea, though there seems to be no way by which these can be distinguished from similar features made by other free-living worms.

PHYLUM NEMATHELMINTHES (ROUNDWORMS)

The Phylum Nemathelminthes (Gr. *nematos*, thread, + *helminthos*, worm), or Nematoda, comprises unsegmented roundworms having slender cylindrical bodies more or less tapered at each end, a resistant cuticle, and a complete and permanent digestive tract. They are a distinctive group of worms, numbering some 4,500 species, and seem to be little related to other phyla, although the Gordiacea and Acanthocephala are commonly grouped with them in the same phylum.

The nematodes differ from typical flatworms in shape (slender, cylindrical, and tapered at both ends), in lacking cilia and suckers, in having a complete digestive tract and a body cavity, and in being dioecious (*i.e.*, having the sexes separate). Food moves along the digestive tract without successive meals becoming mixed, and the undigested residue is expelled through the anus; hence the digestive system of the nematode differs markedly from the branched tract of the flatworm and is suggestive of the more elaborate alimentary tract of the higher invertebrates.

Nematodes, although similar in general appearance and form, are quite different among themselves in structural details and natural history. They are generally small, rarely more than a few centimeters in length, and are present in many environments in astronomic numbers, in this respect rivaling the insects among multicellular animals. More than 2,000 species are free-living in soil and water, and an additional 2,500 are parasitic in both animals and plants.[1] An example of the latter is *Trichina*, the so-called pork worm, which is an endoparasite in swine and commonly is so abundant that a single ounce of infested pork may contain as many as 80,000 individual worms. *Trichina*, which produces **trichinosis** in humans, is but one of a large group of parasitic nematodes which cause dangerous diseases in man. Nematodes are very destructive to certain plants, among which are the pineapple, peach, apple, and okra. They bore into the feeding roots of the plants, producing large knots (root knots), and the plants die of starvation.

[1] FILIPJEV, I. N., 1934, The classification of the free-living nematodes and their relation to the parasitic nematodes, *Smith, Misc. Coll.*, vol. 89, no. 6, 63 pp.

Nematodes have no skeleton but do have a tough and resistant external cuticle, and some parasitic species have tough hooks, teeth, or cutting plates in the mouth (Fig. 6-3). The cuticle itself commonly bears minute bristles, spines, and scales. While it would seem from the preceding statements that the nematodes have certain hard parts that could be preserved as fossils, no

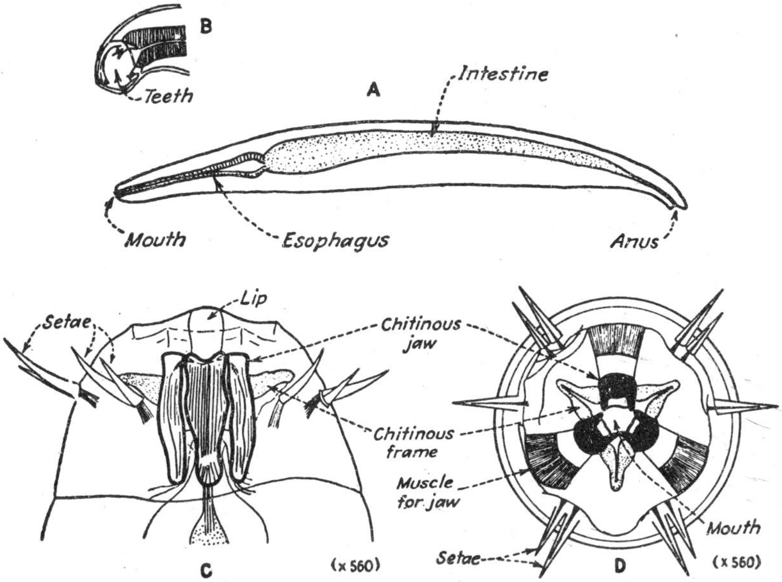

FIG. 6-3. Phylum NEMATHELMINTHES—Class NEMATODA. *A*. Simplified diagram of a nematode showing shape and size of individual and nature of digestive cavity. Other systems of organs are omitted. *B*. Enlarged portion of head of another genus, showing the mouth capsule with the three sets of chitinous teeth. *C–D*. *Enoplus communis: C*, view of head of female from left subventral side, showing nature and position of jaws; *D*, view looking downward into mouth and esophagus, showing the circumoral chitinous frame, the three chitinous jaws, and the muscles that move them. (*A–B generalized from Hatschek*, 1891; *C–D generalized from de Man*, 1886.)

fossil nematodes have been reported thus far. It should not be surprising, however, if they were found; for example, as parasites in refrigerated mammals of Pleistocene age.

PHYLUM GORDIACEA

This small and relatively unimportant phylum of free-living and parasitic worms includes a dozen or more species characterized by a relatively long (as much as 890 mm., or 35 in.) and slender threadlike body, whence the common name "horsehair worm" (Fig. 6-4). Adult worms are often seen wriggling in pools of fresh water, and their presence in such places gave rise to the ancient belief that they originated from a horsehair that had fallen into the water.

The Gordiacea are similar to the Nematoda (Nemathelminthes) in several important respects and commonly are included together with them in the Phylum Nemathelminthes. However, because there are important anatomical differences between the two, most zoologists now seem to prefer a separate phylum for the Gordiacea, and that course is followed here.

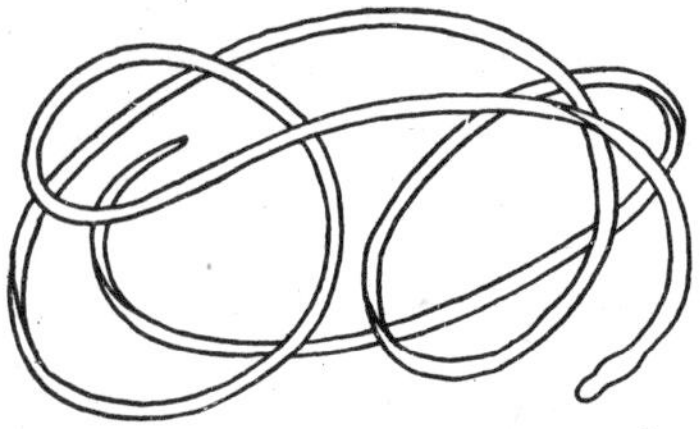

FIG. 6-4. Phylum GORDIACEA. An entire specimen of the common "horsehair worm," *Gordius*, considerably reduced in size. (*Modified after Kükenthal*, 1928–1934.)

Nothing is known of the geologic history of this small phylum, as no fossil forms seem to have been reported, and there is little likelihood that fossil representatives ever will be found, especially if ancient gordiaceans were similar to existing species in number, nature, and habits.

PHYLUM ACANTHOCEPHALA

This phylum comprises some 300 living species of small to moderately long (6 to 650 mm., or ¼ to 26 in.) parasitic worms that are distinguished by an anterior cylindrical **proboscis** bearing rows of recurved spines (Fig. 6-5).

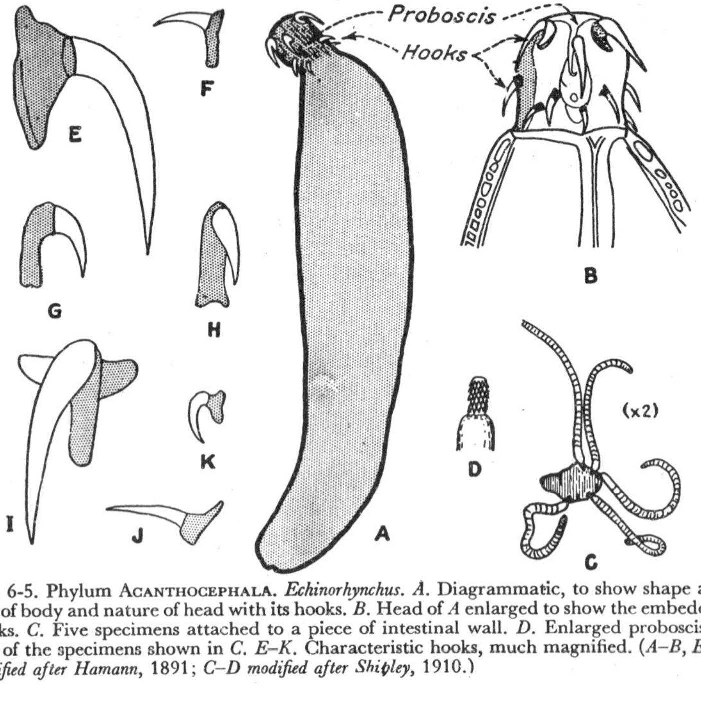

FIG. 6-5. Phylum ACANTHOCEPHALA. *Echinorhynchus*. *A*. Diagrammatic, to show shape and size of body and nature of head with its hooks. *B*. Head of *A* enlarged to show the embedded hooks. *C*. Five specimens attached to a piece of intestinal wall. *D*. Enlarged proboscis of one of the specimens shown in *C*. *E–K*. Characteristic hooks, much magnified. (*A–B, E–K modified after Hamann*, 1891; *C–D modified after Shipley*, 1910.)

(whence the common designation "spiny-headed worms" and the name of the phylum—Gr. *acanthos*, spine, + *kephale*, head). The larvae parasitize arthropods, whereas the adults live in the intestines of vertebrates, attaching themselves to the gut of the host by means of the head spines (Fig. 6-5*C*).

The Acanthocephala show few relationships to other animals and, although formerly grouped with the Trematoda and Gordiacea in the Phylum Nemathelminthes, probably should be considered a separate phylum, as is now common practice among zoologists.

Nothing is known about the geologic history of this phylum. No fossils have been reported, and except for the head spines or hooks, there is nothing in the organism that would have much chance of being preserved except under most unusual conditions, such as those that allowed Pleistocene mammals to be preserved by refrigeration.

PHYLUM KINORHYNCHA

The Phylum Kinorhyncha consists of some 30 species of tiny marine worms that live in bottom muds and sands at all depths. The body is somewhat elongate and has 13 (or 14) rings (Fig. 6-6*A–C*). The two anterior rings form

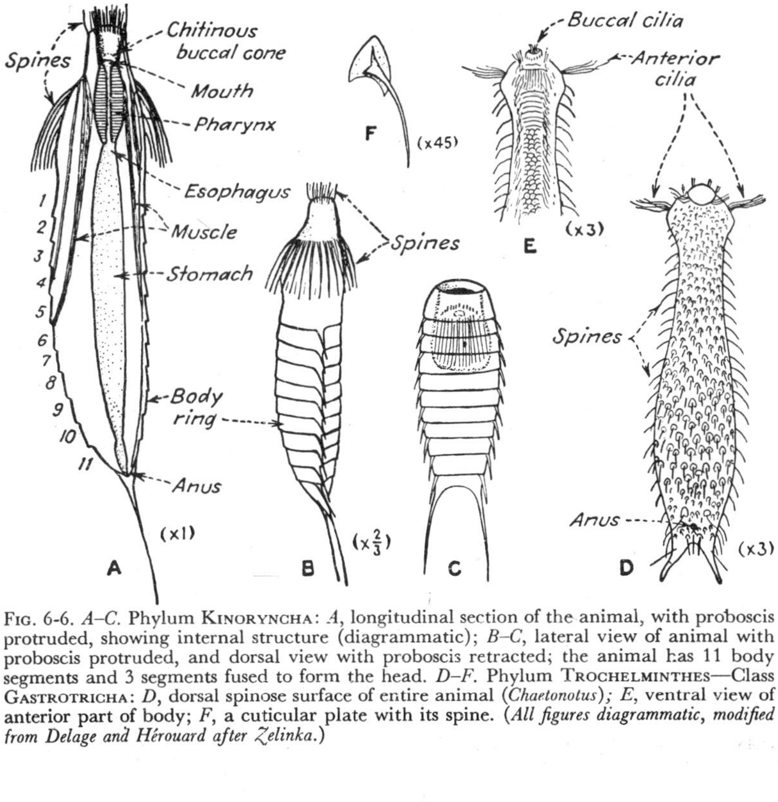

FIG. 6-6. *A–C*. Phylum KINORYNCHA: *A*, longitudinal section of the animal, with proboscis protruded, showing internal structure (diagrammatic); *B–C*, lateral view of animal with proboscis protruded, and dorsal view with proboscis retracted; the animal has 11 body segments and 3 segments fused to form the head. *D–F*. Phylum TROCHELMINTHES—Class GASTROTRICHA: *D*, dorsal spinose surface of entire animal (*Chaetonotus*); *E*, ventral view of anterior part of body; *F*, a cuticular plate with its spine. (*All figures diagrammatic, modified from Delage and Hérouard after Zelinka.*)

the head, which bears a circlet of spines and has a short retractile proboscis. The trunk is surrounded by 11 rings of chitinous cuticle bearing spines, and it may have several prominent posterior spines (Fig. 6-6*B–C*).

Members of the phylum show few relationships with other animals and are best placed in a separate phylum. There is no fossil record; hence nothing is known about the geologic history of the group.

PHYLUM TROCHELMINTHES (ROTIFERS, OR "WHEEL ANIMALCULES")

The Trochelminthes (Gr. *trochos*, wheel, + *helminthos*, worm) are the well-known "wheel animalcules," or rotifers, so named because when moving through the water their beating cilia give them the appearance of tiny rotating wheels. The cilia are disposed in tufts and bands on the anterior apex or on the central circumference of the animal and are responsible for locomotion. The 1,300 species, which are chiefly free-living and microscopic (less than 1 mm. long), are included in two classes—Rotifera and Gastrotricha.

Class Rotifera

The rotifers (Class Rotifera), which include most of the species (1,200) assigned to the Trochelminthes, may be taken as typical of the phylum. They are active microscopic creatures having the appearance of tiny wingless airplanes. They abound in both fresh and stagnant water, especially where there is considerable aquatic vegetation, and also live in marine environments, though in these they are uncommon. Most rotifers are free-living, but some are fixed in protective tubes, or **loricae,** of secreted or agglutinated material, a few are parasitic, and some live in salt water. The organism moves through the water by the beating of the cilia located on the anterior end of the body, and it is the motion of these cilia that suggests the rotation of tiny wheels and is responsible for the phylum and class designations. Rotifers are fascinating to watch under the microscope and have long been favorites with amateur microscopists. As a group they display amazing variety in form and structure and represent a much-differentiated assemblage of organisms that probably have had a long geologic history.

The general internal structure of a rotifer is shown in Fig. 6-7*A*. Of particular interest to paleontologists are the tiny jaws (**trophi**) that lie in a spheroidal muscular expansion of the digestive tract, the **mastax,** between the pharynx and stomach.[1] These jaws are chitinous and show a considerable range in structure (Fig. 6-7*A,C–H*). It would seem that they could be preserved under favorable conditions, but so far no fossil forms seem to have been reported.

[1] At the lower end of the pharynx is a muscular enlargement, the **mastax** or gizzard which contains the masticatory apparatus consisting of hard chitinous chewing organs, the **trophi.** In the ventral wall of the mastax is commonly a median piece, the **fulcrum,** to which are hinged two **rami** (Fig. 6-7*C–H*). The Y formed by these three pieces has been designated the **incus** (anvil). At each side of the gizzard and at a higher level is the hammerlike **malleus** consisting of a long handle, the **manubrium,** and a toothed claw, the **uncus.** These pieces together constitute a relatively powerful and efficient apparatus for chewing and remind one of similar mechanisms in many annelids, to which the term scolecodont has been applied.

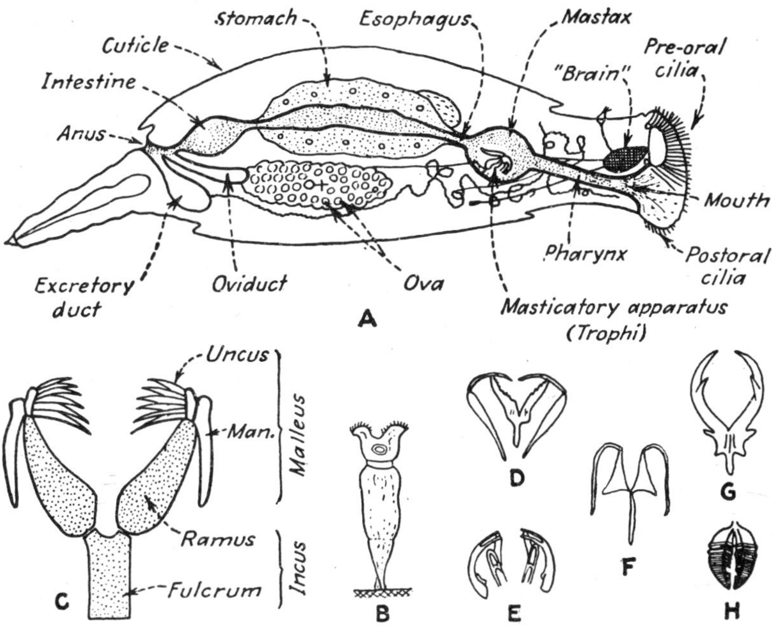

FIG. 6-7. Phylum TROCHELMINTHES—Class ROTIFERA. *A*. Sagittal section of a female rotifer, showing the chief systems of organs. Special attention is called to the chewing apparatus (*trophi*) lying in the muscular mastax. *B*. A rotifer (*Philodina*) fixed by its foot to the substratum. *C*. Diagram showing different parts of the chewing apparatus. *D–H*. Different types of chewing mechanisms. (*Modified from several sources, chiefly after Delage and Hérouard*, 1897.)

The Trochophore Larva

Most worms and many aquatic invertebrates have an early growth stage—a tiny larva that hatches from an egg—that has the general form of an adult rotifer and is referred to as a **trochophore** or **trochosphere.** This larva is a minute, transparent, free-swimming organism shaped like a pear with the several features shown in Fig. 11-2. The larger and upper end has an apical plate with a tuft of cilia and a sense organ, the eyespot. Encircling the organism in the plane of its greatest circumference is a double row of cilia that beat in such a way as to suggest a rotating wheel, whence the term trochophore (Gr. *trochos*, wheel, + *phoros*, bearing). The digestive tract is complete and ciliated internally. It consists of a lateral mouth, a slender esophagus, an enlarged stomach (enteron), a straight intestine, and an adapical anus. The position of the digestive tract determines the median plane of bilateral symmetry. The larva may also contain kidneys (nephridia), muscle and nerve fibers, and a band of mesoderm. Altogether the trochophore larva represents an advanced stage of organic development, and organisms passing through this stage are thought to be related in their evolution.

Trochophore larvae, differing in one way or another from the relatively simple form shown in Fig. 11-2, appear early in the life history of flatworms, nemerteans, bryozoans, phoronids, brachiopods, some annelids, mollusks, and some arthropods. This common larval pattern suggests that the phyla are related in their evolution and that possibly all had some common ancestor in earliest geologic time, but unfortunately there is no known fossil record to corroborate this assumption.

Class Gastrotricha

The Class Gastrotricha is composed of about 100 species of microscopic worms that live among algae and bottom debris in quiet fresh and salt waters. The animal is elongate with an anterior mouth, forked posterior end, simple digestive tract, ciliated ventral surface, and spinose dorsal surface (Fig. 6-6*D–F*). In size and habit it resembles certain ciliated protozoans. There is no larval stage and no fossil record; hence nothing is known about its geologic history.

Geologic History of Trochelminthes

The Trochelminthes represent an interesting and important group of worms, especially with reference to certain problems of invertebrate evolution, and it is unfortunate that there is no fossil record to supplement the evolutionary record still preserved in the trochophore larvae.

PHYLUM CHAETOGNATHA (ARROW WORMS)

The Chaetognatha (Gr. *chaeton*, bristle, + *gnathos*, jaw) are small torpedo-shaped marine worms (Fig. 6-8*A*), some 30 species in all, that commonly are so abundant as to constitute an important part of the marine plankton. They dart through the water in search of microscopic plants and animals and are commonly called "arrow worms" on this account. The name of the phylum calls attention to the bristlelike jaws about the mouth (Fig. 6-9).

The chaetognaths, being unsegmented, lacking cilia on the epidermis and in the digestive tract, and having a postanal tail (a chordate feature), have a combination of characteristics found in no other animals. Their embryonic development somewhat resembles that of many annelids and some chordates. The geologic history of the phylum is obscure, but if *Amiskwia* (Fig. 6-8*B*) from the Middle Cambrian Burgess shale of British Columbia is a fossil representative, as Walcott concluded, then members of the phylum were evolved quite early in the Paleozoic. A fact of interest is that *Amiskwia* is associated with the remains of free-swimming phyllopod crustaceans and is thought to have been an active pelagic swimmer like the living *Sagitta* (Fig. 6-8*A*), which it resembles in some features.

No other fossil chaetognaths have been reported thus far, although it would seem that the tiny chitinous seizing jaws and still smaller teeth (Fig. 6-9) should be capable of preservation.

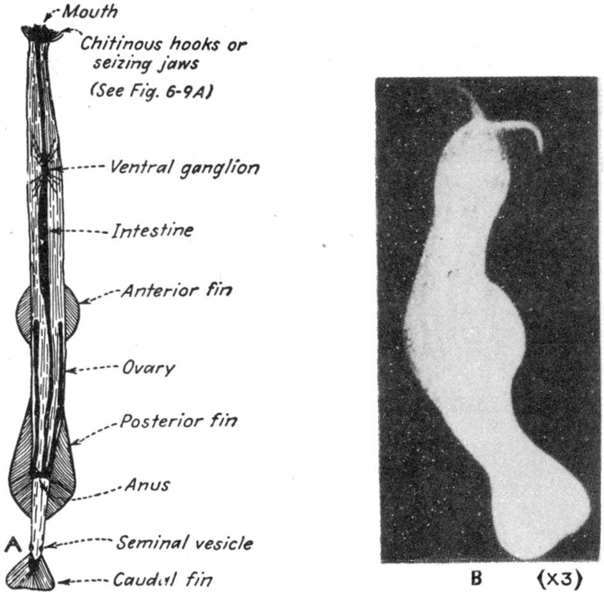

FIG. 6-8. Phylum CHAETOGNATHA. *A. Sagitta*, the common arrow worm; natural size ranges between 20 and 70 mm. *B. Amiskwia sagittiformis*, a flattened specimen from the Middle Cambrian Burgess shale of British Columbia, supposed to be an ancient chaetognath allied to *Sagitta*. (*A after Michael*, 1911; *B after Walcott*, 1911.)

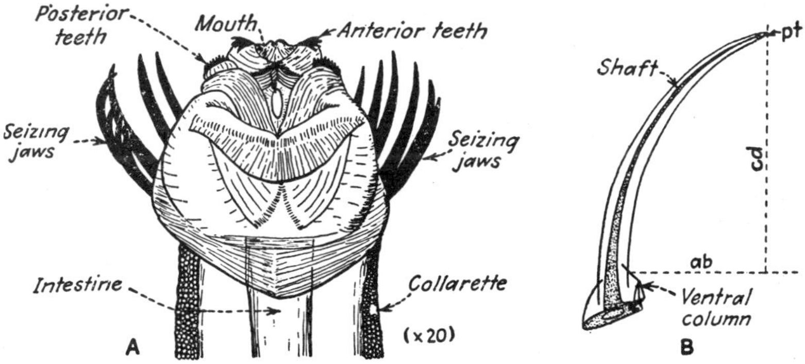

FIG. 6-9. Phylum CHAETOGNATHA. *A*. Head of *Sagitta planktonis*, showing relation of teeth and seizing jaws to the mouth, and system of muscles that manipulate them. *B*. A single seizing jaw, showing general nature and also how dimensions are determined. The line *ab* was drawn perpendicular to the edge of the shaft at its junction with the ventral column; *cd* was then constructed perpendicular to *ab* and including *pt*. The proportion *ab*/*cd* indicates the curvature of the jaw and is of importance in distinguishing the jaws of the different species. Inasmuch as the jaws are resistant structures, they are likely to remain intact in otherwise poorly preserved material. (*After Michael*, 1911.)

REFERENCES

Hyman, L. H. 1951. "The Invertebrates." Vol. 2, "Platyhelminthes and Rhynchocoela; the Acoelomate Bilateria," 550 pp. Vol. 3, "Acanthocephala, Aschelminthes, and Entoprocta; the Pseudocoelomate Bilateria," 572 pp. New York, McGraw-Hill Book Company, Inc.

Michael, E. L 1911. Classification and vertical distribution of the Chaetognatha of the San Diego region. *Univ. Cal. Pub. Zool.*, vol. 8, pp. 21–186.

Storer, T. I. 1951. "General Zoology," 2d ed. McGraw-Hill Book Company, Inc., New York. 832 pp.

CHAPTER 7

PHYLUM BRYOZOA[1] (POLYZOA)[2]

The Bryozoa are small animals that live in close colonial association and build a variety of branched or laminated, chitinous or calcareous structures to a maximum height or thickness of a few centimeters (Fig. 7-1, 7-12, 7-14–7-16). They live attached to the bottom, to objects and organisms on the bottom, and to floating objects or organisms. Bryozoa are dominantly marine, but two small and unimportant groups are confined to fresh water. All ancient Bryozoa that have left a fossil record seem to have been marine, as shown by their association with such salt-water organisms as corals, brachiopods, and crinoids. The fossil record of the phylum is long and extensive. It began in the Late Cambrian and has continued unbroken to the present. About 4,000 fossil species have been described, of which 2,500 are Paleozoic and 1,500 Mesozoic, and more than 2,500 species are now living.

Because of a superficial resemblance to plants, bryozoans were once included in the "zoophytes"—a group of organisms supposedly intermediate between animals and plants. Although the animal nature of this composite group was established early in the eighteenth century (circa 1729) by the Frenchman Peyssonnel, and the animal nature of the hydroids and bryozoans conclusively demonstrated in 1755 by the Englishman Ellis in his classic "Essay on Corallines," it was not until 1830 that Thompson in Ireland showed that each individual bryozoan possessed a complete alimentary tract consisting of a mouth, stomach, and anus. For these organisms, which were totally unlike hydroids, actinozoans, and other animals with which they had been compared, Thompson proposed the name Polyzoa, thereby calling attention to the many individual animals in a single colony—a characteristic not limited to Bryozoa but also exhibited by many nonbryozoan animals. Whether for this reason or for some other, Thompson's name for the phylum has not had wide acceptance, although it clearly has priority over Bryozoa—a term proposed in 1831 by the German Ehrenberg and now used almost universally in continental Europe and in the Americas. Polyzoa continues to be used in England.

GENERAL CHARACTERISTICS

Bryozoan colonies, commonly designated **zoaria** (singular, **zoarium**), vary widely in general shape and architecture.[3] Some are leaflike, bushlike,

[1] Bryozoa—Gr. *bryon*, moss, + *zoon*, animal; referring to the resemblance of certain bryozoan colonies to a tuft of moss.

[2] Polyzoa—Gr. *polys*, many, + *zoon*, animal; referring to the multitude of individuals in a colony.

[3] We use the term **architecture** for the framework of the stony zoarium.

or tuftlike and commonly exhibit a strong resemblance to seaweeds (Fig. 7-1); others are thin incrustations on the surface of stones, shells, living algae, etc. (Figs. 7-25*K*, 7-41*D*); still others, formed by the superposition of many incrustations or layers, are hemispheroidal, globular, nodular, or irregular masses, some of which reach diameters of 10 to 15 cm. (Figs. 7-1, 7-12).

The solid material of the zoarium is membranous, chitinous, or calcareous.

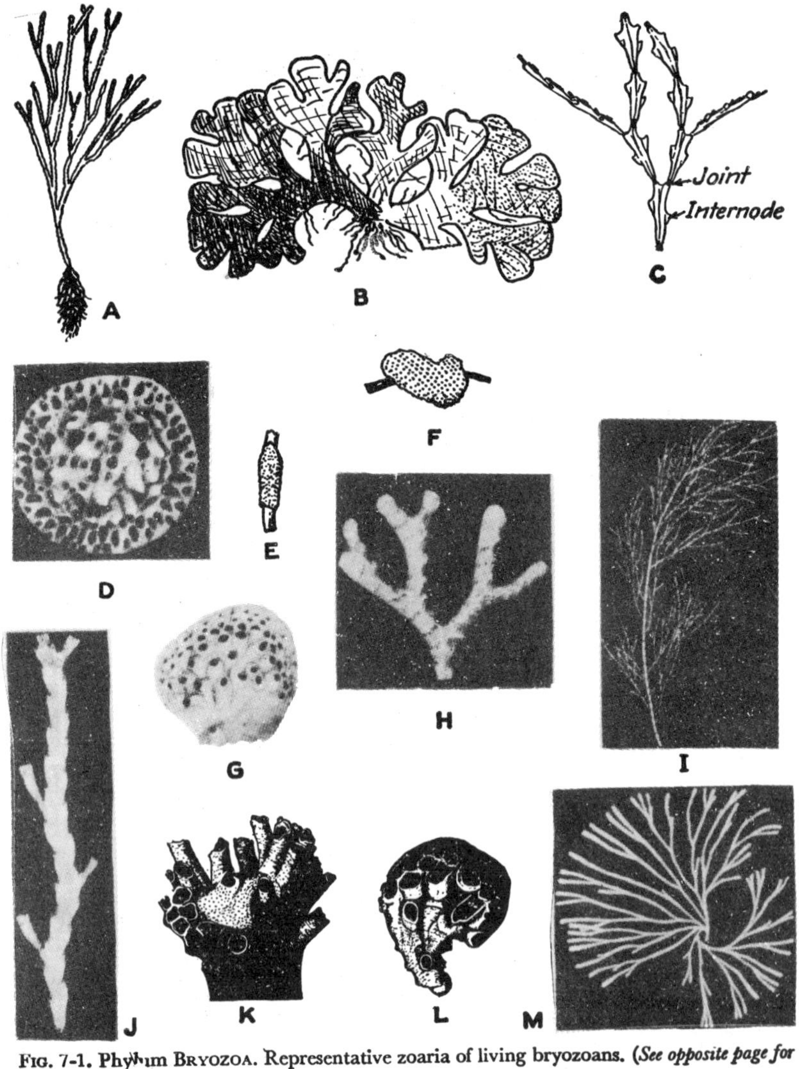

FIG. 7-1. Phylum BRYOZOA. Representative zoaria of living bryozoans. (*See opposite page for descriptions.*)

It is secreted by epidermal cells, and its continuity is broken by several different kinds of pores, perforations, and similar features. Its microscopic structure is likely to be complicated and in many species is so complex as to require special techniques for adequate study. As minute structural details are commonly of the greatest importance in differentiating fossil genera and species of stony bryozoans, polished and etched surfaces and carefully oriented thin sections have to be prepared.

The application of micropaleontology to problems of the petroleum industry has greatly stimulated investigation of Foraminifera, Ostracoda, and certain other microfossils, but not a great deal of attention has been given to the Bryozoa. This perhaps arises from the fact that few paleontologists are trained to use them. Their possible value as microfossils hardly needs emphasis, as even small fragments when properly prepared can generally be readily identified, and since many genera and species have short stratigraphic range and wide geographic distribution, Bryozoa have few peers as index fossils among submacroscopic invertebrates. This use of fragmental zoaria acquires added importance when it is remembered that stony Bryozoa are abundant in many argillaceous and calcareous rocks of the early Paleozoic that do not yield appreciable numbers of significant Foraminifera and other microfossils.

GENERAL MORPHOLOGY

Morphology of Bugula, a Modern Bryozoan

The chief morphological characteristics of a typical bryozoan may be seen in *Bugula* (Fig. 7-2), a living form that builds small tuftlike colonies attached to the bottom in shallow sea water (Figs. 7-1*A*, 7-12*A*). The chief living units of the colony[1] are termed **zooids.** These are small, rarely exceeding 1 mm. in length, and are closely united to form the branches of the zoarium. They commonly show polymorphism, and the several different types of individuals are designated by special terms. **Autozooids** are the normal and most numerous of a colony. Each is a complete individual housed in a chitinous or calcareous tube, the **zooecium.** Other zooids, generally incomplete, modified for a specific purpose and distinctly different from typical autozooids, are designated **heterozooids.**

The autozooid consists of a tubular **visceral sac** inside which lie the chief living parts collectively designated the **polypide.** The anterior end of the

[1] British paleontologists use the term **asty** for a bryozoan colony.

FIG. 7-1. Phylum BRYOZOA. Representative zoaria of living bryozoans, all about two-thirds natural size. *A. Bugula*, a typical dendroidal form with roots (cf. Fig. 7-2). *B. Carbasea*, a flabellate or multifoliate form with numerous branches. *C. Tricellaria*, part of jointed zoarium showing several internodes. *D. Lichenopora*, a young zoarium of discoidal shape. *E–F. Mucronella*, a common incrusting form. *G. Tubulipora*, a typical pyriform zoarium. *H. Idmonea*, a branching form with girdles of zooecia. *I. Gemellaria*, a delicate form resembling a seaweed. *J. Crisia*, a typical internode from a jointed zoarium. *K. Diaperoecia*, tip of branch showing ooecium with the usual semicircular ooeciopore in top center position. *L. Tubulipora*, a young zoarium with ancestroecium. *M. Caberea*, a coarsely branched seaweedlike form. (*A–B, E–F sketched from Busk*, 1884: *C–D, G–M after Osburn*, 1933.)

polypide, the **introvert,** has a tentacle-bearing ridge, the **lophophore,** which surrounds the mouth and is attached to the **aperture** of the zooecium by a flexible chitinous collar. Cilia on the tentacles sweep microscopic organisms toward the mouth, and the tentacles themselves aid in respiration. The introvert can be pulled into the visceral sac completely by the action of **retractor muscles.**

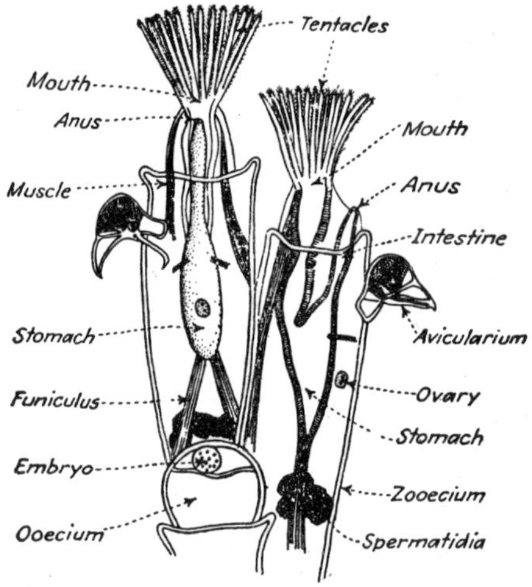

FIG. 7-2. Phylum BRYOZOA. Two autozooids of a modern bryozoan, *Bugula avicularia.* (*Modified after Parker and Haswell*, 1930.)

The larger part of the polypide is the U-shaped, ciliated **digestive tract,** which is suspended in the visceral cavity, or **coelom.** It includes the **mouth** within the tentacular circle, a wide **pharynx,** a slender **esophagus,** a capacious **stomach,** and a slender **intestine** that discharges through the **anus** located outside the lophophore. The stomach is drawn inward by a retractor muscle, the **funiculus.**

As with all other Bryozoa, *Bugula* lacks respiratory, circulatory, and excretory organs. The coelom is lined with a thin peritoneum and is filled with a colorless fluid containing corpuscles. Ectodermal cells form the epidermis of the autozooid and secrete the chitinocalcareous zooecium. Endodermal cells line the digestive tract and perform digestive functions. Mesodermal cells line the coelomic cavity and form muscles, gonads, and the nervous system.

Bugula is hermaphroditic, or monoecious. Eggs develop in a special pouch of the coelomic lining, the **ooecium;** sperm forms in **testes** associated with the muscles. The egg is fertilized in the coelom and develops there into a ciliated trochophore larva, which passes to the outside through a special opening and soon thereafter attaches itself to the bottom. After undergoing metamorphosis it builds a colony by asexual budding.

Autozooids

Although the autozooids of most Bryozoa are monotonously similar in general characteristics, they do vary to some extent in certain features. These variations are discussed in later paragraphs.

The Polypide. The polypide consists of four chief parts (Figs. 7-2, 7-3): (1) the **lophophore,** which is a tentacle-bearing crescentic or circular ridge surrounding the mouth; (2) a **U-shaped digestive tube** that hangs suspended in the visceral cavity with mouth and anus relatively close together on the upper surface of the zooid (Fig. 7-3); (3) certain **muscles** that control the extrusion (evagination) and retraction (invagination) of the introvert or lophophore; and (4) a **nervous system** (not always present) consisting of a single ganglion between the mouth and anus and many nerve fibers that extend to the tentacles and to the upper part of the digestive tract. When the lophophore of some forms is retracted, the tentacles lie in a cavity that is lined

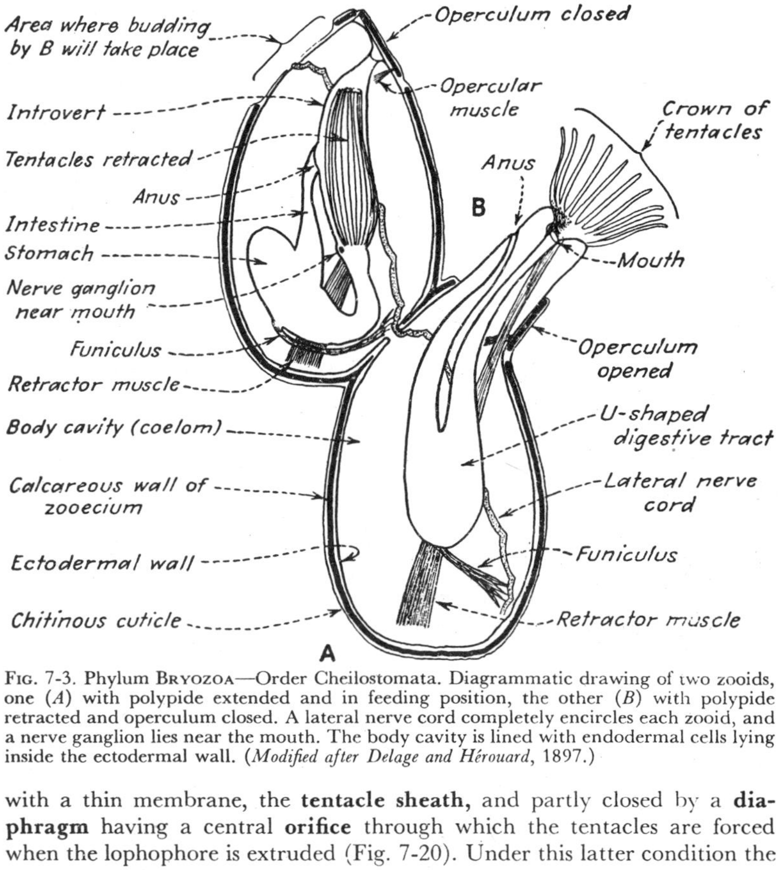

FIG. 7-3. Phylum BRYOZOA—Order Cheilostomata. Diagrammatic drawing of two zooids, one (*A*) with polypide extended and in feeding position, the other (*B*) with polypide retracted and operculum closed. A lateral nerve cord completely encircles each zooid, and a nerve ganglion lies near the mouth. The body cavity is lined with endodermal cells lying inside the ectodermal wall. (*Modified after Delage and Hérouard*, 1897.)

with a thin membrane, the **tentacle sheath,** and partly closed by a **diaphragm** having a central **orifice** through which the tentacles are forced when the lophophore is extruded (Fig. 7-20). Under this latter condition the tentacle sheath constitutes a circular collar connecting the expanded tentacles with the wall of the sac (Fig. 7-3).

Visceral Sac. The sac or body wall of the autozooid encloses the visceral cavity, or true **coelom,** which is filled with colorless fluid containing numerous floating leucocytes. There is neither heart nor vascular system in the Bryozoa, and no nephridial organ. Furthermore, there are no special sense organs and no nervous system other than the simple nerve ganglion present near the mouth in the polypide. It is the polypide that fills the major part of the visceral sac, especially when retracted, and several sets of muscles extend from parts of the polypide across the visceral cavity to its wall (Figs. 7-2, 7-3, 7-40).

The wall of the autozooidal sac is composed of two fleshy layers: (1) an outer layer of ectodermal cells, and (2) an inner layer of mesodermal cells (Fig. 7-4). In some forms (Class Phylactolaemata) a double muscular layer is developed between the ectoderm and endoderm. The mesoderm commonly has the form of a loose network of cells and connecting strands, and the latter may extend across the vascular cavity to the polypide. The ovary is generally attached to the mesodermal wall; the testes to certain muscles (**funiculus,** Fig. 7-2). The ova and spermatozoa, when ripe, break off and float freely in the coelomic fluid, where the fertilized ovum develops into a ciliated trochophore larva before being liberated to the outside.

The Zooecium

In most familiar Bryozoa the autozooid lives in a chitinous or calcareous tube, the zooecium, which is in reality an external envelope, though commonly included by zoologists as an integral part of the wall of the zooid. Inasmuch as the zooecia constitute most of a zoarium, and are therefore likely to be the only part of a bryozoan colony preserved, paleontologists have centered their attention on the characteristics and relationships of these durable structures.

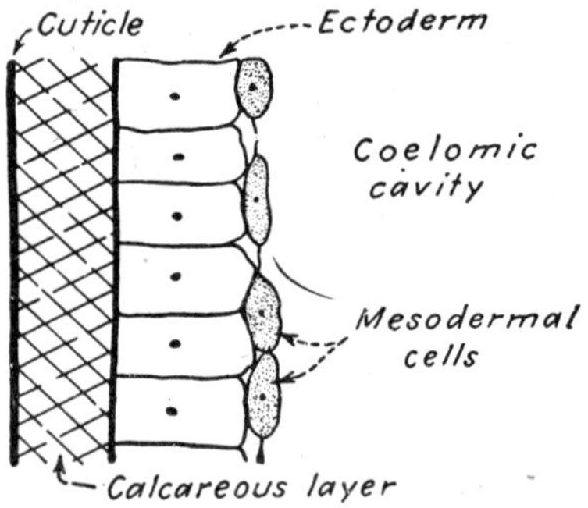

FIG. 7-4. Phylum BRYOZOA. Diagram showing structure of body wall and of enclosing chitinocalcareous zooecium.

The wall of the zooecium lies in direct contact with the outer surface of the ectoderm of the zooid and is secreted by the ectodermal epithelium (Fig. 7-4). It consists typically of two layers: (1) a thick calcareous layer next to the ectoderm, and (2) a thin chitinous cuticle covering the outer surface of the calcareous layer. The cuticle is almost invariably missing in fossil Bryozoa. Both cuticular and calcareous material are secreted by the ectodermal epithelium, the former as a primary layer, the latter, which is missing in some forms, as a secondary layer. The calcium carbonate of the calcareous layer seems to be deposited in the cuticle, and under a microscope appears as a dense laminated network of exceedingly fine threads that pass into the cuticular material. Owing to the fact that calcium carbonate continues to be secreted from the adjacent epithelium throughout the life of the zooid, the zooecial wall tends to be thicker in the older parts of the zoarium. Likewise, the outer or later part of the zooecium in many fossil forms (Order Trepostomata) is likely to have a thicker wall than the inner or earlier part because of the decrease in the rate of forward growth of the zooecium (Fig. 7-27).

Zooecia are always quite small, rarely exceeding a millimeter in diameter though commonly attaining a length of many millimeters, but the zoaria they make may reach dimensions as great as 50 cm. (20+ in.) in some incrusting forms that spread over considerable areas. They exhibit great variety in form, being conical, urn-shaped, prismatic, or casketlike, and in transverse section are circular, oval, elliptical, or polygonal (Figs. 7-25, 7-26, 7-40, 7-41, 7-42). However, the general shape of the zoöecia in a single species remains constant; hence this characteristic has proved to be quite useful in classifying both fossil and living species.

The chief opening into the zooecium, generally referred to as the **aperture** or **orifice,** is essentially as large as the cross section in tubular and prismatic forms, but in boxlike and ovoidal zooecia it is generally small, variously modified, and commonly equipped with a lid, or **operculum.** Its position is terminal in the simpler forms of zooecia but variable in location on the upper face of the more complex forms (Figs. 7-5, 7-23, 7-41, etc.).

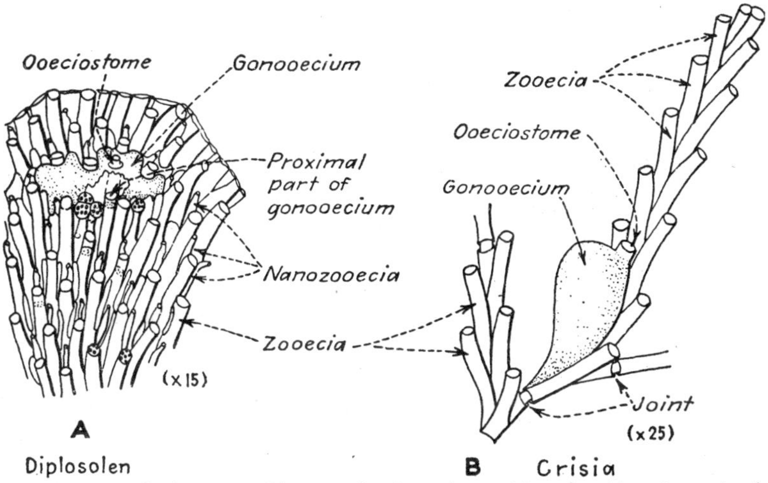

FIG. 7-5. Order Cyclostomata. Diagrams showing polymorphism of zooids and zooecia. *A.* Part of zoarium of *Diplosolen obelia*, viewed from the frontal side. The roof of the gonooecium, or brood chamber, which is stippled, is partly omitted to show the narrow proximal part. Nanozooids are about as numerous as autozooids. *B.* Part of a zoarium of *Crisia eburnea*, showing numerous zooecia and one typical gonooecium with its small ooeciostome. (*A after Borg*, 1926; *B after Harmer*, 1940.)

In the simpler and looser types of zoaria, the zooecia are free except where attached to the stolonal tube or earlier zooecium from which they sprang. In compact zoaria, by contrast, they are in actual contact, being either loosely or tightly packed together, and in some forms the walls of contacting zooecia become amalgamated. In colonies of tightly packed zooecia, the zooids are connected by fleshy tubes and strands that traverse the walls through pores (Fig. 7-3). This interconnection of zooids is immediately apparent if the calcareous zoarium of certain living species is decalcified by the use of an appropriate acid.

Internally the zooecia of many species, particularly fossil ones, are divided into differently shaped compartments by transverse partitions. Some of the partitions extend all the way aross the zooecium, either horizontal or inclined, and are designated **diaphragms,** or **tabulae;** others, which are commonly curved, being arched inward or upward more typically than otherwise, are known as **cystiphragms** (Figs. 7-23, 7-27). These structures are of great importance in the microscopic identification of many fossil genera and species.

The external surface of the zooecia of most living and Cenozoic Bryozoa is commonly a thing of surprising beauty and amazing structure. The orifices, here closed by an operculum and there appearing as semielliptical slits, where the lid is tipped, may be variously modified and are commonly flanked by short, stiff spines (Fig. 7-41*B–C*).

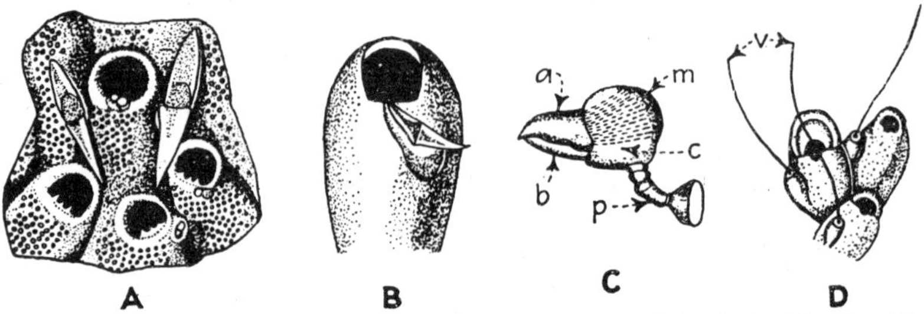

FIG. 7-6. Phylum BRYOZOA. Avicularia and vibracula. *A. Cellopera honoluluensis* with immersed avicularia. Zooecial apertures are shown as solid black. *B.* Apertural part of a single zooecium of *Adeonella* with one conspicuous sessile avicularium. *C.* Diagram of a pedunculate avicularium (*a* = mandible; *b* = beak; *c* = chamber of avicularium; *m* = muscles; *p* = peduncle). *D.* Three zooecia of *Mastigophora*, with long vibracula (*v*). (*All figures from Simpson*, 1897, *after Hincks*, 1880, *and Busk*, 1884.)

Some zooecia bear on their outer side a most singular body called an **avicularium** because of its resemblance to a bird's head (Figs. 7-2, 7-6, 7-36). This little body, which is attached to the wall of the zooecium by a short and flexible neck, can open its lower jaw through an arc of about 180° and from this position can snap it back quickly, if need be, to seize any passing organism that comes within reach. Just why it engages in this activity is uncertain, for the organisms thus seized are not passed on to the food-gathering mechanism, and even if they were, they are rarely the kind of food accepted by the animal.

In addition to the motion of the ciliated tentacles, the trap-door action of opercula, and the snapping of the avicularia, an observer's eye may often note the waving about of tiny bristlelike appendages known as **vibracula** (Fig. 7-38). These are more or less constantly in motion, often in rhythmic sweeps, but their function, like that of the avicularia, is uncertain, although their activity may prevent larvae and noxious matter from settling on the colony. It has been suggested that certain minute tubular spines, the **acanthopores** (Fig. 7-26*A*), common on the zoaria of the Paleozoic Trepostomata may have been places of attachment for structures similar to vibracula and avicularia, but this suggestion cannot be tested in the absence of any known fossils of these tiny structures.

Heterozooids

In the colonies of many living Bryozoa are certain zooids which, because they differ distinctly from the normal individuals (*i.e.*, autozooids) of a colony, have been given the name heterozooids. Of these perhaps the most important are the ones that have become specialized for reproductive purposes (*i.e.*, **gonozooids**). Certain other heterozooids are characteristic of the Order Cyclostomata and are discussed in later paragraphs.

Gonozooids. In modern Cyclostomata the female cells are developed in special **brood chambers** that appear among the zooecia as conspicuously inflated vesicles (Fig. 7-5). These chambers, which have been called **ooecia** or **ovicells** by some investigators and **gonooecia** by others, house certain specialized zooids, the **gonozooids,** which produce large numbers of embryos. Brood chambers are also developed in the Order Cheilostomata, but they are not generally homologous with those of the Order Cyclostomata (Borg, 1926).

REPRODUCTION, GROWTH OF INDIVIDUAL (ONTOGENY), AND COLONIAL DEVELOPMENT (ASTOGENY)

The typical bryozoan is monoecious, *i.e.*, both male and female cells are present in the same individual. The **ovary,** which contains the eggs, is generally developed in some part of the coelomic wall; the **spermary,** around the funiculus (Fig. 7-2).

The eggs are fertilized in the coelom, in some forms in a special brood pouch or **gonooecium** (Figs. 7-35, 7-40, 7-41*B*, 7-42), and develop into trochophore larvae, which then leave the coelom through a special opening to the outside. In the Cyclostomata the egg does not develop into a single individual; instead, many embryos arise by successive fissions of the primary embryo, a process known as **polyembryony.** Most fresh-water Bryozoa encase the egg in a chitinous **statoblast** (Fig. 7-18) which can resist a wide range of unfavorable conditions until it is ready to hatch, generally in the spring, whereupon the chitinous shell splits open equatorially and a young externally ciliated colony of several polypides emerges.

The bryozoan larva, is free-swimming and of the trochophore type, though in many groups it is greatly modified, particularly in lacking a functional alimentary tract. The typical bryozoan trochophore, found in all Entoprocta, but in only certain Ectoprocta, and referred to as *Cyphonautes*,[1] is shown in Fig. 7-7. It has the shape of a bell and possesses a bivalve shell consisting of two triangular pieces of chitin secreted by the ectoderm. The **apical plate** and **ciliated ring** project from between the valves. Internally there are several characteristic organs, including a simple alimentary tract.

[1] Silén (1944, p. 88) concludes that "*Cyphonautes* is a larval form which has appeared secondarily and rather late in the Bryozoa." If true, this would mean "that no phylogenetic conclusions can be drawn from the presence of that larval form as was generally true before."

From this stage of embryology on, several striking differences between the Ectoprocta and Entoprocta appear.

After a short free-swimming period, usually less than 24 hours,[1] the

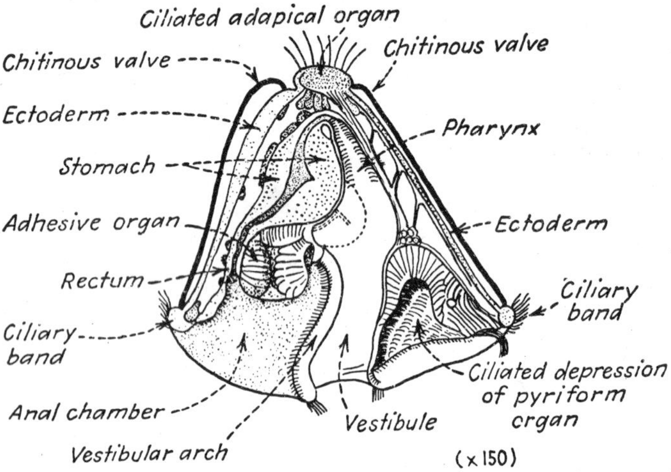

FIG. 7-7. Phylum BRYOZOA. Section through a completely developed *Cyphonautes* larva, showing the adapical organ that ultimately will develop into the first polypide of the colony, the two chitinous valves that fold down over the larva when it attaches itself to the bottom (cf. Fig. 7-8), the adhesive organ used in attachment, and the prominent vestibule that opens into a complete alimentary tract. (*After Prouho*, 1892, *with modifications*.)

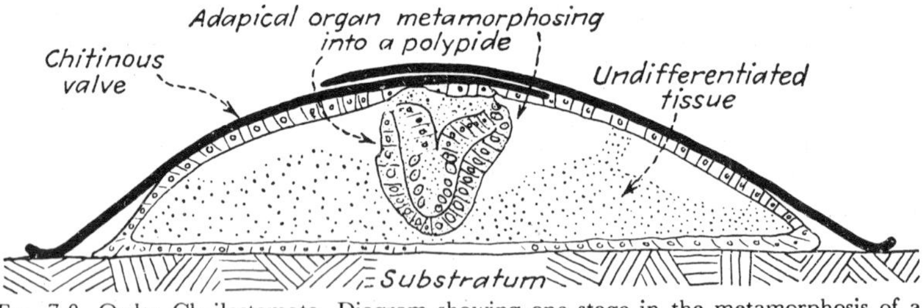

FIG. 7-8. Order Cheilostomata. Diagram showing one stage in the metamorphosis of a *Cyphonautes* larva. The larva is attached to the substratum, and the two valves have flattened out and now overlap one another, thereby making a protective cover for the organism. The adapical organ (apical plate) is developing into a polypide which later will become part of the zooid that will occupy the ancestroecium. Most of the remaining organs of the larva have disintegrated into undifferentiated tissue as a result of the metamorphosis. (*Modified after Kupelweiser*, 1906.)

ectoproct larva fixes itself to some object on the bottom. It then undergoes a remarkable metamorphosis in which all the organs, except the apical plate, disintegrate into an indefinite mass, after which the cells are reorganized and the first individual of the colony is formed from a polypide bud. It is the

[1] Because many ectoproct larvae have no functional alimentary tract, they have to depend for nutrition on the accumulated store of the yolk, hence the period during which they can swim about freely is necessarily short.

ectoderm of the apical plate that gives rise to the new polypide bud (*i.e.*, the tentacles and tentacle sheath, the nerve ganglion, and the alimentary canal). During the metamorphosis a chitinocalcareous covering, the **protoecium,** comes into existence. In the *Cyphonautes* larva this consists of the basal disk and the two primitive valves that flatten out and overlap (Fig. 7-8); in other ectoproct larvae it is a continuous structure and has the form of a tiny moccasin or shoe (Figs. 7-9, 7-19).

The newly formed zooid, the first individual of the colony, now builds about itself a chitinous or chitinocalcareous tube, the **ancestroecium** or **ancestrula,** which is an upward continuation of the protoecium (Fig. 7-9).

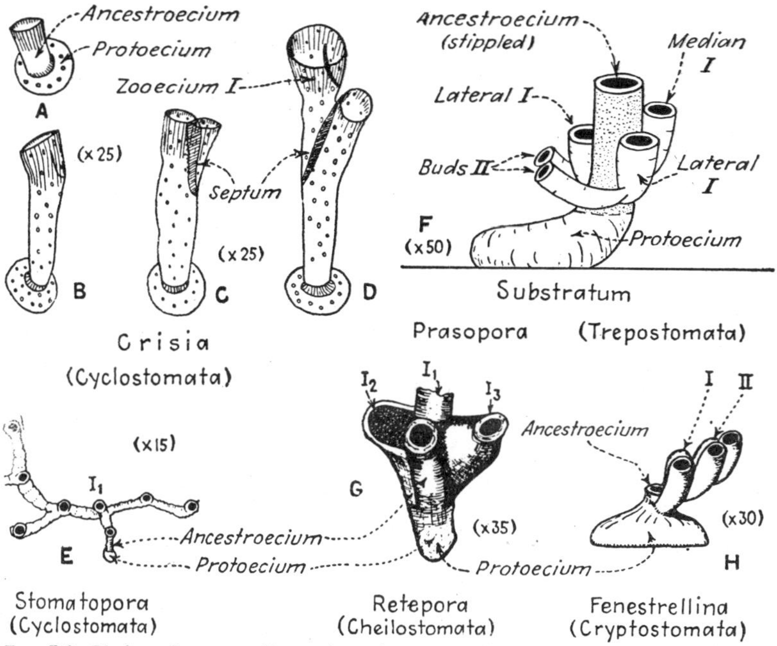

FIG. 7-9. Phylum BRYOZOA. Protoecia and ancestroecia. Diagrams showing the early astogeny of representatives of three bryozoan orders. Generations are indicated by roman numerals. It is to be noted that the first generation of buds arises from the ancestroecium, whereas all later ones grow from preexisting zooecia. *A–D*. Early growth stages of *Crisia eburnea*. In the earliest stages of the colony the primary disk (protoecium) is formed by calcification of the body wall of the metamorphosed larva after its fixation. From the disk grows a tube, the ancestroecium, which has a terminal orifice closed by a membrane. The zooid in this tube is the common bud or first zooid of the colony, and from it is given off the first generation of buds. A new zooecium is indicated by development of an oblique calcareous septum, which grows into the body cavity from the future basal wall of the colony. *E*. Early part of zoarium of *Stomatopora parvipora*. *F*. Diagram of a trepostomatous bryozoan (*Prasopora*) showing a prominent protoecium and ancestroecium, with three buds arising from the ancestroecium. Second-generation buds arise from the lateral zooecia of the first generation. *G*. *Retepora phoenicea* with three first-generation buds. *H*. *Fenestrellina*, with a relatively large protoecium and small ancestroecium. (*A–D after Borg*, 1926; *E modified after Bassler*, 1922; *F–H based on Cumings*, 1904–1915.)

In due time buds form in different positions around the first zooid, and as these develop into mature individuals (autozooids) they build the first zooecia of the zoarium. In general, all Bryozoa, whether recent or fossil, conform to a fundamental plan of primary budding in which an ancestroecium grows from the protoecium and then gives off one or two lateral buds (in some cases a median bud arises somewhat later) (Fig. 7-9). When these buds grow into mature zooids they in turn continue the colony by giving off the next generation, etc. Hence by repeated budding, first from the zooid of the ancestroecium and thereafter from existing autozooids, the colonial structure, which shows great variation in form and architecture, comes into existence. All zooids of a colony remain connected by means of fleshy strands that pass through communication pores in the zooecial walls (Figs. 7-3, 7-28, 7-40).

By contrast with the Ectoprocta, the entoproct larva, after fixing itself to the bottom by its oral surface, undergoes metamorphosis into an adult without degeneration of the alimentary tract and with the mouth rotating upward. No true coelom appears, the tentacles are nonretractile, and other structures develop to make the mature entoproct quite different from the mature ectoproct (see table on page 211).

Degeneration and Brown Bodies

A phenomenon unique to Bryozoa, though not exhibited by some forms, is the periodic degeneration of the polypide, followed by the formation of a compact spherical or ovoid **brown body,** and simultaneously by the development of a polypide bud in the zooid (Fig. 7-10). In this process the polypide has only limited duration, whereas the visceral sac, together with its liga-

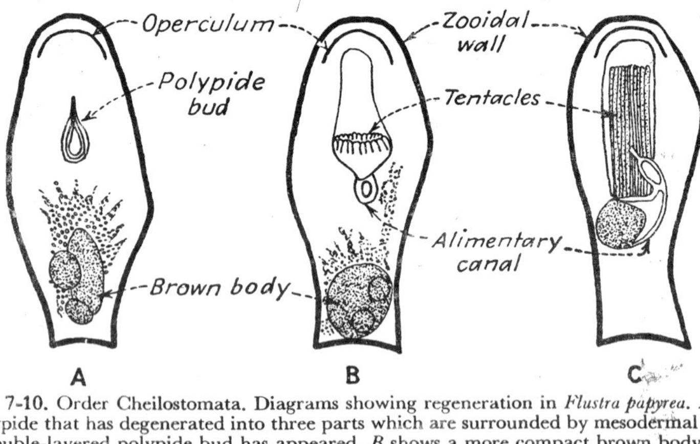

FIG. 7-10. Order Cheilostomata. Diagrams showing regeneration in *Flustra papyrea*. *A* is a polypide that has degenerated into three parts which are surrounded by mesodermal cells; a double-layered polypide bud has appeared. *B* shows a more compact brown body, and the development of tentacles and an alimentary canal in the polypide bud. *C* shows the brown body passing into the stomach of the new polypide. More than one degeneration can take place in a zooid, and successive degenerations are commonly indicated by small residual brown bodies. This phenomenon, which is characteristic of most living Bryozoa, is generally regarded as an excretory process, in the absence of specific excretory organs. (*Adapted from Harmer*, 1930.)

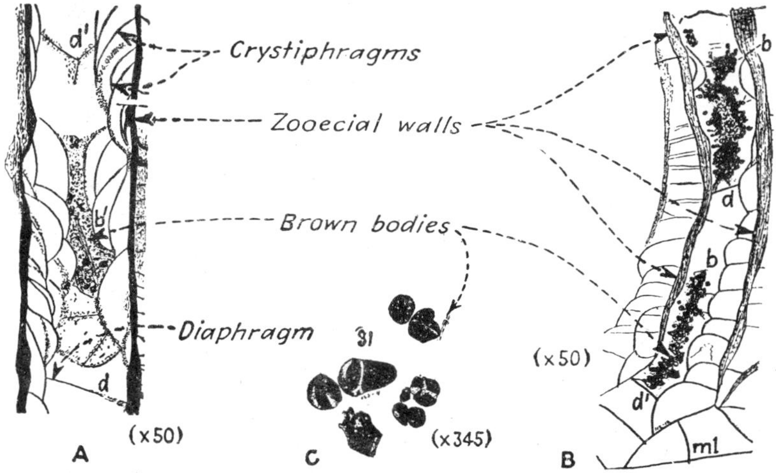

FIG. 7-11. Order Trepostomata. Fossil brown bodies. *A*. Diagrammatic longitudinal section of a zooecium of *Prasopora simulatrix*, showing one well-defined brown mass (*b*), many peripheral cystiphragms, and two prominent diaphragms (*d*, *d'*). *B*. Zooecium of *Peronopora vera*, showing two brown masses, each confined by prominent diaphragms (*d*) and cystiphragms. *C*. A few nodular grains of brown material from a cyst of *Heterotrypa subramosa*. The brown material illustrated in the figures is thought to have developed originally from successive degenerations and regenerations of polypides, as in modern Bryozoa. It is now largely an iron compound that replaced the original organic material left in the zooecium after the death of the polypide. The associated cystiphragms and diaphragms seem to have served the purpose of restricting the intrazooecial space. (*After Cumings and Galloway*, 1915.)

ments and extensor muscles and the vestibule, persists essentially unchanged through successive degenerations. The brown body either remains in the zooid as an inert mass or is evacuated. Degeneration commonly takes place several times during the life of a zooid, and each brown body retained as a residue in the zooid records a degeneration.

The formation and evacuation of brown bodies is probably to be regarded as a process of excretion, since the zooid lacks any specific excretory organs. In certain Entoprocta it may represent sexual rejuvenation (Harmer, 1930). In any event, the loss of the polypide seems to have no permanent effect on the zooid, because soon after its degeneration a bud forms inside the zooid and quickly develops into a new polypide, which lives on in the visceral sac and enclosing zooecium inherited from its parent.

The phenomenon of periodic degeneration with accompanying formation of brown bodies seems to have been characteristic of ancient Bryozoa, as certain Ordovician trepostomes show numerous small brown nodular bodies surrounded by cysts within the same zooecium (Fig. 7-11) (Cumings and Galloway, 1915).

Zoaria

Bryozoan zoaria exhibit great range in shape, size, and architecture (Figs. 7-1, 7-12). Incrusting forms are common, and because they grow rapidly, once they gain a foothold, they may quickly cover large areas. Commonly

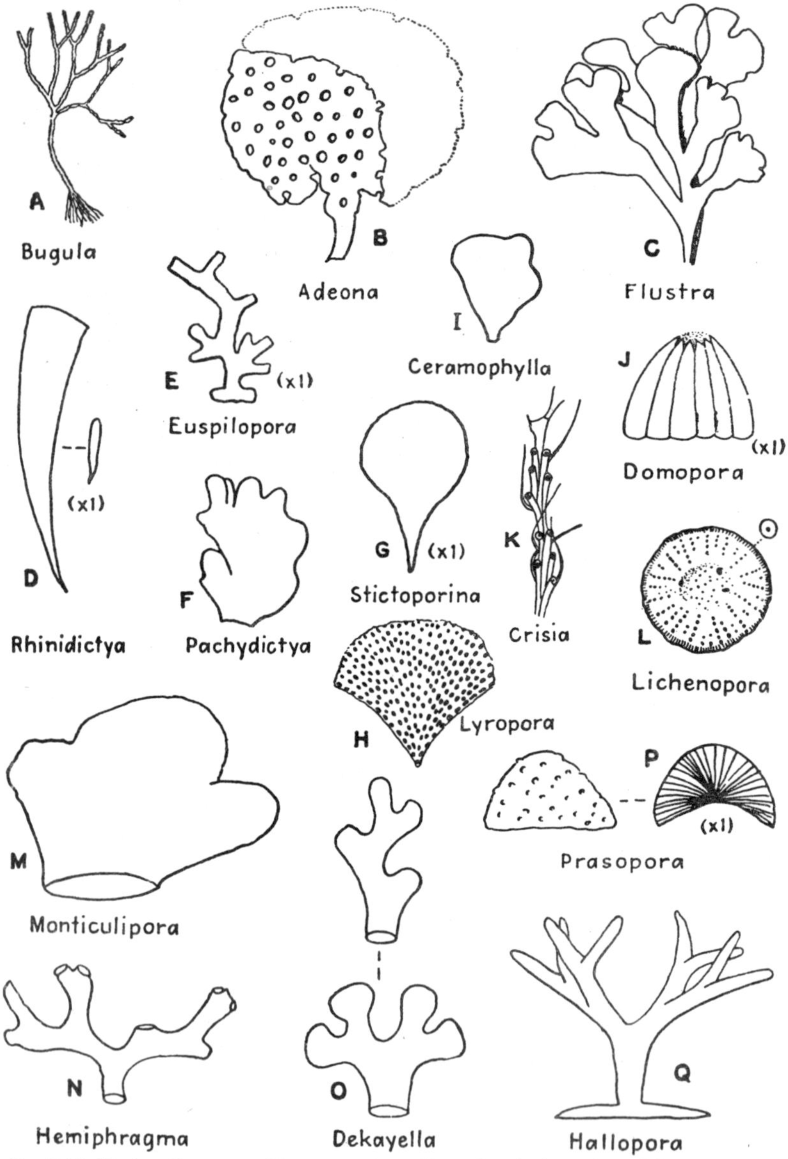

FIG. 7-12 Phylum BRYOZOA. Diagrammatic outlines of typical zoaria, ×½ unless otherwise indicated. *A–C.* Cheilostomata. *D–H.* Cryptostomata. *I–L.* Cyclostomata. *M–Q.* Trepostomata. (*Modified from numerous authors.*)

some take the form of delicate networks and thin laminar expansions. More massive zoaria develop as successive layers are added, and some of these may ultimately become hemispheroidal. Common among the Cryptostomata are zoaria resembling trelliswork, lace, and netlike fan- or funnel-shaped structures. Some zoaria are composed of jointed stems in which the zooecia are arranged radially around the axis in a number of longitudinal rows. Lamellar, leaflike, and dendritic zoaria are especially characteristic of the trepostomatous bryozoans. Lamellar forms are **unilaminar** (**unifoliate**), with all zooecial openings on one side, or **bilaminar** (**bifoliate**), with zooecial openings on both sides. In general, the shape or form of a zoarium is not constant even in the same species—a condition quite the opposite of the zooecia, which do tend to be constant.

The zoarial surface (**frontal side**) commonly has characteristic features of taxonomic importance. Small **mesopores** are present between the zooecia in some species, and they may be concentrated here and there to form stellate clusters (*e.g.*, *Constellaria*, Fig. 7-25*B–D*). Low domes, or **monticules**, and flat or depressed areas called **maculae** mark small groups of zooecia that are abnormal in size or have elevated apertures (Fig. 7-25). **Spines** and **nodes** are present on the surface of some zoaria, and attention is called elsewhere to the different reproductive structures that characterize many of the more geologically recent species.

CLASSIFICATION

The classification of the Bryozoa is now in a state of flux because of the conflicting ideas of different investigators, particularly those who have been studying recent forms. Taking into account the work of Borg, Canu and Bassler, Cori, Harmer, Marcus, Prouho, Silén, Ulrich, and Waters, we prefer the following classification, partly old and partly new, for the reasons given in following paragraphs.

Phylum Bryozoa (Polyzoa).

Subphylum Entoprocta. Relatively simple and archaic bryozoans in which a circular, nonretractile lophophore encloses both mouth and anal opening (Fig. 7-13); no durable hard parts and no fossil record. *Recent.*

Subphylum Ectoprocta. Simple to highly specialized bryozoans in which the anus invariably lies outside the lophophore, which not only is retractile but also has two distinctly different forms, one circular, the other U-shaped (Fig. 7-13); the hard parts show a great range of variation in structure and there is a long and extensive fossil record. *Upper Cambrian to Recent.*

Class Phylactolaemata. Almost exclusively fresh-water bryozoans with a U-shaped lophophore, an overhanging lip around the mouth, a unique type of internal bud (statoblast), and no hard parts. No certain fossil record. *Recent.*

Class Stenolaemata. Marine bryozoans with zooids having a membranous sac; zooecia cylindrical, tapering in the proximal part, and with a terminal aperture; degeneration of polypides and polyembryony in living forms; circular lophophore in living forms; zooecia and zoaria calcified; extensive fossil record.

Order 1. Cyclostomata. *Upper Cambrian to Recent.*

Order 2. Trepostomata. *Ordovician to Permian.*

Class Gymnolaemata. Almost exclusively marine bryozoans with anus lying outside a circular lophophore as in the Stenolaemata (Fig. 7-13); zooecia boxlike with modified and distally placed aperture and with unique appendages; zooecia and zooaria stony and exquisitely ornamented in one group, nondurable in the other; extensive fossil record.

Order 3. Ctenostomata. *Ordovician to Recent.*

Order 4. Cheilostomata. *Cretaceous to Recent.*

Order 5. Cryptostomata. *Upper Cambrian to Permian.*

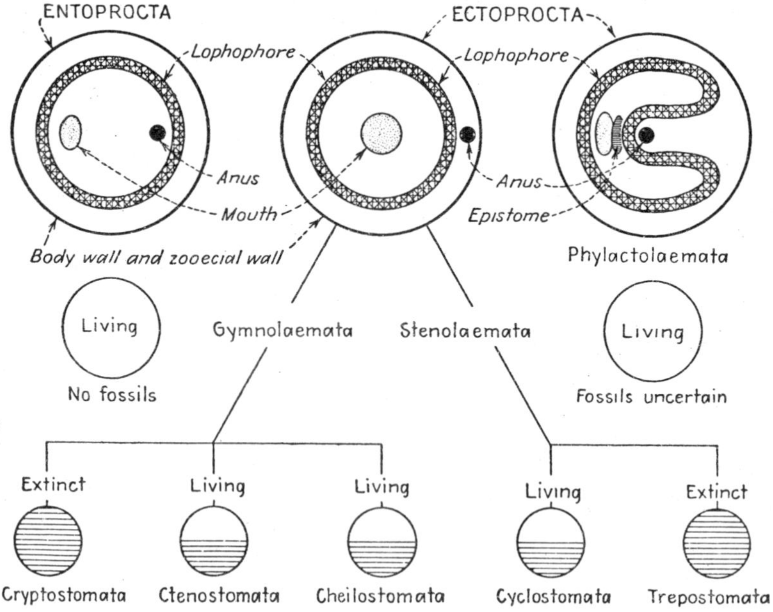

FIG. 7-13. Phylum BRYOZOA. Schematic diagram illustrating the essential differences between the major subdivisions of the phylum. The position of the anus with reference to the lophophore and the ground plan of the lophophore itself are important differentiating characteristics. Entoprocts have about 18 tentacles (not differentiated in our figures) in the lophophore, whereas the number ranges from a dozen or so in the ectoprocta with circular lophophores to more than 60 in some phylactolaematous species. Of the five important orders of Ectoprocta, two are extinct and three are living. All have a fossil record.

Many zoologists and a few paleontologists consider the Entoprocta and Ectoprocta sufficiently distinct from one another to be made separate phyla. While there are some excellent arguments for this separation, as shown in Table 7-1, we prefer to consider the two groups as subphyla of the single Phylum Bryozoa, an admittedly conservative course, following the conclusions of Harmer (1928), who points out that both groups are similar in the following important respects:

1. Embryological development (Prouho, 1892).
2. Mode of attachment of the larvae.
3. Method of budding.
4. Form of the alimentary tract.
5. Position of ciliated tentacles.
6. Position of nerve ganglion.

Furthermore, because there is no fossil record of the Entoprocta, the general problem of the biological affinities of this group, as well as of the Ectoprocta, had best be left to the zoologists.

Recognizing the Phylactolaemata, Stenolaemata, and Gymnolaemata as classes, *i.e.*, as primary subdivisions of the Ectoprocta, makes it possible to take cognizance of much significant work of recent date emphasizing the considerable differences between these major groups. The Stenolaemata are based on the living Cyclostomata, and there is no certainty that the long extinct Trepostomata are properly placed in this class, as almost nothing is known about the soft parts of the animal. However, this extinct group does not fit as well elsewhere, hence its tentative assignment to the Stenolaemata. Some paleontologists would undoubtedly prefer to ignore the Stenolaemata altogether and include all five bryozoan orders with a fossil record in the single class, Gymnolaemata. Admittedly this is a simpler classification, but it is hardly in line with recent bryozoological investigations.

TABLE 7-1. COMPARISON OF CERTAIN MORPHOLOGICAL FEATURES OF ENTOPROCTOUS AND ECTOPROCTOUS BRYOZOA

Entoprocta	*Ectoprocta*
1. Lophophore circular with anus inside circle; with sphincter muscle	1. Lophophore circular in one group, U-shaped in another, with anus outside in both
2. No coelom; cavity between polypide and body wall filled with parenchymatous tissue	2. Coelom present, and filled with a colorless fluid
3. Tentacles not retractile, but can be covered by a circular flap of the body wall	3. Tentacles retractile into an introvert, or tentacle sheath
4. A pair of protonephridia ending in flame cells	4. No nephridia or definite excretory organs
5. Gonads with a duct of their own	5. Intertentacular organ serves for escape of genital products
6. Larva trochospheric; becomes attached by its oral surface and metamorphoses into an adult without degeneration of the alimentary tract	6. Larva trochospheric, much modified; degenerates into brown body, with regeneration of new polypide; periodic degeneration followed by regeneration
7. Zooid without durable hard parts	7. Zooid with durable chitinous or calcareous zooecium and several kinds of heterozooecia

For ready reference and comparison, the more important features of the five bryozoan orders having a fossil record are assembled in Table 7-2. Some of the characteristics noted in this table can also be used as a basis for differentiating taxonomic divisions of lower rank (*e.g.*, families, genera, and even species).

SUBPHYLUM ENTOPROCTA[1] (KAMPTOZOA)[2]

The small group of organisms constituting this subphylum has commonly been included in the Bryozoa, and we prefer such an assignment, in this

[1] Entoprocta—Gr. *entos*, within, + *proktos*, anus; referring to the fact that the anus lies within the ring of tentacles surrounding the mouth (Fig. 7-13).

[2] Kamptozoa—Gr. *kamptos*, flexible, + *zoon*, animal; referring to the flexibility of the zooecia, which allows the branch to bend over (Figs. 7-14, 7-15).

TABLE 7-2. ESSENTIAL CHARACTERISTICS OF THE

Class	Order	Composition of zooecium	Shape of zooecium	Nature of aperture
Stenolaemata	Cyclostomata	Calcareous	Cylindrical	Terminal and round
	Trepostomata	Calcareous	Prismatic or cylindrical	Round or polygonal
Gymnolaemata	Ctenostomata	Horny or membranous, rarely calcified in part	Tubular or conical	Terminal and round
	Cheilostomata	Membranous, chitinous, calcareous	Conical, tubular, prismatic, urn-shaped, etc.	More or less anterior, commonly modified
	Cryptostomata	Calcareous	Cylindrical	Round and concealed

respect following Prouho (1892), Bassler (1922), Harmer (1930, 1931), and Storer (1943) for reasons given on page 210. However, many specialists, mostly zoologists, recognize enough differences between the Entoprocta and the true bryozoans (*i.e.*, Ectoprocta) to justify separating them from the Bryozoa as a separate phylum (Borg, 1926; Marcus, 1926, 1926*a*; Cori, 1929, 1941; Hyman, 1940; and Silén, 1944).[1] Arguments presented by this latter group are as follows:

1. Although the Entoprocta have a bilaterally symmetrical body with well-developed mesoderm of endodermal origin, and a crown of tentacles around the mouth, as in the ectoproctous Bryozoa, the space between the body wall and the U-shaped digestive tract is not a true coelom but a remnant of an early embryonic stage.
2. Although the digestive tract is U-shaped, as in all other Bryozoa, it is so looped that both mouth and anus lie *inside* the crown of tentacles (Fig. 7-13). In the ectoproctous Bryozoa the anus lies *outside* the crown (Fig. 7-13).
3. The construction of the body wall and of the digestive tract is simpler than in ectoproctous Bryozoa in that the lining typical of true coelomate animals is lacking.

[1] This course is followed in the 6th edition of Parker and Haswell's "A Textbook of Zoology" (London, Macmillan & Co., Ltd., 1940), in which Lowenstein recognizes two phyla—Bryozoa (= Ectoprocta) and Calyssozoa (= Entoprocta). Storer (1951) also recognizes the Entoprocta as a separate phylum in the second edition of his "General Zoology," although he considered the group as a class of Phylum Bryozoa in the first edition (1943).

FIVE ORDERS OF BRYOZOA WITH A FOSSIL RECORD

Operculum	Diaphragms	Brood chambers and appendicular organs	Mesopores	Nature of zoarium	Range
None	Rarely present	None	Present or absent	Usually delicate	U. Camb.–Recent
Present	Abundantly present	Probably present	Present	Variable, commonly massive	Ord.–Perm.
Present	None	None	None	Small, delicate, stolonoid	Ord.–Recent
Present	None	Present	None	Delicate, or massive, usually not large	Cret.–Recent
Present	Present or absent	None	Usually absent	Delicate, but may be several centimeters high	U. Camb.–Perm.

4. Finally, the entoproct cannot invaginate its tentacular crown, as it lacks the necessary muscles, and can only double up its ciliated tentacles when disturbed (Fig. 7-14*B*).

5. In contrast to the differences listed above, it is recognized that both entoproctous and ectoproctous bryozoans have a trochophore larva; hence, they, together with the Phoronida and Brachiopoda, are assumed to have evolved from some sort of simple acoelomate worm.

Morphology

The chief features of an entoproct can be seen in the well-known North American fresh-water genus *Urnatella* (Figs. 7-14, 7-15*B*), which, as generally observed, consists of two stems attached to a basal disk with each stem ending in a single zooid having ciliated tentacles. The colonies are minute, rarely exceeding 5 mm. in length, and live attached to stones, pieces of wood, bits of shell, and surfaces of living plants and animals.

The zooid, which is ovoid in form, bears a circlet of short, ciliated tentacles that can be extended or doubled up but not retracted as in the Ectoprocta. Inside the tentacular crown are the anal opening, at the very periphery of the circlet, and a large funnel-shaped depression, the **vestibule,** which has the mouth along the outer margin diametrically opposite the anus. The U-shaped digestive tract, consisting of mouth, esophagus, stomach, intestine, and anus, hangs free in the general body cavity without attachment to the body wall (Fig. 7-14). There is no true coelom; instead the cavity between the alimentary tract and body wall is filled with parenchymatous tissue.

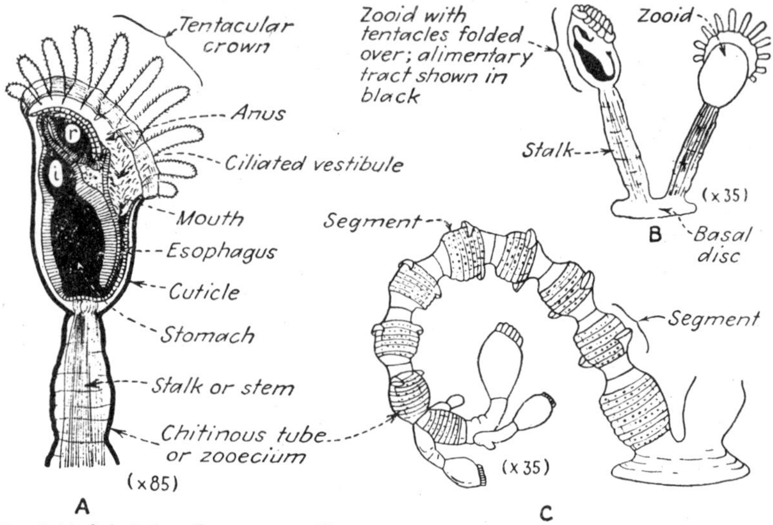

FIG. 7-14. Subphylum ENTOPROCTA. *Urnatella gracilis*, a North American fresh-water entoproctous bryozoan. *A*. Longitudinal section through a zooid with the tentacles extended. Food moves downward into the ciliated vestibule, thence to the mouth and into the alimentary tract, which is shown in black (*i* = intestine; *r* = rectum). Waste leaves the rectum through the anus, which lies just above the upper rim of the vestibule. There is no true coelom, the small body cavity being filled with parenchymatous cells. A layer of thick epithelial cells forms the wall of the alimentary tract. *B*. A young colony, showing the two branches that are typical, one with its zooid having the tentacles extended, the other with tentacles folded over and alimentary tract shown in black. *C*. One branch and several buds of a colony, much magnified to show some of the detailed features of the segmented tube or zooecium. (*Modified after Leidy*, 1884–1895.)

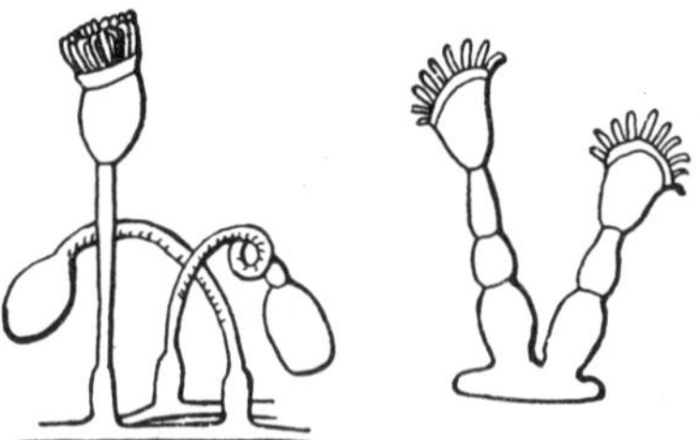

FIG. 7-15. Subphylum ENTOPROCTA. *A*. A colony of the marine *Pedicellina*, showing three individuals rising from stolons. *B*. A young fresh-water *Urnatella* showing the two branches characteristic of the genus, one with three segments, the other with two; both stems rise from a common basal expansion and each has a single zooid at its terminus. (*Modified after Leidy*, 1884–1895.)

The peduncle or stalk is a beaded chitinous tube consisting of as many as 18 segments. It commonly has several lateral branches, each of which terminates in a zooid. Because of its flexibility, the peduncle allows the zooid to bend over like a wilted flower (whence the term Kamptozoa; see footnote, page 211).

The colony begins with the appearance of the two chief branches that spring from the primary zooid and continues to increase by budding and accompanying development of lateral branches.

Pedicellina, a marine form included in the Entoprocta, closely resembles *Urnatella* in several respects as indicated in Fig. 7-15*A*. Of particular interest is the periodic loss and replacement of the complete zooid, a phenomenon similar to the regeneration in the Ectoprocta, though in *Pedicellina* it has been interpreted as a normal process of rejuvenation when the activities of the gonads have been exhausted (Harmer, 1930).

Geologic History

Nothing is known about the geologic history of the Entoprocta, as no fossil representatives have been reported.[1]

SUBPHYLUM ECTOPROCTA[2]

The chief characteristics of the Ectoprocta are listed in Table 7-1 where they are contrasted with those of the Entoprocta. Here it suffices to point out that in the former group the anus invariably lies outside the crown of tentacles (Fig. 7-2, 7-3, 7-13), the tentacles themselves are retractile, there is a true coelom, the polypide degenerates to brown bodies periodically with subsequent regeneration of a new polypide, and the zooid generally builds about itself some sort of tubular or boxlike zooecium that is chitinous or calcareous in most species.

The Ectoprocta have a long and extensive fossil record from the Upper Cambrian to the Recent. Three classes are recognized: (1) Phylactolaemata, a small fresh-water group without fossils; (2) Stenolaemata, with two large and important orders, one with living representatives (Cyclostomata) and the other extinct (Trepostomata); and (3) Gymnolaemata, with two living orders (Ctenostomata and Cheilostomata) and one important extinct order (Cryptostomata).

Class Phylactolaemata[3]

The Bryozoa of this class constitute a small group, with few genera and species, and live exclusively in fresh water.[4] They are readily recognized by the fact that the lophophore has the shape of a horseshoe or crescent, rather than of a circlet as in all other Bryozoa[5] (Fig. 7-13).

[1] Twitchell (1934) has suggested that the Trepostomata are ancient entoprocts and that *Urnatella* is a living survivor of that order. His suggestion, however, has not met with favor among serious students of the Bryozoa.

[2] Ectoprocta—Gr. *ektos*, outside or without, + *proktos*, anus; referring to the fact that the anus lies outside the tentacular ring (Fig. 7-13).

[3] Phylactolaemata—Gr. *phylassein*, to guard, + *laimos*, gullet; referring to the liplike epistome (Fig. 7-17) that covers the mouth.

[4] It is not to be assumed, however, that all fresh-water Bryozoa belong to the Phylactolaemata. Two or three genera of the Class Gymnolaemata may also be found in the same habitat (Harmer, 1930).

[5] A single exception is found in *Fredericella*, which has a circular lophophore and for this reason has been considered a connecting link between the two main groups of bryozoans. This suggestion seems to rest on insecure ground, however, because the genus is not primitive, and particularly because the lophophore of young buds seems to be constructed on the usual phylactolaematous plan (Harmer, 1930).

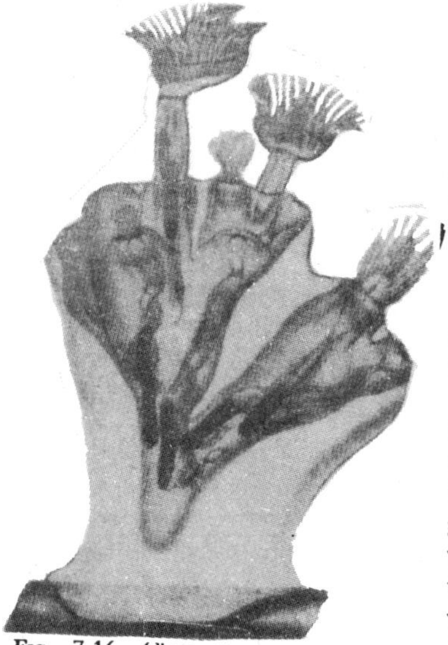

FIG. 7-16. Class PHYLACTOLAEMATA. Adult specimen of *Lophopus crystallinus* (×5), attached to a plant stem (*Lemna*). Two polypides are fully extended, three are partly retracted, and five are fully retracted. There are no hard parts in the zoarium. (*After Allman*, 1856.)

In addition to the usual method of reproduction by eggs, the Phylactolaemata also have the unique **statoblasts** (Fig. 7-18) which are regarded as internal buds. These have a strong chitinous investment, which makes them well adapted to withstand unfavorable conditions while awaiting the time for :rmination. One type of statoblast has wide equatorial ring of chitinous cells containing gas, so that it rises to the surface and floats on being liberated rom the zooid. A second type lacks the ring of gas-filled cells; hence they cannot float but instead remain attached to the substratum where the colony is growing. The floating statoblasts provide a means of distributing a species throughout a continuous body of fresh water, and sessile statoblasts, which have been found in mud adhering to certain aquatic birds, may well be carried in this way from one body of fresh water to another. It may be noted as a point of further interest that the few species of the Gymnolaemata adapted to life in fresh water have developed **hibernacula,** which, though not homologous with statoblasts, nevertheless play a similar role in the general economy of the species (Harmer, 1930).

In view of the fact that most fresh-water Bryozoa do not survive a winter, the statoblasts provide a means of ensuring the continued existence of a

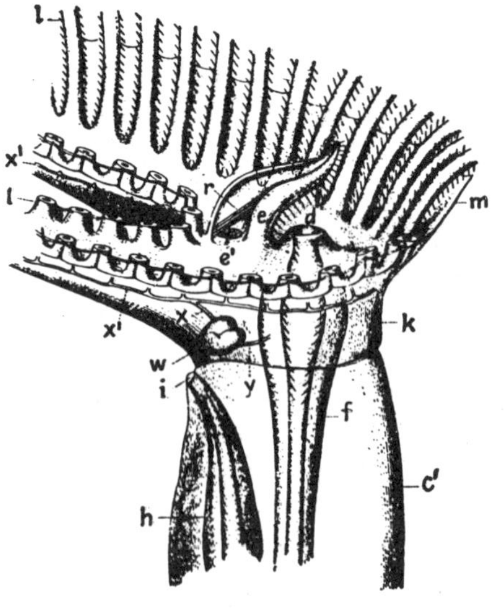

FIG. 7-17. Class PHYLACTOLAEMATA. Diagram of distal part of one zooid of *Lophopus* (×25) (cf. Fig. 7-16), showing nature of tentacular crown, position of mouth, epistome, anus, etc. (*d* = mouth; *e* = epistome; *f* = esophagus; *h* = intestine; *i* = anus; *k* = lophophore; *l* = tentacular row; *w* = nerve ganglion; *x*,*x'*,*y* = nerve filaments.) (*After Allman*, 1856.)

species, even under adverse conditions other than those associated with the winter season (*e.g.*, desiccation). The durable chitinous investment can persist for long periods under favorable conditions, and the question naturally arises, have statoblasts been preserved in ancient fresh-water deposits? Possibly they have, but no fossil statoblasts are known to have been reported.[1] They might well show up in the insoluble residues of Pleistocene and more ancient marls and marlstones.

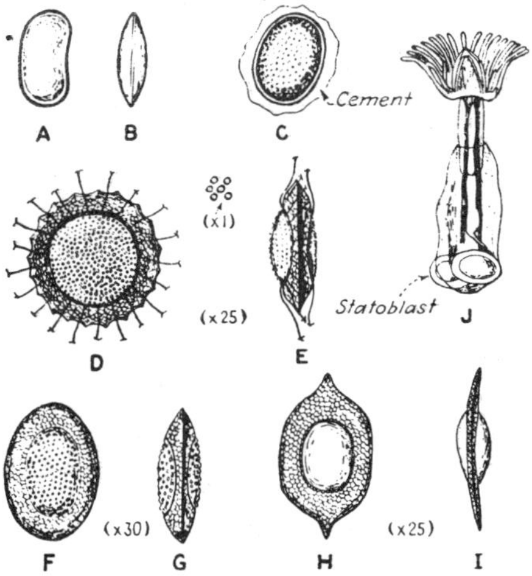

FIG. 7-18. Class PHYLACTOLAEMATA. Statoblasts of fresh-water Bryozoa. *A–B.* Two views of a statoblast of *Fredericella sultana*, greatly magnified; no annulus is developed. *C.* Adherent statoblast of *Alcyonella benedeni*, greatly magnified; it is surrounded with some of the gelatinous cement by which it adheres to the cell wall. *D–E.* Spinose statoblast of *Cristatella mucedo;* a few mature statoblasts are shown natural size between the figures. *F–G.* Statoblasts of *Alcyonella fungosa. H–I.* Statoblasts of *Lophopus crystallinus. J.* Statoblast of *Plumatella repens*, with the young zooid escaping from it. The two parts have separated but are seen still adhering to the posterior part of the zooid which floats through the surrounding water. The polypide is shown as extended. (*Modified after Allman,* 1856.)

The Phylactolaemata have a flexible body wall that is uncalcified. As there are no durable hard parts, preservation is not likely, and no certain fossil record is known. A supposed fossil form, *Plumatellites*, found incrusting a Cretaceous *Unio* may constitute an ancient record of this class, but the poorly preserved nature of the incrustation makes its identification uncertain.

Without a fossil record, there is little evidence to indicate the origin and geologic history of the phylactolaematous Bryozoa. The embryology and ontogeny of living species do not furnish a definite answer as to how the Phylactolaemata are genetically related to other Bryozoa (Fig. 7-45).

Living genera representative of the Class Phylactolaemata are *Plumatella* (Fig. 7-18*J*) and *Lophopus* (Figs. 7-16, 7-17, 7-18*H*,*I*).

[1] Prof. L. R. Wilson of the University of Massachusetts advises us that he has found many types of microscopic chitinoid "cysts" in rocks ranging in age from Ordovician to Pleistocene, and he believes that some of those found in fresh-water, brackish-water, and marine deposits may well be bryozoan statoblasts.

Class Stenolaemata[1]

All living representatives of this class belong to the Order Cyclostomata; essentially nothing is known about the animal in the extinct Order Trepostomata. The class is characterized by narrow, cylindrical, proximally tapering zooids having a calcareous zooecium with a terminal opening. The zooecia are clustered or bundled into loosely knit or compact zoaria (Figs. 7-5, 7-23, 7-25, 7-27). Details of the zooidal morphology are given in the following diagnosis of the Order Cyclostomata.

Order 1. Cyclostomata.[2] *Morphology and Astogeny.* The zoaria of this order are, in general, branching or lobose, more or less flattened, and commonly creeping, either attached to the substratum throughout or erect in the distal parts. The zooecial apertures are confined to one side of the zoarium in some families, arranged in bilaminar disks in another, and scattered around an axis in still another group. It is common to speak of the underside, or non-apertural side, of the zoarium as the **basal side** or **basal lamina** and of the side bearing the zooecial apertures as the **frontal side.** In one family, the Crisiidae, the zoaria are jointed with the sections (**internodes**) between joints held together by flexible chitinous tubes (Figs. 7-1*C*,*J*, 7-21*C*). When the colony dies the internodes are likely to become separated (Fig. 7-1*J*).

The zoarium, which generally consists of many zooecia, develops from a primary disk, or protoecium, made by the larva after it has become attached to the bottom. After the protoecium has been formed the larva metamorphoses into the primary polypide, and zooid, of the future colony. This primary zooid secretes a tubular ancestroecium and then gives off one or more buds which mature into the first generation of zooids and build the first zooecia of the zoarium. All subsequent budding is from the first or from a later generation, because the primary zooid gives rise to only one generation of buds (Figs. 7-9*A–E*, 7-19). The ultimate architecture of the zoarium is determined by the plan of budding.

Zooecia in the Cyclostomata are simple calcareous tubes with thin, minutely porous walls and terminal apertures. The apertures are generally round and somewhat expanded and are not equipped with opercula. Internally the zooecia usually lack the several kinds of transverse partitions so common in trepostomatous and cryptostomatous bryozoans, hence it is generally possible to identify species without preparing thin sections.

Zooids. The cyclostomatous zooids are polymorphic, taking at least three different forms. The first zooid of a colony differs from all others in that it develops directly from the metamorphosis of the larva, is attached to the primary disk, and secretes the ancestroecium. Most of the individuals of a typical zoarium are autozooids and these make the zooecia. Detailed structures of a typical autozooid are shown in Fig. 7-20. There are also other zooids (hetero-

[1] Stenolaemata—Gr. *stenos*, narrow or little, + *laimos*, gullet. Stenolaemata should mean narrow gullet; in proposing the name, however, Borg (1926) emphasized the narrowness of the zooids.

[2] Cyclostomata—Gr. *kyklos* (= *cyclos*), circle, + *stoma*, *stomatos*, mouth; referring to the round terminal aperture of the zooecia.

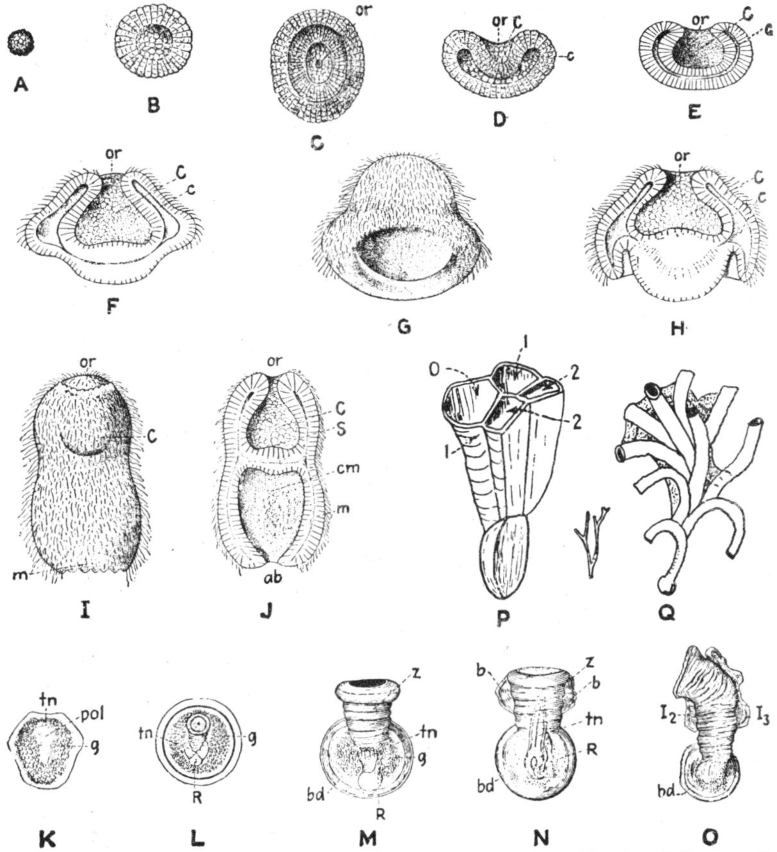

FIG. 7-19. Order Cyclostomata. Development of *Phalangella* (= *Tubulipora*) from early embryo to young colony. *A–F*. Early embryonic stages: *B–F* show invagination of the adhesive organ. *F–J*. Further stages of the free-swimming larva, with *I* and *J* the completed larva. *H* is a longitudinal section of *G*, and *J* of *I*. *K–O*. Formation of polypide and protoecium: *K*, an early sedentary stage, showing rudimentary polypide; *L*, beginning of tentacles and of protoecium; *M–O*, formation of first mature individual (autozooid) and primary buds, I_2 and I_3. *P*. A young zoarium with first and second generations of zooecia. *Q*. A young zoarium somewhat more advanced than *P*, showing prominent bulbous protoecium and several generations of zooecia. (*A–O after Cumings*, 1904, *from Barrois*, 1877; *P after Barrois*, 1877; *Q after Hincks*, 1880.)

zooids) which differ from the normal individuals in being incomplete and having specialized functions. Of these, the most important are gonozooids, which are modified for the production of embryos and which make conspicuously inflated zooecia, the gonooecia (also commonly called ovicells) that serve as brood chambers (Fig. 7-5). The brood chambers have proved important for classification purposes, and the nature of the distal opening, the

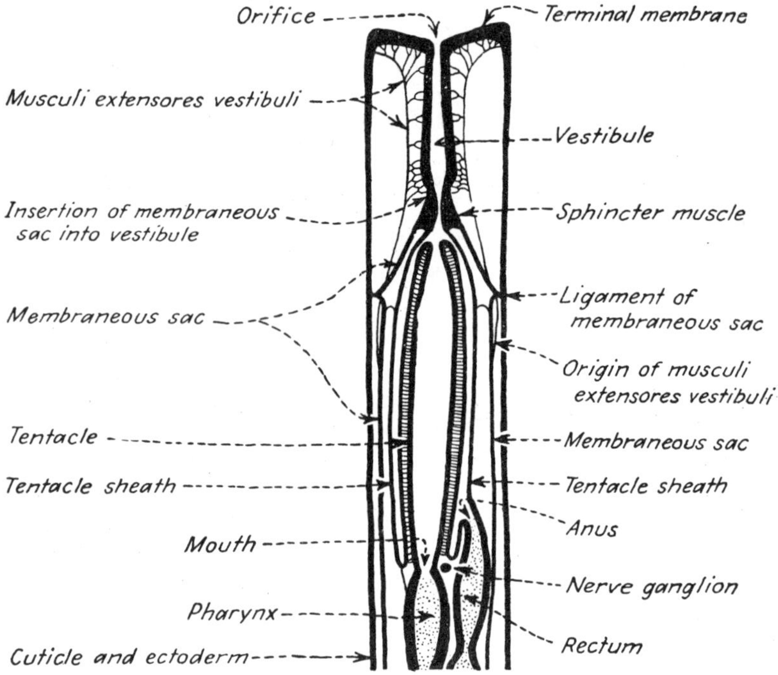

FIG. 7-20. Phylum BRYOZOA—Order Cyclostomata. Diagram of a longitudinal section through the distal half of an autozooid in which the crown of tentacles (crosshatched) is retracted and is surrounded by the tentacle sheath. Only the upper part of the U-shaped digestive tract (stippled) is shown. There is no inorganic zooecium, the ectoderm being surrounded by an external chitinous cuticle. (*Modified after Borg*, 1926.)

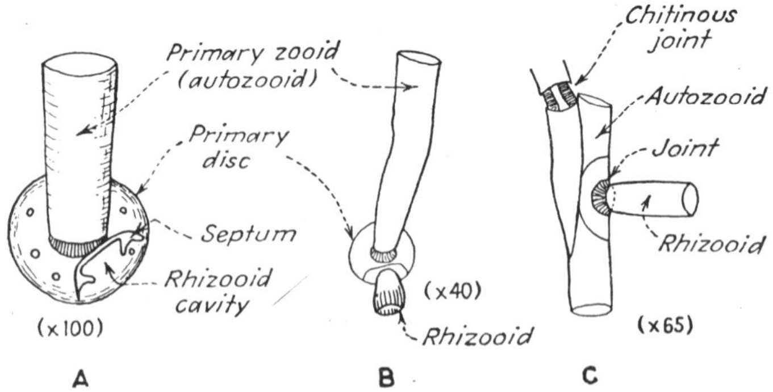

FIG. 7-21. Order Cyclostomata—Family Crisiidae. *A. Crisiella producta;* primary disk (protoecium) and early part of ancestroecium, which is occupied by the primary zooid. The septum surrounds a small cavity from which a rhizooid would have grown later. *B. Crisia eburnea;* primary disk, from which a rhizooid has grown, and the proximal part of the ancestroecium. *C. Crisia aculeata;* an autozooid from which a rhizooid has grown. (*Modified after Borg*, 1926.)

ooeçiostome, is specially important. It has been emphasized repeatedly, however, that the Cyclostomata cannot be subdivided satisfactorily by using any single feature such as these chambers.

In addition to the gonozooids, at least two other kinds of heterozooids are developed in some Cyclostomata. These are **nanozooids** and **kenozooids.** Nanozooids are smaller than normal individuals and, though clearly homologous with autozooids, are incomplete in their polypide structure. They are commonly as numerous as the autozooids, and the small tubular **nanozooecia** that they make end in tiny openings between the zooecial apertures (Fig. 7-5*A*). The function of these miniature autozooids is uncertain.

Kenozooids are long, narrow, cylindrical structures of rhizooid nature that rise from the primary disk, the primary zooid, and the nearest following internodes in Family Crisiidae (Fig. 7-21). They have a strong zooecial wall and a fleshy body wall similar to those of the autozooids, but lack a polypide. Instead of a polypide they have a single body cavity with a network of mesenchymatous cells. Some fossil species of the bryozoan Family Tubiporidae seem to have been provided with kenozooids (Borg, 1926).

Geologic History. One of the oldest known bryozoans is a simple form, *Archaeotrypa* (Fig. 7-22*A–B*), reported from the Upper Cambrian of Alberta (Fritz, 1947). It is assigned to the Cyclostomata because of certain resemblances to *Crepipora* (Fig. 7-22*C–E*).

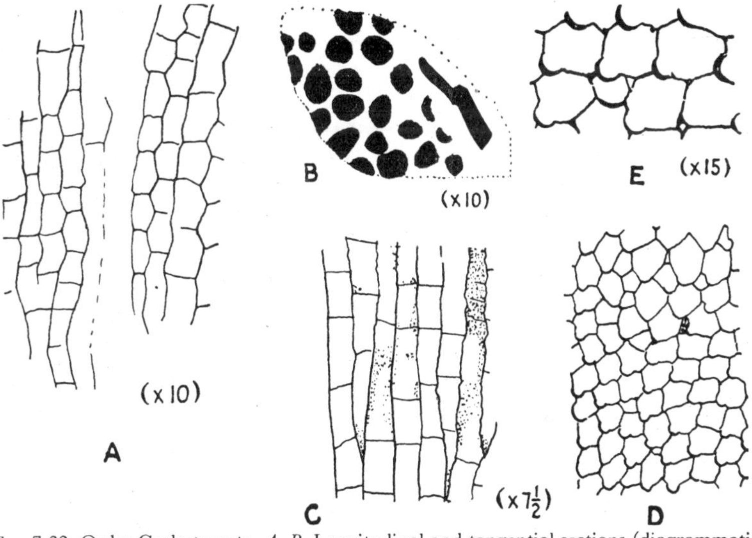

FIG. 7-22. Order Cyclostomata. *A–B*. Longitudinal and tangential sections (diagrammatic from thin sections of two different species) of *Archaeotrypa*, a cyclostome-like bryozoan from the Upper Cambrian of western Canada. *C–E*. Longitudinal and transverse sections of *Crepipora perampla*, an Ordovician cyclostomatous genus from Minnesota. The zooecial walls commonly show crenulation, and lunaria are characteristic. *Archaeotrypa* lacks crenulated zooecial walls but has suggestions of lunaria and has long thin-walled zooecia with numerous diaphragms as in *Crepipora*. The two genera seem to be related, and *Archaetorypa* seems best assigned to the Cyclostomata, at least until it can be studied further. (*A–B adapted from Fritz*, 1947; *C–E after Ulrich*, 1895.)

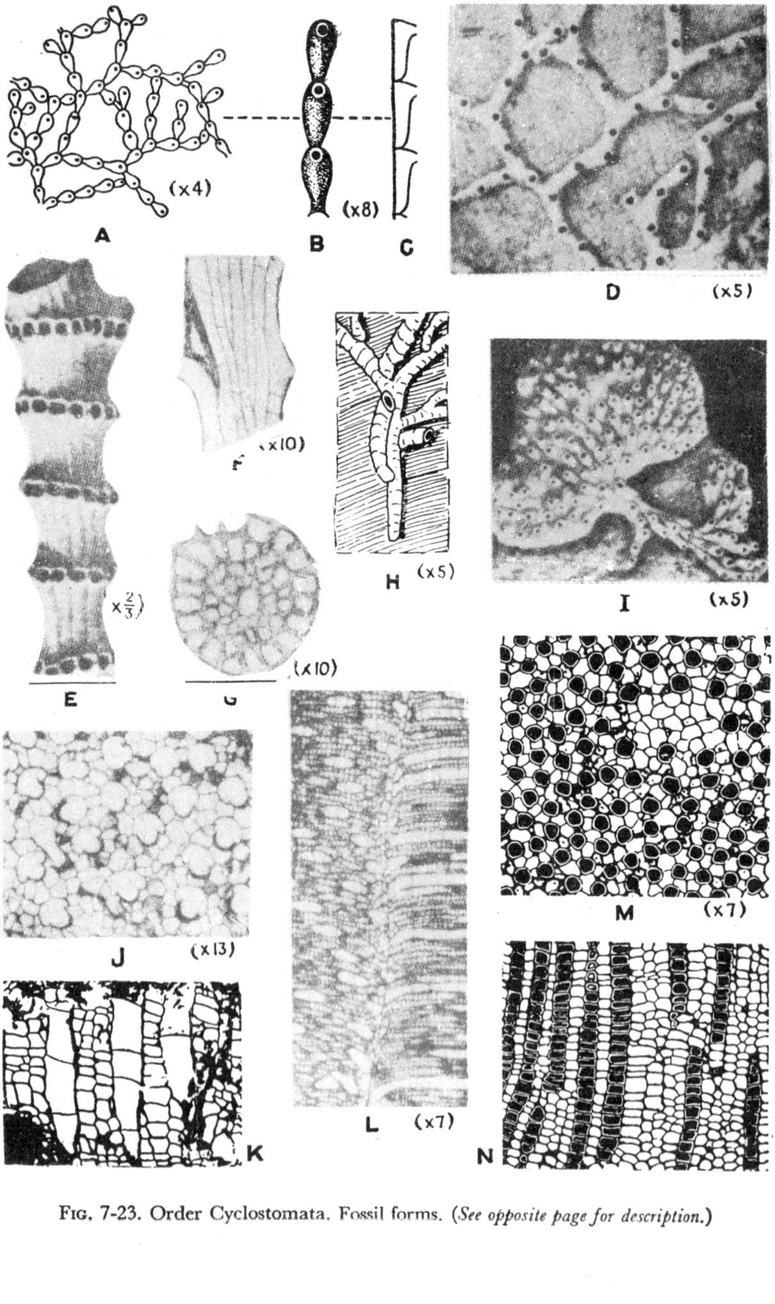

FIG. 7-23. Order Cyclostomata. Fossil forms. (*See opposite page for description.*)

Some authors (*e.g.*, Swinnerton, 1946) believe the Cyclostomata to be primitive and possibly to include the ancestral stocks from which some or all of the other orders were derived. Other investigators (*e.g.*, Borg, 1926), however, consider the order to be monophyletic and to have evolved, as the other orders, from a common ancestral or probryozoan stock (Fig. 7-45).

Whatever the evolutionary history, the Cyclostomata, though second in importance to the Trepostomata during the Ordovician, and to both this order and the Cryptostomata for most of the Paleozoic, were even at that early date a much evolved group, showing great range in both zooecial form and zoarial architecture. A gradual decline began in the Silurian, and the order held a subordinate position until the Jurassic, though development of the fistuliporoid types (Fig. 7-23*J–N*) produced an important group of species during the Late Paleozoic. In the Jurassic the order underwent an expansion which continued into the Cretaceous with the production of many species and individuals. This was short-lived, however, for decline again set in before the close of the Cretaceous, and the order almost immediately became subordinate to the rapidly evolving Cheilostomata, a position it still holds. There are only about a hundred known species now living, but many of these have been intensely studied, and the order as a whole has been widely investigated (see references to the work of Bassler, Borg, Busk, Canu and Bassler, Gregory, Harmer, Hincks, Marcus, Silén, and Waters).

Bassler (1935) recognizes at least 28 families of Cyclostomata which are included in the following four suborders, of which the geological range and typical representatives are indicated.

Tubuliporoidea (Ord.–Recent): *Corynotrypa* (Fig. 7-23*A–C*), *Crisia* (Figs. 7-1*J*, 7-9*A–D*), *Diaperoecia* (Figs. 7-1*K*, 7-23*I*), *Stomatopora* (Figs. 7-9*E*, 7-23*D*). Cerioporoidea (Trias.–Recent): *Lichenopora* (Figs. 7-1*D*; 7-12*L*). Ceramoporoidea (Ord.–Perm.): *Archaeotrypa* (Fig. 7-22*A–B*), *Crepipora* (Fig. 7-22*C–E*), *Cyclotrypa* (Fig. 7-23*M–N*). Hederelloidea (Sil.–Penn.): *Hederella* (Fig. 7-23*H*).

Order 2. Trepostomata.[1] The Trepostomata are an extinct order of exclu-

[1] Trepostomata—Gr. *trepos*, change, + *stoma*, *stomatos*, mouth; referring to the change in the nature of the zooecia from thin wall and simple structure in the early or distal part to a thicker wall and a more complicated structure in the later part.

Inclusion of the Order Trepostomata in the Class Stenolaemata is provisional and is followed because representatives of the order seem closer to the Cyclostomata than to any other known Bryozoa, either living or extinct. The assignment is provisional because there is no known evidence that the Trepostomata had a membranous sac or developed by embryonic fission.

FIG. 7-23. Order Cyclostomata. Fossil forms. *A–C. Corynotrypa inflata* from the Ohio Ordovician: *A*, small portion of a large zoarium incrusting the valve of *Rafinesquina; B*, three zooecia enlarged; *C*, same sectioned longitudinally. *D. Stomatopora contracta* from the Eocene (Midway) of Alabama. *E–G. Spiropora majuscula* from the Eocene (Jackson) of South Carolina: *E*, a fragmental zoarium; *F*, longitudinal section; *G*, transverse section. *H. Hederella thedfordensis*, from the Middle Devonian of Ontario, showing ancestroecial part of a zoarium incrusting a *Tropidoleptus*. *I. Diaperoecia lobulata* from the Eocene (Jackson) of Georgia. *J–K*. Tangential and longitudinal sections of *Triphyllotrypa patentis* from the Permian (Leonard) of Texas. *L*. Longitudinal section of *Meekopora opima* with long erect zooecial tubes, many mesopores, and common diaphragms, from Kansas Pennsylvanian. *M–N*. Tangential and longitudinal sections of *Cyclotrypa abdita*, from the Kansas Permian (Virgil). (*A–C after Ulrich*, 1890; *D-G, I after Canu and Bassler*, 1920; *H after Bassler*, 1939: *J–N after Moore and Dudley*, 1944.)

sively Paleozoic bryozoans that flourished abundantly during the Ordovician, Silurian, and Devonian and then died out by the close of the era without leaving any known descendants (Fig. 7-45). They built calcareous zoaria of variable shapes and sizes (Figs. 7-12, 7-25), some as much as 60 cm. (2 ft.) across, and these now constitute large portions of many calcareous strata. The zoaria consist of long, closely packed, prismatic or cylindrical zooecia with a simple terminal aperture, which in some species was covered with a perforated operculum. Typical zooecia are a fraction of a millimeter across but several millimeters long. Internally the zooecia are divided into compartments by cystiphragms and diaphragms (Figs. 7-25*K*, 7-27).

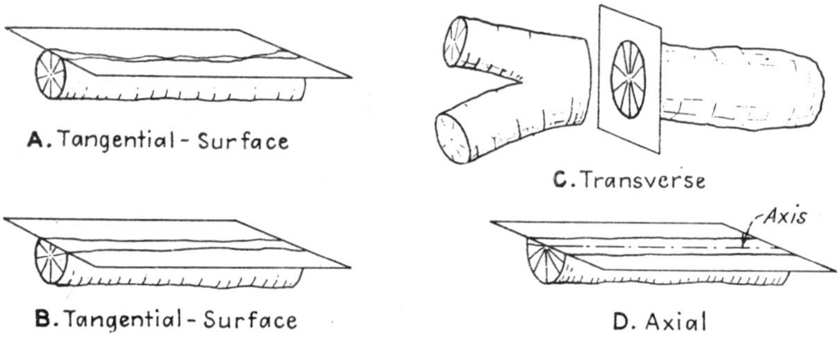

FIG. 7-24. Phylum BRYOZOA. Diagrams illustrating the kinds of sections used in studying bryozoans. *A*. Tangential and quite near the surface, so that the surface or near-surface aspects of the zooecia are shown. *B*. Tangential but somewhat deeper than *A*, so that subsurface features of the zooecia are revealed. *C*. Transverse section directly across axis of zoarial stem, and showing zooecia radiating from axis. *D*. Axial section essentially splitting the zoarial stem lengthwise, and showing mature and immature regions.

Trepostomatous zoaria even in the same species show considerable variation in shape and size, but the zooecia and associated structures are constant. Differentiation of species and genera, therefore, depends on the minute structural details of the zooecial and interzooecial parts of a zoarium, and in order to study these successfully, thin sections must be prepared[1] and examined microscopically.

Zoaria. Trepostomatous zoaria are dendroid, tabular, frondlike or leaflike, globular, hemispherical, conoidal, lenticular, or irregular bifoliate expansions (Fig. 7-12). Some habitually formed incrustations over all sorts of objects on the bottom. The more massive types of zoaria were possible because successive layers of zooecia could be superimposed on earlier layers. Bushlike zoaria were especially efficient in causing deposition of sediments around themselves,

least two, and in many cases preferably three, sections of the zoarium should be made. One should be tangential to the zoarium and transverse to the zooecia in the peripheral zone; one should be an axial section, *i.e.*, it should include the axis of the stem or branch and cut the zooecia longitudinally; and a third should directly transect the stem (Fig. 7-24). Detailed instructions for making thin sections, as well as for preparing loose specimens, are given by Bassler (1922) in his very useful pamphlet, "The Bryozoa, or Moss Animals."

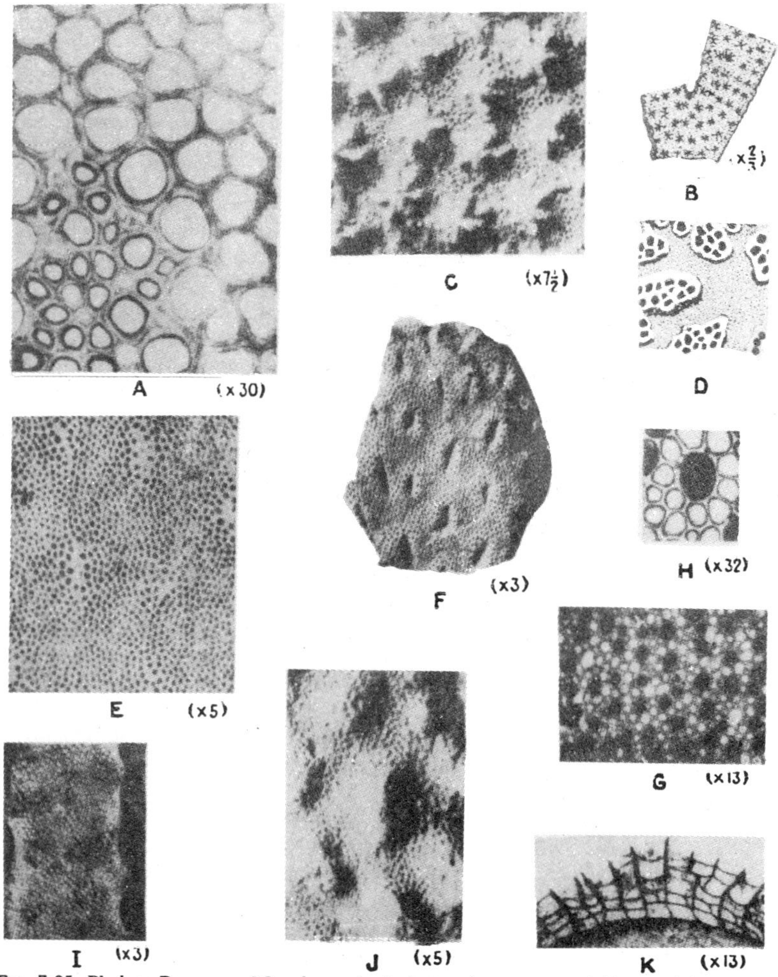

FIG. 7-25. Phylum BRYOZOA. Maculae, monticules, and mesopores. *A*. Tangential section of *Amplexopora*, showing thick zooecial walls of the integrata type and a macula of mesopores in the lower left quarter. *B–D*. *Constellaria florida* from the Upper Ordovician, characterized by strongly developed stellate maculae; *B*, fragmental zoarium; *C*, portion of zoarial surface much enlarged; *D*, tangential section showing arrangement of zooecia in a macula. *E*. Portion of an unweathered surface of *Homotrypa alta*, showing several prominent maculae of larger zooecia. *F*. Fragmental zoarium of *Ceramella casei*, showing prominent elongate maculae. *G–H*. *Lioclema alpenense* from the Michigan Devonian: *G*, tangential section showing relative size of zooecia and mesopores; *H*, a zooecium and surrounding mesopores, showing amalgamate walls and several small acanthopores. *I–J*. Conoidal monticules on a Devonian species of *Anomalotoechus* and an Ordovician species of *Atactopora*, respectively. *K*. Longitudinal section of part of an incrusting zoarium of *Anomalotoechus* (same as *I*) showing distribution of cystiphragms and diaphragms. (*A*, *E*, *J* *after Cumings and Galloway*, 1913; *B*, *D* *after Ulrich*, 1895; *C* *after Bassler*, 1932; *F* *after McNair*, 1937; *G–I*, *K* *after Duncan*, 1939.)

so that they were directly concerned with the accumulation of the sediments in which they were ultimately buried.

The surfaces of many zoaria have the regular pattern of zooecial openings interrupted by elevated and depressed areas containing zooecia larger or smaller than normal. **Maculae** are spotlike areas of abnormally large or small cells that rise above or are level with the zoarial surface, or are depressed somewhat below it (Fig. 7-25). In *Constellaria* (Fig. 7-25*B–D*), for example, the maculae are conspicuously star-shaped and are responsible for the generic name. **Monticules** are celluliferous elevations above the zoarial surface that have the form of small, sharp tubercles, rounded nodes, or elevated rings that completely girdle the zoarium (Fig. 7-25*I–J*). The size, shape, distribution, and other characteristics of maculae and monticules are useful in differentiating species when used together with zooecial structures.

Zooecial Structure. The zooecium of the Trepostomata is typically a long, straight or curved, cylindrical or prismatic tube with a simple terminal aperture (Fig. 7-27). It is generally composed of an earlier (**immature**), or axial, part characterized by thin walls and distantly spaced transverse partitions and a later (**mature**), or outer, part ending in the aperture and having thickened walls and crowded cystiphragms and diaphragms. The immature part typically constitutes from two-thirds to three-fourths of the length of the zooecium, and the latter is prismatic with a polygonal cross section. In the mature region the zooecia are prismatic or tubular, with polygonal or circular cross section, and in many species are separated by small, closely tabulated tubes, or **mesopores** (Fig. 7-25*G–H*). The function of the mesopores is uncertain. Certainly they fill up the space between zooecia, thus giving rigidity to the zoarium, and they may have housed specialized zooids.

It is assumed that each zooecium was extended and the structural features built as successive polypides (Fig. 7-11) degenerated and regenerated within a single persistent zooid.

The walls of the later part of many zooecia have tiny spinelike projections, the **acanthopores**, which can be seen in thin section to be minute tubes included in the wall substance but with a definite structure of their own (Fig. 7-26). Their function is unknown. They traverse the mature region only, and it has been suggested that they may have functioned like the avicularial or vibracular pores of the Order Cheilostomata.

Many Trepostomata, including most of the so-called monticuliporoids, once regarded as corals, were proved to be bryozoans by Cumings (1912), who demonstrated that the colonial development (astogeny) and budding plan of certain typical genera were exactly the same as in living Cyclostomata (see following discussion under Astogeny). After their recognition as bryozoans, but before thin sections were used in identification, most species and genera of Trepostomata were based upon the form of the zoarium, the shape of the zooecia, and surface features such as maculae and monticules. Now, however, all modern students of fossil Bryozoa use thin sections and base identifications on internal and external microscopic features. The fact that accurate identifications can be made from tiny fragments gives to fossil Bryozoa, particularly the Cyclostomata, Trepostomata, and Cryptostomata, unique value for micropaleontological investigations of subsurface geology.

Astogeny. The trepostomatous zoarium began as a minute circular disk, the protoecium, which was constructed by the larva after it settled to the bottom. Soon thereafter the first zooid of the future colony, supposedly arising from the metamorphosis of the larva as in living bryozoans, built the primary zooecium of the future zoarium. This first zooecium has been designated the **ancestrula,** or better, the **ancestroecium.** Next, the primary zooid produced several primary buds, and these built zooecia, the first generation of the zoarium, that spring from the base of the tubular ancestroecium (Fig. 7-9*F*)

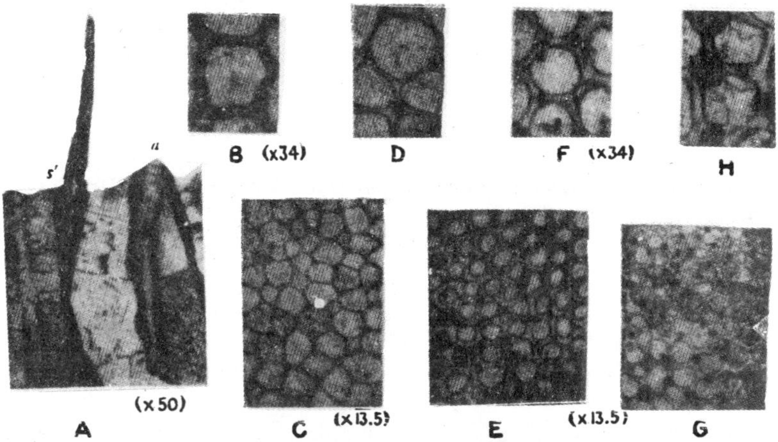

FIG. 7-26. Order Trepostomata. Acanthopores. *A.* Longitudinal section of the peripheral region of a zoarium of *Dekayia maculata*, showing how the acanthopores are related to the zooecial wall. Acanthopores are hollow spines with laminated walls; an exceptionally well developed spine (*s*) is shown rising from the zoarial surface at *s'*. The usual appearance of acanthopores in longitudinal section is shown at *a*. *B.* A zooecium of *Eostenopora picta*, showing small granular acanthopores surrounding zooecia and larger acanthopores at junctions of zooecial walls. *C–D. Anomalotoechus typicus: C*, tangential section showing thin walls, numerous acanthopores at junctions of zooecial walls, and range of size among zooecia; *D*, a few zooecia more enlarged, showing solid acanthopores. *E–F. Stereotoechus typicus.* Tangential section showing different-sized subcircular zooecia and numerous acanthopores, and several more enlarged zooecia showing nature and position of acanthopores. *G–H. Dyoidophragma typicale.* Tangential section showing range of size among zooecia and extremely large acanthopores in monticule (area of larger zooecia in lower left-hand corner), and several zooecia much enlarged to show extremely enlarged monticular acanthopores. (*A after Cumings and Galloway*, 1915; *C–H after Duncan*, 1939.)

and remain adjacent to it. These early structures—the protoecium, ancestroecium, and primary zooecia arising from the latter—are separated from the rest of the zoarium by considerable thickening of their posterior walls. After the first generation there was no further budding from the ancestral zooid in the ancestroecium.

The second generation of buds formed on the zooids occupying the primary zooecia; hence the zooecia that were constructed by the individuals maturing from these buds sprang from the first generation of zooecia. Subsequently the zoarium increased by successive generations of buds that arose from the zooids of the immediately preceding generation, as illustrated in Fig. 7-9*F*.

This astogeny, beautifully demonstrated by Cumings (1912), leaves no doubt about the bryozoan affinities of the Trepostomata, and strongly suggests that this extinct order has its closest relatives among living Cyclostomata.

Geologic History. The oldest known trepostome came from Ordovician rocks. The order appeared in great abundance in this period and continued to be important during the Silurian and Devonian, but thereafter rapidly declined and was extinct by the end of the Paleozoic (Fig. 7-45). Certain Mesozoic bryozoans have been referred to the Trepostomata, but since these have minutely porous walls not found in Paleozoic Trepostomata but characteristic of Cyclostomata, they are better assigned to the latter order. So far as known the Trepostomata did not give rise to any post-Paleozoic descendants.

Classification. The Trepostomata may be divided into the following families, typical genera of which are indicated.

Atactotoechidae (Dev.): *Anomalotoechus* (Figs. 7-25*I,K*, 7-26*C–D*). Prasoporidae (Ord.–Dev.): *Homotrypa* (Fig. 7-25*E*), *Monticuliporella* (Fig. 7-27*B–C*), *Peronopora* (Fig. 7-11*B*), *Prasopora* (Figs. 7-9*F*, 7-11*A*, 7-12*P*). Heterotrypidae (Ord.–Dev.): *Atactopora* (Fig. 7-25*J*), *Dekayia* (Fig. 7-26*A*), *Heterotrypa* (Fig. 7-11*C*, 7-27*A*). Constellaridae (Ord.–Sil.): *Constellaria* (Fig. 7-25*B–D*). Batostomellidae (Ord.–Perm.): *Lioclema* (Fig. 7-25*G–H*). Amplexoporidae (Ord.–Dev.): *Amplexopora* (Fig. 7-25*A*). Halloporidae (Ord.–Dev.): *Hallopora* (Fig. 7-12*Q*). Trematoporidae (Ord.–Dev.): *Hemiphragma* (Figs. 7-12*N*, 7-27*D*), *Monotrypa* (Fig. 7-27*E–F*).

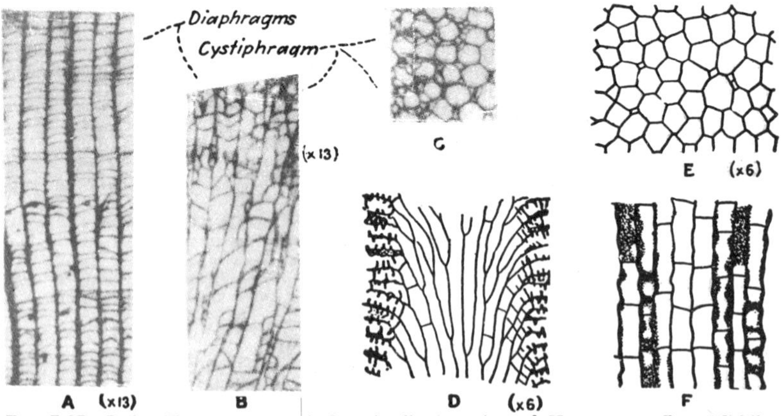

FIG. 7-27. Order Trepostomata. *A*. Longitudinal section of *Heterotrypa taffi*, a Middle Ordovician form, showing excellently developed diaphragms. *B–C*. Longitudinal and tangential sections of *Monticuliporella peculiaris*, showing well-developed diaphragms and cystiphragms. *D*. Longitudinal section of *Hemiphragma irrasum* from the Middle Ordovician, showing complete mature and immature regions and characteristic incomplete diaphragms in the former. *E–F*. Tangential and longitudinal sections of *Monotrypa magna* from the Illinois Middle Ordovician. (*A–C after Loeblich*, 1942; *D–F after Ulrich*, 1890.)

Class Gymnolaemata[1]

The Class Gymnolaemata, which is almost exclusively marine and includes most of the known living Bryozoa, comprises a large and varied group of advanced bryozoans characterized by a circular row of tentacles surrounding the mouth (Fig. 7-13) and by a wide range of structure, architecture, and other characteristics in the zooecia and zoaria. The fundamental structure of an advanced gymnolaematous bryozoan is shown in Fig. 7-28. The Gymnolaemata resemble the Stenolaemata in having a circular ectoproctous

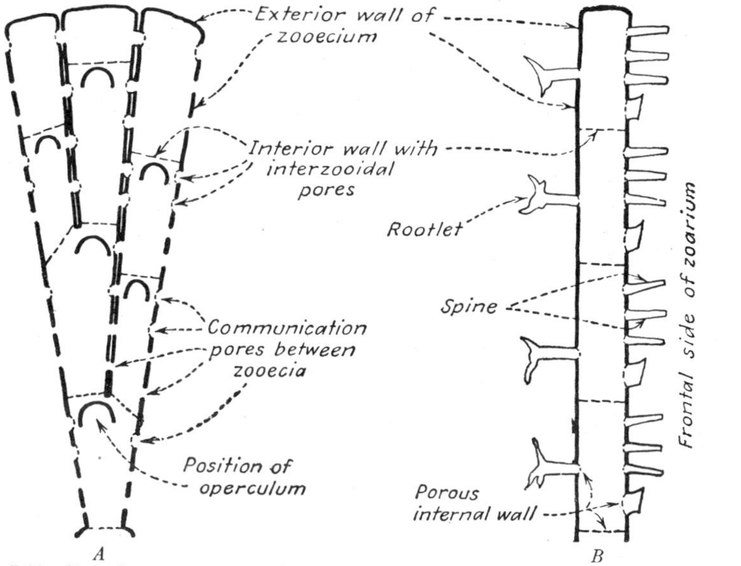

Fig. 7-28. Class Gymnolaemata. Diagrams showing the fundamental structure of the gymnolaematous zoarium. *A*. Horizontal section of part of a zoarium with the positions of opercula indicated. *B*. Longitudinal and vertical section of part of a zoarium, showing individual zooecia, heterozooids of several types (rootlets, avicularia, and spines), and porous internal walls. The frontal side is to the right. (*Modified after Silén*, 1944.)

lophophore, but they differ markedly from that class in having much more complicated zooecia, a different mode of colonial development, and strikingly different zoaria.

As used here the class includes three orders, two extant and one extinct, as summarized below:

Order 1. Ctenostomata. Relatively simple bryozoans with incrusting zoaria composed of tiny stolons from which rise simple tubular zooecia of gelatinous or chitinous composition. The zooecial aperture is closed by folds of the body wall, and the latter is not calcified. *Ordovician to Recent.*

Order 2. Cheilostomata. Advanced bryozoans with chitinous and calcareous zoaria com-

[1] Gr. *gymnos*, naked, + *laimos*, gullet; referring to the fact that the mouth is uncovered, because of the lack of an epistome such as is present in the Phylactolaemata.

posed of boxlike zooecia equipped with avicularia, vibracula, blisterlike brood chambers, and spines. The zooid is protruded through an anterior aperture which has an operculum. *Ordovician? Jurassic* to *Recent.*

Order 3. Cryptostomata. This extinct order comprises massive and fenestrate calcareous zoaria in which the zooecia are short and have the true aperture at the base of a vestibulum opening to the surface. *Upper Cambrian* to *Permian.*

Some earlier authors considered the Ctenostomata to be the ancestors of the Cheilostomata, basing their opinion on the assumption that an operculated aperture of the cheilostomatous type must have developed from a simple aperture of the ctenostomatous type. Some later workers have suggested that the Ctenostomata evolved from primitive cheilostomatous types. Other recent investigators emphasize the close similarity of many zooidal features in the two orders, and favor recognizing them as a single group, differing in many important respects from stenolaematous Cyclostomata.

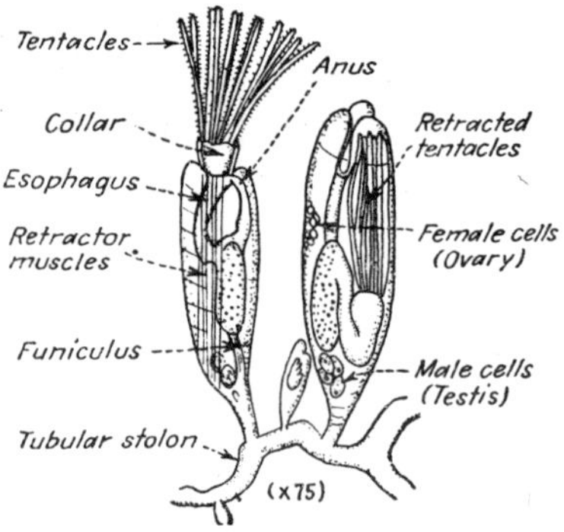

FIG. 7-29. Order Ctenostomata. Three zooids of *Farrella rapens*, rising from a stolon: one has its tentacles expanded, one has them retracted, and a third is in the young stage. (*Modified after Van Beneden*, 1845.)

We favor continued recognition of Ctenostomata and Cheilostomata as distinct orders, especially because of the great difference in the nature of the zooecia and zoaria, and prefer considering them as having evolved from a common probryozoan ancestral stock. We include with these two living gymnolaematous orders, the extinct Cryptostomata which, because of their many resemblances to the cheilostomes, have been called Paleozoic Cheilostomata.[1] Sufficient evidence does not seem to be available to state definitely how the three gymnolaematous orders differentiated from the postulated probryozoan ancestral stock. In Fig. 7-45 the several orders and classes of Bryozoa are all shown as arising independently from a common stock.

Order 1. Ctenostomata.[2] The zoaria of this order are tiny delicate traceries of threadlike **stolons** and tufts of hairlike **stems** (Figs. 7-29, 7-30). Most incrust or penetrate shells, stones, and other bottom objects, but a few are erect. Because the stolons and stems, as well as the zooecia rising from them, are seldom calcified, few ctenostomes have left a fossil record, and when they have, the record is generally in the form of a reticulate network of linear and ovoidal excavations in or on the surface of the substratum (Fig. 7-30).

[1] Ulrich (1900) advanced the view that the Cryptostomata are the Paleozoic representatives of the Cheilostomata, and Cumings (1904) considered that his studies strengthened this view, because the early budding stages in the cryptostomatous genera *Polypora* and *Fenestrellina* are precisely like those in the cheilostomatous genus *Retepora* (Fig. 7-9*G–H*).

[2] Gr. *kteis*, *ktenos*, comb, + *stoma*, *stomatos*, mouth; referring to the fact that when the polypide is retracted the zooecial aperture is closed by an operculum of setae resembling a comb.

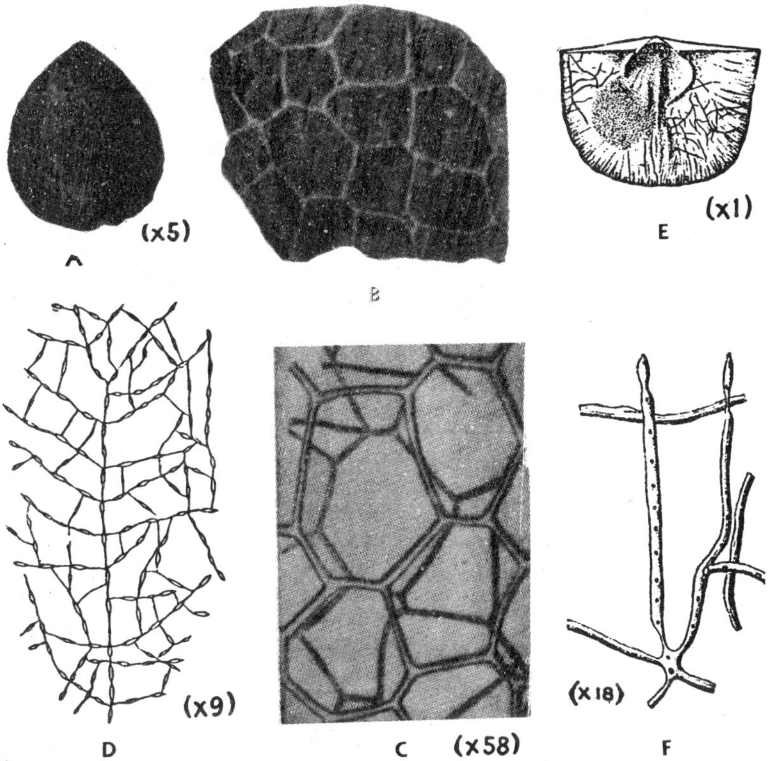

FIG. 7-30. Order Ctenostomata. *A–C. Heteronema priscum*, one of the earliest known bryozoans, from the Lower Ordovician of Estonia: *A*, a zoarium attached to a valve of *Obolus; B*, a small portion of *A* enlarged to show method of branching; *C*, parts of two zoaria with one growing over the other. *D. Rhopalonaria venosa* from the Upper Ordovician of Ohio. *E–F. Vinella repens* from the Middle Ordovician of Minnesota: *E*, two zoaria attached to the inner surface of a pedicle valve of *Strophomena; F*, portion of one of the zoaria magnified to show a nucleus with five divisions of the tubular stolon radiating from it. (*A–C after Bassler*, 1911; *D after Ulrich and Bassler*, 1904; *E–F after Ulrich*, 1893.)

Excavations supposed to be of this type have been found in rocks as old as the Ordovician, attesting the great age of this bryozoan order.

The zooecia, which are commonly isolated, spring from the internodes of the tubular stolon or stem and are formed by autozooids that bud from preexisting zooids (Fig. 7-29). They have soft and uncalcified walls and the terminal aperture is closed during retraction of the polypide by an operculum of setae resembling a comb (whence the name of the order; see footnote on page 230).

The chief features of ctenostomatous zooids are shown in Fig. 7-29. By comparison with Figs. 7-2, 7-3, and 7-40 they may seem to be somewhat similar to typical cheilostomatous zooids, but they lack avicularia, vibracula, and brood chambers. Furthermore, the zooecial wall is soft or flexible (*i.e.*, chitinous), not calcified.

Living Ctenostomata are typically marine, though a few forms live in estuaries, and ancient species seem to have lived in a similar habitat, judging from the fact that their dissolved traceries and ramifications are found on brachiopod shells, crinoid fragments, and similar hard parts of other marine animals.

The oldest fossil ctenostome yet reported, *Heteronema priscum* (Fig. 7-30*A–C*), came from an Estonian sandstone (Unguliten sandstein) of Lower Ordovician age in which it is present on the smooth shells of *Obolus*. Some half dozen genera are known from the Paleozoic, and fewer than that from the Mesozoic and Cenozoic. The thirty or more living genera constitute a relatively insignificant part of the existing bryozoan fauna, and their ancient relatives seem to have been similarly unimportant in the geologic past.

Vinella (Fig. 7-30*E–F*), with radial stolons, and *Rhopalonaria* (Fig. 7-30*D*), with pinnately arranged stolons usually represented by minute excavations in shells and corals, are representative Paleozoic genera. *Farrella* (Fig. 7-29) is a typical living genus.

Order 2. Cheilostomata.[1] The Cheilostomata exhibit the highest type of development and the greatest complexity of zooecial structure found among both living and extinct Bryozoa. Outstanding characteristics are (1) the aperture of the boxlike zooecium is closed by an operculum; (2) the frontal side of the zooid is commonly calcified, so that the zooecial wall has a great range of microscopic structure; and (3) heterozooids of several different types are generally developed. Many cheilostomatous zoaria are objects of great beauty because of the designs made by zooecial boundaries, surface spines, opercular outlines, perforated and ribbed frontal walls, and a wide variety of brood chambers, avicularia, and vibracula. Little wonder, then, that they have long excited the interest and stimulated the curiosity of laymen and scientists alike.

General Morphology. An excellent idea of a typical cheilostome can be gained by examining a zooid of the Family Membraniporidae, which is generally considered to be the least specialized of the numerous cheilostomatous families (Fig. 7-41).

The zooid, and the enclosing zooecium, have the shape of a box or of an old-fashioned coffin with a basal, a frontal, and four side or vertical walls. All zooidal walls except the frontal have pores of connection that perforate the calcareous walls of the enclosing zooecium (Figs. 7-3, 7-28, 7-40). Much of the frontal wall of the zooecium is calcified, but in the distal part is a membranous area, the **apertural field,** around the aperture, which itself is closed by a chitinous operculum. A typical cheilostome brood chamber (**ovicell** or **ooecium**) commonly appears as a prominent blister on the external surface of the zoarium. The internal features, especially the structure of the polypide, are essentially similar to those in all other Cheilostomata and Ctenostomata.

Characteristic structural features of other groups of Cheilostomata are merely additions to the type of zooid just described. Thus, in two groups there is an extensively developed internal calcareous wall, the **cryptocyst** (Fig. 7-31*E–F*); in another there is a frontal shield formed by united spines; and in

[1] Gr. *cheilos*, lip, + *stoma*, *stomatos*, mouth; referring to the chitinous lip or operculum that closes the aperture when the polypide retracts.

still another there is an internal balloonlike sac, the **compensation sac** or **ascus,** which aids in extrusion of the polypide. These additional features are considered briefly in following paragraphs.

Protrusion of Polypide. The way in which the tentacles of a bryozoan are brought into action is of great importance, especially in the Ectoprocta, because of the influence it has exerted on the structural features of the frontal wall of the zooid and zooecium and on the evolution of the zooecium in general (Harmer, 1930).

In the Ectoprocta the tentacles are not merely opened and closed, as in the Entoprocta; they can also be retracted into an **introvert,** the **tentacle sheath,** which lies in the body cavity of the zooid (Figs. 7-3, 7-40). When the tentacles are protruded and outstretched, much of the contents of the body cavity passes from the inside of the zooid to the outside (Fig. 7-3). Since the body

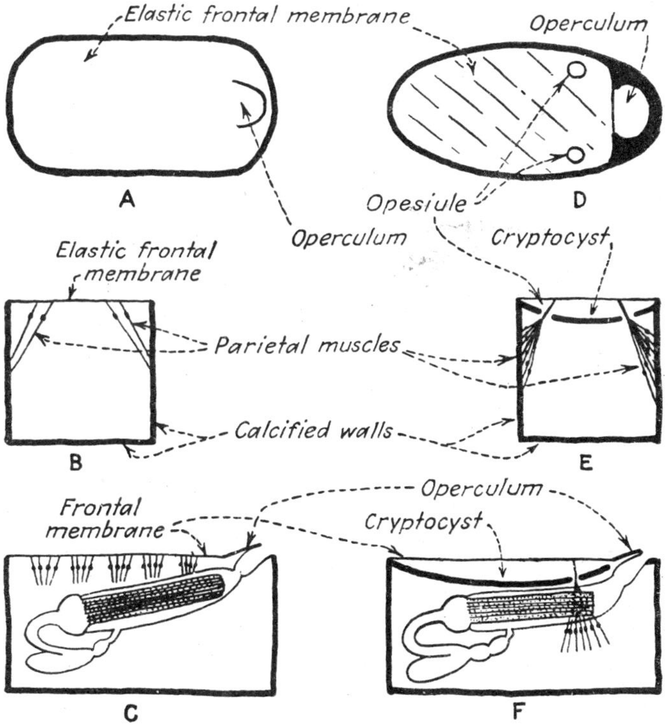

FIG. 7-31. Order Cheilostomata. Diagrams illustrating protrusion of the polypide in the Suborder Anasca (cheilostomatous bryozoans without a compensation sac, or compensatrix). *A–C.* Frontal, end-on, and side views of *Membranipora*, which has an elastic uncalcified frontal membrane attached to the side walls by pairs of parietal muscles. *D–F.* Frontal, end-on, and side views of *Micropora*, which has a single pair of parietal (depressor) muscles. These two groups or pairs of muscles, which are attached to the side walls, extend outward to the calcified cryptocyst, where each becomes a single tendon and passes through a pore (opesiule) to the elastic frontal membrane where they are attached. In the Anasca the parietal muscles pull the frontal membrane downward as the polypide is protruded through the operculated aperture. (*Modified after Harmer,* 1930.)

cavity is a closed space, the fluid it contains cannot be altered appreciably in volume; hence some way or ways had to be developed to maintain the volume of the fluid during protrusion or retraction of the polypide. This has been achieved in two quite different ways.

In one cheilostomatous group, the Suborder Anasca, the zooidal wall contracts during protrusion, so that the volume of the fluid is maintained as the polypide is pushed outward. In these bryozoans the body wall is flexible, in whole or in part. In typical Anasca the frontal wall of the zooid is a flexible membrane to which the parietal (depressor) muscles are attached. It is drawn inward by the action of these muscles when the polypide is protruded (Fig. 7-31*A–C*). It commonly becomes calcified to some extent laterally and peripherally, so that the actual area of flexible membrane may be small. In some anascan species an internal, calcareous second wall, the **cryptocyst,** is developed, presumably to give additional protection to the polypide when retracted. If the cryptocyst is fully developed, as shown in Fig. 7-31*E–F*, the depressor muscles are modified to a single pair that pass through the cryptocyst in two notches or two pores, the **opesiules** (Fig. 7-31*D–F*).

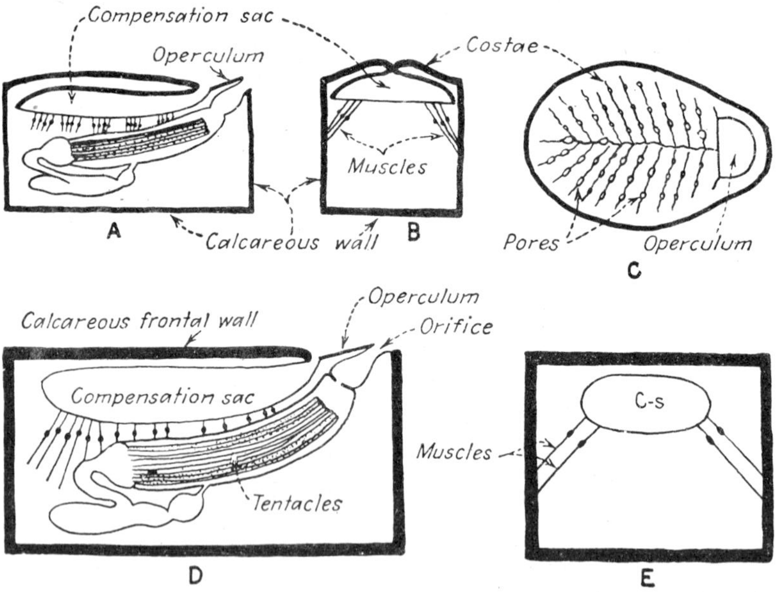

FIG. 7-32. Order Cheilostomata. Diagrams illustrating protrusion of the polypide in the Suborder Ascophora (cheilostomatous bryozoans with a compensation sac). *A–C*. Protrusion of the polypide in the Cribrilinidae. The uncalcified frontal membrane is overarched by a roof composed of costae that have fused with one another except at the pores shown n *C*. *D–E*. Lateral and end-on views of a typical ascophoran autozooid, showing how the compensating sac, which opens to the exterior at the base of the operculum, is attached to the side walls by numerous parietal muscles. In the Ascophora the compensating sac is enlarged by the contraction of these muscles, thus forcing the polypide out of the zooecium through the orifice. (*Modified after Harmer*, 1930.)

In the Suborder Ascophora, the frontal wall is entirely calcified except for two distal openings, the aperture, or orifice, and a much smaller **median pore** opening into a sacklike vessel, the **compensation sac** (Figs. 7-32, 7-40). Inasmuch as the typical ascophoran zooid is surrounded by a rigid boxlike zooecium, the polypide can emerge only if an equal volume of water is somehow introduced to compensate for the protruded organs. The compensation sac has been developed for this purpose. Flow of water into and out of this organ allows protrusion and retraction of the polypide to take place, as shown in Figs. 7-32 and 7-40. Passage of water through the median pore is controlled by the operculum, the base of which passes into the floor of the sac. In many species, modified marginal spines, or **costae,** arch over the frontal membrane and finally unite with one another to form a secondary external calcareous wall, leaving slits or a row of pores (**lacunae**) between adjacent spines. In these forms the compensating sac is the cavity between this secondary wall and the frontal membrane. The muscles operating the sac are shown in Fig. 7-32*A–C*.

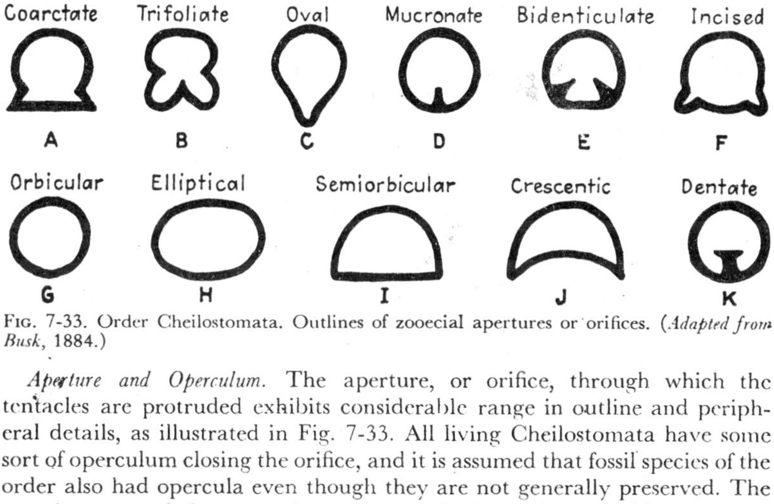

FIG. 7-33. Order Cheilostomata. Outlines of zooecial apertures or orifices. (*Adapted from Busk*, 1884.)

Aperture and Operculum. The aperture, or orifice, through which the tentacles are protruded exhibits considerable range in outline and peripheral details, as illustrated in Fig. 7-33. All living Cheilostomata have some sort of operculum closing the orifice, and it is assumed that fossil species of the order also had opercula even though they are not generally preserved. The anterior part of the operculum, or **anter,** closes the polypide aperture; whereas the posterior part, or **poster,** closes the median pore, or the opening into the compensation sac. Hence the shape of the operculum, which is constant for species, genera, and families, is of considerable taxonomic importance because it reveals the presence of a median pore and indicates the nature of this pore as well as of the orifice. All opercula are more or less circular in outline with the lower border variously modified, as indicated in Fig. 7-34. Some consist of rigid chitin but most are composed of a membrane that is supported or strengthened by rods and other processes or by a complicated framework.

Brood Chambers, or Ooecia. The fertilized eggs of bryozoans develop into embryos and these in turn into larvae inside special chambers of incubation

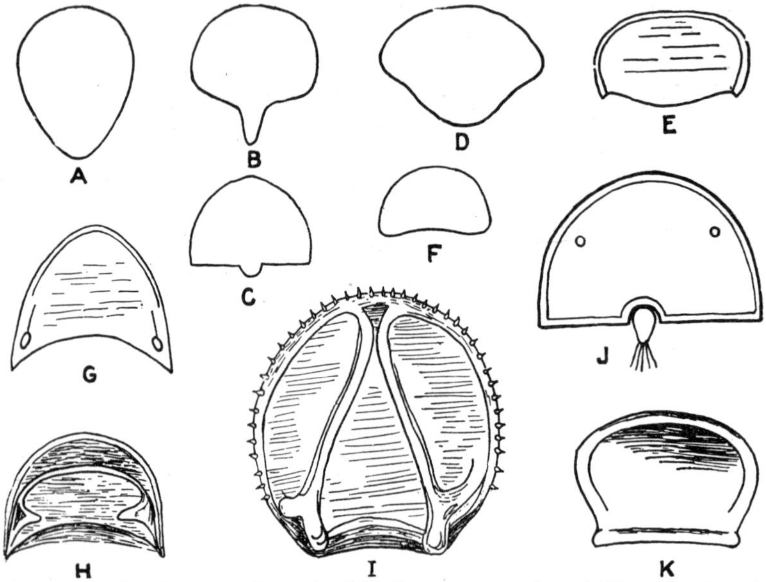

FIG. 7-34. Phylum BRYOZOA. Opercula of Cheilostomata, greatly and differently enlarged. *A. Cellepora. B–E.* Zooecial (*B–C*) and ooecial (*D–E*) opercula of *Adeonella. F. Retepora. G-Salicornaria. H. Vincularia. I.* Complicated ooecial operculum of *Steganoporella. J. Schizoporella. K. Flustramorpha.* (*Modified after Busk,* 1884.)

known as brood chambers, etc. In the Cheilostomata this chamber, which is known as an **ovicell** or **ooecium,** is generally visible on the frontal surface of the zooecium as a prominent blisterlike structure (Fig. 7-40). The nature and position of the ooecium, and its relation to the operculum, have proved useful in the classification of both fossil and living species and genera. Several different types of ooecial structure are shown in Figs. 7-35, 7-41, and 7-42.

It should be pointed out that although the ooecia of the Cheilostomata perform the same function as the gonooecia of the Cyclostomata, they are not homologous because they are not formed in the same way. They are essentially external brood pouches in that they occupy spaces between the distal wall of a fertile zooecium and the frontal wall of its distal successor (Harmer, 1930). Cyclostomatous gonozooecia, by contrast, are modified zooecia that are normal at first but later expand and become specialized (Fig. 7-5*B*).

Appendicular Organs (*Heterozooids*). The avicularia characteristic of many cheilostomatous bryozoans, and discussed briefly on page 202, consist of three parts: (1) a long or short peduncle arising from a normal zooecium and carrying the head; (2) the **head** with a strong pointed **beak**, and (3) the **mandible,** which is a mobile under or lower jaw that conforms with the beak and is operated by powerful muscles (Fig. 7-36*A*). The peduncle and head are calcareous, the beak and mandible chitinous.

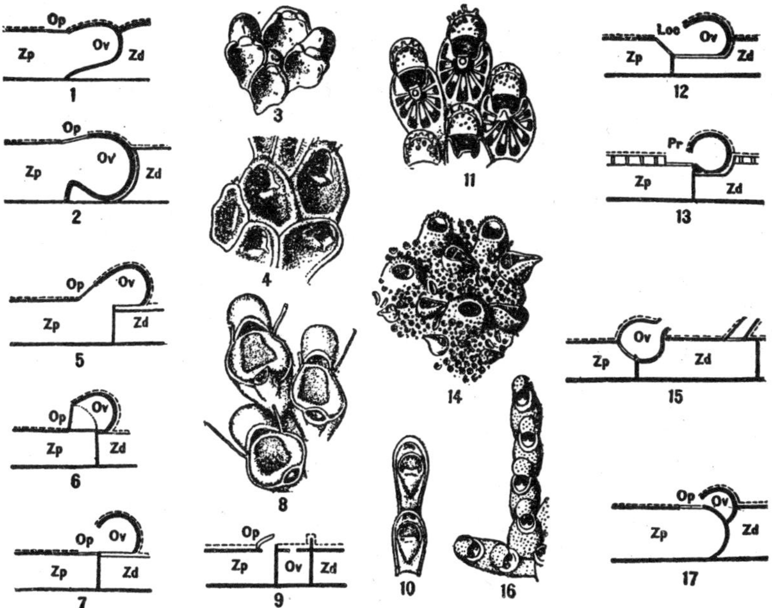

FIG. 7-35. Order Cheilostomata. Ovicell structure. (*Op* = operculum; *Ov* = ovicell; *Zd* = distal zooecium; *Zp* = proximal zooecium; *Loc* = locella; *Pr* = peristomie or tube developed by growth of peristome.) The thin broken line indicates the membraneous ectocyst, while the thin double line represents the operculum. 1–2. Longitudinal sections through zooecia with an endozooecial ovicell. The ovicell is within the zooecium itself, and the operculum closes both the zooecium and the ovicell. In 2 a fold of the zooecial wall separates the ovicell from the zooecium. 3. *Micropora coriacea* Esper. A group of zooecia, ×25, with two showing the endozooecial ovicell and the operculum closing the ovicell as well as the zooecia. 4. *Velumella levinseni* Canu and Bassler. Zooecia, ×40, with the two uppermost bearing the small endozooecial ovicell. 5–7. Sections showing three types of of hyperstomial ovicell in which the ovicell is placed on the distal zooecium; in 5, the ovicell opens below the operculum, and there is thus only one aperture; in 6, there are two apertures, and the operculum in opening closes the ovicell; in 7, the ovicell opens above the operculum. 8. Three ovicelled zooecia of *Ramphonotus minax* Busk, ×50, illustrating the hyperstomial form of ovicell. 9. Sketch of endotoichal ovicell in which the ovicell is completely separated from the zooecium and its orifice is removed from the aperture and placed in the same plane. 10. Two zooecia, ×50, of *Cellaria sinuosa* Hassall, showing the apertures of the small endozooecial ovicell in advance but on the same plane as the large zooecial apertures. 11. Ovicelled zooecia, ×36, of *Umbonula verrucosa* Esper with hyperstomial ovicell opening largely above the aperture. 12–13. Hyperstomial ovicells: in 12, the ovicell is placed in a deep cavity of a distal zooecium; the operculum is very oblique and operates in a special chamber or locella; 13 represents a special type in which the ovicell opens above the operculum in the peristomie or tube formed by the growth of the peristome. 14. A group of zooecia, ×23, of *Tubiporella magnirostris* MacGillivray, with two peristomial ovicells. 15. Diagram of a peristomial ovicell showing its formation by an enlargement of the peristomie. 16. Typical example (*Phylactella labrosa* Buck) of the recumbent ovicell, ×30, in which the ovicell is placed on the distal wall of the zooecium itself. 17. A sketch of a recumbent ovicell showing its relations to the zooecia and operculum. (*After Bassler*, 1922.)

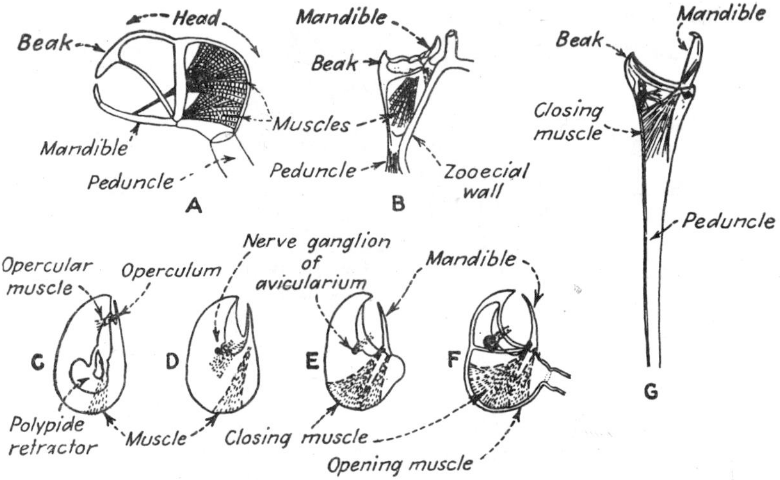

FIG. 7-36. Order Cheilostomata. Avicularia. *A*. A typical pedunculate avicularium. *B*. Sessile avicularium of *Scrupocellaria*. *C–F*. Diagrams showing supposed evolution of a typical avicularium from a normal operculated zooid. It is to be noted that the operculum was modified to become the mandible and that the muscles operating the mandible developed from the polypide retractors. *G*. Avicularium of *Bicellaria*, with conspicuously elongated peduncle. (*A–B after Hincks*, 1880; *C–F after Delage and Hérouard*, 1897; *G after Busk*, 1884; *all diagrams somewhat modified*.)

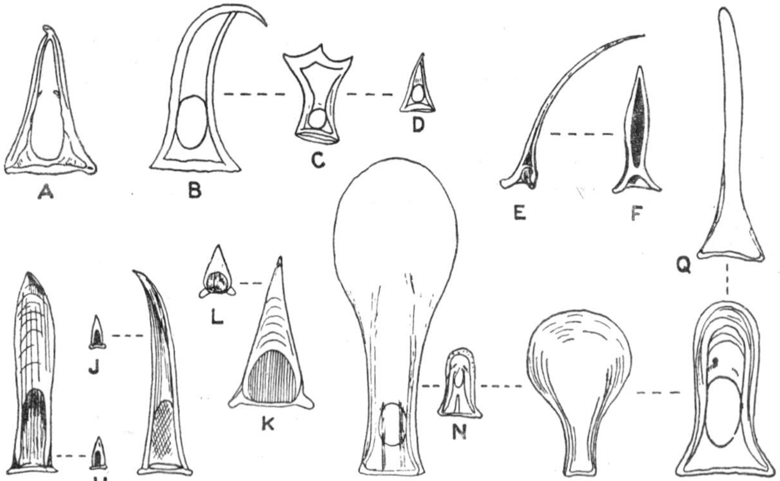

FIG. 7-37. Order Cheilostomata. Chitinous mandibles greatly enlarged from the avicularia of different genera. *A*. *Salicornaria*. *B–D*. Different types of mandibles from species of *Retepora*. *E–F*. Mandibles from two different species of *Flustramorpha*. *G*, *I* and *H*, *J*. Lateral and frontal avicularia, respectively, of *Adeonella*. *K–L*. *Adeona*. *M–Q*. Mandibles from several different species of *Cellepora*. (*Modified after Busk*, 1884.)

In life the avicularium contains a rudimentary polypide and is capable of a vigorous snapping action if disturbed. It is thought by some to have evolved from a normal zooid by gradual loss of polypide structure, modification of distal and frontal walls, specialization of the operculum, and development of a peduncle (Fig. 7-36). Its function is somewhat uncertain, but observers have noted that the snapping action discourages invading microorganisms and larvae from settling on the surrounding zooecial surface.

The mandibles, greatly modified opercula, are symmetrical objects similar to the opercula of normal zooecia. They show considerable variation in shape and other features and, like normal opercula, are of value for classification. A few types are shown in Fig. 7-37.

Vibracula, like avicularia, are highly modified rudimentary heterozooids, but they differ in having a long **seta,** or **lash,** in place of the mandible and in lacking the great variation of structure shown by the avicularia (Fig. 7-38*A*).

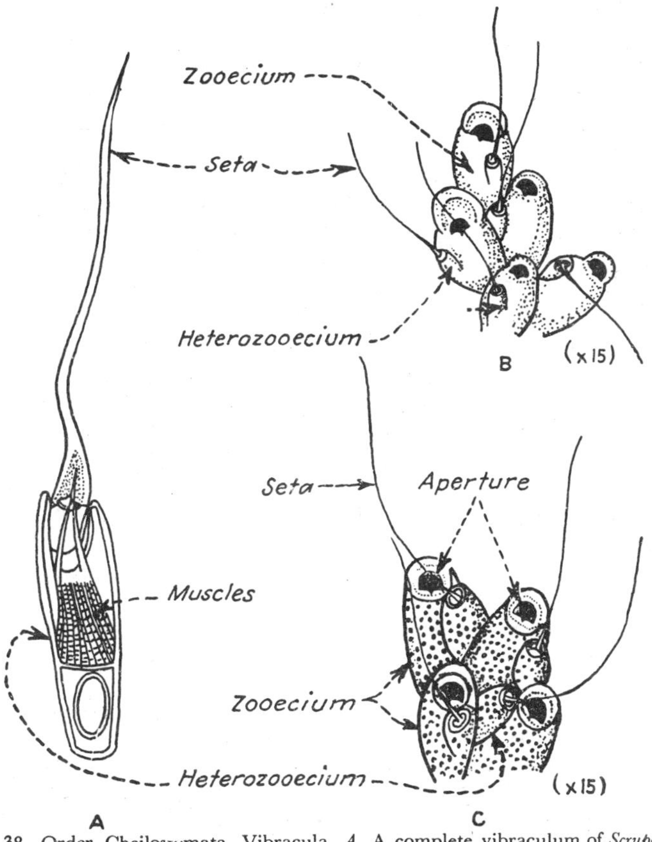

FIG. 7-38. Order Cheilostomata. Vibracula. *A*. A complete vibraculum of *Scrupocellaria* greatly magnified. *B–C*. Groups of zooecia from different species of *Mastigophora*, showing prominent vibracular heterozooids closely associated with autozooidal zooecia. (*Modified after Hincks*, 1880.)

On a living bryozoan the vibraculum is located on the frontal surface of the zooecium (Fig. 7-38*B–C*), and the mobile seta, which at rest lies upon the surface, sweeps slowly to and fro over the zoarial surface discouraging intruders and preventing accumulation of debris around the orifice. In the most specialized form, the vibracula become long and strong serrated lashes that have a locomotory function in certain free-living species.

Avicularia and vibracula, so far as known, are unique to cheilostomatous Bryozoa. They illustrate how a normal individual of a colony, in this case presumably an autozooid originally, can become so greatly modified and highly specialized that its original characteristics are largely obliterated.

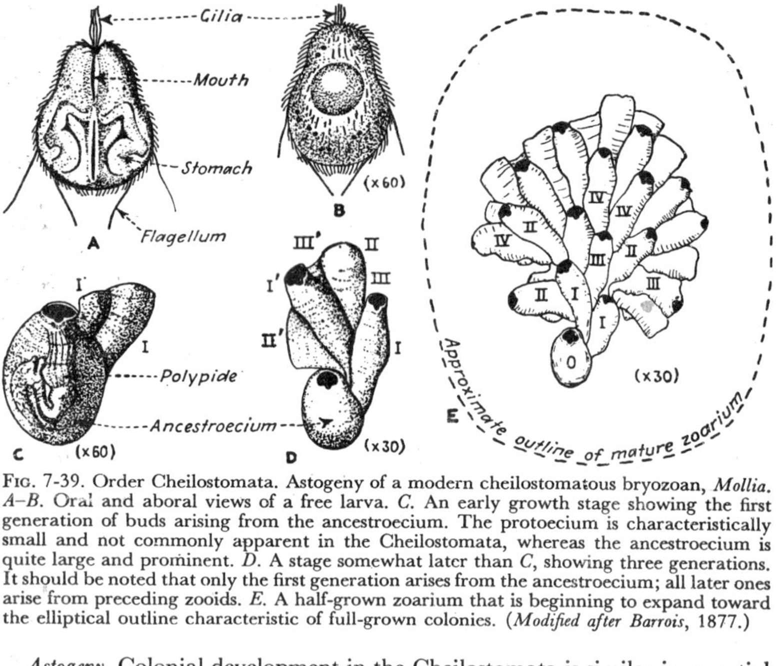

FIG. 7-39. Order Cheilostomata. Astogeny of a modern cheilostomatous bryozoan, *Mollia*. *A–B*. Oral and aboral views of a free larva. *C*. An early growth stage showing the first generation of buds arising from the ancestroecium. The protoecium is characteristically small and not commonly apparent in the Cheilostomata, whereas the ancestroecium is quite large and prominent. *D*. A stage somewhat later than *C*, showing three generations. It should be noted that only the first generation arises from the ancestroecium; all later ones arise from preceding zooids. *E*. A half-grown zoarium that is beginning to expand toward the elliptical outline characteristic of full-grown colonies. (*Modified after Barrois*, 1877.)

Astogeny. Colonial development in the Cheilostomata is similar in essential details to that outlined for the ectoproctous Bryozoa. However, the protoecium tends to be smaller than in other orders, and the ancestroecium differs from the usual cylindrical form in being inflated to such an extent as to resemble an adult zooecium (Figs. 7-9*G*, 7-39).

Zoaria and Zooecia. The zoaria of cheilostomatous Bryozoa range from thin, fragile incrustations to delicate bushlike forms and massive fronds of lobes and leaflike branches (Figs. 7-1, 7-12). Some are wholly calcareous, some are partly so, and many are either chitinous or membranous. In general the better preservation is shown by the more calcified forms.

The zooecia are boxlike or baglike in form and oval or polygonal in outline. They are generally packed together rather loosely, and adjacent zooecia are connected by pores, the **septulae** (Fig. 7-40). The distally placed aperture is characteristically smaller than the diameter of the zooecium and in many species has spines around the margin (Fig. 7-41*B–C*). Typical apertural outlines are shown in Fig. 7-33. The frontal membrane shows great variation in size, shape, and structure. It is spinose in some forms, latticed in others, and fully calcified in still others (Figs. 7-41, 7-42). In the fully calcified species the surface of the zoarium is commonly striking because of the zooecial features and associated brood chambers and appendicular organs (Figs. 7-41, 7-42).

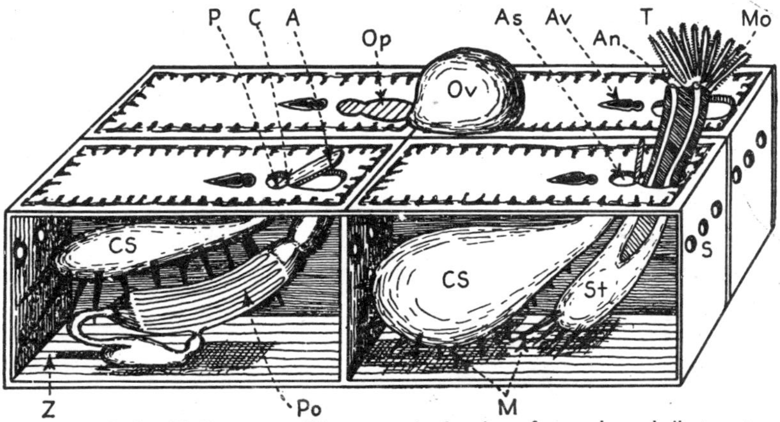

Fig. 7-40. Order Cheilostomata. Diagrammatic drawing of a modern cheilostomatous bryozoan showing four zooecia (*Z*), with the polypides withdrawn in three and protruding from one. The ornamental calcareous wall is equipped with avicularia (*Av*) and ooecia, or ovicells (*Ov*). The posterior part (*P*) of the operculum (*Op*), which operates on a hinge or cardelle (*C*), covers the opening (*As*) into the compensating sac (*CS*), and the anterior part (*A*) closes the orifice of the polypide (*Po*). Adjacent zooecia communicate through small pores, or septulae (*S*). The compensating sac when filled with water forces the polypide out through the orifice, and the animal is then in a feeding position. The operculum is removed in one zooecium, closed in a second, slightly open in a third, and fully open in the fourth. (*An* = anus; *As* = opening into compensation sac; *M* = muscles; *Mo* = mouth; *St* = stomach; *T* = tentacles.) (*Adapted from Bassler*, 1922.)

Classification. Classification of the Cheilostomata has been difficult because of disagreement among specialists as to the relative importance of soft parts and hard structures, and of developmental features.[1] Consequently, we do not list the many families that have been proposed by different authors, but

[1] At one time or another the following characteristics have been used as a basis for subdividing the Cheilostomata:

1. Method of growth (long since discarded).
2. Shape of zooecial aperture, and nature of operculum.
3. Structure of frontal wall.
4. Nature of ooecia, and relation to operculum.
5. Nature of embryo.
6. Appendicular organs.
7. Mode of extrusion of polypide.
8. Nature of zooecia and zoaria.

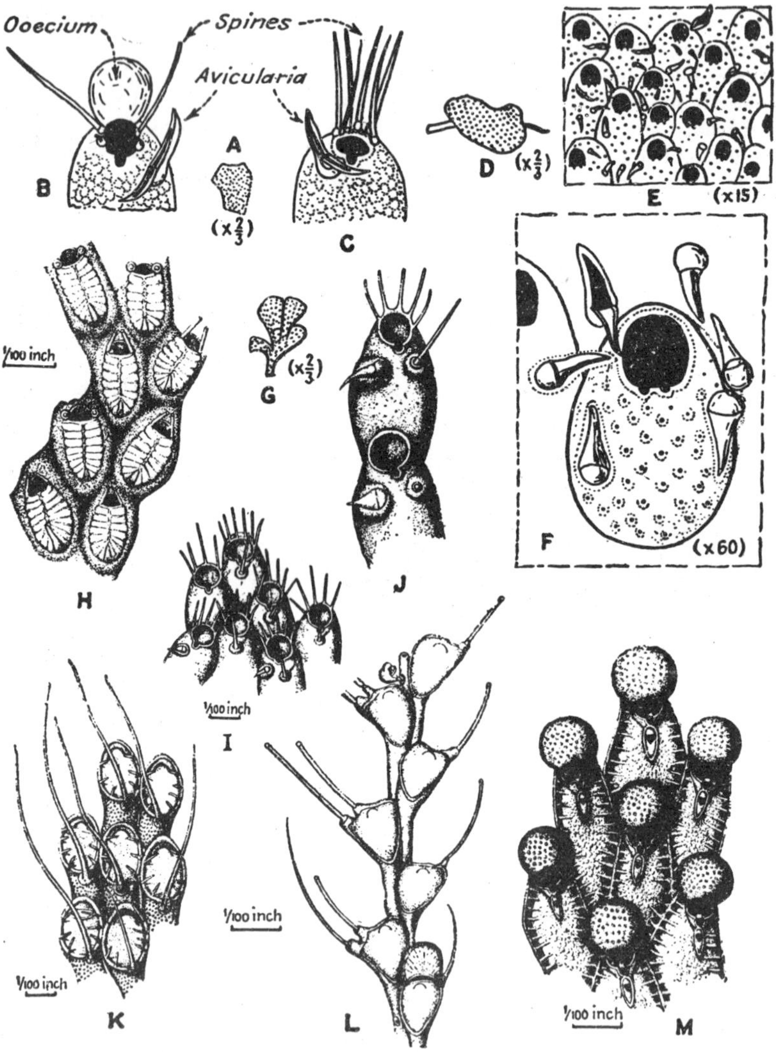

FIG. 7-41. Order Cheilostomata. Frontal features of living forms. *A–C. Schizoporella longispinata: A*, fragmental zoarium; *B*, single zooecium with prominent ooecium, apertural spines, and an avicularium; *C*, normal zooecium with an avicularium and several apertural spines. *D–F. Mucronella bisinuata: D*, zoarium slightly reduced; *E*, portion of surface showing numerous avicularia; *F*, single zooecium much enlarged to show nature and distribution of avicularia around the bisinuate aperture. *G–H. Membraniporella melolontha: G*, zoarium; *H*, a few zooecia, showing frontal structure. *I–J. Schizoporella spinifera;* zooecia showing avicularia, vibracula, and apertural spines. *K. Membranipora pilosa*, with unusually long vibracula and numerous frontal spines. *L. Bicellaria alderi*, with an ooecium and several double spines. *M. Smittia reticulata*, showing well-developed ooecia. (*A–F after Busk*, 1884; *G–M after Hincks*, 1880.)

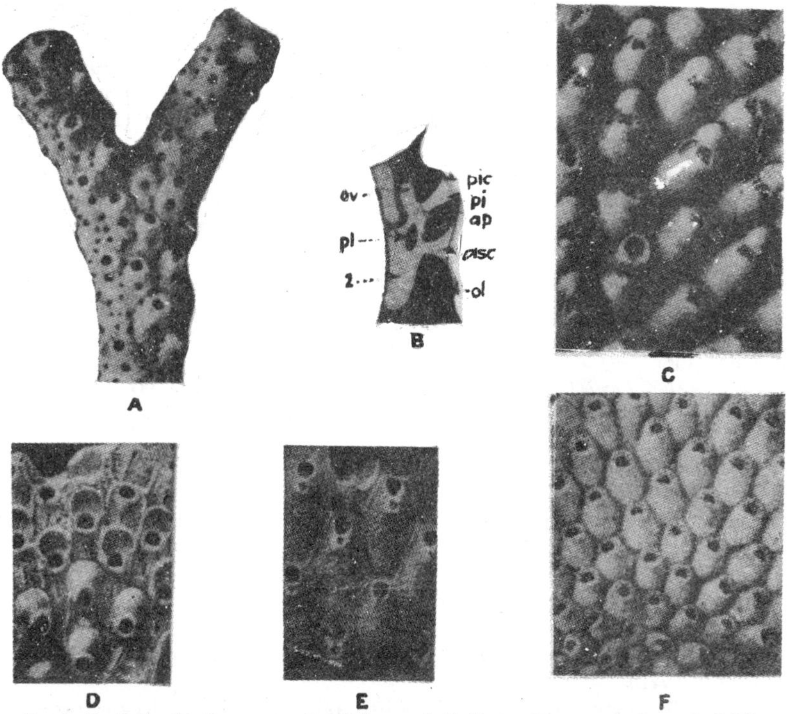

FIG. 7-42. Order Cheilostomata. Fossil forms. *A–B. Gastropella ventricosa* from the Midway (Paleoc.) of Arkansas: *A*, a forked zoarium showing large, swollen zooecia and prominent ascopores; *B*, longitudinal section through a zooecium with ooecium (*ap* = aperture; *asc* = ascopore; *al* = interzooecial substance; *ov* = ooecium or ovicell; *pi* = peristomial tube leading from submerged aperture to surface where it opens as the peristomice, *pic*; *pl* = support in zooecium; *z* = zooecium). *C. Peristomella falcifera* from the Jackson (Eoc.) of North Carolina. A well-preserved zoarium of this incrusting species, with both ordinary and ooecia-bearing zooecia and conspicuous long, curved avicularia. *D–E. Metroperiella* from Jackson of Georgia and North Carolina, respectively: *D*, *M. latipora* with both kinds of zooecia; *E*, *M. biplanata* showing wide polygonal zooecia. *F. Perigastrella oscitans* from the Jackson (Eoc.) of Mississippi showing ancestroecium (solid white) and neighboring zooecia. (*All figures except B are* ×15, *and all are after Canu and Bassler*, 1920.)

instead characterize briefly the two generally recognized suborders, Anasca and Ascophora, which are based primarily on the absence or presence of a compensating sac.[1]

Suborder 1. *Anasca.* Compensation sac lacking; frontal wall membranous or calcareous, generally depressed and surrounded by raised margins; opercular and subopercular areas not separated by a calcareous bar as in the Ascophora. *Ordovician?*[2] *Cretaceous to Recent.*

[1] Vigneaux (1949) recently published a classification of the Cheilostomata which includes 10 superfamilies and 69 families. Many of the latter are new or do not agree with those of Bassler (1935).

[2] Bassler (1935) assigns the genus *Paleschara* (*Ord.—Dev.*) to this suborder.

The following genera are typical: *Bicellaria* (Figs. 7-36*G*, 7-41*L*), *Bugula* (Figs. 7-1*A*, 7-2, 7-12*A*), *Caberea* (Fig. 7-1*M*), *Cellaria* (Fig. 7-35[10]), *Flustra* (Figs. 7-10, 7-12*C*), *Membranipora* (Figs. 7-31*A–C*, 7-41*K*), *Micropora* (Figs. 7-31*D–F*, 7-35[3]), *Mollia* (Fig. 7-39), *Ramphonotus* (Fig. 7-35[8]), *Scrupocellaria* (Figs. 7-36*B*, 7-38*A*), *Steganoporella* (Fig. 7-34*I*), *Tricellaria* (Fig. 7-1*C*), *Velumella* (Fig. 7-35[4]).

Suborder 2. *Ascophora*. Compensation sac present, generally opening on the proximal side of the aperture but rarely also through a median pore (**ascopore**); a calcified transverse bar divides the operculum into opercular and subopercular areas. *Cretaceous to Recent*.

The following genera are typical: *Adeona* (Figs. 7-12*B*, 7-37*K–L*), *Adeonella* (Figs. 7-6*B*, 7-34*B–E*, 7-37*G–J*), *Cellepora* (Figs. 7-6*A*, 7-34*A*, 7-37*M–Q*), *Gastropella* (Fig. 7-42*A–B*), *Mastigophora* (Figs. 7-6*C*, 7-38*B–C*), *Membraniporella* (Fig. 7-41*G–H*), *Mucronella* (Figs. 7-1*E–F*, 7-41*D–F*), *Perigastrella* (Fig. 7-42*F*), *Phylactella* (Fig. 7-35[16]), *Retepora* (Figs. 7-9*G*, 7-34*F*, 7-37*B–D*), *Schizoporella* (Figs. 7-34*J*, 7-41*A–C*,*I–J*), *Smittia* (Fig. 7-41*M*), *Tubiporella* (Fig. 7-35[14]), *Umbonula* (Fig. 7-35[11]).

Geologic History. The Cheilostomata appeared early in the Cretaceous, quickly became a prominent group, and gained dominance over all other Bryozoa by Tertiary time, a position they continue to have in modern seas. Some authors (*e.g.*, Swinnerton, 1946) would derive the order from cyclostomatous ancestors; some, from the Ctenostomata; whereas others (*e.g.*, Borg, 1926) believe they descended from Paleozoic Cryptostomata. If the Ordovician to Devonian *Paleschara* is really a cheilostome, then perhaps there is some reason for considering the Cryptostomata as Paleozoic Cheilostomata (Bassler, 1922).

Order 3. Cryptostomata.[1] This extinct and exclusively Paleozoic order is characterized by delicate lacelike fronds and dendroidal zoaria with the zooecia arranged in different patterns on the branches (Fig. 7-44). There are three general arrangements of zooecia. In one the zoarium is composed of a double layer, with the zooecia of one layer lying base to base with those of the other, and it takes the form of a sword, ribbon, or fan (Fig. 7-44*A–E*). In another the zoaria are lacelike expansions with only a single layer of zooecia, all opening in the same direction, and with the nonzooecial surface covered by a dense calcareous layer of striated or finely granular material. In a third type of architecture the zoaria are ramose and the zooecia, as well as zooecial expansions, arise from the axis of a cylindrical stem. In this group some zoaria have articulating segments (Fig. 7-44*F–G*).

An unusual type of zoarium characteristic of the genus *Archimedes* (Fig. 7-44*I–K*) consists of a screwlike axis from which an ascending lacy and celluliferous sheet is given off like a spiral stairway. It has been suggested that this fossil is actually an animal-plant consortium, *i.e.*, a bryozoan-algal pair that lived together (Condra and Elias, 1944). This hypothesis seems doubtful.

The zoaria are composed of short, straight or curved tubular zooecia that

[1] Gr. *kryptos*, hidden, + *stoma*, *stomatos*, mouth; referring to the fact that the zooecial aperture lies hidden below the zooarial surface at the base of a pit or tubular shaft (Fig. 7-43).

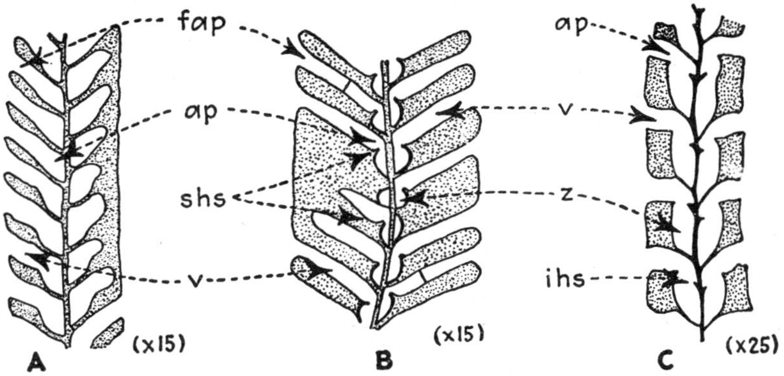

FIG. 7-43. Order Cryptostomata. Longitudinal sections of *Rhinidictya grandis* (*A*), *R. fidelis* (*B*), and *Arthropora simplex* (*C*), showing internal structural features. (*ap* = true aperture at bottom of vestibulum, *v*; *fap* = false aperture; *ihs* = inferior hemiseptum; *shs* = superior hemiseptum; *v* = vestibulum; *z* = zooecium.) (*Modified after Ulrich*, 1895.)

have the true orifice concealed at the base of a pit or shaft, the **vestibulum** (Fig. 7-43). The vestibulum is surrounded by a calcareous deposit and opens to the exterior through a variously sized pore, the **false aperture.** At its base it is partly closed by a **hemiseptum** that restricts the true orifice and extends downward to some extent into the primitive zooecium. In a few species a second plate rises from the bottom of the zooecium and is referred to as the **inferior hemiseptum** to distinguish it from the upper one, which is then designated the **superior hemiseptum.** The function of the hemisepta is unknown, but it has been suggested that they may have served as supports for movable opercula.

The symmetry of zooecial arrangement and the architecture of many zoaria give to the Cryptostomata a beauty only second to that of the Cheilostomata. Both aspects of the cryptostomes have been used for identification of species and genera, but in most modern work the internal structure as revealed in thin sections has become the basis of classification.

The chief characteristics of the order are shown in Table 7-2, where they can be compared with those of the other bryozoan orders. The Cryptostomata have been called Paleozoic Cheilostomata because of similar astogeny and zooecial structure, but they differ from that order in the following respects:

1. Lack appendicular organs (avicularia and vibracula).
2. Seemingly lack ooecia or brood chambers.[1]
3. Lack the different spinose structures so common on the frontal wall and around the aperture in Cheilostomata.[2]

[1] McNair (1937) reports enlarged bulb-shaped zooecia on the older parts of the zoaria of two Devonian cryptostomatous genera, *Semicoscinium* and *Sulcoretepora*, and questions if these supposedly specialized zooecia may have been brood chambers or ovicells.

[2] A possible exception may be certain Devonian species of three genera (*Fenestrellina*, *Polypora*, and *Semicoscinium*), which have been reported to have stellate zooecial apertures because of centrally directed spines that project inward from the walls of the vestibules.

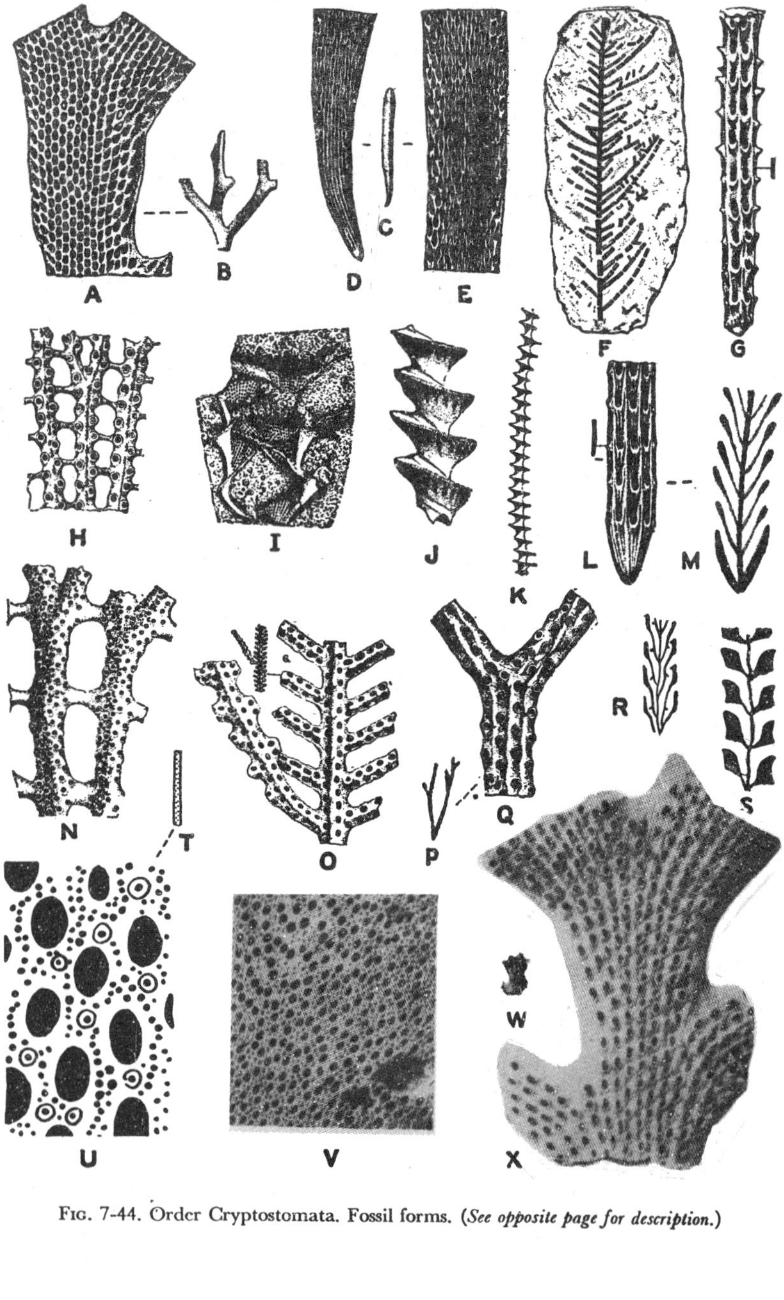

FIG. 7-44. Order Cryptostomata. Fossil forms. (*See opposite page for description.*)

4. Have much thicker and more extensive deposits of calcareous material on the outer surface of the zooecia.

5. Commonly have a long tubular zooecium, supposedly formed by successive generations of polypides.

6. Zoaria with an uninterrupted width of more than 8 mm. have clusters of cells that differ somewhat in size and elevation from the average zooecia.

The last two features cited above suggest an affinity with the Trepostomata, but the Cryptostomata differ from them in having a hemiseptum, a different ratio of the mature and immature regions of the zooecia, and numerous other features listed in Table 7-2.

About 120 genera of Cryptostomata, with many hundreds of species, have been classified in a dozen or more families, some of which are not too well defined. Genera typical of the order are *Acanthoclema* (Fig. 7-44*R*), *Archimedes* (Fig. 7-44*I–K*), *Arthroclema* (Fig. 7-44*F–G*), *Arthropora* (Fig. 7-43*C*), *Ceramella* (Fig. 7-25*F*), *Escharopora* (Fig. 7-44*C–E*), *Euspilopora* (Fig. 7-12*E*), *Fenestrellina* (Figs. 7-9*H*, 7-44*H*), *Helopora* (Fig. 7-44*L–M*), *Intrapora* (Fig. 7-44*V*), *Lyropora* (Fig. 7-12*H*), *Pachydictya* (Fig. 7-12*F*), *Penniretopora* (Fig. 7-44*O*), *Polypora* (Fig. 7-44*N*), *Rhinidictya* (Figs. 7-12*D*, 7-43*A*, 7-44*A–B*), *Rhombopora* (Fig. 7-44*T–U*), *Stictoporina* (Fig. 7-12*G*), *Sulcoretepora* (Fig. 7-44*S*,*W–X*), and *Thamniscus* (Fig. 7-44*P–Q*).

Geologic History. The earliest known representative of the order is an undescribed species supposedly from the *Cedaria* zone of the Cap Mountain member of the Upper Cambrian Riley limestone of central Texas.[1] During the Ordovician and Silurian the order was overshadowed by the Trepostomata,

[1] The specimen, which has been identified by Dr. Helen Duncan as a primitive cryptostome, came from weathered material lying upon the Cambrian limestone. It is not certainly established that the specimen actually came from the Cap Mountain; it may have come from later rocks of Ordovician or even Silurian age—a possibility suggested by the fact that a well-preserved foraminiferal specimen of *Ammodiscus* is associated with the bryozoan fragment.

FIG. 7-44. Order Cryptostomata. Fossil forms. *A–B. Rhinidictya mutabilis* (Ord.). Enlarged surface (×6) showing arrangement of zooecia, and a typical zoarium (×⅔). *C–E. Escharopora recta* (Ord.). A complete zoarium (×⅔), basal part and central portion both enlarged (×6). *F–G. Arthroclema* (Ord.). *A. billingsi*, an almost complete zoarium (×⅔) showing numerous jointed branches, and one small branch segment of *A. armatum* greatly enlarged (×12). *H.* Celluliferous side of *Fenestrellina mimica* (Penn.) (×6) showing double row of zooecia. *I–K. Archimedes* (Miss.–Penn.): *I*, *A. sublaxus* (×⅔) showing fenestrate flange spiraling around axial screw; *J*, *A. wortheni; K*, *A. communis; J*, and *K* are ×⅔ and are fragments of the axial screw. *L–M.* External view and longitudinal section of *Helopora spiniformia* (Ord.; ×12). *N.* A fragment of *Polypora submarginata* (Penn.; ×6), showing numerous zooecia on the main branches and lack of zooecia on the cross bars. *O.* Fragmental zoarium (×⅔ and ×2) of *Penniretopora conferta* (Miss.). *P–Q.* Fragmental zoarium (×⅔) and enlarged fragment (×12) of *Thamniscus furcillatus* (Miss.). *R.* Longitudinal section of *Acanthoclema ohioense* (Dev.; ×7) showing inferior and superior hemisepta. *S.* Longitudinal section of *Sulcoretopora oblique* (Dev.; ×10). *T–U.* Fragmental zoarium (×1) and much magnified surface (×15) of *Rhombopora lepidodendroides* (Penn.) showing scattered large acanthopores and many small acanthopores around the zooecia, which are shown in solid black. *V.* Portion of surface of *Intrapora irregularis* (Dev.; ×7), showing shape and arrangement of apertures and distribution of mesopores. *W–X.* Fragmental zoarium (×⅔ and ×7) of *Sulcoretopora alternata* (Dev.). (*A–Q after Ulrich,* 1890 *and* 1893; *R–S, V–X after McNair,* 1937.)

but in the Devonian it became equally important, and thereafter, until extinction in the Permian, it was the dominant bryozoan order. Fenestrate forms are especially abundant in later Paleozoic limestones and calcareous shales—some parts of the well-known Salem limestone (Indiana building stone) are little more than a felted mass of fragmental zoaria.

BRYOZOAN ECOLOGY AND PALEOECOLOGY

Living Bryozoa inhabit both fresh and salt waters. The Stenolaemata and Gymnolaemata, with stony zoarial structures, are dominantly marine; whereas the Phylactolaemata, lacking durable hard parts, are confined to fresh-water environments. Almost nothing is known of the aquatic adaptations of ancient bryozoans, but from the nature of the fossils with which their remains are associated, it would seem that all fossil Bryozoa thus far reported were marine species.[1]

The sedentary habit makes the bryozoan dependent on the closely surrounding water for its food, which consists largely of diatoms and other microscopic plants, protozoans, and certain other minute animals. These are swept down the funnel formed by the tentacles and into the mouth, whence by continuous ciliary action they pass through the alimentary tract, digestible portions being retained and undesired parts ultimately being rejected through the anal opening.

Like many other sedentary organisms, Bryozoa have constantly to fend off numerous predatory animals such as gastropods, worms, and crustaceans, and larvae and young forms of incrusting animals searching for a surface of attachment. Inasmuch as the snapping and seizing actions of the avicularia and the constant sweeping motion of the vibracula rather effectively prevent such invaders from settling down, it may well be that those two unusual organs are thus performing one of their chief functions.

Bryozoans now live in both shallow and deep marine waters, as well as in shallow fresh waters, and can tolerate turbid waters. That ancient species were similarly adapted is indicated by their abundance as fossils in limestones and shales deposited in both shallow and moderately deep waters. Fossil bryozoans are seldom found in coarse siltstones and in sandstones, and one does not expect to find them indigenous in sandy environments of existing seas.

Living marine Bryozoa attach themselves to seaweeds, to shells of both living and dead invertebrates, to stony colonial structures of hydrozoans, corals, other bryozoans, etc., and to stones, hard sandy bottoms, and rock bottoms. In the past they behaved similarly, judging from the many fossil brachiopod and molluscan shells, coralline masses, and bryozoan zoaria that are incrusted with bryozoan remains.

Incrusting forms can grow rapidly. *Schizoporella*, for example, may grow as much as a millimeter in diameter per day for the first month, after which

[1] A possible exception is the supposed bryozoan genus *Plumatellites* from the fresh-water Upper Cretaceous of Bohemia. The fact that the specimen incrusts a *Unio* strongly indicates that it was made by a fresh-water animal, but its imperfect state of preservation leaves the identification as a bryozoan uncertain (Bassler in Eastman-Zittel, 1927).

growth slows down as the colony reaches its average size of 50 to 70 mm. A tuftlike colony, such as that of *Bugula*, may reach a height of 75 mm. in 3 months. This rapidity of growth makes bryozoans bothersome fouling organisms. Once they are established on a surface, they spread rapidly not only over the original surface but also over other organisms, which they may smother.

Of special interest is the fouling of pipes by certain fresh-water Bryozoa. They form thick crusts inside the pipes, and the dead colonies commonly break loose, become fragmented, and clog smaller pipes and meters. Because they reproduce by statoblasts, which give protection to the eggs and ensure continuance of the species, fresh-water Bryozoa can persist under a wide variety of unfavorable conditions. The most effective method of eliminating them from pipes seems to be to remove their food supply by filtering the water in which they live.

Since their first appearance in the Late Cambrian, Bryozoa have made substantial contributions to calcareous strata in every geologic period. These contributions are the calcareous zoaria that they built and inhabited until death. Most calcareous rocks of Ordovician or younger age are likely to contain some zoaria, and certain limestones are composed of little else. In existing seas Bryozoa are among the more important denizens of coral reefs, and they seem to have held a similar position on ancient reefs or bioherms. The nodular zoaria of *Batostoma* and other monticuliporoids form thin biostromal layers in the Upper Ordovician shales of the Cincinnati region, and fenestrate zoaria are among the more common fossils in the reef rock of many Silurian bioherms of the Upper Mississippi Valley.

Indirectly, the bushy zoaria must have served an important role in deposition on certain bottoms by checking currents and causing deposition of materials in suspension or in solution. On the steep peripheral slopes of ancient coral reefs, where they appear to have grown in great profusion, they almost certainly played an important role in intercepting the reef mud and sand, as these moved down the slopes, and very likely aided in the construction of some of the unusually high initial inclinations observed on many reef structures. They seem also to have lived in great profusion on muddy and calcareous bottoms, where they attached themselves to the bottom directly or incrusted different kinds of objects.

Most large zoaria do not seem to have been molested greatly by scavengers, for they rarely show fragmentation that could be ascribed to such action. With the delicate and fragile forms, however, the situation was probably different, and it seems likely that predators and scavengers may have been responsible for much of the fragmentation that these zoaria so commonly show. Both large and small massive zoaria, as well as fenestrate and lacelike forms, were probably often torn loose by strong waves and currents and transported into other environments, some of which may have been unfavorable for life.

Certain gymnolaematous bryozoans can etch pits, linear grooves, and internal galleries in calcareous shells, and it seems entirely possible that similar traceries now exist as fossils on and in ancient shells (Silén, 1948).

GEOLOGIC HISTORY

The earliest known bryozoans are a simple cyclostomatous representative (*Archaeotrypa*, Fig. 7-22*A–B*) from an Upper Cambrian limestone of Alberta and a primitive cryptostomatous species which is supposedly from the Upper Cambrian Riley limestone of Texas. By early Ordovician time the Ctenostomata had developed, and a simple form, *Heteronema priscum* (Fig. 7-30*A–C*), has been described from the basal Ordovician Unguliten sandstein of Estonia. Representatives of the Trepostomata appeared and spread rapidly during the Ordovician and by the end of the period were the dominant bryozoan order. If *Paleschara* is correctly identified as a cheilostomatous genus, then the fifth and last order of the Ectoprocta was also represented at this early date. In any event, the first four orders—Cyclostomata, Cryptostomata, Ctenostomata, and Trepostomata—were well established by Ordovician time, and all persisted through the Paleozoic. At the end of this era the Cryptostomata and Trepostomata died out, but the Cyclostomata and Ctenostomata lived on and have continued to the present time, though rarely of any importance as compared to the Cheilostomata, which arose to a dominant position in the Cretaceous and have maintained this position to the present.

The Ctenostomata seem never to have been important, as the fossil record is meager and living species few in number. Cyclostomata were common and of some importance during the Paleozoic, with the Ceramoporidae as the most abundant family. Subsequently they held a steady position until the Jurassic and Early Cretaceous, when they became quite important and abundant. They have continued since then to the present as a relatively abundant group but have had to take second position to the Cheilostomata since the Cretaceous. The Trepostomata surpassed all other Bryozoa in the Early Paleozoic but relinquished the position of dominance to the Cryptostomata in the later part of the era. There seems to be good evidence that the Cheilostomata, which made their appearance in the Cretaceous (or possibly in the Ordovician), are a continuation of or a derivative from the Cryptostomata.

EVOLUTION OF THE BRYOZOA

The evolution of the Bryozoa is uncertain and has been a much-discussed subject during the past 25 years. The geologic evidence is stated in the foregoing sections—*i.e.*, (1) no fossil Entoprocta; (2) no certain fossil Phylactolaemata; (3) simple Cyclostomata and primitive Cryptostomata in the Late Cambrian; (4) abundant Trepostomata, and possibly a representative of the Cheilostomata, by Ordovician time; and (5) Cheilostomata first appearing in the Cretaceous (except as noted in 4) and dominant from then to the present. Any evolutionary or developmental sequence must therefore take account of and be consistent with this fossil record.

The trochophore type of larval stage common to all Bryozoa as defined in this work suggests that the phylum had its origin in some wormlike ancestral stock. Some authors hold that the Bryozoa are closely related to the Brachiopoda and Phoronida as well as to several different phyla of worms. Others, however, question or refuse to accept such suggested relationships. Still others

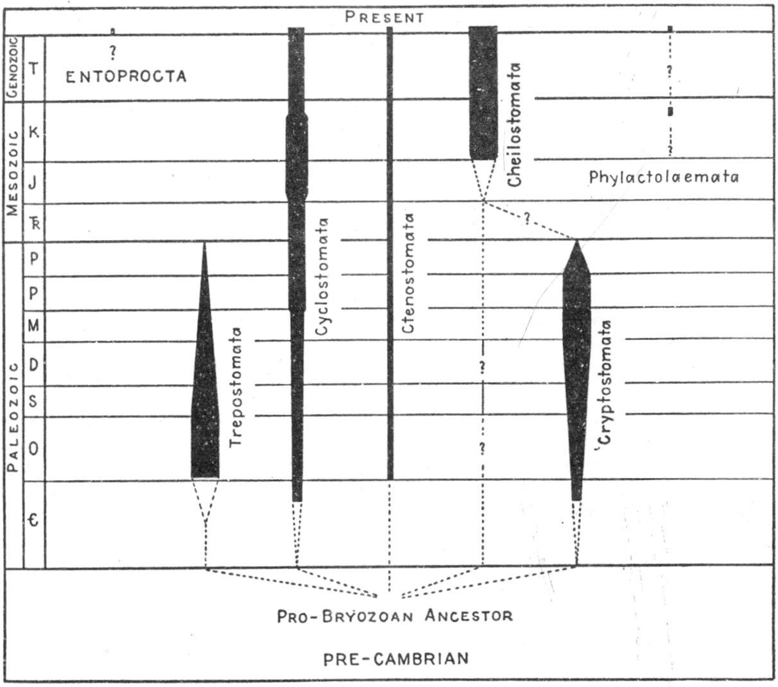

FIG. 7-45. Phylum BRYOZOA. Chart showing the relative importance of the major subdivisions of the phylum throughout geologic time. The width of the line indicates in a general way the relative importance of the division. No fossil Entoprocta are known; Phylactolaemata are extended back to the Cretaceous on a questionable form; and Cheilostomata are extended back to the Ordovician on the assumption that *Paleschara* is a representative of the order. The Cheilostomata are thought by some investigators to have been derived from the Cryptostomata.

would separate the Entoprocta from the Ectoprocta and assign each the rank of phylum (see footnote on page 212). No attempt is made here to evaluate these and other taxonomic questions. Rather, we propose to make a few speculative suggestions.

Whatever the origin of the animals that we include in the Bryozoa, it seems certain that the earliest forms evolved during the Cambrian, and they may well have been differentiated before the Cambrian. The first were probably solitary individuals with only membranous body walls. Colonial forms came later when buds developed and remained attached to the primary zooid.[1] With colonial development went the formation of chitinous and finally calcareous zooecia and zoaria. We are inclined to agree with the opinion of recent bryozoologists (Borg, Silén, etc.) that the Entoprocta, Phylactolaemata, and the five ectoproctous orders with a fossil record all developed independently from a probryozoan ancestral stock. This hypothetical evolutionary history is diagrammed on Fig. 7-45. Other evolutionary possibilities are indicated on this figure by the use of dashed lines, and these are discussed in several places in the preceding text.

REFERENCES[2]

ALLMAN, G. J. 1856. "A Monograph of the Fresh-water Polyzoa." London, Ray Society.

BARROIS, J. 1877. "Recherches sur l'embryologie des Bryozoaires." Lille. 303 pp.

BASSLER, R. S. 1922. The Bryozoa, or moss animals. *Smith Inst.*, *Ann. Rept.* 1920, pp. 339–380.

———. 1935. Bryozoa (Generum et Genotypum Index et Bibliographica). Fossilium Catalogous 1. Animalia, pars 67, pp. 1–229. Berlin.

———. 1939. The Hederelloidea, a suborder of Paleozoic cyclostomatous Bryozoa. *Proc. U.S. Nat. Mus.*, vol. 87, pp. 25–91.

BORG, F. 1926. Studies on Recent cyclostomatous Bryozoa. *Zool. Bidrag*, Uppsala, vol. 10, pp. 181–507.

BUSK, G. 1852. 1875. "Catalogue of Marine Polyzoa in the Collection of the British Museum." London, British Museum (Nat. Hist.).

———. 1884. Report on the Polyzoa collected by *H.M.S. Challenger*, during the years 1873–1876. Pt. 1. The Cheilostomata. *Zool.*, vol. 10, pt. 30.

———. 1886. *Ibid.*, Pt. 2. The Cyclostomata, Ctenostomata, and Pedicellinea. *Zool.*, vol. 17, pt. 50.

CANU, F., and BASSLER, R. S. 1919. Fossil Bryozoa from West Indies. *Carnegie Inst. Wash.*, *Pub.* 291, pp. 73–102.

——— and ———. 1920. North American early Tertiary Bryozoa. *U.S. Nat. Mus.*, *Bull.* 106, 879 pp.

——— and ———. 1922. Studies on cyclostomatous Bryozoa. *Proc. U.S. Nat. Mus.*, vol. 61, 160 pp.

——— and ———. 1923. North American later Tertiary and Quarternary Bryozoa. *U.S. Nat. Mus.*, *Bull.* 125, 302 pp.

[1] Borg (1926, p. 490) concludes: "We may probably have to imagine the Pro-bryozoa as erect sessile animals, more or less cylindrical in shape, combining besides in their organization primitive characters from different orders of Bryozoa now existing. Sooner or later, there may have taken place a process of budding in the upper portion of the body-wall of the animal. One can imagine that the budding zone has been annular, buds thus arising all round the primary individual. This stage is actually represented to a certain extent in the Calyptrostega (Cyclostomata), and possibly it was so also in the Trepostomata. A restriction of the budding to the anal half of the zone or to a limited part of this enables us to form a conception of the origin of, respectively, the ordinary stenolaematous and gymnolaematous zoarium, whereas its restriction to the oral half of the zone gives us a possibility of conceiving the origin of the phylactolaematous type of zoarium. *But these are merely suppositions.*" (Italics are ours.)

[2] This list contains only a few selected references, most of which have extensive bibliographies that will lead the interested student to the great bulk of bryozoan literature.

——— and ———. 1926. Studies on cyclostomatous Bryozoa. *Proc. U.S. Nat. Mus.*, vol. 67, 124 pp.

——— and ———. 1927. Classification of the cheilostomatous Bryozoa. *Ibid.*, vol. 69, 42 pp.

——— and ———. 1928. Fossil and Recent Bryozoa of Gulf of Mexico region. *Ibid.*, vol. 72, 199 pp.

Condra, G. E., and Elias, M. K. 1944. Carboniferous and Permian ctenostomatous Bryozoa. *Bull. Geol. Soc. Amer.*, vol. 55, pp. 517–568.

——— and ———. 1944. Study and revision of *Archimedes* (Hall). *Geol. Soc. Amer., Spec. Paper* 53, 243 pp.

Cori, C. J. 1929. Kamptozoa (Bryozoa entoprocta) *in* Kükenthal and Krumbach's "Handbuch der Zoologie," vol. 2, parts 4–5, pp. 1–64. Berlin and Leipzig, Walter de Gruyter and Co. (This article contains an excellent list of references.)

———. 1936. Kamptozoa *in* Bronn's "Klassen und Ordnungen des Tierreichs," vol. 4, book 4, 119 pp. (This article contains a comprehensive bibliography.)

———. 1941. Bryozoa *in* Kükenthal and Krumbach's "Handbuch der Zoologie," vol. 3, pt. 2, pp. 263–502. Berlin and Leipzig, Walter de Gruyter and Co. (This article contains a comprehensive bibliography.)

Cumings, E. R. 1904. Development of some Paleozoic Bryozoa. *Amer. Jour. Sci.*, vol. 17, pp. 49–78.

———. 1905. Development of *Fenestella. Ibid.*, vol. 20, pp. 169–177.

———. 1912. Development and systematic position of the Monticuliporoids. *Bull. Geol. Soc. Amer.*, vol. 23, pp. 357–370.

——— and Galloway, J. J. 1915. Studies of the morphology and histology of the Trepostomata or monticuliporoids. *Ibid.*, vol. 26, pp. 349–374.

Deiss, C. F. 1932. A description and stratigraphic correlation of the Fenestellidae from the Devonian of Michigan. *Univ. Mich., Contr. Mus. Pal.*, vol. 3, pp. 233–275.

Duncan, H. 1939. Trepostomatous Bryozoa from the Traverse group of Michigan. *Ibid.*, vol. 5, pp. 171–170.

Fritz, M. A. 1932. Permian Bryozoa from Vancouver Island. *Trans. Roy. Soc. Can.*, 3d ser., sec. 4, vol. 26, pp. 93–109.

———. 1947. Cambrian Bryozoa. *Jour. Paleontology*, vol. 21, pp. 434–435.

Grassé, P. P. 195–. "Traité de zoologie," vol. 5. Brachiopodes, Bryozoaires, Annelides, Mollusques. (In press.) Paris, Masson et Cie.

Gregory, J. W. 1896–1909. "Catalogue of the Fossil Bryozoa in the Department of Geology." London, British Museum (Nat. His.).

———. 1896. "The Jurassic Bryozoa." *Ibid.*

———. 1899, 1909. "The Cretaceous Bryozoa." *Ibid.*, vols. I, II.

Harmer, S. F. 1910. Polyzoa *in* Cambridge Natural History, vol. 2, pp. 463–533. New York, Cambridge University Press., Ltd. This author published many important articles on Polyzoa during the twenty years preceding this work, *i.e.*, 1890–1910.

———. 1930. Polyzoa (Presidential Address, Linnean Society). *Proc. Linn. Soc. London*, 1928–1929, pp. 68–118. An excellent summary of the Polyzoa.

———. 1931. Recent work on Polyzoa (Presidential Address). *Ibid.*, 1930–1931, pp. 113–168.

Hincks, T. 1880. "British Marine Polyzoa," 2 vols. London, John van Voorst.

McNair, A. H. 1937. Specialized zooecia in cryptostomatous Bryozoa (abstract). *Geol. Soc. Amer. Proc.* 1936, p. 361.

———. 1937. Cryptostomatous Bryozoa from the Middle Devonian Traverse Group of Michigan. *Univ. Mich., Contr. Mus. Pal.*, vol. 5, pp. 103–170.

———. 1938. Stellate apertures in Bryozoa (abstract). *Geol. Soc. Amer. Proc.* 1937, pp. 284–285.

Marcus, E. 1937–1939. Bryozoarios marinhos brasileiros. I. *Univ. S. Paulo, Bol. Fac.*

Phil. Sc. Letr., vol. I, *Zool.*, vol. 1, pp. 1–224, (1937); II. *Ibid.*, vol. 2, pp. 1–196, (1938); III. *Ibid.*, vol. 3, pp. 111–354 (1939). (This work contains an excellent bibliography.)

MOORE, R. C., and DUDLEY, R. M. 1944. Cheilotrypid bryozoans from Pennsylvanian and Permian rocks of the Midcontinent region. *Univ. Kan. Pub., Bull.* 52, pt. 6, pp. 229–408.

OSBURN, R. C. 1950. Bryozoa of the Pacific Coast of America. Allan Hancock Pacific Expedition, vol. 14, pp. 1–269.

PROUHO, H. 1892. Contribution à l'histoire des Bryozoaires. *Arch. zool. exp.* 2, ser. 10, pp. 557–656.

RAYMOND, P. E. 1941. Invertebrate Paleontology *in* Fiftieth Anniversary Volume, Geological Society of America, pp. 73–103. (Bryozoa, p. 77).

ROBERTSON, A. 1908. The incrusting chilostomatous Bryozoa of the West Coast of North America. *Univ. Cal. Pub. Zool.*, vol. 4, pp. 253–344.

———. 1910. The cyclostomatous Bryozoa of the West Coast of North America. *Ibid.*, vol. 6, pp. 225–284.

SHIMER, H. W., and SHROCK, R. R. 1944. "Index Fossils of North America." New York, John Wiley & Sons, Inc.; Cambridge, Mass., The Technology Press. Chap. VIII, Phylum Bryozoa (Polyzoa), pp. 247–276. (This chapter has a long list of references to works in which North American bryozoans are described.)

SILÉN, L. 1942. Origin and development of the cheilo-ctenostomatous stem of Bryozoa. *Zool. Bidrag.*, Uppsala, vol. 22, pp. 1–59.

———. 1942*a*. On the formation of the interzoidal communications of the Bryozoa. *Ibid.*, vol. 22, pp. 433–488.

———. 1948. On the anatomy and biology of Penetratiidae and Immergentiidae. *Arkiv. Zool.*, Stockholm, vol. 40, pp. 1–48.

TWITCHELL, G. B. 1934. *Urnatella gracilis* Leidy, a living trepostomatous bryozoan. *Amer. Midland Naturalist*, vol. 15, pp. 629–661.

ULRICH, E. O. 1890. Paleozoic Bryozoa. *Illinois Geol. Surv.*, vol. 8, pp. 283–688.

———. 1895. On Lower Silurian [Ordovician] Bryozoa of Minnesota. *Minn. Geol. Surv., Final Rept.* 3, pt. 1, 1895, pp. 96–332.

———. 1911. Bearing of the Paleozoic Bryozoa on paleogeography. *Bull. Geol. Soc. Amer.*, vol. 22, pp. 252–257.

——— and BASSLER, R. S. 1904. A revision of the Paleozoic Bryozoa. Pt. 1, On genera and species of Ctenostomata. *Smith. Misc. Coll.*, vol. 45, pp. 256–294.

——— and ———. 1904*a*. A revision of the Paleozoic Bryozoa. Pt. 2, On genera and species of Trepostomata. *Ibid.*, vol. 47, pp. 15–55.

VAN BENEDEN, P. J. 1845. "Recherches sur l'anatomie, la physiologie, et l'embryogénie des Bryozoaires." Brussels.

VIGNEAUX, M. 1949. Révision des Bryozoaires néogènes du Bassin d'Aquitaine et essai de la classification. *Mém. soc. géol. France*, vol. 28, pp. 1–155.

WATERS, A. W. 1884. Fossil cyclostomatous Bryozoa from Australia. *Quart. Jour Geol. Soc. London*, vol. 40, pp. 674–697.

———. 1887. On Tertiary cyclostomatous Bryozoa from New Zealand. *Ibid.*, vol. 43, pp. 337–350.

———. 1924. The Ancestrula of *Membranipora pilosa* L. and of other cheilostomatous Bryozoa. *Ann. Mag. Nat. Hist.*, (9), vol. 14, pp 594–612.

———. 1925. Ancestrulae of cheilostomatous Bryozoa. *Ibid.*, vol. 15, pp. 341–352. (This author has published many important reports on living and fossil Bryozoa.)

CHAPTER 8

PHYLUM PHORONIDA

The Phoronida are slender worm-like marine organisms that live a solitary and sedentary existence in burrows excavated in the bottom sediments of shallow seas. Most of the 15 described species belong to the genus *Phoronis* (Fig. 8-1*A*). The relatively long cylindrical body, which reaches a length as great as 150 mm. (6 in.), is unsegmented and is crowned with a tufted scroll-shaped structure, the **lophophore,** which bears ciliated tentacles and surrounds the mouth (Fig. 8-1*A–B*). Internally a simple, complete U-shaped

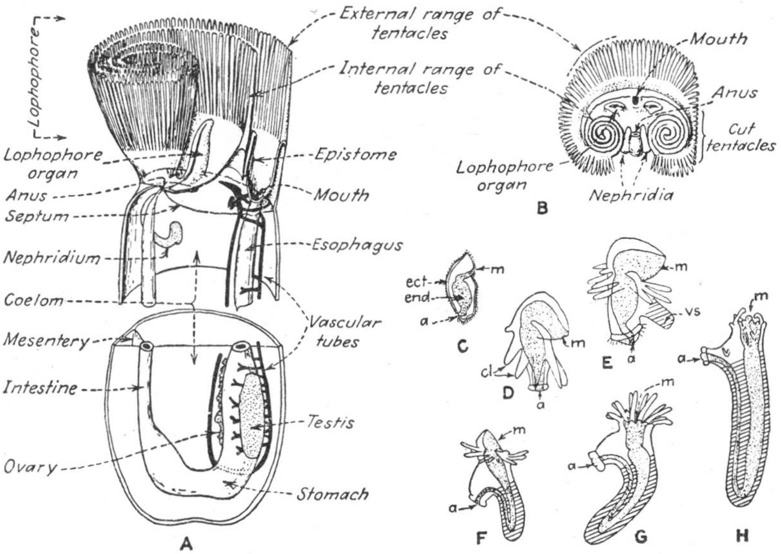

FIG. 8-1. Phylum PHORONIDA. *A.* Schematic diagram of *Phoronis*, with middle part missing, showing chief internal features. Not all the six mesenteries are shown. *B.* Anterior view of *Phoronis*, with inner range of tentacles and right portion of outer range cut away. The nephridial openings are at the upper end of the organs. *C–H. Actinotrocha* and its metamorphosis into an adult *Phoronis*. (*a* = anus; *cl* = ciliary lobes; *ect* = ectoderm; *end* = endoderm; *m* = mouth· *vs* = visceral sac, which is crosslined in *E–H*.) (*Modified after Delage and Hérouard and others*, 1897.)

digestive tract is suspended in the compartmented body cavity, or **coelom,** with the mouth between and at the base of the two arcs of the lophophore and the anus diametrically opposite near the body wall. The organism is hermaphroditic (*i.e.*, monoecious), has a vascular system with hemoglobin, a relatively simple nervous system and two nephridia, and a body wall consisting of cuticle, epidermis, muscle fibers, and coelomic lining.

Phoronids are gregarious, living in groups on the bottoms of shallow seas where their aggregations may cover several square meters (Fig. 8-4). In these assemblages, which might easily be mistaken for colonies, each individual is distinct and has no organic connection with any other member, although the tubes housing the individual phoronids are commonly intertwined in a tangled mass (Fig. 8-2). When disturbed the organism quickly withdraws into the tube, from which after a time it again emerges slowly, not expanding its crown of tentacles until the body is completely extended.

The early life history of the phoronid is a series of remarkable changes, with certain growth stages of a trochophore type (Fig. 8-1*C–H*). The fertilized ova undergo their early developmental stages amidst the tentacles of the parent. The larva resulting from this period of development leaves the parental tentacles early in its growth and swims about actively in the open sea, where it undergoes certain changes to become the distinctive phoronid larva known as the *Actinotrocha* (Fig. 8-1*C–H*). This larva in turn sinks to the sea bottom,

FIG. 8-2. Phylum PHORONIDA. A portion of a tangled colony of *Phoronis kowalevskii*, slightly magnified, showing several tentacular crowns protruding from the tubes. The feltwork made by the tubes attains a thickness of 5 to 8 cm. (2 to 3 in.) and is commonly a foundation on which different animals settle. The tubes themselves, which are generally much longer than the animal, are rendered opaque by excreta and usually have only a little granular detritus attached to the surface. (*Modified after Shipley*, 1910.)

where it undergoes a most remarkable metamorphosis ending in the adult organism.

The place of the Phoronida in the animal kingdom is uncertain. Many zoologists group the phylum with the Bryozoa and Brachiopoda, because of certain resemblances (lophophore, etc.); some, however, emphasizing certain likenesses to *Cephalodiscus*, a hemichordate, prefer to consider the phoronids more closely related to that group of invertebrates. Unfortunately, the lack of a significant fossil record leaves us without any knowledge of the geologic history of this small but interesting phylum.

In life the individual phoronid secretes about itself a tube of organic substance that ultimately becomes opaque and has a leathery or membranous texture. These tubes are intricately intertwined in certain species that live on piles, and they form a dense feltwork upon which other organisms commonly attach themselves (Fig. 8-2). In certain species, sand particles, shell fragments, sponge spicules, and other bottom debris adhere to the outer surface of the tubes, giving them a characteristic appearance not unlike those of some tubicolous annelids (Fig. 8-3).

The individuals of still other species build upright or vertical tubes in the sands and sandy muds of the littoral zone. These tubes consist of a chitinoid

FIG. 8-3. Phylum PHORONIDA. A portion of a colony of *Phoronis psammophila*, slightly magnified, with the tentacular head protruding from several of the tubes. The tubes, which are composed of a tough organic substance, are armored with particles of sand, small shells, shell fragments, and other marine objects. (*Modified after Shipley*, 1910.)

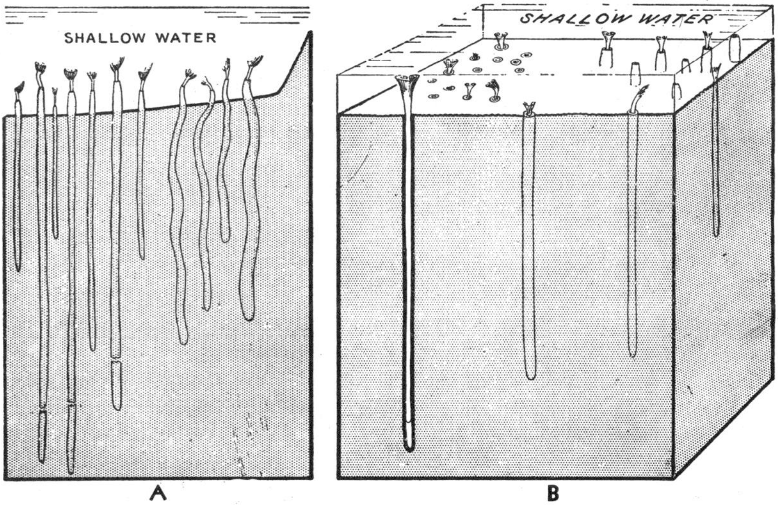

FIG. 8-4. Phylum PHORONIDA. *A*. Straight and sinuous tubes built by *Phoronopsis* in the sands of Elkhorn Slough, Calif. The tubes range in diameter from 3 to 4 mm. and in length from 100 to 150 mm. (4 to 6 in.). They consist of a chitinoid membrane to which sand grains are cemented. *B*. Diagram showing phoronid tubes with apertures at the level of the bottom or slightly below (on the left side of the block) and others which stand several millimeters above the bottom. The longest tube, at the extreme left, is about 150 mm. (6 in.) long and is sectioned longitudinally to show that the body of the phoronid may occupy most of the tube. (*A adapted from Fenton and Fenton*, 1934.)

membrane to which sand generally is cemented. Commonly they are straight and a centimeter or more apart, but they may also be flexuous and crowded (Fig. 8-4*A–B*). They extend downward into the sand as deep as 150 mm. (6 in.), and the upper part of the tube may end a few millimeters below the sand surface or may rise above the bottom as much as 15 mm. In some places along the California coast sand bars are held in place by the great abundance of such phoronid tubes that bind the sand together (Fenton and Fenton, 1934, p. 345).

Present-day phoronids live under conditions and in environments where they would seem to have little chance of leaving a fossil record. If it is assumed that they were similarly adapted in the past—and there is little evidence to the contrary—then it is not difficult to understand why ancient phoronids

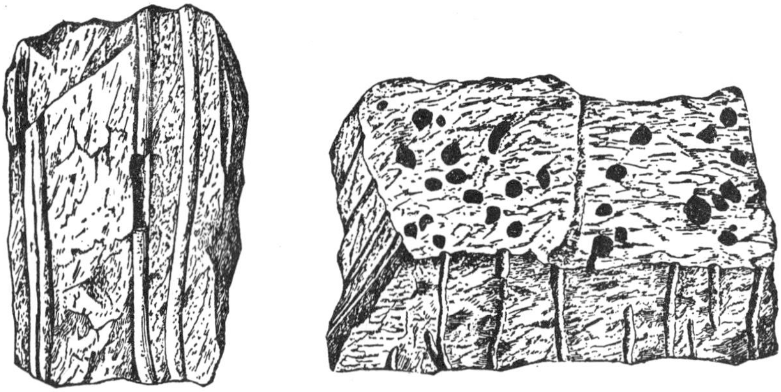

A B

FIG. 8-5. *Scolithus;* borings or tubes ascribed by some to wormlike organisms, by others to phoronids. *A* shows several tubes largely filled with sand: *B* shows longitudinal and transverse sections of numerous tubes. Both specimens are sandstone blocks and are slightly reduced. Scolithid tubes with flaring or funnel-shaped apertures have been referred to the genus *Monocraterion* (L. Camb.–M. Sil.). (*After James*, 1892.)

seem to have left no evidence of their existence. That the phoronids are an ancient group is suggested by their complex embryology and general life history, as well as their present world-wide distribution.

Some years ago the Fentons presented strong arguments for considering upright tubes of a certain type common in Early Paleozoic sandstones as dwelling places of ancient phoronids. These tubes, which are generally referred to the form genus *Scolithus* (Fig. 8-5) and assigned under this designation to ancient burrowing wormlike organisms, closely resemble tubes that are being made today by certain species of *Phoronis* and *Phoronopsis*,[1] as described in the foregoing paragraph. If these fossil tubes were made by

[1] Fenton and Fenton (1934, p. 348) state: "In shape, structure, size, and spacing, tubes built by sand-dwelling phoronids (*Phoronopsis* spp. and *Phoronis architecta* Andrews) correspond closely to those indicated by *Scolithus*. There also are significant, though not convincing, parallels in distribution and sedimentary relationships."

ancient phoronids, (an origin respecting which we reserve opinion), then the Phylum Phoronida was well represented during the earlier part of the Paleozoic, as *Scolithus* has been reported from rocks ranging in age from Cambrian to Devonian or younger. They are comparatively rare in rocks younger than Silurian and do not seem to have been reported from Mesozoic or Cenozoic strata, though certain Cretaceous tubes might be referred to them.

REFERENCES

CORI, C. J. 1937. Phoronidea *in* Kükenthal and Krumbach's "Handbuch der Zoologie," vol. 3, pt. 2, pp. 71–134. Berlin and Leipzig, Walter de Gruyter and Co.

DELAGE, Y., and HÉROUARD, E. 1897. "Traité de zoologie concrète," vol. 5, pp. 156–164. Paris, Schleicher Frères.

FENTON, M. A., and FENTON, C. L. 1934. *Scolithus* as a fossil phoronid. *Pan-Amer. Geol.*, vol. 61, pp. 341–348.

JAMES, J. F. 1892. Studies in problematical organisms; the genus *Scolithus*. *Bull. Geol. Soc. Amer.*, vol. 3, pp. 32–44.

SHIPLEY, A. E. 1910. Phoronis *in* Cambridge Natural History, vol. 2, pp. 450–462. New York, Cambridge University Press.

CHAPTER 9

PHYLUM BRACHIOPODA[1]

INTRODUCTION

Brachiopods are bottom-dwelling bivalve marine animals of generally small size. The **shell,** or **test,** of an adult brachiopod consists of two bilaterally symmetrical **valves,** more or less dissimilar in shape (Fig. 9-1), which enclose the soft parts of the animal, and to the interior of which the body is firmly attached. In one division of the phylum, the Articulata, the two valves are united with one another along a **cardinal margin,** or **hinge line,** by means of **teeth** and **sockets** (Figs. 9-1*A'*, 9-11*I–L*, 9-14*B–C*). Representatives of this division are commonly designated hinged brachiopods, or simply **articulates.** In the other division, the Inarticulata, the two valves lack articulation in the same sense that it obtains in the articulates, and are kept in apposition primarily and in many cases solely by muscles (Fig. 9-1*A–D*, 9-11*A–H*). Shells of this division are commonly spoken of as **inarticulates** (Thomson, 1927).[2]

[1] Brachiopoda—Gr. *brachion*, arm, + *pous*, *podos*, foot; referring to the brachia which early investigators mistakenly thought were homologous with the foot of mollusks.

Early zoologists and paleontologists generally grouped the Bryozoa, Phoronida, and Brachiopoda together in the Phylum Molluscoidea. The reasons for this association were that (1) the embryos of the three groups pass through a somewhat similar trochophore-like stage; (2) the organisms in each group have a lophophore throughout life, although this structure varies in morphology in the three groups; and (3) the soft parts, particularly the digestive tract and nervous system, are similar.

It is now almost universal practice to consider the Bryozoa, Phoronida, and Brachiopoda as distinct and separate phyla and to drop the designation Molluscoidea altogether. This practice is based on the facts that (1) the adults in the three groups differ considerably in morphology and otherwise, the differences beginning to appear soon after the larvae have settled down; (2) there are fundamental differences in the shelly structures within which the fleshy parts of the animal are housed; and (3) there are great differences in the habits and habitats of the adult organisms. The individual brachiopod is solitary, always reproduces sexually, and has a bivalve shell that bears no resemblance to the tubular zooecium of a bryozoan or the burrow of a phoronid, though it is interesting to note that the common brachiopod *Lingula* lives in a burrow excavated in sandy bottoms. Bryozoans, in contrast to solitary brachiopods and phoronids, are colonial and build a wide range of calcareous and chitinoid structures (zoaria) composed of numerous individual tubular or prismatic zooecia, each of which houses an individual bryozoan. Whereas the adult brachiopod shell is commonly a centimeter or more across, the typical bryozoan zooecium is seldom more than a millimeter or two in diameter, and the whole colonial structure, or zoarium, is commonly no larger than a single brachiopod shell. The phoronid lacks hard parts altogether.

[2] The interested reader will find an excellent discussion of brachiopods in J. A. Thomson's "Brachiopod Morphology and Genera (Recent and Tertiary)," published in 1927 as Manual 7 of the New Zealand Board of Science and Art. This work has served as a model for much of the present discussion.

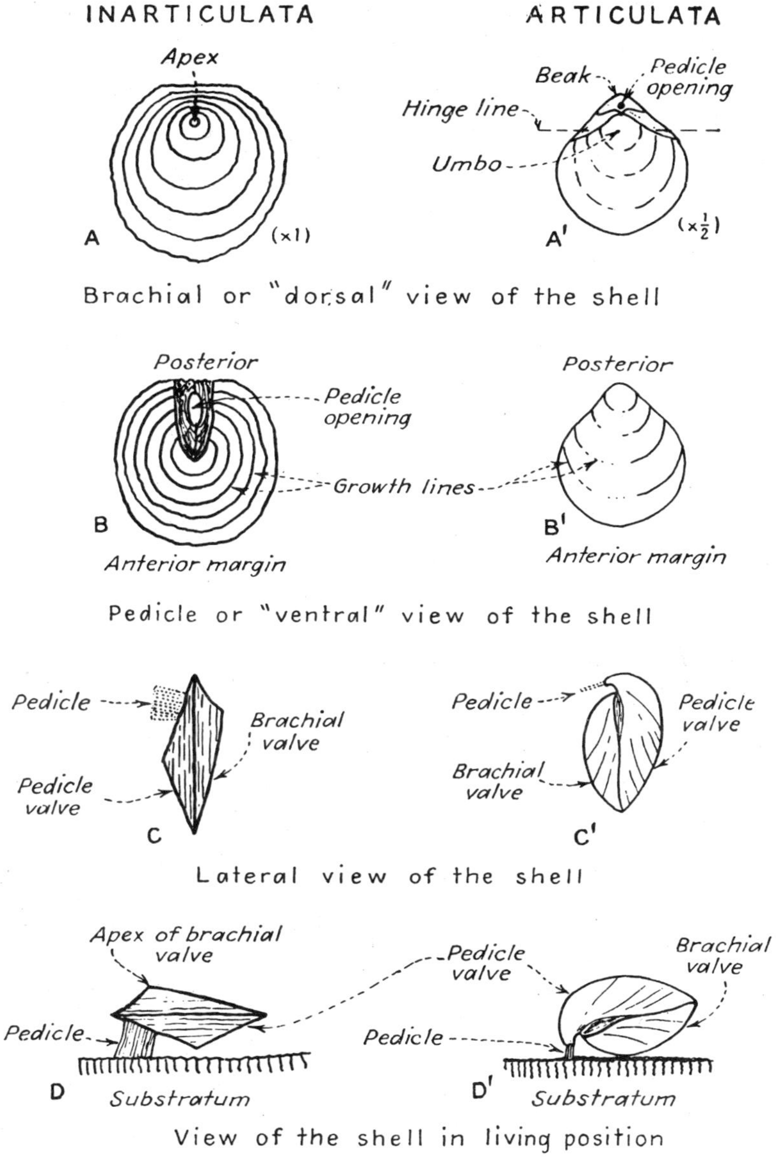

FIG. 9-1. Phylum BRACHIOPODA. Diagrams showing typical representatives of the two chief divisions of the phylum. The views of the inarticulate are based on the common living neotremate genus *Discinisca*, which has a chitinophosphatic shell. The views of the articulate are based on the well-known telotremate genus *Magellania*, which has a calcareous shell.

Living brachiopods are dominantly bottom-dwelling animals. They are normally attached to rocks, shells, or other objects on the bottom by a flexible stalk, the **pedicle** or **peduncle** (Fig. 9-1*D–D'*). The pedicle is attached to the inner surface of the larger valve by muscles within the shell and protrudes either between the valves, which is relatively uncommon, or through a **pedicle opening** restricted to the posterior part of the larger or **pedicle valve** (Fig. 9-1*A'*, *B*). In a few brachiopods the pedicle is atrophied, and the shell either lies free on the bottom, becomes anchored by spines, or is cemented by one of the valves to some object on the bottom (Fig. 9-10). Members of one family, the Lingulidae, can move about over the bottom, though characteristically they anchor themselves in sandy and muddy sediments by encasing the end of the long pedicle in a sand tube (Fig. 9-10*S–W*).

Embryological studies show that the two valves of a brachiopod are morphogenetically ventral and dorsal in position, in contrast to the bivalve molluscan Pelecypoda in which the valves are right and left.

Although it has long been common practice to designate the two valves as ventral and dorsal, many brachiopod specialists prefer **pedicle** for ventral and **brachial** for dorsal, inasmuch as these terms more accurately describe the shells and are less ambiguous.[1] We also prefer these designations.

The pedicle valve of an articulate is longer than the brachial, carries the teeth, and in most species has the pedicle opening wholly or partly restricted to its posterior portion. The brachial valve is the shorter, bears the sockets, and in many species has an internal structure, the **brachidium,** that acts as a support for the fleshy **brachia** and associated soft parts (Fig. 9-4).

The body of a brachiopod, as well as the two valves of the shell, is bilaterally symmetrical, with the plane of symmetry passing through the pointed posterior **beaks** and rounded anterior margins in such position as to divide the animal and the two enclosing valves into essentially similar halves (Figs. 9-1, 9-2, 9-3). It should be noted that although certain brachiopods and pelecypods have somewhat similar shapes, the plane of symmetry lies *between* the valves in the latter but longitudinally transects those of the former (Fig. 9-2).

In the study, description, and illustration of a brachiopod shell it is customary to show the valves with the pointed part directed toward the top of the page (Figs. 9-1, 9-3), regardless of whether the valve is viewed externally or internally. Several different positions in which shells and valves can be viewed are shown in Fig. 9-3. The beak or pointed part of a valve is posterior, the opposite rounded margin is anterior, and the line along which the valves are in contact is the **commissure.** In those shells of Neotremata that have no true beak the posterior part of the shell is generally more or less straight,

[1] Percival (1944, p. 1), from embryological studies, has shown that, for some brachiopods at least (*e.g.*, *Terebratella*), "the current terms of dorsal and ventral, as applied to the adult, must be reversed, what was called dorsal is primarily ventral as it is related to the blastopore, and vice versa." Cloud (1948) has discussed the paleontological implications of Percival's discoveries, and the interested reader will find additional pertinent suggestions on brachiopod evolution in his article.

whereas the anterior margin is rounded. If these same shells have an apex, it is posterior (Fig. 9-1*A–D*).

If there is doubt as to which valve of a brachiopod shell is the pedicle or which the dorsal, or if the posterior end cannot be determined from observation of the shell's exterior, these questions can be answered definitely by studying internal features, such as muscle scars, pallial markings, and teeth or sockets (see detailed discussion in later paragraphs).

The latest census of living and extinct brachiopods differs somewhat from

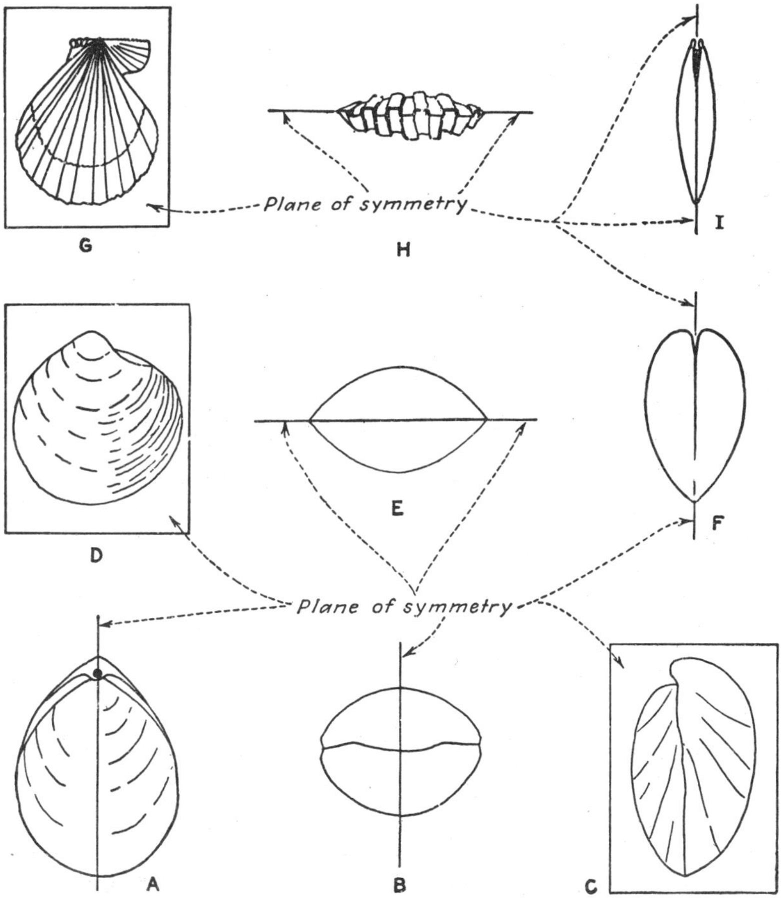

FIG. 9-2. Brachiopoda and Pelecypoda compared, with reference to plane of bilateral symmetry. In the brachiopod shell (*A–C*) the valves are unequal but equilateral, so that the plane of symmetry passes through the beaks and anterior margins of the valves. In pelecypod shells (*D–I*) the valves are generally equal but inequilateral, hence the plane of symmetry passes between the valves.

previous estimates and from one specialist to another. Recent estimates indicate between 200 and 225 described living species, which are distributed among some 61 or so genera, and about 30,000 described fossil species included in 1,250 to 1,400 genera, of which the majority are Paleozoic. These figures, even though only approximate, suffice to demonstrate the major role of brachiopods in the past as compared to their minor role in present marine faunas.

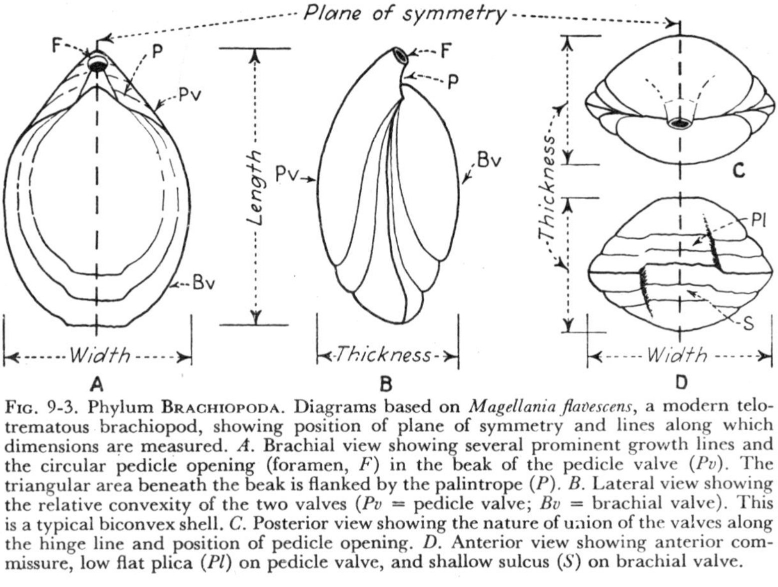

FIG. 9-3. Phylum BRACHIOPODA. Diagrams based on *Magellania flavescens*, a modern telotrematous brachiopod, showing position of plane of symmetry and lines along which dimensions are measured. *A*. Brachial view showing several prominent growth lines and the circular pedicle opening (foramen, *F*) in the beak of the pedicle valve (*Pv*). The triangular area beneath the beak is flanked by the palintrope (*P*). *B*. Lateral view showing the relative convexity of the two valves (*Pv* = pedicle valve; *Bv* = brachial valve). This is a typical biconvex shell. *C*. Posterior view showing the nature of union of the valves along the hinge line and position of pedicle opening. *D*. Anterior view showing anterior commissure, low flat plica (*Pl*) on pedicle valve, and shallow sulcus (*S*) on brachial valve.

THE ANIMAL

The soft parts of the brachiopod animal are contained between the two valves of the shell (Fig. 9-4). For purposes of description they may be divided into the following five parts: (1) the **body,** containing the viscera, and traversed by the muscles; (2) the **mantle;** (3) the **lophophore;** (4) the **pedicle;** and (5) the **muscles.**

The Body

The actual body is confined to the posterior third or so of the space between the valves. It is surrounded by the **body wall** that consists of an external epidermis, connective tissue, and a ciliated coelomic lining. The space thus surrounded is termed the **visceral cavity,** or **coelom,** and in it lie the chief organs of the animal.

Two folds of the body wall extend anteriorly as the **mantle lobes.** These are in contact with the two valves and send fine papillae into the shell sub-

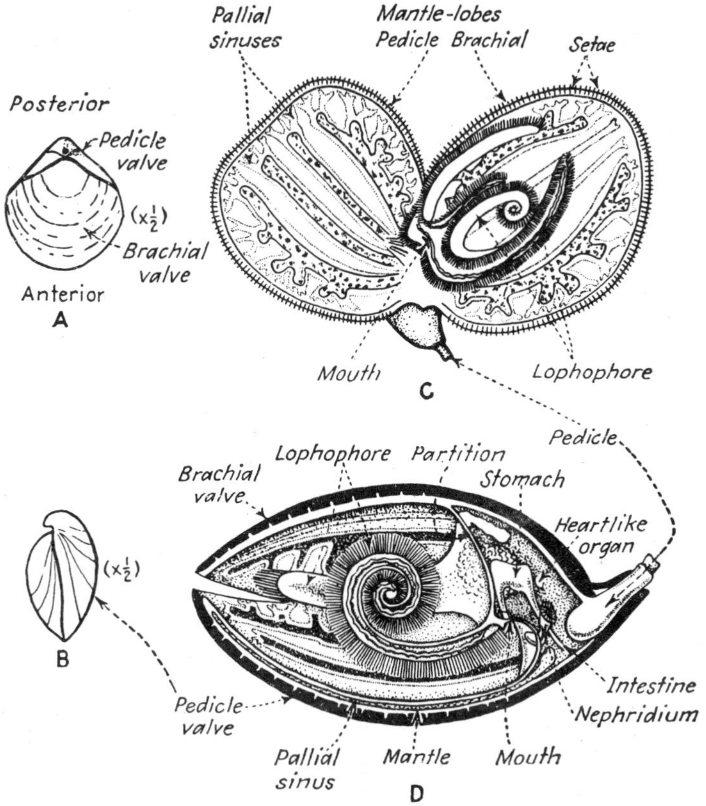

FIG. 9-4. Phylum BRACHIOPODA. Structure of a modern articulate brachiopod, *Magellania lenticularis*. *A–B*. Brachial and lateral views, respectively, of the whole shell. *C*. The body removed from the shell and with the mantle lobes spread apart to show internal features. *D*. Section of animal and shell along the plane of symmetry. Both figures are diagrammatic, and the lophophore is represented as smaller proportionately than in the actual animal. (*A–B adapted from Davidson*, 1886–1888; *C–D modified after Parker and Haswell*, 1928.)

stance (Fig. 9-5). Anteriorly, the body wall forms an obliquely transverse partition that separates the coelom from the anterior **mantle cavity.**

Extending forward from this partition and occupying most of the space between the mantle lobes (*i.e.*, the mantle cavity) is the lophophore. At the middle of the area along which the lophophore joins the wall is a simple oral opening, the so-called **mouth,** through which food-bearing currents enter the alimentary tract.

The mouth opens directly into the digestive tube, which occupies much of the visceral cavity. The tube consists of a short **gullet,** a pouchlike **stomach** with paired **digestive glands,** and a long slender **intestine.** The intestine of **articulate** brachiopods ends blindly, as there is no anus. By contrast, the

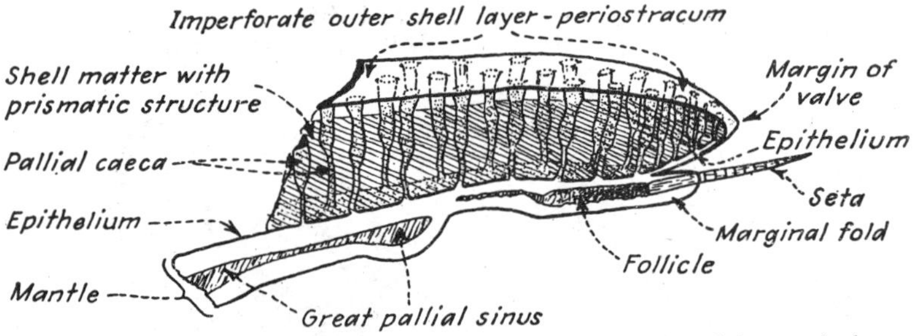

FIG. 9-5. Phylum BRACHIOPODA. Diagram of a longitudinal section of the marginal part of one valve of *Magellania australis*, a modern telotremate, showing relation of fleshy mantle to shell substance. The mantle gives off trumpet-shaped caeca which penetrate the prismatic calcareous layer but end blindly against the imperforate chitinous periostracum. The epithelium of the mantle seems to be continuous with the periostracum along the margin of the shell. A prominent marginal fold of the mantle is beset with setae. The shell is said to be endopunctate since the tubes, or puncta, occupied by the caeca are visible only on the inner surface of the valve. (*Adapted from Hancock*, 1859.)

intestine of an inarticulate makes three or four convolutions and leads through a **rectum** to an **anus,** which opens into the mantle cavity.

In addition to the alimentary tract, the fluid-filled coelom contains the internal organs, which are supported on mesenteries. Branches of the coelomic cavity extend into the mantle lobes as **pallial sinuses** and into the lophophore. In the lingulids there is also a tubular extension into the pedicle. A digestive gland, the so-called **liver**, envelops the stomach and discharges its secretions into the stomach through special canals. The **blood** of a brachiopod contains corpuscles, leucocytes, and spindle bodies, and circulates through a complicated system of ciliated **blood vessels.** A small **heartlike organ** that beats weakly may aid slightly in the circulation of the blood, but the cilia lining the vessels seem to be almost entirely responsible for maintaining circulation. A large trumpet-shaped "kidney," or **nephridium,** lies on each side of the intestine. These two excretory organs open into the coelom at their larger end; at the smaller end they join the transverse partition of the body wall and open through it into the mantle cavity. A **nerve ring** surrounds the gullet and sends extensions to several of the organs, but no special sense organs are present.

Brachiopods are dioecious, *i.e.*, the sexes are separate. Each animal has two **gonads,** one ventral and the other dorsal, and the nephridia serve as reproductive ducts, discharging eggs and sperm to the exterior through the mantle cavity. The fertilized egg grows into a free-swimming trochophore-like larva, which ultimately attaches itself by means of a special structure that becomes the pedicle (see Embryology).

The Mantle

The **mantle,** or **pallium,** is a thin double-layered membrane that lines the interior of the shell anteriorly from the body (Fig. 9-4*C–D*). As previously stated, it consists of two anterior extensions of the body wall; one of these

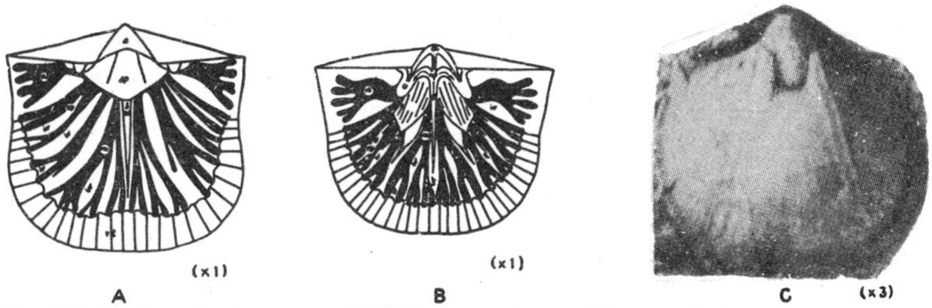

FIG. 9-6. Phylum BRACHIOPODA. Pallial or vascular markings. *A–B*. Pedicle and brachial interiors, respectively, of *Estlandia marginata*, showing in solid black a reconstruction of the vascular markings. It is to be noted that the main trunks split into smaller ones toward the shell margins, and that the system of markings is essentially bilaterally symmetrical. *C*. An impression of the interior of the pedicle valve of an Upper Cambrian species, *Billingsella plicatella*, made by pressing molding clay into the valve. (*A–B after* Öpik, 1934; *C original from Bell; also Bell*, 1941.)

lines the pedicle valve, the other lines the brachial valve. The outer layer of the membrane is in contact with the shell and sends minute **papillae,** or **caeca,** into all shell layers except the outermost, the **periostracum** (Fig. 9-5). The tiny perforations left in the shell after death of the animal are termed **puncta**[1] (singular, **punctum**) and are discussed more fully under shell structure. The inner layer of the membrane represents an extension of the anterior and lateral surfaces of the body wall.

The two layers are firmly united over most of the mantle, but in certain places they are separated to accommodate slender tubular extensions of the visceral cavity known as **pallial** or **vascular sinuses** (Figs. 9-4*C*, 9-5). These tubes have the form of large trunks, entering the mantle lobes from the coelom, and generally they branch toward the shell margins. The sinuses exhibit considerable variation in nature and pattern in different groups of brachiopods.

The presence and disposition of the sinuses are commonly indicated in fossil shells by **pallial** or **vascular markings** (Fig. 9-6). In some brachiopods the shell substance is thinner opposite the main trunks, so that the latter can be followed either by translucence from the exterior or by slight depres-

[1] It has been pointed out by several writers that the form *punctum* (plural, *puncta*) is preferable to *puncta* (plural, *punctae*) for the tiny holes in brachiopod shells, arthropod exoskeletons, etc. We are indebted to Rev. J. D. Gauthier, S.J., of Boston College for the information that both *punctum* (neuter) and *puncta* (feminine) come from the perfect participle of *pungere*, to prick or puncture, hence should have identical meanings. *Puncta* (feminine), however, is extremely rare, although it is found once in a fourth-century writer, Vegetius. *Punctum* (neuter), by contrast, is the traditional form of the word used for the past three centuries and is the one defined in Webster's New International Dictionary and in the large Oxford English Dictionary. Even though some authors may wish to rationalize the use of *puncta* (plural, *punctae*), on the grounds that there is such a word, rare though it be, we prefer to follow traditional practice as established by more than three centuries of usage and to adopt *punctum* (plural, *puncta*) in this and other chapters.

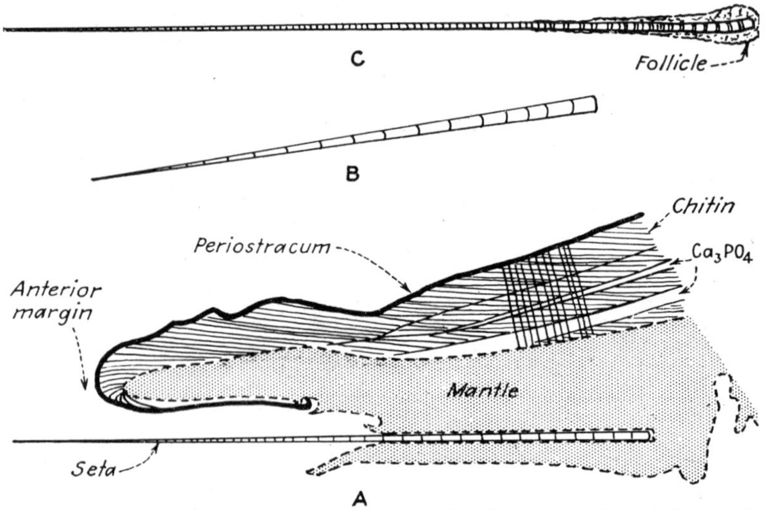

FIG. 9-7. Phylum BRACHIOPODA. Chitinous setae. *A*. Idealized section through the anterior margin of the mantle and shell of *Lingula*, showing the relations of seta, periostracum, and successive shell layers to ectodermal surface of mantle. *B*. One of the setae from the mantle margin of *Waldheimia australis*, highly magnified. *C*. A marginal seta of *Terebratulina caput-serpentis* greatly enlarged and showing the follicle from which it originates. (*A after Blochmann*, 1900; *B–C after Hancock*, 1859.)

sions on the inner surface of the valves. On the surfaces of internal fillings (cores, or steinkerns) the markings commonly stand out in a way similar to the blood vessels of a person's hand.

The external epithelium of the mantle seems to be continuous with the horny periostracum of the shell and serves as the shell-secreting part of the animal (Figs. 9-5, 9-7*A*). In those brachiopods having a perforate or punctate shell, the papillae or caeca of the mantle fill the pores and end blindly against the periostracum. The function of the papillae is uncertain. Some authors suppose that they aid in nourishing the organic tissue of the shell (Thomson), whereas others have assigned them a sensory function (Sollas).

Slightly within the edge of the shell the inner layer of the mantle is thickened into a fold that is free anteriorly and capable of extension and contraction. This fold generally carries a row of stout chitinous **setae,** which project beyond the shell margin (Figs. 9-5, 9-7) and vary in form in different genera. Extended setae have also been noted in dead brachiopods and traces of them have been found in a few fossils (*e.g.*, *Dictyonina pannula* from the Middle Cambrian Burgess shale of British Columbia; Walcott, 1912, Fig. 32, page 362).

The Lophophore

The **lophophore** is a variously shaped appendage attached to the anterior surface of the body and with the mouth at the middle in the plane of bilateral symmetry (Figs. 9-4*C–D*; 9-8). It occupies a large part of the anterior two-thirds of the space between the mantle lobes, and for this reason the mantle cavity is commonly also called the **brachial cavity.** It creates within this cavity currents of sea water that not only enable the mantle lobes to perform their respiratory function but also bring food to the mouth and carry away rejected substances, waste materials, and sexual products.

The lophophore exhibits considerable morphological variation in different families and genera (Fig. 9-8). In some brachiopods it consists of a simple or simply lobed disk, in others of two elongate **brachia,** or arms, that are coiled or folded. The edge of the disk, or of the brachia, bears long tentaclelike **cirri,** at the base of which lies a trough, the **brachial groove,** leading to the mouth. **Cilia** are disposed in two series on the cirri; they line the brachial grooves, and cover the body of the lophophore and the mantle lobes. These cilia by a lashing motion produce currents in the brachial cavity—an incoming current on each side, and an outgoing current in the middle.

When the shell is open, the coiled or folded brachia can be unwound and extended beyond the anterior margin of the shell (Fig. 9-9). Calcareous sup-

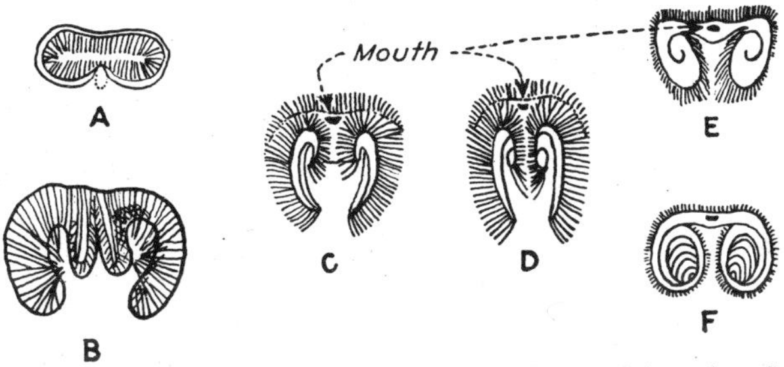

FIG. 9-8. Phylum BRACHIOPODA. Brachia. *A*. Adult lophophore of *Argyrotheca*. *B*. *Lacazella*. *C*. *Magellania*. *D*. *Cancellothyris*. *E*. *Lingula*. *F*. *Hemithyris*. All are from adult individuals and are variously enlarged or reduced. (*Adapted from Beecher*, 1897, *after several different authors*.)

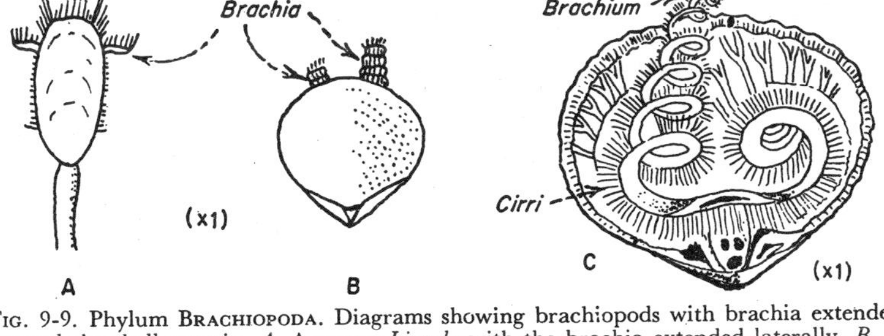

FIG. 9-9. Phylum BRACHIOPODA. Diagrams showing brachiopods with brachia extended beyond the shell margin. *A*. A young *Lingula* with the brachia extended laterally. *B*. A rhynchonellid with the brachia protruded. Specimens have been observed with the brachia extended beyond the anterior margin as much as 4 cm. (twice the length of the shell). The brachia themselves were sluggish, but the cirri were constantly in motion. *C*. *Hemithyris*, with one brachium extended and the other enrolled. (*A–B modified after Morse*, 1902 and 1869, *respectively; C after Hall and Clarke*, 1894.)

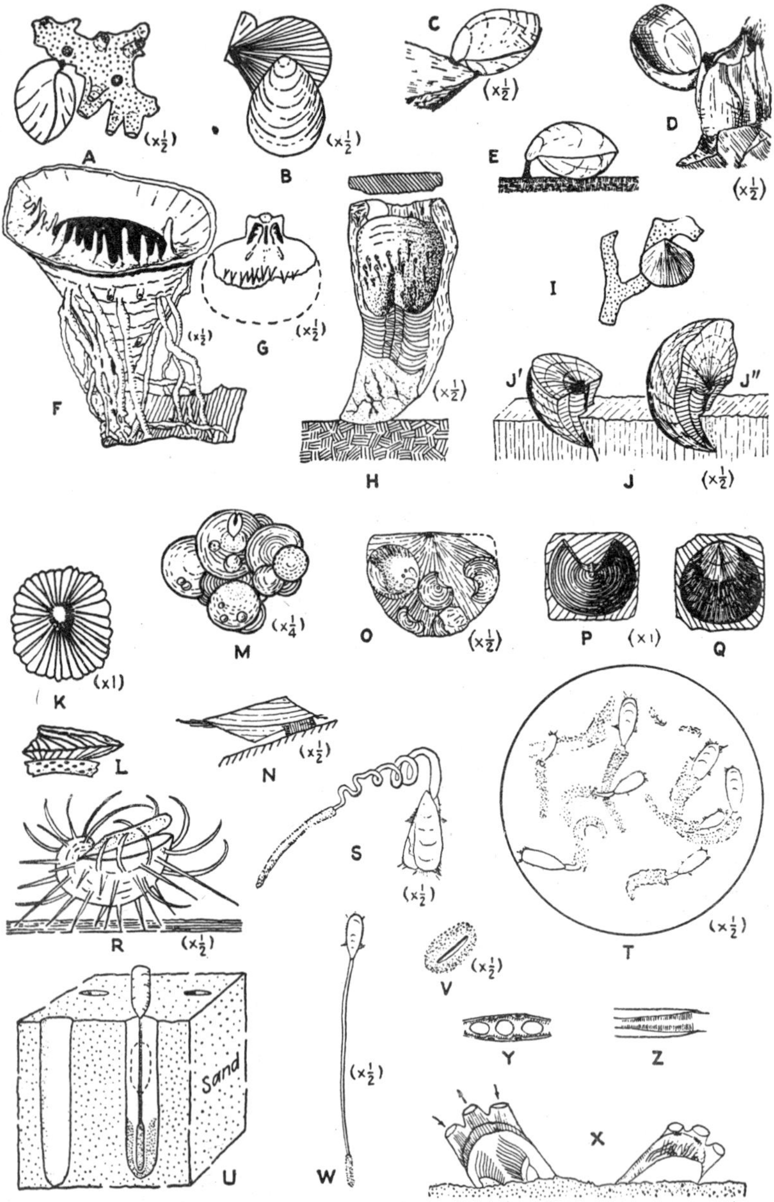

FIG. 9-10. Phylum BRACHIOPODA. Modes of attachment. (*See opposite page for detailed description.*)

ports for the fleshy brachia are present in the shells of some living brachiopods and are characteristic of many fossil shells representing more highly specialized brachiopods of the past (Fig. 9-35). They range from mere spikelike processes to elaborate spires and loops. All are grouped under the general term **brachidia** (singular, **brachidium**). They are discussed more fully in connection with internal shell structures.

The Pedicle or Peduncle

The **pedicle,** or **peduncle,** is the organ by which most brachiopods attach themselves to the substratum. It is a tough, flexible cylinder that is attached to the interior of the so-called pedicle valve and protrudes from the posterior part of the shell. It may be so short as to appear lacking or as long as several centimeters. In *Lingula* and *Glottidia* (Fig. 9-10*S–W*), for example, it is a conspicuous structure with the distal end modified for burrowing. In these genera it is contractile and secretes mucus that serves to attach it after a fashion. In certain inarticulates the pedicle becomes nonfunctional (atrophied) and the pedicle valve is cemented directly to the substratum. Some extinct brachiopods seem to have lacked a pedicle and to have lain free on the bottom or to have attached themselves by spines (Fig. 9-10*F*,*R*).

It is generally assumed that the brachiopods of the past attached themselves in much the same ways as do modern forms. As many of the earliest brachiopods were liguloids, it may well be that many of them had burrowing habits

FIG. 9-10. Phylum BRACHIOPODA. Modes of attachment. *A–E.* Telotrematous brachiopods attached to a piece of coral (*A*, *Liothyris*), a pecten (*B*, *Laqueus*), a piece of rock and a barnacle, respectively, (*C–D*, *Terebratalia* in Pacific waters), and a firm bottom (*E*). *F–G. Prorichthofenia*, a much specialized Permian productid: *F*, the pedicle valve of an old shell attached to a shell fragment by spines; the lidlike brachial valve lies deep in the conical pedicle valve where the dark shadow is indicated; *G*, inner surface of a fragmental brachial valve. *H. Richthofenia lawrenciana* from the Permian (Salt Range) of India. The conical pedicle valve, shown in longitudinal section, is cemented directly to the substratum; the flat and reduced brachial valve acts as a lid. The internal structure of the pedicle valve is somewhat analogous to that of certain corals. *I.* A small telotrematous brachiopod (*Kraussina*) attached to a branching coral. *J.* Two specimens of the European *Clitambonites* in living position with beaks buried deeply in the substratum. *J′* is a young individual that still possesses a short pedicle, whereas *J″*, an adult, has lost the pedicle. *K–N. Crania ignabergensis*, a Cretaceous neotremate: *K*, apical view of brachial valve; *L*, lateral view of shell cemented to a coral fragment by the pedicle valve. *M–N. Discinisca lamellosa: M*, a cluster of young and old shells, illustrating the gregarious habit of the species; *N*, lateral view showing the shell attached to the substratum by a short thick pedicle. *O–Q. Schizocrania filosa*, a common Ordovician neotremate, cemented to a valve of *Rafinesquina alternata: O*, a group of valves; *P*, the pedicle valve seen on the extreme right of *O*; *Q*, exterior of a brachial valve. *R.* Schematic and hypothetical drawing of a spinose productid supported above the bottom, brachial valve up, by means of long spines. *S–Z.* Lingulids—*Lingula* and *Glottidia: S*, *G. pyramidata*, showing convoluted vermiform pedicle with posterior end enclosed in sand tube; *T*, eight young specimens of *G. pyramidata*, which attached themselves to the bottom of a bowl by means of rude sand burrows made of the sand placed in the bowl; *U*, diagram of *L. anatina* in its burrow; *V*, the opening into the burrow of *L. lepidula; W*, *G. pyramidata* with outstretched pedicle that is five times as long as the shell and has the posterior part enclosed in sand; *X–Z*, *L. lepidula*, showing attitude in sand and formation of setal tube openings (*X*), front view showing anterior pallial folds and smaller setal tube openings (*Y*), and side view with lateral seta vertical and meeting (*Z*). (*All diagrams more or less diagrammatic and simplified. A–B, I, M after Davidson*, 1886–1888; *C–D after Fenton and Fenton*, 1932; *F based on original photograph by Cooper; G after Cooper*, 1944; *H after Waagen*, 1882–1885; *J after Opik*, 1930; *K–L after Nicholson*, 1872; *N, S–T, V–Z after Morse*, 1902; *O–Q after Hall and Clarke*, 1892.)

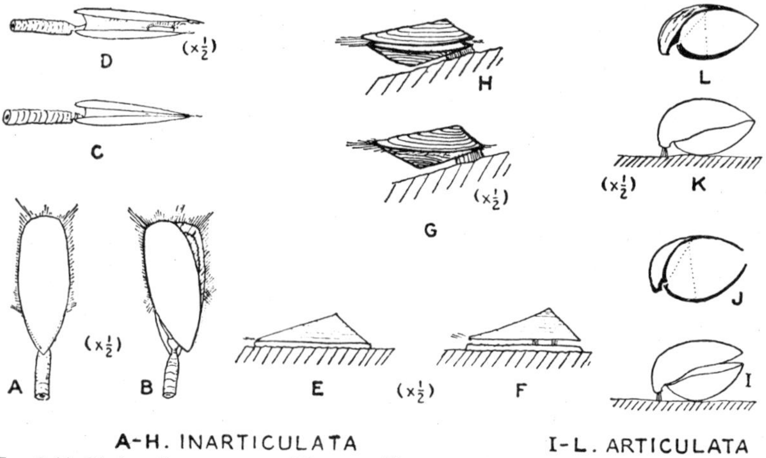

FIG. 9-11. Phylum BRACHIOPODA. Diagrams illustrating how inarticulates and articulates differ in opening and closing their shells. *A–D. Glottidia pyramidata*, a burrowing atremate that can rotate the brachial valve (*A–B*), and can also push the valves apart (*C–D*). *E–F.* A neotremate cemented to the substratum, with brachial valve pulled down so that shell is closed (*E*), and with valve pushed up (*F*). *G–H.* A pedunculate neotremate, *Discinisca lamellosa*, with shell closed (*G*), and open (*H*). In the latter condition the brachial valve is simply pushed upward. *I–L.* A typical articulate, *Magellania lenticularis*, with shell open (*I*), and as seen in longitudinal section (*J*) showing teeth and sockets and muscles (dotted line); and with shell closed (*K–L*). In the articulate shell the valves cannot be pushed apart because of the articulation along the hinge line. They merely gape along the anterior and lateral margins (cf. Fig. 9-14B–C). (*A–G adapted from Morse*, 1902.)

similar to those of modern *Lingula* and *Glottidia* (Fig. 9-10*U*,*X*). Others seem to have used the pedicle for attachment, or they cemented the pedicle valve directly to some object on the bottom (*e.g.*, *Schizocrania* and *Petrocrania* on *Rafinesquina;* Fig. 9-10*O*; 9-42*D–E*), whereas the later Paleozoic spiny genera (*e.g.*, *Prorichthofenia* and *Productus*) used elaborate spines for anchorage or to keep the shell well above the bottom (Fig. 9-10*F*,*R*).

Morphologically and embryologically a fundamental distinction can be drawn between the pedicles of the inarticulates and those of the articulates. Because of the striking difference in the way the pedicle originates in these two groups, and also because the pedicles themselves are morphologically different, Thomson (1927) proposed that the two generally accepted divisions Inarticulata and Articulata be renamed Gastrocaulia[1] and Pygocaulia,[2] respectively, with the nature and origin of the pedicle, rather than the absence or presence of articulation, as the basis of subdivision. Fortunately the orders and superfamilies remain the same in the new divisions as in the old, so that

[1] Gastrocaulia—Gr. *gaster*, *gastros*, stomach, + *kaulos*, stem; referring to the development of the pedicle from the mantle lobe.

[2] Pygocaulia—Gr. *pyge*, rump, + *kaulos*, stem; referring to the development of the pedicle from the posterior or caudal segment.

we may continue to use the familiar terms **inarticulate** and **articulate** without doing violence to Gastrocaulia and Pygocaulia, respectively.

The pedicle of living lingulids (atremates) is exceptionally long—four to six times the length of the shell in *Lingula*, and nine times in *Glottidia* (Fig. 9-10*U*, *W*)—and is a flexible tube capable of muscular movements (*i.e.*, contraction and extension). It has a central cavity which communicates with the coelom by a narrow canal and in life contains corpuscle-bearing fluid (or blood) similar to that filling the coelom. The pedicle of the discinids (neotremates) is extremely short but relatively large in diameter (Fig. 9-10*N*, 9-11*G–H*) and is similar to that of the lingulids in most respects. As previously stated, the pedicle of the lingulids is encased within a tube of agglutinated sand, whereas that of the neotremates is attached to some object on the sea bottom (Fig. 9-10).

The pedicle of *Lingula* and *Pelagodiscus* develops relatively late in the larval period, during the free-swimming stage after the protegulum is completely formed. It originates from the ventral mantle lobe and is not protruded between the valves until fixation is about to take place (Fig. 9-16).

By contrast, the pedicle of the articulates (Pygocaulia) is generally short and stout, with the protruded part covered with a thick horny sheath, and it has no communication with the coelom. It originates early from the caudal segment of the embryo (Figs. 9-17, 9-18, 9-19).

The Muscles

The muscles by which brachiopods open and close their shells and adjust them on the pedicle traverse the body. They are attached to the inner surfaces of the valves, and their places of attachment, marked by smooth and sculptured depressions or elevated areas, are referred to as **muscle scars** or **muscle markings** (Fig. 9-14*A*).

The system of muscles in the inarticulate brachiopods is much more complex than in the articulates because of the greater number of relative motions that are given to the two valves. As a consequence, fossil inarticulates have a more complicated set of muscle scars than articulates, as can be seen by comparing Figs. 9-12 to 9-15.

Living inarticulate brachiopods are known to be able to move the brachial valve from right to left with a rotary and sliding motion, and to move the valves toward and away from one another (Fig. 9-11). The muscles of a typical atremate (*Lingula*) and neotremate (*Discinisca*) are compared in Figs. 9-12 and 9-13.

The muscular system of a representative modern articulate, *Magellania flavescens*, is described in Table 9-1 and illustrated in Fig. 9-14, and that of another, *Gryphus vitreus*, is illustrated in Fig. 9-15. Two pairs of **diductor muscles** open the shell by contraction. These are attached at one extremity to the inner surface of the pedicle valve, where the larger and more anterior pair of scars is made by the **principal diductors,** and the smaller and posterior pair by the **accessory diductors** (not always preserved on fossil shells). The other ends of the diductors are attached to the **cardinal plate** or **hinge plates,** or to a special structure, the **cardinal process** (Fig. 9-14), near the

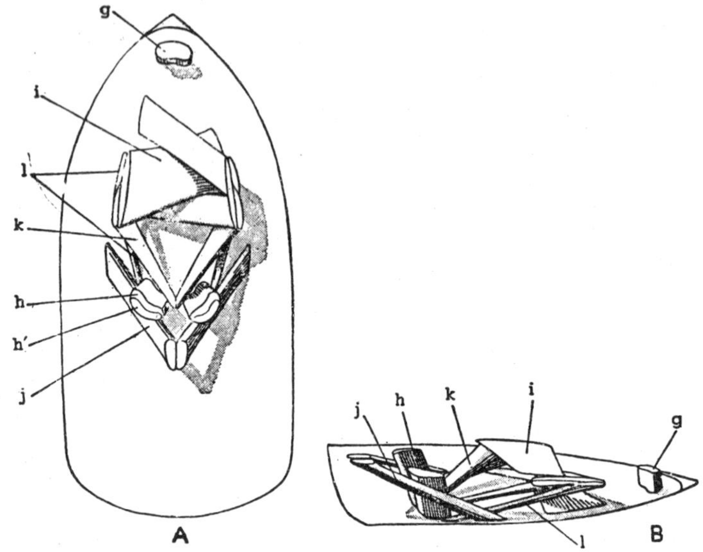

FIG. 9-12. Order Atremata. Muscle system of *Lingula*, viewed from above (*A*) and from the side (*B*), after removal of the brachial valve. (*g* = umbonal muscle; *h,h'* = central muscles; *i* = transmedians; *j* = anterior laterals; *k* = middle laterals; *l* = outside laterals.) (*From Bulman*, 1939, *mainly after Hancock*, 1859, *and Blochmann*, 1900.)

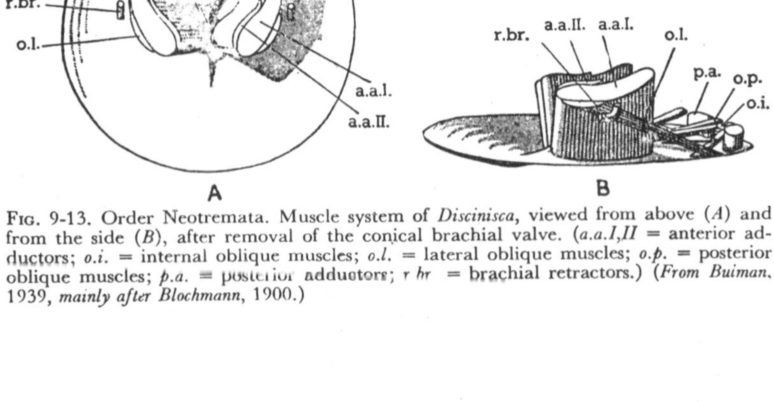

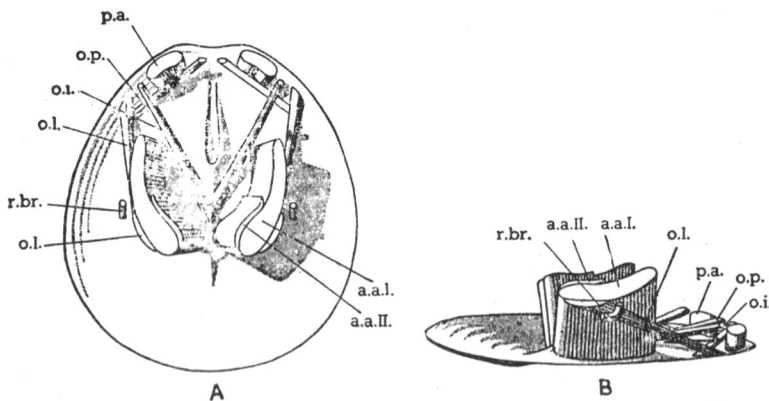

FIG. 9-13. Order Neotremata. Muscle system of *Discinisca*, viewed from above (*A*) and from the side (*B*), after removal of the conical brachial valve. (*a.a.I,II* = anterior adductors; *o.i.* = internal oblique muscles; *o.l.* = lateral oblique muscles; *o.p.* = posterior oblique muscles; *p.a.* = posterior adductors; *r br* = brachial retractors.) (*From Buiman*, 1939, *mainly after Blochmann*, 1900.)

posterior part (umbo or beak) of the brachial valve. The principal diductors are attached to the anterior part of the process, the accessory diductors to the posterior part. Both pairs pass between the two hinge teeth.

In most shells the diductor muscles utilize the principle of a lever of the first order, the axis of rotation being the line joining the hinge teeth, as shown in Fig. 9-14*B*–*C*. Since the diductors are attached anteriorly to this line in the pedicle valve and posteriorly to it on the brachial (*i.e.*, to the cardinal process), contraction of the muscles pulls the brachial umbo downward toward the center of the pedicle valve, thus forcing the valves apart. These muscles are quite powerful because they have only a short lever arm on which to exert their pull.

The two **adductor muscles,** which cross the body directly more or less perpendicularly to the plane of the valves, close the shell by contraction (Fig. 9-14*B*). At one end they are attached to the inner surface of the pedicle valve, one on each side of the median line, where, being close together, they make a heart-shaped impression. About midway across the body cavity each muscle splits so that two pairs are attached to the inner surface of the brachial valve—an anterior pair fairly close together and a posterior pair farther apart. Hence two pairs of adductor muscle scars are left on the inner surface of the brachial valve.

The **pedicle muscles,** five in number, serve partly to attach the pedicle to the larger valve and partly to adjust the shell on the pedicle. They consist of a single muscle, the **protractor,** and a pair each of **ventral adjustors** and **dorsal adjustors.** The protractor joins the pedicle to the pedicle valve and leaves an elongate scar posteriorly to the other scars. The dorsal adjustors are attached to the posterior part of the brachial valve, whereas the ventral adjustors form two scars that lie outside the adductor scars. The pedicle muscles are relatively weak; hence their scars are not prominent and are not likely to be as obvious as those of the adductors and diductors.

In some specialized articulates the diductors utilize a lever of the third order (*e.g.*, *Platidia*) and in others (Thecideidae) the arrangement of the muscles is quite different from that described above.

The shells of recently dead brachiopods and of many fossil forms are likely to be closed unless the valves were actually disarticulated or torn apart. This situation is the reverse of that found in most pelecypods, because in the latter the mechanics of the muscles and hinge structures is such that the valves are forced apart on death.

ONTOGENY

The development (ontogeny) of only a few genera of brachiopods is known in detail. Those best known are *Lingula* (Yatsu, 1902), an inarticulate; and *Argyrotheca* (Kowalevsky, 1873; Shipley, 1883), *Lacazella* (Kowalevsky, 1873), *Terebratulina* (Morse, 1871, 1873; Conklin, 1902), and *Terebratella* (Percival, 1944), all articulates.

From the standpoints of embryology and physiology, to say nothing specifically of the shell differences, the Phylum Brachiopoda consists of two sharply differentiated groups which, because they show profound differences in development, are considered by Percival (1944) to be an example of **con-**

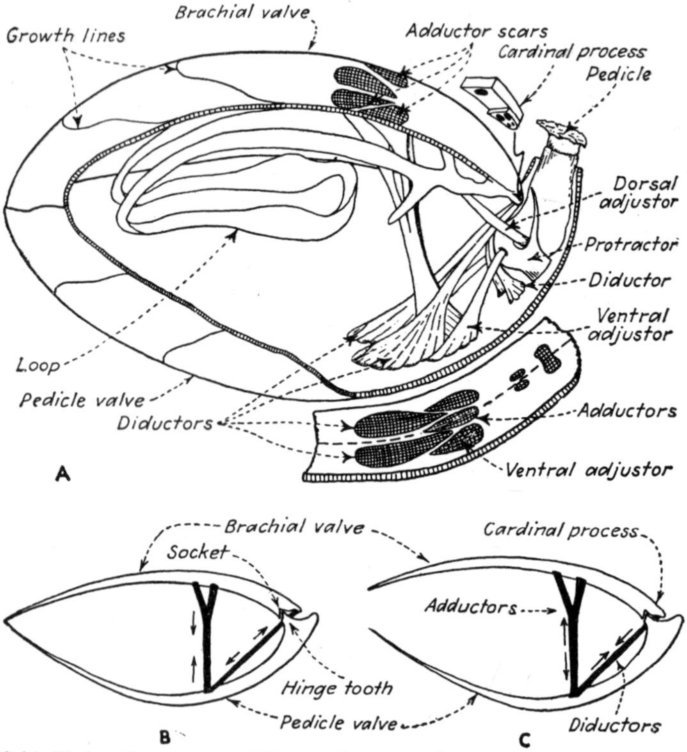

FIG. 9-14. Phylum BRACHIOPODA. The muscle system of a modern articulate brachiopod, *Magellania flavescens*, diagrammatized to show the relations of the different muscles to shell structures. *A*. Left lateral view of the shell, with part removed. The muscle scars of the brachial valve are shown as though they could be seen through a transparent shell, and the scars of the pedicle valve are shown on the imaginary mirror surface directly beneath the shell. *B–C*. Diagrams showing the actions of the muscles when the valves are pulled together and apart, respectively. (*A modified from Hancock*, 1859.)

vergence of two quite distinct groups (possibly different phyla) rather than of **divergence** from a common ancestor. This question of the twofold major division of the Brachiopoda is considered further under Classification.

Embryology

The coelomic mesoderm of all brachiopods whose embryology is known seems to begin with similar proliferations of initially endodermal cells (Cloud, 1948). From this stage on, however, the coelom differs in its development. According to available descriptions, *Lingula*, and probably *Lacazella* (a "protrematous" articulate), develop by the mechanical process termed **schizocoely** (a perivisceral cavity, the schizocoel, arises from the hollowing out of

TABLE 9-1. EXPLANATORY CHART FOR FIG. 9-14
Number, names, and functions of muscles in an articulate brachiopod

Name of muscle	Number	Function
Adductors	2	Close the shell by contraction; they are stretched when the shell is open, relaxed when it is closed
Diductors:		
Principal diductors	2	Open the shell by contraction; they are relaxed when the shell is closed
Accessory diductors (or divaricators)	2	
Pedicle muscles:		
Ventral adjustors	2	Adjust the shell up and down on the pedicle, or twist on the pedicle
Dorsal adjustors	2	
Protractor (also a ventral adjustor)	1	

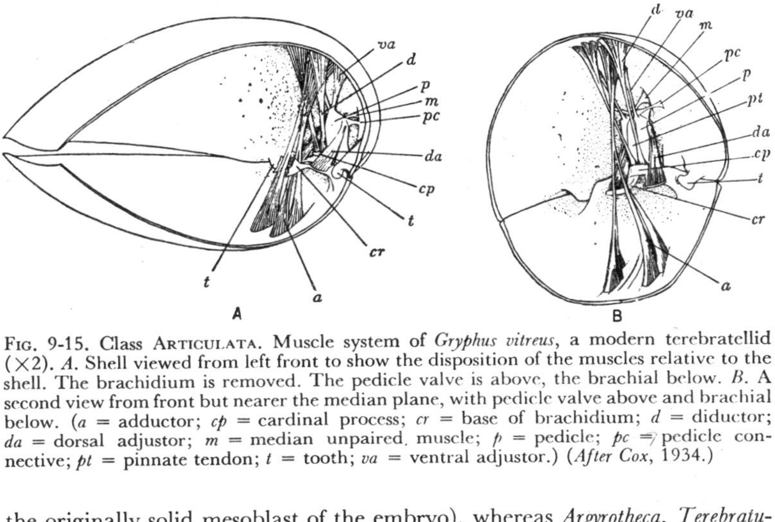

FIG. 9-15. Class ARTICULATA. Muscle system of *Gryphus vitreus*, a modern terebratellid (×2). *A*. Shell viewed from left front to show the disposition of the muscles relative to the shell. The brachidium is removed. The pedicle valve is above, the brachial below. *B*. A second view from front but nearer the median plane, with pedicle valve above and brachial below. (*a* = adductor; *cp* = cardinal process; *cr* = base of brachidium; *d* = diductor; *da* = dorsal adjustor; *m* = median unpaired muscle; *p* = pedicle; *pc* = pedicle connective; *pt* = pinnate tendon; *t* = tooth; *va* = ventral adjustor.) (*After Cox*, 1934.)

the originally solid mesoblast of the embryo), whereas *Argyrotheca*, *Terebratulina*, and *Terebratella*, all "telotrematous" articulates, develop by **enterocoely** (a perivisceral cavity arises through outpouching of mesodermal cells and is hollow from the time these cells differentiate from the primary endoderm). The following comparison of the ontogenies of *Lingula*, *Argyrotheca*, *Terebratella*, and *Lacazella* shows how profoundly they differ.

Lingula has a relatively long embryonic life (Fig. 9-16)[1] and does not emerge from the egg membrane until about the time the first pair of cirri appear.

[1] The development of *Lingula* has been studied in detail by Yatsu (1902), and the interested reader is referred to his work for further details.

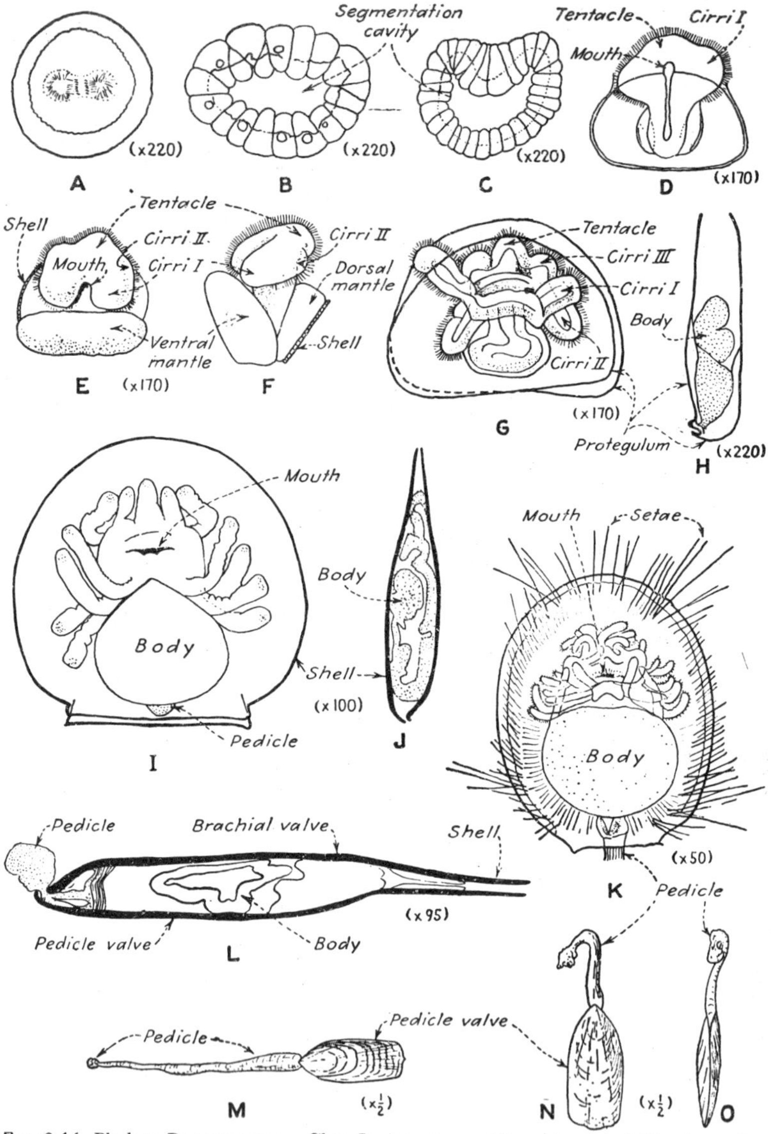

Fig. 9-16. Phylum Brachiopoda—Class Inarticulata. The ontogeny of *Lingula anatina*. (*See opposite page for detailed description.*)

Segmentation gives rise to a blastula consisting of two apposed layers, each of which has 16 cells (incidentally, a similar condition occurs in phylactolaematous Bryozoa). The mesoderm arises as two lateral proliferations from the endoderm, and the coelomic spaces are schizocoeles. The mouth, morphogenetically ventral in position, arises on the site of the closed blastopore, the lophophore is extraoral in origin, and the alimentary tube continues backward, downward, and forward to open by an anus. The mantle begins as a posterior circular flange which later becomes bilobed. The pedicle appears late and has quite a different kind of origin and constitution from that in *Terebratella* and *Terebratulina*. Two almost equidimensional valves develop, of which the smaller is dorsal and the larger ventral. The pedicle develops from the ventral mantle lobe and is related to the large valve. Thus, the orientation of an adult *Lingula* is essentially that of the embryo, since this stage is achieved without metamorphosis (Percival, 1944).

In *Terebratella* and its allies (*Terebratulina* and *Argyrotheca*), segmentation of the fertilized egg, where known, is generally similar and begins with the gradual growth of a spherical blastula (Fig. 9-17*A–C*). The gastrula forms by invagination, develops a blastopore, and soon becomes ciliated. It changes its form rapidly as growth proceeds, and ultimately the body has two fairly well defined regions—a thickened, ciliated anterior end, with a downward bulge and an anterodorsal slope, and an unciliated more slender posterior part that gives rise to the pedicle.

At this stage, a mantle fold, slightly higher dorsally and laterally than ventrally, completely surrounds the body. The mantle rudiment grows backward as a sheath enclosing a continually narrowing peduncular rudiment, until, finally, only the tip of the pedicle is visible (Fig. 9-17*F*, 9-19*A–B*). During elaboration of mantle and pedicle, the latter changes to a slender tapering structure. At this stage of development the embryo is said to have attained the **cephalula** stage. It is now ready to leave the parent.

When the mature larvae leave the parental mantle cavity, they are about 0.2 mm. long and about 0.14 mm. broad at the widest part through the base of the mantle rudiment. The apical tuft, so conspicuous at earlier stages, is lost before the larva leaves the parent (Fig. 9-17*F*).

After about thirty hours of free-swimming life the larva becomes loosely stuck to the substratum by the tip of the pedicle (Fig. 9-17*G*). After some

FIG. 9-16. Phylum BRACHIOPODA—Class INARTICULATA. The ontogeny of *Lingula anatina*, the well-known burrowing atremate. *A*. Fertilized egg undergoing segmentation mitosis. *B*. Median section through a blastula. *C*. Median section through an early gastrula. *D*. Early embryo with the first pair of cirri and beginning of tentacle. *E–F*. Anterior and left lateral views of an embryo with two pairs of cirri, and the beginning of the protegulum on the dorsal mantle. *G–H*. An embryo with three pairs of cirri and a fully formed protegulum. The protegulum is still a single plate but is folded at the posterior part. The embryo can now swim and will start adding shell matter to the protegulum. *I–J*. Embryo with six pairs of cirri, viewed ventrally and laterally in a median sagittal section. The pedicle has begun to form and the protegulum has split into the two valves of the shell. *K–L*. Larva at the stage of 15 pairs of cirri. The pedicle, which has a coelomic extension, now protrudes from between the valves but clearly belongs to the larger valve, as indicated in the median sagittal section. *M*. An adult with the pedicle fully outstretched. *N–O*. Two different views of an adult with the pedicle slightly crooked. (*A, K adapted from Yatsu*, 1902; *M–O modified after Davidson*, 1886–1888.)

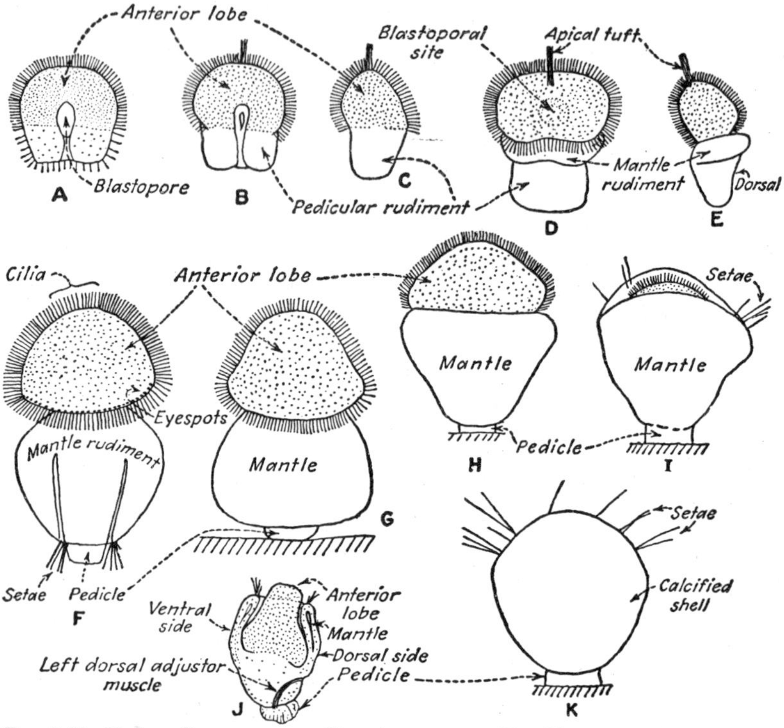

FIG. 9-17. Phylum BRACHIOPODA—Class ARTICULATA. The life history of *Terebratella inconspicua*, a modern telotremate. *A*. Late gastrula stage, showing disappearance of posterior cilia, differentiation into regions of anterior lobe and mantle and stalk, and appearance of ventral groove from blastopore. *B–C*. Views of a somewhat more advanced stage, showing development of apical tuft, differentiation of anterior lobe, and development of pedicular rudiment. *D–E*. Closure of blastopore, appearance of mantle rudiment and of rudiment of pedicle. *F*. Mature larva on emergence from parental mantle cavity, with eyespots and setae. *G*. Larva after two hours' attachment by the pedicle. The mantle has not yet begun to reverse. *H*. After somewhat longer attachment, the mantle is beginning to reverse. *I*. Mantle reversion almost completed, with only small part of anterior lobe still visible. *J*. Sagittal section of larva just after reversion of the mantle and before shell formation has begun. *K*. Enclosure of anterior lobe is now complete, and the shell is beginning to be calcified. All figures much magnified. (*Adapted from Percival*, 1944.)

hours the mantle is suddenly reversed, at first only partly enclosing the anterior lobe but ultimately surrounding it completely as shown in Figs. 9-17*H–J* and 9-19*C–D*.

During the later period of enclosure, when some dorsoventral flattening has taken place, the outer surface of the mantle becomes glistening white and smooth, and shell secretion has begun. It seems that the mantle ectoderm first produces a thin cuticular shell, the periostracum of the complete shell, to which it then adds a calcareous layer, which comes ultimately to lie between the mantle ectoderm and the outer cuticular layer.

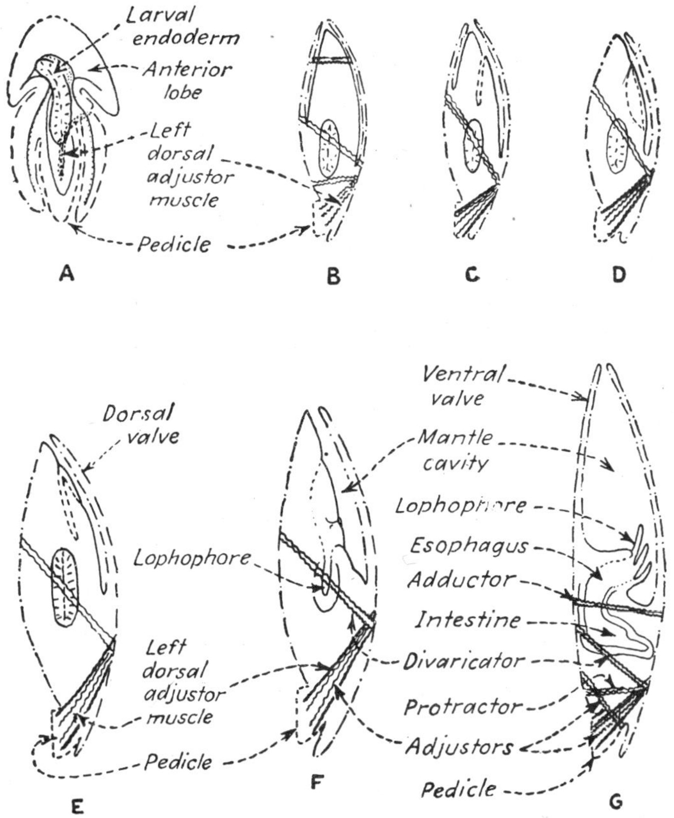

FIG. 9-18. Phylum BRACHIOPODA—Class ARTICULATA. Diagrams showing transformation from larva to young adult in *Terebratella inconspicua*, as seen in sections viewed from the left side. *A*. A larva after leaving parent but before attachment and reversal of the mantle. *B–G*. Diagrams showing particularly the elaboration of the digestive system from the larval endoderm and the relation between dorsal valve and dorsal adjustor muscle. It is important to note that the larger valve, which is also the pedicle valve, has a dorsal position morphologically, hence the usual practice of designating this valve the ventral would be incorrect. (*Adapted from Percival*, 1944.)

During this early period the animal sits upright on the pedicle but soon the body tends to tilt. It becomes evident that one of the valves is longer than the other and that the pedicle belongs to the larger valve (Fig. 9-18). It is pertinent to emphasize here that the larger valve is morphogenetically *dorsal* and the smaller *ventral*—the direct opposite of what has always been accepted—a relation which, as Percival (1944) and Cloud (1948) point out, is justification for dropping the use of the terms ventral and dorsal for the valves of brachiopod shells and for adopting the more accurate terms pedicle and brachial.

Shell growth is fairly rapid, with additions being made along the lateral and anterior margins at the mantle edge. A specimen grown under laboratory conditions increased its original length by a half in the three weeks following attachment.

During this same period, while the valves are being secreted around the animal, the anterior lobe undergoes a series of profound changes resulting in the formation of the lophophore, alimentary tube, coelom, and other adult structures.

With the development of the mouth, stomodaeum (anterior part of alimentary canal), and stomach, the main outlines of metamorphosis have been laid down. There now exists a small and quite simple brachiopod with pedicle, pedicle area, unequal valves, and an elementary gut or digestive tube. Later changes lead mainly to the elaboration of these structures, further develop-

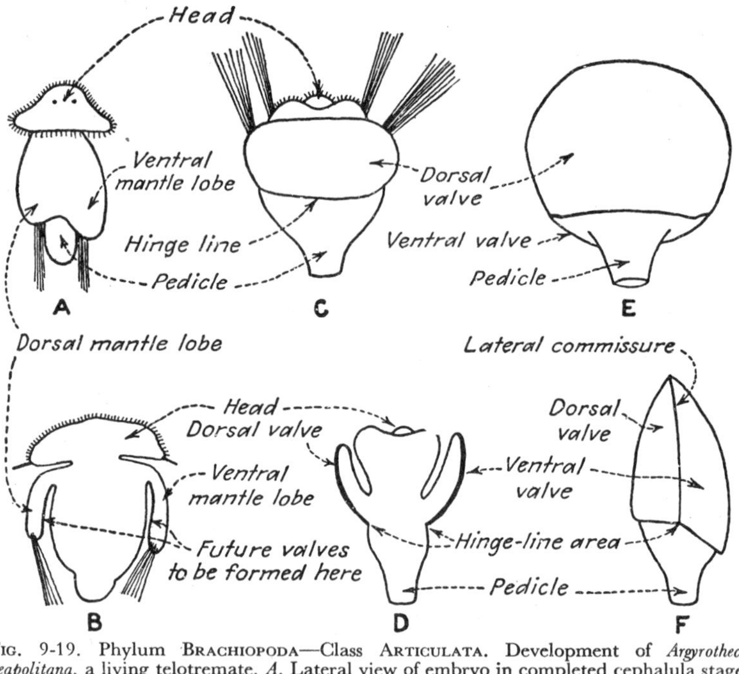

FIG. 9-19. Phylum BRACHIOPODA—Class ARTICULATA. Development of *Argyrotheca neapolitana*, a living telotremate. *A*. Lateral view of embryo in completed cephalula stage, with three distinct segments. *B*. Dorsoventral longitudinal section (internal structures omitted) of *A*, showing posteriorly extended mantle lobes, on the inner surfaces of which the dorsal (brachial) and ventral (pedicle) valves will form. *C*. Dorsal view of embryo after mantle has been reversed. *D*. Dorsoventral section of *C* in outline, showing the two mantle lobes directed anteriorly, thus bringing the shell-secreting surfaces and the valves themselves into an exterior position. Ultimately the valves grow large enough to enclose the entire animal excepting the pedicle, which emerges between the valves posteriorly as shown in *E* and *F*. *E–F*. Dorsal (brachial) and side views of the earliest shell (protegulum), showing large posterior opening between the valves. (*A–B after Shipley*, 1884; *C–F after Kowalevsky*, 1874, *and Beecher*, 1892; *all more or less modified and variously magnified.*)

ment of the coelom and enlargement of the shell. It is significant that at this stage there is no indication of the calcareous loop that is developed in mature shells as a support for the lophophore.

The development of *Argyrotheca neapolitana*, long ago worked out by Kowalevsky (1873) and Shipley (1833), is shown diagrammatically in Fig. 9-19 for comparison with the recent investigations of *Terebratella* by Percival (1944).

Lacazella, a protrematous articulate, differs in several important respects from the telotrematous genera *Terebratella* and *Argyrotheca* (Figs. 9-17, 9-18, 9-19). The embryo first divides transversely into a **cephalic** and a **caudal segment,** but soon the former divides with the production of a median **thoracic segment** (Fig. 9-20*A*). Next the body wall of the thoracic segment develops an encircling fold, which consists of a small ventral mantle lobe and a prominent dorsal mantle lobe (Fig. 9-20*B*). Meanwhile a head with four "eyespots" is differentiated at the anterior end of the cephalic segment, and the embryo, now in the **cephalula** stage (Fig. 9-20*A–B*), leaves the brood pouch of the mother and becomes free-swimming.

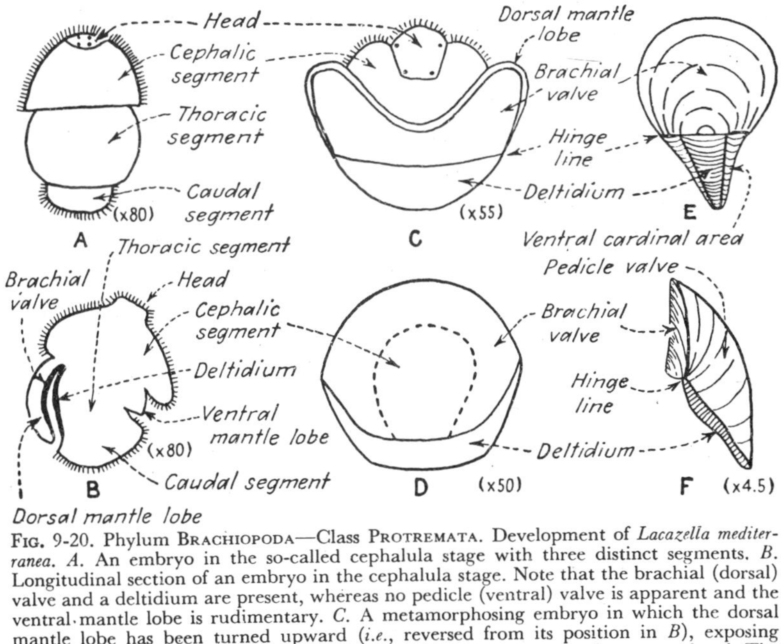

FIG. 9-20. Phylum BRACHIOPODA—Class PROTREMATA. Development of *Lacazella mediterranea*. *A*. An embryo in the so-called cephalula stage with three distinct segments. *B*. Longitudinal section of an embryo in the cephalula stage. Note that the brachial (dorsal) valve and a deltidium are present, whereas no pedicle (ventral) valve is apparent and the ventral mantle lobe is rudimentary. *C*. A metamorphosing embryo in which the dorsal mantle lobe has been turned upward (*i.e.*, reversed from its position in *B*), exposing brachial valve and deltidium. *D*. Embryo after metamorphosis has been completed. The cephalic lobe (dotted) is now covered by the dorsal mantle lobe. *E*. Adult shell showing brachial valve, and deltidium enclosed by the cardinal area of the pedicle valve. Note position of hinge line and lack of a pedicle opening. (*Adapted from Arber*, 1942, *Beecher*, 1892, *Kowalevsky*, 1874, *and Davidson*, 1887.)

During the cephalula stages the first rudiments of the shell appear as a chitinous layer on the dorsal surface of the trunk segment and on the inner surface of the dorsal mantle lobe (Fig. 9-20*B*). After a brief free-swimming period the embryo attaches itself to the substratum by the caudal segment, now serving as a pedicle, and changes form by reversing the mantle—the dorsal mantle lobe and the lateral parts of the general mantle fold are lifted up, the incipient ventral mantle lobe having by this time disappeared altogether. The first shell rudiment is now external in position. The part on the dorsal lobe develops into the dorsal valve, whereas the plate on the thoracic segment becomes a **deltidium.**[1] The origin of the ventral valve has not been observed, but it is assumed to be secreted by some part of the mantle lobe of the thoracic segment, inasmuch as the ventral lobe is atrophied. As the shell grows, the cardinal area comes into existence as the posterior margin of each valve expands laterally. In the middle line, the cardinal area has a large gap that is filled by the thoracic segment which bears the deltidium on its dorsal surface (Fig. 9-20*E*). The dorsal valve comes to be separated from the thoracic segment by the ventral valve, so that the hinge line lies to the dorsal side of the thoracic segment. This is quite different from telotrematous articulates (*e.g.*, *Argyrotheca*), in which the two valves are formed separately, one on each side of the thoracic segment, thus giving the hinge line a median position. The complete protegulum of *Lacazella* consists of *three* distinct plates—dorsal and ventral valves secreted by the circumthoracic mantle fold, and a third plate, the deltidium, secreted by the dorsal surface of the thoracic segment. No true pedicle seems to develop in *Lacazella*, but in the ancient strophomenids (*e.g.*, *Leptaena*) a small pedicle is thought to have emerged through a tiny apical or supra-apical foramen in the ventral valve anterior to the deltidium (Fig. 9-21*A–B*,*F–G*).

On the basis of restudy of the original descriptions of the embryology *Lacazella*, Arber (1940, page 186) was led to the conclusion that the deltidium in the embryo of the extinct strophomenids (*e.g.*, *Leptaena;* Fig. 9-21) was probably borne on the thoracic mantle ring in direct lateral continuity with the ventral valve—a relation exactly opposite the previously supposed funda-

[1] Arber (1939, p. 82) has proposed the adoption of Bronn's (1862) original term **pseudodeltidium** for this embryonic plate, which Schuchert (1897) called the **prodeltidium** arguing that it was included in his definition. Cloud (1942, p. 18), on the other hand, has pointed out that Bronn used the term for several distinctly different structures, and that Ting (1936) has further confused the issue by still another application of the term. He proposed, therefore, that pseudodeltidium be dropped altogether, and that either deltidium, as defined by American authors (see Table 9-2), be adopted, or, better, that deltidium be retained as a nongenetic general term, and a new term **xenidium** be used. We approve Cloud's suggestion that deltidium be retained as a general term for any shelly plate or group of such plates restricting the delthyrial opening, regardless of how the plates originated. There is a real advantage in having such a nongenetic term available for designating a plate or plates which because of poor preservation or for other reasons do not reveal their mode of formation. Although a particular deltidium may be a primary shell structure (*i.e.*, it began as a primary component of the first shell), we prefer to use the term deltidium in a collective nongenetic sense.

TABLE 9-2. COMPARISON OF USES OF THE TERMS FOR PLATES RESTRICTING OR CLOSING THE DELTHYRIUM (See Fig. 9-30)

Definition of feature	Von Buch, 1835; Bronn, 1862; Woods, 1937; Arber, 1939–1942	Most American authors; Thomson, 1927	Cloud, 1942; Shrock and Twenhofel, 1952
Single plate closing the delthyrium in the Protremata (*e.g.*, Strophomenacea; *Lacazella-Leptaena*)	Pseudodeltidium (Bronn 1862, Arber 1939–1942) Deltidium (Von Buch 1835)	Deltidium, and prodeltidium of Schuchert	Deltidium (S + T) Xenidium (C)
Two plates partly or completely closing the delthyrium in the Telotremata (*e.g.*, Terebratulacea)	Pseudodeltidium (Bronn 1862) Deltidium (Von Buch 1835) Composed of two deltidial plates (Arber 1939–1942)	Deltidial plates and pseudodeltitium	Conjugate or discrete Deltidial plates
A single secondary plate closing part or all of the delthyrium along the hinge line, in addition to two deltidial plates that may be obsolete			Henidium (Cloud 1942)
Any shelly plate or group of such plates restricting or closing the delthyrium, regardless of time and mode of formation	Deltidium (Davidson)		Deltidium

mental distinction between these two components of the protegulum (Schuchert, 1897). If this is true, it follows that neither the mode of development of the valves nor the nature of the delthyrium and deltidium in protrematous *Lacazella* and *Leptaena* is homologous with similar features in the Telotremata, and that there is, therefore, a fundamental difference between the protremate and telotremate divisions of articulate brachiopods.

The Protegulum

While the larva grows and differentiates into the organs and other fleshy structures that ultimately constitute the completely developed brachiopod, the mantle increases in size and its ectoderm begins to secrete shell material. In the earliest stages, the embryonic shell, which is designated the **protegulum** (N.L. *pro*, before, + *tegulum*, little covering), is merely a thin horny covering, having the form of a single plate in *Lingula* (Fig. 9-16*H*), a tiny bivalve shell in *Argyrotheca* (Fig. 9-19*E–F*) and its allies, and a three-piece covering in *Lacazella* (Fig. 9-20*E–F*). As growth proceeds, calcium phosphate

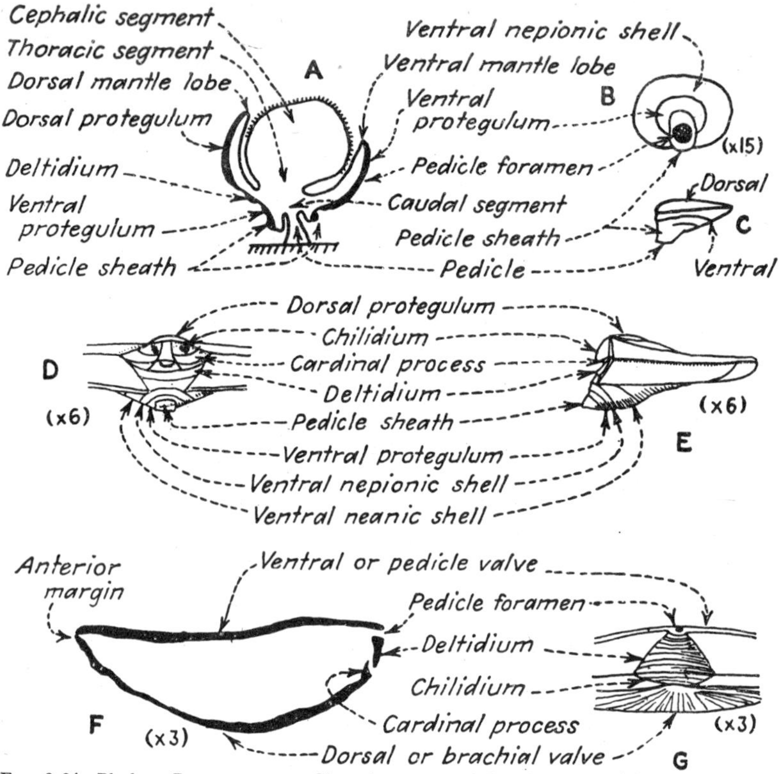

FIG. 9-21. Phylum BRACHIOPODA—Class ARTICULATA. Development of *Leptaena* and its bearing on the evolution of the Strophomenacea. *A.* Diagrammatic dorsoventral longitudinal section of the hypothetical typembryo of *Leptaena*. Arber (1942, p. 186) recently concluded that the deltidium of the strophomenids was borne on the mantle ring, not directly on the dorsal body wall as shown in this figure. *B–C.* Ventral and side views of a nepionic infant or larval shell of *L. rhomboidalis* from the Lower Devonian of New York, showing the conspicuous pedicle sheath rising above the ventral protegulum. *D–E.* Side and posterior views of an adult shell of *L. emarginata* from the Upper Silurian of Poland, showing the several important and significant embryonic and larval shell structures in the posterior region. *F.* Median section of *Strophomena planoconvexa*, showing open apical foramen and its position relative to the deltidium. *G.* Cardinal view of umbonal region of *Strophomena incurvata*, showing scar of supra-apical foramen entirely enclosed within the larval or nepionic pedicle (ventral) valve. (*Modified after Arber*, 1939 *and* 1940.)

or calcium carbonate is added to the horny cuticle of most forms, chiefly along the anterior and lateral margins, and ultimately the shell begins to take on its adult shape, which in many brachiopods is quite different from that of the protegulum (Fig. 9-22). Because of its delicate nature and minute size, the protegulum is not generally present on mature living shells and is almost never preserved on fossil shells, though its form and other characteristics may be determined from impressions remaining in the calcareous shell.

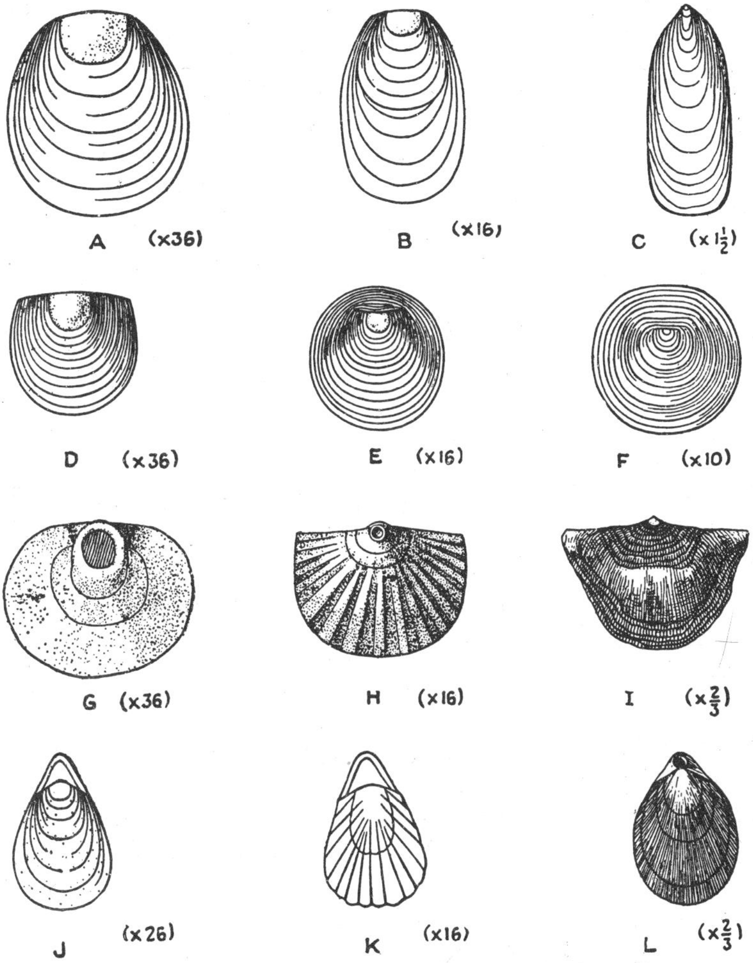

FIG. 9-22. Phylum BRACHIOPODA. Growth stages from protegulum to adult shell. *A–C. Glottidia albida*, a modern atremate: *A*, young shell, with protegulum stippled; *B*, a somewhat older shell which has changed from an oboloid to a linguloid outline; *C*, an adult shell. *D–F. Orbiculoidea minuta*, an extinct neotremate; *D*, early shell, with protegulum stippled, showing hemiperipheral growth; *E*, older shell with some holoperipheral growth; *F*, an adult shell. *G–I. Leptaena rhomboidalis*, an extinct protremate: *G*, early shell with short hinge and prominent sheath around pedicle opening; *H*, partly grown shell with radiating costae; *I*, adult shell. *J–L. Terebratulina septentrionalis*, a modern telotremate: *J*, a young shell with open delthyrium; *K*, partly grown shell with radiating costellae; *L*, adult shell. (*From Beecher*, 1892, *after several authors.*)

The protegulum shows some striking differences in its development as indicated in Figs. 9-18, 9-19, 9-20, and 9-22. In *Lingula* it appears early, even before the free-swimming stage, when the embryo has only one or two pairs of cirri, and has the form of a single circular cuticular plate folded over the posterior part of the mantle into two semicircular valves bound together by a connecting fold of cuticle (Thomson, 1927) (Fig. 9-16*H*). As growth proceeds during the free-swimming stage, the fold breaks and the two valves become free of each other (Fig. 9-16*J*). At a somewhat later stage the pedicle develops within the valves but is not protruded until fixation is about to take place (Fig. 9-16*L*). It is pertinent to reemphasize that the pedicle of *Lingula* arises from the mantle lobe and has a cavity connected with the coelom; hence it cannot be regarded, either morphologically or embryologically, as homologous with the pedicle of *Argyrotheca* and its allies, because in the latter the pedicle develops from the caudal segment of the larva and has no open connection with the coelom.

In *Argyrotheca*, *Terebratulina*, and *Terebratella*, the protegulum first appears as two distinct, elliptical valves separated by the whole width of the caudal segment at first and later by the pedicle. At this stage, the so-called cephalula stage (Fig. 9-17*F*, 9-19*D*), the valves are not in contact, but a little later the construction of the protegulum is completed as the two valves come into contact on the right and left sides of the body where teeth and sockets are developed. The smaller or brachial valve of the youthful shell develops a nearly straight posterior or **cardinal margin,** but the larger or pedicle valve has its cardinal margin interrupted by a triangular aperture, the **delthyrium,** through which the thick pedicle emerges (except in *Lacazella*, which does not develop a pedicle). The delthyrium of the adult *Argyrotheca* remains little modified, but in many telotrematous genera it becomes partly or wholly closed by shelly plates, the **deltidial plates,** which grow inward from its lateral edges and ultimately close most or all of the delthyrial opening (Fig. 9-30). Since these plates are secreted by the mantle lining the pedicle valve, they are to be considered an integral part of that valve.

Summary

It is clear from the preceding discussion that much more needs to be known about the development of both soft parts and shell plates before ontogenetic distinctions can take their proper place as one of the bases for major subdivision of the Brachiopoda. At the present time it may be said that embryologically the Brachiopoda can be subdivided into two major groups: (1) the *Gastrocaulia* (Inarticulata), in which the pedicle develops from the mantle lobe, and (2) the *Pygocaulia* (Articulata), in which the pedicle develops from the caudal segment of the larva. How valid these two groups really are, and whether or not they are the only major groups, cannot be determined until more is known about the embryology of the neotrematous and protrematous brachiopods. We prefer, therefore, to retain the terms Inarticulata and Articulata, which concern shell features rather than soft parts, for the two major divisions of the Brachiopoda. The chief differences between these two major divisions are summarized in Table 9-3.

TABLE 9-3. DIFFERENCES BETWEEN INARTICULATE AND ARTICULATE BRACHIOPODA

Inarticulata (*Gastrocaulia*)	*Articulata* (*Pygocaulia*)
1. Pedicle produced from the mantle lobe at a late stage in development	1. Pedicle arises as a primary lobe of the body on the embryologically dorsal side
2. Does not rotate mantle lobe	2. Rotates or reverses mantle lobes after fixation
3. Development of all organs direct	3. Metamorphosis
4. Musculature complex	4. Musculature relatively less complex
5. Intestinal tract has anus	5. Intestinal tract ends blindly
6. Chitinous or chitinophosphatic shells characteristic; calcareous shells rare	6. Shell calcareous, with chitinous periostracum
7. Shell interior relatively simple, and without any shelly supports for the brachia	7. Shell interior complicated by numerous structures; calcareous supports for brachia variously developed

THE SHELL

The embryonic shell, or protegulum, of a brachiopod is a minute structure ranging in size from 0.05 to 0.60 mm. in diameter and seems always to have been horny and impunctate (Fig. 9-22). Although it represents the first stage of shell formation, it seldom persists throughout the full life of an individual. It is soon strengthened or succeeded by calcareous substances, and although additions are first made at the margins of the protegulum, the shape and size of the protegulum do not necessarily determine the shape of the mature shell. As the shell proper grows and changes, many new features and structures are added. These range in taxonomic importance from ordinal, through familial and generic, to subspecific. The following aspects of the mature shell deserve consideration:

General morphology of the shell.
- Shape.
- Dimensions.
- Homoeomorphy.

External morphology of the valves.
- Development of outline and form.
- Types of cardinal margin.
- Palintropes and interareas.
- Nature of commissure and types of folding.
- Surface sculpture.
- Morphology of the pedicle opening.
- Curvature of the beak.

Internal morphology of the valves.
- Articulation.
- Cardinalia.
- Brachial supports (brachidia).
- Pallial markings.
- Muscle marks.

General Morphology of the Shell

Shape. The shell of a brachiopod is bivalve and bilaterally symmetrical, with the plane of symmetry bisecting the valves into similar halves by passing through the beaks of the valves posteriorly and through the middle of the anterior margin (Figs. 9-2, 9-3). It is to be noted that with few exceptions brachiopod shells are placed with the beaks upward when illustrated, although the shell may have any of several different living orientations with reference to the substratum.

The valves of a brachiopod shell are equilateral but of different sizes and shapes. The larger valve, which always has the pedicle if that structure is present, is typically convex, in some genera conspicuously so (*e.g.*, productids), and in most forms its beak projects or arches over the rounded umbo or posterior part of the smaller (brachial) valve. Although brachiopod shells tend to be biconvex, they show considerable variation in profile[1] depending upon the curvature of both valves, as illustrated in Fig. 9-23. In describing the different types of profiles the first term refers to the curvature of the brachial valve, the second to that of the pedicle valve. This is purely arbitrary, and some authors prefer the reverse.

Biconvex shells have both valves rather strongly convex. **Dorsibiconvex** shells have the brachial (generally dorsal) valve more convex than the pedicle. Shells with convex pedicle and plane brachial valves are described as **plano-convex,** whereas those with convex pedicle and concave brachial valves are said to be **concavo-convex.** If the shell is concavo-convex in its early growth stages but reverses to convexo-concave in maturity—as is the case with *Strophomena* and some of its allies—it is **resupinate.** Shells that are convexo-concave, without the incipient flat stage in the brachial valve, a condition well illustrated by many orthids such as *Valcourea*, may be designated **pseudo-resupinate** or simply **convexo-concave.**

Dimensions. The dimensions of brachiopod shells are measured as indicated in Fig. 9-3. **Length,** or **height,** is measured in the plane of symmetry from point of beak to anterior margin; **width,** or **breadth,** perpendicular to the plane where the shell is broadest; and **thickness,** in the plane roughly at right angles to the valves.

Brachiopod shells vary considerably in size. Excluding dwarfed forms, adult shells range from a few millimeters to many centimeters wide or long. Shells of *Conchidium* (Sil.) and *Scacchinella* (Perm.) exceed 150 mm. (6 in.) in length. *Isogramma* (Penn.) was more than 150 mm. (6 in.) wide, and the giant of all brachiopods, appropriately named *Gigantoproductus giganteus* (formerly *Gigantella gigantea*), from the Viséan (Miss.), reached a width of 37.5 cm. (15 in.) and a length of more than 25.0 cm. (10 in.). On the average, however, most brachiopod shells will fit in a box about 50 by 50 by 100 mm. (2 by 2 by 4 in.).

Homoeomorphy. It has long been known that among certain invertebrate groups, notably brachiopods and ammonites, external characters and shell form are not reliable for generic differentiation and phyletic studies because of a phenomenon known as **homoeomorphy. Homoeomorphs** are species that are almost alike so far as superficial appearance goes but are unlike if particular structural details are closely examined (Buckman, 1901). Species

[1] Shell shape and convexity can be used to a certain extent in classifying brachiopods, and McEwan (1939) has included these in a comprehensive key to the chief shell features of 22 articulate families.

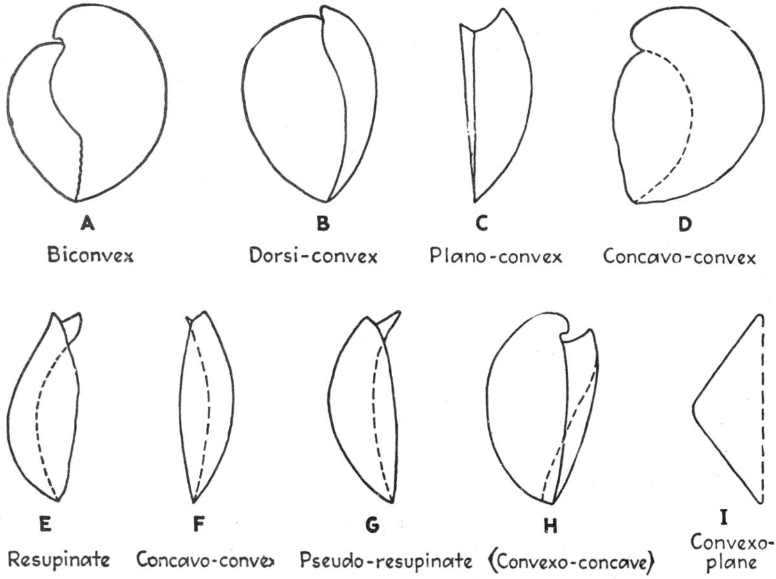

FIG. 9-23. Phylum BRACHIOPODA. Lateral views of brachiopods to show convexity and concavity of valves. All figures are semidiagrammatic. The left valve is brachial; the right, pedicle. Dashed lines indicate the position of the concealed valve along the plane of symmetry. *A*. Biconvex shell represented by *Conchidium*. *B*. Dorsi-biconvex shell illustrated by *Atrypa*. *C*. Plano-convex shell common among the orthids. *D*. Concavo-convex shell, a form characteristic of the productids. *E*. Resupinate shell well illustrated by *Strophomena*. *F*. A thin concavo-convex shell such as that found in *Rafinesquina*. *G–H*. Pseudoresupinate or convexo-concave shells: *G*, a form found in *Valcourea*; *H*, a shape characteristic of *Hebertella*. *I*. Convexo-plane or conical shell characteristic of many neotremate shells that are cemented by much of the surface of the pedicle valve.

of different stocks may display homoeomorphy contemporaneously, in which case they are designated **isochronous homoeomorphs,** or a later form may simulate an earlier one, in which case the organisms concerned are said to be **heterochronous homoeomorphs.**

The prevalence and importance of homoeomorphy among the Brachiopoda has been demonstrated by Buckman (1901, etc.), Cooper (1930), Cloud (1942, 1948), and Wilson (1944, 1945), among others, and the phenomenon should be kept in mind by those paleontologists who must identify old genera and species and describe new ones.

External Morphology of the Valves

The completed protegulum consists of essentially similar pedicle and brachial valves that have a straight posterior margin and a semicircular outline (Figs. 9-16, 9-17*K*, 9-19*E–F*, 9-20*E–F*, 9-22). The ultimate outline of the valves depends upon how and where additions are made to the margins of the plates of the protegulum. Other important changes that take place as the shell grows produce variations in the cardinal margins, along the commissure, in the surface sculpture, and in the pedicle opening and beaks.

Development of Outline and Form. Shell increments made after the protegulum is completed are secreted by the outer surface and growing margin of the mantle lobes and body walls. These additions are made to the inner surface of the shell, thus strengthening the protegulum, and as a band beyond the margins of the two protegular plates. Marginal increments are indicated by **growth lines,** which are spaced on the valve surface at irregular intervals and seem to indicate periods of cessation or marked retardation of shell growth.

In **hemiperipheral** growth, successive shell increments extend the lateral and anterior margins, but not the posterior margin. The protegulum marking the apex of the valve remains at the middle of the posterior margin, and the valve tends to retain the outline of the protegulum (Fig. 9-22). Broader increments laterally result in a shell that is **transverse,** *i.e.*, wide relative to its length; whereas broader additions anteriorly result in an elongate shell.

In **holoperipheral** growth, shell matter is added along the entire valve margin, and the apex, which bears one plate of the bivalve protegulum, recedes from the posterior margin to some point between it and the center of the valve (Fig. 9-22). Shells showing holoperipheral growth tend to have circular or oval outlines and conical valves, with growth lines concentric to the beaks. Many neotremate genera illustrate this type of growth.

In **mixoperipheral growth,** successive shell increments are added as usual to the lateral and anterior margins, but on the posterior margin they are added in such manner that the shell surface turns inward and back on itself resulting in a small shelflike extension, the **palintrope,** that faces the opposite valve (see later discussion; also Fig. 9-25*N–O*).

One form of growth may follow another, and the two valves of the same shell commonly have different types of growth. These variations are responsible for the differences in outline and form seen in the valves of most adult brachiopods.

Types of Cardinal Margin. In articulate brachiopod shells in which mixoperipheral growth of at least the pedicle valve is the general rule, the posterior margin bears the teeth and is designated the **cardinal margin** or **hinge line**[1] (Fig. 9-24). The hinge line shows considerable variation, being straight and long in some transverse shells (*e.g.*, the spiriferids), short and curved in others (*e.g.*, the terebratulids), etc. The five following terms have been proposed for the different types of hinge-line development (Fig. 9-24).

Spiriferid—hinge line quite long and straight, or nearly so; commonly drawn out well beyond lateral margins in **alate** shells (Fig. 9-24*C*). **Megathyrid**—hinge line long and straight (Fig. 9-24*E,G*). **Submegathyrid**—hinge line slightly less than the greatest width of the shell, and straight or nearly straight (Fig. 9-24*A–B*). **Subterebratulid**—hinge line much less than the greatest shell width, but not strongly curved (Fig. 9-24*F*). **Terebratulid**—hinge line much less than the greatest shell width, and strongly curved (Fig. 9-24*D,H*)

[1] The cardinal margin and hinge line are different features of the posterior margin, though they may be coincident in some shells, as pointed out by Cloud (1942).

As a general rule the spiriferid and megathyrid stages are associated with a short pedicle, whereas the terebratulid stage characterizes shells with a long pedicle.

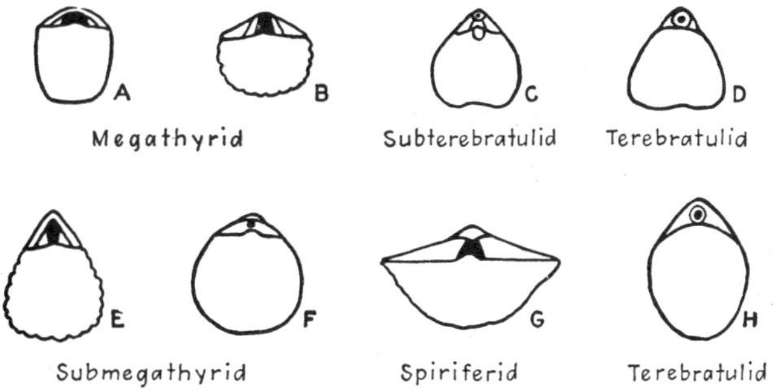

FIG. 9-24. Class ARTICULATA. Diagrams showing shell outlines and types of cardinal margins. *A. Kraussina cognata. B. Megathyris detruncata. C. Kraussina bisum. D. Dallina floridana. E. Eucalathis tuberata. F. Clidonophora incerta. G. Spirifer increbescens. H. Gryphus vitreus.* (*All except G modified after Thomson,* 1927.)

Palintropes and Interareas. Between the hinge line and beak ridges of many articulate shells is an anteriorly curved shelflike extension of the posterior part of each valve, to which the term **palintrope** (Fig. 9-26) has been applied (see page 300). In some forms the palintrope is a simple curved shelf broken only by a delthyrium in the median plane; in many others (*e.g.*, the spiriferoids, Fig. 9-25*A*) the lateral margins are first slightly curved but soon give way to a prominent flat shelf that extends from hinge line to beak and generally has a roughly triangular shape. The flat surface of this type of palintrope is generally designated the **interarea,** or **cardinal area,** and may be so large that it is coincident with the outline of the palintrope itself. In such a case the term palintrope applies to the shelflike structure as a whole and the cardinal area (or simply interarea) refers to the outside or external surface of the palintrope.

The interarea on the pedicle valve is almost invariably wider than that on the brachial—in some shells (*e.g.*, *Cyrtina*) conspicuously so—and is generally triangular in shape (Fig. 9-25*A*,*G*,*I*). Interareas commonly have growth lines parallel to the hinge line, and in some cases other lines essentially perpendicular to the hinge line. Although it may be said that true interareas are found only in articulate shells, nevertheless somewhat similar surfaces are present on a few of the Inarticulata. These surfaces are designated **pseudointerareas,** or **pseudocardinal areas.**

Although the exact function of the palintropes is not certain, they seem to have kept the beaks apart so that the shell could be opened more widely. The beak of each valve generally overhangs its associated palintrope, with the pedicle beak showing the greater degree of overhang. As a matter of fact the pedicle beak of some brachiopods with small and narrow palintropes overhangs to such an extent that it conceals the entire brachial interarea.

Nature of Commissure and Types of Folding. The line along which the

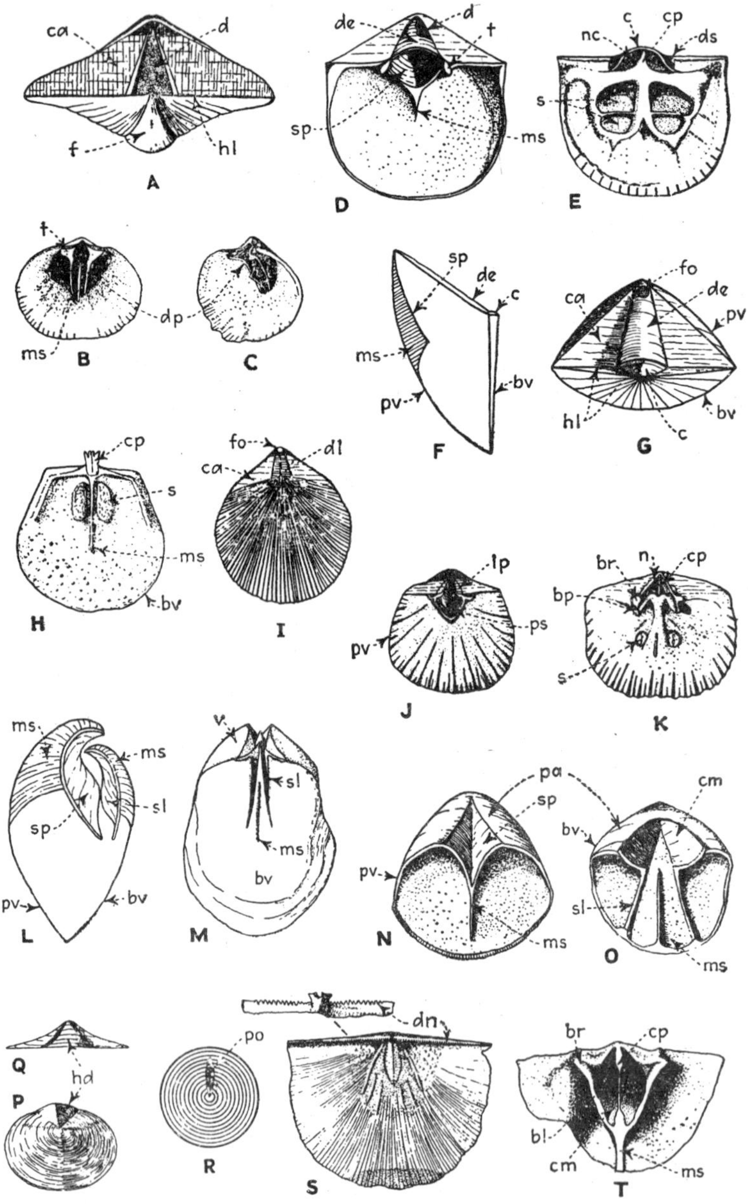

FIG. 9-25. Phylum BRACHIOPODA. Hinge-line structures. (*See opposite page for detailed descriptions.*)

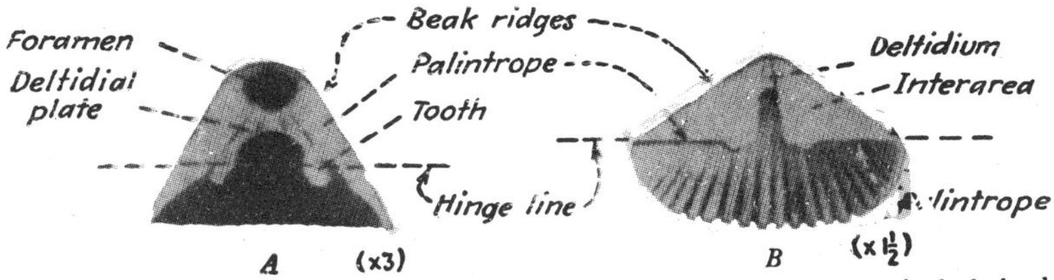

FIG. 9-26. Phylum BRACHIOPODA. Structures in posterior part of valves, particularly beak ridges, palintropes, interareas, and hinge lines. *A. Campages. B. Hesperorthis.* (*Modified after Cooper*, 1944.)

valve margins touch is designated the **commissure** (Fig. 9-1), and according to the development of the margins it may lie in a plane, or be wavy or serrate. Because of the way brachiopod shells develop, the greatest modifications of the commissure are made along the anterior part, where the valves may become conspicuously corrugated. Some of the complications common in anterior commissures are shown in Fig. 9-27.

The most primitive form of shell as regards convexity is the biconvex shell with smooth valves (Fig. 9-1, 9-27*A*). The anterior commissure of such a shell is perfectly plane or **rectimarginate.** Many shells depart widely from this primitive condition by developing prominent radial corrugations that involve the full thickness of the valve. The ridge of such corrugations is designated a **fold, or plica** (plural, **plicae**). The trough is termed a **sulcus** (plural, **sulci**).

FIG. 9-25. *A.* Posterior view of a spiriferid, showing long and straight hinge line, high and wide pedicle interarea, large triangular delthyrium, and prominent fold in brachial valve. *B–C.* Interior views of pedicle valve of *Enteletes*, showing teeth, dental plates, and prominent median septum. *D–G.* Views of *Vellamo: D*, interior of pedicle valve; *E*, interior of brachial valve, showing particularly the chilidium arching over the notothyrial cavity and cardinal process; *F*, longitudinal section, showing relation of chilidium and deltidium; *G*, posterior view. *H.* Brachial interior of *Juresania*, showing prominent trifid cardinal process. *I.* Brachial view of *Trigonosemus*, showing deltidial plates and foramen. *J.* Interior of pedicle valve of *Glossorthis*, showing lateral plates and pseudospondylium. *K.* Brachial interior of *Dolerorthis*, showing triangular notothyrium, notothyrial cavity with median cardinal process, brachiophores, and brachiophore processes. *L–M. Pentamerus: L*, median longitudinal section of shell, showing prominent median septa, spondylium, and septal plate; *M*, steinkern, or filling, of shell, showing impressions left by the several posterior structures. *N–O. Conchidium: N*, posterior part of pedicle valve, showing spondylium and supporting median septum; *O*, posterior part of brachial valve, showing septal plates and cruralium. *P–Q.* Apical and posterior views of *Paterina*, showing well-developed **homoeodeltidium.** *R.* Apical view of the pedicle valve of *Orbiculoidea*, showing concentric growth lines and slitlike pedicle opening. *S.* Interior of pedicle valve of *Stropheodonta*, showing prominent denticulation along the hinge line. Part of hinge of dorsal valve, considerably enlarged, is added to show the denticulation and nature of cardinal process. *T.* Brachial interior of *Linoporella*, showing a cruralium formed by union of the brachiophore plates with the floor of the valve. [*br* = brachiophore; *bl* = brachiophore plate; *bp* = brachiophore process; *c* = childium; *ca* = cardinal area (interarea); *cp* = cardinal process; *cr* = crural plates; *cm* = cruralium; *d* = delthyrium; *dc* = delthyrial cavity; *de* = deltidium; *dl* = deltidial plates; *dn* = denticle; *do* = brachial (dorsal) valve; *dp* = dental plate; *ds* = dental socket; *f* = fold; *fo* = foramen; *hd* = homoeodeltidium; *hl* = hinge line; *lp* = lateral plate; *ms* = median septum; *n* = notothyrium; *nc* = notothyrial cavity; *pa* = palintrope; *po* = pedicle opening; *ps* = pseudospondylium; *s* = muscle scar; *sl* = septal plate; *sp* = spondylium; *t* = tooth; *v* = pedicle (ventral) valve.] (*B–C after Dunbar and Condra*, 1932; *J–L, N–O, T adapted from Schuchert and Cooper*, 1932; *P–Q after Walcott*, 1912; *S after Hall and Clarke*, 1894.)

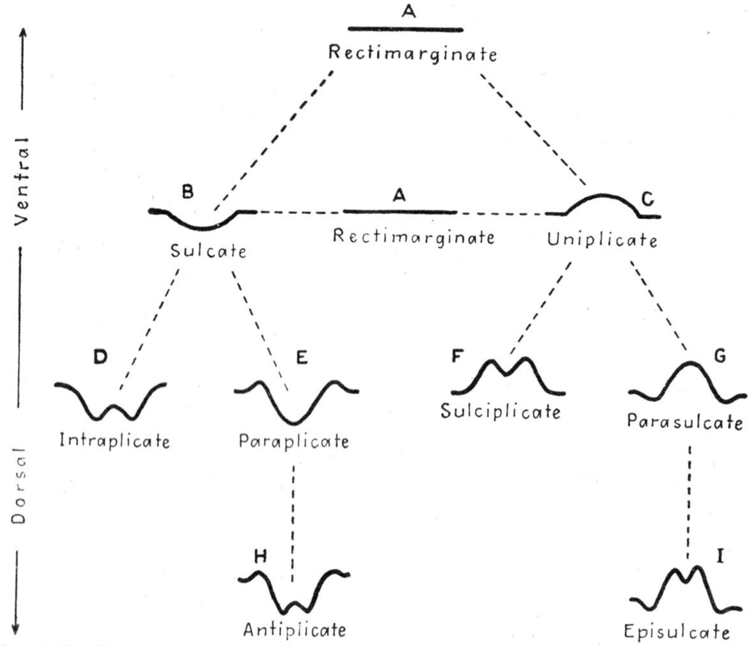

FIG. 9-27. Class ARTICULATA. Diagram of anterior commissures, illustrating types of folding in their several lines of development from a supposed originally **rectimarginate** condition (*A*). The **sulcate** form (*B*) has a single sulcus in the brachial valve, opposed by a plica in the pedicle valve; **intraplicate** (*D*) has a plica raised in the middle of the brachial sulcus, and a corresponding sulcus in the fold on the pedicle valve; **paraplicate** (*E*) has plicae formed along the lateral margins of a persistent brachial sulcus; and **antiplicate** (*H*) is further complicated by a small brachial plica in the prominent brachial sulcus. In the **uniplicate** commissure (*C*) there is a single plica in the brachial valve opposing a sulcus in the pedicle valve; **sulciplicate** (*F*) shows the uniplicate fold broken by a median sulcus, with a corresponding fold in the middle of the pedicle sulcus—the reversed counterpart of intraplicate; **parasulcate** (*G*) preserves the uniplica unbroken, but a sulcus is formed on each side of it to produce the reversed counterpart of paraplicate; **episulcate** (*I*), the reversed counterpart of antiplicate, results from modification of parasulcate by imposition of a sulcus upon the median plica. **Multiplicate** shells develop when the simpler types of commissure are modified, *e.g.*, by repeated splitting (sulcation) of plicae in a folded shell. (*After Thomson*, 1927, *slightly modified.*)

It is thought that the more complicated types of commissure evolved from the primitive rectimarginate type as indicated in Fig. 9-27.

Surface Sculpture. The shells of most primitive brachiopods are smooth on the exterior except for concentric growth lines, and even these may be only barely visible. The shells of young individuals of all species also tend to be smooth, but with increased age and added growth there is usually development of additional sculpture. This ornamentation takes many different orms—fine or minute lines and ridges, coarser ribs, fine granules and coarser ubercles, and several kinds of spines (Fig. 9-28). These features may be arranged concentrically, following the lines of growth, or radially, more or less perpendicular to the growth lines.

Radial sculpture consists of fine ridges, **fila** (singular, **filum**) or **costellae,** also commonly called **striae** in previous works, and more prominent ribs, or **costae.** As a comparison of size, there are about 40 costellae in the width of 1 cm. (slightly less than ½ in.) in a certain species of *Terebratulina*, whereas multicostate species of *Magellania* and *Terebratella* have only 5 to 10 costae in the same space. As the shell grows, both costellae and costae generally increase in number, either by bifurcation of ridges already present or by intercalation or implantation of new ones.

Concentric sculpture takes the form of growth lines, lamellar outgrowths as in *Discinisca* and *Athyris* (Fig. 9-28*J*), or corrugations (**rugae**) as in *Magellania joubini*. Rugae are transverse ridges crossing the costae on the older part of the shell and are responsible for the reticulated or cancellated appearance of many productids; *e.g.*, *Dictyoclostus americanus* (= *Productus semireticulatus*, in part) and *Marginifera lasallensis* (Fig. 9-50*C–D*).

If both radial and concentric sculpture are present on the same shell, the valve surface has a **reticulated** or checkered appearance. Where growth lines cross costellae or costae, tubercles may develop or the growth lamellae may become raised at the intersections, making the ribs **squamose** or **imbricate.**

Spines may also develop, not only at the intersections of growth lines and radial features, but elsewhere over the shell surface, and even along the hinge line, as in *Chonetes* (Fig. 9-29*A–B*). Spines and rugae are particularly characteristic of the productids. The spines are slender pointed tubes rising from the outer layers of the shell, either at a high angle to the surface, in which case they are described as **erect,** or at an angle approaching tangency, hence **oblique** (Fig. 9-29). The oblique spines taper more rapidly than the erect and are commonly small and relatively short.

Since the spines are hollow, they leave a tiny hole in the shell when broken off, and this characteristic led many earlier paleontologists to describe the productids as punctate. However, the productid shell lacks puncta altogether and is now described as pseudopunctate. Dunbar and Condra (1932) have demonstrated the importance of spines in discriminating between Pennsylvanian productids.

Growth lines, radial ridges of all kinds, granules, tubercles, pustules, spines, and rugosities of many types are variously combined on brachiopod shells to give them a great variety of surface sculpture. This sculpture is an important character for specific discrimination, and in a few cases has even been used for generic differentiation (Dunbar and Condra, 1932; Bond, 1941). In the case of growth lines an added use is provided in that study of them may make it possible to determine the shape of the shell during successive growth stages (Fig. 9-22); such study can be of importance in working out the morphogeny of a species.

Morphology of the Pedicle Opening. The opening through which the fleshy stalk protrudes is broadly referred to as the **pedicle opening.** In some early brachiopods it is a simple gap along the posterior margin in its middle

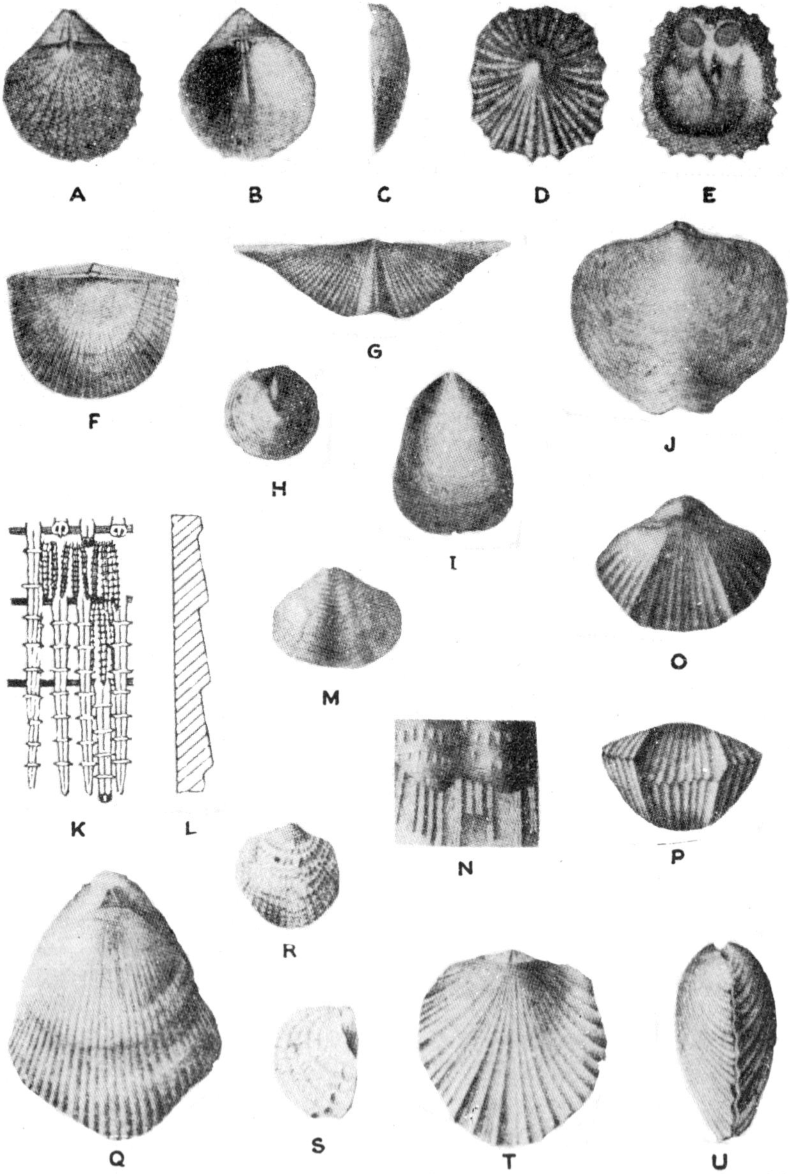

FIG. 9-28. Phylum BRACHIOPODA. Surface ornamentation. (*See opposite page for detailed descriptions.*)

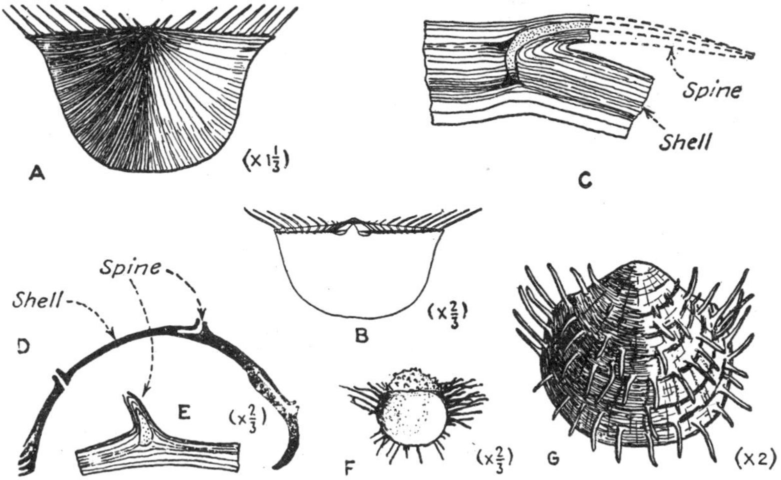

FIG. 9-29. Phylum BRACHIOPODA. Surface spines. *A*. Restoration of the pedicle valve of *Chonetes*, showing prominent spines rising from the cardinal margin. *B*. A drawing of the cardinal spines of *Chonetes granulifer*, showing the courses of the tubular spine bases through the palintrope of the shell. *C*. Longitudinal section of a spine base of *Echinoconchus*. The matrix is stippled, and the spine is hypothetically restored by dashed lines. It is to be noted that this spine tends to parallel the shell surface (cf. *D* and *E*). *D*. Longitudinal section of the pedicle valve of *Marginifera muricatina* (beak at the right), showing the courses of the hollow spines in the shell. These spines were erect. *E*. Oblique section through the base of a spine in *Linoproductus*. *F*. A pedicle valve of *Avonia*, showing strong development of surface spines. *G*. Exterior of the pedicle valve of a productid, showing strong spines rising from the surface of the valve, commonly along growth lines. (*B–E after Dunbar and Condra*, 1932; *F adapted from Cooper*, 1944; *G adapted from Davidson*, 1886–1888.)

FIG. 9-28. *A–C*. Brachial exterior, pedicle interior, and lateral view of *Thecidia papillata*, which has a pronounced papillate surface. *D–E*. Brachial exterior and interior of *Crania ignabergensis*, a Cretaceous species, with heavy radial costae. *F*. *Strophomena antiquata*, brachial exterior, showing larger costellae separated by groups of much finer costellae. This shell could be described as multicostellate. *G*. *Mucrospirifer mucronata*, with strong angular costae having lamellose structure, and with a prominent pedicle sulcus carrying a small median plica. *H*. Pedicle valve of *Discina*, showing concentric growth lines produced by holoperipheral growth. *I*. Pedicle valve of *Lingula* having many fine growth lines produced by hemiperipheral growth. *J*. *Athyris spiriferoides*, brachial exterior, showing many irregular lamellose growth lines. *K–L*. Surface spines of *Phricodothyris perplexa*. The larger are double-barreled. A section of the same bit of shell is shown in profile in *L* to indicate the relation of the terracelike bands to the rows of spines. *M–N*. *Elytha fimbriata*. Pedicle exterior, and a small part of the exterior much enlarged, showing double-barreled spines and general fimbriate nature of the shell surface. *O–P*. Brachial and anterior views of *Sieberella*, showing the relatively few strong costae and the prominent sulcus in the brachial valve. *Q*. Brachial exterior of *Conchidium nysius*, a heavy shell with many strong rounded costae, all of equal size. *R–S*. Pedicle exterior and lateral view of different species of *Atrypa*, showing concentric rugosities produced by spinose and crinkled growth lamellae. *T–U*. Dorsal exterior and lateral view of *Stricklandinia anticostiensis*, which is characterized by angular costae that multiply by implantation. Most figures are about two-thirds natural size; *K–L* and *N* are magnified. (*A–I after Nicholson*, 1872; *J*, *M–U after Hall and Clarke*, 1893; *K–L after Dunbar and Condra*, 1932.)

part. In other forms it is a restricted opening shared by both valves (delthyrium in the pedicle; notothyrium in the brachial), or a specialized perforation restricted to the pedicle valve (**foramen**). In no known brachiopod is the pedicle opening restricted to the brachial valve, but in a few forms it is shared with that valve, at least in maturity.

Because the pedicle plays such an important part in the life of a brachiopod, it is to be expected that the opening through which it protrudes has undergone extensive modification during the evolution of the phylum. So important did these modifications appear to Beecher (1891, 1893) that he based his four orders—Atremata, Neotremata, Protremata, and Telotremata—on characteristics of the opening. Although there is now grave doubt about the validity and homogeneity of these orders, no one has yet proposed a better grouping, and we simply combine the last two in order to have a subdivision to which can be assigned the many protrematous and telotrematous superfamilies (see later discussion under Classification).

In the early stages, the pedicle of the young brachiopod is large enough to occupy the entire delthyrium (Fig. 9-19*F*), but as growth continues the opening gets too large for the pedicle and the delthyrium is partly or wholly closed by secondary shell matter. Closure takes place differently in the several orders of brachiopods.[1]

In one group (Protremata) an impunctate triangular plate partly or completely fills the delthyrium. It is set off from the remainder of the palintrope by grooves and represents one type of **deltidium.**[2]

In many Telotremata the mantle adds calcareous material along the sides of the delthyrium and on the hinge-line side of the pedicle. The two plates thus formed from the lateral growth are designated the **deltidial plates,** and the single plate that grows forward from the hinge line as a secondary development has been designated the **henidium** (Cloud, 1942, page 18). These plates can be combined in several ways, as indicated in Fig. 9-30 and Table 9-2. As with the deltidium, they also are set off from the remainder of the palintrope by grooves and are punctate in punctate shells.

In many rynchonellids the deltidial plates have simply coalesced along the middle line but with the line of junction quite visible; in contrast, the same plates in many terebratulids have become so completely fused that the cover of the delthyrium is a single plate with transverse rugosities. To this plate Buckman (1918) applied the term **symphytium** (Gr. *symphytos,* grown together, healed, of a wound).

In cemented articulates, and also in certain shells that seem to have lain free on the bottom, the delthyrial opening is completely closed by a deltidium of some sort and the foramen is plugged.

Protremata and Telotremata with a palintrope on the brachial valve have in this valve a counterpart of the delthyrium. This also is a triangular reen-

[1] There has been much disagreement and confusion over the terms used for the several kinds of plates developed in connection with closure of the delthyrium. The problem has been discussed recently by Arber (1939–1942) and Cloud (1942), and conflicting uses are shown on Table 9-2. For further discussion of the problem see p. 284.

[2] See footnote on p. 284.

trant of the posterior margin and is designated the **notothyrium** (Fig. 9-25*K*). It is closed by two discrete plates, the **chilidial plates,** or by a single plate, the **chilidium** (Fig. 9-25*E–G*), which is punctate in punctate shells.

The most primitive type of pedicle opening known is that of a young *Lingula* immediately after the protrusion of the pedicle. It is a simple broad gap in the posterior part of the shell made by the outward bending of the posterior margins of both valves (Fig. 9-16*K–L*). As the shell grows, additions are made around the margin of this gap, and the reentrant in one valve gradually closes, whereas that in the other, the pedicle valve, deepens into what may be called the **pedicle notch.** In neotrematous brachiopods, which generally have holoperipheral growth in both valves, the pedicle is confined to a foramen slightly posterior to the apex of the valve.

A somewhat different line of development is found in the Lingulacea and Siphonotretacea which have hemiperipheral growth succeeded by mixoperipheral in each valve. While the brachial palintrope remains intact, the pedicle palintrope is interrupted by a pedicle notch. In the adult *Lingula* the main part of the pedicle lies in a deep groove, or more correctly a peri-

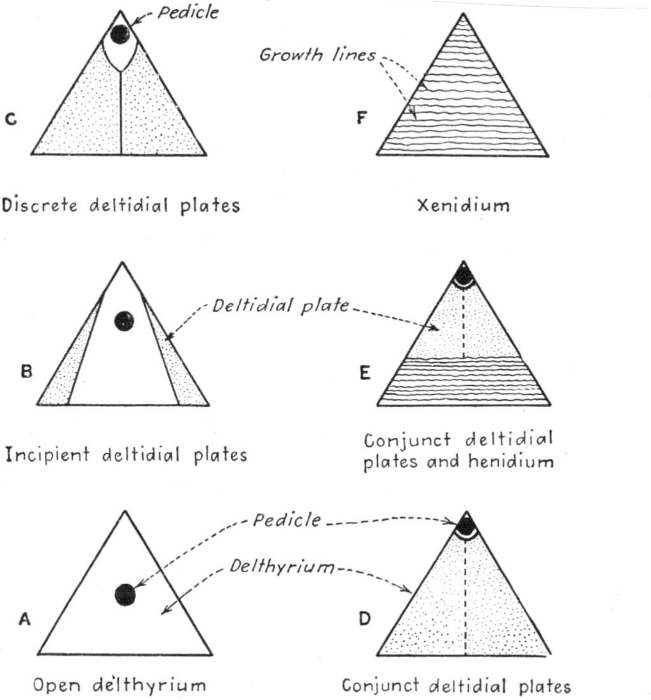

FIG. 9-30. Class ARTICULATA. Highly generalized and diagrammatic figures showing the different modifications of the delthyrial opening in telotrematous brachiopods. (*Based on discussions by Thomson,* 1927, *and Cloud,* 1942.)

ostracum-lined depression, in the pedicle palintrope. In certain genera of Siphonotretacea (*e.g.*, *Schizambon*), the pedicle opening migrates through the apex and finally attains a position anterior to the apex on the pedicle valve (Fig. 9-41*C*).

A third line of development is shown by a large number of protrematous and telotrematous forms which have articulated calcareous shells developed by mixoperipheral growth. In these the posterior gap of the valves consists of an open delthyrium and notothyrium. In the early growth stages of most Telotremata there is a large open delthyrium through which the pedicle protrudes, but this condition does not persist for long. The pedicle becomes confined to a small part of the delthyrium, because of the growth of deltidial plates (Fig. 9-30), or shifts its position partly or wholly out of the delthyrium. The general rule in the Telotremata is for the pedicle opening to move away from the brachial umbo toward or even completely through the apex of the pedicle valve. A nomenclature has been proposed for the different positions occupied by the pedicle, as illustrated in Fig. 9-31.

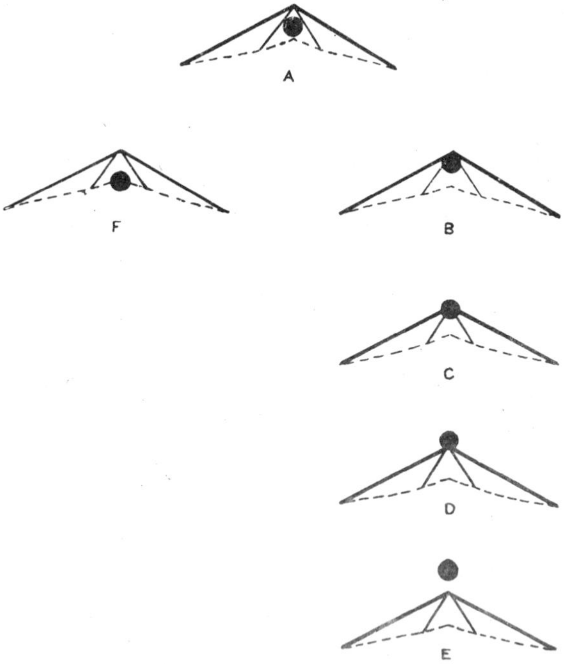

FIG. 9-31. Class ARTICULATA. Diagrams showing different positions of the foramen. The upper heavy lines represent the line of the beak ridges; the lower broken lines mark the posterior margin of the brachial valve; the inner thin lines represent the margins of the delthyrium; and the black circular spots mark the site of the foramen. *A*. Hypothyrid. *B*. Submesothyrid. *C*. Mesothyrid. *D*. Permesothyrid. *E*. Epithyrid. *F*. Amphithyrid. (*Adapted from Thomson*, 1927.)

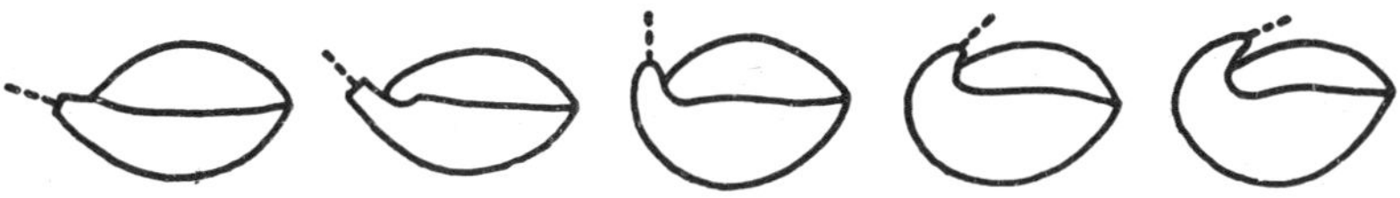

FIG. 9-32. Class ARTICULATA. Diagrams showing incurvature of the beak of the pedicle valve and the terminology used to describe it. The pedicle is indicated by a heavy dashed line. (*Adapted from Thomson*, 1927.)

While much attention has been paid to the development of the delthyrium and the migration of the pedicle, both involving the pedicle valve, less attention has been directed to concomitant changes in the brachial valve. Generally the notothyrium becomes closed by a chilidium, but there are many details concerning the development of the posterior part of the brachial valve that remain to be determined.

Curvature of the Beak. The curvature of the beak of the pedicle valve toward the brachial valve is a conspicuous feature of most brachiopod shells, and authors vary in the terms used to describe the relation. Thomson's (1927) suggestions, as illustrated in Fig. 9-32, seem definite and acceptable.

Internal Morphology of the Valves

The internal features of the valves of a brachiopod, particularly those of the brachial valve, are of the greatest importance in differentiating major divisions and in detecting homoeomorphy (see page 290). Attention is directed here to four kinds of internal features: (1) articulation; (2) cardinalia; (3) brachial supports; and (4) muscular impressions.

Articulation. The hinge line has already been defined as the line along which articulation takes place, and it is also the line along which the valves of articulate shells join posteriorly. It is divided into two equal parts by the median pedicle opening consisting of the delthyrium and notothyrium. In protrematous and telotrematous shells there are two prominent **hinge teeth** on the pedicle valve, one at each corner of the delthyrium along the hinge line (Figs. 9-25*D*, 9-33*B*,*D*), and corresponding **sockets** (**dental sockets**) in the brachial valve at the basal corners of the notothyrium (Figs. 9-25*E*, 9-33*A*,*C*). In some shells the teeth are supported by vertical **dental plates** or **dental lamellae**[1] (Fig. 9-25*B*–*C*) extending below to the base of the beak cavity and back toward the apex. These plates divide the beak cavity beneath the palintrope into a large central **delthyrial cavity** and two smaller lateral cavities. In a few genera the dental plates are joined by a callous deposit extending across the floor of the beak cavity, and the structure thus formed resembles the spondylium of the Pentameracea (see pages 309–310).

The teeth and sockets function chiefly for articulation, acting as a fulcrum as well as a device for preventing lateral slipping of the valves, but in a few genera they share these functions with supplementary structures, some of which are almost equally important in articulation.

[1] Bond (1941) has used dental lamellae to differentiate species of *Schizophoria*.

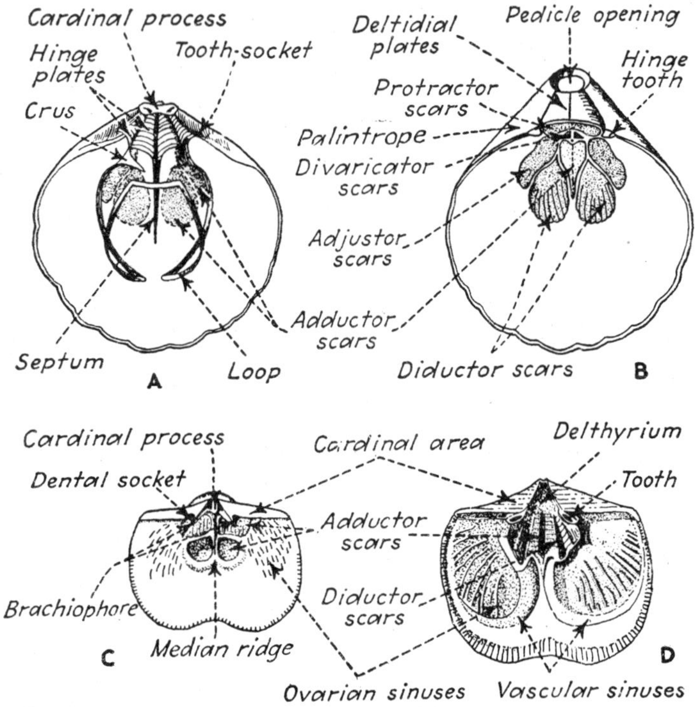

FIG. 9-33. Class ARTICULATA. Internal structures of brachiopod shells. *A–B. Magellania flavescens*, a modern telotremate: *A*, brachial interior with all flesh removed; *B*, interior of pedicle valve, showing muscle scars and deltidial plates. *C. Hebertella*, an extinct protremate common in Upper Ordovician strata; brachial interior of *H. occidentalis*. *D*, Interior of pedicle valve of *Glyptorthis insculpta*. (*A–B after Davidson*, 1886–1888; *C–D after Winchell and Schuchert*, 1895.)

Cardinalia. Near the posterior (cardinal) margin in the interior of the brachial valve is a complex of structures that are connected with articulation, muscular attachment, and attachment of brachial supports. The elements of this complex are collectively designated the **cardinalia,** and they consist of (1) the articulating apparatus; (2) the brachial supports (brachidia) and associated features; and (3) the cardinal process. They may be intimately fused together, but commonly they can be resolved into distinct structures with definite functions. Closely associated with the cardinalia in many genera is a **median septum** or **ridge,** which may be free or attached to the cardinalia as a buttress (Figs. 9-25, 9-33*A*,*C*).

The cardinalia are **weak** if they consist of thin platelike processes and **strong** if they are composed of stout rounded processes more or less fused together. Strong cardinalia form along the cardinal margin a bench or step that descends steeply to the floor of the valve in front. The hinge sockets lie on the outer edges of this bench and may be bounded by **inner** and **outer**

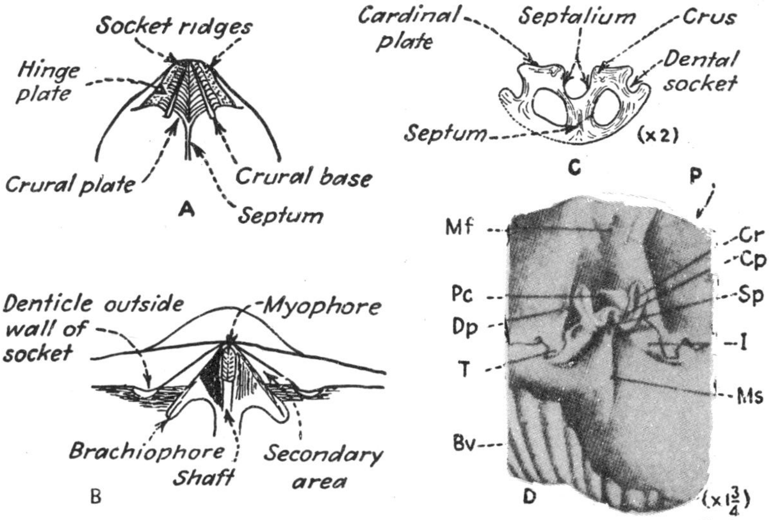

FIG. 9-34. Class ARTICULATA. Features of the cardinalia. *A*. Posterior part of brachial interior of *Dallina floridana*, showing particularly well how the crural plates are conjoined below to form a median septum. *B*. Diagram showing cardinalia with the cardinal process composed of a myophore and a shaft. *C–D*. *Septaliphoria astieriana*: *C*, transverse section of apical part of brachial valve, showing structure of the septalium and its relation to the septum; *D*, posterior part of articulated valves seen from the interior, showing relation of septalium to associated features. (*Br* = brachial valve; *Cp* = cardinal plate; *Cr* = crus; *Dp* = dental plate; *I* = internal wall of dental socket; *Mf* = muscle field; *Ms* = median septum; *Pc* = pedicle collar; *Pv* = pedicle valve; *Sp* = septalium; *T* = tooth.) (*A after Thomson*, 1927; *B after Schuchert and Cooper*, 1932; *C–D after Wisniewska*, 1932; *all somewhat modified*.)

socket ridges (Fig. 9-34*A*). In certain Telotremata two **crura** (singular, **crus**), the basal parts of the brachial supports, arise near the brachial umbo and traverse the cardinalia on the inner sides of the socket ridges, projecting slightly forward beyond the latter (Fig. 9-34*A*). The parts of the crura that unite with the cardinalia are designated **crural bases. Hinge plates** are more or less horizontal plates extending toward the median line from the socket ridges and separated anteriorly from the floor of the valve by a cavity. They may remain free or become coalesced along the median line, and in some forms they descend to join the median septum.

The designation **hinge trough** is used for those cardinalia in which a trough-shaped structure is formed by the posterior bifurcation of a strong septum, with the forks of the septum rising to join the inner anterior corners of the combined socket ridges and crural bases.[1] If the area between the

[1] Students of Mesozoic telotrematous brachiopods use the term **septalium** for a somewhat similar Y-shaped structure in the posterior part of the brachial valve beneath the crura (Fig. 9-34*C–D*). It is a trough formed by plates extending from the cardinal plate to a median septum with which they are united. The structure is not the result of the bifurcation of the septum, as it might seem to be, but is formed by the posterior reflection of the cardinal plate (see Fig. 9-34*C*).

socket ridges is occupied by a solid filling of shell matter, the designation **hinge platform** is applied to the cardinalia.

Beneath and between the margins of the notothyrium is the **notothyrial cavity** (Fig. 9-25*E*), the floor of which may become built up into a **notothyrial platform.**

At the posterior extremity of the brachial valve, in a median position, commonly rising from the notothyrial platform and projecting anteriorly over hinge plates, hinge trough, or hinge platform, is an unpaired but commonly forked structure, the **cardinal process,** which serves as a base for the attachment of the diductor muscles (Figs. 9-25*E,H,K,T*, 9-33*A*). This process, as would be expected from its constant use in the opening of the shell, varies greatly in different species in its size, shape, structure, and ornamentation. Complex cardinal processes have an anterior part, the **shaft,** and a posterior roughened area, the **myophore,** where the diductor muscles are attached (Fig. 9-34*B*). Nine different types of cardinal processes have been observed in the Orthacea, and others are present in other articulates (Schuchert and Cooper, 1932). The cardinal process is one of the more important internal structural features of articulate brachiopod shells.

Brachial Supports (Brachidia). The fleshy lophophores of many protrematous and all telotrematous brachiopods are supported by calcareous processes attached to the brachial valve—either to the cardinalia, or to a septum, or to both. The term **brachidium** is used in a collective sense for the supporting structure, and more specific terms are applied to the several different parts of the more complex brachidia.

In the primitive protrematous orthids the brachia were not supported by a brachidium, but certain structures associated with brachial support were developed as part of the cardinalia. First of these is the **brachiophore** (Figs. 9-25*K*, *T*, 9-34*B*), consisting of the structures on either side of the notothyrium that bound the dental sockets. Attached to the brachiophore are elongate brachiophore processes (Fig. 9-25*K*) to which the lophophore was attached. In some genera the brachiophores are supported by more or less vertical **brachiophore plates** or **supports** (Fig. 9-25*T*). These structures, which did not actually support the fleshy brachia, as true brachidia do, but were rather a base of attachment for the lophophore, are not commonly well preserved but are of great evolutionary importance.

The simplest brachidium is found in the Rhynchonellacea. It consists of two rodlike crura or small **crural plates** (Fig. 9-51*H*) that diverge from the cardinalia. These can support the lophophore only at its proximal end near the mouth. If the crural plates join beneath the brachial interarea, or if they unite with the floor in the same position, a **cruralium** (Fig. 9-25*T*) is formed.

In the Spiriferacea the crura are produced anteriorly as a pair of spirally enrolled calcareous ribbons, or **spiralia** (Fig. 9-35*G–K*), and in a few highly specialized brachidia of this type a secondary spire, or **diplospire,** was developed inside the spiralium. In some genera the two spiralia are separated, but more commonly they are united medially by **jugal processes** (Fig. 9-35*C–F,I*) which together constitute a **jugum** (Fig. 9-35*J–K*).

A brachidium in which spiralia make up the major portion of the structure

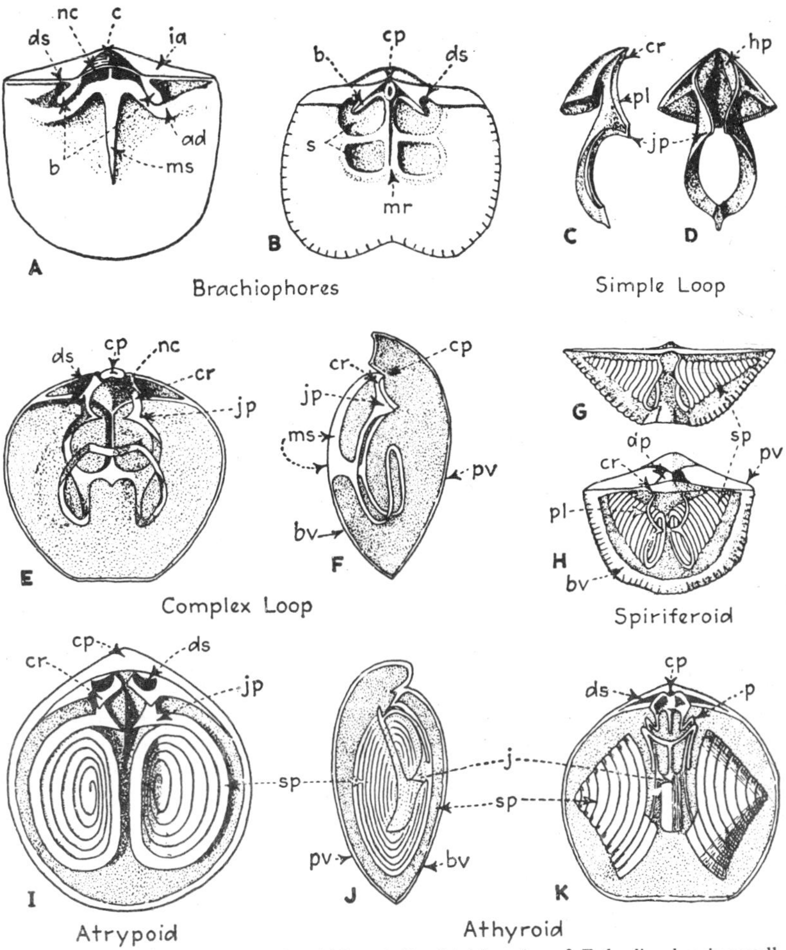

FIG. 9-35. Class ARTICULATA. Brachidia. *A*. Brachial interior of *Estlandia*, showing well-developed brachiophores and a chilidium roofing the notothyrial cavity. *B*. Brachial interior of *Hebertella* (cf. Fig. 9-33*C*), showing brachiophores and prominent muscle scars. *C–D*. Lateral and dorsal views of the brachidium of *Centronella*, showing a simple loop arising from a divided hinge plate. *E–F*. Interior and lateral views of brachial valve of *Magellania*, showing a complex loop attached to the inner surface of the valve by a median septum. The jugal processes are not united. *G–H*. Interior views of a spiriferid, with the brachial valve partly removed, showing spiriferoid spiralia with discrete jugal processes. *I*. Brachial interior of *Atrypa*, somewhat generalized. *J–K*. Lateral and interior views of the brachial valve of *Athyris*, showing athyroid spiralia joined by a jugum. The posterior reflection of the primary lamellae is conspicuous. [*a* = cardinal area (or interarea); *ad* = adventitious shell matter; *b* = brachiophore; *bv* = brachial valve; *c* = chilidium; *cp* = cardinal process; *cr* = crus, crura; *dp* = deltidial plates; *ds* = dental socket; *hp* = hinge plate; *j* = jugum; *jp* = jugal process; *mr* = median ridge; *nc* = notothyrial cavity; *pl* = primary lamella; *pv* = pedicle valve; *s* = muscle scar; *sp* = spiralium.] (*A after Schuchert and Cooper*, 1932; *B after Winchell and Schuchert*, 1895; *C–D, G–H after Hall and Clarke*, 1892; *E–F, J–K after Davidson*, 1886–1888.)

can be divided into three distinct parts (Fig. 9-35*C*)—the **primary lamellae,** by which the spiralia are joined to the brachial interarea; the jugum, by which the spiralia are joined together; and the spiralia themselves. Of the last named there are three general types. In one, the **atrypoid type** (Fig. 9-35*I*), the primary lamellae extend anteriorly, following the lateral margin of the valve, and give way to spiralia whose apices are directed toward the plane of symmetry or in a dorsal or ventral direction. In the second group, the **spiriferoid** type (Fig. 9-35*G–H*), the primary lamellae extend directly anteriorly to the central part of the shell where the spiralia arise and point laterally. The primary lamellae in the third group extend directly anteriorly for a short distance and then are deflected backward upon themselves. The spiralia begin following the deflection, and the apices are usually directed laterally. Such a brachidium belongs to the **athyroid** type (Fig. 9-35*J–K*). Shells with atrypoid and athyroid brachidia usually have short hinge lines; those with spiriferoid brachidia, long ones.

In the Terebratulacea the brachidium has the form of a calcareous **loop** the open ends of which are attached to the cardinalia by the crura. Projecting anteriorly from the margin of the hinge plates (Fig. 9-35*D*) are two short crura to which the loop proper is attached. At the distal ends of the crura are two pointed processes, the **crural processes,** that project obliquely inward and upward toward the pedicle valve. The long-looped brachidia typical of the higher genera of the Terebratellacea attain their final form after a complex series of changes, during which the loop remains attached to a median septum (Fig. 9-35*E–F*).

Brachidia of the loop-bearing type exhibit many different forms which are not designated by specific terms, but instead are usually indicated by the adjectival names derived from typical genera (*e.g.*, **Terebratelliform** type).

Brachial supports of all kinds are of great importance, not only for the identification of many genera and species but also because of their evolutionary significance. Unfortunately they are seldom preserved. If present, the character in some cases can be ascertained by making serial sections of the shell starting at the beak and from these constructing a three-dimensional restoration. Rarely the brachidia are silicified with the remainder of the shell still calcareous. In such cases the calcareous shell can be removed by acid treatment, and the brachial process obtained intact. Rare indeed is the case, however, in which the brachidium is not incomplete in one way or another.

Pallial Markings. Parts or all of the mantle are commonly folded in such a way as to produce tubular extensions (**pallial sinuses**) of the coelom which contain body fluids and in some cases parts of genital organs (Fig. 9-4). Such sinuses commonly leave **pallial markings** on the inner surfaces of the valves (Figs. 9-6, 9-33*C–D*). These markings usually have a dendritic appearance, with the smaller and more numerous branches situated laterally and anteriorly. They are of considerable importance in the definition of certain families, but unfortunately are usually not found well preserved.

Muscle Marks. The places of attachment of the different muscles are marked by pits depressed below the general inner surface of the valves, by roughened areas elevated slightly above the surface, or by areas on platforms of different kinds. These **muscle scars** and structures for attachment are of considerable importance in classification.

The term **muscle mark** is used by some authors for any mark that indicates muscular attachment; the term **muscle track,** for the path down the valve taken by successive muscular attachments. When these terms are employed, muscle scar is then applied to a well-defined area representing the place of ultimate muscle attachment.

Brachiopod shells without definite articulation have complex sets of muscles by which the valves are pushed apart and pulled together, or slipped about with respect to one another in a variety of ways. The scars of such a system are illustrated in Figs. 9-12 and 9-13.

Articulate shells possess a somewhat simpler muscular system, as the shell opens and closes on only one axis. The valves are prevented from moving laterally by the several articulating devices along the hinge line and additionally in some forms by crenulation of the lateral and anterior shell margins. The adductor muscles leave two scars on the pedicle valve and four on the brachial. The diductors leave two main scars on the pedicle valve and two on the cardinal process. The adjustors may leave small scars on both valves. and secondary diductors leave faint impressions which are seldom preserved in fossil shells (Fig. 9-14, 9-15).

Supplementary processes for the attachment of muscles are developed in some brachiopods. In certain Atremata (*e.g.*, *Trimerella* and *Lingulasma*) each valve has an elevated area, the **platform,** to which muscles are attached (Fig. 9-39*E–F*). This structure is solid in some shells, whereas in others it is a plate supported by a median septum. It serves to elevate the place of muscle attachment above the floor of the valve. An analogous structure, the **spondylium,** is developed in the pedicle valve of many articulates. This is a cuplike or spoon-shaped platform lying directly anterior to the delthyrium and is supported by a median septum (Fig. 9-25*D*,*N*).

The different types of muscle platforms in articulate shells may be characterized as follows:

1. Dental lamellae converge only slightly and extend directly to the floor without ever becoming united. This type of muscle cavity is common in orthid shells and has been designated a **spondylium discretum,** though it is not a true spondylium.

2. In the mature shells of some primitive Protremata (*e.g.*, *Glossorthis* and *Finkelnburgia*) the anterior part of the muscular area is elevated on a callosity produced by shell thickening within the delthyrial cavity, and the dental lamellae appear to be united, but they are really discrete, resting on the floor of the valve. For this type of muscle platform the term **pseudospondylium** is used (Fig. 9-25).

3. A **spondyloid** is a structure simulating a spondylium in form but produced by deposition of secondary shell substance on and about the dental lamellae, swelling them laterally until union is effected. It is well developed in *Porambonites* and has been mistaken for a true spondylium.

4. In a few rare cases the dental plates converge to form a spoonlike muscle platform that is unsupported by a median septum, hanging freely suspended in the pedicle valve (*e.g.*, *Protorthis*). This structure has been designated a **free spondylium.**

5. The term spondylium, as previously stated, is generally applied to the spoon-shaped muscle platform produced by the convergence and union of the dental lamellae with a

median septum, regardless of the structure of the septum. It has been determined, however, that the spondylium arises in several different ways; hence Kozlowski has defined several kinds. One type, well displayed in *Vellamo* (Fig. 9-25*D*), is formed by the union of the dental plates with a single simple median septum, and the entire structure is one piece. Kozlowski designated this a **spondylium simplex,** but because it is the simplest type of supported spondylium, Schuchert and Cooper (1932) have suggested that the limiting word simplex be dropped. Quite in contrast to the simple spondylium of *Vellamo* is the **spondylium duplex** typical of the Pentameracea. In this type the supporting septum is actually double, as it is composed of two united vertical plates (Fig. 9-25*M*).

It is possible that the cruralium of some genera served as a place for muscle attachment in the brachial valve, just as the spondylium had this function in the pedicle valve. A second feature in the brachial valves of certain articulates is a thickened ridge along the median plane on the floor of the valve between the adductor muscle scars. This has the form of a low septum or partition in a few shells (Figs. 9-33*C*, 9-35*B*).

Composition and Structure of the Shell

Composition, Structure, and Punctation. Unaltered brachiopod shells, so far as they have been chemically analyzed, fall into two quite different groups on the basis of inorganic composition:

1. **Chitinophosphatic** shells, such as *Lingula*, which contain chitin, calcium phosphate, and lesser quantities of magnesium carbonate and sulphate and calcium carbonate.
2. **Calcareous** shells, such as *Laqueus*, which consist almost entirely of calcium carbonate (98 + per cent) with traces of a few other salts.

It is assumed that there was a purely **chitinous** shell among the most primitive brachiopods, because of the fact that the protegula of all species are chitinous.

In addition to the inorganic substances that make up the "hard" part of the shell, there is also, in living shells, a variable amount of organic substance, partly in the form of a chitinous periostracum and partly pervading the shell substance in a membranous form.

Table 9-4 shows analyses of atremate, neotremate, protremate, and telotremate shells of living species.

Some brachiopod shells have been described as **horny** (**corneous**) or **calcareocorneous,** but these terms are hardly applicable, because no shells are known to be composed dominantly of either substance. It is possible that there are chitinocalcareous and calcareophosphatic shells, but the existence of these does not yet seem to have been proved (Thomson, 1927).

The chitinophosphatic shells are structurally of two types, exemplified by *Discinisca* and *Lingula*. In *Discinisca* the calcium phosphate and accessory mineral salts are distributed uniformly throughout the chitinous material, whereas in *Lingula* chitinous and phosphatic layers alternate (Fig. 9-36*A*–*B*).

The shell of *Discinisca* consists of lamellae that make a slight angle with the surface and overlap one another like shingles. These lamellae, which are of uniform composition and structure, have a chitinous foundation containing a large content of calcium phosphate. The shell layer is traversed through its complete thickness by fine, essentially parallel tubuli disposed perpendicularly to the surface and commonly branching toward the inner surface. The lamellar punctate layer of the shell is covered by a dense nonpunctate concentrically striated periostracum (Fig. 9-36*C*).

TABLE 9-4. ANALYSES (CLARKE AND WHEELER, 1915) OF THE SHELLS OF SOME RECENT BRACHIOPOD SPECIES

Inorganic matter stated separately, with mineral salts recalculated to 100 per cent

	Chitinophosphatic shells			Calcareous shells				
	(1)	(2)	(3)	(4)	(5)	(6)	(7)	(8)
SiO_2	0.91	0.50	0.85	0.22	0.15	0.06	0.52	0.18
$(Al,Fe)_2O_3$	0.54	0.29	0.58	0.27	0.23	0.04	0.15	0.48
$MgCO_3$	2.70	0.79	6.68	8.63	0.49	0.93	1.37	0.68
$CaCO_3$	1.18	4.25	8.35	88.59	98.20	98.61	96.78	98.30
$CaSO_4$	2.93	4.18	8.37	1.72	0.55	0.36	1.18	0.36
$Ca_3P_2O_8$	91.74	89.99	75.17	0.57	0.38	Trace	Trace	Trace
	100.00	100.00	100.00	100.00	100.00	100.00	100.00	100.00
Organic matter...	40.00	39.50	25.00	3.52	2.65	0.93	4.73	1.55

(1) *Lingula "anatina"* (Gmelin), Japan. (2) *Lingula unquis* (Linné), Philippine Islands. (3) *Discinisca lamellosa* (Brod.), Peru. (4) *Crania anomala* (Müller), Norway. (5) *Hemithyris psittacea* (Gmelin), Shetland Islands. (6) *Gryphus cubensis* (Pourtalès), Florida. (7) *Terebratulina septentrionalis* (Gray), Maine. (8) *Laqueus californicus* (Koch), California.

In the shell of *Lingula*, the alternating chitinous and phosphatic lamellae are covered by a smooth periostracum (Fig. 9-36*A*) and are traversed by tubules that in life contain fine papillae of the epithelial layer of the mantle lobes. The chitinous layers have an oblique fibrous structure whereas the phosphatic layers are transversely prismatic (Fig. 9-36*A*–*B*).

Calcareous shells have a much more complex structure than those that are chitinophosphatic. The modern articulate *Laqueus californicus* has a three-layered shell (Fig. 9-36*M*). The outermost layer, which is entirely organic, is a thin periostracum. The middle layer, which is best described as the **outer carbonate layer,** is composed of exceedingly fine fibers oriented more or less perpendicularly to the surface. The **inner carbonate layer** is composed of somewhat coarser fibers oriented obliquely to the inner shell surface at a rather constant angle. The fibers of the inner layer are flattened and elongated rods arranged parallel to one another and trending obliquely forward (*i.e.*, posteriorly) from the outer surface to the inner at a constant angle. The ends of the fibers trace an imbricated mosaic, like the scales of a snake, on the interior of the shell (Fig. 9-36*H*), and it has been suggested that variations in pattern of this mosaic may be useful for specific discrimination (Hobbs and Cloud, in Cloud 1942).

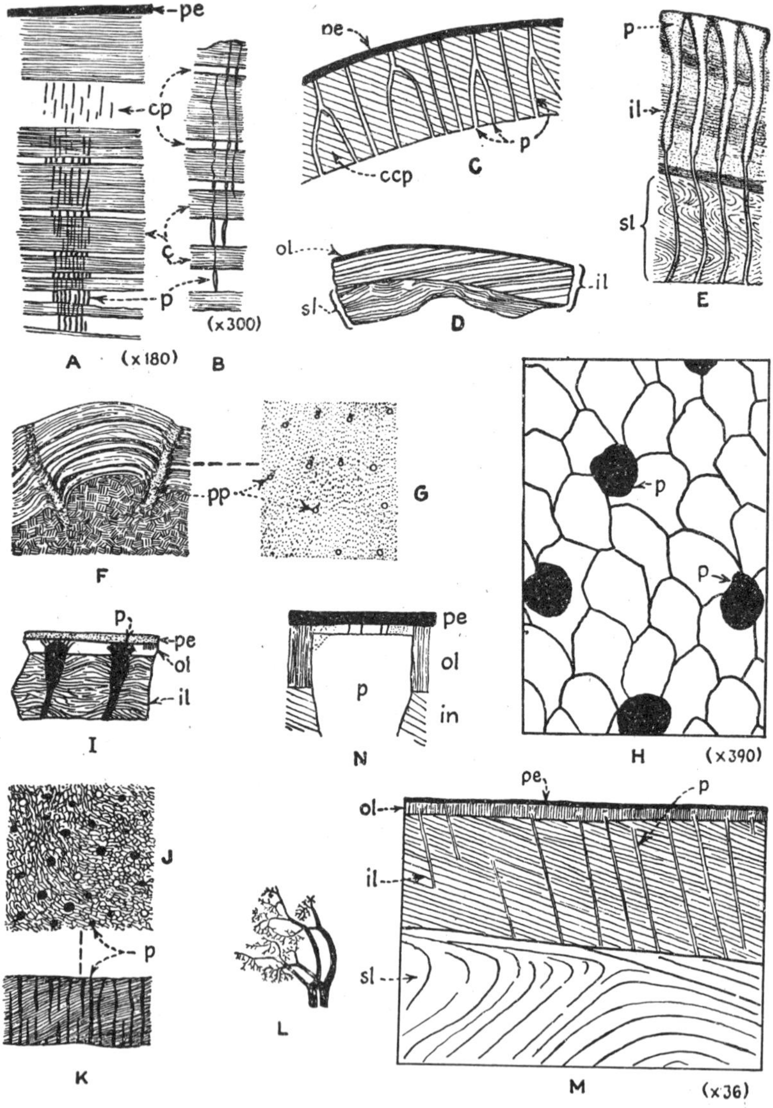

Fig. 9-36. Phylum Brachiopoda. Microscopic shell structure. (*See opposite page for detailed description.*)

The puncta in *Laqueus* are irregular tubes that begin at the inner surface of the shell, completely penetrate the inner carbonate layer, and end just before reaching the outer surface of the outer carbonate layer. Tiny tubules, arising from the abruptly flattened ends of some of the puncta, seem to continue through the rest of the outer carbonate layer, but whether they enter the periostracum is unknown (Fig. 9-36*M–N*).

If secondary shell substance is added to the inner surfaces of the valves, it is punctate except where added to impunctate median septa, hinge plates, dental plates, and hinge teeth. In these secondary deposits the lamellae are arranged in random fashion, seemingly depending on the relation of the mantle lobe to the shell at the time of deposition (Fig. 9-36*D,M*).

A three-layered shell seems also to have been present in many ancient articulates, although, as would be expected, the periostracum is not preserved. In the Superfamily Orthacea, for example, the shell has a thin outer non-fibrous layer and a much thicker inner fibrous layer. The outer **laminated** or **lamellar** layer, against which the inclined fibers of the inner **prismatic** layer end, grows only at the margin of the shell, determines the outline and convexity of the valves, as well as the corrugations (folds and sulci), and is, therefore, the primary shell layer when present. The prismatic layer, which is composed of fibers of calcite directed forward and inward with respect to the whole shell (as in *Lingula* also), is deposited by the mantle proper upon

FIG. 9-36. *A–B. Lingula murphiana.* Transverse sections of the chitinophosphatic shell, showing periostracum and alternating layers of chitin and calcium phosphate. The tubuli, or puncta, are much broader in the calcareous layers, as shown in the more magnified diagram (*B*). They do not enter the periostracum. *C.* Diagrammatic section of the shell of *Discinisca*, a neotremate, showing puncta that fork toward the inner surface. They do not enter the periostracum. *D.* Median longitudinal section through the central part of a ventral valve of *Lingulella acutangula*, showing the thin, structureless outer layer (*ol*), the obliquely laminated inner layer (*il*), and the crinkled innermost layer (*sl*) produced by secondary thickening of the shell. *E.* Highly magnified transverse section of a small part of the shell of *Magellania australis*, showing the trumpet-shaped puncta typically developed in the primary lamellar layer but contracted in the secondary innermost layer. *F–G.* Transverse and tangential sections of the shell of *Strophomena aculeata*, showing minute pseudopuncta and the lamellar nature of the calcareous shell substance. *H.* Tangential section of brachial valve of *Laqueus californicus*, showing shell mosaic. The large black spots represent puncta filled with opaque material. *I.* Shell structure of a punctate articulate, showing inner prismatic layer, outer laminated layer, periostracum, and trumpet-shaped puncta with radiating grooves about the outer extremity (cf. *N*). *J–K.* Tangential and transverse sections of the shell of *Cyrtina hamiltonensis*, a Devonian telotremate, showing puncta in prismatic inner layer. *L.* Branching tubuli in the shell of *Crania anomala*, a modern neotremate. These differ from similar features in articulate shells in that they are largest at the inner margin of the shell and branch dendritically toward the outer margin. The shell substance is calcium carbonate. *M–N.* Shell structure of *Laqueus californicus: M*, a longitudinal section through part of a pedicle valve, front of shell to right, showing differences of orientation of primary (*il* and *ol*) and secondary (*sl*) shell layers; puncta transect the inner layer and almost but not quite penetrate the outer layer, ending instead just before reaching the periostracum, as shown more clearly in *N*; *N*, a much magnified bit of *M*, showing enlargement of a punctum in the outer layer, and the several small tubules that extend from its outer boundary toward the periostracum. All figures are more or less diagrammatic and have been adapted from the original illustrations. (*c* = chitin; *ccp* = chitinophosphatic; *cp* = calcium phosphate; *il* = inner shell layer; *ol* = outer shell layer; *p* = punctum or puncta; *pe* = periostracum; *pp* = pseudopunctum or pseudopuncta; *sl* = secondary layer.) (*A–B after Blochmann*, 1900; *C based on Thomson*, 1927; *D after Walcott*, 1912; *E–G after Davidson*, 1886–1888; *H*, *M–N after Hobbs and Cloud in Cloud*, 1942; *I after King*, 1870; *J–K after Nicholson and Lydekker*, 1889; *L after Joubin*, 1906.)

the entire inner surface of the shell, and as it thickens it tends to obliterate corrugations and other irregularites on the inner surfaces of the valves and to enlarge articulating and brachiophore processes (Figs. 9-35*A*, 9-36*D*,*M*).

The tubules, or puncta, in the orthids and many other articulates are of the following two types:

1. **Exopuncta.** These puncta, visible on the outer valve surface only, lie mainly in the lamellar layer and do not penetrate far into the inner or prismatic layer. *Paurorthis*, *Hebertella*, and *Plectorthis* are examples.

2. **Endopuncta.** These are pores found only in the inner fibrous prismatic layer. They are common over the whole inner surface of the shell but are not visible on the outer shell surface if the outer lamellar layer is present.

Punctation is an important shell feature, and Cooper (1944) has suggested that calcareous shells can be subdivided into the following three broad groups on the basis of the punctation:

1. **Punctate** brachiopods, having shells that are characteristically endopunctate.

2. **Impunctate** brachiopods, in which the shell substance is dense and lacking in puncta. These are by far the most numerous of all brachiopods.

3. **Pseudopunctate** brachiopods, like *Rafinesquina* and *Juresania*, in which the fibrous layer surrounds and commonly covers internal calcareous spicules. In worn and exfoliated specimens the shell layers tear away from the spicules and leave coarse pits that may be mistaken for puncta. Broken hollow spines also cause holes on the shell's outer surface, and these have been mistakenly identified as puncta by some previous investigators.

Secondary shell material may be added to the interior of the valves as growth proceeds, making these progressively thicker, particularly in the posterior or older part (Figs. 9-35*A*, 9-36*D*). On the other hand, it is possible for the valves to be reduced in thickness by resorption. Boring animals and other predaceous organisms commonly reduce the thickness of the valves, whereas sedentary incrusting forms such as corals, worms, bryozoans, small brachiopods, and barnacles commonly attach themselves to living or dead shells and subsequently add much bulk to them. The normal structure of the shell substance in some individuals shows interruptions caused by injuries that were later healed.

Spicules. In addition to the bivalve shell and its several accessory structures (plates, brachidia, etc.), articulate brachiopods also have small calcareous platelike spicules lying free in the mantle, lophophore, and body wall. These lie below the epithelium in the connective tissue and each is surrounded by a membrane by which it was secreted.

The spicules are tiny single crystals of clear calcite generally flattened parallel to the basal plane (111), so that they give a good uniaxial interference figure. Some of the flatter spicules are perforated and, because of the growth of lateral processes, have a stellate form. Others are adorned with tiny spines and form antlerlike bodies (Fig. 9-37). Spicules are not likely to be preserved but have been reported in fossil Thecideidae. The function of the spicules is unknown. They could serve to stiffen the surrounding flesh if they were united into an internal skeleton. In *Lacazella*, for example, they take the form of tiny plates that consitute a vault in the pedicle mantle over the sinuses containing the genital organs.

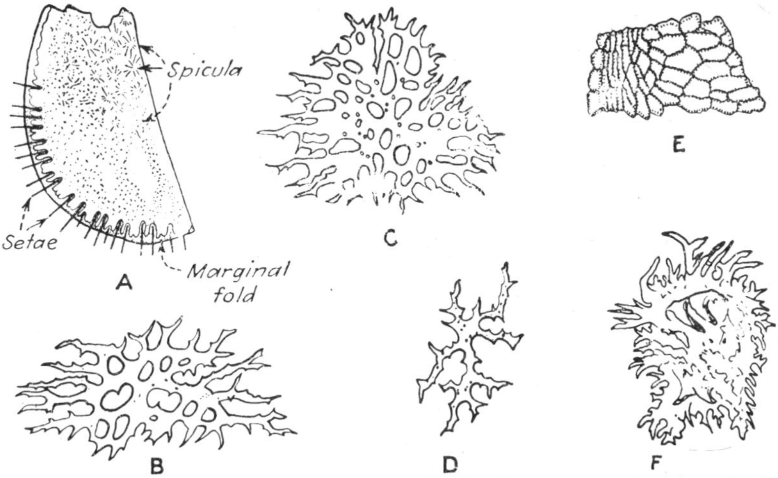

FIG. 9-37. Phylum BRACHIOPODA. Spicules. *A–D*. *Terebratulina caput-serpentis*. *A*, lateral half of pedicle mantle lobe, showing arrangement of calcareous spicula and border of chitinous setae; *B–D*, three of the flat calcareous spicula highly magnified. *E–F*. Enlarged portion of the mantle of *Megerlia truncata*, showing the closely packed spicula undisturbed; and one of the flattened spicula highly magnified. (*Modified after Hancock*, 1859.)

CLASSIFICATION

The classification of the Phylum Brachiopoda is now in a state of flux, particularly with respect to ordinal divisions in the Articulata. Two major subdivisions have been recognized for more than a century, although they have not always been based on the same features. Of these probably Inarticulata and Articulata are now most familiar. Other names for the same groups are shown in Table 9-5.

TABLE 9-5. HISTORICAL DEVELOPMENT OF BRACHIOPOD CLASSIFICATION

<table>
<tr><th>Owen, 1858</th><th>Bronn, 1862</th><th>Huxley, 1869</th><th>Beecher, 1891</th><th>Thomson, 1927</th><th>Cooper, 1944</th></tr>
<tr><td rowspan="2">Lyopomata</td><td rowspan="2">Ecardines (without hinge)
Pleuropygia (with anus)</td><td rowspan="2">Inarticulata
Valves without articulating devices</td><td rowspan="2">Atremata
Neotremata</td><td>Gastrocaulia</td><td>Inarticulata</td></tr>
<tr><td colspan="2">Atremata
Neotremata</td></tr>
<tr><td rowspan="2">Arthropomata</td><td rowspan="2">Testicardines (hinged shell)
Apygia (without anus)</td><td rowspan="2">Articulata
Valves with articulating devices</td><td rowspan="2">Protremata
Telotremata</td><td>Pygocaulia</td><td>Articulata</td></tr>
<tr><td colspan="2">Palaeotremata
Protremata
Telotremata</td></tr>
</table>

The provisional classification adopted in the present work is based on the works of Thomson (1927), Schuchert and LeVene (1929), and Cooper (1944). The two latter classifications are given here in a footnote[1] for comparison with ours, because they are familiar to most American paleontologists and have served a useful purpose in brachiopod study. No existing classification of Brachiopoda should be considered as final. Most specialists feel that much more needs to be known about both living and fossil brachiopods before a completely satisfactory classification can be devised.

The more important features and characteristics of brachiopods that are used for classification are:

1. Embryonic development (ontogeny).
2. Shell development and nature of the union of the valves.
3. Nature, position, and modification of the pedicle opening.
4. Nature and form of the brachia and brachial supports.
5. Numerous internal structures such as cardinalia, muscle scars, muscle platforms, septa, and pallial markings.
6. Microscopic structure of shell (*i.e.*, presence or absence of and nature of puncta).
7. Size, shape, and surface sculpture of shell, and curvature (convexity or concavity) of valves.

---

[1] Comparison data for the study of Brachiopoda:

Schuchert and Le Vene (1929)

Class Brachiopoda.
- Order 1. Palaeotremata.
  - Superfamily 1. Paterinacea.
  - Superfamily 2. Rustellacea.
  - Superfamily 3. Kutorginacea.
- Order 2. Atremata.
  - Superfamily 1. Obolacea.
  - Superfamily 2. Lingulacea.
  - Superfamily 3. Trimerellacea.
- Order 3. Neotremata.
  - Superfamily 1. Siphonotretacea.
  - Superfamily 2. Acrotretacea.
  - Superfamily 3. Discinacea.
  - Superfamily 4. Craniacea.
- Order 4. Protremata.
  - Superfamily 1. Orthacea.
  - Superfamily 2. Clitambonacea.
  - Superfamily 3. Dalmanellacea.
  - Superfamily 4. Syntrophiacea.
  - Superfamily 5. Pentameracea.
  - Superfamily 6. Strophomenacea.
- Order 5. Telotremata.
  - Superfamily 1. Rhynchonellacea.
  - Superfamily 2. Atrypacea.
  - Superfamily 3. Spiriferacea.
  - Superfamily 4. Rostrospiracea.
  - Superfamily 5. Terebratulacea.

Cooper (1944)

Class Inarticulata.
- Order Atremata.
  - Superfamily Obolacea.
  - Superfamily Trimerellacea.
- Order Neotremata.
  - Superfamily Paterinacea.
  - Superfamily Siphonotretacea.
  - Superfamily Acrotretacea.
  - Superfamily Discinacea.
  - Superfamily Craniacea.

Class Articulata
- Order Palaeotremata
- Impunctate Articulata
  - Superfamily Orthacea (including Clitambonacea).
  - Superfamily Syntrophiacea.
  - Superfamily Pentameracea.
  - Superfamily Triplesiacea.
  - Superfamily Rhynchonellacea.
  - Superfamily Spiriferacea (including Atrypacea, Spiriferacea and Rostrospiracea).
- Pseudopunctate Articulata.
  - Superfamily Strophomenacea.
  - Superfamily Productacea.
- Punctate Articulata.
  - Superfamily Dalmanellacea.
  - Superfamily Terebratulacea.
  - Superfamily Punctospiracea.

In the classification that follows, the two classes of Brachiopoda are based largely on pedicle development (Gastrocaulia and Pygocaulia) and the presence (Articulata) or absence (Inarticulata) of articulating processes. Orders are based on the nature of the method of valve growth, position of pedicle opening, and modification of the latter by accessory plates. Superfamilies and families are determined by characters found principally[1] in the brachial interior (*e.g.*, brachial supports), in the region around the pedicle beak, and on the inner surface of both valves (*e.g.*, pallial markings). Genera are based on smaller but consistent modifications of internal features, especially those in the posterior of the brachial valve, and to a less extent on external characteristics.

The classification used in the present discussion follows:[2]

Phylum Brachiopoda.
- Class Inarticulata (Gastrocaulia).
  - Order Atremata.
    - Superfamily Lingulacea.
    - Superfamily Trimerellacea.
  - Order Neotremata.
    - Superfamily Acrotretacea.
    - Superfamily Siphonotretacea.
    - Superfamily Discinacea.
    - Superfamily Craniacea.
- Class Articulata (Pygocaulia).
  - Order Palaeotremata.[2]
    - Superfamily Rustellacea.[2]
  - Orders Protremata-Telotremata (undifferentiated)[3]
    - Superfamily Kutorginacea.
    - Superfamily Orthacea.
    - Superfamily Clitambonacea.
    - Superfamily Dalmanellacea.
    - Superfamily Triplesiacea.
    - Superfamily Syntrophiacea.
    - Superfamily Plectambonitacea.
    - Superfamily Pentameracea.
    - Superfamily Strophomenacea.
    - Superfamily Chonetacea.
    - Superfamily Productacea.
    - Superfamily Rhynchonellacea.
    - Superfamily Stenoscismacea.
    - Superfamily Atrypacea.
    - Superfamily Spiriferacea.
    - Superfamily Rostrospiracea.
    - Superfamily Punctospiracea.
    - Superfamily Terebratulacea.
    - Superfamily Terebratellacea.

[1] McEwan (1939) has constructed a key to 22 articulate families in which convexity is one of the more important familial characters. Also see footnote, p. 290.

[2] The Superfamily Paterinacea has been removed from the Order Palaeotremata and has been omitted altogether from the classification used here because of its uncertain taxonomic position. Representatives of the superfamily are considered by some specialists to be among the most primitive brachiopods known and to belong to a group that cannot be included in any of the conventional five orders. *Paterina* (Figs. 9-25*P–Q*) and *Micromitra* (Fig. 9-38) illustrate the chief features of the superfamily, which is limited to the *Cambrian* (Fig. 9-58).

The Superfamily Rustellacea is believed to include the simplest of all articulates, hence the Order Palaeotremata, which is based on the single rustellacean genus *Rustella* (Fig. 9-43*A*), is placed in the Class Articulata.

The Superfamily Kutorginacea, formerly included in the Order Palaeotremata, is now considered to include protrematous forms resembling the orthacean genus *Nisusia;* hence it is placed directly before the Superfamily Orthacea.

A few genera of unquestioned brachiopods are so different that they cannot yet be placed in any recognized family. Among these are *Dictyonella*, *Eichwaldia*, and *Isogramma.*

[3] Thomson (1927), Cooper (1944), and other brachiopod specialists now hold that, in the light of the discoveries of the past two decades, Beecher's (1891–1892) Orders Pro-

Class Inarticulata (Gastrocaulia)

Inarticulate brachiopods have shells composed of conical or tongue-shaped valves that lack articulation and are held in apposition by muscles alone. The shell matter is chitinophosphatic or calcareous, and growth of the shell is either holoperipheral or mixoperipheral.

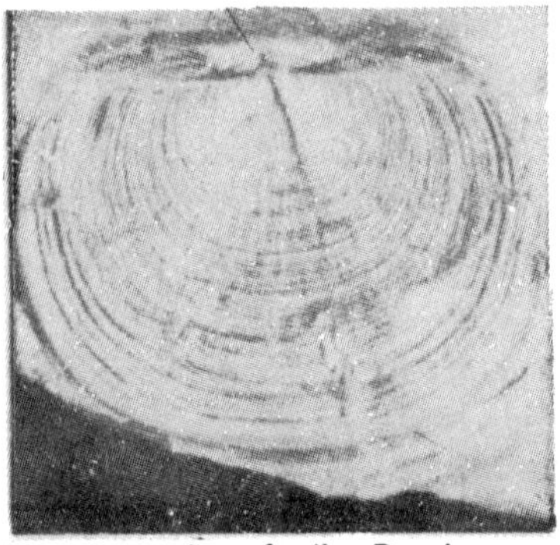

Fig. 9-38. Superfamily Paterinacea. *Micromitra sculptilis* (×8). Apical view of brachial valve. (*After Bell*, 1941.)

In the embryo the mantle lobes develop directly without revolution, and the pedicle develops during the free-swimming stage within the valves of the protegulum from the ventral mantle lobe. At a later stage the pedicle, which clearly is attached to the larger (pedicle) valve, is protruded and used for fixation (Fig. 9-16).

Most inarticulate shells are circular or oval in outline and asymmetrically conical in profile, but some have a tongue-shaped outline and a flat lenticular

---

tremata and Telotremata are no longer acceptable as originally proposed and that the superfamilies and families formerly assigned to them must be grouped in some other way. As Cooper (1944, p. 284) points out:

"The Protremata are described as having delthyria in both valves, which are often more or less closed by a pseudodeltidium and chilidium. The brachial supports are rudimentary. The Telotremata, on the other hand, are described as having the delthyrium more or less closed by deltidial plates and the lophophore supported by crura, loops or spires. As a matter of fact these simple statements are misleading; in actuality the delthyria are variously modified.

"Discovery of deltidial plates in the Protremata and what appears to be a pseudodeltidium in the Telotremata leads to confusion of the two orders. Add to these complications the presence of a loop in the impunctate, protremate *Enantiosphen* and the confusion becomes still greater. It seems best, therefore, in order to make separations easier, not to use the ordinal characters of the Protremata and Telotremata as defined but to arrange the superfamilies falling in these orders according to the three types of shell structure into which the Articulata are divisible. The arrangement above is not at present proposed as a genetic classification; it would require the demonstration that loops and spires were developed independently in the punctate and impunctate brachiopods. This is a possibility, but present knowledge does not permit such a demonstration. Nevertheless, it seems clear that the punctate and pseudo-punctate shells were derived from the impunctate Orthacea."

In a more recent classification, the Termiers (1949) propose that the Articulata be divided into two groups based on the nature of the lophophore—"1, Brachiopodes à lophophore spiralé, and 2, Brachiopodes à lophophore en bandelette," *i.e.*, spire-bearers and loop-bearers, respectively.

Because no ordinal division of the Articulata satisfactory to the specialists has yet been devised, we simply list the superfamilies once assigned to the Protremata and Telotremata in approximately the order that they have been listed previously under these two subdivisions.

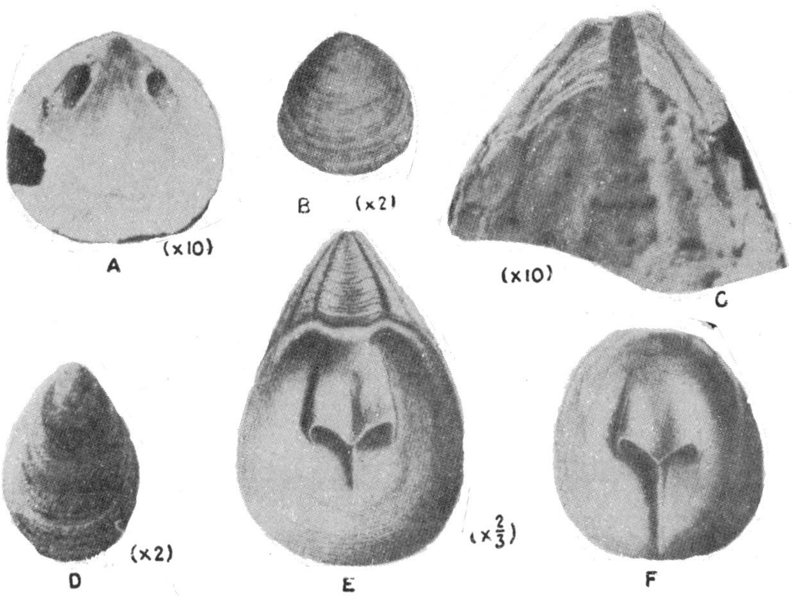

FIG. 9-39. Order Atremata—Superfamilies Lingulacea (*A–D*) and Trimerellacea (*E–F*). ***A–B. Dicellomus***, a common Upper Cambrian atremate: *A*, brachial interior of *D. nanus;* ***B***, pedicle exterior of *D. occidentalis. C–D. Lingulella. C*, fragmental pedicle interior of *L. acutangula; D*, pedicle exterior of *L. ampla. E–F. Trimerella grandis* from the Upper Silurian (Guelph) of Ontario. *E*, pedicle interior, showing prominent muscle platform; *F*, brachial interior. (*A–C after Bell*, 1941, *or original; E–F after Davidson and King*, 1872.)

profile. The conical forms, excepting the cemented Craniidae, are fixed by a short pedicle (Fig. 9-10*N*), whereas the lingulids have a long flexible pedicle and live in burrows with their setiferous anterior margin protruding (Fig. 9-10*X*–Z). The development and migration of the pedicle are complex and have been discussed on foregoing pages. The complicated muscle system leaves an equally complicated set of muscle marks.

The Inarticulata range from Lower Cambrian to Recent, but they have never been important as a group and are rarely abundant except locally. While few of the genera have any biostratigraphic value, except in the Cambrian, the class as a whole is interesting and important for the possible light its species throw on the problems of brachiopod evolution.

Order 1. Atremata.[1] The Atremata are inarticulate, chitinophosphatic-shelled brachiopods of subtriangular, oval to subrounded, or tonguelike outline having the pedicle attached to the larger (or pedicle) valve, in which it occupies a groove. Specialized forms have heavy calcareous shells with internal platforms for muscle attachment. *Lower Cambrian to Recent.*

[1] Atremata—Gr. *a*, without, + *trema, trematos*, perforation; referring to the absence of a strictly defined pedicle opening, the opening being shared by both valves.

Superfamily Lingulacea. Elongate chitinophosphatic and thin-shelled atremates. They have highly differentiated muscles (Fig. 9-12) and a wormlike tubular and flexible pedicle, both of which aid the animal in its free-living and burrowing habit. Modern *Lingula* (Figs. 9-10, 9-16, 9-28*I*), a typical genus of the superfamily, has a long and ancient history. It appeared first in the Ordovician, as one of the earliest representatives of the superfamily, and has persisted to the present with little evident change. *Lower Cambrian to Recent.*

Lingula (Fig. 9-10) and *Glottidia* (Figs. 9-10, 9-22) are the best known living genera of the superfamily, whereas *Lingulella* (L. Camb.-L. Ord.) and *Dicellomus* (U. Camb.) are representative fossil genera (Fig. 9-39).

Superfamily Trimerellacea. Thick-shelled and strongly inequivalved calcareous brachiopods, seemingly divergent from the Lingulacea. The shells have posterior platforms to which certain muscles were attached. Some advanced forms have rudimentary articulation and a more or less prominent pedicle pseudointerarea that is triangular and transversely striated. *Middle Ordovician through Silurian.*

Trimerella (Fig. 9-39*E–F*), from the Silurian, is representative of the superfamily and is an important index fossil.

Order 2. Neotremata.[1] The Neotremata are specialized and, to a certain extent, degenerate brachiopods having small chitinous, chitinophosphatic (or, rarely, supposedly calcareophosphatic[2]), or calcareous shells consisting typically of high or flattened conical valves. The pedicle when present emerges through a perforation or sheath, or a triangular cleft, and in maturity in certain forms may be lost when the pedicle valve is cemented to the substratum. The protegulum is semicircular or semielliptical, and the adult shells are circular or elliptical because shell growth is holoperipheral (Fig. 9-22*D–F*).

The earliest neotremates are found in Lower Cambrian rocks, and living forms are common but not particularly abundant in present seas. The Neotremata as a group seem at no time to have been highly varied, but they are extremely abundant locally in certain beds. *Lower Cambrian to Recent.*

Superfamily Acrotretacea. Shells circular in outline and conical in form with a minute circular pedicle opening behind the apex of the pedicle valve; long pseudointerarea in some cases modified by an intertrough (Fig. 9-40*A–B*). *Lower Cambrian through Ordovician.*

Prototreta (Fig. 9-40*C–F*), from the Cambrian, is a representative genus.

Superfamily Siphonotretacea. Shells with pedicle emerging through a ventral sheath. The pedicle opening is circular and unmodified, tends to be axial, may be posterior or anterior to protegulum, and tends to be elongate if it has migrated through the protegulum. *Cambrian and Ordovician.*

[1] Neotremata—Gr. *neos*, young, + *trema*, *trematos*, perforation; referring to the fact that the pedicle opening is present in young shells but commonly is closed and nonfunctional in those adult forms that cement their shell to the substratum.

[2] Bell (1948) has reported a Cambrian acrotretoid from Texas that is composed of hydroxy apatite.

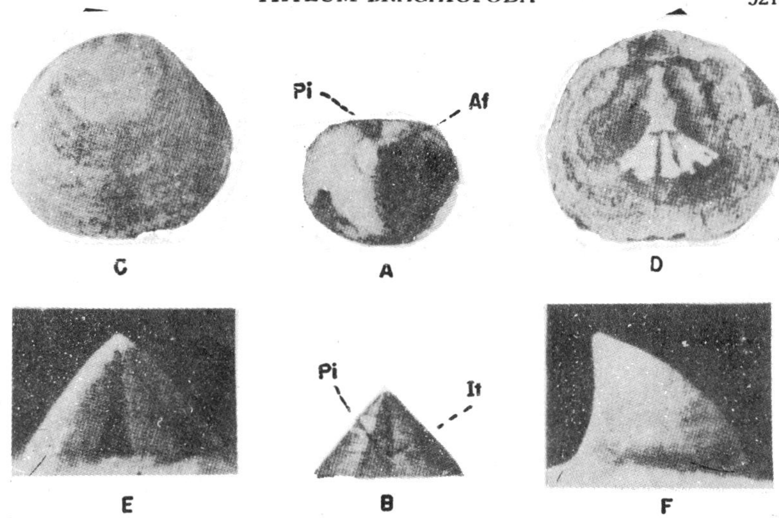

FIG. 9-40. Superfamily Acrotretacea. *A–B*. The pedicle valve of an acrotretid, viewed apically and posteriorly, showing holoperipheral growth, apical foramen (*Af*), pseudointerarea (*Pi*), and furrow or intertrough (*It*). *C–F*. *Prototreta trapeza* (×10), an Upper Cambrian species: *C*, brachial exterior; *D*, brachial interior, showing expanded median septum characteristic of the genus; *E*, posterior view of pedicle valve (cf. *B*); *F*, lateral view of pedicle valve. (*A–B after Walcott*, 1912; *C–F after Bell*, 1941.)

Siphonotreta (Ord.) and *Schizambon* (Camb.–Ord.) are typical genera (Fig. 9-41).

Superfamily Discinacea. Neotremates with circular and flattened conical shells in which the foramen lies in a modified pedicle slit. *Ordovician to Recent.*

Schizocrania (Figs. 9-10*O–Q*, 9-42*D*) (Ord.–Dev.), *Orbiculoidea* (Figs. 9-22, 9-25*R*, 9-42) (Ord.–Cret.), and *Discinisca* (Figs. 9-1, 9-10*M–N*, 9-13) (Tert.–Recent) are representative genera.

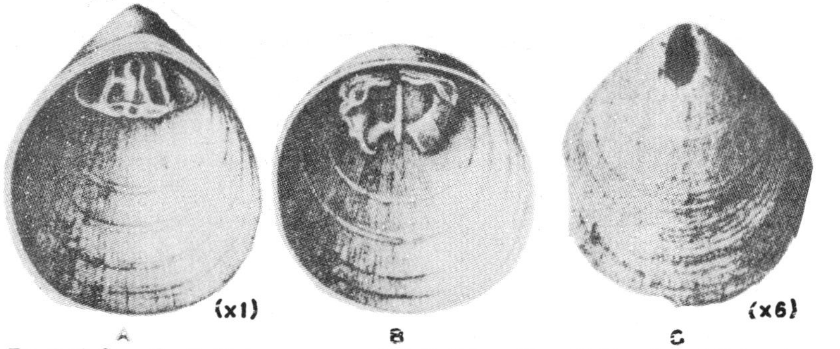

FIG. 9-41. Superfamily Siphonotretacea. *A–B*. Pedicle and brachial interiors of *Siphonotreta unguiculata*. *C*. Pedicle exterior of *Schizambon pennsylvanicum*, from the Lower Ordovician, showing prominent pedicle foramen anterior to the beak. (*A–B after Davidson*, 1851–1855; *C after Ulrich and Cooper*, 1938.)

Superfamily Craniacea. Neotremates with flattened conical calcareous shells lacking a pedicle opening and usually cemented to some object by the pedicle valve. *Ordovician to Recent.*

Petrocrania (Fig. 9-42*E*), a familiar genus with a long stratigraphic range (M. Ord.–Perm.?), is commonly cemented to other brachiopods. *Crania* (Fig. 9-10*K–L*, 9-28*D–E*, 9-42*C*), with a much longer record (Ord.–Recent), is a second representative genus.

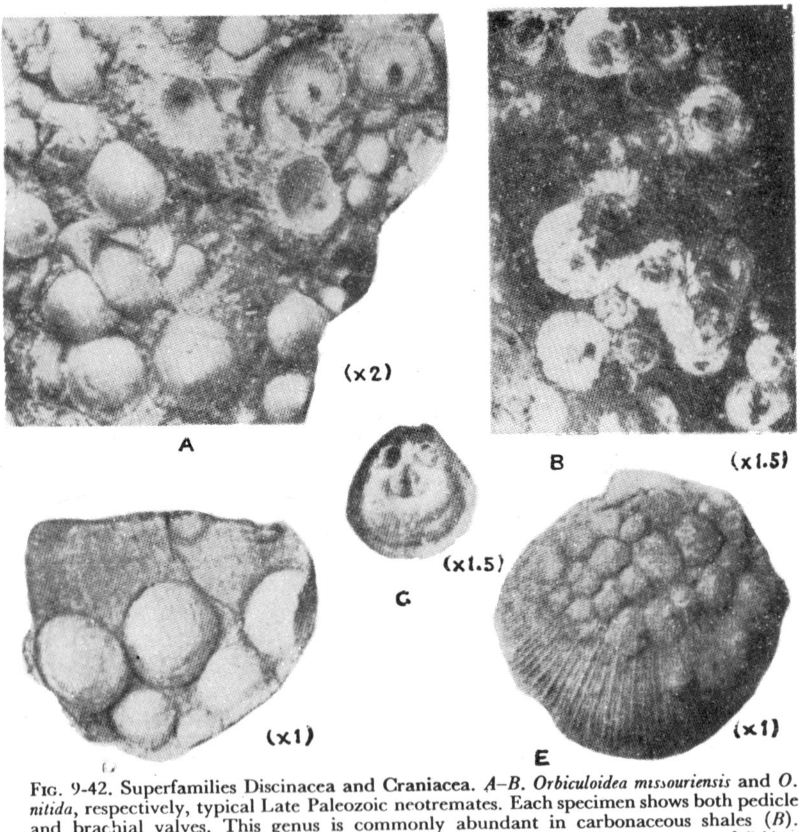

FIG. 9-42. Superfamilies Discinacea and Craniacea. *A–B. Orbiculoidea missouriensis* and *O. nitida*, respectively, typical Late Paleozoic neotremates. Each specimen shows both pedicle and brachial valves. This genus is commonly abundant in carbonaceous shales (*B*). *C.* Interior of brachial valve of *Crania antiqua* from the Cretaceous. *D.* A group of *Schizocrania filosa* on *Rafinesquina*. *E.* A cluster of valves of *Petrocrania scabiosa* on *Rafinesquina*. *D* and *E* are common Ordovician neotremates. (*A, D–E after Cooper*, 1944.)

CLASS ARTICULATA (PYGOCAULIA)

Articulate brachiopods have oval and transverse calcareous shells composed of two typically convex but unequal-sized valves that are held together along the posterior hinge line by means of articulating devices. The shell is opened and closed by the action of specialized muscles, and the valves, like the leaves of a book, have only one motion, that of simple opening and closing on a single axis. Each valve has a triangular opening in the palintrope. The opening on the pedicle palintrope is the delthyrium, whereas that on the brachial is the notothyrium; both tend to be modified. The shell can be adjusted on the pedicle, and most articulates have the pedicle opening more or less modified. The shell matter is dominantly calcium carbonate with fibrous prismatic structure, and growth is largely hemiperipheral.

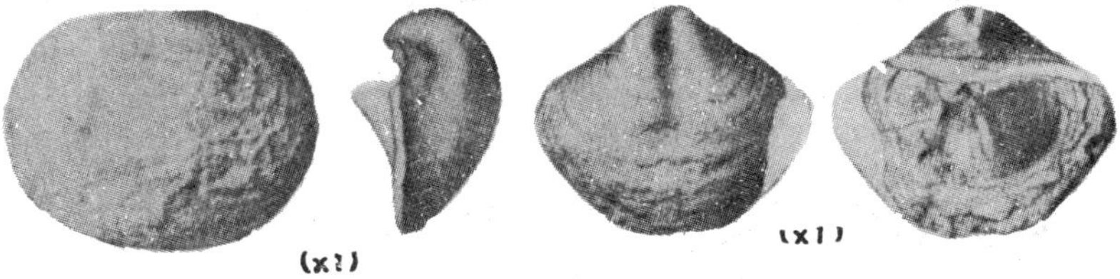

A B C D

FIG. 9-43. Superfamilies Rustellacea (*A*) Order Palaeotremata and Kutorginacea (*B–D*). *A*. Exterior of a compressed pedicle valve of *Rustella edsoni* from the Lower Cambrian. *B–D*. Lateral, pedicle, and brachial views of an almost complete shell of *Kutorgina cingulata* from the Lower Cambrian of Vermont. (*After Walcott*, 1912.)

In the development of the embryo the mantle lobes are revolved or reversed from a posterior to an anterior position (Figs. 9-17, 9-19, 9-20). The pedicle is developed from the caudal part of the embryo and is never enclosed within the shell as in the Inarticulata. It is attached by muscles to both valves, but it belongs to the larger or pedicle valve. In no known articulate brachiopod does the pedicle lie in the brachial valve. Although the intestine lacks an anus in all living representatives, it may have had an anal opening in some ancient and extinct genera.

Most articulates show modifications of the delthyria and notothyria and development of many kinds of calcareous supports for the brachia. They also exhibit much variation in surface sculpture. All these features, as well as several others, have been described on preceding pages.

The Articulata range from Lower Cambrian (Palaeotremata) to Recent. They had their greatest development in the Paleozoic, and since the close of the Tertiary have greatly diminished in importance until now they are no longer an important component of modern marine faunas. Fossil articulates are important index fossils in many formations and have been used extensively for correlation.

Order 1. Palaeotremata.[1] The Order Palaeotremata includes the single genus *Rustella*, which has a calcareous shell without fully developed articulation or delthyria. *Rustella* is considered to be the earliest and most primitive representative of the Articulata, though probably not the direct ancestor of any later brachiopods (Thomson, 1927). The order is limited to the Lower Cambrian.

[1] Palaeotremata—Gr. *palaios*, ancient, + *trema*, *trematos*, perforation; referring to the presence of a pedicle opening in the most ancient shells.

Superfamily Rustellacea. Primitive, relatively thick-shelled brachiopods having chitinous or calcareophosphatic shells in which the pedicle groove is rudimentary. *Rustella* (Fig. 9-43*A*), the genus on which the superfamily is based, is limited to the Lower Cambrian.

Orders 2 and 3. Protremata[1] and Telotremata[2] (Undifferentiated).[3] The brachiopods referred to these two orders, which no longer have good standing,[4] have biconvex, concavo-convex, plano-convex, and resupinate shells that are smooth, costellate, costate, or plicate and commonly have prominent concentric growth lines. All the shells are completely articulate and have well-developed interareas. The pedicle opening (delthyrium) may be open or may be closed by one or two plates. Brachidia show a wide range of structure, becoming much complicated in the more highly specialized genera, and spondylia of several types are developed. Many shells are impunctate, others are pseudopunctate, and some are punctate.

The simpler of these brachiopods seem to have appeared first in the Cambrian and to have become almost extinct by the close of the Paleozoic. Only two families are extant. The more advanced forms—those generally included in the Telotremata—constitute a large assemblage with more than 200 genera in the Paleozoic, 175 or more in the Mesozoic, and perhaps as many as 75 genera in the Cenozoic and Recent.

The superfamilies are considered roughly in the order in which they are usually listed under the Protremata and Telotremata, but with some additions.

Superfamily Kutorginacea. Thick-shelled brachiopods tending to be wider than long and having rudimentary articulation, deltidia, and more or less rudimentary cardinal areas. The system of muscle scars is prophetic of that found in later protrematous genera, and for this reason in part some investigators regard *Kutorgina* (Fig. 9-43*B–D*), the genus on which the superfamily is based, as the most primitive genus of the Order Protremata. (*Lower Cambrian.*)

Superfamily Orthacea. Orthid shells are impunctate, typically costate and costellate, and commonly plicate and sulcate, with wide hinges, well-developed interareas on both valves, and commonly a simple cardinal process. A deltidium and a chilidium are present in some genera. Pseudospondylia are common, but true spondylia are rare. The superfamily arose in the Early Cambrian, became a prolific group in the Ordovician, and then declined to extinction in the Devonian. *Lower Cambrian to Middle Devonian.*

[1] Protremata—Gr. *pro*, early, + *trema*, *trematos*, perforation; referring to the nature of the pedicle opening.

[2] Telotremata—Gr. *telos*, last or end, + *trema*, *trematos*, perforation; referring to the fact that telotrematous brachiopods have the last or most highly modified pedicle opening.

[3] Because the homogeneity of both Protremata and Telotremata is now in doubt (Cooper, 1944; Cloud, 1942; Arber, 1942; etc.), we treat the orders together by considering the widely recognized superfamilies without giving taxonomic significance to the sequence in which they appear. (See footnote on p. 317.)

[4] See footnote on p. 318.

Nisusia, Billingsella (Fig. 9-6*C*), *Eoorthis, Hesperorthis* (Fig. 9-26*B*), *Hebertella* (Figs. 9-33*C–D*, 9-35*B*), *Platystrophia, Dolerorthis* (Fig. 9-25*K*), and *Glossorthis* (Fig. 9-25*J*) are representatives of this large and important superfamily (Fig. 9-44).

Superfamily Clitambonacea. Clitambonaceans are specialized impunctate orthoids with a prominent spondylium simplex and anchorlike cardinalia. *Ordovician and Silurian.*

Clitambonites (Fig. 9-10*J*, 9-45*A*), *Clinambon* (Fig. 9-45*B*), and *Estlandia* (Figs. 9-6, 9-35*A*) are common European genera; *Vellamo* (Fig. 9-25*D–G*) is the common North American clitambonitid.

Superfamily Dalmanellacea. Dalmanellids are endopunctate orthoids with a lobed (usually trilobed) cardinal process. The superfamily is thought to have been derived from the Orthacea. Representatives appeared first in the Middle Ordovician; they spread widely and diversified greatly during the Silurian and Devonian; and the superfamily finally became extinct during the Permian. *Middle Ordovician through Permian.*

Resserella, Bilobites, Rhipidomella, Schizophoria, and *Enteletes* (Fig. 9-25*B–C*) are typical genera (Fig. 9-46).

Superfamily Triplesiacea.[1] Triplesiids are impunctate, generally biconvex brachiopods having a flat deltidium, minute apical foramen, long forked cardinal process, and short, divergent brachial processes. *Middle Ordovician through Silurian.*

Triplesia and *Oxoplecia* are typical genera (Fig. 9-47).

Superfamily Syntrophiacea. Syntrophiids are impunctate orthoids with generally smooth shells having strongly developed pedicle plica and brachial sulcus, a spondylium simplex, and variable cardinalia. *Middle Cambrian into Lower Devonian.*

Syntrophopsis (Fig. 9-47*F–G*) is a representative genus.

Superfamily Plectambonitacea. Concavo-convex to gently biconvex; pseudopunctate shells having a simple cardinal process commonly fused with chilidial plates and brachial interior commonly multiseptate (one to three septa). *Lower Ordovician to Upper Devonian.*

Sowerbyella (Fig. 9-49*H–J*), *Sowerbyites* (Fig. 9-46*J–K*) and *Plectodonta* are representative genera.

Superfamily Pentameracea. The pentamerids are typically large biconvex impunctate shells with a spondylium duplex and usually a cruralium or two parallel vertical plates in the brachial valve. The hinge line is narrow, interareas small, and surface smooth or costate. *Middle Ordovician through Devonian.*

Pentamerus (Figs. 9-25*L–M*, 9-48*B–C*), *Conchidium* (Figs. 9-25*N–O*, 9-48*A,D*) and *Stricklandinia* (Figs. 9-28*T–U*), all excellent index fossils of the Silurian, are typical genera.

Superfamily Strophomenacea. Strophomenids have flat transverse costellate shells with a profile that may be concavo-convex, convexo-concave, or resupinate. The shell is pseudopunctate. The pedicle opening, if present, is an

[1] See Ulrich, E. O., and Cooper. G. A., 1936, New Silurian brachiopods of the Family Triplesiidae, *Jour. Paleontology*, vol. 10, pp. 331–347.

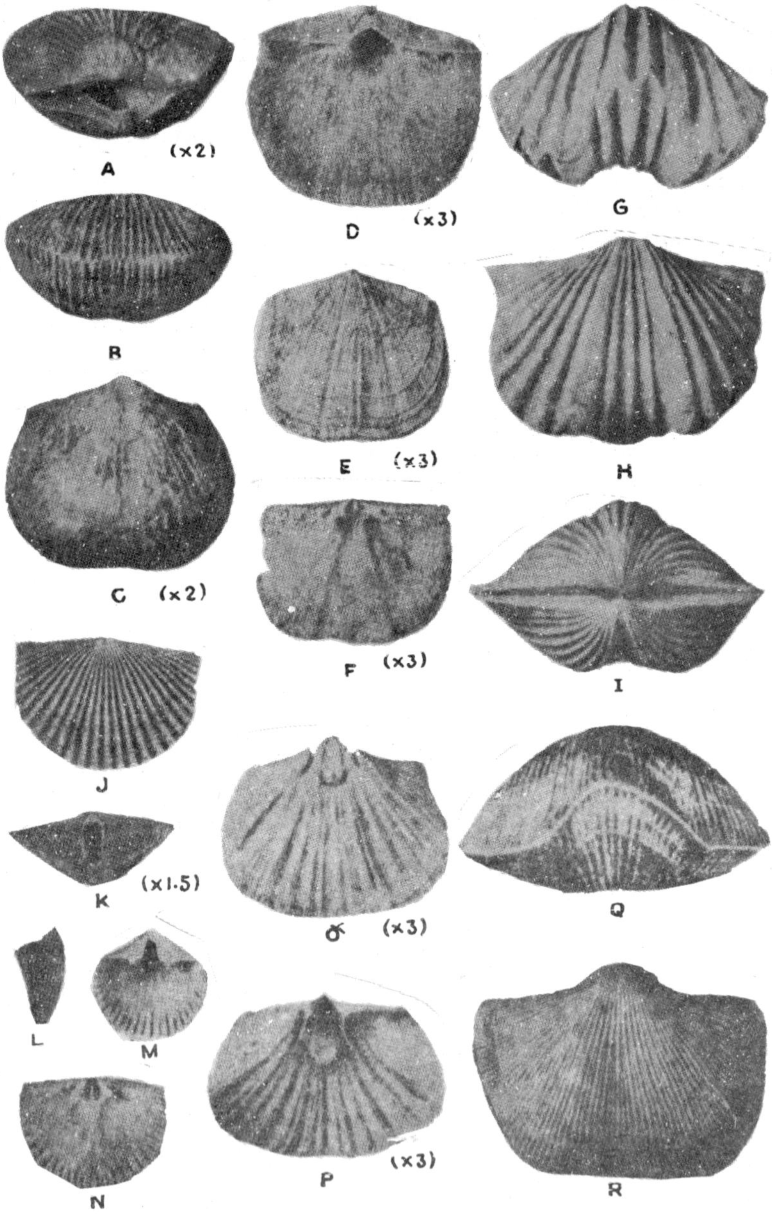

Fig. 9-44. Superfamily Orthacea. *(See opposite page for detailed description.)*

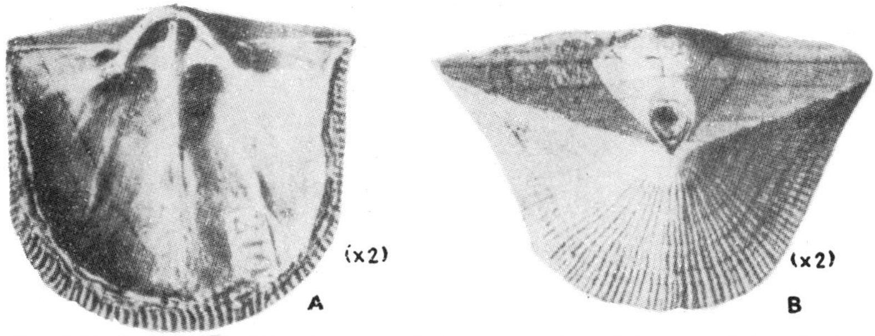

FIG. 9-45. Superfamily Clitambonacea. *A.* Interior of brachial valve of *Clitambonites*, showing the prominent interarea, arched chilidium, strong median cardinal process beneath chilidium, heavy median ridge, and other features of the cardinalia. *B.* A young shell of *Clinambon*, viewed from directly above the beak of the pedicle valve, showing the prominent and high interarea, the large and strongly curved deltidium which fits against the smaller chilidium, the prominent subapical foramen, and the numerous costae. (*After Öpik*, 1934.)

apical foramen in adult shells, but the pedicle seems never to have emerged through the delthyrium because in young shells it protruded through a tiny supra-apical foramen, surrounded by a sheath and anteriorly situated with reference to the deltidium (Fig. 9-21) (Arber, 1942). Commonly the pedicle atrophied and the shell lay free on the bottom or was attached by part of the surface of the pedicle valve. A deltidium and a chilidium were usually well developed, and short brachiophores supported the lophophore. The Strophomenacea were an important and prolific group during the Paleozoic, but since then have gradually declined until now only two living genera remain. *Lower Ordovician to Recent.*

Strophomena (Fig. 9-28*F*), *Rafinesquina* (Figs. 9-10*O*, 9-42*D–E*), *Stropheodonta* (Fig. 9-25*S*), and *Leptaena* (Figs. 9-21, 9-22*G–I*) are well known and typical Paleozoic genera (Fig. 9-49). *Leptaena* (Fig. 9-49*G*) (M. Ord.–Miss.) is one of the most widespread and longest ranging extinct brachiopods known. *Thecid'a* (Fig. 9-28*A–C*) is a Cretaceous genus, and *Lacazella* (Fig. 9-20) is one of the two living genera of the superfamily.

Superfamily Chonetacea. Flattish shells with concavo-convex profile, pseudo-

FIG. 9-44. Superfamily Orthacea. *A–C.* Posterior, anterior, and pedicle exterior of *Nisusia deissi*, from Middle Cambrian. *D–F.* Interior and exterior of pedicle valve, and interior of brachial valve of *Billingsella exasperata*, a common Upper Cambrian genus. The pedicle valve has a prominent deltidium and a small subapical foramen; the brachial valve has a tiny cardinal process rising from the floor of the notothyrial cavity. *G–I.* Anterior, pedicle exterior, and posterior views of *Platystrophia ponderosa*, showing prominent sulcus in the pedicle valve and alternate plica on the brachial valve. This shell, common in the Upper Ordovician, has sharply angular costae which give the anterior commissure a zigzag appearance. *J.* Brachial exterior of *Dolerorthis flabellites*. *K–N. Hesperorthis tricenaria*, a common Mid-Ordovician orthid: *K*, posterior view; *L*, lateral view; *M*, interior of pedicle valve, with high interareas and relatively large delthyrium; *N*, brachial interior. *O–P. Eoorthis remnicha*, an Upper Cambrian orthid: *O*, filling of a pedicle valve; *P*, clay squeeze made from *O*, showing features of the valve interior. *Q–R.* Anterior view and brachial exterior of *Hebertella sinuata*, showing many costae and prominent sulcus on the brachial valve and smooth curvature of the pedicle valve. All figures are ×1 unless otherwise indicated. (*A–F, O–P after Bell*, 1941; *G–H, J–N, Q–R after Cooper*, 1944.)

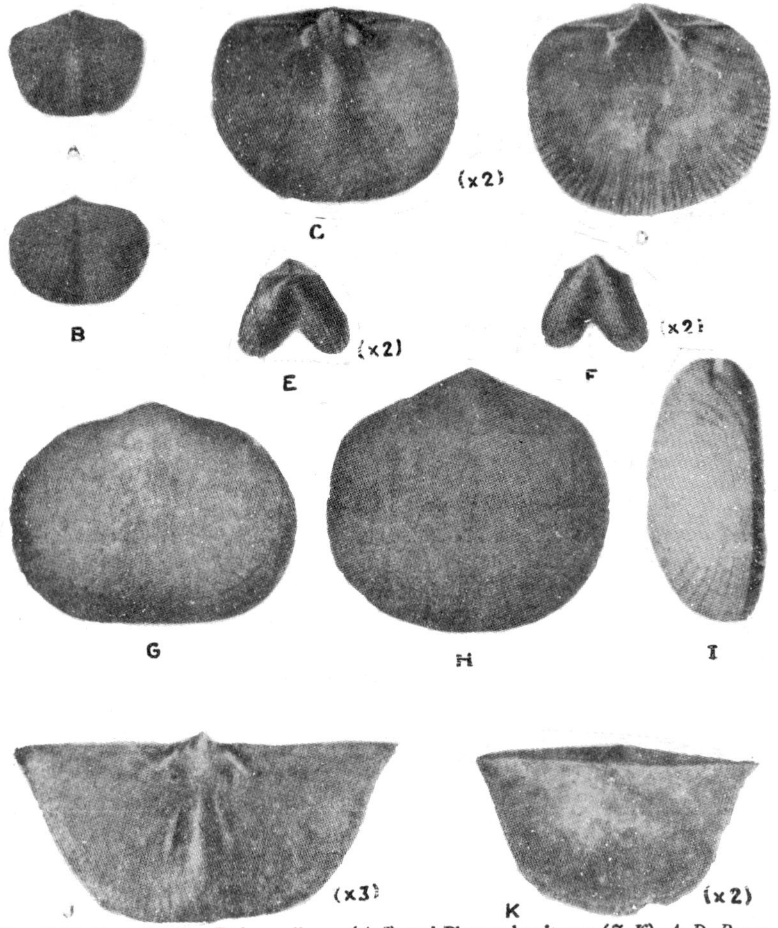

FIG. 9-46. Superfamilies Dalmanellacea (*A–I*) and Plectambonitacea (*J–K*). *A–D. Resserella meeki*, a common Upper Ordovician species: *A*, pedicle valve; *B*, brachial view; *C*, brachial interior; *D*, interior of pedicle valve. *E–F.* Brachial and pedicle exteriors of *Bilobites bilobus*, an index brachiopod of the Middle Silurian. *G.* Pedicle valve of *Schizophoria nevadensis*, a multicostellate dalmanellid. *H–I.* Pedicle exterior and side view of *Rhipidomella musculosa*, a common Lower Devonian species. *J–K. Sowerbyites triseptatus: J*, brachial interior (×3); *K*, brachial exterior (×2). (*A–D, G, J–K after Cooper*, 1944; *E–F after Hall*, 1892; *H–I after Schuchert and Maynard*, 1913.)

punctate, and with spines along posterior margin of pedicle valve. *Upper Ordovician through Permian.*

Chonetes (Fig. 9-29*A–B*) is a representative genus.

Superfamily Productacea. Members of this superfamily, the productids, are pseudopunctate brachiopods, typically plano- or concavo-convex, with conspicuous spines on the entire exterior of the shell. They are particularly

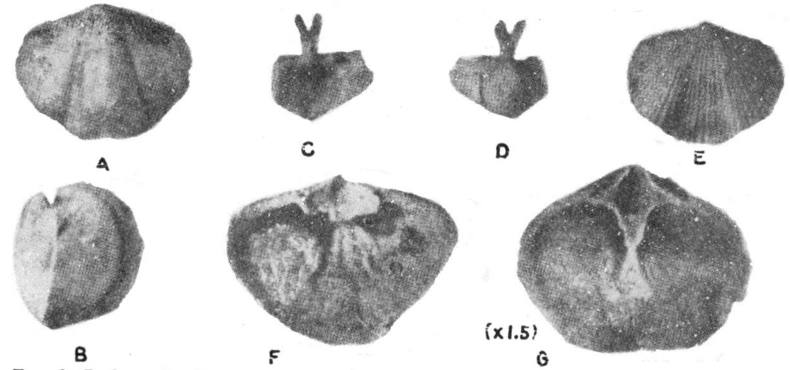

FIG. 9-47. Superfamilies Triplesiacea (*A–E*) and Syntrophiacea (*F–G*). *A–B*. Brachial exterior and side view of *Triplesia cuspidata*. *C–D*. Two views of the forked cardinal process of *Triplesia ortoni*. *E*. Exterior of pedicle valve of *Oxoplecia gouldi*, showing numerous costellae and broad sulcus. *F–G*. Interiors of brachial and pedicle valves of *Syntrophopsis magna*. The pedicle valve has a well-developed spondylium. (*A–E after Cooper*, 1944; *F–G after Ulrich and Cooper*, 1938.)

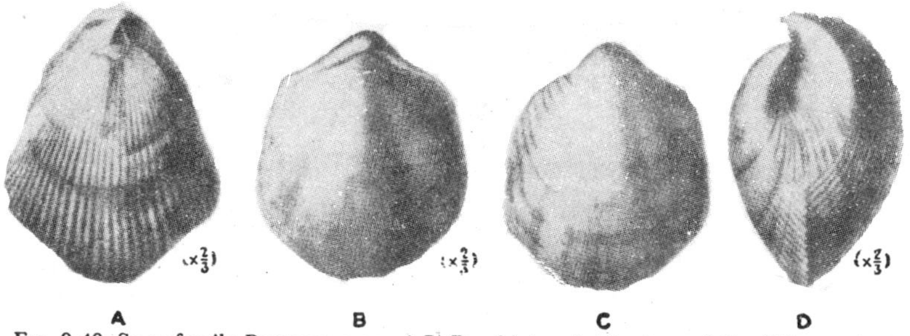

FIG. 9-48. Superfamily Pentameracea. *A,D*. Brachial and side views of *Conchidium nysius*, a strongly costate pentamerid from Kentucky. *B–C*. Brachial view and exterior of pedicle valve of *Pentamerus laevis*, a smooth pentamerid. (*After Hall and Clarke*, 1894.)

abundant in Pennsylvanian rocks the world over, but range from Lower Devonian to the end of the Paleozoic. *Lower Devonian through Permian.*

Dictyoclostus, *Marginifera*, *Juresania*, *Avonia* (Fig. 9-29*F*), *Richthofenia* (Fig. 9-10*H*), and *Prorichthofenia* (Fig. 9-10*F–G*) are typical genera of this important later Paleozoic superfamily (Fig. 9-50).

Superfamily Rhynchonellacea. The rhynchonellids are impunctate[1] subtriangular, rostrate shells, characteristically costate, and the delthyrium is usually closed by deltidial plates. The small foramen lies just anterior to the beak. The lophophore is supported by crura. The Rhynchonellacea are supposedly the earliest and simplest telotrematous brachiopods. *Middle Ordovician to Recent.*

Lepidocyclus (= *Rhynchotrema* of authors), *Rhynchotreta*, *Camarotoechia*, and *Leiorhynchus* are typical rhynchonellids (Fig. 9-51).

[1] Except for *Rhynchopora*, which is considered by many to be a punctate rhynchonellid.

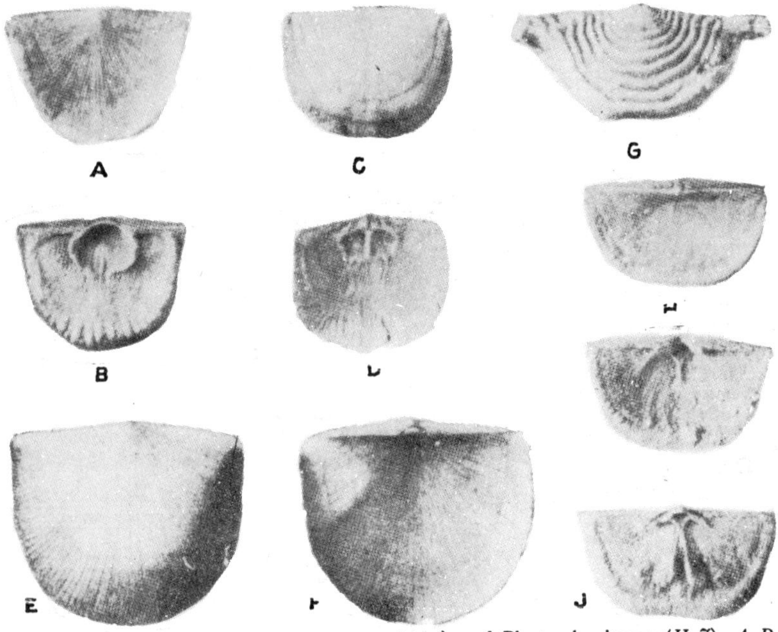

Fig. 9-49. Superfamilies Strophomenacea (*A–G*) and Plectambonitacea (*H–J*). *A–D*. *Strophomena planumbona*, a common Mid-Ordovician resupinate strophomenid: *A–B*, exterior and interior of pedicle valve; *C–D*, exterior and interior of brachial valve. *E–F*. Exterior of pedicle valve and brachial view of *Rafinesquina alternata*, a common Upper Ordovician strophomenid. *G*. Exterior of pedicle valve of *Leptaena richmondensis*, showing characteristic concentric growth rugosities. *H–J*. *Sowerbyella punctostriata*: *H*, brachial exterior; *I–J* interiors of pedicle and brachial valves, respectively. (*A–D*, *H–J after Cooper*, 1944; *E–F after Whitfield.*)

Superfamily Stenoscismacea. Rhynchonelloid in outline and profile but pedicle interior with spondylium and brachial interior with elongated spoon-shaped cruralium in addition to long delicate crura. *Middle Devonian through Permian.*

Stenoscisma (Figs. 9-50*H–J*, 9-51*H*) is a representative genus.

Superfamily Atrypacea. Atrypids are subquadrate spiriferoids with a complex brachidium. The primary lamellae follow the margins of the shell, and the spiralia are directed inward or toward the floor of the brachial valve. *Middle Ordovician to Upper Devonian, with a single specimen said to have come from the Lower Mississippian.*

Atrypa (Figs. 9-28*R–S*, 9-35*I*, 9-52) is the best known representative of the superfamily.

Superfamily Spiriferacea. Spiriferids, or spirifers as they are most commonly called, are typically transverse shells having calcareous spiralia of the spiriferoid type. The interareas are commonly well developed, and the delthyrium is modified by a deltidium or by deltidial plates. The shell may be smooth, costellate or costate, and in many genera it is also plicate and sulcate. *Ordovician through Triassic.*

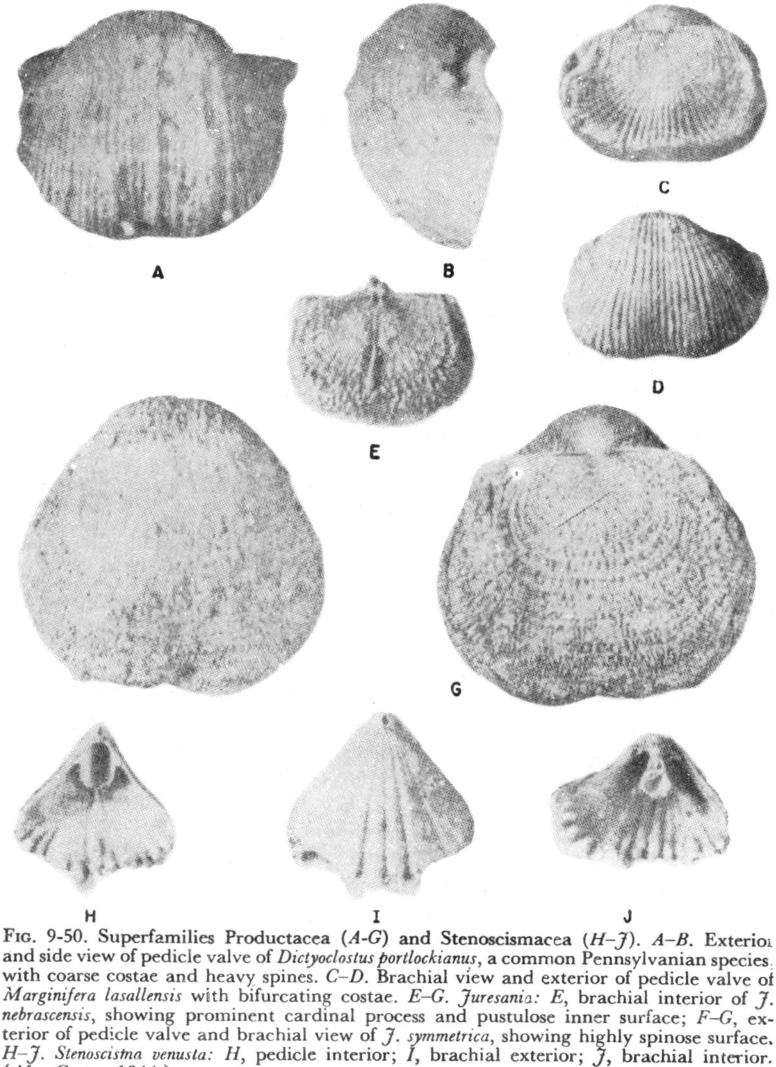

FIG. 9-50. Superfamilies Productacea (*A-G*) and Stenoscismacea (*H–J*). *A–B*. Exterior and side view of pedicle valve of *Dictyoclostus portlockianus*, a common Pennsylvanian species with coarse costae and heavy spines. *C–D*. Brachial view and exterior of pedicle valve of *Marginifera lasallensis* with bifurcating costae. *E–G*. *Juresania*: *E*, brachial interior of *J. nebrascensis*, showing prominent cardinal process and pustulose inner surface; *F–G*, exterior of pedicle valve and brachial view of *J. symmetrica*, showing highly spinose surface. *H–J*. *Stenoscisma venusta*: *H*, pedicle interior; *I*, brachial exterior; *J*, brachial interior. (*After Cooper*, 1944.)

Eospirifer, *Cyrtospirifer*, *Mucrospirifer* (Fig. 9-28*G*), *Platyrachella*, *Spirifer* (Fig. 9-24*G*), *Neospirifer*, *Elytha* (Fig. 9-28*M–N*), and *Phricodothyris* (Fig. 9-28*K–L*) are representative of this important superfamily (Fig. 9-53).

Superfamily Rostrospiracea. Rostrospirids are impunctate, rostrate, spiriferoids with an athyroid brachidium. *Ordovician through Triassic.*

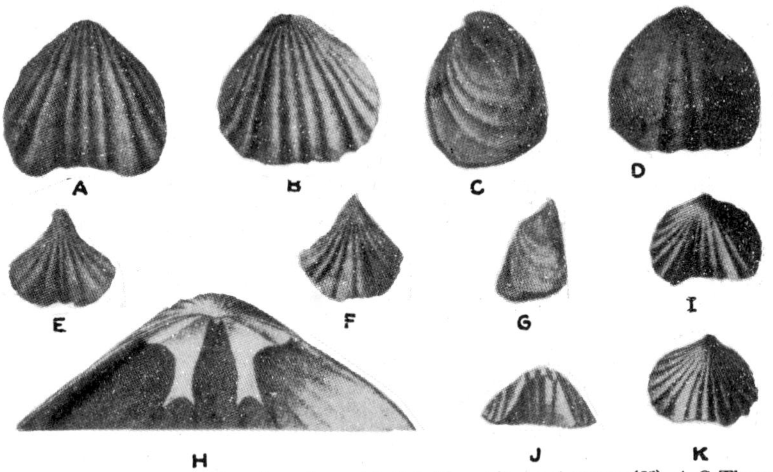

FIG. 9-51. Superfamilies Rhynchonellacea (*A–G, I–K*) and Stenoscismacea (*H*). *A–C*. Three views of *Lepidocyclus* ("*Rhynchotrema*") *capax*, a common Upper Ordovician rynchonellid that has relatively few coarse costae, which show herringbone structure due to growth lamellae. *D*. Brachial view of *Leiorhynchus rockymontanum*, a widespread mid-continent Pennsylvanian rynchonellid. *E–G*. Pedicle exterior, brachial, and side views of *Rhynchotreta americana*. *H*. Posterior part of the brachial valve of *Stenoscisma formosa*, from the New York Devonian, showing the minute cardinal process and well-developed crura. *I–K*. Three views of *Camarotoechia contracta*, a widespread American index brachiopod of the Upper Devonian. (*A–G after Cooper*, 1944; *H–K after Hall and Clarke*, 1894.)

Hyattidina, *Meristella*, *Athyris* (Figs. 9-28*J*, 9-35*J–K*), and *Composita* are typical of the superfamily (Fig. 9-54).

Superfamily Punctospiracea. Punctospirids are punctate spiriferoids having spiriferoid brachidium and complicated cardinal process. *Lower Silurian through Jurassic.*

Punctospirifer and *Homeospira* are representative genera (Fig. 9-55).

Superfamily Terebratulacea. Terebratulaceans are endopunctate articulates with a looplike brachidium, with hinge plates or a cardinal plate in the brachial valve, without a notothyrium or chilidium, and with the delthyrium usually more or less restricted by deltidial plates (Cloud, 1942). *Lower Silurian to Recent.*

Centronella (Fig. 9-35*C–D*), *Rensselaeria*, *Rensselaerina*, *Beachia*, and *Cryptonella* are typical Paleozoic genera (Fig. 9-56); *Chlidonophora* (Fig. 9-24*F*), *Eucalathis* (Fig. 9-24*E*), and *Gryphus* (Figs. 9-15, 9-24*H*) are modern representatives.

Superfamily Terebratellacea. Terebratellids are terebratuloids in which the loop in the higher genera develops from both the cardinalia and the median septum; the loop generally becomes free from the median septum as the latter is partly or wholly resorbed. *Mesozoic to Recent.*

Kingena, *Trigonosemus*, and *Choristothyris*, all from the Cretaceous, are three fossil terebratellids (Fig. 9-57); *Argyrotheca* (Fig. 9-19), *Megathyris* (Fig. 9-24*B*), *Kraussina* (Figs. 9-10*I*, 9-24*A,C*), *Terebratalia* (Figs. 9-10*C–D*, 9-57*A–B*) *Dallina* (Fig. 9-24*D*), *Laqueus* (Fig. 9-10*B*), *Terebratella* (Figs. 9-17, 9-18), and

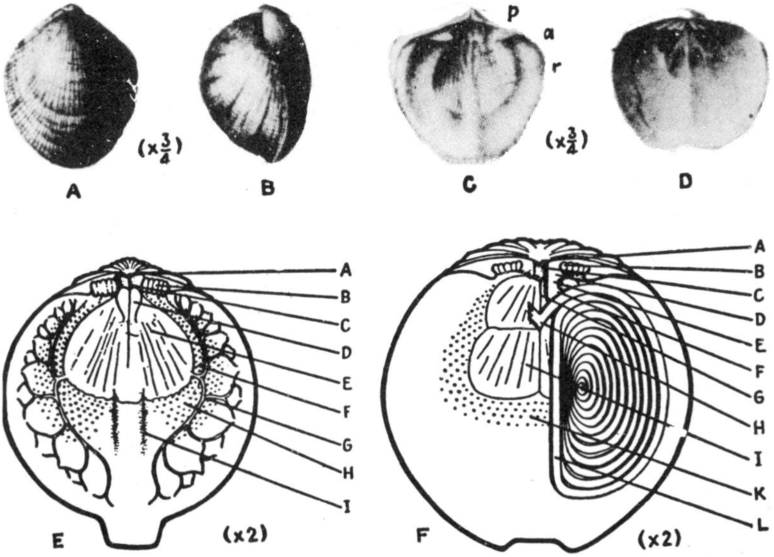

FIG. 9-52. Superfamily Atrypacea. *Atrypa reticularis*, a catchall species. *A–B*. Two views of a Devonian form. *C*. Interior of a pedicle valve, showing the broad pedicle cavity; widely separated teeth; pedicle (*p*), adductor (*a*), and diductor (*r*) muscle scars; and crenulated margins. *D*. Interior of a brachial valve, showing the structure of the hinge plate, and the several muscle scars. *E*. Reconstruction of the pedicle valve of *A. reticularis* (Linnaeus) (*A* = umbo; *B* = tooth; *C* = pseudoseptum; *D* = adductor-muscle impression; *E* = diductor-muscle impression; *F* = muscle callosity; *G* = vascular impression; *H* = genital impression; *I* = groove for the accommodation of the first ascending lamella). *F*. Reconstruction of brachial valve with only one spire drawn (*A* = umbo; *B* = socket; *C* = callosity for the attachment of the diductor muscles; *D* = fused crural base and inner socket wall; *E* = crus; *F* = primary lamella; *G* = pseudoseptum; *H* = jugum; *I* = adductor-muscle impression; *K* = genital impression; *L* = first ascending lamella). (*A–B after Schuchert and Maynard*, 1913; *C–D after Hall and Clarke*, 1894; *E–F after Alexander*, 1948.)

Magellania (Figs. 9-1–9-4, 9-14, 9-33*A–B*, 9-35*E–F*) are genera with living representatives. Some of these latter genera also have Mesozoic and Cenozoic species.

ECOLOGY AND PALEOECOLOGY[1]

Very little has yet been written on the ecology of living brachiopods, and even less on the paleoecology of the fossil forms, and, as Cooper (1948*a*) points

[1] This discussion is based on a 1911 article by Schuchert (Paleogeographic and geologic significance of Recent Brachiopoda, *Bull. Geol. Soc. Amer.*, vol. 22, pp. 258–280) and the following recent summaries by Cooper: 1937, Brachiopod ecology and paleoecology, *Nat. Res. Council, Rept. Comm. Paleoecol.* 1936–1937, pp. 26–53. 1942, Ecology of some Permian brachiopods, *Nat. Res. Council, Rept. Comm. Marine Ecology as Related to Paleontology*, 1941–1942, pp. 36–37. 1948, Annotated bibliography of brachiopod ecology, *Nat. Res. Council, Rept. Comm. Treatise Marine Ecology and Paleoecology*, 1946–1947, pp. 38–44 1948*a*, Annotated bibliography of brachiopod paleoecology, *Ibid.*, pp. 45–53.

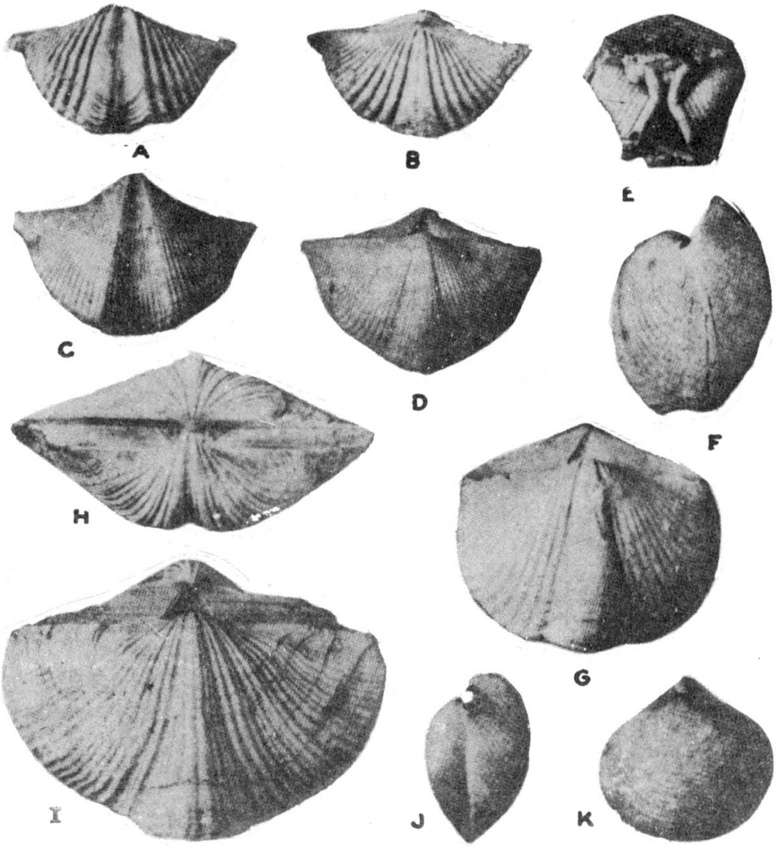

FIG. 9-53. Superfamily Spiriferacea. *A–B. Mucrospirifer consobrinus: A*, exterior of pedicle valve, showing prominent sulcus; *B*, brachial view, showing alternate plica. *C–D. Cyrtospirifer whitneyi: C*, pedicle exterior; *D*, brachial view. *E–G. Platyrachella oweni*, the well-known silicified Devonian spiriferid from southern Indiana: *E*, posterior fragment of brachial valve, showing the two spiralia, swollen in the process of fossilization; *F*, side view; *G*, brachial view. *H–I. Neospirifer cameratus*, a common Pennsylvanian spirifer: *H*, posterior view; *I*, brachial view, showing costae of different sizes and the large delthyrium. *J–K. Phricodothyris perplexa: J*, side view; *K*, brachial view (also cf. Fig. 9-28*M–N*). (*A–I after Cooper*, 1944; *J–K after Dunbar and Condra*, 1932.)

out, probably many significant bits of paleoecological data lie undiscovered in the wealth of literature on ancient brachiopods.

Brachiopods seem always to have been marine benthonic[1] organisms. A few species can live in waters less saline than that of the open ocean, but the

[1] Some Paleozoic forms are thought by Ruedemann (1934) to have been planktonic, living attached to seaweeds, and Allan (1937) has suggested that brachiopods might be transported by attachment to the shells of migratory pectens. (See Cooper, 1948, 1948*a*, and Fig. 9-10*B*.)

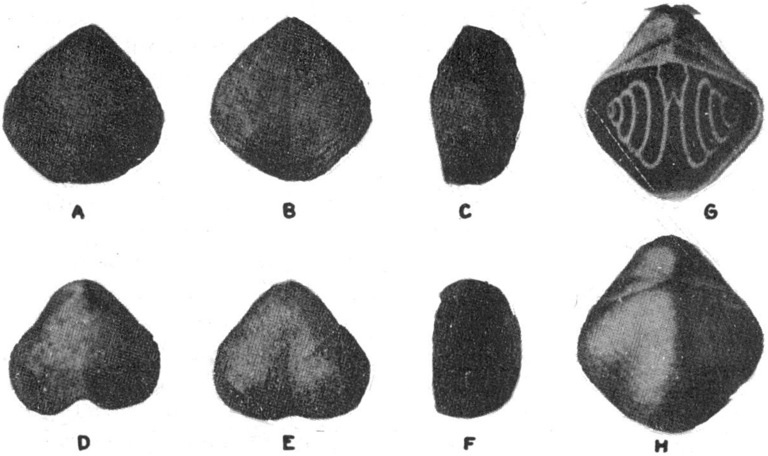

FIG. 9-54. Superfamily Rostrospiracea. *A–C. Composita subtilita*, a common Pennsylvanian index fossil: *A*, brachial view; *B*, pedicle exterior; *C*, side view. *D–F. Meristella atoka*, a Lower Devonian species: *D*, brachial view; *E*, pedicle exterior; *F*, side view. *G–H. Hyattidina congesta: G*, view showing part of brachial valve removed so that brachidium is visible; *H*, brachial view. (*A–F after Cooper* 1944; *G–H after Hall and Clarke*, 1894.)

vast majority of modern species live in normal marine habitats and ancient species probably were similarly adapted.

Living brachiopods have been taken at depths ranging from the high-tide line to 2,900 fathoms (5650 m. or 17,500 ft.).[1] *Lingula*, which lives only in tropical and warm temperate seas, seldom inhabits bottoms at depths greater than 20 fathoms (40 m., or 120 ft.), whereas *Pelagodiscus* is generally found in depths greater than 1,000 fathoms (1950 m., or 6,000 ft.), though its range is known to be from 200 to 2,737 fathoms (1,200 to 16,422 ft.). Because so little is known about the depth ranges of living brachiopods, and also because

[1] Cooper (1937) gives the following bathymetric data for 177 of the slightly more than 200 known living species (25 are excluded because they lack any data):

Fifty-nine species (33 per cent) are confined to shallow water; *i.e.*, depths less than 100 fathoms (200 m., or 600 ft.). *Lingula* and *Glottidia* are typical genera.

Sixty-three species (36 per cent) live in water deeper than 100 fathoms. Examples are *Pelagodiscus* (200 to 2,737 fathoms) and *Abyssothyris*, taken from 2,900 fathoms.

Fifty-five species (31 per cent) range from the tide zone to the deeps, hence can endure a great range in pressure and temperature. Examples are *Crania*, *Argyrotheca*, and *Terebratulina*.

From the above data it can be seen that, whereas about 60 per cent of living brachiopod species for which data are available can live in shallow water, only about 33 per cent are actually confined to shallow water. Hence, excepting *Lingula* and *Glottidia*, which are known to be confined to depths less than 100 fathoms, no generalizations on bathymetric distribution can be safely drawn.

As stated elsewhere in the text, it is probably not safe to draw inferences about depths at which extinct brachiopods lived, except for the lingulids, if the inferences are based only on the brachiopods themselves.

FIG. 9-55. Superfamily Punctospiracea. *A*. Brachial view of *Punctospirifer pulcher*, a Permian species, showing smooth plica flanked by strong costae. *B–C*. Brachial and side views of *Homoeospira evax*, a Mid-Silurian species. (*After Cooper*, 1944.)

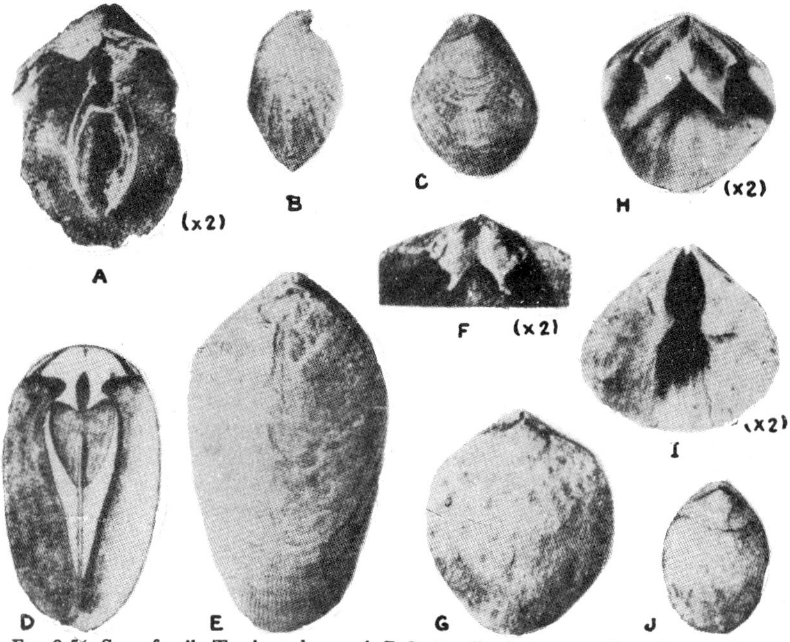

FIG. 9-56. Superfamily Terebratulacea. *A–C. Cryptonella planirostra*, a Devonian species: *A*, view of brachidium, with pedicle valve removed; *B*, side view; *C*, brachial view. *D–E. Rensselaeria marylandica: D*, restoration of a loop with a long vertical plate; *E*, brachial view, showing numerous costellae. *F–G. Beachia suessana: F*, a typical cardinal plate, in the posterior part of the brachial valve; the brachidium is broken off, but was attached to the anteriorly projecting processes; *G*, brachial view of an adult shell. *H–J. Rensselaerina medioplicata: H*, cardinal plate with brachidium broken off; *I*, view of interior of pedicle valve; *J*, brachial view of an immature shell. (*After Cloud*, 1942.)

some species show an unusually great range (as much as 2,537 fathoms, or 15,222 ft.), it is probably not correct to infer depths for most fossil brachiopods if the inference is based entirely on the brachiopod itself. If the inference is drawn from the nature of the sedimentary materials enclosing the shells—and this too is fraught with difficulty—then it must be proved that the organisms

FIG. 9-57. Superfamily Terebratellacea. *A–B*. *Terebratalia transversa*, a living terebratellid: *A*, brachial view; *B*, exterior of pedicle valve. *C*. Brachial view of *Choristothyris plicata* from the Upper Cretaceous of New Jersey. *D*. Brachial view of *Trigonosemus palissa* from the Upper Cretaceous of Holland, with cardinal margin, deltidial plates, and apical foramen indicated in white. *E*. Brachial view of *Kingena wacoensis* from the Lower Cretaceous of Texas. (*A–C, E after Cooper*, 1944.)

lived where their shells were buried. An outstanding exception is *Lingula*, which seems on the basis of both ecological and paleoecological data to have been throughout its long geological history a dweller in the littoral waters of warm shallow seas (less than 25 fathoms deep).

Modern brachiopods as a group are world-wide in their distribution and are found in the cold waters of both polar regions, as well as in the warmer waters of all oceans. If this was the case with ancient brachiopods—and there is no good reason to think otherwise—then little or no temperature significance can be assigned to extinct species.

Judging from the nature of the sedimentary rocks in which fossil brachiopod shells are preserved, it may be inferred that extinct brachiopods could live on bottoms having a considerable range of environmental conditions. Some species seem to have flourished only where the waters were clear; others seem to have been able to live in muddy waters; and a few, including *Lingula* and *Glottidia*, developed the habit of living in burrows excavated in sandy and muddy bottoms. Certain atremates and neotremates are abundant in black shales, but their relations with the original black muds are uncertain.[1] Many brachiopod shells were quite clearly buried on bottoms removed from the sites where the organism lived, and under such circumstances the valves are likely to have been separated and broken during transport. Such death assemblages are termed **thanatocoenosic.**

Brachiopods commonly make up a large part of fossil assemblages in

[1] Some have argued that the presence of *Lingula* in a black shale indicates that the sediment accumulated in a shallow water environment close to shore; others hold that the shells were transported into deep water where the black muds were being deposited. (See Cooper, 1937, p. 37.)

Paleozoic rocks (*e.g.*, Upper Ordovician rocks of the Cincinnati region; Silurian beds of New York, Gotland, etc.; and Pennsylvanian shales and limestones the world over). In a few cases a single genus was responsible for extensive biostromes (*e.g.*, *Pentamerus*, *Conchidium*, and certain productids), but brachiopods do not seem to have been generally common or abundant on many ancient bioherms. However, assemblages of brachiopods were extremely common in interstices between coral colonies that formed the Silurian reefs of Anticosti and Gotland, and Cooper (1937) mentions a great variety of brachiopods in and about the *Prismatophyllum* bioherms in the Devonian Alpena limestone of Michigan.

Because brachiopod shells usually remain closed after death, they are like hollow pebbles on the bottom, and they serve excellently as objects of attachment for worms, bryozoans, corals, other brachiopods, mollusks, and barnacles. It is not uncommon to find Recent shells so incrusted that little of the surface sculpture is visible, and many fossil shells are similarly incrusted, particularly with bryozoans, worm tubes, and cemented neotremates (Fig. 9-42*D–E*). An occasional shell may also preserve evidence of some boring organism.

All living brachiopods except the free-moving burrowing forms attach their shell to some object on the bottom or to the bottom directly. Species with a fully developed pedicle use it for attachment and live with the umbo of the brachial valve resting on the substratum (Fig. 9-10*C,E*). Some use the pedicle only during the early growth stages; in maturity the pedicle valve is cemented to the substratum and there is no further use for the pedicle. In still others, although the pedicle atrophies, the shell is not cemented but lies free on the bottom, generally with the larger and heavier valve (*i.e.*, the pedicle valve) down. All these modes of attachment seem to have been used by extinct brachiopods,[1] and in addition, certain of the latter, particularly the productids, developed spines of several different kinds that served to keep the shell above the bottom regardless of the position of the valves with reference to the bottom (Fig. 9-10*R*). At least two spinose types were developed. One group of shells, typified by *Productella* (provisionally called "*Avonia*") (Fig. 9-29*F*), developed long straight spines extending obliquely from the cardinal extremities and long body spines curving over the shell margins. These spines kept the shell off a firm bottom and permitted the brachial valve to be opened regardless of the orientation of the shell. In another group typified by *Prorichthofenia*, the pedicle valve developed into a long cone that was attached to the substratum and held in an upright position by many anchoring and supporting spines, and the brachial valve decreased in size until it became only a small lid for the cone (Fig. 9-10*F–G*). Cooper (1942, 1950) has etched from Texas Permian limestones some beautiful tangled clusters of *Prorichthofenia* that show how the shells not only were interconnected by long spines but also were attached as a mass to the substratum.

[1] *Derbyia*, a common Late Paleozoic protremate, lost its pedicle and cemented the beak of the pedicle valve to the substratum. *Stropheodonta* lost the pedicle in maturity, and thereafter lay unattached. Most of the Productacea utilized spines for attachment, anchorage (cementation), or stability.

He also found large reeflike clusters of *Scacchinella* in which each shell was cemented to some part of the pedicle valve of its neighbor.

It is possible that a few species may have lived with the pedicle beak buried in the soft bottom, as Öpik has suggested for *Clitambonites* (Fig. 9-10*J*), but it is also possible that many shells found in this position assumed it after death because the posterior part of the shell was heavier.

The living habits of the free-moving genera typified by the oft-mentioned *Glottidia* and *Lingula*, probably the most studied and best known of all brachiopods, are fairly well known. *Lingula* is widespread in the warm shallow waters of the shore zone at localities widely distributed throughout the world, and it can be fairly safely inferred, from the nature of the rocks in which fossil Lingulas are preserved, that the genus probably was always confined to the littoral zone and shallower waters of the neritic zone. Today it is particularly well represented in Japanese waters. In certain bays and estuaries of southern Japan individuals live in burrows a few inches (50 to 300 mm., or 2 to 12 in.) long and move up and down the flattened tubular holes by means of the highly contractile pedicle. Ordinarily *Lingula* lives fastened in a burrow by its pedicle. If it is removed and placed on the sand, it will again bury itself by lateral movements of the brachial valve, movements of the lateral setae, and a wiggling motion of the pedicle. It fixes itself in its burrow by secreting mucus that agglutinates sand grains or mud to form a tube.

When in the burrow, the individual lingulid lies with the posterior margin down and the setaceous anterior margin protruding (Fig. 9-10*X–Z*). The front margins of the mantle lobes are so folded as to produce three tubular gaps. Water is drawn in through the side openings and expelled through the middle one.

Lingula does not seem to be affected by brackish water or by water so foul from decomposing organic matter that burrowing mollusks are unable to survive. It may be exposed on intertidal flats for hours with no apparent injury. The genus is characteristic of waters less than 25 fathoms deep and of either sandy or muddy bottoms.

In one living genus, *Terebratalia obsoleta*, the form of the shell reflects environmental conditions in a significant way. Shells taken from 30 to 50 fathoms varied from transverse to round and almost smooth forms. Shorter and more rotund shells developed where the water was roughest, whereas spirifer-like shells were found in quieter water. Cooper (1937) concluded that adaptations such as folding, lobation, plication, cementation, alation and mucronation, elongation, compressed form, resupination, reversion, spinescence, and geniculation all helped the brachiopod keep the front margin out of the mud.

As brachiopods are sessile benthonic organisms throughout most of life, they can be dispersed only during the brief embryonic period when the larvae are free-swimming and can be carried by currents. Since the young growth stages are known in detail in only five genera—*Lingula* (Atremata), *Lacazella* (Strophomenacea), *Terebratulina* (Terebratulacea), and *Argyrotheca* and *Terebratella* (Terebratellacea)—the available data are far too sparse for much generalization on dispersion during the embryonic period.

Certain obstacles tend to limit or prevent dispersal. The free-swimming larval stage in the articulates is short—10 to 12 days in *Terebratulina*, and less in *Terebratella*—and the known larvae do not swim after the development of a functioning stomach. Time, therefore, seems to limit dispersal among the articulates. By contrast, the larvae of inarticulates have a mouth and functioning stomach, which allow them to live for some time in the free-swimming stage. However, inarticulates live chiefly in warm shallow water; hence wide deep waters and cold waters should constitute a barrier to their dispersal. Whether they do in fact constitute such a barrier seems questionable in the face of a report that a larva of *Lingula* was taken from a depth of 2,200 fathoms (13,200 ft.) some 200 miles off the coast of Ceylon, and another report that free-swimming larvae of *Pelagodiscus*, which in maturity generally lives in depths greater than 1,000 fathoms (6,000 ft.), were taken from 40 fathoms in the Indian Ocean. Furthermore, there is a possibility, as suggested by Allan (1937), that brachiopods may attach themselves to free-swimming pectens (*e.g.*, *Chlamys*) and thereby get a free ride to far places. It is also possible that young brachiopods may have become attached to floating seaweeds. Furthermore, the extremely wide distribution of such species as *Pentamerus oblongus* and *Coelospira hemispherica* shows that some forms overcame factors limiting migration.

It is probably too early to say whether the world-wide distribution of brachiopods through geologic time was achieved by migration limited to the shallow waters along continental margins, or whether the larvae were able by some as yet undiscovered means to travel directly across the deeper oceans.

GEOLOGIC HISTORY

The relative abundance of genera in the different superfamilies through geologic time is shown diagrammatically in Fig. 9-58. From this figure the relative importance of the different superfamilies during any geologic period can be determined. The significance of this stratigraphic distribution was recently discussed by Cooper and Williams (1952).

There are several doubtful records of alleged brachiopods of Pre-Cambrian age, but these are all questionable for one reason or another. The earliest brachiopods thus far reported are alleged atrematous shells from the supposed Pre-Cambrian Vindhyan rocks of India and an alleged *Lingulella* from the Belt series of Montana (generally correlated with the Pre-Cambrian). Chapman (1935) reported from India more than 100 carbonized specimens, which he interpreted as primitive chitinous protegula and as the possible ancestors of Lower Cambrian Atremata. Sahni (1936), however, after a careful examination of the specimens, doubted that they were brachiopod remains and concluded that there was insufficient evidence to assign a definite systematic position to the remains.

Fenton and Fenton (1933, 1936) found a dozen or so specimens, all pedicle valves, of a *Lingulella* in the Newland argillaceous limestone of the Beltian series, supposedly of Pre-Cambrian age. These Montana specimens seem to be brachiopods, to judge from the published photographs of them; they are

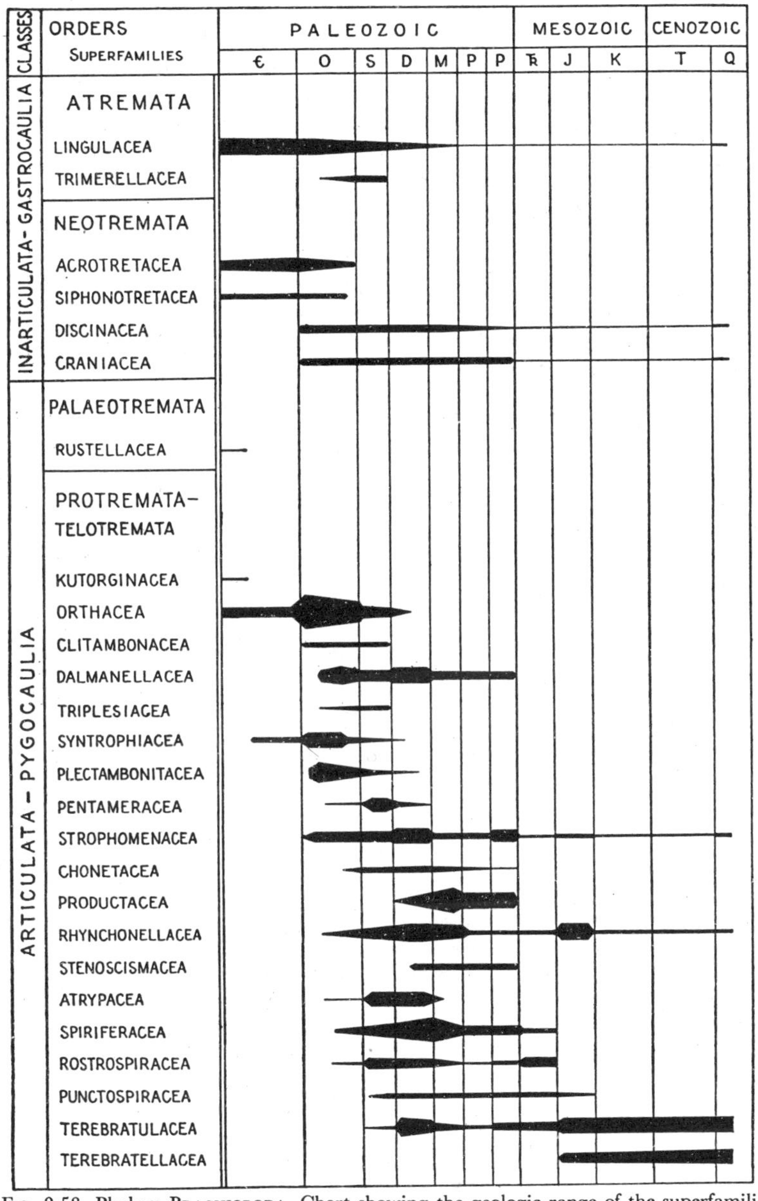

FIG. 9-58. Phylum BRACHIOPODA. Chart showing the geologic range of the superfamilies of the phylum. The width of the time band is roughly proportional to the numerical importance of the superfamily. (*Based on an original chart by Cooper and Williams*, 1950, *plus additional data from G. A. Cooper. See also Cooper and Williams*, 1952.)

therefore the oldest ones thus far reported if the Newland limestone is actually of Pre-Cambrian (Algonkian) age. Not all geologists, however, agree that the rocks are Pre-Cambrian; hence the true age of the Fentons' *Lingulella* remains debatable. There are good reasons, however, to assume that primitive brachiopods were present in Pre-Cambrian seas and that they had chitinous shells (Thomson, 1927), because the fossil record of Lower Cambrian rocks in many parts of the world leaves no doubt about the abundance and variety of brachiopods in the earliest Cambrian seas.

The earliest Lower Cambrian brachiopods are generally of primitive type, but some show rather high specialization along certain lines. None of them is composed wholly of chitin. It is thought that the earliest and most primitive shell substance had given way long before to chitinophosphatic and calcareous materials. Most of the Early Cambrian brachiopods seem to have been inarticulates (*e.g.*, *Lingulella*), but there were also a few primitive articulates (*e.g.*, *Nisusia*, *Rustella*, *Wimanella*, *Diraphora*, and *Kutorgina*).

By Late Cambrian time the Rustellacea and Kutorginacea had disappeared, whereas the Lingulacea, Acrotretacea, Siphonotretacea, Orthacea, and Syntrophiacea had become firmly established. It was during the Ordovician, however, that the brachiopods underwent their greatest differentiation with the appearance of 14 new superfamilies. Some of these were short-lived, but eight persisted into the Permian and four have living representatives. The brachiopods reached their greatest development during the Silurian and Devonian, when two new superfamilies were added in each period to bring the total to 19 in the Silurian and 18 in the Devonian. By Carboniferous time the number of superfamilies had decreased to 14, but 13 of these persisted into the Permian, and 9 lived through the Triassic, 7 through the Jurassic, and 6 into the Recent. After the Devonian, only a single new superfamily appeared—the Terebratellacea in the Early Mesozoic.

During the Carboniferous and Permian the Productacea and Spiriferacea reached a position of dominance over all other brachiopods, and in Pennsylvanian and Permian seas the productids reigned supreme, with many highly spinose genera, some of bizarre type (*e.g.*, *Prorichthofenia*). With the end of the Permian almost all Paleozoic genera became extinct except for a few stragglers that lived on into the Mesozoic. The Spiriferacea and Rostrospiracea were gone by the end of the Triassic and the Punctospiracea by the end of the Jurassic. Shortly after the disappearance of these three groups, the Terebratellacea appeared and the Terebratulacea and Rhynchonellacea seemingly took on new vigor. With the decline of the rhynchonellids at the close of the Jurassic, the first two superfamilies assumed the leading role and have kept it to the present time. The decline of the brachiopods in the later Mesozoic and Tertiary has been ascribed to their constitutional inaptitude to compete successfully with the Mollusca under changing conditions (Elliott, 1947).

It now seems that the first brachiopods were tiny atremates with simple chitinous shells. These are believed to have been present in Pre-Cambrian seas, although no unquestioned fossils have yet been found to prove their existence. From these simple forms evolved two main branches—the Gastrocaulia, characterized by an embryonic development in which the pedicle forms late from the ventral mantle lobe; and the Pygocaulia, characterized by complex embryonic development and metamorphosis during which the pedicle develops early from the caudal segment. These two primary and fundamentally different groups differentiated quite early—possibly in the late Pre-Cambrian—into the Atremata and Neotremata on one hand and the Palaeotremata, Protremata, and Telotremata on the other. This postulated evolution is presented schematically in Table 9-6.

NATURE AND STRATIGRAPHIC USE OF FOSSIL BRACHIOPODS

Fossil brachiopods are commonly well preserved because of little alteration of shell substance and no evident change in form or structure of the shell. Calcareophosphatic shells, particularly if contained in argillaceous rocks, are likely to be well preserved, and those of many Cambrian species seem to be little altered from the condition in which they were buried.

Because the shells of articulate brachiopods are generally closed at the death of the animal, complete and undistorted shells are common in many superfamilies (*e.g.*, Orthacea, Productacea). Separate valves are also common in many rocks and are most useful in the study of certain internal shell features.

The two valves of inarticulates are likely to become separated when the animal dies, and in rocks specimens of one valve may be present almost to the exclusion of the other. Commonly, however, the two valves are mixed together in random fashion.

Brachiopod shells are commonly replaced by calcite, silica, or iron sulphide. Some such replaced shells have been so greatly altered that few of the original structures remain and little more than the shape of the shell is left to indicate that the fossil is a brachiopod. On the other hand many silicified shells have even the finest structural features beautifully preserved, and the magnificent specimens that Dr. G. Arthur Cooper of the U.S. National Museum has etched from Paleozoic limestones are well known to all students of brachiopods.

In many cases the original shell has been dissolved away completely, but generally there remain an impression of the exterior, from which shape and surface sculpture can be determined, and a filling of the interior (*i.e.*, a **core,** or **steinkern**), which bears on its surface impressions of the different internal features of the shell (*e.g.*, muscle scars and pallial markings). By pressing some molding substance into these impressions or around the cores, excellent replicas of the features present in the original shell can be obtained.

Certain features of brachiopod shells have been used widely and successfully for stratigraphic purposes. Ulrich and Cooper (1938), in one of the most carefully prepared and documented brachiopod studies in recent years, describe the species found in North American rocks of latest Cambrian and earliest Ordovician age, and a similar work by Cooper, soon to be published, will show the great value of the brachiopods in identifying and correlating Middle Ordovician strata. Cumings (1903), by studying shell shape, plica-

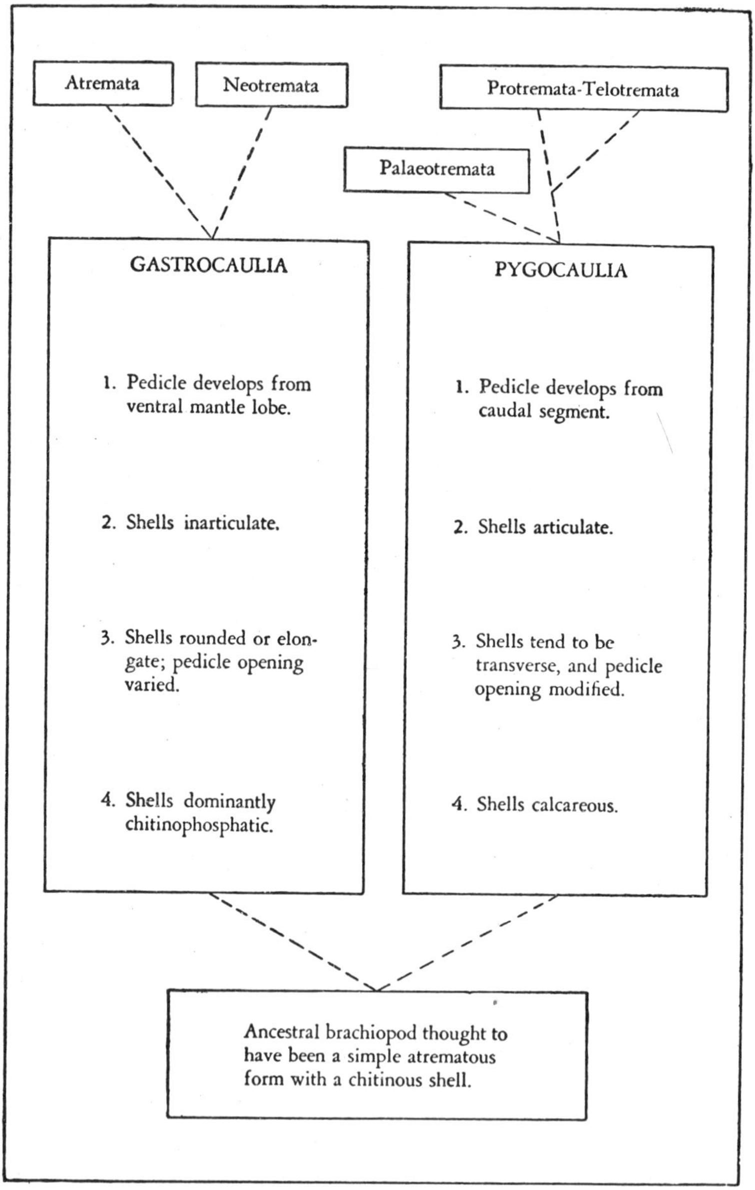

TABLE 9-6. DIAGRAM SHOWING HYPOTHETICAL EVOLUTION OF THE BRACHIOPODA

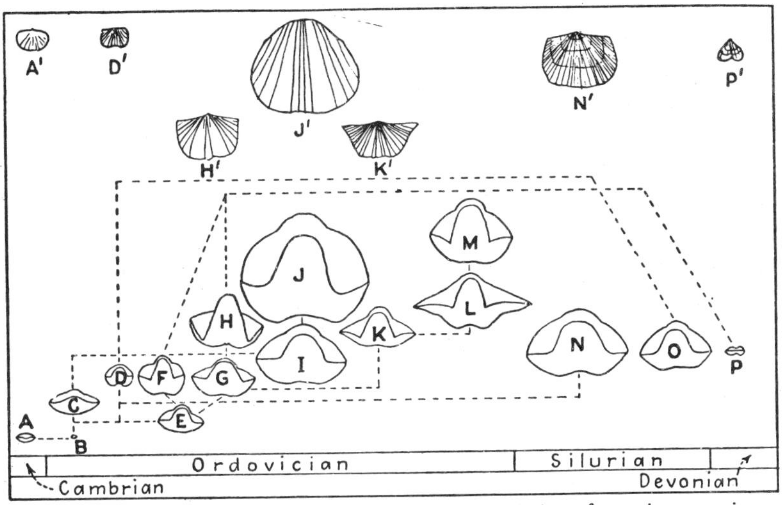

FIG. 9-59. Phylum BRACHIOPODA. Diagrams showing variation of anterior commissure and shell costation in *Platystrophia* (*sensu lato*) from its beginning in the Middle Ordovician (Trenton) to its end in the Middle Silurian (Niagaran). The dashed lines represent supposed lines of derivation or descent. The Cambrian genus *A* is "*Orthis*" *lenticularis;* the Lower Devonian genus *P* is *Bilobites varicus*. This chart illustrates how variation of anterior commissure, shell shape, and ornamentation can be used stratigraphically. (*Adapted from Cumings*, 1903.)

tion, and costation, demonstrated how the species of *Platystrophia* can be used for zoning strata from Ordovician to Devonian age (Fig. 9-59). Salmon (1942) has shown how the 50 species of "*Rafinesquina*," which is now divided into *Rafinesquina* and *Öpikina*, have stratigraphic significance if account be taken of shell structure and certain interior characters, and the Fentons (1931) found *Atrypa* useful as a horizon marker and index of faunal shifting in the Iowa Devonian when they studied detailed shell features. Gill (1945) zoned the Ordovician to Permian rocks of Victoria (Australia) on the basis of the chonetids, Sutton (1938) found the productids similarly useful in identifying Mississippian strata, and Dunbar and Condra (1932) showed several generic series (*e.g.*, the Orthotetinae) to have zoning value in their study of Pennsylvanian brachiopods (Fig. 9-60). Buckman's classic work on the British Jurassic brachiopods is familiar to all serious paleontologists, and Crickmay (1933) has suggested that the North American Jurassic can also be zoned, and possibly thermal conditions in Jurassic seas determined, by studying beak details, muscle marks, and other shell features of the telotrematous genera. For the most recent rocks Crespin and Chapman (1927) have used the brachiopods to zone the Tertiary rocks of Australia, Allan (1939, 1940*a*) has recorded the genera from the Tertiary of both Australia and New Zealand, and Thomson (1927), in his indispensable manual, repeatedly mentions the stratigraphic value of Tertiary genera and calls attention to aspects of living species that might be used in the study of fossil forms.

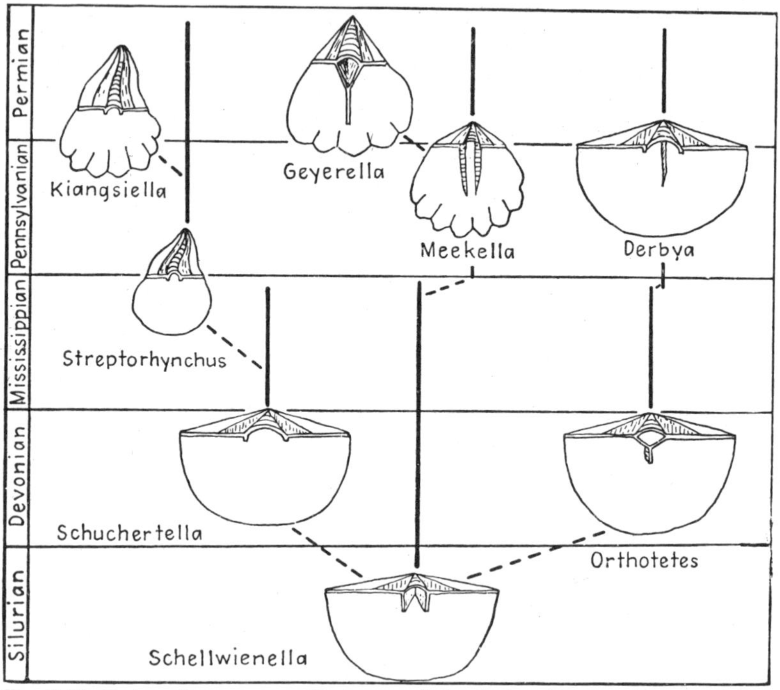

FIG. 9-60. Phylum BRACHIOPODA. Phylogeny of the Orthotetinae, a subfamily of strophomenids. This diagram illustrates how the variations in the cardinal area and in shell shape and ornamentation among closely related genera can be used for stratigraphic purposes. (*Modified after Dunbar and Condra*, 1932.)

REFERENCES

ALEXANDER, F. E. S. 1949. A revision of the brachiopod species *Anomia reticularis* Linneaus genolectotype of *Atrypa* Dalman. *Quart. Jour. Geol. Soc. London*, vol. 104, pp. 207–220.

ALLAN, R. S. 1938, 1940. Studies on the Recent and Tertiary Brachiopoda of Australia and New Zealand. *Rec. Cant. Mus.* (N.Z.), vol. 4, pp. 231–248 (1939), 227–297 (1940).

———. 1940*a*. A revision of the classification of the terebratelloid Brachiopoda. *Ibid.*, vol. 4, pp. 267–275.

ARBER, M. A. 1939. The nature and significance of the pedicle foramen of *Leptaena* Dalman. *Geol. Mag.*, vol. 76, pp. 82–92.

———. 1940. The relation of the valves to the pedicle in the strophomenid brachiopods. *Ibid.*, vol. 77, pp. 161–174.

———. 1942. The pseudodeltidium of the strophomenid brachiopods. *Ibid.*, vol. 79, pp. 179–187.

BEECHER, C. E. 1891–1892. Development of the Brachiopoda. *Amer. Jour. Sci.*, vol. 41, pp. 343–357 (1891); vol. 44, pp. 133–155 (1892).

———. 1893. Some correlations of the ontogeny and phylogeny in the Brachiopoda. *Amer. Nat.*, vol. 27, pp. 599–604.

BELL, W. C. 1941. Cambrian Brachiopoda from Montana. *Jour. Paleontology*, vol. 15, pp. 193–255.

———. 1944. Early Upper Cambrian brachiopods. *Geol. Soc. Amer., Spec. Paper* 54, pp. 144–155.

———. 1948. Acetic acid etching technique applied to Cambrian brachiopods. *Jour. Paleontology*, vol. 22, pp. 101–102.

Blochmann, F. 1892, 1900. "Untersuchungen ueber den Bau der Brachiopoden." Jena. Pt. I, 1892; Pt. II, 1900.

Bond, G. 1941. Species and variation in British and Belgian Carboniferous Schizophoriidae. *Proc. Geol. Assoc.*, vol. 52, pp. 285–303.

Buckman, S. S. 1901. Homoeomorphy among Jurassic Brachiopoda. *Proc. Cotteswold Nat. Field Club*, vol. 13, pp. 231–290.

———. 1918. The Brachiopoda of the Namyan Beds, Northern Shan States, Burma. *Pal. Indica* (n.s.), vol. 3, *Mem.* no. 2, 299 pp.

Bulman, O. M. B. 1939. Muscle systems of some inarticulate brachiopods. *Geol. Mag.*, vol. 76, pp. 434–444.

Chapman, F. 1935. Primitive fossils, possibly atrematous and neotrematous Brachiopoda from the Vindhyans of India. *Rec. Ind. Geol.*, vol. 69, pp. 109–120.

Clarke, F. W., and Wheeler, W. C. 1915. The composition of brachiopod shells. *Proc. Nat. Acad. Sci.*, vol 1, pp. 262–266.

Cloud, P. E., Jr. 1942. Terebratuloid Brachiopoda of the Silurian and Devonian. *Geol. Soc. Amer., Spec. Paper* 38, pp. 1–182.

———. 1948. Notes on Recent brachiopods. *Amer. Jour. Sci.*, vol. 246, pp. 241–250.

Conklin, E. G. 1902. The embryology of a brachiopod, *Terebratulina septentrionalis*, Couthouy. *Proc. Amer. Phil. Soc.*, vol. 41, pp. 41–76.

Cooper, G. A. 1930. The brachiopod genus *Pionodema* and its homoeomorphs. *Jour. Paleontology*, vol. 4, pp. 369–382.

———. 1937. Brachiopod ecology and paleoecology. *Nat. Res. Council, Rept. Comm. Ecology*, 1936–1937, pp. 26–53.

———. 1942. Ecology of some Permian brachiopods. *Nat. Res. Council, Rept. Comm. Marine Ecology as Related to Paleontology*, 1941–1942, pp. 36–37.

———. 1944. Phylum Brachiopoda *in* Shimer and Shrock's "Index Fossils of North America" Chap. 9, pp. 277–365. New York, John Wiley & Sons, Inc., Cambridge, Mass., The Technology Press. (This reference has a carefully selected bibliography.)

———. 1948. Annotated bibliography of brachiopod ecology. *Nat. Res. Council, Rept. Comm. Treatise Marine Ecology and Paleoecology*, 1946–1947, pp. 38–44.

———. 1948*a*. Annotated bibliography of brachiopod paleoecology. *Ibid.*, pp. 45–53.

———. 1950. Permian faunas of the Glass Mountains, Texas, and their environment. *Trans. N.Y. Acad. Sci.*, vol. 12, pp. 80–81.

Cooper, G. A., and Williams, A. 1952. Significance of the stratigraphic distribution of the brachiopods. *Jour. Paleontology*, vol. 26, pp. 326–337.

Cox, I. 1934. Notes on the shell musculature of *Gryphus vitreus* (Born). *Geol. Mag.*, vol. 71, pp. 226–230.

Crespin, I., and Chapman, F. 1927. Range in Time and Distribution of the Australian Tertiary Brachiopods *in* Thomson's "Brachiopod Morphology and Genera (Recent and Tertiary)," pp. 298–302.

Crickmay, C. H. 1933. Attempt to zone the North American Jurassic on the basis of its brachiopods. *Bull. Geol. Soc. Amer.*, vol. 44, pp. 871–894.

Cumings, E. R. 1903. The morphogenesis of *Platystrophia;* a study of the evolution of a Paleozoic brachiopod. *Amer. Jour. Sci.*, vol. 15, pp. 1–48, 121–136.

Dall, W. H. 1920. Annotated list of Recent Brachiopoda. *Proc. U.S. Nat. Mus.*, vol. 57, pp. 261–377.

Davidson, T. 1858–1880. "Monograph of British Fossil Brachiopoda." *Palaeontographical Society*, vols. 2 and 4.

———. 1886–1888. A monograph of Recent Brachiopoda. *Trans. Linn. Soc. London*, vol. 4, pp. 1–248.

DUNBAR, C. O., and CONDRA, G. E. 1932. Brachiopods of the Pennsylvanian system of Nebraksa. *Neb. Geol. Surv. Bull.* 5, 377 pp.

ELLIOTT, G. F. 1947. Distribution des Brachiopodes en espèces et ses causes illustrée par les Térébratulinides de l'Eocene de l'Europe occidentale. *France soc. géol. Bull.*, 5th sér., vol. 17, pp. 67–80.

———. 1947*a*. The development of a British Aptian brachiopod. *Quart. Jour. Geol. Soc. London*, vol. 58, pp. 144–159.

FENTON, C. L. 1931. Studies of evolution in the genus *Spirifer*. *Wagner Free Inst. Sci. Pub.*, vol. 2, 436 pp.

FENTON, C. L., and FENTON, M. A. 1931. *Atrypa* as horizon marker. *Bull. Geol. Soc. Amer.*, vol. 42, pp. 352–353.

——— and ———. 1932. Orientation and injury in the genus *Atrypa*. *Amer. Midland Naturalist*, vol. 13, pp. 63–74.

——— and ———. Oboloid brachiopods in the Belt Series of Montana (abstract). *Bull. Geol. Soc. Amer.*, vol. 44, p. 190.

——— and ———. 1936. Walcott's "Pre-Cambrian Algonkian algal flora" and associated animals. *Ibid.*, vol. 47, pp. 609–621.

GILL, E. D. 1945. Chonetidae from the Paleozoic rocks of Victoria and their stratigraphical significance. *Proc. Roy. Soc. Victoria*, vol. 57 (n.s.), pts. I–II, pp. 125-151.

GRABAU, A. W. 1931–1932. The Brachiopoda (studies for students). *Sci. Quart., Nat. Univ. Peking*, vol. 2, pp. 235–254 (pt. 1); 397–422 (pt. 2); vol. 3, no. 2, pp. 75–112 (pt. 3); no. 4, pp. 85–117 (pt. 4).

HALL, J., and CLARKE, J. M. 1892, 1894. An introduction to the study of the Brachiopoda, intended for a handbook for the use of students. *N.Y. State Geol. Surv., An. Rept.* 11, pp. 133–223; 1892; pt. II, *An. Rept.* 13, pp. 945–1137, 1894.

——— and ———. 1892, 1893. An introduction to the study of the genera of Paleozoic Brachiopoda. *N.Y. State Geol. Paleontology*, pt. 1, 367 pp., 1892; pt. 2, 394 pp., 1893.

HANCOCK, A. 1858. On the organization of the Brachiopoda. *Phil. Trans. Roy. Soc. London*, vol. 148, pp. 791–869.

HATAI, K. M. 1938. On some Cenozoic Brachiopoda from the North American region. *Amer. Midland Naturalist*, vol. 19, pp. 706–722.

KOWALEVSKY, M. 1883. Observations sur le developpement des brachiopodes. Analysis by Oehlert and Deniker. *Arch. zool. exp.* ser. 2, vol. 1, pp. 57–76.

MCEWAN, E. D. 1939. Convexity of articulate brachiopods as an aid in identification. *Jour. Paleontology*, vol. 13, pp. 617–620.

MORSE, E. S. 1869. Note on the extension of the coiled arms in *Rhynchonella*. *Amer. Jour. Sci. Arts*, vol. 17, p. 257.

———. 1871. On the early stages of *Terebratulina septentrionalis*. *Mem. Boston Soc. Nat. Hist.*, vol. 2, pp. 29–39.

———. 1873. On the embryology of *Terebratulina*. *Ibid.*, pp. 249–264.

———. 1902. Observations on living Brachiopoda. *Ibid.*, vol. 5, pp. 313–386.

ÖPIK, A. 1930. Brachiopoda Protremata der Estländischen Ordovizischen Kukruse-Stufe. *Pub. Geol. Inst., Univ. Tartu* (Estonia), vol. 20, pp. 1–261.

———. 1934. Ueber Klitamboniten. *Ibid.*, no. 39, 239 pp.

PERCIVAL, E. 1944. A contribution to the life history of the brachiopod *Terebratella inconspicua* Sowerby. *Roy. Soc. New Zealand, Trans.* 74, pp. 1–23.

RAYMOND, P. E. 1941. Invertebrate Paleontology *in* Fiftieth Anniversary Volume, Geological Society America, pp. 73–103. (Brachiopoda, pp. 81–82.)

REED, F. R. C. 1895. Brachiopoda. Pt. II. Palaeontology of the Brachiopoda. Cambridge Natural History, vol. 3, pp. 489–512.

———. 1943. Brachiopoda and Mollusca from the Productus limestones of the Salt Range. *Pal. Indica*, n.s., vol. 23, *Mem.* no. 2.

Sahni, M. R. 1936. *Fermoria minima.* A revised classification of the organic remains from the Vindhyans of India. *Rec. Ind. Geol.*, vol. 69, pp. 458–468.

St. Joseph, J. K. S. 1935. A description of *Eospirifer radiatus* (J. de C. Sowerby). *Geol. Mag.*, vol. 72, pp. 316–327.

———. 1938. The Pentameracea of the Oslo region. *Norsk Geol. Tidsskrift*, vol. 17, pp. 225–336.

———. 1941. The brachiopod family Parastrophinidae. *Geol. Mag.*, vol. 78, pp. 317–401.

Salmon, E. S. 1942. Mohawkian *Rafinesquina. Jour. Paleontology*, vol. 16, pp. 564–603.

Schuchert, C. 1897. A synopsis of American fossil Brachiopods. *U.S. Geol. Surv.*, *Bull.* 87, 464 pp.

———. 1922. Paleogeographic and geologic significance of Recent Brachiopods. *Bull. Geol. Soc. Amer.*, vol. 22, pp. 258–275.

——— and Cooper, G. A. 1931. Synopsis of the brachiopod genera of the suborders Orthoidea and Pentameroidea, with notes on the Telotremata. *Amer. Jour. Sci.*, vol. 22, pp. 241–251.

——— and ———. 1932. Brachiopod genera of the suborders Orthoidea and Pentameroidea. *Peabody Mus. Nat. Hist.* (Yale), *Mems.*, vol. 4, pt. 1, 270 pp.

——— and Le Vene, C. M. 1929. Fossilium Catalogus I. Animalia, pars. 42, Brachiopoda, 140 pp. Berlin.

Shipley, A. E. 1883. On the structure and development of *Argiope. Mitth. Zool. Sta. Neapel*, vol. 4, pp. 494–520.

———. 1895. Brachiopoda. Pt. 1, Recent Brachiopoda. Cambridge Natural History, vol. 3, pp. 461–488. New York, Cambridge University Press.

Sutton, A. H. 1938. Taxonomy of Mississippian Productidae. *Jour. Paleontology*, vol. 12, pp. 537–569.

——— and Summerson, C. H. 1943. Cardinal process of Productidae. *Ibid.*, 17, pp. 323–330.

Termier, H., and Termier, G. 1949. Sur la classification des brachiopodes. *Bull. soc. hist. nat. de l'Afrique du Nord*, vol. 40, pp. 51–63.

Thomson, J. A. 1927. "Brachiopod Morphology and Genera (Recent and Tertiary)." Wellington, N.Z., *New Zealand Board of Science and Art*, Manual No. 7, 338 pp.

Ulrich, E. O., and Cooper, G. A. 1936. New Silurian brachiopods of the family Triplesiidae. *Jour. Paleontology*, vol. 10, pp. 331–347.

——— and ———. 1938. Ozarkian and Canadian Brachiopoda. *Geol. Soc. Amer.*, *Spec. Paper* 13, pp. 1–323. (Reviewed by Schuchert, *Amer. Jour. Sci.*, vol. 237, pp. 135 *ff.*, 1939.)

Walcott, C. D. 1912. Cambrian Brachiopoda. *U.S. Geol. Surv.*, *Mon.* 51, 2 vols.

Wang, Y. 1949. Maquoketa Brachiopoda of Iowa. *Geol. Soc. Amer.*, *Mem.* 42, pp. 1–55.

Wilson, A. E. 1944. *Rafinesquina* and its homomorphs *Öpikina* and *Öpikinella. Trans. Roy. Soc. Can.*, vol. 38, pp. 145–203.

———. 1945. *Strophomena* and its homomorphs *Trigrammaria* and *Microtypa. Ibid.*, vol. 39, pp. 121–150.

Wisniewska, M. 1932. Les rhynchonellidés du Jurassique sup. de Pologne. *Pal. Polonica*, (Warsaw), vol. II, 71 pp.

Yatsu, N. 1902. On the development of *Lingula anatina. Jour. Coll. Sci. Imp. Univ. Tokyo*, vol. 17, pp. 1–112.

CHAPTER 10

PHYLUM MOLLUSCA[1]

GENERAL CONSIDERATIONS

The Mollusca, widely known as **mollusks** and containing many of the so-called shellfish,[2] constitute a large group of invertebrates that are alike in fundamental morphology but unlike with respect to the shell enclosing the soft parts. Included in this large and important phylum, which is represented by more than 150,000 living and many thousands of fossil species, are such well-known animals as the chitons (*Amphineura*); tooth shells (*Scaphopoda*); oysters, mussels, and clams (*Pelecypoda*); slugs, snails, whelks, and limpets (*Gastropoda*); and nautiluses (or nautili), cuttlefish, and octopuses (or octopi) (*Cephalopoda*). Mollusks are of great interest to man because of their extensive use as food, because they are intermediate hosts in the life cycle of many parasitic worms that are responsible for a variety of human and animal diseases (*e.g.*, schistosomiasis), because some are destructive wood borers, and because the shells, which are mostly calcareous, are widely used for certain commercial purposes (*e.g.*, manufacture of pearl buttons and ornaments) and are prized by collectors the world over.

Mollusks have a soft unsegmented body that is generally protected by an external calcareous shell of one or more pieces secreted by the **mantle**—a modified part of the body wall so designated because it generally is a flap or fold enveloping a part or the whole of the body. Part of the ventral surface of the body is usually modified into a muscular organ, the **foot,** which in the different groups is used for creeping, digging, burrowing, swimming, etc. If appendages are present they are invariably unsegmented The body proper consists of a relatively simple alimentary tract, generally arcuate but twisted in the gastropods, a well-developed heart and vascular system, one or more pairs of gills or, in the absence of these, of other breathing organs, a nervous system composed of three pairs of ganglia with longitudinal and transverse commissures, and commonly certain sense organs (*e.g.*, eyes and hearing and tactile organs). The sexes may or may not be separate. In all classes except the cephalopods, the fertilized egg develops into a trochophore larva, which passes through a veliger stage (discussed later) to become an adult, and the adult mollusk is invariably solitary, though individuals commonly congregate into clusters.

Since their appearance in the Early Cambrian, the Mollusca as a group have deployed into every type of aquatic habitat and have successfully in-

[1] Mollusca—L. *mollusca*, a kind of soft nut with a thin shell. The typical mollusk is a soft-bodied animal inside a shell of some sort.

[2] Other shellfish that are eaten by man include brachiopods, crabs and lobsters, and certain echinoderms.

vaded the land. Representatives of the phylum now live in all kinds of water and may be floaters (plankton), swimmers (nekton), or bottom-dwellers (vagrant and sessile benthos). Certain gastropods also live on the land under environmental conditions ranging from moist places to the desert. They climb trees and cliffs, and some species are abundant over grasslands. Fossil mollusks seem to have been adapted to a similarly broad range of environmental conditions. Because of their wide distribution and extensive development, in the present as well as during the long geologic past, mollusks constitute one of the most successful of all animal phyla.

Individual mollusks range in size from tiny adult snails scarcely a millimeter long to giant squids of the Atlantic Ocean that span more than 15 m. (50 ft.) from the tip of the tail to the end of the longest tentacle. The molluscan shell ranges in size from minute forms less than a millimeter in greatest dimension to the giant straight-coned shells of Ordovician nautiloids, some of which are as much as 4½ m. (15 ft.) long, the somewhat smaller coiled ammonites of the Mesozoic, which reach 2 m. (6 ft.) in diameter,[1] and the huge pelecypod *Tridacna* of the present seas, which weighs more than 225 kg. (500 lb.), and whose bivalve shell is commonly large enough even when closed to hold a small child.

The Mollusca have left an abundant fossil record in the rocks of every geologic period since the beginning of the Cambrian, and many genera are reliable index fossils.

The Animal

The relations of soft parts to the hard shells in the different classes of Mollusca are shown diagrammatically in Fig. 10-2. If the hard external shell is disregarded for the moment, the molluscan animal is a soft, relatively short, unsegmented, generally bilaterally symmetrical body with an anterior mouth, a posterior anus, a ventral foot, and a dorsal mantle. Inside the body are the vital organs—alimentary tract, heart and vascular system, nervous system, digestive and excretory organs, and certain other soft parts (Fig. 10-1). Respiration is carried on by special organs (**gills,** or a **lung**) lying in the **mantle cavity** (the space between the mantle and the body proper) and also under certain conditions by the mantle itself or by the epidermis.

The digestive tract is usually a simple, straight or U-shaped tube with an anterior **mouth,** a short **esophagus,** an expanded median **stomach,** a looped or coiled **intestine** and a posterior[2] **anus** opening into the mantle cavity. Certain specialized glands secrete digestive fluids. A dorsally situated heart pumps blood through a system of vessels and tubes, some of which ramify the breathing organs.

All mollusks except the pelecypods have the front or anterior part of the body modified into a **head.** This part of the body usually bears sensory appendages or external sense organs and contains the mouth portion (**buccal**

[1] The largest ammonite shells if uncoiled would be as much as 11 m. (35 ft.) long.

[2] In many gastropods the soft parts are twisted in such a way that the digestive tract is looped and the anus lies directly above the head in an anterior position. Other deviations from the simple or gently arcuate digestive tract are shown in Fig 10-2.

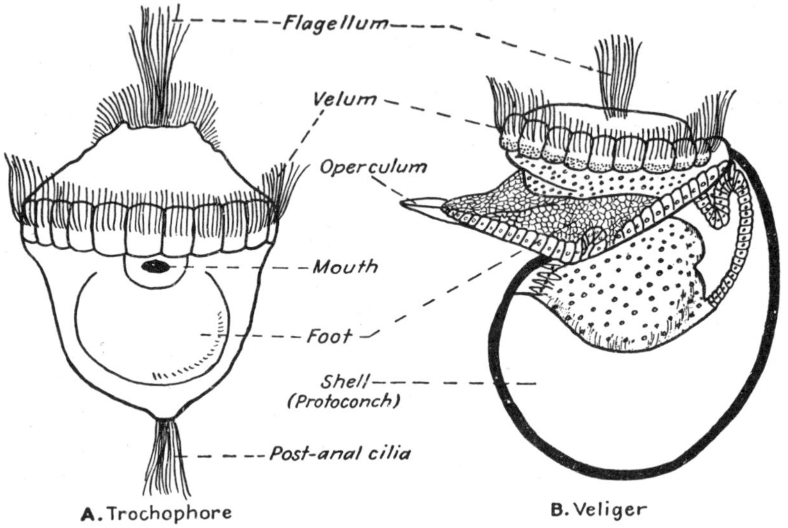

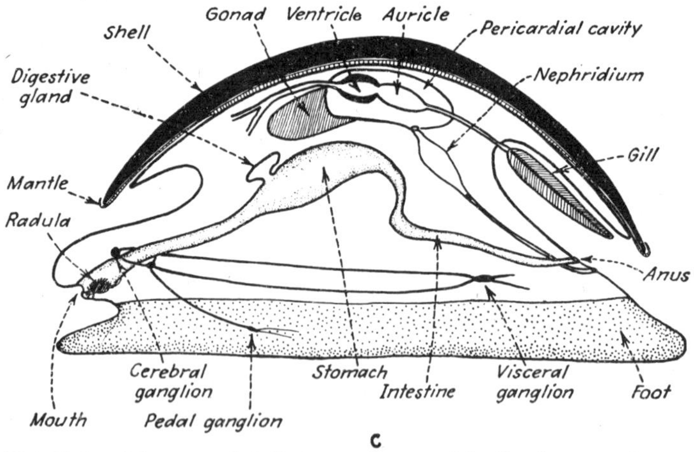

FIG. 10-1. Phylum MOLLUSCA. *A–B*. Early growth stages of *Patella vulgata*, a modern marine gastropod with a simple orthoconic shell. The trochophore stage (*A*) is the earliest larval stage and resembles the larvae of many annelid worms. During the next or veliger stage (*B*) the organism secretes the calcareoconchiolinitic protoconch. The adult shell when completed is a simple orthocone. *C*. Diagram of a hypothetical or archetypal mollusk that might have been ancestral to the five classes of Mollusca. A radula is shown, although it is lacking in the Pelecypoda. The shell is shown as a solid and continuous curved plate, but in the Scaphopoda and in some Gastropoda its apex is perforated. (*By Turner; adapted from several authors.*)

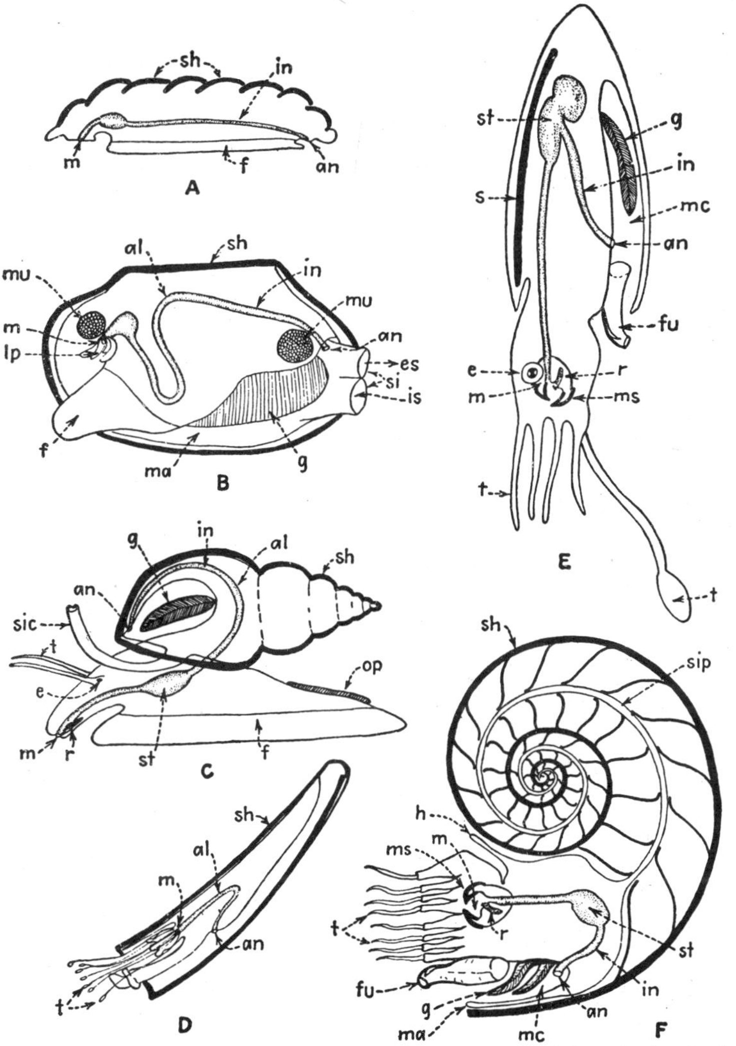

Fig. 10-2. Phylum Mollusca. Diagrams illustrating the relation of the animal to its shell in the five classes of Mollusca. *A*. Amphineura. *B*. Pelecypoda. *C*. Gastropoda. *D*. Scaphopoda. *E–F*. Cephalopoda. (*al* = alimentary canal or digestive tube; *an* = anus; *e* = eye; *es* = exhalant or excurrent siphon; *f* = foot; *fu* = hyponomic funnel; *h* = hood; *in* = intestine; *is* = inhalant or incurrent siphon; *lp* = labial palp; *m* = mouth; *ma* = mantle; *mc* = mantle cavity; *ms* = horny mandibles or jaws; *mu* = muscle; *r* = buccal cavity containing odontophore and radula; *s* = sepion, or internal shell; *sh* = shell; *si* = siphon; *sic* = siphonal canal; *sip* = siphuncle; *st* = stomach; *t* = tentacle.) (*By Turner; adpated from numerous authors.*)

cavity) of the digestive tract. The buccal cavity generally contains horny structures of some sort that aid in seizing and masticating food. These are sharp **beaks** in some forms and platelike chewing structures in others; in addition, most mollusks other than bivalves have a tonguelike rasp, the **radula** (Fig. 10-3). The beaks tend to be situated quite near the front of the mouth, whereas the radula lies in a radular sac directly behind and below the mouth opening and, if unusually long, may be coiled up on the floor of the buccal cavity. The typical radula consists of a tough, flexible muscular band, the **odontophore,** beset with many microscopic teeth composed of **conchiolin,** a horny shell substance. It can be protruded from the mouth if the animal so desires (Fig. 10-3). The radula has the dual function of cutting up the food upon which the animal feeds and of aiding in rasping activities.

All but a few marine mollusks have the sexes separate, but most land snails are hermaphroditic. The reproductive organs are usually in pairs and lie in the walls of the body cavity. The eggs are generally discharged and fertilized externally in marine forms; in a few cases, however, they are fertilized within the body of the female. Some mollusks prepare egg cases within which the young hatch and live for a short time; others retain the eggs in special pouches in the mantle cavities. Those that make no preparations for the care of the young, as the oysters, produce immense numbers of eggs to assure a new generation.

In all classes except the Cephalopoda, segmentation of the fertilized egg is followed by a trochophore larval stage (Fig. 10-1*A*) and this in turn by a **veliger** larva (Fig. 10-1*B*) which ultimately develops into an adult mollusk. Both trochophore and veliger larvae are generally free-swimming. Cephalopods lack a veliger stage and differ in other ways from typical Mollusca. Their embryology is discussed more fully in later paragraphs.

Most mollusks live only a few years, three or four being a normal lifetime, but an extreme age of 20 years has been reported, and it is possible that some pelecypods, as *Tridacna*, may exceed this by many years. Although a few aquatic mollusks are planktonic or pelagic, by far the majority are free-moving or attached bottom dwellers. The phylum as a whole has attained world-wide distribution, and representatives live from 10,450 m. (34,000+ ft.)[1] below sea level to 5,480 m. (18,000 ft.) above it on the highest mountains. Gastropods have attained the greatest elevation, though pelecypods also live thousands of feet above sea level in favorable lakes and streams.

The Shell

Most mollusks have an external or internal shell of some kind (Fig. 10-2). This always has a framework of conchiolin[2] and generally an external layer of the same substance, but in most shells the framework is an inconspicuous part of the two fundamental calcareous layers. Shell substance, both conchio-

[1] The Danish research vessel *Galathea* is reported to have dredged two bivalves from a depth of more than 34,000 feet (approximately 10,450 m.) in the Mindanao Deep off the Philippine Islands (*New York Times*, July 25, 1951).

[2] A nitrogenous substance constituting the organic basis of most molluscan shells. Its macrostructure can be studied by dissolving the associated shelly material.

lin and the calcareous materials, is secreted largely by cells of the epidermal epithelium of the mantle.

The generalized molluscan shell has three fundamental layers (Figs. 10-19*E*, 10-46*A*): ostracum, hypostracum, and periostracum.

1. The **ostracum** is the middle layer in three-layered shells and the outer of the two calcareous layers. It bears a growth record in the form of growth lines. It constitutes most of the shell and consists of several layers of different structure in some pelecypods, whereas in the coleoid cephalopods it is little developed. It is invariably calcareous and may be composed of calcite or aragonite, a combination of both minerals, calcified conchiolin, or calcareous material with interlaminated conchiolin.
2. The **hypostracum** is the inner calcareous shell layer, the laminae of which lie at an angle to those of the ostracum. In the gastropods, for example, the hypostracum is secreted by the general epithelium of the mantle, whereas the ostracum is secreted by certain gland cells on the supramarginal ridge of the mantle.[1]
3. The **periostracum,** which constitutes the outer layer in most living molluscan shells, is a rather thin cuticular layer of conchiolin. It may become thickened and calcified as suggested by Naef (1922), who homologized with it the conchiolinitic and calcareous rostra of squid shells.

As would be expected in a group as old and diversified as the Mollusca, shell structure and composition show wide variation, so that the three fundamental shell layers are likely to vary considerably from the simple description given above.

The shells of marine mollusks are composed of both aragonite and calcite, but those of fresh-water and terrestrial habitats seem generally to be of aragonite. The beautiful color patterns of tropical shells are developed in either or both calcareous layers, and may also be present in the periostracum.

No general statements can adequately characterize molluscan shells, because of the great variation they display in size, architecture, and sculpture in the different classes, nor can one generalize about the relations of soft parts to the shelly hard parts. In the Amphineura, for example, the animal has an elliptical outline, with a large flat ventral foot and a dorsal shelly cover of eight overlapping calcareous plates (Fig. 10-2*A*). The Scaphopoda are dorsoventrally elongated mollusks with the soft parts encased in a slightly curved, slowly expanding conical shell open at both ends (Fig. 10-2*D*). Gastropoda generally have a prominent flat foot, in many forms considerably larger than the shell, a distinct head with tentacles and eyes, and a moundlike body that is protected by a simple conical shell into which the animal can withdraw completely (Fig. 10-2*C*). The simplest gastropod shell is a flat cone; more complex shells result from coiling of a slowly expanding cone, and it is interesting to note that in a spiral shell coiling may be either clockwise or counterclockwise. Most Pelecypoda are compressed laterally, have a small anteroventral foot, lack a head, and are enclosed within two lateral dorsally hinged valves, one right and one left. (Figs. 10-2*B*, 10-9). Cephalopoda have a large head, which bears conspicuous lateral eyes and a circlet of fleshy arms sur-

[1] Newell (1937, p. 16) uses the term hypostracum in a restricted sense for "a filmlike calcareous secretion of the muscles, deposited between the muscle and shell proper, coextensive with the muscle impressions." (Fig. 10-19*C*.)

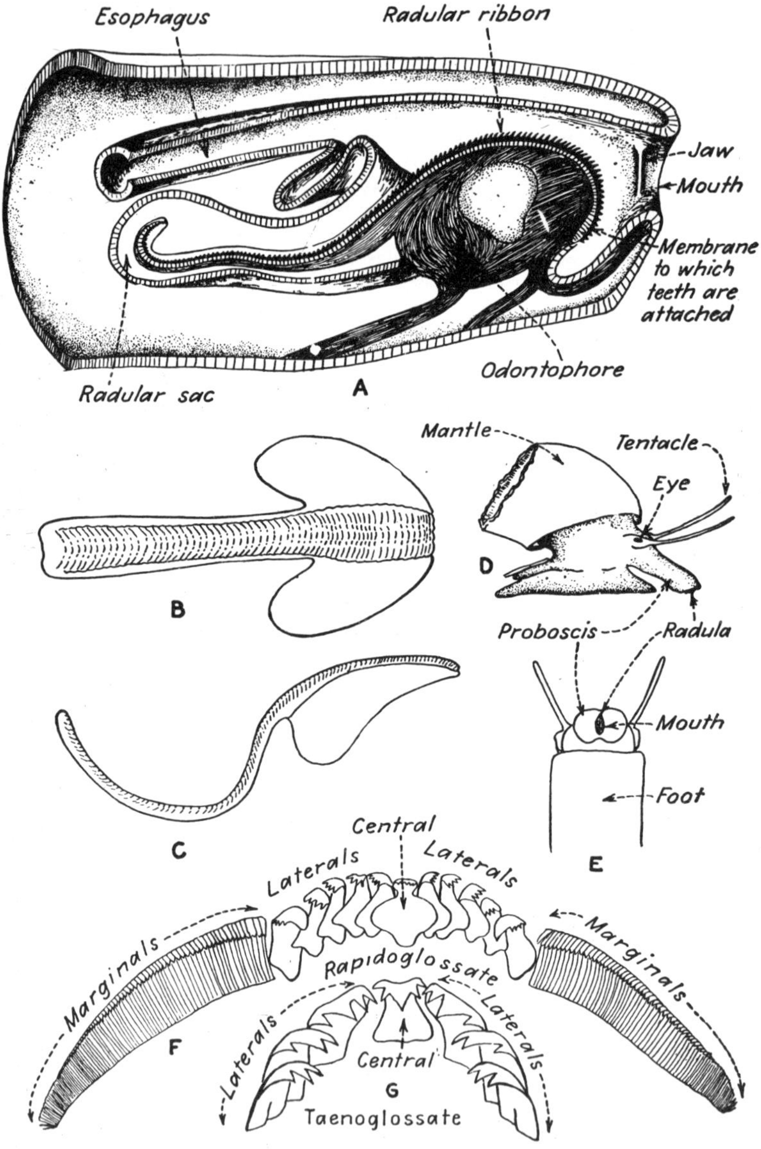

FIG. 10-3. Class GASTROPODA. Radulae. (*See opposite page for detailed dercriptions.*)

rounding the mouth, and a bulbous or elongate body. The body lacks a hard external shell in all living forms except *Nautilus* (Figs. 10-2*F*, 10-55) and *Argonauta* (Fig. 10-83), has an internal vestigial penlike or coiled shell in most forms, and lacks any shell whatsoever in a few genera.

CLASSIFICATION

The Phylum Mollusca is generally divided into the following five classes, which are based primarily on the nature of the foot and certain other soft parts:

Class 1. Amphineura. Chitons, or "coat-of-mail" shells. The fossil record is sparse. *Cambrian to Recent.*

Class 2. Scaphopoda. Tooth shells or tusk shells. Fossils are uncommon in rocks older than Mesozoic.

Class 3. Pelecypoda. Oysters, clams, mussels, scallops, etc. The fossil record, which extends from Lower Ordovician to Recent, includes many genera and species.

Class 4. Gastropoda. Snails, slugs, whelks, limpets, etc. Fossils are abundant and the record extends from Cambrian to Recent.

Class 5. Cephalopoda. *Nautilus;* squids, octupuses, etc. Fossil cephalopods are common in earlier Paleozoic rocks, abundant in middle and later Paleozoic rocks, and exceedingly abundant in Mesozoic rocks. In Tertiary and younger rocks, by contrast, few genera are represented.

Subclasses, orders, and lower taxonomic divisions are based on the following characteristics:

1. Number and nature of breathing organs.
2. Structure and type of genitalia.
3. Nature of nervous system.
4. Structure and type of radulae, if present.
5. Number and nature of shell muscles.
6. Shell architecture, including general form of shell, hinge dentition, apertural armature, structures for muscle attachment, and certain internal structures.

Of the five classes listed above, the Amphineura and Scaphopoda are of least importance as to number, variety, fossil record, and stratigraphic value. Pelecypods are numerous at the present time and seem to have been a fairly important molluscan group since the Early Paleozoic. The Gastropoda now seem to be near the peak of their greatest development, both in numbers and variety, though they have been an important component of bottom faunas since the Early Paleozoic. Although of little present importance, the shell-bearing Cephalopoda were extremely abundant and varied at certain times during the Paleozoic (Ord.–Dev.) and again during the Mesozoic, and their shells are among the most reliable of index fossils. Considered as a group, the

FIG. 10-3. *A.* Median longitudinal diagrammatic section through the proboscis of a prosobranch showing jaw, radula, and associated musculature. *B–C.* Top and side views of radular ribbon on *Oncomelania*, the notorious intermediate snail host of schistosomiasis; anterior end to right. *D–E.* Shell and animal of *Oncomelania quadrasi: D*, side view, showing radula being applied to a plant fragment; *E*, ventral view of head and portion of foot. *F.* A rapidoglossate radula from *Trochus cinerarius*. *G.* A taenoglossate radula. (*By Turner; A adapted from Lang*, 1900; *B–E adapted from Abbott*, 1945; *F–G adapted from Simroth in Bronn*, 1896–1907.)

Mollusca constitute one of the most important of the invertebrate phyla for the paleontologist because of the long and extensive fossil record.

Fossil Mollusca, particularly the Gastropoda and Cephalopoda, have been much studied because of the evidence for evolution recorded in the structures of the shells. Of unusual interest in this respect are the ammonite cephalopods. The shells of these mollusks, if broken back progressively from aperture to embryonic shell, reveal step by step, but in reverse, the successive changes that took place in the features of the shell as it grew from a tiny beginning to its adult form. The embryonic shells, **nuclear whorls** or **protoconchs,** of Gastropoda have also been studied extensively for the light they throw on the evolution of that class.

Living Mollusca have been investigated far more intensively than fossil forms because of their economic importance and because they are an ideal group for morphologic and ecologic study. The literature of the phylum, as a consequence, is enormous, and the following discussion, therefore, is merely a brief summary of a few of the more paleontologically important aspects of the group.

Class Amphineura[1]

The Class Amphineura has two orders:

Order 1. Polyplacophora. Represented by the familiar chitons and characterized by a dorsal shield of eight shelly plates.

Order 2. Aplacophora. A group of small wormlike mollusks that lack any evidence of shell except tiny calcareous spicules in the mantle.

The chitons have left a few scattered fossils, the oldest in Cambrian rocks, but the Aplacophora seem to lack a fossil record altogether.

The amphineurans are generally considered to be the most primitive of all Mollusca, although they do not appear as fossils so early as some of the more advanced classes.

Order 1. Polyplacophora (Fig. 10-4). The chiton has an elongate elliptical body that bears on its convex dorsal surface a flexible shield of eight overlapping calcareous **valves,** or **plates.** These plates are held together and are partly or entirely covered by a thick fleshy part of the mantle, the **girdle,** which is studded with bristles, spines, or scales (Fig. 10-5). The mantle covers the dorsal and lateral surfaces of the body, whereas a prominent flat foot constitutes most of the ventral surface. Beneath and behind the front margin of the girdle is the small head, which contains the mouth but lacks eyes or tentacles. The buccal cavity has a long radula with many transverse rows of tiny teeth. The simple alimentary tract consists of the mouth, a large stomach and a long slender simply looped intestine that terminates in a posterior anus (Fig. 10-2*A*). The simple nervous system consists of a circumesophageal ring to which are connected two pairs of ventral longitudinal nerve cords. The gills lie in the **pallial groove**—the space between the foot and mantle—and number 6 to 80 depending on the species.

[1] Amphineura—Gr. *amphi*, on both sides, + *neuron*, sinew; referring to the paired organs which lie one on each side of the band connecting the shelly plates of the dorsal shield.

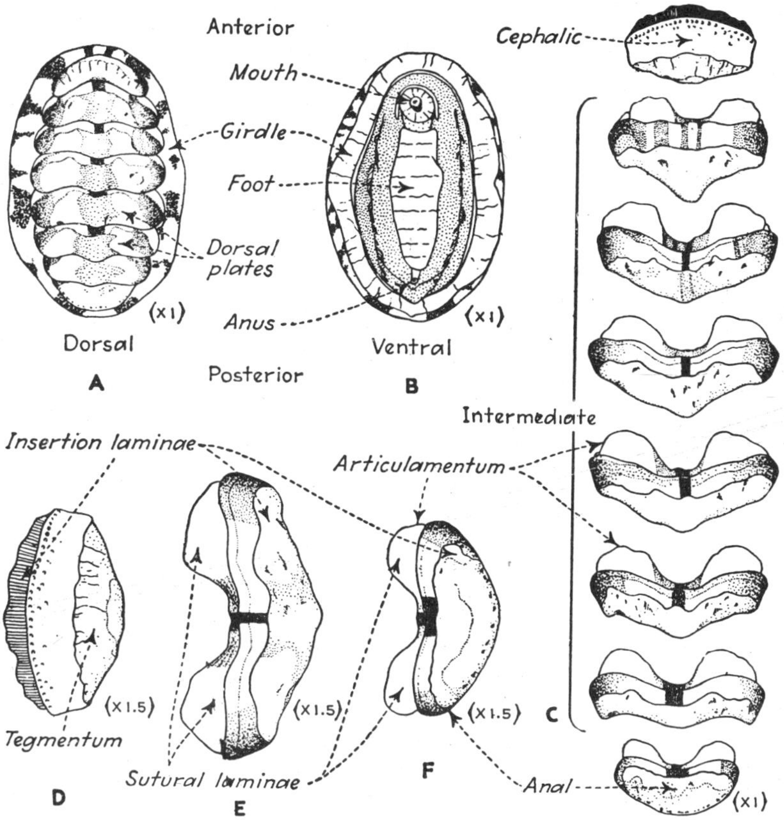

FIG. 10-4. Class AMPHINEURA—Order Polyplacophora. A common living polyplacophoran, *Chiton*. *A*. Dorsal view showing the eight imbricating plates, or valves, embedded in the girdle. *B*. Ventral view. *C*. The eight valves separated to show their different characteristics. *D*. The cephalic valve somewhat enlarged to show the prominent insertion lamina which is concealed in *E* and *F*. *E*. An intermediate, or median, valve with conspicuous sutural laminae. *F*. The anal valve, slightly enlarged.

Sexes are separate, and fertilization of eggs is external. The eggs are discharged individually or as long strings and may number as many as 200,000 in some females.

The eight valves of the chiton shell are arranged in a single continuous series, with the anterior, or **cephalic,** valve and the posterior, or **anal,** valve always conspicuously different from one another and from the six **intermediate,** or **median,** valves that are essentially alike (Fig. 10-4). Each valve has two layers—an upper, the **tegmentum,** consisting largely of conchiolin impregnated with calcium carbonate, more or less porous, and with the surface sculptured: and an under nonporous layer, the **articulamentum,** which is entirely calcareous. The tegmentum has minute pores of two sizes that are

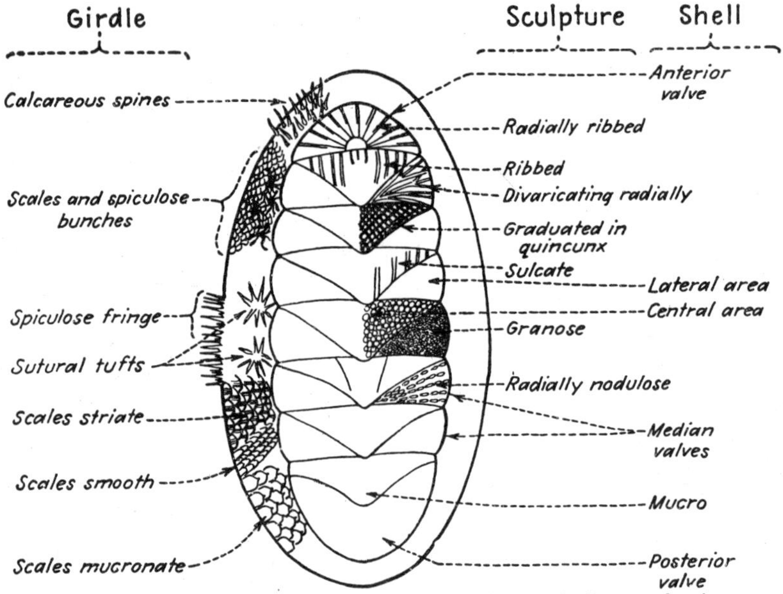

FIG. 10-5. Class AMPHINEURA—Order Polyplacophora. Schematic diagram showing types of *Chiton* sculpture and variations in nature of girdle. (*By Turner; slightly modified after Cotton and Godfrey*, 1940.)

connected by tiny canals, and the canals themselves contain nerve fibers, the tips of which extend into the pores and act as "eyes" on the outer surface of the valve. The articulamentum is generally larger than the tegmentum, in which case some of it is covered by the thick fleshy girdle, which is part of the mantle, or by parts of adjacent valves (Fig. 10-4). Extensions of the articulamentum on the anterior edge of the cephalic valve, on the posterior margin of the anal valve, and on the posterolateral edges of the intermediate valves are designated **insertion laminae.** These are embedded in and concealed by the girdle. Anterior extensions of the articulamentum on the intermediate and anal valves are termed **sutural laminae.** These are covered by the valve in front. Because of the way the valves overlap and are jointed to one another, the entire dorsal shield is flexible and the animal can roll up like a "pill bug."

Chitons are exclusively marine and are represented by about 600 living species, among which are forms ranging from ½ in. (12 mm.) to as much as 10 in. (250 mm.) in length. They live mainly on rocky bottoms, mostly in relatively shallow water, but living specimens have been taken from depths as great as 4,200 m. (13,800 ft.). They cling to rocks and other solid objects by means of the foot or creep slowly over the bottom. If dislodged or disturbed the animal slowly rolls up into a ball and remains so for some time before slowly unrolling.

The fossil record of the Polyplacophora is a meager one. No complete assemblage of valves has been described, but isolated valves of about 100

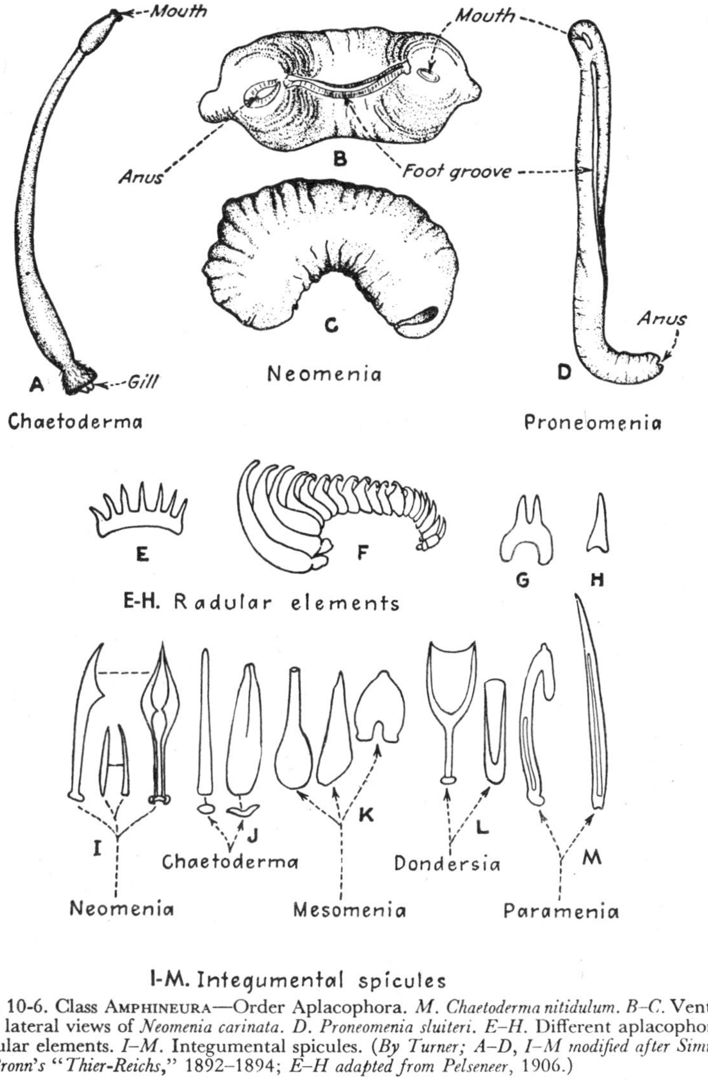

FIG. 10-6. Class AMPHINEURA—Order Aplacophora. *M. Chaetoderma nitidulum. B–C.* Ventral and lateral views of *Neomenia carinata. D. Proneomenia sluiteri. E–H.* Different aplacophoran radular elements. *I–M.* Integumental spicules. (*By Turner; A–D, I–M modified after Simroth in Bronn's "Thier-Reichs,"* 1892–1894; *E–H adapted from Pelseneer,* 1906.)

species have been reported from strata ranging in age from Cambrian to Recent. The earliest of these valves are present in the Upper Cambrian Trempealeau beds of Wisconsin and Eminence limestone of Missouri.[1] Paleozoic valves differ from later ones in lacking well-developed insertion and sutural laminae.

[1] Personal communication from J. Brookes Knight, 1949.

Order 2. Aplacophora (Fig. 10-6). Aplacophorans are small wormlike animals in which the mantle covers the entire body. The foot is either lacking altogether or is a ciliated ridge in a ventral groove extending from mouth to anus (Fig. 10-6*B*,*D*). No shell is developed; but the mantle contains tiny calcareous spicules (Fig. 10-6*I*–*M*). Aplacophorans are seldom found in shallow water and never alive in the littoral zone. A few burrow in mud, but most forms live on or within corals and hydroids in moderately deep water (2,280 m., or 7,500 ft.). Less than a dozen genera and only about 50 species of living aplacophorans have been described; no fossil remains at all seem to have been reported.

Class Scaphopoda[1]

Scaphopods are marine mollusks with shell and mantle together forming a slightly curved and gradually expanding tube open at both ends (Figs. 10-2*D*, 10-7). The concave side of the animal is dorsal, the smaller end posterior, and the larger anterior. Muscles near the posterior end attach the body to the shell. At the opposite or anterior end is a long conical or spadelike foot that can be protruded beyond the shell margin. The mouth is situated on a snoutlike projection of the pharynx at the base of the foot and is flanked by small bunches of ciliated and contractile tentaclelike filaments, the **captacula,** which are both sensory and prehensile and which help to catch the microscopic organisms used as food. No head is developed, but the buccal mass of the alimentary tract has a radula. Gills are lacking, and the mantle serves for respiration.

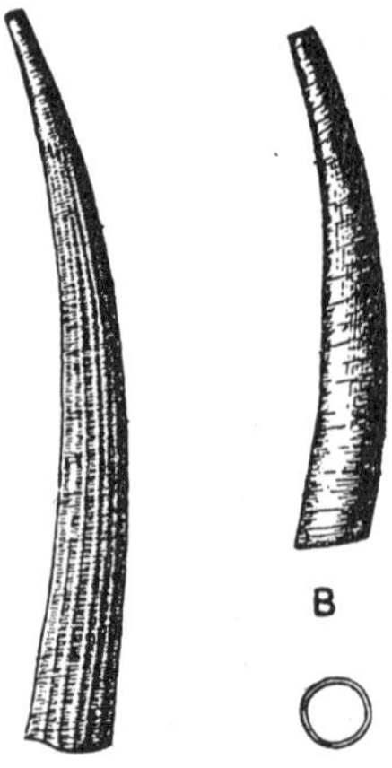

Fig. 10-7. Class Scaphopoda. Two Miocene species of *Dentalium*. *A* shows a longitudinally ribbed shell; *B* and *C* illustrate a smooth shell. (*After Dall*, 1904.)

They live in marine waters at nearly all depths from shoals to bottoms as deep as 4,570 m. (15,000 ft.). In life the animal is partly buried in the mud or sand, with the shell obliquely oriented and with the smaller or posterior end above the bottom, as shown in Fig. 10-2*D*.

The scaphopod shell consists of aragonite and grows by successive additions to the larger end. The posterior end may be crenulated or fissured, and commonly some of this part of the shell is lost through breakage, wear, or solution. The largest of the shells rarely attain a length of 15 cm. (6 in.);[2]

[1] Scaphopoda—Gr. *scaphe*, a boat, + *pous*, *podos*, foot.

[2] The largest living scaphopods usually attain a maximum length of about 133 mm. (5¼ in.) and a corresponding maximum diameter of about 15 mm. (⅗ in.) (Pilsbry and Sharp, 1897–1898, pp. 65, 80), but Miller (1949) has recently reported a giant fossil scaphopod from the Pennsylvanian of Texas that is almost double these dimensions—*i.e.*, somewhat more than 252 mm. (10 in.) long and about 30 to 35 mm. (1⅖ in.) in apertural diameter—and also cites a few other shells that may even have exceeded these dimensions.

most are 25 to 50 mm. (1 to 2 in.) long. The outer surface is smooth in some forms and transversely or longitudinally ribbed in others.

The scaphopod shell differs from the shells of most gastropods in being open at both ends; from those of all shell-bearing cephalopods in being unchambered, as well as open at both ends; from the calcareous tubes of tubicolar annelids in having three instead of two shell layers; and from the conularids (Class Scyphozoa?) in being ribbed or fissured longitudinally.

About 200 living species of scaphopods are known, and more than 300 fossil species have been described. Most of these, both living and fossil, have been referred to the widespread living *Dentalium* (Fig. 10-7), the oldest known scaphopod genus, which first appeared in the Ordovician. Fossil scaphopods are uncommon in rocks older than Cretaceous.

Class Pelecypoda[1] (Lamellibranchia)[2]

The Pelecypoda, or Lamellibranchia, constitute the second largest group of mollusks and are particularly interesting because representatives of the class show such a wide range of adaptation. They are familiar to the layman because of their economic importance. Oysters, clams, and scallops are an important source of food; certain pelecypods produce pearls of gem quality and are cultivated for that purpose, especially in Japan; and the shells of fresh-water bivalves are cut into pearl buttons. On the negative side, the boring depredations of the shipworms (*Teredo* and *Bankia*) have long been a serious problem in maintaining seaside and seagoing wooden structures (wharves, ships, etc.).

The typical pelecypod is bilaterally symmetrical in both soft and hard parts and can generally be distinguished from all other Mollusca by its bivalve shell,[3] which is calcareous and has two valves that are alike (except along the hinge line) but reversed one to the other. The organism itself is laterally compressed, and has a prominent ventral foot. When the shell is closed all soft parts are usually completely enclosed within the two valves.

Throughout their long geologic history, beginning almost certainly in the Cambrian, most pelecypods have been free-moving bottom dwellers. Today many live partly or completely buried in the bottom mud or sand, others bore into wood and rock, a few attach their shells loosely by a tuft of horny threads (byssus), and a few others become permanently cemented to objects on the bottom (Fig. 10-8). It is known from the fossil record that ancient pelecypods were similarly adapted. The more common adaptations of modern pelecypods are shown in Fig. 10-8.

Most marine bivalves live in shallow waters from the intertidal zone to a

[1] Pelecypoda—Gr. *pelekys*, a hatchet, + *pous*, *podos*, foot; referring to the resemblance of the foot to the blade of an ax or hatchet.

[2] Lamellibranchia—L. *lamella*, a small plate or leaf, + *branchia*, gills; referring to the leaflike nature of the gills.

[3] It is interesting to note that the embryonic shell of the Scaphopoda consists of two calcareous laminae, but these subsequently unite to form the typical tusk-shaped tubular shell.

few hundred fathoms of depths, but many deep-water forms are also known. Certain pelecypods can also live in brackish water, and the members of three families (Unionidae, Mutelidae, and Sphaeridae) flourish in lakes, streams, and other bodies of fresh water.

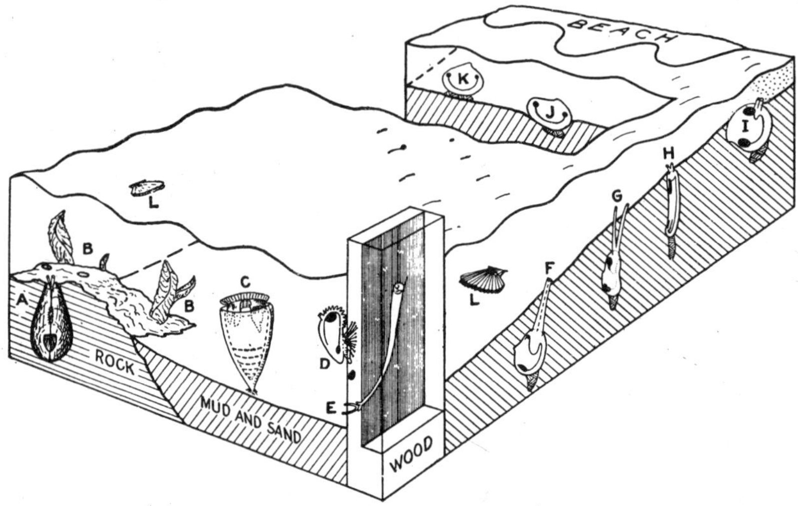

FIG. 10-8. Class PELECYPODA. Diagrams (not all to same scale) showing adaptations among pelecypods. Only one valve is shown in *D* and *F–K*, and the muscles are indicated by black spots. The foot is shown by complex crosshatching. A few of these forms may be exposed at low tide. *A. Pholas*, the common rock borer. *B. Ostrea*, the familiar edible oyster. *C. Hippurites*, an unusual extinct pelecypod in which one valve, highly modified to a conical shape, was attached to the bottom whereas the other, also much modified, served as a lid. *D. Mytilus*, the black sea mussel, attached to the surface of a wooden pier by a byssus. *E. Teredo*, the shipworm, which bores into wooden piers and similar structures. The pier is sectioned vertically to show the path of the boring. *F. Mya*, the mud clam, a burrowing form with a long siphon. *G. Tagelus*, a burrowing form with a long double siphon. *H. Ensis*, the razor clam, with a short double siphon. *I. Venus*, the familiar quahog, almost completely buried in bottom sediment. *J. Crassatellites*, nestling in a depression in the sand. *K. Nucula*, a form which moves over the surface of the sand. *L. Pecten*, shown swimming (at left) and lying on the bottom (at right). (*Adapted from Berry*, 1929.)

The Animal

The soft parts of the pelecypod are enclosed within a shell composed of two hinged calcareous valves (hence the common term "bivalve"), to the inner surfaces of which the soft body is attached (Fig. 10-9*A*). One or two transverse **adductor muscles** pull the valves together, whereas the **ligament** and **resilium** force them apart. The body proper consists of four parts: (1) the visceral mass; (2) the foot; (3) the mantle lobes; and (4) the gills.

Visceral Mass. The visceral mass lies chiefly in the dorsal part of the body and is attached dorsally to the inner surfaces of the two valves. The alimentary tract consists of a **mouth, esophagus, stomach** (into which two digestive glands discharge), a coiled **intestine,** and an **anus.** The mouth is anterior and lacks the radula that is present in all other classes of Mollusca. The anus is

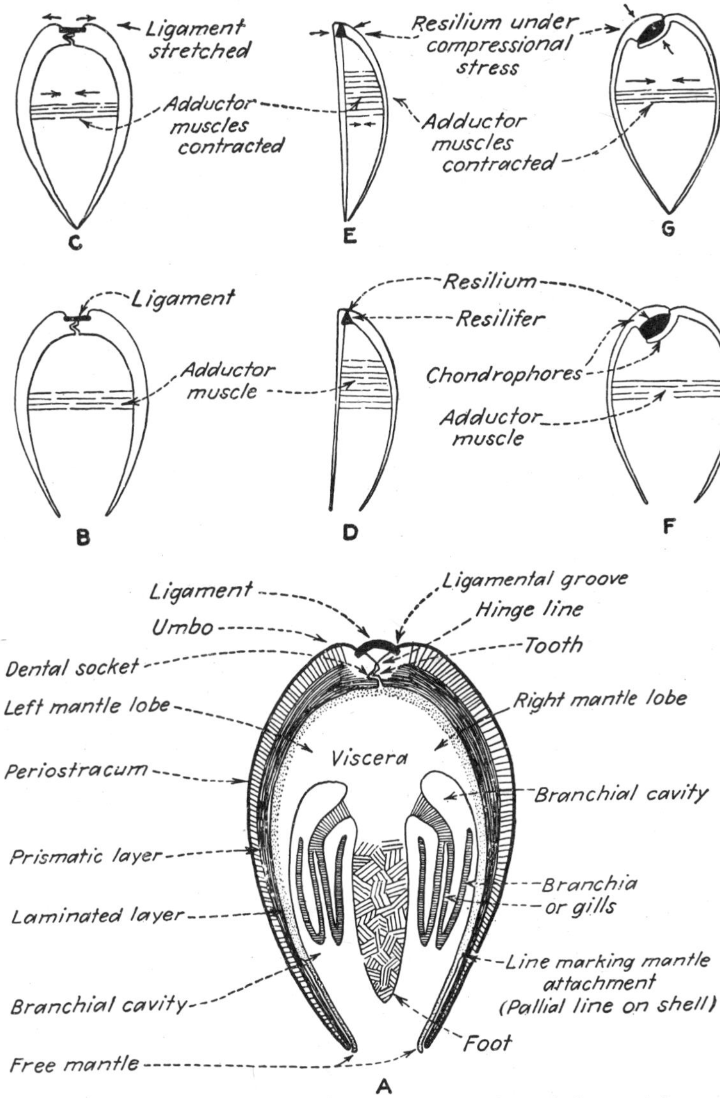

FIG. 10-9. Class PELECYPODA. Diagrams illustrating internal morphology and the opening and closing of shells. *A*. Diagrammatic median section through animal and shell, at right angles to the plane of symmetry, showing the general nature of both and the relations of the different soft parts to the shell. *B–C*. An open shell (*e.g.*, *Venus*) with ligament and adductor muscles relaxed, and the same shell closed with both structures under stress. *D–E*. An open and closed shell, respectively, of *Pecten*. When the shell is closed the resilium is compressed in a nutcracker action. *F–G*. An open and closed shell, respectively, of *Mya*, showing a modified resilium and the enclosing chondrophores. In *B–F* the adductor muscles are indicated by broken lines, as they do not lie in the plane of the section except in *Pecten* (*D–E*).

posterior and opens directly to the outside or in more advanced forms near the excurrent siphon (see later discussion). The **circulatory system** consists of a heart and a system of tubes through which blood is pumped to the different organs. The **nervous system** includes three pairs of ganglia with interconnecting fibers, light-sensitive organs on the mantle, and certain other sensory structures. **Kidneys** (**nephridia**) discharge nitrogenous wastes, and **gonads** produce the sex products. In many species the sexes are separate.

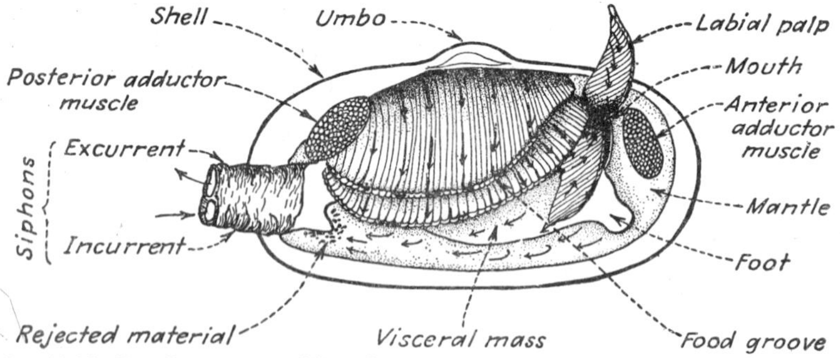

FIG. 10-10. Class PELECYPODA. Ciliary feeding and respiration in bivalves as illustrated by *Mya arenaria*, sketched from life. The right valve and right mantle lobe are removed, and one labial palp is turned upward to show inner surface and to expose mouth. The arrows indicate the effective direction of cilia-activated currents of water. Water enters the mantle cavity through the incurrent siphon, moves dorsally, and is strained through the gills, after which it passes downward over the outer surface of the gills. Strings of mucus containing minute particles of food and sediment are carried ventrally across the surface to the marginal food groove, thence along the food groove to the labial palps and finally into the mouth. Large or heavy particles are rejected all along the food grooves and from the margins of the labial palps. These are carried posteriorly by the cilia lining the mantle cavity and are collected at the base of the incurrent siphon. Periodically the animal discharges these particles through the incurrent siphon by temporarily reversing the direction of flow. Gas exchange takes place as the water passes over and through the gill lamellae. Deoxygenated water containing normal waste products is eliminated through the excurrent siphon. Interested readers are referred for further details to Riderwood's (1903) discussion. (*By Turner.*)

Foot. The foot is an anteroventral[1] extension of the visceral mass and is a muscular organ that can be extended beyond the shell margin, when the valves are apart, and used for locomotion and burrowing. It lies in the median plane, which is also the plane of bilateral symmetry, and divides the **mantle cavity** (*i.e.*, the space enclosed within the mantle lobes) into a right and a left part. It is typically wedge-shaped (whence the name of the class; see footnote on page 363), but in some species it is cylindrical; in sedentary forms it is essentially functionless. In a few species it has a gland that secretes horny threads, collectively designated the **byssus,** which are used for temporary or permanent attachment (Figs. 10-8*D*, 10-13*E*–*F*). Many species use byssal threads for anchorage in their younger growth stages, abandoning them in early maturity, and *Anomia* becomes permanently attached by a byssus impregnated with calcite.

[1] In *Tridacna*, which lives with the dorsal part of the shell on the susbtratum and the valves agape upward, the foot is bent dorsally to aid in attachment.

Mantle Lobes. The mantle hangs freely from the visceral mass and consists of two thin fleshy sheets, the mantle lobes, each adhering to the inner surface of a valve (Fig. 10-9*A*). The marginal edges of the mantle are free of the shell, and the line along which they become free and which marks the marginal limit of mantle attachment is indicated on the shell by a prominent groove, the **pallial line,** paralleling the shell margin. The mantle edges are muscular and can be brought together to enclose the mantle cavity within.

The free edges of the mantle secrete the two outermost layers of the three-layered shell, whereas the outer surfaces of the mantle lobes deposit the innermost layer.

The living pelecypod maintains an incoming current[1] of water to bring oxygen to the gills and food to the mouth and an outgoing current to carry away rejected material and waste products[2] (Fig. 10-10). These currents enter and leave the mantle cavity by passing between the posterior margins of the right and left mantle lobes, and they are kept separate so that the incoming food- and oxygen-bearing waters are not fouled by the outgoing waste. The posterior edges of the mantle lobes have been modified in several different ways to keep the currents separated, as shown in Figs. 10-11 and 10-12.

In the more primitive pelecypods the posterior edges of the mantle lobes are so folded as to produce an upper (dorsal) exhalant and a lower (ventral) inhalant channel. The opposite edges of the lobes touch but are not united (Fig. 10-11*B*). A somewhat more advanced modification is present in many fresh-water clams, in which the posterior edges of the mantle lobes are united at one place, above which the mantle edges are drawn out slightly to form an **exhalant siphon** (Fig. 10-11*C*). Below, the edges remain separate and are merely held together to restrict the entrance of the incoming current. In the next stage of modification the edges of the mantle lobes unite at two places, so as to form two restricted openings, and extend posteriorly to form an upper exhalant and a lower **inhalant siphon** (Fig. 10-11*D*). These two siphons, either separate or united to one another along one side, may grow to considerable lengths, particularly in burrowing forms that commonly have siphons two or three times as long as the shell (Figs. 10-11, 10-12). In some siphonate bivalves the edges of the mantle lobes in front of the siphons are united completely except for a small ventral slit, the **pedal gape,** through which the foot protrudes (Fig. 10-11*E*). Such a mantle is described as **closed.**

Siphons in most forms can be partly or entirely drawn within the shell by retractor muscles. These muscles are attached to the inner surfaces of the valves directly below the posterior adductor muscle, and they represent enlargement of the muscles by which the mantle is attached along the pallial line.

In the folding of the mantle lobes to produce extensible siphons, the mantle

[1] Circulation of water is produced by the action of cilia that cover the surfaces within the mantle cavity (Fig. 10-10).

[2] As an example of the amount of water circulated by a pelecypod, it has been reported that an oyster filters a barrel every day (*Science News Letter*, Jan. 11, 1936, p. 30).

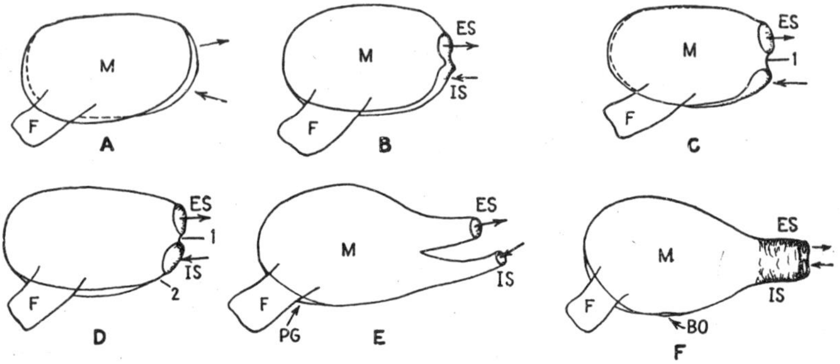

Fig. 10-11. Class Pelecypoda. Development of siphons. Diagrams showing modification of mantle in production of siphons and pedal gape (shell removed). *A*. Mantle completely open. *B*. Rudiments of siphons, with mantle still completely open. *C*. Mantle joined at one place. (1). *D*. Mantle joined at two places (1 and 2), with formation of complete siphonal openings. *E*. Mantle closed posteroventrally and tubular siphons well formed. *F*. Tubular siphons adherent and encased in muscular sheath with mantle almost completely closed ventrally, except for a small opening (*BO*) through which the byssus may protrude. (*BO* = byssal opening; *ES* = excurrent or exhalant siphon; *M* = mantle; *IS* = incurrent or inhalant siphon; *F* = foot; *PG* = pedal gape.) (*By Turner; adapted from Lang*, 1900.)

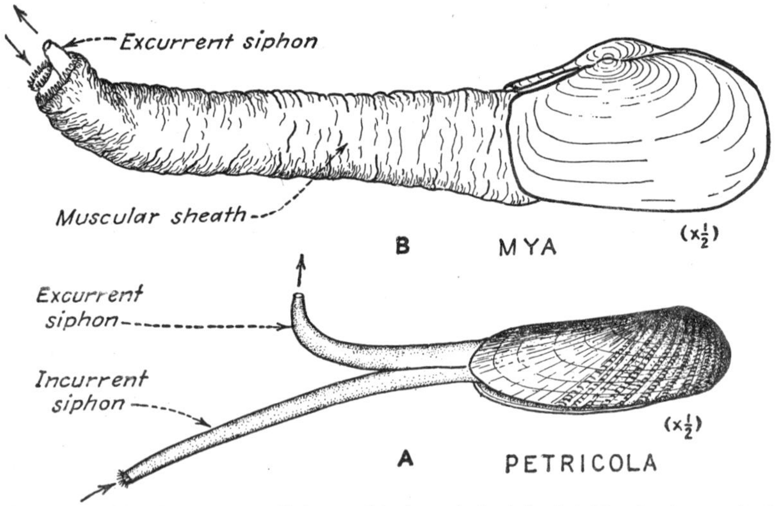

Fig. 10-12. Class Pelecypoda. Siphonate bivalves. *A*. *Petricola pholadiformis*, characterized by separate siphons. *B*. *Mya truncata*, with two siphons encased in a strongly muscular sheath. The siphons are extensible in both species. *Mya* is a mud burrower; *Petricola* burrows in mud, clay, and peat. (*By Turner*.)

itself is pulled away from the shell to some extent so that the pallial line has a rounded reentrant marking the position of the siphons. This reentrant is the **pallial sinus** (Fig. 10-13*B*) and, if present on fossil shells, indicates that the **animal had extensible siphons.** Furthermore, since extensible siphons are best

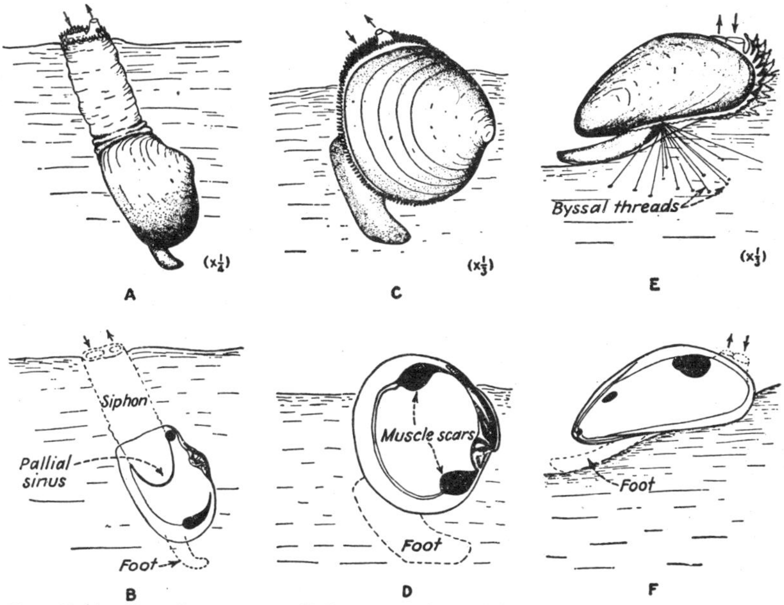

FIG. 10-13. Class PELECYPODA. Siphonate and nonsiphonate pelecypods. *A–B*. *Mya truncata*, a typical burrowing form with the two siphons encased in a tubular sheath: *A*, animal in living position; *B*, interior of left valve and trace of siphonal sheath to show conspicuous pallial sinus. *C–D*. *Arctica* (*Cyprina*) *islandica*, a siphonate form with a nonextensible siphon, hence without a pallial sinus: *C*, animal in living position; *D*, interior of left valve. *E–F*. *Mytilus edulis*, a byssal-bearing form that lives on the surface of the bottom, attached temporarily by means of the byssal strands: *E*, animal in living position; *F*, interior of right valve. (*By Turner; C–D adapted from Mobius*, 1872.)

developed in burrowing bivalves, fossil shells with pallial sinuses may be assumed to have belonged to such forms. Pelecypods with nonextensible siphons do not burrow deeply, and their shells lack a pallial sinus.

Gills. Most modern pelecypods have two pairs of gills (also termed **branchia,** or **ctenidia**), but the number may have been different in extinct species. The gills arise from ridges on each side of the body and lie in the two **gill cavities,** which are the spaces between the foot and the two mantle lobes (Fig. 10-9*A*). They exhibit considerable variety in morphology, structure, and arrangement and are of great importance for classification of living species (Fig. 10-14).

The most primitive gill type consists of free, short, wide, and flat filaments (Fig. 10-14*A*) and is designated **aspidobranchiate,** or **protobranchiate.** A second type, referred to as **filibranchiate,** is composed of long, slender filaments that are kept from becoming tangled by interlocking patches of cilia or bars of connective tissue (Fig. 10-14*B*). The gills may become plaited so as to exhibit vertical folds, in which case they are described as **pseudolamellibranchiate.** The filaments of the filibranchiate type may become interconnected so as to produce a perforated, leaflike variety designated **eulamellibranchiate** (Fig. 10-14*C*), and this type may be further modified into a perforated, muscular partition that is termed **septibranchiate** (Fig. 10-14*D*).

One of the more common earlier classifications of the Pelecypoda was based on gill structure and morphology and consisted of the following five orders: Protobranchia, Filibranchia, Pseudolamellibranchia, Eulamellibranchia, and Septibranchia. This classification cannot be used by paleontologists, however, because fossil bivalve shells yield no evidence as to the nature of the gills possessed by the animals that produced them.

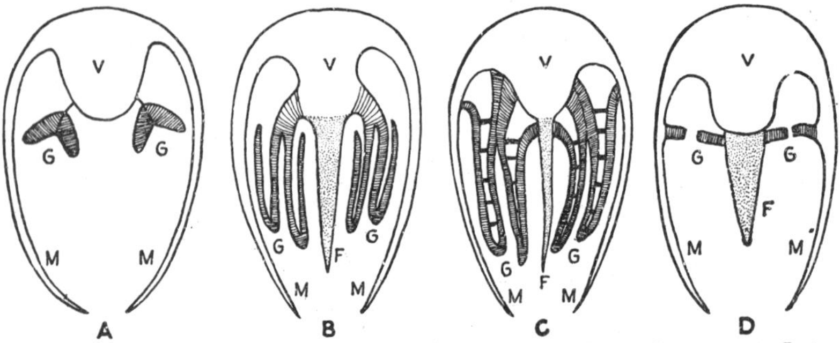

FIG. 10-14. Class PELECYPODA. Morphology of the branchia, shown diagrammatically in transverse section. *A*. **Protobranchiate** type, exemplified by *Nucula*, consisting of two simple double-leaved gills. The foot is not shown, as it lies anterior to the plane of the section. *B*. **Filibranchiate** type, in which each gill has two reflected plates; present in the Arcidae, Mytilidae, etc. If the reflected parts of the gill become completely united and the leaves crumpled, as in the Pectinidae and Ostreidae, the branchia are described as **pseudolamellibranchiate.** *C*. **Eulamellibranchiate** gills; the leaves of each gill plate are interconnected and the adjacent filaments intimately connected. *D*. **Septibranchiate** variety, in which the gills are more or less muscular perforated partitions that divide the pallial cavity into two distinct parts. (*F* = foot; *G* = gill; *M* = mantle; *V* = viscera.) (*By Turner; adapted from Lang*, 1900.)

The Shell

General Considerations. Pelecypod shells consist essentially of two convex calcareous valves that articulate along a dorsal hinge line. The two valves, designated **right** and **left,** respectively, because they are attached to the right and left sides of the animal, are generally equal in size and asymmetrical in outline (Fig. 10-16). In shells cemented to other objects by one valve, the attached valve tends to be the larger, and the smaller acts as a lid (Fig. 10-8*C*). The earliest part of each valve is the pointed **beak,** which usually curves toward the front (anterior) end of the shell. The rounded and elevated part of each valve directly adjacent to the beak is the **umbo.**

The typical pelecypod is bilaterally symmetrical, with the plane of symmetry passing between the valves, but many exceptions are found among attached shells. It is customary in orienting the shell to place the beaks uppermost and pointing forward or away from the observer. In this view, which is along the plane of symmetry, the valve on the right hand is the right valve; that on the left, the left one. The front of the shell. *i.e.,* the way the beaks

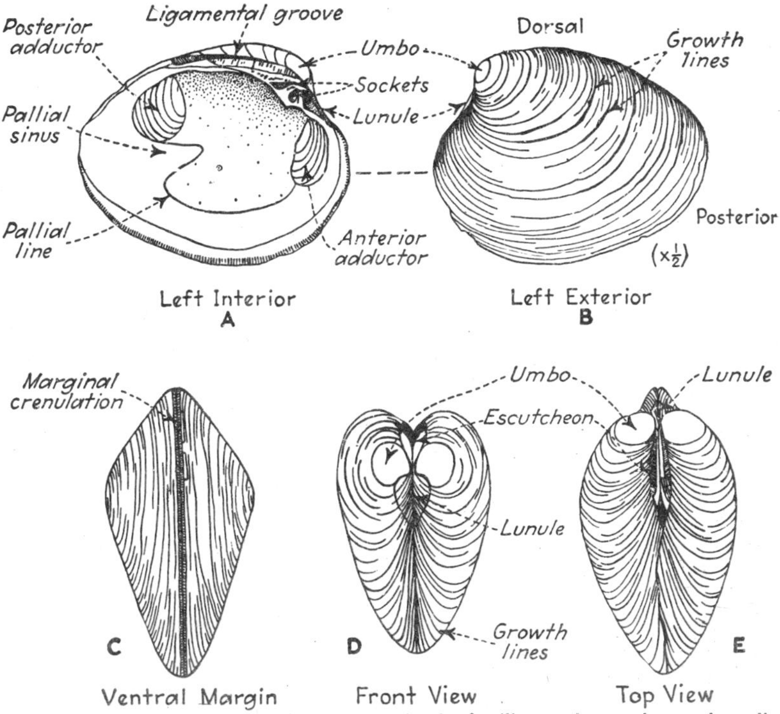

FIG. 10-15. Class PELECYPODA. *Venus mercenaria*, the familiar quahog and a modern dimyarian pelecypod.

TABLE 10-1. COMPARISON OF PELECYPODA AND BRACHIOPODA

Pelecypoda	*Brachiopoda*
1. Valves inequilateral	1. Valves equilateral
2. Generally equivalved	2. Inequivalved
3. Valves, right and left	3. Valves, pedicle and brachial
4. Plane of symmetry between valves and beaks including the hinge line	4. Plane of symmetry across valves and hinge line through the beaks
5. No pedicle; hence no pedicle opening	5. Pedicle opening present, except in primitive and mature cemented forms
6. Teeth and sockets, if present, in each valve	6. Teeth and sockets in opposite valves, except in the Inarticulata, which lack teeth altogether
7. Shells opened by ligament, resilium, or both; closed by muscles	7. Shells opened and closed by muscles; no ligament present
8. Shells typically three-layered—periostracum, prismatic layer, laminated layer (or periostracum, ostracum, and hypostracum, respectively)	8. Shells three-layered or multilayered—periostracum, laminated layer, prismatic layer; also periostracum and alternating layers of chitin and calcium phosphate

are directed, is anterior and the part nearest the observer, posterior. If a pallial sinus is present, it interrupts the posterior part of the pallial line, and if only a single adductor muscle scar is present, it has posterior position on

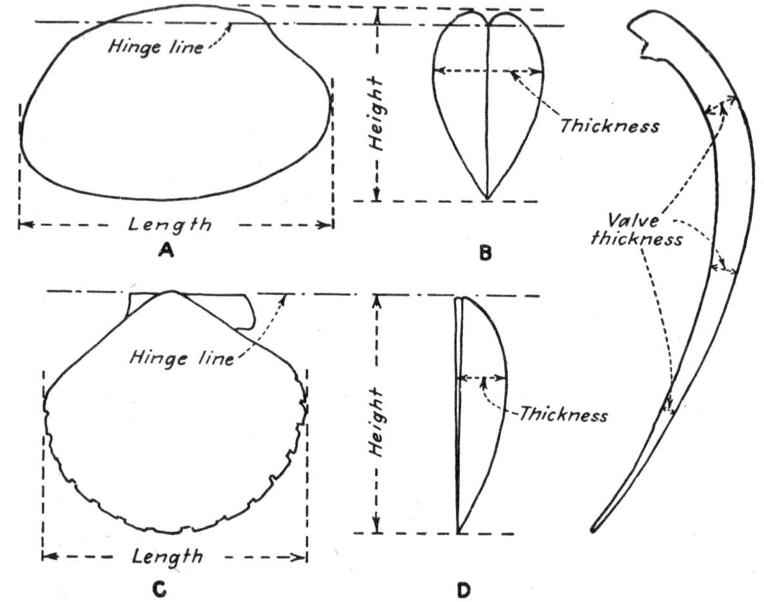

FIG. 10-16. Class PELECYPODA. Diagrams illustrating how dimensions of pelecypod shells and of the valves of such shells are measured.

the valve's inner surface. The upper or hinge-line part of the shell is dorsal; the lower, ventral. In all but a few pelecypods the foot is ventral.[1]

Pelecypods and brachiopods resemble one another in having external, bivalve calcareous shells, though they differ strikingly in most other respects. The more important features that can be used to differentiate the shells of the two groups are compared in Table 10-1 (also see Fig. 9-2).

The **height** of a pelecypod shell is measured along a line in the plane of symmetry from umbo to ventral margin (Fig. 10-16). The **length** is the greatest distance between the anterior and posterior margins measured in the plane of symmetry. The **thickness** is the greatest distance through the shell when it is closed and is measured in a line perpendicular to the plane of symmetry. The thickness of a valve is the distance directly across the valve from inner to outer surface.

Pelecypod shells exhibit great range in size, from minute forms less than 3 mm. long to the giant *Tridacna*, which reaches a meter in length. Although shape varies to some extent, most shells tend to be laterally compressed (biconvex) and somewhat elongated longitudinally, with the dorsal margin straight or gently curved and the ventral rounded.

The typical pelecypod shell in its complete development is composed of

[1] Exceptions to a ventral foot are found in *Tridacna*, in which the foot is dorsal, and in *Pecten*, *Lima*, and a few others, in which it is really internal (*i.e.*, reduced and inside the shell).

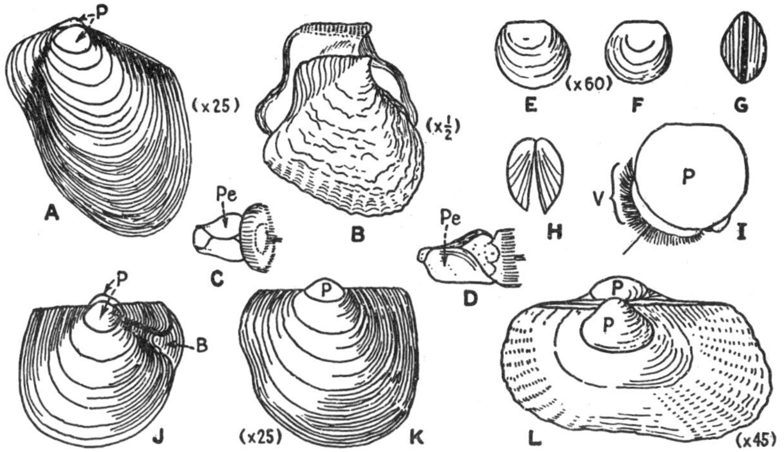

FIG. 10-17. Classes PELECYPODA and SCAPHOPODA. Embryonic and nepionic shells. *A–B. Perna ephippium: A*, young form showing prodissoconch (*P*) and succeeding growth; *B*, adult shell. Note striking difference between nepionic and adult shells. *C–D.* Young specimen of *Dentalium*, a scaphopod, showing two views of the periconch (*Pe*). *E–H.* Four views of the prodissoconch of *Ostrea edulis* from the gill chamber of the parent. Note straight hinge line and symmetrical outline. *I.* Young embryo of *Cardium exiguum* with prodissoconch, which is symmetrical and has a straight hinge line. Note the prominent swimming organ, or velum (*V*), which characterizes the free-swimming larval stage. *J–K.* Young *Pecten irradians*, viewed from right and left sides, respectively, showing prodissoconch (*P*), and subsequent growth. *B* is a byssal notch. *L.* Young nepionic shell of *Arca pexata*, showing prodissoconch (*P*) and subsequent growth. (*All figures adapted from Jackson*, 1890; *C–D are adapted from Lacaze-Duthiers.*)

three distinct layers (Fig. 10-19)—an outer **periostracum** that is composed of conchiolin; a middle **prismatic layer** (the ostracum) consisting of an intergrowth of conchiolin and calcareous prisms; and an inner **laminated layer** (the hypostracum) of closely spaced and alternating lamellae of conchiolin and calcium carbonate.

Shell Development.[1] The first rudiment of the pelecypod shell appears along the middorsal line of the veliger larva in a position corresponding to the hinge part of the adult shell. It consists of two tiny horny valves, one on each side of the line, which are equal, almost symmetrical, and connected along a straight hinge line by an elastic band composed of conchiolin. As the shell grows by successive additions, chiefly along the ventral margin, growth lines develop concentrically about the umbo, which lies near the dorsal margin, and during the same period the hinge line becomes arched. This embryonic shell, which is seldom preserved in fossil bivalves, is the **prodissoconch** (Fig. 10-17).

When the veliger loses its swimming apparatus and settles to the bottom, it becomes a young pelecypod ready to adopt a benthonic mode of life. Thus far, the mantle has secreted the shell; now, however, the shell-secreting areas are extended into the two mantle lobes, the edges of which continue to secrete

[1] Swinnerton, 1947.

the periostracum, meanwhile also adding prisms of calcareous material in a matrix of conchiolin. Soon the entire surface of the mantle begins to add laminae to the inner surface of the prismatic layer, and the shell now has its three fundamental layers—periostracum, outer ostracum (prismatic layer), and inner hypostracum (laminated layer).

Along the hinge line little or no calcareous material is deposited, and as a consequence the conchiolin of the prodissoconch retains its elasticity and functions as a hinge between the valves. As the shell grows, the dorsal margins of the valves become thickened and denticulate, as a result of which the valves fit more perfectly together. The teeth and sockets developed along this margin determine the hinge line and guide the valves as the shell closes, so that their edges fit accurately one against the other. The elastic band of conchiolin (*i.e.*, the ligament) tends to keep the valves apart, but this tendency is counteracted by one or two internal transverse muscles, the **adductors,** that can pull the valves together. It is important to point out that the pelecypod shell is merely a bivalve shield that serves as a protective cover. Because it is not closely intergrown with the soft parts of the body, it tends to remain a relatively simple structure. However, it is quite sensitive to environmental influences, hence likely to show wide variation in form.

Modification in Shape of Shell. Although the majority of pelecypod shells have one valve the mirror image of the other (except for teeth and sockets along the hinge line), many others have dissimilar valves because of adaptation to unusual modes of life. A few pelecypods cement one valve to some object on the bottom and use the other as a lid (Fig. 10-8*B–C*). As a result of such attachment the two valves become different. In many oysters the difference is not extreme, but in the extinct *Exogyra* (Fig. 10-29*A–C*) and *Gryphaea* (Fig. 10-29*I*) the upper valve is distinctly different from the lower and much smaller, and in the extinct rudistids (*e.g.*, *Radiolites* and *Hippurites*), the valves are markedly different, with the lower a coral-like conical structure and the upper a peculiarly toothed lid[1] (Fig. 10-8*C*). Some pelecypods attach by the right valve (*e.g.*, *Spondylus*, *Plicatula*, and the rudistids); others by the left (*e.g.*, *Exogyra*, *Gryphaea*, and *Ostrea*). It is to be noted, however, that upon attachment many bivalves change their usual right-and-left orientation to one that is top-and-bottom, or even a position that is somewhat indifferent (*e.g.*, *Ostrea*). A similar orientation is shown by several free-swimming forms such as *Pecten* and *Lima*. Certain attenuated shells (*e.g.*, *Mytilus* and *Pinna*) are held to the substratum by a bundle of horny threads (the byssus), arising from a gland on the foot, but these have undergone little or no valve differentiation so that the valves are essentially similar and equal.

In addition to the differences between valves in the same shell, many shells with essentially similar valves show wide variation in the shape of the shell as a whole (Fig. 10-18), depending on whether the animal swims, ploughs through bottom sediments, burrows in sand and mud, or bores in wood and rock (see further discussion on page 397). The more common shell shapes as related to mode of life are shown in Fig. 10-8.

[1] This modification in shell shape is analogous to that among cemented protremate brachiopods as illustrated by the *Derbyia—Protorichthofenia—Richthofenia* sequence.

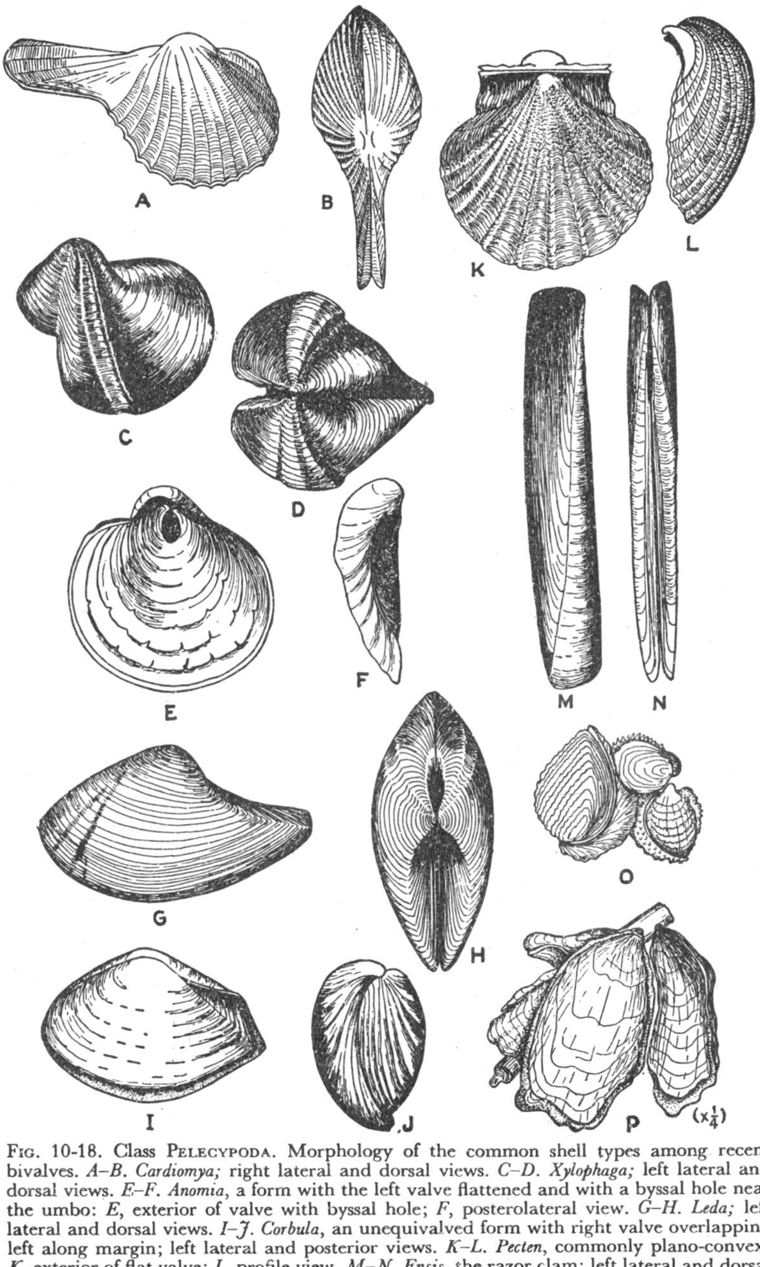

FIG. 10-18. Class PELECYPODA. Morphology of the common shell types among recent bivalves. *A–B. Cardiomya;* right lateral and dorsal views. *C–D. Xylophaga;* left lateral and dorsal views. *E–F. Anomia*, a form with the left valve flattened and with a byssal hole near the umbo: *E*, exterior of valve with byssal hole; *F*, posterolateral view. *G–H. Leda;* left lateral and dorsal views. *I–J. Corbula*, an unequivalved form with right valve overlapping left along margin; left lateral and posterior views. *K–L. Pecten*, commonly plano-convex: *K*, exterior of flat valve; *L*, profile view. *M–N. Ensis*, the razor clam; left lateral and dorsal views. *O–P. Chama* and *Ostrea*, respectively, shells of highly variable shape. All shells except *P* are about half natural size. (*By Turner.*)

Shell Structure and Composition. The external periostracum, which is a thin horny covering secreted by the free edges of the mantle, protects the underlying shell from solution. It is commonly lacking in the umbonal region of old shells, because of erosion, and is rarely if ever preserved in fossil shells. The middle or prismatic layer is composed of closely packed, polygonal prisms of calcium carbonate separated by films of conchiolin like that in the periostracum and disposed roughly perpendicular to the outer surface of the valve (Fig. 10-19). The prisms[1] are secreted by the edge of the mantle, hence they are added only along the valve margins. The prismatic layer constitutes the outer layer in all fossil pelecypod shells because of loss of the periostracum. The inner shell layer is composed of thin laminae of calcite or aragonite disposed roughly parallel to the inner surface of the valve. This calcareous lining, commonly termed the **laminated, nacreous,** or **mother-of-pearl** layer, is secreted by the entire surface of the mantle, hence it grows continuously during life and is thickest in the oldest part of the shell. The laminae are crumpled in some shells and when exposed to light display iridescence; in others they are thick and not crumpled, and the valve surface then has a porcelaneous appearance.

The term **ostracum** has been used by some authors for the shell proper (*i.e.*, the calcareous part) regardless of the number of calcareous layers. In those pelecypod shells with a two-layered ostracum, the outer prismatic layer constitutes the **outer ostracum** and the inner laminated covering is termed the **inner ostracum.** In many shells a local filmlike calcareous secretion of the muscles, the **hypostracum,** is deposited between the muscle and shell proper, coextensive with the muscle impressions (Newell, 1937) (Fig. 10-19*C*).

The microstructure of ancient fossil pelecypod shells is little known, partly because original shell substance was dissolved or destroyed during fossilization, making well-preserved specimens rare, and partly because little study has been devoted to the subject. The interested reader will find an excellent review and discussion of this subject by Newell (1937).

Nacre, the most common material constituting the innermost layer of the pelecypod shell, consists of thin, uniform lamellae of aragonite, all of the same thickness, separated by equally thin leaves of organic material, probably conchiolin. The lamellae average slightly less than 1 μ in thickness and are always parallel to the lines of shell growth and never irregular like the calcitic lamellae of an oyster or the foliaceous structure of a pecten. Nacreous structure is unique to mollusks and is well exemplified by the "mother-of-pearl" part of the shell in many living genera. The organic lamellae are generally destroyed in fossilization and the pearly luster is also lost in most of the older fossils.

Another type of shell structure, the **crossed lamellar,** is present in all mollusks except cephalopods and is unknown in any other phylum. This structure, which is more commonly developed in aragonite than in calcite, consists of an intergrowth of first-order lamellae which are themselves com-

[1] The prisms are extremely well developed in some shells (*e.g.*, *Inoceramus* and *Pinna*), in which they reach diameters of a millimeter and lengths of 5 or 6 mm. Such large prisms are useful in some regions for identification and correlation of the containing strata.

posed of minute second-order lamellae. The microscopic structure produced is shown in Fig. 10-19*A–B*.

Nacreous and crossed lamellar structure are both characteristic of the inner ostracum but are not found together in the same shell. This mutual exclusion seems to be true of both fossil and modern shells.

Certain pelecypods are commercially important because of the **pearls** they produce. These are small, concentrically laminated calcareous bodies secreted by the mantle in response to some type of irritation. Free pearls are usually spherical and are produced mainly in the mantle tissue. They are later liberated within the mantle cavity, where additional layers are added. Irregu-

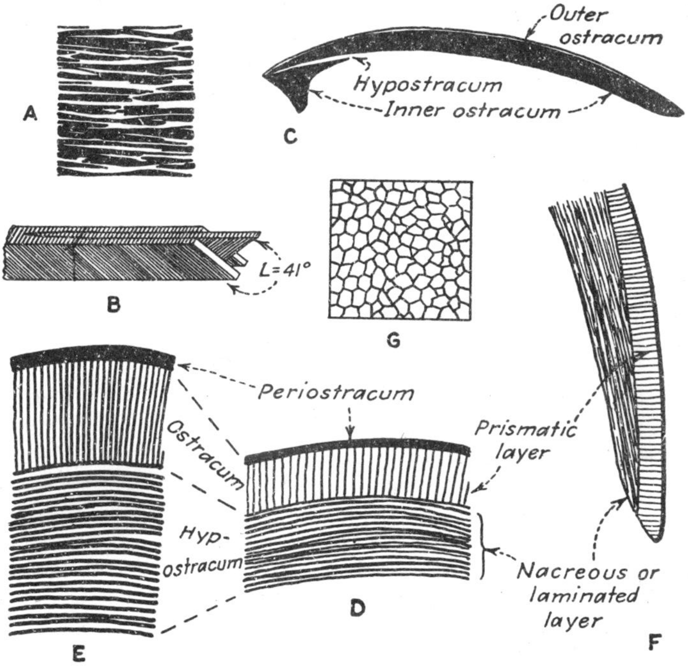

FIG. 10-19. Class PELECYPODA. Microscopic shell structure. *A–B*. Diagrams illustrating cross-lamellar structure: *A*, section of shell cut parallel to shell surface, showing edges (white and black) of branching and wedging first-order lamellae; *B*, three first-order lamellae much magnified, composed of second-order lamellae inclined 41° in alternately different directions in alternate lamellae of the first order. Section *A* was taken in the top plane of *B*; the thickness of a first-order lamella is equal to the vertical line in *B*. *C*. Diagrammatic longitudinal section through a Paleozoic pectinoid shell, showing three fundamental shell constituents. The outermost constituent, the periostracum, is lacking. The outer ostracum is composed of calcite; the hypostracum and inner ostracum, of aragonite. *D–G*. Diagrams illustrating the three fundamental shell layers: *D–E*, sections transverse to shell; conchiolinitic portions are shown in black, inorganic substances by lines; *F*, marginal part of a valve, showing how laminated layer thins toward edge and thickens in older part of shell; *G*, transverse section of prismatic layer, showing crowded prisms of calcite separated by films of conchiolin. (*A–C slightly modified after Newell*, 1937; *E–G, highly generalized and diagrammatic.*)

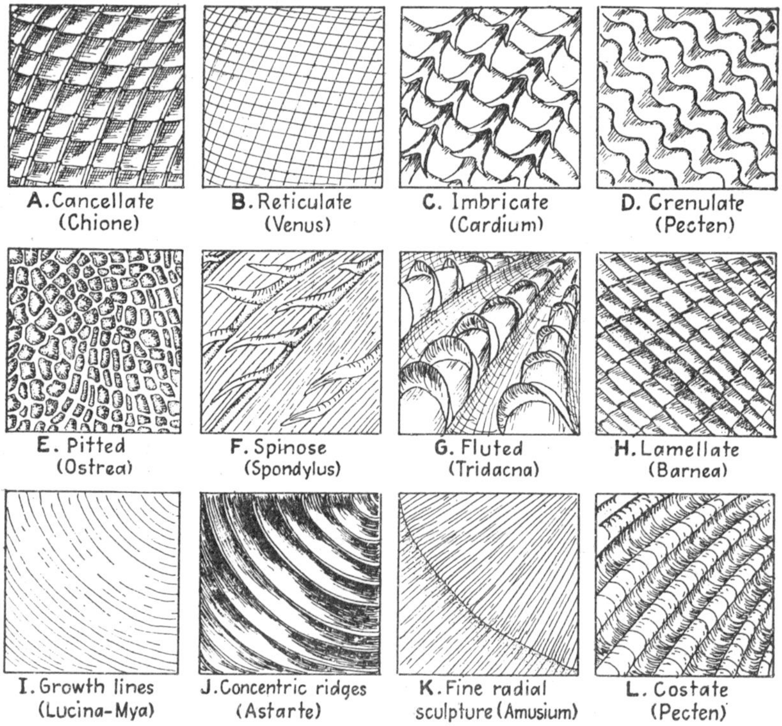

FIG. 10-20. Class PELECYPODA. Diagrams illustrating different types of surface sculpture. The lower name in parentheses is that of a genus which has the particular type of sculpture developed in some of its species. (*By Turner.*)

lar pearls are commonly present in the muscle tissue or even in other soft parts of the animal. Attached pearls, usually termed blister pearls, may be produced between the mantle and the nacreous layer. All common types of pearls have been preserved as fossils. The earliest are from Jurassic rocks and are ascribed to *Gryphaea*. Many Cretaceous pearls have been reported, and these were produced by several species of *Inoceramus*, which might well be dubbed the "pearl oyster" of the Cretaceous. Other examples of fossil pearls, among them an unusually large blister pearl in a Miocene *Panope*, have been discussed by Berry (1936).

The mineralogical composition of the shell material differs among species. In one group of oysters both prismatic and nacreous layers are composed of calcite, whereas in another group they are wholly of aragonite. In *Pinna* the outer prismatic layer is composed of calcite and the inner of aragonite.

Since aragonite is more readily dissolved than calcite, the mineral composition of a calcareous shell largely determines how it is preserved. As an example, shells composed originally of aragonite alone are likely to be represented by external impressions and internal fillings only, due to the complete solution of the original shell substance. Fossil shells having the outer layer of cal-

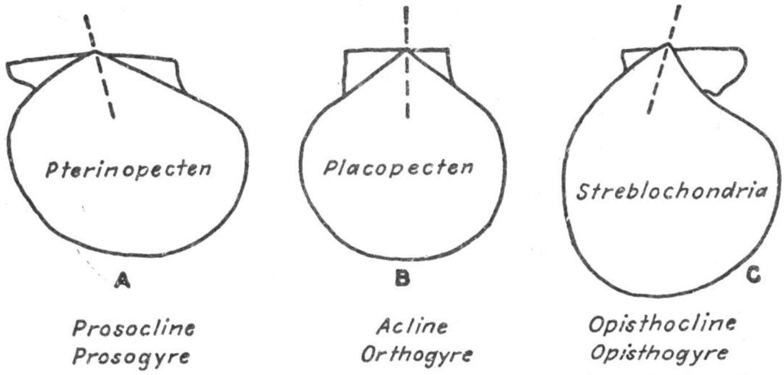

FIG. 10-21. Class PELECYPODA. Diagrams illustrating shell obliquity and orientation of the beaks. The dashed line approximately bisects the umbonal angle, and its inclination with reference to the hinge line, which is horizontal, indicates the obliquity. (*A and B outlined from Shimer and Shrock*, 1944; *C from Newell*, 1937.)

cite and the inner of aragonite commonly show the former intact and the latter partly or entirely dissolved away.

Surface Sculpture. The exterior of most pelecypod shells is relatively smooth, but in many species it is somewhat sculptured and in a few it is exceedingly rough because of crenulations, ribs, spines, and other surface features (Fig. 10-20). One of the commonest types of sculpture,[1] designated **concentric,** consists of lines of growth, or **growth lines,** that are concentric about the umbos. These may be fine or coarse and may be closely or distantly spaced. Many pelecypod shells have no other sculpture. A second type of sculpture is **radial** and consists of elements that radiate from the beaks. These may be fine grooves and ridges (**costellae**), coarser rounded and angular ridges termed **costae,** prominent **ribs,** and rows of protuberances and spines. **Cancellated sculpture** is present on those shells having both concentric and radial elements about equally developed. A few shells, exemplified by *Tridacna* (Fig. 10-20), have the entire valves crenulated into a series of broad and rounded radial folds and troughs which interlock along the ventral margin when the shell is closed. Curved shelly flanges also rise from the folds and extend some distance outward from the shell surface. Although some pelecypod shells are spiny, *e.g.*, *Spondylus*, spinosity is less conspicuous among the Pelecypoda than among the Gastropoda. In fact, living pelecypods tend in general to have only limited sculpture, and this also seems to have been the case with extinct species.

Hinge-line Structures. The valves of most pelecypods interlock dorsally by means of **teeth** and **sockets** that lie along a straight or curved **hinge line.**

[1] The word **sculpture** is used widely as a collective term for the surface features of molluscan shells, and we adopt this practice. **Ornament** and **ornamentation** have also been used in a similar sense, though not so acceptable. Recently the term **prosopon** was proposed as a substitute for ornament (GILL, E. D., 1949, *Prosopon*, a term proposed to replace the biologically erroneous term ornament, *Jour. Paleontology*, vol. 23, p. 572), but there is little likelihood that it will meet with approval, particularly since the word has long been the name of a genus of Brachyura (WRIGHT, C. W., 1950, Comment on proposal to substitute *prosopon* for ornament, *Jour. Paleontology*, vol. 24, p. 497).

Teeth are present in both valves and fit into corresponding sockets in the opposite valves. In most shells the valves can be separated without difficulty, but in a few they cannot be disarticulated without breaking some of the teeth. The teeth, sockets, and certain other closely associated structures are collectively termed the **dentition,** and shells lacking dentition are described as **edentulous.** The number, nature, and arrangement of the components of the dentition are important for discrimination of both fossil and modern shells, and the more important types are discussed in a subsequent section.

Most pelecypod shells have a flat or slightly curved **cardinal area** on each valve between the beak and hinge line (Fig. 10-15). This has sculpture different from that on the remainder of the shell and in some species is divided into two parts—an anterior heart-shaped area, the **lunule;** and a posterior elongate depression, the **escutcheon**[1] (Fig. 10-15).

Shells in which the beaks are directed toward one another are described as **orthogyre** (Fig. 10-21); those with anteriorly directed beaks, as *Venus* (Fig. 10-15), are **prosogyre;** and those with beaks directed posteriorly are **opisthogyre.** The **obliquity** of a shell refers to the inclination of the mid-umbonal line (Fig. 10-21). Shells with forward obliquity are **prosocline;** upright shells, like most Pectinidae, are **acline;** and shells with backward obliquity are **opisthocline.** It is known from ontogeny that prosocline shells appeared first and gave rise to acline forms which in turn gave rise to opisthocline types.

Most pelecypod shells have specialized structures along the hinge line that cause the shell to open when the adductor muscles are relaxed and certain other structures associated closely with the muscles themselves. The **ligament** is an elastic band, rod, or hemicylinder of conchiolin passing from one valve to the other above the hinge line and so disposed that it is bent or stretched when the shell is closed (Figs. 10-9, 10-22). It consists of lamellar conchiolin secreted by the mantle and is continuous with the periostracum. Its lateral edges are inserted into **ligamental grooves** situated in the cardinal area of each valve. The most primitive type of ligament, described as **amphidetic,** extends along and above the hinge line with part in front of the beaks and part behind. A few shells have a **prosodetic** ligament, which lies wholly in front of the beaks, but by far the majority have the ligament situated behind the beaks, in which cases it is described as **opisthodetic** (Fig. 10-22*D–F*). An **alivincular** ligament consists of a single cordlike strand, **multivincular** ligaments are composed of a bundle of strands, and a **parivincular** ligament is a hemicylindrical band (Fig. 10-22*A–C*).

A second type of hinge structure, similar to the ligament in origin and customarily designated a **resilium** (Figs. 10-9, 10-22*I*), is a triangular pad residing in a central pit along and below (*i.e.*, ventral to) the hinge line or

[1] It has been pointed out by some authors that the terms lunule and escutcheon are nondescriptive of the features to which they are applied, and recently it has been suggested that the terms be dropped in favor of **sigilla** for lunule and **vallis** for escutcheon (Dodge, H., 1950, Suggested substitutes for the terms "lunule" and "escutcheon" in Pelecypoda, *Jour. Paleontology*, vol. 24, pp. 500–501). Even if the proposed terms were outstandingly appropriate, it is doubtful if they could replace ones so long in use as lunule and escutcheon.

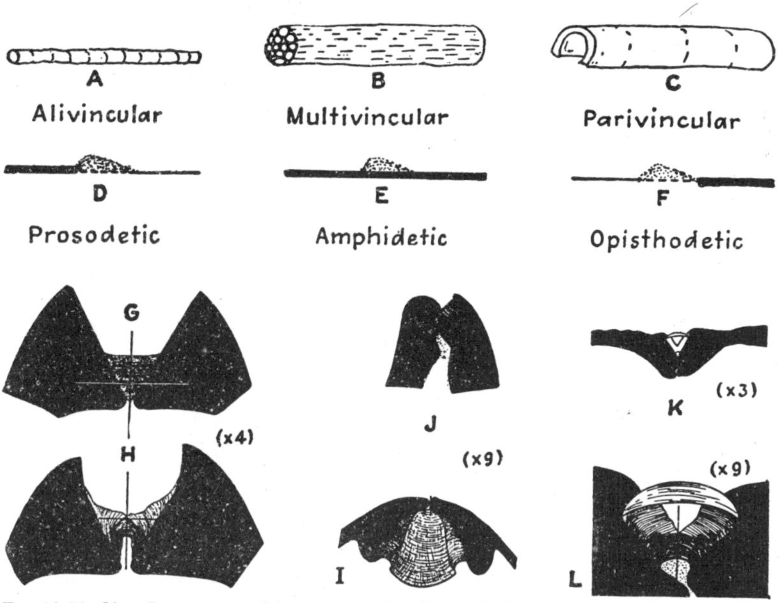

FIG. 10-22. Class PELECYPODA. Ligaments and resilia. *A–C.* Diagrams of the three structural types of ligaments. *D–F.* Diagrams showing positions of ligament (in black) with reference to the beak (stippled), which points anteriorly to the left. *G–H. Pinctada savignyi: G,* section through the lamellar elastic ligament just back of the resilifer; *H,* section through fibrous compressional ligament. *I–J. Pecten (Chlamys) islandicus: I,* section through resilium, showing the bell-shaped lamellar ligament flanked on both sides by the calcareous vestigial fibrous ligament; *J,* section through the lamellar ligament behind the beaks. *K–L. Arca pexata.* Drawings showing the single elastic, lamellar ligament band above and the relatively massive fibrous ligament below. In drawings *G–L* the valves are in tight apposition; the inorganic shell is black; the stippled parts are extensions of the mantle that secrete the teeth and ligament; and the hinge or cardinal axis is indicated by a cross. (*G–L after Newell,* 1937.)

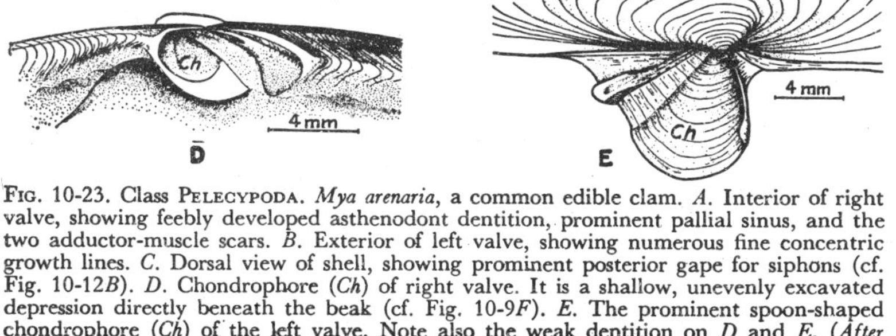

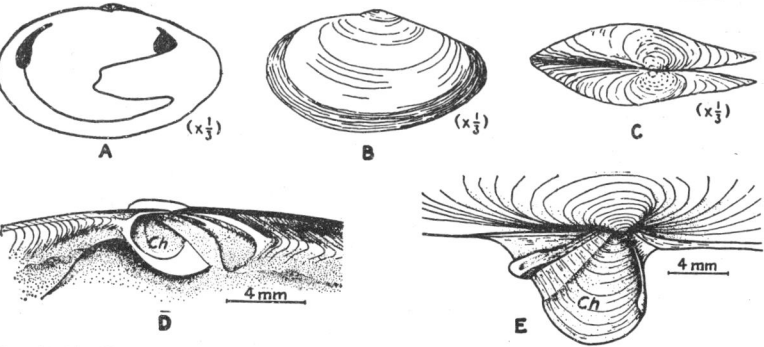

FIG. 10-23. Class PELECYPODA. *Mya arenaria,* a common edible clam. *A.* Interior of right valve, showing feebly developed asthenodont dentition, prominent pallial sinus, and the two adductor-muscle scars. *B.* Exterior of left valve, showing numerous fine concentric growth lines. *C.* Dorsal view of shell, showing prominent posterior gape for siphons (cf. Fig. 10-12*B*). *D.* Chondrophore (*Ch*) of right valve. It is a shallow, unevenly excavated depression directly beneath the beak (cf. Fig. 10-9*F*). *E.* The prominent spoon-shaped chondrophore (*Ch*) of the left valve. Note also the weak dentition on *D* and *E.* (*After Foster,* 1946.)

dorsal margin. The two pits, one on each valve, that receive the resilium together constitute the **resilifer** (Figs. 10-9*D–G*, 10-22*I*), but this term is also applied to either pit singly. The resilium is composed of fibrous conchiolin impregnated with calcareous spicules and is an organ of compression; *i.e.*, it is elastic chiefly under compressional stress. In certain shells, well exemplified by *Mya*, a common edible and burrowing clam, the resilifers are greatly modified and are designated **chondrophores** (Fig. 10-23). The chondrophore of the right valve is a shallow, unevenly sculptured pit with its lower portion flattened where the ventral margin fuses with the shell. That of the left valve is a prominent, spoon-shaped structure with its anterior edge projecting sharply from the hinge line at about a right angle. Generally, shells with a resilium have only one of these structures, but a few forms (*e.g.*, *Perna*) have more than one resilium and resilifer.

As indicated in Fig. 10-22, the ligament is under tensional stress when the valves are tightly apposed, whereas the resilium is compressed under the same conditions. Pelecypods with well-developed ligaments are not likely to have a resilium, and those with large resilia generally have relatively weak ligaments.

Dentition. The dorsal, or cardinal, margin of the valves of denticulate pelecypods is generally thickened[1] and bears numerous tubercles and corresponding pits, ridges and troughs, or teeth and sockets. As stated previously, the term dentition applies collectively to the number, nature, and arrangement of these irregularities, whereas **hinge** (or **dental apparatus**) is customarily used as a designation for the whole interlocking mechanism. It has been suggested that the first hinge teeth developed from crenulations and irregularities on the dorsal margins and that they served primarily for preventing the valves from shearing over one another.

As early as the Ordovician a dental apparatus had been developed, and it consisted of many small teeth and alternating sockets. In the earliest stages, the dentition was little more than a series of slightly modified irregularities of the cardinal margin. Soon, however, two quite different patterns of tooth arrangement evolved. In the **taxodont,** or **ctenodont,** type the teeth tended to converge from the hinge line toward the center of the valve, as illustrated by *Ctenodonta* and *Nucula* (Fig. 10-24). In the **actinodont** type the teeth radiated downward from the umbo, as illustrated by *Actinodonta* and *Lyrodesma* (Fig. 10-24). Taxodont dentition does not seem to have given rise to other patterns although it has persisted to the present and is characteristic of many modern species and genera, which constitute the Order Taxodonta. Actinodont dentition, by contrast, provided an arrangement of teeth and sockets that was readily modified, and from this early type arose a long series of modifications.

In the early actinodont dentition the centrally located teeth were short and likely to be heavy, whereas the outer, or posterior, ones were elongated and slender (Fig. 10-24). Modification tended to follow two general lines: (1) reduction in the number of teeth; and (2) increase in size and projection

[1] In some species the cardinal margin is reinforced by a thick vertical **hinge plate** (Fig. 10-26*G*), which if present carries the teeth and sockets.

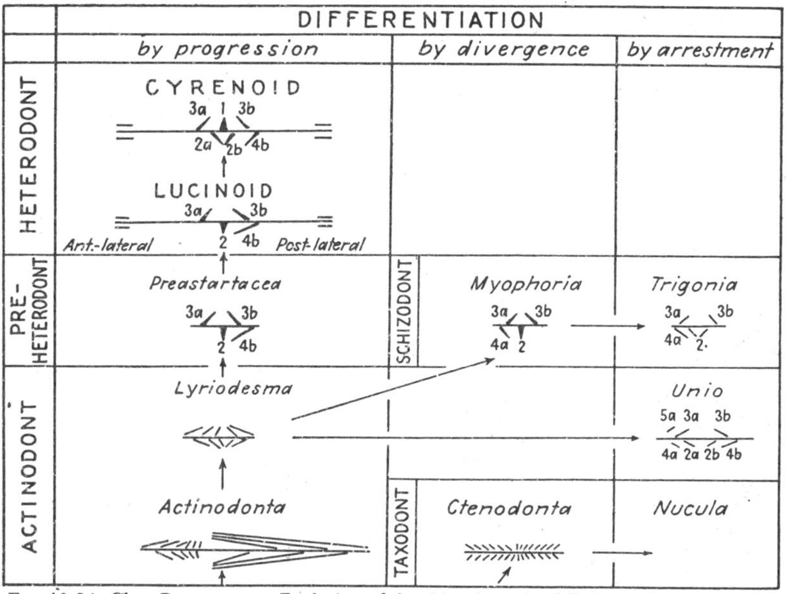

FIG. 10-24. Class PELECYPODA. Evolution of dentition in normal Pelecypoda as suggested by Swinnerton (1947). This diagram is essentially an adaptation of Swinnerton's illustration. Consult Fig. 10-25 for significance of the numbers and letters by which the teeth are designated. In each diagram the horizontal line separating the groups of teeth is the hinge line.

of those remaining. By progressive differentiation a dentition evolved that consisted characteristically of a few of the older teeth as strong **cardinals** underlying the ligament and a group of new teeth, the **laterals,** lying beyond the limits of the ligament. This type of dentition, termed **heterodont** on post-Paleozoic shells and **preheterodont** on the Paleozoic ancestral forms, is characteristic of the vast majority of pelecypods, both fossil and living, and is well exemplified in the common *Venus* (Fig. 10-15). These progressive changes are indicated diagrammatically in Fig. 10-24, along with certain other nonprogressive modifications that produced another common dentition, the **schizodont,** characteristic of modern *Myophoria* and the Trigoniacea.

In describing the actinodont-heterodont series of dentitions, it is customary to use a system of notation for the hinge teeth devised by the French paleontologists Bernard[1] and Munier-Chalmas. Numbers are used for the principal teeth, and the numbering starts with the lowest figure for the centrally placed teeth. Odd numbers are used for the teeth of the right valve, even for those on the left. The letter *a* following a number indicates an anterior tooth, *b* a posterior one. Sockets are indicated by a dash (—) between numbers, and lateral teeth are indicated by short dashes parallel to the hinge

[1] BERNARD, F., 1895–1897, Note sur le développement et la morphologie de la coquille chez les Lamellibranches, *Bull. soc. géol. France,* 3d ser., vol. 23, pp. 104–154; vol. 24, pp. 54–82, 412–449; vol. 25, pp. 559–566.

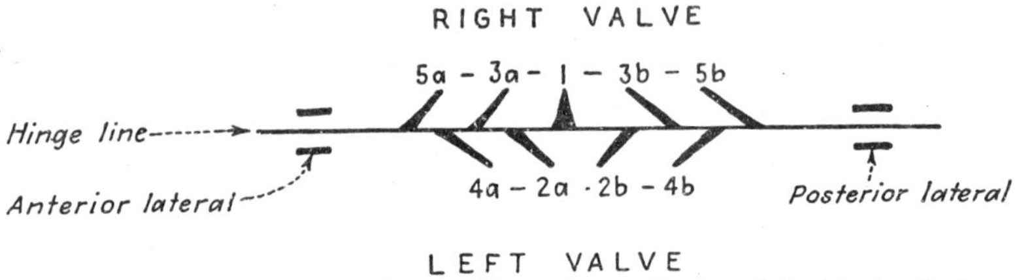

FIG. 10-25. Class PELECYPODA. Diagram illustrating application of the Munier-Chalmas notation to the dentition of pelecypods.

line. By this notation the dentition as a whole can be represented by a formula such as that shown in Fig. 10-25.

The dentition of some pelecypod shells is so complex that the notation described in the foregoing paragraph cannot be applied satisfactorily.[1] Specialists are not in agreement as to how many types of dentition should be recognized and what notation should be used. Thus far no general system of notation has been devised that is applicable to all denticulate bivalve shells.

The terms widely used by malacologists for the three common types of dentition are taxodont, heterodont, and schizodont. Other terms are used for certain specialized types. Paleontologists are generally familiar with a somewhat larger number of types, because some extinct pelecypods had dentition unlike any present in living forms. Ten different types of dentition were defined, and most of them illustrated, in our first edition. Here we include them as a footnote[2] (also see Fig. 10-26).

[1] Transposition of dentition has been reported in a few pelecypod shells (Popenoe and Findlay, 1934). In such shells the interior of a valve with completely transposed dentition presents the appearance of a mirror image of the interior of a normal opposite valve of the species.

[2] **Taxodont** dentition consists of a series of many similar alternating teeth and sockets, as many as 35 of each, and appears first in Ordovician shells (*e.g.*, *Ctenodonta*). *Nucula* is a typical modern example; *Arca* (Fig. 10-26*A*), another modern genus, has secondary taxodont dentition.

Teleodont (heterodont) represents the highest development of dentition among the pelecypods and is characterized by a few large cardinal teeth with or without laterals. Roughened lateral areas and accessory lamellae are also present in some shells. The more primitive examples merge into or resemble diagenodont forms. *Venus* (Figs. 10-15, 10-26*E*), with three cardinals in the right valve and two in the left, is a modern example.

Schizodont dentition is not well defined and is supposed to be a divergent or arrested derivative from the early actinodont type (Fig. 10-24). Generally the teeth are variable in shape and size and are more or less divisible into subumbonal, pseudocardinal, and posterolateral groups. Irregular development is exemplified in the shells of the fresh-water Unionidae (Fig. 10-26*D*); regular development, in *Myophoria* and *Trigonia* (Fig. 10-24).

Isodont, as the term suggests, consists of about equally developed teeth. Each valve has two main teeth, and these are so arranged that two sockets and an anterior and a posterior lateral are near the middle of the hinge line of one valve, and two teeth with one anterior and one posterior socket lie near the middle of the hinge line of the other valve. The teeth are slightly curved and fit into correspondingly curved sockets. In shells with

(*Footnote continued on page 386.*)

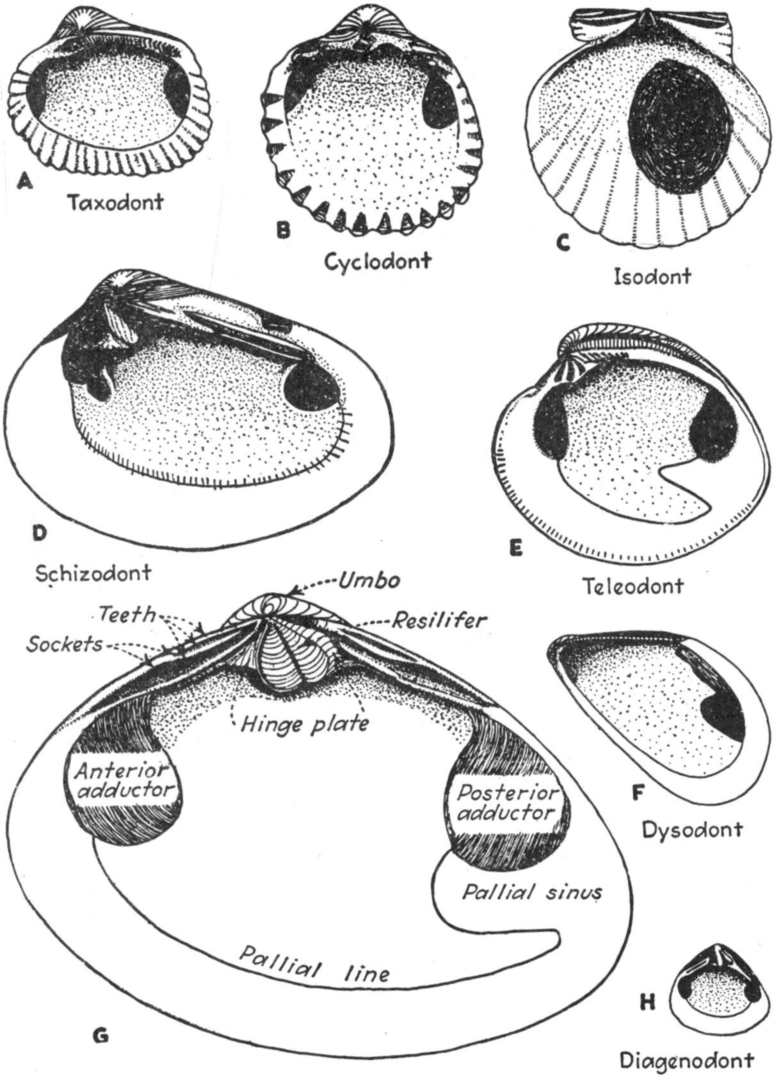

FIG. 10-26. Class PELECYPODA. Interior view of right valves, showing the different types of dentition and differences in the muscle scars. *A. Arca*, taxodont and isomyarian. *B. Cardium*, cyclodont and isomyarian. *C. Pecten*, isodont and monomyarian. *D. "Unio,"* schizodont and anisomyarian. *E. Venus*, teleodont and isomyarian. *F. Mytilus*, dysodont and anisomyarian. *G. Mactra*, with a well-developed hinge plate, a prominent double-lobed resilifer, a pallial sinus and isomyarian muscle scars. *H. Astarte*, diagenodont and isomyarian. (All drawings are about one-half natural size.)

Musculature. The pelecypod shell is closed by the contraction of adductor muscles extending between the valves and attached to each valve. At the site of attachment each muscle leaves a distinct **muscle scar** (Fig. 10-26), which is generally shallow in thin shells and deep in thick shells. Typically there are two adductor muscles, one anterior and one posterior, and these leave a scar on each valve at each end of the hinge line. Shells with valves having two such scars are termed **dimyarian** (Fig. 10-27*A–B,E*). If the two muscles are about equal in size, the shell is **isomyarian** (Figs. 10-26*G*; 10-27*A–B,E*). If the anterior adductor is conspicuously smaller, the shell is **anisomyarian** or **heteromyarian** (Fig. 10-27*D*); and if it is completely lacking, as in Family Pectinidae, in which case the posterior adductor is unusually large, the shell is described as **monomyarian** (Fig. 10-27*C*).

Other less conspicuous muscle scars are present on some shells. The pedal muscles leave a **protractor** and a **retractor** scar in the anterior part of each valve and a second retractor scar behind the protractor in some forms. Siphonated shells have muscles to retract the siphon, and these may leave a scar on each valve directly below the posterior adductor scar.

The musculature is an important feature for classification of both living and fossil pelecypods. In fossil shells the adductor muscle scars are commonly well marked and provide a ready means for differentiating forms as dimyarian, isomyarian, heteromyarian, and monomyarian.

Classification

Classification of the Pelecypoda has always been a difficult problem for paleontologists, because most existing classifications were proposed by zoologists who based the major divisions on the soft parts of the animal. We prefer the classification of Thiele (1935), which has three orders:

---

extreme development of this interlocking relation, as in *Spondylus*, it may not be possible to separate the valves one from the other without breaking some of the teeth. In more typical development, as in *Pecten* (Fig. 10-26*C*), the teeth are slender ridges.

Dysodont dentition, illustrated by *Mytilus* (Fig. 10-26*F*), is a fully developed type and seems to have evolved from external sculpture across the cardinal area from the beak.

Pantodont, which is restricted to Ordovician and Silurian shells (*e.g.*, *Allodesma* and *Orthodonticus*), seems to be a stage of development that possibly antedated true teleodont dentition. The teeth are differentiated into grooved cardinals and groups of more than two laterals.

Diagenodont (preheterodont) dentition consists normally of three or less cardinals and one or two laterals in any group. *Astarte* (Fig. 10-26*H*) is an example.

Cyclodont dentition is characterized by the lack of a flat hinge plate and extreme bending of the teeth. *Cardium* (Fig. 10-26*B*) and *Tridacna* have this kind of dentition.

Asthenodont dentition, marked by obsolete teeth, is present in burrowing and boring bivalves, such as *Mya* (Fig 10-23) and *Pholas* (Fig 10-8*A*).

Anomalodont dentition, present in *Allorisma*, a primitive borer, is characterized by teeth that are quite small or wanting altogether.

Edentulous bivalves are those lacking teeth of any kind. A group of ancient pelecypods (Paleoconcha), of which *Acharax* is the only surviving genus, have edentulous shells. *Grammysia* and *Solenopsis* are well-known fossil representatives of the group. An edentulous condition also exists in the recent genus *Anodonta* of the Family Unionidae.

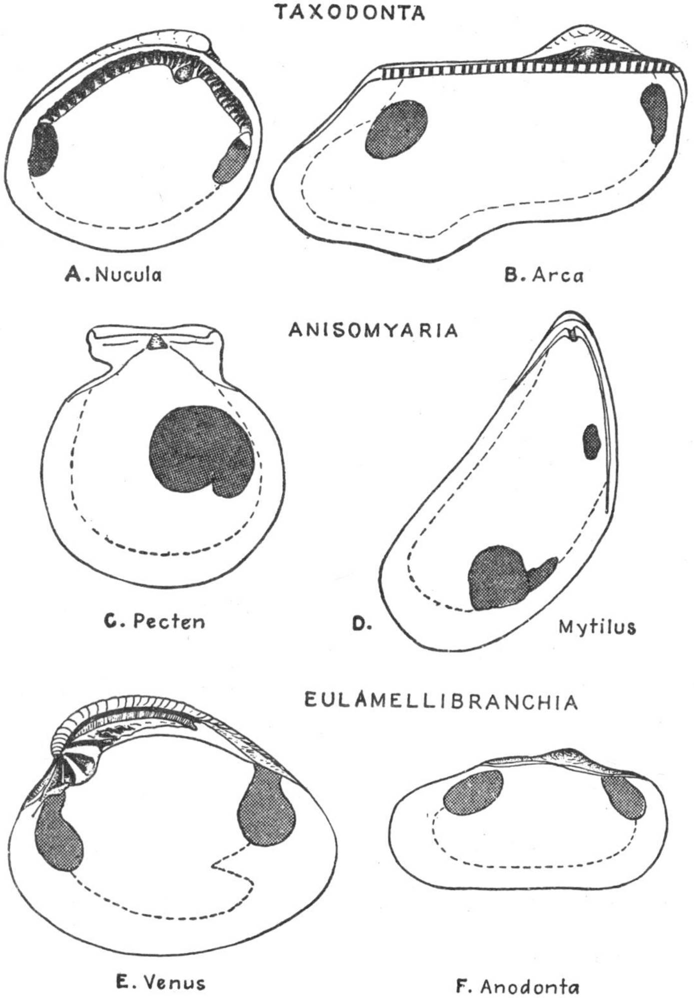

FIG. 10-27. Class PELECYPODA. Diagrams illustrating Thiele's three orders, which are based largely on hinge dentition and muscle scars. *A–B*. Order Taxodonta: *A*, *Arca*, taxodont dentition, two essentially equal muscle scars (isomyarian); *B*, *Nucula*. *C–D*. Order Anisomyaria: *C*, *Pecten*, with isodont dentition and a large single muscle scar (monomyarian); *D*, *Mytilus*, with isodont dentition and two unequal muscle scars (anisomyarian): *E–F*. Order Eulamellibranchia: *E*, *Venus*, with teleodont dentition and two equal muscle scars (isomyarian); *F*, *Anodonta*, edentulous and with two equal muscle scars. (*By Turner*.)

Order 1. Taxodonta.
Order 2. Anisomyaria.
Order 3. Eulamellibranchia.

These are discussed later, and the chief features of the shells in each order are shown in Fig. 10-27.

The features that have been most used for ordinal[1] classification of the Pelecypoda are:

1. Nature of gills and associated soft parts. Five orders based on gill morphology are recognized by many zoologists. These are Protobranchia, Filibranchia, Pseudolamellibranchia, Eulamellibranchia, and Septibranchia (see discussion of these on page 370 and also Fig. 10-14). This classification is not acceptable to paleontologists because the type of gill that was present in most fossil pelecypods cannot be determined.

2. Nature of hinge dentition. We recognize as fundamental the types of dentition diagrammed in Fig. 10-24—*i.e.*, taxodont, actinodont, schizodont, and heterodont; we also define in a footnote a few other specialized and uncommon types (see page 384 and Fig. 10-26). Neumayr (1884, 1891) was one of the first to propose a classification based on shell features (chiefly dentition), and this has served as the basis of the classifications now widely accepted by paleontologists. Of these classifications, American paleontologists have generally accepted that of Dall (1890–1903), which is based largely on the nature of the dentition. Three orders of denticulate shells and one separate subordinal group of nondenticulate or edentulous shells were recognized by Dall, as follows: Prionodesmacea which are forms with taxodont, schizodont, isodont, and dysodont types of dentition; Anomalodesmacea, characterized by anomalodont dentition; Teleodesmacea, which have teleodont, pantodont, diagenodont, cyclodont, and asthenodont dentition; and Paleoconcha, of which the shells have no dentition of any kind. This classification, well known to paleontologists through its adoption in the English edition of Zittel's "Text-book of Palaeontology," is not altogether satisfactory for several reasons, but chiefly because it requires unnatural grouping of many forms.

3. Number and nature of adductor muscles. Two general categories are recognizable: (1) monomyarian, and (2) dimyarian (isomyarian and anisomyarian). No classification based on musculature alone has been proposed, but most classifications take account of the chief muscles.

4. Evolution. A natural or evolutionary classification was outlined by Douvillé (1912), but it has not been widely accepted in spite of its merits. It is based on the concept of adaptative radiation, and assumes a primary radiation into the three following divisions, adapted to the three chief modes of life of pelecypods: (1) normal benthonic existence, either active and moving in an upright position (**orthoconchs**) or passive and lying on one side (**pleuroconchs**); (2) a fixed life on the bottom, with the shell suspended from some foreign body by the byssus or directly cemented by shelly matter to some object on the bottom; (3) burrowing in soft sediments or boring in wood and rocks.

We adopt Theile's classification because it seems to be the best grouping that can be made on the composite basis of dentition, musculature, and gill structure. The geologic ranges of the many superfamilies are indicated in Fig. 10-31.

Order 1. Taxodonta.[2] Taxodont pelecypod shells have many similar teeth

[1] Among the more important shell features used in differentiating subordinal groups (families and genera) are the following:

1. Features of the musculature and of the shell-opening devices.
2. Type of hinge dentition.
3. Nature of surface sculpture.
4. Shell microstructure.

[2] Taxodonta—Gr. *taxis*, an arrangement, or order, + *odous*, *odontos*, tooth; referring to the orderly arrangement of the teeth.

and sockets along the hinge margin (*i.e.*, taxodont dentition, whence the name) and usually are dimyarian. The earliest known representatives are from Ordovician rocks. *Ordovician to Recent.*

Ctenodonta (Ord.–Sil.), *Nucula* (Sil.–Recent), and *Nuculana* (Sil.–Recent) are typical of the Superfamily Nuculacea. *Vanuxemia* (Ord.), *Cyrtodonta* (Sil.), *Glycymeris* (Cret.–Recent), *Breviarca* (U. Cret.–Paleoc.), and *Arca* (Jura.–Recent) are representative of the Superfamily Arcacea[1] (Fig. 10-28).[2]

The following genera of the Taxodonta are illustrated: Nuculacea: *Nucula* (Figs. 10-8*K*, 10-14*A*, 10-27*A*, 10-28*A–C*), *Ctenodonta* (Figs. 10-28*F*, 10-34*B–C*), and *Leda* (Figs. 10-18*G–H*, 10-28*D–E*). Arcacea: *Arca* (Figs. 10-26*A*, 10-27*B*), *Glyptarca* (Fig. 10-34*A*), *Cyrtodonta* (Fig. 10-28*H–I*), and *Glycymeris* (Fig. 10-28*G*).

Order 2. Anisomyaria.[3] In this order the anterior adductor muscle is much reduced or lacking altogether, and the posterior adductor is a large powerful muscle lying near but slightly behind the center of the shell. The more primitive shells had small nodes or special tooth structures, whereas later forms (*e.g.*, Ostreacea) developed a type of rudimentary schizodont dentition. The embryonic shell has a grooved margin. The mantle is open and without siphons. The gill lamellae are flat with similar filaments or plicate with dissimilar filaments. The earliest known fossil anisomyarians are from Ordovician rocks. *Ordovician to Recent.*

Five superfamilies have been proposed. These, together with their geologic range, are included in Fig. 10-31.

Ambonychia (Ord.), *Pterinea* (Dev.–Recent), and *Inoceramus* (Jura.–Cret.) are examples of the Pteriacea. *Mytilus* (Trias.–Recent) is representative of the Mytilacea. *Ostrea* (Trias.–Recent), *Gryphaea* (Jura.–Eoc.), and *Exogyra* (Jura.–Cret.) are typical of the Ostreacea. *Pecten* (Miss.–Recent), with its many subgenera, exemplifies the Pectinacea. *Anomia* (Jura.–Recent) typifies the Anomiacea (Fig. 10-29). Many other anisomyarian genera are described and illustrated in the standard references cited in the footnote on this page.

The following genera of the Anisomyaria are illustrated: Mytilacea: *Mytilus* (Figs. 10-8*D*, 10-13*E–F*, 10-26*F*, 10-27*D*) and *Modiolopsis* (Fig. 10-34*D*). Pteriacea: *Ambonychia* (Fig. 10-29*H*), *Pterinea* (Fig. 10-29*G*), and

[1] The Arcacea are now regarded by some investigators (MacNeil, 1937; Nicols, 1950) as more closely related to certain groups of the Anisomyaria because the hinge structure is of a later type and may possibly be the result of convergence. It may well be, therefore, that this superfamily should be transferred to the Order Anisomyaria or used as the basis for a new order.

[2] It is beyond the scope of the present work to illustrate more than a few fossil or living pelecypods. The interested reader can find most of the common genera described and illustrated in standard reference works, among which may be mentioned the following: "Text-book of Palaeontology" (Zittel), "Index Fossils of North America" (Shimer and Shrock), "Tertiary Faunas" (Davies), "Catalogue of the Marine Pliocene and Pleistocene Mollusca of California" (Grant and Gale), "Handbuch der systematischen Weichtierkunde" (Thiele), and "Tertiary Fauna of Florida" (Dall).

[3] Anisomyaria—Gr. *anisos*, unequal, + *mys*, *myos*, muscle; referring to the fact that the two adductor muscles are not equal in size.

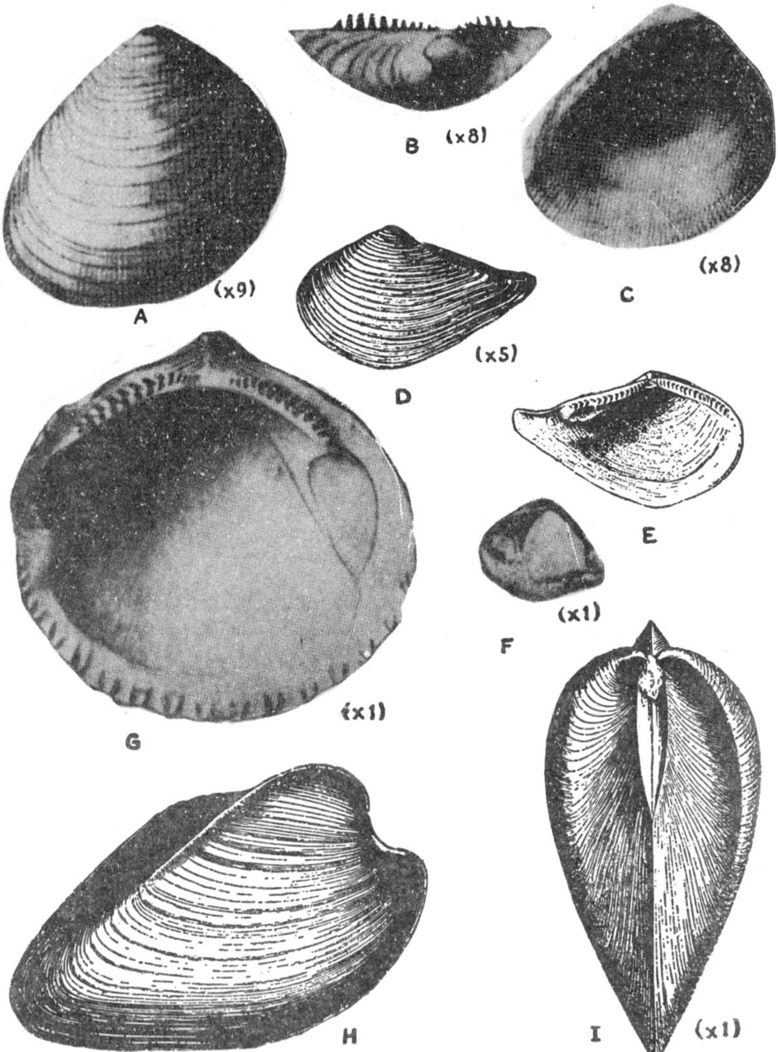

FIG. 10-28. Class PELECYPODA—Order Taxodonta. *A–F.* Nuculacea: *A–C*, exterior of right valve, hinge view of right valve, and interior of left valve of *Nucula defuniak*, a Miocene species; *D–E*, exterior and interior of left valve of *Leda acuta*, a Miocene species; *F*, interior of left valve of *Ctenodonta gibberula*, an Ordovician form. *G–I.* Arcacea: *G*, interior of left valve of *Glycymeris subovata*, a common Miocene species; *H–I*, exterior of right valve and dorsal view of complete shell of *Cyrtodonta hindi* from the Ordovician. (*A–E, G after Gardner*, 1926; *F after Ulrich*, 1897; *H–I after Nicholson*, 1872.)

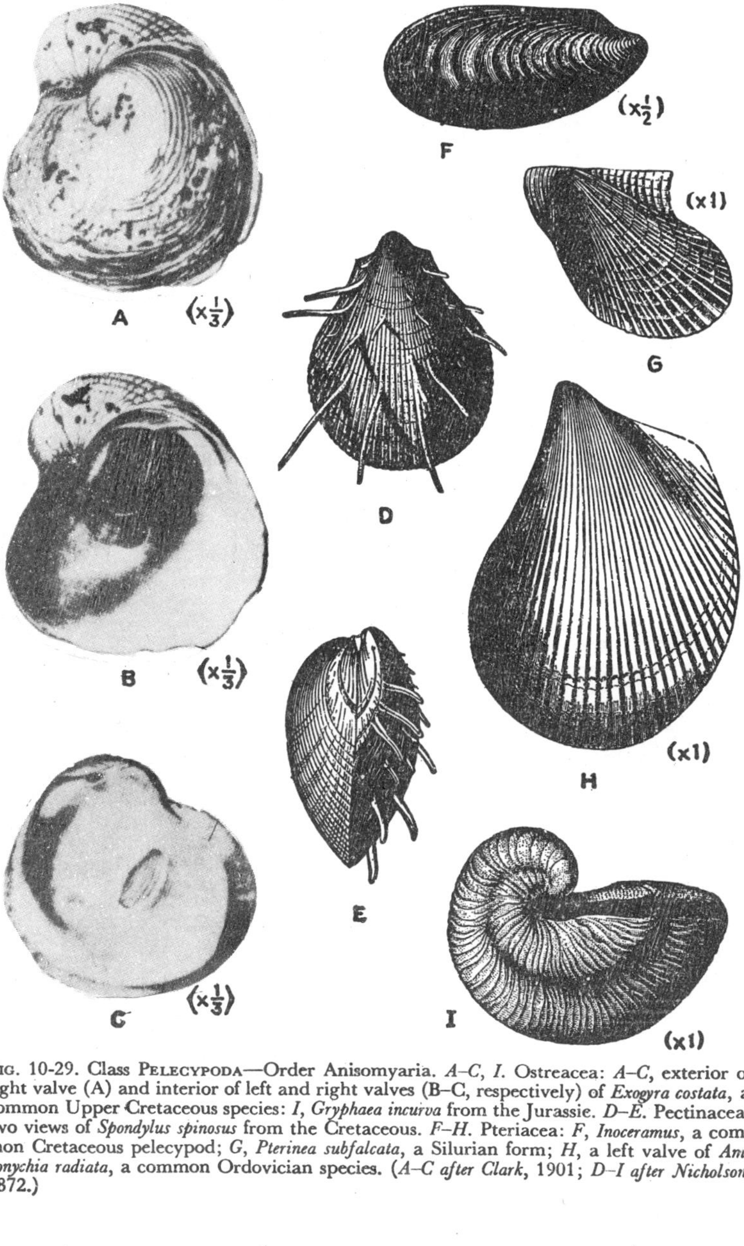

Fig. 10-29. Class Pelecypoda—Order Anisomyaria. *A–C, I.* Ostreacea: *A–C,* exterior of right valve (A) and interior of left and right valves (B–C, respectively) of *Exogyra costata,* a common Upper Cretaceous species: *I, Gryphaea incurva* from the Jurassie. *D–E.* Pectinacea; two views of *Spondylus spinosus* from the Cretaceous. *F–H.* Pteriacea: *F, Inoceramus,* a common Cretaceous pelecypod; *G, Pterinea subfalcata,* a Silurian form; *H,* a left valve of *Ambonychia radiata,* a common Ordovician species. (*A–C after Clark,* 1901; *D–I after Nicholson,* 1872.)

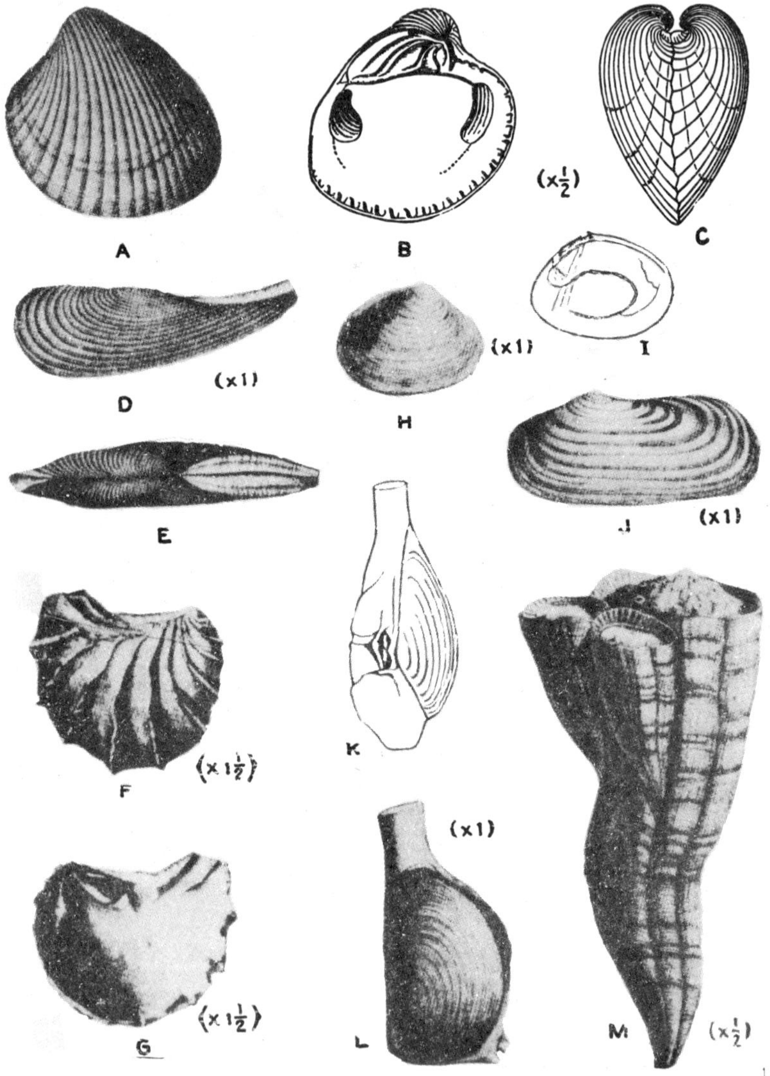

FIG. 10-30. Class PELECYPODA—Order Eulamellibranchia. *A–C.* Three views of *Venericardia planicosta* (Carditacea), a world-wide index fossil of the Eocene. *D–E. Laternula* (=*Anatina*) *spatulata* (Pandoracea) from the Upper Jurassic. *F–G. Trigonia eufalensis*, exterior and interior of right valve, from the Upper Cretaceous. *H–I.* Interior and exterior of the right valve of *Tellina proxima* (Tellinacea), a post-Pliocene species. *J.* Left valve of *Saxicava rugosa* (Saxicavacea), a post-Pliocene and Recent form. *K–L. Clavagella cretacea* (Clavagellacea) from the Cretaceous. *M. Hippurites toucasiana* (Rudistacea), a large individual, with two smaller ones attached to it; an extinct genus confined to the Cretaceous. (*A–E, H–M* after *Nicholson*, 1872; *F–G after Clark*, 1916.)

Inoceramus (Fig. 10–29*F*). Pectinacea: *Pecten* (Figs. 10-8*L*, 10-9*D–E*; 10-18*K–L*, 10-26*C*, 10-27*C*), *Spondylus* (Fig. 10-29*D–E*), *Pterinopecten* (Fig. 10-21*A*), *Placopecten* (Fig. 10-21*B*), and *Streblochondria* (Fig. 10-21*C*). Anomiacea: *Anomia* (Fig. 10-18*E–F*). Ostreacea: *Ostrea* (Fig. 10-8*B*; 10-18*P*), *Exogyra* (Figs. 10-29*A–C*, 10-33), and *Gryphaea* (Fig. 10-29*I*).

Order 3. Eulamellibranchia.[1] The Eulamellibranchia is by far the largest and most important order of pelecypods, and to its 25 superfamilies are assigned most living and many extinct genera.

The gill filaments are fused to form perforated leaflike plates (eulamellibranchiate type, whence the name of the order), and the mantle is commonly produced into posterior siphons. The anterior muscle may be somewhat smaller than the posterior, but in many forms the muscles are essentially equal. The hinge teeth are usually small in number in both valves and the dentition is of the heterodont type or one of its derivatives. A few cardinals and usually some laterals constitute the hinge apparatus, but in some shells teeth are lacking altogether and are replaced by irregular dental processes. The earliest known fossil representative came from Silurian rocks, but the order did not become important until the Mesozoic. The ranges of the 25 superfamilies are shown in Fig. 10-31. *Silurian to Recent.*

The following genera of the Eulamellibranchia are illustrated: Trigoniacea: *Trigonia* (Jura.–Recent; Fig. 10-30*F–G*). Unioniacea: *Anodonta* (Tert.–Recent; Fig. 10-27*F*). Astartacea: *Astarte* (Trias.–Recent; Fig. 10-26*H*) and *Crassatellites* (U. Cret.–Recent; Fig. 10-8*J*). Carditacea: *Venericardia* (Fig. 10-30*A–C*). Cyprinacea: *Arctica* (*Cyprina*) (Jura.–Recent; Fig. 10-13*C–D*). Chamacea: *Chama* (Cret.–Recent; Fig. 10-18*O*). Rudistacea: *Hippurites* (Cret.; Fig. 10-8*C*, 10-30*M*). Cardiacea: *Cardium* (Trias.–Recent; Fig. 10-26*B*). Veneracea: *Venus* (Jura.–Recent; Fig. 10-8*I*, 10-9*B–C*, 10-15, 10-26*E*, 10-27*E*) and *Petricola* (?–Recent; Fig. 10-12*A*). Mactracea: *Mactra* (U. Cret.–Recent; Fig. 10-26*G*). Tellinacea: *Tellina* (Jura.–Recent; Fig. 10-30*H–I*) and *Tagelus* (Cret.–Recent; Fig. 10-8*G*). Solenacea: *Ensis* (Tert.–Recent; Fig. 10-18*M–N*). Saxicavacea: *Saxicava* (Tert.–Recent; Fig. 10-30*J*). Myacea: *Mya* (Tert.–Recent; Figs. 10-8*F*, 10-9*F–G*, 10-10, 10-12*B*, 10-13*A–B*, 10-23) and *Corbula* (Trias.–Recent; Fig. 10-18*I–J*); Adesmacea: *Pholas* (Jura.–Recent; Fig. 10-8*A*), *Teredo* (Jura.–Recent; (Fig. 10-8*E*), *Bankia* (?–Recent; Fig. 10-32), and *Xylophaga* (?–Recent; Fig. 10-18*C–D*). Pandoracea: *Laternula* (Jura.–Recent; Fig. 10-30*D–E*). Clavagellacea: *Clavagella* (Cret.–Recent; Fig. 10-30*K–L*). Poromyacea: *Cardiomya* (?–Recent; Fig. 10-18*A–B*).

Ontogeny and Evolution

With few exceptions[2] the fertilized pelecypod egg hatches into a free-swimming trochophore-like larva, which soon develops a bell-shaped swimming organ, the **velum,** by enlargement of the ciliated ring, and undergoes certain other morphological changes to become a veliger larva. Before the

[1] Eulamellibranchia—Gr. *eu*, well, + L. *lamella*, a thin plate, leaf, or layer, + Gr. *branchia*, gills; referring to the well-developed leaflike gills characteristic of the order.

[2] Unionidae, some (if not all) Sphaeridae, and probably a few marine genera.

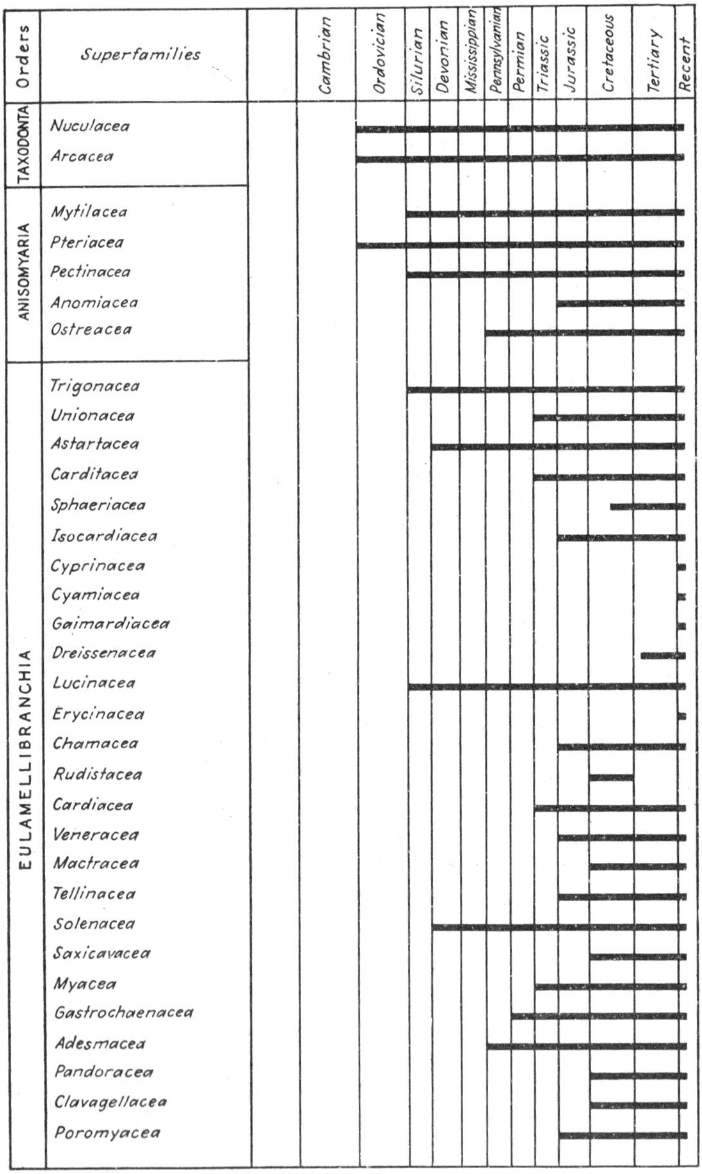

FIG. 10-31. Class PELECYPODA. Chart showing the geological range of the 33 superfamilies of pelecypods.

veliger settles to the bottom it forms a simple univalve shell of conchiolin over the middorsal part. This embryonic shell, termed the **prodissoconch** (Fig. 10-17), consists of two lateral flaps, which later become the right and left valves, and a dorsal band continuous between the two. At about this stage in development the veliger settles to the bottom and begins the series of changes leading to the adult animal and its completely developed shell.

The prodissoconch of modern bivalves is a thin, fragile bivalve shell consisting of two valves of conchiolin joined along the dorsal margin by a flexible band of conchiolin essentially like that of the valves except for the calcareous spicules of the latter. Although little is known about the prodissoconchs of the more ancient pelecypods, it is assumed that they were of similar nature. Such a fragile structure was of little use to the young pelecypod, unless it could be modified in several ways to make it a stronger and more efficient protection. Several of these modifications merit brief consideration.

Calcareous material was secreted by the mantle, and as the shell grew, the two valves came to be composed largely of that substance, with only a minor volume of conchiolin in the form of periostracum and of a network of fibers ramifying the calcareous shell. Along the dorsal margin at first the shell received only minor calcareous additions, and the elastic band continued to function as a rude hinge. Soon, however, irregularities developed along the dorsal margin of each valve, and from these arose teeth and sockets, at first simple and similar but later complex and of several varieties, and these served to interlock the valves firmly along the hinge line while allowing them to open along the lateral and ventral margins without sliding over one another. Marginal crenulations, which developed much later, also served to interlock the valves when the shell was closed.

The valves themselves were strengthened in several ways. Some were greatly thickened (veneraceans), some were conspicuously arched (arcaceans) and ribbed (pectens), and still others developed strong rounded crenulations (tridacnids). Accompanying these modifications went the development of less conspicuous features of sculpture, such as nodes, spines, and rugosities.

Many changes were made in the shell as the animals adapted themselves to different bottom conditions and assumed different modes of life. Those that continued to move freely over the bottom sediments, or to plough through them to a shallow depth, did not greatly change the shape of the shell from that of the prodissoconch. Those that adopted the burrowing habit, however, became elongated and developed extensible siphons, and the shell likewise took on an elongate form with the umbo situated well forward. Boring pelecypods went even farther in shell modification, as exemplified by *Pholas* (a rock borer) and *Bankia* (Fig. 10-32) and *Teredo* (wood borers).

One group of pelecypods acquired the habit of attaching their shells temporarily or permanently to some object on or above the bottom by byssal threads (*e.g.*, *Mytilus* and *Pinctada*). Swinnerton (1947), in his excellent discussion of shell modification among the pelecypods, points out the several important changes produced in the shell, as well as in the soft parts, by adaptation to the byssal mode of attachment. Most important of these are change in shell form from flattened ellipsoidal (elliptical outline) to flattened conoidal (triangular outline); rotation of shell with reference to the substratum, with consequent rotation of the hinge line from horizontality to a position perpendicular to the substratum; gradual shrinking of anterior part of shell; and disappearance of the anterior adductor muscle accompanied by enlargement and centering of the posterior adductor.

Some pelecypods assumed the habit of lying on one valve—the left in some cases, the right in others—and these soon began to attach the under valve to the substratum by byssal threads or by direct cementation. The fixed valve underwent great changes in size, shape, and structure, whereas the free valve, in some cases acting as little more than a lid, remained less changed and has proved the more reliable for discriminating genera and tracing ancestry (Fig. 10-8*B–C*).

In such forms as the chamids and rudistids, which arose from free ancestors, the fixed valve came to resemble a solitary coral in general appearance. In the rudistids the attached valve became a cone, in which coralloid structure was developed, and the free valve was modified into a lid that carried on its underside long teeth fitting into deep sockets in the internal calcareous framework of the fixed valve (Fig. 10-8*C*). These pelecypods lived in great profusion and were widely distributed in the clear, shallow Cretaceous seas of the Mediterranean and Texas-Mexican regions.[1] Locally they formed bioherms (reefs) and biostromes.

In those forms, like the oysters, which arose from ancestors that already used byssal attachment, excessive shell growth was confined to the ventral margin and resulted in a shell that tended to become spiral in one plane or another. *Ostrea* illustrates the wide range in shell modification that is brought about by this kind of attachment, and its two ancient relatives, *Gryphaea* (Fig. 10-29*I*) and *Exogyra* (Figs. 10-29*A–C*, 10-33) exemplify forms with heavy spiral valves.

Ecology[2] and Natural History

Living Pelecypoda, other than certain Sphaerids,[3] are exclusively aquatic and are adapted to life in fresh, brackish and normal marine waters, in any one of which individuals may be exceedingly abundant. They are characteristically bottom dwellers, but within the limits of bottom environments they have acquired a wide range of adaptation.[4] Some move freely over

[1] Douvillé, H., 1935, Les Rudistes et leur évolution, *Bull. soc. géol. France*, vol. 5, pp. 319–358.

[2] The interested reader will find excellent discussions of pelecypod ecology in the following articles: Bartsch, P., 1937, An ecological cross section of the lower part of Florida based largely upon its Molluscan fauna, *Nat. Res. Council, Rept. Comm. Paleoecology*, 1936–1937, pp. 11–25. Stenzel, H. B., 1945, Paleoecology of some oysters, *Nat. Res. Council, Comm. Marine Ecology as Related to Paleontology*, 1944–1945, pp. 37–46.

[3] A few species in Family Sphaeridae will leave their normal fresh-water habitat and invade the land for a short distance, working in between wet leaves with their muscular foot.

[4] Since pelecypods are filter feeders and cannot, therefore, be either scavengers or

bottom sediments, ploughing through soft mud and sand by means of the foot and leaving a typical furrow as a trail; some burrow into soft bottoms, in extreme cases burying the shell completely and maintaining circulation by means of long siphons, and others bore holes into rocks on the bottom or into wooden structures rising from the bottom; others attach themselves temporarily or semipermanently by byssal threads, or permanently by cementation; and a few adults are swimmers. Most living genera are confined to marine environments, where they exhibit a wide range of adaptation to depth and other conditions. Some live in the ever-changing littoral zone,[1] others are spread widely over the continental shelves and slopes, and deep-sea forms have been taken from depths as great as 10,450 m. (34,000 ft.). Over the continental shelves and slopes, in places where deposition is slow, individuals of some species are widespread and prodigiously numerous.[2] One such area, situated on the south end of Dogger Bank in the North Sea, has an extent of 700 sq. miles and an estimated population of 4,500,000,000,000 clams. The oyster beds of many bottoms with their enormous populations are another well-known example. Among fossil pelecypods, an outstanding example of a similar nature is *Exogyra cancellata*, from the Upper Cretaceous, which has been traced with interruptions from New Jersey to Mexico (Fig. 10-33), a distance of over 4,000 km. (2,500 miles), and in one place the down-dip spread of the fossiliferous layers is as much as 185 km. (115 miles) (Stephenson, 1933).

Pelecypods are dispersed chiefly by currents that transport the free-swimming larvae. Fish carry certain parasitic larvae of the fresh-water Unionidae, and a wading fowl is known to have transported a living pelecypod which clamped its shell onto one of the bird's feet.

Burrowing and boring pelecypods swallow large quantities of mud, sand,

---

predators, they are generally found in limited ecological niches. Their chief adaptations, therefore, are to the substratum.

[1] The tidal zone and the shallow waters directly adjacent may be densely populated by species of *Mya* (Fig. 10-23), a burrowing genus, whose presence may commonly be detected by the water that is squirted from the exhalant siphon as one walks near the buried animal. Also common in the tidal zone and ajdacent shallow waters are *Mytilus*, the black mussel, *Pholas*, the well-known rock borer, and *Bankia*, *Teredo*, and *Martesia*, the notorious wood borers. *Mytilus* may locally carpet rocky bottoms by the millions. The rocks of some coasts are fairly riddled with holes excavated by *Pholas*, shells of which may be found trapped in the hole that was progressively enlarged as the animal grew. The boring activities of *Teredo* and *Bankia*, the so-called "shipworms," are of great concern to persons having the responsibility of maintaining wooden shore installations in marine waters. These particular pelecypods have undergone spectacular modification in adapting to wood-boring activities, as illustrated in Fig. 10-32. The interested reader will find an excellent discussion of the natural history and of the depredations of *Bankia* in the following report: Clench, W. J., and Turner, R. D., 1946, The genus *Bankia* in the Western Atlantic, *Johnsonia* (M.C.Z., Harvard), vol. 2, no. 19, pp. 1–28.

[2] Many modern pelecypods are gregarious in habit, living together by the thousands where biological and physical environmental conditions are congenial. However, at other places on the same bottom and at the same depth, the same forms may be rare or absent, possibly because of unfavorable conditions when the spat settled.

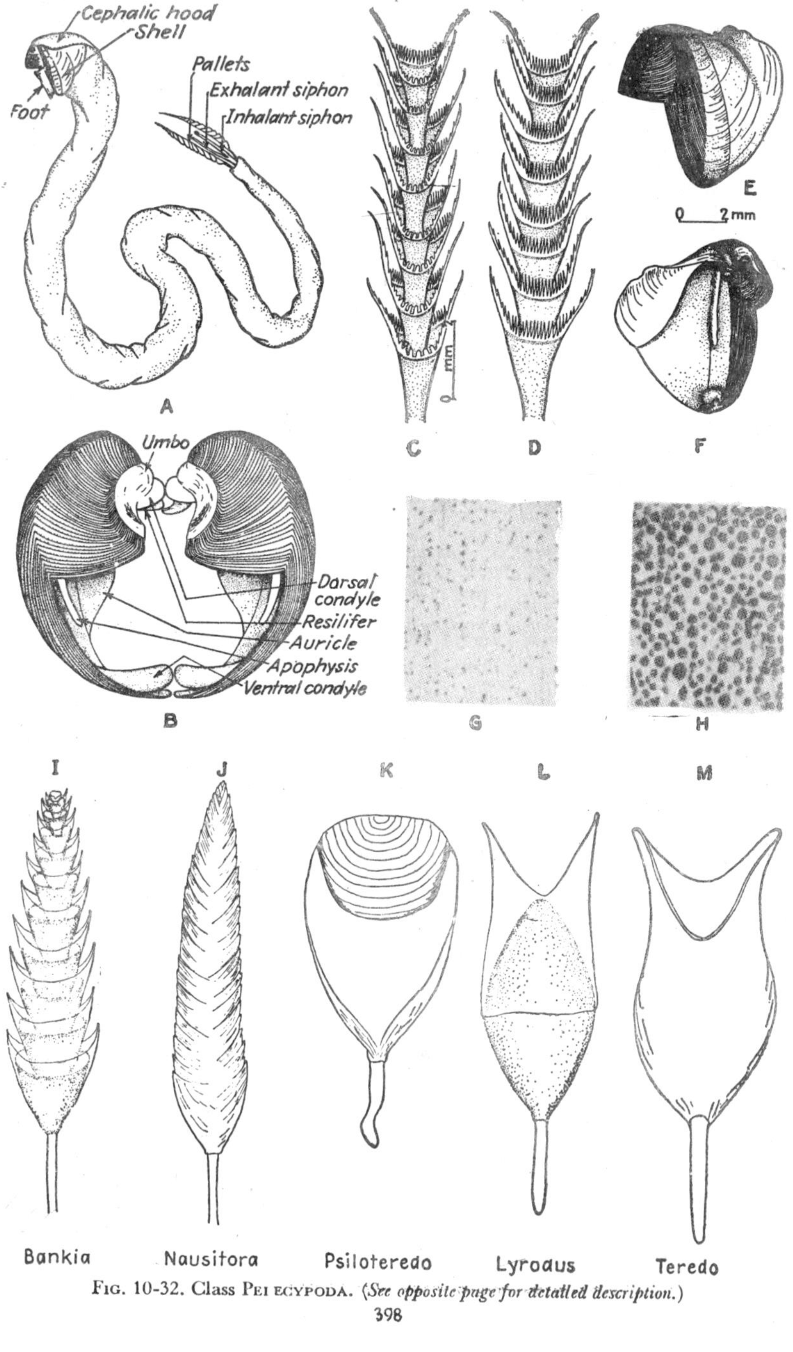

FIG. 10-32. Class PELECYPODA. (*See opposite page for detailed description.*)

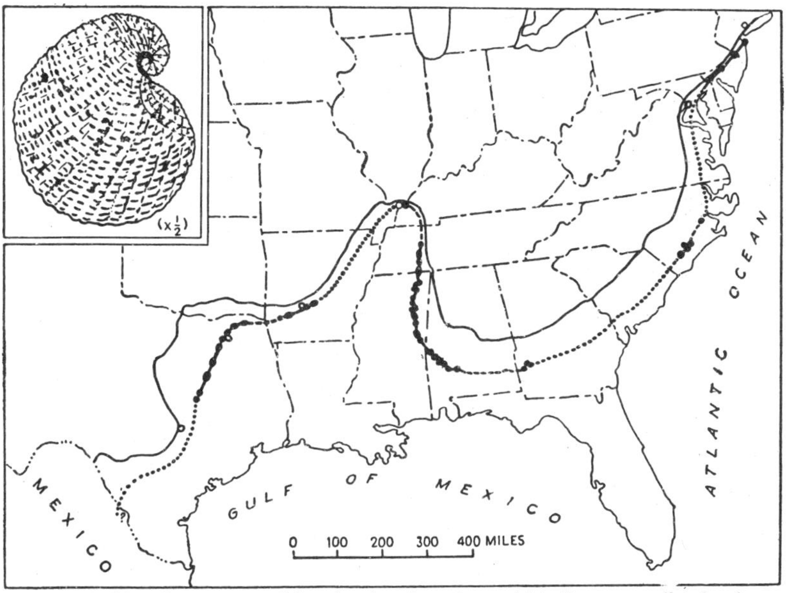

FIG. 10-33. Class PELECYPODA. Map showing line along which *Exogyra cancellata* has been collected, including down-dip occurrences on salt-dome structures as well. Spots indicate localities; dashed lines indicate parts of outcrop zone along which no specimens were collected; dotted lines indicate stretches along which zone is overlapped by younger formations. Inset shows sketch of left valve. (*Map adapted from Stephenson*, 1933.)

pulverized rock, and shredded wood, and these materials are certain to undergo both physical and chemical change during passage through the animal's alimentary tract.

Commensalism, the habit of sharing habitation and food without being parasitic, is common among pelecypods. Some live in the burrows of crustaceans and worms, others live on echinoderms, and still others live embedded in the tests of Ascidians or in the tissue of sponges. Rarely a species is parasitic

FIG. 10-32. Class PELECYPODA. *Bankia*, the notorious and destructive wood-boring pelecypod. *A*. Diagrammatic drawing of an entire specimen, showing the highly modified shell at the anterior end, and at the posterior extremity the peculiar pallets, which are used for pushing comminuted wood out of the boring and also for plugging up the small entrance to the boring. Pallets are composed of a calcareous base covered with a thin periostracum and consist of two parts—the **blade,** composed of numerous conelike elements, and the **stalk** which has an axial position, is nearly circular in cross section, and may be visible through the cones. *B*. Anterior view (pedal gape) of a typical teredinid shell, showing the two highly specialized valves and the apposition of the condyles on which they rotate. *C–D*. Outer and inner surfaces, respectively, of one pallet of *Bankia fosteri*. *E–F*. External and internal views, respectively, of one valve of *B. fosteri*. *G*. Outside of a piece of board bored by *B. gouldi*. The smaller holes lead to the internal borings of individuals that were alive when the board was removed; the larger holes lead to tubes in which the animals were dead. *H*. Inside of same board as *G*, showing tunnels after young *Bankia* had grown to a length of 12 mm. (½ in.) *I–M*. A series of pallets showing variation in the Family Teredinidae, all members of which are wood borers (shipworms). These pallets are largely calcareous with a chitinous cover, and range from *I*, which consists of a series of cones threaded on an axial stem, to *M*, which is a single cone. (*All by Turner*, 1946.)

(*e.g.*, parasitic larvae of the Unionidae infest certain fish, and *Entovalva* lives in holothurians).

The surface of a pelecypod shell, regardless of whether the shell is empty or occupied by a live animal, offers an excellent place for all sorts of benthonic organisms to become attached; hence algal mats, worm tubes, bryozoan zoaria, small oysters and other cementing pelecypods, and barnacle shells commonly incrust much or all of the shell exterior.

Pelecypods are known to have been an important source of food for man since an early stage in his evolution, and the shell heaps accumulated by certain groups of ancient men emphasize the importance of pelecypods as food in early human history. Oysters, scallops, marine mussels, and marine clams have been most commonly used for food by man, but pelecypods are also eaten by other animals. Starfish feed over shell beds, especially over oyster banks, and kill their prey by more or less completely enveloping the whole animal in the protruded digestive system. Many gastropods drill holes into living shells to get at the flesh within, and along some coasts the majority of dead pelecypod shells of certain species have a neat borehole near the beak. Certain carnivorous gastropods have devastated commercial oyster beds. Small pelecypod shells are swallowed by certain fish, and by rooting crustaceans and other animals.

The shells of certain fresh-water clams have long been exploited for their mother-of-pearl, which has been used the world over for buttons, knife handles, inlay work, and ornaments. Pearls have been prized as gems since antiquity, and the Japanese have developed an important industry based on pearl culture.

The damage done by shipworms to wooden structures in marine waters has been a constant concern to man, as pointed out in earlier paragraphs (see footnote on page 397), and in some cases has involved great losses. As an example, it has been estimated that in four years' time (1917–1921) shipworms caused 25 million dollars damage to piers and other wooden structures in San Francisco Bay.

Geologic History

The earliest known fossil pelecypods have come from Lower Ordovician rocks. They are small simple shells that have been assigned to *Ctenodonta*, *Glyptarca*, *Modiolopsis*, etc.[1] (Fig. 10-34). Before the close of the Ordovician many genera and species had appeared and the pelecypods were on the way to establishing themselves as an important component in benthonic faunas, a position that they have held with certain interruptions to the present time. Large pelecypod faunas have been collected from Silurian and Devonian rocks. Some Devonian species lived in fresh water. Brackish-water forms appear in great numbers in later Paleozoic rocks.

The Pelecypoda underwent great change at the close of the Paleozoic and during the Triassic as many ancient genera became extinct and many new genera, some of which have persisted to the present, made their ap-

[1] Hicks, H. H., 1873, On the Tremadoc rocks in the neighbourhood of St. David's, South Wales, and their fossil contents, *Quart. Jour. Geol. Soc. London*, vol. 29, pp. 39–52.

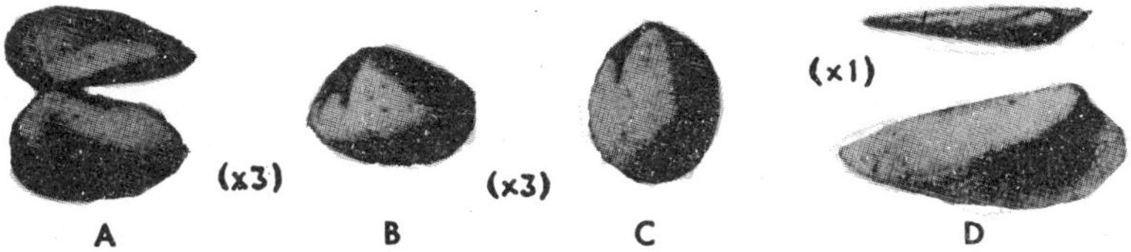

FIG. 10-34. Class PELECYPODA. Earliest known pelecypod shells, all from Tremadoc (L. Ord.) strata of South Wales, England. *A. Glyptarca primaea. B–C. Ctenodonta menapiensis. D. Modiolopsis ramseyensis.* (*After Hicks*, 1873.)

pearance. Since the Triassic the pelecypod fauna of the world has become increasingly more modern.

The number of new families appearing in each geologic period is approximately as follows (also see Fig. 10-31):

Ordovician	10	Jurassic	15
Silurian	10	Cretaceous	18
Devonian	9	Eocene	15
Carboniferous	3	Miocene-Pliocene	2
Permian	1	Pleistocene-Recent	3
Triassic	14		

At certain times in the geologic past pelecypods have contributed importantly to the construction of calcareous rocks. Most Early Paleozoic shells, although adding to the enclosing sediments at the time of burial, are not now preserved because of removal by solution owing, possibly, to the shells having been composed of aragonite. In Pennsylvanian rocks, however, shell matter is commonly preserved, and in later rocks preservation is likely to be good. During the Cretaceous the chamaceans and rudistids built bioherms, and in Colorado during Pierre time the shells of *Lucina occidentalis ventricosa* piled up into heaps referred to as "tepee buttes."[1] Many genera were locally abundant during the Mesozoic, and their shells now constitute a large part of the rock. *Gryphaea*, *Exogyra*, and *Inoceramus* were such genera in the Jurassic and Cretaceous, and *Ostrea* has been an important contributor of shells from the Cretaceous to the present time. Reference has already been made to the 700-sq.-mile accumulation that is building on Dogger Bank.

The shells of most pre-Pennsylvanian pelecypods are not generally preserved, the record of their existence consisting largely of impressions and fillings of the interior (steinkerns, cores); in contrast, Pennsylvanian and later shells are likely to be preserved. The explanation for this difference in preservation may be that the Early Paleozoic pelecypods constructed a dominantly aragonitic shell, whereas those of Pennsylvanian and later time built shells that were dominantly calcitic. The shells of aragonite, being composed of a relatively unstable form of calcium carbonate, were easily destroyed by solution, whereas those of calcite could persist through the changes leading to fossilization.

[1] These shell mounds, or "tepee buttes," average about 4½ m. (15 ft.) in diameter and are generally higher than wide. They mark sites where the pelecypods continued to live on in spite of muddy conditions.

The shells of most dead pelecypods are open, and commonly the valves are either spread apart more widely than the muscles would have permitted in life or are completely separated. This condition, which is the opposite of that characteristic of dead brachiopod shells, is due to the action of the ligament or resilium or both, which force the valves apart when death causes the adductors to relax.

The stony structures made by algae, hydrocorallines and corals, worms, bryozoans, brachiopods, other pelecypods, and barnacles commonly incrust the valves and whole shells of fossil pelecypods, just as they do on the shells of living forms in modern seas. Holes bored by predatory gastropods indicate the manner in which some pelecypods met death (Fig. 10-54), and repaired injuries indicate accidents of one kind or another in the seas of long ago.

In general, pelecypods are not among the best index fossils, chiefly because poor preservation often makes accurate generic or specific determinations uncertain or impossible. Especially is this true of pre-Devonian forms. In Pennsylvanian and later rocks the better preserved shells provide many useful index fossils, among which may be mentioned *Myalina*, *Aviculopecten*, *Pernopecten*, *Pseudomonotis*, *Streblochondria*, *Exogyra*, *Gryphaea*, *Inoceramus*, *Hippurites*, *Protocardium*, *Radiolites*, *Venericardia*, and *Glycymeris*. (Figs. 10-28, 10-29, 10-30). The value of nonmarine pelecypods for correlation of Pennsylvanian rocks has been discussed by Dix and Trueman (1937) in a comprehensive article that also contains a lengthy bibliography, and Hessland (1945) has shown the importance of *Mya* in the European Quaternary.

Class Gastropoda[1]

General Considerations

Typical of the Class Gastropoda are the land and fresh-water snails and the myriads of marine conchs, whelks, limpets, periwinkles, etc. Many have shells that are beautiful in both architecture and coloration, and certain of the shells have long been prized by collectors.

The animal has a well-developed head region, a ventral foot specialized for creeping, a simple or twisted alimentary tract, and ctenidia or specialized structures for breathing. A radula is present in all but a few parasitic forms. Fertilization is usually internal, and the sexes are united or separate, depending on the species. The embryo develops into a trochophore-like larva that usually passes through a veliger stage during which it undergoes torsion before becoming an adult.

The typical gastropod has a univalve calcareous shell in the form of a simple unchambered cone that is generally coiled in some manner. In all but a few forms that have the shell reduced and internal or lacking altogether (land slugs), or have a shell in only the earliest growth stages (nudibranchs), the adult shell is external and is borne dorsally. It serves as a refuge into which the animal can withdraw for protection. Most gastropods can withdraw completely into the shell, and many bear on the foot a horny or cal-

[1] Gastropoda—Gr. *gaster*, *gastros*, stomach, + *pous*, *podos*, foot; referring to the position of the foot on the ventral part of the animal.

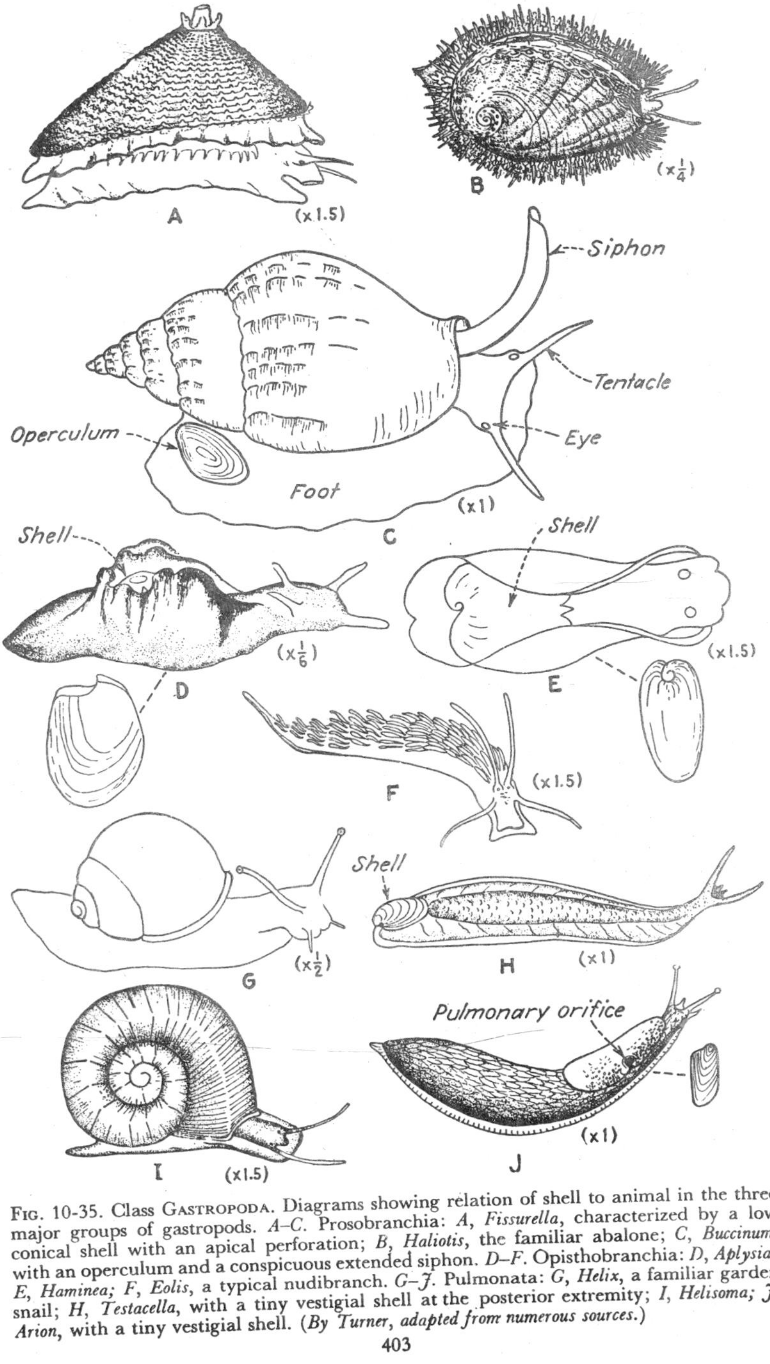

FIG. 10-35. Class GASTROPODA. Diagrams showing relation of shell to animal in the three major groups of gastropods. *A–C*. Prosobranchia: *A*, *Fissurella*, characterized by a low conical shell with an apical perforation; *B*, *Haliotis*, the familiar abalone; *C*, *Buccinum*, with an operculum and a conspicuous extended siphon. *D–F*. Opisthobranchia: *D*, *Aplysia*; *E*, *Haminea*; *F*, *Eolis*, a typical nudibranch. *G–J*. Pulmonata: *G*, *Helix*, a familiar garden snail; *H*, *Testacella*, with a tiny vestigial shell at the posterior extremity; *I*, *Helisoma*; *J*, *Arion*, with a tiny vestigial shell. (*By Turner, adapted from numerous sources.*)

careous plate, the operculum, that closes the aperture of the shell when the animal completely withdraws into the shell.

The majority of gastropods are aquatic, and most species live in shallow marine waters. Some, however, have become adapted to brackish- and fresh-water environments, and two groups, by modifying the breathing apparatus, have successfully invaded the land. Today gastropods may be found in a variety of marine environments, in brackish-water bays and lagoons, along lake shores and in streams, and in a variety of land habitats commonly far from permanent water bodies. They are among the most adaptable of invertebrates with reference to such conditions as depth below and height above sea level, water and air pressure, temperature and salinity of water, and temperature and humidity of air. Although it is generally assumed that ancient and extinct species were adapted in ways similar to those of living forms, there is little evidence from the fossil record either to support or to refute this assumption.

At the present time gastropods excel all other mollusks in variety of shell form and importance as a faunal component. More than 50,000 species have been described, of which about 15,000 are fossil and 35,000 living. Their geologic history begins in the early Cambrian and extends unbroken to the present, and they are now probably as numerous and varied as they have ever been.

Morphology of Soft Parts

The body of a gastropod is easily divisible into three parts—**head, foot,** and **visceral hump** (Fig. 10-35). The hump is more or less completely covered with a membranous **mantle,** which secretes and lines the dorsally borne shell. In many adult gastropods the body is asymmetrical as a result of torsion of certain of the soft parts, but the fundamental symmetry, as in all Mollusca, is bilateral and is evident in the early growth stages.

The well-developed head has a ventrally situated **mouth,** a pair of **eyes** that are commonly stalked, and one or two pairs of retractile **tentacles** that act as sensory organs (Figs. 10-35, 10-37). The mouth opens into the **buccal cavity,** or **pharynx,** which contains the radula (Fig. 10-3*A*). The typical gastropod radula is a chitinous band beset with transversely arranged teeth supported on a cartilaginous odontophore. When moved back and forth the radula acts as an effective rasp or file. The number, arrangement, and shape of the tiny teeth are so constant that radulae are important in classifying living gastropods.[1] The number of teeth ranges from 16 to as many as 750,000 and the teeth themselves are usually arranged in two symmetrical sets, one set on either side of a median tooth, as shown in Fig. 10-3. The individual teeth exhibit considerable variety in form and in certain details bear some resemblance to the chitinophosphatic denticles collectively termed **conodonts** (Chap. 15). A few paleontologists have suggested that conodonts

[1] Gastropod radulae show constant and systematic variations of such nature that they can be divided into several different types, which have been used by many zoologists as a basis for ordinal and superfamilial divisions. The characteristics of the several types are illustrated in Fig. 10-36.

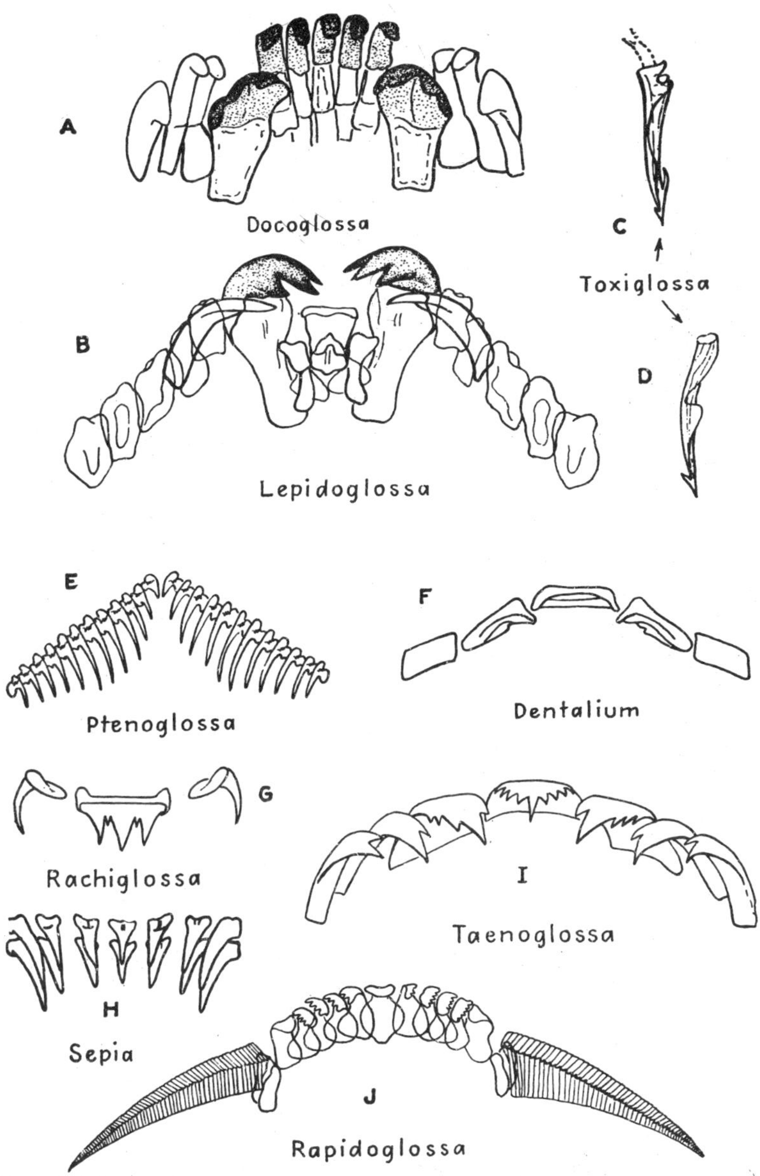

FIG. 10-36. Phylum MOLLUSCA. Radulae. Diagrams showing the different types of molluscan radulae; in most figures only one transverse row of radular elements is shown. *A*. Docoglossate radula from *Ancistromesus*, an amphineuran. *B*. Lepidoglossate radula from *Cryptoconchus*, a marine gastropod. *C–D*. Toxiglossate radulae from *Conus* and *Terebra*, respectively, familiar marine gastropods. *E*, Ptenoglossate type from *Actaeon*, a marine gastropod. *F*. Radula of *Dentalium*, a common scaphopod. *G*. Rachiglossate radula of *Murex*, a marine gastropod. *H*. Two transverse rows from the radula of *Sepia*, a common sepioid squid. *I*. Taenoglossate radula of *Lanistes*, a fresh-water gastropod. *J*. Rapidoglossate radula of *Clanculus*, a marine gastropod. (*By Turner, adapted from several authors.*)

are actually the radular teeth of extinct gastropods but the suggestion is not acceptable for reasons discussed in Chap. 15.

In some gastropods there are also horny **jaws** (**mandibles**) in the dorsal part of the buccal cavity directly opposite the ventrally situated radula (Fig. 10-3*A*). The radula and associated dental plates probably function most often as tearing and grasping organs, and they are particularly useful to boring forms. Common boring gastropods are the whelks (*e.g.*, *Purpura*) and the naticoids (*e.g.*, *Natica*). These bore through the shells of pelecypods, other gastropods, and brachiopods largely by chemical means—*i.e.*, by using acids[1] as solvents—then insert the proboscis into the borehole and employ the radula to seize and macerate the flesh of the prey. Many shells on modern beaches have boreholes made by gastropods, and a few fossil shells as old as Devonian have similar holes that are generally ascribed to boring gastropods (Fig. 10-54*C*–*E*). Boring gastropods are particularly destructive on many oyster banks.

The foot is typically a flat creeping organ back of the head on the underside of the body. Bottom-dwelling gastropods move slowly over the substratum by causing the surface of the foot to be rhythmically rippled. In the pelagic Heteropoda the foot is modified into a vertical laterally compressed fin, and in the Pteropoda it consists of a pair of winglike membranes near the head. Members of both groups swim by means of the modified foot.

The intestinal tract in young forms is a nearly straight tube with an anterior mouth and a posterior anus. In the adult stage of many forms, however, it is looped in such a way that the anal opening is situated above the head (Fig. 10-37*A*).

The liver, kidneys, and certain glands are normally clustered about the posterior part of the alimentary tract, but because of torsion they vary from this position.[2] The heart has one auricle and a ventricle in all but the most primitive forms, and the blood is circulated through a much-branched system of vessels. Respiration is carried on in a few forms by the mantle, but most gastropods have gills or lungs. Fundamentally, the gills are paired, but owing to distortion of the body the right gill is lost in many groups. The gills are lamellar, plumelike, or featherlike and occupy the mantle cavity. In a few specialized forms (the nudibranchs) supplementary pseudogills of considerable variety have been developed, usually following loss of true gills (ctenidia). These supplementary organs may be exposed on the back or on the sides. Gastropods that have adapted themselves to life on the land (*i.e.*, the Pulmonata) have lost their ctenidia, and the mantle cavity, lined with blood vessels, functions as a lung. Some pulmonates have returned to an aqueous habitat, but having a lung instead of gills, they must come to the surface at intervals to breathe air or to absorb oxygen directly through the integument.

In order to supply the gills with uncontaminated water and to flush impurities out of the mantle cavity, the gastropod must maintain a current of

[1] Sulphuric acid has been reported in some gastropods. This would serve as an effective solvent.

[2] Primitively the gastropod had pairs of most organs, but after torsion only the left organ of a pair was left when the right one became atrophied as a result of the torsion.

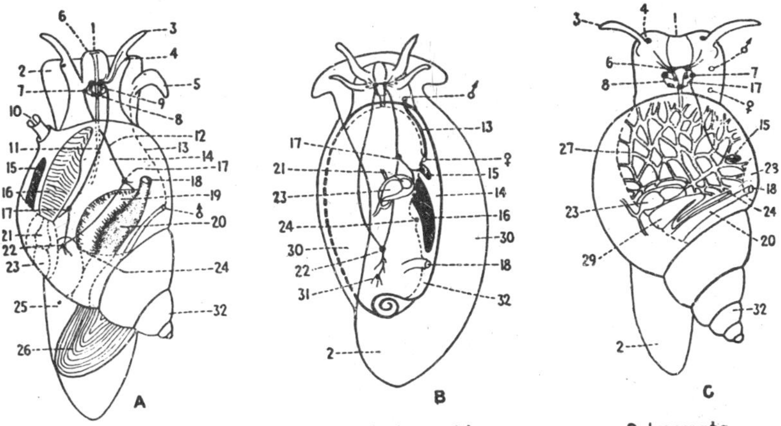

FIG. 10-37. Class GASTROPODA. Diagrams showing the morphological characteristics on which the three orders of Gastropoda are based. *A*. Idealized diagram of a representative of the Prosobranchia, showing twisted nerve commissures, single pair of tentacles, gills anterior to heart, single opening for monoecious sex products, spiral shell, and operculum. *B*. Representative of the Opisthobranchia, characterized by gills posterior to heart, untwisted nerve commissures, two pairs of tentacles, combined sexes, and nonoperculate partly internal shell. *C*. A typical representative of the Pulmonata, having a pulmonary sac, untwisted nerve commissures that are concentrated around the esophagus, two openings in the head for escape of sex products, two tentacles, and a spiral shell that is nonoperculate except in the Family Amphibolidae. (1 = mouth; 2 = foot; 3 = tentacle; 4 = eye; 5 = penis; 6 = cerebral ganglion; 7 = pleural ganglion; 8 = pedal ganglion; 9 = statocyst; 10 = siphon; 11 = supraintestinal connective; 12 = subintestinal connective; 13 = sperm channel; 14 = mantle cavity; 15 = osphradium; 16 = ctenidium; 17 = parietal ganglion; 18 = anus; 19 = hypobranchial gland; 20 = rectum; 21 = auricle; 22 = visceral ganglion; 23 = heart cavity; 24 = outlet of nephridium; 25 = foot; 26 = operculum; 27 = vascular network of the lung; 28 = breathing aperture; 29 = kidney; 30 = parapodial lobe; 31 = genital ganglion; 32 = shell.) (*By Turner; modified after Lang*, 1900; *numbered as in Wenz*, 1938.)

water over the gills and through the mantle cavity. Normally, water enters under the anterior part of the mantle cavity and flows so as to bathe the gills, after which it cleanses the cavity of excreted and secreted impurities before passing to the exterior (Fig. 10-45*B–D*). Many specializations have been developed to separate clean water from foul; some advanced forms, for example, have developed in the anterior part of the mantle a tubular fold, the **siphon,** through which water is drawn into the mantle cavity. The gastropod siphon is in no way analogous to or homologous with the "siphon," or hyponome, of the cephalopods (Fig. 10-55), but it is similar to the siphon of pelecypods. Siphonated gastropods have several shell features that are closely associated with the siphon (Figs. 10-2*C*, 10-35*C*).

In most marine gastropods the sexes are separate, whereas terrestrial forms are typically hermaphroditic (*i.e.*, the reproductive organs of both sexes are present in the same individual). In the simplest type of reproduction, sperm and unfertilized eggs are released to the exterior, where fertilization takes place. The fertilized egg then develops into a trochophore larva, and this in turn passes through the veliger stage unique to the Mollusca before the developing animal attains adult characteristics. In many forms, however, the fertilized eggs are held in chitinized capsules, or egg cases, and in these the embryonic stages are usually completed and the young hatched out. The young commonly remain in the capsule for a short time before abandoning it, so that when they finally leave they are young gastropods with completely formed shells (Fig. 10-38*A–G*). In a few cases the parent retains the capsules within the body and gives birth to young individuals. Such gastropods are described as **ovoviviparous** in contrast to **oviparous** forms, which bring forth eggs from within the body.

One of the most interesting aspects of gastropod development is the torsion that takes place in certain of the soft parts. The rudimentary alimentary tract of the larva becomes bent upon itself as growth proceeds, and the visceral hump begins to rotate around its main axis in a movement that carries the mantle cavity and anal opening to a position on the front side of the hump, so that the mantle cavity opens forward. The loop of nerve commissures serving the visceral hump has its distal part rotated during the torsion just described and comes to have the form of a figure 8. Most modern gastropods have the loop twisted; the pulmonates, however, have developed a symmetrical system, and the nudibranchs have undergone detorsion (Fig. 10-37).

Zoologists have long based the major divisions of the Class Gastropoda on the nature of the nervous system and the relationships of the more important soft parts. Unfortunately for the student of fossil gastropods, only the shell is available for study, and it carries little or no record of the profound anatomical changes that can be observed to take place in living forms. Consequently, the classification adopted in the present work is based largely on the nature of certain soft parts (particularly the gills) of living forms, and the fossils are fitted into the classification as well as possible.

The Shell

A complete gastropod shell consists of a tiny apical portion, the **nucleus,** composed of the **nuclear whorls** of the embryonic shell, and the shell proper, or **conch** (Figs. 10-38, 10-39). The nuclear whorls are usually smooth and commonly differ strikingly from those of the conch. They are collectively termed the **protoconch**[1] by some investigators, and that useful designation is adopted in the present work.

In those Gastropoda with the more primitive mode of reproduction (*i.e.*, with a free-swimming larval stage), the first rudimentary shell appears on the free-swimming larva as a cap-shaped calcareous plate. Additions are made

[1] Some malacologists and gastropod specialists use the designation nucleus or nuclear whorls for the complete embryonic shell and restrict the term protoconch to the simple cap-shaped plate that constitutes the first shell rudiment (Knight, 1941). Inasmuch as the first plate commonly passes insensibly into the nuclear whorl, so that its termination is not detectable, we prefer to use the term protoconch in a general sense to apply to the fully formed embryonic shell

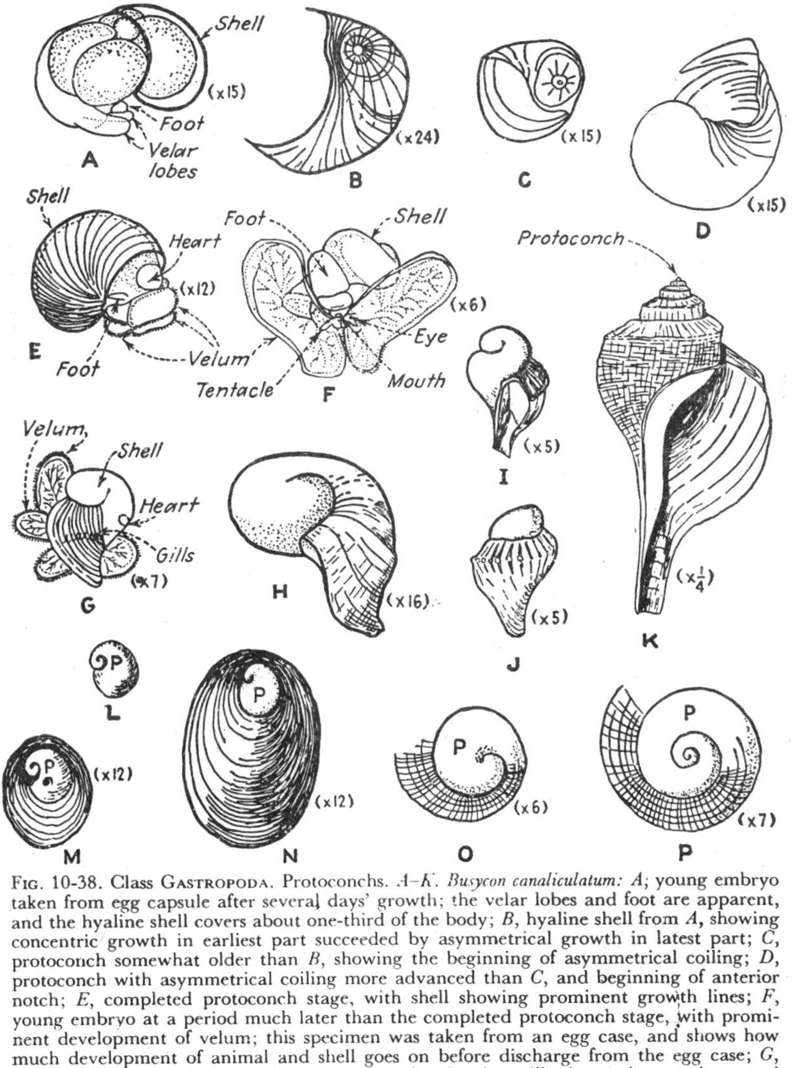

FIG. 10-38. Class GASTROPODA. Protoconchs. *A–K. Busycon canaliculatum: A;* young embryo taken from egg capsule after several days' growth; the velar lobes and foot are apparent, and the hyaline shell covers about one-third of the body; *B*, hyaline shell from *A*, showing concentric growth in earliest part succeeded by asymmetrical growth in latest part; *C*, protoconch somewhat older than *B*, showing the beginning of asymmetrical coiling; *D*, protoconch with asymmetrical coiling more advanced than *C*, and beginning of anterior notch; *E*, completed protoconch stage, with shell showing prominent growth lines; *F*, young embryo at a period much later than the completed protoconch stage, with prominent development of velum; this specimen was taken from an egg case, and shows how much development of animal and shell goes on before discharge from the egg case; *G*, embryo just before emerging from egg capsule, showing gills, heart, large velum, and complete protoconch with numerous growth lines; *H*, a complete protoconch; *I–J*, young protoconchs as taken from an egg capsule, showing some surface sculpture; *K*, an adult shell. *L–N. Crepidula fornicata: L*, a complete protoconch; *M*, protoconch with earliest part of conch; *N*, a young shell, somewhat older than *M*. *O–P. Ficus: O, F. papyratia*, a protoconch occupying about one whorl; *P, F. filia*, a protoconch consisting of about two and one-half whorls. (*P* = protoconch.) (*A–D, G–J after Grabau*, 1903, 1910; *E–F, L–N after Jackson*, 1890; *O–P after Smith*, 1945.)

to this plate along its entire free margin, but the protorsional posterior edge grows faster than the anterior so that the shell becomes coiled in a plane. Except in the bellerophonts, this planispiral coiling is of short duration, however, and soon changes to the helicoid type of coiling that characterizes the Gastropoda as a class. Thus, the embryonic shell, or protoconch, passes through three distinct stages characterized by (1) a simple cap-shaped plate; (2) planispiral mode of coiling; and (3) helicoid type of coiling. In its complete development, therefore, the protoconch is generally a smooth, helically coiled shell with a small hollow cone, the **umbilicus,** along the axis of coiling (Fig. 10-38*H–J*). Certain Lower Cambrian genera (*e.g., Scenella*) believed to be primitive gastropods have noncoiled cap-shaped adult shells somewhat like the first stage of the typical protoconch of modern coiled gastropods.

In those more advanced Gastropoda that protect the eggs by capsules, the embryonic and early nepionic[1] stages are passed within the capsule, so that by the time the young gastropod emerges from the capsule, it has built a young shell that consists of a complete protoconch and one or more whorls of the conch (Fig. 10-38*H–J*). In most species the protoconch merges into the early conch stage (nepionic) without any visible line of demarcation (*e.g., Busycon carica*), but in some (*e.g., Ficus papvratia* and *F. filia*) the conch begins abruptly and the whorl has conspicuously different surface sculpture (Fig. 10-38*O–P*).

The ultimate shape of the adult shell depends upon the way in which shell matter is added to the margin of the nuclear whorl Elongation of the protorsional posterior margin of the whorl, with coiling about a horizontal axis that is perpendicular to the plane of coiling, produces an adult planispiral shell, as illustrated by *Knightites* (Fig. 10-45*B–D*), in which the coil has right and left sides. More commonly, however, the planispiral coiling in the protoconch gives way to helicoid coiling, and the adult shell, as a consequence, is spirally coiled. As noted in a preceding paragraph, the nuclear whorls of all species of which the embryonic development is known are helically coiled regardless of the shape of the adult shell. It does not necessarily follow, however, that this has always been the case. It is possible that the cap-shaped gastropod shells (*e.g., Scenella*) of the Cambrian represent the first condition of the protoconch; the planispirals (*e.g., Owenella*) of Cambrian, Ordovician, and later periods, the second stage of protoconch development; and the helicoid spirals (*e.g., Lophospira*), the third. Certain of the later cap-shaped shells (*Crepidula;* Fig. 10-38*L–N*) do not represent a primitive stage, however, because they have a helicoid protoconch.

Except for the protoconch, which is quite distinct from the remainder of the shell (*i.e.*, the conch) and generally missing from adult specimens, the typical gastropod shell is essentially a simple unchambered cone open at the larger end and closed at the smaller except in a few apically perforate genera.

The simplest type of gastropod shell, architecturally considered, is the noncoiled conical shell characteristic of certain Cambrian forms (*e.g., Scenella*) and of modern **limpets** (Figs. 10-1; 10-35*A*). In coiled shells the

[1] Hyatt (1894) suggested the following terms for ontogenetic stages: **nepionic,** or babyhood: **neanic,** or youthful; **ephebic,** or adult; and **gerontic,** or senile.

fundamental cone may be coiled planispirally, helically, or irregularly and the successive volutions may be free or in contact (Fig. 10-40). One complete volution about the axis of coiling is a **whorl,** and the line marking the trace of the area of contact between adjacent whorls is the **suture** (Fig. 10-39). All the whorls except the last in fully developed shells together constitute the **spire.** The last may be conspicuously larger than those of the spire (Figs. 10-41*A*, 10-42). If the whorls are not in actual contact, or touch along a limited interface, (Fig. 10-40), they are completely visible, and their cross section tends to be circular or elliptical. Commonly each whorl overlaps part of the preceding whorl, and in extreme cases (*e.g.*, *Bulla;* Fig. 10-51*C*)

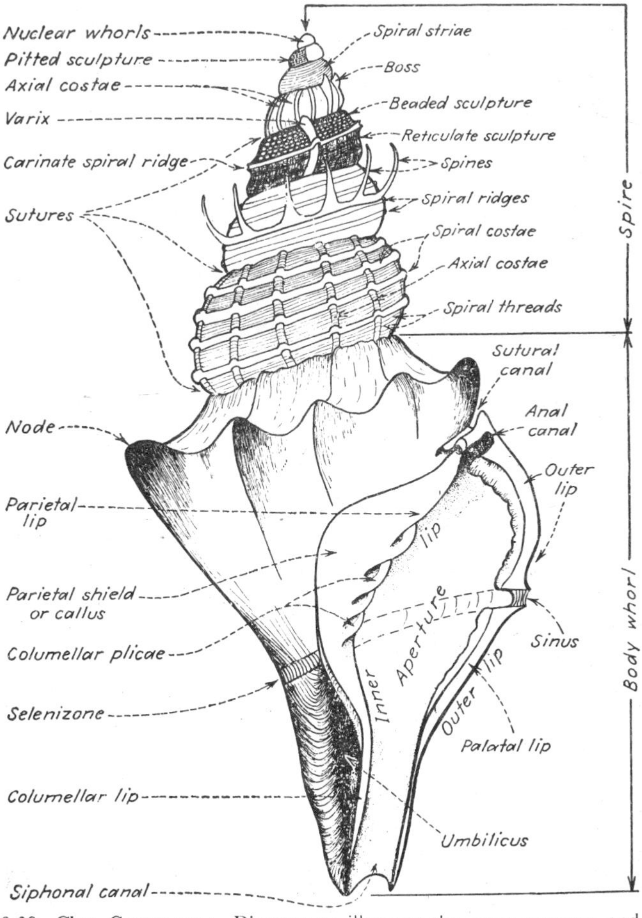

FIG. 10-39. Class GASTROPODA. Diagram to illustrate the more common and typical features of a gastropod shell. (*By Turner.*)

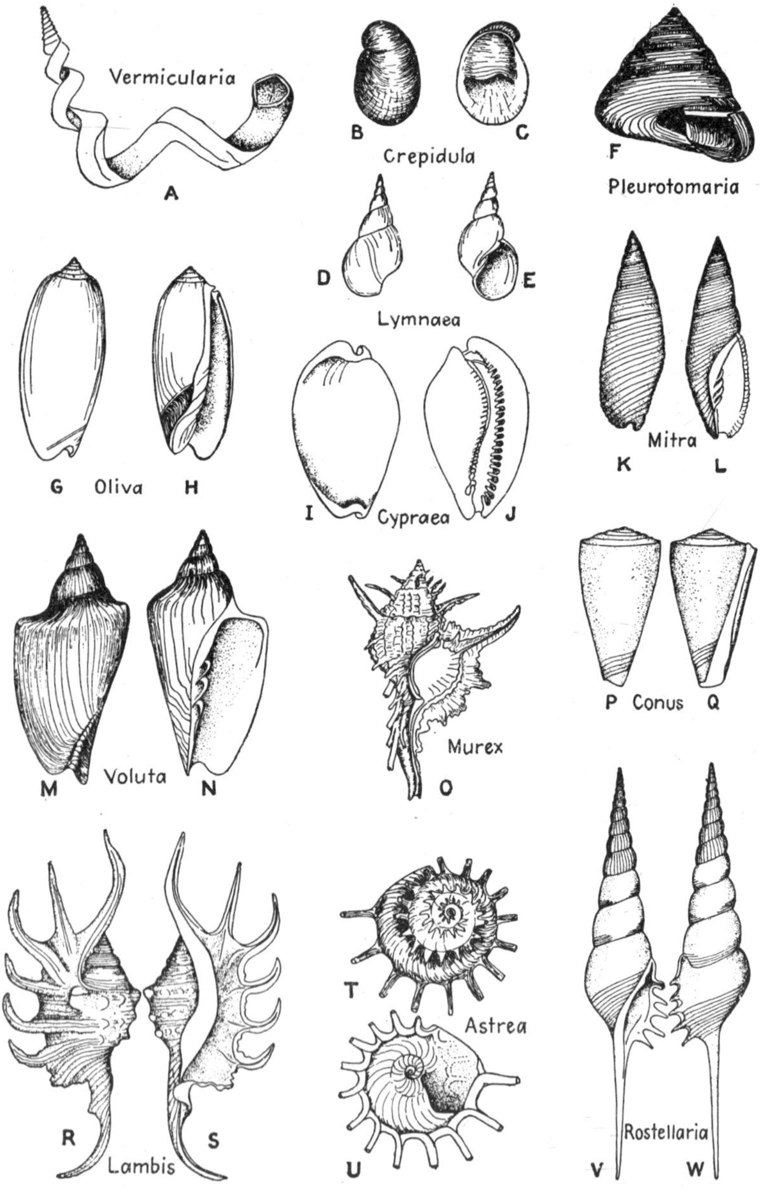

FIG. 10-40. Class Gastropoda. (*See opposite page for detailed description.*)

the overlap is so complete that only the last whorl is visible.[1] This type of coiling, in which the inner whorls are more or less completely concealed by the outer whorl or whorls, is described as **convolute** or **involute,** and the shell involved is similarly designated.

By far the majority of gastropod shells have the form[2] of a helicoid spiral in which the fundamental cone is coiled loosely or tightly around an imaginary axis. Shells in which the whorls are coiled around an imaginary axial cone, or **umbilicus** (Figs. 10-39, 10-41*C–E*), in such a way that a portion of each whorl is visible in the adult shell are described as **umbilicate** or **perforate.** In some shells the opening into the umbilicus is partly or completely closed by a shelly **callus**[3] (Figs. 10-39; 10-45*A*). **Imperforate** shells are those in which the whorls are so tightly coiled that no umbilicus exists or in which the umbilical opening has been closed by a reflection of the parietal shield or by a columellar fold. Some shells that are perforate in earlier growth stages become imperforate in the adult.

The whorls of coiled shells, when viewed in cross section, show more or less modification depending upon the tightness and nature of coiling. The upper side of the whorl tends to become differentiated into distinct areas and to vary from flat to convexly or concavely curved. The spire then has a contour consonant with the nature of the whorls. Flat upper surfaces give the spire a terraced aspect, as in *Trochonema* (Fig. 10-48*C–D*), or a uniformly inclined surface, as in *Eotomaria* (Fig. 10-48*Q*).

[1] *Cypraea* (Fig. 10-40*I–J*) has the last whorl covering quite a different appearing earlier shell.

[2] Knight (1941) used three terms to describe the form of Paleozoic gastropod shells: (1) **spiral** for those in which the whorls are coiled symmetrically in one plane, so that each flank of the coil is the mirror image of the other, one being the left and the other the right side (*e.g.*, *Bellerophon*); (2) **discoidal** for shells coiled enough out of plane that the flanks of a whorl have a top and a bottom (*e.g.*, *Euomphalus*); and (3) **helicoidal** for shells in which the whorls follow a helix (*e.g.*, *Helix* and *Trochus*).

[3] Knight (1941) recognized four types of shells with respect to the absence or presence of an umbilicus and the nature of the axis or cone of coiling:

1. **Anomphalous.** Structurally without an umbilicus; the axis of coiling falls within the shell wall or even within the whorl.
2. **Phaneromphalous.** Umbilicus free from callus and axis of coiling outside shell wall proper. The sides of an umbilicus may diverge up to 180°.
3. **Hemiomphalous.** Umbilicus partly filled with callus and axis of coiling outside shell wall proper.
4. **Cryptomphalous.** With a structural umbilicus completely filled with callus; the axis of coiling lies outside the shell wall proper.

FIG. 10-40. Class GASTROPODA. Architectural types of gastropod shells. *A. Vermicularia*, a shell that is irregularly uncoiled in the adult stage. *B–C. Crepidula*, the slipper shell. *D–E. Lymnaea*, a fresh-water form. *F. Pleurotomaria*, characterized by a prominent slit band. *G–H. Oliva*. *I–J. Cypraea*, the familiar ear shell. *K–L. Mitra*, with a dentate inner lip. *M–N. Voluta*, with a flaring outer lip. *O. Murex*, one of the most rugose and spinose of all gastropod shells. *P–Q. Conus*, the familiar cone shell. *R–S. Lambis* (=*Pterocera*), a shell that mimics a crab. *T–U. Astrea*, with many marginal spines. *V–W. Rostellaria*, with a conspicuous siphonal canal and denticulate outer lip. All shells are about one-third natural size. (*By Turner, from specimens and after numerous authors.*)

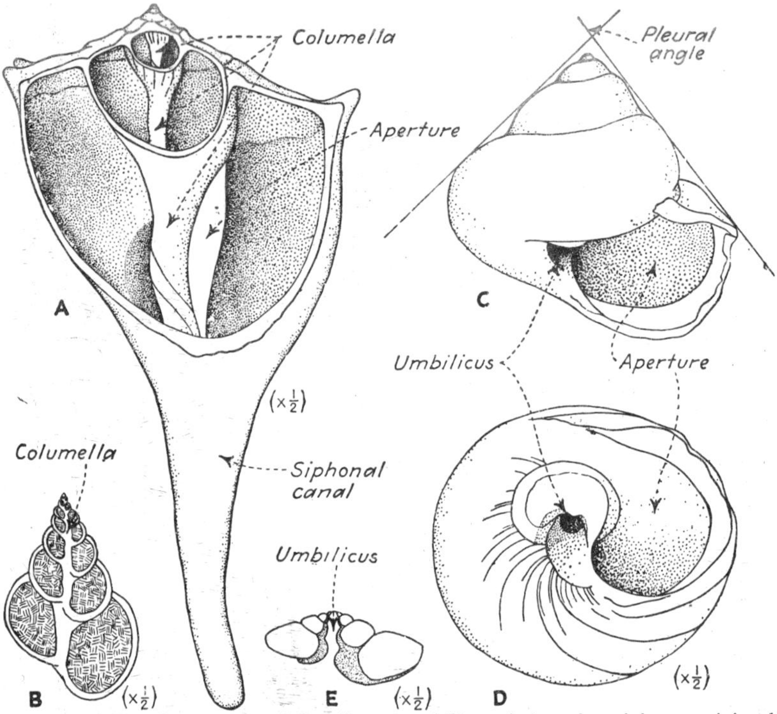

FIG. 10-41. Class GASTROPODA. Columella and umbilicus. *A*. A modern siphonate, sinistral gastropod shell (*Busycon*) sectioned to show the twisted columella. *B*. Columellar section of *Auriptygma virgatum*, from the Pennsylvanian of Missouri, showing the prominent columella and the hollow whorls filled with sediment. *C–D*. Apertural and umbilical views of a modern umbilicate gastropod. *E*. Vertical section of *Liospira angustata*, from the Middle Ordovician of Minnesota, showing a prominent umbilicus and thickening of shell along the umbilicus. (*B modified after Knight*, 1931; *E after Ulrich and Scofield*, 1897.)

Most shells have regular coiling, either **dextral** (clockwise, or right-handed) or **sinistral** (counterclockwise, or left-handed) when viewed from above (Figs. 10-40, 10-42), and most species tend to coil in only one direction, though some exhibit both types within the same species. Most spirally coiled shells are dextral.[1] Still another variant is *Maclurites* (Fig. 10-48*A–B*), which is actually dextral but seems to be sinistral because of the slight spire and is therefore described as **hyperstrophic.** A few adult shells with regular coiling in the earlier part unwind regularly or irregularly in the later part [*e.g.*, *Loxoplocus* (Fig. 10-48*I*) and *Vermicularia* (Fig. 10-40*A*)].

To orient a gastropod shell it is customary[2] to place or hold it so that the

[1] Sinistral shells are rare. Such coiling is inconstant in some species but constant in others.

[2] This is the conventional orientation now used by most paleontologists and zoologists, but formerly some European investigators oriented gastropod shells in exactly opposite position; *i.e.*, with the spire pointing downward.

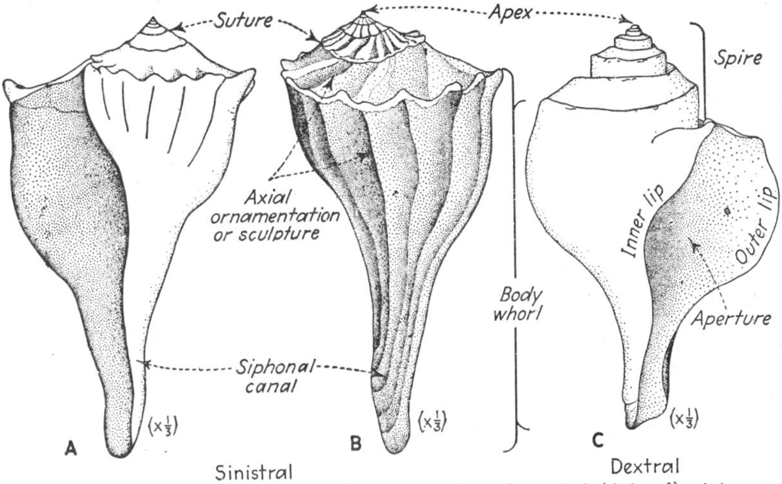

FIG. 10-42. Class GASTROPODA. *A–B*. Two views of a left-handed (sinistral), siphonate Recent gastropod (*Busycon*), showing prominent axial sculpture. *C*. A right-handed (dextral) specimen with a siphonal canal somewhat shorter than in *A* and *B*.

apex of the spire points upward and the aperture is visible (Figs. 10-39, 10-40). If the aperture is on the right side the shell is dextral; if it is on the left, sinistral. Planispiral shells of the bellerophontid type are customarily oriented with the aperture below (Fig. 10-40).

The pointedness of a shell is indicated by the **pleural,** or **apical, angle,** which is the angle included between two straight lines each tangent to the last two whorls on opposite sides (Fig. 10-41*C*). A low-spired shell has a large obtuse pleural angle that may be as great as 180°, as in *Maclurites*, whereas a high-spired form has a small acute pleural angle.

The opening, or aperture, of many gastropod shells can be closed by means of a horny or calcareous plate, the **operculum,** that is carried on the posterior part of the foot (Figs. 10-35*C*, 10-43). It comes into a closing position when the animal withdraws completely into the shell (Fig. 10-43*A*,*L*). The initial part of the operculum, made in the earlier growth stages, may be central or to one side on the complete operculum, and subsequent growth additions may be concentric or spirally arranged around it (Fig. 10-44). Some opercula are composed entirely of conchiolin, some have this substance impregnated with calcium carbonate, and others are largely calcareous. Some are smooth, but others are ornamented to some degree. Many species of the Architectonidae have a conical operculum that is covered externally with many spiral laminae. If the operculum is twisted, the twist, viewed from the exterior, is opposite to that of the shell, viewed from above. Thus a dextral shell, which spirals in clockwise direction, has an operculum that is twisted in a counterclockwise direction, when viewed from the outside. Many gastropods, including nearly all Pulmonates, lack an operculum. Fossil opercula are rare, and only calcareous ones have been reported from the Paleozoic. The earliest

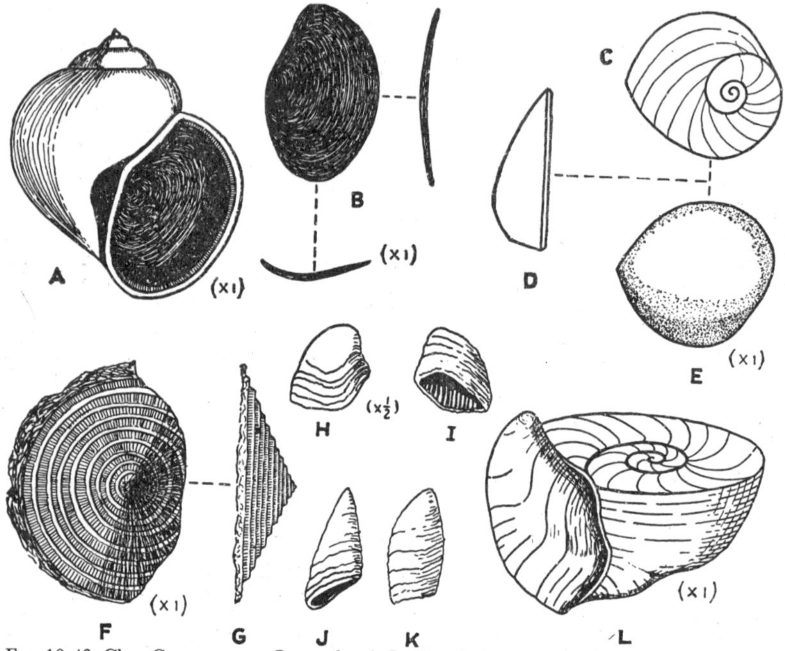

FIG. 10-43. Class GASTROPODA. Opercula. *A–B. Ampullaria*, a modern fresh-water gastropod with a chitinous operculum: *A*, apertural view of shell showing operculum in place; *B*, three views of the thin, slightly arched operculum. *C–E*. Interior, profile, and exterior views of a calcareous operculum of a modern Pacific gastropod. *F–G*. Supposed operculum of a Silurian gastropod from northern Indiana. *H–L. Ceratopea*, an operculated Lower Ordovician gastropod from Tennessee: *H–K*, views of two different silicified opercula; *L*, the operculum in place on the hyperstrophic shell (shown inverted). (*F–G after Kindle and Breger*, 1904; *H–K based on original photographs by J. Bridge; L after Oder*, 1932.)

come from Lower Ordovician beds (Fig. 10-43*H–L*). The opercula of some living gastropod groups are sufficiently different one from another that they can be used for discriminating genera and species (Bartsch, 1946) (Fig. 10-44).

The apertural outline may be circular, elliptical, oval, crescentic, or slit-like (Fig. 10-40). The margin of the aperture, generally termed the **peristome,** may be divided into an **inner lip,** which is further divided into a **columellar** and a **parietal** part, and an **outer lip,** which may be divided into an upper, an outer, and a lower part (Figs. 10-39, 10-45*A*). The inner lip in closely coiled forms is modified by the wall of the last whorl and may bear spiral ridges which in some forms extend to the apex (Fig. 10-45*A*).

In siphonate shells the lower part of the outer lip is interrupted by a **siphonal notch** and may be produced into a prominent shelly trough, or **canal,** which serves as a protective roof over the incurrent siphon. This canal may be straight or curved, and in some shells it is longer than the greatest apertural diameter (Figs. 10-40, 10-42).

The outer lip of the peristome may be thin or thick, curved outward (**reflected**) or inward (**inflected**), or produced into winglike or fingerlike processes (Fig. 10-40*R*–*S*). In many gastropods with two gills, as the Pleurotomaridae and Bellerophontidae, the outer lip is interrupted by a parallel-sided notch, or slit, that serves for the egress of water and feces from the mantle cavity, and for this reason is commonly termed the **anal notch** or **anal mark.** This reentrant is progressively closed as the shell grows, and the resulting lunulate band is termed the **selenizone** or **slit band** (Fig. 10-45).

The gastropod shell is secreted by the mantle. It consists of an organic base of conchiolin mineralized and made rigid with calcite or aragonite. The organic base can be revealed by dissolving away the calcareous matter with a dilute solution of hydrochloric acid.

A typical shell is composed of three layers: (1) a thin outer layer of conchiolin, which is easily destroyed and almost never preserved in fossil gastropods; (2) an outer prismatic layer, usually of calcite; and (3) an inner lamellar layer, generally of aragonite (Fig. 10-46*A*). These correspond to the three fundamental layers of all molluscan shells—periostracum, ostracum, and hypostracum, respectively. A layer of lamellar shell matter may be secondarily deposited over the outer surface of a shell, usually within the aperture where the outer surface is encroached upon by the succeeding whorl, and in rare cases may cover the entire shell exterior if the entire shell is enveloped in the mantle. This deposit is termed the **inductura** (or **callus**) and when present may be thin or quite thick. That part of the deposit laid over the parietal wall by the advancing mantle is the **parietal inductura** or **parietal shield** (Fig. 10-45*A*, 10-46).

The sculpture of gastropod shells is varied and often complex. It is made along the apertural margin by the advancing mantle, and consists of two sets of differently oriented features—revolving (spiral) and transverse (axial). In revolving sculpture the features run with the whorl; in transverse sculpture they pass across the whorl and are likely to be parallel with the growth lines. The chief features of sculpture may be defined briefly as follows (also see Fig. 10-39):

Carina, -ae—prominent angular or rounded ridges, generally revolving. **Costa, -ae**—coarse raised threads that may be transverse, revolving, or both. **Growth lines**—transverse markings that indicate former positions of the shell margin in the growth of the shell. **Lira, -ae**—fine raised threads that may be transverse, revolving, or both. **Stria, -ae**—very fine grooves that may be transverse, revolving, or both. **Varix, varices**—keels, flanges, or series of spines transverse to the whorl. A varix marks a stage during which there was little growth at the shell margin.

In addition to these, there are many other less regular features that appear on only a few shells. Obvious features, such as ribs, nodes, spines, and cancellated areas are not defined.

Sculpture might be expected to be simple on primitive gastropods and complex on advanced forms, but such is not always the case. Some advanced forms have smooth shells and some simple forms have much-sculptured shells. The complexity of sculpture, therefore, is not an index to the position of a

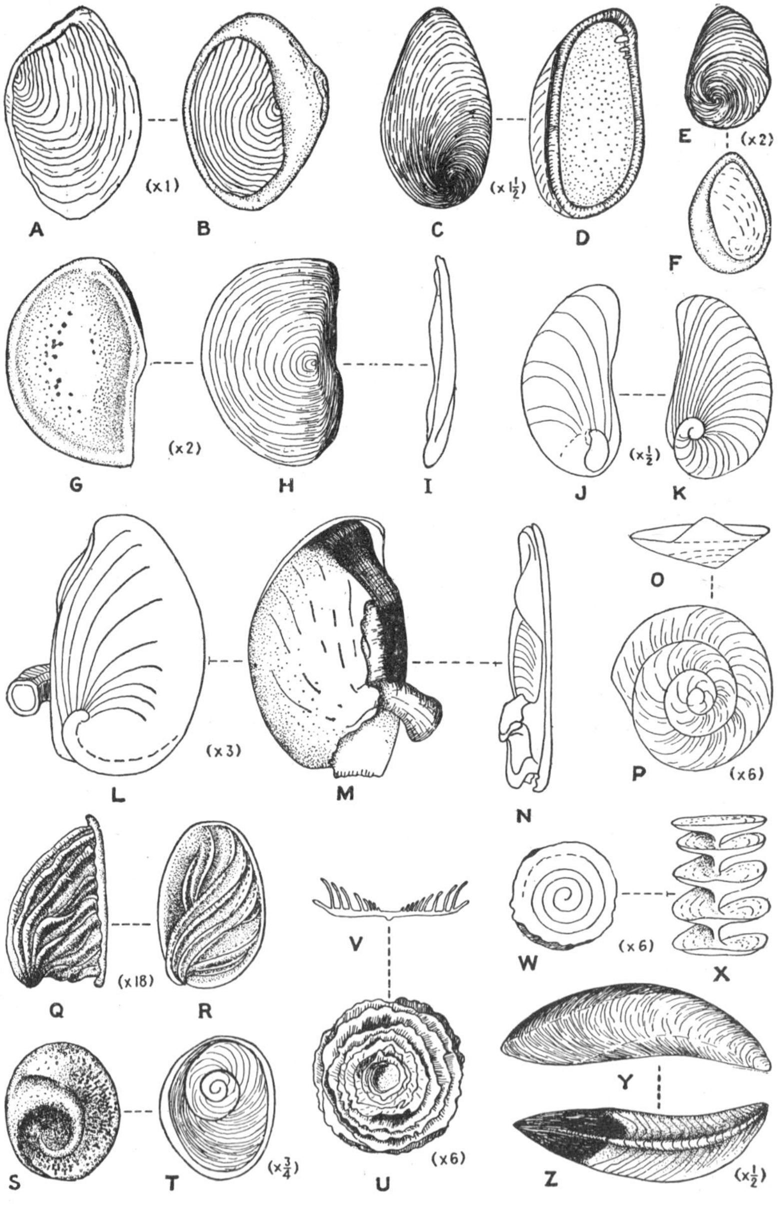

FIG. 10-44. Class GASTROPODA. Opercula. (*See opposite page for detailed descriptions.*)

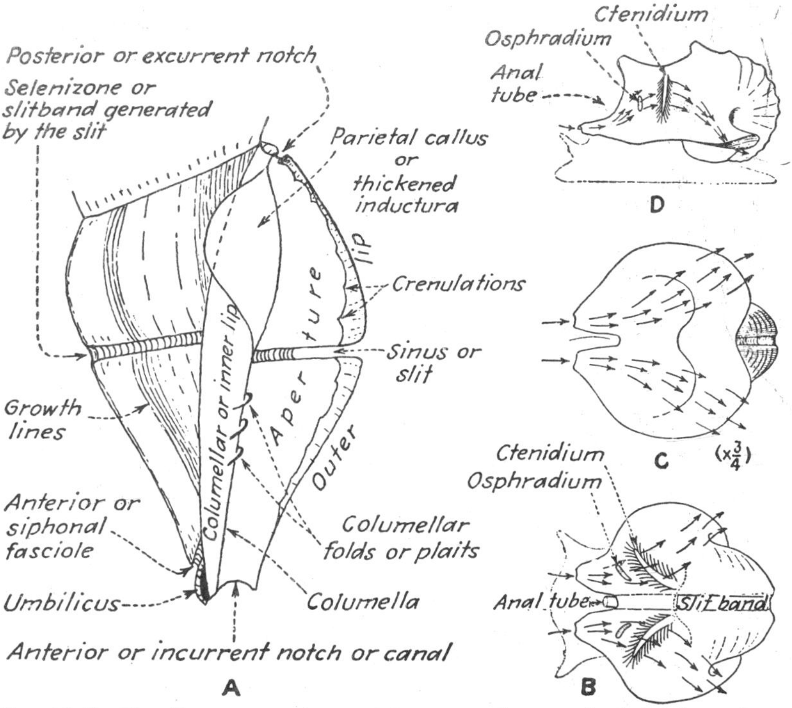

FIG. 10-45. Class GASTROPODA. Apertural features and water circulation in ancient gastropods. *A*. Diagram showing the terms customarily used by paleontologists in describing fossil gastropod shells (cf. Fig. 10-39). *B–D*. Diagrams showing inferred relations of soft parts to apertural features, and inferred circulation, in *Knightites*, a Pennsylvanian bellerophontid gastropod: *B*, top view of shell (represented as transparent), with water currents indicated by arrows; *C*, apertural view of empty shell, showing water currents; *D*, side view of shell (represented as transparent), and inferred relations of soft parts and water currents. (*B–D after Knight in Moore*, 1941.)

species in an evolutionary series. Gerontic shells may show highly complex sculpture on the last whorl or may have the sculpture suppressed. Suppression of sculpture in the later whorls is a useful criterion for demonstrating that tiny shells are adult forms (*i.e.*, some specimens of a tiny species commonly show suppressed sculpture, thus indicating that the shells were made by adult animals).

FIG. 10-44. Opercula of modern gastropods; some or all species belonging to the genera mentioned below have opercula like those shown. Unless otherwise stated, the genus is marine and the first figure of a series is a view of the outer face. Dominantly calcareous opercula are indicated as (*Ca*); chitinous, as (*Ch*). *A–B*. *Murex* (*Ch*). *C–D*. *Oocorys* (*Ch*). *E–F*. *Pleurocera* (*Ch*), fresh-water. *G–I*. *Viana* (*Ca*), a land form: *I*, edge view, with inner face to left and outer to right. *J–K*. The inner and outer face of *Natica* (*Ca*). *L–N*. *Nerita* (*Ca*): N, edge view, with inner face to left and outer to right. *O–P*. Edge view and outer face of *Cirsotrema* (*Ch*). *Q–R*. *Taheitia* (*Ca*), a land form. *S–T*. *Turbo* (*Ca*). *U–V*. Outer face and cross section of *Fijipoma* (*Ca*), a land form. *W–X*. Outer face and pulled-out operculum of *Kanapa* (*Ch*), a land form. *Y–Z*. *Strombus* (*Ch*). (*By Turner*.)

Vivid colors, both regularly and irregularly distributed, are conspicuous on many gastropod shells and commonly produce attractive designs. The color pattern may not be obvious on certain shells because of being concealed beneath the periostracum. Color is rarely preserved in fossil shells, but the design made by the colors may be faintly outlined on unusually well preserved shells.

Gastropod shells range in height, the distance from apex to lower margin of aperture, from less than a millimeter in *Pupa* to more than ½ m. (24 in.) in *Megalotractus aruanus*. The shape may be conoidal, cylindrical, ellipsoidal, spheroidal, turbinate, or fusiform (Fig. 10-40). Ordinarily shape has no more than generic value, and in many cases unrelated shells have quite similar shapes.

The fleshy part of the gastropod is attached to the shell by muscles. By contracting these muscles the animal can pull itself into the shell and close the aperture behind itself with the operculum, if one is present. Most gastropods have an external shell. A few, however, have so completely surrounded the shell with soft body tissue that it no longer functions for protection and has degenerated into a small plate completely unlike the usual shell. In extreme cases the shell has disappeared altogether, and the animal is naked—a condition existing in the nudibranchs, in some pteropods, and among many of the air-breathing (pulmonate) forms (Fig. 10-35).

Classification

The classification of modern Gastropoda is based almost entirely on the nature of the soft parts, among which those considered most important are (1) respiratory organs, (2) nervous system; (3) heart; (4) reproductive organs; (5) foot; and (6) radular complex. Only among lower taxonomic categories (families, genera, etc.) are shell characteristics considered important for classification. Most treatises on Gastropoda recognize two subclasses

TABLE 10-2. GEOLOGIC RANGE OF THE ORDINAL AND SUPRAORDINAL DIVISIONS OF THE CLASS GASTROPODA

Thiele, 1931; Wenz, 1938; Knight, 1944

Division	*Range*
Subclass Protogastropoda	
Order 1. Cynostraca	L. Camb.–Carb.; ?Perm.
Order 2. Cochliostraca	L. Camb.–Ord.
Subclass Prosobranchia	
Order 1. Archaeogastropoda (Aspidobranchia)	U. Camb.–Recent
Order 2. Mesogastropoda (Taenioglossa)	L. Ord.–Recent
Order 3. Neogastropoda (Stenoglossa)	Ord.–Recent
Subclass Opisthobranchia	
Order 1. Pleurocoela	Carb.–Recent
Order 2. Pteropoda	?L. Camb.–?Perm.; Cret.–Recent
Order 3. Sacoglossa	Recent
Order 4. Acoela	Eoc.–Recent
Subclass Pulmonata	
Order 1. Basommatophora	Penn.–Recent
Order 2. Stylommatophora	U. Cret.–Recent

based on the nature of the nerve commissures—Streptoneura, in which the visceral nerve commissures have the form of a figure 8, and Euthyneura, in which the commissures have been uncrossed by secondary detorsion. This classification was used in our first edition, but we now prefer to abandon it and use the one proposed by Thiele (1931). Thiele's classification, which has been followed by Wenz (1938) and Knight (1944) and seems to be gaining wide acceptance among both malacologists and paleontologists, is based partly on the nature and position of the gills and partly on other anatomical features. This classification, slightly enlarged to include one extinct subclass, is shown in Table 10-2. (The relationship of animal to shell and of soft parts to one another in each of the three subclasses with living representatives is shown in Fig. 10-35.) The geologic ranges of the orders are only approximate because of differences of opinion as to the orders to which certain families and genera should be assigned.

Subclass Protogastropoda[1]

This extinct subclass includes the earliest known gastropod shells, which are conoidal or planispirally coiled. These early gastropods presumably had not undergone torsion, hence may be regarded as close to the archetypal mollusk. *Lower Cambrian to Carboniferous, ?Permian.*

Order 1. Cynostraca.[2] This order includes simple conical and cornucopia-shaped shells that are smooth or sculptured with more or less prominent growth lines and concentric or transverse ridges. Members of the order are considered to be the most primitive of all gastropods and probably close to the prototypical mollusk from which the class descended. The earliest representatives appear in Lower Cambrian rocks, and the order seems to have become extinct sometime during the Carboniferous or Permian.

Scenella and *Palaeacmaea* are representative genera (Fig. 10-47).

Order 2. Cochliostraca.[3] This order includes the earliest known coiled gastropod shells. The first representatives are from Lower Cambrian rocks, the last from Ordovician. *Pelagiella* and *Matherella* are typical genera (Fig. 10-47).

Subclass Prosobranchia[4]

The Subclass Prosobranchia includes chiefly marine gastropods that are characterized by torsion of the visceral hump in which the gills and anus are brought forward to an anterior position. In the two most primitive superfamilies (Pleurotomariacea and probably Bellerophontacea), the gills are paired; in all other prosobranchs, however, the right-hand gill, along with

[1] Protogastropoda—Gr. *protos*, first, + *gaster*, *gastros*, stomach, + *pous*, *podos*, foot; referring to the fact that these were the first or earliest gastropods.

[2] Cynostraca—Gr. *kyon*, *kynos*, dog, + *ostracon*, shell; probably referring to the resemblance of the shell to a dog tooth.

[3] Cochliostraca—Gr. *kochlos*, spiral shell, + *ostracon*, shell; referring to the spiral coiling of the shells.

[4] Prosobranchia—Gr. *proso*, forward, + *branchia*, gills; referring to the position of the gills anterior to the heart.

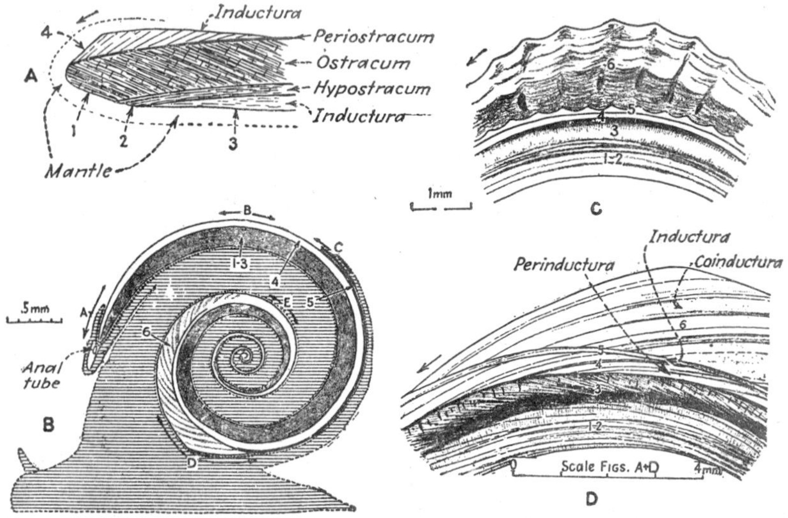

FIG. 10-46. Class GASTROPODA. Shell growth and structure. *A*. Diagram of outer apertural margin, showing places where mantle adds shell substance. At 1, inclined laminae of the ostracum are laid down, with prisms of calcite roughly perpendicular to the laminae. At 2, thin parallel laminae of aragonite are deposited to form the hypostracum. At 3 and 4, lamellar shell matter is secondarily deposited to make the inductura. The periostracum, which is conchiolinitic and thin, is not likely to be preserved in fossil shells. *B*. Diagram of *Euphemites*, a Pennsylvanian bellerophontid gastropod, showing three inductural layers (4-6) added to the fundamental shell (1-3). The large dots at *A*, *C*, and *D* indicate the chief areas of shell formation. *C–D*. Small portions of the shell of *B*, showing microscopic shell structure in greater detail. Layers 1-3 correspond to the ostracum and hypostracum in *A*. (*B–D after Moore*, 1941.)

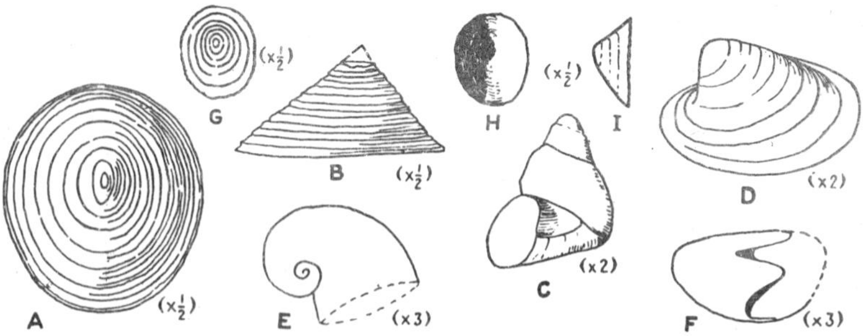

FIG. 10-47. Subclass PROTOGASTROPODA. *A–B*. Top and side views of *Palaeacmaea*, a Late Cambrian and Ordovician genus with a flat conical shell. *C*. *Matherella*, one of the earliest known coiled gastropods, from the Upper Cambrian. *D*. *Scenella*, from the Lower Cambrian. *E–F*. Top and apertural views of an incomplete specimen of *Pelagiella*, the earliest known coiled gastropod, from the Lower Cambrian. *G–I*. Top, apertural, and side views of the shell of modern *Acmaea*, which, although architecturally simple like *Palaeacmaea* (*A–B*), belongs to a highly specialized gastropod (introduced for comparison). (*A–F outlined from illustrations in Shimer and Shrock's "Index Fossils of North America,"* 1944.)

the other right-hand organs, has been lost as a result of torsion of the visceral hump, so that only the left-hand gill remains. If gills are lacking altogether, the animal respires by means of a pallial growth, as in the Patellidae, or a pulmonary chamber, as in the Helicinidae and Hydrocenidae. The nervous system is streptoneurous (*i.e.*, the visceral nerve commissures are twisted), there is only one pair of tentacles, and the sexes are separate.

Most shells are dextral and coiled in a helicoid spiral, but a few are cup-shaped, caplike, or cornucopia-shaped. The more primitive are nonsiphonate; the more advanced are siphonate, and some have a canaliculate apertural margin. A horny or calcareous operculum is usually present.

The Prosobranchia constitute by far the largest subclass of gastropods and include most of the typical shell-bearing forms. Their geologic history begins in later Cambrian time, and they have left a good fossil record in the rocks of every geologic period. *Upper Cambrian to Recent.*

Order 1. Archaeogastropoda (Aspidobranchia). This is the most primitive order of the Prosobranchia, and to it most Paleozoic gastropods seem to belong. The shells are not distinctive, because of wide variation in architecture, but they are characterized by a selenizone (slit band). An operculum may be present or lacking. The oldest shells come from Upper Cambrian rocks. The order, which is dominantly marine, is well represented in modern marine faunas, and there are also a few fresh-water and land members.

It is customary to recognize two suborders of living prosobranchs, based on certain features of the radula—Docoglossa and Rhipidoglossa (Fig. 10-36). The Docoglossa, or limpets, are open-conical shells, with spiral coiling confined to the nucleus. *Acmaea* (Fig. 10-47*G–I*) is a typical modern limpet. The Rhipidoglossa include limpet-like or spiral shells that are operculate. All are exclusively marine except three families (Neritidae,[1] Helicinidae, and Proserpinidae).

Because the radula and associated jaws are never preserved in the more ancient gastropods, and rarely if ever in younger fossil forms, it is not possible to assign fossil genera to the suborders Docoglossa and Rhipidoglossa on the basis of radular characteristics. Shell features, therefore, have to be used, and among the many hundreds of genera referred to the Archeogastropoda, the following are representative: Superfamily Bellerophontacea: *Cyrtolites* (Ord. Fig. 10-48*L–M*), *Salpingostoma* (Ord.–Dev.; Fig. 10-48*F–G*), and *Knightites* (Penn.; Fig. 10-45*B–D*). Superfamily Pleurotomariacea: *Rhacopea* (U. Camb.–L. Ord.; Fig. 10-48*R–T*), *Ophileta* (Ord.; Fig. 10-48*U*), *Loxoplocus* (*Lophospira*) (Ord.; Fig. 10-48*H–I*), *Pleurotomaria* (Sil.–Recent; Fig. 10-40*F*), *Haliotis* (Cret.–Recent; Fig. 10-35*B*), and *Fissurella* (Recent; Fig. 10-35*A*). Superfamily Trochacea: *Astrea* (Mioc.–Recent; Fig. 10-40*T–U*). Superfamily Euomphalacea: *Ceratopea* (Ord.; Fig. 10-43*H–L*), *Maclurites* (Ord. Fig. 10-48*A–B*), *Lytospira* (Ord.–Sil.; Fig. 10-48*E*), and *Straparolus* (Sil. Jura.; Fig. 10-48*J–K*). Superfamily Trochonematacea: *Trochonema* (Ord. Dev.; Fig. 10-48*C–D*), *Platyostoma* (Sil.–Dev.; Fig. 10-48*O–P*), and *Orthonychia* (Sil.–Perm.; Fig. 10-48*N*).

[1] Some Neritidae are marine: others live in fresh water; and a few have invaded a land habitat.

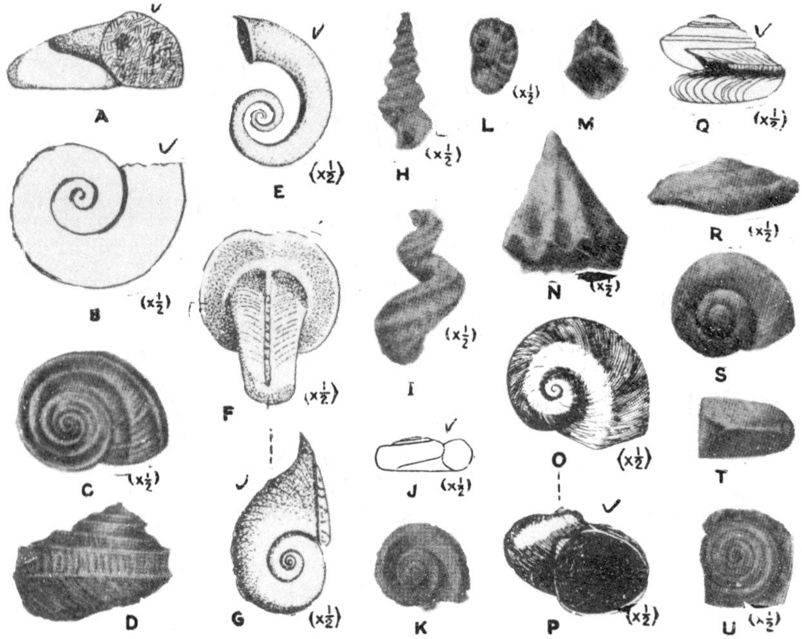

FIG. 10-48. Order Archaeogastropoda. *A–B. Maclurites*, a hyperstrophic shell, viewed aperturally and from below. This Ordovician genus commonly has a heavy calcareous operculum. *C–D. Trochonema*, a terraced Ordovician form. *E. Lytospira*, a loosely coiled Lower Ordovician genus. *F–G*. Filling of interior of *Salpingostoma: F*, back view, showing riblike keel, the filling of the slit, and traces of external markings; *G*, side view. *H–I*. Two different species of the common Ordovician genus *Loxoplocus* (*Lophospira*). *J–K*. Apertural view in outline and top view of *Straparolus* (*Euomphalus*), a flat spiral form. *L–M*. *Cyrtolites*, a prominently sculptured shell from the Ordovician. *N. Platyceras* (*Orthonychia*), species of which are commonly found attached to the tegmen of a crinoid. *O–P. Platystoma*, a common Silurian genus. *Q. Eotomaria*, from the Ordovician, showing a slit band and prominent growth lines. *R–T. Rhacopea*, an Upper Cambrian and Lower Ordovician genus with a prominent slit band. *U. Ophileta*, a flat-spired multiwhorled Ordovician form. (*C–M, Q after Ulrich and Scofield*, 1897; *N–P after Hall*, 1897; *R–U after Ulrich and Bridge*, 1931.)

Order 2. Mesogastropoda. This is a large and important order of prosobranch gastropods, including Suborders Gymnoglossata, Ptenoglossa, Taenioglossa and composed of many families of both fossil and living genera, among which are marine, fresh-water, and land forms. The shell is in no case nacreous, and it is usually helicoid but may be discoidal or cap-shaped (Fig. 10-49). The apertural margin is entire; *i.e.*, it lacks a siphonal notch or canal.

The breathing organ is the single left gill in the mantle cavity, the heart has one auricle, there is only one kidney (the left one of the original pair), the nervous system is not concentrated, the sexes are separate, there is an **osphradium** (supposedly a sensory organ), and the radula is usually taenioglossate. The earliest fossil representatives come from Ordovician rocks. *Lower Ordovician to Recent.*

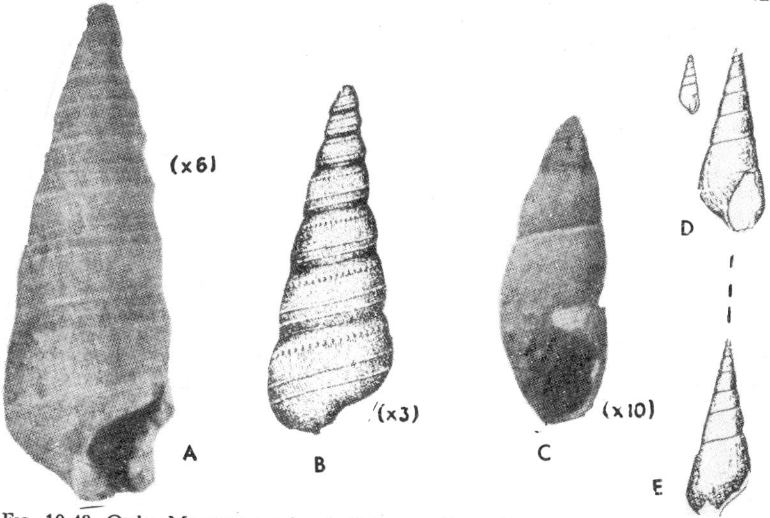

FIG. 10-49. Order Mesogastropoda. *A. Orthonema* (Penn.–Perm.). *B. Acanthonema* (Dev.). *C. Girtyspira* (Miss.–Penn.). *D–E. Meekospira* (Ord.–Perm.). (*A, C original after Knight; B after Meek*, 1873; *D–E after Whitfield*.)

Representative Paleozoic genera are *Meekospira* (Fig. 10-49*D–E*), *Acanthonema* (Fig. 10-49*B*), *Girtyospira* (Fig. 10-49*C*), and *Orthonema* (Fig. 10-49*A*). Typical Mesozoic and Cenozoic genera with living species are *Vermicularia* (Fig. 10-40*A*), *Crepidula* (Fig. 10-40*B–C*), *Lambis* (Fig. 10-40*R–S*), *Rostellaria* (Fig. 10-40*V–W*), *Cypraea* (Fig. 10-40*I–J*), *Ampullaria* (Fig. 10-43*A–B*), and *Littorina*.

Order 3. Neogastropoda (Stenoglossa). The neogastropods, which include a large number of familiar living genera (*e.g., Murex, Voluta, Oliva, Conus*, and *Terebra*), are operculate gastropods in which the apertural margin of the shell has a siphonal notch or canal. The nervous system is concentrated, the sexes are separate with the male having a large copulatory organ on the right side, the proboscis is well developed, the esophagus has an unpaired gland, and there is a double sensory organ (osphradium) situated near the gills that supposedly serves an olfactory function or tests the purity of the water passing to the gills. The operculum is horny and paucispiral. The radula is narrow, and the transverse rows have only three or less teeth each.

The order, which seemingly began in the Early Ordovician, is divided into the four following superfamilies of living forms (not all fossil genera are included in these):

Muricacea (U. Cret.–Recent). *Sargana, Rapana, Purpura, Murex* (Fig. 10-40*O*), and *Urosalpinx* (Fig. 10-50*D–E*) are familiar genera.

Buccinacea (U. Cret.–Recent). *Pyrene, Buccinum, Busycon* (Figs. 10-41, 10-42), and *Fusus* are representative genera.

Volutacea (U. Cret.–Recent). *Oliva* (Fig. 10-40*G–H*), *Harpa, Voluta* (Fig. 10-40*M–N*), *Cymbium, Mitra* (Fig. 10-40*K–L*), and *Cancellaria* (Fig. 10-50*F*) are typical genera.

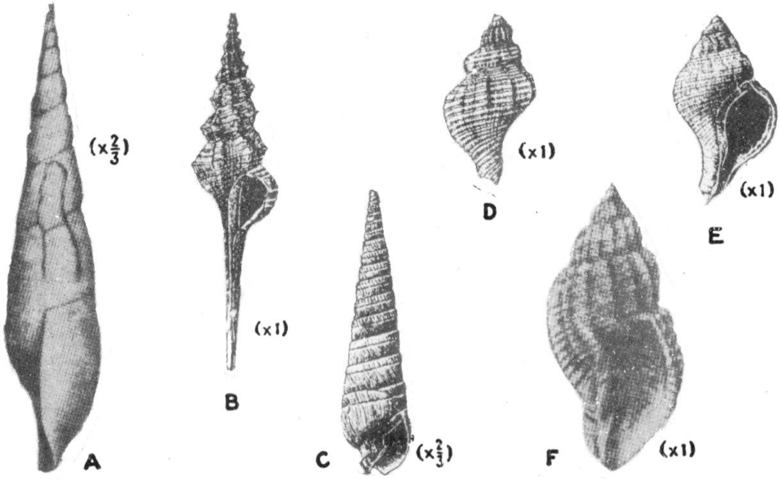

FIG. 10-50. Order Neogastropoda. *A*. An Ordovician species of *Subulites*, a widespread Ordovician and Silurian genus. *B*. *Falsifusus*, an Early Tertiary genus. *C*. A Miocene species of the familiar existing *Terebra* (Tert.–Recent). *D–E*. *Urosalpinx* (Tert.–Recent). *F*. *Cancellaria* (Tert.–Recent). (*A after Ulrich*, 1897; *B after Harris*, 1899; *C–E after Martin*, 1904; *F after Dall*, 1891.)

Conacea (Toxoglossa) (U. Cret.–Recent). *Drillia*, *Conus* (Fig. 10-40*P–Q*), and *Terebra* (Fig. 10-50*C*) are well-known living genera.

Subulites (Ord.–Sil.; Fig. 10-50*A*) is a typical Paleozoic representative, and *Falsifusus* (Fig. 10-50*B*) is an early Tertiary genus.

Subclass Opisthobranchia[1]

The Opisthobranchia usually have but a single gill, nephridium, and auricle (Fig. 10-37*B*). The breathing organs, if present, are posterior to the heart; if they are lacking, the animal uses the external surface of the mantle for respiration. The nervous system is almost invariably euthyneurous;[2] *i.e.*, the visceral nerve commissures are untwisted so as to have a primitive appearance in plan and are concentrated around the esophagus. All are hermaphroditic and marine.

The shells, if present, are small and of relatively simple architecture, and the adults except in the Actaeonidae and Limacinidae are without an operculum. Four orders are recognized by Thiele (1931) and Wenz (1938)—Pleurocoela, Pteropoda, Sacoglossa, and Acoela. The subclass has only a sparse fossil record, which probably does not go back as far as the Cambrian, but is well represented in modern seas by a wide variety of genera.

[1] Opisthobranchia—Gr. *opisthen*, behind, + *branchia*, gills; referring to the position of the gills posterior to the heart.

[2] *Actaeon* (Fig. 10-51*F*), the most primitive opisthobranch, retains the streptoneurous type of nervous system, but all other opisthobranchs seem to be euthyneurous. From this it would seem that the Opisthobranchia descended from a prosobranch ancestor.

The shells, if present, fall into the following groups:

1. Normal spiral shells with more or less ear-shaped aperture, and commonly with columellar folds (Pleurocoela). An operculum is lacking except in the Actaeonidae and Limacinidae.

2. Involute, hemiellipsoidal cowrie-like shells.

3. Limpet-like shells.

4. Shells of forms adapted to pelagic life and generally with a secondary bilateral symmetry (Pteropoda).

5. Much reduced internal shells with little resemblance to typical gastropod shells (Acoela).

6. No shell except for the nuclear whorls which appear only in the larval stage and are then lost (Nudibranchia); this is the case with many living opisthobranchs.

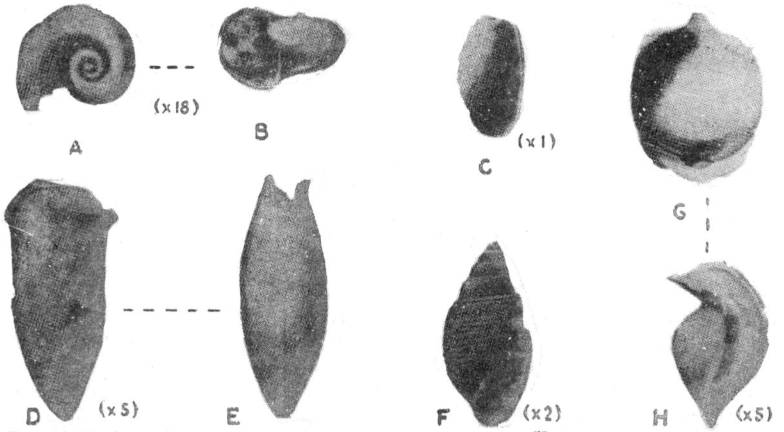

FIG. 10-51. Class GASTROPODA—Subclass OPISTHOBRANCHIA. A group of Miocene species belonging to existing genera. *A–B. Limacina* (Tert.–Recent). *C. Bulla* (Jura.–Recent). *D–E. Vaginella* (U. Cret.–Recent). *F. Actaeon* (Cret.–Recent). *G–H. Cavolina* (Mioc.–Pleist.). (*A–B, D–E, G–H after Collins*, 1934; *C after Woodring*, 1928; *F after Martin*, 1904.)

Order 1. Pleurocoela (= **Tectibranchia** = **Steganobranchia**). This order includes opisthobranchs that have a large head, a broad cephalic hood or disk, sessile eyes, and a mantle cavity that opens to the right if present. If a shell is present it is usually thin and commonly rudimentary and is largely or entirely enveloped by a fold of the mantle and foot (the **epipodium** or **parapodium**). The oldest fossil tectibranchs come from Carboniferous rocks, but the order did not become well established until the Late Mesozoic. *Actaeon* (Cret.–Recent; Fig. 10-51*F*), *Bulla* (Jura.–Recent; Fig. 10-51*C*), *Aplysia* (Recent; Fig. 10-35*D*), and *Haminea* (Recent; Fig. 10-35*E*) are representative genera.

Order 2. Pteropoda.[1] The pteropods are pelagic swimming opisthobranchs with the foot modified into a pair of winglike lobes or fins, the epipodia or parapodia. They are hermaphroditic, usually lack gills, and may

[1] Pteropoda—Gr. *pteron*, wing, + *pous*, *podos*, foot; referring to the winglike lobes of the foot.

or may not have a shell. They are small in size but commonly so abundant as to cover the surface of the sea for miles, and their tiny calcareous shells have accumulated locally in prodigious quantities on some ocean bottoms to form **pteropod ooze.** One of the two suborders, Gymnosomata, is distinguished by having no shell or mantle in the adult stage. Members of the other suborder, Thecosomata, have tiny calcareous shells of several types, of which *Vaginella* (U. Cret.–Olig.), *Limacina* (Tert.–Recent), and *Cavolina* (Mioc.–Pleist.) are representative (Fig. 10-51).

Coleoloides (L. Camb.), *Hyolithes* (Camb.–Perm.), *Tentaculites* (Ord.–Dev.), and *Styliolina* (Dev.) are typical of a group of Paleozoic fossils of uncertain affinities that have commonly been assigned to the Pteropoda, or more generally to the Mollusca. They are illustrated in "Index Fossils of North America" and need no further discussion, except for emphasis on the statement that their zoological affinities are uncertain.[1]

The established geologic range of the order, therefore, is Cretaceous to Recent.

Order 3. Sacoglossa. This order consists of a small group of opisthobranchs characterized by thin, colorless external shells with small whorls and more or less wide aperture. The shell covers only a small part of the body. No fossil forms seem to have been reported.

Order 4. Acoela (Nudibranchia). The shell, if present, is small, shaped like a bowl or ear, and usually internal; it is commonly lacking altogether, in which case the dorsal region may be protected by a wide **notum.** The mantle cavity is shallow or entirely lacking, and the gills, if present, are variable in size and position. They may be on the right, in the region of the anus, under the edge of the notum, or lacking. Only one family, the Umbraculidae, seems to have left a fossil record, which begins in the Eocene. *Eolis* (Fig. 10-35*F*) is a typical genus. The Acoela as a whole, and particularly the Suborder Nudibranchia with its 35 families, constitute a large and important group of living mollusks, and many genera have been differentiated on the basis of the radular complex, which is quite varied.

Subclass Pulmonata

The Pulmonata are air- and water-breathing gastropods distinguished by lacking gills and by usually having a lung or respiratory sac.[2] The nervous system is consolidated (Fig. 10-37*C*), and all members of the subclass are hermaphroditic. The subclass, which is a large one, includes most land snails and slugs and many fresh-water snails. Adults lack an operculum except in the Amphibolidae, and the shell is vestigial or lacking altogether in a few groups.

The Pulmonata constitute the second largest group of gastropods, and the

[1] The Termiers (1947) create a new molluscan class, the Eopteropoda, to include the Hyolithidae and Tentaculitidae along with the Conulariidae, but it is doubtful if such disposition of the three families is an improvement over earlier assignments.

[2] The interior of the mantle cavity is traversed by a system of branching blood vessels and thus constitutes a pulmonary organ adapted for breathing air. It opens to the exterior through a contractile orifice that can be entirely closed.

number of described species approximates 7,000, of which 6,300 are living and 700 are fossil. The earliest supposed pulmonate came from Pennsylvanian rocks; fossils are rare in rocks older than Upper Cretaceous.

Two orders are customarily recognized:

Order 1. Basommatophora
Order 2. Stylommatophora

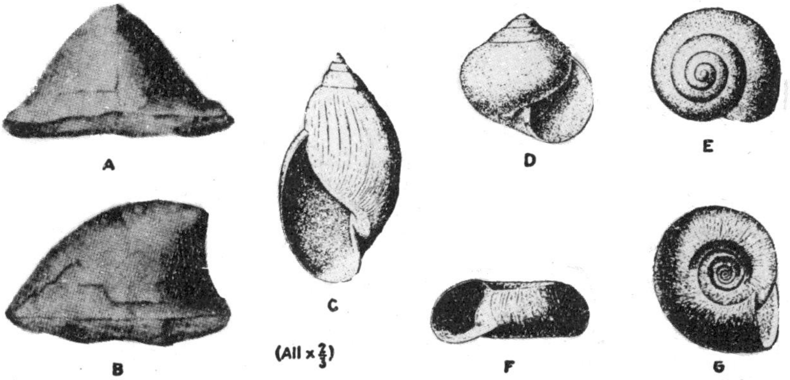

FIG. 10-52. Class GASTROPODA—Subclass PULMONATA. *A–B*. *Anisomyon*, a patelliform pulmonate, from the Upper Cretaceous. *C*. A Pliocene species of *Physa*, a fresh-water genus, which ranges from Jurassic to Recent. *D–E*. An Oligocene species of *Helix*, a terrestrial genus ranging from Eocene to Recent. *F–G*. A Pliocene species of *Planorbis*, a fresh-water genus, which ranges from Lower Jurassic to Recent. (*A–B*, *F–G after Meek*, 1876; *C–E after Dall*, 1890–1892.)

Order 1. Basommatophora. Representatives of this order have the eyes at the base of a single pair of nonretractile tentacles, and the genital orifices are generally separated. They are exclusively aquatic and can live in fresh, brackish, or salt water. The shell is helicoid or cap-shaped. Genera typical of the order (Penn.–Recent) are *Physa* (Fig. 10-52*C*), *Lymnaea* (Fig. 10-40*D–E*), *Helisoma* (Fig. 10-35*I*), and *Planorbis* (Fig. 10-52*F–G*). *Anisomyon* (Fig. 10-52*A–B*) is an extinct genus.

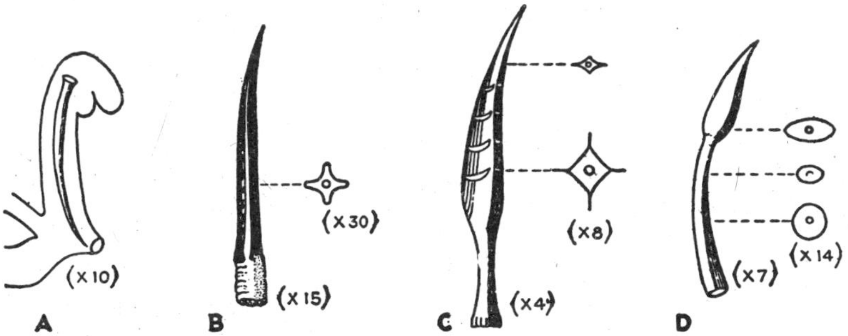

FIG. 10-53. Class GASTROPODA. Calcareous darts, unique to the Family Helicidae, are found in females and are used during mating. *A*. A dart in the dart sac. *B–D*. Darts from different species of *Helix*. (*After Ashford*, 1883–1885.)

Order 2. Stylommatophora. In the Stylommatophora the eyes are situated at the tips of one of the two pairs of contractile tentacles, and the genital orifices are generally united. The animal usually has a helicoid shell, but this may be greatly reduced or lacking. Members of the order are largely land snails. A genus typical of the order (U. Cret.–Recent) is *Helix* (Figs. 10-35*G*, 10-52*D–E*). It is representative of the Family Helicidae, females of which are unique in having small calcareous darts associated with the reproductive organs (Fig. 10-53). These darts, though calcareous, have not yet been reported as fossils. Other typical living genera are *Arion* (Fig. 10-35*J*) and *Testacella* (Fig. 10-35*H*).

Geologic History of Gastropoda

The geologic ranges of the numerous superfamilies of the Class Gastropoda are shown in Table 10-2. From this table it can be seen that the gastropods have a long and extensive fossil record and are today seemingly at or near the peak of development.

Development. The oldest known fossil gastropods (*Scenella, Helcionella, Pelagiella*) were collected from Lower Cambrian rocks, but judging from the nature of these earliest genera, it seems likely that the Class Gastropoda came into existence before the Cambrian. If account be taken of the fact that they are associated with typical marine fossils (*e.g.*, trilobites and brachiopods), it would seem that the first gastropods originated in marine waters. The oldest known pulmonate gastropods are from Pennsylvanian rocks, and some of these may have been land snails.

In modern gastropods the first shell to appear in the embryo is a tiny cap-shaped plate, which, as shell matter is added, tends first to coil in a planispiral fashion and then to change to a helicoid type of coiling, which in most forms persists into adulthood. It is interesting to note that among the Cambrian gastropods are adult shells that show each of the three stages just mentioned. *Palacmaea* is a simple cap-shaped form; *Helcionella* and *Scenella* are cornucopia-shaped shells less than one complete planispiral coil in length; *Pelagiella* is a small shell of a few planispiral coils; and *Scaevogyra* and *Matherella* are helicoidally coiled. It would seem, therefore, that the earliest gastropod had a cap-shaped shell and probably was not greatly unlike the supposed archetypal mollusk discussed on page 350 and illustrated in Fig. 10-1. From such a simple and primitive form of shell could be developed the three chief types that have characterized Gastropoda throughout their long history—the orthoconic or cyrtoconic cap-shaped shell; the rotund shell with planispiral whorls; and the snail-like form with whorls following a helicoid spiral (Figs. 10-35, 10-40).

It has been suggested that the Gastropoda have possibly reached their zenith, but there is no satisfactory way of proving this. Certainly it can be said that they now constitute one of the largest, most widely distributed, and most successfully adapted groups of animals that has ever left a fossil record.

Ecology and Paleoecology. Considered as a whole the Gastropoda have world-wide distribution and show an amazing range in adaptation. They live in the sea, in brackish waters, in fresh-water environments, and on the land. The largest populations are found in the shallow well-lighted marine waters of the continental shelves, but some species live at depths of more than 5,300 m. (17,400 ft.),[1] and on the land both terrestrial and fresh-water forms have been found at elevations as great as 5,480 m. (18,000 ft.)

Marine gastropods show a wide range of adaptation. A few swim (*e.g.*, Heteropoda and Pteropoda), but most dwell on the bottom, where they may move freely over the mud and sand (commonly leaving a characteristic trail; Fig. 10-54*A*), adhere to rocks, shells, and plants on the bottom, or burrow in the soft substratum. The limpets and limpet-like forms adhere to rocks (whence the common designation "rock clingers") by means of the foot, and in some cases the suction is so strong that the animal can be detached only with difficulty. As an example, the large abalone (*Haliotis*) of the California coast clings so tenaciously that it is difficult to tear the animal away from the substratum, and this must be done with an iron bar or hook. Assemblages of gastropods commonly show remarkable persistence in the habitation of certain kinds of bottoms.

Most marine gastropods soon die if placed in fresh water, and fresh-water species suffer a similar fate when put in sea water. A few exceptional fresh-water genera (*e.g.*, *Melania* and *Neritina*) can tolerate brackish water, and the marine *Littorina* invades the mouths of small streams where they enter the sea. In general, however, aquatic gastropods tend to be limited to a single habitat.

Pulmonate gastropods attain world-wide distribution and show great vertical range. Fresh-water forms have been taken from lakes in The Himalaya more than 5,480 m. (18,000 ft.) above sea level and at lower elevations elsewhere in lakes, ponds, swamps, rivers, and hot springs. Land forms live on the moist and plant-covered surfaces of cliffs and crags and particularly in woodlands and other places where there is a plant cover. Locally they may be prodigiously abundant. Although widely adapted to land habitats, gastropods do not seem to be as tolerant of external conditions as many terrestrial invertebrates. Extremes of temperature do not seem to limit distribution so much as lack of moisture and low content of lime in the soil. Lack of shade and places of concealment are also important. Not many genera are found in arid habitats, but individuals of some species may be extremely abundant under such conditions. Land snails and slugs inhabit areas of considerable vegetation, such as woods, pastures, hedges, garden areas, and plant-covered rock surfaces. A few forms burrow underground, some climb trees, and some

[1] A study by Durham (1947) of 12,000 occurrence records representing 964 "generic" units from about 1,600 different localities from all oceans showed that the greatest concentration of genera of gastropods is in depths less than 200 meters. Four genera have been recorded from depths between 5,000 and 5,500 m., but none deeper; there seems to be a secondary population concentration between 1,100 and 1,400 m. of depth; and about 70 genera seem to be commonly indicative of depths less than 10 m. Of the 964 units recorded, 26 are confined to the Arctic Zone, 211 to the Temperate Zone, and 384 to the Tropical Zone (zones were based on surface temperatures).

even invade houses. More than 6,000 species of land gastropods have been described.

Most gastropods are herbivorous, both in the sea and on the land, and feed upon various plants that they cut up by means of the radula. Some are unusually voracious, as many a gardener can attest. A few eat living and

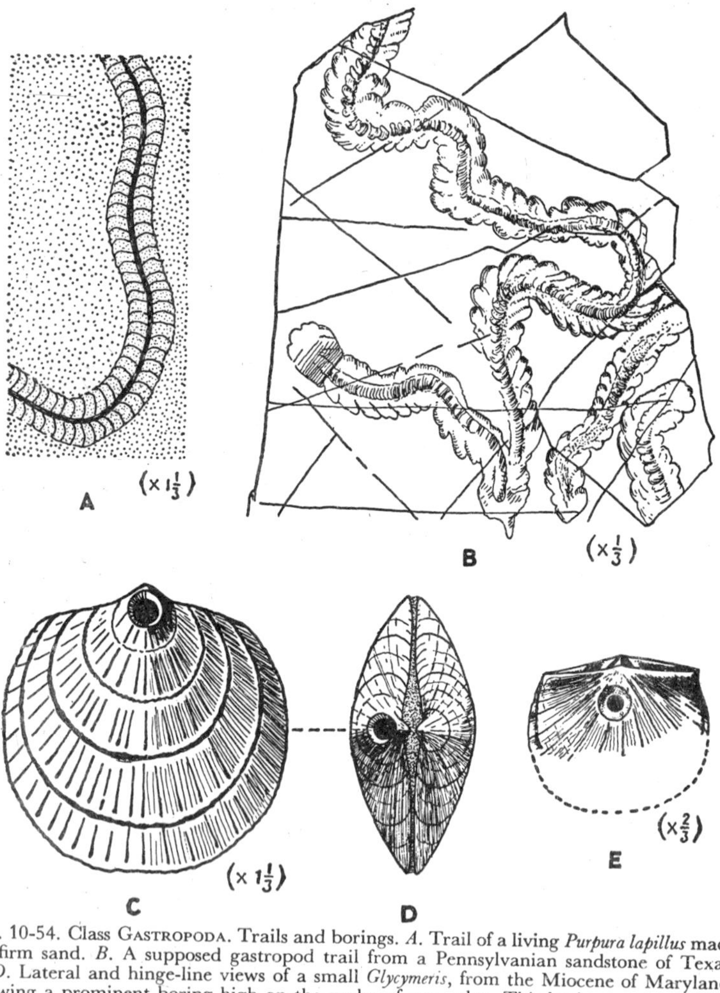

FIG. 10-54. Class GASTROPODA. Trails and borings. *A*. Trail of a living *Purpura lapillus* made on firm sand. *B*. A supposed gastropod trail from a Pennsylvanian sandstone of Texas. *C–D*. Lateral and hinge-line views of a small *Glycymeris*, from the Miocene of Maryland, showing a prominent boring high on the umbo of one valve. This boring may well have been made by a gastropod, as it closely resembles borings of certain living species. *E*. Shell of *Rafinesquina*, from the Upper Ordovician of Indiana, showing a supposed gastropod boring. (*A modified after Nicholson and Lydekker*, 1889; *B after Powers*, 1922; *E modified after C. L. and M. A. Fenton*, 1931.)

decomposed animal matter, and some (*e.g.*, *Natica*, *Purpura*, and *Murex*) are predators that drill holes in the shells of other mollusks to get at and feed upon the soft parts within. On many coasts the majority of the dead shells of certain pelecypod species have gastropod borings, and similarly bored fossil shells have been reported (Fig. 10-54*C–D*).

Gastropods are preyed upon by birds, fish, crustaceans, and other gastropods, as well as by the usual parasitic population. Usually the shell is sufficient for protection, but some predators, such as certain carnivorous beetles, thrust their head into the aperture of land shells and feed on the soft parts of the snail. Commensalism is not common among gastropods; one of the best known examples is that of *Platyceras* (*Orthonychia; Igoceras*), which lived on certain Mississippian crinoids. The gastropod attached its cap-shaped shell to the tegmen near the anal opening and fed on the waste products expelled by the crinoid.

Gastropods are of considerable importance to man. Snails are used extensively for food in some parts of the world, were highly considered by the Romans, and seem to have been an important item in the diet of Paleolithic man, judging from the size and extent of shell heaps discovered in the debris of his camping sites. Certain species are hosts for parasites that are responsible for human and other animal diseases. One of the most notorious of these is *Oncomelania*, a fresh-water snail, some species of which serve as a host for the Oriental blood fluke (*Schistosoma*), which when it infests man causes the dread disease of **schistosomiasis.**[1] In past ages certain gastropods were exploited for dye materials. The "Tyrian purple" of ancient Mediterranean civilizations was made from certain glands of *Murex trunculus* and *M. brandarus*, and in Mexico *Purpura patula pansa* is similarly exploited for dye. The beauty of form and the pattern of coloration among gastropod shells have long made them attractive to collectors, and as a consequence they are probably more assiduously collected by both amateurs and professionals than almost any other group of invertebrates. In some coastal and island regions certain shells have in the past been so highly prized as to be used for currency, and today many kinds of shell ornaments are being manufactured.

Stratigraphic Importance. In Early Cambrian time the gastropods were represented by several genera that had small, smooth cap-shaped shells, such as *Helcionella* and *Scenella*, and by a few that had a simple helically coiled shell, as illustrated by *Pelagiella*. By Late Cambrian time the planispirally coiled bellerophontids (*e.g.*, *Owenella*) and several helicoid genera, such as *Matherella* and *Scaevogyra*, had appeared. Many families came into existence during the Ordovician, and new families have continued to be added from then to Recent times.

Present-day gastropods constitute the largest class of the Phylum Mollusca, and of the 35,000 described living species about 29,000 breathe by means of gills and 6,000 by means of lungs. In addition to the 35,000 living species, something of the order of 15,000 fossil and extinct species have been described.

[1] It is interesting to note that *Schistosoma* infests a prosobranch (*Oncomelania*) in the Philippines but prefers pulmonates (*Bulinus* and *Physopsis*) in Africa. The reason for this specificity of host is unknown.

Nature of Fossil Gastropods. Shell material of Paleozoic gastropods is not generally preserved except in the later rocks; the fossil record consists largely of external impressions and internal fillings (or steinkerns). Quite commonly fossil gastropods from the earlier Paleozoic rocks are too poorly preserved for complete description and accurate identification, either specifically or generically. It has been suggested that this failure of preservation is to be ascribed to the fact that much of the shell of most early gastropods was composed of aragonite. The few well-preserved shells, typified by those of the cyclonemids (Ord.–Sil.) and the platycerids (Sil.–Perm.), are thought to have been partly calcitic. The shells of many Mesozoic and Cenozoic gastropods are commonly well preserved and were probably largely calcitic like their living descendants.

Gastropods make a variety of trails in soft mud and sand and are thought to have been responsible for certain traillike markings that are preserved in ancient rocks[1] (Fig. 10-54*B*). These markings, which have been given various names, are of debatable origin, but some of them do resemble the trails of living gastropods and may well have been made by extinct members of that class. Some fossil shells have boreholes that may have been made by boring gastropods (Fig. 10-54*C–E*).

CLASS CEPHALOPODA[2]

General Considerations

The Cephalopoda, including the ancient nautiloids and ammonoids and the modern squids, cuttlefishes, and octopuses, constitute the most morphologically and anatomically advanced class of the Mollusca. They are bilaterally symmetrical animals with a highly developed head surrounded by a crown of elongated mobile and muscular tentacles, which are usually furnished with prehensile suckers or hooks. The head has large well-organized eyes, which are similar to the vertebrate eye in some respects, well-developed hearing organs, and in most forms a cartilaginous brain case. The mouth is provided with powerful horny beaklike jaws and a radula. Water from the mantle cavity is expelled through a muscular funnel beneath the head, and the animal can swim rapidly when the expulsion is violent. Most existing forms (*Nautilus* excepted) have a bag filled with an inklike fluid that can be ejected into the water before it is expelled through the funnel. The clouded water thus expelled provides a screen behind which the cephalopod may escape from its enemies. One or two pairs of breathing organs, the ctenidia, occupy the mantle cavity, and the class is commonly divided on the basis of the number of gills into two orders: Tetrabranchiata with four gills, and Dibranchiata with two gills. The tetrabranchiates, now represented by only

[1] POWERS, S., 1922, Gastropod trails in Pennsylvanian sandstones in Texas, *Amer. Jour. Sci.*, vol. 53, pp. 101–107.

[2] Cephalopoda—Gr. *kephale*, head, + *pous*, *podos*, foot; referring to the fact that the animal has its foot quite near the head, and hence seems to stand on its head.

a single genus, the well-known *Nautilus*, have an external calcareous shell in the form of a chambered and planispirally coiled cone. Most of the myriads of fossil cephalopods are thought to have been tetrabranchiates, but this cannot be definitely proved for many of them (see later discussion of classification). The dibranchiates have either some sort of modified or vestigial internal shell or none at all. Sexes are separate; the eggs are heavily yoked; and development is direct without the trochophore and veliger larval stages characteristic of all other Mollusca.

The cephalopods are an old and seemingly waning group of highly specialized mollusks that have long since passed their zenith. The 400 species (150 to 170 genera) that remain in existing seas are but a small fraction of those that have lived. The few species with external shells, all belonging to the single genus *Nautilus*, are but a mere remnant of a once great race that filled the seas of the past more than once. Many of the ancient cephalopods had beautifully sculptured shells, some of which reached gigantic proportions for invertebrate structures. At least 600 genera and more than 10,000 species of fossil cephalopods have been described. The earliest of these come from Upper Cambrian rocks, and the class as a whole seems to have had two periods of great development—one during the earlier Paleozoic (Ord.–Sil.), when the nautiloids reached their peak in diversity and number of individuals, and again during the Mesozoic, when the ammonoids were the dominant group. It is possible that the shell-less forms and those with internal shells are more advanced now than ever before, and they may be moving toward a third peak.

Fossil cephalopods are common in marine formations of Ordovician to Tertiary age, and judging from the nature of the fossils with which they are associated, cephalopods seem always to have been exclusively marine, as they are today.

Because the majority of cephalopod genera are extinct, and also because there is uncertainty as to the number of gills many of these genera had, paleontologists customarily use a classification that is somewhat different from that in general use by zoologists.[1] The classification that is used in the present work is as follows:

[1] It seems probable that the ancestral cephalopod had four gills and that the dibranchiates, therefore, have lost two gills sometime during their evolution. Because the time of this reduction is unknown and since gill structures leave no impressions on the shell, it is impossible to determine whether many fossil cephalopods were tetrabranchiate or dibranchiate. Paleontologists, therefore, have abandoned the number of gills as a basis of classification and have turned to shell features, on the basis of which they customarily recognize the following three subclasses: the Nautiloidea, with simple-sutured shells; the Ammonoidea, with complex sutures and sculpture; and the Coleoidea, in which the shell has become internal and vestigial or secondarily lost. This classification has distinct merit with present knowledge, as it is not yet certain whether the earliest of the coleoids (*i.e.*, the belemnites) evolved from a simple straight-coned member of the Nautiloidea or whether, as is suggested by certain shell features, they were derived from a simple representative of the Ammonoidea (*e.g.*, *Bactrites*). (Flower, 1946).

Subclass 1. Nautiloidea. Straight and coiled chambered shells with gently curved transverse septa, simple lines of contact (**sutures**) along which the septa meet the wall of the shell, and generally simple surface sculpture. The only living representative of this important subclass is the chambered *Nautilus*, which is tetrabranchiate. It is assumed that the many extinct representatives also had four gills. The earliest fossil nautiloids come from Upper Cambrian rocks; hence the range of the subclass is from Upper Cambrian to Recent.

Subclass 2. Ammonoidea. Coiled shells with wrinkled septa, complex sutures, and more or less highly sculptured exteriors. Members of the subclass, which has been extinct since the end of the Cretaceous, may have been tetrabranchiate, but this has not yet been demonstrated; hence it seems preferable to consider the ammonoids as a separate subclass equal in rank to the Nautiloidea. The oldest known ammonoid was collected from Upper Silurian rocks, and the range of the subclass is Upper Silurian through Upper Cretaceous.

Subclass 3. Coleoidea (Dibranchia). Dibranchiate cephalopods with 8 or 10 arms, or tentacles, and with an internal calcareous vestigial shell or none at all. The subclass includes the living *Spirula* and extinct *Belemnites* (Belemnoidea), the squids and cuttlefish (Sepioidea), and *Argonauta* and other octopuses (Octopoda). The oldest fossil representative is *Eobelemnites* from the Mississippian (Chester), and the range of the subclass is from Upper Mississippian to Recent.

Essentially all that is known about the soft parts of the Nautiloidea is based on the only living genus, *Nautilus*. Almost nothing is known about the animal in the Ammonoidea, which became extinct in the Cretaceous, although paleontologists have generally assumed that the ammonoid was more or less like the nautiloid. By contrast the Coleoidea with many living representatives have been much studied so that the morphology of their soft parts is generally well known.

Before discussing the three subclasses briefly characterized in preceding paragraphs, it seems appropriate to compare typical living representatives of the Nautiloidea and Coleoidea and to describe the several different types of shells made by the animal. *Nautilus* will serve this purpose for the Nautiloidea, and *Sepia* is a typical example of a coleoid.

Morphology of Nautilus and Sepia

As the last survivor of a large and varied group of cephalopods, which remained coiled throughout the Mesozoic and Cenozoic, *Nautilus* may be expected to have undergone specializations in tissue organization and body pattern and to be somewhat different, therefore, from the individuals that made the cephalopod shells of the Paleozoic (Fig. 10-57). Flower (1946) has ably discussed this matter in considerable detail.

Figure 10-55 shows a diagrammatic sagittal section of *Nautilus* that is designed to portray the general body plan of the animal and the relations of soft parts to shell structures. Figure 10-57 compares a restoration of the animal that made the first nautiloid shells in the Ordovician seas with modern *Nautilus*. The ancient animal is supposed to have had four gills, numerous tentacles, and general body morphology like that of *Nautilus*.

The shell of *Nautilus* is a chambered cone that is coiled planispirally. The animal occupies the terminal **living chamber** and is cemented to the shell by conchiolin on each side. As the animal grows, it moves forward in the shell and periodically shuts off a chamber by secreting a septum. The septa are perforated in the middle and are traversed by the **siphon,** which is a slender tubular prolongation of the visceral hump that extends backward to the very beginning of the shell. The siphon has blood vessels and in ancient nautiloids is thought to have deposited certain secondary shell structures. All the chambers except the last contain a gas similar to air except for somewhat more nitrogen and less oxygen. This gas obviously increases the buoyancy of the shell, thereby making it easier for the animal to swim.

A thin **mantle** covers the entire animal and adheres to the shell; it cannot, therefore, participate in respiration or locomotion. The protruding **head** consists of an elaborate system of retractile and adhesive **tentacles** and a pair of laterally situated **eyes** which are simpler than those of squids, in that they lack a crystalline lens. The tentacles, which may number as many as 90,[1] are arranged in two circles around the mouth. The anterior part of the head, where it touches the part of the shell that is coiled forward over the neck of the animal, is much thickened to form the **hood,** and when the animal withdraws into the living chamber the hood acts as an operculum. Directly beneath the head is the much modified **foot,** which has the form of an elongated **funnel** and is commonly termed the **hyponome.** This opens directly into the capacious **anal chamber** or **mantle cavity,** which contains four leaflike gills. *Nautilus* lacks an ink sac, and there is no evidence that the sac was developed in any of the extinct forms included in the Subclass Nautiloidea.

The front of the mouth is equipped with a pair of horny or calcareous jaws which together resemble the beak of a parrot (Fig. 10-55). Behind this beak are two prelingual processes, a radula, and a tongue. Connecting mouth and crop is a short tubular esophagus which is surrounded by a complex ring of nerve tissue, the dorsal part of which constitutes a brain, whereas the remaining parts are a fusion of the dorsal ganglia. The crop becomes constricted posteriorly and connects by a small tube with the stomach, which has thick muscular walls. Behind the stomach, a large **liver** and certain other digestive glands discharge into the alimentary canal. A simple tubular intestine turns back anteriorly from the stomach and empties through a simple anus into the anal chamber. The nephridia and heart lie near the intestine, and the former together with the reproductive system also discharge their products into the anal chamber. Oxygen-bearing water enters the chamber, or **ctenidial cavity,** through an opening between the relaxed mantle and the adjacent wall of the hyponome (Fig. 10-56*E*). The water bathes the gills and when expelled carries to the outside the waste from metabolism and the products of the sex organs. During normal expulsion the mantle is pressed

[1] Extinct Paleozoic nautiloids may have had relatively few tentacles of large size, as suggested by certain discrete impressions (as few as 10 in number) that are preserved in Upper Ordovician rocks and that have been interpreted as impressions made by the tentacles of a *Treptoceras* (Flower, 1946).

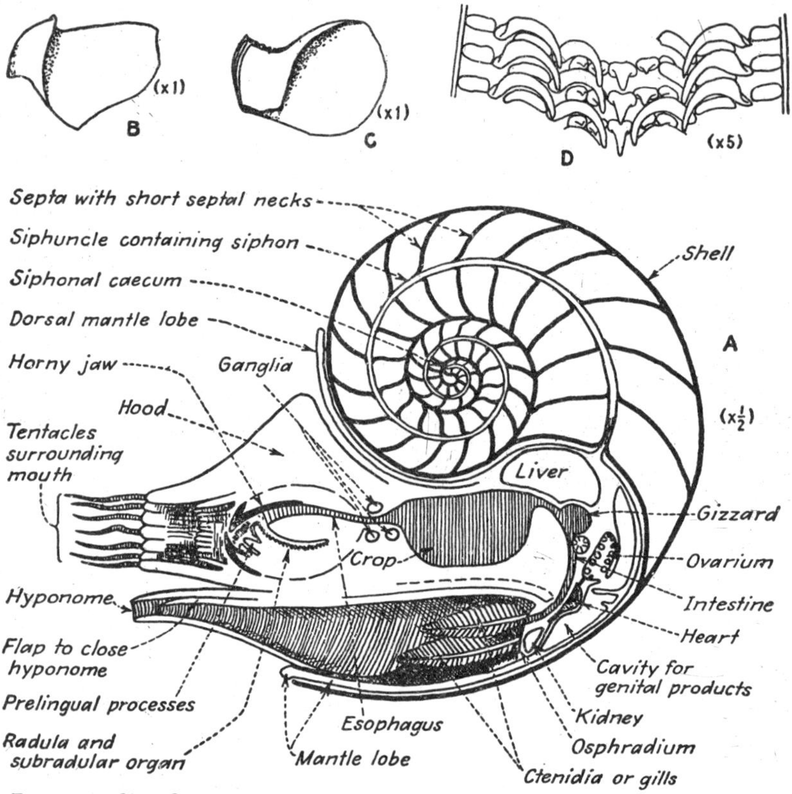

FIG. 10-55. Class CEPHALOPODA. *A*. Diagrammatical sagittal section of *Nautilus*, showing chief organs and relation of body to shell. *B–D*. Jaws and radular elements of *Nautilus pompilius: B*, side view of upper jaw; *C*, inner surface of lower jaw; *D*, three complete rows of radular elements. (*A, D adapted from Naef*, 1928; *B–C modified after Griffin*, 1900.)

against the hypónome, and the valve of the latter flat against its wall (Fig. 10-56*A*). If water is forcibly ejected, it acts as a jet to propel the animal forward or backward, depending on the way the hyponome is directed.

The three generally recognized species[1] of *Nautilus* inhabit moderately shallow water (rarely more than 700 m. deep) along the coasts and around coral reefs in a large area of the southwest Pacific from the Malay region eastward to Australia, New Caledonia, and the Fiji Islands, and thence northward to the Philippines. Empty shells have been reported from a much

[1] Of the 10 supposed living species, excepting Iredale's (1944) recently created ones about which there is question, only three are generally recognized—the involute non-umbilicate *N. pompilius*, and the two umbilicate species, *N. umbilicatus* and *N. macromphalus*. The problem of how many species of *Nautilus* are now living will probably not be solved without careful ecological, morphological, life-history, and statistical studies of the different forms.

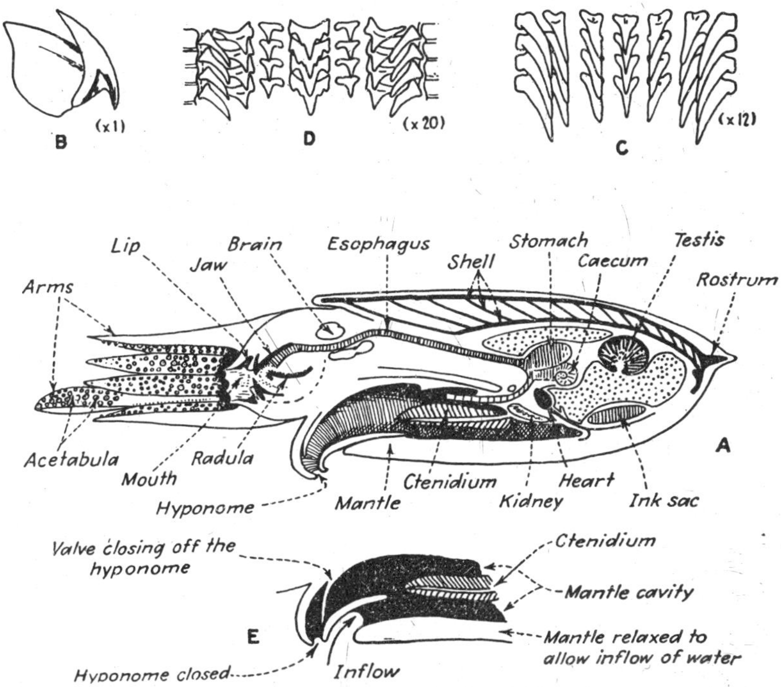

FIG. 10-56. Class CEPHALOPODA. *A*. Anatomy of a dibranchiate, *Sepia*, as revealed in a sagittal section. *B–C*. *Sepia officinalis*: *B*, jaw; *C*, portion of radula. *D*. Portion of radula of *Octopus vulgaris*. *E*. Small part of *A*, showing the inhaling phase when water is taken into the mantle cavity. The mantle is relaxed to permit the inflow of water as indicated by the arrow, and the valve of the hyponome is closed so as to prevent escape of water through that organ. *Sepia* (*A*) is shown in the expiratory condition with the mantle pressed against the hyponome, and the valve of the latter flat against the wall. (*Modified after Naef*, 1921; 1928.)

larger area and have drifted to such remote places as Japan, India, and Africa—a condition that should be borne in mind by those who are concerned with interpreting the distribution of fossil remains.

The gas in the chambers of the shell serves to buoy the animal so that it can hover near the bottom or swim by using the hyponome. In either case the gas-filled chambers must be situated above the living chamber. Although it is commonly stated that the animal crawls over the bottom by means of its tentacles, observations by competent investigators do not support this idea. *Nautilus* is a rapid swimmer. It moves by jet propulsion, and the more violent movements probably result when the animal suddenly withdraws into the shell, thus compressing the anal chamber and causing a jetlike expulsion of water from the hyponome.

Although a living specimen of *Nautilus* has been captured in water less than 4 m. deep, individuals are not abundant in water less than 100 m. deep, and the best catches are made in water about 500 to 700 m. deep. How much deeper *Nautilus* can actually go is uncertain because little is known about its life habits. In fact, some of the deeper records may be based on dead individuals.[1]

The shell of *Nautilus* is a chambered cone that is coiled planispirally and in such manner that only the last of the several volutions is wholly visible. The shell has a blunt rounded apex with a sacklike shelly structure, the **siphuncular caecum,** extending almost if not entirely to the terminal wall, as shown in Fig. 10-75*G*. Septa, convex toward the apex of the shell (*i.e.*, the adapical end), divide the cone into chambers, or **camerae,** and are themselves perforated in the center by an opening through which the fleshy siphon passes. The siphon extends from the posterior portion of the visceral hump adapically through all chambers to the first, where it terminates in the sacklike siphuncular caecum previously mentioned. The surface of the siphon is composed of mantle tissue and this tissue can secrete calcium carbonate. Consequently, the siphon is generally encased in a continuous or discontinuous calcareous tube, the **siphuncle,** which consists of two distinctly different structures, the septal necks and the connecting rings. The **septal neck** is a short tube that extends posteriorly from the convex side of a septum and is continuous with it. It is generally too short to reach back to the earlier septum, but may be joined to that septum by a thin, porous cylinder, the **connecting ring.** Each connecting ring fits over the outside of a septal neck at one end and into the earlier neck at the other end. Connecting rings are thin fragile structures, hence are easily broken and are commonly missing in both recent and fossil nautiloids. When fully developed, septal necks and connecting rings together make a continuous calcareous tube around the siphon, and in fossil nautiloids that have such a continuous siphuncle it is not always possible to determine where a septal neck ends and a connecting ring begins.

The **septa** are transverse partitions secreted by the posterior mantle and added periodically as the animal grows forward in the living chamber. Where they fuse with the inner surface of the shell, they tend to turn forward, particularly along the dorsal margin, and make a narrow flange. A typical septum, therefore, consists of three main regions; the mural part just described, the septum proper, or free part, that acts as a transverse partition, and the septal neck, which serves as a partial sheath around the siphon. The line along which the septum leaves the wall of the shell, or where the free part of the septum grades into the mural part, is termed the **suture.** This line is an important feature in fossil shells because it indicates the extent of corrugation of a septum and also the positions of successive septa. As will be pointed out later, the sutures became amazingly complicated in the more advanced ammonoids. On the internal fillings of many ancient nautiloids a narrow **septal furrow** marks a middorsal zone where the mural part of the septum was lacking (Fig. 10-65).

The first half of the initial whorl of the shell is somewhat different from

[1] For an excellent summary of ecological and paleoecological data on *Nautilus*, see Stenzel (1948, 1948*a*).

the remainder of the shell. It is not closely coiled, so that there is a minute **umbilical perforation** (Fig. 10-75*F–G*). A similar perforation is present in most fossil coiled nautiloid shells, but it is commonly concealed by the later whorls. The first septum differs from all others in that it lacks a perforation but has a prominent central depression, the siphuncular caecum, in which the siphon ends. The terminal chamber represents the first calcareous camera of the nepionic or youthful shell and supposedly succeeded an initial horny embryonic shell, the so-called **protoconch.** A **scar,** or **cicatrix,** is present on the rounded end of the first camera and is thought to mark the place of attachment of the protoconch. However, no such supposed horny protoconch has been demonstrated to be present in *Nautilus*. To explain its absence in-

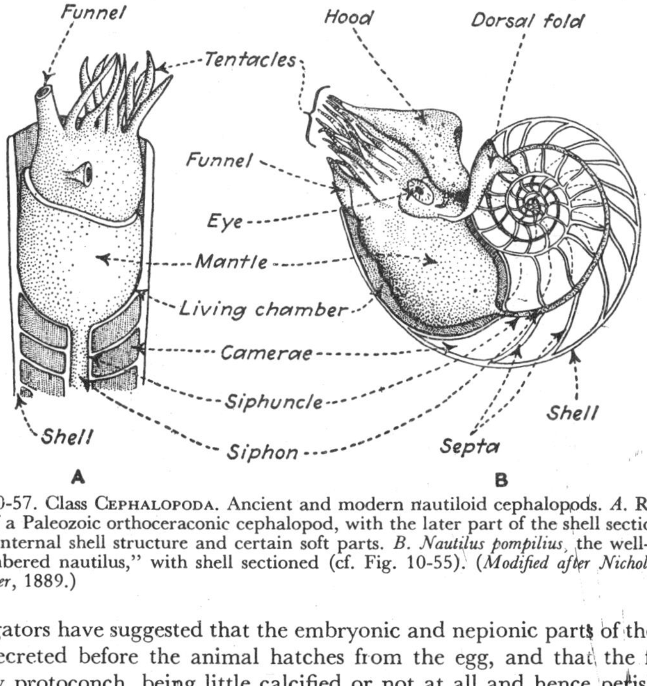

FIG. 10-57. Class CEPHALOPODA. Ancient and modern nautiloid cephalopods. *A*. Restoration of a Paleozoic orthoceraconic cephalopod, with the later part of the shell sectioned to show internal shell structure and certain soft parts. *B*. *Nautilus pompilius*, the well-known "chambered nautilus," with shell sectioned (cf. Fig. 10-55). (*Modified after Nicholson and Lydekker*, 1889.)

vestigators have suggested that the embryonic and nepionic parts of the shell are secreted before the animal hatches from the egg, and that the fragile horny protoconch, being little calcified or not at all and hence perishable, is soon destroyed, leaving only its scar of attachment as evidence of its former existence. In this absence of knowledge about the embryonic and nepionic stages of *Nautilus*, it seems impossible to state definitely whether or not it has a protoconch. As regards extinct nautiloids, it would seem from the investigations of Clarke (1893, 1897), Ruedemann (1905), and Flower (1946) that some definitely had a protoconch whereas in others no protoconch as such was developed.

The wall of the shell is composed of two layers, both aragonitic: an outer porcelaneous layer and an inner nacreous layer. In addition, the entire inner surface of the shell, including the septa, is coated with an opaque calcareous film, the exact mineralogical composition of which seems not to have been determined. Recently taken shells generally have a film of black organic material above the aperture where the hood contacts the surface, but this substance is easily destroyed and hence would not be likely to survive fossilization.[1]

Sepia (Fig. 10-56) is typical of a large group of dibranchiate cephalopods in which the shell has become internal and vestigial. The general shape of the animal and disposition of the organs are much as they would be if the animal inhabited the living chamber of a shell like that of *Nautilus*. As a whole the body is cylindrical, with the visceral hump constituting the bluntly rounded posterior part and the head with its numerous tentacles forming the anterior part. Conspicuously lacking is the siphon that is present in *Nautilus*. The head has a central mouth and two relatively large lateral eyes, which are more complex than those of a vertebrate. Encircling the mouth are 10 prominent tentacles for seizing prey. Four pairs of these are short and stout and are covered with suckers on their inner surfaces. The tentacles of the fifth pair, which is fourth in position counting from the dorsal surface, are long and can be retracted into large basal pits. These longer tentacles have suckers only at their free ends. Internally, the organs of *Sepia* are much like those of *Nautilus*, as may be seen by comparing Figs. 10-55 and 10-56, except that the gills number two instead of four. The mantle, which surrounds all the body except the head, contains a complex of strong muscles, which by interaction compress and expand the mantle cavity. During normal respiration there is a gentle rhythmic intake and expulsion of water through the hyponomic funnel. If the animal is suddenly disturbed, the muscular action becomes greatly accelerated and spasmodic, and the jetlike ejection of water through the funnel causes the animal to dart quickly through the water. By directing the funnel the animal can move either forward or backward.

The shell of *Sepia* is internal and is a rather strong plate that acts as an endoskeleton. It is considered to be the vestige of an ancestral chambered shell that underwent profound modification as indicated in Fig. 10-80. The several types of vestigial internal shells characteristic of the Coleoidea are shown in Fig. 10-80 and are discussed more fully on following pages.

Architecture and Structure of Cephalopod Shells

Many of the more important structures of the cephalopod shell are common to both Nautiloidea and Ammonoidea, and a few are common to all three subclasses. It seems appropriate, therefore, to discuss the shell in a

[1] Little is known about the structure and composition of the shell material of ancient nautiloids, owing to alteration or destruction during fossilization. According to a recent report by Fischer and Finley (1949), certain Pennsylvanian nautiloid shells preserved with little or no alteration in asphaltic rock show nacreous luster and microstructure of the aragonitic shell material. Shell wall, septa, protoconch, and cameral deposits are all preserved, and the authors conclude that the cameral deposits are of organic origin (rather than being secondary and inorganic) and are the product of cameral mantle tissues (also see p. 457).

general way as a preliminary to considering its special characteristics in the three major taxonomic divisions.

General Considerations. The external cephalopod shell,[1] which is confined to the Nautiloidea and Ammonoidea, is termed the **conch.** In its simplest form it is a chambered shelly cone open at the larger end and closed at the smaller (Fig. 10-58). It is divisible into a large anterior **living chamber,** which generally constitutes from a fourth to a third of the conch by volume, and a posterior chambered **phragmocone,** which is partitioned into intercommunicating chambers, or camerae, by singly perforated transverse septa. A discontinuous or continuous shelly tube, the siphuncle, begins at the very apex of the shell, extends from one septal perforation to the next, and ends in the perforation of the latest septum, which forms the posterior wall of the living chamber. Some shells are straight cones, some are gently curved, and many are loosely or tightly coiled, chiefly planispirally but also rarely helicoidally (Figs. 10-59, 10-60). A delicate and fragile embryonic shell, the **protoconch,** constitutes the initial part of the conch, but it is almost never preserved in straight fossil shells and rarely in coiled ones.

A large vocabulary of technical terms has been created for use in describing the many aspects and structures of cephalopod shells. Most of the more important of these are defined in later paragraphs, but a few need to be defined here as they are used generally throughout the discussion.

Adapertural, or **adoral**—toward the aperture or anterior end of the shell. **Adapical**—toward the apex of the shell. **Apicad**—toward the apex from a point of reference. **Centrad**—toward the center of the shell. **Distad**—toward the periphery or tip; away from base or center. **Dorsad**—toward the dorsal side of the shell. **Dorsum**—the dorsal side of the shell. **Mural**—pertaining to the wall of the shell. **Venter**—the underside or ventral side of the shell, indicated generally by the hyponomic sinus and conchial furrow. **Ventrad**—toward the ventral side of the shell.

In life it is supposed that an animal basically like modern *Nautilus* occupied the living chamber and had a fleshy siphon that extended back to the first chamber of the shell, passing through successive septal perforations as in *Nautilus* (Fig. 10-55).

The nature of the first cephalopod shell is unknown. Formerly it was thought that the earliest shell must have been a simple straight chambered cone with a centrally situated siphuncle (Fig. 10-58) and that from this simple shell by progressive coiling and later uncoiling were developed the many types of coiled shells that are characteristic of both Nautiloidea and Ammonoidea. Unfortunately, several facts make this simple supposition untenable. The earliest nautiloids (which are also the oldest cephalopods) are small curved shells with marginal siphuncles, and the supposedly simple straight shells with circular section are actually not primitive. Furthermore, some nautiloid shells had a protoconch (Clarke, 1893, 1897), whereas in several large and important orders of straight shells (Actinoceratida and

[1] Not considered as true external shells are the shells of *Spirula*, a coleoid, and the egg cases of the octopod *Argonauta*, which cases superficially resemble a coiled shell but lack any internal structures.

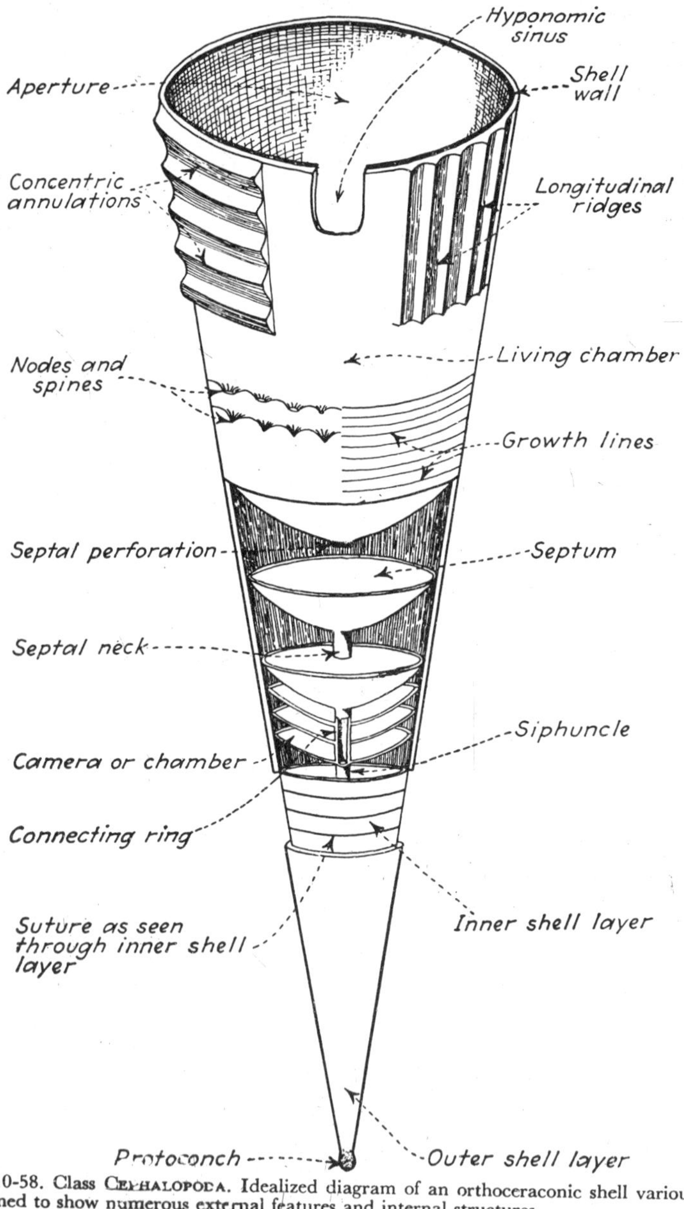

Fig. 10-58. Class Cephalopoda. Idealized diagram of an orthoceraconic shell variously sectioned to show numerous external features and internal structures.

Endoceratida), no protoconch as such seems to have been developed. It is now generally accepted that the concept of gradual coiling and later uncoiling of the cephalopod shell is not a valid basis for subdividing the class and its orders into successive evolutionary stages. It has been demonstrated, for example, that progressive coiling of the shell has taken place more than once and that shells almost exactly similar in external aspects are in actuality greatly different in important internal features. Consequently, internal shell structure is now considered to be the most reliable basis for subdividing the Nautiloidea into its numerous orders, and this aspect of shell structure is considered later in the discussion of classification (page 462).

Coiling and Shell Form. The more typical forms of nautiloid and ammonoid shells are shown diagrammatically in Figs. 10-59 and 10-60. It was long customary to assign all coiled shells of similar type to a single genus, so that there was developed through the years a series of descriptive terms that were derived from the generic names. As these terms are both appropriate and in wide use, they are listed and defined in Table 10-3. It is convenient to recog-

TABLE 10-3. TERMINOLOGY OF SHELL FORM IN NAUTILOIDEA AND AMMONOIDEA
Largely after Flower, 1946. Ammonoid terms are italicized

Shell form	Portion of shell	Entire shell
Straight	Orthocone Orthoconic	Orthoceracone Orthoceraconic Orthoceran *Bactriticone*
Curved	Cyrtocone Cyrtoconic	Cyrtoceracone Cyrtoceraconic Cyrtoceran
Loosely coiled	Gyrocone Gyroconic	Gyroceracone Gyroceraconic Gyroceran *Gyroceratiticone*
Whorls in contact	Tarphycone Tarphyconic	Tarphyceracone Tarphyceraconic Tarphyceran *Dactylioceracone*
Involute coil		Nautilicone Nautiliconic Nautilian *Ammoniticone*
Eccentric coil		Trochoceracone Trochoceraconic Trochoceran *Turriliticone*
Short gibbous shells		Brevicone Breviconic Gomphoceroid
Secondarily straight shells due to uncoiling		Lituiticone Lituiticonic *Baculiticone*

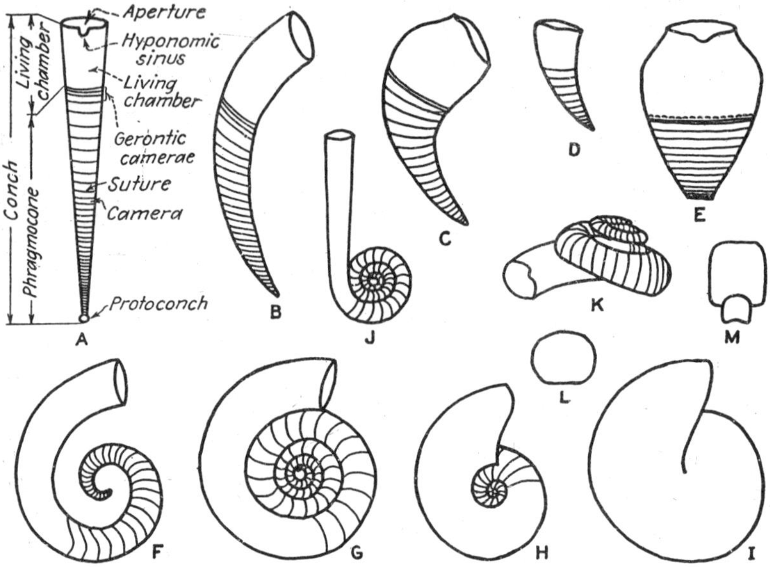

FIG. 10-59. Class CEPHALOPODA—Subclass NAUTILOIDEA. Shell forms and coiling. *A*. An orthoceracone, illustrative of genera like *Michelinoceras*, with the principal shell features indicated. This figure serves as a general key for the other shells. *B*. Cytoceracone. *C*. An adult cyrtoceraconic shell which is breviconic in its latest growth stages (cf. *E*). *D*. Cyrtoconic young stage of the same form as *C*. *E*. Brevicone. *F*. Gyroceracone. *G*. Tarphyceracone. *H–I*. Convolute and involute nautilicones, respectively. *J*. Lituiticone; nautiliconic in the earlier stages, orthoconic in the latest part. *K*. Trochoceracone; also described as trochoceroid. *L*. Cross section of a whorl of a tarphyceracone (of *G*), with no true impressed zone but only a slight flattening of the dorsum. *M*. Cross section of two whorls of a *Tarphyceras*, showing slight development of impressed zone for reception of the earlier whorl. (*Modified after Flower*, 1946.)

nize two groups of terms—those which may be applied to the shell as a whole, and those which may be applied only to a portion of a shell. The types of shells to which these terms are applicable are shown in Figs. 10-59 and 10-60

The simplest form of cephalopod shell is a straight cone, termed an **orthoceracone** (*Orthoceras*) among the Nautiloidea and a **bactriticone**[1] (*Bactrites*) among the Ammonoidea. Slightly curved shells, illustrated by the common nautiloid genus *Cyrtoceras*, are **cyrtoceracones;** no ammonoid shells of this type are known. Loosely coiled nautiloid shells in which the coils do not touch are designated **gyroceracones,** after *Gyroceras*, and ammonoid shells of similar nature are **gyroceratiticones** (*Gyroceratites*). If the whorls of a coiled shell are in contact, like a coiled rope on the deck of a ship, the nautiloid form is a **tarphyceracone** (*Tarphyceras*), and the ammonoid, which may show impressing of earlier coils on later, a **dactylioceracone** (*Dactylioceras*). Tighter

[1] If *Bactrites* is a nautiloid, as some contend, then the term bactriticone should be suppressed. For secondarily straightened shells Miller (1938, p. 12) has suggested **lobobactriticone** or **baculiticone.**

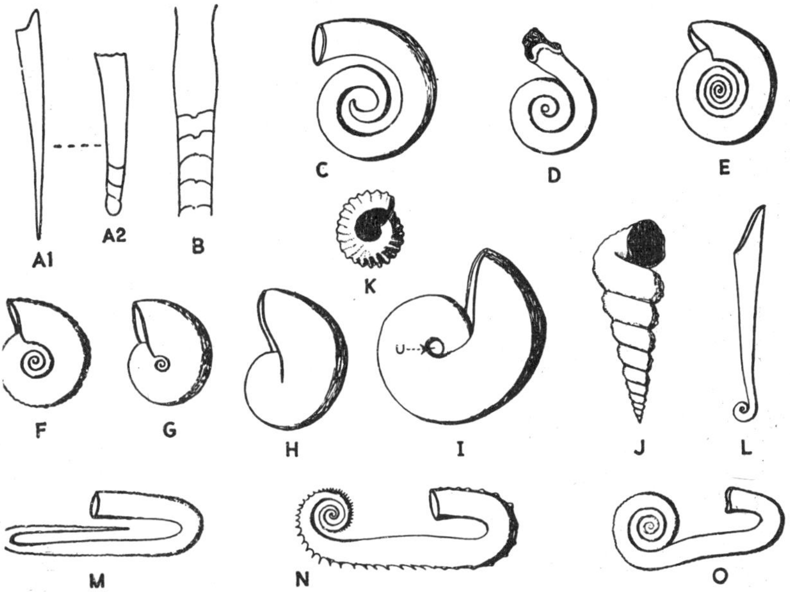

FIG. 10-60. Class CEPHALOPODA. Coiling in ammonoid shells. *A–B*. Bactriticones characteristic of *Bactrites: A*1, a complete conch; *A*2, a supposed young shell, *B*, an incomplete mature shell viewed ventrally and showing living chamber with a few of the latest camerae. *C*. Gyroceratiticone. *D*. Dactylioceracone. *E–H*. Ammoniticones illustrative of *Clymenia* (*E*), *Trachyceras* (*F*), *Hoplites* (*G*), and *Placenticeras* (*H*), showing progressive encroachment of last whorl on preceding whorls until the shell is involute as in *H*. *I*. An ammoniticone with an umbilicus (*u*), illustrated by *Scaphites*. *J*. A turriliticone illustrative of *Turrilites*. *K*. View of larger end of *J* looking along the axis of coiling and into the umbilicus (black circle). *L*. A modified ammoniticone—a baculiticone—showing uncoiling and straightening of the shell in maturity as illustrated by *Baculites*. *M–O*. Modified ammoniticones coiled in different ways: *M*, *Hamites; N*, *Ancycloceras; O*, *Scaphites*. (*Adapted from figures by numerous authors.*)

planispiral coiling causes more and more impressing of earlier coils on later ones, with progressively less of the earlier whorl remaining visible; such shells are described as **convolute.** Planispiral coiling reaches its limit when the last whorl covers all preceding ones; such shells are **involute.** Convolute and involute nautiloid shells are generally designated **nautilicones** (*Nautilus*) and similar ammonoid shells, **ammoniticones** (*Ammonites*), both with appropriate adjectival modifiers (*e.g.*, involute nautilicone; convolute ammoniticone). A few cephalopod shells are eccentrically coiled in such fashion that the whorls follow a helix, developing more or less of a spire; such nautiloid shells are termed **trochoceracones** (*Trochoceras*) and the ammonoids, **turriliticones** (*Turrilites*).

In some shells coiling ceased in the adult stage and the whorl continued to grow forward in a straight or slightly curved line, so that it became free from the earlier coiled part. Such uncoiling is present in a few nautiloids

(*Lituites*) and in numerous ammonite genera (*e.g.*, *Baculites* and *Scaphites*). Such shells may be designated **lituiticones** and **baculiticones,** respectively. In *Scaphites* and a few other genera the latest part of the shell started to coil back on itself so that the complete shell has somewhat the profile appearance of a boat, whence the generic name *Scaphites* (Gr. *scaphe*, boat).

A small group of nautiloid cephalopods developed short gibbous shells illustrated by *Brevicoceras* and *Gomphoceras*. Such shells have been described as **breviconic** and **gomphoceroid,** respectively. Similar forms do not seem to have developed among the Ammonoidea.

Another set of terms included in Table 10-3 are merely descriptive of a part of a shell and are not applied to complete shells. These are **orthocone, cyrtocone, gyrocone,** and **tarphycone.** Their meaning can be inferred readily from the data listed in the column headed "Shell form."

Orientation. The mid-ventral part of an orthoceraconic shell is indicated by the hyponomic sinus and the conchial furrow, and in a few shells preserved as internal fillings a narrow septal furrow may mark the middorsal region (Fig. 10-65). The siphuncle of orthoceraconic and bactriticonic shells may be central or eccentric.

Most curved and coiled shells have the ventral side, or venter, on the outside of the curve and the dorsum on the inside (Fig. 10-61*C–D*). Such shells are **exogastric.** A considerable number of nautiloids, particularly crytoceraconic genera, are known or believed to have curved their shells in the opposite direction, so that the dorsum is on the outside of the curve

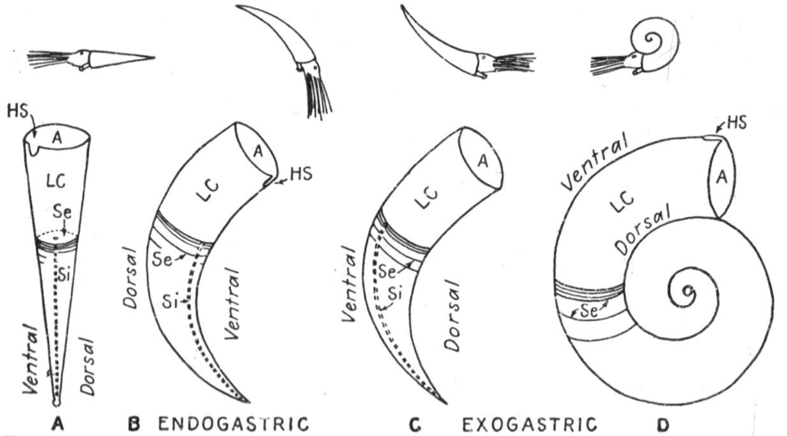

FIG. 10-61. Class CEPHALOPODA. Diagrams showing orientation of cephalopod shells. *A*. A typical orthoceracone with prominent hyponomic sinus and central siphuncle. *B*. An endogastric cyrtoceracone. *C*. An exogastric cyrtoceracone. *D*. A typical tarphyceracone which could only be exogastric. The small figures in the upper row indicate how the animal and shell may have been related; they are obviously conjectural. Only a few of the septa are shown in the lower figures. (*A* = aperture; *HS* = hyponomic sinus; *LC* = living chamber; *Se* = septum; *Si* = siphuncle.)

and the venter on the inside. Such shells are **endogastric** (Fig. 10-61*B*). Recognition of a shell as endogastric has been based on several different criteria, not all of which are reliable. The most reliable criterion seems to be the presence of a hyponomic sinus on the concave side of the shell. The Silurian *Phragmoceras* and its allies have such a sinus and also have the siphuncle near the concave side; hence they are truly endogastric. Many other cyrtoceraconic genera, which show no hyponomic sinus, have, however, been described as endogastric because the siphuncle is close to the concave side of the shell. That position of siphuncle alone is not a reliable criterion for classing a shell as endogastric is demonstrated by the fact that *Archiacoceras* from the German Devonian has an exogastrically curved shell with a dorsal siphuncle! It is quite possible, therefore, that genera originally described as endogastric on the basis of a dorsal siphuncle alone may actually be exogastric (Flower, 1943; 1946). The problem of orientation is of particular interest with respect to the oldest cephalopods in that most of them have the siphuncle in dorsal position, hence have been considered endogastric.

The several types of internal cephalopod shells are shown in Fig. 10-80 and are discussed more fully on following pages. Since they are vestigial and hence incomplete, they can be understood better once a typical complete shell has been fully described.

Shape and Size of Shell. External cephalopod shells show wide variation in shape and size. Straight shells vary from slender cones to short robust pear-shaped structures and range in size from tiny forms less than 25 mm. (1 in.) long to the giant endoceroid cones of the Ordovician with a maximum length of 4½ m. (15 ft.) and a maximum diameter across the living chamber of about 30 cm. (1 ft.). Coiled shells may be spheroidal, as many Pennsylvanian nautiloids, thickly discoidal, as *Nautilus*, or thinly discoidal, as *Placenticeras*. Some coiled shells are shaped like a hook (*Hamites*) or a boat (*Scaphites*). The largest known coiled shell is that of *Pachydiscus seppenradensis* from the Cretaceous of Westphalia, which when complete is supposed to have been about 2.55 m. (8 ft. 6 in.) in diameter and would be more than 10 m. (35 ft.) long if uncoiled.[1] From this extreme, shells range in dimension to tiny forms less than 25 mm. (1 in.) in diameter.

Surface Sculpture and Coloration. Except for growth lines, many cephalopod shells are smooth—a condition well shown by the living *Nautilus*, by many extinct nautiloids (particularly orthoceraconic forms), and by earlier and some later ammonoids. Many others, however, have spiral and transverse ridges and ribs, or nodes and spines similarly arranged. Sculpture was developed early in the history of the cephalopods and gradually increased in complexity to a climax in the late Mesozoic. Especially rapid and varied was the development of sculpture during the Mesozoic, when many bizarre and specialized forms evolved. The general extinction of cephalopods in the

[1] In 1895 Landois illustrated a giant specimen that is 1.8 m. (5 ft. 11 in.) in diameter, and he estimated that if the living chamber of this shell were complete its maximum diameter would be about 2.55 m. (8 ft. 4 in.). Other records of giant ammonoid shells are summarized by Miller and Youngquist (1946).

late Cretaceous eliminated all the highly sculptured forms so that only a few simply sculptured types remained and on these coloration was more conspicuous than sculpture.

The sculpture of many Mesozoic shells is so varied and yet so constant within the smaller taxonomic divisions that it has been used widely and successfully in differentiating genera and species.

Shell coloration in living cephalopods is limited to *Nautilus*, the only surviving genus with a true external shell. The calcareous egg case of *Argonauta*, which superficially resembles an external shell, and the small, spirally coiled internal shell of *Spirula* lack coloration, though the former does have sculpture (Fig. 10-83*C*). The coloration of *Nautilus* is quite prominent. Patterns of color were also undoubtedly present on the shells of many ancient cephalopods, though coloration is not commonly preserved on fossils. If preserved, the color pattern may be present on all sides of a shell or on only a part,

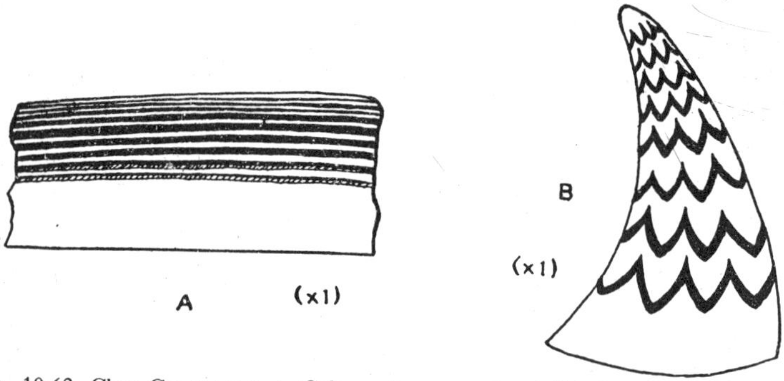

Fig. 10-62. Class Cephalopoda. Color patterns on the shells of ancient cephalopods. *A*. Lateral view of portion of conch of *Geisonoceras tenuitextum* (Ordovician) showing numerous rectilinear color bands parallel with the shell axis. *B*. A breviconic shell of *Cyrtoceras parvulum* (Silurian) showing transverse zigzag color bands that completely encircle the conch. (*After Ruedemann*, 1921.)

and may consist of spots or of annular, undulating or zigzag bands.[1] Such markings have been found on shells of *Cyrtoceras* (Fig. 10-62*B*) as old as Silurian. Some shells, as *Kionoceras angulatum*, have a series of zigzag color bands on the dorsal side only, and the shells of *Geisonoceras tenuitextum* (Fig. 10-62*A*) have bands on the dorsal side that are parallel to the long axis of the shell. It is presumed that cephalopods with dorsally ornamented shells moved with the dorsal side of the shell always upward. Shells with uniform coloration on all sides are believed to have been maintained in a vertical position during life, and the animal is assumed to have floated or swum with the pointed extremity of the shell upward.

[1] The interested reader will find an excellent discussion of shell coloration in the following articles: Foerste, A. F. 1930. The color patterns of fossil cephalopods . . . , etc. *Univ. Mich., Contr. Mus. Pal.*, vol. 3, pp. 109–150. Miller, A. K. 1947. Tertiary nautiloids of the Americas. *Geol. Soc. Amer., Mem.* 23, pp. 13–20.

The Aperture. The terminal opening, or aperture, of the cephalopod shell may be circular, elliptical, oval, or variously restricted and modified as shown in Fig. 10-63. The apertural margin is characteristically sinuate. Marginal convexities that project forward are termed **crests,** and concavities that form posteriorly directed reentrants are designated **sinuses.** Most shells have a **hyponomic sinus,** which is invariably ventral, a lateral crest on each side, and a dorsal sinus. The absence of a hyponomic sinus has been interpreted by many investigators to mean that the hyponome was wanting or greatly reduced, but such a view is difficult to maintain.

The aperture of mature shells of some groups is restricted in some way, and may be so greatly modified as to have the shape of the letter T or to be little more than a slit (Fig. 10-63). Restricted apertures are particularly characteristic of certain nautiloid families (Oncoceratidae, Poterioceratidae, Trimeroceratidae, and Phragmoceratidae), but they were also developed in ammonoid shells. The aperture of *Mandaloceras* is small and shaped like the letter T, with the hyponomic sinus at the base of the T; that of *Phragmoceras* is constricted laterally, so that there is a rather large ventral part, a smaller dorsal part, and a narrow slit connecting the two.

Restriction of the apertural margin has been variously interpreted. It probably aided in keeping the animal in the shell, and as some shells with restricted apertures are thought to have floated in the water with the aperture downward, the restricting structure may well have supported the body, as well as protecting part of it, and thus have been of value to the animal. On the other hand, restriction of the aperture limited motion, almost certainly interfered with locomotion, and may have forced the animal to use food of smaller dimensions. Restriction of the aperture seems to have taken place only after the shell had attained maturity.

Nautilus lacks any sort of plate for closing the aperture when the animal withdraws into the body chamber, and no extinct nautiloids are known to have had such a plate. By contrast, many ammonoids did develop a single or double plate for this purpose, and these plates have been preserved in the living chambers of some shells (Fig. 10-64). Some ammonoids had a single horny plate, the **anaptychus;** others had two equal and similar calcareous plates, collectively designated the **aptychus.**[1]

Septa and Septal Structures. The number of septa varies greatly among different species, and the distance between adjacent septa is likewise variable, but both number and spacing are fairly constant in the same species. Septa tend to be relatively far apart in the earlier portion of the shell and quite close together in the part of the shell directly adjacent to the living chamber (Fig. 10-59). The septa of most nautiloid shells are single curved plates convex toward the shell apex and concave toward the aperture (Fig. 10-58). The earlier and more primitive shells (*i.e.*, the Nautiloidea) have smoothly curved, bowllike septa that join the shell wall along a circular or elliptical suture. In the more advanced forms, which constitute the Ammo-

[1] The interested reader will find a valuable summary of reports of aptychi and anaptychi in the following article: O'CONNELL, M., 1921, New species of ammonite opercula from the Mesozoic rocks of Cuba, *Amer. Mus. Nov.*, no. 28, pp. 1–15.

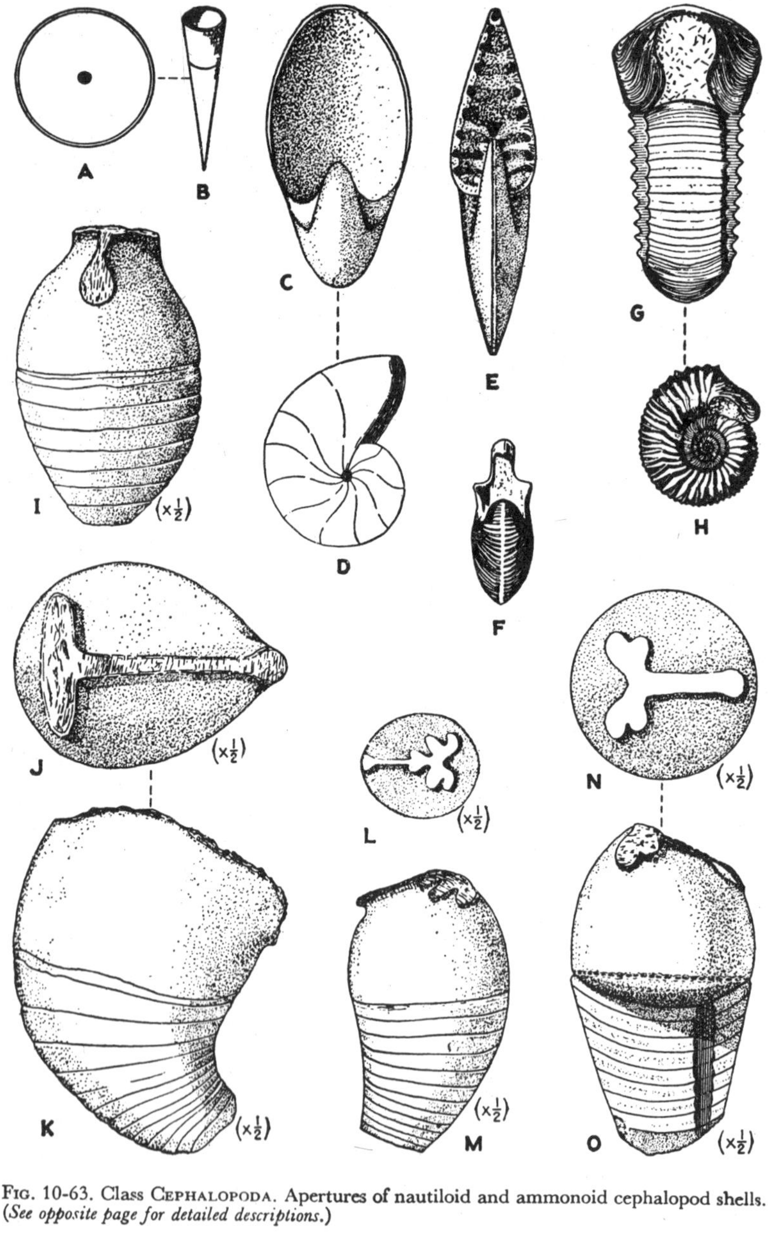

FIG. 10-63. Class CEPHALOPODA. Apertures of nautiloid and ammonoid cephalopod shells. (*See opposite page for detailed descriptions.*)

noidea, the septa are much inflected along the periphery, and the suture, as a consequence, is a more or less intricately convoluted line or tracery. On the middorsal region of many nautiloid shells is a narrow region, the septal furrow (Figs. 10-65, 10-67*E*), where the mural part of the septum is lacking. This line is commonly prominent on internal fillings.

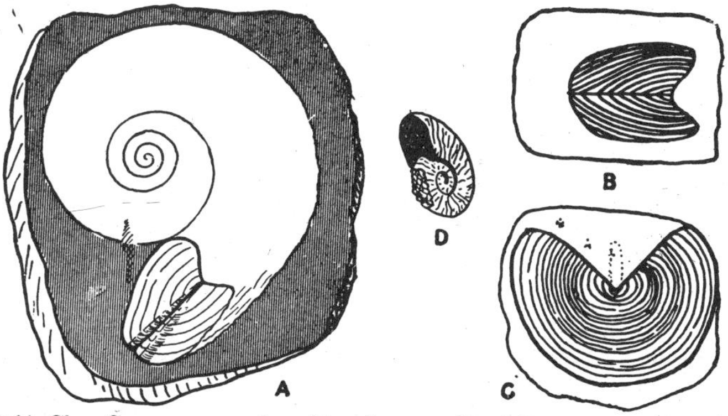

FIG. 10-64. Class CEPHALOPODA. Aptychi and anaptychi. *A*. Impression of an ammonite shell with an aptychus in place in the living chamber, from Jurassic of Solenhofen, Bavaria. *B*. Outside of an aptychus similar to that shown in *A*. *C*. An anaptychus, which seems to be a single plate from the Upper Devonian of Germany. *D*. A Jurassic ammonite with an aptychus in place in the aperture of the shell. (*A–B redrawn from Rüppel*, 1829; *C–D redrawn from Woodward*, 1885.)

The Siphuncle. By definition the siphuncle is the discontinuous or continuous shelly tube that sheathes the fleshy siphon in modern *Nautilus* and supposedly sheathed the siphonal tube in ancient cephalopods. It consists of two quite different structures—septal necks and connecting rings (Figs. 10-65, 10-67). In certain nautiloids (*e.g.*, Actinoceratida) additional features were developed within and around the siphuncle, so that it became a highly complex structure of great taxonomic importance (Figs. 10-66, 10-67, 10-71).

The septal neck is a short tubular extension of the septum surrounding

---

FIG. 10-63. Class CEPHALOPODA. *A–B*. "*Orthoceras*": *A*, showing circular aperture with central siphuncle; *B*, entire shell showing size of living chamber as compared to remainder of shell. *C–D*. *Nautilus*: *C*, showing interrupted elliptical aperture; *D*, side view of shell showing its involute nature. *E*. Apertural (septal) view of *Placenticeras placenta*, from the Cretaceous of Western United States, showing the sagittate aperture, one of the crenulated septa, and the ventral siphuncle. *F*. Apertural view of *Scaphites refractus*, showing a modified aperture. *G–H*. Apertural and lateral views, respectively, of a Jurassic *Stephanoceras*, showing a modified aperture. *I*. *Mandaloceras hawthornensis* from the Illinois Silurian showing the constricted aperture and several of the larger chambers. *J–K*. Aperture and side views, respectively, of *Phragmoceras rex* from the Bohemian Silurian, showing a T-shaped aperture. *L–M*. Apertural and side views of different species of *Hexameroceras* from the Illinois Silurian: *L*, *H. byronense*, apertural view; *M*, *H. jolietense*, side view. *N–O*. *Gomphoceras deshayesi*, from the Silurian of Bohemia: *N*, apertural view; *O*, side view, showing complete living chamber and several chambers longitudinally sectioned to show the position and nature of the siphuncle. (*Adapted from numerous sources. E after Meek; F–H after Wright; I, L-M after Foerste; J–K, O after Barrande.*)

the septal perforation, or **foramen,** and generally extending toward the apex of the shell (Figs. 10-58, 10-65). Its form varies considerably among the different groups of nautiloids and ammonoids, and upon its form and structure were largely based the major divisions of nautiloids proposed by Hyatt (1900) and used generally for the past 50 years. Its chief features are shown in Figs. 10-65, 10-66, and 10-67.

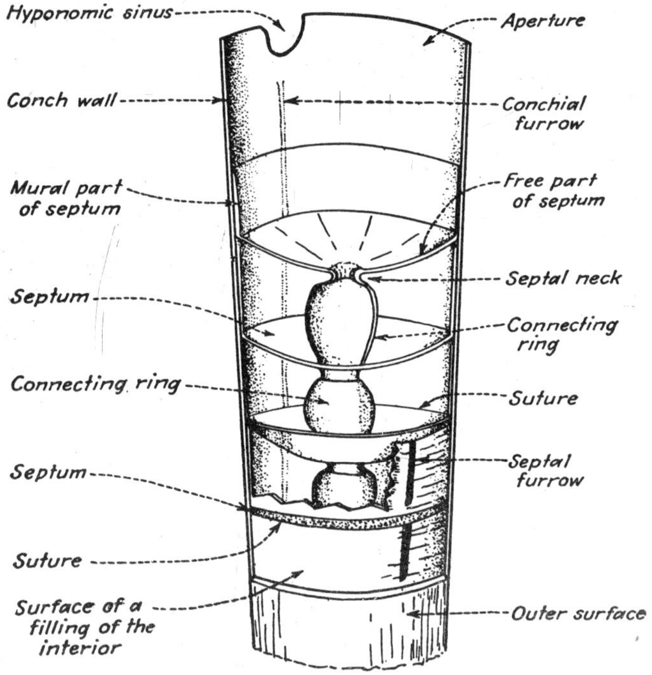

FIG. 10-65. Class CEPHALOPODA. A variously sectioned orthoceracone, showing internal structure. The hypnomic sinus and conchial furrow mark the mid-ventral part of the shell. Adorally the shell is cut to the center. The first segment of the siphuncle is cut to the center and is seen to consist of septal neck and connecting ring. The next septum and siphuncular segment are complete. (*Modified after Flower*, 1946.)

The connecting ring is a shelly tubular structure which typically extends from the tip of one septal neck to the tip of the next one. It is not secreted from a specialized area of ectodermal epithelium, as are true shell parts; instead, it is secreted within the tissue of the siphon and is calcitic in contrast to the aragonitic septum. In *Nautilus* the ring consists of minute calcareous spicules held together partly by organic matter. As a consequence of its fragility, it is commonly broken or lacking altogether in sectioned shells of living *Nautilus*. Connecting rings probably were equally fragile structures in the shells of many ancient cephalopods, as shown by their not having been found; on the other hand, many fossil shells have well-preserved con-

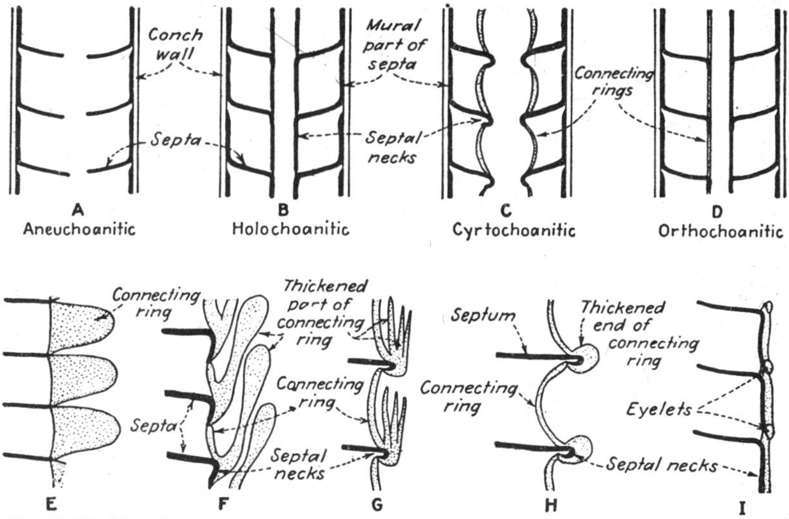

FIG. 10-66. Class CEPHALOPODA. Siphuncles, showing different types and role and modification of connecting rings. Septa are shown in solid black; connecting rings and associated secondary deposits are stippled. *A–D*. The four common types of siphuncle in the Nautiloidea. Types *C* and *D* have been collectively designated ellipochoanitic. *E*. Aneuchoanitic necks and thickened rings with obscure marginal layers. *F*. Ellipochoanitic necks with part of neck thickened into a forward projecting lobose process. *G*. Actinosiphonate structure showing development of handlike process from thickening of connecting ring. *H*. Connecting rings thickened around ends of cyrtochoanitic necks. *I*. Portion of a holochoanitic endoceroid with thick connecting rings, each terminating in an eyelet. (*A–D diagrammatic; E–I modified after Flower*, 1946.)

necting rings, and in several groups the rings came to be quite highly specialized, as shown in Figs. 10-66 and 10-67.

The following types of siphuncles, which are defined largely by the nature of the septal neck and connecting rings, may be recognized (Flower, 1946) (Fig. 10-67):

Holochoanitic. Septal necks extend apicad for the length of one camera; connecting rings generally present (see ellipochoantic); now restricted to Endoceratida.

Orthochoanitic. Siphuncle cylindrical and septal necks straight and simple.

Cyrtochoanitic. Siphuncular segments expand within the camerae, and the septal necks are bent outward or distad at their tips.

Aneuchoanitic. Siphuncle in which septum forms only a vestigial neck, being scarcely bent apicad; found only in earliest nautiloids.

Ellipochoanitic. Septal necks relatively short and supplemented by connecting rings, as opposed to the old concept of holochoanitic (see above), in which connecting rings were supposed to be lacking; includes both orthochoanitic and cyrtochoanitic types.

It has been customary to consider the complete siphuncle as consisting of two distinct parts: (1) the **ectosiphuncle,** composed of the outer wall formed by the septal necks and connecting rings; and (2) the **endosiphuncle, con-**

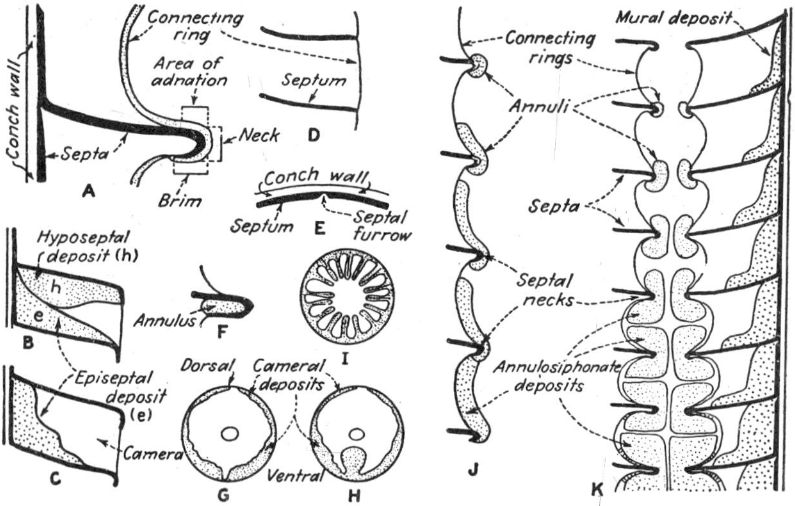

FIG. 10-67. Class CEPHALOPODA—Subclass NAUTILOIDEA. Cameral and siphuncular structures. *A*. Portion of phragmocone, showing essential features. The septum is continuous from the tip of its mural part to the end of the septal neck. *B–C*. Portions of camerae, showing cameral deposits and orthochoanitic siphuncle. *D*. Aneuchoanitic siphuncle (cf. Fig. 10-66*A*). *E*. Restored section of conch wall, showing mural part of a septum interrupted on the middorsal area by the septal furrow. *F*. Cyrtochoanitic septal neck of an actinoceroid reinforced by an annulus. *G–H*. Cross sections of camerae of two different genera, showing circumferential cameral deposits. *I*. Cross section of a typical actinosiphonate siphuncle. *J*. Longitudinal section through one side of a pseudorthoceroid siphuncle, showing gradually enlarged annuli. *K*. Diagrammatic section showing the spatial relationships of the different cameral and siphonal deposits. (*Modified after Flower*, 1946.)

sisting of all structures within the ectosiphuncle. These latter structures, which were for the most part added after the ectosiphuncle was formed, include dissepiments, endocones, annuli, vertical plates converging toward the center of the siphonal cavity, spokelike radial rods, and several types of irregular processes (Fig. 10-67).

Tabulae, diaphragms, and dissepiments cross the cavity of the siphuncle in a few early Paleozoic nautiloids but do not seem to have persisted in later shells. In one large group of nautiloids, the Endoceratida, a series of solid cones, the **endocones,** lying one within another, fill the siphuncle (Fig. 10-71*A*5). The tips of the endocones continue in flat or circular tubes, the **endosiphotubes,** and these may be divided by minute tabulae. Endocones were also developed independently in several other orders of nautiloids.

Calcareous rings resembling doughnuts formed around the septal foramens of some shells. These are termed **annuli,** or **annulosiphonate deposits,** and show considerable variation in structure. In some shells of the Actinoceratida they occupy most of the cavity of the siphuncle (Fig. 10-67*K*). The siphuncles of some shells have vertical rays extending inward from the wall toward the center. These are termed **actinosiphonate deposits** (Figs. 10-67*I*, 10-71*R*) and are much complicated in some of the more advanced nautiloid shells. In addition to the several endosiphuncular structures just described, there are several other less important structures that form parts of the characteristic pattern of deposits in those groups of cephalopod shells in which deposits are developed. These are defined and illustrated in the many special works on the Nautiloidea, and are not considered further in the present work.

Cameral Deposits (Fig. 10-67). The camerae of many ancient nautiloid shells have deposits that are considered true shell secretions laid down by a supposed cameral mantle (Flower, 1946; Fischer and Finley, 1949). Such a mantle, or possibly a membrane like that found in living *Nautilus*, supposedly kept in contact with the main part of the animal by the exchange of materials with the blood system of the siphon through the porous connecting ring (Flower, 1946). These deposits, which are not to be confused with deposits in the camerae made in other ways, are of considerable taxonomic importance in some groups and have been divided into several types for which special terms have been proposed (Teichert, 1933, 1935; Flower, 1946).

Sutures. The suture is the line along which the septum leaves the wall of the shell. It is also, therefore, the line of transition from the mural to the free part of the septum (Figs. 10-58, 10-65). The simplest sutures are circular and are normal to the axis of the shell. The more complex sutures are lobed and some are amazingly convoluted. Undulations or lobes convex toward the aperture are termed **saddles;** those convex toward the apex, **lobes** (Fig. 10-68). Lobes and saddles are described as **lateral** if they are on the sides of the shell; **dorsal** (also **columellar** or **antisiphonal** in older works on the Ammonoidea) if dorsal in position; and **ventral** (**siphonal** in older reports on the Ammonoidea) if ventral in position. In general, the suture pattern is much simpler in nautiloids than in ammonoids, though in some of the more specialized coiled nautiloids it is more complicated than in the older and simpler of the ammonoids.

It is almost a universal practice to show the suture as it would appear if transferred to a plane surface by rotating the whorl through 360°, with the axis of rotation perpendicular to the plane of the suture. The ends of the suture line thus mark the median dorsal line of the shell, and the mid-point of the line, which is customarily marked by an arrow pointing toward the aperture, marks the position of the median ventral line.[1] If the suture line is unusually long, only one half of it is illustrated. This half shows the mid-dorsal and mid-ventral lines as its extremities.

The suture of whorls that are free or merely touching can be determined directly by observing the free part of the septum. This is the case in all shells except those in which the whorls are so tightly coiled that there is an impressed zone where the later whorl conceals a part of the next earlier whorl, the ventral portion of which is impressed into the dorsal portion of the later whorl along the line of involution. In such tightly coiled shells, only part of the suture, the **external** part, can be seen; the **internal** part can only be observed when the covering whorl is broken away. The sutures

[1] Furnish and Unklesbay (1940) discuss the best means of taking sutures from ammonoid shells. Also see Elias (1938).

of such shells are customarily divided into external and internal parts with the division line, or **umbilical seam,** along the edge of the umbilicus or impressed zone. The umbilical seam is indicated on the suture by a single curved line or by two short, closely spaced parallel lines between the mid-dorsal and mid-ventral arrows (Fig. 10-68).

The typical ammonoid suture has several lobes and saddles on the external

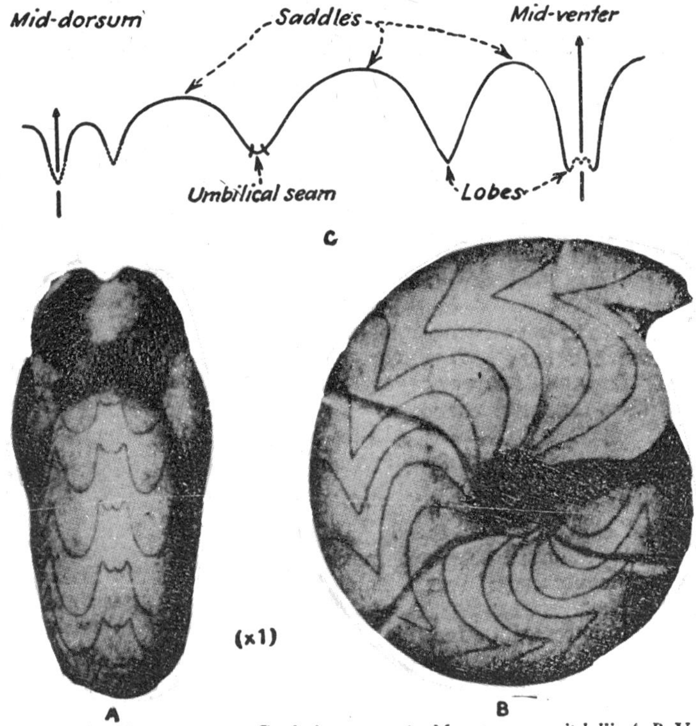

FIG. 10-68. Class CEPHALOPODA. Goniatite sutures in *Muensteroceras mitchelli*. *A–B*. Ventral and lateral views of complete conch, somewhat restored. *C*. Diagrammatic representation of one of the adoral sutures. Note that the saddles are rounded and the lobes angular. The arrows point adorally. The right two-thirds of the suture is external; the left one-third, internal. (*After Enich in Miller*, 1935.)

suture and commonly a corresponding number on the internal suture, as shown in Fig. 10-68. The first suture in both nautiloids and ammonoids is a simple circle, or an undulating closed line composed of a slight ventral saddle and corresponding shallow lateral lobes. In all but the most primitive ammonoids the second suture has a marked **ventral lobe** and a less conspicuous **dorsal lobe,** either or both of which may become crenulated in later sutures. After these two lobes have appeared, there develops between them a pair of broad lateral lobes that are largely external in position. These lateral lobes subsequently become divided by a pair of saddles that arise in the region of the umbilicus, so that four lateral lobes (*i.e.*, two pairs of them) come into existence. The ventral pair consists of the **first lateral lobes of the external suture,** whereas the dorsal includes the **first lateral lobes of the internal suture.** The saddles that separate the first lateral lobes of the external suture from the ventral lobe are designated the **first lateral saddles of the external suture,** and the pair of saddles separating the first lateral lobes of the internal sutures from the dorsal lobe should be termed the **first lateral saddles of the internal sutures.**

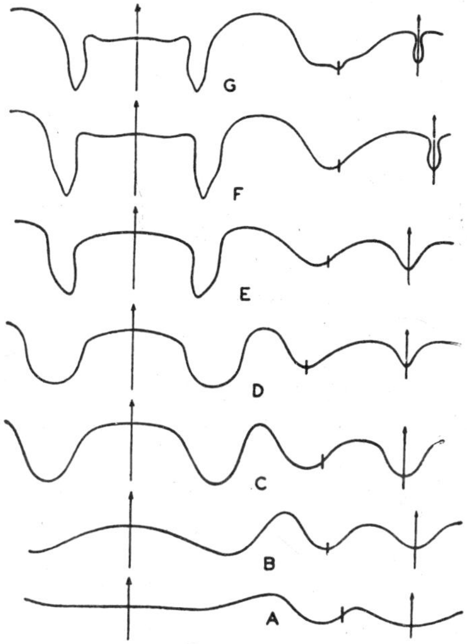

FIG. 10-69. Class CEPHALOPODA. Developmental series of nautiloid sutures, showing progressive changes in ventral and dorsal lobes. *A. Cimomia* (Paleoc.). *B.* A primitive *Hercoglossa* (Lower Eoc.). *C.* An advanced *Hercoglossa* (Lower Eoc.). *D.* The first suture of a typical *Aturia* (Upper Eoc.). *E. Aturoidea* (Lower Eoc.). *F.* A typical Upper Eocene *Aturia*. *G.* A more advanced *Aturia* (Mioc.). *Aturia* became extinct in the Pliocene. (*After Miller and Furnish*, 1938.)

In the next stage of sutural development the saddles on the umbilical zones are subdivided by a pair of lobes, and the four saddles thus resulting are designated the **second lateral saddles** of the external and internal sutures, except in certain forms where all saddles become components of the external sutures. In subsequent stages additional inflections may develop successively near the umbilical seam and then migrate ventrad and dorsad from the seam as the whorls enlarge with growth. If the external suture develops a second pair of lateral saddles immediately dorsad of the first

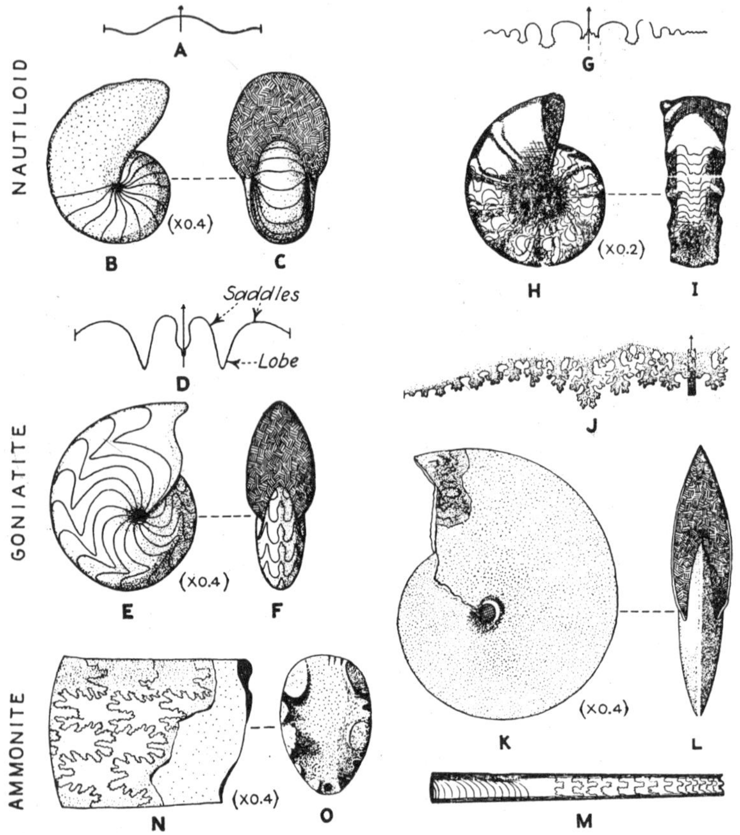

FIG. 10-70. Class CEPHALOPODA. Sutures. *A–C. Nautilus undulatus* from the Cretaceous of England: *A*, a complete suture with the extremities marking the axis of coiling; *B–C*, side and apertural views, respectively, of an internal filling of the shell. *D–F. Muensteroceras rotarius*, a typical goniatite, from the Mississippian of the United States: *D*, a complete suture with rounded saddles and angular lobes; *E–F*, side and apertural views, respectively, of an internal filling. *G–I. Ceratites nodosus* from the Triassic of Germany: *G*, the complete visible suture (part of the suture is hidden on the dorsal side), with rounded saddles and crenulated lobes; *H–I*, side and back views respectively, of an internal filling. *J–L. Placenticeras lenticularis* from the Cretaceous of western United States: *J*, the left half of the complex suture; *K–L*, side and apertural views, respectively, of a shell slightly smaller than that represented by the suture shown in *J*. A small portion of the shell in *K* has been removed to reveal the sutures. *M–O. Baculites*, a common Cretaceous ammonoid in many parts of the world: *M*, *B. anceps*, from England, showing the later part of the conch; *N*, a small fragment of *B. ovatus* with most of shell removed to show sutures; *O*, view of a septum showing the lobes and saddles and the siphuncle. (*A–C after Sharpe; J–L and N–O after Meek; M after Wright.*)

lateral lobes, these are termed the **second lateral saddles** of the external suture, and the lobes immediately dorsad of them are the **second lateral lobes** of the external suture. All other inflections that are added between the second lateral lobes and the umbilical seam are termed **auxiliaries.** Corresponding lobes and saddles of the internal sutures are similarly designated. Prominent lobes that arise in some shells from subdivision of the first lateral saddles or lobes are termed **adventitious lobes.**

Four general types of sutures are customarily recognized. These, in order of complexity, are orthoceratite, or nautiloid; goniatite; ceratite; and ammonite, including a degenerate form which, because it resembles a ceratitic suture, has been designated pseudoceratite.

The **orthoceratite,** or **nautiloid,** suture is either a circle (a straight line when projected), as in *Michelinoceras;* or an undulating line with gently rounded lobes and saddles, as in *Nautilus* (Fig. 10-70*A–C*) and a group of extinct Tertiary genera (Fig. 10-69). Shells with this simple type of suture appeared first in the late Cambrian.

Typical **goniatite** sutures are exhibited by the Mississippian genus *Muensteroceras,* shown in Figs. 10-68 and 10-70*D–F*. The goniatite suture has rounded saddles and somewhat angular lateral lobes. Shells with goniatite sutures appear first in Devonian rocks and are not found in beds younger than Permian.

Ceratite sutures are characterized by smooth rounded saddles and much crinkled lobes. The designation comes from the Triassic genus *Ceratites* (Fig. 10-70*G–I*), in which the suture is typically developed. Ceratite sutures first appear on the adult whorls of shells of Mississippian age (*e.g.*, *Prodromites*) and do not persist after the Triassic, though a somewhat similar type of suture again reappears in certain Late Mesozoic shells, in which case, however, it is termed **pseudoceratite** because it developed from an ammonitic type rather than from a nautiloid type.

Ammonite sutures, so termed because they characterize so many of the more specialized ammonoids, have both lobes and saddles minutely crinkled, in some species so intensely as to give to the suture a dendritic or mosslike appearance (Fig. 10-70*J–O*). Shells with ammonite sutures appear first in the Permian (*e.g.*, *Perrinites*) and disappear with the extinction of the Ammonoidea at the close of the Cretaceous.

Sutures constitute one of the most important features of ammonoid shells for both generic differentiation and stratigraphic identification. They have also proved useful in determining evolutionary series because in the same shell the earliest sutures are relatively simple whereas the later ones become progressively more complicated. Just as successive sutures in a single shell become more and more complicated, so adult sutures of the genera of an evolutionary series likewise become progressively more complicated (Fig. 10-69). Because of these variations, it is considered possible to determine the ontogeny of a single species and to work out the genetic relationships between closely related genera.

Internal Shells. Thus far the discussion of the cephalopod shell has been concerned almost entirely with external shells, since these constitute the overwhelming majority of fossil cephalopods and hence are of greatest interest and importance to paleontologists. However, there is one subclass, the Coleoidea, in which the shell has become internal and vestigial, or has been lost altogether, and the evolution of this group of cephalopods, most of which have living representatives, has proved of great interest and evolutionary importance. These shells are considered in connection with the Subclass Coleoidea.

Summary. Cephalopod shells stand almost alone among invertebrate hard parts as objects illustrating many of the phenomena of evolution. They have been much studied for this reason because a single shell has preserved in itself the successive shell modifications that were made as the animal passed through successive growth stages from embryo and larva to mature and senile adult. Variation in shell features affected shell form (shape and coiling), aperture, surface sculpture (particularly among the Mesozoic ammonoids), siphuncle and camerae, and septa and sutures. From the study of the evolution of cephalopod shells has come the formulation of many of the supposed "laws" of evolution, which, through the years, have greatly stimulated evolutionary studies of other invertebrate groups. Raymond (1941) reviews these studies for the past 50 years, and Swinnerton (1947) summarizes some of the more spectacular examples. Reading the works of these two authors and of the investigators they cite will carry the interested reader into one of the great compilations of descriptive and theoretical writing concerning the general subject of the evolution of the invertebrates.

Classification

Classification of the Cephalopoda has been in a state of flux for the past several decades, particularly at ordinal and familial levels, and as more and more new species and genera have been discovered, it has become obvious that older classifications, as good as could be devised at the time, are no longer adequate because of new facts. In the present work, therefore, we abandon most of the older divisions and introduce in their stead a somewhat more elaborate classification based on numerous recent works. In adopting these newer taxonomic divisions we follow modern investigators in recognizing the importance of siphuncular structures, septal and sutural variations, surface sculpture, and general shell morphology.

The three widely recognized subclasses—Nautiloidea, Ammonoidea, and Coleoidea—are briefly characterized in a preceding section on page 436. Each subclass is considered at some length on the following pages.

Subclass Nautiloidea

This subclass includes at least 300 genera and more than 2,500 species, all of which are fossil except the lone genus *Nautilus* with its few living species. Almost nothing definite is known about the soft bodies of the earlier nautiloids, but they are assumed to have been essentially like modern *Nautilus* (Figs. 10-55, 10-57), whose characteristics are discussed on preceding pages.

Since 1900, when Hyatt proposed the classification[1] introduced in the Zittel-Eastman "Text-book of Palaeontology," there has been no attempt until quite recently to regroup the rapidly increasing number of genera into a classification more in accord with growing knowledge. In 1950 Flower and Kümmel proposed the following ordinal grouping adopted in the present work:

Order 1. Ellesmeroceratida.
Order 2. Endoceratida.
Order 3. Actinoceratida.
Order 4. Michelinoceratida.
Order 5. Ascoceratida.
Order 6. Bassleroceratida.
Order 7. Oncoceratida.
Order 8. Discosorida.
Order 9. Tarphyceratida.
Order 10. Barrandeoceratida.
Order 11. Rutoceratida.
Order 12. Centroceratida.
Order 13. Solenochilida.
Order 14. Nautilida.

The interested reader will find in their article the reasons for the regrouping and descriptions of the detailed characteristics of the 14 orders, which are briefly described in the following paragraphs.

Order 1. Ellesmeroceratida (Fig. 10-71*D–E*). The shells of this order, which supposedly includes the earliest and most primitive forms, are small and closely septate; the siphuncle is tubular and marginal; septal necks are extremely short, and connecting rings, when present, are relatively thick and commonly of complex structure; and the siphuncle lacks endocones and annuli but may have a few diaphragms. Most of the straight and endogastric genera that dominate the earliest (Lower Canadian) cephalopod associations

[1] Hyatt (1900) proposed the following major divisions of the Nautiloidea which he based largely on the structure of the siphuncle, *i.e.*, the septal necks, connecting rings, and accessory features:

Holochoanites. Septal necks reaching from septum of origin to next septum apicad or even beyond to plane of second septum.

Mixochoanites. Shells in which are marked changes from orthochoanitic to cyrtochoanitic siphuncles.

Schistochoanites. Septal necks more or less imperfect, present on the internal side, and absent or split on the outer side. (This division, based on *Conoceras*, has been suppressed.)

Orthochoanites. Siphuncular segments slightly nummuloidal, fusiform, or tubular, but never markedly nummuloidal, nor are the septal necks bent sharply outward as in Cyrtochoanites.

Cyrtochoanites. Siphuncle highly variable, but septal necks generally short and bent outward or crumpled.

Modern students of the Nautiloidea are finding Hyatt's classification unsatisfactory, hence we abandon it and adopt in its place a more detailed classification recently proposed by Flower and Kümmel (1950).

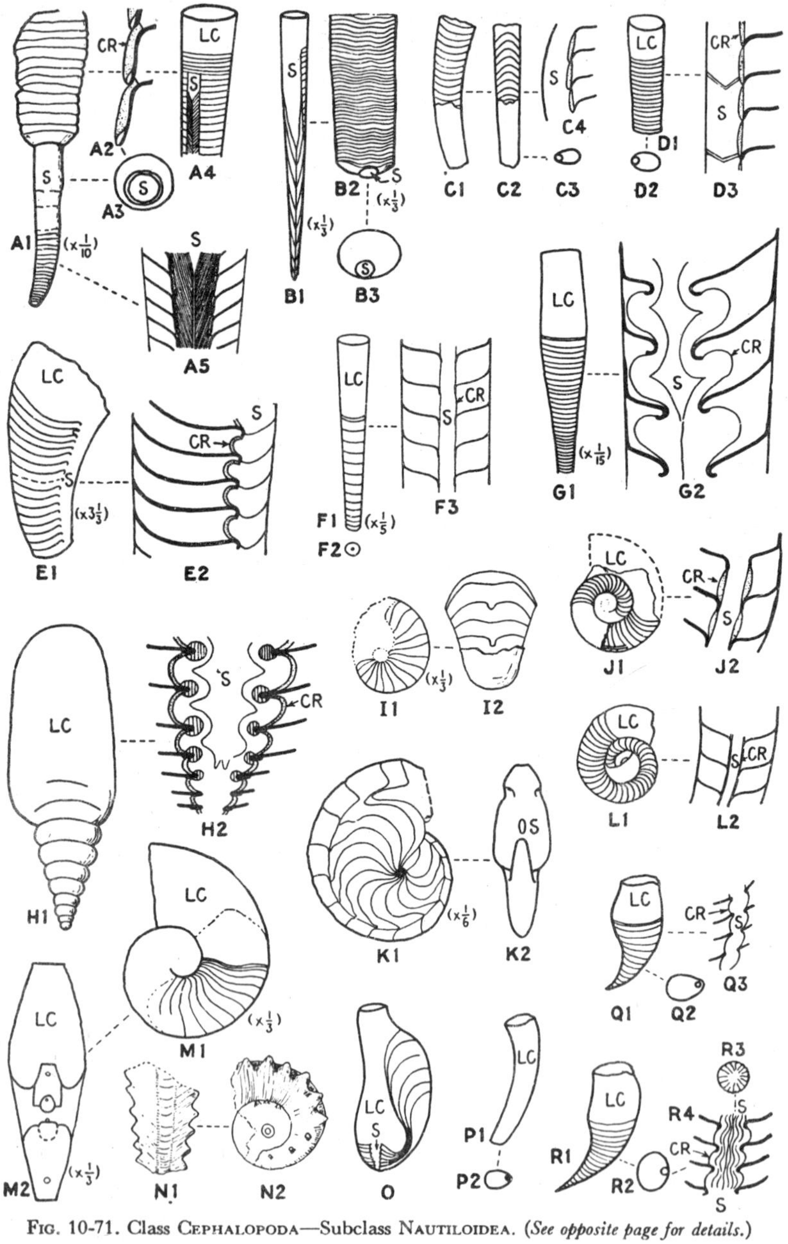

FIG. 10-71. Class CEPHALOPODA—Subclass NAUTILOIDEA. (*See opposite page for details.*)

are included in this order. The earliest representative of the order, and supposedly the earliest unquestioned cephalopod, is *Plectronoceras* (Fig. 10-71*E*) from the Upper Cambrian of China. The range and stratigraphic importance of the order are shown in Fig. 10-72. *Upper Cambrian to Lower Silurian.* *Ellesmeroceras* (Fig. 10-71*D*) is a typical genus.

Order 2. Endoceratida (Fig. 10-71*A–B*). The endoceratids are long orthoceraconic shells with relatively large, tubular to conical siphuncles containing well-developed endocones. This order, which had only a short duration, includes the longest orthoceracones known, some of which exceed 4½ m. (15 ft.) in length. *Proterocameroceras* (Fig. 10-71*B*) and *Endoceras* (Fig. 10-71*A*) are typical of the order, which is limited to the Ordovician (Fig. 10-72).

Order 3. Actinoceratida (Fig. 10-71*G*). The shells included in this large, important order are dominantly orthoconic and have large, broadly expanded siphuncles that are generally occupied by large pendant annulosiphonate deposits. These deposits typically grew to such size as to fill the

FIG. 10-71. Highly diagrammatized figures illustrating typical representatives of the 14 orders of the subclass NAUTILOIDEA (see Fig. 10-72 for the stratigraphic range of the different orders). *A*1–*A*5. *Endoceras* (Endoceratida): *A*1, incomplete conch, with siphuncle and numerous camerae preserved; *A*2, part of siphuncular wall, showing long septal necks and thick connecting rings; *A*3, transverse section of conch, showing size and position of siphuncle; *A*4, later part of a conch, showing part of siphuncle with its prominent endocones; *A*5, later part of siphuncle showing the prominent endocones, the tips of which form the endosiphotubes along the axis. *B*1–*B*3. *Proterocameroceras* (Endoceratida): *B*1, early part of conch, showing endocones and earliest septa; *B*2–*B*3, view of siphonal side of multicamerate conch, and transverse section showing size and position of siphuncle. *C*1–*C*4. *Bassleroceras* (Bassleroceratida): *C*1, side view; *C*2, ventral view; *C*3, cross section; *C*4, longitudinal section of a small part of the siphuncle. *D*1–*D*3. *Ellesmeroceras* (Ellesmeroceratida): *D*1, later part of conch; *D*2, cross section showing marginal position of siphuncle; *D*3, longitudinal section of siphuncle, showing short septal necks, relatively thick connecting rings, and two diaphragms across the siphuncle. *E*1–*E*2. *Plectronoceras* (Ellesmeroceratida), the oldest known cephalopod: longitudinal section of later part of conch (E1) and of a small part of the siphuncle (*E*2), showing short septal necks, strongly curved connecting rings, and diaphragms. *F*1–*F*3. *Michelinoceras* (Michelinoceratida): *F*1, later part of conch; *F*2–*F*3, cross section and longitudinal section, showing orthochoanitic siphuncle. *G*1–*G*2. *Actinoceras* (Actinoceratida): later part of conch and longitudinal section of the crytochoanitic siphuncle with endosiphuncular structure. *H*1–*H*2. *Discosorus* (Discosorida): *H*1, ventral view, showing lower part of living chamber and nummuloidal siphuncle; *H*2, longitudinal section of part of siphuncle, showing characteristic structures. The septal necks are recumbent, the connecting rings strongly curved and heavy, and endosiphuncular structures complicated. *I*1–*I*2. *Solenochilus* (Solenochilida): side and ventral views of an incomplete conch. *J*1–*J*2. *Tarphyceras* (Tarphyceratida): *J*1, variously sectioned incomplete conch, showing small part of ventral siphuncle; *J*2, longitudinal section of siphuncle, showing thick connecting rings. *K*1–*K*2. *Aturoidea* (Nautilida): side and apertural views of a conch lacking most of the living chamber; the siphuncle is eccentric and the sutures undulatory. *L*1–*L*2. *Barrandeoceras* (Barrandeoceratida): *L*1, side view of an incomplete specimen, *L*2, longitudinal section of siphuncle, showing thin connecting rings. *M*1–*M*2. *Domatoceras* (Centroceratida): *M*1, side view of an incomplete conch (restored); *M*2, section of another specimen, showing ventral siphuncle and quadrate nature of whorls. *N*1–*N*2. *Temnocheilus* (Rutoceratida): ventral and lateral views of an incomplete conch, showing characteristic nodes. *O*. *Ascoceras* (Ascosceratida), a mixochoanitic genus, with unusual septal development. *P–R*. Genera of Oncoceratida: *P*1–*P*2. *Oonoceras; Q*1–*Q*3, *Oncoceras; Q*1, a complete conch; *Q*2, cross section; *Q*3, longitudinal section of the siphuncle; *R*1–*R*4, *Poteriocerina; R*1, a complete conch; *R*2, cross section of conch; *R*3, cross section of siphuncle; *R*4, transverse section of siphuncle, showing radial endosiphuncular septa. (*CR* = connecting rings; *LC* = living chamber; *S* = siphuncle; septa are shown in several diagrams as heavy black lines.) (*All figures more or less diagrammatic and based on many authors; most figures of siphuncular structure based on original sketches by Flower.*)

earlier part of the siphuncle except for the siphonal vascular system, a central canal, and radial canals that terminate in a hollow space, the **perispatium,** near the connecting ring (Fig. 10-67). *Lower Ordovician through Mississippian* (Fig. 10-72).

Actinoceras (Fig. 10-71*G*), a widespread Mid-Ordovician genus with a complicated siphuncle, is typical of the order.

Order 4. Michelinoceratida (Fig. 10-71*F*). This long-ranging order consists dominantly of orthoceracones that are subcircular in section and have a relatively central siphuncle. The siphuncle is orthochoanitic or cyrtochoanitic, with well-developed septal necks and thin homogeneous connecting rings. No representatives of the order have been found in rocks older than Chazyan, but the great variety of forms in beds of that age indicates a long period of development of which no record has yet been discovered. *Lower Ordovician through Triassic* (Fig. 10-72).

Michelinoceras (Fig. 10-71*F*), Ordovician to Mississippian, a genus commonly identified in the past as *Orthoceras*, is typical of the order.

Order 5. Ascoceratida (Fig. 10-71*O*). The ascoceratids are primitively orthoceracones and cyrtocones with a subcentral siphuncle. The adoral segments of the siphuncle are broadly expanded; the mature part of the shell is much inflated; and the septa rise adorally on the dorsum, becoming sigmoidal in the last to be added. *Ordovician to Silurian* (Fig. 10-72).

Ascoceras (Fig. 10-71*O*) is a typical genus.

Order 6. Bassleroceratida (Fig. 10-71*C*). This order includes exogastric cyrtoceracones with a ventral tubular siphuncle. The order arose and flourished in the earlier part of the Ordovician and was extinct by the end of the period. *Ordovician* (Fig. 10-72).

Bassleroceras (Fig. 10-71*C*) is a typical genus.

Order 7. Oncoceratida (Fig. 10-71*P–R*). The oncoceratids are primitively exogastric cyrtoceracones and brevicones that are compressed in section. The siphuncle, which is ventral in position, is suborthochoanitic in the early part but cyrtochoanitic in the mature portion. Most of the cyrtoconic and breviconic shells of the Paleozoic belong in this order, and as they exhibit wide variation in form and structure, it is not possible to define the order accurately without an extended description. *Middle Ordovician to Lower Pennsylvanian* (Fig. 10-72).

Oncoceras (Fig. 10-71*Q*), *Oonoceras* (Fig. 10-71*P*), and *Poteriocerina* (Fig. 10-71*R*) are typical of this large and important order.

Order 8. Discosorida (Fig. 10-71*H*). The Discosorida constitute a genetic line of straight to curved shells that are characterized by a relatively large, broadly expanded siphuncle in which the siphuncular segments are broadly expanded from the earliest growth stage. They resemble Actinoceratida so far as the general aspects of the siphuncle are concerned but differ from that group in the thickening of the connecting rings, in lacking any evidence of an endosiphuncular vascular system, and in the quite different form of the annulosiphonate deposits when they are present. The order had its greatest development in the later Ordovician and earlier Silurian. *Middle Ordovician to Upper Devonian* (Fig. 10-72).

Discosorus (Fig. 10-71*H*), a widespread Silurian genus with a prominent nummuloidal siphuncle resembling a string of large beads, is typical of the order.

Order 9. Tarphyceratida (Fig. 10-71*J*). The tarphyceratids are a small group of dominantly Lower Ordovician coiled shells with persistently thick connecting rings as in the Bassleroceratida. The siphuncle is ventral in young shells, but may be central or dorsal in adult shells. *Lower Ordovician to Middle Silurian* (Fig. 10-72).

Tarphyceras (Fig. 10-71*J*) is a typical genus.

Order 10. Barrandeoceratida (Fig. 10-71*L*). This order, of which the limits are not yet sharply defined, includes Early Ordovician nautilicones that appear similar to some members of the Family Tarphyceratidae (Order Tarphyceratida) but have thin homogeneous connecting rings. *Middle Ordovician to Upper Silurian* (Fig. 10-72).

Barrandeoceras (Fig. 10-71*L*) is a typical representative of the order.

Order 11. Rutoceratida (Fig. 10-71*N*). The rutoceratids are a fairly large group of gyroconic shells particularly characterized by varices of growth in the form of frills, nodes, spines, or lirae. Later forms are more closely coiled, and the whorls vary from angular and nodose to smooth and rounded. The order is chiefly later Paleozoic. *Lower Devonian to Middle Jurassic* (Fig. 10-72).

Temnocheilus (Fig. 10-71*N*) is a typical genus.

Order 12. Centroceratida (Fig. 10-71*M*). The centroceratids are gyroconic to nautiliconic, with primitively quadrate whorls in which the venter is narrow and the dorsum broad. Whorls tend to become more rounded in later shells, and lateral lobes deeper. *Lower Devonian to Upper Jurassic* (Fig. 10-72).

Domatoceras (Fig. 10-71*M*), a common and widespread Late Paleozoic genus, is characteristic of the order.

Order 13. Solenochilida (Fig. 10-71*I*). This is a small group of smooth rapidly expanding globose nautilicones with a ventral siphuncle and simple sutures. The order ranges from Lower Mississippian through the Permian (Fig. 10-72) and is typified by *Solenochilus* (Fig. 10-71*I*) on which the single Family Solenochilidae is based.

Order 14. Nautilida. The Nautilida, which is the only nautiloid order with living representatives, includes a relatively small group of involute nautilicones in which the siphuncle is central (except for *Aturia*), the sutures straight to quite sinuous ("goniatitic" in the Family Hercoglossidae), and the shell surface smooth or with sinuous plications. *Middle Devonian to Recent* (Fig. 10-72).

Nautilus (Figs. 10-55, 10-57), the sole living survivor of the order, and *Aturia* (Fig. 10-71*K*), a widespread extinct genus (U. Cret.–Plioc.), are typical of the order.

The stratigraphic record of the Nautiloidea, as shown in Fig. 10-72, is a long and interesting one with some orders having quite a long range and others being limited to a single period. The genetic relationships of the 14 orders and 75 families are most uncertain in the present state of knowledge, partly because many valid genera cannot yet be fitted into the orders just

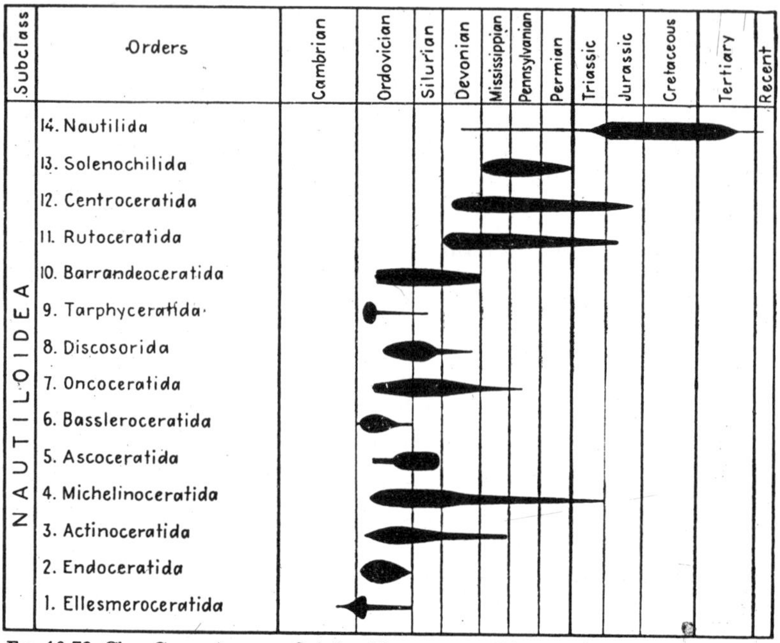

FIG. 10-72. Class CEPHALOPODA—Subclass NAUTILOIDEA. Diagram showing geologic range of the orders of the subclass. The width of the band is roughly proportional to the abundance and variety of species within each order, but the bands are not comparable from one order to another. (*Original from R. Flower*, 1950.)

described and partly because of insufficient data on many genera of uncertain validity. Figure 10-73 is a highly speculative diagram suggesting possible relationships between most of the orders. Much careful and detailed work will have to be done, however, before the many questionable connections can be validated or disproved, whichever the case may be.

Subclass Ammonoidea

The Ammonoidea constitute the largest and in some aspects the most important major subdivision of the Cephalopoda. Little is known about the soft parts of the animal, because the subclass became extinct at the close of the Cretaceous. However, the myriads of shells preserved in rocks ranging in age from Upper Silurian to Cretaceous exhibit such rapid and varied changes in morphology and structure that they are important index fossils in the later Paleozoic rocks and excel all other invertebrates in this respect in Mesozoic rocks. They have been widely and intensively studied on every continent, and the literature concerning them is enormous; so great, in fact, that most modern cephalopod specialists are content to study only certain taxonomic or stratigraphic groups. No satisfactory comprehensive classification of the Ammonoidea seems yet to have been worked out, and the taxonomy of some

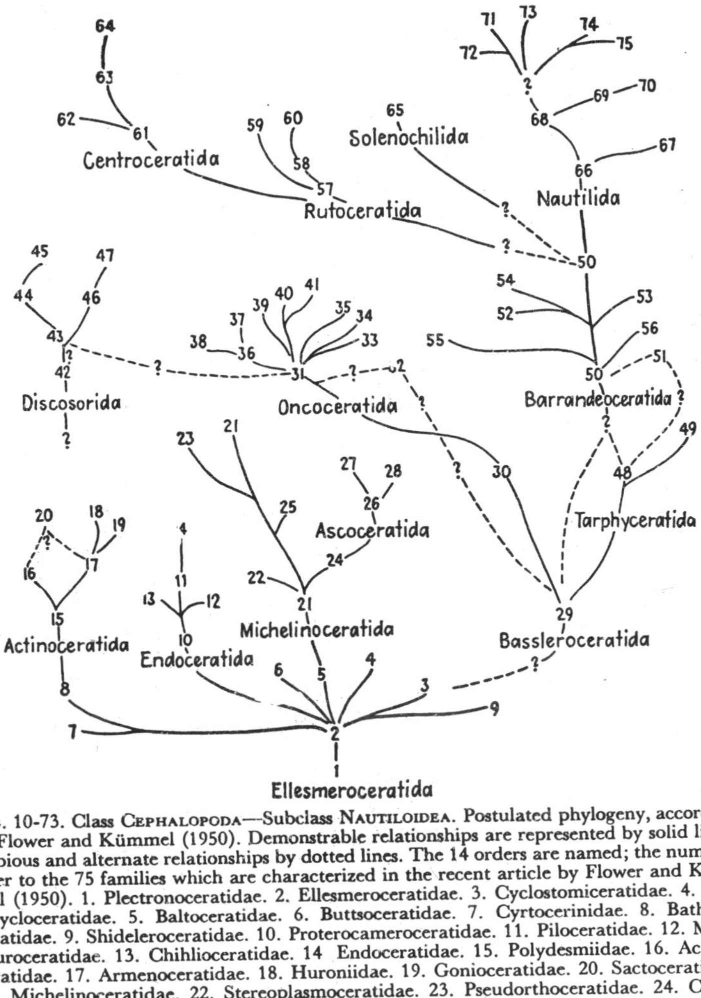

Fig. 10-73. Class Cephalopoda—Subclass Nautiloidea. Postulated phylogeny, according to Flower and Kümmel (1950). Demonstrable relationships are represented by solid lines; dubious and alternate relationships by dotted lines. The 14 orders are named; the numbers refer to the 75 families which are characterized in the recent article by Flower and Kümmel (1950). 1. Plectronoceratidae. 2. Ellesmeroceratidae. 3. Cyclostomiceratidae. 4. Protocycloceratidae. 5. Baltoceratidae. 6. Buttsoceratidae. 7. Cyrtocerinidae. 8. Bathmoceratidae. 9. Shideleroceratidae. 10. Proterocameroceratidae. 11. Piloceratidae. 12. Manchuroceratidae. 13. Chihlioceratidae. 14 Endoceratidae. 15. Polydesmiidae. 16. Actinoceratidae. 17. Armenoceratidae. 18. Huroniidae. 19. Gonioceratidae. 20. Sactoceratidae. 21. Michelinoceratidae. 22. Stereoplasmoceratidae. 23. Pseudorthoceratidae. 24. Clinoceratidae. 25. Paraphragmitidae. 26. Hebetoceratidae. 27. Choanoceratidae. 28. Ascoceratidae. 29. Bassleroceratidae. 30. Graciloceratidae. 31. Oncoceratidae. 32. Allumettoceratidae. 33. Hemiphragmoceratidae. 34. Trimeroceratidae. 35. Brevicoceratidae. 36. Valcouroceratidae. 37. Jovelaniidae. 38. Diestoceratidae. 39. Nothoceratidae. 40. Acleistoceratidae. 41. Archiacoceratidae. 42. Ruedemannoceratidae. 43. Westonoceratidae. 44. Lowoceratidae. 45. Discoscoridae. 46. Cyrtogomphoceratidae. 47. Phragmoceratidae. 48. Tarphyceratidae. 49. Trocholitidae. 50. Barrandeoceratidae. 51. Plectoceratidae. 52. Uranoceratidae. 53. Lechritrochoceratidae. 54. Rhadinoceratidae. 55. Apsidoceratidae. 56. Lituitidae. 57. Rutoceratidae. 58. Tetragoneratidae. 59. Koninckinoceratidae. 60. Tainoceratidae; 61. Certroceratidae. 62. Triboloceratidae. 63. Domatoceratidae. 64. Syringonautilidae. 65. Solenocheilidae. 66. Liroceratidae. 67. Ephippioceratidae. 68. Paranautilidae. 69. Clydonautilidae. 70. Gonionautilidae. 71. Nautilidae. 72. Paracenoceratidae. 73. Cymatoceratidae. 74. Hercoglossidae. 75. Aturidae.

groups is in an almost hopeless state of confusion because of the uncertainty regarding the validity of many of the described genera.[1] We do not, therefore, attempt an ordinal classification of the subclass, as is done for the Nautiloidea; instead, we discuss the more important characteristics of the subclass and cite a few genera that are typical of some of the hundred or more[2] families that have been proposed. The reader who is interested in the evolution and geological history of the Ammonoidea is referred to Swinnerton (1947) for an excellent discussion of evolution, and to the works of Hyatt (1900), A. K. Miller *et al.*, J. P. Smith, Arkell (1950), and Wright (1952) for descriptions of genera and classification of taxonomic and stratigraphic groups. Extensive lists of references to North American cephalopods, in general, and to Nautiloidea and Ammonoidea, in particular, are given in "Index Fossils of North America" (Shimer and Shrock, 1944), pp. 527–531 and 563. For a recent paper on Triassic ammonoids, see Kümmel (1952).

The nature and developmental history of ammonoid shells can perhaps be best understood by briefly considering the development of a typical shell. For this purpose *Dactylioceras commune* (Fig. 10-74) from the Jurassic will serve excellently. The adult shell is thinly discoidal, is sculptured with costae and bifurcating ribs, and consists of many whorls coiled planispirally in such fashion that they embrace one another only slightly, so that all inner whorls remain visible. When the outer whorl of the shell is peeled off progressively backward, it shows a succession of changes in ornamentation and in the shape of its cross section. In this succession growth stages in the development of the shell can be recognized. The first, or embryonic, stage is represented by the protoconch, which is an ovoidal calcareous chamber with its long axis perpendicular to the plane of coiling of the shell. The first three or four whorls, which are formed in infancy and constitute the **nepionic,** or **brephic,** stage, are smooth and flattened and increase rapidly in width from coil to coil. The next few whorls, which constitute the adolescent, or **neanic,** stage, show a series of changes and gradually take on the features characteristic of the adult, or **ephebic,** stage. In old or senile shells certain changes commonly appear, particularly in the sculpture; these constitute a final, or **gerontic,** stage.

Internally the adult shell is divided into a large living chamber that may constitute most or all of the last whorl, preceded adapically by 125 to 130 chambers separated by singly perforated septa. A small ventrally situated

[1] Kümmel (1952), Wright (1952), and Arkell (1950) have recently proposed classifications of the Triassic, Cretaceous, and Jurassic ammonoids, respectively, as bases for extended treatments of these groups in the proposed "Treatise on Invertebrate Paleontology" initiated by the Paleontological Society of America, and it is expected that the "Treatise" will include a complete classification of all Ammonoidea.

[2] We do not know of any recently compiled list of all ammonoid families, but the number of such families must now exceed several hundred. Wright (1952) recently listed 54 families of Cretaceous ammonoids, of which 9 also occur in Jurassic rocks; Arkell (1950) listed some 45 families of Jurassic ammonoids; Plummer and Scott (1937) listed 28 families of Mississippian to Permian genera; and Hyatt, as long ago as 1900 in Zittel's "Textbook of Palaeontology," had recognized almost a hundred families of Ammonoidea.

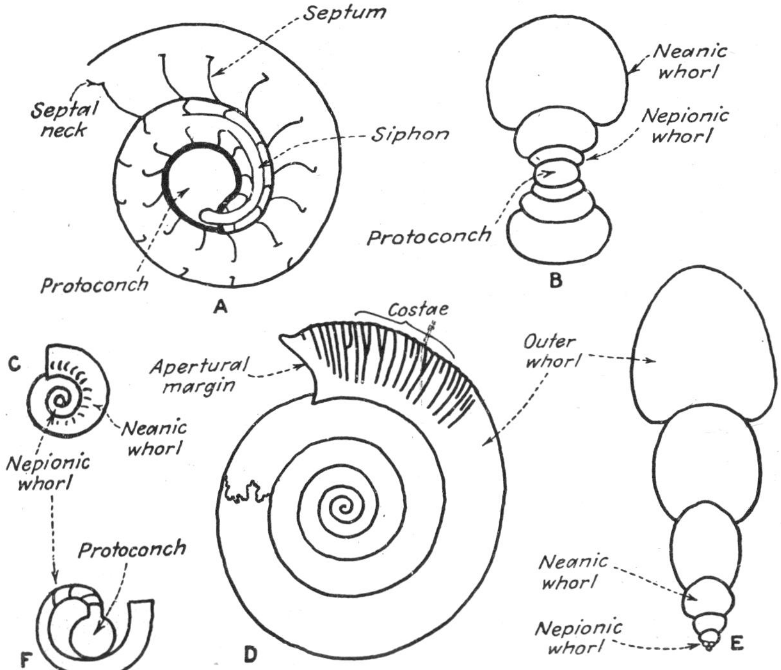

FIG. 10-74. Class CEPHALOPODA. Growth stages of *Dactylioceras commune* (*A–E*), a Mesozoic ammonite, and "*Goniatites*" *canaliculatus* (*F*), a Paleozoic goniatite. *A*. Median section of innermost whorls, much enlarged, showing large protoconch, and succeeding nepionic and neanic whorls. *B*. Transverse section of innermost whorls. *C*. Side view of first few coils, much enlarged. *D–E*. Side view and transverse section, respectively, of an adult shell, somewhat reduced. *F*. Globular protoconch succeeded by nepionic whorl with the earliest septa. (*A–E modified after Swinnerton*, 1947; *F from Jackson*, 1890, *and Hyatt*, 1894, *after Sandberger*.)

siphuncle starts at the latest septum, which is the posterior wall of the living chamber, and extends back to the nepionic whorls, in which it shifts gradually to a central position in the septum, and ends blindly against the first septum in a bulbous extension (caecum) of that septum into the protoconch.[1]

[1] It is assumed that the embryo built the simple hemispheroidal protoconch, and that the conch began when the infant cephalopod constructed a living chamber in the form of a tubular collar rising from the margin of the permanent aperture of the protoconch. After building this chamber the animal rested and secreted the first septum, which closed off the protoconch completely. The siphonal caecum was also incorporated in the first septum, which differs from all later ones in being imperforate and lacking a septal neck of any sort. The caecum, which projects into the protoconch as a bulbous extension of the first septum, is connected with the internal surface of the protoconch by calcareous strands that are collectively termed the **prosiphon.** These strands are thought to have served as a support for the caecum. A somewhat similar structure is present in the apical part of nautiloid and coleoid shells (Fig. 10-75).

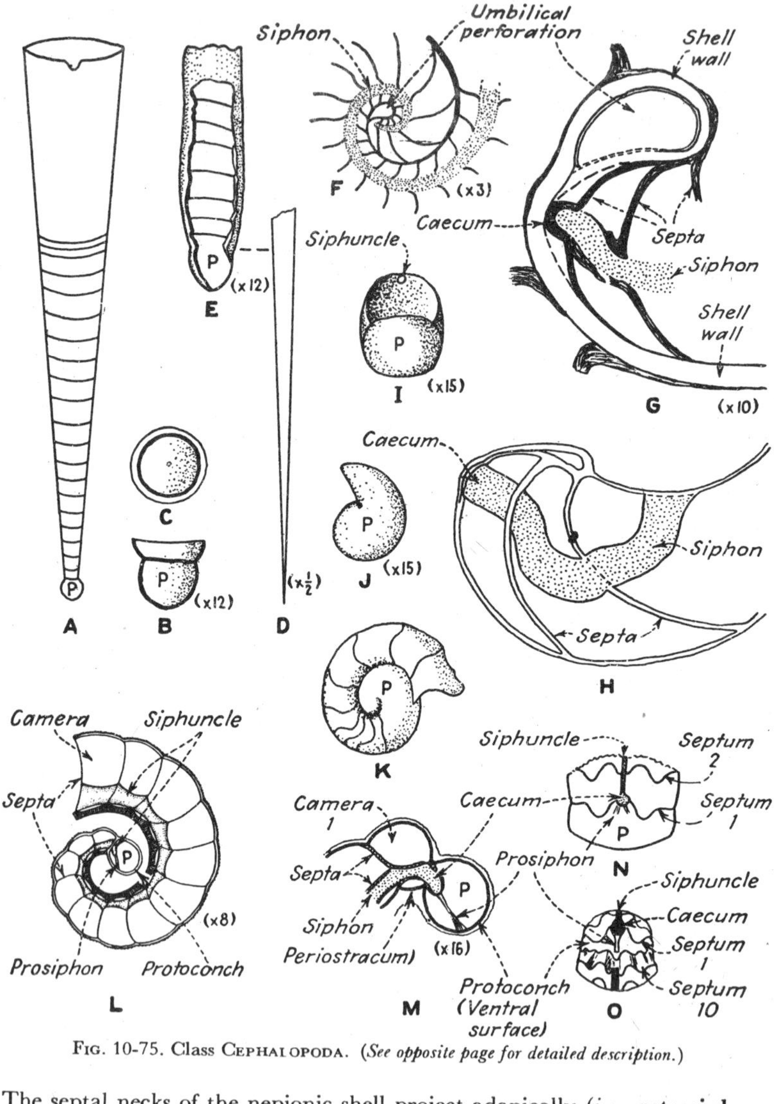

FIG. 10-75. Class CEPHALOPODA. (*See opposite page for detailed description.*)

The septal necks of the nepionic shell project adapically (*i.e.*, **retrosiphonate**), as in the Nautiloidea, but in the neanic and ephebic stages they are reversed and project adaperturally (*i.e.*, **prosiphonate**). The sutures of the adult whorls are typically ammonitic, but as they are followed backward they become less and less complex until they are simple and smooth curves on the nepionic whorls. It is noteworthy that the first suture, which marks the line of attachment of the first septum, differs from the second and all later ones by having a large ventral saddle.

From the data thus obtained by breaking back the adult shell to its protoconch, it is possible to reconstruct the successive growth stages and to study the changes made in the shell as the ammonoid animal grew from an embryo to an adult cephalopod. Studies of this kind have been made of many genera, and from these investigations has grown a large body of knowledge that has gradually been systematized and integrated into numerous "laws" or "principles" of evolution. Alphaeus Hyatt, an American paleontologist of the last century, was one of the outstanding leaders in this kind of cephalopod study and he formulated many of the principles that today are given such formidable names as **tachygenesis, lipopalingenesis,** and **bradygenesis.** The interested reader will find an excellent discussion of these and other aspects of ammonoid evolution in Swinnerton's "Outlines of Palaeontology" (1947).

In their evolution the Ammonoidea underwent great diversification in shell morphology and sculpture, mode of coiling, shape of whorls, nature of septa and siphuncles, and complexity of sutures. Certain changes of shell morphology and mode of coiling recur frequently and may reflect adaptations to constantly recurring modes of life. These changes are not reliable for working out the evolution of a group. On the other hand the sutures have been found to have great taxonomic and evolutionary importance and are now used extensively for this purpose. The siphuncle, which has such great taxonomic importance in the Nautiloidea, lacks importance as a shell feature in the Ammonoidea, although it was formerly used as a basis for subdividing the latter into the Intrasiphonata, with a dorsal siphuncle, and the Extrasiphonata, with a ventral siphuncle. Since only a few dozen genera of ammonoid shells, the compact clymenid group (*e.g., Clymenia;* Fig. 10-76*D*), have dorsal siphuncles, all the remainder of the 5,000 species

---

FIG. 10-75. Class CEPHALOPODA. Protoconchs and nepionic shells. *A.* A generalized orthoceracone, viewed from the ventral side, with the wall of the conch almost completely removed so that the septa are visible (see Fig. 10-58 for labeling of parts). *B–C.* Side and apical views of a protoconch of a supposed orthoceracone. *D–E. Bactrites,* a primitive ammonoid: *D,* a nearly entire shell; *E,* nepionic shell, with part of wall stripped away, showing small protoconch, nepionic expansion and contraction, and first few septa. *F.* Nepionic part of *Heminautilus etheringtoni,* showing nature of early septa, and caecal ending of siphon before it reached the apical wall of the initial chamber. *G.* A median longitudinal section of the adapical portion of the conch of *Nautilus pompilius,* showing the siphuncular caecum and the umbilical perforation; the umbilicus persists only during the early growth stages. *H.* Section through the first three camerae of *Eutrephoceras dekayi,* showing siphon ending in a caecum. The shell has been somewhat distorted in fossilization (cf. *G*). *I–K. Manticoceras fasciculatum: I–J,* apertural and side views of protoconch, showing first septum and ventral siphuncle; *K,* nepionic shell, showing large protoconch succeeded by the early nepionic camerae. *L–M. Spirula: L,* adult shell; *M,* protoconch and earliest nepionic septa and camerae. *N.* Protoconch and earliest nepionic structures of *Lytoceras alamedense.* The prosiphon is a flattened spreading web as compared to the hour-glass shape of the specimen shown in *O.* (*P* = protoconch.) *O.* A transparent protoconch and early nepionic camerae of "*Ammonites*" *breweri,* showing a conspicuous prosiphon extending from the wall of the proconch to the siphuncular caecum. (*B–E, I–K, H after Clarke,* 1899; *F after Durham,* 1947; *G after Miller, Dunbar, and Condra,* 1933; *L after Munier-Chalmas,* 1873; *M after Naef,* 1922; *N–O after Crickmay,* 1925.)

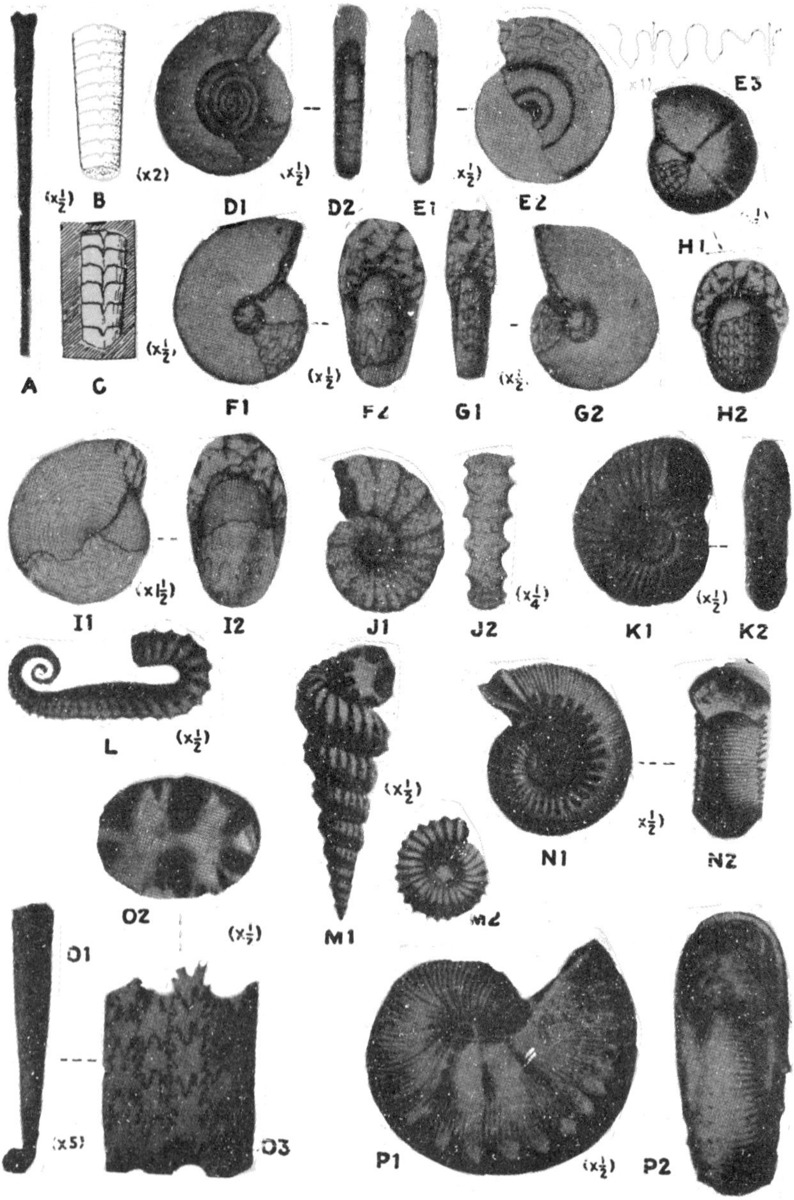

Fig. 10-76. Class Cephalopoda—Subclass Ammonoidea. (*See opposite page for detailed descriptions.*)

having ventral siphuncles, it follows that position of the siphuncle is not a very satisfactory basis for subdividing the Ammonoidea.

Sutures have thus far provided the most useful basis for separating ammonoid genera into large generalized groups, but these groups are of questionable taxonomic homogeneity and it would seem that a natural classification of the Ammonoidea at the ordinal level remains to be proposed. For the present, therefore, we follow the customary practice of recognizing three general groups of ammonoids based on the nature of the suture: goniatites, or goniatitoids; ceratites, or ceratitoids; and ammonites, or ammonitoids. *Muensteroceras* (Fig. 10-68) is a representative goniatite; *Ceratites* (Fig. 10-70*G–I*) is a typical ceratite; and *Placenticeras* (Fig. 10-70*J–L*) is typical of the ammonites. A few other typical genera are shown in Figs. 10-76 and 10-85.

Ammonoid shells run the full gamut of shell forms—bactriticones, mimoceracones, dactylioceracones, ammoniticones, and turriliticones (Fig. 10-60). They are typically more highly sculptured than nautiloid shells, and the sculpture of large groups of Mesozoic genera has great taxonomic and stratigraphic importance. The aperture in many genera has a prominent ventral **rostrum** (*e.g.*, *Dipoloceras*), and this has lateral crests and lappets in Late Mesozoic forms. The presence of a rostrum supposedly indicates the lack of a hyponome and has been cited as evidence in support of the contention that rostrate forms were active crawlers rather than swimmers.

Single and double plates have been found *in situ* in the aperture of many Mesozoic ammonoids (Fig. 10-64) and have generally been interpreted as opercula that closed the aperture and protected the animal when it withdrew into the shell. The single plates, to which the term anaptychus has been applied, are carbonaceous when preserved and are thought to have been horny in the living animal. They are rare in Paleozoic ammonoids but have been found in numerous Mesozoic shells. The double plate, or aptychus, is invariably calcareous and has a three-layered microscopic structure. Many of these plates have been found, some separately and others *in situ*,

---

FIG. 10-76. *A*. Ventral view of *Lobobactrites*, a Devonian orthoconic ammonoid. *B*. Ventral view of small part of *Bactrites* from the Middle Devonian, showing small ventral lobe. *C*. Ventral aspect of a fragmental conch of *Eobactrites*, from the Ordovician, the oldest known ammonoid. *D*1–*D*2. Side and apertural views of *Clymenia*, from the Devonian an ammonoid that is unique in having a dorsal siphuncle. *E*1–*E*3. Ventral and lateral views and diagrammatic suture of *Protocanites* from the Mississippian. *F*1–*F*2. Lateral and ventral views of *Schistoceras*, a common Pennsylvanian ammonoid. *G*1–*G*2. Two views of *Proudenites*, a widespread Upper Pennsylvanian genus. *H*1–*H*2. Two views of *Peritrochia*, a world-wide genus in rocks from Lower Pennsylvanian to Middle Permian age. *I*1–*I*2. Two views of the Permian genus *Agathiceras*. *J*1–*J*2. Two views of *Ceratites nodosus*, the well-known Triassic genus. *K*1–*K*2. Two views of *Idoceras*, from the Jurassic. *L*. *Ancyloceras*, from the Lower Cretaceous. *M*1–*M*2. Two views of *Turrilites*, from the Lower Cretaceous. *N*1–*N*2. Two views of a highly sculptured ammonoid from the Middle Jurassic. *O*1–*O*3. A young specimen and two views of a filling of the interior of *Baculites*, the familiar and widespread Upper Cretaceous genus. *P*1–*P*2. Two views of *Scaphites*, one of the commonest and most widespread of Upper Cretaceous genera. (*A after Hall*, 1879; *B after Whiteaves*, 1898; *C adapted from Miller*, 1938, *after Schindewolf; D, J, L–N after Nicholson*, 1872; *E after Miller and Butts*, 1936; *F–I after Miller and Furnish*, 1939 *and* 1944; *K after Imlay*, 1939; *O after Reeside*, 1927; *P after Meek*, 1876.)

and they are sufficiently different one from another that some specialists have classified them into several distinct groups (see footnote, page 451).

As stated previously, the siphuncle of ammonoid shells varies in structure from shell to shell and also in different parts of the same shell. If it consists entirely of adapically directed septal necks, it is **monochoanitic;** if it consists of both septal necks and adaperturally directed tubular **collars,** it is **diplochoanitic;** and if the septal necks are lacking, but the collars persist, it is **cloiochoanitic.** Earlier ammonoids have monochoanitic siphuncles; diplochoanitic or transitional types appear in specialized Carboniferous genera; and cloiochoanitic types are present in most Triassic and all Jurassic and Cretaceous genera. The siphuncle becomes reduced in size in the more advanced and specialized ammonoids, and siphuncular and cameral deposits are not common as in the Nautiloidea.

The earliest ammonoid seems to be *Eobactrites* (Fig. 10-76*C*) from the Ordovician (Miller, 1938), but the subclass does not seem to have become well established until the Devonian, when some 90 or more genera appeared. Many more genera appeared in succeeding periods of the Paleozoic, but it was not until the Mesozoic that the ammonoids attained their greatest development. An entirely new group, those with ceratite sutures, made their appearance in the Mississippian and arose to a maximum in the Triassic, so that the Triassic has commonly been called the "Age of Ceratites," just as the later Paleozoic has sometimes been called the "Age of Goniatites." In the Jurassic there was another great surge in ammonoid evolution, this time by the complex ammonites, which had appeared earlier in the Permian, and these persisted, though with gradually diminishing vigor, to the close of the Cretaceous when the entire Subclass Ammonoidea suddenly and inexplicably became extinct.

Figure 10-76 shows genera that are typical of a few of the more important families or larger groups of ammonoids. Space unfortunately prevents listing and illustrating more than these few representatives of one of the greatest of all fossil groups.

Subclass Coleoidea

General Considerations. The Coleoidea include all living cephalopods except the tetrabranchiate nautiloid, *Nautilus*, and are widely familiar through the notoriety of the octopus and the squid. We adopt the classification of Naef (1922), which consists of four orders, one extinct and three living, as follows: Belemnoidea (extinct), Sepioidea, Teuthoidea, and Octopoda.

Coleoids have only two gills (whence the name Dibranchiata, commonly used in older works), and 8 or 10 arms that bear hooks and sucker disks (**acetabula;** Fig. 10-77). The Octopoda have only eight arms, all of which are equally long and covered on one side with many fleshy acetabula. In the other three orders there are eight sucker-bearing sessile arms of essentially equal length and two much longer retractile tentacles that have a terminal clublike expansion armed with tiny horny hooks and partly covered with clusters of suction disks. These 10-armed coleoids are commonly grouped

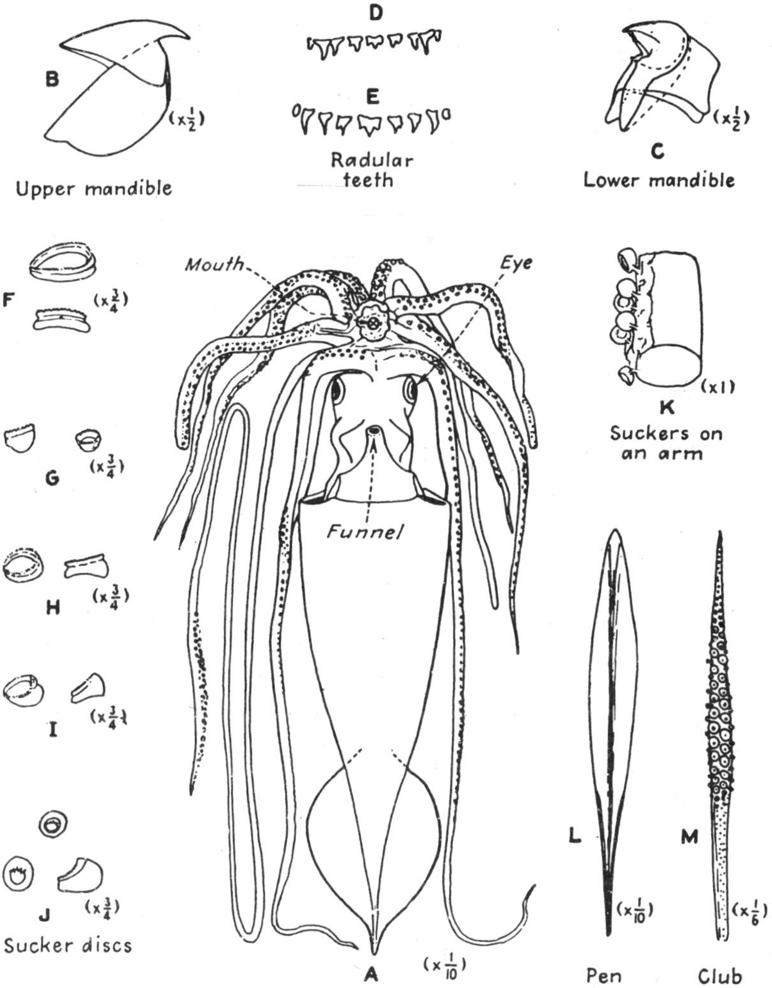
D
E
Radular
teeth
B
(x½)
Upper mandible
C
(x½)
Lower mandible
F
(x¾)
G
(x¾)
H
(x¾)
I
(x¾)
J
(x¾)
Sucker discs
Mouth
Eye
Funnel
A
(x 1/10)
K
(x1)
Suckers on
an arm
L
(x 1/10)
Pen
M
(x 1/6)
Club

FIG. 10-77. Class CEPHALOPODA—Subclass COLEOIDEA. *Architeuthis*, a living representative of the Order Teuthoidea, with numerous parts of the animal shown separately. *A*. The animal, which is shown in ventral aspect, lacks most of one of the two long club-bearing tentacles and the tips of several of the shorter sessile arms as a result of being buffeted about for several days by storm waves. It probably is a young specimen. *B–C*. Upper and lower jaws, or mandibles, respectively. These are shown in place in the mouth. *D–E*, Radular teeth: *D* shows a row of teeth from the part of the radula near the mouth opening; *E* shows another row farther back on the radula, which is about 5 mm. wide. *F–J*. Horny denticulate rings from the sucker disks on the different arms. *K*. Small part of a sessile arm, showing the pedunculate sucker disks as they are attached to the surface of the arm *L*. Internal shell, or pen. *M*. Sucker-bearing club of the long tentacle. (*Modified after Mitsukuri and Ikeda*, 1895.)

together as the Decapoda, but as that designation is also widely used for certain crustaceans, it is not adopted here.

The hyponome forms a complete tube and functions as a powerful jetlike swimming organ. This feature, along with the conspicuous streamlining, makes the coleoids probably the most efficient and versatile swimmers among the invertebrates. An ink sac is present in certain living coleoids and is known from many well-preserved specimens to have been present in some ancient forms as early as the Jurassic. The contents of the ink sac can be discharged into the gill cavity, where the inky substance blackens the water, and the darkened water when expelled through the hyponome provides a protective screen under cover of which the coleoid can dart away to safety.

The mouth is provided with a pair of powerful horny beaks, or mandibles, and a multidenticulate radula is present (Figs. 10-56*B–D*, 10-77). The radula is sufficiently constant in its different forms that it has been used to subdivide living octopods into four taxonomic groups.

The shell is either internal or lacking altogether, with the single exception of a delicate external shell unique to *Argonauta* and not homologous with other molluscan shells. The internal shells exhibit wide variation in form and structure, with the general tendency being toward more highly specialized (or possibly degenerate) shells in the geologically younger forms.

Hard Parts and the Fossil Record. The fossil record of the Coleoidea may consist of one or more of the following components (Fig. 10-77):

1. Internal Shells. Internal shells of great variety characterize the four orders of Coleoidea. The belemnoids had cigar-shaped shells of calcite; sepioid shells are highly modified chambered cones, to which the term **sepion** has been applied; teuthoid shells have been termed a **pen** or **gladius** (L. *gladius*, sword); and rare octopod shells have irregular shape or resemble the teuthoid gladius (Fig. 10-84).

2. External Shells. A true external shell is unknown among the Coleoidea, but in one octopod genus, *Argonauta* (Fig. 10-83*A–D*), the large females secrete a delicate nonseptate planispirally coiled calcareous shell that is used as a nesting place and egg case. This shell, although superficially resembling certain ammoniticones, is not homologous with true cephalopod shells and is unique among the Mollusca. It is known fossil from rocks as old as the Miocene (Naef, 1922, page 294).

3. Jaws. Coleoids typically have a pair of powerful horny beaks in the mouth, and these have been found as fossils in Mesozoic and Cenozoic rocks. Usually they are best preserved in specimens that consist of an impression of the entire body (Figs. 10-78; 10-84), but isolated jaws have also been reported.

4. Radular Elements. The radula is well developed in living coleoids, and it may be assumed to have been present in extinct species. The radular elements are tiny horny hooks and pointed plates (Figs. 10-56*C–D*, 10-77*D–E*) and should be preserved under favorable conditions, but no records of fossil cephalopod radulae are known to us.

5. Hooks and Sucker Disks (Figs. 10-56, 10-77, 10-82). The arms and tentacles of coleoids are partly covered with horny hooks and muscular suction cups, or acetabula, that commonly have a horny denticulate rim. Both hooks and cups are susceptible to fossilization, but they have seldom been preserved.

6. Compressions and Impressions (Figs. 10-78, 10-84). The most complete preservation of ancient coleoids is illustrated by many well preserved compressions and impressions of whole animals found particularly in the well-known Jurassic lithographic limestone of Solenhofen, Bavaria. Some of these specimens are so well and completely preserved that the jaws, arms, tentacles, ink sac, body outline, and internal shell are easily discernible.

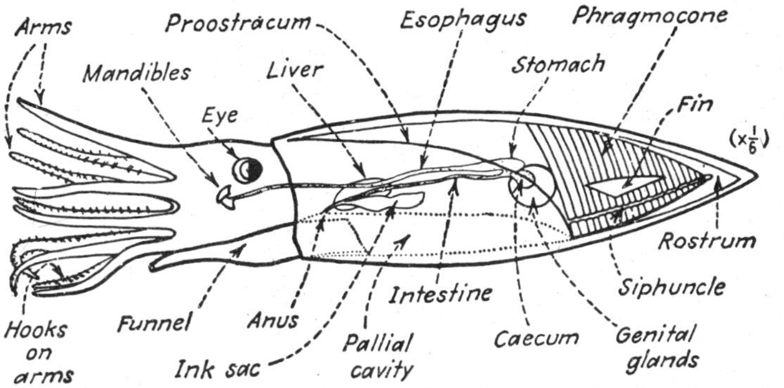

FIG. 10-78. Class CEPHALOPODA—Order Belemoidea. Reconstruction of an Upper Cretaceous belemnoid, *Belemnoteuthis syrica*, with an internal shell consisting of phragmocone, proostracum, and rostrum. Most of the soft parts are indicated on the impression, the ink sac and canal leading to the intestine are preserved, and mandibles indicate the position of the mouth. (*Modified after Roger*, 1945.)

The Internal Shell. The internal shells of extinct and living coleoids present one of the most interesting and challenging arrays of shell modification to be found in any group of Invertebrata, and though there are many gaps, certain definite trends in structural modification seem evident, as illustrated in Fig. 10-80. In general the trend has been to reduce the shell.

The shell of *Belemnites* (Figs. 10-80*B*, 10-81*E–G*), an extinct genus that was abundant the world over during the Jurassic and Cretaceous, shows the chief morphological features characteristic of coleoids as a group. (Figure 10-78 is a restoration of a closely related genus *Belemnoteuthis*, showing how the shell supposedly was carried inside.) The shell is composed of three fundamental parts—rostrum, phragmocone, and pro-ostracum.[1] The **rostrum,** or **guard,** is a solid, conoidal or cigar-shaped structure constituting the apical part of the shell, and commonly is the only part of the shell that survives fossilization. When sectioned transversely it is seen to consist of prisms of calcite that radiate from an axial line eccentrically located slightly nearer the ventral surface. Fitting into the hollow conical **alveolus,** in the blunt end of the guard, is the **phragmocone,** which is divided into chambers by septa, each of which is perforated by the ventrally situated siphuncle. The **pro-ostracum** is a bladelike plate, supposedly composed of imperfectly calcified conchiolin that extends the dorsal part of the shell forward from the aperture of the phragmocone. In life it probably acted as a dorsal protection for the viscera.

The phragmocone is homologous with the conch of the Nautiloidea and Ammonoidea, and it indicates that the Coleoidea bear close kinship with

[1] Naef (1922) homologizes these three shell parts with the three fundamental molluscan shell layers as follows:

Coleoid Shell	*Fundamental Layer*
Rostrum; commonly calcified (*e.g.*, *Belemnites*).	Periostracum.
Conotheca—sheath around phragmocone plus pro-ostracum.	Ostracum.
Phragmocone.	Hypostracum.

both of these groups. It resembles the shells of certain cyrtoconic nautiloids in its simple septa and ventral siphuncle[1] but differs in having a prominent protoconch. It differs from typical ammonoids in its simple sutures and from both Nautiloidea and Ammonoidea in having the dorsal margin of the phragmocone wall, or conotheca, extended forward as the pro-ostracum.

The **conotheca** (Fig. 10-79), which consists of the pro-ostracum and the thin wall investing the phragmocone, is composed of three thin laminae, the outer of which bears certain peculiar markings, the **conothecal striae** (Fig. 10-79*C*), made by the membranes that invested it in life. These striae are of

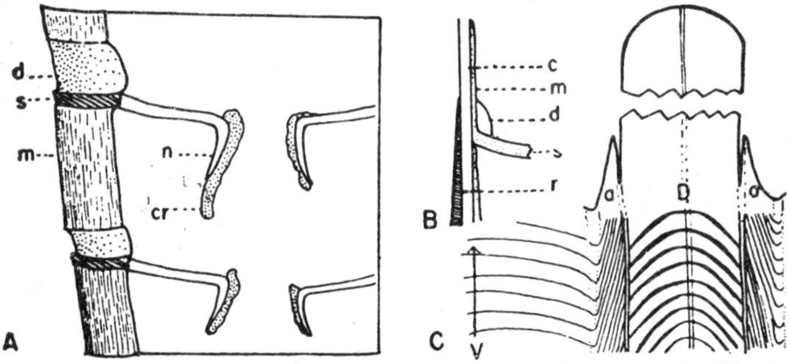

FIG. 10-79. Class CEPHALOPODA. Details of the morphology of *Eobelemnites*, the oldest known belemnite. *A*. Part of phragmocone, showing natural exfoliated surface on left, artificially ground section on right, exposing siphuncle (*d* = nonstriated dull zone representing mold of cameral deposit; *s* = septum, showing broken edge at surface of specimen; *n* = septal neck; *cr* = connecting ring; *m* = polished and finely striated surface of internal mold of mural part of septum). *B*. Reconstructed cross section through wall of phragmocone (*c* = conotheca or conch; *r* = rostrum; *m* = mural part of septum; *s* = free part of septum; *d* = cameral deposit). *C*. Projection of conothecal striae (*V* = venter; *D* = dorsum). The relative thickness of the lines is designed to indicate their relative clarity (*a* = asymptotes). Above the dorsum the outline of the rostrum is reconstructed. This is broken, since it is impossible to estimate its actual length. The reentrants dorsad of the asymptotes are dubious. (*After Flower*, 1945.)

some importance because they commonly make it possible to reconstruct the outlines of the missing pro-ostracum and to determine other shell features. Of particular interest in this regard are dorsal, lateral, and ventral bands of transverse conothecal striae present on many belemnoids (Fig. 10-79). The dorsal area is bounded by two straight narrow bands, the **asymptotes** (Fig. 10-79*C*), which extend from the apex of the cone as far as the aperture.

The oldest fossil coleoid shell, *Eobelemnites*, was found in a boulder of supposed Mississippian age. It belongs to the extinct Order Belemnoidea, and

[1] A recently described Triassic belemnoid, *Choanoteuthis mulleri* (Fischer, 1951), has a most unusual siphuncular structure, which differs markedly from any heretofore recorded for cephalopods of any sort. The siphuncle is holochoanitic and is composed of invaginated funnels (septal necks?), each of which extends through slightly more than two chambers. The structure somewhat resembles that of the holochoanitic nautiloids and of the Jurassic belemnoid *Megateuthis quinquesulcatus*. It would seem that the belemnoids, like the nautiloids, developed holochoanitic forms at the time of their first great morphologic radiation and their rise to abundance (Fischer, 1951).

though the earliest known coleoid, it is not a primitive form, as can be seen from Fig. 10-79. As a matter of fact it is so specialized as already to have the features characteristic of the Coleoidea—*i.e.*, rostrum and pro-ostracum, in addition to the phragmocone. The morphology of the phragmocone and pro-ostracum is of particular interest because of the light it throws on the general problem of coleoid origin. It is now generally believed that the Coleoidea developed from a group of cephalopods that had straight external shells, but the exact nature of the ancestral form or forms is a moot question (Flower, 1945).

In the evolution of the Coleoidea, it seems that the postulated external shell of the ancestral form was gradually enveloped by the mantle until finally it became entirely internal, and that during this envelopment the shell became greatly modified and tended to lose bulk or weight until in extreme cases, as in the octopods, it was lost altogether. Some of the stages in this shell modification are shown diagrammatically in Fig. 10-80.

It may be hypothecated that during the later Paleozoic and earlier Mesozoic some straight-shelled cephalopods (possibly like *Michelinoceras*) added a conical sheath to the apical part of the septate conch to form the

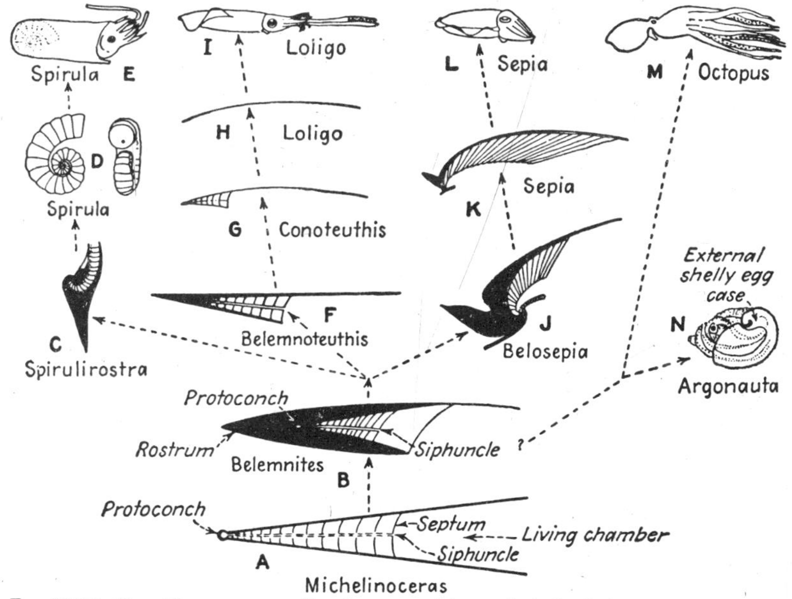

FIG. 10-80. Class CEPHALOPODA. Development of internal shells. It is postulated that a Paleozoic orthoceraconic form of the type of *Michelinoceras* grew around its shell, finally became a belemnitid with a heavy rostrum, a relatively small phragmocone, and a conspicuous pro-ostracum. At least three lines of shell modification are believed to have taken place, as indicated by the arrows. The origin of *Octopus*, which lacks a shell or has only the merest vestige of one, and *Argonauta*, the female of which secretes a thin external shell-like egg case, is uncertain. (*Adapted from numerous authors, including Borradaile et al.*, 1935, *Naef*, 1922 *and* 1923, *Roger*, 1947, *and Swinnerton*, 1947.)

rostrum and extended anteriorly the dorsal part (and possibly also the ventral at first) of the conothecal layer surrounding the phragmocone to form the pro-ostracum. *Eobelemnites* (Miss.), *Metabelemnites* (U. Trias.), and *Belemnites* (Jura.–Cret.) fall in this line of modification.

With the opening of the Tertiary began acceleration of four general trends of shell modification that had already been established in the Mesozoic. The genera developed constitute the fossil and living members of the three orders, Sepioidea, Teuthoidea, and Octopoda. Whatever the other changes in the three trends, there was in all three the common tendency to reduce the weight of the shell.

In one sepioid line (*Belemnites—Belosepia—Sepia*) the rostrum was progressively reduced until it became little more than a point at the posterior end of the shell; the septa became oblique and the siphuncle wide open; and the pro-ostracum came to form a roof over the forwardly projecting septa (Fig. 10-80).

In a second sepioid line (*Belemnites—Spirulirostra—Spirula*) the phragmocone changed in shape from a short to a long and slowly expanding cone, which then became cylindrical in its adult whorls (with the septa becoming more widely spaced) and curved through an arc to the gyroconic coiling seen in *Spirula*. The rostrum was progressively reduced until it became lost altogether, hence the only part of the shell that remained was the phragmocone, which meanwhile had become coiled, as in *Spirula* (Fig. 10-81*A–D*).

In the teuthoid line (*Belemnites—Belemnoteuthis—Conoteuthis—Loligo*) there was early loss of the rostrum, gradual decrease in size of the conical phragmocone, and development of a long penlike or lanceolate pro-ostracum of conchiolin. At the extreme end of this line is common *Loligo* (Fig. 10-82), which has only a thin spatulate pro-ostracum.

The stages of shell modification leading to complete loss of internal shell, as exemplified in the octopods, are as yet unknown.

Classification. The classifications of the Coleoidea that have been proposed during the period since 1885 (Zittel) differ greatly because of differences of opinion as to the taxonomic significance of many of the morphological features involved. Flower (1945) reviews this situation fully for those who may be interested.

As previously stated, the classification proposed by Naef in 1922 is adopted in the present work, because it seems to take greatest cognizance of shell morphology, structure, and evolution. Four orders are included in the Subclass Coleoidea:

Order 1. Belemnoidea (extinct).
Order 2. Sepioidea.
Order 3. Teuthoidea.
Order 4. Octopoda.

These are characterized briefly in the following paragraphs.

Order 1. Belemnoidea. The belemnoids are supposedly the oldest and most primitive of the dibranchiate cephalopods and are the only fossil coleoids of much paleontologic importance. The oldest representative, *Eobelemnites*

Fig. 10-79) from the Mississippian,[1] is already so specialized, in having a rostrum and pro-ostracum in addition to a phragmocone, as to make it obvious that the ancestors of the Order Belemnoidea, as well as of the Subclass Coleoidea, must be sought in rocks of Early Mississippian or pre-Mississippian age. The last of the true belemnoids became extinct, along with their cousins, the ammonoids, at the close of the Cretaceous, but it is highly probable that the general belemnoid stock lives on in the modern orders of Coleoidea. *Upper Mississippian through Cretaceous.*

Eobelemnites (U. Miss.), *Metabelemnites* (U. Trias.), and *Belemnites* (Jura.–Cret.) are representative genera (Figs. 10-78, 10-79, 10-80). *Belemnites* is an excellent index fossil of Jurassic and Cretaceous rocks the world over, and in this respect is scarcely less important than many ammonite genera.

Order 2. Sepioidea. The sepioids include the coleoids with highly modified belemnitoid shells in which the rostrum has been greatly reduced or lost, the pro-ostracum[2] greatly modified, and the phragmocone persistent to rudimentary, with a tendency to curve or coil. The oldest fossil sepioid, *Voltzia palmeri*, which is an almost complete pro-ostracum with a fragmental phragmocone, was recently reported as coming from an Upper Jurassic (Oxfordian) limestone of Cuba (Schevill, 1950). Fossil sepioids are not common but have been reported from several Tertiary formations. Living sepioids are abundant in existing seas. *Upper Jurassic to Recent.*

Sepia, *Belosepia*, *Spirulirostra*, and *Spirula* constitute a group of typical sepioids in which two quite different trends of shell modification are shown, as illustrated in Fig. 10-80. The shell, or sepion, of *Sepia officinalis* is the familiar cuttlebone of commerce, so widely used to provide a source of lime for caged birds, and the small shiny coiled shells of *Spirula* are widely used in paleontological laboratories to acquaint beginning students with the fundamental structures of the septate cephalopod shell.

The beautiful little shell of *Spirula* (Fig. 10-81*A–D*) is a gyroconically and endogastrically coiled cone divided into 30 to 40 chambers by simple septa that are convex toward the apex and perforated by a ventral siphuncle composed of long septal necks. The initial chamber is a spherical protoconch, which is separated from the first camera of the conch by a constriction where the imperforate first septum closes its original aperture (Figs. 10-75*L*, 10-81*A*). It is calcified and contains a prosiphon, which, as in other cephalopods, ends at the first septum. The closely spaced septa have a tiny ventral perforation through which the siphuncle extends from the siphuncular caecum of the first septum to the apertural end of the shell. In life the shell is carried at the posterior extremity of the animal, with a small part exposed

[1] *Eobelemnites caneyense* (Flower, 1950) was collected from a boulder, supposedly of Mississippian age, in the Caney shale of Chester (Upper Misisissippian) age.

[2] The pro-ostracum (commonly termed **sepion**) of the sepioids has the form of a spatulate pen or an oarlike paddle and acts as a stiffening axis in the dorsal part of the body (Fig. 10-82). Some pro-ostraca have a short pointed **mucro,** or **thorn,** at the small end, from which the broader part of the shell extends. The anterior part of the pro-ostracum has numerous shelly laminae that are considered homologous with the septa of the phragmocone of other coleoids.

and part internal (Fig. 10-81*D*). Unlike *Nautilus*, which it resembles in some respects, the shell of *Spirula* has no true living chamber and no lobation of the simple circular apertural margin. It also has a calcified protoconch—a feature which is missing in *Nautilus*, supposedly because it is chitinous and is lost soon after the formation of the apical chamber of the calcareous conch. *Spirula* supposedly represents the product of a trend of shell modification in which both rostrum and pro-ostracum were lost, whereas the shell of

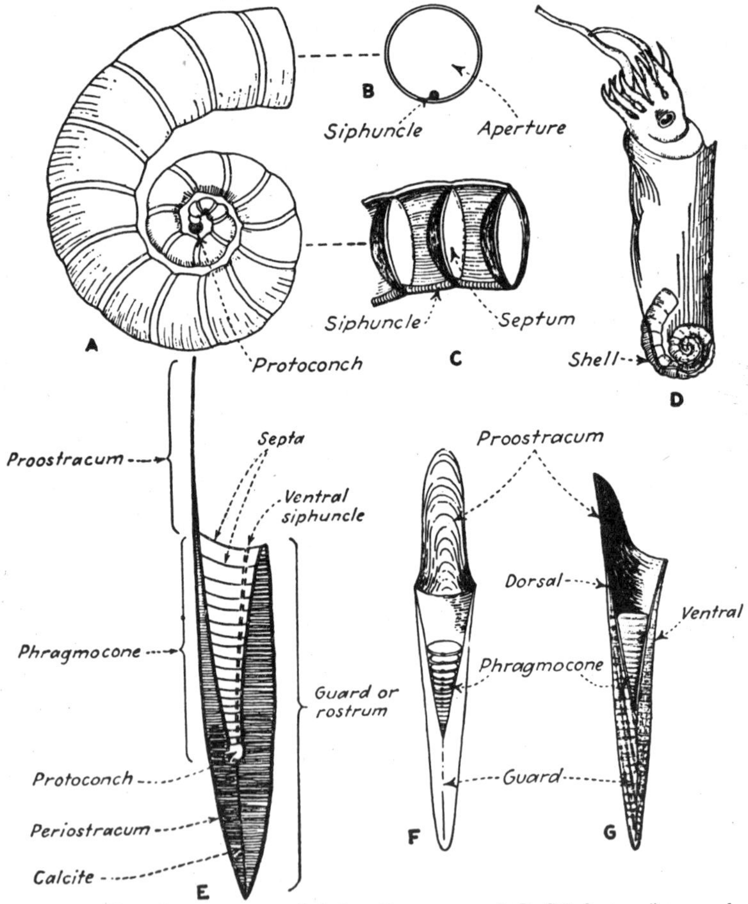

FIG. 10-81. Class CEPHALOPODA—Subclass COLEOIDEA. *A–C. Spirula peronii*, a modern sepioid from the Atlantic Ocean: *A*, the complete shell, a gyroceracone, which is internal as shown in *D*; *B*, apertural view showing ventral position of siphuncle; *C*, small portion of shell sectioned to show the septa, which are convex adapically, and the ventral siphuncle. *D. Spirula australis*, a modern sepioid showing position of shell in relation to remainder of organism. *E*. Diagram of a belemnitid shell, showing the several important features. *F–G*. Two views of a *Belemnites* (Jurassic) showing the several characteristic structures. (*D after Wright; F–G after Phillips*.)

Sepia is the result of a different trend in which the phragmocone is almost unrecognizedly modified, and the pro-ostracum and rostrum reduced to vestiges. The conspicuous periostracum is heavily calcified.

Order 3. Teuthoidea. Teuthoids are 10-armed coleoids in which the rostrum is undeveloped, the phragmocone becomes rudimentary or lost, and the pro-ostracum has the form of an elongated pen, or gladius (Fig.

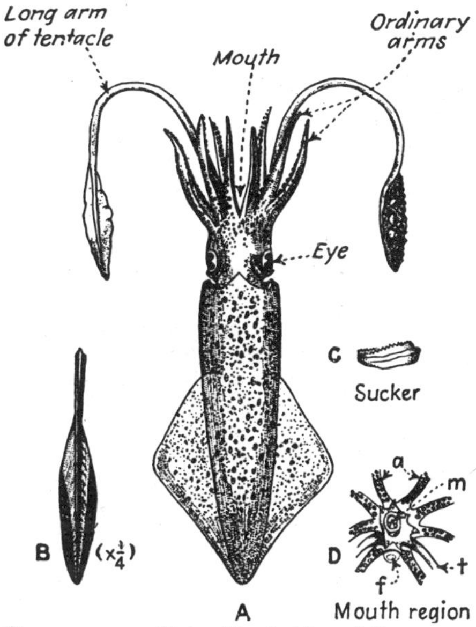

FIG. 10-82. Class CEPHALOPODA—Order Teuthoidea. *Loligo*, a modern cuttlefish, or teuthoid. *A*. Dorsal view of the whole animal. *B*. The internal shell, or pen. *C*. Lateral view of one of the small suckers from one of the arms, showing the serrated margin made by the horny hooks. *D*. View of the mouth region, showing the mouth (*m*), funnel or hyponome (*f*), bases of the arms (*a*), and one of the two tentacles (*t*). (*A, C–D modified after Nicholson and Lydekker*, 1889; *B after Woodward.*)

10-82*B*). The oldest known fossil teuthoids seem to be Jurassic in age, although specialists believe the order was in existence much earlier. *Jurassic to Recent.*

Plesioteuthis, an extinct genus (Jura.–Cret.) with a penlike gladius, and *Ommastrephes* and *Loligo* (Fig. 10-82), modern squids with a thin paddlelike pro-ostracum, are typical representatives of the order.

Order 4. Octopoda. The Octopoda include the notorious octopuses or devilfish of existing seas. The baglike body is shorter and rounder than in other coleoids and has eight subequally long arms that bear suckers without horny rims. The body generally lacks the prominent lateral finlike structures so characteristic of the sepioids, although some forms have bluntly rounded lateral expansions.

In spite of the common statement that all octopods are shell-less, some living forms actually have a small vestigial internal shell, and the oldest fossil representative, *Palaeoctopus newboldi* (Fig. 10-84) from the Upper Cretaceous, had a small but distinct saddle-shaped internal shell. Attention has also been called to the peculiar teuthoid gladius of some of the Vampyro-

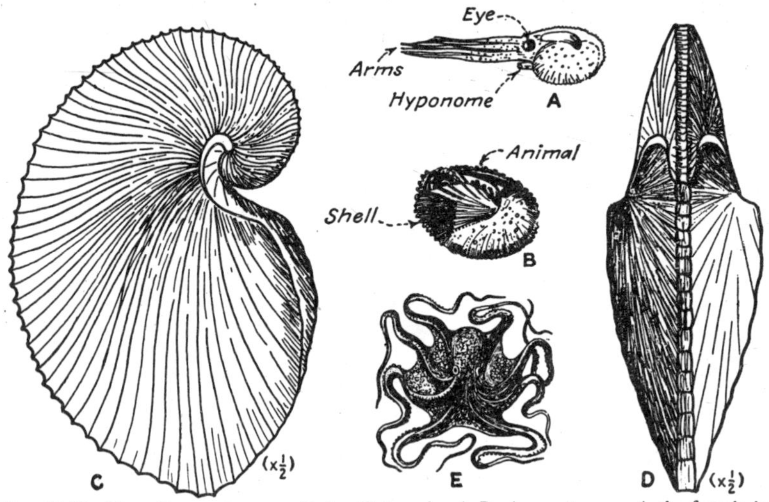

FIG. 10-83. Class CEPHALOPODA—Order Octopoda. *A–D. Argonauta argo: A*, the female in swimming position, with the shell that serves as a brood pouch; *B*, female withdrawn into shell; *C–D*, side and apertural views, respectively, of the paper-thin, unchambered calcareous shell. *A* and *B* are greatly reduced. *E. Octopus vulgaris*, showing nature of body and the eight sucker-bearing arms. (*A–B, D after Wright.*)

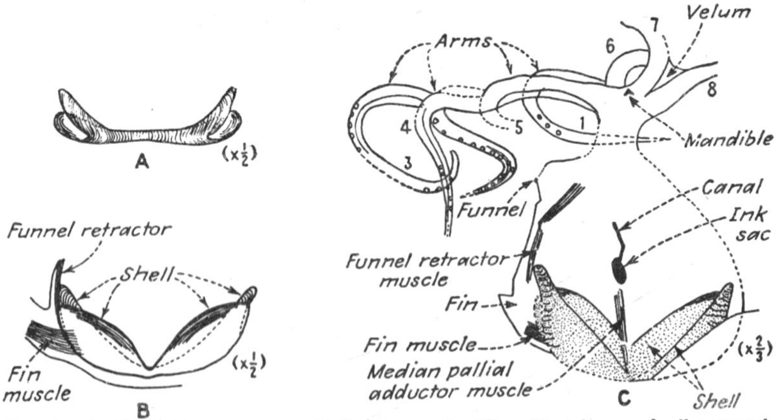

FIG. 10-84. Class CEPHALOPODA—Order Octopoda. The oldest known fossil octopod, *Palaeoctopus newboldi*, from the Upper Cretaceous fish beds of Syria. *A–B*. Posterior and dorsal views of the saddle-shaped vestigial shell. *C*. Diagrammatic sketch of whole body, showing arms, body outline, ink sac and canal, shell, and muscles associated with the shell. The shell in *C* has been stippled so as to be more conspicuous. (*Modified after Roger*, 1946.)

morpha,[1] a small group of living coleoids generally included in the Octopoda.

Of particular interest in connection with shell development is the fact that the females of *Argonauta*, which are many times as large as the males, secrete a delicate external shell (Fig. 10-83*A–D*) that serves as a nest and egg sac. This shell, which is secreted by special expansions of the dorsal arms, is an unchambered, rapidly expanding planispirally coiled cone with surface sculpture consisting of folds and tubercles and two nodose ventral keels.

The oldest known fossil octopod is *Palaeoctopus newboldi* (Fig. 10-84) from the Upper Cretaceous fish beds of Lebanon (Syria).[2] This early representative is particularly interesting because it has the outline of the soft body and of the ink sac preserved, and also because it has a vestigial saddle-shaped internal shell. As would be expected of soft-bodied organisms, however, fossil octopods are extremely uncommon. *Upper Cretaceous to Recent.*

Palaeoctopus (Fig. 10-84), previously cited, is a fossil representative, and *Octopus* (Fig. 10-83*E*) and *Argonauta* (Fig. 10-83*A–D*) are typical living genera.

Geologic History of Cephalopoda

The evolutionary history of the Cephalopoda has been intensively investigated because the shell contains a complete record of successive growth stages from the embryonic protoconch to the adult or senile conch. By starting at the aperture of external shells and breaking away the shell wall at successively older or earlier septa, the succession of growth stages passed through by the shell is revealed. As progress is made toward the apex, the septa, sutures, siphuncle, sculpture, and whorl shape either change from complexity toward simplicity or show subsequent modifications caused by secondary shell structures. In many shells each change corresponds to a stage not only in the shell itself but also in the developmental history of the race to which it belongs. Hence, it has been said that in such shells **ontogeny recapitulates phylogeny,** *i.e.*, the changes that take place in a single shell during its lifetime epitomize the changes in its race through the ages. This **biogenetic law,** as it has been called, is not universally applicable, even to cephalopod shells, and has been seriously challenged by many investigators, particularly zoologists. Cephalopod specialists recognize that the law has imperfections and exceptions but continue to find it valid for many specific and generic series.

Ontogeny. Not a great deal is known about the ontogeny of most cephalopods. Coleoids lay their large-yolked eggs either singly or in clusters on objects projecting above the sea bottom (*e.g.*, rocks and coral heads). The eggs hatch and develop directly into adults without passing through the

[1] A small group of somewhat peculiar living octopods, the Vampyromorpha (Pickford, 1946–1949) have, in addition to the eight regular arms, two retractile filaments, which have been interpreted as two additional arms. A further peculiarity is a well-developed shell remarkably like an uncalcified teuthoid gladius. These and other pecuilarities have led Pickford (1946) to propose that the Vampyromorpha be recognized as a separate order of Coleoidea equal in rank to the true Octopoda.

[2] ROGER, J., 1946, Les invertébrés des couches a poissons du Crétacé supérieur du Liban, *Mem. soc. géol. France*, vol. 23 (Mem. 51), 92 pp.

usual molluscan trochophore and veliger larval stages. A protoconch appears early and soon gives way to the initial chambers of the adult shell. *Argonauta* is unique among the coleoids in protecting the eggs by retaining them inside its paper-thin external shell (Fig. 10-83).

Although *Nautilus* has been much studied and many times discussed, almost nothing is known of its life history except the fact that the eggs are discharged to the exterior through the hyponome. The actual embryology of *Nautilus* does not yet seem to have been worked out. Nothing is definitely known of the developmental history of the soft animal in the extinct genera of Nautiloidea, in the Ammonoidea, and in the Belemnoidea, although certain inferences can be drawn from the succession of changes recorded in the shell, as discussed in a preceding paragraph.

The embryonic shell of cephalopods is termed the protoconch. It is supposedly horny or conchiolinitic in *Nautilus*, and is thought to have been of the same character in extinct nautiloids, but it seems generally to have been lost quite early, as it is seldom preserved. It is supposed to leave a scar in the posterior wall of the initial camera showing its place of attachment. Among the Ammonoidea and Coleoidea the protoconch is a large and conspicuous ovoidal or spheroidal calcareous chamber containing a prosiphon (Figs. 10-75, 10-81*A*). The initial camera of the postembryonic shell, or conch, has as its apical wall the first septum of the conch, and this septum completely closes the original aperture of the protoconch. To its convex apical surface is attached the prosiphon that extends across the protoconch, and in shells lacking a protoconch the cicatrix on the apical surface marks the place where the protoconch was attached to the septum. On its concave adoral side the first septum has a conspicuous tubular depression, the siphuncular caecum, which receives the terminus of the fleshy siphon. This caecum, as the term implies, is completely closed off and presumably never opened into the protoconch. The second septum of camerate shells generally has a conspicuous septal neck reaching back part or all of the way to the first septum, and in certain genera is somewhat different from later septa. All septa after the first or second are essentially alike, and they are added to the shell at the posterior of the living chamber with more or less constant spacing, except that in old shells they commonly become crowded near the living chamber.

Ecology and Paleoecology.[1] Less than 200 living species of the Cephalopoda remain out of a great assemblage that includes more than 10,000 fossil forms. Excepting the two or three species of *Nautilus*, all living cephalopods are coleoids. Included in the 10,000 fossil species are about 7,500 ammonoids, more than 2,500 nautiloids, and a few hundred coleoids. Living cephalopods are exclusively marine, and extinct forms, judging from the fossils with which they are commonly associated, seem also to have lived only in marine environments.

Living species of *Nautilus* have been reported from widely scattered areas in the Southwest Pacific (Melanesia, northern Australia, East Indian region,

[1] The reader interested in the ecology and paleoecology of the Cephalopoda is referred to recent summaries by Miller and Furnish (1937), Scott (1940), and Flower (1942).

and the Philippines) and seem to be limited to bottoms of less than 700 m. (2,300 ft.) in depth, though some investigators have questioned whether this depth was not too great. They come into shallow water at night but stay in deeper water during the day. The fossil record of the Nautiloidea indicates that as a group they were dwellers in relatively shallow water, probably limited largely to depths like those prevailing on existing continental shelves and in shallow gulfs and seas.

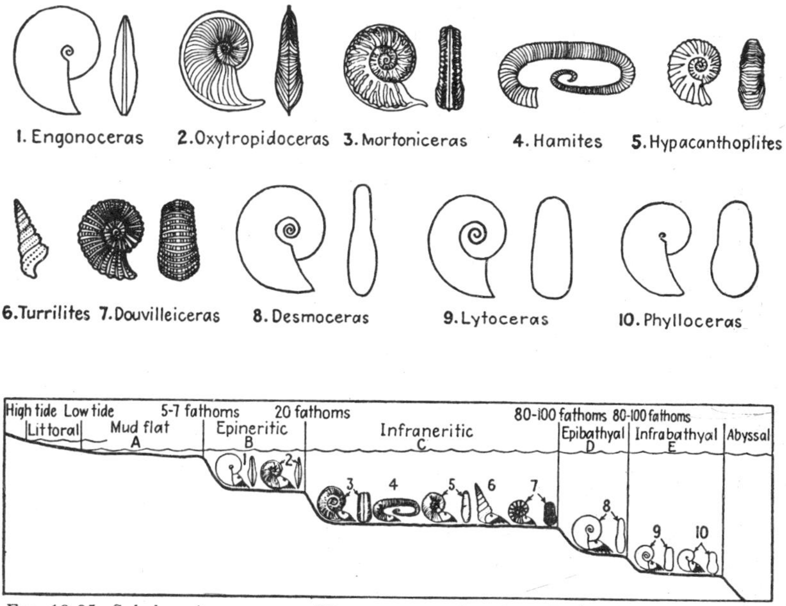

FIG. 10-85. Subclass AMMONOIDEA. Diagram suggesting the nektobenthonic habitat and bathymetric distribution of the principal types of Texas Cretaceous ammonoids, with enlarged sketches of the different genera. The diagram is not to scale, and the terraced arrangement is used merely to emphasize the groupings. There are also numerous gradations from one zone to another, but these are not shown. *Lytoceras* and *Phylloceras* are not found in Texas. (*Adapted and modified from Scott*, 1940.)

Although nothing is definitely known about the soft parts of the Ammonoidea, considerable knowledge of where and under what conditions these extinct cephalopods lived has been deduced from the study of the shells themselves and of the other fossils with which the shells are associated. As an example, Scott (1940) has shown how different genera of Texas Cretaceous ammonites seem to have been limited in bathymetric distribution, as shown diagrammatically in Fig. 10-85.

The larger squids are pelagic, but some smaller forms (*e.g.*, *Sepia* and *Loligo*) may live near shore. The ancient squids (Belemnoidea) seem also to have lived in both shallow and deep waters. Modern squids are gregarious and often swim in vast numbers in the open seas or shore waters where they attract whales and certain fish that feed upon them. The cod fishermen of the North Atlantic recognize this relationship by the laconic expression, "Plenty squeed, plenty cod." They use the common squid of the Gulf of St. Lawrence as bait, and their catch is likely to be proportional to the quantity of squid they can obtain. Storms occasionally cause extensive destruction of squids. In July, 1929, storm waves of only ordinary magnitude along the Labrador coast of Quebec washed squids ashore in Betchewan Bay in such large numbers that fishermen gathered them by boatloads. The great abundance of the cigar-shaped *Belemnites* in thin layers of Jurassic and Cretaceous rocks is evidence that ancient squids were also present in enormous numbers.

Squids and octopuses also provide food for man in some parts of the world. In the Mediterranean, in the China seas and around Japan, and in Polynesia and Melanesia they are caught in great numbers for this purpose. Indians dwelling along the Pacific coast in Canada and Alaska are reported to eat the devilfish on many occasions. A rather odd and unexpected use of fossil ammonoids is that practiced by certain Plains Indians of Wyoming and Montana, who live on a terrane of Cretaceous strata. The medicine men collect the beautifully preserved shells, many of which have most of the nacreous layer intact, and carry them in their bags as "medicine."

The ink sac and the use to which its contents are put are unique to the squids, which can, if necessary, produce inky black water in their wake by discharging the contents of the ink sac into the terminal part of the intestine, whence it enters the mantle cavity and from there is expelled to the exterior. Not only, then, did the cephalopods develop jet propulsion in using the hyponomic funnel for swimming, but they also developed the smoke screen.

Besides having developed the first smoke screen and the first organ for jet propulsion, cephalopods have also been credited with the discovery and use of a third principle, that illustrated by the submarine when it sinks or rises through a change in content of certain chambers. Crediting this discovery to the cephalopods presupposes, of course, that they could and can replace the gas of the camerae with water, or vice versa, and thereby sink or rise. However, it has not yet been demonstrated that living *Nautilus* can replace the gas of its shell chambers with water, and the same statement applies to extinct nautiloids as well as to the ammonoids. It is quite possible that many ancient cephalopods increased the specific gravity of the shell so that it was equal to that of water by partially filling the chambers with secondary calcareous matter. With such a shell the animal could then rise or sink by simply adjusting itself to the living chamber.

The octopus, or devilfish, and the squids have played a somewhat exaggerated rolê in popular literature, and stories are current describing the horrors of attack on small ships by giant squids and octopuses. One such story pictures a giant squid dragging a small ship beneath the waves and grasping a helpless sailor in its long arms. While their reputation for ferocity is not unmerited, it is rather unlikely that serious damage has ever been done to ships of any size by them. The giant squid of the north Atlantic, *Architeuthis princeps*, with a length exceeding 15 m. (52 ft.), might conceivably damage a small boat and make conditions uncomfortable for those on board, but no record and authenticated report of such an event is known to exist. It is altogether possible, however, that individuals swimming in marine waters might be attacked by octopuses.

The chambered *Nautilus* has been described in poetry by Oliver Wendell Holmes as follows:

This is the ship of pearl, which, poets feign,
Sails the unshadowed main,—
The venturous bark that flings
On the sweet summer wind its purpled wings
In gulfs enchanted, where the Siren sings,
And coral reefs lie bare,
Where the cold sea-maids rise to sun their streaming hair.

The same romantic picture of *Nautilus* rising to the surface of the sea, spreading out some part of its body, and sailing away across the waves appears in Pope's "Essay on Man," in the lines

Learn of the little Nautilus to sail,
Spread the thin oar, and catch the driving gale.

These pictures are only imaginary, for no one has ever seen a *Nautilus* perform in the fashion painted by the poets.

It is to be expected that in a class of animals as large and varied as the cephalopods. and over a period of time as long as 500,000,000 years, different members of the class would have become adapted to every marine environment during every period. All other dominant groups of animals whose history is fully known made such adaptations, and it does not seem unreasonable, therefore, to assume that during every period there were planktonic, nektonic, nektonic-planktonic, and benthonic cephalopods. Only one form of marine adaptation seems to have been omitted, that of sessile benthonic life, and even that may have been acquired by the peculiar *Nipponites* from the Upper Cretaceous of Japan. It is almost certain that the foot became adapted for swimming, thus augmenting or displacing the funnel in that function; that it was modified for crawling on the bottom; and finally that it became vestigial in some forms. Some cephalopods seem to have learned to swim with the shell held in a horizontal position. This is suggested by the calcareous fillings in the empty chambers of *Actinoceras*. Some appear also to have held the shell in an oblique position, and still others carried the shell with the aperture downward. Some forms seem to have dragged the shell along the bottom, while others carried the structure upon their backs. Shells with streamline contours (*Placenticeras*) were probably swimmers, whereas such clumsy forms as *Mandaloceras* and *Phragmoceras*, both of which had restricted apertures, were almost certainly nektonic-planktonic or entirely planktonic, more or less drifting with the shell in a vertical position and with the apex upward. The slender orthocones, such as *Actinoceras*, seem to have been swimmers; and *Endoceras*, with its large, partly filled siphuncle, may well have had a similar habit. It was probably easier for the latter to float than to drag its large shell over the bottom. Color bands preserved on some small cyrtoceracones (Fig. 10-62*B*) strongly suggest that the shell was carried with the apex pointed upward.

Stratigraphic Range. The oldest known cephalopod now generally accepted by specialists is the cyrtoconic nautiloid genus *Plectronoceras* from the Upper Cambrian of Manchuria (Flower and Kümmel, 1950).[1] The shell is an endogastric cyrtoceracone with short septal necks that have the lower margins curved strongly outward.

Many nautiloid genera appeared during the Ordovician, and by the close of the period most orders of Nautiloidea had been differentiated (Fig. 10-72). The majority of Ordovician cephalopods seem to have had straight or only slightly curved shells, and some of the straight shells had a length of 4.5 m. (15 ft.), a length never again attained by an invertebrate shell. Most large nautiloids were gone by Silurian time, but smaller forms continued to appear during the Silurian and Devonian, though by the Devonian the subclass was definitely on the decline and continued to decline to the end of the Paleozoic. Straight-shelled forms were extinct by the close of the Triassic, and only half a dozen coiled genera persisted to this time. Of these five, two remained in the Jurassic and one lived on into the Tertiary before dying out. The single living genus *Nautilus* appeared in the Jurassic.

The ammonoids seem to have made their appearance in the Ordovician in the goniatitic genus *Eobactrites*, but they did not diversify extensively until the Devonian, after which they increased in importance until they became the dominant subclass of cephalopods before the close of the Paleozoic. Shells with ceratite sutures first appeared in the Mississippian but did not become common until the Permian and Triassic. True ammonitic shells appeared in the Permian and rapidly gained dominance over the Triassic ceratites to become the overwhelmingly dominant group of cephalopods in the Jurassic and Cretaceous. Why this greatly diversified group of ammonoids suddenly vanished at the close of the Cretaceous is one of the great unanswered questions of organic evolution.

The Paleozoic goniatites are small and relatively smooth; ceratites are considerably larger, and some are highly sculptured; and earlier ammonites are smooth and rotund. Jurassic shells, however, tend to be flattened laterally, and some are highly sculptured. Extreme sculpture characterizes many Cretaceous shells, some of which reached gigantic proportions, as exemplified by *Pachydiscus seppenradensis* which has a maximum diameter of 2.55 m. (8 ft. 6 in.)

The earliest known Coleoidea appeared in the Late Mississippian in the genus *Eobelemnites*, but the subclass does not seem to have been well represented until the belemnoids appeared in great numbers in the Jurassic seas. With the extinction of this order in the Cretaceous, however, the coleoids began to decline, and though the sepioids, teuthoids, and octopods are now the dominant group of cephalopods, they have left only a meager fossil

[1] *Volborthella*, a peculiar septate orthocone from the Lower Cambrian of the Baltic region, was formerly considered to be the oldest known cephalopod, but specialists are now uncertain as to the exact affinities of these tiny shells.

record since their appearance in the Jurassic (sepioids and teuthoids) and Cretaceous (octopods).

Nature of Fossil Record of the Cephalopoda. The external shells of Cephalopoda, particularly those of the Ammonoidea, are excellent index fossils and are probably the most satisfactory of all fossils for zoning the marine beds of the Mesozoic. In identifying fossil cephalopods and discriminating genera, the following shell features have proved most useful:

1. Nature of septa and sutures (Nautiloidea and Ammonoidea).
2. Position and structure of siphuncle (Nautiloidea).
3. Siphuncular and cameral deposits (Nautiloidea).
4. Length of body chamber and nature of apertural modifications (Nautiloidea and Ammonoidea).
5. Nature of protoconch and first few chambers.
6. Shell morphology and surface sculpture (Ammonoidea).
7. Microstructure, composition, and interrelationships of the different parts of the shell (Coleoidea).

Kinship between genera and families usually cannot be demonstrated without taking into consideration several of the shell aspects listed above, and preservation commonly determines whether or not a given fossil can be generically identified.

Earlier Paleozoic cephalopods are usually preserved as fillings of the interior, as impressions, or as much replaced and permineralized specimens, and more often than not they are incomplete. Preservation of later Paleozoic forms is somewhat better, but the best preserved specimens come from Mesozoic strata. Exceptionally well preserved shells have the original shell matter essentially unaltered and intact, and from these much can be learned about the ontogeny by longitudinally sectioning the fossil or by breaking away successive camerae from the aperture backward to the initial chambers.

The shells of the Nautiloidea are rarely well enough preserved to have retained original color patterns or nacreous shell substance, but a few specimens have been reported in which patterns of color can be detected, and others in which the actual shell substance is essentially intact. By contrast, ammonoid shells are commonly exquisitely preserved and as a result have been intensively studied by specialists. Coleoid shells are abundant and may be well preserved in some rocks, but the fragile pro-ostracum is likely to be badly crushed or missing altogether. The best preserved specimens, as those from the Jurassic beds of Solenhofen and the Upper Cretaceous fish beds of Syria, are compressions of the whole organism in which soft as well as hard parts may be outlined and partly or entirely preserved (Fig. 10-84). On such specimens it may be possible to detect mandibles, hooks on the arms, and particularly the ink sac, commonly with its contents preserved as a finely divided carbonaceous material. This carbonized substance has been converted into an ink and used to label fossil specimens for museum display.

GEOLOGIC HISTORY OF THE MOLLUSCA

The geologic history of the Phylum Mollusca is long, complex and varied, and richly recorded by an abundance of fossils in the rocks of every geologic period beginning with the Cambrian. Some large divisions (*e.g.*, Protogastropoda, Paleoconcha, Ammonoidea, and Belemnoidea) have been extinct for a long time; others, once important, have only a few living representatives (*e.g.*, Nautiloidea); and still others (*e.g.*, Coleoidea) are the intriguing end members of many separate branches of molluscan evolution. Ascendant in present seas are the gastropods, but this has not always been so because the pelecypods and cephalopods have also had their day, as indicated in Fig. 10-86.

Evolution of the Phylum. The Phylum Mollusca is thought to have evolved from a Pre-Cambrian annelid-like ancestor. This belief is based on the facts that fossil mollusks are present in the earliest Cambrian rocks and that all living Mollusca except the Cephalopoda have a trochophore type of larva somewhat like that of the Annelida. Unique to the Mollusca is the peculiar veliger larva that follows the trochophore stage and develops into the adult mollusk. The Cephalopoda, although typical mollusks in most respects, have a heavily yolked egg that develops into the adult directly without either the trochophore or veliger larval stages.

The first shell seems to have been a hollow dorsally borne conchiolinitic hemisphere that ultimately became calcified. This primitive and archetypal shell became highly modified in the several major groups of mollusks as they evolved. In the aplacophorus Amphineura it is little more than a horny dorsal shield studded with spicules and spines, whereas in the polyplacophorus forms it has become a conspicuous dorsal shield of eight articulating calcareous plates. The shell of the Pelecypoda is bivalve and that of the Scaphopoda a hollow tapering cone open at both ends, but significantly in both these classes the young shell is at first a single caplike plate. In the Gastropoda and Cephalopoda the shell begins as a bulbous protoconch, which then grows forward as a straight or coiled cone that remains unilocular in the Gastropoda but becomes chambered in the Cephalopoda. A tendency to extend the mantle over the shell, and ultimately to surround it completely, in which case it becomes an internal shell, was developed in each of the three major molluscan classes.

The five classes of Mollusca must have diverged from the ancestral stock far back in the Pre-Cambrian, because the Pelecypoda, Gastropoda, and Cephalopoda were quite diversified by Late Cambrian time and do not seem to have been directly ancestral to either the Amphineura or Scaphopoda, which themselves must go back to some Early Cambrian or Pre-Cambrian ancestor, as suggested in Fig. 10-86.

Ecology. No other group of Invertebrata except the ubiquitous Arthropoda has ever successfully invaded the wide variety of habitats now occupied by mollusks. Two classes sent invaders into the fresh-water habitat, and the gastropods continued on to invade the land. In the sea, mollusks of one kind or another live at all depths and in nearly all latitudes; they swim, float, crawl over the bottom, attach themselves to objects on the bottom, or bore and burrow into bottom materials. The Amphineura lead a sluggish existence on the sea bottom, and the scaphopods and many pelecypods have become adapted to living partly or completely buried in mud and sand. By contrast gastropods and cephalopods have always maintained more freedom of movement, and living cephalopods, excepting *Nautilus*, are probably the best swimmers ever developed among the invertebrates.

Throughout geologic time the Mollusca have been dominantly marine, though brackish- and fresh-water pelecypods developed as early as Mid-Paleozoic. Neither brackish-water nor fresh-water molluscan faunas are ever

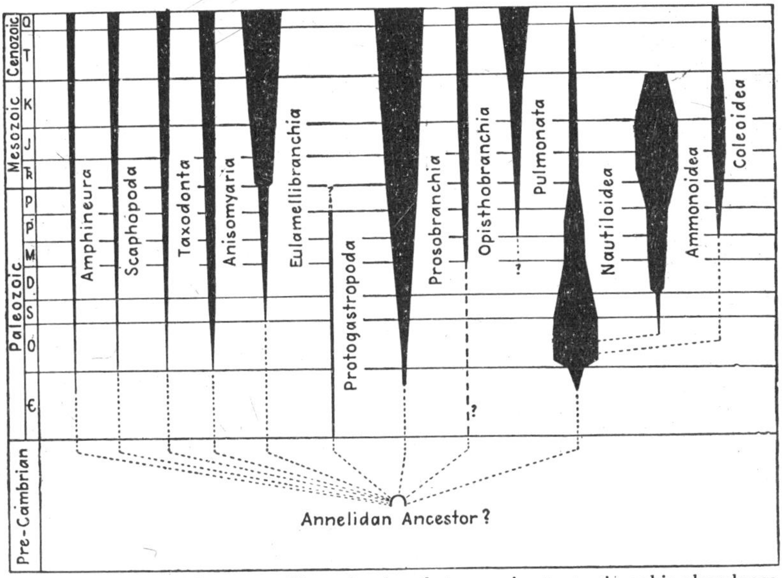

FIG. 10-86. Phylum MOLLUSCA. Chart showing the approximate stratigraphic abundance or importance of the major divisions of the Mollusca, based in part on the number of families listed in the latest edition of Zittel's "Textbook of Palaeontology" (Eastman, 1927). More detailed charts of this kind appear on earlier pages (Figs. 10-31, 10-72, 10-80). The actual time of beginning of each group is uncertain, but all are believed to go back to a primitive ancestral molluscan that developed from some sort of annelid worm sometime in the Pre-Cambrian.

large, and land-dwelling mollusks (*i.e.*, gastropods), as would be expected, have seldom been preserved. One of the most interesting fossil faunas of land-dwelling mollusks is the group of pulmonate gastropods found in the loess associated with Pleistocene glacial deposits (Baker, 1931). Land snails should also be expected in local pockets in ancient and recent deposits of residual soils derived from and lying on limestones (Shrock, 1948).

Some living species of land snails have been collected as far north as 73° 30′ N. lat. in Siberia, where the mean annual temperature is −10°F. and where the range is from 40°F. in July to −30°F. in January. Individuals may be frozen in ice and still live. Snails also live in the Algerian desert, where the temperature rises in excess of 120°F., and they have also been reported from spring waters as hot as 122°F. Certain snails and pelecypods have been taken from water at an elevation of 5,480 m. (18,000 ft.) in The Himalaya. The range in pressure tolerated by members of the phylum is from 6½ lb. per sq. in. at the highest altitudes to 9,000 lb. (4½ tons) per sq. in. at the greatest depths, a factorial difference of more than 1,400 times.

The life span of most mollusks is only a relatively few years. Pulmonate gastropods generally live less than a year, though individuals have reached an age of 7 years. Marine streptoneurous gastropods live for several years, but marine nudibranchs seldom live more than a year. Pelecypods average 3 or 4 years, but certain forms are known to have lived 20 years. Cephalopods, so far as known, have an average life span of 4 to 5 years, though it is quite possible that some of the gigantic nautiloids and ammonites of past geological ages lived much longer.

Stratigraphic Range. The stratigraphic range and relative abundance of the major divisions of Mollusca are shown on Fig. 10-86. Amphineura appear first as fossils in Upper Cambrian rocks, but in no rocks are fossil amphineurans of any stratigraphic importance.

Marine pelecypods are believed to have appeared first in Early Cambrian seas; fresh-water species seem to have been in existence as early as the Devonian; and brackish-water species left shells in Late Paleozoic strata. Throughout geologic time pelecypods have left an important fossil record, and during the Mesozoic the shells of *Ostrea*, *Exogyra*, *Inoceramus*, *Chama*, and the rudistids were so abundant on some bottoms as to form extensive shell beds or biostromes.

Marine gastropods made their appearance in the Early Cambrian, but fresh-water forms are not known fossil before the Jurassic, although they must have been in existence much earlier to have given rise to the land pulmonates that left fossils in Pennsylvanian rocks. Fossil gastropods are not especially abundant in Paleozoic rocks, and though much more numerous in Mesozoic and Cenozoic rocks, they were seldom important rock builders. On the bottoms of present tropical and subtropical seas, however, the tiny calcareous shells of the pteropods, which are specialized swimming opisthobranchs, are so numerous that locally they constitute as much as half of the bottom ooze. Scaphopods, though having a considerable geologic range, have little stratigraphic importance.

Among the Cephalopoda, the Nautiloidea dominate the Early and Middle Paleozoic; the Ammonoidea dominate the Late Paleozoic and reign supreme over all cephalopods throughout the entire Mesozoic; and the Coleoidea, with all the ammonoids gone and only a few nautiloids straggling on, naturally have a position of dominance in present seas.

Only locally did the Paleozoic cephalopods play any important part in rock building, but during the Mesozoic the ammonoids, with their large shells, contributed importantly to many calcareous deposits. The internal shells of *Belemnites* and other closely related belemnoids are locally abundant enough to constitute an important part of the rock, but more commonly they are sporadically scattered through the rock.

REFERENCES[1]

General

BAKER, F. C. 1920. The life of the Pleistocene or Glacial period. *Univ. Illinois, Contr.* 7, *Mus. Nat. Hist.*, 476 pp.

———. 1937. Pleistocene land and fresh water Mollusca as indicators of time and ecological conditions *in* G. G. MacCurdy's "Early Man," pp. 67–74. (Abstract) *Pan-Amer. Geol.*, vol. 67, pp. 317–319.

CLARK, A. 1934. Environment of marine Mollusca living off Long Beach, California, and its bearing on Pleistocene correlations (abstract). *Geol. Soc. Amer. Proc.*, 1933, p. 393.

CLARKE, F. W., and WHEELER, W. C. 1922. The inorganic constituents of marine invertebrates. *U.S. Geol. Surv., Prof. Paper* 124, 62 pp. (Data on the composition of many molluscan shells.)

COOKE, A. H. 1895. Molluscs *in* Cambridge Natural History, vol. 3, pp. 6–459. New York, Cambridge University Press.

COTTON, B. C., and GODFREY, F. K. 1938–1940. "The Molluscs of South Australia." pt. 1, 1938; pt. 2, 1940. Handbooks of the flora and fauna of South Australia, issued by the British Science Guild, South Australian branch. Adelaide.

DALL, W. H. 1889. A preliminary catalogue of the shell-bearing marine mollusks and brachiopods of the southeastern coast of the United States. *U.S. Nat. Mus., Bull.* 37, pp. 1–221.

———. 1890–1903. Tertiary fauna of Florida. *Wagner Free Inst. Sci., Philadelphia, Trans.* 3, pts. 1–6, 1654 pp.

———. 1921. Marine shell-bearing mollusks of the northwest coast of America. *U.S. Nat. Mus., Bull.* 112, pp. 1–217.

GARDNER, J. 1926–1929. The Molluscan fauna of the Alum Bluff group of Florida. *U.S Geol. Surv., Prof. Paper* 142, parts I–VIII.

GRANT, U. S., IV, and GALE, H. R. 1931. Catalogue of the marine Pliocene and Pleistocene Mollusca of California. *San Diego Soc. Nat. Hist., Mem.*, vol. 1, 1036 pp.

HENDERSON, J. 1935. Fossil non-marine Mollusca of North America. *Geol. Soc. Amer., Spec. Paper* 3, 313 pp.

HYATT, A. 1894. Phylogeny of an acquired characteristic. *Proc. Amer. Phil. Soc.*, vol. 32, pp. 349–647.

JOHNSON, C. W. 1934. List of marine mollusca of the Atlantic coast from Labrador to Texas. *Boston Soc. Nat. Hist., Proc.* 40, pp. 1–203.

Johnsonia (M.C.Z., Harvard). 1941 to date. Monographs of Western Atlantic mollusks. Harvard University, Cambridge, Mass.

[1] The literature of fossil and living Mollusca is enormous, and only a few of the larger and more comprehensive works are listed, along with numerous shorter works that are cited because of special paleontological interest. Extensive lists of European references to fossil Mollusca are given in the well-known editions of Zittel's "Text-book of Palaeontology," and similar lists of North American references are given in Shimer and Shrock's "Index Fossils of North America" (1944). The several major works on living Mollusca listed in the present work contain long lists of references that will lead the interested reader to the important literature of the many different groups.

KEEN, A. M. 1937. "An Abridged Check List and Bibliography of West North American Marine Mollusca." Stanford University, Calif., Stanford University Press, 87 pp.

——— and BENTSON, H. 1944. Check list of California Tertiary marine Mollusca, *Geol. Soc. Amer., Spec. Paper* 56, 280 pp.

OLDROYD, I. S. 1924–1927. "The Marine Shells of the West Coast of North America." Vol. I (Pelecypoda); vol. II (Scaphopoda and Gastropoda). Stanford University, Calif., Stanford Univ. Press.

PALMER, K. E. H. 1937. The Claibornian Scaphopoda, Gastropoda, and dibranchiate Cephalopoda of the southern United States. *Bull. Amer. Pal.*, vol. 7, no. 32, in 2 parts, 730 pp.

PELSENEER, P. 1906. Mollusca *in* Lankester's "A Treatise on Zoology," pt. 5. London, A. and C. Black.

PILSBRY, H. A. (see Tryon).

———. 1939–1948. The land Mollusca of North America. *Acad. Nat. Sci. Philadelphia, Mon.* 3, vol. 1 (1939–1940), 994 pp.; vol. 2 (1946–1948), 1113 pp.

RAYMOND, P. E. 1941. Invertebrate Paleontology *in* Fiftieth Anniversary Volume, Geological Society of America (Mollusca, pp. 83–86). An excellent review of progress in the study of fossil Mollusca during the past 50 years.

ROGER, J. 1946. Les invertébrés des couches à poissons du Crétacé supérieur du Liban. *Soc. géol. France, Mém.* 51, vol. 23, 92 pp.

ROGERS, J. E. 1908. "The Shell Book." New York, Doubleday & Company, Inc., 485 pp.

SIMROTH, H. 1892–1894. Mollusca *in* Bronn's "Klassen und Ordnungen des Tier-Reichs," vol. 3, pt. 1.

SWINNERTON, H. H. 1947. "Outlines of Palaeontology," 3d ed. London, Edward Arnold & Co., 393 pp.

THIELE, J. 1929–1935. "Handbuch der systematischen Weichtierkunde." Jena, Gustav Fischer. Pt. 1 (1929): Loricata [Amphineura], pp. 1–22; Gastropoda, (Prosobranchia), pp. 23–376. Pt. 2 (1931): Gastropoda (Opisthobranchia, Pulmonata), pp. 377–378. Pt. 3 (1934): Scaphopoda, pp. 779–782; Bivalvia [Pelecypoda], pp. 782–948; Cephalopoda, pp. 948–995; Addenda, pp. 995–1022. Pt. 4 (1935): Grundlagen des natürlichen Systems der Mollusken, etc., pp. 1023–1154.

TROSCHEL, F. H. 1856–1893. "Das Gebiss der Schnecken, etc." 2 vols. Berlin.

TRYON, G. W. (continued by H. A. Pilsbry). 1879–1935. "Manual of Conchology" 1st ser., vols. 1–17, 1878–1898; 2d ser., vols. 1–28, 1885–1935.

YONGE, C. M. 1949. "The Sea Shore." London, Collins, 311 pp.

ZITTEL, K. A. VON. 1900. "Text-book of Palaeontology," translated by C. R. Eastman, vol. 1. New York, The Macmillan Company; London, Macmillan & Co., Ltd. 839 pp. (Mollusca, pp. 344–604.)

Amphineura—Scaphopoda

HENDERSON, J. B. 1920. A monograph of the East American scaphopod mollusks. *U.S. Nat. Mus., Bull.* 111, pp. 1–177.

MILLER, A. K. 1949. A giant scaphopod from the Pennsylvanian of Texas. *Jour. Paleontology*, vol. 23, pp. 387–391.

Pelecypoda

BERRY, C. T. 1936. A Miocene pearl. *Amer. Midland. Naturalist*, vol. 17, pp. 464–470.

DIX, E., and TRUEMAN, A. E. 1937. The value of non-marine lamellibranchs for the correlation of the Upper Carboniferous. *IIième Congres pour l'advancement des études de stratigraphie Carbonifère, Heerlen*, 1935, *Compte Rendu* (W. Jongmans), vol. 1, pp. 185–201.

DOUVILLÉ, H. 1912. Classification des Lamellibranches *Bull. soc. géol. France.* 4th ser., vol. 12, pp. 419–467.

HESSLAND, I. 1945. On the Quaternary *Mya* period in Europe. *Arkiv. Zool.*, vol. 37A, no. 8, 51 pp.

JACKSON, R. T. 1890. Phylogeny of the Pelecypoda. *Mem. Boston Soc. Nat. Hist.*, vol. 4, no. 8, pp. 277–400.

KEEN, A. M., and FRIZZELL, D. L. 1939. "Illustrated Key to West North America Pelecypod Genera." Stanford University, Calif., Stanford University Press, 28 pp.

MACNEIL, F. S. 1937. The systematic position of the pelecypod genus *Trinacria*. *Jour. Wash. Acad. Sci.*, vol. 27, pp. 452–458.

NEUMAYR, M. 1884. Zur Morphologie des Bivalvenschlosses. *Sitz. k. Akad. Wiss. Wien (Math.-nat. Kl)*, vol. 88, pp. 385–418. (Also see Beiträge zur einer morphologischen Eintheilung der Bivalven, *Denkschr. k. Akad. Wiss. Wien.*, vol. 58, pp. 701–801.)

NEWELL, N. D. 1937. Late Paleozoic pelecypods: Pectinacea. *State Geol. Surv. Kansas*, vol. 10, pp. 1–123.

NICOL, D. 1950. Origin of the pelecypod family Glycymeridae. *Jour. Paleontology*, vol. 24, pp. 89–98.

POPENOE, W. P., and FINDLAY, W. A. 1934. Transposed hinge-structures in Lamellibranchs (abstract). *Geol. Soc. Amer. Proc.*, 1933, p. 387.

REINHART, P. W. 1935. Classification of the pelecypod family Arcidae. *Mus. roy. d'hist. nat. Belgique, Bull.*, vol. 11, no. 13, 68 pp.

———. 1943. Mesozoic and Cenozoic Arcidae from the Pacific slope of North America. *Geol. Soc. Amer., Spec. Paper* 47, 117 pp.

RIDEWOOD, W. G. 1903. On the structure of the gills of the Lamellibranchia. *Phil. Trans. Roy. Soc. London*, vol. 195, pp. 147–284.

SCHENCK, H. G. 1934. Literature on the shell structure of pelecypods. *Mus. roy. d'hist. nat Belgique, Bull.*, vol. 10, 20 pp.

——— and KEEN, A. M. 1936. Bathymetric Distribution of Marine Pelecypoda (abstract). *Geol. Soc. Amer. Proc.*, 1935, p. 367.

STEPHENSON, L. W. 1933. The zone of *Exogyra cancellata* traced twenty-five hundred miles. *Bull. Amer. Assoc. Petrol. Geol.*, vol. 17, pp. 1351–1361.

WRIGLEY, A. 1946. Observations on the structure of lamellibranch shells. *Proc. Malacol. Soc. London*, vol. 27, pp. 7–19.

Gastropoda

ASHFORD, C. 1883–1885. The darts of British Helicidae. *Jour. Conch.*, vol. 4, pp. 69–270, with interruptions.

BAKER, F. C. 1931. Pulmonate Mollusca peculiar to the Pleistocene period, particularly the loess deposits. *Jour. Paleontology*, vol. 5, pp. 270–292.

BARTSCH, P. 1946. The operculate land mollusks of the family Annulariidae of the Island of Hispaniola and the Bahaman Archipelago. *U.S. Nat. Mus., Bull.* 192, pp. 1–264.

COLLINS, R. L. 1934. A monograph of the American Tertiary pteropod mollusks. *Johns Hopkins Univ., Stud. Geol.* 11, pp. 137–324.

DURHAM, J. W. 1947. Bathymetric distribution of gastropod genera (abstract). *Bull. Geol. Soc. Amer.*, vol. 58, p. 1260.

EALES, N. B. 1950. Torsion in Gastropoda. *Proc. Malacol. Soc. London*, vol. 28, pp. 53–61.

FENTON, C. L., and FENTON, M. A. 1931. Apparent gastropod trails in the Lower Cambrian (Ross Lake, British Columbia). *Amer. Midland Naturalist*, vol. 12, pp. 401–405.

——— and ———. 1931*a*. Some snail borings of Paleozoic age. *Ibid.*, vol. 12, pp. 522–528.

——— and ———. 1937. *Archaeonassa*, Cambrian snail trails and burrows. *Ibid.*, vol. 18, pp. 454–456.

GRABAU, A. W. 1902–1910. Studies of Gastropoda: I. *Amer. Nat.* vol. 36: pp. 917–945, 1902. II. *Fulgur* and *Sycotypus*. *Ibid.*, vol. 37, pp. 515–516, 539, 1903. III. On orthogenetic variation in Gastropoda. *Ibid.*, vol. 41, pp. 607–646, 1907. IV. Value of the protoconch and early conch stages in the classification of Gastropoda. *Int. Zool. Cong.* VII, Boston, 1907, Proc., pp. 753–766, 1910.

———. 1904. Phlogeny of *Fusus* and its allies. *Smith. Misc. Coll.*, vol. 44, pp. 1–157.

KNIGHT, J. B. 1930–1934. The gastropods of the St. Louis, Mo., Pennsylvanian outlier. *Jour. Paleontology*, vols. 4–8.

———. 1939. Review of Wenz's Gastropoda *in* Schindewolf's "Handbuch der Palaeozoologie." *Ibid.*, vol. 13, pp. 230–232.

———. 1941. Paleozoic gastropod genotypes. *Geol. Soc. Amer., Spec. Paper* 32, 510 pp.

———. 1944. Paleozoic Gastropoda *in* Shimer and Shrock's "Index Fossils of North America," pp. 437–479. New York, John Wiley & Sons, Inc.; Cambridge, Mass., The Technology Press.

ODER, C. R. L. 1932. Fossil opercula from the Knox dolomite. *Amer. Midland Naturalist*, vol. 13, pp. 133–152.

SHROCK, R. R. 1952. Fossils in laterite and bauxite (abstract). *Bull. Geol. Soc. Amer.*, vol. 58, pp. 1227–1228.

SMITH, B. 1945–1946. Observations on gastropod protoconchs. *Pal. Americana*, vol. 3, nos. 19 and 21.

WENZ, W. 1938–1941. Gastropoda in Schindewolf's "Handbuch der Palaeozoologie." Berlin, Gebrüder Borntraeger. Pt. 1 (1938): General, pp. 1–84; Prosobranchia, pp. 85–240. Pt. 2 (1938): Prosobranchia, pp 241–480. Pt. 3 (1939): Prosobranchia, pp. 481–720. Pt. 5 (1941): Prosobranchia, pp. 721–1200. (Reviewed by Knight in *Jour. Paleontology*, vol. 13, pp. 230–232; 1939.)

Cephalopoda

ARKELL, W. J. 1950. A classification of the Jurassic ammonites. *Jour. Paleontology*, vol. 24, pp. 354–364.

BRUNN, A. F. 1943. The biology of *Spirula spirula* (L). *Carlsberg Found. Oceanog. Exped.*, 1928–1930, Dana-Rept. 24, 46 pp.

CLARKE, J. M. 1893. The protoconch of *Orthoceras*. *Amer. Geol.*, vol. 12, pp. 112–115.

———. 1897. The Lower Silurian [Ordovician] Cephalopoda of Minnesota. *Minn. Geol. Surv., Final Rept.*, vol. 3, pt. 2, pp. 761–812.

DEAN, B. 1901. Notes on living *Nautilus*. *Amer. Nat.*, vol. 35, pp. 819–837.

ELIAS, M. K. 1938. Studies of the late Paleozoic ammonoids: 1, Methods of drawing sutures, bibliography. *Jour. Paleontology*, vol. 12, pp. 86–90.

FISCHER, A. G. 1951. A new belemnoid from the Triassic of Nevada. *Amer. Jour. Sci.*, vol. 249, pp. 385–393.

FISCHER, A. G., and FINLEY, R., JR. 1949. Microstructure of some Pennsylvanian nautiloids (abstract). *Bull. Geol. Soc. Amer.*, vol. 60, p. 1887.

FLOWER, R. H. 1940. The apical end of *Actinoceras*. *Jour. Paleontology*, vol. 14, pp. 436–442.

———. 1942. Environment of early Paleozoic nautiloids. *Nat. Res. Council, Comm. Marine Ecology as Related to Paleontology*, 1941–1942, pp. 37–40.

———. 1943. Studies of Paleozoic Nautiloidea (pts. 1–7). *Bull. Amer. Pal.*, vol. 28, no. 109, 140 pp.

———. 1945. A belemnite from a Mississippian boulder of the Caney shale. *Jour. Paleontology*, vol. 19, pp. 490–503.

———. 1946. Ordovician cephalopods of the Cincinnati region. *Bull. Amer. Pal.*, vol. 29, no. 116, 657 pp.

——— and KÜMMEL, B. JR., 1950. A classification of Nautiloidea. *Jour. Paleontology*, vol. 24, pp. 604–616.

FURNISH, W. M., and UNKLESBAY, A. G. 1940. Diagrammatic representation of ammonoid sutures. *Jour. Paleontology*, vol. 14, pp. 598–602.

GRIFFIN, L. E. 1900. The anatomy of *Nautilis pompilius*. *Nat. Acad. Sci.*, vol. 8, mem. 5, pp. 103–232.

HYATT, A. 1900. Cephalopoda *in* Zittel-Eastman's "Textbook of Palaeontology," vol. 1, pp. 502–604. New York, The Macmillan Company; London, Macmillan & Co., Ltd.

IREDALE, T. 1944. Australian pearly *Nautilus*. *Australian Zool.*, vol. 10, pp. 294–298.

KERR, J. G. 1931. Notes upon the Dana specimens of *Spirula* and upon certain problems of cephalopod morphology. *Danish "Dana" Exped.* 1920–1922, Rept. 8, pp. 1–34.

KOBAYASHI, T. 1935. On the phylogeny of the primitive nautiloids, with descriptions of *Plectronoceras liatolungense*, etc. *Japanese Jour. Geol. Geogr.*, vol. 12, pp. 17–26.

KÜMMEL, B. 1952. A classification of the Triassic ammonoids. *Jour Paleontology*, vol. 26, pp. 847–853.

MILLER, A. K. 1938. Devonian ammonoids of America. *Geol. Soc. Amer.*, *Spec. Paper* 14, 262 pp.

———. 1943. Cambro-Ordovician cephalopods. *Biol. Revs.*, vol. 18, pp. 98–104.

———. 1947. Tertiary nautiloids of the Americas. *Geol. Soc. Amer.*, *Mem.* 23, 234 pp.

———. 1949. American Permian nautiloids. *Ibid.*, *Mem.* 41, 218 pp.

——— and Furnish, W. M. 1937. Paleoecology of the Paleozoic cephalopods. *Nat. Res. Council*, *Rept. Comm. Paleoecology*, 1936–1937, pp. 54–63.

——— and UNKELSBAY, A. G. 1943. The siphuncle of Late Paleozoic ammonoids. *Jour. Paleontology*, vol. 17, pp. 1–25.

——— and YOUNGQUIST, W. 1946. A giant ammonite from the Cretaceous of Montana. *Ibid.*, vol. 20, pp. 479–484.

NAEF, A. 1921–1928. Die Cephalopodon. *Flora e Fauna del Golfo di Napoli*, *Mon.* 35. Pt. 1, vol. 1, 863 pp. (1923); Pt. 1, vol. 2, 357 pp. (1928).

———. 1922. "Die fossilen Tintenfische." Jena, Gustav Fischer, 322 pp.

PICKFORD, G. E. 1946–1949. *Vampyroteuthis infernalis* Chun, an archaic dibranchiate cephalopod. *Carlsberg Found. Oceanog. Exped.* 1928–1930, Dana-Repts. 29 (1946) and 32 (1949).

PLUMMER, F., and SCOTT, G. 1937. Upper Paleozoic ammonites in Texas. *Univ. Texas Bull.* 3701, 516 pp.

ROGER, J. 1944. *Acanthoteuthis* (*Belemnoteuthis*) *syrica*, n. sp., cephalopode dibranche du Crétacé supérieur de Syrie. *Bull. soc. géol. France*, 5th ser., vol. 14, fasc. 1–3, pp. 3–10.

———. 1947. Découverte d'une coquille de *Sepia* . . . dans la Vindobonien supérieur de Sanbrigues . . . et histoire paléontologique des Sepiidae. *Bull. soc. géol. France*, 5th ser., vol. 17, pp. 225–232.

RUEDEMANN, R. 1905. The structure of some primitive cephalopods. *N.Y. State Mus.*, *Bull.* 80, pp. 296–341.

———. 1921. Observation on the mode of life of primitive cephalopods. *Bull. Geol. Soc. Amer.*, vol. 32, pp. 315–320.

RUEPPELL, E. 1829. "Abbildung und Beschreibung eininger neuen oder wenig gekannten Versteinerungen aus der Kalkschieferformation von Solenholfen." Frankfurt am Main, H. L. Brönner.

SCHEVILL, W. E. 1950. An Upper Jurassic sepioid from Cuba. *Jour. Paleontology*, vol. 24, pp. 99–101.

SCOTT, G. 1940. Paleoecological factors controlling the distribution and mode of life of Cretaceous ammonoids in the Texas area. *Jour. Paleontology*, vol. 14, pp. 1164–1203.

STENZEL, H. B. 1948. Ecology of living nautiloids. *Nat. Res. Council*, *Rept. Comm. Treatise Marine Ecology and Paleoecology*, 1947–1948, pp. 84–90.

———. 1948*a*. Paleoecology of Tertiary nautiloids. *Ibid.*, pp. 96–97.

TEICHERT, C. 1933. Der Bau der actinoceroiden Cephalopoden. *Palaeontographica*, vol. 78, Abt. B.

———. 1935. Structures and phylogeny of actinoceroid cephalopods. *Amer. Jour. Sci.*, vol. 29, pp. 1–23.

ULRICH, E. O., FOERSTE, A. F., MILLER, A. K., *et al.* 1942–1944. Ozarkian and Canadian cephalopods Pt. I. Nautilicones: *Geol. Soc. Amer., Spec. Paper* 37 (1942), 157 pp. Pt. II. Brevicones: *Ibid., Spec. Paper* 49 (1943), 240 pp. Pt. III. Longicones and Summary: *Ibid., Spec. Paper* 58 (1944), 226 pp.

VERRILL, A. E. 1882. Report on the cephalopods of the northeastern coast of America. *Ann. Rept. Comm. Fish and Fisheries* 1879, pp 1–244.

WOODWARD, H. 1885. On some Palaeozoic phyllopod-shields, etc. *Geol. Mag.*, dec. 3, vol. 2, pp. 345–352.

WRIGHT, C. W. 1952. A classification of the Cretaceous ammonites. *Jour. Paleontology*, vol 26, pp. 213–222.

CHAPTER 11

PHYLUM ANNELIDA (SEGMENTED WORMS)

The strongly segmented earthworms, sandworms, and leeches, which compose the Phylum Annelida, stand in striking contrast to the many groups of unsegmented worms discussed in Chap. 6. The typical annelid most nearly fits the popular conception of a worm, although some members of the phylum differ considerably from that conception. The body is fundamentally an elongated, cylindrical or flattened, and segmented tube with a mouth at one end, an anus at the other, tiny legs, or **parapodia,** on the ventral side, and a smooth dorsal surface (Fig. 11-1). The ringlike segments, or **somites,** of the bilaterally symmetrical annelid body are essentially similar, and the segmentation is generally evident both externally and internally.

The annelid, besides its characteristic segmentation, has (1) a body cavity, or **coelom,** surrounding the complete digestive tract; (2) a single preoral segment; (3) a muscular body wall with an external circular layer and an internal longitudinal layer; (4) a central nervous system with a pair of preoral ganglia, ventral cords extending the length of the body, and dendritic branches of the cord in each segment; (5) a closed circulatory system of main and lateral branches with hemoglobin-bearing blood; (6) nephridia and associated ducts; and (7) a larva, when developed, of the trochophore type (Fig. 11-2). In addition to the foregoing characteristics, which are common to all annelids, some representatives of the phylum have a definite but nonchitinous cuticle, and chitinous bristles, or **setae** (**chaetae**), that lie embedded in the ectoderm (Figs. 11-8, 11-9). Certain of the typical annelid features are lacking, or greatly suppressed, in representatives of the phylum that have adapted themselves to special conditions. Reproduction is varied; in some annelids the sexes are united and development is direct (earthworms and leeches), in others the sexes are separate and there is a trochophore larval stage (polychaetes and archiannelids), and in a few earthworms and polychaetes, reproduction is asexual and by budding.

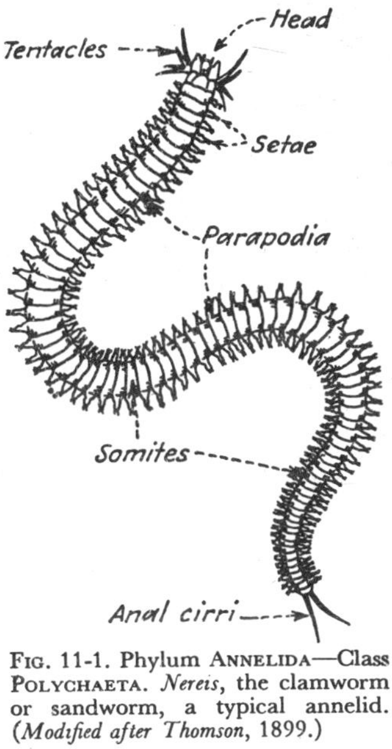

FIG. 11-1. Phylum ANNELIDA—Class POLYCHAETA. *Nereis,* the clamworm or sandworm, a typical annelid. (*Modified after Thomson,* 1899.)

The phylum may be divided into the following classes:

Class Archiannelida. Small, marine annelids with primitive characteristics and a trochophore larva; believed to have been derived from the Polychaeta by a process of simplification. No fossil record.

Class Polychaeta (sandworms and tube worms). Strongly segmented, predominantly marine worms with numerous somites and lateral parapodia that bear bristles (Fig. 11-1). The fossil record, which is considerable for worms, includes representatives from rocks of many ages, the oldest at least Cambrian.

Class Oligochaeta (earthworms). Strongly segmented annelids without a head or parapodia, and with few setae (Figs. 11-17, 11-18). These dwell chiefly in fresh water and moist soil. The fossil record begins with Silurian representatives.

Class Hirudinea (leeches). Body with prominent posterior sucker; without tentacles, parapodia, or setae; 34 somites; dwell in fresh or salt water and on land. No certain fossil record is known.

Class Gephyrea. Body sausage-shaped, without segmentation or parapodia; anterior end with retractile proboscis or crown of tentacles; marine dwellers, burrowing in sand and mud, chiefly in shallow waters. The fossil record includes representatives from rocks as old as Middle Cambrian.

Class Archiannelida

This class of annelids is composed of a few tens of species of small marine worms of simplified character included in five or six genera. Members of the simplest genera are more like a larval polychaete than a full-grown worm, and for this reason many zoologists prefer to include all archiannelids in the Polychaeta. Inasmuch, however, as the less simple genera show characteristics setting them apart from the Polychaeta, it seems better to recognize the Archiannelida as a distinct class, whose members, as they evolved from a polychaete-like ancestor, became simplified by retaining certain youthful characteristics and losing certain distinctive features, such as setae and parapodia.

Polygordius, whose trochophore larva is shown in Fig. 11-2, is a typical archiannelid. No fossil Archiannelida seem yet to have been reported, though Walcott 40 years ago wrote that he would not be surprised if they were represented in the well-known Burgess fauna.

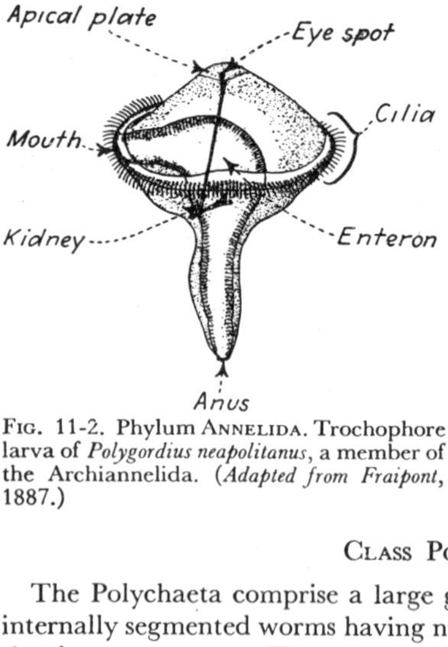

FIG. 11-2. Phylum ANNELIDA. Trochophore larva of *Polygordius neapolitanus*, a member of the Archiannelida. (*Adapted from Fraipont*, 1887.)

Class Polychaeta

The Polychaeta comprise a large group (3,500 species) of externally and internally segmented worms having numerous somites with lateral parapodia that bear many setae. There is a head region with tentacles, and the anterior end of the digestive tract is generally equipped with tiny chitinous hooks or teeth. There is a trochophore larval stage in many species, but a few species reproduce asexually by budding. Members of the class are predominantly marine and are adapted to a wide range of environments. Some are pelagic, others are free-living on the bottom, still others live in burrows, and certain species build free or attached tubes. Living and fossil forms with these habits are discussed in a following paragraph. The fossil record of the Polychaeta consists of tiny jaws (scolecodonts), burrows, tubes, and supposed trails.

The following three orders of Polychaeta, two existing and one extinct, are generally recognized:

Order 1. Errantia.
Order 2. Sedentaria.
Order 3. Miskoa.

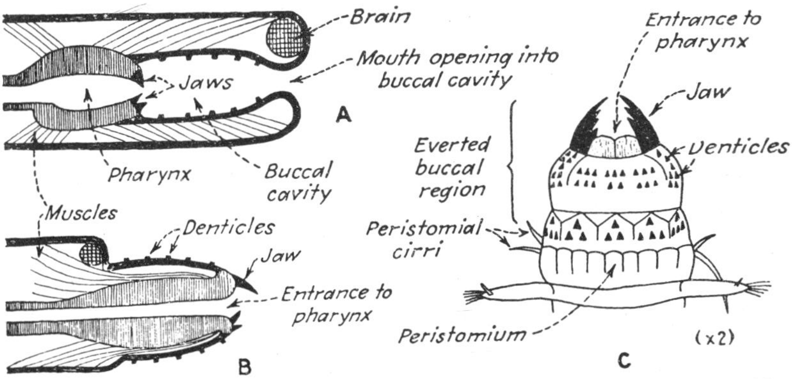

FIG. 11-3. Class POLYCHAETA. Diagrams illustrating the jaw- and denticle-bearing eversible buccal apparatus characteristic of the Polychaeta. *A* shows apparatus at rest; in *B* the pharynx has been thrust forward and the buccal cavity turned inside out, so that the jaws now lie some distance in front of the head, which is represented by the brain, and the denticle-bearing surface of the buccal cavity is now on the exterior. *C*. Head of *Nereis*, with buccal region everted, as seen in ventral view. (*A–B modified from Benham after Lang*, 1910; *C modified from Benham*, 1910.)

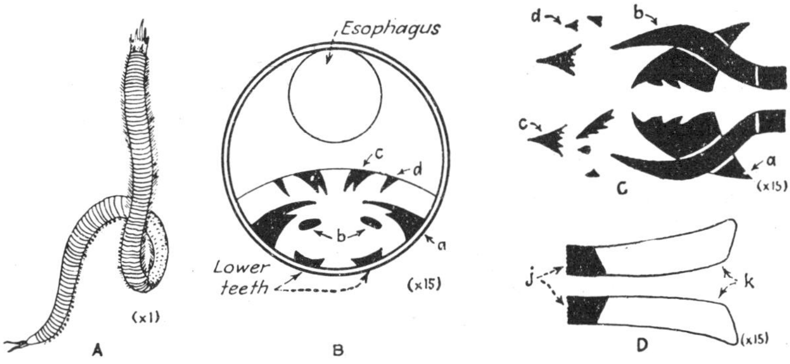

FIG. 11-4. Class POLYCHAETA—Genus *Eunice*. *A*. A complete individual. *B*. Transverse section of anterior end of head portion, showing the several different parts of the masticatory mechanism in a special chamber below the esophagus, much enlarged. *C*. Upper series of denticles. *D*. Lower pair of denticles. (*a* = grinders; *b* = forceps; *c* = rasping plates; *d* = grater; *j* = tooth; *k* = base into which muscles are inserted.) (*A after Ehlers*, 1887; *B–D modified after Benham*, 1910.)

Order 1. Errantia. Free-moving, predaceous polychaetes with well-marked head having an eversible buccal structure armed with powerful chitinous jaws (Fig. 11-3), and well-developed parapodia bearing setae that are used for locomotion. The fossil record, consisting of carbonaceous films, chitinous jaws and denticles, excrements or castings, burrows, tubes, and supposed trails, reaches back as far as the Early Cambrian, and possibly into the Pre-Cambrian. *Nereis* (Figs. 11-1, 11-3), the sandworm; *Arenicola* (Fig. 11-12), the lugworm (lobworm); and *Eunice*, a worm with an excellently developed set of jaws (Fig. 11-4), are typical living representatives. *Nereites*, *Arenicolites*, and *Eunicites* are typical fossil genera (Figs. 11-11, 11-12).

Order 2. Sedentaria. Predominantly sedentary polychaetes with indistinctly separated head, short and generally not eversible buccal structure without jaws, and short parapodia that are not used for swimming. Typical Sedentaria inhabit more or less firm tubes, which they secrete or construct of extraneous particles, and live chiefly on phytoplankton. The fossil record, which begins with the Ordovician, consists of tubes of many kinds.

Serpula and *Spirorbis* (Fig. 11-15) are well-known and widespread genera whose members secrete calcareous tubes. Both are also common Paleozoic fossils, the earliest fossil Sedentaria being an Ordovician *Spirorbis*.

Order 3. Miskoa. This is an extinct, Early Paleozoic group of supposed marine polychaetous worms with similar somites and parapodia throughout the length of the body, retractile proboscis, straight digestive tract, and no distinct specialization of somites.

This order is based upon a group of remarkably preserved annelids from the Middle Cambrian Burgess shale of British Columbia, found and described by Walcott early in the present century. *Miskoia* (Fig. 11-5*C*), *Canadia* (Fig. 11-5*B*), and *Wiwaxia* (Fig. 11-10*C–D*) are typical of the three families.

Certain Ordovician and Silurian genera (*Protoscolex*,[1] *Eotrophonia*, and *Bertiella*) have been assigned to this order by some paleontologists, but there is too much present uncertainty about the affinities of these to use them as a basis for extending the range of the order to the Late Silurian. As a matter of fact, they more probably belong to other polychaete orders.

Living and Fossil Polychaeta. Although living polychaetes are divided into free-swimming—or, more correctly, free-moving—forms (*Errantia*) and sedentary forms (*Sedentaria*), this classification is misleading because only a comparatively few species are free-swimming throughout life. By far the majority live in some sort of tube or burrow. Furthermore, certain errant species secrete a tube or make one of mud, whereas certain sedentary species carry their tube with them, leave it and swim about freely, or build no tube at all. It seems more practicable, therefore, to consider the class as a whole rather than to discuss the two existing orders separately.

A typical member of the Polychaeta is the common sandworm, *Nereis* (Fig. 11-1), which reaches a length of 150 mm. (6 in.) It is composed of a

[1] E. O. Ulrich (Observations on fossil Annelids, and descriptions of some new forms, *Jour. Cin. Soc. Nat. Hist.*, vol. 1, pp. 87–91, 1878–1879) erected *Protoscolex* and *Eotrophonia* to include tiny segmented worms preserved in the shales of the Upper Ordovician at Cincinnati, Ohio.

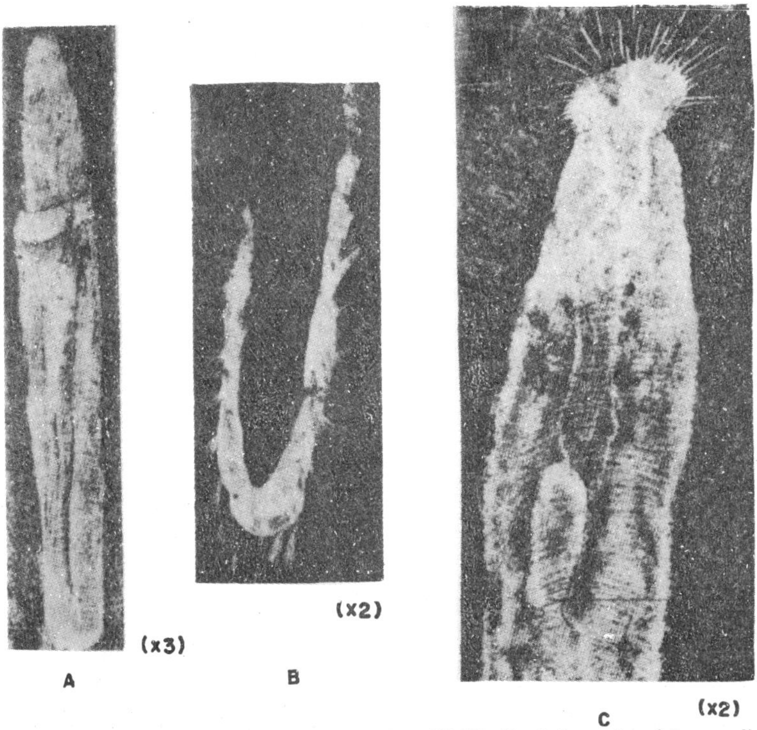

FIG. 11-5. Phylum ANNELIDA. Miskoa, an extinct Middle Cambrian order of the annelid Class Polychaeta. *A. Selkirkia major*, a flattened tube with the anterior portion of the worm projecting from it. *B. Canadia setigera*, a slender flattened specimen showing bundles of setae and one of the tentacles of the head at the extreme left end. *C. Miskoia preciosa*, anterior end of a specimen 9 cm. (3.6 in.) in length in which proboscis seems to have been withdrawn. Circumoral setae, annular lines, segments, and part of the digestive tract (*e*) are shown. (*After Walcott*, 1911.)

considerable number of segments which are essentially similar throughout the length of the body but modified at the anterior end to make a head and at the posterior end to make a tail. A typical body segment has on each side a muscular parapodium that bears bundles of bristles (setae or chaetae) (Figs. 11-8, 11-9) and filamentous sensory organs known as **cirri.** The setae of each bundle project from a large sack and each seta arises from a single cell at the bottom of the sack (Fig. 11-9). The parapodium is bilobed and each lobe has a strong needlelike internal bristle, the **aciculum,** to which are attached the muscles that operate the whole bundle of setae. The chitinous acicula, therefore, serve as an internal skeleton to the parapodium (Fig. 11-8).

The **head** consists of modified segments bearing eyes, sensory tentacles, and palps. Below these structures is the anterior end of the digestive tract consisting of the **buccal** region or mouth, which is lined with tiny chitinous

denticles, and the **pharynx,** which carries a set of powerful chitinous jaws. The buccal region can be everted or retracted at will, so that when the worm grasps food it can bring the powerful jaws into direct contact with the object and, having seized it, can then draw it toward the esophagus (Fig. 11-3). The jaws and denticles, being chitinous and offering considerable resistance to many common solvents, should be capable of preservation under favorable conditions, and certain fossil "teeth" in ancient rocks have been interpreted as polychaete jaws (see later discussion of **scolecodonts;** Figs. 11-6, 11-7).

Nereis is common in shallow water along the New England coast and elsewhere around the Atlantic and is known as the sandworm because of its habit of burrowing in the sand and mud between tide levels. It secretes a mucilaginous substance that cements the sand grains together, thus enabling the burrow to remain open. The animal is predatory and also eats some of the organic debris of the bottom. Excremental material is ejected into the burrow below the animal and gradually fills the burrow. The filling thus made, being slightly different from the surrounding sand or mud in composition and texture, commonly stands out on a weathered surface, especially by its lighter color.

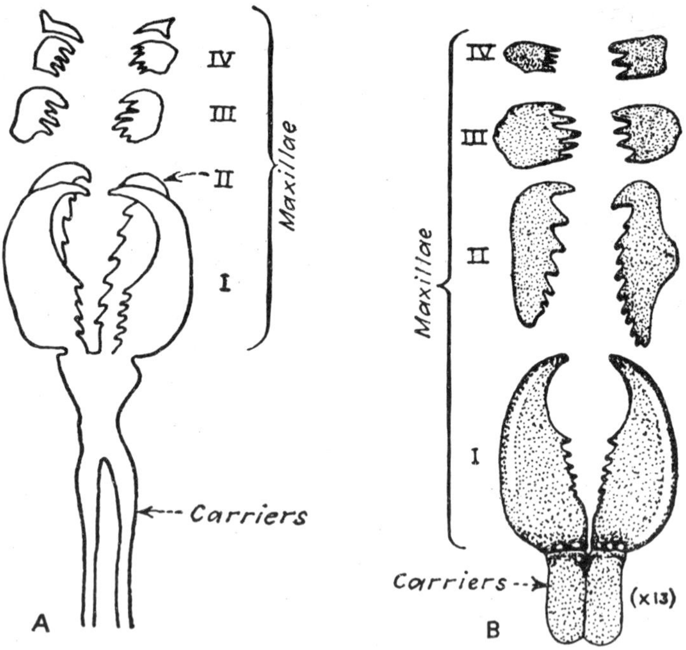

FIG. 11-6. Class POLYCHAETA. Scolecodonts. *A*. The jaw apparatus of a living worm, *Arabella setosa*, showing morphology and arrangement of the maxillae. *B*. Schematic reconstruction of the jaw apparatus of a Devonian genus, *Arabellites*, based on isolated scolecodonts. (*After Eller*, 1934.)

Polychaeta are typically free-moving, but comparatively rarely free-swimming, predaceous and mud-loving worms with a well-developed head, a long body of many similar segments, and a small tail. The head is provided with an eversible buccal structure that generally has a variety of tiny denticles and larger powerful jaws, all chitinous (Fig. 11-3). Each typical body segment has a pair of bilobed parapodia that bear bundles of setae and have a pair of internal bristles to which the setae-moving muscles are attached (Fig. 11-9).

Polychaetes are largely restricted to marine waters, though a few species live in fresh- and brackish-water environments. They are generally found between tide levels and in shallow waters of less than 40 m. (120 ft.) depth, but they live at all depths and some have been dredged from depths of more than 5,500 m. (18,000 ft., or 3,000 fathoms). In general, they are most abundant where the bottom sediments consist of sand admixed with considerable mud containing decaying organic matter. These sediments not only provide food for the worm but also have a consistency that enables the burrows to retain form and remain open. The chitinous denticles and jaws, and possibly the setae and acicula as well, are capable of preservation if conditions are favorable.

Scolecodonts. The jaws and denticles that are present in the mouth of many polychaetes are of great interest to paleontologists because fossils of similar nature are present in sedimentary rocks of all ages from Ordovician

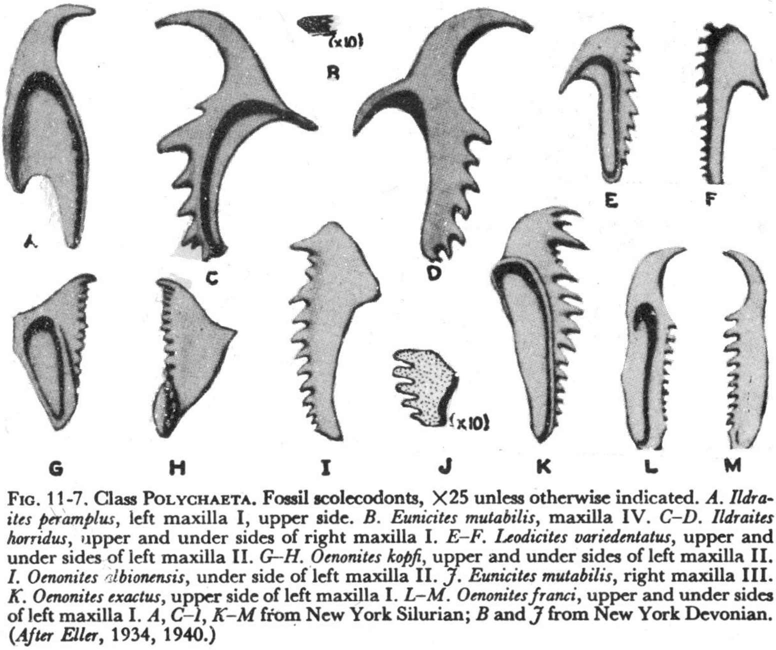

FIG. 11-7. Class POLYCHAETA. Fossil scolecodonts, ×25 unless otherwise indicated. *A. Ildraites peramplus*, left maxilla I, upper side. *B. Eunicites mutabilis*, maxilla IV. *C–D. Ildraites horridus*, upper and under sides of right maxilla I. *E–F. Leodicites variedentatus*, upper and under sides of left maxilla II. *G–H. Oenonites kopfi*, upper and under sides of left maxilla II. *I. Oenonites albionensis*, under side of left maxilla II. *J. Eunicites mutabilis*, right maxilla III. *K. Oenonites exactus*, upper side of left maxilla I. *L–M. Oenonites franci*, upper and under sides of left maxilla I. *A*, *C–I*, *K–M* from New York Silurian; *B* and *J* from New York Devonian. (*After Eller*, 1934, 1940.)

to Recent. They are now commonly called **scolecodonts**[1] (Gr. *skolex*, worm, + *odous*, tooth) to distinguish them from superficially similar but quite different toothlike objects known as **conodonts**, with which they have been confused. The latter are discussed in Chap. 16.

Scolecodonts show considerable range in size, shape, structure, and number, in both living and fossil species, and they present a troublesome problem to the paleontologist in one important respect. Unless he has a complete assemblage of the different parts of a jaw system, he can err by assigning to different genera or species isolated scolecodonts that actually belong to a single species or individual.[2]

In living polychaetes there may be a single stabbing tooth or a circle of such denticles (Fig. 11-4), but more commonly large grasping and tearing jaws are present, as in *Nereis* (Fig. 11-3) and *Arabella* (Fig. 11-6). *Eunice* carries its numerous denticles and jaws in a special pouch below the digestive tract (Fig. 11-4), and certain Paleozoic scolecodonts, assigned to the extinct genus *Eunicites* (Fig. 11-7*B*,*J*), may well have been part of a similar assemblage.

Scolecodonts are common in many shales and impure limestones, considered to be near-shore deposits, and are known to increase in abundance around ancient Silurian bioherms just as living polychaetes are common in the vicinity of Pacific coral reefs. From these relations Eller[3] concludes that the Silurian seas in which the scolecodont-bearing dolomitic limestones were deposited were warm, shallow, well aerated, and of relatively low salinity—an interesting example of how fossils may furnish information about ancient oceanographic conditions.

Scolecodonts may be expected as black microscopic or near-microscopic objects in shales and impure limestones, from which they can be removed by digesting the rock in appropriate solvents. They are composed of about 50 per cent volatile matter—presumably largely carbon, as typical scolecodonts are coal black—and more than 45 per cent silica.[4] Generally, they lie isolated in the rock or on a bedding surface, commonly in great numbers, and about a dozen complete assemblages of mouth parts have been found (Lange, 1947–1949). These latter are of great importance because they show what parts belong together in a single individual. Typical fossil scolecodonts are shown in Fig. 11-7.

Setae (Chaetae) and Acicula. The bristles which are so characteristic of the polychaetous annelids, and to which the terms setae and chaetae have

[1] CRONEIS, C., and SCOTT, H. W., 1933, Scolecodonts (abstract), *Bull. Geol. Soc. Amer.*, vol. 44, p. 207.

[2] Croneis (1941) has suggested an alternate method of classifying isolated scolecodonts. It is a military classification, hence nonbiological, with numerous categories that can be considered roughly equivalent to individual, species, etc.

[3] ELLER, E. R., 1944, Scolecodonts of the Silurian Manitoulin dolomite of New York and Ontario, *Amer. Midland Naturalist*, vol. 32, p. 749. For additional comments on the paleoecology of ancient scolecodont-bearing worms, see the same author, pp. 4–5 in *Nat. Res. Council, Rept Comm. Marine Ecology as Related to Paleontology*, 1941–1942.

[4] CRONEIS, C., Jan. 8, 1935, personal communication.

been applied, are of three types: (1) simple (Fig. 11-8); (2) jointed; and (3) **uncini** (Fig. 11-9). Each seta, which is a solid chitinous bristle (excepting those of *Euphrosyne* that are hollow and calcareous), is the product of a single formative cell that lies in a cyst at the base of the **setae sack** (Fig. 11-9*A*). Although setae are used mainly in locomotion, it is possible that some of the stronger serrated types may also be used in offensive or defensive actions.

Isolated fossil setae seem to have been reported only from Upper Ordovician shales,[1] but clusters of setae are beautifully preserved in the numerous polychaetous worms of the Middle Cambrian Burgess shale and of the Jurassic Solenhofen lithographic limestones of Bavaria (Figs. 11-5, 11-11).

Although no isolated fossil acicula seem to have been recognized, they are preserved essentially in place in the Jurassic *Eunicites* (Fig. 11-11).

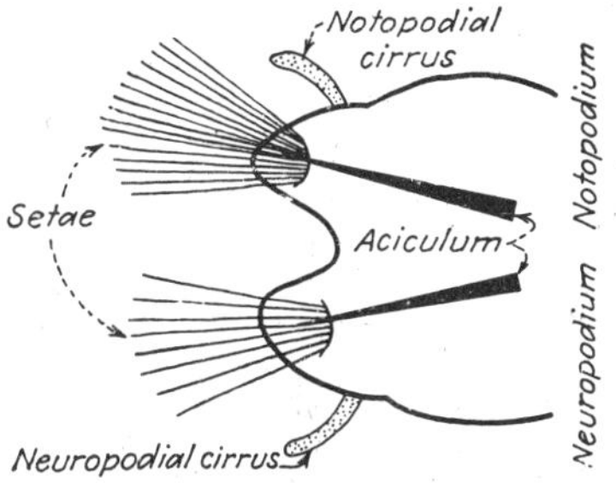

FIG. 11-8. Phylum ANNELIDA—Class POLYCHAETA. Generalized diagram of a double-lobed polychaete parapodium showing relation of acicula to bundles of bristles. Some parapodia have two bundles on each lobe. Not all parapodia have the two sensory cirri, and some have them greatly modified from the simple form shown.

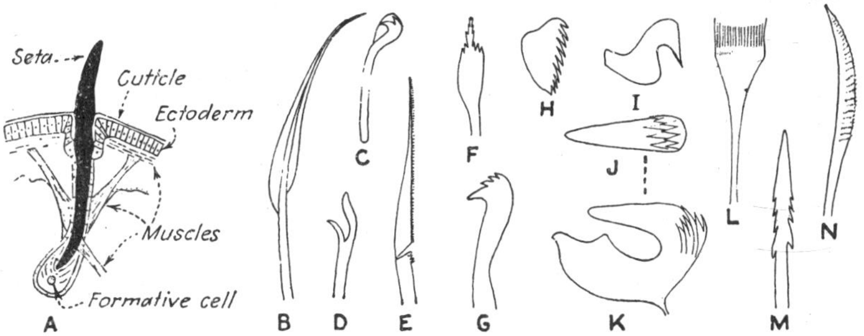

FIG. 11-9. Phylum ANNELIDA. Setae of annelid worms. *A*. Diagram of one seta and the cell that made it in the body wall of *Lumbricus*, the common earthworm. *B–N*. Setae from several different polychaete worms. All are greatly magnified, some more than others. The forked seta *D* is hollow and calcareous; all others are solid and chitinous, which is the usual condition. (*A modified after Borradaile and Potts*, 1935; *B–D modified after Benham*, 1910.)

Scales. One small family of Polychaeta, the Aphroditidae, are unique in having flattened dorsal cirri in the form of horny scales carried on alternate segments of the body (Fig. 11-10*A–B*). The well-preserved scales and spines of the Middle Cambrian genus *Wiwaxia* (Fig. 11-10*C–D*) are probably of similar nature.

Impressions and Carbonaceous Films. Under unusually favorable conditions worms can be buried intact and persist long enough to leave in the enclosing sediment impressions of their surfaces and carbonaceous films

[1] ULRICH, *op. cit.*

representing the residues of their body substances. Preservations of this sort among the Polychaeta have been discovered in the Middle Cambrian Burgess shale (Walcott), in a Mazon Creek concretion from the Illinois Pennsylvanian (Croneis), and in the famous Jurassic lithographic limestone of Bavaria. The latter two cases include specimens that have jaw parts and acicula preserved (Fig. 11-11).

Burrows and Borings. Most polychaetes live a large part of life, and some all of it, in tubes or burrows. Some burrow through soft sediments, but leave no tubes that may be preserved. Others excavate a permanent burrow in which they spend most of the time. A well-known example is *Arenicola marina*, the common lugworm (lobworm) (Fig. 11-12*B–C*). It burrows to a depth of 0.45 to 0.60 m. (18 to 24 in.) and throws up a considerable quantity of sediment as **castings** or **sand ropes,** which are a common sight on muddy and sandy bottoms (Davison, 1891). The burrow is U-shaped or L-shaped (Wells, 1945), and the castings are piled about one end of the burrow, the other being represented by a nearby hole (Fig. 11-12*C*). Fossil burrows and castings, supposedly of the same origin and hence given a somewhat similar name (*Arenicolites*, Fig. 11-12*D*), have been found in rocks as old as Devonian. Supposed burrows known collectively as *Scolithus* (Fig.

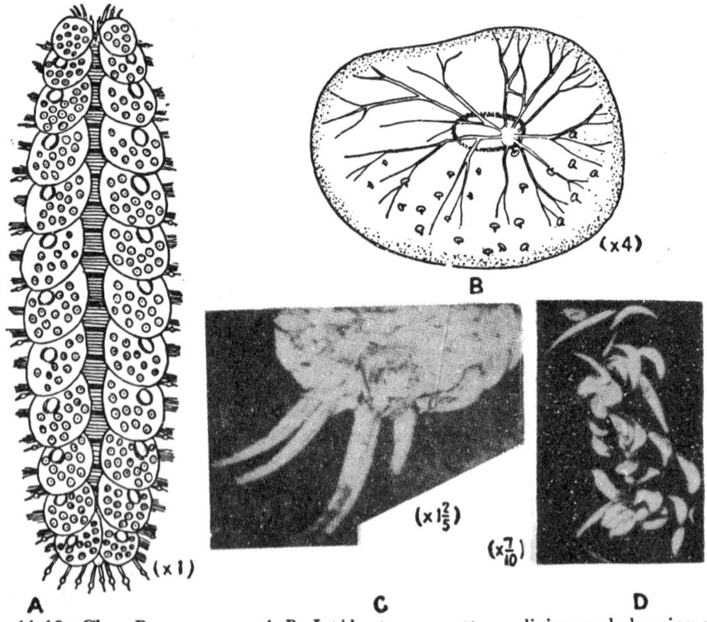

FIG. 11-10. Class POLYCHAETA. *A–B. Lepidonotus gymnonotus*, a living scale-bearing polychaete: *A*, dorsal view of a complete specimen; *B*, a single scale or elytron, which is somewhat horny in texture. *C–D. Wiwaxia corrugata*, a scale-bearing and spinose polychaete from the Middle Cambrian Burgess shale of British Columbia: *C*, portion of the dorsal surface with numerous scales and five of the dorsal spines; *D*, scattered scales and spines on surface of shale. (*A–B after M'Intosh*, 1885; *C–D after Walcott*, 1911.)

11-12*E*) have long been considered of worm origin, but recently it has been suggested that they were made by phoronids.[1] They are especially abundant in Upper Cambrian sandstones but are common throughout Paleozoic rocks, and if they do represent the burrows of worms, then burrowing worms were present in the earliest Cambrian seas.[2]

More or less vertical tubes, like those of *Scolithus* but differing in having funnel-shaped apertures, have been reported from rocks as old as Lower Cambrian and are believed to have been made by ancient burrowing annelids.[3]

Most ancient of all supposed worm burrows are those from the Pre-Cambrian (Algonkian) Greyson shale of Montana, to which the name *Planolites corrugatus* (Fig. 11-12*A*) has been applied. These fossils are supposed to be fillings of a burrow made by a wormlike creature; whether or not that is what they are, will probably remain an intriguing but unanswerable question. Recently described and unnamed "fossil burrows" from the Middle Huronian Ajibik quartzite of Michigan,[4] may possibly have been made by worms, inasmuch as

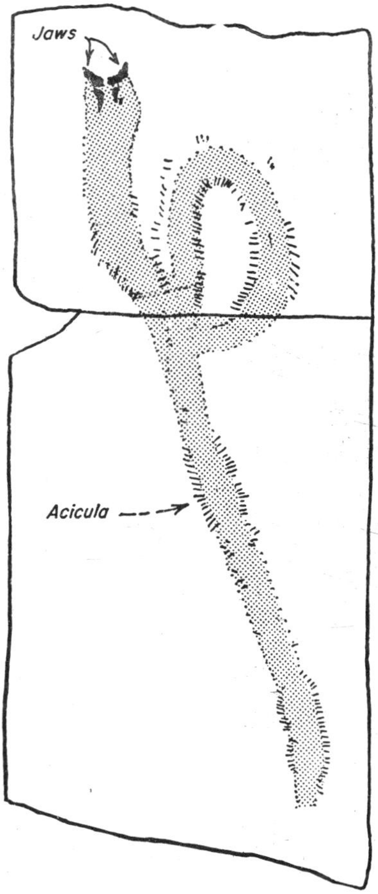

FIG. 11-11. Phylum ANNELIDA. A fossil polychaete, *Eunicites avitus*, from the Jurassic lithographic limestone of Solenhofen, Bavaria. The jaws are seen at the anterior end and the acicula along each side. The shape of the body is clearly outlined, though indistinctly in some parts. This unusual fossil represents preservation of hard parts along with an impression of the soft body. (*Sketch based on figure by Ehlers*, 1868.)

[1] FENTON, C. L., and FENTON, M. A., 1934, *Scolithus* as a fossil phoronid, *Pan-Amer. Geol.*, vol. 61, pp. 341–348.

[2] The puzzling *Arthrophycus* and *Daedalus* from the Lower Silurian (Medinan) have been interpreted as probable worm burrows (SARLE, C. J., 1906, The burrow origin of *Arthrophycus* and *Daedalus* [*Vexillum*], *Rochester Acad. Sci., Proc.* 4, pp. 203–210), but alternative interpretations that seem more acceptable have also been suggested.

[3] HOWELL, B. F., 1946, Silurian *Monocraterion clintonense* burrows showing the aperture, *Bull. Wag. Free Inst. Sci.*, vol. 21, p. 33.

[4] FAUL, H., 1949, Fossil burrows from the Pre-Cambrian Ajibik quartzite of Michigan, *Nature*, vol. 164, p. 32.

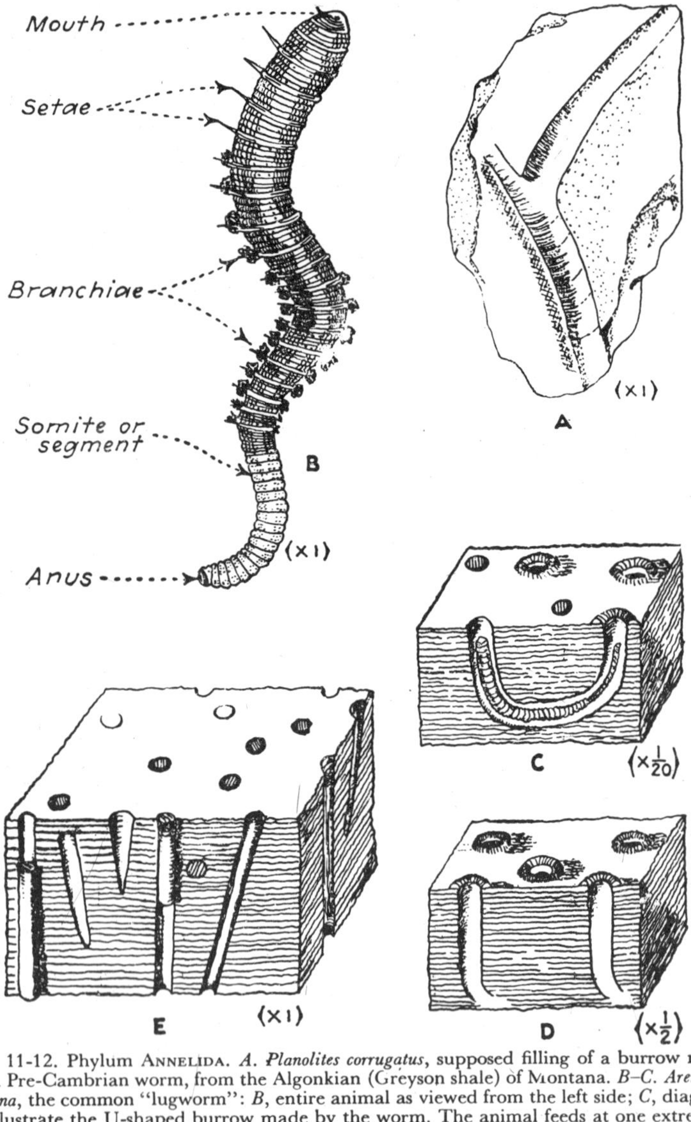

FIG. 11-12. Phylum ANNELIDA. *A. Planolites corrugatus*, supposed filling of a burrow made by a Pre-Cambrian worm, from the Algonkian (Greyson shale) of Montana. *B–C. Arenicola marina*, the common "lugworm": *B*, entire animal as viewed from the left side; *C*, diagram to illustrate the U-shaped burrow made by the worm. The animal feeds at one extremity and casts off excreta at the other. These burrows may extend downward into sand for several feet. *D. Arenicolites chemungensis*, a burrow from the Devonian (Chemung) of New York, which may have been made by a worm similar to the living *Arenicola*. *E. Scolithus*, the common Early Paleozoic burrow usually ascribed to some type of worm. The hollow or filled tubular burrows are usually vertical or slightly inclined. These have also been interpreted as the burrows of phoronids. (*A after Walcott*, 1898; *B after Thomson*, 1895; *D after Whitfield*, 1904.)

they are described as "uniform, sinuous ropes of sand following the uneven and ripple-marked bedding surfaces" for as much as 0.6 m. (2 ft.) (Fig. 11-16*A–B*).

A few polychaetes have the power to bore holes and even galleries in hard rocks and shells.[1] *Polydora*, for example, makes a U-shaped burrow in chalk, limestone, shells, and even shale, and *Leucodore* can excavate burrows in almost any kind of rock. They seem to bore mechanically, possibly by means of their strong setae, rather than by employing chemical substances, as has so commonly been supposed. No fossil worm borings of the type just described seem to have been reported. They probably would be difficult to distinguish from similar holes made by a variety of other animals.

Castings (Coprolites). Many burrowing worms, as the common marine lugworm, *Arenicola marina* (Fig. 11-12*B–C*), and the well-known earthworm, *Lumbricus terrestris* (Fig. 11-17), swallow enormous quantities of mud, sand, and soil as they burrow. This material, losing its organic content and being otherwise affected chemically and also physically as it passes through the digestive tract of the worm, is ultimately cast off in the form of pellets or cylindrical and ropelike masses. These **castings** are a common sight on muddy and sandy bottoms between tide levels and on the surface of the ground. They are usually closely associated with the opening of the burrow from which they were expelled. Fossil castings are generally called **coprolites.**

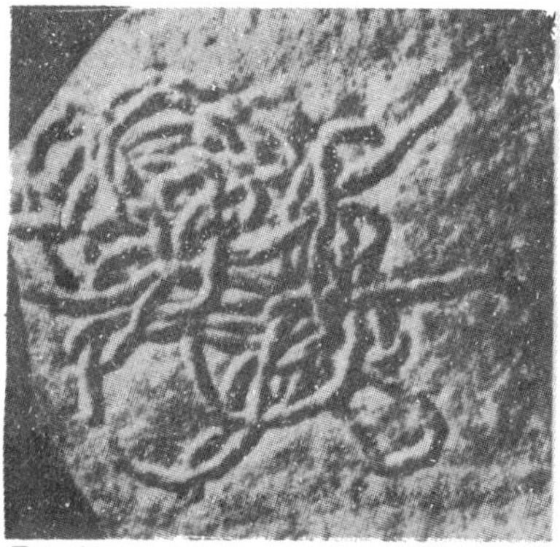

FIG. 11-13. *Lumbricaria*, natural size, from the Solenhofen lithographic limestone (Jurassic) of Bavaria. This fossil is considered a worm casting by some paleontologists, a holothuroid casting by others. The mass of tangled ropelike strands lies on the upper surface of the layer.

Modern sea-bottom sediments contain great numbers of **excremental** or **fecal pellets,** of which some may be of worm origin. Somewhat similar objects in ancient rocks may have had the same origin. Castings and trails of supposed annelids have been reported from Lower Cambrian rocks of South Australia.[2] Castings are associated with U- and L-shaped tubes in certain New York Devonian beds (Fig. 11-12*D*). *Lumbricaria* (Fig. 11-13), the well-known ropelike fossil from the Jurassic limestone of Bavaria, usually has been considered a worm casting, though it has also been interpreted as the casting of an ancient holothuroid, an interpretation for which there is some

[1] M'INTOSH, W. C. 1868. On the boring of certain annelids. *Ann. Mag. Nat. Hist.*, ser. 4, vol. 2, pp. 276–295.

[2] MADIGAN, C. T., 1926, Organic remains below the Archaeocyathinae limestone at Myponga Jetty, South Australia, *Trans. Roy. Soc. S. Australia*, vol. 1, pp. 31–35.

basis,[1] and is an example of a large group of similar fossils which can be assigned a worm origin only with uncertainty.[2]

Tubes. Polychaete worms build a great variety of tubes, some so soft and perishable as to vanish along with the flesh when the animal dies but others so hard and durable as to form rock masses and even reefs. Naturalists of 50 to 100 years ago extensively studied the tube-forming annelids, and their observations need little modification today.[3] At least three different groups of worms of this type can be recognized, if the nature and composition of the tube be used as a basis of division.

One group includes several forms that secrete a fluid substance that may remain soft and gelatinous or harden to such an extent that the tube is firm and tough enough to be fashioned into pens for writing. Some of the organic substances of these secreted tubes are highly complex compounds whose composition is hidden in such terms as **onuphin, spirographin, mucin,** and **conchiolin.**

Supposed fossil tubes of the sort just described present an interesting and intriguing collection of objects. Walcott (1911) found "chitinous or parchment-like" tubes with part of the worm protruded from the open end in the Middle Cambrian Burgess shale, and named the worm *Selkirkia major* (Fig. 11-5*A*). With it he also illustrated an excellently preserved specimen of *Hyolithes carinatus*, which genus he suggested should be removed from the Pteropoda and included in the Annelida. Howell (1949) believes his new Ordovician genus *Tubulelloides* may have had an organic tube of some sort, but there is a question about its being a worm at all.

A second worm group, consisting of a large number of species, is characterized by **agglutinated tubes.** The individuals of this group use the secretion of their surface glands as a mortar to cement together a wide variety of bottom materials. The agglutinated tubes thus formed consist of sand and gravel particles, sponge spicules, shells and shell fragments, and similar materials. On bottoms that are muddy this substance is used exclusively, being strengthened by an internal coating of secreted matter, or is mixed with coarser particles of the type just mentioned.

Agglutinated worm tubes show considerable variety in form and structure. Some are simple cones of mud (*Eupista*, Fig. 11-14*C–D*) or of sand grains, etc. (*Amphictene*, Fig. 11-14*G–K*); others are cylindrical tubes of many-sized particles commonly coated with shells (*Nothria*, Fig. 11-14*A,E*);

[1] Fenton, C. L., and Fenton, M. A., 1934, *Lumbricaria:* a holothuroid casting? *Pan-Amer. Geol.*, vol. 61, pp. 291–292.

[2] In the first edition of the present work certain peculiar strings of spheroids and cylindrical rods appearing in the Mississippian Salem limestone, the well-known Indiana building stone, were interpreted as "probable castings of large worms" (p. 138 and Fig. 43*A*). These may be of some other origin, possibly algal, as no scolecodonts seem to be associated with them.

[3] Excellent descriptions will be found in the following references: Dalyell, J. G., 1853, The powers of the Creator displayed in the creation, vol. 2, London. Etheridge, R., Jr., 1880, A contribution to the study of the British Carboniferous tubicular Annelida, *Geol. Mag.* vol. 7, pp. 109–369 (with interruptions). M'Intosh, W. C., 1894, On certain homes or tubes formed by Annelids, *Ann. Mag. Nat. Hist.*, ser. 6, vol. 13, pp. 1–18.

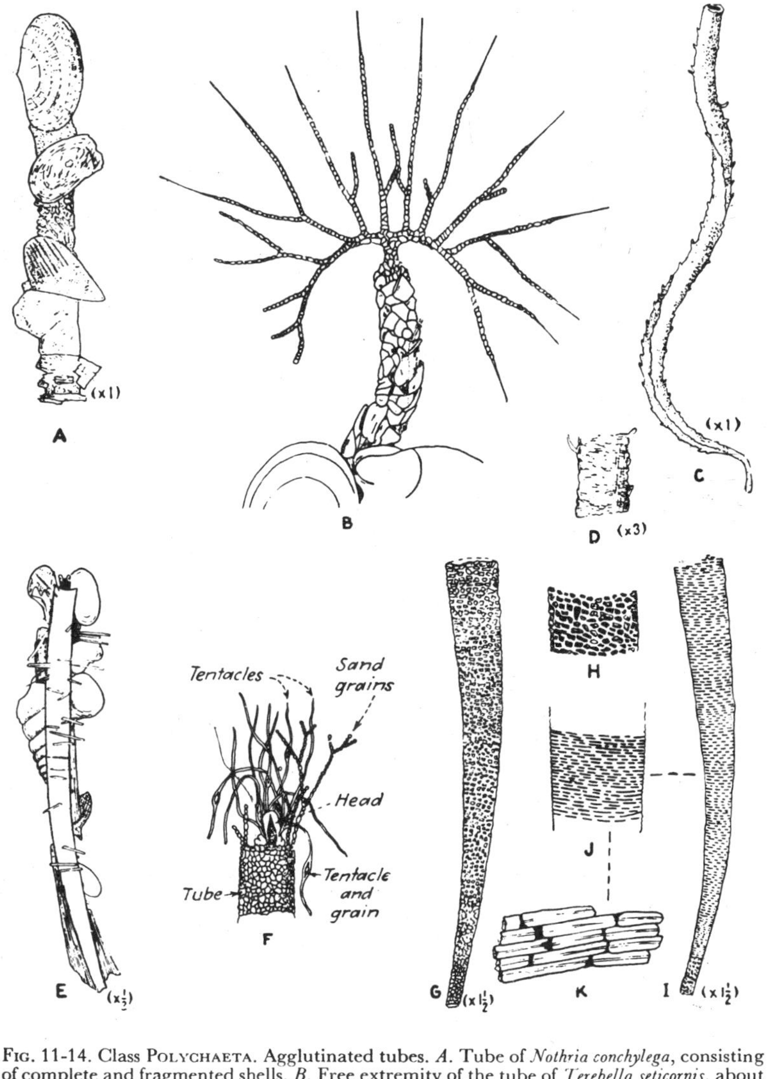

FIG. 11-14. Class POLYCHAETA. Agglutinated tubes. *A.* Tube of *Nothria conchylega*, consisting of complete and fragmented shells. *B.* Free extremity of the tube of *Terebella seticornis*, about natural size. *C–D.* Mud tube of *Eupista grubei: C*, a complete tube with longitudinal and spiral spinous ridges; *D*, a small part of tube enlarged to show scalelike arrangement of mud particles. *E.* Tube of *Nothria abranchiata*, showing shells and shell fragments adhering to main tube. *F.* Upper end of tube of *Terebella conchilega*, showing head of worm, with sand grains in its mouth. The tentacles have collected sand grains in their grooves. *G–H.* Tube of *Amphictene auricoma*, composed of sand grains. *I–K.* Tube of *A. auricoma*, composed of sponge spicules. (*A, G–I after M'Intosh*, 1894; *B–E after M'Intosh*, 1885; *F after Benham*, 1910.)

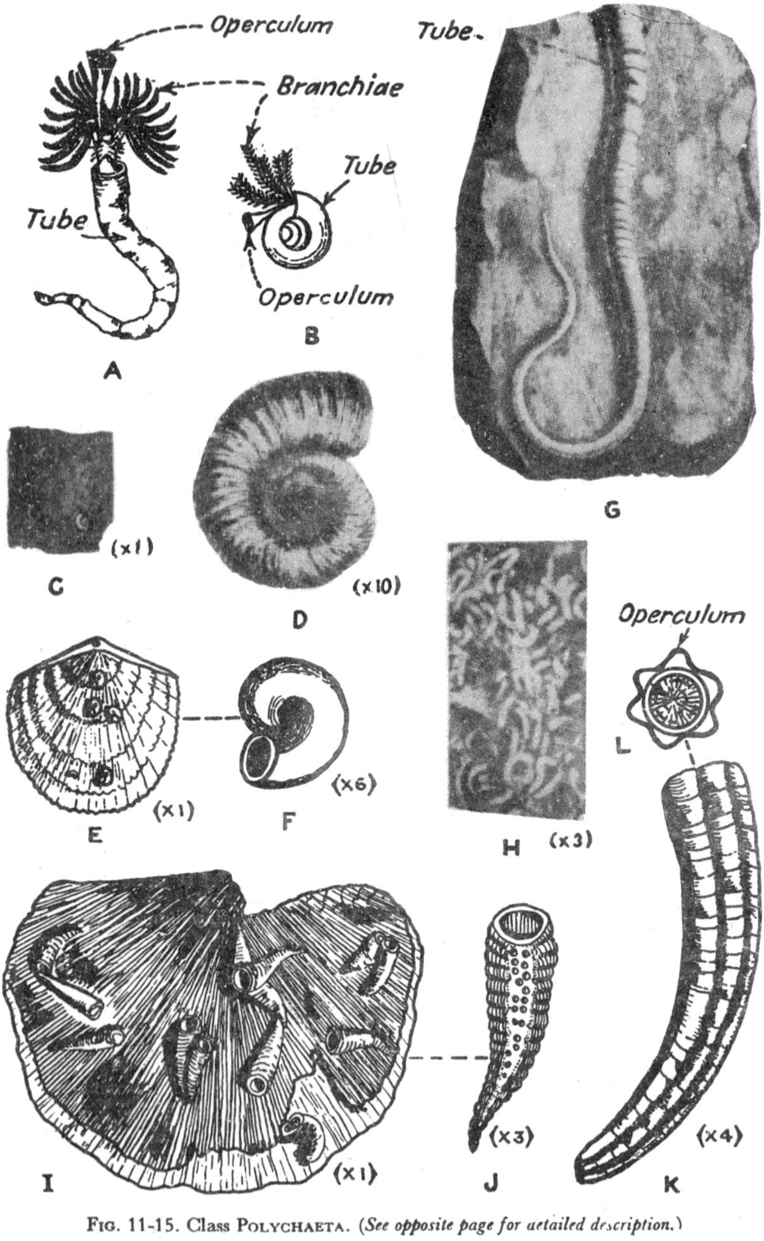

FIG. 11-15. Class POLYCHAETA. (*See opposite page for detailed description.*)

whereas still others have the tube crowned with a mass of radiating slender tubes (*Terebella*, Fig. 11-14*B,F*). The worm selects the particles to be used and then by means of the tentacles conveys them to the mouth for a coating of some sort after which they are carried to the tube by means of the tentacles (Fig. 11-14*F*).

Some agglutinated tubes are attached to the bottom or to objects on the bottom, others simply lie in the soft mud, and still others are carried about by the worm. An unusual tube is built by a specialized polychaete, *Chaetopterus*. This worm lives in shallow water (30 m., or 90+ ft.), in a parchmentlike U-shaped tube embedded in soft mud and sand, with the two open ends just above the surface of the bottom. Such a tube would, if preserved, bear close resemblance to U-shaped burrows, from which it might be distinguished by the absence of castings around either opening.

So far as is known to the authors, no fossil worm tubes of the agglutinated type have been reported, but it is reasonable that they may be expected, as foraminiferal tests of similar size and construction have been found.

In the third group of tube-forming polychaetes the animal constructs a tube of calcium carbonate on a base either of secreted mucin or of magnesium phosphate and an organic substance known as onuphin. Such tubes show considerable variety in general architecture and have been taken from depths of more than 6,000 m. (3,125 fathoms, or 18,000+ ft.). They range from simple curved cones to irregularly or regularly coiled tubes, both types commonly having ornamented lids, or **opercula** (Fig. 11-15).

Serpula (Fig. 11-15*A,G,H*) is one of the best known living genera and is common in modern seas. On some bottoms the calcareous tubes are so numerous that they form tangled biohermal masses. Such structures are present along the coast of Bermuda, where they are known as *Serpula atolls*, and in the Pleistocene of California. The separated tubes of *Serpula* are known from rocks as old as Cretaceous. Another serpulid genus is *Hamulus* (Fig. 11-15*K*), which is known from Paleozoic and younger rocks.

One of the most common and widespread of the tubicolous polychaetes is *Spirorbis* (Fig. 11-15*B–F*), which builds a tiny coiled calcareous tube that has the appearance of a small gastropod shell. These are commonly seen adhering to shell fragments, stones, algae, etc., and are found as fossils adherent to brachiopod shells and bryozoan zoaria in rocks as old as Ordovician. They can easily be overlooked because of their small size. *Cornulites*

FIG. 11-15. Class POLYCHAETA, tubicolous worms. *A, G–H. Serpula*, a common polychaete with a calcareous tube: *A, S. contortuplicata*, a living tubicolar worm with an operculated tube; *G, S. flagellum*, a Jurassic serpulid: *H, S. pervermiformis*, an Upper Cretaceous serpulid with intertwined tubes. *B–F. Spirorbis*, a common tubicolous polychaete with a coiled and operculated calcareous tube: *B, S. communis*, considerably enlarged; *C, S. carbonarius*, a Pennsylvanian species, showing several tubes attached to a fragment of a fossil plant; *D*, a single specimen from *C*, magnified. *E–F*, tubes of *Spirorbis* cemented to the brachial valve of a Devonian *Atrypa*, and a single tube somewhat enlarged. *I–J. Cornulites conica*, calcareous tubes of a tubicolar polychaete cemented to the valve of an Ordovician *Rafinesquina*, and a single tube considerably enlarged. *K–L. Hamulus onyx*, a common operculated Upper Cretaceous tube. (*A–B, G after Nicholson*, 1872; *C–D after Dawson*, 1868; *H, K–L after Wade*, 1926; *I–J after Nicholson*, 1879.)

(Ord.–Penn.) is an extinct genus whose cornucopia-shaped calcareous tubes are commonly found incrusting brachiopod shells (Fig. 11-15*I*–*J*).

Trails. Those worms that crawl over soft mud and sand commonly leave shallow sinuous trails that may be flanked by tiny pits made by the parapodia. There seems no good reason why some of these trails might not have been preserved under favorable conditions, but such a possibility is hardly justification for ascribing every such marking to worms—a practice that has

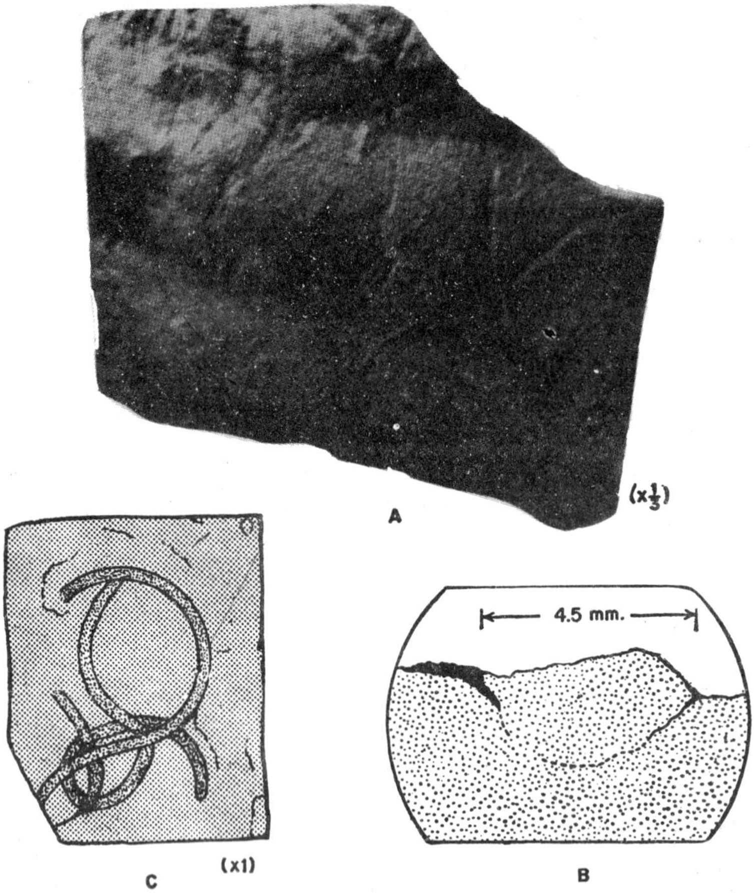

Fig. 11-16. Probiematical Pre-Cambrian worm trails. *A–B*. Trails on the ripple-marked surface of a layer of Ajibik (Middle Huronian) quartzite from Michigan: *A* shows the underside of the layer; *B* is a diagrammatic transverse section through the supposed "burrow" filling, the sand of the filling being identical with the surrounding sediment. *C*. *Helminthoidichnites meeki*, a supposed worm trail from the Algonkian of Montana. The trails here illustrated may have been made by a mollusk, arthropod, or some other invertebrate organism; actually their origin is problematical, though it seems reasonable to consider them organic. (*A–B after Faul*, 1949; *C after Walcott*, 1898.)

been prevalent for many years. Other organisms in no way related to worms make similar markings. An example of a questionable trail is *Helminthoidichnites meeki*, from the Pre-Cambrian (Algonkian) of Montana (Fig. 11-16), which bears a generic name suggesting its similarity to a worm trail. Another Pre-Cambrian trail (or burrow) (Fig. 11-16*A–B*) recently reported from the Middle Huronian Ajibik quartzite of Michigan may have been made by an ancient worm (see footnote, page 513).

Class Myzostoma

This class consists of a few species of small oval polychaetous worms that are ectoparasites on living crinoids or asteroids. Inasmuch as the myzostomarians pass through a trochophore larval stage similar to that of the Polychaeta, they are considered by many zoologists to have degenerated from some polychaete ancestor to their present form in adapting to their peculiar ectoparasitic life.

Little is known about the geologic history of this class. The tendency toward parasitism seems to have been established by Early Mesozoic time, as supposed myzostomarians are reported to have infested the columnals of Jurassic crinoids. There is no other known fossil record of the class.

Class Oligochaeta

The Class Oligochaeta includes annelids with few setae per segment (Gr. *oligos*, few, + *chaete*, spine). The most familiar representative is the common earthworm, *Lumbricus terrestris*, which may be as much as 300 mm. (12 in.) long and 10 mm. (3/8 in.) in diameter.[1]

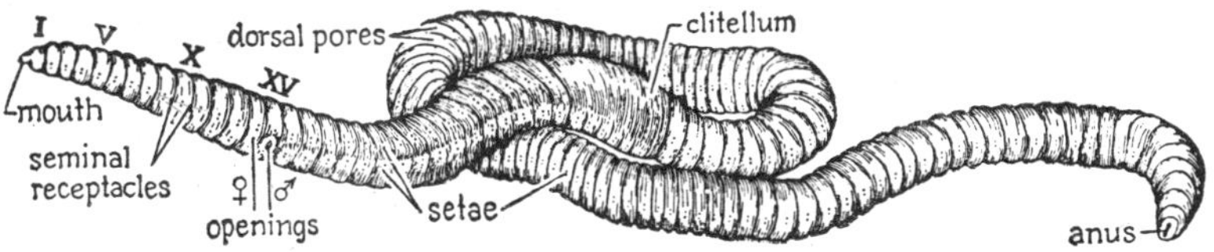

Fig. 11-17. Class Oligochaeta. The earthworm, *Lumbricus terrestris*, with chief external features indicated. (*After Storer*, 1943.)

The oligochaete body is elongate and cylindrical, strongly segmented, bluntly tapered at each end, and flattened on the ventral side (Fig. 11-17). There is no distinct head. The body of an adult oligochaete consists of more than 100 (to as many as 200) somites that are separated by transverse grooves. The mouth is in the first somite, the anus in the last, and between somites 31 to 37 is a conspicuous swelling, the **clitellum,** which secretes a substance forming cocoons to contain eggs.

There are no parapodia or cirri as in the Polychaeta, but each somite, excepting the first three and the last, bears four pairs of tiny setae that pro-

[1] Most oligochaetes are smaller than *Lumbricus* and are of the order of 1 to 100 mm. long, but notable exceptions are the giant earthworms of Ecuador and Australia that reach lengths greater than 2 m. (7 ft.) and a diameter of 25 mm. (1 in.).

ject slightly on the under and lateral sides. These setae, which are chitinous bristles, can be extended or retracted and they serve as grappling devices when the worm is moving over the ground or in its burrow.

The specialized part of the digestive tract is shown in Fig. 11-18. The **pharynx** is lined with cuticle but jaws and denticles are lacking. The **crop** is used for storing food until it can be ground up in the **gizzard,** which usually contains sand grains that aid in the grinding process. Different chemical substances are secreted by the digestive tract, and these act on the food as it passes along from mouth to anus. There is, therefore, little mechanical action on the passing material except in the gizzard.

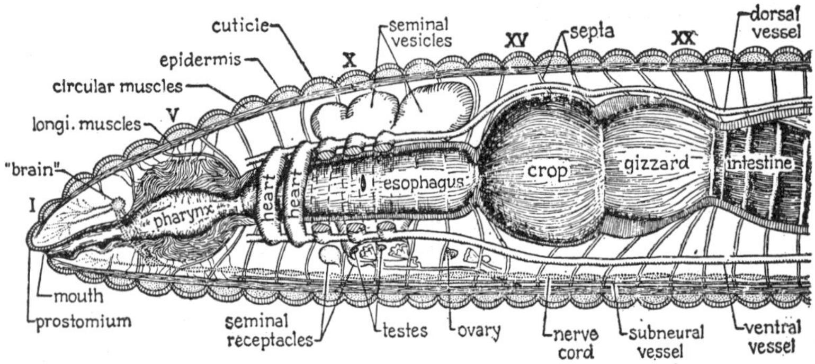

FIG. 11-18. Class OLIGOCHAETA. Internal structure of anterior portion of an earthworm from the left side; body wall and digestive tract cut in median section. (I,X,XV,XX = somites.) (*After Storer*, 1943.)

The majority of the 2,400 species of oligochaetes are earthworms living in soil, under stones, or in moist banks along shores of ponds, lakes, and streams. A few live deep in lakes, some live under stones and seaweeds along the seashore, and a few species live in shallow marine waters. Most oligochaetes are vegetarians and lack the jaws and denticles so characteristic of the polychaetes. The only firm structure in the head region is the cuticular lining of the pharynx, but sand grains are usually found in the gizzard, which is an enlargement of the digestive tract just in front of the opening into the intestine (Fig. 11-18). The food, which consists of leaves, grasses, and other vegetation, and of both plant and animal debris under some circumstances, is first moistened by a secretion from the mouth region and then drawn into the digestive tract, along which it travels to the gizzard, undergoing chemical alteration on the way. In the gizzard it is ground up by the muscular action of the walls, aided by sand grains, and is then passed into the intestine, where digested material is absorbed and residues pushed along to be expelled through the anus. If the oligochaete is ingesting soil that contains organic matter, the latter is extracted and the modified soil expelled as castings. Such castings are a common and familiar sight on the ground where earthworms are present.

The great importance of earthworms to soil conditioning was long ago pointed out by Charles Darwin (1881), who estimated that in a single year on favorable sites earthworms can bring as much as 18 tons of castings to the surface on every acre. The burrows they excavate in this process furnish easy avenues of entry into the soil for air and water, and the soil cast on the surface has undergone certain chemical changes as a result of having passed through the worm.

Almost nothing is known about the geologic history of the Oligochaeta, though it is generally supposed that the class descended from some marine polychaete or polychaete-like ancestor which adapted itself to life on land. About the only record an oligochaete can leave is a carbonaceous film, or an impression of its body; its burrow, so ephemeral that there seems little likelihood it would ever be preserved; and its castings and trails. The fossil record of the class is meager and questionable.

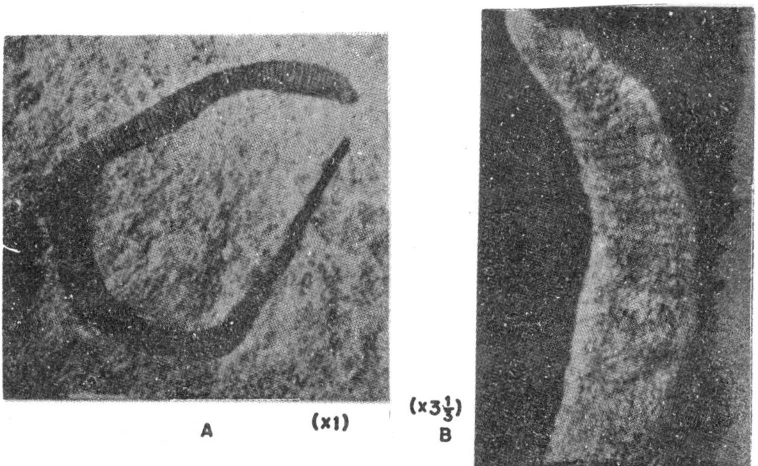

FIG. 11-19. Class OLIGOCHAETA. Fossil forms. *A. Protoscolex batheri*, a complete worm with a clitellum, from the Silurian Lockport limestone of New York. *B.* Questionable fossil earthworm—actually a sand filling of a burrow—from the Paleocene of Wyoming. (*A after Ruedemann*, 1925; *B after Hazen*, 1937.)

The oldest fossil worms that seem to have been assigned to the Oligochaeta are some half dozen species of *Protoscolex* ranging in age from Upper Ordovician to Upper Silurian. All these are marine forms, a fact that is not out of harmony with the supposition that the terrestrial oligochaetes were derived from marine ancestors. *Protoscolex batheri* (Fig. 11-19*A*) from the Silurian (Lockport) of New York is a typical but relatively large form. It is interesting to note that it was found in a limestone that was deposited in a shallow marine bay or lagoon; hence the waters may have been somewhat brackish at times.[1]

The only other supposed fossil earthworm that has come to our attention

[1] RUEDEMANN, R., 1925, Some Silurian (Ontarian) faunas of New York, *N.Y. State Mus. Bull.* 265, pp. 7, 40–45.

is a burrow with a segmented sand filling (Fig. 11-19*B*), reported from a Paleocene stream deposit (Fort Union) of Wyoming.[1]

Class Hirudinea

This class contains the familiar leeches, of which almost 300 species have been described. The body of a typical leech, which ranges in length from 10 to 200 mm., has 34 somites subdivided externally into many annuli, a large posterior sucker, and in some forms a smaller anterior one. Tentacles, parapodia, and setae are lacking, and the organism differs in several other respects from most annelids.

Leeches are predatory or parasitic annelids living in a wide range of habitats—fresh-water, marine and terrestrial. Lacking parapodia they travel like a "measuring worm" on a surface and swim by undulations of the body. They attach themselves by the suckers and are notorious for their blood-sucking habit.

There seems to be no certain fossil record of the Hirudinea; hence nothing is known about their geologic history.

Class Gephyrea

This class is an assemblage of three distinct groups of burrowing marine worms having sausage-shaped bodies that show little or no evidence of segmentation. Although some zoologists consider them to be an independent phylum, or even three separate and distinct phyla, others see in them enough annelid features to include them as a class in the Phylum Annelida.[2] Here the three groups are considered as orders:

Order 1. Echiurida.
Order 2. Sipunculida.
Order 3. Priapulida.

Order 1. Echiurida. The members of this order have a gourd-shaped body bearing a long, extensible proboscis that has a ciliated groove leading to the mouth. The anus is posterior and lies at the end of a long intestine. There is a single pair of setae at the posterior end. Of possible interest to paleontologists is *Urechis caupo*, a species reaching a length of 480 mm. (19 in.) that lives in U-shaped burrows along the California coast. *Echiurus* (Fig. 11-20*A*) is another common coastal genus in the Northern Hemisphere.

[1] Hazen, B. M., 1937, A fossil earthworm(?) from the Paleocene of Wyoming, *Jour. Paleontology*, vol. 11, p. 250. Professor B. F. Howell has furnished the following additional details on this interesting fossil, a photograph of which is shown in Fig. 11-19*B*: "The worm was collected from the Mantua lentil of the Early Paleocene Polecat Bench Formation at Section 31, T. 57 N., R. 9 W., Park County, Wyoming . . . The specimen has never yet been given a scientific name; but Professor J. Percy Moore, who is a specialist on living worms and who has seen the specimen, considers it to be undoubtedly an earthworm." (Personal communication, Nov. 11, 1949.)

[2] The possession of paired nephridia; a nervous system with an unsegmented ventral nerve cord, dorsal ganglia, and an esophageal ring; separated sexes with one or two gonads; setae (in only one order); and a trochophore larva are thought to indicate relationships with the Annelida.

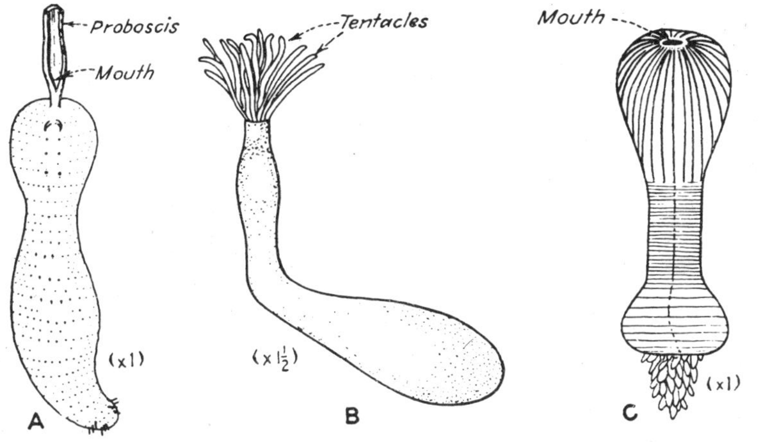

FIG. 11-20. Class GEPHYREA. *A. Echiurus*, a typical representative of Order Echiurida. *B. Dendrostoma*, Order Sipunculida. *C. Priapulus*, Order Priapulida. Some priapulids have small spines around the mouth. (*Diagrammatic after Delage and Hérouard*, 1897.)

Nothing is known of the geologic record of this group, as no fossil echiurids seem yet to have been definitely identified. The U-shaped burrows, however, present a possible feature for preservation under favorable conditions.

Order 2. Sipunculida. *Dendrostoma* (Fig. 11-20*B*), a typical sipunculid, has a crown of retractile tentacles or lobes around the mouth, a twisted intestine, and the anus located anteriorly in a middorsal position. There is a pair of nephridia, which serve as sex ducts, and the gonads are paired.

The geologic history of this order is indefinite. It has been suggested that the burrows assigned to the form genus *Scolithus* (L. Camb.–Mesozoic) were possibly made by ancient sipunculids, but this is admittedly an uncertain matter. Walcott (1911) called attention to the fact that his *Ottoia* (Fig. 11-21*A*) shows certain similarities to the Sipunculida, but he did not go so far as to assign the genus to any specific order of the Gephyrea. No other supposed fossil sipunculids have come to our attention.

Order 3. Priapulida. *Priapulus* (Fig. 11-20*C*), a representative genus of this order, has a gourd-shaped body with a spiny proboscis but without tentacles. There is a mouth cavity with chitinous teeth and a straight digestive tract ending in a posterior anus. A vascular system, nephridia, and sense organs are lacking in the adult, and the gonads, which have ducts and also serve for excretion, are paired. The lack of any fossil record leaves us in complete ignorance of the geologic history of the order.

Fossil Gephyrea. As stated in the foregoing paragraphs, no fossils have been assigned specifically to any of the three recognized orders of living gephyreans. However, Walcott tentatively assigned *Ottoia* and *Banffia* (Fig. 11-21*B*) to the Class Gephyrea, basing the assignment on the presence of an anterior introvertible proboscis and the elongate cylindrical body. He also called attention to the segmentation of *Ottoia*, as well as the superficial resemblance to leeches, but did not consider this feature important enough to assign the genus to the Class Hirudinea.

As for *Scolithus*, which has been assigned to this class, nothing definite can be said about its true origin. Some of the burrows included in this catchall *may* have been made by ancient gephyrean worms, but there seems no convincing way of proving it.

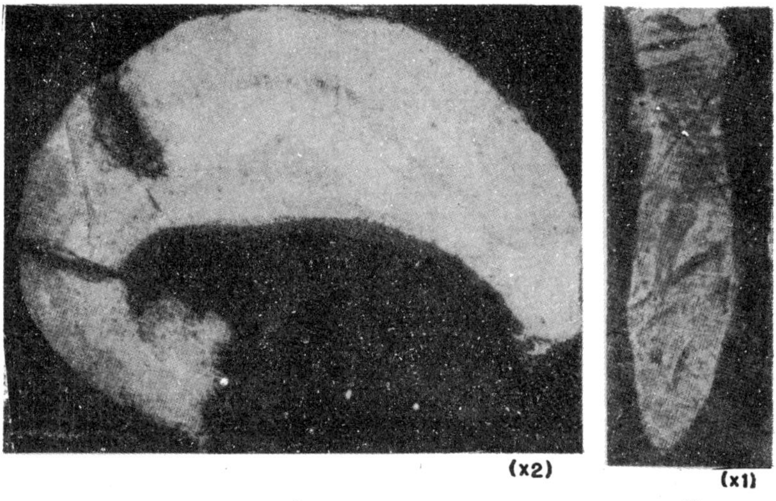

FIG. 11-21. Class GEPHYREA. Fossil Gephyrea, from the Middle Cambrian Burgess shale of British Columbia. *A. Ottoia prolifica*, a small specimen showing the hooks about the mouth (right end), proboscis, posterior end, and the distinct annular lines and segments. *B. Banffia constricta*, anterior section, constricted portion, and part of posterior section. (*After Walcott*, 1911.)

GEOLOGIC HISTORY OF THE ANNELIDA

Segmented worms have had a long geologic history. They are now believed to have evolved in Pre-Cambrian time and to have been the general ancestral stock of the present Annelida, Onychophora, Arthropoda, and Mollusca,[1] as well as the forerunners of several extinct groups in these same phyla (Fig. 12-3).

Modern Annelida resemble the Arthropoda in having a segmented body covered with a cuticle (not chitinous, however) that is secreted by the epidermis, in the general structure of the nervous system, and in certain other respects. The Annelida differ strongly, however, in having a large coelom, only simple unjointed appendages if these are present at all, almost no specialization of somites, and no succession of larval stages (**molts**).

The Annelida and Onychophora have several features in common, as dis-

[1] SNODGRASS, R. E., 1938, Evolution of the Annelida, Onychophora, and Arthropoda, *Smith. Misc. Coll.*, vol. 97, pp. 1–159.

cussed more fully in the next chapter, and are thought to have descended from the same ancestral form (Fig. 12-3).

The embryology of the annelids resembles that of certain other worms, of the Onychophora and Arthropoda, and of the Mollusca in certain features, and the presence of a trochophore larva in these several phyla suggests an evolutionary relationship of some sort.

When, where, and from what kind of ancestor the first segmented worms evolved is one of the unanswered questions of animal evolution. That annelids appeared early in geologic time is proved by their excellently preserved representatives in the Middle Cambrian Burgess shale of British Columbia. This fauna contains 10 genera of worms, of which all except the single chaetognathan genus *Amiskwia* (Fig. 6-8*B*) belong to widely separated and only distantly related extinct families of Annelida. Inasmuch as these worms are highly evolved creatures, with some of the fundamental characteristics of the several existing annelid classes (Polychaeta, Gephyrea), it follows that the Phylum Annelida must have been well differentiated by the beginning of the Paleozoic. If this view is accepted, then it leads to the almost inescapable conclusion that the annelids and their closest relatives (*i.e.*, the arthropods and onychophorans) must have had a long Pre-Cambrian history. Unfortunately, no significant record of this early developmental period has yet been reported. The only fossil record of annelids older than the Cambrian consists of certain burrows and trails (Fig. 11-16) considered by their discoverers possibly to have been made by annelid worms, but whether or not these vague and scattered fossils were actually made by annelids—or by any other worms, for that matter—must await confirmation. Certainly they should stimulate further search, and it should come as no surprise if more substantial and definite remains of Pre-Cambrian segmented worms are some day discovered.

The fossil record of the annelid worms is meager if we except the remarkably preserved specimens of the Middle Cambrian Burgess shale and the less spectacular fossils in the Jurassic limestones of Solenhofen.

The few trails and burrows thus far reported from Pre-Cambrian rocks are of uncertain origin; they *may* have been made by worms, but until scolecodonts, acicula, setae, or impressions are found in close association with them, their worm identity must remain in doubt.

Annelid evolution had progressed far by the Middle Cambrian, as proved by the wealth of fossil evidence in the Burgess shale. Castings, trails, and burrows of supposed worms have been found in Cambrian rocks of all ages, but again their worm origin is unproved.

By Ordovician time annelid worms were adapted to life on planktonic organisms,[1] and *Spirorbis*, *Cornulites*, and other forms had developed the

[1] That worms of several different kinds lived with the graptolites and other planktonic organisms in the Early Paleozoic seas has been emphasized by Ruedemann's studies (RUEDEMANN, R., 1934, Paleozoic plankton of North America, *Geol. Soc. Amer., Mem.* 2, pp. 11, 36, 86–90). Some of these worms seem to have been segmented and to have borne setae on the segments. As would be expected, their relationships with existing worms can only be surmised.

ability to secrete calcareous tubes in which to live. There were almost certainly some worms that built chitinous tubes earlier than the Ordovician, but of these there seems to be no certain record.

Fossils of many kinds ascribed to worms have been reported from almost every major subdivision of the geologic column since the Ordovician. Scolecodonts, calcareous tubes, burrows with their associated castings, and carbonaceous films or impressions leave no doubt that the Phylum Annelida has been well represented in marine faunas since Early Cambrian time, even if there is no extensive fossil record to fill the gaps. Although it seems reasonable to suppose that some of the many trails, tubes, borings, burrows, and other less definite structures illustrated in paleontological literature were made by ancient worms, it is unfortunately true that this will generally be difficult or impossible to prove.

GEOLOGIC WORK OF THE ANNELID WORMS

Throughout their existence, since their probable appearance in Pre-Cambrian times, annelid worms have affected to an important extent the sediments of sea bottoms and the soil mantle on the lands. The vast quantities of sediments that have passed through their digestive tracts have been modified both physically and chemically. The grinding and churning actions to which the sedimentary material is subjected, not only in the pharynx and gizzard but also in the intestine as well, reduce the particles to a smaller size. The chemical actions of the different digestive secretions bring about successive compositional changes in the sedimentary material as it moves through the digestive tract. The castings, therefore, generally differ from the material originally taken into the mouth in showing a somewhat finer texture, in lacking most or all of the organic matter originally contained, and in having a reduced content of certain soluble salts. For these reasons fossil castings should be expected to differ somewhat from the surrounding rock.[1]

At the present time earthworms are important to agriculture because of their action on surface and subsurface soils. Darwin (1881), who wrote so interestingly and in such detail on this aspect of annelid habits, states that some garden areas may have as many as 50,000 earthworms per acre, though the average for all soil areas, he thinks, is more likely about half that figure. From the castings over a square yard ejected during 45 days in a valley area, he calculated that the worms transported about 18 tons of soil to the surface during a year. If spread out evenly over the surface, the castings would make an annual layer about 5 mm. (⅕ in.) thick. In addition to the substantial transport of material by the worms, there is also the production of innumerable tubes and burrows along which air and water can penetrate the soil mantle for considerable distances.

Boring and burrowing marine annelids carry on important activities on many mud and sand bottoms. The lugworms, or lobworms (*Arenicola marina*,

[1] It seems likely that all marine and lacustrine sediments of slow deposition have made many passages through the digestive tracts of annelid worms as of other sediment-eating animals.

Fig. 11-12), which inhabit the sand flats of the littoral zone along the Northumberland coast, have been investigated by Davison (1891), Wells (1945), and others, and they prove to be far more effective than earthworms in moving sediments from depth to a surface. These annelids eat the mud and sand through which they burrow and cast the rejected sediment onto the surface around one opening of the U-shaped burrow. Davison (1891) estimated that the individual castings would average more than 84,000 per acre at any specified time and about 50,000,000 per square mile. Where the worms are most abundant, he estimated that they brought on the average more than 1,900 tons of sand to the surface of every acre every year—136 times the weight of soil brought up by earthworms over an equal area in the same time! Inasmuch as the worms burrow to a depth of 0.6 m. (2 ft.) or more, it follows that most of the material in the upper ½ m. would ultimately be involved in the slow churning process, and Davison concluded that the entire 2-ft. layer would pass through the worms once in about two years. Taking geologic time into account, and considering the physical and chemical changes undergone by the sediments during their passage through the digestive tracts of the lugworms, it would seem obvious that these and similar annelids perform important geological work. That this work is of long duration is indicated by the presence of castings and associated burrows ascribed to arenicoloid worms present in rocks as old as the Devonian (Fig. 11-12*D*).

REFERENCES

BUCHSBAUM, R. 1940. "Animals without Backbones." Chicago, University of Chicago Press. Chap. 19, pp. 207–234.

Cambridge Natural History, vol. 2. 1910. London, Macmillan & Co., Ltd. (Polychaete Worms by W. B. Benham; Earthworms and Leeches by F. E. Beddard; Gephyrea by A. E. Shipley.)

CRONEIS, C. 1934. Paleoecology of the worms (abstract). *Geol. Soc. Amer. Proc.* 1933, p. 361.

———. 1941. Micropaleontology—past and future. *Bull. Amer. Assoc. Petrol. Geol.*, vol. 25, 1208–1255. (Scolecodonts, pp. 1244—1247.)

DARWIN, C. R. 1881. "The Formation of Vegetable Mould, through the Action of Worms, with Observations on their Habits." London, John Murray, 326 pp.

DAVISON, C. 1891. On the amount of sand brought up by lobworms to the surface. *Geol. Mag.*, dec. 3, vol. 8, pp. 489–493.

DELAGE, Y., and HÉROUARD, E. 1897. "Traité de zoologie concrète." Paris, Schleicher Frères. (Worms, pp. 4–46, 182–251.)

ELLER, E. R. 1946. New scolecodonts from the Kagawong (Ordovician) of Manitoulin Island, Ontario. *Penn. Acad. Sci. Proc.*, vol. 20, pp. 71–75. (This author has published a number of important articles on scolecodonts.)

HOWELL, B. F. 1949. New hydrozoan and brachiopod and new genus of worms from the Ordovician Schenectady formation of New York. *Bull. Wag. Free Inst. Sci.*, vol. 24, pp. 1–11.

LANGE, F. W. 1947–1949. Anelídeos poliquetos dos folhelhos Devonianos de Paraná [Brazil]. *Mus. Paranese, Arquivos*, vol. 6, pp. 161–230 (1947). Polychaete Annelids from the Devonian of Parana, Brazil. *Bull. Amer. Pal.*, vol. 33, No. 134, 102 pp. (1949). [See reviews by Caster in *Jour. Paleontology*, vol. 22, pp. 647–648 (1948); *Amer. Jour. Sci.*, vol. 246, pp. 724–725 (1948); and *Bull. Amer. Assoc. Petrol. Geol.*, vol. 33, pp. 1771–1772 (1949).]

RICHTER, R. 1928. Die fossilen Fährten und Bauten der Würmer, ein Überlick uber ihre biologischen Grundformen und deren geologische Bedeutung. *Pal. Zeitsch.*, vol. 9, pp. 193–204.

SHIMER, H. W., and SHROCK, R. R. 1944. "Index Fossils of North America." New York, John Wiley & Sons, Inc.; Cambridge, Mass., The Technology Press. Chap. VI, pp. 228–234. (This work contains a list of most of the important references to fossil annelids in North America.)

STORER, T. I. 1943. "General Zoology." New York, McGraw-Hill Book Company, Inc. Chap. 21, pp. 412–435. 2d ed., 1951, Chap. 17, pp. 438–460.

WALCOTT, C. D. 1911. Middle Cambrian annelids. *Smith. Misc. Coll.*, vol. 57, pp. 107–145.

WELLS, G. P. 1945. The mode of life of *Arenicola marina* Linnaeus. *Jour. Marine Biol. Assoc. United Kingdom*, vol. 26, pp. 107–207.

CHAPTER 12

PHYLUM ONYCHOPHORA

The Phylum Onychophora consists of 70 living species of a wormlike creature, all belonging to the single genus *Peripatus* (with eight or nine subgenera) (Fig. 12-1), and two supposed fossil genera, of which the better known is the remarkable *Aysheaia pedunculata* (Fig. 12-2*B*–*C*) from the Middle Cambrian Burgess shale of British Columbia. Because of its superficial wormlike appearance, *Peripatus* was commonly classified with slugs, annelid worms, or myriapods by early zoologists. When its anatomy and life history became known, however, it was immediately recognized as a somewhat advanced and unusual invertebrate closely allied to, but distinct from, the Annelida on one hand and the Arthropoda on the other. The mixture of annelidan and arthopodan characteristics, together with certain unique features of its own, is considered by many present zoologists a valid reason for recognizing the Onychophora as a separate phylum.

A typical *Peripatus* (Fig. 12-1*A*) resembles an ordinary caterpillar in general appearance, though it is not, of course, closely related. The long cylindrical body, which has a velvety skin, or cuticle, lacks the external segmentation of typical annelids and arthropods, but it has many (17 to 43) pairs of short legs, the **parapodia,** a pair for each of the internal segments of the body. Each leg terminates in a pair of claws, and all legs are exactly alike (Fig. 12-1*A*,*C*–*D*).

The head, which is not marked off from the remainder of the body, consists of three fused segments. One is **preoral,** bearing a pair of antennae; two are **postoral,** one bearing jaws and the other bearing oral papillae (Fig. 12-1*D*). The mouth lies at the posterior end of a ventral depression, the **buccal cavity,** and is surrounded by a raised lip. Within the buccal cavity, on each side of the mouth, are the two jaws, which are stumplike muscular structures with a pair of chitinous cutting blades or claws at their free extremities (Fig. 12-1*D*–*F*). These claws are quite similar to those borne by the legs (Fig. 12-1*C*) and seem to be thickenings of the cuticle. They are directed posteriorly with the serrated cutting edge turned backward and seem to be used for tearing the food held by the swollen annular lip surrounding the mouth. At the front of the buccal cavity is a median tonguelike muscular protuberance that is attached to the dorsal wall of the mouth (Fig. 12-1*D*). It bears a row of tiny chitinous teeth and somewhat resembles the radula of certain mollusks.

The fusion of the three anterior segments to form the head places *Peripatus* midway between the annelids, which lack such fusion, and the arthropods, which have a head of six fused segments.

The remaining segments of the body, varying in number according to the species, are all alike, and each bears a pair of short cylindrical parapodia.

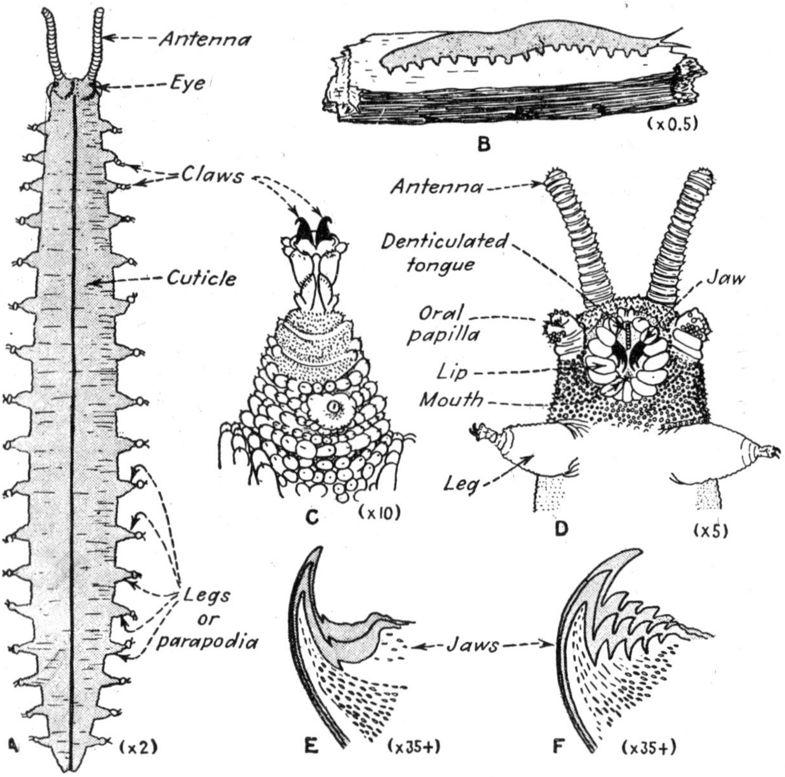

FIG. 12-1. Phylum ONYCHOPHORA. *Peripatus* (*Peripatopsis*) *capensis*, a typical onychophoran. *A*. Dorsal view of an entire specimen. *B*. Lateral view of live specimen crawling on a piece of wood. *C*. The last (seventeenth) left leg of a male viewed from the ventral side, showing the pair of terminal claws. *D*. Ventral view of head and oral region, showing the pair of jaws in the buccal cavity. *E–F*. Outer and inner jaws, respectively. *C–F* are much enlarged. (*A*, *C–F after Balfour*, 1883; *B after Sedgwick*, 1887.)

Each parapodium ends in a pair of chitinous claws (whence the name of the phylum: Onychophora—Gr. *onyx*, *onychos*, claw + *phoros*, bearing).

The body wall is composed of a thin chitinous cuticle, an epidermis, and a thin plexus of muscles, certain of which operate the legs and the several parts of the head region. Enclosed by the body wall is an undivided cavity, the **hemocoel,** containing the simple tubular digestive tract, the circulatory and nerve systems, and the excretory and reproductive organs. Tubular ingrowths of the body wall, designated **tracheae,** seem to serve for respiration.

The Onychophora resemble the Annelida in the following respects: (1) structure of the eyes; (2) paired nephridia in all but the first two segments; (3) ciliated reproductive ducts; and (4) simple digestive tract. They more closely resemble the Arthropoda in having (1) the jaws derived from

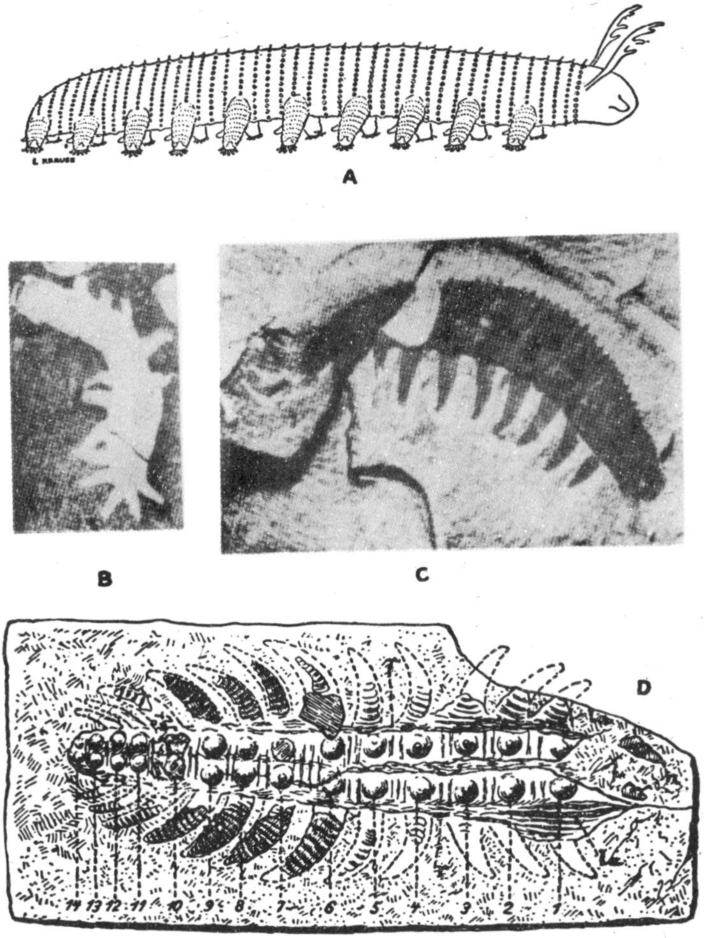

FIG. 12-2. Phylum ONYCHOPHORA. Fossil forms. *A–C. Aysheaia pedunculata*, a supposed onychophoran from the Middle Cambrian Burgess shale of British Columbia: *A*, conjectural restoration; *B–C*, unretouched photographs of two different specimens (×1.5). *D. Xenusion auerswaldae*, a supposed onychophoran (incompletely restored) from a Pre-Cambrian (Algonkian) or Early Cambrian quartzite glacial boulder of Ostpriegnitz (×1). (*A–C after Hutchinson*, 1930; *D after Pompeckj*, 1927.)

modified appendages; (2) a hemocoelic body cavity; (3) a dorsal "heart" with ostia; (4) the coelom reduced to cavities (nephridia and reproductive ducts); (5) similar structure of reproductive organs; and (6) tracheae resembling those of insects. They are unique in having (1) a single pair of jaws; (2) poorly developed segmentation externally; (3) peculiar arrangement of tracheal apertures; (4) velvety nature of skin; (5) separate nerve cords without true ganglia; and (6) simplicity and similarity of all body segments behind the head.

In summary then, the Onychophora seem to be an autonomous and unique group of organisms occupying a position somewhere between the annelids and arthropods. They do not seem to have descended from annelids or to have been ancestral to any known arthropods. More likely, they share with these a common ancestor which evolved from a trochophore larva long before the Cambrian (Fig. 12-3).

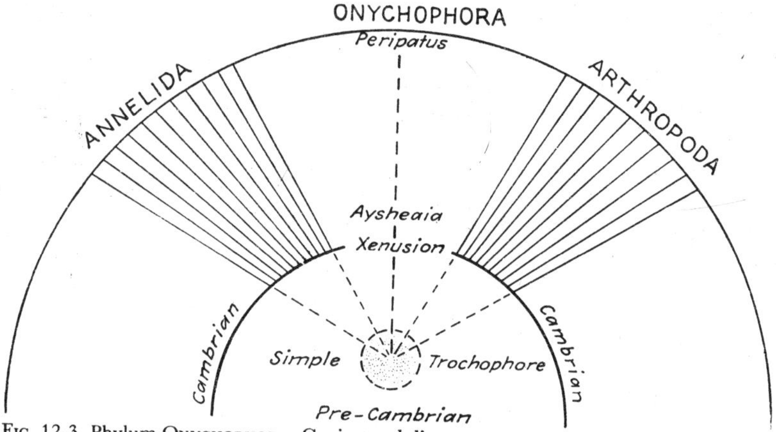

FIG. 12-3. Phylum ONYCHOPHORA. Conjectural diagram suggesting that *Peripatus*, *Aysheaia*, and *Xenusion* may be connecting links of the Onychophora, which, like the Annelida and Arthropoda, evolved from a simple Pre-Cambrian trochophoric ancestor common to all of them.

That the Onychophora became differentiated early in geologic time is suggested by the discovery of supposed fossil representatives in the Algonkian (or Lower Cambrian) of Europe[1] and in the Middle Cambrian Burgess shale of British Columbia. Eleven specimens are known from the latter area, and these indicate an organism that reached a length of 50 mm. (2 in.) and possessed at least 11 pairs of appendages, of which 10 pairs were legs and the eleventh pair were anterior branched structures. A conjectural restoration of this organism, to which the name *Aysheaia pedunculata* has been given, is shown in Fig. 12-2*A*.

[1] The relationship of *Xenusion auerswaldae* Pompeckj (Fig. 12-2*D*) to the Onychophora, Annelida, and Arthropoda is uncertain. This supposedly Algonkian or Early Cambrian animal seems to have been a segmented wormlike creature with annulated uniramous appendages, but its exact taxonomic relationships are unknown. In discussing it Heymons (1928) suggests that Onychophora, Tardigrada, and Pentastomida—the latter two groups without a fossil record and not discussed in this work—are merely relics of a large and important assemblage of Paleozoic animals. Perhaps *Xenusion* and *Aysheaia* give us a faint suggestion of the ancestral forms of some of the few surviving members of that fauna (Fig. 12-2).

If *Aysheaia* is really a fossil onychophoran, and the preponderance of evidence now seems to support this identification,[1] then several interesting questions immediately arise. All known existing onychophorans (*i.e.*, the species of *Peripatus*) are terrestrial and live only in moist places under logs in the tropical forests of widely separated regions throughout the world (Fig. 12-1*B*). *Aysheaia*, on the contrary, was almost certainly marine, as the remains of all 11 individuals have been found in association with typical marine organisms. Hutchinson (1930) and others have grappled with this baffling question, and about the best generalization that can be made is that *Aysheaia* represents an extinct marine organism belonging to an ancient group from which the living *Peripatus* may have evolved, meanwhile adapting itself to a terrestrial habitat and spreading over the tropical regions of the entire world. Today, its restriction to local regions in such widely separated areas as Australia, Africa, Asia, South America, and Central America suggests that, once more successful and widespread, it is now gradually shrinking in its distribution.

Onychophorans have no hard parts likely to be preserved except the tiny claws and jaws; hence it is not surprising that their fossil record is so meager. Only under unusually favorable conditions, such as those that prevailed for a short time in the Middle Cambrian seas of British Columbia, could a soft perishable body like that of an onychophoran be preserved. It is possible, however, that some of the minute chitinous hooks and jaws of Paleozoic rocks may be the remains of ancient and extinct representatives of the phylum.

It would seem then, that the *Aysheaia-Peripatus* couplet may be another combination of closely related organisms widely separated in time, like that represented by *Mackenzia-Edwardsia* among the coelenterates. In these combinations, the fossil forms can hardly be called "missing links," but surely they are suggestive "connecting links," in that they do throw faint shafts of light into that dark abyss of Pre-Cambrian time when the ancestors of all living things were slowly differentiating from primordial life.

REFERENCES

CUÉNOT, L. 1949. Les Onychophores *in* Grassé's "Traité de zoologie," vol. 6, pp. 1–37. Paris, Masson et Cie.

HEYMONS, R. 1928. Ueber Morphologie und Verwandschaftliche Beziehungen des *Xenusion auerswaldae* Pompeckj aus dem Algonkium. *Zeitschr. Morphol. Ökol. Tiere*, vol. 10, p. 307.

HUTCHINSON, G. E. 1930. Restudy of some Burgess shale fossils. *Proc. U.S. Nat. Museum*, vol. 78, art. 11, pp. 1–24.

POMPECKJ, J. F. 1927. Ein neues Zeugnis uralten Lebens. *Pal. Zeitschr.*, vol. 9, pp. 287–313.

SEDGWICK, A. 1883. A monograph of the genus *Peripatus*. *Quart. Jour. Micros. Sci.*, vol. 28, pp. 431–493.

———. 1895. *Peripatus in* Cambridge Natural History, vol. 5, pp. 1–26.

WALCOTT, C. D. 1931. Addenda to descriptions of Burgess shale fossils. *Smith. Misc. Coll.*, vol. 85, no. 3, 46 pp.

ZACHER, F. 1933. Onychophora *in* Kükenthal und Krumbach's "Handbuch der Zoologie," vol. 3. Berlin and Leipzig, Walter de Gruyter and Co.

[1] Resser, in a posthumous paper by Walcott (1931), cites letters sent to Walcott by several authorities who raised the question of the possible onychophoran affinities of *Aysheaia*.

CHAPTER 13

PHYLUM ARTHROPODA[1]

INTRODUCTION

The Phylum Arthropoda comprises an unusually large and varied group of highly developed invertebrates whose long and extensive history reaches back almost certainly into the Pre-Cambrian (Fig. 13-58). The typical arthropod is an elongate, segmented, and bilaterally symmetrical animal with mouth and anus at opposite ends of the body. The body is encased in a jointed chitinous[2] or calcareochitinous exoskeleton or is attached to a calcareous shell, and most arthropods have numerous jointed appendages that are modified for several different functions. Members of the phylum show great variation in size, ranging from tiny insects less than 0.25 mm. in length to large trilobites more than 60 cm. (24 in.) long, eurypterids exceeding 150 cm. (5 ft.) in length, and giant crabs that can span 3.4 m. (11 ft.) with their outstretched claws. Fossil arthropods are more or less common in sedimentary rocks of all ages from earliest Cambrian to Recent, and living forms are nearly everywhere abundant in the seas, on land, and in the air. Latest estimates place the total number of species at more than 1,000,000, and some of these are represented by astronomic numbers of individuals (*e.g.*, ants and flies). During the long geologic history of the phylum, representatives have invaded every life habitat at one time or another, and arthropods excel all other animals, with the possible exception of the vertebrates, in the success with which they have adapted themselves. A few of the more familiar members of the phylum are crabs, lobsters, crayfishes, barnacles, shrimps, centipedes, millepedes, spiders, scorpions, and insects. Two extinct groups, the trilobites and eurypterids, were important during the Early and Middle Paleozoic, but they did not survive the end of that era.

The Phylum Arthropoda has been variously subdivided because of differences of opinion as to how subclass and ordinal lines should be drawn. The classification adopted in this book is based largely on the following:

[1] Arthropoda—Gr. *arthron*, joint, + *pous*, *podus*, foot; referring to the jointed or segmented nature of the appendages.

[2] Chitin is a complex noncellular organic substance that is present in most arthropods and also is found in sponges, hydroid coelenterates, annelid worms, and other phyla, but never in protozoans and vertebrates. It is a nitrogenous polysaccharide, $(C_{32}H_{54}N_4O_{21})_x$, that is insoluble in water, alkalis, dilute acids, or the digestive juices of many animals. It has a fibrous structure and is commonly adorned with surface pits, spines, ridges, and irregular protuberances or pierced by tiny perforations. It is quite stable, and once buried in sediments, it is highly resistant to decomposition and destruction by ordinary ground water. Therefore, it is commonly found even in the most ancient fossiliferous rocks in an excellent state of preservation. It is often hardened by inorganic salts, of which calcium carbonate is probably the most common. In the arthropods, as well as in other invertebrates, it is secreted by the epidermis and is a strong skeletal material and protective cover. (Also see Richards, 1946 and 1951.)

1. Nature of body segmentation.
2. Structure and number of appendages.
3. Nature and position of respiratory processes.

The classification is as follows:

Phylum Arthropoda.
 Class Crustacea. Crabs, shrimps, barnacles.
  Subclass Branchiopoda. Fossil and living phyllopods. *Cambrian?; Silurian?; Devonian to Recent.*
  Subclass Ostracoda. Fossil and living ostracodes. *Upper Cambrian to Recent.*
  Subclass Copepoda. Living copepods; no fossil forms known.
  Subclass Cirripedia. Fossil and living barnacles. *Silurian?; Devonian?; Cretaceous to Recent.*
  Subclass Malacostraca. Fossil and living crabs, crayfish, etc. *Silurian?; Devonian to Recent.*
 Class Arachnoidea. Scorpions and spiders.
  Subclass Merostomata. Fossil and living water-breathing arachnoids. *Middle Cambrian to Recent.*
  Subclass Arachnida (Embolobranchiata). Fossil and living air-breathing arachnoids. *Silurian to Recent.*
 Class Trilobita. Extinct trilobites. *Lower Cambrian to Permian.*
 Class Chilopoda. Fossil and living centipedes. *Pennsylvanian to Recent.*
 Class Diplopoda. Fossil and living millepedes. *Devonian to Recent.*
 Class Symphyla. Living symphylans; no fossils known.
 Class Insecta. Fossil and living insects. *Lower Pennsylvanian to Recent.*

The trilobites are treated as a class because there is no present consensus as to how closely these ancient and extinct arthropods are related to the other classes. The centipedes (Class Chilopoda), millepedes (Class Diplopoda), and symphylans (Class Symphyla) are recognized as separate classes, though they commonly have been included together with the Pauropoda in a higher taxonomic subdivision, the Myriapoda. Although this latter division is not recognized in the following discussion, there is frequent reference to myriapods and myriapodan characteristics with these terms having general group significance rather than strict taxonomic meaning.

THE ANIMAL AND ITS EXOSKELETON

The body of an arthropod consists of many segments, or **somites,** which may be alike or different and which constitute three primary body divisions—head, thorax, and abdomen. In all arthropods the head somites are fused together, and in many of them other additional somites are also fused. Each somite of the simpler arthropods has one or two pairs of jointed appendages, which are made to articulate by the action of special muscles. In more specialized arthropods most or all of the posterior appendages are lost, so that only the somites of the front or anterior part of the body have appendages. The appendages show a wide range of modification and have been used for interpreting the phylogenetic relationships of the different groups. They function for locomotion (crawling and swimming), respiration, grasping, mastication, and oviposition, and as sensory organs.

The well-developed nervous system consists of paired dorsal ganglia over the mouth and connectives to a pair of ventral nerve cords, with a ganglion in each somite. In highly specialized forms the ganglia are concentrated. Sensory organs include antennae, simple and compound eyes, and auditory organs.

An efficient circulatory system consists of a dorsal heart and numerous arteries through which blood is distributed to organs and tissues. The blood returns to the heart through body spaces (**hemocoeles**). Respiration is carried on by gills (external **book gills** and internal **book lungs**), **branchial appendages,** and internal air ducts (**tracheae**), or by the general body surface.

The more highly specialized an arthropod becomes, the more complex are the nervous, circulatory, and respiratory systems, the sense organs, and the appendages.

The soft body is encased in an exoskeleton of some sort, which may be a chitinous or calcareochitinous dorsal shield (**carapace**) of numerous segments or plates (*e.g.*, lobsters, crabs), a calcareous tentlike structure of overlapping pieces (*e.g.*, barnacles), or a bivalve shell of two convex calcareous pieces (*e.g.*, ostracodes). A few arthropods lack an exoskeleton (*e.g.*, myriapods), but even these usually have the skin hardened and strengthened (*i.e.*, scleritized) by chitin. The arthropod exoskeleton functions as a true skeleton in that it provides places for the attachment of muscles, some of which move the segments of the body covering and the jointed appendages. Impermeability of the encasing material no doubt facilitates transition from aquatic to terrestrial life—a transition successfully accomplished by several groups of arthropods, of which the insects are probably the most notable.

Since the body and appendages of most arthropods are enclosed in a continuous and rigid exoskeleton, which cannot be extended or expanded by muscular action, the animal cannot increase in size except during periods when the hard covering is shed, or molted. In **molting,** or **ecdysis,** which takes place periodically, the integument separates at some place around the head or along the back, and the animal crawls out of its old covering. Then follows a short period during which the animal is without any protective cover. By the end of this period, during which the animal may expand, it has secreted a new exoskeleton which hardens only when the expansion is completed. In general, molting arthropods increase in size with successive molts, but some do not change dimensions appreciably, and if the food supply happens to be curtailed the animal may actually shrink in size. A few parasitic crustaceans grow without molting; in these the external integument seems to increase by interstitial growth.

Growth in some arthropods seems to continue at a diminishing rate from youth to old age; hence the intervals between moltings increase in length with age.[1] In the winged insects, however, growth ceases at the end of the larval stage, and the adult insect no longer molts.

The exoskeleton of an arthropod is intimately related to the enclosed soft

[1] Smith (1935) reported that lobsters molt from two to five times annually for the first few years, but after the fifth to eleventh year they generally molt only once a year. Storer (1951) states that most arthropods molt from four to seven times.

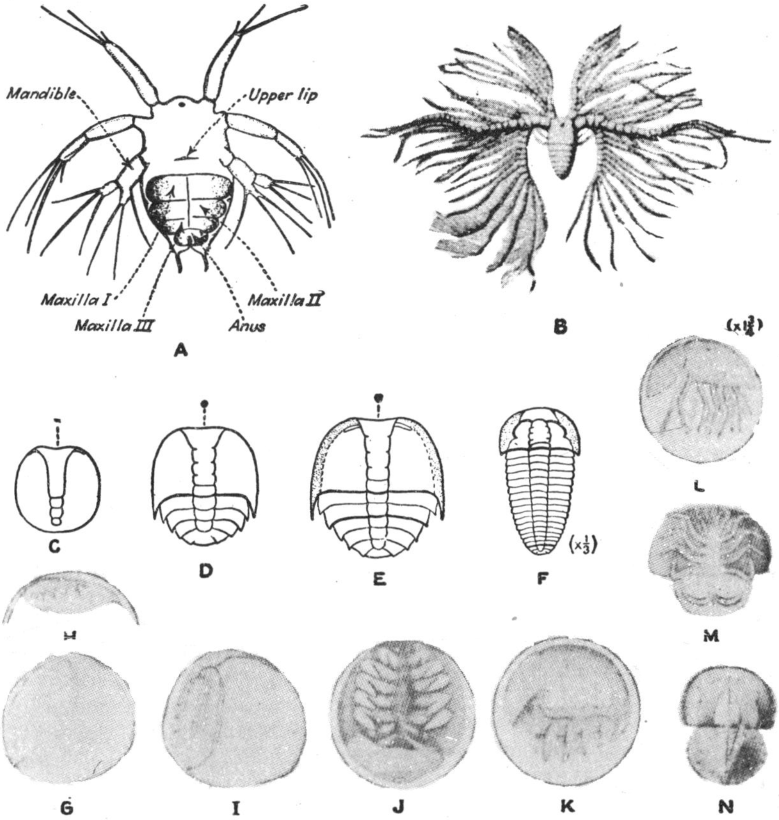

FIG. 13-1. Phylum ARTHROPODA. Larvae and early growth stages. *A*. Larval stage (much enlarged) of *Euphausia*, a modern marine crustacean (Subclass Malacostraca—Order Euphausiacea). This larva is a typical nauplius. *B*. Restoration of a bizarre supposed primitive crustacean, *Marria walcotti*, from the Middle Cambrian Burgess shale of British Columbia. It is unlike all other known Cambrian or later crustacean genera, but possesses certain features found in the larval stages of several orders of modern Crustacea. It may represent the larval stage of some crustacean, but at present its taxonomic position can only be conjectured. *C–F*. Stages in the development of the opisthoparian trilobite, *Sao hirsuta*, from the Cambrian of Bohemia: *C*, protaspis stage, with cephalon and pygidium undifferentiated; possibly at this stage the larva had five cephalic appendages, although these have not been found; *D*, meraspid stage, with cephalon and pygidium well differentiated but without any thoracic segments; *E*, later meraspid stage, showing two thoracic segments and narrow free cheeks; *F*, an adult (holaspid stage) with 16 free thoracic segments. The free cheeks are stippled in *E* and *F*. *G–N*. Growth stages of the modern king crab, *Limulus polyphemus* (Class Arachnoidea—Order Xiphosura): *G–H*, front view and partial profile of the embryo as it first appears inside the egg; the mouth (*M*) and six pairs of appendages are apparent; *I*, the embryo, farther advanced than *G*, showing the fingerlike rudimentary appendages; *J–K*, a more advanced stage than *I*, with the abdomen differentiated from the head, and the segments well marked; the five lobes of the liver in *K* indicate the five posterior cephalic segments; *L–M*, two views of the embryo just before hatching, a stage designated the "trilobitic stage" by some authors; note that there are six pairs of legs; *N*, a freshly hatched young larva. (*A modified after Metschnikoff; B after Ruedemann*, 1913; *C–F modified after Barrande*, 1852, *and Beecher*, 1895; *G–N after Packard*, 1872.)

parts of the body—probably more so than in any other invertebrate. The body of the animal grows by adding successive somites, which are introduced in regular order from front to rear, with the latest one always added directly in front of a posterior abdominal region. The protective chitinous integument develops around the somite as the latter grows to full size.

In many of the more advanced arthropods partial or complete fusion of adjacent exoskeletal elements takes place, so that the complete exoskeleton is readily divisible into cephalic, thoracic, and abdominal regions. Fusion may even go so far as to result in a **cephalothorax** and a single abdominal shield, both showing only obscurely the original segmentation (*e.g.*, *Limulus*, Fig. 13-13).

Arthropods have several striking features in common with the annelid worms, the most conspicuous of which is the segmentation in the body, muscles, and nervous system. Because of this segmentation, which is shown by no other invertebrate phyla, many investigators favor the view that the Phylum Arthropoda arose from a primitive aquatic annelid sometime during the later Pre-Cambrian. In contrast, worms (also the Onychophora) have only a cuticle, whereas arthropods (except for the myriapods) have an exoskeleton.

Presumably developed from some aquatic annelid-like ancestor in the Pre-Cambrian, and probably persisting for a long period of time in the soft condition before the development of the exoskeletal armor, the arthropods were in an advanced stage of evolution by the beginning of the Paleozoic.

The Crustacea, Arachnoidea, and Trilobita, at least, had been differentiated by the beginning of the Cambrian, and the Chilopoda, Diplopoda, and Insecta were to make their appearance before or during the Carboniferous. The crustaceans still continue to live in water, as did the extinct trilobites, but the arachnoids, chilopods, and diplopods, and many of the insects have adapted themselves to terrestrial habitats, and some of the insects have modified their bodies for prolonged flight.

Class Crustacea[1]

General Considerations

The Class Crustacea is a large group of aquatic arthropods which have the soft parts encased in a chitinous, calcareochitinous or totally calcareous exoskeleton that protects the animal and serves as a place of attachment for muscles. Typical and familiar members of the class are shrimps, water fleas, pill bugs, barnacles, crayfishes, crabs, and lobsters, the latter three of which are widely sought for food. Crustaceans are chiefly herbivorous, but many familiar forms (*e.g.*, decapods) are carnivorous or scavenging. The food passes from the mouth into a large stomach and then travels through a straight intestinal tube to the anal opening at the posterior extremity of the body The anterior and posterior parts of the digestive tube are lined with a chitinous substance that is continuous with the exoskeleton; hence only the

[1] Crustacea—L. *crusta*, crust; referring to the crustlike nature of the chitinous integument of most crustaceans.

middle part of the canal can secrete digestive fluids and absorb nutrients from the passing food. The sexes are separate, and reproduction is ordinarily oviparous. Most lower Crustacea hatch from the eggs as a free-swimming larva termed a **nauplius** (Fig. 13-1*A*). At this nauplius stage of development the animal has an unsegmented body, a single median eye, and three pairs of appendages. By successive molts and attendant modifications, the animal adds somites in the process of developing into an adult. In more advanced Crustacea the nauplius stage is passed in the egg, and when the young organism leaves the egg it already has the form of the adult, although it is much smaller.

Each somite of a typical, relatively unspecialized crustacean has a pair of

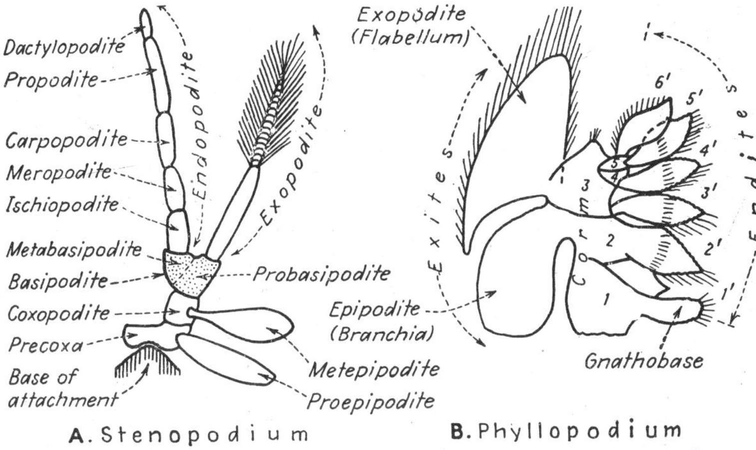

FIG. 13-2. Class CRUSTACEA. Highly generalized diagrams comparing the stenopodous and phyllopodous types of appendages. The protopodite, not indicated, would include the precoxa, coxa, and basipodite; the basipodite, shown by stippling, includes the probasipodite and metabasipodite if these two parts are differentiated. (*Adapted from Borradaile*, 1926.)

two-branched, jointed appendages (Fig. 13-2), which contain opposed sets of internal muscles for their movement. The appendages on different somites consist of parts that differ in number, structure, and function, but all are referable to a common fundamental form, the **biramous appendage;** *i.e.*, all appendages, however much modified, can be homologized with the supposed primitive and biramous appendage of the hypothetical ancestral crustacean (Figs. 13-2, 13-8).

Crustacean appendages, because of the remarkable variations they exhibit in living forms, are of great interest to both zoologists and paleontologists—to the former because of the modifications they have undergone in becoming adapted to different functions, and to the latter because of the light they throw, or fail to throw, on the relationships of several extinct groups of arthropods (*e.g.*, Trilobita; Eurypterida).

The first pair of appendages, designated **antennules,** differ from all the others in being uniramous and of simple structure.[1] All remaining appendages are of one or the other of two types—stenopodia or phyllopodia—both of which, however, are but variants of a fundamentally biramous plan of construction that is believed to have been characteristic of the appendages of the archetypal crustacean.

In the **stenopodium**[2] (Fig. 13-2), which has commonly been called "biramous," the two rami—an inner **endopodite** and an outer **exopodite**—arise independently from a common base, the **protopodite.** In addition to the rami, the protopodite in many cases also bears, on its outer side, one or more processes designated **epipodites.** Generally the endopodite is the larger of the two rami; in many cases it forms with the protopodite an axis, the **corm,** from which the exopodite extends laterally.

The protopodite may consist of as many as three pieces (or four depending on the nomenclature), but most commonly it consists of two—a proximal **coxopodite** and a distal **basipodite.** In a few forms a basal piece, the **precoxa,** precedes the coxopodite, and the basipodite itself may be divided into two parts—the **probasipodite** (which then usually bears the exopodite) and the **metabasipodite,**[3] or **preischiopodite.**[1] Epipodites, if present, arise from either the precoxa as **proepipodites** or from the coxopodite as **metepipodites.**

The endopodite is typically slender and segmented and consists of a maximum of five pieces, or **joints** (six if the preischiopodite is included), which, if present as in the crayfish, are designated, in order from the base distally, the **ischiopodite, meropodite, carpopodite, propodite,** and **dactylopodite.**

The exopodite is unsegmented in many forms and does not have a fixed number of joints. It is more likely to be reduced or lacking than is the endopodite. Either ramus may have a slender many-jointed terminal **flagellum.**

A composite diagram of a stenopodium is shown in Fig. 13-2*A* so that it can be compared with a phyllopodium and other arthropodan appendages.

The typical **phyllopodium** (Fig. 13-2*B*) is a broader and flatter appendage than most stenopodia, and its axial part, or corm, bears on the outer side one or more lobes, the **exites,** and on the median side a row of lobes, the **endites** (endites are rare in stenopodia). The more distal exite, which is usually about opposite the third or fourth endite from the base, is designated the **flabellum** and is homologous with the exopodite of the stenopodium or biramous appendage. Exites proximal to the flabellum are epipodites. The endite standing at the base of the phyllopodium is usually somewhat different in form from the rest and is designated a **gnathobase.** The corm consists of the usual five joints of the normal appendage.

[1] Many malacostracans have biramous antennules, but the rami are probably not homologous with those of the other appendages.

[2] Stenopodium—Gr. *stenos*, narrow, + *pous*, *podos*, foot; referring to the slender form of the appendage as compared to the foliose phyllopodium.

[3] If the term metabasipodite is used, it is implied that the component is a part of the basipodite, thus adding a possible fourth piece to a complete protopodite. If, on the other hand, the term preischiopodite is employed, it is implied that the component is considered to be the first segment of the endopodite; in which case a complete protopodite would have only three components.

Either a stenopodium or a phyllopodium may, because of modification, differ considerably from the ideal form just described. Even on the same animal, homologous joints may be quite different in structure.

Which of the two types—stenopodium or phyllopodium—more closely resembles the supposed appendage of the archetypal crustacean? This is a moot question, but an important one because it bears on the relationships of the Crustacea, Trilobita, and Annelida. Strong support can be cited for considering either the more primitive, and the interested reader will find our general question fully discussed by Borradaile (1926, 1926*a*, 1935), Störmer (1939), and Heegaard[1] (1945). The question is also considered briefly in our discussion of the phylogeny of the Trilobita.

The appendages of most crustaceans are modified in some way to perform a special function. The appendages of *Astacus* (Fig. 13-3), the familiar crayfish, may be taken as typical of advanced Crustacea. The first pair are termed **antennules,** and each has two whiplike lashes (the endopodite and the exopodite) that bear chitinous setae at their tips. In the second pair, designated **antennae,** the endopodite is a long whiplike process whereas the exopodite has the form of a blade-shaped plate. Both antennules and antennae have sensory functions. The third pair, designated **mandibles,** are modified into strongly chitinized teeth that aid in chewing. The fourth and fifth pairs, referred to respectively as first and second **maxillae,** lie behind the mandibles and form a sort of lower lip, which assists in the manipulation of food. All five pairs of appendages thus far described—*i.e.*, antennules, antennae, man-

[1] Since many paleontologists consider the trilobites as early crustaceans because their appendages are similar to those of some Crustacea, it is perhaps pertinent to comment briefly on one aspect of arthropod appendages. Heegaard (1945) has suggested that the primitive arthropod limb be regarded as a biramous appendage with an unjointed basal segment, the **sympodite** (protopodite of most authors; precoxa and coxa of Störmer, 1939), which is furnished with two branches, an endopodite and an exopodite (Fig. 13-2). In the ontogeny of certain crustaceans, the appendage at first begins as a bud, which at a very early stage becomes bifurcate before any articulation begins. Subsequently, two branches are divided off from the sympodite, which remains unjointed, and these form an independent endopodite and exopodite respectively. It is at this stage that the nauplians of all Crustacea begin. In some crustaceans the appendages remain at this stage throughout life or may not even progress to this stage, as with certain phyllopods in which either the endopodite alone or both endopodite and exopodite remain connected with the sympodite without any articulation or suture (Heegaard, 1945). More typically, however, both branches of the appendage undergo many modifications and commonly the exopodite is greatly reduced or wanting altogether, as in *Astacus*, the appendages of which are described in following paragraphs. In addition to reduction or loss of the exopodite, the endopodite in Crustacea is also commonly reduced, particularly in the circumoral appendages.

It would seem from Heegaard's (1942) arguments, therefore, that both Crustacea and Trilobita have primitive arthropod appendages and probably descended from a common ancestor.

As previously stated, we prefer to consider the Crustacea and Trilobita as separate and distinct classes which are probably rather closely related in their origin and ancestry.

In contrast Störmer (1939; 1944; 1951) does not consider the two branches of a crustacean appendage homologous with the two branches of a trilobite appendage, hence he would not support the contention that the Crustacea and Trilobita are closely related.

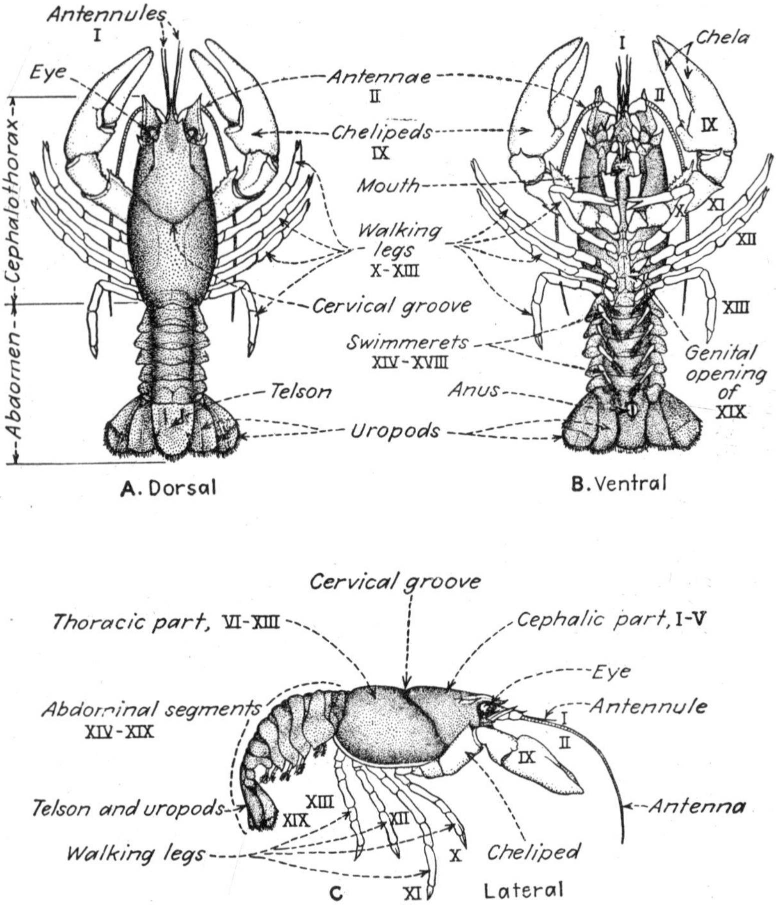

FIG. 13-3. Class CRUSTACEA. Diagrams showing three views of a common crayfish. Of the 19 appendages, the following are not designated on the diagrams: III, mandibles; IV and V, maxillae; and VI–VIII, maxillipeds. The five cephalic and the eight thoracic somites are covered by a continuous rigid carapace, or cephalothorax, and their division is marked by the transverse cervical groove. The six abdominal somites are covered by jointed abdominal segments.

dibles, and maxillae—originate from the cephalic part of the body. The thorax bears eight pairs of appendages, the first three (the **maxillipeds**) modified for food manipulation and the other five (the **walking legs**) modified for locomotion. The abdomen bears five pairs of **swimmerets** and a terminal swimming paddle consisting of a pair of broad **uropods.**

Appendages with distal segments modified into claws (**chelae;** singular, **chela**) are designated **chelipeds** and are described as **chelate.** The modifications shown by the thoracic and abdominal segments of *Astacus* (Fig. 13-3) illustrate the variations that may be expected in other Crustacea. The last segment, or **telson,** is flanked by a pair of flipper-like uropods.

The number of body segments varies widely among the Crustacea. As a rule, earlier and more primitive groups have larger numbers, whereas later and more specialized forms have fewer somites, suggesting that evolution has been accompanied by reduction in the number of somites through fusion.

The exoskeleton of crustaceans is composed either wholly of chitin or of that substance impregnated with calcium carbonate or calcium phosphate, $Ca_3(PO_4)_2$. It is typically divided into three parts—cephalic, thoracic, and abdominal. In many Crustacea, *e.g.*, lobsters and crayfishes, the cephalic and abdominal exoskeletal segments are fused into a **cephalothorax,** and the single rigid unjointed shield or bivalve shell covering this part of the animal is termed the carapace (Fig. 13-3). In the Ostracoda the exoskeleton has the form of a calcareous bivalve shell, and in the Cirripedia it is composed of more than a dozen calcareous plates.

Classification

Classification of the Crustacea has long been a subject of much discussion and divided opinion. In this work we use the widely accepted subclasses and orders of living crustaceans that Storer describes and illustrates in the 1951 edition of his "General Zoology" (McGraw-Hill Book Company, Inc.). The treatment of extinct subclasses and orders is as follows: (1) the Trilobita are removed from the Crustacea and considered as a separate class with their discussion postponed until both Crustacea and Arachnoidea have been considered; (2) we accept the three extinct[1] subclasses—Homopoda, Xenopoda, and Archaeostraca—but consider them as Arthropoda Incertae Sedis; their discussion is postponed until the subclasses of Crustacea and Arachnida that have living representatives have been considered.

The subclasses of Crustacea are considered in the following order: Branchiopoda, Ostracoda, Copepoda, Cirripedia, and Malacostraca.

Subclass Branchiopoda[2]

General Considerations. The Branchiopoda constitute a specialized group of living crustaceans that have a long geologic history reaching back possibly to the Cambrian, and certainly to the Silurian or Devonian. Some are naked; others have a chitinous or calcareochitinous covering or a calcareous bivalve shell. The somites vary in number, and the appendages, generally leaflike and lobed rather than leglike, are not primitive. Two **caudal furcae** are

[1] The well-known Middle Cambrian fauna from the Burgess shale of British Columbia, first described and illustrated by Walcott (1910–1931), contains an interesting group of unusual arthropods that possibly represent long extinct crustaceans or crustacean-like animals. As these fossils cannot be directly related to living genera by intermediate forms, they have been used as a basis for three extinct subclasses—Homopoda, Xenopoda, and Archaeostraca—which are to be regarded largely as provisional groups of present taxonomic convenience.

[2] Branchiopoda—Gr. *branchia*, gills + *pous, podos*, foot; referring to the breathing organs carried on certain of the appendages.

present in some species on the posterior extremity of the abdomen. There is considerable diversity of appearance among so-called branchiopods, and if only the exoskeletons were considered, it would seem that unlike organisms were being grouped together. The ontogeny of widely different forms, however, proves that they are closely related in their ancestry. The essential features of the several different kinds of exoskeletons are shown in Fig. 13-4.

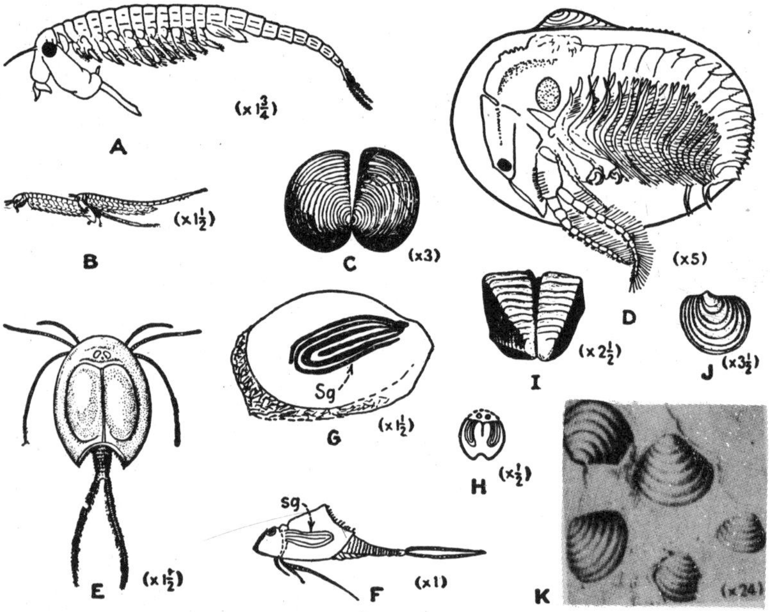

Fig. 13-4. Class Crustacea—Subclass Branchiopoda. *A. Branchipus vernalis*, a fresh-water branchiopod without a carapace. *B.* A pair of *Artemia gracilis* from Great Salt Lake, Utah; the male is clasping the female, with the claspers (*c*) in front of the ovisac (*e*), as they swim. *C. Pseudestheria ovata*, a complete carapace from the Triassic. *D.* "*Estheria*" *belfragei;* a male, much enlarged, showing features of organism and relation of shell and organism. *E–H. Apus: E, A. cancriformis*, a modern species; *F, A. aequalis*, from Colorado, a modern species showing shell gland (*Sg*) on left side; *G–H*, fossil impression of *A. beedei*, showing nature of shell gland (*Sg*) on right side and posterior emargination of carapace and flat projection of outline of carapace, from Permian of Oklahoma. *I. Leaia leidyi*, from the Mississippian. *J–K. Rhabdostichus: J, R. buchoti* from marine Silurian or Devonian of France; *K, R. pulex* from New York marine Devonian. (*A–B, D after Packard*, 1883; *C after Nicholson*, 1872; *E after Nicholson and Lydekker*, 1889; *F–H after Ruedemann*, 1922; *I after Jones in Nicholson*, 1872; *J after Raymond*, 1946; *K after Hall and Clarke*, 1888.)

Branchiopods are widespread and abundant in all kinds of waters, even those that are extremely salty. It seems probable that they had similar distribution during the past, though this assumption cannot be proved because fossil representatives are not common or abundant.

Classification. The Subclass Branchiopoda is generally subdivided into four orders:

Order 1. Anostracea.
Order 2. Notostraca.
Order 3. Conchostraca.
Order 4. Cladocera.

The first three orders constitute the so-called Phyllopoda and have a fossil record that extends from Silurian or Devonian to Recent. The Cladocera, or water fleas, are not known as fossils before the Pleistocene.

Order 1. Anostraca.[1] The Anostraca are elongate crustaceans that have a distinct head, which is provided with stalked eyes; 11 to 19 pairs of thoracic appendages; and no carapace. The order includes two doubtful genera from Carboniferous rocks,[2] one genus (*Artemia*) from the Eocene, and two families of living genera that include forms inhabiting salt lakes and brine pools as well as the less saline waters of the oceans. *Carboniferous?; Eocene to Recent.*

Order 2. Notostraca.[3] Notostracans have low oval carapaces, sessile eyes, and 40 to 60 somites. Middle Cambrian genera first referred to this order by Walcott (1912, 1931) have been assigned to the extinct Order Pseudonotostraca, but they are believed to be linked with modern notostracans, possibly through such a form as the Mid-Devonian *Lepidocaris rhyniensis* (see Raymond 1935, page 218).

The fossil record of the order is uncertain. *Apus* (Fig. 13-4*E–H*), which ranges from Permian to Recent and is abundant in modern seas, seems to be the oldest known undoubted notostracan.[4] *Permian to Recent.*

Order 3. Conchostraca.[5] The Conchostraca are dominantly fresh-water crustaceans characterized by a compressed body of 10 to 27 somites enclosed within a bivalve carapace. The eyes are sessile, the appendages are numerous and phyllopodous, and the caudal furcae have clawlike branches. The carapace is commonly difficult to distinguish from ostracode and certain molluscan shells. Most fossil conchostracans consist only of the carapace, because the appendages and other anatomical features were destroyed or lost before or during fossilization. Consequently, families, genera, and species have to be based entirely on the characteristic features of the carapace—*e.g.*, outline of valves, curvature of hinge, position of umbo, nature of muscle scar, and nature of surface sculpture.

[1] Anostraca—Gr. *an*, without, + *ostracon*, shell of a testacean; referring to the lack of a shell.

[2] The Middle Cambrian genera *Leanchoilia*, *Opabinia*, and *Yohoia* reported from the Burgess shale by Walcott (1912), and referred by him to the Anostraca, are now generally included in the extinct Order Pseudanostraca of the extinct Subclass Homopoda (Raymond, 1935). *Lepidocaris rhyniensis* (Scourfield, 1926), from the Devonian, resembles an anostracan in some details, but is different enough that Scourfield based a new order, Lipostraca, on it alone. This order is not included in the present work.

[3] Notostraca—Gr. *not(o)*, the back part, + *ostracon*, shell of a testacean; referring to the shell on the back of the animal.

[4] Howell and Kobayashi (1936) reported a new notostracan genus from the Ordovician of Siberia, and stated that other supposed notostracans of Upper Cambrian age had been found. The exact affinities of these Siberian arthropods are uncertain.

[5] Conchostraca—Gr. *konche* (L. *concha*), shell, + *ostracon*, shell of a testacean; referring to the fact that the covering of the animal is a shell.

According to Raymond (1946), who recently classified all known genera of fossil Conchostraca, the age of the oldest conchostracan cannot be stated definitely. *Rhabdostichus* is the oldest genus thus far reported if it is truly Silurian; however, there is a possibility that the containing rock may be Middle Devonian, in which case it would not be the oldest form. Several genera have been reported from the Lower Devonian of Belgium and the Eifel district of Germany, and other genera are found in Middle Devonian and younger rocks in North America. True[1] conchostracans are believed to have descended from some bivalve crustacean of Cambrian or Ordovician time and, with the exception of the marine *Rhabdostichus*, to have invaded fresh-water environments in Late Silurian or Early Devonian times. *Silurian?, Lower Devonian to Recent.*

Rhabdostichus (Fig. 13-4) is a Mid-Paleozoic (Sil. or Dev.) genus and *Leaia* (Fig. 13-4) a common later Paleozoic genus. Many species formerly assigned to *Estheria* have had to be reassigned to several new genera because that old and familiar generic name was preoccupied (Raymond, 1946).

Order 4. Cladocera.[2] This order includes the minute water fleas which move jerkily through the water by using the somewhat enlarged second antennae. The animal has a tiny bivalve shell that encloses the body but generally leaves the head exposed, paired sessile eyes that are fused and medially situated, and four to six pairs of trunk appendages. No fossil representatives are certainly known. A doubtful specimen (*Lynceites*) from the Carboniferous of Europe may belong in this order, and some egg cases found in Pleistocene glacial deposits of Germany may have been made by cladocerans. *Carboniferous?, Pleistocene?, Recent.*

Subclass Ostracoda[3]

General Considerations. Ostracodes are minute, lentil-shaped crustaceans having a bivalve carapace that completely encloses the indistinctly segmented body (Fig. 13-5). The appendages are few and much modified. The carapaces, or shells, range in size from less than 1 mm. in length to large specimens more than 20 mm. long and are chitinous or calcareous. Ostracodes live in all aquatic environments but are most abundant in marine waters where they often appear in great numbers, either swimming at or near the surface or creeping over the bottom. They live in shallow waters and have scavenging habits. So voracious are many forms that they will quickly remove all flesh from a dead animal placed in their midst.

The body of the animal is attached to the enclosing carapace by means of adductor muscles that pull the valves together. The body is smaller than the

[1] Numerous Cambrian genera referred to the Order Conchostraca by Ulrich and Bassler (1931) are considered by Raymond (1946) to be not true conchostracans but rather an extinct order of bivalve crustaceans, some of which may have been ancestors of the Ostracoda or Conchostraca. Raymond created a new archaeostracan order, Bradorina, to include these ancient forms.

[2] Cladocera—Gr. *clados*, a branch, + *keras*, a horn; referring to the branched antennae.

[3] Ostracoda—Gr. *ostracodes*, testaceous, from *ostracon*, shell of a testacean; referring to the fact that the typical ostracode has a calcareous bivalve shell.

carapace and is only indistinctly segmented. The seven pairs of appendages consist of two pairs of antennae (first and second; or antennules and antennae, respectively), one pair of mandibles, two pairs of maxillae, and two pairs of slender trunk appendages, or legs. In fresh-water[1] forms there are three pairs of legs and only one pair of maxillae. The rudimentary abdomen ends in a single or bifurcated spine, which is used mainly for clearing out foreign matter that comes between the valves. The typical ostracode has a small **median eye** and a pair of large **lateral eyes**. **Eye tubercules,** or **ocular spots,** on the shell exterior may indicate the positions of the lateral eyes.

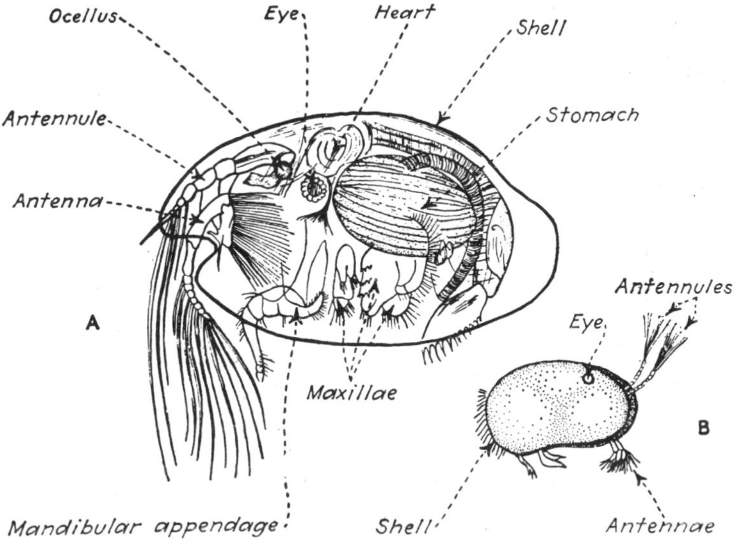

FIG. 13-5. Subclass OSTRACODA. *A. Cypridina messinensis*, viewed from the side and greatly enlarged, with one valve of shell removed so as to reveal the different parts of the body. The appendages designated maxillae are probably more correctly considered legs, particularly the back two. *B. Cypris fusca*, viewed from the side, with the two valves slightly agape and with several of the appendages protruding. (*Modified after Nicholson and Lydekker*, 1889.)

The sexes are separate, but the shells of the two are essentially alike except for size and shape (*e.g.*, the male shell is the larger in *Candona*, whereas it is the smaller in *Cypris*). Some members of the Paleozoic Beyrichidae have shells with a strongly inflated posterior part or with a swollen pouch in a posterior position. Shells with such **brood pouches,** as they have been termed, are assumed to have belonged to the female of the species (Fig. 13-6*H*).

The Carapace, or Shell. Students of fossil Ostracoda customarily refer to the carapace as the shell, and that term is used in the present discussion. The typical ostracode shell consists of two valves, a right and a left, which

[1] The interested reader will find excellent illustrations of fresh-water Ostracoda in the following article: EURTOS, N. C., 1935, Fresh-water Ostracoda from Florida and North Carolina, *Amer. Midland Naturalist*, vol. 17, pp. 471–522.

articulate along the dorsal edge, commonly designated the hinge line, and meet or overlap along the ventral, dorsal, or entire margin. The valves are pulled together by a subcentral adductor muscle that commonly leaves a scar of some kind (tubercle, pit, or cluster of spots) on the inner surface of each valve. The outer surface of the valves may be smooth and glossy, granulose, pitted, striate, or reticulate. In addition to having these small

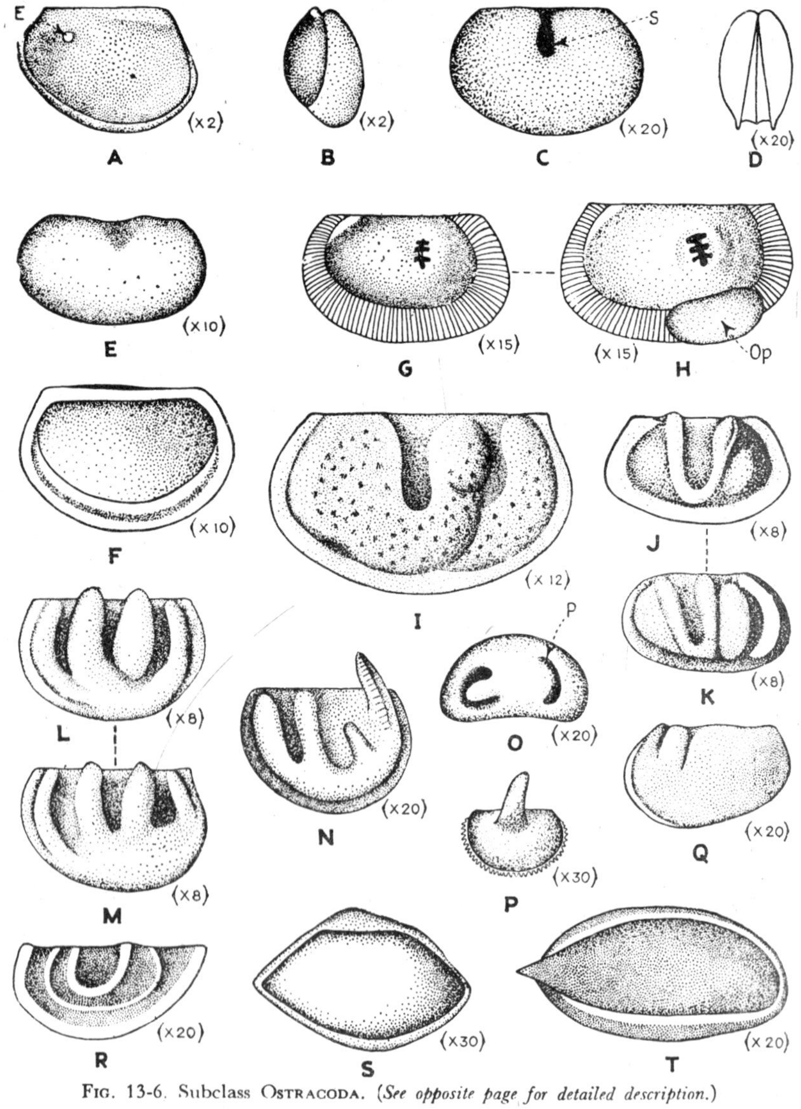

FIG. 13-6. Subclass OSTRACODA. (*See opposite page for detailed description.*)

sculptural details, the valves of many fossil ostracodes are strongly lobate, sulcate, or nodose. These more prominent surface features are taxonomically significant in some species and hence of interest.

Classification. The classification of living Ostracoda is based largely on the nature of the appendages, but since these structures are almost never preserved in fossil ostracodes, identification and classification of fossil forms must be based entirely on shell features. The more important of these features are:

1. Shape or outline of valve as seen in side view; valve convexity; location of greatest shell thickness.
2. Nature of surface sculpture.
3. Contact relations of valves.
4. Nature of brood pouches, if present.
5. Nature and position of adductor muscle scars.
6. Microstructure of shell material.

Geologic History and Stratigraphic Range. The earliest ostracode thus far described was found in Upper Cambrian rocks and has been assigned to the Family Leperditiidae. Other Cambrian genera once assigned to the Ostracoda are now included in other divisions of the Crustacea. Ostracodes have been relatively abundant in every geologic period since the Early Ordovician to the present time, and in a few cases their shells constitute a large part of the rocks in which they are preserved. Paleozoic and Cenozoic ostracodes have been studied extensively and have been widely used for stratigraphic correlation; Mesozoic species, on the other hand, are not so well known. More than 250 genera and 2,500 species of fossil and living ostracodes have been described. Many of these are illustrated in "Index Fossils of North America" (Shimer and Shrock, 1944). *Upper Cambrian to Recent.*

Representatives of a few of the families are illustrated in Fig. 13-6.

Subclass Copepoda[1]

The copepods are minute crustaceans without any sort of distinct carapace, which are present in enormous numbers in modern seas (also less numerous

[1] Copepoda—Gr. *cope*, an oar, + *pous*, *podos*, foot; referring to the oarlike legs or appendages.

FIG. 13-6. *A–B. Leperditia fabulites* from Ordovician of Minnesota: *A*, left side of entire shell, showing eyespot (*e*) and characteristic overlapping of the larger right valve; *B*, posterior view. *C–D. Euprimitia sanctipauli* from Ordovician of Minnesota, showing right valve and end view of entire carapace. *E–F. Leperditella inflata* from Ordovician of Kentucky, showing exterior and interior views of two left valves. *G–H. Chilobolbina dentifera* from Ordovician of Estonia, showing male and female, respectively. *I.* Male left valve of *Zygobeyrichia ventribunctata* from Silurian of West Virginia. *J–K.* Left valves of *Zygosella: J*, *Z. vallata* from Silurian of West Virginia, a male left valve; *K*, *Z. macra* from same horizon in Virginia, a female left valve showing the narrow, ridgelike brood pouch paralleling the posterior border. *L–M. Drepanellina clarki*, from the Silurian of Maryland, showing male and female valves, respectively. *N. Ceratopsis chambersi*, left valve from Ordovician of Minnesota. *O. Thlipsura v-scripta discreta*, a left valve from Silurian of Gotland. *P. Aechmina bovina*, a left valve from Silurian (Wenlock) of England. *Q. Kloedenella obliqua*, right valve from Silurian of Maryland. *R. Strepula concentrica*, a right valve from Silurian of England. *S. Bairdia beedei*, a completes hell from Pennsylvanian of Kansas. *T. Krausella inaequalis*, right side of a complete shell from Ordovician of Illinois. (*E* = eyespot; *Op* = brood pouch; *P* = pit; *S* = sulcus.) (*All figures redrawn from Ulrich and Bassler*, 1923.)

in fresh waters). They do not seem to have left any fossil record. The typical copepod has nine free trunk somites, of which only the first five have appendages. The antennae and antennules are usually well developed, and a caudal furca is present in most forms. Most copepods are free-living, but some are commensal or parasitic and these are likely to show considerable modification in number of somites and nature of appendages. *Recent.*

Subclass Cirripedia[1]

General Considerations. The Cirripedia, popularly known as **barnacles,** are a group of greatly modified exclusively marine crustaceans which, because they closely resemble certain mollusks, were referred to the Phylum Mollusca until 1830, when study of their early life history showed them to be true arthropods.

The body is enclosed within a membranous mantle developed from the carapace, and in many forms this mantle is covered with calcareous plates. The adult barnacle, except for parasitic forms, is attached by the anterior extremity of the head (actually by secretions from a special gland on the first antenna). The adult body is either unsegmented or only obscurely segmented, and the six or less pairs of slender, feathery biramous appendages are situated behind the mouth and are used in food gathering.

A cirriped egg hatches into a free-swimming larva, or nauplius, which molts several times before it secretes a bivalve shell like that of the ostracodes. Ultimately this larva attaches itself to some object on the bottom and grows into an adult animal.

Parasitic cirripeds lack a shell, and this condition may also have prevailed in some ancient species, though no fossils have been found to prove such a suggestion. Those provided with calcareous shells attach themselves to some object on the bottom during the larval stage, and the multiplated shell develops as the young barnacle grows to adulthood. Rocky coasts and the boulders of gravel-covered bottoms that are not too strongly swept by currents and waves are commonly literally plastered with the shells of the familiar acorn barnacle. Along many rocky coasts the high-tide line is strikingly indicated by the sharp upper limit of the "barnacle zone." When the tide goes out, the barnacle simply draws in its appendages and closes the shell by means of special plates.

Cirripeds also live attached to floating or stranded plant materials, to living shelled invertebrates and testaceous arthropods and vertebrates, and to all sorts of dead shells and skeletal fragments. Certain species are an expensive nuisance in ship operations because they build calcareous masses of such extent on the ship's hull that they impede progress through the water and must, therefore, be removed from time to time. Some cirripeds are also parasitic on or within certain animals. These, of course, have left no fossil record.

The goose barnacle, which is considered more primitive than the sessile forms, is attached by a thick, flexible stalk (**peduncle**) that is naked in some

[1] Cirripedia—L. *cirrus*, a curl, + *pes*, *pedis*, foot; referring to the fringelike appearance of the appendages when extended beyond the edges of the shell plates.

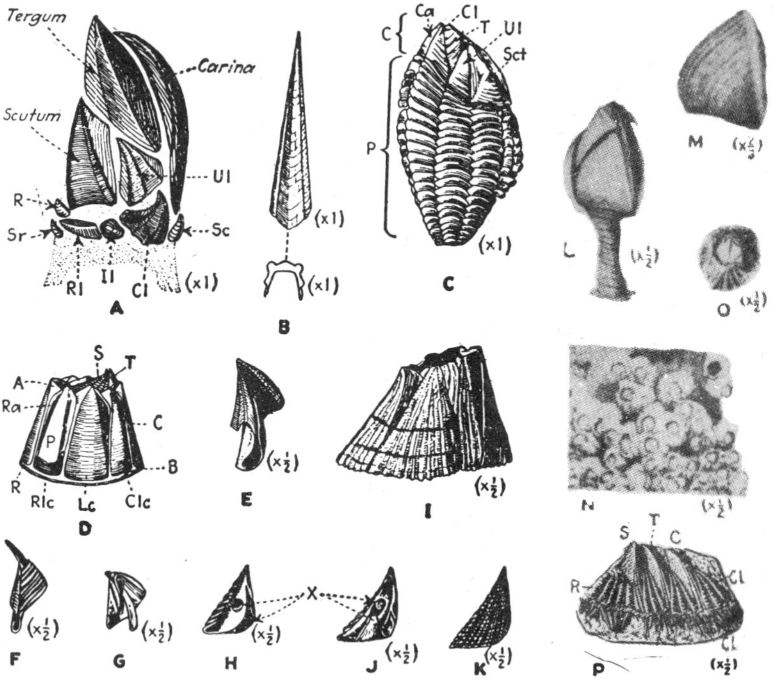

FIG. 13-7. Subclass CIRRIPEDIA. Living and fossil barnacles. *A*. Capitulum of *Scalpellum fossulum*, a pedunculate cirriped from the Upper Cretaceous of England. The peduncle is stippled. *B*. Dorsal view of the carina from *A*. *C*. *Loricula darwini*, a pedunculate individual found attached to an ammonite from the Cretaceous of England. *D*. Diagram of the shell of modern *Balanus*, a sessile barnacle. *E*. Tergum of *B. concavus* from the Tertiary of England. *F–G*. Exterior and interior of tergum of *Balanus*. *H*. Interior of scutum of *Balanus* showing the muscle scars (*X*). *I*. Shell of *B. concavus* (cf. *E*). *J–K*. Interior and exterior of the scutum of *I*. *L–M*. *Anatifa lepas*, a modern pedunculate cirriped popularly known as a "goose barnacle": *L*, complete scalpellum and peduncle; *M*, a detached scutum. *N–O*. *Balanus balanoides*, a modern acorn barnacle: *N*, cluster of shells attached to a pile; *O*, a detached shell of conical form. Note that the terga and scuta are in place in both *N* and *O*. *P*. *Hercolepas signatus*, a Silurian sessile barnacle from the island of Gotland. (*A* = alae; *B* = basis; *C* = capitulum; *Ca* = carina; *Cl* = carinolatus; *Clc* = carinolateral compartment; *Il* = inframedian latus; *L* = lateral; *P* = peduncle; *Pa* = paries; *R* = rostrum; *Ra* = radii; *Rl* = rostral latus; *Rlc* = rostrolateral compartment; *Sc* = subcarina; *Sct*, *S* = scutum; *Sr* = subrostrum; *T* = tergum; *Ul* = upper latus; *X* = muscle scars.) (*A–B, D–K after Darwin*, 1854; *C after Woodward*, 1865; *L–N after Nicholson*, 1872; *N–O after Pilsbry*, 1916; *P after Aurivillius in Withers*, 1915.)

forms (*e.g.*, *Anatifa*, Fig. 13-7*L–M*, and *Scalpellum*, Fig. 13-7*A*). At its free extremity the peduncle bears the **capitulum.** This corresponds to the shell of the sessile balanoids and is composed of numerous separate and free-moving calcareous plates that form a shield about the body and its appendages (Fig. 13-7). The several kinds of plates are illustrated and named in Fig. 13-7 and may be homologized with similar plates on an acorn barnacle as shown in the same figure.

The acorn barnacle, typified by *Balanus* (Fig. 13-7), has a well-organized shell composed of four to ten calcareous plates more or less fused together along the sides and bottom to form a truncated cone that is cemented by its base to any one of a wide variety of objects (Fig. 13-7). The opening into the shell is covered by two pairs of plates, the **terga** (singular, **tergum**) and **scuta** (singular, **scutum**). These four plates are hinged, and when retracted they protect the animal within from drying out when the shell is exposed during tidal change.

Geologic History and Stratigraphic Range. The fossil record of the Cirripedia is so sparse that little of its pre-Mesozoic history can be deduced from the few scattered fragments that have been found. Certain fossils from Cambrian strata have been provisionally referred to the Cirripedia, but the taxonomic relationships of these are too uncertain to warrant the statement that the cirripeds appeared in the Cambrian. Serious doubt has also been cast on supposed fossil barnacles from the Lower Ordovician because these are isolated plates that have never been found in a complete assemblage. It is generally stated that the oldest undoubted fossil barnacle is the Silurian genus *Hercolepas* (Fig. 13-7) which Withers (1915) considered to be a sessile form of the balanoid type but unlike any other known living or fossil cirriped. The true nature of other supposed fossil barnacles from Paleozoic rocks is uncertain.[1] The next oldest undoubted fossil cirripeds come from Jurassic and Cretaceous rocks, and these, interestingly, are pedunculate forms (*e.g.*, *Loricula*, Fig. 13-7; *Scalpellum*, Fig. 13-7). Beginning with the Cretaceous, well-preserved pedunculate and sessile barnacles are known from both Cretaceous and Tertiary rocks. It is difficult to explain why the sessile barnacles appear fossil so much earlier than the supposedly more primitive pedunculate group from which they are generally assumed to have developed. In this connection, it is also to be noted that there is no acceptable record of sessile barnacles between the Silurian and Upper Cretaceous. These data have led some investigators to doubt the validity of all Paleozoic and Early Mesozoic Cirripedia, and to postulate that the subclass did not originate until later Mesozoic time.

The Cirripedia are thought to have evolved from an ancestral crustacean and to have developed their unique features in part, at least, as they became adapted to a sessile habit. In their evolution they became hermaphroditic, developed numerous calcareous plates on the exterior of the mantle, simplified certain parts of the body, and lost other parts altogether. Although the Cirripedia are much modified and highly specialized animals, their crustacean affinities are clearly indicated by their embryology.

At least 1,000 species of cirripeds have been described. Of these somewhat more than 800 are living and about 200 extinct. Many living species have no hard parts at all. *Anatifa lepas* (Fig. 13-7*L–M*) typifies living pedunculate

[1] *Eobalanus* from the Upper Ordovician (Ruedemann, 1924) and *Protobalanus* from the Devonian (Withers, 1915; Van Name, 1926) are examples of supposed ancestral balanoids, but the cirriped nature of these fossils is uncertain. Likewise, several supposed pedunculate barnacles (*e.g.*, *Turrilepas*, *Plumulites*, and *Strobilepas*) based on isolated plates or incomplete specimens from Ordovician to Carboniferous rocks are of uncertain relationships (Withers, 1915).

barnacles, and *Scalpellum* (Fig. 13-7) and *Loricula* (Fig. 13-7), both from the Cretaceous, are typical of supposed fossil pedunculate forms.

Balanus (Fig. 13-7) from modern seas is an example of the acorn or sessile barnacles and *Hercolepas* (Fig. 13-7), from the Silurian, may be an ancient representative of this group.

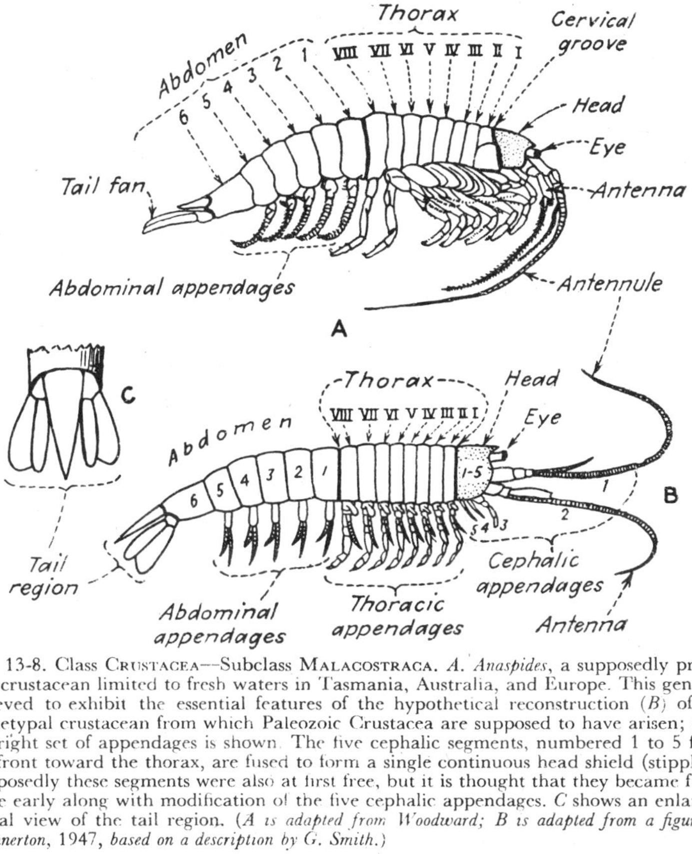

FIG. 13-8. Class CRUSTACEA—Subclass MALACOSTRACA. *A. Anaspides*, a supposedly primitive crustacean limited to fresh waters in Tasmania, Australia, and Europe. This genus is believed to exhibit the essential features of the hypothetical reconstruction (*B*) of the archetypal crustacean from which Paleozoic Crustacea are supposed to have arisen; only the right set of appendages is shown. The five cephalic segments, numbered 1 to 5 from the front toward the thorax, are fused to form a single continuous head shield (stippled). Supposedly these segments were also at first free, but it is thought that they became fused quite early along with modification of the five cephalic appendages. *C* shows an enlarged dorsal view of the tail region. (*A is adapted from Woodward; B is adapted from a figure by Swinnerton,* 1947, *based on a description by G. Smith.*)

Subclass Malacostraca[1]

General Considerations. The Subclass Malacostraca is a large, compact group of diverse but closely related crustaceans among which are the familiar crabs, crayfishes, lobsters, and shrimps. Malacostraca have been much studied by zoologists, but paleontologists have paid little attention to the subclass because of the rarity of fossils, particularly of forms older than the Mesozoic.

The archetypal crustacean, which is believed to have been present in

[1] Malacostraca—Gr. *malacos*, soft, + *ostracon*, shell of a testacean; referring to the soft-shelled nature of many members of this subclass.

earliest Cambrian seas and probably also in Pre-Cambrian seas, is believed to have had 19 body segments divided as follows: 5 cephalic segments fused with one another to make the head; 8 thoracic segments, all free from one another; and 6 free abdominal segments, with a flattened telson (Fig. 13-8).

Supposedly the eyes were carried upon movable stalks, the antennules and antennae were biramous, the antennae had the outer branch scalelike, and the other cephalic appendages were adapted to perform mastication. The eight thoracic appendages were all alike and were typically crustacean; *i.e.*, they had a basal part, the protopodite, with two divisions, the endopodite and exopodite, respectively. The endopodite had six divisions and served entirely for walking. The exopodite was adapted for swimming. The five abdominal appendages were biramous and all alike except for the last, which

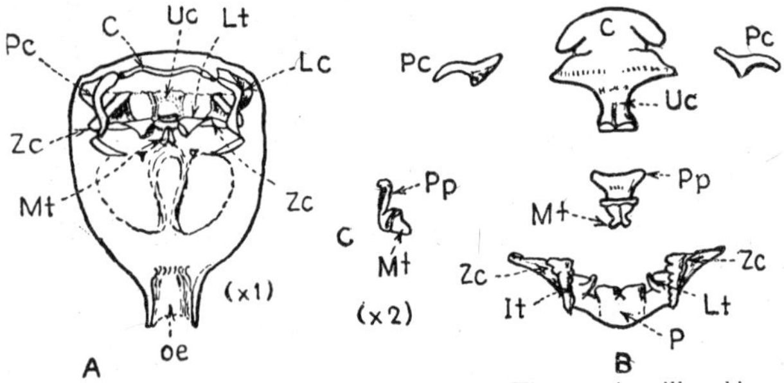

FIG. 13-9. Class CRUSTACEA—Subclass MALACOSTRACA. The gastric mill and its components in the common crayfish, *Astacus fluviatilis*. *A*. The stomach, viewed from the front, after removal of the anterior wall The components are seen in place. *B*. The ossicles of the gastric mill separated from one another. *C*. The prepyloric ossicle and median tooth, seen from the right side. The several components of the gastric mill are calcified cuticle. (*C* = cardiac ossicle; *Lt* = lateral tooth; *Mt* = median tooth; *Oe* = esophagus; *P* = pyloric ossicle; *Pc* = pterocardiac ossicle; *Pp* = prepyloric ossicle; *Uc* = urocardiac process; *Zc* = zygocardiac ossicle.) (*Adapted from Huxley*, 1880.)

was much flattened and together with the telson formed a tail fin. The essential exoskeletal features of this hypothetical ancestral crustacean are shown in Fig. 13-8. It is believed that the earliest malacostracans came from this sort of ancestor.

Modern Malacostraca have the body typically composed of 19 somites distributed the same as in the postulated ancestral form—5 cephalic, 8 thoracic, and 6 abdominal.[1] The head is fused to one or more thoracic somites; a carapace is commonly present; and the abdominal somites have appendages. The newly born malacostracan has progressed beyond the nauplius stage and in many forms is quite similar in appearance to its parents except for size.

The Malacostraca are an important source of food for man, and every year millions of individual crabs, lobsters, and shrimps are taken. They are also important scavengers on the sea bottom, where they eat everything of

[1] Possum shrimps in the embryonic stage and modern Leptostraca (*e.g.*, *Nebalia*) throughout life have seven abdominal segments, and it may be that the archetypal malacostracan also had seven instead of six.

organic nature irrespective of the degree of decomposition. Many have a **gastric mill** (Fig. 13-9) in the intestinal tract in which shells that have been swallowed are broken into bits or ground into powder. Some of the larger species break open shells with their claws in order to obtain the animals inside.

Geologic History and Classification. Although malacostracans are extremely abundant in modern seas and probably have been so since the middle of the Mesozoic, they are rarely abundant as fossils. Perhaps the best-preserved individuals come from the famous lithographic limestone of Solenhofen, Bavaria—a rock that was deposited as a lime mud. Almost equally good specimens have been found in the well-known concretions from the Pennsylvanian shales along Mazon Creek in Illinois. Many specimens of later time have also been found in concretions.

The Subclass Malacostraca is generally divided into the primitive Leptostraca and the more highly specialized Eumalacostraca. Subdivisions are as follows:

Series 1. Leptostraca. The carapace is bivalve; abdomen has seven somites; all forms are small and have wide distribution in existing marine waters; and the group seems to have evolved during the Middle Paleozoic.[1]

Order 1. Nebaliacea. Fossil forms are questionable. *Nebalia* (Fig. 13-10*A*), a living genus, is an example.

Series 2. Eumalacostraca. The abdomen has six or less somites, and most representatives have developed since the beginning of the Mesozoic.

Order 1. Anaspidacea. The carapace is lacking and the first thoracic somite is fused with the head, or defined therefrom by a groove. The order, which seems to have appeared in the Carboniferous, has only four living genera, all of which live in fresh-water habitats (Australia and Tasmania). A typical subterranean aquatic form also lives in Europe. *Carboniferous to Recent.*

Palaeocaris (Fig. 13-10*C*), from the Pennsylvanian, is a fossil representative, and *Anaspides* (Fig. 13-8*A*) is a typical recent form.

Order 2. Mysidacea (Prawns). These shrimplike forms, which have a carapace over much of the thorax and a tail fan of uropods, are mostly marine and are represented by some 300 living species. *Mississippian to Recent.*

Anthrapalaemon (Fig. 13-10*B*), from the Pennsylvanian, has been referred to this order.

Order 3. Cumacea. These small eumalacostracans have a carapace with two anterior plates commonly jointed over the head. They are mostly marine and burrow in soft sediments. No fossils are known.

Order 4. Tanaidacea. These are tiny marine forms having a carapace and a chelate second thoracic appendage. They live in mud or in tubes and have been found at depths as great as 2,000 fathoms (12,000 ft.). No fossils are known.

Order 5. Isopoda. These are the familiar terrestrial pill bugs and their marine relatives. *Devonian to Recent.*

Cyclosphaeroma (Jura.) (Fig. 13-10*F*) and *Sphaeroma* (Recent) (Fig. 13-10*E*) are representatives of the order.

Order 6. Amphipoda. These are the sand hoppers, which lack a carapace and commonly have a laterally compressed body; they live in both fresh and salt waters.

[1] Many of the forms once included in the Phyllocarida and here assigned to the Archaeostraca (Arthropoda Incertae Sedis) may actually be Leptostracans.

The first undoubted fossil amphipods come from Tertiary rocks. Supposed fossil amphipods from earlier rocks (*e.g.*, *Necrogammarus* from the Silurian) are questionable. *Tertiary to Recent.*

Order 7. Stomatopoda. These are the mantis shrimps, which have a head with two movable anterior somites bearing eyes and antennules. The 200 living species are marine and dwell on the bottom in sand or crevices. The oldest fossil stomatopod yet reported is *Squillites spinosus*, based on two fragmental specimens from the Mississippian Heath shale of Montana (Scott, 1938). Supposed fossls from Pennsylvanian rocks are of uncertain affinities, but well-preserved forms (*e.g.*, *Sculda*) have been found in the Jurassic beds of Solenhofen, Bavaria. *Mississippian to Recent.*

Order 8. Euphausiacea. These are marine forms with all thoracic segments biramous. A nauplius of the common genus *Euphausia* is shown in Fig. 13-1. No undoubted fossil specimens seem to have been reported.

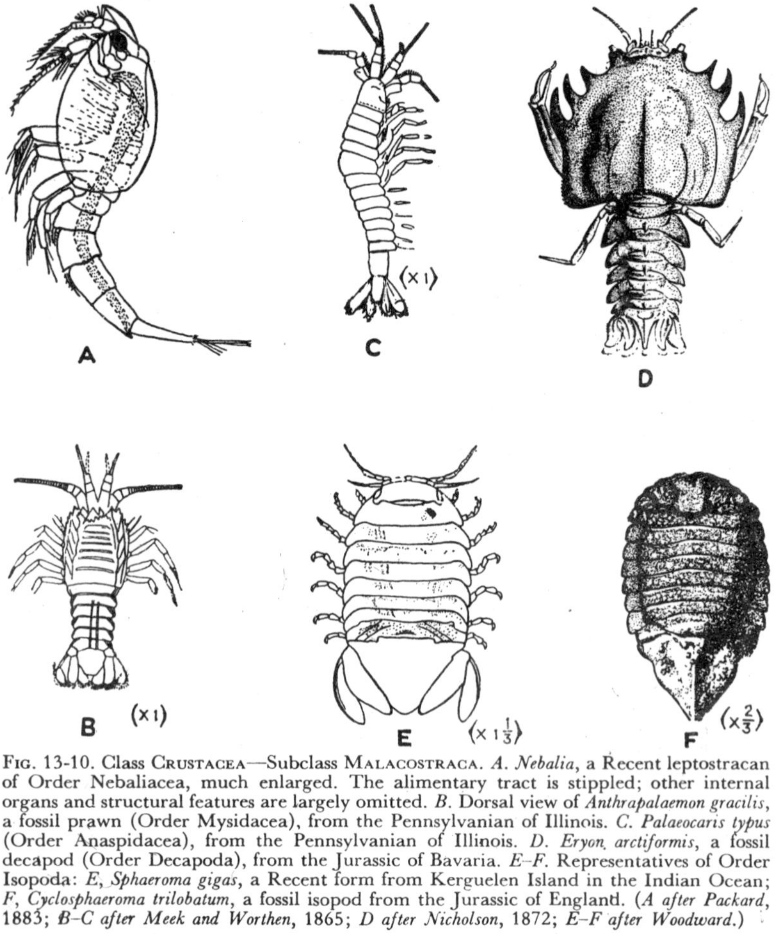

FIG. 13-10. Class CRUSTACEA—Subclass MALACOSTRACA. *A. Nebalia*, a Recent leptostracan of Order Nebaliacea, much enlarged. The alimentary tract is stippled; other internal organs and structural features are largely omitted. *B.* Dorsal view of *Anthrapalaemon gracilis*, a fossil prawn (Order Mysidacea), from the Pennsylvanian of Illinois. *C. Palaeocaris typus* (Order Anaspidacea), from the Pennsylvanian of Illinois. *D. Eryon arctiformis*, a fossil decapod (Order Decapoda), from the Jurassic of Bavaria. *E–F.* Representatives of Order Isopoda: *E*, *Sphaeroma gigas*, a Recent form from Kerguelen Island in the Indian Ocean; *F*, *Cyclosphaeroma trilobatum*, a fossil isopod from the Jurassic of England. (*A after Packard*, 1883; *B–C after Meek and Worthen*, 1865; *D after Nicholson*, 1872; *E–F after Woodward.*)

Order 9. Decapoda. This important order, which is represented by fossils as old as Triassic, includes the familiar shrimps, crabs, crayfishes, and lobsters. The thoracic appendages are mostly uniramous and five pairs of walking legs are characteristic. Most of the 8,000 species live in salt water, but some dwell in streams and fresh-water lakes and ponds, and a few are terrestrial. *Triassic to Recent.*

Astacus (Fig. 13-3), the familiar fresh-water crayfish, is a typical decapod. Fossil decapods (*e.g.*, *Eryon;* Fig. 13-10*D*) are especially well preserved in the Jurassic limestone of Solenhofen, Bavaria.

Order 10. Nahecarida. This is an extinct Lower Devonian order based on the genus *Nahecaris* (Fig. 13-11), which has a pair of biramous antennules and a pair of biramous antennae. The other cephalic segments are unknown. The trunk consists of eight thoracic and eight abdominal segments. The eight pairs of thoracic appendages are slender and biramous, with the exopodite weaker than the endopodite. The front five abdominal segments bore biramous pleopods. The carapace was all in one piece.

Nahecaris closely resembles certain genera of the Order Decapoda and it may be that the Decapoda were derived from some *Nahecaris*-like ancestor (Raymond, 1935).

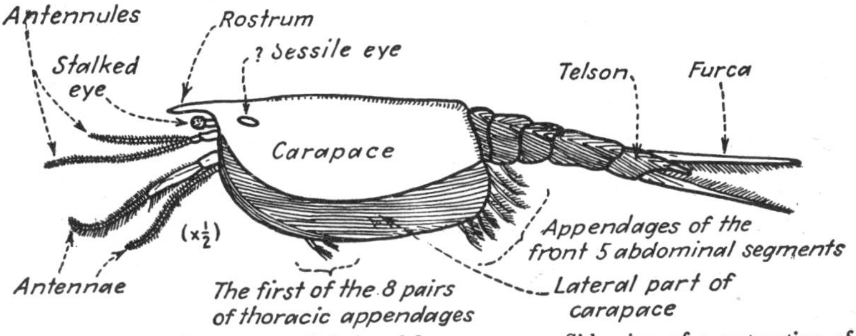

FIG. 13-11. Class CRUSTACEA—Subclass MALACOSTRACA. Side view of a restoration of *Nahecaris stuertzi* from the Lower Devonian black shales of Bundenbach, Germany. The Order Nahecarida is based on this extinct genus. The animal is supposed to have had eight thoracic and eight abdominal segments (the last abdominal segment includes the telson). The first eight pairs of appendages were slender and biramous; five segments of the abdomen bore biramous pleopods, with the endopodites and exopodites about equal in strength. In the restoration most of the cephalic and thoracic segments are missing. (*Modified after Broili*, 1929.)

CLASS ARACHNOIDEA[1]

General Considerations

The Class Arachnoidea comprises a large and varied group of specialized arthropods among which spiders, mites, ticks, scorpions, and king crabs are living examples. One large group of aquatic members, the Eurypterida and allies, are extinct, but *Limulus*, the king crab, is a living descendant.

Most arachnoids are small, free-living land animals that are most numerous in the warmer and drier regions of the world. Many have poison glands and poison claws by which they kill the prey on which they feed, and a few

[1] Arachnoidea—Gr. *arachne*, spider, + *-oid*, like; referring to the spiderlike characteristics of the class.

have bites or stings that may cause man serious illness and even death. The majority, however, are quite harmless, and most spiders are actually beneficial to man because they eat certain insects. The lacy webs of the spiders have long been admired for their beauty.

Arachnoids vary greatly in body form and nature of their appendages. The body of a typical arachnoid (except in the Acarina) is distinctly divisible into a cephalothorax, or **prosoma,** and an abdomen (Figs. 13-15, 13-18). The cephalothoracic somites are coalesced, but those of the abdomen may be fused or free. Antennules, antennae, and mandibles are never present. Typically there are six pairs of jointed appendages, all on the cephalothorax; these consist of an anterior pair of **chelicerae,** a pair of large chelate **pedipalpi,** and four pairs of **walking legs.** All appendages are uniramous, and those around the mouth are not greatly modified as jaws, but all have an inner prolonged part, the **gnathobase,** similar to that in the Trilobita and used for crushing food.

In the scorpions the abdomen is divided into a posterior (**metasoma**) and an anterior region (**mesosoma**) (Fig. 13-18); the abdomen is less clearly divided in the extinct eurypterids (Fig. 13-15) and is consolidated in the spiders and limuli. Body segmentation is not apparent in living mites and ticks. Some arachnoids have a posterior terminal spine.

Arachnoidea respire by means of several special breathing structures. The ancient aquatic forms (Eurypterida) and modern *Limulus* have book gills, scorpions have book lungs, and the later terrestrial arachnids are equipped with air tubes, or tracheae, similar to those of the Insecta. As previously stated, the majority of arachnoids now live on land, and this mode of life is probably the reason for the sparse fossil record of ancient terrestrial forms.

Some arachnoids lay eggs, whereas in others (*e.g.*, scorpions and mites) the young are hatched within the body of the mother. In either case the embryonic arachnoid has reached a mature stage of form and structure by the time it is hatched or born and differs little from its parents except in size (Fig. 13-1*G–N*). Consequently there are no larval stages comparable to those in the Insecta and no nauplius stage or other free-swimming larval stage like that found in the Crustacea.

The Arachnoidea represent a persistent evolutionary stock that is believed to have arisen from a primitive arthropod group in the later Pre-Cambrian and to have differentiated into several orders by the end of the Cambrian. This differentiation followed a definite structural plan that persists in living groups. They differ from the Crustacea in lacking gills (except Merostomata), biramous appendages, and anterior sensory structures (antennules and antennae). They differ greatly from the Chilopoda and Diplopoda, particularly in body form and in having few appendages. Terrestrial arachnoids resemble insects in having tracheae and certain other tubular structures, but they differ in lacking wings, a distinct head, and a larval stage in all but two orders, and their mouth parts and sex openings are different. They bear certain resemblances to the extinct Trilobita, but there is uncertainty as to how closely they are related to this ancient class (see later discussion on the relationships of the Trilobita, page 598). Some investigators believe that the arachnoids are descendants of very early trilobites, whereas others would reverse the relationship and have the trilobites derived from early arachnoids.

The oldest Arachnoidea seem to have been marine. The terrestrial habitat does not seem to have been invaded until Late Ordovician or Early Silurian, but since the Carboniferous, terrestrial forms have constituted the principal group of arachnoids. Some investigators (Gaskall, 1908; Patten, 1912) have suggested that an arachnoid may have been ancestral to the vertebrates, but this suggestion is not at all acceptable to modern vertebrate paleontologists (Romer, 1941; 1949).

Classification

The Class Arachnoidea is customarily divided into two subclasses[1]—Merostomata, a water-breathing division; and Arachnida (or Embolobran-

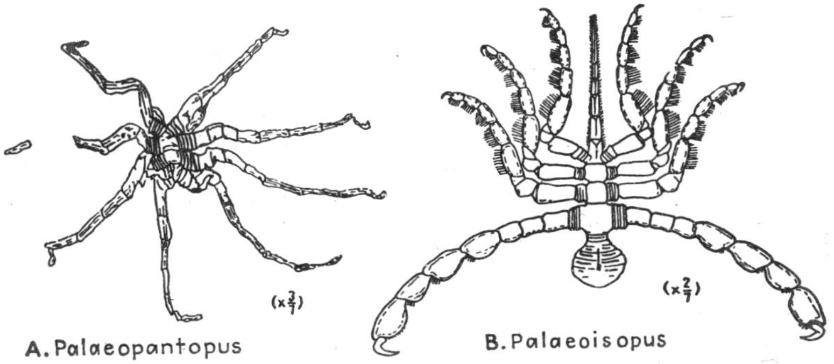

FIG. 13-12. Phylum ARTHROPODA—Pycnogonida. *A.* An almost complete specimen of *Palaeopantopus maucheri*. *B.* Reconstruction of a specimen of *Palaeoisopus problematicus* in ventral view. Both figures represent ancient pycnogonids from the famous Lower Devonian black shale at Bundenbach, Germany. (*Redrawn from Broili,* 1929 *and* 1933.)

chiata), an air-breathing group. The merostomes were the more primitive, lived in aquatic habitats, and breathed by means of specialized gills. The arachnids replaced the lung gills with tracheae and developed other structures necessary for life on land.

[1] Some authors include in the Arachnoidea three small but zoologically interesting groups of uncertain systematic position: Pycnogonida (sea spiders), Pentastomida (soft, unsegmented worm-like arthropods), and Tardigrada (water bears). Of these, only the pycnogonids have a possible fossil record; a brief description of this group follows. The organism has a short thin body with rudimentary abdomen. The cephalothorax has five somites with seven pairs of appendages, including a proboscis with suctorial mouth, a pair of 10-jointed egg-bearing legs, and four or six pairs of long, slender jointed walking legs. In some there is a four-legged larva with metamorphosis, whereas in others development is direct. About 400 species are known, and all are marine. The only known fossils thus far assigned to the pycnogonids are the two Lower Devonian genera, *Palaeopantopus* and *Palaeoisopus* (Fig. 13-12). Both were described from the Rhineland by Broili, and there is considerable doubt as to whether these can be related to any known living arthropods (Fage *in* Grassé, 1949).

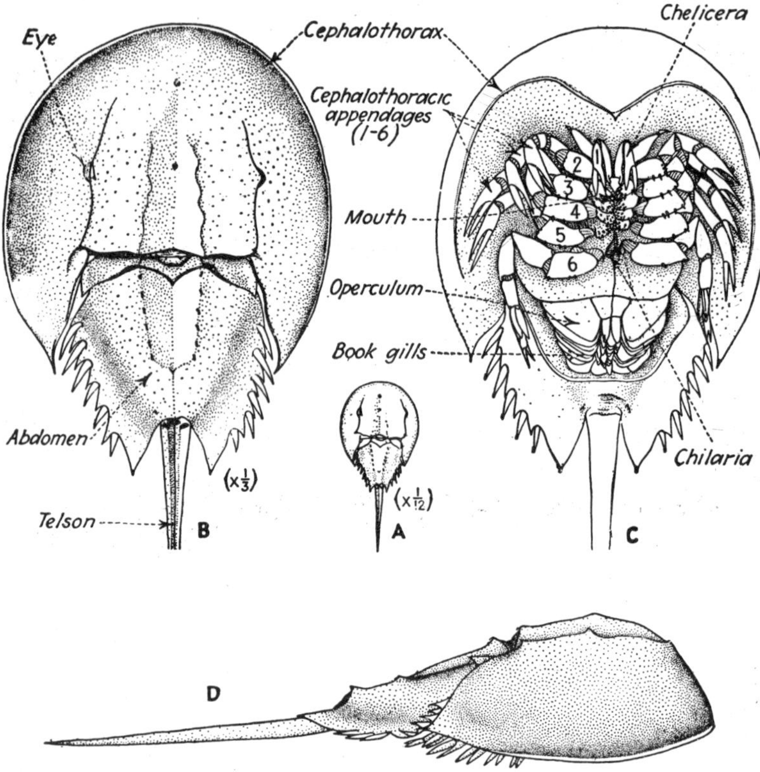

FIG. 13-13. Order Xiphosura. *Limulus polyphemus*, the king crab of the Atlantic Ocean and the only surviving genus of the order. *A*. Entire animal viewed dorsally. *B*. Dorsal view showing how exoskeleton is divided into two prominent shields—the carapace over the cephalothorax and the abdominal shield over the abdomen. *C*. Ventral view, showing number and nature of the appendages, all cephalothoracic, and nature and position of book gills. *D*. Side view of entire animal.

Subclass Merostomata[1]

General Considerations. The Subclass Merostomata is an ancient race of marine arachnoids that made its appearance in the Cambrian, flourished during the Middle Paleozoic, and declined through the Mesozoic to its present state of a single surviving genus, the familiar king crab, *Limulus* (Fig. 13-13), with four living species. The Merostomata include the largest known arachnoids and for this reason have been called the *Gigantostraca* (gigantic shells). Individual eurypterids attained a length of almost 3 m. (10 ft.) and were, so far as is known, the largest invertebrates of their time.

[1] Merostomata—Gr. *meros*, thigh, + *stoma*, *stomatos*, mouth; referring to the position of the mouth.

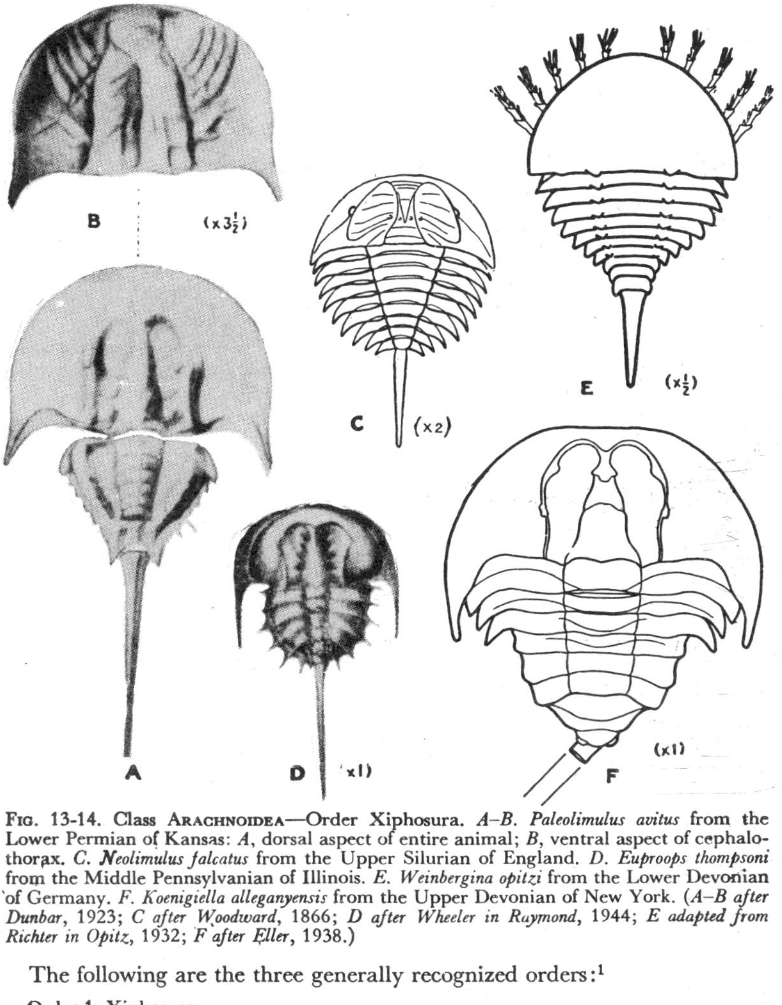

FIG. 13-14. Class ARACHNOIDEA—Order Xiphosura. *A–B. Paleolimulus avitus* from the Lower Permian of Kansas: *A*, dorsal aspect of entire animal; *B*, ventral aspect of cephalothorax. *C. Neolimulus falcatus* from the Upper Silurian of England. *D. Euproops thompsoni* from the Middle Pennsylvanian of Illinois. *E. Weinbergina opitzi* from the Lower Devonian of Germany. *F. Koenigiella alleganyensis* from the Upper Devonian of New York. (*A–B after Dunbar*, 1923; *C after Woodward*, 1866; *D after Wheeler in Raymond*, 1944; *E adapted from Richter in Opitz*, 1932; *F after Eller*, 1938.)

The following are the three generally recognized orders:[1]

Order 1. Xiphosura.
Order 2. Eurypterida.
Order 3. Aglaspida.

The distinguishing features of the Merostomata are emphasized in the following discussion of these three orders.

Order 1. Xiphosura. The xiphosurans include both ancient and modern king crabs and their close allies, and the sole surviving genus, *Limulus*, may be taken as typical of the order.

[1] In an article published too late to be used in the present work Störmer (1952) proposes a new classification of the xiphosura and related merostomes

Limulus (Fig. 13-13) is common along the Atlantic shore from Maine to Yucatan and in the Pacific from Japan to Sumatra—typically on muddy and sandy bottoms, where it ploughs through the soft sediments in search of worms and other small animals. Over its body *Limulus* bears an arched chitinous carapace of horseshoe shape that is joined to a broad abdominal shield, or abdomen, which carries numerous movable spines on its lateral margins and a long bayonetlike telson attached to its posterior extremity. The telson, which is articulated with the abdominal shield, is used to assist in ploughing through soft sediments and also as a lever by which the animal can regain its normal position when accidentally turned on its back.

Limulus has six pairs of segmented appendages on the cephalothorax and six pairs of broad thin abdominal appendages that are joined medially. On the posterior surface of the last five pairs of abdominal appendages[1] are the book gills, specialized respiratory structures consisting of 150 to 200 leaves containing blood passages. The upper surface of the carapace bears two medial simple eyes and two lateral compound eyes. Other features of the exoskeleton are shown in Fig. 13-13.

The earliest fossil Xiphosura come from Lower Ordovician rocks (Caster, 1944), and *Limulus*, with its four living species, dates from the Triassic. *Neolimulus* (Fig. 13-14*C*) from the Upper Silurian, and *Weinbergina* (Fig. 13-14*E*) from the Lower Devonian, are typical of the Suborder Synziphosura (xiphosurans with all true abdominal segments movable); *Koenigiella* (Fig. 13-14*F*), from the Upper Devonian, *Euproops* (Fig. 13-14*D*), a common Pennsylvanian genus, *Paleolimulus* (Fig. 13-14*A*–*B*), from the Permian, and *Limulus* (Fig. 13-13), ranging from the Triassic to Recent, are representatives of the Suborder Limulada[2] (xiphosurans with some or all trunk segments fused and prosoma with eye ridges, at least at the posterior margin).

Order 2. Eurypterida.[3] The eurypterids are extinct aquatic arachnoids which are of interest to paleontologists because of their structural plan (Fig. 13-15) and excellent preservation, and also because of the light that they throw on both arachnoid and arthropod evolution. They were especially abundant during the Silurian and Devonian, and some of them are the largest of known arthropods. *Pterygotus* (Fig. 13-16*C*) and *Stylonurus* (Fig. 13-16*B*), for example, were commonly between 2 and 3 m. long from front of head to tip of telson.

The elongate body was covered with a thin, chitinous integument with tubercles or fine scalelike markings. The body segmentation and appendage development typical of the eurypterids are well shown in *Eurypterus*, illustrated in Fig. 13-15.

The comparatively small prosoma consists of the first six somites (1–6 in Fig. 13-15) which are completely fused and covered with a single shield, or carapace. Dorsally, the prosoma bears a pair of small, median eyes (**ocelli**)

[1] Störmer (1939) sees in these appendages evidence of a primitive biramous condition.

[2] Not to be confused with Walcott's Order Limulava, which is included in the arthropodan Subclass Xenopoda.

[3] Eurypterida—Gr. *eurys*, broad, + *pteron*, wing; referring to the broad, winglike appendages, which are developed by the modification of one or more of the pairs of appendages.

and two large, crescentic, lateral eyes. On the ventral side are six pairs of appendages that vary somewhat in different genera. In *Eurypterus* the first pair are chelicerae; the second, third, and fourth pairs are walking legs; the fifth pair are balancing legs; and the sixth pair are paddlelike swimming legs.

The abdomen consists of 13 free segments; the anterior six (7–12 in Fig. 13-15) constitute the mesosoma; the posterior six plus a telson (13–18 in

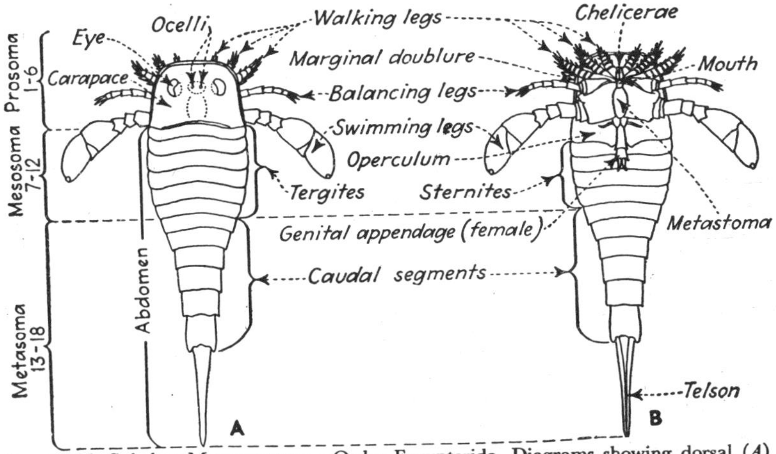

FIG. 13-15. Subclass MEROSTOMATA—Order Eurypterida. Diagrams showing dorsal (*A*) and ventral (*B*) aspects of a typical eurypterid, with **the chief** features indicated. (*Adapted from Clarke and Ruedemann*, 1912.)

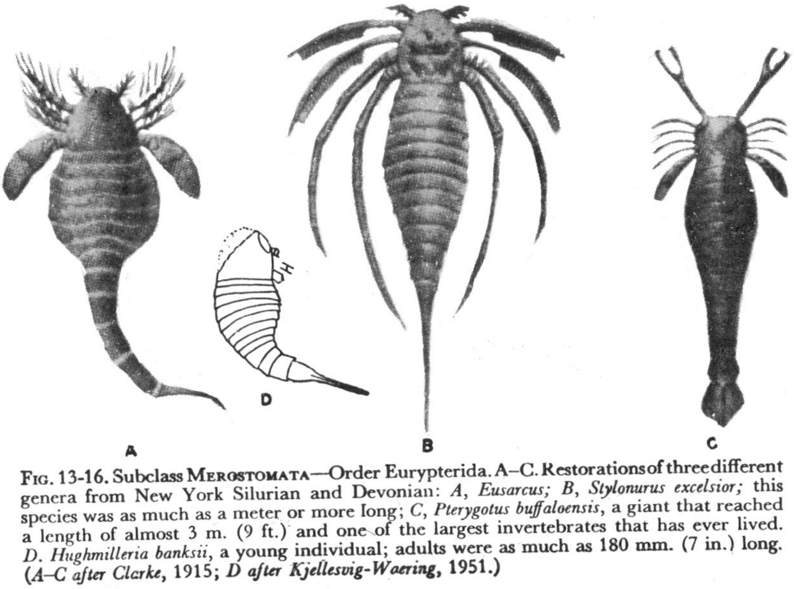

FIG. 13-16. Subclass MEROSTOMATA—Order Eurypterida. A–C. Restorations of three different genera from New York Silurian and Devonian: *A*, *Eusarcus; B*, *Stylonurus excelsior;* this species was as much as a meter or more long; *C*, *Pterygotus buffaloensis*, a giant that reached a length of almost 3 m. (9 ft.) and one of the largest invertebrates that has ever lived. *D. Hughmilleria banksii*, a young individual; adults were as much as 180 mm. (7 in.) long. (*A–C after Clarke*, 1915; *D after Kjellesvig-Waering*, 1951.)

Fig. 13-15), the metasoma. Each mesosomatic segment except the first (7 in Fig. 13-15), which carries the genital operculum, bears a pair of broad, leaf-like appendages which have generally been assumed to have had something to do with respiration. Although it is commonly stated that the six mesosomatic appendages of an eurypterid correspond to the operculum and branchial appendages of a xiphosuran (*e.g.*, *Limulus*), Moore (1941) concludes that they are not homologous with the gill-bearing appendages of *Limulus*, as there is no conclusive evidence of their having had lamellae attached to them. He would homologize the six mesosomatic segments in Xiphosura, Eurypterida, and Scorpionida as follows:

Segment	Eurypterida	Xiphosura	Scorpionida
Pregenital	Metastoma	Chilaria	Embryonic
Mesosoma 1	Operculum gills	Operculum	Operculum
Mesosoma 2	Appendage + gills	Appendage + gills	Pectines
Mesosoma 3	Appendage + gills	Appendage + gills	Lung
Mesosoma 4	Appendage + gills	Appendage + gills	Lung
Mesosoma 5	Appendage + gills	Appendage + gills	Lung
Mesosoma 6	Appendage + gills	Appendage + gills	Lung

The metasoma consists of six free, annular, gradually tapering segments together with a long, bladelike telson. Metasomatic segments have no appendages of any kind.

All known species of Eurypterida were aquatic, but whether they lived in marine, brackish, or fresh water is a moot question. That they lived on and near the bottom as vagrant benthos seems reasonably certain,[1] but the development of two large, paddlelike appendages in many forms strongly suggests that these species also had a swimming habit. McConnell (1916), among others (also see Störmer, 1944), has suggested that the eurypterids originated and lived in streams, from which they were carried into brackish and saline lagoons or even to the open sea to be buried in lagoonal or marine deposits. The distribution of their fossil remains in sedimentary rocks suggests that eurypterids did not live in normal marine waters, except possibly in the very beginning of their history.

Known growth stages show that when the young eurypterid was hatched it was similar to the adult except that it had a relatively larger cephalothorax and relatively larger eyes, fewer segments in the body, and a less marked difference between mesosoma and metasoma.

[1] It is generally held that the Eurypterida were sluggish bottom dwellers or poor swimmers. They are thought to have spent most of their life crawling about sluggishly over muddy bottoms, but the presence of swimming appendages in some forms suggests a swimming habit. All were probably carnivorous, and the giant forms may well have been predatory. The general absence of eurypterid remains in marine deposits suggests that they were confined to rather narrowly restricted environments, and there is much to be said for the assumption that they lived in fresh-water streams and lakes and possibly in coastal lagoons. Ruedemann (1934), however, believes that the earlier eurypterids, particularly those of the Ordovician and Silurian, lived in marine habitats.

The oldest known eurypterids have been found in Lower Ordovician rocks (Deepkill graptolite shales of New York), but the remains are fragmentary and poorly preserved. Excellently and completely preserved specimens are locally abundant in certain Silurian and Devonian strata, and it would seem that the order attained its zenith during these two periods. Later Paleozoic forms are rare, and the order became extinct sometime during the Permian (Decker, 1938).

Most fossil eurypterids have come from rocks that were deposited under exceptional environmental conditions, and some incomplete specimens may be parts of molted exoskeletons that drifted from the habitat of life into an environment where the animal itself could not have lived. It seems probable, therefore, that the known fossil record, which includes about 200 species, does not accurately indicate their actual abundance.

The Order Eurypterida may be divided, on the basis of the development of the last pair of prosomal legs, into at least eight families (Störmer, 1951). The following genera illustrate the differences between the more important families: *Pterygotus* (Pterygotidae; Fig. 13-16*C*), *Eurypterus* (Eurypteridae; Figs. 13-15, 13-16), *Eusarcus* (Carcinosomidae; Fig. 13-16*A*), *Stylonurus* (Stylonuridae; Fig. 13-16*B*), and *Hughmilleria* (Hughmilleridae; Fig. 13-16*D*).

Order 3. Aglaspida. The aglaspids are Early Paleozoic Merostomata characterized by a chitinophosphatic exoskeleton consisting of a longitudinally trilobed dorsal shield and a longitudinally cleft postventral plate. The dorsal shield consists of a semicircular carapace and 12 abdominal segments. In Family Aglaspidae (Upper Cambrian) the segments are all freely articulating, and the twelfth bears a long sharp telson spine. In another family, the Beckwithiidae, based on the single Mid-Cambrian genus *Beckwithia*, some of the posterior segments are coalesced.

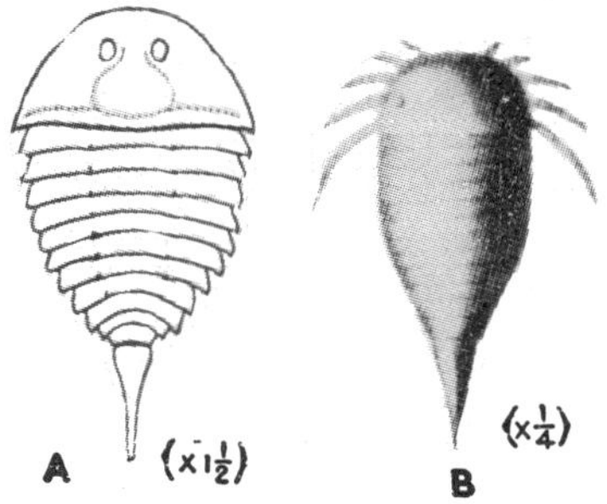

FIG. 13-17. Subclass MEROSTOMATA—Order Aglaspida. *A*. Generalized diagram of the dorsal view of a typical aglaspid, based on the holotype of *Aglaspella eatoni* and several other specimens from the Upper Cambrian of Wisconsin. *B*. *Strabops thatcheri* from the Upper Cambrian of Missouri. (*B after Clarke and Ruedemann*, 1912.)

The cephalothorax bears a pair of compound eyes dorsally and six pairs of uniramous jointed appendages ventrally. The first pair of appendages are strongly chelate; the remaining five pairs are simple and unspecialized. The abdomen is divisible into a mesosoma of nine segments and a metasoma of two or three segments. At least the first seven mesosomal segments have each a pair of simple appendages like the posterior five pairs under the cephalothorax. The metasomal segments lack appendages altogether.

Aglaspella (Fig. 13-17*A*) and *Strabops* (Fig. 13-17*B*) are representatives of the order, which, so far as is known, is limited to the Cambrian and Ordovician.[1]

[1] Caster and Macke (1952) recently described *Neostrabops* from the Upper Ordovician of Ohio.

Subclass Arachnida[1] *(Embolobranchiata*[2]*)*

General Considerations. The Arachnida comprise the air-breathing arachnoids, among which are included the familiar spiders, scorpions, mites, and ticks. So far as is known, all the more than 30,000 described living species are terrestrial and breathe by means of book lungs or tracheal tubes. It is assumed that the fossil species did likewise.

The arachnid body, and likewise the exoskeleton, typically exemplified in the ancient scorpion *Palaeophonus* (Fig. 13-18), is divided into prosoma, mesosoma, and metasoma. The prosoma bears two large median dorsal eyes and numerous simple eyes at the lateral edges of the frontal margin. Ventrally it carries six pairs of jointed uniramous appendages—a pair of anterior chelicerae, a pair of long and powerful chelate pedipalps, and four pairs of seven-jointed walking legs tipped with three small claws. The gnathobases of the first three pairs of appendages are arranged about the mouth and were probably used in eating.

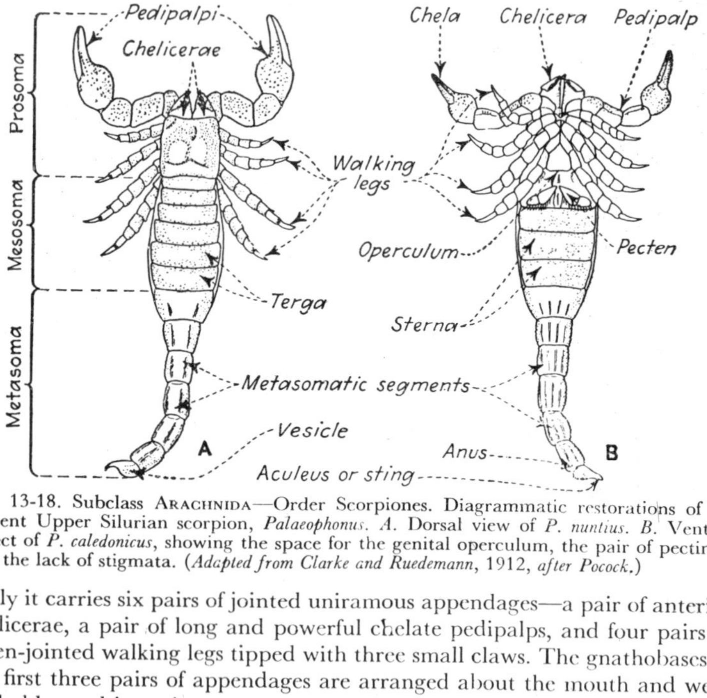

FIG. 13-18. Subclass ARACHNIDA—Order Scorpiones. Diagrammatic restorations of an ancient Upper Silurian scorpion, *Palaeophonus*. *A*. Dorsal view of *P. nuntius*. *B*. Ventral aspect of *P. caledonicus*, showing the space for the genital operculum, the pair of pectines, and the lack of stigmata. (*Adapted from Clarke and Ruedemann*, 1912, *after Pocock*.)

The mesosomatic somites are enclosed in six segments, each of which is composed of a dorsal piece, or **tergum** (plural, **terga**), and a ventral piece, or **sternum** (plural, **sterna**), which supposedly were united laterally by soft skin. The first mesosomatic segment bears a ventral genital opening; the second carries a pair of comblike appendages, the **pectines** (singular, **pecten**),

[1] Arachnida—Gr. *arachne*, spider; referring to the fact that this subclass includes the spiders.

[2] Embolobranchiata—Gr. *embolos*, anything pointed, + *branchia*, gills; referring to the nature of the respiratory organs.

which are tactile organs; and the third to the sixth segments have broad, flat sternal plates that are perforated by oblique slits along the lateral margin. These slits, or **stigmata,** open into the tracheae.

The seventh to twelfth segments constitute the metasoma (also designated **postabdomen** in some works). The first segment of this series is wide like the preceding ones but differs from them in having the tergum and sternum more or less united and in lacking stigmata. Consequently, it is considered a metasomatic segment, though it is somewhat larger and of different shape than the succeeding ringlike segments. The last, or twelfth, segment carries a

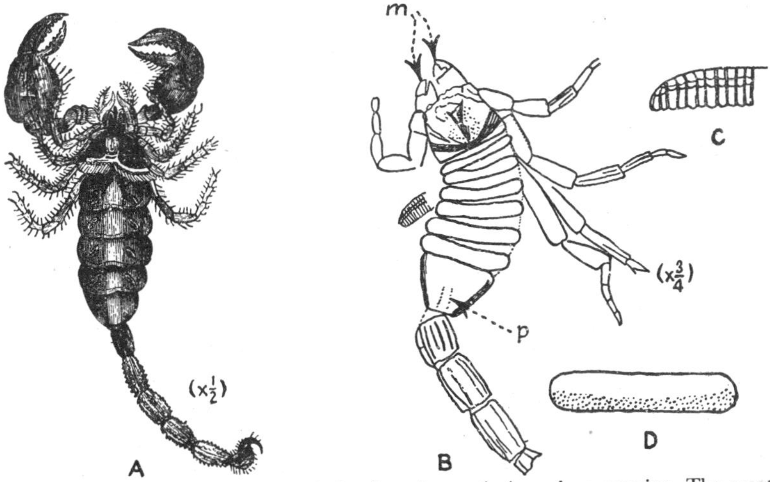

FIG. 13-19. Subclass ARACHNIDA—Order Scorpiones. *A*. A modern scorpion. The great nipping claws are not legs, but are a development of organs belonging to the mouth. *B–D*. *Eoscorpius carbonarius* from the Pennsylvanian of Illinois: *B*, fragmental specimen, with one of the pectines lying detached from the body; *C*, pecten enlarged; *D*, one of the body segments enlarged to show the surface granules (*m*, mandibles somewhat crushed and distorted; *p*, small pits). (*A after Nicholson*, 1872; *B–D after Meek and Worthen*, 1868.)

spinelike stinging structure composed of a basal expansion, the **vesicle,** which contains poisonous fluid and a pointed stinger, the **aculeus.**

Scorpions such as the Silurian genus *Palaeophonus*, described in preceding paragraphs, are the most primitive arachnids known and probably are nearest the ancestral stock in general appearance, whereas the ticks seem to be the most highly specialized. The arachnids may well be the first animals that adapted their bodies to terrestrial life. The changes necessary should not have been too difficult, as only slight modification of the respiratory apparatus had to be made. It may well be that the book gills of the early invader developed a covering and became book lungs and that the branchia were similarly modified into tracheae.

The earliest arachnid known is *Palaeophonus* (Fig. 13-18) from the Upper Silurian of Europe and North America. Representatives of three orders have been discovered in the Scottish Devonian, and nine additional orders left

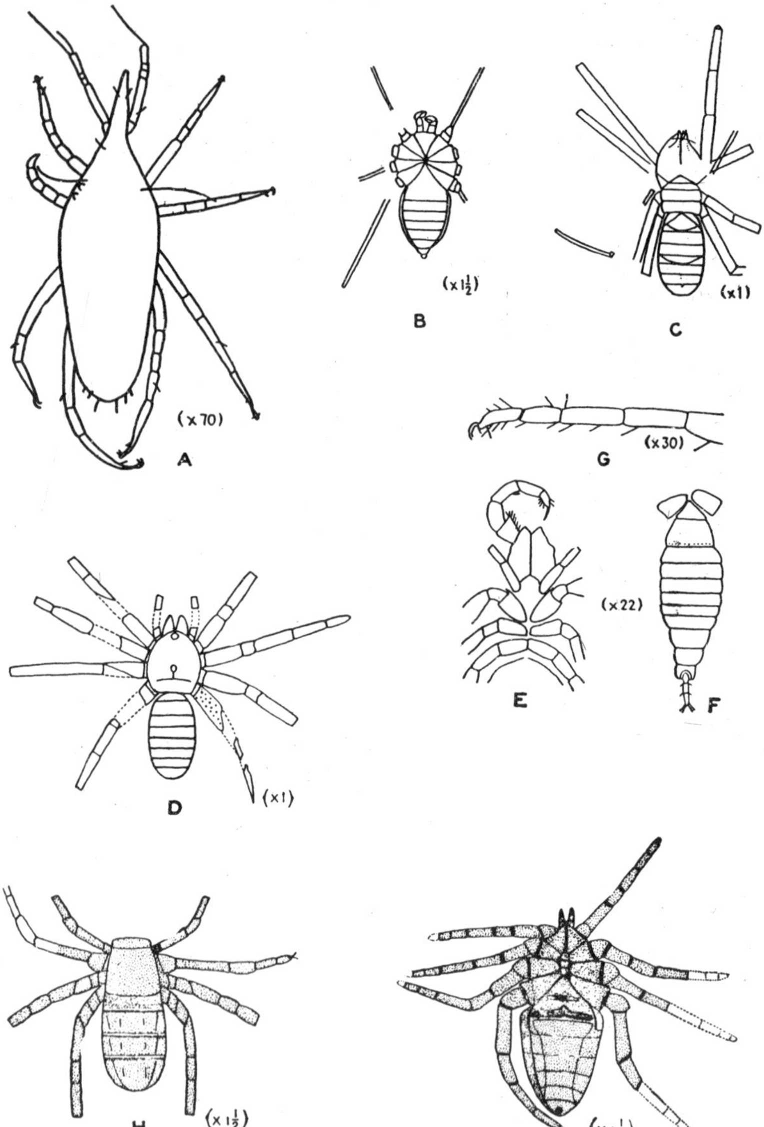

Fig. 13-20. Subclass Arachnida. Fossil arachnids; orders indicated in parentheses; all except *A* and *E–G* from Pennsylvanian of Illinois. *A. Bdella vetusta* (Acari), a mite from the Cretaceous(?) amber of Manitoba. *B. Kustarachne tenuipes* (Kustarachnae). *C. Protosolpuga carbonaria* (Solifugae). *D. Arthrolycosa antiqua* (Araneae). *E–G. Calcitro fisheri* (Schizomida), a fossil whip scorpion from supposed Miocene onyx of Arizona: *E*, ventral aspect showing legs: *F*, ventral aspect of an abdomen; *G*, the fourth leg. *H. Polychera punctulata* (Ricinulei). *I. Architarbus rotundatus* (Architarbi). (*A after Ewing in Carpenter et al.*, 1937; *B–D, H–I after Petrunkevitch*, 1913; *E–G after Petrunkevitch*, 1945.)

representatives in the North American Pennsylvanian. Other orders are known from Mesozoic and Cenozoic rocks, and future discoveries may show that some of these were also represented in the Paleozoic. Petrunkevitch (1949), long the outstanding leader in the study of Paleozoic Arachnida, believes that "there can be scarcely any doubt as to the existence of all sixteen orders [of Arachnida] at least in the late Paleozoic," even though fossil representatives of some of the orders have not yet been found except in later rocks or not at all.

Many investigators have emphasized the similarities between the earliest arachnids and eurypterids, and some have proposed that the latter were ancestral to the primitive scorpions. While it is true that the Arachnida resemble the Merostomata in having a single pair of chelate appendages in front of the mouth, and five additional cephalothoracic appendages, they differ in having book lungs and tracheal tubes instead of book gills, in lacking compound eyes, in the structure of the carapace, and in certain other external and internal features. On the other hand, the Merostomata possess features not present in any arachnid. It would seem, therefore, that there is considerable evidence in support of the contention that both Arachnida and Merostomata developed independently from an older and more primitive arthropod stock (Petrunkevitch, 1949)

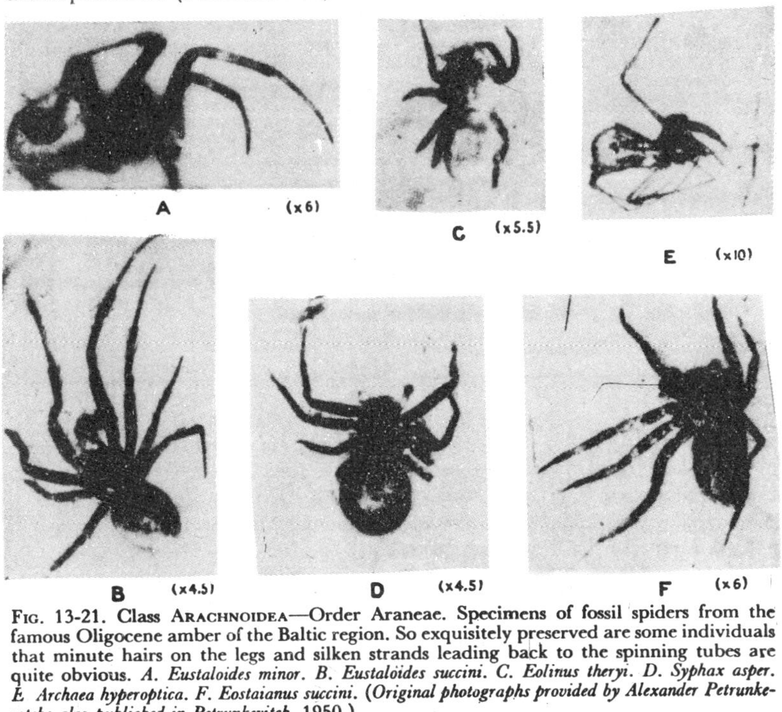

FIG. 13-21. Class ARACHNOIDEA—Order Araneae. Specimens of fossil spiders from the famous Oligocene amber of the Baltic region. So exquisitely preserved are some individuals that minute hairs on the legs and silken strands leading back to the spinning tubes are quite obvious. *A. Eustaloides minor. B. Eustaloides succini. C. Eolinus theryi. D. Syphax asper. E. Archaea hyperoptica. F. Eostaianus succini.* (*Original photographs provided by Alexander Petrunkevitch; also published in Petrunkevitch,* 1950.)

TABLE 13-1. GEOLOGIC RANGES OF THE ORDERS of SUBCLASSES ARACHNIDA AND MEROSTOMATA.

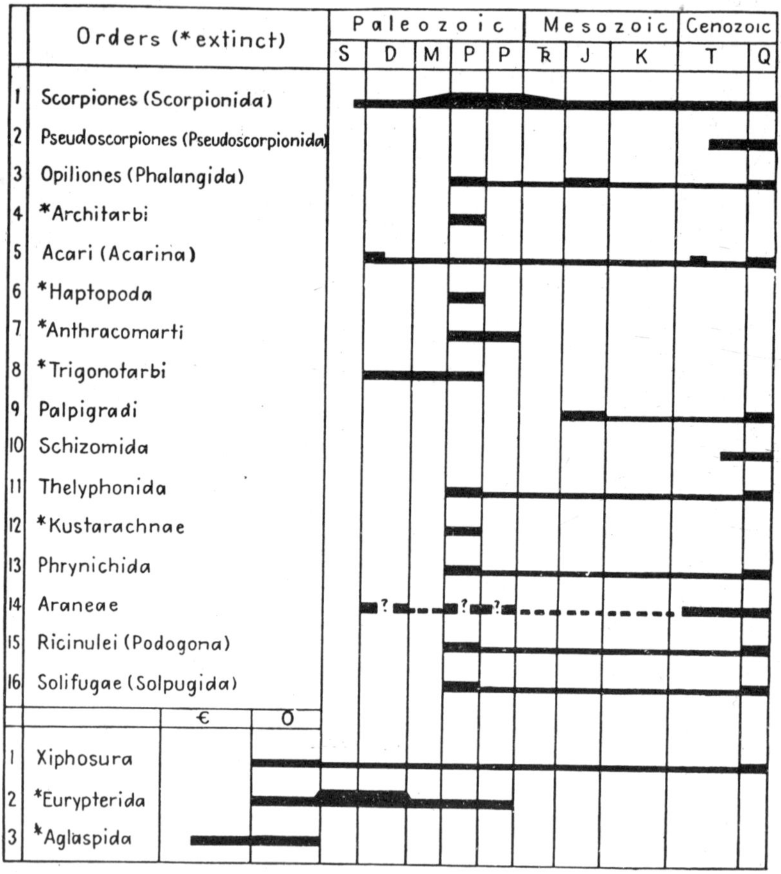

Classification. As the fossil record of the Subclass Arachnida is quite incomplete and based on only about 200 species (as compared to more than 30,000 living species), it will suffice merely to list below the 16 generally recognized orders, with a few pertinent comments, and to indicate the known geologic ranges of the orders in Table 13-1.

Order 1. Scorpiones (Scorpions). The telson is modified to form a poison gland. The scorpions first appeared in the Silurian, seem to have reached their zenith in the later Paleozoic, and are still living. *Upper Silurian to Recent.*

Palaeophonus (Fig. 13-18) from the Silurian and *Eoscorpius* (Fig. 13-19*B–D*) from the Pennsylvanian are ancient and extinct representatives, which differ considerably from a modern scorpion (Fig. 13-19*A*).

Order 2. Pseudoscorpiones (Book Scorpions). *Lower Oligocene to Recent.*

Order 3. Opiliones (= Phalangida). These are the familiar daddy longlegs. *Pennsylvanian to Recent.*

Order 4. Architarbi. This is an extinct order limited to the *Pennsylvanian*, and exemplified by *Architarbus* (Fig. 13-20*I*).

Order 5. Acari (= Acarina). These are the mites and ticks. The only Paleozoic species yet reported is *Protocarus crani* from the Devonian Rhynie chert of Scotland. The next oldest specimens come from supposed Cretaceous amber (Fig. 13-20*A*) and from the Eocene Green River shale. *Devonian to Recent.*

Order 6. Haptopoda. An extinct "Upper Carboniferous" (*Penn.*) order based on a single species, *Plesiosiro madeleyi* (Fig. 13-22*D*).

Order 7. Anthracomarti. An extinct Upper Paleozoic order based on seven closely related genera of which *Cryptomartus* (Fig. 13-22*C*) is typical. *Pennsylvanian, Permian.*

Order 8. Trigonotarbi. An extinct later Paleozoic order exemplified by *Trigonotarbus* (Fig. 13-22*F*). *Devonian to Pennsylvanian.*

Order 9. Palpigradi. Minute arachnids with only a single fossil species, *Sternarthron zittelli*, from the Jurassic of Europe. *Jurassic to Recent.*

Order 10. Schizomida (= Pedipalpi in Part). A modern order of whip scorpions with a single fossil species, *Calcitro fisheri* (Fig. 13-20*E–G*), from Late Cenozoic onyx in Arizona. *Miocene to Recent.*

Order 11. Thelyphonida (= Pedipalpi in Part). A modern order of whip scorpions, with a single Paleozoic genus, *Geralinura* (Fig. 13-22*B*), from the "Upper Carboniferous." *Pennsylvanian to Recent.*

Order 12. Kustarachnae. An extinct North American order based on the single genus *Kustarachne* (Fig. 13-20*B*). *Pennsylvanian.*

Order 13. Phrynichida (= Pedipalpi in Part). A modern order with three fossil genera from the "Upper Carboniferous." *Pennsylvanian to Recent.*

Order 14. Araneae (Spiders). These are the familiar spiders with spinnerets in both sexes and with copulatory organs on the palpi of the males. Tertiary spiders had these features, but they have not been demonstrated for older supposed fossil Araneae. Consequently, the Devonian species *Palaeocteniza crassipes* is a questionable spider, and there is likewise doubt about "Upper Carboniferous" and Permian Araneae. No Mesozoic spiders are known, but Eocene and Oligocene forms are not uncommon. *Devonian?, Pennsylvanian to Permian?, Cenozoic to Recent.*

Arthrolycosa (Fig. 13-20*D*) is a supposed fossil spider from the Illinois Pennsylvanian. Several spiders from the famous Oligocene amber of the Baltic region are shown in Figs. 13-21 and 13-22.

Order 15. Ricinulei (= Podogona). Rare tropical arachnids which have a unique movable plate attached to the front of the carapace and turned under ventrally to protect the mouth parts when these are not in use. All fossil forms have come from "Upper Carboniferous" (Pennsylvanian) rocks of Europe and North America. *Pennsylvanian to Recent.*

Polyochera (Fig. 13-20*H*) is a fossil genus from the Lower Pennsylvanian of Illinois.

Order 16. Solifugae (= Solpugida). A modern order with a single fossil species from the Pennsylvanian of Illinois, *Protosolpuga carbonaria* (Fig. 13-20*C*). *Pennsylvanian to Recent.*

Geologic History. The extremely fragmentary record of the Arachnida, as shown in Table 13-1, is misleading to the extent that arachnids were almost certainly much more varied and abundantly represented than the fossil record suggests. Because they seem to have lived in environments where chances of preservation were small, it is rather surprising that so many fossil forms have survived. Perhaps this is why only about 200 fossil species are

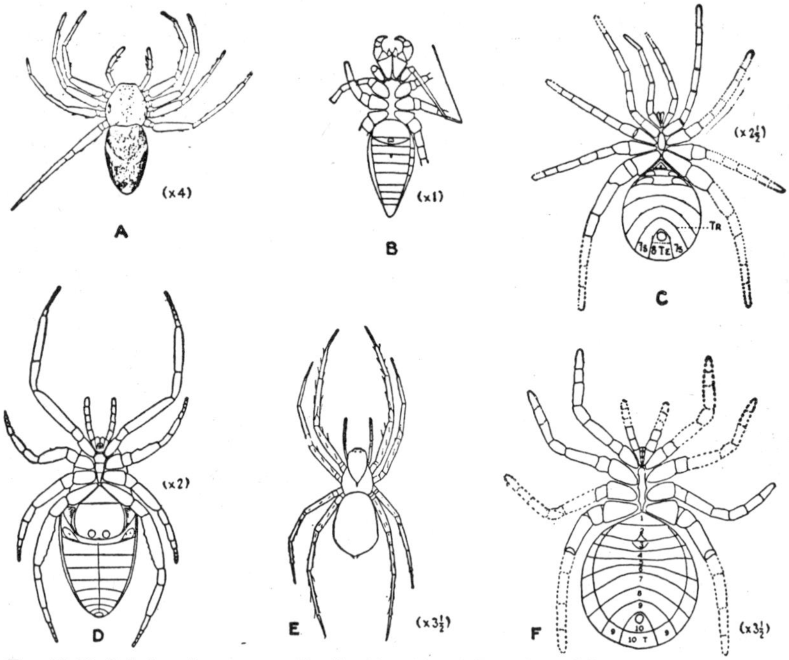

FIG. 13-22. Subclass ARACHNIDA. Fossil spiders *A* and *E* are from Oligocene Baltic amber; all others are from "Upper Carboniferous" of Great Britain. Orders are indicated within parentheses. *A*. *Caduceator quadrimaculatus* (Araneae). *B*. *Geralinura priesti* (Thelyphonida). *C*. *Cryptomartus priesti* (Anthracomarti). *D*. *Plesiosiro madeleyi* (Haptopoda). *E*. *Eocrypoeca distincta* (Araneae). *F*. *Trigonotarbus johnsoni* (Trigonotarbi). (*All after Petrunkevitch: A*, 1950; *B*, *E*, 1913; *C–D*, *F*, 1949.)

known as against more than 30,000 described living species. Fossil species have to be based entirely on external characters, because internal organs are almost never preserved well enough to be of taxonomic use.

ARTHROPODA INCERTAE SEDIS

Included in the Phylum Arthropoda, but of uncertain taxonomic position within the phylum, is a heterogeneous group of intriguing fossils that have been shifted about from one class to another by different investigators. Because there does not seem to be a present consensus as to how these fossils should be classified, they are placed here so that their chief characteristics may be compared with those of the Crustacea and Arachnoidea, and to anticipate later discussion of the Trilobita. The following provisional classification follows that proposed by Raymond (1935). All divisions are extinct.

Subclass Homopoda (Cambrian).
- Order 1. Marrellina (Middle Cambrian).
- Order 2. Pseudanostraca (Middle Cambrian).

Order 3. Pseudonotostraca (Lower and Middle Cambrian).
Order 4. Hymenocarina (Cambrian).
Subclass Xenopoda (Middle Cambrian).
Order 1. Limulava. (Middle Cambrian).
Subclass Archaeostraca (Cambrian to Triassic).
Order 1. Bradorina (Cambrian).
Order 2. Ceratiocarina (Cambrian—Pennsylvanian).
Order 3. Rhinocarina (Devonian—Pennsylvanian).
Order 4. Discinocarina (Ordovician—Triassic).

Subclass Homopoda[1]

Raymond (1920, 1935) created this subclass to include a diversified group of extinct Cambrian shrimp-like arthropods, which have two pairs of uniramous tactile organs and the other appendages biramous (*i.e.*, of the trilobitan type), and which have or lack a carapace. Because the forms included in this subclass are extinct and cannot be directly connected by intermediate forms with animals of much later date that they closely resemble, it seems advisable to follow Raymond's suggestion that the several orders involved be included in a separate subclass. The homopods, therefore, may be regarded as ancient and extinct relatives of modern Crustacea, but not necessarily direct ancestors of any living crustaceans. (*Cambrian.*)

Order 1. Marrellina. These are homopods with a trilobite-like form in which the pleural lobes are reduced. All segments behind the antennal, which have uniramous antennae, bear biramous appendages. The order is based on the single Mid-Cambrian genus, *Marrella* (Fig. 13-23*A*). *Middle Cambrian.*

Order 2. Pseudanostraca.[2] Members of this extinct order are homopods with two pairs of tactile organs and with appendages on all segments or absent from one or more of the posterior ones. Not all appendages are biramous, and there is no carapace. The order is extinct and includes the well-known Middle Cambrian genera *Opabinia*, *Leanchoilia*, and *Yohoia* and possibly certain other questionable genera (Fig. 13-23). So far as is known the order is limited to the Middle Cambrian, and it may have led to the Anostraca (Raymond, 1935).

Order 3. Pseudonotostraca.[3] Members of this extinct order are homopods with two pairs of tactile organs, the other appendages trilobitan; biramous appendages on front segments but lacking on some posterior segments; carapace depressed; believed to lead to the notostracans. *Lower and Middle Cambrian.*

Protocaris (Fig. 13-23*E*), Lower Cambrian, and *Burgessia* (Fig. 13-23*D*) and *Waptia* (Fig. 13-23*F–G*), both Middle Cambrian, are included in the order.

Order 4. Hymenocarina. These are homopods with a compressed carapace in which the rostrum is free or attached. Biramous appendages are

[1] Homopoda—Gr. *homos*, the same, + *pous*, *podos*, foot; referring to the similarity of the different pairs of limbs or appendages.

[2] Pseudanostraca—Gr. *pseudo*, false, + *an*, without, + *ostracon*, shell of a testacean; referring to the fact that members of the order are not to be falsely taken as anostracans.

[3] Pseudonotostraca—Gr. *pseudo*, false, + *noton*, back, + *ostracon*, shell of a testacean; referring to the fact that members of the order are not to be falsely taken as notostracans.

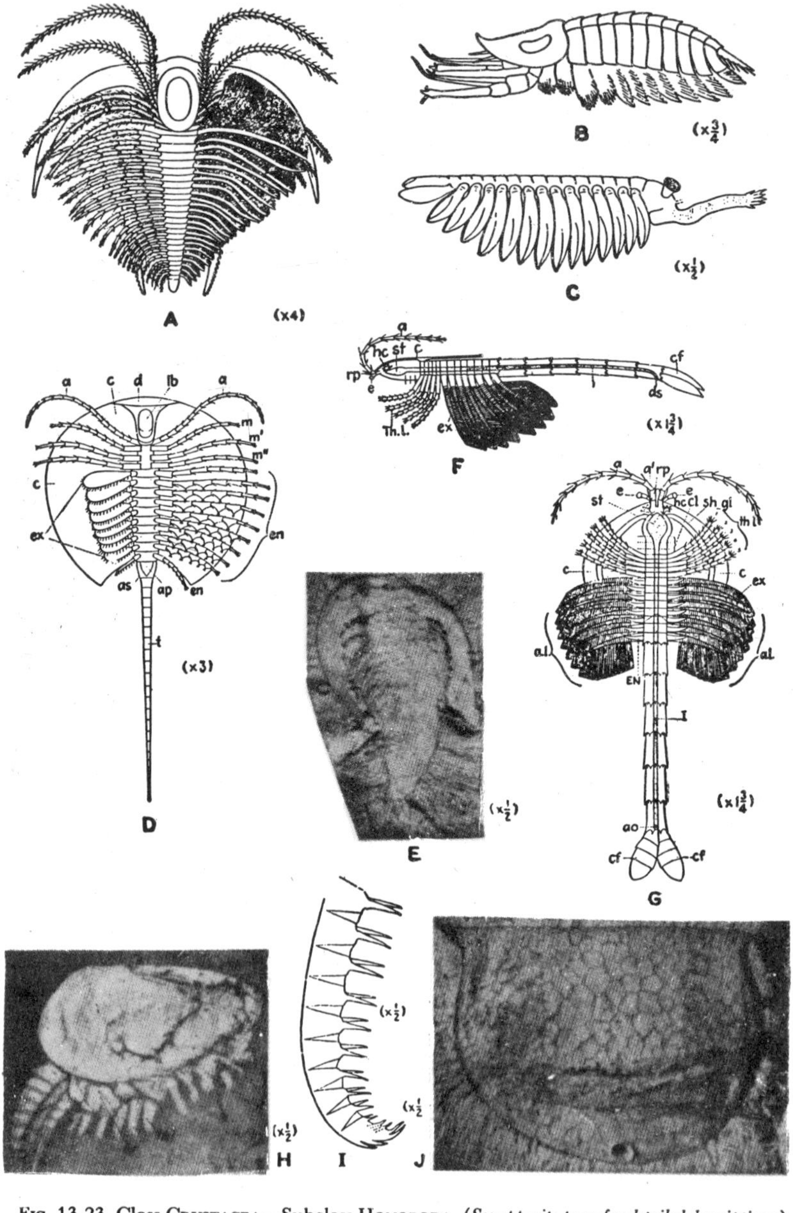

FIG. 13-23. Class CRUSTACEA—Subclass HOMOPODA. (*See opposite page for detailed descriptions.*)

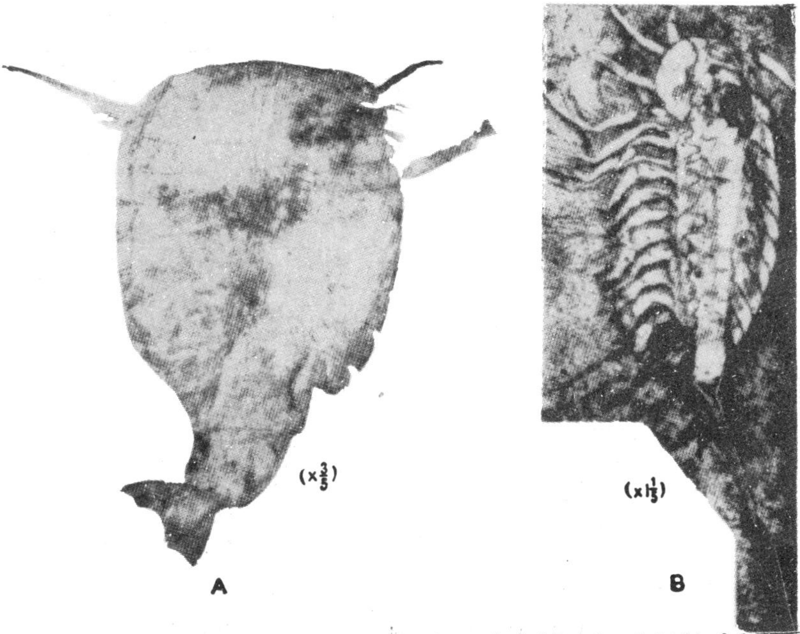

Fig. 13-24. Subclass Xenopoda—Order Limulava. *A.* A large dorsal shield of *Sidneyia inexpectans*. *B.* A flattened specimen of *Emeraldella brocki*, giving a partial view of the head and body and a fine profile view of the abdomen and long telson. Both specimens are from the Middle Cambrian Burgess shale of British Columbia. (*After Walcott*, 1911 *and* 1912.)

present on the anterior portion of the trunk. The order, which seems to be limited to the Cambrian, includes the genera *Hymenocaris*, *Anomalocaris*, and *Tuzoia*, and possibly several other genera of uncertain relationships (*e.g.*, *Carnarvonia*, *Fieldia*, and *Hurdia*). (Fig. 13-23.)

Subclass Xenopoda[1]

This subclass embraces Arthropoda with more or less eurypterid-like form, one pair of uniramous antennae, modified appendages on cephalic part, and

[1] Xenopoda—Gr. *xeons*, strange or foreign, + *pous*, *podos*, foot; referring to certain peculiarities of the appendages.

Fig. 13-23. Ancient and extinct arthropods of uncertain affinities from the Middle Cambrian Burgess shale of British Columbia. *A.* Restoration of *Marrella splendens* (Order Marrellina), showing ventral aspect. Although all the limbs seem to be biramous, only endopodites are shown on one side and exopodites on the other, for the sake of greater clearness. *B–C.* Order Pseudanostraca: *B*, restoration of *Leanchoilia superlata; C*, conjectural restoration of *Opabinia regalis*. *D–G.* Order Pseudonostraca: *D*, *Burgessia bella*, showing diagrammatic outline of ventral side with appendages partly indicated; *E*, *Protocaris pretiosa*, with shell mostly removed to show remainder of body; *F–G*, side and ventral aspects of a restoration of *Waptia fieldensis*. *H–J.* Order Hymenocarina: *H*, side view of right valve of *Hymenocaris perfecta; I*, outline of a specimen of *Anomalocaris canadensis*, in which nine abdominal segments and the caudal segment are preserved; *J*, left valve of *Tuzoia retifera*. (*A after Wood in Raymond*, 1920; *B after Raymond*, 1935; *C after Krause in Hutchinson*, 1930; *D–H, J after Walcott, several different dates; I after Whiteaves*, 1884.)

biramous appendages on anterior part of trunk. Some xenopod may be ancestral to the Merostomata (Raymond, 1920, 1935). The subclass is based on three Middle Cambrian genera, *Sidneyia* (Fig. 13-24*A*), *Amiella*, and *Emeraldella* (Fig. 13–24*B*), all of which are placed in the Order Limulava (Raymond, 1920).

Subclass Archaeostraca[1]

This subclass was created by Stromer for a small group of extinct Paleozoic Arthropoda which had a compressed or, more rarely, a depressed carapace, and of which nothing is known about the appendages. In addition to the genera assigned to the following orders, a dozen or more other genera of uncertain relationships have been referred to this admittedly "catchall" subclass at one time or another (*e.g.*, *Eopteria*, *Euchasma*, *Ischyrina*, *Ribeiria*, *Ribeirella*, and *Technophorus*).

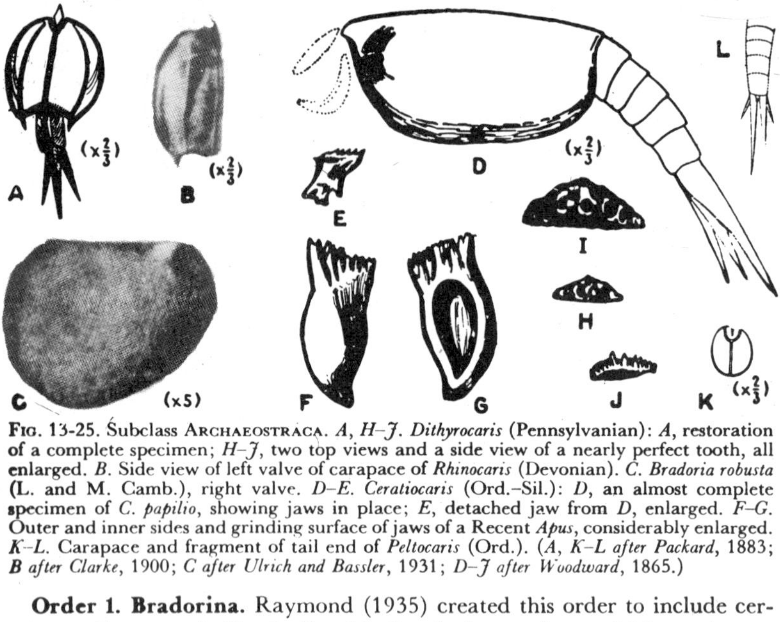

FIG. 13-25. Subclass ARCHAEOSTRACA. *A*, *H–J*. *Dithyrocaris* (Pennsylvanian): *A*, restoration of a complete specimen; *H–J*, two top views and a side view of a nearly perfect tooth, all enlarged. *B*. Side view of left valve of carapace of *Rhinocaris* (Devonian). *C*. *Bradoria robusta* (L. and M. Camb.), right valve. *D–E*. *Ceratiocaris* (Ord.–Sil.): *D*, an almost complete specimen of *C. papilio*, showing jaws in place; *E*, detached jaw from *D*, enlarged. *F–G*. Outer and inner sides and grinding surface of jaws of a Recent *Apus*, considerably enlarged. *K–L*. Carapace and fragment of tail end of *Peltocaris* (Ord.). (*A*, *K–L after Packard*, 1883; *B after Clarke*, 1900; *C after Ulrich and Bassler*, 1931; *D–J after Woodward*, 1865.)

Order 1. Bradorina. Raymond (1935) created this order to include certain small, ostracode-like fossils with the single muscle scar high on the anterior of the bivalve shell. These forms, included in the families Bradoriidae, Beyrichonidae, and Indianidae, may have been ancestral to the Ostracoda, from which they differ, however, in having a less calcareous shell (Raymond, 1935, 1946). The order seems to be limited to the *Cambrian*.

Bradoria (Fig. 13-25*C*) is representative of the Family Bradoriidae.

Order 2. Ceratiocarina. These are extinct shrimp-like arthropods with a bivalve carapace strongly coalesced medially and having a free rostrum. Almost nothing is known of the appendages. *Cambrian to Pennsylvanian.*

[1] Archaeostraca—Gr. *archaios*, ancient or primitive, + *ostracon*, shell of a testacean; referring to the fact that the subclass includes ancient or primitive shelled animals.

Ceratiocaris (Fig. 13-25*D–E*), from Ordovician and Silurian strata, is a representative of the order.

Order 3. Rhinocarina. These are extinct shrimp-like arthropods in which the bivalve carapace has a free rostrum and a narrow median dorsal plate separated from the valves by a straight or slightly curving hinge at each side. *Devonian to Pennsylvanian.*

Rhinocaris (Fig. 13-25*B*) and *Dithyrocaris* (Fig. 13-25*A,H–J*) are representative genera.

Order 4. Discinocarina. These extinct arthropods have a chitinous carapace in the form of subcircular or oval shields with a triangular rostrum filling an anterior notch. Surface sculpture consists of raised concentric lines. *Ordovician into Triassic.*

Peltocaris (Fig. 13-25*K–L*), from the Ordovician of Great Britain, is a representative of this order.

Class Trilobita[1]

General Considerations

The trilobites constitute an extinct group of exclusively Paleozoic arthropods in which the body was divided into a variable number of somites and partly encased in a supposedly chitinous integument. This integument consisted of a dorsal covering of mineralized chitin, a turned-under ventral doublure, one or more anteroventral plates (hypostome;[2] epistome), and a ventral membrane that is never preserved and probably consisted of nonchitinous and nonmineralized fleshy substance.

The dorsal part of the integument, which is the part of the trilobite exoskeleton most commonly preserved, typically has a prominent transverse trilobation, a characteristic that gives the name to the class. Ventrally, the living trilobite had a pair of biramous appendages on each somite and a pair of long delicate antennules extending anteriorly from beneath the head.[3] Only under exceptionally favorable conditions, however, were these delicate appendages and antennules preserved (Fig. 13-26).

The average trilobite was a small creature, usually 50 to 75 mm. (2 to 3 in.) in length, but some were quite tiny (less than 10 mm.) and a few giants (*e.g.*, *Terataspis;* frontispiece) attained a length of more than half a meter (67.5 cm., or 27 in.).[4]

Trilobites seem to have been exclusively marine, since their remains are always associated with those of typical salt-water animals (*e.g.*, corals, crinoids, brachiopods, and cephalopods), and they are supposed to have been

[1] Trilobita—L. *tri*, three, + *lobus*, lobe; referring to the transverse as well as the longitudinal trilobation of the exoskeleton.

[2] Two spellings have been used for the following terms, as indicated; epistome (epistoma), hypostome (hypostoma), and metastome (metastoma). The first alternative is preferred in the present work.

[3] One genus, *Neolenus*, has posterior appendages that may have served a sensory function.

[4] Among the giants are *Terataspis grandis* from the Devonian (675 mm., or 27 in.), *Isotelus gigas* from the Ordovician (450 mm., or 18 in.), *Paradoxides harlani* from the Cambrian (450 mm., or 18 in.), and *Dalmanites myrmecocophorus* and *Homalonotus major*, both from the Devonian (375 mm., or 15 in.). The average trilobite, however, has a length of about 25 mm. (1 in.) and a width of about half that dimension.

largely vagrant bottom dwellers, although some delicate spiny forms (*e.g.*, odontopleurids largely) are thought to have been planktonic. Some were probably scavengers; others may have been filter feeders.

Appearing as widely differentiated and highly developed forms in earliest Cambrian seas, they were dominant among the invertebrates until the cephalopods displaced them in the Late Ordovician. Thereafter they declined gradually to the close of the Paleozoic when they became extinct. They attained world-wide distribution and are of great value as index fossils for local, continental, and intercontinental correlation. More than 1,500[1] genera have been described, and these have been variously grouped into orders, superfamilies, etc.

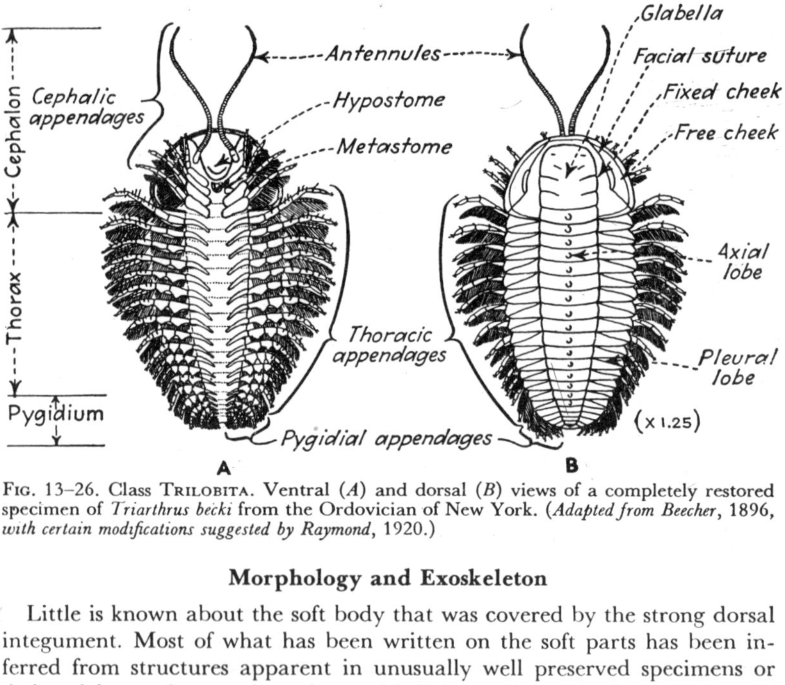

FIG. 13–26. Class TRILOBITA. Ventral (*A*) and dorsal (*B*) views of a completely restored specimen of *Triarthrus becki* from the Ordovician of New York. (*Adapted from Beecher*, 1896, *with certain modifications suggested by Raymond*, 1920.)

Morphology and Exoskeleton

Little is known about the soft body that was covered by the strong dorsal integument. Most of what has been written on the soft parts has been inferred from structures apparent in unusually well preserved specimens or deduced by analogy or homology with living arthropods. Some aspects of the soft parts of *Ceraurus pleurexanthemus*, a Middle Ordovician species, are shown in Fig. 13-27. Because of the paucity of information on the soft parts of the trilobite, chief emphasis is laid on the exoskeleton in the following discussion.

In general, the parts of the exoskeleton where no movement took place were thickened and strengthened with calcium carbonate; elsewhere the in-

[1] Prof. J. Marvin Weller writes (Jan. 7, 1952) that his card file includes more than 2,500 trilobite generic names; of these about 1,000 are spelling variants, homonyms, or obviously unnecessary synonyms.

tegument was thin and flexible. It has been assumed that these thinner places acted as joints for movement and some as places where adjacent exoskeletal parts could separate during molting.

In life the trilobite exoskeleton consisted of a trilobate dorsal shield; a small, ventral liplike plate, the **hypostome,** similar to the labrum of the Crustacea; a thin **ventral membrane** (never found preserved); and numerous biramous, jointed appendages (Figs. 13-26, 13-30).

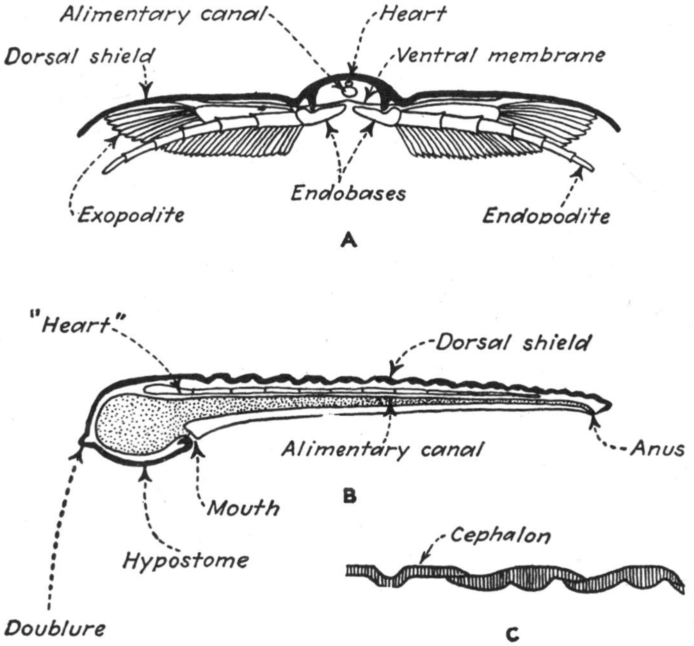

FIG. 13-27. Class TRILOBITA. Structure of *Ceraurus pleurexanthemus*, a Mid-Ordovician species. *A*. Transverse section of thorax. *B*. Longitudinal section along the axis or axial lobe. *C*. Diagram showing the posterior part of the cephalon and the first two thoracic segments, with their typical imbricating relationship. (*Adapted from Raymond*, 1920.)

In most typical trilobites the dorsal shield is divided by two axial furrows into a prominent convex axial region, or **axial lobe,** flanked on each side by a pleural region, or **pleural lobe.** When viewed in transverse section (Fig. 13-27*A*), the trilobate profile of the dorsal shield is quite apparent. It should be noted, however, that the three regions are not freely separated, as the integument is continuous across the entire exoskeleton.

Longitudinally the dorsal shield is divided into three distinct parts—the head, or **cephalon;** the flexible middle part, or **thorax;** and the abdominal part, or **pygidium.** Each of these exhibits transverse trilobation with a strongly convex axial part and more flattened pleural areas (Fig. 13-26).

The axial part of the cephalon is termed the **glabella** and is generally arched prominently above the two lateral **cheeks.** The cheek areas of most trilobites are divided along a **facial suture** into outer **free cheeks,** which commonly become detached, and inner **fixed cheeks,** which are marginally continuous with the glabella. If eyes are present, they lie along the facial suture on the free cheeks. The thorax is composed of a series of imbricating segments that are supposed to have been united during life by thin flexible integument. Each segment, though divisible transversely into axial and pleural regions, is a single rigid plate. The pygidium is a segmented or smooth plate that consists of several posterior segments more or less completely fused together. In some species the posterior marginal area of the pygidium is prolonged into a **caudal spine.**

Few fossil trilobites are completely enough preserved to show the ventral side; hence knowledge respecting appendages and other ventral structures is lacking in all but a few genera. In a few, however (*e.g.*, *Triarthrus*, Fig. 13-26; *Cryptolithus*, *Ceraurus*, and *Neolenus*), the ventral side is preserved, and the number and nature of appendages have been determined from these specimens. In these there are five cephalic appendages—an anterior pair modified into antennules, followed by four pairs of essentially similar, biramous appendages. So far as known each thoracic and pygidial segment was also provided with a pair of biramous limbs, regardless of whether the dorsal shield segment is separate or fused with adjacent segments. The evolutionary importance of these biramous appendages is discussed later.

Students of trilobites have developed a complex nomenclature for designating the many different parts of the exoskeleton and for describing the wide variations shown by these parts. As yet there seems to be no universal agreement as to the use of many terms, but recent efforts at standardizing the nomenclature give hope that ultimately general agreement may be reached. An example is the recent attempt to standardize the nomenclature for Cambrian trilobites (Howell *et al.*, 1947). In the following discussion only a few of the more general terms are used (see Fig. 13-28 for illustrations of the more common terms), but even these may seem to the beginning student to be too numerous.

The Cephalon. The cephalon of most trilobites is a more or less rigid plate formed from the fusion of the five to seven anterior cephalic segments. It is typically somewhat arched in longitudinal profile, and in transverse profile shows the trilobation characteristic of the entire dorsal shield. Along the side and front margins it is ventrally reflexed to a greater or less degree, and the turned-under portion is designated the **doublure** (Fig. 13-29). In outline the cephalon ranges from semicircular or semielliptical to triangular, and the posterior margin is usually straight or but slightly curved. The angle included between the posterior and lateral margins is designated the **genal angle,** and if this part of the cephalon is prolonged posteriorly, the prolongation is the **genal spine.** The angle may be acute or obtuse, and the spine short and broad or long and pointed (Fig. 13-28).

The axial portion of the cephalon constitutes the **glabella.** This may be segmented or smooth. It is divided by transverse furrows into an **anterior lobe,** one to three pairs of **lateral lobes,** and, if the **lateral furrows** do not

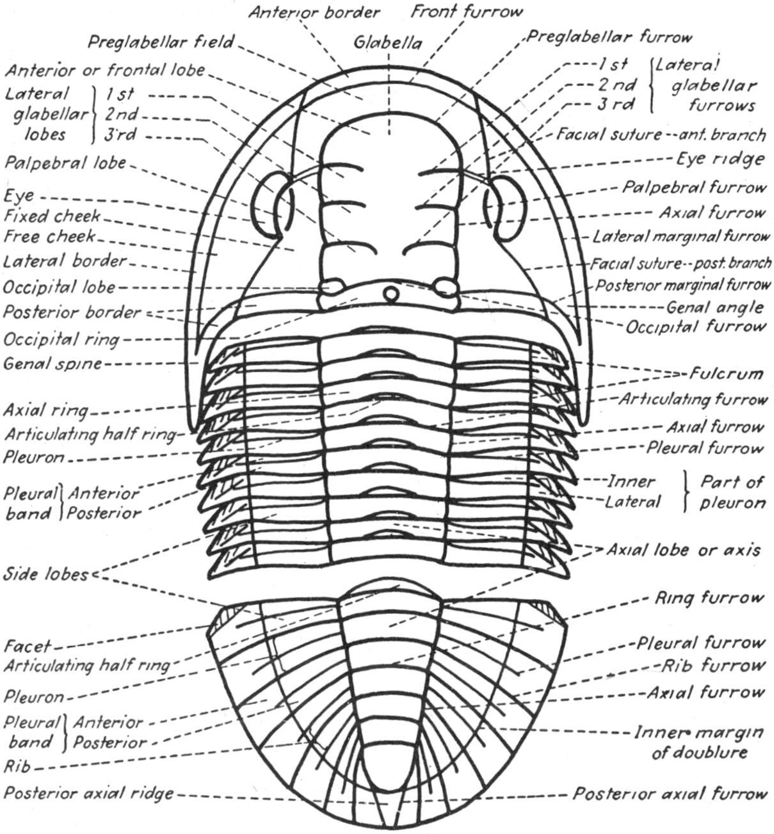

FIG. 13-28. Class TRILOBITA. Diagram showing the dorsal aspect of an opisthoparian trilobite with all the more important features labeled. (*Modified slightly after Warburg*, 1925.)

join across the median line, a **median lobe,** and an **occipital furrow** separating the **occipital ring** (Fig. 13-28). It has been suggested that the five glabellar somites represent integumental plates covering the five anterior somites of the trilobite ancestor, and that they became fused to form the glabella (Fig. 13-33).

Furrow development and lobe shape exhibit considerable variation. In some species the glabella is quite small, whereas in others it constitutes most of the cephalon. It may be only a fraction of the length of the cephalon. It may expand forward, be parallel-sided, or narrow forward; it may extend to or even slightly beyond the anterior border. The furrows may be quite distinct, indistinct, or absent altogether, and they may extend entirely across the glabella or appear as notches on the sides. In some forms they join longitudinally, in which case they separate small lobes from the side of the glabella.

The occipital furrow is almost always well defined, even if the others are indistinct.

The glabella is flanked laterally by the fixed cheeks, from which it is set off by the longitudinal **axial furrows.** It is also separated from the frontal margin of the cephalon by a part of the **marginal furrow.**

The two fixed cheeks together with the glabella constitute the **cranidium.** The fixed cheeks may be large or small depending upon the position of the facial suture. In *Conocoryphe* (Fig. 13-36) they constitute more than half the cranidium, whereas in *Albertella* (Fig. 13-36) they are quite small. The axial or dorsal furrows are well defined in some genera (*e.g.*, *Calymene*, Fig. 13-36), faint in others (*e.g.*, *Isotelus*, Fig. 13-36), and almost invisible in still others (*e.g.*, *Bumastus*, Fig. 13-36). The fixed cheeks do not include the genal angles or spines unless the facial suture cuts the lateral margins of the cephalon. About midway along the lateral margin of each fixed cheek is a bean-shaped elevation, the **palpebral lobe.** This may be connected with the frontal lobe of the glabella, and it lies upward and inward from the eye itself which, if present, is situated on the adjacent inner margin of the free cheek.

Sutures of several different kinds mark on the cephalon lines along which component parts are joined. These **cephalic sutures** are of considerable taxonomic and ontogenetic interest, and are illustrated in Fig. 13-29. The most conspicuous of these, and the ones generally most easily seen, are the **facial sutures.** These consist of two lateral branches that start symmetrically on the posterior margin or lateral doublures, cross the border and then take symmetrical courses forward over the dorsal surface of the cephalon to the anterior border where they unite on the dorsal surface or along the anterior margin, or continue ventrally across the doublure as the **connective sutures.** In some cases the two branches meet the anterolateral angles of a special median rostral shield, or **rostrum** (also designated the **epistome**), which interrupts the doublure (Fig. 13-29) and is typically ventral in position. The

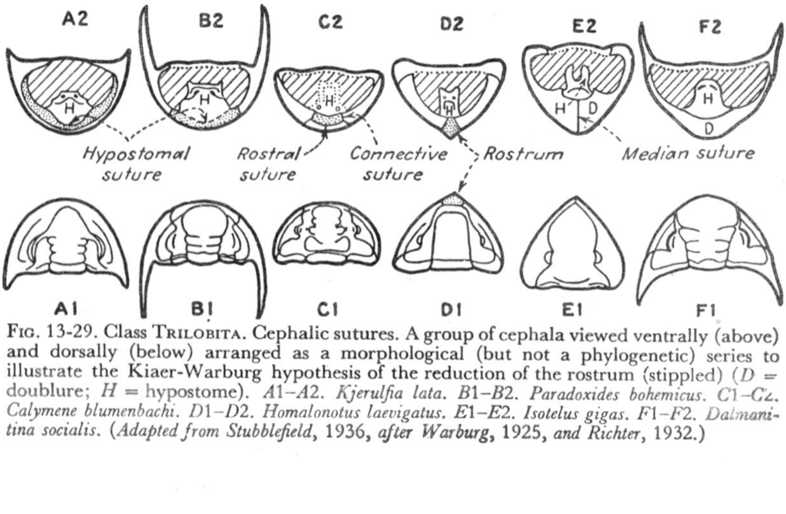

FIG. 13-29. Class TRILOBITA. Cephalic sutures. A group of cephala viewed ventrally (above) and dorsally (below) arranged as a morphological (but not a phylogenetic) series to illustrate the Kiaer-Warburg hypothesis of the reduction of the rostrum (stippled) (*D* = doublure; *H* = hypostome). *A1–A2. Kjerulfia lata. B1–B2. Paradoxides bohemicus. C1–C2. Calymene blumenbachi. D1–D2. Homalonotus laevigatus. E1–E2. Isotelus gigas. F1–F2. Dalmanitina socialis.* (*Adapted from Stubblefield*, 1936, *after Warburg*, 1925, *and Richter*, 1932.)

anterior suture bounding the rostrum is the **rostral suture.** It is typical for the hypostome to be separated anteriorly from the cephalic doublure or the rostrum by a transverse cleft, the **hypostomal suture.** In many Cambrian forms, however, this suture is closed and marked only by an impressed line, so that the hypostome is continuous with the doublure. In many trilobites (*e.g.*, Cyclopygidae; Phacopidae) the facial sutures supposedly join anteroventrally and are not united with the hypostomal suture, so that no rostrum is delineated and the doublure is continuous across the anteroventral part of the cephalon. In other forms, however, the facial sutures are joined to the hypostomal suture by the connective sutures. There may, however, be only a single axial suture, as in the Asaphidae, in which case the designation **median suture** is used. The reader interested in suture evolution is referred to the excellent discussions by Raw (1927) and Stubblefield (1936).

The facial suture separates the cheek into an inner fixed cheek and an outer free cheek, and passes on the axial side of the eye which, if present, lies on the free cheek. The nature and position of facial sutures were once considered of great taxonomic importance, and trilobites were divided into three orders based upon whether the suture is ventral or marginal (Hypoparia), whether it cuts the posterior margin (Opisthoparia), in which case the free cheeks bear the genal angles or spines, or whether it cuts the lateral margin (Proparia), in which case the fixed cheeks carry the genal angles or spines. This classification has had to be modified because the assumptions on which it was based have been shown to be invalid in important particulars. The two terms **opisthoparian** and **proparian,** however, are still useful for describing the facial suture.

The free cheeks separate easily from the cranidium in many genera (*e.g.*, *Ogygopsis*, Fig. 13-36*G*), but in some (*e.g.*, *Olenellus*, Fig. 13-35) they are completely fused with the cranidium, and in a few others their presence is doubtful. It is thought that the facial sutures served as places along which parts of the cephalon could separate in molting.

The eyes of trilobites are complex organs that vary greatly in size, position, and structure. Some trilobites were eyeless; some (*e.g.*, *Cryptolithus; Harpes*) had a tubercle with one or two facets; but by far the majority (*e.g.*, *Phacops*) had a multiple-faceted compound eye like that common among many modern arthropods. The eye surface was convex outward, hence the trilobite had a wide range of vision. A few trilobites (*e.g.*, *Encrinurus*, Fig. 13-37*B*, and *Encrinuroides*, Fig. 13-31*A*) had the eyes on short stalks.

Compound trilobite eyes are of two kinds. In one group the entire eye is covered with a transparent chitinous covering, the **cornea.** This covering may be quite smooth and give no indication of the compound nature of the eye beneath it, or it may be granular, thereby reflecting the underlying eye facets. Eyes of this type, which are described as **holochroal,** are the ones found on most trilobites. The second type of eye, the **schizochroal,** has a separate cornea for each individual eye facet. The corneas may be circular or polygonal. Facets range in size from less than 0.10 mm. to as much as 0.50 mm. and in number from 14 to 600 among schizochroal eyes to more than 15,000 in some holochroal types. In most eyes the facets are regularly arranged with greatest economy of space (Fig. 13-30).

Eye development of unusual nature is shown in the Ordovician *Telephus* (Fig. 13-30). The two eyes are quite large and are situated on small free

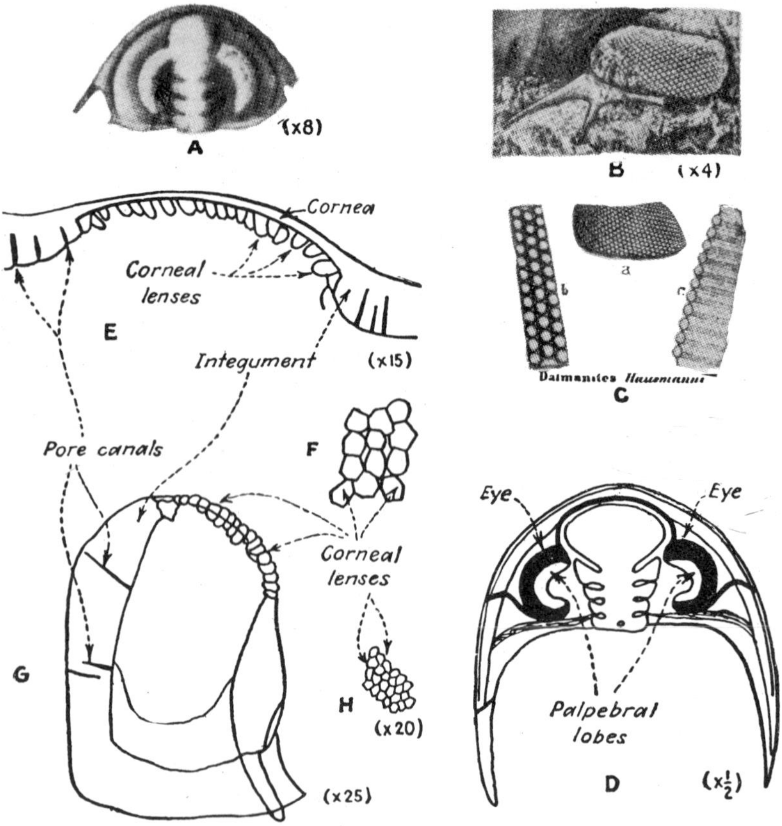

FIG. 13-30. Phylum ARTHROPODA. Arthropod eyes (*A–D* and *G–H*, Trilobita; *E–F*, Arachnoidea). *A*. Cephalon of *Wanneria halli*, from Lower Cambrian of Alabama, showing large crescentic eye lobes joining anteriorly with the hemicylindrical glabella. *B*. Side view of eye and spinose free cheek of *Telephus*, from the Tennessee Ordovician. *C*. Eye of *Dalmanites hausmanni: a*, complete eye, with lenses preserved; *b*, three vertical rows of lenses, each lens of which is surrounded by six small granules disposed in hexagonal manner; *c*, vertical section along one of the vertical rows. *D*. Cephalon of *Dalmanites*, showing in black the crescentic eyes which lie on the lateral surfaces of the two curved palpebral lobes. The facial suture, indicated by heavy line, passes around both ends of each eye and between the outer margin of the palpebral lobe and the inner margin of the eye, so that the eye lies on the free cheek. *E–F*. Eye of the arachnoid *Limulus*, introduced for comparison: *E*, cross section, showing the hard parts; *F*, tangential section of a few lenses. *G–H*. Eye of *Isotelus gigas: G*, cross section; *H*, tangential section of the lenses to show how compactly they are arranged. (*A after Walcott*, 1910; *B after Ulrich*, 1930; *C–D after Barrande*, 1852; *E–H after Packard*, 1880.)

cheeks along the lateral margins of the cephalon. The eye surface is hemi-ellipsoidal; hence it is assumed that the animal could see what was going on in every direction along or above the bottom. The eyes of many trilobites (*e.g.*, *Nevadia; Wanneria*) are situated on the ends of eye ridges that extend posteriorly over the free cheeks from the frontal lobe of the glabella.

The border of the cephalon is commonly set apart from the remainder of the shield, with which it is continuous, by the marginal furrow. Along the outer margin the integument is turned under to form the doublure (Figs. 13-27*B*, 13-31), which may be interrupted along the anterior part by the median rostrum (Fig. 13-29). The underlip, or hypostome, is a subfrontal plate lying directly in front of the mouth. In front of the hypostome, in some forms, is the rostrum or epistome, isolated by sutures. A third ventral plate, the **metastome,** is situated directly behind the mouth (Fig. 13-26*A*). The mouth itself faces backward, that is, toward the pygidium, an arrangement which

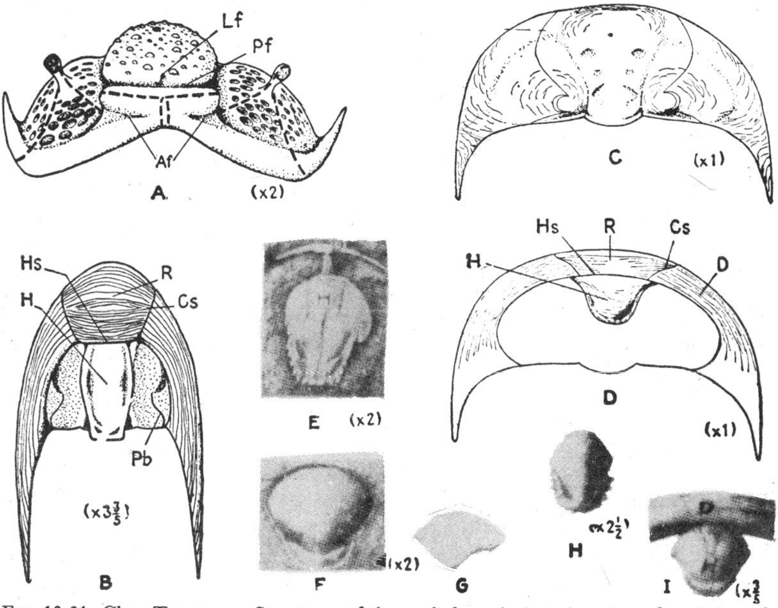

FIG. 13-31. Class TRILOBITA. Structures of the cephalon. *A*. Anterior view of cephalon of *Encrinuroides sexcostatus*, an Ordovician proparian trilobite with prominently stalked eyes. The course of the facial sutures is shown by the dashed line. *B*. Ventral view of reconstructed cephalon of *Phillipsinella parabola*, from the Upper Ordovician, showing prominent rostrum and elongate hypostome. *C–D*. Dorsal and ventral views, respectively, of *Bronteopsis scotica*, a Middle Ordovician proparian with a relatively plain cephalon. *E*. *Paedeumias transitans* (L. Camb.); hypostome attached to doublure by a narrow median support. *F–G*. *Olenellus thompsoni* (L. Camb.); top and side views of hypostome. *H*. *Elliptocephala asaphoides* (L. Camb.); hypostome. *I*. *Callavia bröggeri* (L. Camb.); hypostome attached to doublure. (*Af* = anterior furrow; *Cs* = connective suture; *C* = doublure; *H* = hypostome; *Hs* = hypostomal suture; *Lf* = longitudinal furrow; *Pb* = posterior branch of facial suture; *Pf* = preglabellar furrow; *R* = rostrum.) (*A–D after Whittington*, 1950; *E–I after Walcott*, 1910.)

is in harmony with the assumption that the food was brought from that direction.

The Thorax. The thoracic part of the trilobite exoskeleton consists of a variable number of trilobed segments (from two in *Agnostus*, Fig. 13-34, to 44 in *Paedeumias robsonensis*), which articulate with one another (Fig. 13-27*C*) and thereby allow enrollment. Transversely the segments are a single piece but divisible into an axial and two pleural lobes, the latter of which may be elongated into posterior spines, or **pleura** (singular, **pleuron**) (Fig. 13-26*B*). Each segment has a frontal extension of the axial portion that is inserted beneath the posterior margin of the segment directly in front (Fig. 13-27*C*). These flanges permitted articulation of the segments and also protected the thoracic part of the body when the exoskeleton was enrolled. The axial part of each segment has a furrow at the anterior margin, and **pleural furrows** may also be present.

There seems to be a significant relation between number of thoracic segments and size of pygidium. Forms with few segments have relatively large pygidia (*e.g.*, *Agnostus*, *Eodiscus*, and *Isotelus;* Figs. 13-35, 13-36), whereas those with many tend to have relatively small pygidia (*e.g.*, *Paradoxides*, with 16 to 20, and *Harpes*, with 29; Fig. 13-36). As the evolutionary trend seems to have been toward reduction in number of thoracic segments and increase in the number fused together to make the pygidium, it has been suggested that such forms as *Agnostus* and *Eodiscus* are actually advanced and specialized rather than primitive.

As with other arthropods, each new thoracic segment was added between the pygidium and the last segment of the thorax.

The Pygidium. The pygidium constitutes the third and posterior part of the exoskeleton, and in life covered the abdominal part of the body. It is a single transversely trilobate shield composed of a variable number (2 to 29) of segments firmly fused together. The individual segments may or may not be distinct. In some species it closely resembles the cephalon in shape and size, whereas in others it may differ greatly in these respects. *Olenellus* (Fig. 13-35*D*) seems to have lacked a pygidium altogether, whereas the pygidia and cephalons of some forms (*e.g.*, agnostids) are so much alike that it is difficult to tell one from the other.

The axial lobe may extend the full length of the pygidium, as in *Calymene;* may range to only a fraction of it as in *Scutellum* (Fig. 13-36*J*); or may merge so completely with the remainder of the pygidium that its identity as a distinct feature is lost, as in *Bumastus* (Fig. 13-36*I*). Segmentation may be conspicuous, indistinct, or lacking altogether on the axial lobe, and may be present or absent on the pleural lobes. The pygidial border may be continuously smooth or variously frilled with spines. There is usually a marginal furrow and the marginal border is reflexed to form a doublure, which is commonly of considerable width. Caudal and marginal spines are characteristic of certain groups.

The Ventral Side. The ventral side of a trilobite, except for the parts covered by the doublures and the plates around the mouth, seems to have been covered in life by nothing more than a soft epidermis or membrane, which is never preserved. An **axial groove** extends from the posterior margin to the mouth and is bordered on each side by the appendages. Inward extensions of the basal segment of the appendages may have aided in conveying food to the mouth and in mastication (Fig. 13-27*A*). Appendages are known in only a few trilobites, either because they were not preserved or because they are not visible in the fossil specimens. In some specimens (*e.g.*, *Triarthrus*) they were revealed only after careful removal of the hard rock in which the underside of the trilobite was embedded. Figure 13-26 shows such a form. Ventral structures are now well known in *Calymene* (*Flexicalymene*), *Ceraurus*, *Isotelus*, *Neolenus*, and *Triarthrus*, and imperfectly in a few other genera, but it seems likely that similar structures in other genera will come to light with further collecting. Reference is made in an earlier paragraph to the hypostome, epistome, and metastome, all situated around the mouth.

Appendages. Specimens preserving the ventral side of a trilobite show five pairs of appendages on the cephalon (Fig. 13-33) and a pair of biramous appendages on each thoracic and pygidial segment (Fig. 13-26).

The typical trilobitan appendage (Fig. 13-32) is biramous and consists of

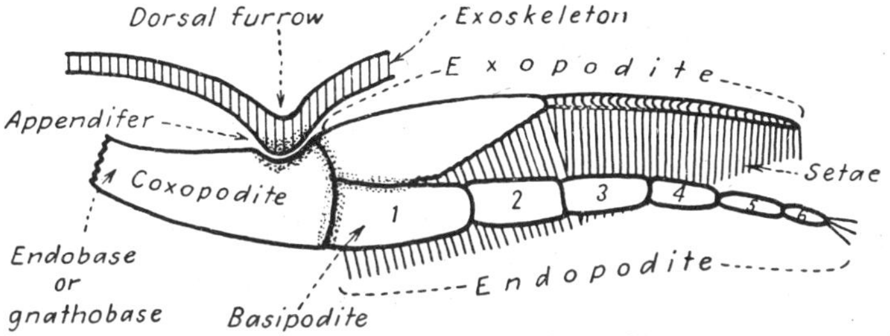

FIG. 13-32. Class TRILOBITA. Generalized diagram of a trilobitan appendage. The four areas of articulation are indicated by stippling. These are (1) between appendifer and coxopodite; (2) between coxopodite and the large basal shaft of the exopodite; (3) between coxopodite and basipodite; and (4) between basipodite and the basal plate of the exopodite. The marginal setae of the exopodite are partly hidden by the proximal segments of the endopodite. (*Adapted from Raymond*, 1920.)

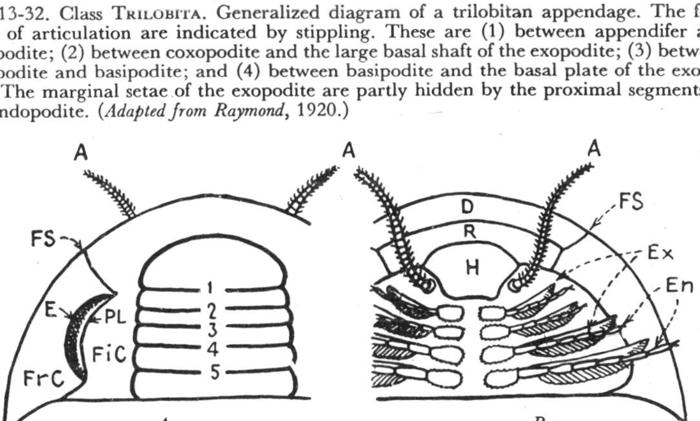

FIG. 13-33. Class TRILOBITA. *A–B*. Diagrams of dorsal and ventral aspects, respectively, of an opisthoparian trilobite, showing essential features. In *A* the glabellar furrows are numbered; in *B* the front pair of biramous appendages have been designated mandibles, the next two pairs maxillae, and the fourth pair maxillipeds. (*A* = antennule; *D* = doublure; *E* = eye; *En* = endopodite; *Ex* = exopodite; *FiC* = fixed cheek; *FrC* = free cheek; *FS* = facial suture; *H* = hypostome; *PL* = pleural lobe; *R* = ı trum.) (*Adapted from Richter*, 1933.)

two strikingly different branches, the **endopodite** and **exopodite,** which spring from the outer extremity of a large basal segment, the **coxopodite.** The coxopodite is prolonged inwardly as an **endobase** (**gnathobase** on the four cephalic appendages) and is articulated with a special process, the **appendifer,** formed by invagination of the dorsal shield at the extremities of the transverse furrows of the axial lobe (Fig. 13-32).

The endopodite is a long slender jointed appendage composed of six segments, the outermost of which bears three bristles. The innermost or basal segment, which is termed the **basipodite,** articulates with the coxopodite along its inner extremity and partly supports the exopodite with which it is articulated along part of one side. The endopodite seems to have been used for walking.

The exopodite, though somewhat variable in structure, typically consists of a flattened shaft with closely spaced bristles, or setae, along one margin. It articulates with the coxopodite, as well as with part of the basal segment (basipodite) of the endopodite, and in life it is probable that there was little individual freedom of movement between the two branches of the appendage. The exopodite is believed to have been used for both respiration and swimming.

The appendages of the thorax and pygidium are all essentially alike except for size, but the five pairs of cephalic appendages are somewhat modified and deserve some discussion. A point of possible evolutionary significance is the fact that the trilobite seems to have had only five cephalic somites and five pairs of appendages, whereas arachnoids have six of each.

The anterior pair of cephalic appendages are uniramous, jointed antennules (Fig. 13-33). These are attached ventrally to the axial furrows on each side of the anterior glabellar lobe and are directed forward beyond the front margin of the cephalon. The other four pairs, which are postoral in position, are biramous and of the same fundamental structural plan as the thoracic and abdominal appendages, except that they were more or less modified for mastication.

The whole appendage articulated with the appendifer and must have been controlled in life by powerful muscles. The inner extensions of the four pairs of cephalic appendages are termed gnathobases because they are thought to have functioned in mastication; similar extensions on the thoracic and pygidial segments (endobases) possibly aided in moving food forward along the axial groove toward the mouth.

Classification

In the different classifications of the Subclass Trilobita, the following characteristics, all exoskeletal, have been recognized as more or less important in differentiating major divisions:

1. Ontogeny as revealed in series of growth stages.
2. Nature and position of facial sutures.
3. Number of thoracic segments.
4. Nature of cephalon or pygidium or both.
5. Absence or presence of eyes, and their structure.

Of these the first is the most important by far because the series of growth stages indicate the characteristics of the very young trilobites, which in many cases are quite different from the same features in adults. For example, the young of *Calymene* have indisputable proparian facial sutures, whereas in the adults the facial suture may bisect the genal angle, thus leaving the investigator who is dealing with only adult exoskeletons in doubt as to whether he should call it proparian or opisthoparian.

For the first third of the present century the classification proposed by Beecher (1897) was widely used. On the basis of his classic studies, which in part consisted of the laborious excavation and preparation of the ventral sides of many specimens of *Triarthrus becki* (Fig. 13-26), he concluded that the trend and orientation of the facial suture and its position with respect to the genal angle, together with the position of the free cheeks and the nature of the eyes, afforded a fundamental basis for classification. Accordingly he subdivided the Trilobita into the following three orders:

Order Hypoparia. The free cheeks form a continuous marginal ventral plate on the cephalon, extending in some forms onto the dorsal side at the genal angles, and the facial suture ranges from marginal to submarginal.

Order Opisthoparia. The facial suture cuts the posterior margin of the cephalon, and the free cheeks are limited to the dorsal side and include the genal angles or spines.

Order Proparia. The facial suture ends on the lateral margins, the free cheeks are dorsal in position, and the fixed cheeks bear the genal angles or spines.

Although Beecher's classification was the best that could have been made at the time and with the material available, it has had to be considerably modified and expanded in the light of important new discoveries.[1] Of particular evolutionary importance among these discoveries are the exquisitely preserved specimens from the Middle Cambrian Burgess shale and the critically important early growth stages that are being released from

[1] Beecher (1901) thought that the free cheeks were at first ventral and the facial suture marginal, and that there were only eye lines and eye cells (ocelli) rather than true eyes. This primitive condition represents his hypoparian stage. The facial suture is then supposed to have migrated from the margin onto the upper surface of the cephalon, with the genal angles on the free cheeks, and eyes are supposed to have appeared first along the margin of the cephalon and then to have moved inward toward the facial suture on the dorsal surface of the free cheeks. This stage represents Beecher's opisthoparian condition. Finally the facial suture migrated laterally and then anteriorly until it cut the posterolateral border of the cephalon in front of the genal angles, which then became a part of the fixed cheek. This condition he supposed to be the most advanced, and he designated it proparian.

Subsequent investigators, having the advantage of many excellently preserved specimens discovered since Beecher's time, have found that some of the Hypoparia (Eodiscidae) had descendants with proparian cheeks, whereas others (Cryptolithidae) are thought to have had an ancestor with an opisthoparian suture. These and other discoveries have necessitated abandoning the Order Hypoparia, with the result that genera formerly included in this group have been assigned to several new orders. Likewise, the ontogeny of *Leptoplastus salteri* shows a proparian suture in youthful stages but an opisthoparian condition in adults, and this led Raw (1925; 1927) to suggest that trilobites with proparian sutures in the adult stage are examples of arrested development.

Ordovician limestones by acid treatment (Lalicker, 1935; Whittington, 1941, 1941*a*, 1947; Ross, 1951).

The following classification, which is based on the discoveries and suggestions of many students of trilobite evolution and taxonomy, assumes that the Class Trilobita is a natural group of rather closely related ancient arthropods, some of which may have been ancestral to other classes of Arthropoda:

Order 1. Agnostida.
Order 2. Eodiscida.
Order 3. Olenellida.
Order 4. Opisthoparia.
Order 5. Proparia.

It is to be noted particularly that Beecher's orders Opisthoparia and Proparia are retained, his order Hypoparia is abandoned, and three additional orders are included—Agnostida, Eodiscida, and Olenellida. The interested reader is referred to Rasetti (1948) for a recent discussion of these orders and for pertinent references to the literature.

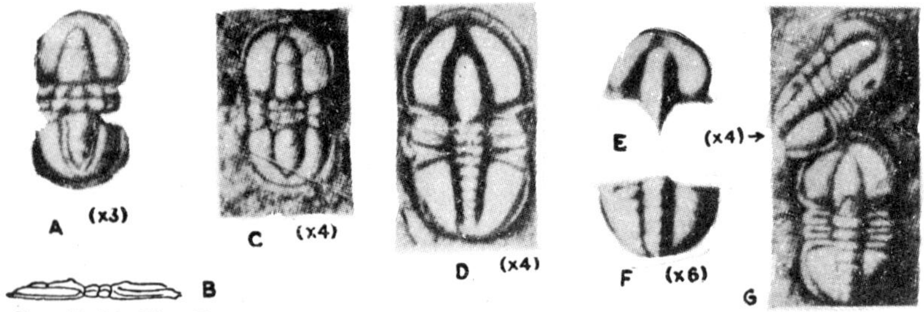

FIG. 13–34. Class TRILOBITA. Middle Cambrian species. *A–C. Agnostus* (Order Agnostida): *A–B, A. interstrictus; C, A. montis. D–G.* Representatives of Order Eodiscida: *D, Eodiscus punctatus; E–G, Pagetia clytia,* cephalon (note missing free cheeks), pygidium, and two almost complete specimens. (*A–B after Walcott,* 1886; *C–D, photographs by C. E. Resser; E–G, after Walcott,* 1916.)

Order 1. Agnostida. The agnostids are quite small eyeless[1] trilobites characterized by strong similarity of cephalon and pygidium, constant number (two) of thoracic segments, complete lack of facial sutures and free cheeks, and highly specialized pygidium consisting supposedly of a constant number of greatly differentiated segments fused into a single rigid plate. Agnostids differ most strikingly from all other trilobites in the small and constant number of thoracic segments and highly specialized pygidium. The last-named feature represents the same tendency toward fusion of segments that is characteristic of the cephalons of all trilobites, and it is taken as proof that the agnostids are a highly specialized group that were differentiated from the ancestral trilobitan stock far back in the Pre-Cambrian, possibly before they acquired a hard test (Whitehouse, 1939). *Lower Cambrian through Ordovician.*

Agnostus (Fig. 13-34) is typical of the order.

[1] Some students of trilobites now believe that the agnostids are highly specialized trilobites that became blind secondarily.

Order 2. Eodiscida. These are quite small trilobites characterized by equal-sized but conspicuously dissimilar cephalic and pygidial shields, small number of thoracic segments (two or three, as contrasted with 5 to 42 among other trilobites, except for the agnostids which invariably have two), and either total lack of facial sutures or small free cheeks of proparian type (*Pagetia;* Fig. 13-34). Some were blind, whereas others had eyes.

The cephalon has a tapered glabella, which never has the typical agnostid bilobation, and the pygidium is typically trilobitan with numerous segments apparent in the axial lobe and in a few forms on the pleural lobes. The eodiscids do not seem to have had any post-Cambrian descendants. The North American species of the Family Eodiscidae were recently revised by Rasetti (1952). *Lower and Middle Cambrian.*

Eodiscus and *Pagetia* are examples of this small order (Fig. 13-34).

Order 3. Olenellida (also called Mesonacida). The olenellids are typical trilobites in most respects, but differ in lacking distinguishable facial sutures (Rasetti would not include any genera with well-developed facial sutures). The cephalon is large, the pygidium small and simple, and the thorax of many segments (13 to 27). The eyes are large and the prominent curved palpebral lobes extend to the glabella. The olenellids are among the oldest of trilobites. They are believed to have differentiated from the common trilobitan ancestral stock long before the beginning of the Cambrian, and possibly to have given rise to both the Crustacea and the Arachnoidea. *Lower Cambrian.*

Mesonacis (Fig. 13-35), with the third thoracic segment enlarged and the pleura extended as prominent spines, and with the fifteenth segment having a large spine on the axial lobe, is typical of the order, which includes among others the following familiar genera: *Elliptocephala*, *Paedeumias*, *Olenellus*, *Holmia*, *Wanneria*, and *Callavia* (Fig. 13-35).

Order 4. Opisthoparia. This is the largest by far of all trilobite orders, and it includes among its members some of the oldest (*e.g.*, *Bonnia;* L. Cambrian) and all the latest (*e.g.*, *Neogriffithides;* Permian) genera. All members of the order, which as presently constituted is almost certainly not a natural taxonomic group, share in common an adult opisthoparian facial suture. The eyes are holochroal and are situated on the free cheeks, which commonly bear prominent genal spines. *Lower Cambrian through Permian.*

More than two dozen families of opisthoparian trilobites have been proposed, and these have been grouped into half a dozen superfamilies by different authors (Swinnerton, 1915; Richter, 1933; Rasetti, 1948). Genera illustrating the variations characteristic of the order are shown in Fig. 13-36.

Order 5. Proparia. In proparian trilobites the facial sutures extend from the lateral margins of the cephalon in front of the genal angles, inward and forward, cutting the front margin separately, or uniting in front of the glabella; in the first case the two free cheeks come free from the cephalon as separate pieces, whereas in the second they form a single yokelike piece. In all proparian genera the genal angle or genal spine is a part of the fixed cheek. Eyes are carried on the free cheeks of most forms, but a few of the more primitive genera seem to have been blind. The Proparia appeared in

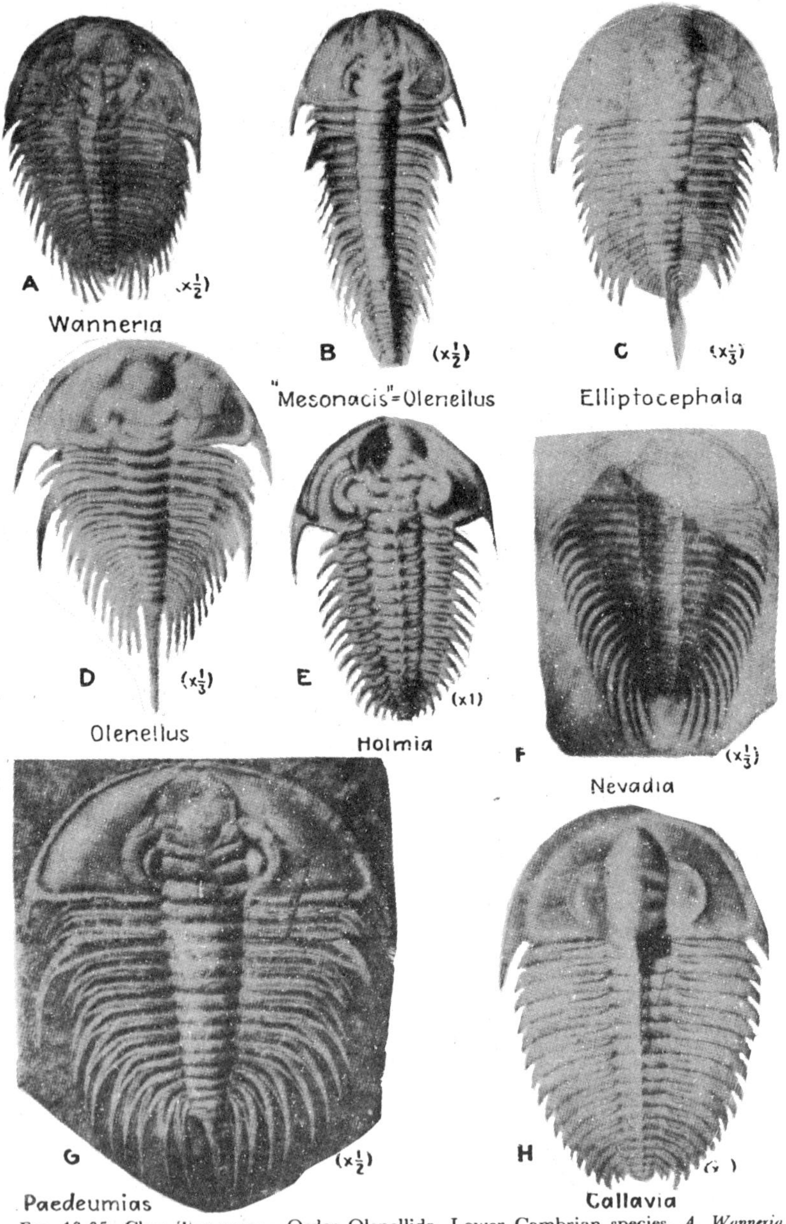

FIG. 13-35. Class TRILOBITA—Order Olenellida. Lower Cambrian species. *A. Wanneria walcottana. B. Olenellus vermontanus* (formerly referred to *Mesonacis*). *C. Elliptocephala asaphoides. D. Olenellus thompsoni. E. Holmia kjerulfi;* the left part of the glabella has been cut away so as to reveal part of the outline of the hypostome. *F. Nevadia weeksi. G. Paedeumias transitans. H. Callavia broggeri.* (*All from Walcott*, 1910, *with E after Holm*, 1887.)

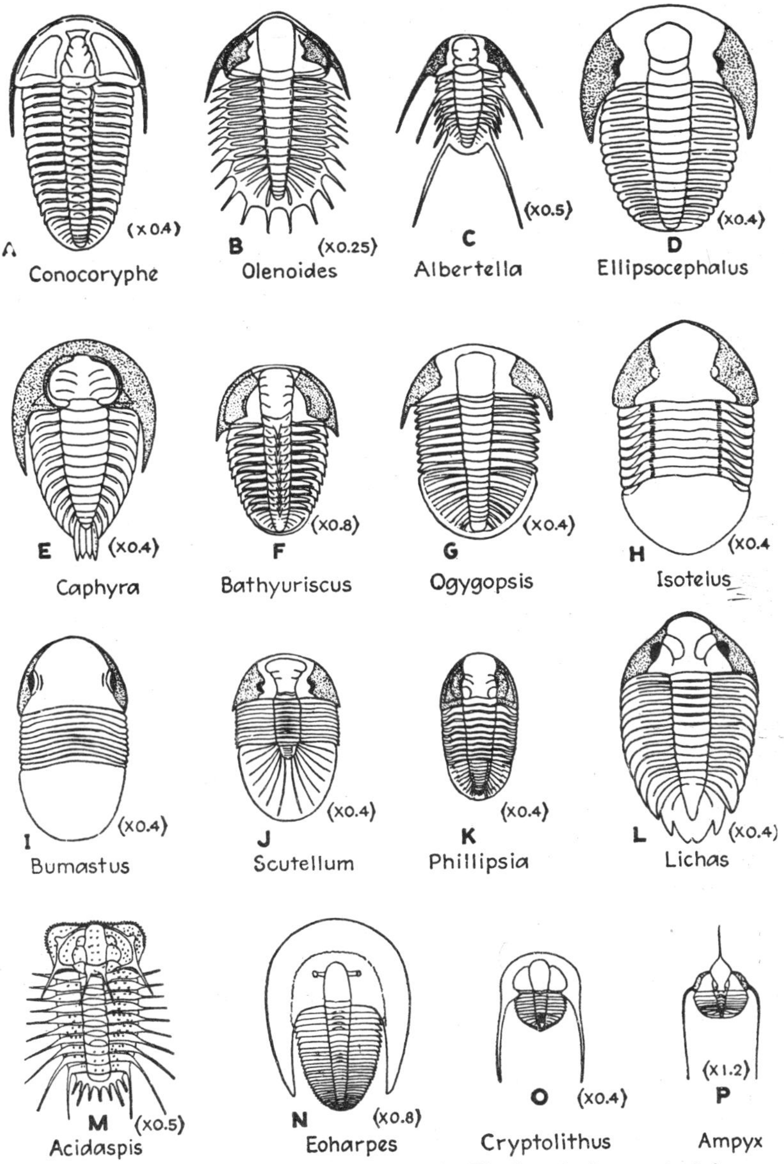

FIG. 13-36. Class TRILOBITA—Order Opisthoparia. The free cheeks are stippled except where the facial suture is marginal or submarginal, or, as in *Acidaspis*, where it is not indicated. (*Diagrams based on figures by the following authors: A, D–E, J, L–P after Barrande,* 1852; *B–C after Walcott,* 1908; *H, K after Nicholson and Lydekker,* 1889.)

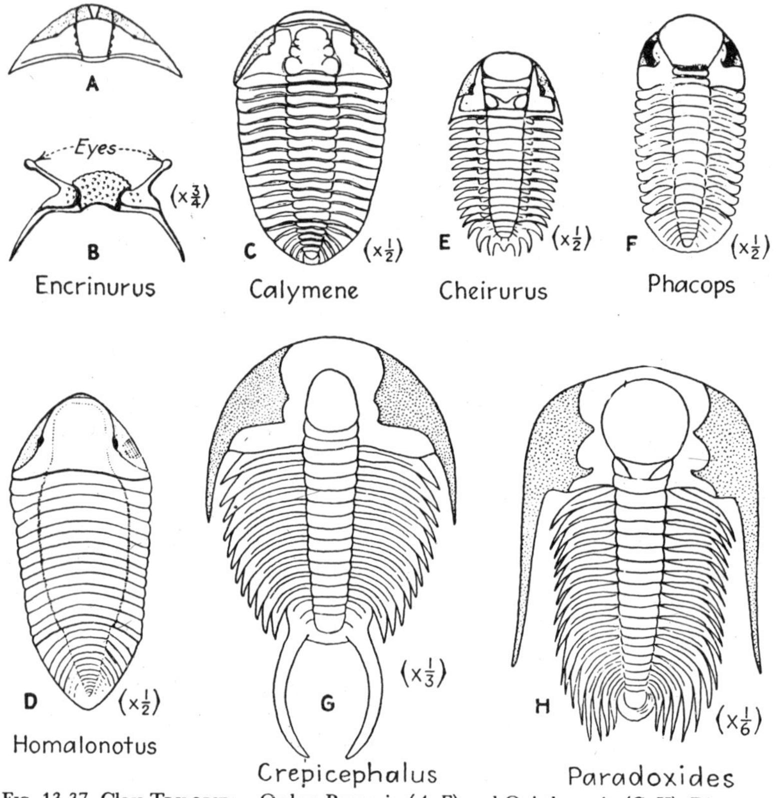

FIG. 13-37. Class TRILOBITA—Orders Proparia (*A–F*) and Opisthoparia (*G–H*). Diagrams illustrating representative genera. Free cheeks are stippled. (*Adapted from figures by the following authors: A after Beecher*, 1897; *B after S. Weller*, 1907; *C, E–F after Barrande*, 1852; *D after Hall*, 1861; *G–H after Walcott*, 1886, 1916.)

earliest Ordovician seas, differentiated rapidly during the Silurian and Devonian, and became extinct before the beginning of the Mississippian. A few Cambrian genera with proparian facial sutures (*e.g.*, *Pagetia*, Fig. 13-34; *Pagetides*) have been referred to this order, but they are so different from typical Proparia that they have been retained in the Eodiscida, with which they seem to have closer affinities. *Lower Ordovician through Devonian.*

A few representatives of the Proparia are illustrated in Figs. 13-36 and 13-37.

Ontogeny and Phylogeny

Trilobites are assumed to have laid eggs, but this has not been definitely established (Walcott, 1879). Certain ovoidal bodies associated with trilobite remains have been interpreted as fossil eggs, but their true origin is unknown. It would seem that the trilobite egg hatched out at an early stage in the ontogeny, because the youngest known larval stages differ greatly from the

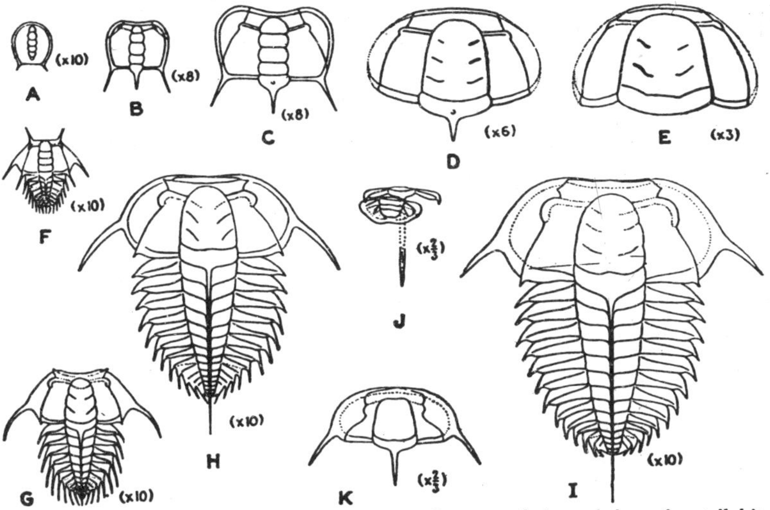

FIG. 13-38. Class Trilobita. *A–E.* Developmental stages of the opisthoparian trilobite *Peltura scarabaeoides*, from the Upper Cambrian of Bornholm: *A*, protaspis stage; complete shield with the last or posterior axial ring belonging to the pygidial somite; *B*, cephalon of early meraspid (nepionic) stage with proparian facial suture, prominent genal spines, and spine on the fifth or occipital ring; *C*, a later nepionic stage, showing prominent posterior migration of facial suture; *D*, neanic (late meraspid or early holaspid) stage with opisthoparian facial suture; the free cheeks are restored in perspective; *E*, adult cephalon of holaspid stage; free cheeks restored in perspective. *F–K.* Stages of development of the opisthoparian trilobite, *Leptoplastus salteri: F–I*, meraspid stages; *J*, pygidium and last thoracic segment bearing its median spine; *K*, large cephalon, restored. *J* and *K* are parts of adult (holaspid) specimens. (*A–E from Stubblefield*, 1936, *after Poulsen; F–I after Raw*, 1925 *and* 1927.)

adult. Having attained the first larval stage, the young trilobite, designated a **protaspis,** then passed through a succession of growth stages ending finally in the adult. Fortunately, specimens representing the different stages have been preserved, and it has been possible to reconstruct the ontogeny of a number of species. In such ontogenetic studies, it has become customary to recognize three general periods of growth—protaspid, meraspid, and holaspid[1]—which mark, respectively, the beginning and the completion of the thorax (Fig. 13-38).

The **protaspid period** includes all the earliest growth stages from the hatching to the appearance in the dorsal shield, or protaspis, of the first transverse suture, which divided the protaspis into two separate parts, the cephalon and **transitory pygidium.** The cephalon was complete with its five fundamental segments in the youngest protaspid stage; postcephalic segments constituting what has been designated a transitory or rudimentary pygidium appeared in later protaspid stages.

The **meraspid period** began with the separation of the cephalon and transitory pygidium by the transverse suture and ended with the completion

[1] Beecher (1895) first proposed the term protaspid, and Raw (1925) proposed the use of protaspid, meraspid, and holaspid as designations for three periods of trilobite development.

of the last segment of the thorax. It consists, therefore, of a definite number of stages, or **degrees,** each of which was preceded by a molt and marked by the appearance of a new segment. The degrees are designated as 0 to $(n - 1)$, in which sequence 0 marks the beginning stage, when the cephalon and transitory pygidium were separated but before the first thoracic segment had appeared, and n represents the number of segments in the completed thorax. Each new thoracic segment was added from the anterior border of the pygidium, and if the trilobite molted before forming the segment, it follows that the number of molts during the period would be the same as the number of segments in the complete thorax.

With completion of the thorax the larval stage of development ended and the trilobite, now an adult in all essential particulars, entered the **holaspid period.** During this period its most obvious change was great increase in size, although it also underwent further development in form, culminating in the completely developed adult individual. It is assumed that the trilobite continued to molt throughout the holaspid period, but it is not possible to determine exactly how many molts followed completion of the thorax. Raw (1927) estimates that an individual of *Leptoplastus salteri* molted 3 times during the protaspid period, 12 times during the meraspid, and 14 times[1] during the holaspid, for a total of 29 molts.

Although the three periods just described are defined by the development of the thorax, other important changes also took place during growth. Of special interest in this regard are the changes shown by the cephalon. *Peltura scarabaeoides*, an opisthoparian, may be taken as an example to illustrate these changes. Figure 13-38 shows two protaspid stages (*A*,*B*), two meraspid cephala (*C*,*D*), one transitional meraspid-holaspid cephalon (*E*), and an adult cephalon of the holaspid stage. It is important to note that this species of *Peltura* has a proparian suture in the larval stages (*i.e.*, during the meraspid period), but an opisthoparian suture in the adult stage. On the basis of this kind of development, Stubblefield (1936) has suggested that the proparian condition may be regarded as arrested development.

Two fundamentally different views of the significance of trilobite ontogeny have developed during the past half century. One view holds that the larval stages of development fairly closely represent the ancestry (*i.e.*, that the ontogeny recapitulates the phylogeny); the other considers them as secondary adaptations to the larval conditions of life. Raw (1927) probably is not far wrong when he suggests that the truth lies between the two and that the position differs greatly in different cases. It might be added that many more ontogenies need to be worked out before the fundamental divisions of the Trilobita can be discriminated satisfactorily.

There arises from the preceding discussion the moot question, what is the ancestry and what are the descendants, if any, of the Trilobita? There now seems to be a wide consensus that the trilobites descended from a simple archetypal arthropod, which in turn evolved from an annelidan ancestor, and Raw (1927) concludes that the protaspid is predominantly embryonic and larval and that the paucity of its segments is ascribable partly to its small

[1] Raw based his estimate of holaspid molts on the proportional increase in size during the period.

size and partly to inheritance from the larval stage of the ancestral annelid. Although the protaspis is comparable with the annelid larva, it is quite different from the youngest larvae of all classes of arthropods except the Arachnoidea; it is strikingly different in some respects from the nauplius larva (Fig. 13-1) typical of the Crustacea, for example, but it is somewhat similar to the embryo of *Limulus*, the sole surviving genus of the Merostomata. This latter resemblance has led some investigators to suggest that the earliest arachnoids descended from a trilobitan ancestor (the Olenellida) (Raw, 1927).

Störmer has long been a proponent of the arachnoidean affinities of the Trilobita, basing his conclusion on the nature of the appendages in the two classes, and his views and classification have been presented in several recent publications (Störmer, 1933; 1939; 1942; 1944; 1949; 1951). In the restudy of specimens investigated earlier by Beecher (1895), Raymond (1920), and Walcott (1921, etc.) Störmer (1939) claimed to have found at the base of the appendage a "pre-coxal segment" to which the branchial ramus (*i.e.*, the exopodite) is said to be attached. The branchial ramus is regarded as a pro-epipodite, and hence not homologous with the exopodite of the crustacean appendage. Extending this concept to the Trilobita in general, and also considering the nature of the arachnoidean appendages, he reached the conclusion that "the trilobite limb probably is to be regarded as a prototype of the appendages in the Arachnomorpha [= Merostomata + Arachnida + Trilobita, etc.,]"[1] and only remotely resembles the postulated biramous

[1] On the basis of this conclusion, along with many related considerations, Störmer (1944), in an excellent review of the Arthropoda, proposed the following classification (which is compared with that being used in Grassé's "Traité de zoologie"):

Störmer, 1944			Grassé, vol. VI, 1949	
Myriapoda-Insecta	Progoneata	Symphyla	Myriapoda	Mandibulata
		Pauropoda		
		Diplopoda		
	Chilopoda	Chilopoda		
	Insecta	Apterygota	Insecta	
		Pterygota		
Crustacea		Branchiopoda	Crustacea	
		Ostracoda		
		Copepoda		
		Cirripedia		
		Archaeostraca		
		Malacostraca		
Arachnomorpha	Trilobitomorpha	Trilobita	Trilobitomorpha	
		Merostomoidea		
		Marrellomorpha		
		Pseudocrustacea		
		?Arthropleurida		
	Chelicerata	Merostomata	Chelicerata	
		Arachnida		
		Pycnogonida		
Pantopoda		Palaeopantopoda		
		Pantopoda		

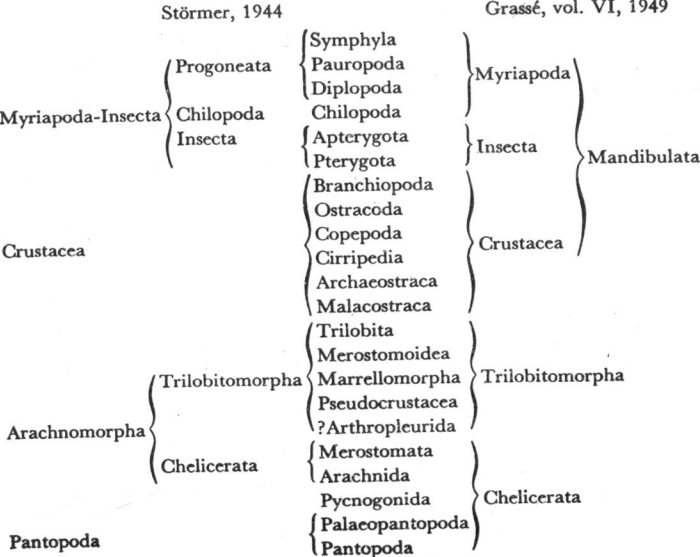

appendage of the archetype crustacean. The evidence and reasoning of Störmer (1939; 1944) in developing his classification have been strongly questioned, particularly by Garstang (1940) and Heegaard (1945), and it would seem that other groupings of the numerous arthropodan orders will probably have to be made (*e.g.*, Grasse in "Traité de zoologie" is following quite a different grouping, as indicated in the footnote on page 599). Also see Störmer (1951) and Henningsmoen (1951).

One final question often raised in connection with discussions of trilobite phylogeny is, did the Arthropoda have a single ancestral form in common (*i.e.*, were they of *monophyletic origin*) or did they arise as independent branches from different but closely related ancestors (*i.e.*, were they of *polyphyletic* origin)? Perhaps the best way to emphasize the controversy is to quote briefly from two of the most recent students.

For the single ancestor, Heegaard (1945, page 13) states: "The *Arthropods* should presumably be regarded as having a *monophyletic origin* which from *Trilobites* or more probably *trilobite-like* forms in one direction tend towards *Crustaceans* and *Myriapoda-Insecta*, in the other towards *Xiphosura*, *Gigantostraca* and *Arachnida*." (See Fig. 13-57.)

For several ancestors, Störmer (1944, page 148) concludes: "To the present author a polyphyletic origin of the Arthropoda seems to be most in accordance with the fossil record."

For the confused reader who wishes to learn more about the evidence and reasoning used in support of the two opposed postulates, there is a voluminous literature that can be consulted. The more important references are listed by Snodgrass (1938), Störmer (1944; 1951), and Heegaard (1945).

Paleoecology[1]

Inasmuch as trilobites are extinct, nothing can be directly determined about their habits and habitats by observation. Much about their paleoecology, however, can be inferred from their faunal associates, their methods of burial and preservation, and their different exoskeletal structures. In a group of invertebrates as large and varied as the Trilobita, it is to be expected that there must have been many adaptations to every marine environment. If trilobites ever lived in fresh waters or upon the land, either they left no record of that existence, or the forms that lived in those habitats have not yet been recognized as trilobites. In the sea there were certainly some forms adapted to both soft and hard bottoms (*i.e.*, mud and sand on the one hand and solid rock on the other). Others lived along the strand line, where they wandered over the surface and burrowed in the soft sediments as do some modern crabs. Still others were almost certainly swimmers or floaters. Some of these probably lived near the bottom and spent more or less time on the bottom, whereas others, particularly spiny forms, may well have been planktonic. Modified and specialized structures of different trilobites indicate that species were adapted to many environments.

[1] For brief discussions of the paleoecology of the trilobites see Howell and Resser (1934), Raymond (1936), and Schevill (1936).

Certain types of trilobites lived in great numbers on Paleozoic bioherms, and in reef environments that have been studied extensively the same genera are rarely found both in the bioherm and in the interbiohermal beds. Likewise, trilobites seem to have shown preference for different kinds of depositional environments, but the critical evidence for determining the nature and extent of this preference is only now beginning to be assembled.

The food of trilobites is unknown and their mode of feeding can only be conjectured. It is known, however, that trilobites lived in close association with simple plants (algae), protozoans, sponges, worms of many kinds, bryozoans, branchiopods, and mollusks, and it is assumed that some forms selected their food from among their living associates. Some were quite likely herbivorous, feeding on the simple plants of the earlier Paleozoic seas. Others may have been scavangers, cleaning up organic debris included in the soft sediments through which they ploughed and burrowed.

Trilobites almost certainly served as food for other animals, and it may well be that this was one of the important factors leading to their extinction. Among possible enemies were cephalopods, early chordates, sharks, and possibly even cannibalistic members of their own class. It seems more than a mere coincidence that trilobites underwent a marked decline in the Late Silurian and in the Devonian with the advent of sharks and other early fishes, and it may well be that these new invaders soon swept the seas almost bare of trilobites, particularly of those forms thought to have been denizens of clear and open waters.

The original development of the exoskeleton was possibly a response to an unfavorable environment, though there may be other equally plausible reasons. The development of spines has also been ascribed to environmental conditions. Some investigators have suggested that the earliest trilobites had no exoskeleton and that cannibalistic forms may have attacked one another, since no other enemies are known. Such attacks would have eliminated those without protective structures and would have put a premium on those that developed a protective armor. The armor would have served well until the sharks and other fish appeared in the Silurian and Devonian. The appearance of these early chordates may well have caused, or at least hastened, the ultimate extinction of trilobites, since thereafter trilobite remains are not common. Nektonic forms probably were the first to go, as these would fall an easy prey to the new enemy that moved swiftly through the water with mouth wide open to capture the swimming trilobites. There were also probably bottom-browsing fish which rooted and ploughed through the soft sediments in search of food just as modern carp still do. To such fishes the mud-dwelling trilobites probably provided a good source of food. Many trilobites could enroll, thereby protecting their undersides, but this practice, ironically enough, instead of saving the animal may actually have aided the fish in capturing and swallowing it.

The ability to enroll seems to have been developed early, as a few Cambrian forms are known to have had this ability. It became common in the Ordovician, Silurian, and Devonian, but was not acquired by all contemporary trilobites. In enrollment the pygidium was opposed to the cephalon in such a way as to cover the entire ventral surface in the manner of an armadillo. The doublures adjusted themselves to fit one another and were so perfectly opposed that no animal could reach the ventral part of the trilobite. It has been suggested, but not proved, that enrollment was developed chiefly for protection against animal enemies.

Trilobites, like many other arthropods, molted periodically, but the number of times of molting is unknown. As a consequence of molting, exoskeletons greatly outnumbered actual individual organisms on the sea bottom, and it is likely, therefore, that the parts or the whole of several exoskeletons in a faunal assemblage may have been made and shed by the same individual. The many free cheeks, cranidia, thoracic segments, and pygidia scattered sporadically through the sedimentary rocks should probably be regarded in large part as fragments of molts rather than as parts of exoskeletons to which the trilobites were attached at the time of death and burial. The molts were no doubt of low specific gravity, with large surface area relative to volume, and hence could be transported long distances by slowly moving waters of low competency. In this way the remains might well have been buried far from the places where the animals lived or died and in localities where trilobites did not live and could not have lived. Such transportation of molts must be given serious consideration in efforts to interpret the environments of trilobite life. Entire specimens (*i.e.*, with the ventral as well as dorsal structures preserved) indicate that the animal was inside at the time of burial and, further, that it probably lived and died quite near the place of burial.

Fossil Record and Stratigraphic Range

A complete trilobite exoskeleton is one of the rarest of fossils and when found is likely to be more or less altered. Entire dorsal shields are relatively common, but antennules and appendages are rarely preserved. The usual finds are separate cephala, cranidia, free cheeks, hypostomes, thoracic segments, and pygidia (Fig. 13-39). Such remains probably represent fragments of molted exoskeletons that came apart after being abandoned.[1] Certain ovoidal bodies associated with trilobite remains have been interpreted as eggs; peculiar markings have been questionably identified as "nests," and some tracks and trails have been ascribed to trilobites; but all these enigmatic fossils are of uncertain origin.

The fossil record of trilobites is commonly in the form of chitinoid fragments or, if the actual exoskeletal substance is missing, impressions of the upper or under sides of the exoskeletal parts. In many cases the general integument is so well preserved that surface markings are still apparent, and in a few cases most of the original structure and substance of the eyes are preserved.

The oldest known trilobite remains have been collected from Lower

[1] It has been suggested that the trilobites molted quite frequently, perhaps more often than existing marine Crustacea, and it is probable that many fragmental fossils represent molted exoskeletons (Raw, 1927).

Cambrian rocks, and these represent animals of such an advanced stage of development that it seems certain the Trilobita had a long Pre-Cambrian history, possibly as soft-bodied creatures without much of an integument. By the end of the Cambrian hundreds of genera and thousands of species had come and gone; many, however, lived on in Ordovician seas and gave rise to additional new groups. Soon after the close of the Ordovician, however, trilobites began to decline in kinds and numbers and they steadily decreased

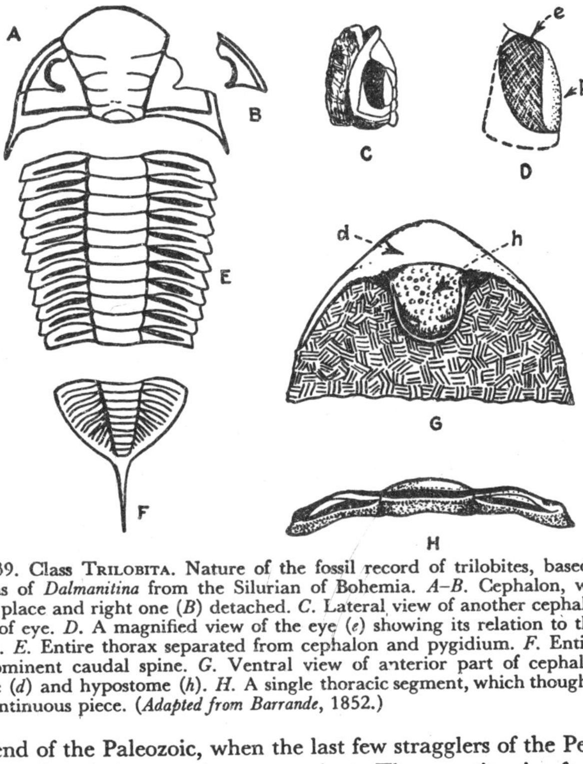

FIG. 13-39. Class TRILOBITA. Nature of the fossil record of trilobites, based mainly on specimens of *Dalmanitina* from the Silurian of Bohemia. *A–B*. Cephalon, with left free cheek in place and right one (*B*) detached. *C*. Lateral view of another cephalon, showing position of eye. *D*. A magnified view of the eye (*e*) showing its relation to the palpebral lobe (*pl*). *E*. Entire thorax separated from cephalon and pygidium. *F*. Entire pygidium with prominent caudal spine. *G*. Ventral view of anterior part of cephalon, showing doublure (*d*) and hypostome (*h*). *H*. A single thoracic segment, which though trilobed is a single continuous piece. (*Adapted from Barrande*, 1852.)

to the end of the Paleozoic, when the last few stragglers of the Permian seas disappeared and the race became extinct. The question is often asked of paleontologists, why did a vigorous and varied group of animals like the trilobites, which had a protective covering, wide range of adaptation, and large representation in types and numbers of individuals, decline so rapidly after such extensive development in Cambrian and Ordovician seas? No satisfactory answer has yet been given this question, but it should be noted that they were adapted to a biologic environment in which agile forms like the fish were not present. Cannibalistic relatives may have assisted in the decline. Perhaps the giant cephalopods of the Ordovician were also a factor. However, the sharp decline in genera and species with the advent of fish strongly suggests a causal relationship. Certainly after the Devonian, trilobites were rare, and only a few (less than a dozen genera) rather generalized forms persisted through the Carboniferous into the Permian.

Trilobites had world-wide distribution and many are excellent index fossils. Cosmopolitan genera like *Agnostus*, *Eodiscus*, *Olenellus*, *Paradoxides*, *Cryptolithus*, *Bumastus*, *Phacops*, *Flexicalymene*, and *Dalmanites* are useful for intercontinental correlation, and more restricted genera are equally valuable for continental and provincial correlation. Some formations locally contain large numbers of well-preserved remains (*e.g.*, Burgess shale, Lodi siltstone, Utica shale, Niagaran limestone).

INTRODUCTION TO CHILOPODA, DIPLOPODA, SYMPHYLA, AND PAUROPODA

Many authors, especially in the past, have included four relatively small groups of tracheate noninsectan arthropods in Class Myriapoda, a name which calls attention to the numerous legs in some of the forms. This class is usually divided as follows on the basis of the position of the genital aperture (**gonopore**):

Subclass Opisthogoneata. Gonopore posterior (Fig. 13-40).
  Order 1. Chilopoda (centipedes).
Subclass Progoneata. Gonopore anterior (Figs. 13-41, 13-56).
  Order 1. Pauropoda.
  Order 2. Diplopoda (millepedes).
  Order 3. Symphyla (garden centipedes).

It has become common practice among zoologists to drop Myriapoda, Opisthogoneata, and Progoneata as definite taxonomic divisions, though retaining the terms in adjectival form, and to consider each of the orders listed above a distinct class. This practice is adopted in the present work. The Chilopoda and Diplopoda are considered briefly because both classes have left a meager though important fossil record. The Symphyla, although without a known fossil record, are also discussed because of their importance in insect evolution. The Pauropoda are not considered further because they have no known fossil representatives and are of little evolutionary importance.

CLASS CHILOPODA[1]

The Class Chilopoda includes the familiar centipedes, which have a slender, elongate, segmented body that is flattened dorsoventrally (Fig. 13-40). The head has a pair of long, jointed, uniramous antennae, a pair of mandibles, and two pairs of maxillae. Body somites range in number from 15 to 173. The first somite has a pair of poison claws; all others except the last two have a pair of small, uniramous, seven-jointed appendages, or **walking legs.** Centipedes breathe by means of tracheae and are well adapted to life on land, where they scurry about at night in search of food. Approximately 1,700 living species have been described, but only a few fossil forms have been found, and the fossil record is consequently of little importance. Living chilopods are divided into five orders, three of which have fossil

[1] Chilopoda—Gr. *cheilos*, lip, + *pous*, *podos*, foot.

representatives in Tertiary amber and fresh-water deposits. The oldest fossils, which come from the Pennsylvanian of Illinois, as well as certain younger specimens, cannot be definitely assigned to any of the five existing orders. These fossils have been made the basis of two extinct families.

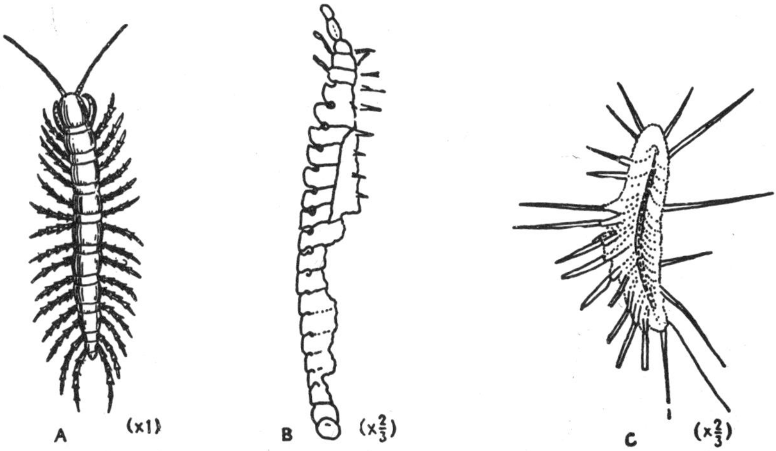

FIG. 13-40. Class CHILOPODA. Recent and fossil centipedes. *A. Lithobius forificatus*, a Recent species, about natural size. *B–C*. Incomplete centipedes from the Pennsylvanian of Illinois: *B*, *Latzelia primordialis*; *C*, *Palenarthrus impressus*. (*A after Nicholson and Lydekker*, 1889; *B–C outlined from drawings by Emerton in Scudder*, 1890.)

CLASS DIPLOPODA[1]

This class includes the millepedes, which are typically elongate, segmented creatures with a cylindrical body, the wall of which is hardened by deposits of calcium carbonate. The small head bears a pair each of antennae, mandibles, and maxillae and has two clusters of simple eyes (Fig. 13-41*A*). The short thorax consists of four single somites, each with one pair of legs. The long abdomen has from 20 to more than 100 double somites, each of which has two pairs of special breathing organs and of nerve ganglia, and two pairs of seven-jointed legs. The tracheal system is extensive and complicated. The millepedes live in damp dark places and rarely come out into the light. Unlike the agile centipedes, they move about slowly, testing the path ahead with the short antennae, and some species roll up in a spiral when disturbed.

Of the 6,300 described species only a few are fossil. Living species are generally distributed among eight orders, and those fossil forms that cannot be assigned to any of the existing orders are included in one or the other of two extinct orders (Protosygnatha and Archipolypoda).

At least five modern families have fossil representatives that have been found in Tertiary amber and in fresh-water sediments of the same general age (*e.g.*, Eocene Green River shales of Wyoming and Miocene Florissant beds of Colorado). All pre-Tertiary forms have been referred to one or the

[1] Diplopoda—Gr. *diplos*, double, + *pous*, *podos*, foot; referring to the two pairs of legs on each of the abdominal segments.

other of the two extinct orders previously mentioned. The Order Protosygnatha is based on a Pennsylvanian genus, *Palaeocampa*. The Order Archipolypoda includes several Devonian and Carboniferous genera (*e.g.*, *Euphoberia;* Fig. 13-41*B*) and possibly one Mesozoic species from Greenland. No fossil millepedes older than Devonian are known to have been described.

As with the centipedes, the millepedes have a sparse fossil record, and although this record is of considerable evolutionary interest, it is of little stratigraphic value.

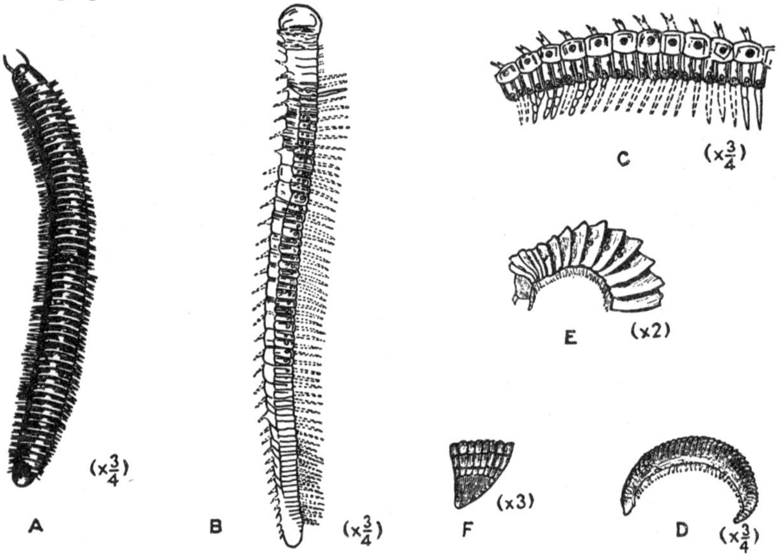

FIG. 13-41. Class DIPLOPODA. *A. Julus maximus*, a modern millepede. Note that each body segment has two pairs of legs, whence the name of the class. *B–C. Euphoberia armigera*, a fossil millepede from the Pennsylvanian of Illinois: *B*, an entire specimen, with the head preserved; the small dorsal spines and longer legs (on the ventral side) are dotted in because they are not well preserved in the actual specimen; *C*, a small part of an individual, showing two kinds of segments. *D–F. Xylobius sigillariae*, a fossil millepede from the Pennsylvanian of Nova Scotia: *D*, an entire individual; *E*, anterior portion; *F*, posterior portion. (*A after Nicholson and Lydekker*, 1889; *B–C after Meek and Worthen*, 1868; *D–F after Dawson*, 1868.)

CLASS SYMPHYLA[1]

The Symphyla are eyeless terrestrial arthropods (Fig. 13-56) that live in damp places with humus. Certain species are known as garden centipedes, although they are not, of course, true chilopods. They have antennae, jaws, and two pairs of maxillae, and in the adult have 12 pairs of legs. They have no known fossil record, and their taxonomic position is uncertain, because, as the name implies, they possess a singular combination of characters that indicate close relationships with insects on the one hand and myriapods on the other. In recent years the embryology of certain of the Symphyla has

[1] Symphyla—Gr. *syn*, with, + *phyle*, a clan; referring to the combination of myriapodous, insectan, and thysanurous characters.

been worked out (Tiegs, 1947; 1949), and it is now quite apparent that, while they display obvious affinity with the lower grades of myriapods (and even with the onychophoran, *Peripatus*) on the one hand, on the other they reveal an organization in advance of these, and in that respect they foreshadow the insects. It is not to be assumed, however, that the earliest insects evolved from a symphylan, but rather that they and the symphylans arose from more generalized creatures to which Imms (1947) has applied the designation Protosymphyla. This question of insect origin is further discussed in following paragraphs.

Class Insecta[1]

General Considerations

The typical adult insect is a six-legged, air-breathing arthropod with an elongate, bilaterally symmetrical body distinctly divided into **head, thorax,** and **abdomen** and bearing a pair of **wings** on the second and third thoracic segments (Fig. 13-42). A few insects seem never to have developed wings; these constitute the Subclass Apterygota. By far the majority, however, developed wings; these are embraced in the Subclass Pterygota. The earliest winged insects, the Palaeoptera, could move their four wings only up and down. Later forms, included in the Neoptera, could flex the wings and fold them back over the abdomen when at rest. Some representatives of every order of these neopterous insects became secondarily wingless, but it is interesting to note that, so far as known, no palaeopterous insect ever became similarly wingless (see later discussion of wing evolution). The numerous orders of living and extinct insects are shown in Fig. 13-55, and it is obvious from this tabulation that the winged insects have been the dominating group since wings first appeared.

Those competent to judge have estimated that approximately 800,000[2] species of living insects have been described and named, and they consider it probable that this enormous number may in fact not represent more than one-half of those that actually exist upon the earth at the present time. Standing in strong contrast, and emphasizing the exceedingly incomplete fossil representation of ancient insects, are the 12,000 extinct species distributed among 24 living and 10 extinct orders.

Insects live in every ecological niche, from the coldest polar region to the warmest and most humid tropical areas, and wherever found are always represented by some forms that make life miserable for other animals. In many parts of Canada, for example, the "black flies" of the "bush" are at times so abundant that humans cannot long endure without some protection against their vicious attack, and in the swampy areas of the same and other

[1] Insecta—L. *in*, into, + *secare*, to cut; referring to the strong constrictions of the insect body.

The authors gratefully acknowledge the assistance of Prof. Frank M. Carpenter (Harvard University), who critically read this discussion of the Insecta and provided several of the illustrations for this section.

[2] Also see Calvert, P. P., 1923, The number of living insects, *Entomol. News*, vol. 34, p. 122.

regions mosquitoes make life a burden for any animal entering their domain. Many insects are responsibl or the spread of dangerous diseases, and every schoolboy knows the story of malaria and the *Anopheles* mosquito.

It costs hundreds of millions of dollars annually to keep certain insect pests under control, and many governments are doing their utmost to prevent introduction of harmful species by imposing quarantines and other restrictive measures. Some of the most extensive damage is done by termites, which enter houses and tunnel through the timbers, in some cases reducing the woodwork to a mere shell. Other familiar destructive insects are grasshoppers, locusts, chinch bugs, fruit flies, and the cotton boll weevil. The larvae of certain forms are voracious feeders (*e.g.*, caterpillars) and they may strip a plant of foliage in a very short time.

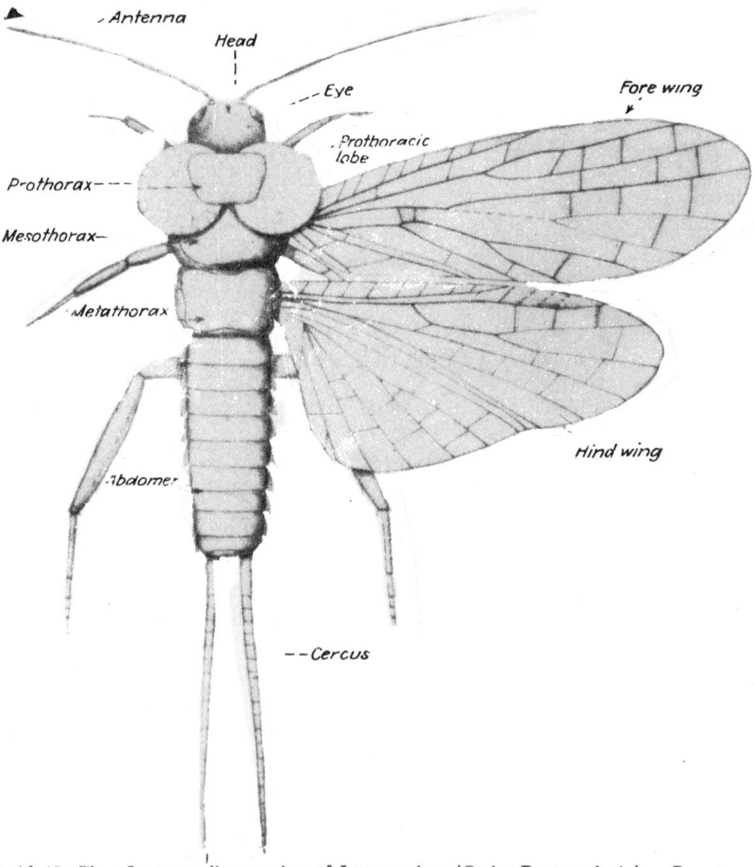

FIG. 13-42. Class INSECTA. Restoration of *Lemmatophora* (Order Protoperlaria), a Permian insect from the Wellington formation of Kansas. The two left wings are omitted in order to reveal the legs. (*By F. M. Carpenter.*)

The reproductive ability of insects is almost beyond imagination. It has been stated that if the common housefly could multiply unimpeded, with all individuals preserved, the progeny of a single female would soon bury the surface features of the earth. A somewhat less fanciful estimate of similar nature has been made for the common cabbage aphid (Herrick, 1926). It has been estimated that in a period of 135 days the 12 generations starting with a single mother would number 564 quadrillion (564,000,000,000,000,-000, or 564×10^{15}) individuals, which, if they all survived and were present at the end of the period, would have an aggregate weight of more than 800,000,000 tons, or the equivalent of some 8,000,000,000 men averaging 200 pounds apiece! It is certainly true that the reproductive ability of the Insecta is of astronomic proportions.

Of the animals inhabiting the earth, no group is more numerous and varied than the insects. In one form or another they are adapted to live anywhere and to eat almost anything organic, and were it not for the fact that they serve as food for numerous animals (*e.g.*, birds and even other insects), it would not be long before they would make the earth uninhabitable for other creatures. It has often been stated that man's most serious competitors for mastery of the earth are the Insecta. He makes constant war upon them in one form or another to prevent them from destroying his wooden buildings, killing the animals and plants he intends to eat, eating the clothes that he wears, and even infecting his own body with disease-producing germs.

Imaginative writers have pictured the earth inhabited by giant insects with powers thousands of times those of men, but nature seems to have decreed that insects must forever remain relatively small, for there does not seem to be an evolutionary trend toward gigantism among living forms. Although it is true that a few insects have attained wingspreads of 77 cm. (2½ ft.), the great majority are measurable in terms of a few millimeters.

General Morphology

The insect head,[1] which, as in most other arthropods, consists of six fused somites, is devoid of external segmentation, generally carries a pair of compound dorsal eyes, and has three pairs of more or less modified ventral appendages. The eye is composed of many individual facets, of which there are 4,000 in the housefly and 28,000 in the dragonfly, and in some forms the eyes are a conspicuous part of the head. The anterior pair of appendages are antennae; the next pair are mandibles; the third pair are maxillae; and the fourth pair are fused to form a **labium**. The appendages exhibit considerable variation from order to order.

The thorax is composed of three segments, each of which bears a pair of jointed legs, and the second and third of which typically bear wings on the dorsolateral surface in adult members of the Pterygota. Some of the Pterygota (*e.g.*, Diptera), however, have only a single pair of wings, these being situated on the second thoracic segment; some (*e.g.*, beetles) have the front

[1] Snodgrass (1951) gives an excellent review of the morphology of the head of mandibulate arthropods.

pair modified into coriaceous covers (elytra); and some have become secondarily wingless. The abdomen always has eleven somites in all postembryonic stages, and these lack walking appendages in the adult individual.

The typical insect breathes air, and the respiratory organs have the form of tiny breathing tubes, or tracheae, through which air is conveyed to all parts of the body. The nervous system and sense organs are quite complex, and most insects are exceedingly sensitive to surrounding environmental conditions.

Reproduction and Development

The sexes are separate, and development of an individual typically proceeds through a definite series of growth stages beginning with a fertilized egg and ending with an adult. In one group, those characterized by **incomplete metamorphosis** (Fig. 13-43*A–E*), a miniature of the adult hatches directly from the egg. This young insect, or **nymph,** is much smaller than its parents and lacks wings, but after a few molts (ecdyses) it develops wings and by this time has all the essential characteristics of an adult. A point of ecological interest is the fact that in general nymphs live in the same environment and eat the same food as their parents. Grasshoppers and cockroaches are examples of such insects.

In the insects with **complete metamorphosis** (Fig. 13-43*F–K*), by contrast, an ellipsoidal larva hatches from the egg and immediately begins to consume food in large amount,[1] meanwhile increasing in size. This stage, which may be passed in the ground, in water, in wood, in flesh, or on vegetation, is exemplified by the maggots of the housefly and the caterpillar of the moth or butterfly. The larval stage may be long or short, depending on the species. At its completion, the full-grown larva stops eating and in some forms encases itself in a silken capsule (*e.g.*, cocoon of the "silkworm"). Then follows a resting period, the **pupal stage,** during which the adult characteristics develop from the larval. When the young insect emerges from its pupal case, at which stage it is designated an **imago** (Fig. 13-43*I*), it is an adult in all essential respects.

Although two quite distinct series of changes are outlined in the preceding paragraphs, as a matter of fact the course of developmental stages varies greatly in the different orders of insects, with some stages dropping out completely in certain of the groups. Among the several more important features that show interesting variations are the wings, and as these are a much-used organ, they may be expected to have undergone extensive modifications. A brief consideration of wing development, therefore, seems in order, particularly because wings are the only fossil remains of many species and because they probably represent the most striking feature of insect evolution.

[1] Many insects have larvae that bore into wood, and timbers left outdoors for several years are likely to be honeycombed with small tubes made by larvae as they chewed their way through the wood. Many a person has been dismayed to find woolen clothes in his closet perforated with small holes cut by moth larvae, and it sometimes happens that carpets and rugs on the floors of dark rooms are partly or totally destroyed by these and other larvae.

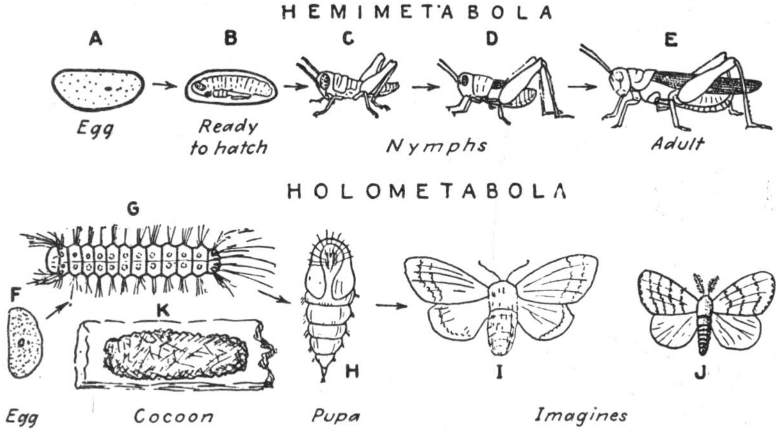

FIG. 13-43. Class INSECTA. Metamorphosis. *A–E.* Successive stages in the development of a grasshopper, illustrating incomplete metamorphosis. Such insects are included in the Hemimetabola. *F–J.* Complete metamorphosis as illustrated by the gypsy moth. Such insects belong to the Holometabola. Some holometabolous insects (*e.g.*, moths) make silken pupal cases or cocoons, such as that shown in *K*. (The different figures are not to the same scale; they are highly generalized and diagrammatic. (*Adapted from several authors.*)

Wing Development

The insect wing is a marvel of structural design and is conspicuous for lightness and strength. The basic framework consists of a double membrane strengthened by subparallel cuticular strands radiating from the base of attachment to the outer margin (Fig. 13-44). These rods, or **veins,** are customarily designated by some system of terms and numbers. The system shown in Fig. 13-44 is now in general use and is based upon a terminology proposed by Comstock (1918) and modified by Lameere (1922). Wings of somewhat different type from two extinct Permian species are illustrated in Figs. 13-45 and 13-46 to show how closely these ancient structures resemble the wings of modern insects. It should be emphasized that although wings are referred to as appendages, they are not really appendages at all, but rather dorsolateral outgrowths of the integument surrounding the thoracic somites.

The earliest insects all had four wings. Those of the two palaeopterous orders, Palaeodictyoptera (Fig. 13-44*B*) and Protodonata, could move the wings only up and down (Figs. 13-47 and 13-48), whereas those of the earliest neopteran order, Protorthoptera, even though contemporaneous with palaeopterous forms (so far as the fossil record goes), could flex their wings and fold them back over the abdomen. There is no doubt that the Neoptera descended from the Palaeoptera (Martynov, 1925); hence the latter must have lived earlier than the Pennsylvanian, in the rocks of which period the oldest fossil insects have been found.

The origin of insect wings is a moot question, but some light may be thrown on the problem by examining the earliest known winged forms. For this purpose the Permian protoperlarian *Lemmatophora* (Fig. 13-42) is useful. Of special interest are the two prominent membranous lobes on the prothoracic segment. These seem to be homologous with the functional wings of the mesothoracic and metathoracic segments, and many students of insects believe that wings developed from similar lobes on the two posterior segments of the thorax. In support of this view is the additional fact that prothoracic lobes are prominent on all true Palaeodictyoptera and on other generalized insects.

It is postulated, therefore, that the first insect wings were developed from

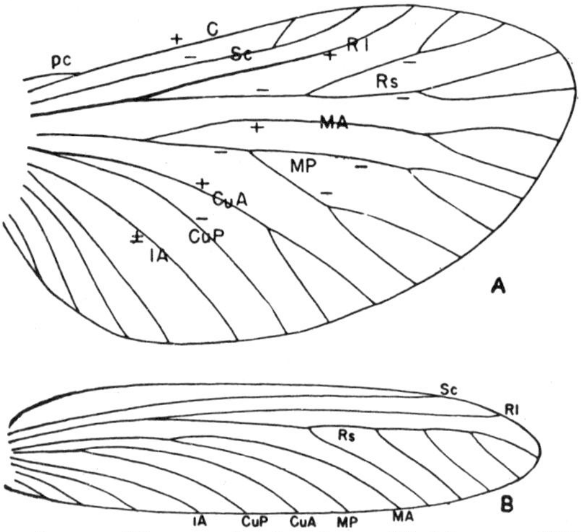

FIG. 13-44. Class INSECTA. Diagrams showing the modified Lameere (1922) terminology of wing venation now in common use. *A*. Archetype of wing venation—a generalized diagram of the forewing of all insects. *B*. A forewing of *Stenodictya*, a member of the extinct Order Palaeodictyoptera, from the Carboniferous of France. This wing, which belonged to an early palaeopterous insect, has relatively simple structure. (*C* = costa; *Sc* = subcosta; *RI* = radius; *Rs* = radial sector; *MA* = anterior media; *MP* = posterior media; *CuA* = anterior cubitus; *CuP* = posterior cubitus; *IA* = first anal; *pc* = precostal area; + = convex vein; – = concave vein; ± = flat, convex, or concave vein.) (*Furnished by F. M. Carpenter.*)

dorsal thoracic lobes of some sort, and that these wings could be moved only up and down. Insects having such wings, *i.e.*, the Palaeoptera, could not seek refuge beneath stones and pieces of wood, or in crevices and holes, yet some have persisted to the present time, as exemplified by modern dragonflies (Odonata) and May flies (Ephemerida).

The second important step in wing evolution came when the insect developed the ability to flex the wings and fold them back over the abdomen when at rest. This great forward step not only changed flight behavior but also made it possible for the insect to hide under objects and in crevices or to crawl into burrows and tubes of its own making where it could deposit its eggs in safety. It is interesting to note, in passing, that man has followed a somewhat similar course in changing the rigidly attached wings of a land

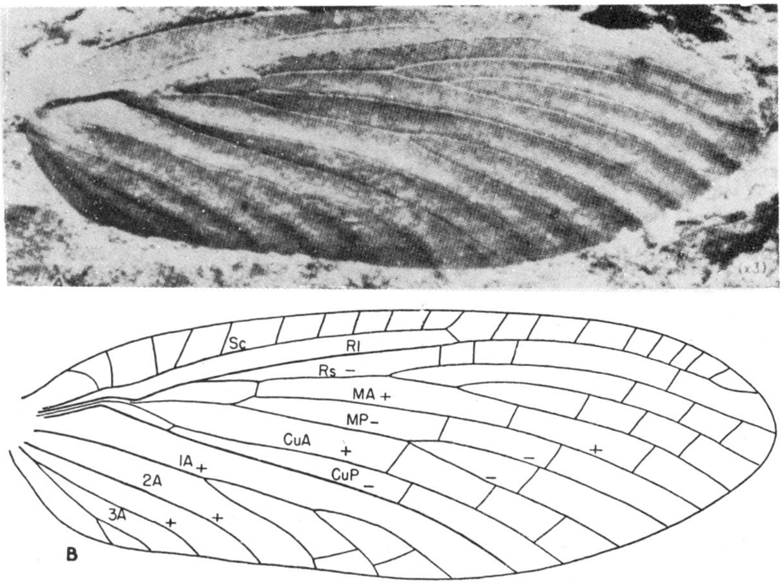

FIG. 13-45. Class INSECTA. Forewing of *Parelmoa revelata* (Order Megasecoptera), from the Permian of Oklahoma. *A*. A complete and splendidly preserved specimen, the holotype, showing the venation. *B*. Drawing of *A* with components of venation indicated. Compare with Fig. 13-44 for meaning of abbreviations. (*Originals from Carpenter*, 1947; *specimen in Museum of Comparative Zoology, Harvard University.*)

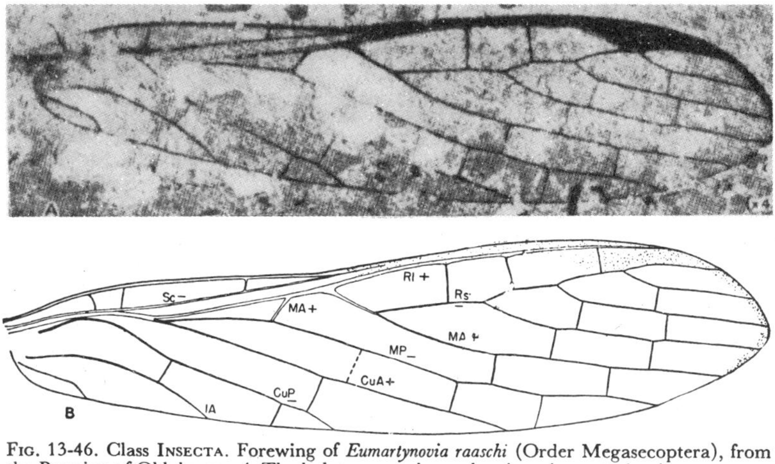

FIG. 13-46. Class INSECTA. Forewing of *Eumartynovia raaschi* (Order Megasecoptera), from the Permian of Oklahoma. *A*. The holotype specimen showing pigmentation in upper part and tip and well-preserved veins. *B*. Drawing of *A*. See Fig. 13-44 for meaning of the abbreviations. (*Originals from Carpenter*, 1947; *specimen in Museum of Comparative Zoology, Harvard University.*)

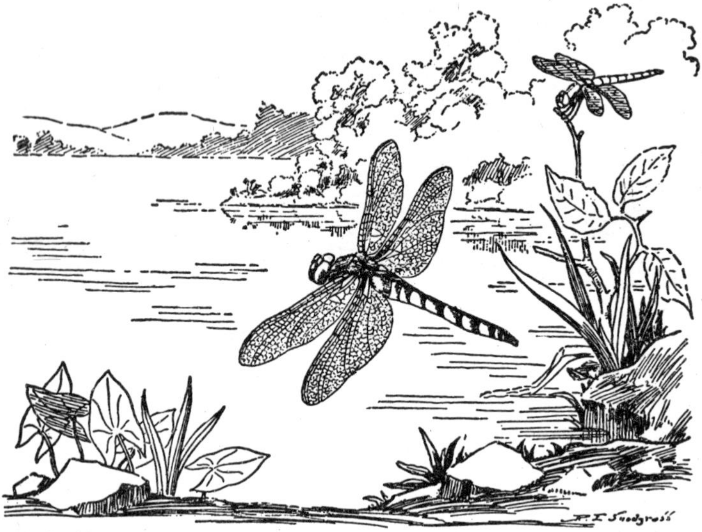

FIG. 13-47. Class INSECTA. Dragonflies (Order Odonata), modern representatives of the Palaeoptera. It is to be noted that when the insect is perched, the wings are held horizontally at right angles to the body and are not flexed at the base so as to rest on the abdomen. (*After Snodgrass*, 1930.)

plane to the articulating backfolding wings of a modern carrier plane, thereby making it possible, as with the insect, to take the machines through relatively small openings and store them in small spaces.

The more advanced stage of wing evolution was marked by the modification of the two forewings into thick chitinous covers, the **elytra** (singular, **elytron**), which rest over the folded hind wings (when the insect is at rest), as in the beetles (Coleoptera) and Protelytroptera (Fig. 13-49). Insects with wings protected by elytra could still further exploit places of safety for themselves and their eggs, and as they adapted themselves to a wider and wider range of surface and subsurface environments, wings became less and less important and finally were lost altogether in the adult stage. With attainment of this secondarily wingless stage, insects had completed the cycle of wing evolution from wingless ancestor, through palaeopterous and neopterous forms, the latter with elytra in certain groups, to secondarily wingless descendants. The existing insect fauna includes many species which as adults have one or the other of the different stages; *i.e.*, some are fundamentally wingless, some have palaeopterous wings, some have neopterous wings with or without elytra, and some neopterous forms are secondarily wingless. These relationships are diagrammed in Fig. 13-50. It is worth emphasizing again that whereas secondarily wingless groups have developed in every known order of living Neoptera, not a single such group is known to have developed among the Palaeoptera.

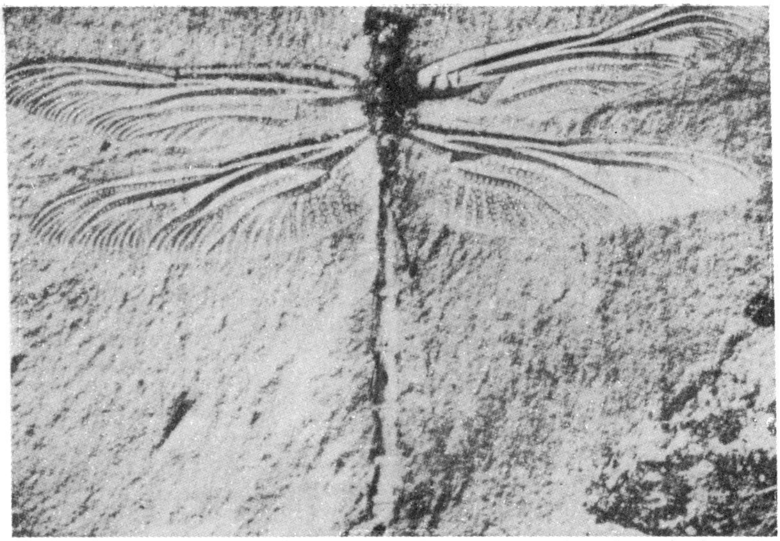

FIG. 13-48. Class INSECTA. *Protolindenia witteri* (Order Odonata), a dragonfly from the Upper Jurassic limestone of Solenhofen, Bavaria. Like modern dragonflies, this ancient form could move its wings only up and down (cf. Fig. 13-47), hence when it was buried the wings remained outstretched as in life. (*Photograph by F. M. Carpenter; specimen in Carnegie Museum.*)

Classification

The taxonomy of a group as numerous and varied as the Insecta presents many difficult problems; hence there is division of opinion as to how many and what orders should be recognized. Thus far 44 extinct orders have been proposed, but only 10 of these probably need be considered, as the remaining 34 are synonyms or so incompletely known that their taxonomic validity is dubious. Of the many orders with living representatives, 24 are now generally recognized. The ordinal classification here used is that followed in the latest works of F. M. Carpenter (1947; 1947*a*; 1950). Extinct orders are indicated by an asterisk.

Subclass Apterygota.
- Order 1. Thysanura.
- Order 2. Entotrophi.

Subclass Pterygota.
- *Order 3. Palaeodictyoptera.
- *Order 4. Megasecoptera.
- *Order 5. Protohemiptera.
- *Order 6. Protodonata.
- Order 7. Odonata.
- *Order 8. Protephemerida.
- Order 9. Ephemerida.
- *Order 10. Protoperlaria.
- Order 11. Perlaria (= Plecoptera).
- *Order 12. Protorthoptera.
- Order 13. Orthoptera.
- *Order 14. Caloneurodea.
- *Order 15. Glosselytrodea.
- Order 16. Blattaria.
- *Order 17. Protelytroptera (= Elytroptera).
- Order 18. Isoptera.
- Order 19. Dermaptera.
- Order 20. Embioptera.

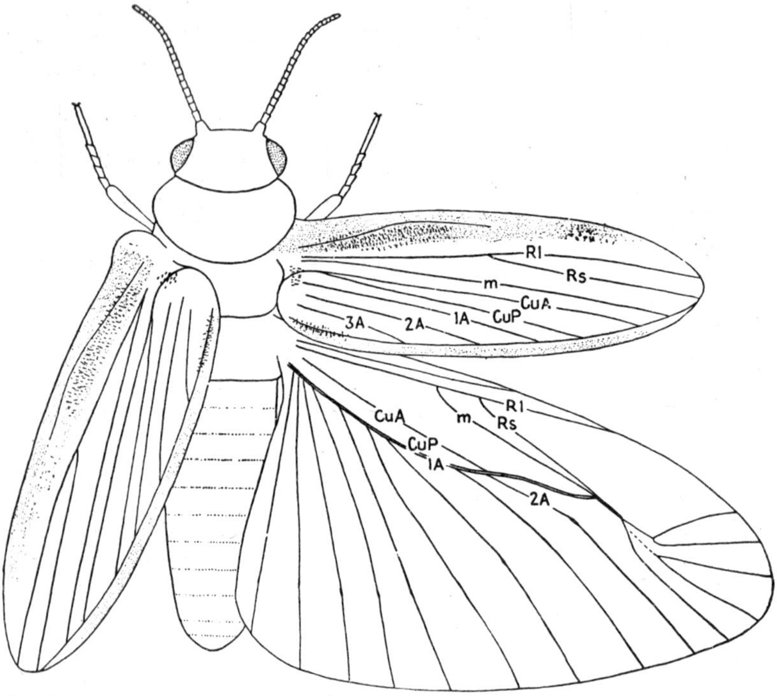

FIG. 13-49. Class INSECTA. A highly modified elytrophorous insect, *Protelytron permianum* (Order Protelytroptera), from the Lower Permian of Kansas. The forewings, though preserving vestiges of venation, have been modified into true elytra. The hind wings were large, with a greatly expanded anal region, and with hinges on the longitudinal veins enabling the wing to be folded up transversely as well as lengthwise like a fan. This insect, therefore, not only had highly modified expansible hind wings, which could be folded back over the abdomen when at rest, but it also could protect these delicate wings by thick convex elytra that could be folded back over the hind wings and abdomen. (*An original restoration by F. M. Carpenter based upon specimens in the Museum of Comparative Zoology, Harvard University.*)

Order 21. Mallophaga.
Order 22. Psocoptera (= Corrodentia).
Order 23. Thysanoptera.
Order 24. Hemiptera.
Order 25. Anoplura.
Order 26. Neuroptera.
Order 27. Mecoptera.
Order 28. Diptera.
Order 29. Trichoptera.
Order 30. Lepidoptera.
Order 31. Siphonaptera (= Aphaniptera).
Order 32. Coleoptera.
Order 33. Strepsiptera.
Order 34. Hymenoptera.

Figure 13-55 shows the geologic range of all generally recognized insect orders, both existing and extinct, and is based upon the most recently reported fossil discoveries. Supraordinal divisions are those in common use by present entomologists.[1]

[1] Some investigators prefer to recognize a superclass, the Hexapoda, in which they

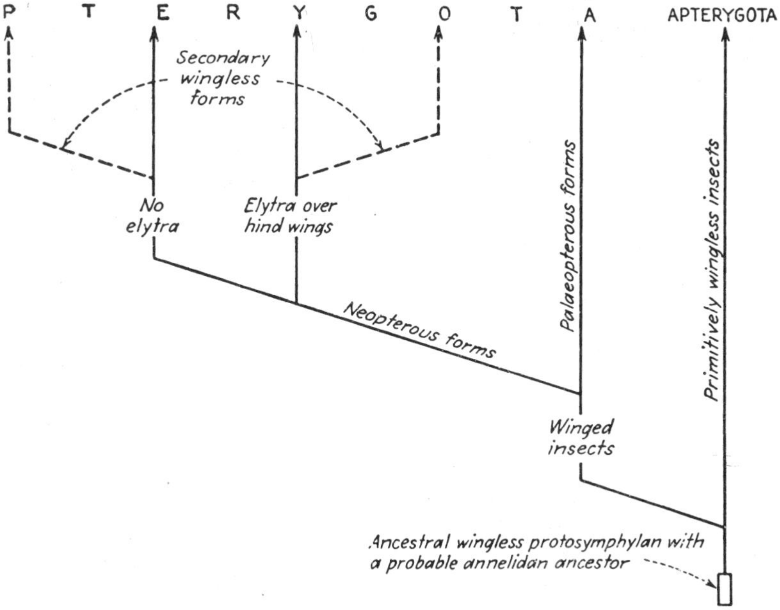

FIG. 13-50. Class INSECTA. Diagram illustrating wing development in the insects. It is worth noting that winged insects of the palaeopterous type are not known to have developed secondarily wingless forms. In contrast, both elytrophorous and nonelytrophorous Neoptera developed secondarily wingless forms.

group the six-legged arthropods into three classes—Collembola, Protura, and Insecta (Fig. 13-57). The members of these classes differ fundamentally in the number and time of appearance of the abdominal segments.

The Insecta typically have 11 abdominal segments in the embryonic stage, and these are retained or reduced in number in the adult.

The Protura have only nine abdominal segments in the embryonic stage but gain three more during postembryonic development, so that the adult has 12 segments. The class seems to be without a fossil record and is represented by approximately 70 living species assigned to seven or eight genera. It is of little paleontological interest and is not considered further.

The Collembola (Fig. 13-51) embrace the familiar springtails, which have six abdominal segments that appear fully developed in the embryonic stage and persist in the adult although in some cases reduced in number. The oldest known fossil Collembola come from the Devonian Rhynie chert of Scotland (Scourfield, 1940). A later species, *Protentombrya walkeri*, has been described from the supposed Cretaceous amber of Manitoba, and some 70 species have been found in the Baltic Oligocene amber (Fig. 13-51*B*). Living Collembola include approximately 2,000 species, of which the common *Podura aquatica* (Fig. 13-51*A*) is representative. The class is of little paleontological importance and is not considered further.

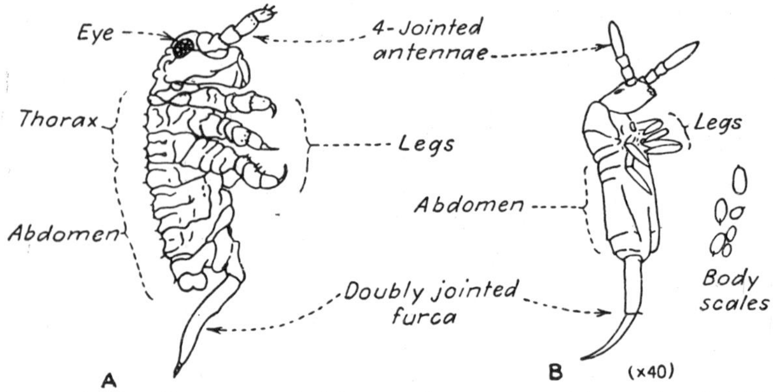

FIG. 13-51. Class COLLEMBOLA. *A. Podura aquatica*, a common living species. *B. Lepidocyrtus ambricus*, an almost complete specimen from the Baltic Oligocene amber. Note the general similarity between the living and the extinct forms. (*A modified after Willem*, 1900; *B redrawn from Handschin*, 1926.)

Subclass Apterygota[1]

The Subclass Apterygota includes all primitively wingless insects. These are characterized by an abdomen of 11 somites with vestigial appendages and a pair of terminal cerci. Two orders are generally recognized[2]—Thysanura and Entotrophi. Both lack metamorphosis.

Order 1. Thysanura. These are the bristletails, which comprise some 300 living species of small, swift, dark-loving insects exemplified by the troublesome "silverfish," *Lepisma saccharina* (Fig. 13-52*A*), which eats the starch in books, papers, and clothes. The earliest known fossil species was recently reported from the Lower Triassic. Numerous species have also been found in the Oligocene Baltic amber. *Triassic to Recent.*

Order 2. Entotrophi. These are small, elongate wingless insects with long slender antennae; three thoracic segments, each with a pair of legs; and 11 abdominal segments, with two cerci. *Campodea* (Fig. 13-52*B*) is a living genus; a fossil species was recently reported from supposed Miocene onyx marble of California (Pierce, 1951).

Subclass Pterygota[3]

The Subclass Pterygota, as the name implies, includes all winged and secondarily wingless insects, and these are characterized by the lack of abdominal appendages with the exception of cerci. The subclass is customarily divided into the more primitive Palaeoptera, which are characterized by two pairs of wings that can only be moved up and down (Fig. 13-53) and not flexed backward, and the more advanced Neoptera, which have backward-flexing wings that can be folded over the abdomen. Some Neoptera

[1] Apterygota—Gr. *a*, no or not, + *pteryx, pterygos*, wing; referring to the absence of wings.

[2] See footnote, p. 617.

[3] Pterygota—Gr. *pteryx, pterygos*, wing; referring to the presence of wings.

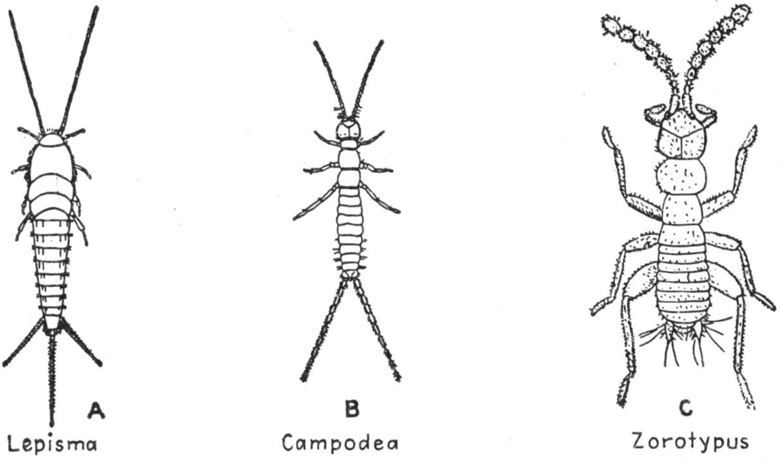

FIG. 13-52. Class INSECTA. *A. Lepisma saccharina* (Subclass Apterygota; Order Thysanura), the familiar "silverfish," enlarged several times. *B. Campodea staphylinus* (Subclass Apterygota; Order Entotrophi), another primitively wingless insect. *C. Zorotypus guineensis* (Subclass Pterygota; Order "Zoraptera"), a secondarily wingless insect. (*A, B adapted from Lameere*, 1935; *C adapted from Silvestri*, 1934.)

have only one pair of wings, some have elytra, and a few representatives of every order have become secondarily wingless. (Fig. 13-52*C*). Those Neoptera which display incomplete metamorphosis constitute the Hemimetabola; those with complete metamorphosis, the Holometabola. The relations of the several supraordinal divisions, together with the geologic ranges of the orders referred to them, are shown in Fig. 13-55. The orders are briefly characterized below; as before, extinct orders are indicated by an asterisk.

***Order 3. Palaeodictyoptera.** These are the earliest of the palaeopterous insects and appear first in Lower Pennsylvanian rocks, but it is believed that they developed even earlier in the Paleozoic. A typical representative has two pairs of wings essentially equal in size and similar in form. The wing venation is primitive (Fig. 13-44*A*), and the wings could not be flexed backward or folded over the abdomen. Two prothoracic lobes are present, and these may represent rudimentary wings. The order is confined to the Pennsylvanian and Permian and is believed to have given rise to several extinct Paleozoic orders as well as to the two existing palaeopterous orders, Ephemerida and Odonata. Most fossil species from Paleozoic rocks were once assigned to this order. Now, however, many of these are referred to other extinct orders, which themselves may have evolved from some palaeodictyopteran ancestor. *Lower Pennsylvanian through Permian.*

***Order 4. Megasecoptera.** (Figs. 13-44, 13-45, 13-46). This is one of the most intriguing of all extinct insect orders because of the range in wing development. Most megasecopterans had palaeopterous wings (Figs. 13-53, 13-54); these constitute the Eumegasecoptera. Some (the Paramegasecoptera) could flex their wings over the abdomen, but they did this by a mecha-

FIG. 13-53. Class INSECTA. An excellently preserved and highly specialized palaeopterous genus, *Mischoptera nigra* (Order Megasecoptera), from the Carboniferous beds of Commentry, France (×⅔). This early insect had short and modified forelegs (the other two pairs of legs are unknown), an armed prothorax, and conspicuously falcate wings. (*Photograph by F. M. Carpenter; also see Carpenter,* 1951.)

nism that seems to have been developed independently of that which led to neopterous wings. Almost 100 species referable to more than 30 genera have now been described, and the order is known to have flourished during the Late Paleozoic. *Upper Pennsylvanian and Permian* (*dominantly*).

***Order 5. Protohemiptera.** These insects are palaeopterous and related to the Palaeodictyoptera. Mouth parts were suctorial. *Upper Pennsylvanian through Permian.*

***Order 6. Protodonata.** (Fig. 13-48). This order is based on several dozen species distributed among three well-delimited families. The mouth parts are mandibulate, and the insect was probably predaceous. The powerful legs were set well forward and probably were used for grasping. The wing venation is basically similar to that in the Odonata. *Lower Pennsylvanian through Permian; ?Triassic* (*possible fragmental wing*).

Order 7. Odonata. These are the familiar dragonflies (Fig. 13-47); they have large bodies, enormous eyes, and four equal-sized diaphanous wings. More than 5,000 living species have been described, and many additional extinct species have been collected from rocks of Lower Permian to Recent age. The Odonata and Protodonata are thought to have come from a common palaeodictyopteran ancestor. *Lower Permian to Recent.*

***Order 8. Protephemerida.** This order is based on a single specimen composed of four wings and parts of the body, which has two cerci and a median filament. The wing venation seems intermediate between that of the

Palaeodictyoptera and Ephemerida, and the Protephemerida have been considered a connecting link between these two orders. *Upper Pennsylvanian.*

Order 9. Ephemerida. These are the May flies which appear periodically in great numbers and live only a few hours or days. There are several thousand existing species. Adults are preserved in many rocks from Lower Permian to Recent age, and they constitute most of the fossil record of the order, as the delicate nymphs are seldom preserved. *Lower Permian to Recent.*

Order 10. Protoperlaria (Fig. 13-42). These ancient relatives of modern stone flies (Perlaria) are known only from Permian rocks, but the fossil record is excellent. *Permian.*

Order 11. Perlaria (= Plecoptera). This order consists of several thousand living species of stone flies. They have never been a large component of any insect fauna and have a fragmentary record that goes back to the Upper Permian. Jurassic and Oligocene representatives are included in the fossil record. *Upper Permian to Recent.*

***Order 12. Protorthoptera.** This large and important order includes the earliest fossil orthopteroid insects, of which more than 100 genera and almost 200 species have been described. The legs are diverse, an ovipositor is commonly developed, and the wing venation shows primitive aspects. The order is believed to have been the ancestor of Recent orthopteroid orders. *Pennsylvanian and Permian.*

Order 13. Orthoptera. The Orthoptera are an ancient and complex group of neopterous insects in which there has been a tendency toward a tegminous wing, reduction of venation, and total loss of wing. Sound-producing organs (strigulatory mechanisms) are present, and these have caused morphological changes in the wings. The order includes the familiar grasshoppers (Fig. 13-43*A–E*), katydids, locusts, crickets, praying mantises, and walking sticks. The first true Orthoptera appeared in the Triassic and are thought to have descended from the Protorthoptera. *Lower Triassic to Recent.*

***Order 14. Caloneurodea.** This small order of about 20 species includes neopterous insects with hind wing and forewing similar; little is known about the body except that there were no jumping legs. *Upper Pennsylvanian through Permian.*

***Order 15. Glosselytrodea.** Only three species of this order are known; they have elytra and a peculiar venation. *Permian into Jurassic.*

Order 16. Blattaria. The Blattaria are the bothersome cockroaches, a group with many fossil as well as modern representatives. More than 300 Pennsylvanian and Permian species are known from North America and an even larger number from Europe. Approximately 80 Jurassic species have been described, 50 or more are known from the Tertiary, and several thousand living species have been described. *Middle Pennsylvanian to Recent.*

***Order 17. Protelytroptera (= Elytroptera).** These forms have the forewing modified into a coriaceous elytron and the back wing much modified (Fig. 13-49). *Permian.*

Order 18. Isoptera. The Isoptera include the notorious termites whose borings may reduce a wooden structure like a house to a mere shell. There are almost 2,000 living species, and possibly as many or more undiscovered

(Snyder, 1949). Termites are thought to have made their appearance in the Early Tertiary. *Eocene to Recent.*

Order 19. Dermaptera. This order of some 900 living species includes the earwigs, which have stout horny forceps, or cerci, at the end of the abdomen. They first appear as fossils in the Middle Jurassic. *Middle Jurassic to Recent.*

Order 20. Embioptera. The embiids have a long straight-sided body. About 150 living species have been described, and a few extinct forms have been found in Tertiary amber and in the Miocene Florissant shales. *Oligocene to Recent.*

Order 21. Mallophaga. These are the biting lice, several thousand species in number, that infest both fowls and mammals. None is known fossil.

Order 22. Psocoptera (= Corrodentia). About 900 species of book lice constitute this order, which has fossil representation as early as the beginning of the Permian. *Lower Permian to Recent.*

Order 23. Thysanoptera. These are the thrips, represented by about 2,000 living species and a few extinct species that go back as far as Late Permian. They are very destructive to many flowering and other plants. *Upper Permian to Recent.*

Order 24. Hemiptera. Included in this large and important order, which is commonly divided into Heteroptera and Homoptera, are more than 50,000 living species of true bugs, cicadas, aphids, and scale insects. The earliest fossils are found in Lower Permian rocks, and other extinct species have been collected in Mesozoic and Cenozoic rocks. *Lower Permian to Recent.*

Order 25. Anoplura. The Anoplura are bloodsucking insects (*e.g.*, "cooties") on many mammals. They were not known fossil until 1948, when an extinct species was described from a Pleistocene ground squirrel found in Russia. *Pleistocene to Recent.*

Order 26. Neuroptera. (Fig. 13-54). This order includes several thousand species of Dobson flies, ant lions, and lacewing flies, and a few tens of extinct species, mostly Tertiary in age. *Lower Permian to Recent.*

Order 27. Mecoptera. These are the scorpion flies, a small group of perhaps 300 living species and a few fossil species, the earliest of which appeared in the Early Permian. It is thought that this order may have been ancestral to the Diptera, Trichoptera, and possibly Lepidoptera. *Lower Permian to Recent.*

Order 28. Diptera. This important order includes nearly 100,000 living and extinct species of true flies (Fig. 13-54), the earliest of which appeared in the Early Jurassic. The order is thought to have been derived from the Mecoptera. *Lower Jurassic to Recent.*

Order 29. Trichoptera. The modern caddis flies, represented by 3,500 or more species, constitute this order. There are a few Mesozoic species and several hundred extinct Tertiary forms. In some rocks (*e.g.*, Indusial limestone of Auvergne) the larval cases of Trichoptera are extremely abundant. *Middle Jurassic to Recent.*

Order 30. Lepidoptera (Fig. 13-43*F–K*). The butterflies and moths that constitute this order are among the most beautiful of insects, and are sought eagerly by many collectors. This is one of the largest of existing orders, with

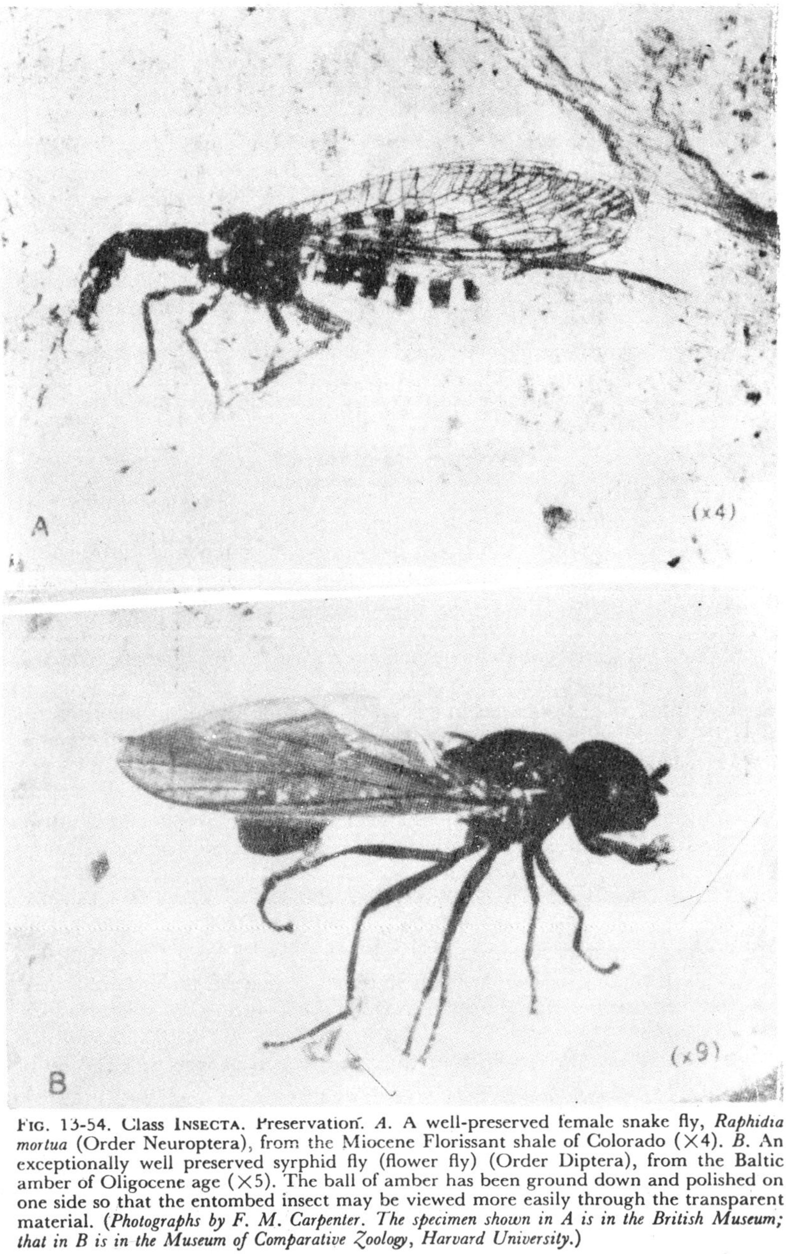

FIG. 13-54. Class INSECTA. Preservation. *A*. A well-preserved female snake fly, *Raphidia mortua* (Order Neuroptera), from the Miocene Florissant shale of Colorado (×4). *B*. An exceptionally well preserved syrphid fly (flower fly) (Order Diptera), from the Baltic amber of Oligocene age (×5). The ball of amber has been ground down and polished on one side so that the entombed insect may be viewed more easily through the transparent material. (*Photographs by F. M. Carpenter. The specimen shown in A is in the British Museum; that in B is in the Museum of Comparative Zoology, Harvard University.*)

more than 120,000 living species. In contrast scarcely 100 fossil species have been described. *Eocene to Recent.*

Order 31. Siphonaptera (= Aphaniptera). This order includes the fleas, perhaps a thousand species in number, which first appear fossil in the Oligocene amber. *Oligocene to Recent.*

Order 32. Coleoptera. The beetles and weevils constituting this order are represented in living faunas by more than 300,000 species. Only a few extinct species are known from the Paleozoic, but 350 or more have been described from Mesozoic rocks, and more than 2,500 from Tertiary formations. *Upper Permian to Recent.*

Order 33. Strepsiptera. Strepsiptera are largely parasitic on Hymenoptera and Orthoptera, and the several hundred living species are supplemented with a few extinct species from Tertiary amber. *Oligocene to Recent.*

Order 34. Hymenoptera. To this order are assigned the ants, bees, and wasps, numbering 100,000 living species, and several hundred extinct forms, the oldest of which are Middle Jurassic. *Middle Jurassic to Recent.*

Preservation of Insects

Insects are not likely to be preserved except under special conditions because of the soft nature of the body and the total lack of inorganic hard parts. Unless burial is prompt, the body parts quickly soften and disintegrate, leaving only the wings, but these, because they are more resistant to decomposition, can be preserved if conditions are at all favorable. Complete insects, therefore, are not likely to be preserved except in materials that quickly covered and protected the delicate body (*e.g.*, tree resins, volcanic ash, and unusually fine-grained muds). Wings, being stronger and more resistant to decomposition, were preserved in a somewhat wider range of rock types because they could withstand some transport and buffeting about without complete destruction. These are the reasons why nearly half the described species of fossil insects are based on wings alone.

In addition to complete specimens and detached wings and elytra, insects may also leave certain kinds of external evidence of their previous existence in the form of fossilized larval chambers (Brown, 1945), pupal cases (Lea, 1925; Fairbridge, 1950), gall impressions (Hoffman, 1932), fecal pellets (Light, 1930), and excavated galleries in fossil wood (Walker, 1938). These fossil materials, though interesting as curiosities, are of little paleontological value, except as they indicate certain activities of ancient insects, and should never be the basis for extending the range of a known species or establishing a new species.

When special conditions favored the preservation of insects, large numbers of individuals were commonly buried, and the deposits thus made are today the ones that have provided the greatest collections of fossil specimens. As examples, the great Baltic Oligocene amber deposits have furnished more than 150,000 specimens (Fig. 13-54); the well-known Miocene ash beds of the extinct lake at Florissant, Colorado, have produced some 60,000 specimens (Fig. 13-54); and the Permian Wellington formation (Elmo limestone member) has yielded more than 18,000 specimens (Fig. 13-53).

Fossil insects have been found at more than a hundred different localities scattered geographically over the entire earth and stratigraphically in beds from Lower Pennsylvanian to Recent age. However, nearly nine-tenths of the specimens have come from only 12 of these localities;[1] the other important localities have yielded only a few specimens. The larger and more important collections of Carboniferous insects have come from the Commentry beds (Stephanian = Upper Pennsylvanian) of France, the Pennsylvanian (and Permian) strata of the Saar Basin in Germany, and the well-known Pennsylvanian concretions of the Mazon Creek area in Illinois. The outstanding contributor of Permian specimens has been the Lower Permian Wellington formation of Kansas and Oklahoma; smaller collections have also been made from the Middle Permian of Russia and the Upper Permian of New South Wales. The only Triassic insects of importance come from Queensland. As with several other groups, the prolific lithographic limestone of Solenhofen, Bavaria, has yielded the best of Jurassic insects, but almost equally important specimens have come from southern England and Russia. Supposed Cretaceous specimens have been obtained from placers of amber fragments in Manitoba, but the actual source of these specimens is uncertain. The most prolific of all fossil insect localities is the well-known Baltic area with its sequence of Lower Oligocene shales. Scarcely less important as sources of Tertiary insects are the Miocene ashstones of Florissant, Colorado, and the Eocene Green River beds of Colorado and Wyoming. References to

[1] One of the best known North American localities for fossil insects is near Florissant, Colo., where an ancient fresh-water lake received deposits of Miocene volcanic ash. This ash in falling carried down insects, other animals, and plant materials from both the air and water and buried them together in the bottom sediments of the lake. Consequently, both terrestrial and aquatic organisms are preserved together, a mixture of forms that could not possibly have lived together.

A second well-known locality is along Mazon Creek in Illinois, where countless thousands of argillaceous concretions in Lower Pennsylvanian sandstones have yielded one of the best known of all Carboniferous biotas (*i.e.*, floras and faunas). At this locality insects, along with crustaceans, arachnids, and other arthropods, were buried in fine-grained silts and muds deposited in or near coal swamps, and later they became the nuclei of the familiar flattened concretions, which happily tend to split in such manner as to reveal the buried fossil. As would be expected, however, fossil arthropods are exceedingly rare in these concretions as compared to plant remains, and a collector should be prepared to split tens of thousands of concretions to find a scrap or two of an ancient insect or other arthropod.

At Solenhofen, Bavaria, insects were entombed in impalpably fine calcareous muds that accumulated on lagoonal bottoms between Jurassic coralline bioherms. Here, too, are found exquisitely preserved remains of many phyla of invertebrates and of the first known bird, *Archaeopteryx*.

Probably the richest of all fossil-insect localities is that of the Baltic region of East Prussia, where tens of thousands of specimens are preserved in amber particles incorporated in a Lower Oligocene shale. (Some authors have pointed out that the insects may actually be somewhat older than the shale containing them, *i.e.*, possibly Eocene, but most writers consider the insects to be essentially contemporaneous with the shale, hence Lower Oligocene). Some of these insects are so exquisitely preserved that when viewed through the transparent amber they seem as though they were still alive. Yet, when such an amber fragment is broken, the insect is seen to consist of only a shell of its former self—all internal parts are missing because of decomposition.

most of these famous localities are included in the few articles listed in the bibliography at the end of this chapter. (Also see footnote on page 625.)

Geologic History, Evolution, and Origin

The oldest unquestionable fossil insects—three single wings representing three different extinct orders—have come from Lower Pennsylvanian rocks (= lower part of "Upper Carboniferous" of Europe). One comes from Czechoslovakia, one from Germany, and one from Pennsylvania. Two of the wings belong to palaeopterous orders (Palaedictyoptera and Protodonata) and the third to a neopterous order (Protorthoptera). Whatever else may be inferred from this fossil record, it is certain that insects with fully developed wings of both palaeopterous and neopterous type existed in Europe and North America at the beginning of Pennsylvanian time. Furthermore, as it is generally assumed that neopterous insects descended from palaeopterous ancestors, it seems logical to conclude that palaeopterous forms appeared earlier than the Pennsylvanian, either in the Mississippian or possibly Devonian. Every fine-grained fresh-water sedimentary rock of these systems should be carefully examined for fossil insects, because any specimen found would be of great interest to students of insect evolution.

From their assumed beginning in Pennsylvanian or somewhat earlier time, insects rapidly evolved in the Carboniferous forests of the world, with the development of five new orders in addition to the three mentioned in the preceding paragraph (Fig. 13-55). During the Permian 12 new orders appeared to swell the insect fauna of that period to 19 orders (one Pennsylvanian order had become extinct by the end of that period), and out of this great fauna soon arose the chief orders of living insects either during the Triassic or Jurassic.

If the fossil record of the Insecta is set aside for a moment, and the developmental history of existing insects considered, morphological studies give conclusive evidence that the first true insects were wingless, like the members of the Thysanura and Entotrophi, which constitute the Subclass Apterygota, and that they combined the generalized characteristics of both of these groups. The development of winged insects (Subclass Pterygota) marked the first great evolutionary step within the Insecta. How wings actually originated is by no means clear, as stated on page 611, but it seems reasonable to suppose that they developed from lateral tergal flaps similar to the prothoracic lobes of the Palaeodictyoptera (Fig. 13-42), and that they were of the palaeopterous type (*i.e.*, they could not be flexed over the abdomen when the insect was resting). The second great evolutionary step was made when the development of wing articulation enabled the insect to flex the wings and fold them back over the abdomen when not in flight, and thus to seek protection beneath objects and in holes and crevices where a palaeopterous form could not go. The earliest of the neopterous insects had a simple or direct type of postembryonic development (Hemimetabola), but quite early in their history the third great step in insect evolution came with the attainment of a more complex type of metamorphosis, with larval and pupal stages, and the homometabolous Neoptera were born. Thus insect

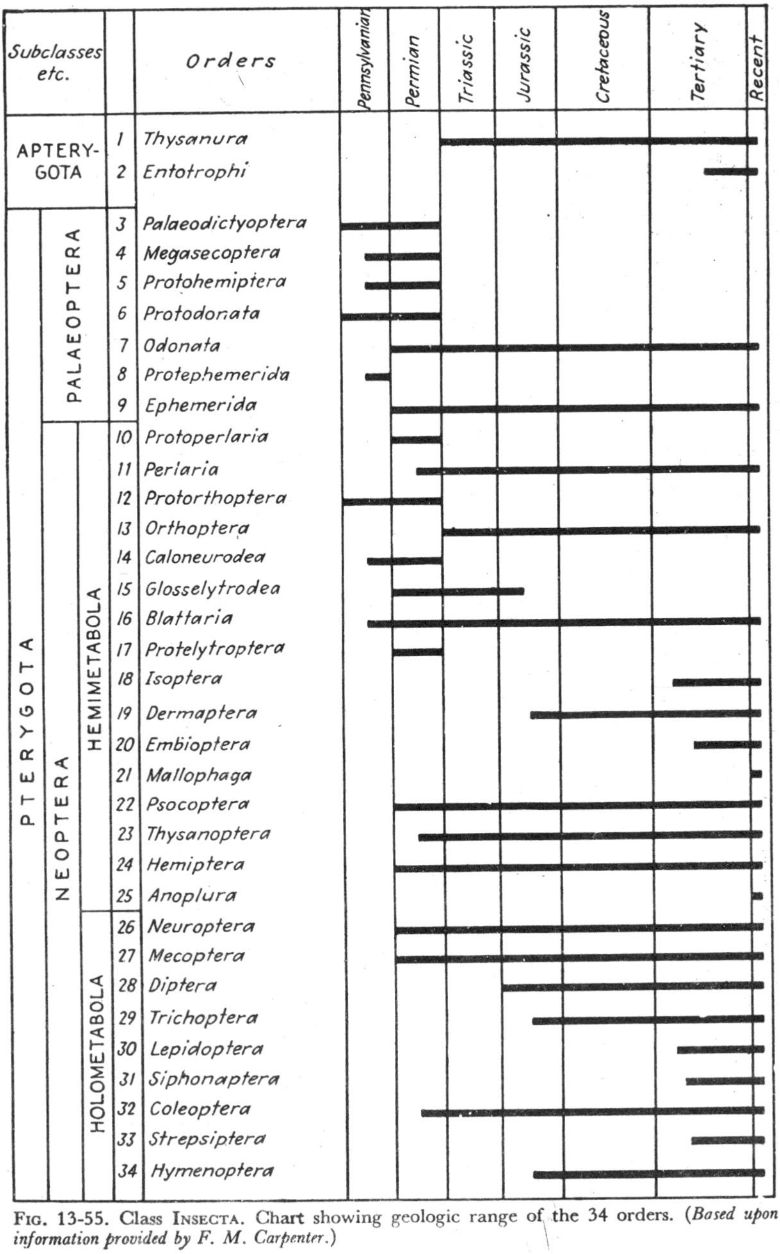

FIG. 13-55. Class INSECTA. Chart showing geologic range of the 34 orders. (*Based upon information provided by F. M. Carpenter.*)

evolution has been marked by three successive steps—acquisition of wings, flexing of wings, and complete metamorphosis. The last of these three steps was made in the Permian, and insects do not seem to have made any later evolutionary changes of comparable importance since then, *i.e.*, during the last 250,000,000 years!

There remains for brief consideration the difficult and challenging question, from what ancestors did the insects come? This has long been a moot question, and only recently does the probable answer seem to have been suggested. The several different theories of insect ancestry that have been expounded—(1) crustacean, (2) trilobitan, and (3) myriapodan—have been ably outlined and evaluated in recent discussions by Tiegs (1947; 1949) and Imms (1947), and interested readers are referred to their papers for both discussion and pertinent references. Suffice it here to present briefly the chief evidence supporting the theory that the insects arose from a primitive and generalized polypodous arthropod of symphylan type (Protosymphyla of Imms, 1947), which itself had evolved from a still more generalized arthropod of the *Peripatus* type (Fig. 12-1) that had already solved the physiological problem of terrestrial existence.

Primitive insects share with existing members of the myriapodan classes Chilopoda, Diplopoda, and Symphyla the following basic structural characteristics:

1. The head and its appendages have a considerable degree of uniformity but are markedly different from those of the Crustacea and Arachnoidea.
2. The trunk appendages are simple walking legs, which at no time either embryonically or postembryonically give any indication of the biramous structure so characteristic of the Crustacea and Trilobita.
3. Specialized structures termed **exsertile vesicles** are present in some groups of both myriapods and primitive insects.
4. There is a straight simple intestine with specialized excretory organs (**Malpighian tubes**), which shows no general resemblance to the arachnoid-crustacean type of intestine.
5. Tracheae and a long tubular heart are present.
6. Embryonic development, which differs in important particulars from that of the Crustacea, is rather uniform.

Most of the lower insects pass through a stage in their embryonic development in which they have a 6-segmented head and 14-segmented trunk, with each of the 20 segments bearing a pair of rudimentary limbs. The Symphyla have the same general type of organization. The prevalence of the polypod condition and the fixed number of segments in both insects and myriapods of primitive nature suggest that the ancestor of both was a polypod creature of somewhat the same nature as living *Peripatus* (Fig. 12-1).

In spite of the several important resemblances between primitive insects and myriapods, however, one great difference has been difficult to rationalize. The Diplopoda, Pauropoda, and Symphyla (Fig. 13-56*B*) all have the gonopore near the front end of the trunk (these are **progoneate**), whereas the Chilopoda have the gonopore at the hind end of the trunk (they are described as **opisthogoneate**) (Fig. 13-56*A*). The Insecta are opisthogoneate, yet otherwise they resemble the progoneate myriapods much more closely. Furthermore, though their nearest relatives among existing myriapods would seem to be the opisthogoneate chilopods, these are ruled out as possible ancestors because of the remarkable specialization of the first abdominal segments as poison jaws. The dilemma was resolved when investigation of the embryonic development of a symphylan showed that it actually was originally opisthogoneate and that it became progoneate as a result of secondary development (Tiegs, 1940; 1945). On the basis of this important discovery, specialists on insect evolution now favor the theory, as stated previously, that the earliest insects and the most primitive myriapods descended from a generalized opisthogoneate symphylan arthropod, which itself may well have developed from a still more primitive and generalized creature somewhat like modern *Peripatus*. Of special interest in the last particular is the fact that an ancient relative of *Peripatus*, the genus *Aysheaia* (Fig. 12-2), was present in Middle Cambrian seas, and some of its relatives may well have migrated onto the land and undergone the changes necessary to terrestrial existence and leading to the earliest of the protosymphylans.

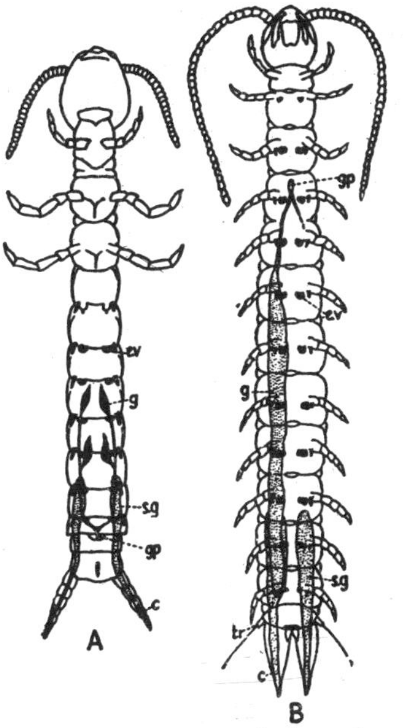

FIG. 13-56. PHYLUM Arthropoda—Classes INSECTA and SYMPHYLA. *A. Anajapyx* (Class Insecta), a female, showing the opisthogoneate condition. *B. Scutigerella* (Class Symphyla), a female, showing the progoneate condition. (*c* = cercus; *ev* = exsertile vesicles; *g* = gonad; *gp* = gonopore; *sg* = spinning gland; *tr* = trichobothrium.) (*After Tiegs*, 1949.)

GEOLOGIC HISTORY OF THE ARTHROPODA

The geologic history of the Arthropoda is varied and complex, and because of its fragmentary nature, it has many gaps which leave in doubt the exact relationships of some extinct groups. Only four aspects of the history are considered here: (1) the evolution and geologic history of the phylum; (2) the ecology of living forms and the paleoecology of ancient ones; (3) distribution through time (*i.e.*, stratigraphic position shown diagrammatically in Fig. 13-58); and (4) nature of the fossil record.

Evolution. The Arthropoda are now generally believed to have evolved during the later Pre-Cambrian from some group of simple and generalized aquatic segmented worms. In their subsequent evolution they have retained the annelid structure of the nervous system, the mode of growth of the body segments, and the distinctive elongated and segmented body, but they have developed one conspicuous feature which sets them apart from all worms—

the hardening of the integumental cuticle. This hardening of the integument resulted in the loss of the flexibility and contractility of their annelidan ancestors and led to the development of telescoping segments. Similarly, the annelidan appendages were encased in chitin and came to be jointed. These two striking advances in early arthropod evolution were followed by other almost equally important changes, and the whole sequence of these changes may be briefly outlined as follows:

1. A chitinous integument was developed over part or all of the ancestral annelid, and the loss of flexibility resulting from such a rigid skin was compensated by telescopic movement between successive body segments when motion was required.

2. The worm appendages (parapodia) were sclerotized and became jointed so that the whole appendage consisted of a series of individually movable parts.

3. Some of the anterior segments of the body were fused to form a head; there was a similar fusion of the posterior segments to form an abdomen; and finally, as in the Trilobita, there was a subdivision of the body and exoskeleton into cephalon, thorax, and abdomen (pygidium). These changes were accompanied by a concentration of the ganglia of the segments to form a brain. Further fusion of the segments ultimately resulted in but two

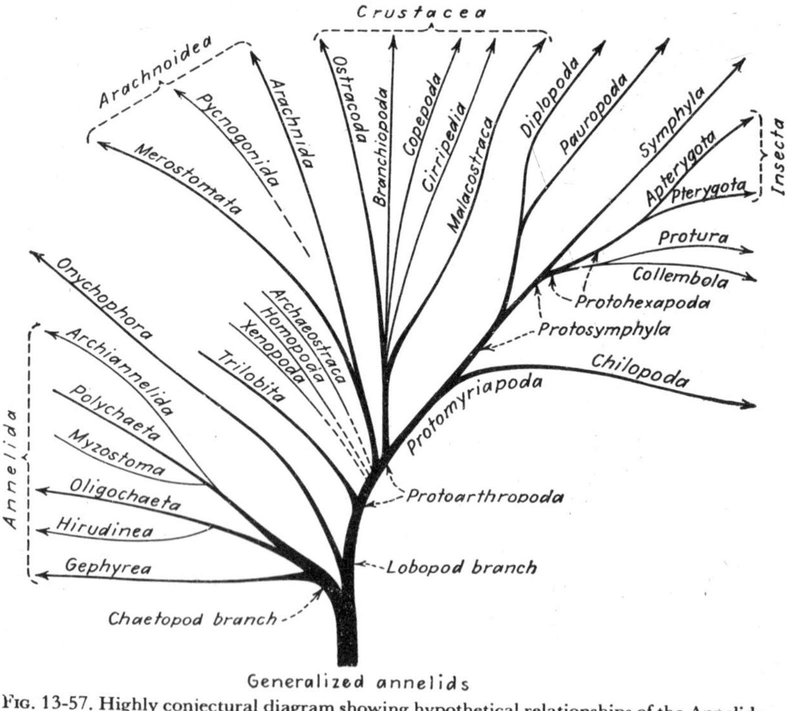

FIG. 13-57. Highly conjectural diagram showing hypothetical relationships of the Annelida, Onychophora, and Arthropoda. (*Adapted in part from Snodgrass*, 1938.)

rigid exoskeletal plates—the cephalothorax (carapace) and the abdomen. Throughout the phylum, regardless of the group concerned, there seems always to have been a tendency to eliminate free segments by fusion with others.

4. The appendages, originally simple and similar parapodia, were modified to sclerotized and jointed structures specialized for a wide range of function. In more highly specialized arthropods there was a strong tendency toward reduction in number of appendages and loss of them completely on some segments (especially those of the abdomen).

5. Mechanisms of respiration changed from simple gills to book gills to book lungs and from branchia to tracheae, thereby permitting a change from water breathing to air breathing and thus making invasion of land habitats possible.

6. Wings developed on the dorsal surface of the thorax, making possible invasion of the air.

The interested reader will find an extended discussion of arthropod evolution in an article by Snodgrass (1938).

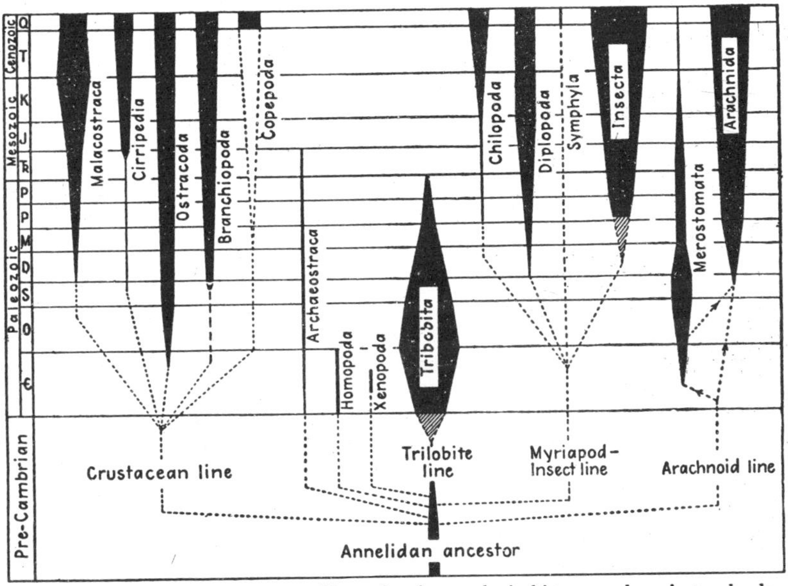

FIG. 13-58. Phylum ARTHROPODA. Chart showing geologic history and conjectural relationships of the different subclasses of the phylum. The width of the dark bands is a relative indication of the importance of the division during its history, but the bands are not comparable one with another. Inclined lining indicates probable existence, though not supported by fossil evidence. Dashed lines indicate only the connection of the different divisions and are not to be interpreted as meaning that all divisions arose at the same time. Four great lines seem to have arisen from a common annelidan ancestor and to have been established in Pre-Cambrian or very early Cambrian time—the Trilobita, Arachnoidea, Crustacea, and Myriapoda-Insecta. There is no present unanimity of opinion as to how these groups are interrelated, hence all are shown as arising independently from the ancestral stock. The relationships of the Archaeostraca, Homopoda, and Xenopoda are uncertain. Some or all of them may have arisen from the Crustacean line. (*Based on data and suggestions of numerous authors.*)

No unquestioned fossil arthropods have yet been found in Pre-Cambrian rocks.[1] By Early Cambrian time, however, arthropods had reached a high level of development, as indicated by the record of the trilobites, and the variety of the arthropodan forms is indicated by the rich Middle Cambrian fauna of the Burgess shale. It seems reasonable to conclude that the Cambrian Arthropoda had a long previous history, although as yet little can be determined as to the nature of this history except by inference. It would seem, therefore, that the remote ancestors of the Arthropoda are certainly to be sought in Pre-Cambrian strata.

It is evident from the fossil record that arthropods had already spread over a wide range of environmental conditions by the early part of the Paleozoic, and that several major divisions of the phylum seem to have appeared about the same time (Fig. 13-58). Whether the phylum, as now defined, had its origin in a single ancestral stock, or in several closely related but distinctly different stocks, is one of the unanswered questions of arthropod phylogeny. In the present state of knowledge, strong evidence can be presented in support of either possibility.

The evolution and geologic history of the several classes of Arthropoda are discussed separately under each class and need not be repeated here. One interpretation of arthropod phylogeny, that of Snodgrass (1938), is shown in Fig. 13-57. The known and inferred geologic history of the phylum is shown in Fig. 13-58.

Ecology and Paleoecology.[2] In a group like the Arthropoda, which is represented by so many different kinds of organisms in such prodigious numbers, it is to be expected that representatives will be found in every possible life habitat from the deepest ocean bottom[3] to the highest mountain top and from the poles to the equator, and it is further to be expected that arthropods probably were similarly adapted during the geologic past. Whether conditions be frigid,[4] temperate, or torrid; whether it be dry, moist, or humid; whether it be on the surface or in the soil; in fresh, brackish, or salt waters; or in the air; representatives of the phylum are sure to be present, commonly in abundance and variety. No other phylum of animals probably has ever attained the wide range of adaptation shown by the Arthropoda.

The Crustacea have been called "the insects of the sea," and quite appro-

[1] Raymond (1935) reviews, evaluates, and then rejects as fossil arthropods all reported specimens except certain pieces designated *Beltina danai* from the Beltian series of western North America. He is inclined to believe that the latter are fragments of some kind of arthropod exoskeleton. Other investigators, however, consider all specimens of *Beltina* as algal in origin and some even doubt that the pieces had an organic origin.

[2] See Raymond (1936).

[3] The Danish research vessel *Galathea* is reported to have dredged a crustacean from a depth greater than 10,360 m. (34,000 ft.) in the Mindanao Deep off the Philippines (*New York Times*, July 25, 1951).

[4] In the *New York Times* for Sunday, May 27, 1951, p. 17, the North American Newspaper Alliance reported that the British-Scandinavian Antarctic Expedition had found two species of mites on Queen Maud Land at 71° 18′ S. lat.

priately so, for existing oceans and seas are teeming with their countless numbers and the fossil record indicates that they were sometimes prodigiously abundant in past seas. They are also commonly abundant in fresh waters, where they are in competition with insects as well as with other creatures, and scarcely a pool, ditch, stream, pond, or lake can be found which does not have its assemblage of crustaceans. On land they are relatively uncommon, although a few species have become adapted to a terrestrial habitat. Some of the Crustacea, notably the crabs, lobsters, and shrimps, have for a long time been used as food by man. It is in the economy of nature, however, that the Crustacea, especially the amphipods, isopods, ostracodes, and copepods, stand supreme. The first three are scavengers that live mainly in the shallower waters of the sea, where they feed on all sorts of organic debris and in turn serve as food for many larger marine animals (*e.g.*, fish). The copepods, which feed on diatoms and other microscopic plankton, are themselves planktonic and are sometimes so abundant as to discolor the surface waters of the ocean for miles and thereby guide fishermen and whalers to the best fishing grounds.

In the Early Paleozoic seas the extinct trilobites and earliest archaeostracans, conchostracans, and ostracodes must have played an important role as scavenging organisms on the sea bottom and in the waters above. It is also probable that the ancient arachnoids may have carried on scavenging activities in some cases.

The Insecta show a wide range of adaptation to terrestrial and aquatic habitats. Their relations with the plants constitute an entrancing story. They not only acquire food from and find shelter in living plants, but they are also the chief destroyers of dead vegetation in many tropical environments. Bees seek the nectar of flowers, and some flowers (*e.g.*, orchids) are dependent on certain insects for pollenizing. In fact, without insects, many plants could not reproduce at all. To a large degree scavenging insects are quite beneficial in facilitating rapid decomposition and removal of decaying organic substances of all kinds. Aquatic insects are found chiefly in fresh waters; there are almost no insects on the open sea and only a few in the shallow coastal waters. Some insects are cannibalistic; some are predatory; and many are either ectoparasitic or endoparasitic. A few live with their relatives and feed at their expense. Social life reaches amazing perfection among ants and bees, and the gregarious habit seems to be fairly typical of the class as a whole.

Stratigraphic Distribution. The general relations and stratigraphic ranges of the major subdivisions of the Arthropoda are shown in Fig. 13-58. The Trilobita appear first in Lower Cambrian rocks, but it is believed that they evolved during the later Pre-Cambrian. They flourished during the Cambrian and Ordovician but started to wane in the Silurian and after the Devonian declined rapidly to extinction in the Permian.

The several subclasses of Crustacea made their appearance at different times during the Paleozoic, expanded and diversified during the Mesozoic, and are today persistent groups of widely adapted organisms.

The millepedes (Diplopoda) appeared first in the Devonian and the centipedes (Chilopoda) in the Pennsylvanian. Lack of a fossil record leaves uncertain the time of appearance of the Pauropoda and Symphyla.

The oldest known fossil insects come from Lower Pennsylvanian rocks, but their advanced state of development clearly indicates a long period of evolution, and it should not come as a surprise if Mississippian or even Devonian fossil insects are ultimately discovered. Although insects are prodigiously abundant today, they left few fossils in rocks older than Tertiary. This discrepancy is believed to be largely ascribable to difficulty of preservation.

The Arachnoidea, like the Trilobita, undoubtedly had a long Pre-Cambrian history, as the Lower Cambrian representatives of the class are complex and rather highly specialized creatures. The Merostomata reached their greatest development during the Paleozoic and since then have rapidly declined until now only a single genus, *Limulus*, survives. The Arachnida, in contrast, have developed gradually since their first appearance in the Silurian and are today a large and varied group of highly specialized arthropods.

Many fossil arthropods are important index fossils, particularly those that attained world-wide distribution, and many others, even though too poorly preserved or with an incomplete fossil record, are nevertheless of great evolutionary interest. In spite of the enormous numbers of individuals that must have lived during the geologic past, there is a real paucity of well-preserved arthropod remains if the phylum as a whole is considered.

Nature of the Fossil Record. Since most arthropod exoskeletons are largely chitin and relatively few are of calcareous substances, most fossil remains are in the form of impressions made by the hard parts or thin carbonaceous films representing what is left of the decomposed chitin.[1] The calcareous structures have an excellent chance of preservation, although in many cases they have been partly or totally destroyed during fossilization. Among the most satisfactory arthropod fossils in some respects are those preserved in amber (Figs. 13-21, 13-54). Many of these retain the original body form undistorted, the coloration and surface markings, and even the most delicate hairs, spines, and other surface features. There is the great disadvantage, however, that the fossils cannot be removed from the amber for study because they are so fragile that they disintegrate immediately upon being freed from the matrix. Although less spectacular than the amber specimens, the exquisitely preserved insect wings from several Carboniferous and Permian shales leave little to be desired. Certainly as tantalizing and challenging as any arthropod remains are the beautifully preserved specimens that have been recovered from the Middle Cambrian Burgess shale of British Columbia. Although actual organic substance is not preserved, outlines of anatomical and morphological features are shown in almost unbelievable detail. So unique and varied is this faunal assemblage that many of the fossils cannot be assigned to existing classes or orders, probably because they represent groups that long ago became extinct. This great discovery not only demonstrated the complexity of the Arthropoda as early as the Middle Cambrian, but also made quite clear that the beginnings of the phylum would have to be sought in Pre-Cambrian rocks. Almost equally spectacular has

[1] For some unusual types of preservation see Lance (1946) and Pierce (1944–1947).

been the discovery in Kansas and Oklahoma of a Permian insect fauna that is throwing much light on the evolution of that class of arthropods (see Carpenter, 1947, 1947a). Finally, the exquisitely preserved growth stages of trilobites that are being recovered from soluble rocks by etching techniques are beginning to provide a reliable basis on which ultimately a new and better classification of the Trilobita can be based (see Ross, 1951; Lalicker, 1935; and Whittington, 1941; etc.).

REFERENCES[1]

Arthropoda (General)

BORRADAILE, L. A., *et al.* 1935. "The Invertebrata," 2d ed. New York, Cambridge University Press. (Arthropoda, pp. 305–542.) The Macmillan Co.

DAVID, T. W. E., and TILLYARD, R. J. 1936. "Memoir on Fossils of the Late Pre-Cambrian (Newer Proterozoic) from the Adelaide Series, South Australia." Sydney, Angus and Robertson, Ltd., in conjunction with the Royal Society of New South Wales. 122 pp.

HEEGAARD, P. 1945. Remarks on the phylogeny of the Arthropoda. *Arkiv. Zool.*, vol. 37A, no. 3, 15 pp.

LANCE, J. F. 1946. Fossil arthropods of California. *S. Calif. Acad. Sci. Bull.*, vol. 45, pp. 21–27.

ÖPITZ, R. 1932. "Bilder aus der Erdgeschichte des Nahe-hunsrück-Landes Birkenfeld." Birkenfeld, Hugo Enke. 224 pp.

PIERCE, W. D. 1944–1947. Fossil arthropods of California. *S. Calif. Acad. Sci. Bull.*, vols. 43, 44, and 45.

RAYMOND, P. E. 1935. *Leanchoilia* and other mid-Cambrian Arthropoda. *Bull. Mus. Comp. Zool. (Harvard)*, vol. 76, pp. 205–230.

———. 1935*a*. Pre-Cambrian life. *Bull. Geol. Soc. Amer.*, vol. 46, pp. 375–392.

———. 1936. Paleoecology of the Arthropoda. *Nat. Res. Council, Rept. Comm. Paleoecology*, 1935–1936, pp. 22–27.

———. 1941. Invertebrate Paleontology *in* Fiftieth Anniversary Volume, Geological Society of America. (Arthropoda, pp. 86–90.)

RICHARDS, A. G. 1946. The organization of arthropod cuticle: a modified interpretation. *Science*, vol. 105, pp. 170–171.

———. 1951. The integument of arthropods. The chemical components and their properties, the anatomy and development, and the permeability. Minneapolis, University of Minnesota Press. 411 pp.

ROMER, A. S. 1941. "Man and the Vertebrates." Chicago, University of Chicago Press. 405 pp.

———. 1949. "The Vertebrate Body." Philadelphia, W. B. Saunders Company. 643 pp.

SCUDDER, S. H. 1873–1890. On Carboniferous Myriapoda from Illinois. *Mem. Boston. Soc. Nat. Hist.*, vols. 2–4.

SCHUCHERT, C. 1916. The earliest fresh-water arthropods. *Proc. Nat. Acad. Sci.*, vol. 2, pp. 726–733.

SNODGRASS, R. E. 1938. Evolution of the Annelida, Onychophora, and Arthropoda. *Smith. Misc. Coll.*, vol. 97, pp. 1–159.

———. 1951. Comparative studies on the head of mandibulate arthropods. Ithaca, Comstock Publishing Company, Inc. 118 pp.

[1] Key references to the voluminous literature on Arthropoda will be found in some of the more comprehensive works listed here.

Störmer, L. 1949. Classes des Merostomoidea, Marellomorpha, et Pseudocrustacea *in* Grassé's "Traité de zoologie," vol. 6, pp. 198–210. Paris, Masson et Cie.

Storer, T. I. 1951. "General Zoology," 2d ed. New York, McGraw-Hill Book Company, Inc. 832 pp.

Vandel, A. 1949. Généralities composition de l'embranchement des Arthropodes *in* Grassé's "Traité de zoologie," vol. 6, pp. 80–158. Paris, Masson et Cie.

Walcott, C. D. 1931. Addenda to descriptions of Burgess shale fossils (edited by C. E. Resser). *Smith. Misc. Coll.*, vol. 85, pp. 1–46.

Crustacea

Agnew, A. F. 1948. Annotated bibliography of Paleozoic ostracodes. *Nat. Res. Council. Rept. Comm. Treatise Marine Ecology and Paleoecology*, 1946–1947, no. 7, pp. 58–67.

Berry, C. T. 1939. A summary of the fossil Crustacea of the Order Stomatopoda. *Amer. Midland Naturalist*, vol. 21, pp. 461–471.

Borradaile, L. A. 1926. Notes upon crustacean limbs. *Ann. Mag. Nat. Hist.*, vol. 17, pp. 193–213.

———. 1926*a*. On the primitive phyllopodium. *Ibid.*, vol. 18, pp. 16–18.

Clarke, J. M. 1896. The structure of certain Paleozoic barnacles. *Amer. Geol.*, vol. 17, pp. 137–143.

Cooper, C. L. 1942. Occurrence and stratigraphic distribution of Paleozoic ostracodes. *Jour. Paleontology*, vol. 16, pp. 764–776; reprinted as *Ill. Geol. Surv. Rept. Inv.* 83.

Frederickson, E. A., Jr. 1946. A Cambrian ostracode from Oklahoma. *Jour. Paleontology*, vol. 20, p. 578.

Howell, B. F., and Kobayashi, T. 1936. A new notostracan genus from the Ordovician of Siberia. *Ann. Carnegie Mus.*, vol. 25, pp. 59–61.

Huxley, T. H. 1880. "The Crayfish." New York, Appleton-Century-Crofts, Inc. 371 pp.

Kesling, R. V. 1951. The morphology of ostracod molt stages. *Ill. Biol. Mem.*, vol. 21, pp. 1–324.

Packard, A. S. 1883. A monograph of the phyllopod Crustacea of North America, with remarks on the Order Phyllocardia. *Rept. U.S. Geol. Surv. Terr.*, vol. 12, pp. 295–592.

Pilsbry, H. A. 1906. The sessile barnacles (Cirripedia) contained in the collections of the U.S. National Museum; including a monograph on the American species. *U.S. Nat. Mus., Bull.* 93, 360 pp.

Rathbun, M. J. 1935. Fossil crustacea of the Atlantic and Gulf Coastal Plain. *Geol. Soc. Amer., Spec. Paper* 2, 160 pp. (Many other papers on Crustacea were written by this author.)

Raymond, P. E. 1946. The genera of fossil Conchostraca—an order of bivalved Crustacea. *Bull. Mus. Comp. Zool.* (Harvard), vol. 96, pp. 218–307.

Ruedemann, R. 1922. On the occurrence of *Apus* in the Permian of Oklahoma. *Jour. Geol.*, vol. 30, pp. 311–318.

———. 1924. An ancestral acorn barnacle. *N.Y. State Mus., Bull.* 251, pp. 93–104.

Scott, H. W. 1938. A stomatopod from the Mississippian of central Montana. *Jour. Paleontology*, vol. 12, pp. 508–510.

Scourfield, D. J. 1926. On a new type of crustacean from the Old Red Sandstone (Rhynie chert bed, Aberdeenshire). *Phil. Trans. Roy. Soc. London*, vol. 214, pp. 153–187.

Smith, G. W. 1909. On the Anaspidacea, living and fossil. *Quart. Jour. Micros. Sci.*, vol. 53, pp. 489–578.

Smith, W. C. 1935. Growth of the young lobster (*Homarus vulgaris*). *Trans. Liverpool Biol. Soc.*, vol. 48, pp. 51–60.

TRIEBEL, T. 1941. Zur Morphologie und Oekologie der fossilen Ostracoden. *Senckenbergiana*, vol. 23, pp. 294–400.

ULRICH, E. O., and BASSLER, R. S. 1923. Paleozoic Ostracoda, their morphology, classification, and occurrence. *Md. Geol. Surv.*, Silurian vol., pp. 271–391.

——— and ———. 1931. Cambrian bivalved Crustacea of the order Conchostraca. *Proc. U.S. Nat. Mus.*, vol. 78, 130 pp.

VAN NAME, W. G. 1925. The supposed Paleozoic barnacle *Protobalanus* and its bearing on the origin and phylogeny of the barnacles. *Amer. Mus. Nov.*, no. 197, 8 pp.

———. 1926. A new specimen of *Protobalanus*, supposed Paleozoic barnacle. *Ibid.*, no. 227, 6 pp.

WALCOTT, C. D. 1912. Middle Cambrian Branchiopoda, Malacostraca, Trilobita, and Merostomata. *Smith. Misc. Coll.*, vol. 57, pp. 145–228.

WITHERS, T. H. 1915. Some Paleozoic fossils referred to the Cirripedia. *Geol. Mag.*, dec. 2 vol. 6, pp. 112–123.

WRIGHT, L. M. 1948. "A Handbook of Ostracoda," 1st ed. Tulsa, Okla., privately printed.

Arachnoidea

BROILI, F. 1928. Crustaceenfunde aus dem rheinischen Unterdevon. 1. Über Extremitätenreste. *Sitz. math.-naturw. Abt. bay. Akad. Wiss., München*, pp. 197–201.

CASTER, K. E. 1941. Trails of *Limulus* and supposed vertebrates from Solenhofen lithographic limestone. *Pan-Amer. Geol.*, vol. 76, pp. 241–258.

———. 1944. Limuloid trails from the Upper Triassic (Chinle) of the Petrified Forest National Monument Arizona. *Amer. Jour. Sci.*, vol. 242, pp. 74–78.

———. 1944*a*. New synziphosuroid merostome from the Lower Ordovician (of Tenn.) (abstract). *Bull. Geol. Soc. Amer.*, vol. 55, pp. 1465–1466.

CLARKE, J. M., and RUEDEMANN, R. 1912. The Eurypterida of New York. *N.Y. St. Mus. Mem.* 14, 2 vols.

DECKER, C. E. 1938. A Permian eurypterid from Oklahoma. *Jour. Paleontology*, vol. 12, pp. 396–397.

FAGE, L. 1949. Classe des Mérostomacés, *in* Grassé's "Traité de zoologie," vol. 6, pp. 219–262. Paris, Masson et Cie.

MCCONNELL, M. 1916. The habit of the Eurypterida. *Bull. Buffalo Soc. Nat. Sci.*, vol. 11, 277 pp.

MOORE, P. F. 1941. On gill-like structures in the Eurypterida. *Geol. Mag.*, vol. 78, pp. 62–70.

PETRUNKEVITCH, A. 1948. Arachnida, Eurypterida, Spiders, Xiphosurans in Encyclopaedia Britannica, 14th ed.

———. 1949. A study of Palaeozoic Arachnida. *Trans. Conn. Acad. Arts Sci.*, vol. 37, pp. 69–315.

———. 1950. Baltic amber spiders in the Museum of Comparative Zoology. *Bull. Mus. Comp. Zool.* (Harvard), vol. 103, pp. 257–337. (This article contains references to earlier works.)

PRANTL, F., and PŘIBYL, A. 1948. "Revision of the Bohemian Silurian Eurypterida." Prague. Pp. 65–116; Czech edition, pp. 1–64.

RAYMOND, P. E. 1944. Late Paleozoic xiphosurans. *Bull. Mus. Comp. Zool.* (Harvard), vol. 94, pp. 477–508.

RUEDEMANN, R. 1934. Eurypterids in graptolite shales. *Amer. Jour. Sci.*, 5th ser., vol. 27, pp. 374–387.

SHARPE, C. F. S. 1932. Eurypterid trail from the Ordovician. *Amer. Jour. Sci.*, 5th ser., vol. 24, pp. 355–361.

STÖRMER, L. 1936. Eurypteriden aus dem Rheinischen Unterdevon. *Abhandl. preuss. geol. Landesanst.*, heft 175, pp. 1–74.

———. 1944. On the relationships and phylogeny of fossil and recent Arachnomorpha. *Skrift. Norske Vid.-Akad., Oslo, Mat.-Nat. Klasse*, vol. 5, 158 pp.

———. 1951. A new eurypterid from the Ordovician of Montgomeryshire, Wales. *Geol. Mag.*, vol. 88, pp. 409–422.

WILLS, L. J. 1947. "A Monograph of British Triassic Scorpions." *Palaeontographical Society*, 137 pp.

Trilobita

BEECHER, C. E. 1893. A larval form of *Triarthrus. Amer. Jour. Sci.*, 3d ser., vol. 46, pp. 142–147.

———. 1897. Outline of a natural classification of the trilobites. *Ibid.*, 4th ser., vol. 3, pp. 181–207.

———. 1901. "Studies in Evolution." New York, Charles. Scribner's Sons. 638 pp.

DELO, D. 1935. Locomotive habits of some trilobites. *Amer. Midland Naturalist*, vol. 16, pp. 406–409.

FENTON, C. L., and FENTON, M. A. 1937. Trilobite "nests" and feeding burrows. *Amer. Midland Naturalist*, vol. 18, pp. 446–451.

GARSTANG, W. 1940. Störmer on the appendages of trilobites. *Ann. Mag. Nat. Hist.*, vol. 6, pp. 59–66.

——— and GURNEY, R. 1938. The descent of Crustacea from trilobites, and their larval relations *in* "Evolution: Essays Presented to Prof. E. S. Goodrich on His 70th birthday." New York, Oxford University Press, pp. 271–285.

GHEYSELINCK, R. F. C. R. 1937. "Permian Trilobites from Timor and Sicily with a Revision of Their Nomenclature and Classification." Amsterdam, Scheltema and Holkema's Boekhandel. 108 pp. (Reviewed in *Jour. Geol.*, vol. 47, pp. 443–444.)

HENNINGSMOEN, G. 1951. Remarks on the classification of trilobites. *Norsk. Geol. Tidsskrift*, vol. 29, pp. 174–217.

HOWELL, B. F., and RESSER, C. E. 1934. Habitats of the agnostian trilobites (abstract). *Geol. Soc. Amer. Proc.*, 1933, pp. 360–361.

——— *et al.* 1947. Terminology for describing Cambrian trilobites. *Jour. Paleontology*, vol. 21, pp. 72–76.

LALICKER, C. G. 1935. Larval stages of trilobites from the Middle Cambrian of Alabama. *Jour. Paleontology*, vol. 9, 394–399.

LINDSTRÖM, G. 1901. The visual organs of trilobites. *Kungl. Svenska Vetenskapsakad. Handlingar*, vol. 34, no. 8, 86 pp.

RASETTI, F. 1948. Lower Cambrian trilobites from the conglomerates of Quebec. *Jour. Paleontology*, vol. 22, pp. 1–24.

———. 1952. Revision of the North American trilobites of the Family Eodiscidae. *Jour. Paleontology*, vol. 26, pp. 434–451.

RAW, F. 1925. The development of *Leptoplastus salteri* and other trilobites. *Quart. Jour. Geol. Soc. London*, vol. 81, pp. 223–324.

———. 1927. The ontogenies of trilobites and their significance. *Amer. Jour. Sci.*, 5th ser., vol. 14, pp. 7–35, 131–149.

———. 1937. Systematic position of the Olenellidae (Mesonacidae). *Jour. Paleontology*, vol. 11, pp. 575–598.

RAYMOND, P. E. 1920. The appendages, anatomy, and relationships of trilobites. *Mem. Conn. Acad. Arts. Sci.*, vol. 7, 169 pp.

———. 1921. Criteria for species, phylogenies and faunas of trilobites. *Bull. Geol. Soc. Amer.*, vol. 32, pp. 349–352.

———. 1935. Protaspides of trilobites. *Jour. Palentology*, vol. 9, pp. 400–401.

Richter, R. 1933. Crustacea (Paläontologie). *in* "Handwörterbuch der Naturwissenschaften," 2d ed., vol. 2, Jena, Gustav Fischer. (Trilobita, pp. 840–856.)

Ross, R. J., Jr. 1951. Stratigraphy of the Garden City formation in eastern Utah, and its trilobite faunas. *Peabody Mus. Nat. Hist. (Yale), Bull.* 6, 161 pp.

Schevill, W. E. 1936. Habits of trilobites [based largely on Richter's observations]. *Nat. Res. Council, Rept. Comm. Paleoecology*, 1935–1936, pp. 29–43.

Störmer, L. 1933. Are the trilobites related to the arachnids? *Amer. Jour. Sci.*, 5th ser., vol. 26, pp. 147–157.

———. 1939. Studies on trilobite morphology. Pt. I. The thoracic appendages and their phylogenetic significance. *Norsk. Geol. Tidsskrift*, vol. 19, pp. 143–273.

———. 1942. *Ibid.*, Pt. II. The larval development, the segmentation and the sutures, and their bearing on trilobite classification. Vol. 21, pp. 49–164.

———. 1944. On the relationships and phylogeny of fossil and recent Arachnomorpha. *Skrift. Norske Vid.-Akad., Oslo, Mat.-Nat. Klasse*, vol. 1, 158 pp.

———. 1949. Classe des trilobites *in* Grassé's "Traité de zoologie," vol. 6, pp. 160–197. Paris, Masson et Cie.

———. 1951. Studies on trilobite morphology. Pt. III. The ventral cephalic structures with remarks on the geologic position of the trilobites. *Norsk. Geol. Tidsskrift*, vol. 29, pp. 108–158.

———. 1952. Phylogeny and taxonomy of fossil horseshoe crabs. *Jour. Paleontology*, vol. 26, pp. 630–639.

Stubblefield, C. J. 1936. Cephalic sutures and their bearing on current classifications of trilobites. *Biol. Revs.*, vol. 11, pp. 407–440.

Swinnerton, H. H. 1915. Suggestions for a revised classification of trilobites. *Geol. Mag.*, dec. 6, vol. 2, pp. 487–496, 538–545.

———. 1919. The facial suture of trilobites. *Ibid.*, vol. 6, pp. 103–110.

Teichert, C. 1944. Permian trilobites from Western Australia. *Jour. Paleontology*, vol. 18, pp. 455–463.

Ulrich, E. O. 1930. Ordovician trilobites of the family Telephidae, etc. *Proc. U.S. Nat. Mus.*, vol. 76, 101 pp.

Walcott, C. D. 1879. Note on the eggs of the trilobites. *N.Y. State Mus., Ann. Rept.* 31, pp. 66–67.

———. 1908–1918. Cambrian trilobites. *Smith. Misc. Coll.*, vols. 53, 57, 64, and 67, and many other papers.

Warburg, E. 1925. The trilobites of the Leptaena limestone in Darlarne. *Bull. Geol. Inst. Uppsala*, vol. 18, 446 pp.

Weller, J. M. 1936. Carboniferous trilobite genera. *Jour. Paleontology*, vol. 10, pp. 704–715.

———. 1937. Evolutionary tendencies in American Carboniferous trilobites. *Ibid.*, vol. 11, pp. 337–346.

———. 1944. Permian trilobite genera. *Ibid.*, vol. 18, pp. 320–327.

Westergård, A. H. 1946. Agnostidea of the Middle Cambrian of Sweden. *Sverig. Geol. Undersokung*, vol. 40, 140 pp.

Whittington, H. B. 1941. The Trinucleidae—with special reference to North American genera and species. *Jour. Paleontology*, vol. 15, pp. 21–41.

———. 1941*a*. Silicified Trenton trilobites. *Ibid.*, pp. 492–522.

———. 1947. Silicified Ordovician trilobites (abstract). *Bull. Geol. Soc. Amer.*, vol. 58, pp. 1239–1240.

———. 1950. "British Trilobites of the Family Harpidae." *Paleontographical Society*, 1949, 55 pp.

———. 1950*a*. Sixteen Ordovician genotype trilobites. *Jour. Paleontology*, vol. 24, pp. 531–565.

Insecta[1]

Bachofen-Echt, A. 1949. "Der Bernstein und seine Einschuesse." Vienna, Springer. 204 pp.

Brown, R. W. 1945. *Celliforma spirifer*, the fossil larval chambers of mining bees. *Jour. Wash. Acad. Sci.*, vol. 24, pp. 532–539.

Brues, C. T., Melander, A. L., and Carpenter, F. M. 1952. Classification of insects [including fossils]. *Bull. Mus. Comp. Zool.* (Harvard), vol. 107 (in press).

Carpenter, F. M., 1947. Early insect life. *Psyche*, vol. 54, pp. 65–85.

———. 1947*a*. Lower Permian insects from Oklahoma. Part 1. Introduction and the orders Megasecoptera, Protodonata, and Odonata. *Proc. Amer. Acad. Arts Sci.*, vol. 76, pp. 25-54. (Also see later papers in same publication by same author.)

———. 1950. The Lower Permian insects of Kansas. Part 10. The Order Protorthoptera: The Family Liomopteridae and its relatives. *Ibid.*, vol. 78, pp. 185–219. (This article contains references to previous papers.)

———. 1951. Studies on Carboniferous insects from Commentry, France: Part II. The Megasecoptera. *Jour. Paleontology*, vol. 25, pp. 336–355.

———. 1952. Fossil insects *in* "The 1952 Yearbook of Agriculture," pp. 14–19.

——— *et al.* 1937. Insects and arachnids from Canadian amber. *Univ. Toronto Studies, Geol. Ser.*, no. 40, pp. 7–62.

Cockerell, F. D. A. 1937. Recollections of a naturalist, V. Fossil insects. *Bios*, vol. 8, pp. 51–56. (A short but interesting autobiographical sketch of a paleoentomologist who wrote many papers on fossil insects.)

Comstock, J. H. 1918. "The Wings of Insects." Ithaca, The Comstock Publishing Company. 430 pp.

Fairbridge, R. W. 1950. The geology and geomorphology of Point Peron, Western Australia. *Jour. Roy. Soc. West. Australia*, vol. 34, pp. 35–72.

Handschin, E. von. 1926. Revision der Collembolen des baltischen Bernsteins. *Ent. Mitteil.*, vol. 25, no. ¾, pp. 211–223; no. ⅚, pp. 330–342.

———. 1926*a*. Die Collembolen des baltischen Bernsteins. *Zool. Anz.*, vol. 65, pp. 179–182.

Handlirsch, A. 1908. "Die fossilen Insekten." Leipzig, W. Engelmann.

———. 1920. Palaeontologie (Insects) *in* Schroeder's "Handbuch der Entomologie," vol. 3, pp. 117–306. Jena, Gustav Fischer.

———. 1922. Fossilium Catalogus I. Animalia, pars 16, Insecta palaeozoica, 230 pp. Berlin.

———. 1937–1939. Neue Untersuchunger über die fossilen Insekten, Teil. I and II. *Ann. natur. Mus.*, Wien, vol. 48, 140 pp.; vol. 49, 240 pp. (Reviewed in *Amer. Midland Naturalist*, vol. 18, pp. 1106–1109, 1937; *Amer. Jour. Sci.*, 5th ser., vol. 35, pp. 308–309.)

Herrick, G. W. 1926. The "ponderable" substance of aphids (Homop.). *Entomol. News.*, vol. 37, pp. 207–210.

Hoffman, A. D. 1932. Miocene insect-gall impressions. *Bot. Gazette*, vol. 93, pp. 341–342.

Imms, A. D. 1937. "Recent Advances in Entomology," 2d ed. Philadelphia, The Blakiston Company. 431 pp. (Chap. IV, pp. 72–102, is an excellent discussion to date of the paleontology of the Insecta.)

———. 1947. The phylogeny of insects. *Tijdschr. Entomol.* (*Festbundel* 1945), vol. 88, pp. 63–66.

[1] According to Professor Frank M. Carpenter of Harvard University, the literature on fossil insects includes about 4,000 titles contributed by almost 700 different authors. The few articles listed here are cited because of their monographic nature, importance in revision, contribution to the geologic history of insects, or extensive lists of more recent references.

LAMEERE, A. 1922. Sur la nervation alaire des insectes. *Bull. classe sci., Acad. roy. Belgique*, pp. 138–149. (Translation by A. M. Brues, On the wing-venation of insects. *Psyche*, vol. 30, pp. 123–132).

———. 1935. "Précis de zoologie," vols. 4 (pp. 57–458) and 5 (pp. 1–536).

LEA, A. M. 1925. Notes on some calcareous insect puparia. *Rec. South. Australian Mus.*, vol. 3, pp. 35–36.

LIGHT, S. F. 1930. Fossil termite pellets from the Seminole Pleistocene (of Florida). *Univ. Cal. Pub., Bull. Dept. Geol. Sci.*, vol. 19, pp. 75–80.

MARTYNOV, A. B. 1925. Ueber zwei Grundtypen der Flügel bei den Insekten und ihre Evolution. *Zeitcher. Morphol. Ökol. Tiere*, vol. 4, pp. 465–501.

———. 1938. Etudes sur l'histoire géologique et de phylogenie des ordres des insects (Pterygota). Part 1. *Trav. de l'inst. paleont.*, vol. 7, 147 pp.

PIERCE, W. D. 1951. Fossil arthropods from onyx-marble. *S. Calif. Acad. Sci. Bull.*, vol. 50, pp. 34–49.

SCOURFIELD, D. J. 1940. The oldest known fossil insect. *Proc. Linn. Soc. London*, 152d Session, pp. 113–131.

SCUDDER, S. H. 1890. "The Fossil Insects of North America; with Notes on Some European Species." 2 vols. New York. (1. The pretertiary insects. 2. The tertiary insects; also published as vol. 13 of *Rept. U.S. Geol. Surv. Terr.*)

———. 1891. Index to the known fossil insects of the world including myriapods and arachnids. *U.S. Geol. Surv., Bull.* 71, 744 pp. (This author wrote many important papers on fossil insects during the period from 1860 to 1900.)

SNODGRASS, R. E. 1928. Morphology and evolution of the insect head and its appendages. *Smith. Misc. Coll.*, vol. 81, no. 3, 158 pp.

———. 1930. How insects fly. *Rept. Smithsonian Inst.* 1929, pp. 383–421.

———. 1938. Evolution of the Annelida, Onycophora, and Arthropoda. *Smith. Misc. Coll.*, vol. 97, no. 6, 159 pp.

SNYDER, T. E. 1949. Catalog of the termites (Isoptera) of the world. *Smith Misc. Coll.*, vol. 112, 490 pp.

TIEGS, O. W. 1940. The embryology and affinities of the Symphyla, based on a study of *Hanseniella agilis*. *Quart. Jour. Micros. Sci.*, London (N.S.), vol. 82, pp. 1–225.

———. 1945. The post-embryonic development of *Hanseniella agilis* (Symphyla). *Ibid.*, vol. 85, pp. 191–328.

———. 1947. The development and affinities of the Pauropoda, based on a study of *Pauropus silvaticus*. *Ibid.*, vol. 88, pp. 165–267, 275–336.

———. 1949. The problem of the origin of insects. *Australian and New Zealand Assoc. Adv. Sci.*, vol. 27, pp. 47–56.

TILLYARD, R. J. 1930. The evolution of the Class Insecta. *Papers Proc. Roy. Soc. Tasmania*, 89 pp. (Also *Amer. Jour. Sci.*, 5th ser., vol. 23, pp. 529–539, 1923.) (This author has published a dozen or more papers on Kansas Permian insects in the *Amer. Jour. Sci.*, beginning in 1924.)

WALKER, M. V. 1938. Evidence of Triassic insects in the Petrified Forest National Monument, Ariz. *Proc. U.S. Nat. Mus.*, vol. 85, pub. 3033, pp. 137–141.

WILLEM, V. 1900. Recherches sur les Collemboles et les Thysanoures. *Mem. cour. mem sav. étr., Acad. roy. Belgique*, vol. 58, 144 pp.

CHAPTER 14

PHYLUM ECHINODERMA[1]

GENERAL CONSIDERATIONS

The echinoderms constitute an exceptionally well defined and sharply limited group of exclusively marine animals that are readily divisible into clearly separated and easily distinguished classes, some of which have representatives that first appeared in the Early Cambrian. Twelve of these classes are now recognized; of these, four have living representatives and eight have long been extinct. The stratigraphic range and relative geological importance of the 12 classes are indicated in Fig. 14-56.

Classes and lower taxonomic divisions with living representatives are designated by names that call attention to some characteristic feature. Thus, there are the feather stars and sea lilies (unstalked and stalked *Crinoidea*, respectively); the starfish (*Stelleroidea*), subdivided into the true starfish (sea stars) (*Asteroidea*) and the brittle stars (*Ophiuroidea*); the sea urchins (*Echinoidea*); and the sea cucumbers (*Holothuroidea*). In Paleozoic seas there were representatives of eight other classes, all now extinct—*Cystoidea*, *Eocrinoidea*, *Paracrinoidea*, *Edrioasteroidea*, *Carpoidea*, *Machaeridia*, *Cyamoidea*, and *Cycloidea* (see Table 14-1).

The Crinoidea are easily recognizable by their featherlike arms and usually by the presence, at the center of the dorsal or aboral side, of a cluster of slender, segmented appendages (**cirri**) or of a stalk that may or may not have cirri. Fossil crinoids are abundant in many Paleozoic and Mesozoic rocks, and the class dates from the Ordovician. Living representatives are mainly confined to the deeper waters, and the class does not now seem to be as important as it has been at times during the geologic past.

The Asteroidea exhibit considerable variety of size and form but are conspicuous by the pentagonal or hexagonal arrangement of the arms, which may be quite long or so short as hardly to be distinguishable from the body. The arms are hollow and contain radial extensions of the body; in no case do they have cirri or are they featherlike. Fossil starfish have been found in rocks as old as Middle Ordovician; they were never preserved in abundance except under unusual conditions of burial.

The Ophiuroidea have solid, slender, noncirrate arms which are more or less segmented and which, in some highly specialized groups, are dendritic or forked. The arms of the so-called "sea spiders" or "basket fish" may have as many as a dozen divisions. The fossil record of the Ophiuroidea, which

[1] Echinoderma—Gr. *echinos*, spiny, + *derma*, skin; referring to the spiny or prickly skin of the animals. The Greeks applied the name *echinos* to the hedgehog as well as to the sea urchin, both of which have a prickly skin. The term *Echinus* continues to be used as a generic name for a certain urchin.

TABLE 14-1. CLASSIFICATION OF PHYLUM ECHINODERMA

(Based on Jaekel, 1918, etc.; Bather, 1900; Whitehouse, 1941; Moore and Landon, 1943; Regnell, 1945; etc. Extinct groups are indicated by an asterisk.)

Subphylum Pelmatozoa (attached forms)
- *Class Cystoidea (M. Ord.–L. Perm.)
  - *Subclass Hydrophoridea (M. Ord.–M. Dev.)
  - *Subclass Blastoidea (M. Ord.–L. Perm.)
- *Class Eocrinoidea (L. Camb.–M. Ord.)
- *Class Paracrinoidea (M. Ord.)
- Class Crinoidea (L. Ord.–Recent)
  - *Subclass Inadunata (M. Ord.–Trias.)
  - *Subclass Flexibilia (M. Ord.–M. Perm.)
  - *Subclass Camerata (L. Ord.–M. Perm.)
  - Subclass Articulata (Trias.–Recent)
- *Class Edrioasteriodea (L. Camb.–Penn.)

*Subphylum Homalozoa (simple, supposedly attached forms)
- *Class Carpoidea (M. Camb.–L. Dev.)
- *Class Machaeridia (Ord.–Dev.)

*Subphylum Haplozoa (simple, supposedly unattached forms)
- *Class Cyamoidea (M. Camb.)
- *Class Cycloidea (M. Camb.)

Subphylum Eleutherozoa (free and vagrant benthos)
- Class Stelleroidea (L. Ord.–Recent)
  - Subclass Asteroidea (M. Ord.–Recent)
  - Subclass Ophiuroidea (Middle Miss.–Recent)
  - *Subclass Auluroidea (L. Ord.–Upper Miss.)
  - *Subclass Somasteroidea (L. Ord.)
- Class Echinoidea (M. Ord.–Recent)
- Class Holothuroidea (M. Camb.?, Ord.–Recent)

begins in the Mississippian, consists largely of the individual plates and segments of the arms; few complete exoskeletons seem to have been preserved, probably because quick burial under special conditions was required to ensure that the delicate arms were not broken.

The Echinoidea, including the familiar sand dollars, heart urchins, etc., are encased in a firm, more or less rigid test that is characteristically spiny and without projecting arms of any kind. The test is typically hemispheroidal or hemiellipsoidal with the oral (lower) side flattened. The spines are much diversified in size and shape, ranging from microscopic needles to club-shaped structures as much as 25 mm. in diameter and 30 cm. in length. Fossil echinoids, which first appear in Ordovician rocks, are commonly abundant and well preserved. Seldom is the complete skeleton, spines and all, preserved, but the complete corona (the hemispheroidal part without the spines) is commonly found, and the spines, both complete and fragmental, are quite abundant in some rocks.

The Holothuroidea differ conspicuously from the other classes in that the skeleton is greatly reduced, being represented by tiny plates in the soft fleshy wall of the ellipsoidal body, and the oral-aboral axis is elongated so that the animal lies on one side with the mouth anterior and the anus posterior. Although there is no obvious external radial symmetry, such as characterizes the other classes thus far discussed, a transverse section through the animal midway between mouth and anus reveals a symmetrical arrangement of body parts similar to that of a regular sea urchin or a pentagonal sea star cut in a similar plane.

The four classes with living representatives are not only sharply defined in their respective patterns of structure and in their morphology, but they also are characterized by an equal and similar diversity in their movements, methods of feeding, and modes of adaptation. These differences were long ago used as a basis for subdividing the Echinoderma into attached forms (Pelmatozoa) and free-moving forms (Eleutherozoa).

Little is known definitely about the soft parts of the animal in the eight extinct classes (and in the extinct stelleroid subclasses *Auluroidea* and *Somasteroidea*), hence discussion of their characteristics, which must be conjectured from the skeletal remains, is postponed to following pages where each group is considered in detail.

Figure 14-1 schematically compares typical representatives of different classes of the Echinoderma, with reference to living position and nature of arms (if present), etc. Figure 14-56 similarly shows the known geologic history of the phylum.

Owing to their relatively sedentary habits, their aversion to fresh or even

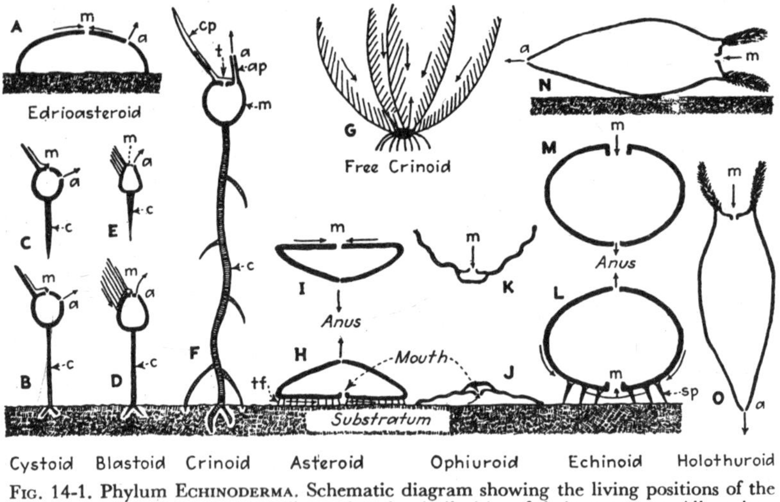

FIG. 14-1. Phylum ECHINODERMA. Schematic diagram showing the living positions of the chief types of echinoderms and illustrating the similarities of their structures. All sections are cut in the plane of symmetry that includes the mouth (*m*) and anus (*a*). The arrows indicate the direction of food-bearing or waste-carrying water currents. The test or integument is shown by a heavy, solid line. *A*. An edrioasteroid cemented to the substratum. *B*. Cystoid attached by stem (*c*) and with one arm shown. *C*. Free cystoid with short stem. *D*. Stemmed blastoid with a few brachioles shown. *E*. Free blastoid with a feebly developed stem. *F*. Stemmed crinoid with one arm, covering plates (*cp*) and tegmen (*t*) shown. The anus is elevated on an anal proboscis (*ap*). *G*. Free-living crinoid with prominent arms and short cirri encircling the base. *H–I*. An asteroid in living position (*H*), held above the bottom by tube feet (*tf*), and inverted (*I*). *J–K*. An ophiuroid in living position (*J*) and inverted (*K*). *L–M*. An echinoid in living position (*L*), held off the bottom by spines, and inverted (*M*). *N–O*. A holothuroid in living position on the bottom (*N*), and turned with the mouth upward (*O*) for comparison. Only two tentacles are shown.

brackish water, the brevity or complete absence of a free-swimming larval life, and the usually small bathymetrical range, echinoderms provide remains that are remarkably suitable as a basis for studying paleoecological and paleogeographical conditions through geologic time (Clark, 1946). The ancient ancestry of the phylum, the characteristic appearance of the skeletons and skeletal fragments, the fair abundance of well-preserved remains in the rocks of every period since the beginning of the Paleozoic, and the great variety and abundance of certain groups in existing seas have encouraged extensive study of both living and fossil forms. Many extinct species have proved to be excellent index fossils.

Morphology

General Considerations. Living echinoderms are readily distinguishable through their possession of radial symmetry, in which pentameral arrangement is most common; a mesodermal skeleton composed of plates of calcite with a characteristic microstructure resembling trelliswork; and a system (the water-vascular system) of sacs, canals, and tubes that carry water through the body, particularly by means of five radial canals from which small branches, the podia, are given off to the exterior. Extinct echinoderms in which significant skeletal features are preserved seem to have had a similar organization.

The echinoderms have a true body cavity (**coelom**) in which there is a distinct **gut** that may be straight, curved, or twisted and coiled. The mouth and anus may be on the same side of the body, as in many crinoids; they may be diametrically opposite, as in some echinoids and most asteroids; or the mouth may be slightly anterior and the anus conspicuously posterior, as in irregular echinoids. Whatever their position, however, the mouth and anus are arbitrarily taken to define a plane of primary bilateral symmetry that is common to all echinoderms. In most forms, however, this original bilateral symmetry is obscured by the well-developed secondary radial symmetry, usually pentamerous, that is such a conspicuous feature of living echinoderms. In recent Echinoidea certain species have developed a second bilateral symmetry that coincides with the original; this necessarily distorts the radial arrangement so characteristic of that class.

The arrangement and position of mouth and anus, as well as of other body features and skeletal structures, change with alteration in the symmetry. Echinoderms that are attached to the bottom generally have radial symmetry, whereas those that are mobile have tended to move the mouth toward the front and the anal opening toward the rear, with consequent development of a second bilateral symmetry.

In most echinoderms the body is divided into five radially disposed extensions. These may have the form of mobile arms, as in the starfishes; fixed radial areas, as in the sea urchins; or internal compartments, as in the holothuroids. Externally, these five extensions are marked by five radial grooves, or bands of porous plates, along which food-bearing water currents move toward the mouth. These grooves, or bands, are designated **ambulacra** (singular, **ambulacrum**) or **rays.** If the skeleton is a continuous structure, as in the echinoids, the ambulacral areas are separated by groups of plates that constitute an **interambulacrum.**

In addition to the water-vascular system, which is unique to the Echinoderma and which was probably first used for respiration, the echinoderm has a distinctly localized nervous system and a crude circulatory system without a heart or regular circulation.

The Water-vascular System. The water-vascular system of the common Atlantic starfish, *Asterias forbesi* (Fig. 14-2), may be taken as typical of many living echinoderms. Its essential parts are shown diagrammatically in Fig.

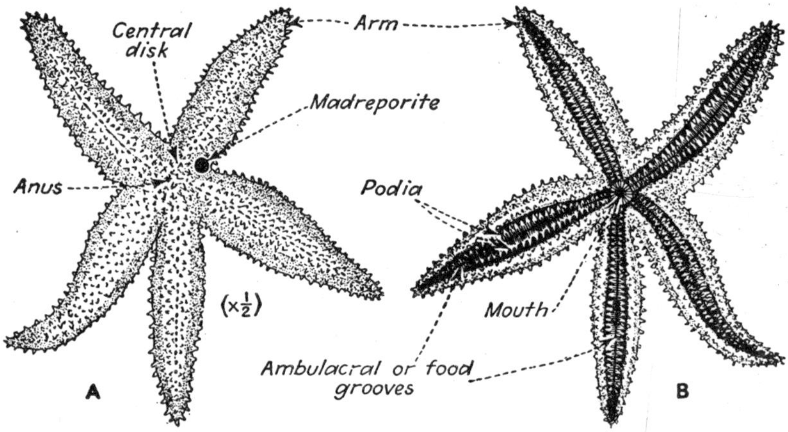

FIG. 14-2. Subclass ASTEROIDEA. *Asterias forbesi*, the common starfish along the Atlantic coast. *A*. Aboral view. *B*. Oral view.

14-3. Water enters the system through a perforated plate, the **madreporite,** situated on the upper surface of the starfish in an interambulacral position, and passes downward through the **stone canal** into the **ring canal** (and its **reservoirs**) surrounding the mouth. The ring canal gives off five **radial vessels** which have ambulacral position. On each side the vessel gives off four rows of **tube feet,** or **podia,** which are small, muscular tubes closed at the outer end and expanded at the inner into balloonlike **ampullae** (singular, **ampulla**). The ampullae lie inside the arm, protected by the skeletal plates, whereas the closed end of the podium extends between the plates and protrudes into the ambulacral groove, ending in a disklike expansion that functions as a suction disk for adherence. Sea water, held in reservoirs between the radial vessels, is admitted to the vessels and thence into an ampulla. By contraction of the ampulla, the water is forced into the corresponding tube foot, which is thus greatly lengthened. When the extended foot touches a solid object, the central part of the terminal expansion adheres to the object, the water is withdrawn, and as the tube foot contracts, the animal pulls itself forward. When water is again forced into the foot by the ampulla, the end is relaxed and loosened, and the foot again lengthens in the direction of locomotion. By repeated and independent action of many tube feet, the starfish slowly moves (about 6 in. a minute) over the surface with a gliding motion, characteristically with the arm to the left of the madreporite in advance.

In addition to its function for locomotion, the water-vascular system also aids in respiration, and it is probable that this was the original function of the system in the earliest echinoderms.

A water-vascular system seems to have been developed in the earliest echinoderms (Lower Cambrian edrioasteroids, Ordovician cystoids, etc.), judging from the internal structure and skeletal arrangement of ambulacral plates. In the primitive cystoid *Aristocystites*, irregularly arranged canals

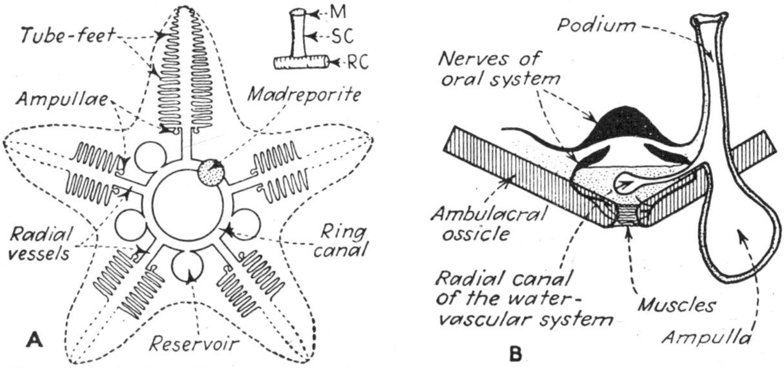

FIG. 14-3. Phylum ECHINODERMA—Subclass ASTEROIDEA. Water-vascular system. *A*. Simplified schematic diagram illustrating the structure of the water-vascular system of a starfish. The madreporite (*M*) communicates with the ring canal (*RC*) through the vertical stone canal (*CS*), as shown in the small diagram at the upper right. Ampullae are shown on only the first pair of tube feet. *B*. Section across an ambulacrum of an asteroid, showing how different parts of the water-vascular system are related to other parts of the starfish. (*B modified after Bather*, 1900, *in Lankester's "A Treatise on Zoology*.")

ramify the plates, and in later and more complex cystoids these become systematized into diplopores, pore rhombs, etc. (see later discussion). In the Blastoidea the pores are confined to the ambulacral areas and connect interiorly into saclike structures (hydrospires) produced by the folding of certain parts of the test (Fig. 14-9*E–F*). The pores are also arranged along the ambulacra in the Crinoidea. The water-vascular system reaches its highest degree of development in the Stelleroidea and Echinoidea.

Ambulacra and Food Grooves. In most echinoderms five (or rarely six) conspicuous ambulacra or ambulacral areas radiate from the mouth. Lying along the axis of each of these is a food groove provided with cilia, which by a rhythmic lashing motion drive water and contained food particles toward the mouth. Elongation and branching of the food grooves greatly increase the food-gathering area, just as the modern conveyor-belt system extends the gathering area from which materials can be transported to a common terminal. One large Jurassic crinoid species belonging to the existing genus *Pentacrinus* had a food-groove system totaling several hundred meters in length.

Ecology and Paleoecology

Attachment. Most living echinoderms are benthonic, and it seems likely that ancient and extinct forms were likewise bottom dwellers. Attached forms have a stem or column that elevates them above the bottom, and this terminates in some sort of anchoring device. Stemmed forms (crinoids, cystoids, and blastoids) were quite abundant during the Paleozoic, but only a few existing crinoids retain a stem throughout life; most lose the stem, which is present in the early growth stages, by the time maturity is attained. It is possible that some Paleozoic pelmatozoans also abandoned their anchorage in maturity; this may partly account for the paucity of calices in many crinoidal limestones that are composed almost exclusively of stem plates.

In some echinoderms the food grooves are open; hence the food-bearing water is exposed at all times. In others they are partly roofed over by spines, and in the cystoids, edrioasteroids, and many extinct crinoids they were completely covered with tiny hinged plates.

Habits and Habitats. Echinoderms are among the most important of bottom scavengers in existing seas, and it has been said that "an ocean without echinoderms might become a putrid cesspool." They attack and devour their living associates, eat the bodies of dead organisms, and ingest vast quantities of bottom sediments for their organic content. Starfish prey on worms, other echinoderms, mollusks, and fish, and upon the oyster banks their depredations may be tremendous. Like certain other echinoderms, starfish can regenerate lost arms; hence, when an angry oyster fisherman tears one of them apart, he may unwittingly increase the number of individuals. The voracity of these creatures is well illustrated by a large specimen of *Luidia clathrata*, with arms 145 mm. (almost 6 in.) long, that was picked up dead on the beach at Sarasota, Fla. This particular starfish had swallowed a sand dollar (*Mellita*) 60 mm. (more than 2 in.) in diameter, and seemingly died when it could not disgorge the test of its prey.

Holothuroids and certain echinoids gorge themselves with mud and sand for the purpose of extracting the contained organic matter. As a result of passing through the alimentary tracts of the animals, these sediments are affected both chemically and physically, and their organic material is partly or wholly removed. It seems likely that similar activities were characteristic of ancient echinoderms, and possibly the seemingly unfossiliferous nature of some marine limestones may be partly ascribable to the comminution or destruction of fragile shells by such scavengers.

Modern echinoderms are strongly gregarious, preferring to live in colonies, assemblages, or "gardens," in which they are likely to be present in great numbers, as illustrated by the multitudes of familiar urchins along certain coasts and the great numbers of starfish in some waters. The *Challenger* expedition dredged 10,000 feather stars (unstalked crinoids) in a single haul from fairly deep water, and brittle stars (ophiuroids) are known to be equally abundant. In regard to the latter, as many as 60 individuals have been counted on 0.25 sq. m. (3.5 sq. ft.) of bottom (depth about 65 ft.) in the Kattegat (Howe, 1942).

Crinoid gardens, like the famous Mississippian bioherm near Crawfordsville, Ind., crinoid rafts such as those of *Uintacrinus* in the Upper Cretaceous of Kansas, and urchin assemblages, such as that of the Mississippian *Melonechinus* at St. Louis, Mo., indicate that ancient echinoderms were also gregarious and assembled locally in great numbers.

As stated previously, all known living echinoderms are marine, and they seem always to have been confined to that environment. Living specimens have been taken at depths from tidemark to 10,360 m. (more than 34,000 ft.)[1] and in practically all latitudes. On the basis of these observations, paleontologists universally consider as marine any sedimentary rock that contains an appreciable volume of echinodermal remains.

Many organisms live with echinoderms in commensal relationship. Protozoans, sponges, annelid worms, mollusks, and arthropods feed upon the excreta of certain species. Similarly in the Mississippian, a small gastropod, *Orthonychia* (= *Igoceras*), sat upon or beside the anal opening on certain crinoids, and in some cases actually cemented its shell to the upper surface of the crinoid skeleton. A little fish, known as *Fierasper*, gains protection by crawling into the mouth of holothuroids or among the tentacles tail first. Hordes of nematodes, trematodes, and other parasitic worms infest the echinoderms, either as ectoparasites or as endoparasites.

Ancestry

The ancestry of the Echinoderma surely goes back into Pre-Cambrian time, because distinctive members of the phylum are preserved in Lower Cambrian rocks, and large and advanced groups were fully evolved by Ordovician time. The embryology of modern forms suggests that some Pre-Cambrian annelid worm may have been the ultimate ancestor of the phylum but that, between the early worm stage and the differentiation into the numerous classes, there was an intermediate stage in the form of a simple bilaterally symmetrical animal. This larva, which is supposed to have been the starting stage of the several classes, has been termed the *Dipleurula* (meaning "little two-sides" and referring to the supposed bilateral symmetry) (Fig. 14-55) and is noted in several parts of the following discussion (see particularly the section on phylogeny of the Echinoderma). A hypothetical evolutionary tree and the relative importance of different classes of the phylum during geologic history are shown in Fig. 14-56.

Stemmed and sessile forms (*i.e.*, pelmatozoans) flourished during the Paleozoic, but most of them were extinct by the end of the era. Vagrant forms (*i.e.*, eleutherozoans), which were subordinate in numbers and variety, rapidly expanded in both respects in the Early Mesozoic, and in existing seas they constitute the major part of the echinoderm fauna.

THE SKELETON

The inorganic components of an echinoderm have the form of plates, ossicles, spines, spicules, etc., and are loosely or firmly bound together into a variety of structures. Inasmuch as the components are formed in the meso-

[1] The Danish research vessel *Galathea* is reported to have dredged 61 sea cucumbers (holothuroids) from a depth of more than 34,000 ft. in the Mindanao Deep just east of the Philippines (*New York Times*, July 25, 1951).

derm, the structure they make is a true skeleton. However, several different terms in common use for the skeleton are introduced and defined in the appropriate places on following pages. For the present discussion three terms are used more or less interchangeably—**skeleton, test,** and **theca.**

The typical echinoderm test is hollow and is pear-shaped, hemispheroidal, hemiellipsoidal, globular, or star-shaped. It is composed of a tough epidermal integument studded with calcareous spicules and plates in the Holothuroidea; in all other classes it is a more or less rigid structure composed of many calcareous plates arranged either asymmetrically or symmetrically. Although the plates are situated in and directly beneath the skin, or epidermis, so that the test of the living animal is covered with dermal tissue, they are secreted in the mesoderm; hence the test is truly of internal origin. Tissue permeates the individual plates, and constitutes more than 62 per cent by volume of certain echinoid spines, the other 38 per cent being the calcite that remained when all organic matter was removed. The surface of many skeletons is covered with articulated spines of the same composition as the test proper (*e.g.*, spiny echinoids), and individual plates or groups of plates bear external grooves, pits, ridges, nodes, tubercles, and other features of sculpture.

A point of some interest is the fact that the skeletal component of echinoderms, whether it be spicule, plate, or spine, is a fine-pored, continuous single crystal of calcite. Consequently, each component breaks throughout with uniformly oriented rhombohedral cleavage—a property which sets the skeletal components of echinoderms apart from those of all other organisms that secrete crystalline inorganic substances. In echinoid spines, for example, the *C* axis of the crystal is parallel to the long axis of the spine, and one plane of cleavage cuts across the spine at an inclined angle. In all other organisms the crystalline hard parts are composed of an aggregate of small discrete crystal grains whose orientation is either random or to some extent parallel along one or more crystallographic axes.

The skeletal plates may be so firmly united as to form a rigid structure capable of retaining its shape upon burial (*e.g.*, many crinoids and echinoids). In many forms, however, the individual plates are held together by fleshy, cartilaginous or dermal matter, so that when the animal dies, the test is crushed in being buried or falls apart if the organic substances are destroyed before burial. In the Holothuroidea the scattered calcareous spicules are rarely united into any sort of continuous structure, hence isolated spicules are the fossils generally left by this group.

Figure 14-1 shows that some echinoderms live with the mouth downward, some with it upward, and the holothuroids with it variously placed with respect to the substratum. It would not always be correct, therefore, to speak of the upper side as dorsal or the underside as ventral; hence other terms are used in designating the different parts of the animal. The side of the animal and test on which the mouth is situated is designated as the **oral** or **actinal** side (also **ventral,** but without implication as to orientation with reference to the substratum); the diametrically opposite side, where the anus may be, but not necessarily is, located, is **aboral, abactinal,** or **dorsal.** In many echinoderms the anal opening is near the mouth, hence on the oral side (as in edrioasteroids, etc.), and in certain echinoids it has moved downward from an aboral position to become posterior or even ventral.

Considered as a whole the echinoderm test exhibits the same radial pentamerous symmetry (or such symmetry distorted by a second development of bilateral symmetry) as that shown by the animal; hence, there is an ambulacral area on the test corresponding to the same area of the body, and an interambulacral area between adjacent ambulacra. Food grooves on the test are underlain and margined by calcareous plates, and in some cases are, or were, covered by special roofing plates. The simplest and earliest echinoderms (*i.e.*, cystoids and edrioasteroids) had five short food grooves lying directly upon the oral surface of the test. The original number of grooves in the ancestral echinoderm is not definitely known, but there are some indications that only three were originally present. The five grooves possessed by most living echinoderms, as well as by most extinct forms, are thought to have developed from the original three by the branching of two of the latter, as shown in Fig. 14-4*A*. Extensions of these grooves beyond the immediate oral area, along with successive branching, which took place quite early, greatly enlarged the food-gathering area for the organism, and in some cystoids the five major food grooves extend downward over the test almost to the stem on the aboral side.

The food grooves of many early cystoids (*e.g.*, blastoids) were lined with small jointed appendages, termed **brachioles** (Fig. 14-9*A*,*E*), which in some cases were themselves equipped with delicate segmented bristlelike **pinnules.** The food grooves in these forms were continued onto the brachioles from the

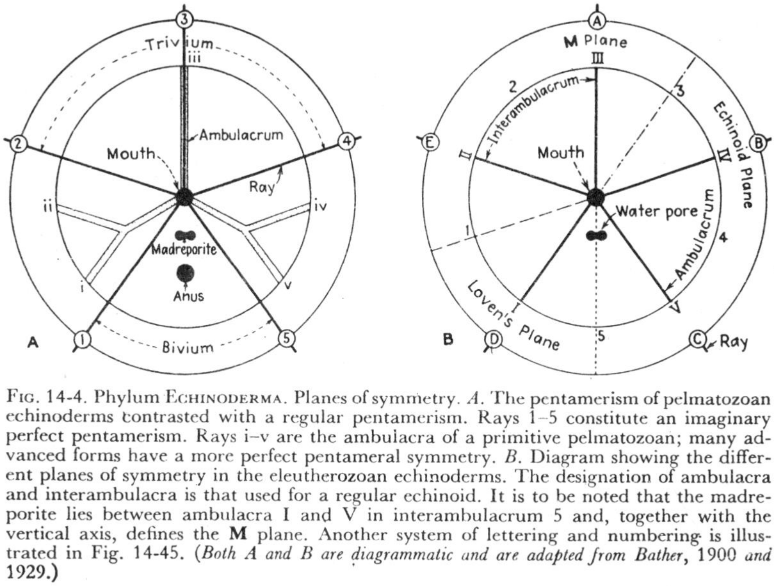

FIG. 14-4. Phylum ECHINODERMA. Planes of symmetry. *A*. The pentamerism of pelmatozoan echinoderms contrasted with a regular pentamerism. Rays 1–5 constitute an imaginary perfect pentamerism. Rays i–v are the ambulacra of a primitive pelmatozoan; many advanced forms have a more perfect pentameral symmetry. *B*. Diagram showing the different planes of symmetry in the eleutherozoan echinoderms. The designation of ambulacra and interambulacra is that used for a regular echinoid. It is to be noted that the madreporite lies between ambulacra I and V in interambulacrum 5 and, together with the vertical axis, defines the **M** plane. Another system of lettering and numbering is illustrated in Fig. 14-45. (*Both A and B are diagrammatic and are adapted from Bather*, 1900 *and* 1929.)

oral surface of the test, but were not extended onto the pinnules. The last step in the development of the food-gathering apparatus came with the development of five (six in some cases) **arms** or **brachia,** which were much larger and more complicated than the brachioles. These arms, particularly characteristic of the crinoids, both ancient and modern, show great variation in structure, nature of branching, number and arrangement of plates, etc., and are more fully considered on following pages.

Although complete or partly complete tests of echinoderms are not uncommon fossils, particularly locally or in certain beds, skeletal fragments are far more common and may be expected in marine shales and sandstones of every age from Cambrian to Recent. Some limestones consist of little else than the skeletal plates and stem fragments of crinoids and cystoids or the biscuitlike tests of echinoids. Echinoderms were common on Paleozoic bioherms and are also present in modern reef environments. In recent years fragmental echinodermal remains have become important microfossils (Smiser, 1933; Geis, 1936; Moore, 1938; Howe, 1942).

CLASSIFICATION

The Phylum Echinoderma is subdivided, as shown in Table 14-1, largely on the basis of the following characteristics, not all of which are of equal importance:

1. Mode of life and attendant modifications—presence or absence of stem or place for attachment on the aboral side of the animal. (The stem generally serves as a means of attachment, though this function is lost in certain forms.)
2. Presence or absence of fixed or movable arms.
3. Nature of the water-vascular system.

A comparison of the different classes of echinoderms, as that in Fig. 14-1, shows that the phylum comprises a closely related group of animals. If the organisms are all so oriented that the mouth is up, then it is seen that in each class the five radially disposed arms or ambulacra radiate from the more or less centrally located mouth. In contrast, the anus varies in position in the different organisms; in some it is about diametrically opposite the mouth, whereas in others it is nearer the mouth. In a general way the arms of the starfish, brittle star, and crinoid are homologous to the ambulacral areas of the cystoid and echinoid. If the starfish is turned over, so that the mouth is up, then the lower aboral side corresponds to the base of the cystoid and crinoid, and the upper side, on which the five ambulacra radiate from the mouth, is similar to the upper surface of these two groups. In order to orient the starfish with the echinoid, the arms of the former must be bent upward until their tips almost meet; then union along the margins of the arms by means of interambulacral plates would produce a skeletal structure like that of an echinoid; in this and the previous case, however, the madreporite of the starfish must also be shifted so as to lie on the upper surface of the animal. The holothuroid differs from the starfish in having the central part of the body elongated into an ellipsoidal form and in having the mouth anterior and the anus posterior. Too little is known about the extinct haplozoans to be certain how they were oriented on the substratum, but a conjectural position is indicated in Fig. 14-30.

SUBPHYLUM PELMATOZOA[1]

General Considerations

This large and paleontologically important subphylum includes five classes, four of which are extinct. The fifth, the Crinoidea, although represented by a few groups in existing faunas, is no longer as important as it once was during the Paleozoic (see Table 14-1). As a consequence, little is actually known about the soft parts of the animal and how they were related to the hard parts. Most of these relationships must be conjectured, but a few can be determined from study of living crinoids.

All pelmatozoans have in common a hollow, modified ellipsoidal skeleton, or test, composed of many calcareous plates and ordinarily, although not always, attached to some foreign object or the bottom by a flexible jointed **stem,** or **column,** or by direct cementation. The stem is composed of small, centrally perforated calcareous disks, the **columnals,** which are placed one above another to form a flexible, essentially solid cylinder, which in the living organism is completely surrounded and permeated by dermal tissue (Fig. 14-17). It is attached at the upper end to a centrodorsal plate of the test, and at the lower it terminates in a pointed plate, an anchor of some sort, or a holdfast system consisting of numerous rootlets similar in structure to the stem itself.

Surmounting the stem is the **crown,** which houses the animal and includes its food-gathering system. This is composed of the cuplike basal part, the **calyx,** which contains the body, and the **arms** or **brachioles,** which rise from the upper margin or upper surface of the calyx and bear extensions of a food groove. Brachioles are slender, delicate structures that characteristically line the food grooves (Fig. 14-6), whereas arms are more robust structures that stand at the ends of the food grooves and rise conspicuously above the surface of the test (Fig. 14-6). From the bases of the arms or brachioles food grooves lead to the centrally located mouth. The food grooves of many cystoids, edrioasteroids, and crinoids are roofed over with tiny calcareous **cover plates** (Figs. 14-9*E*, 14-17*G*), and the oral surface of many crinoid calices is covered with a tough integument or a roof of calcareous plates, either of which constitutes a **tegmen** (Fig. 14-17*B*). If the mouth lies beneath the tegmen, it is described as **subtegminal.** The anal opening is typically situated in a ventrolateral position between the two posterior rays that constitute the bivium of the imaginary pentameral system of all pelmatozoans (Fig. 14-4). In the cystoids and edrioasteroids it is covered with numerous plates, and in some crinoids it lies well above the calyx at the end of an extended anal tube (Fig. 14-18).

[1] Pelmatozoa—Gr. *pelma, pelmatos*, the sole of the foot, + *zoon*, animal; referring to the fact that representatives are attached to the bottom by the under part.

Food was brought to the mouth of the pelmatozoan by currents that moved along the ciliated food grooves of the so-called **subvective**[1] **system,** or food-gathering system. The food grooves, which radiate from the mouth, may lie between the skeletal plates (**endothecal**), extend over the surface of the calyx (**epithecal**), lie on brachioles and arms rising above the surface (**exothecal**), or lie beneath the surface (**hypothecal**).

Table 14-2 shows the chief characteristics and stratigraphic range of the five classes now included in the subphylum.

TABLE 14-2. RANGE AND CHIEF CHARACTERISTICS OF THE PELMATOZOA

Class	Subvective system	Plate arrangement	Pores	Range
Cystoidea	Food grooves on surface of calyx; biserial brachioles	Irregular to regular	Pore plates of several types and hydrospires	M. Ord.–L. Perm.
Eocrinoidea	Biserial brachioles	Regular	Thecal pores lacking	L. Camb.–M. Ord.
Paracrinoidea	Uniserial brachia	Irregular	Subepithecal pores	M. Ord.
Crinoidea	Food grooves extend onto uniserial branched arms	Regular	Outgrowths of radial canals	L. Ord.–Recent
Edrioasteroidea	Curved food grooves; sessile on theca	Irregular	Pores in ambulacra	L. Camb.–Penn.

CLASS CYSTOIDEA[2]

General Considerations

The cystoids are stemmed or stemless echinoderms with a globular, ovoidal, hemispheroidal, or hemiellipsoidal test composed of a variable number (13 to more than 200) of calcareous plates arranged with or without symmetry. The mouth is near, or at the center of, the oral surface and in some forms is covered by five small plates. The anus is typically ventrolateral in position, lying between rays 1 and 5, and is covered by five or more triangular plates, which make a **valvular pyramid** (Fig. 14-6), or by a variable number of small plates. Some cystoids have one or two small openings between the mouth and anus and in the same plane. One of these is supposed to have served as a **genital pore,** and the other may have been an opening for the entrance of water (*i.e.*, a madreporite or **hydropore**).

The Calyx. The calyx plates are tetragonal, pentagonal, hexagonal, or irregularly polygonal and are united by close suture. In earlier and more primitive cystoids the plates may number several hundred, and these lack any symmetry of arrangement. In the more specialized forms (*i.e.*, the

[1] Subvective—L. *subvectio*, conveyance; referring to the conveying of food particles.

[2] Cystoidea—Gr. *cystis*, bladder, + *-oid*, like; referring to the bladderlike appearance of the calyx.

Blastoidea), the plates are relatively few in number (as few as 13 calyx plates excluding those connected with the food grooves) and symmetrically arranged (Figs. 14-6, 14-7, 14-8). It is assumed that calyx development in the Cystoidea proceeded from many plates with irregular shape and arrangement to few plates with regular shape and arrangement.

In addition to the larger openings into the calyx (*i.e.*, mouth, anus, genital pore, and hydropore), certain or all of the calyx plates are perforated by tiny tubes or canals, the so-called **thecal pores** (Fig. 14-5), which are supposed to have functioned in respiration by providing a way for the coelomic fluid to come into osmotic contact with the surrounding sea water. These pores are regarded as an essential cystoid characteristic, and are unknown in any other class of the Pelmatozoa. They are of several distinct types and are used as a basis for dividing the Cystoidea into two subclasses: Hydrophoridea, with haplopores, diplopores, pore rhombs, and pectinate rhombs (pectinirhombs); and Blastoidea, with hydrospires. These several types of pores are described in following paragraphs.

Ambulacral Areas, Brachioles, and Arms. The ambulacral areas and food grooves of cystoids are irregular and branched in the more primitive hydrophorideans but symmetrical and unbranched in the blastoids. Typical cystoids have five food grooves with more or less perfect pentamerous symmetry, but these are supposed to have been developed by the branching of the two posterior ambulacra of an original trio, as suggested in Fig. 14-4*A*.

Arms have the form of biserial brachioles and are pinnulate in a few cases. They may be rather delicate structures on the one hand or robust on the other, and in either case they bear exothecal extensions of the food grooves.

Stratigraphic Range and Classification

The earliest known cystoids, as the class is here delimited, are of Middle Ordovician age. The Hydrophoridea were extinct by the Middle Devonian, whereas the more highly specialized Blastoidea persisted into the early part of the Permian.

Cystoids vary greatly in the morphology, structure, and symmetry of skeletal parts, but they share two features that make them unique among the Pelmatozoa—thecal pores and biserial brachioles. They are subdivided into two[1] subclasses and five orders as follows, largely on the basis of differences in the thecal pores:

[1] In the first edition of the present work the following divisions were recognized:

Class Cystoidea.
  Order *Amphoridea*.
  Order Rhombifera.
  Order Diploporita.
  Order *Aporita*.
Class Blastoidea.

Specialists no longer recognize the two orders in italics, referring the genera formerly included in them to other divisions, and they prefer considering the Blastoidea as a subclass of Cystoidea rather than as a separate class.

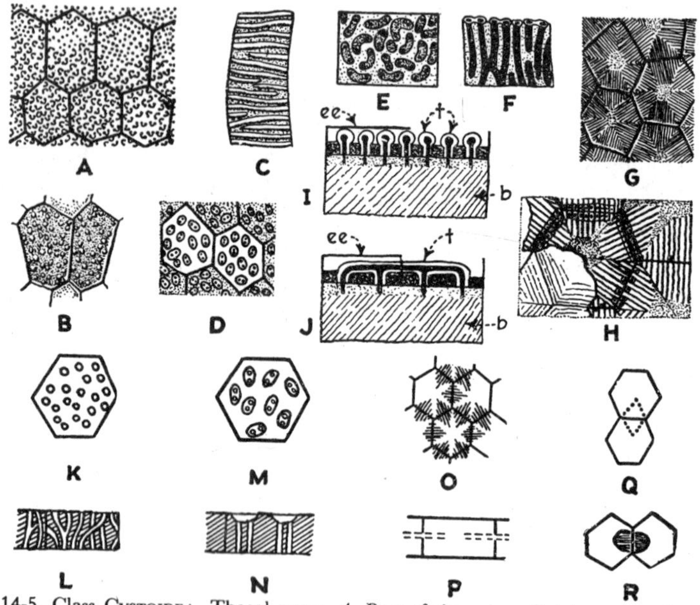

FIG. 14-5. Class CYSTOIDEA. Thecal pores. *A*. Part of the calyx of *Aristocystites*, showing plates perforated by single and horseshoe-shaped pores. *B–C*. *Aristocystites bohemicus* from the Ordovician of Bohemia: *B*, inner surface of two calyx plates, showing simple pores; *C*, vertical section of part of a plate, showing transverse tubes which terminate at both ends in pores. *D*. *Craterina bohemica*, from the Ordovician of Bohemia; plates with external epidermis removed to show diplopores. *E–F*. *Aristocystites bohemicus*: *E*, part of a plate bearing diplopores; *F*, vertical section of part of a plate showing relation of pores and tubes. *G–J*. *Echinosphaerites infaustus* from the Ordovician of Bohemia: *G*, a few plates showing pore rhombs, with tiny surface canals; *H*, plates with external epidermis partly removed to show tubes of pore rhombs; *I*, section transverse to tubes and parallel with the edge of the plate showing parallel tubes (*t*) communicating internally with the body; the external epidermis (*ee*) is shown in the left half of the diagram; *J*, section along the longest tube of the pore rhomb, showing tube crossing suture between adjacent plates and communicating with interior. The left half of the diagram has the external epidermis in place. *K–L*. Haplopores. *M–N*. Diplopores. *O–R*. Pore rhombs: *O–P*, striated; *Q*, porous; *R*, pectinirhomb. (*A–J after Barrande*, 1887; *K–R after Chauvel*, 1941.)

Class Cystoidea.
- Subclass Hydrophoridea.
  - Order 1. Diploporita.
  - Order 2. Rhombifera.
- Subclass Blastoidea.
  - Order 1. Eublastoidea.
  - Order 2. Coronata.
  - Order 3. Parablastoidea.

Subclass Hydrophoridea[1]

The Hydrophoridea include those cystoids that have numerous, small irregular plates arranged with little semblance of definite symmetry and containing all types of thecal pores except hydrospires.

The simplest thecal pores, designated **haplopores** (Fig. 14-5*K–L*), are tiny

[1] Hydrophoridea—Gr. *hydor*, water, + *phoros*, bearing; referring to the development of the thecal pores supposedly for circulation of water fluids.

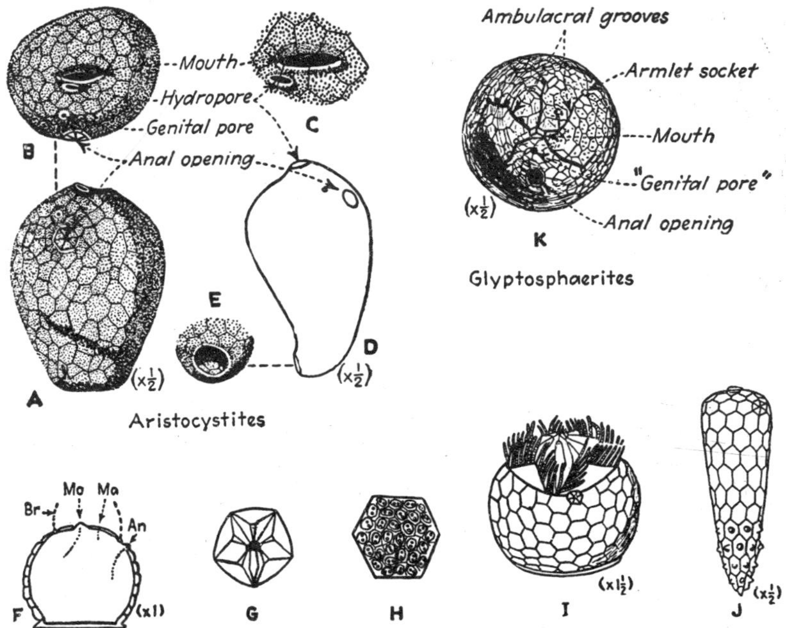

FIG. 14-6. Class CYSTOIDEA—Order Diploporita. *A–E. Aristocystites bohemicus* from the Ordovician of Bohemia: *A*, an almost complete specimen with only the basal termination missing; the anal opening is closed by a valvular pyramid composed of numerous triangular plates; *B*, top view of *A*, showing the four openings into the slightly deformed test; *C*, enlarged view of the mouth region, showing the nature and arrangement of the perforated plates; *D*, outline of a complete specimen with three of the four thecal openings visible; *E*, enlarged view of basal part of test, showing the prominent concavity. *F–I. Tholocystis kolihai* from the Middle Ordovician of Europe: *F*, section through theca, showing position of mouth (*Mo*), madreporite (*Ma*), anus (*An*), and brachioles (*Br*); parts of the alimentary tract and water-vascular system are indicated by dotted lines; *G*, diagram showing the thecal plates that are visible in an oral view; the peristome is shown in black; *H*, a thecal plate, enlarged several times, showing numerous diplopores; *I*, reconstruction of complete test. *J*. Reconstruction of *Calix sedgwicki* from the Middle Ordovician of Bohemia. *K*. *Glyptosphaerites leuchtenbergi* from the Ordovician of Russia. Upper surface of test showing the five ambulacral grooves, the mouth covered by oral plates, the anal opening without the covering plates, the supposed genital pore, and the places of attachment of the armlets. (*A–E after Barrande*, 1887; *F–J after Chauvel*, 1941; *K after Nicholson and Lydekker*, 1889.)

canals that traverse the plate at right angles to its planar dimension. If these are connected in pairs, still perpendicular to the surface, they are designated **diplopores** (Fig. 14-5*E–F,M–N*). In a more complicated arrangement they lie parallel to the greater surface of the plate and perpendicular to the plane of the suture (Fig. 14-5). Those crossing any one suture, from one plate to the next, occupy a rhombic area bisected by the suture line, and, since, in weathered specimens, there seem to be pores at the ends of these canals, the rhombic areas have been designated **pore rhombs** (Fig. 14-5*G–H,O–R*). A highly specialized type, the **pectinate rhomb,** now generally designated **pectinirhomb,** developed from folding in the plate. Inasmuch as haplopores and diplopores are characteristic of cystoids having grooves on the surface of the test, whereas pore rhombs and pectinirhombs are confined to those in which the food grooves are extended onto brachioles or arms, two orders may be recognized: Diploporita and Rhombifera, respectively.

Order 1. Diploporita. These are Hydrophoridea characterized by diplopores; thecal canals seldom traverse the sutures between plates and tend to be restricted to certain areas or plates. Radial symmetry affects the five food grooves and to some extent the calyx plates around the mouth. The food grooves are epithecal and exothecal on the brachioles lining the ambulacra. Included in the order is *Aristocystites* (Fig. 14-6), from the Ordovician of Bohemia, which seems to have been attached directly to the bottom by its aboral surface, as no stem is known. Among other genera assigned to the order, which ranges from Middle Ordovician to Middle Devonian, are *Calix*, *Craterina*, and *Glyptosphaerites* (Ord.; Fig. 14-6); *Sphaeronites*, and *Tholocystis* (Ord.; Fig. 14-6), *Megacystites* (Sil.); and *Carpocystites* (Dev.).

Order 2. Rhombifera. The Rhombifera, as the name implies, have calyx plates that bear pore rhombs and pectinirhombs. The food grooves exhibit radial symmetry and are borne largely on jointed biserial brachioles. Most of the forms seem to have had stems. Typical of the order is the Silurian genus *Caryocrinites* (Fig. 14-7), which has a globular test in which the plates are arranged in cycles with hexamerous symmetry. There are 6 to 13 arms, dis-

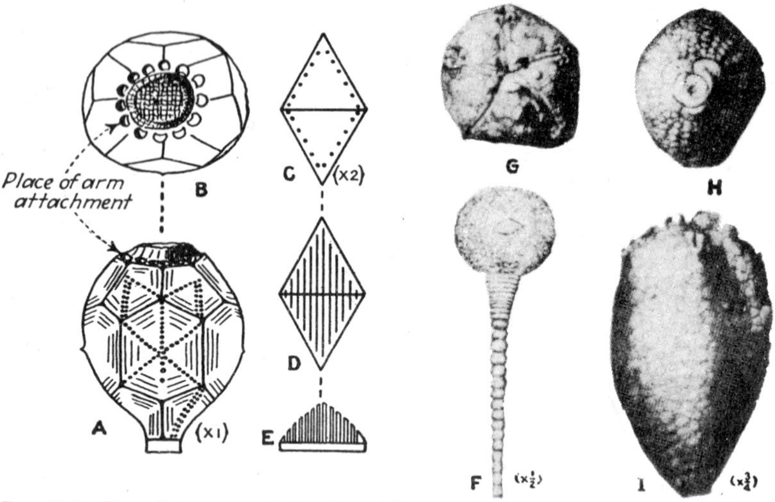

FIG. 14-7. Class CYSTOIDEA—Order Rhombifera. A–E. *Caryocrinites ornatus*, a common Silurian cystoid: *A*, a complete calyx showing nature and arrangement of plates, distribution of pore rhombs, and places where the arms were attached; *B*, top view of *A*; *C*, surface view of a pore rhomb showing the pores; *D*, section parallel to the surface of the plate showing arrangement of the internal tubes; *E*, transverse section of *D*. *F*. *Strobilocystites calvini*, one of the last of the cystoids, from the Devonian of Iowa. *G–I*. *Eumorphocystis multiporata*, from the Ordovician of Oklahoma: *G*, summit view; *H*, basal view; *I*, posterior view. (*C–E after Wachsmuth and Springer*, 1879; *F after Stainbrook*, 1941; *G–I after Branson and Peck*, 1940.)

posed in a circle around the mouth, which, together with the food grooves, is endothecal. The anus is on the outer margin of the ventral surface and has a valvular pyramid. There is a long stem composed of cylindrical segments.

Among genera assigned to the order, which ranges from Middle Ordovician into the Devonian, are *Echinosphaerites* (Ord.; Fig. 14-5*G–J*), *Pleurocystites* (Ord.), *Eumorphocystis* (Ord.; Fig. 14-7*G–I*), and *Glyptocystites* (Ord.); *Callocystites* (Sil.); and *Strobilocystites* (Dev.· Fig. 14-7*F*).

Subclass Blastoidea[1]

The blastoids are extinct, short-stemmed or stemless pelmatozoans with a rigid budlike theca in which the plates have a definite number and arrangement. Five (or rarely four) epithecal food grooves, which in most forms lie on a lancet-shaped plate, radiate symmetrically from a central peristome between five interradial plates, the **deltoids** (Fig. 14-9), and are margined by alternating **side plates** bearing brachioles onto which extend branches of the main food groove. The grooves and peristome are protected by small cover plates, which can open over the grooves; hence in life there was a continuous roof over the entire convective system. The calyx plates, however irregular in some species, are always arranged in at least three cycles, which, ascending from the top plate of the stem, are designated **basals, radials,** and **deltoids** (Fig. 14-9*B*,*G*). In many species, however, other plates are included in the calyx, with the result that the entire calyx, exclusive of ambulacral elements, may contain a hundred or more plates.

The slender uniserial brachioles, in some cases provided with delicate pinnules, were, and in a few fossil specimens still are, attached along the lateral margins of the five main food grooves in single or double rows. True arms, such as those in the Crinoidea, were never developed.

Two distinct lines of development are evident in the Blastoidea as a whole. First is a strong tendency toward greater perfection of pentameral symmetry, accompanied by decrease in number of calyx plates and a more and more definite arrangement of the plates. Second, is the concentration of semiporous structures into the interradial areas, and from this the development of an elaborate system of endothecal folds, the **hydrospires,** which hang into the thecal cavity (Fig. 14-9*E–F*).

The earliest known blastoids are from the Ordovician and the last come from Permian rocks. They were particularly abundant during the Mississippian and are so numerous in certain limestones that they give the name to the rock (*e.g.*, Pentremital limestone, after the genus *Pentremites*). The most primitive blastoids are not greatly different in some important respects from certain hydrophoridean Diploporita, and this resemblance has led some specialists to suggest that the ancestors of the Blastoidea are to be found among members of the Hydrophoridea. However, the most recent researches on the ancestry of the Blastoidea suggest that they were derived from an echinoderm stock that already in the Ordovician had only three plates in the basal series, for all adult blastoids have but three basals, and only three

[1] Blastoidea—Gr. *blastos*, bud, + *-oid*, like; referring to the resemblance of a blastoid calyx to a flower bud.

are present in the earliest growth stages of known blastoid larvae (*Mesoblastus* and *Pentremites*, both Upper Miss.; Figs. 14-8, 14-9). (See further discussion under Class Eocrinoidea.)

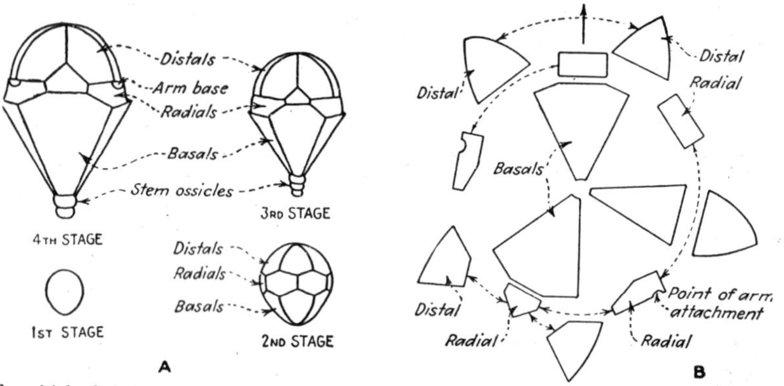

FIG. 14-8. Subclass BLASTOIDEA. Early growth stages of *Mesoblastus glaber*, an Upper Mississippian eublastoid from southern Illinois. *A*. Four stages in the larval development, much enlarged, shown diagrammatically. *B*. Diagrammatic plate analysis of the fourth stage of larval development (cf. *A*). (*After Croneis and Geis*, 1940.)

Order 1. Eublastoidea. The Order Eublastoidea, consisting of those pelmatozoans ordinarily designated **blastoids,** is considered first so that the other two orders of Blastoidea may be compared with it later. The order can be excellently characterized by describing in some detail the well-known Mississippian genus *Pentremites* (Fig. 14-9).

The complete skeleton consists of the **calyx** and its brachioles, the column or stem, and the root system by which it was attached to the substratum (Fig. 14-9*A*). The calyx is ovoidal and the five food grooves radiate from the central mouth with almost perfect radial symmetry. The 13 primary calyx plates are arranged in three cycles—3 basals, 5 radials, and 5 deltoids. Two of the basal plates are of equal size and somewhat larger than the third; the three encircle the top plate of the column. The five large and deeply incised radials are also designated **forked plates** because they have two oral extensions that enclose the ambulacrum. In many blastoids the radials are the largest and most conspicuous calyx plates. The five interradial plates—or deltoids, as they are designated because their triangular outline suggests the Greek letter delta (Δ)—constitute the third cycle of the calyx and lie near the summit in notches at the oral junctions of the radials. In the two blastoids whose ontogeny is relatively well known, the first plate to appear in the interradial position (the "distals" in Fig. 14-8) is a truly larval structure, for it is resorbed or shed at the metamorphosis of the individual, and only later in the development do the deltoids appear. The deltoid cycle is interrupted by the ambulacral areas. In *Pentremites* the deltoids are quite small, but in other blastoids they may be relatively larger.

The conspicuous petaloid ambulacra radiate from the central mouth with almost perfect pentameral symmetry and extend well down on the lateral

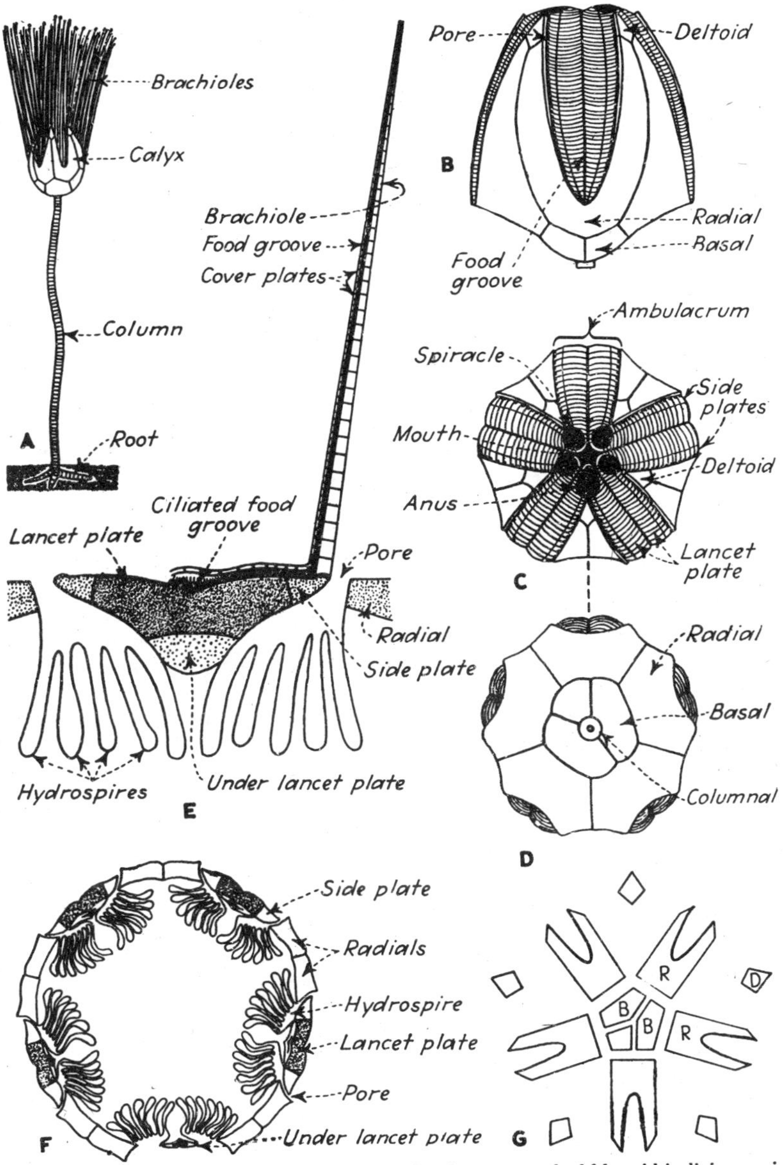

FIG. 14-9. Subclass BLASTOIDEA. *A.* Diagram showing an attached blastoid in living position. *B–G. Pentremites*, the common Mississippian blastoid: *B–D*, lateral, oral, and basal views, respectively, of the calyx; *E*, schematic transverse section of the mid-portion of an ambulacrum as it appeared in life, with a brachiole shown along one margin; *F*, cross section of calyx of *P. pyriformis* at about the mid-point of the ambulacra, with the plates of one ambulacrum removed; *G*, diagram to show analysis of the chief calyx plates (*B* = basal; *R* = radial; *D* = interradial or deltoid). (*F after Wachsmuth and Springer*, 1879.)

slopes of the calyx. Each ambulacrum is floored with a median **lancet plate** and two rows of **side plates,** as shown in Fig. 14-9*E*. In addition, an **under lancet plate** lies beneath the median part of the lancet plate. Blastoids typically have five ambulacra, although one may not be developed in a few forms, and the areas may be broad or narrow, long or short, pointed at one or both ends, and raised above, level with, or depressed below the surface of the calyx.

The main food groove (also designated **ambulacral groove**) lies along the median line of the lancet plate, and from it many minute grooves traverse the two parts of the plate more or less perpendicularly to the margin of the ambulacral area where they end on a side plate near a small dome to which a brachiole was attached. From these points the grooves became exothecal and continued outward along the brachioles. Both ambulacral areas and the grooves on the brachioles were covered with tiny alternating cover plates, so that the entire convective system was roofed over in life. A food particle acquired at the end of a brachiole was swept down the ciliated groove to the base of the brachiole, thence transversely across the ambulacral area to the median food groove, and then on to the mouth on the summit of the calyx—all the way under cover.

Pores of two kinds in the ambulacral areas lead to the internal hydrospires. A single row of small pores lies along each margin of each ambulacrum outside the brachioles (Fig. 14-9*E*). These openings lie between the side plates and the radials in *Pentremites;* in certain other blastoids a second series of plates, the **outer side plates,** is interposed between the regular side plates and the radials, and in these forms the porelike openings lie between the outer side plates and radials. It is important to emphasize here that these marginal openings, although loosely designated pores, are not comparable to the haplopores and diplopores of the Hydrophoridea, to the water pores of the Crinoidea, or to the ambulacral pores of the Echinoidea. Five conspicuous elliptical clefts, the **spiracles,** are arranged symmetrically around the mouth at the points of the deltoids (Fig. 14-9*C*). These represent the summit openings into the hydrospires. One of them, lying between the two posterior rays, is larger than the other four and is divided internally by two partitions. The middle one of the three divisions connects with the internal cavity only and is believed to have served as the anal opening; each of the other two divisions leads to a hydrospire. Each of the four equal-sized spiracles has a single partition, and each of the divisions likewise leads to a hydrospire. There are, therefore, 10 separate hydrospires hanging into the calyx and communicating to the exterior through the marginal pores and the spiracles.

The hydrospire is a structure that is unique to the Blastoidea, and it merits a brief description because it is believed to have served a respiratory function. Viewed as a whole it is a much folded elongate vessel extending into the thecal cavity and hanging suspended from the summit and internal surface of the calyx. It is completely closed except for a large opening in the spiracle at its upper end and the numerous openings in the canal that lies between the side plate and radials along the margin of the ambulacral area.

The structure was first designated a hydrospire because an investigator inferred that in the folded part, because of the great surface area provided by the numerous folds, the coelomic fluids could most efficiently come into osmotic contact with oxygenated sea water that lay on the other side of the hydrospire wall. It is now generally believed that when the blastoid was alive water was admitted into the hydrospire through the marginal openings and expelled through the spiracles. In its passage the oxygenated water presumably bathed the hydrospires so that the whole hydrospire system performed a respiratory function. In *Pentremites*, which is a highly specialized eublastoid, almost all the hydrospire lies inside the calyx wall, but in less specialized genera (*e.g.*, *Codaster*), it is formed by the folding of the thecal wall between the radial plate and the plates of the ambulacrum (Fig. 14-10). Both the

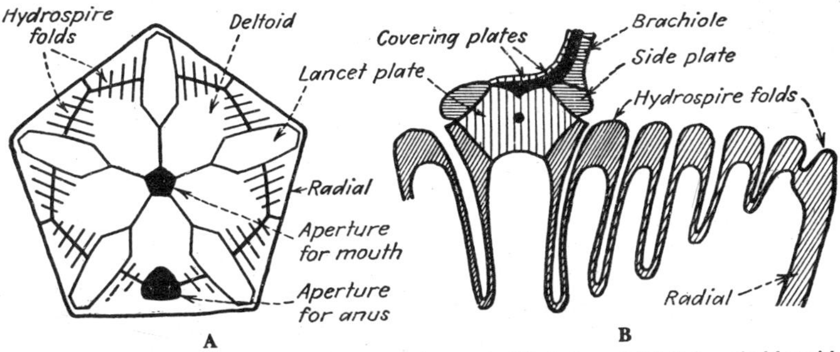

FIG. 14-10. Order Eublastoidea. Structural features of *Codaster*, a Mid-Paleozoic blastoid with simple hydrospire structure. *A*. Cross section of theca a short distance below the oral surface, showing sutures between thecal plates and positions of hydrospire slits. *B*. Slightly restored section across part of a radius. (*Diagrammatic and adapted from Bather*, 1900.)

structure of the hydrospire and the morphology and arrangement of the ambulacral plates are quite diverse and are used in differentiating subordinal divisions.

The living eublastoid quite probably resembled a small brush, held with the bristles up, but it was rarely completely preserved (Fig. 14-9*A*). The slender brachioles and delicate pinnules are not commonly preserved as they existed in life because of the ease with which they and also the cover of the convective system fell apart into the constituent plates when the animal died. Likewise, the complete stem and root system are seldom preserved, but the separated plates are common in many limestones and calcareous shales. It is the calyx that was most often preserved, and because these are both abundant and objects of considerable beauty, they are among the better known fossils of Mississippian rocks.

The Eublastoidea range from Silurian into Lower Permian. *Pentremites* (Fig. 14-9) excellently represents the order, though as previously stated it is a somewhat specialized form. Other typical genera are *Troostocrinus* (Sil.), *Codaster* (Sil.–Miss.; Fig. 14-10), *Nucleocrinus* (Dev.), *Mesoblastus* (Miss.; Fig. 14-8), and *Schizoblastus* (Miss.–Perm.).

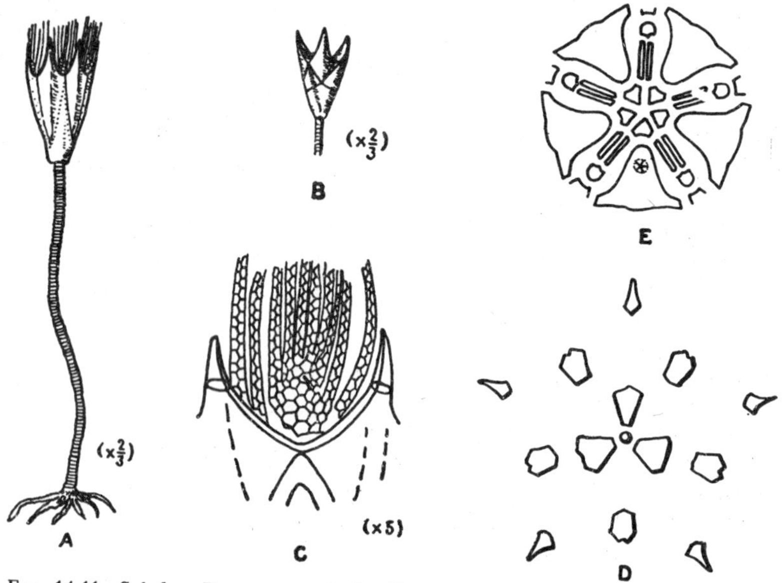

FIG. 14-11. Subclass BLASTOIDEA—Order Coronata. *Stephanocrinus angulatus* from the Middle Silurian of New York. *A*. Reconstruction of complete specimen. *B*. A calyx and part of the stem. *C*. A part of one ambulacral region, showing the structure and arrangement of plates in the brachioles. *D*. Plate analysis of the calyx. *E* Analysis of plates constituting the summit of the theca. (*After Hall*, 1852.)

Order 2. Coronata. This small extinct order is based largely on the following genera: *Tormoblastus* (Ord.) and *Stephanocrinus* (Ord.–Sil.; Fig. 14-11). Other genera are *Mespilocystites* (Ord.), *Stephanoblastus* (Sil.), and possibly *Paracystis* (Ord.). These genera are referred to the Blasto dea because of the nature and development of the skeletal elements, the radially pentamerous arrangement of the ambulacra, and the presence of delicate biserial brachioles. Because they seem to lack hydrospires, or have them almost completely atrophied, and to have the brachioles concentrated at the distal ends of the ambulacra, Jaekel (1918), Wanner (1924), Moore (1940) and Regnell (1945) have concluded that they should be grouped n a separate order of Blastoidea.

Stephanocrinus (Fig. 14-11) illustrates the essential features of the order, which is limited to the Ordovician and Silurian.

Order 3. Parablastoidea. This is based on the Early Ordovician genus *Blastoidocrinus*[1] (Fig. 14-12) which, although clearly a blastoid, differs in such important respects from all other members of that subclass that it has been

[1] *Blastoidocrinus* was formerly included ir Father's (1900) Protoblastoidea, but since it has hydrospires it must be excluded from that group which, by definition, lack "hydrospire folds hanging into the thecal cavity." Bather's Protoblastoidea as a group has now been so completely dismembered that it is no longer used by specialists (Regnell, 1945).

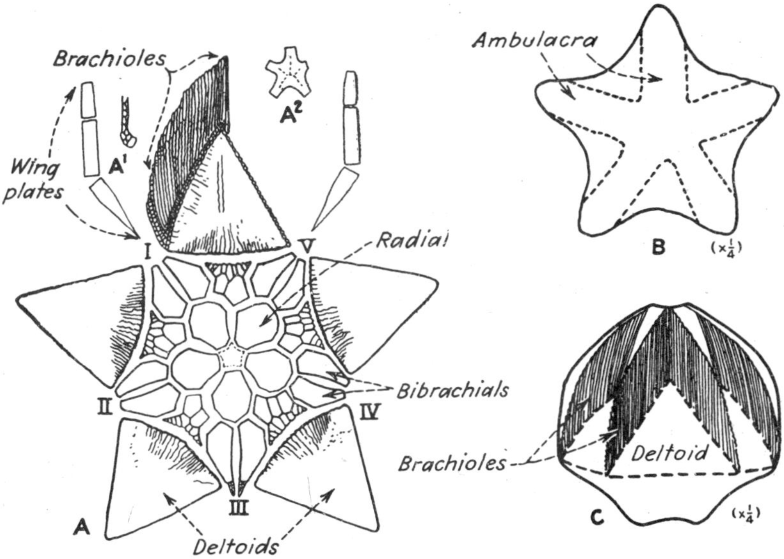

FIG. 14-12. Subclass BLASTOIDEA—Order Parablastoidea. *Blastoidocrinus* from the Ordovician (Chazy) of New York. *A*. Analysis of the theca. The brachioles are partly sketched in for one side of one of the large deltoids; the individual ossicles are omitted in all except the lower three. The wing plates of only two of the radii are shown. The detail of the base of a brachiole is shown in A^1 and the apical piece or central plate in A^2. *B*. Outline of oral view with the five ambulacra indicated by dashed lines. *C*. Reconstructed side view, showing one complete deltoid with brachioles and parts of two other deltoids. (*Adapted from Hudson*, 1907.)

made the basis of a separate order, the Parablastoidea (Hudson, 1907), which may be characterized briefly as follows.

The complete calyx, or theca, consists of several hundred plates and is more or less clearly separable into an aboral and an oral part. The aboral part consists of about 80 plates that are arranged in three or more circlets, the first two of which are basals and radials, as in crinoids. The basals are unknown, but there are five radials, as indicated in Fig. 14-12. The third cycle consists of five pairs of brachial plates (the **bibrachials**), and 13 interbrachials. The plates of the fourth circlet vary between 50 and 60 in number, and though they are horizontally disposed, their continuity as a circlet is broken at each radius by the bibrachials. In the oral part the most conspicuous plates are the five great triangular deltoids, each of which occupies an entire interambulacral area. The five ambulacra lack the lancet plate of the Eublastoidea, but have adambulacrals meeting under the food grooves. They also have cover plates and numerous brachioles. Hydrospires cross one or more interambulacral plates, and it is believed that water flowed down the brachioles into a series of brachiolar cavities from which it moved into the hydrospires through pores passing between the adambulacrals. *Ordovician.*

Blastoidocrinus (Fig. 14-12) is the only genus that is definitely assigned to

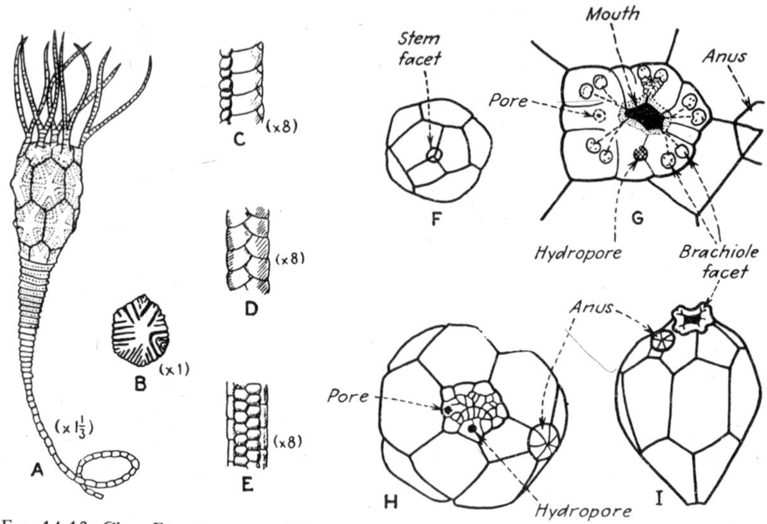

FIG. 14-13. Class EOCRINOIDEA. *A–E. Macrocystella mariae* from the Cambrian of England: *A*, side view of complete specimen; *B*, single thecal plate of a large specimen; *C–E*, side, dorsal, and ventral views, respectively, of a small part of a brachiole. *F–I. Cryptocrinus: F*, aboral view showing place where stem was attached and the three plates of the first circlet; *G*, enlarged view of oral region with the tegminal plates removed; *H*, oral view; *I*, side view. (*Adapted from Bather*, 1900.)

the order, but *Blastocystis*, also from the Ordovician, may belong here (Regnell, 1945).

CLASS EOCRINOIDEA[1]

The Class Eocrinoidea consists of a small group of Cambrian and Ordovician pelmatozoans that have some characteristics of both cystoids and crinoids yet differ from these two classes in such important respects that specialists tend to consider them a separate class. They resemble the Cystoidea in the way the ambulacral grooves branch and in the ventrolateral location of the anus, but they differ in lacking the thecal pores (*i.e.*, hydrospires and diplopores) that are regarded as characteristic of the Cystoidea. They resemble the Crinoidea in lacking these pores and in the tricyclic arrangement of the plates of the calyx, but differ in having biserial brachioles—another cystoidean feature—whereas all crinoid brachia are primitively uniserial. It may well be that they constitute the ancestral stock of all pelmatozoans, as suggested by Regnell (1945) and favored by Moore (1948).

Macrocystella (Fig. 14-13), from the Cambrian, and *Cryptocrinus* (Fig. 14-13), from the Ordovician, typify the class, which ranges from Lower Cambrian to Middle Ordovician.

[1] Eocrinoidea—Gr. *eos*, dawn or early, + *crinoidea*; referring to the fact that members of the class were regarded as early and primitive crinoids.

Macrocystella has an ellipsoidal calyx in which the plates are arranged in three circlets of five each, and a fourth cycle of equal number bearing the brachioles. There are five branched brachioles and a rapidly tapering stem by which the animal may have attached itself to the bottom. The radiating folds of the calyx plates are strongly marked, but thecal pores are wanting.

Cryptocrinus (Fig. 14-13*F–I*) has a small irregularly spheroidal theca composed of four circlets of plates. The basal circlet has but three plates, supposedly the result of fusion of two pairs of the original five. The unfused plate is in the right anterior interradius (see Fig. 14-4). The second cycle consists of five large hexagonal plates which alternate with the five subpentagonal plates of the third circlet. The latter surround a pentagonal area in which lie the tiny plates of the fourth circlet, together with numerous small tegminal plates. Five main food grooves lead from the mouth to facets on the plates of the fourth circlet. Slender brachioles arose from the facets. Other less important features are shown in Fig. 14-13.

In shape, and in number and arrangement of its plates, the larval theca of the blastoid genus *Mesoblastus* somewhat resembles *Cryptocrinus* and, according to Croneis and Geis (1940), does not lose this resemblance entirely until the end of larval development. This similarity has led some investigators to suggest that possibly the Blastoidea arose from some eocrinoid like *Cryptocrinus* which even as early as the Ordovician had only three basal plates.

Class Paracrinoidea[1]

The extinct Class Paracrinoidea, which is limited to the Middle Ordovician, embraces a group of three much-discussed genera—*Comarocystites*, *Amygdalocystites*, and *Canadocystis*—and may also include the genus *Wellerocystis*. These genera collectively have the following characteristics: (1) the plates of the test are not arranged in cycles, as in other pelmatozoans, and show no differentiation into calicinal and tegminal parts, as in the Crinoidea; (2) the appendages are developed as free or recumbent uniserial brachia bearing uniserial pinnules—a crinoid feature; and (3) a system of subepithecal pores is present in typical forms. As with the Eocrinoidea, the Paracrinoidea have some features of both cystoids and crinoids but cannot be assigned to either group; hence it has been proposed that they be recognized as a separate and extinct class (Regnell, 1945), and that course is followed here.

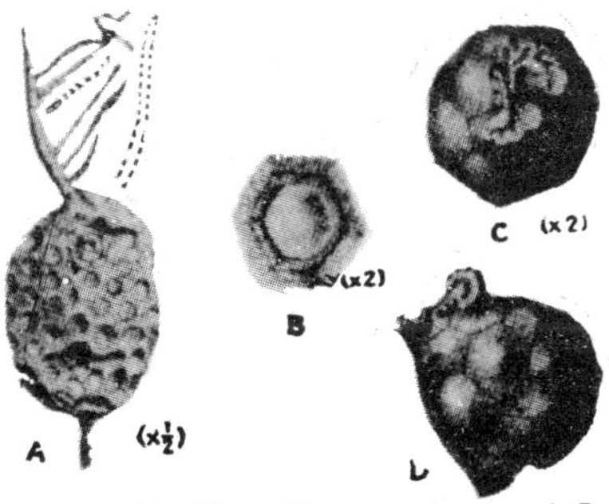

Fig. 14-14. Class Paracrinoidea. *A–B*. *Comarocystites punctatus* from the Ordovician: *A*, anterior view of a large theca with one pinnulated arm partly preserved; *B*, a thecal plate somewhat enlarged. *C–D*. *Canadocystis emmonsi* from the New York Ordovician: *C*, side view of the type; *D*, summit view. (*A–B after Billings*, 1858; *C–D after Hudson*, 1905.)

[1] Paracrinoidea—Gr. *para*, beside or alongside, + *crinoidea*; referring to the supposed relationships of the class with the Crinoidea.

Comarocystites (Fig. 14-14*A–B*) and *Canadocystis* (Fig. 14-14*C–D*) supposedly typify this recently erected class.

CLASS CRINOIDEA[1]

The Crinoidea include the sea lilies and feather stars of existing seas and the so-called stone lilies of the past (Fig. 14-15). Although there has long been a widely held opinion that living crinoids are few in variety and only a mere vestige of a class that was much more important during the past,

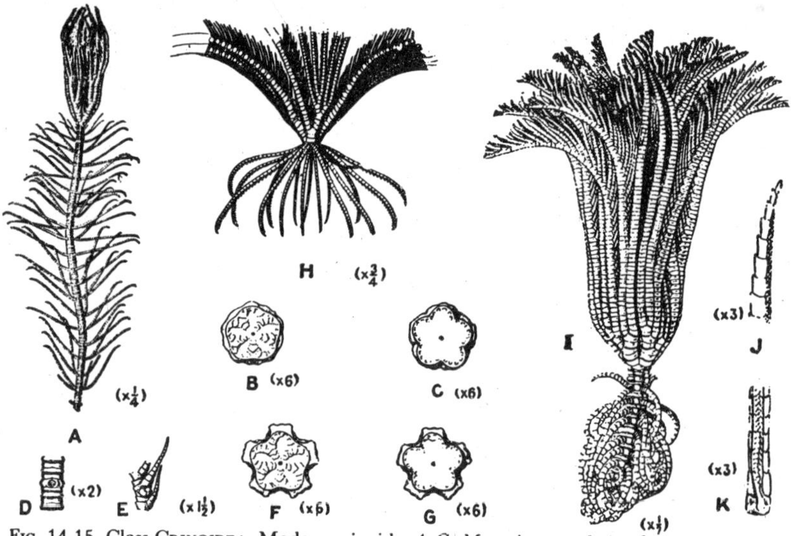

FIG. 14-15. Class CRINOIDEA. Modern crinoids. *A–G. Metacrinus moseleyi*, a form with a long cirrate stem: *A*, a young individual; *B–C*, an internodal and an infranodal joint; *D*, portion of the stem; *E*, portion of an arm; *F–G*, upper and lower faces of a nodal joint. *H, Antedon elegans*, a free-swimming form with a circlet of long cirri around the base. *I–K. Pentacrinus maclearanus*, a form with a short highly cirrate stem: *I*, a complete specimen; *J*, the end of a pinnule from the side; *K*, ventral view of basal part of a pinnule, showing the development of its ambulacral skeleton. (*After Carpenter*, 1884 *and* 1888.)

the facts show a somewhat different situation, as pointed out by Clark (1915) in his monograph on the free-swimming *Comatulids*. The *Challenger* dredged as many as 10,000 unstalked individuals in a single haul, and during the cruise of the *Albatress* in 1906, Clark (1915) reports that he handled tens of thousands of specimens and several times saw the forward deck of the steamer literally buried under several tons of individuals belonging to a species larger than any known fossil form. Crinoids are widely distributed in existing seas, having been found in many places between 52° S. lat. and 81° N. lat., and the known fossil localities indicate a similar world-wide distribution during the geologic past. Living crinoids are gregarious animals tending to live in groups or "gardens" of immense populations, as indicated by the haul of

[1] Crinoidea—Gr. *krinon*, lily, + *-oid*, like; referring to the lilylike appearance of the skeleton.

the *Challenger*, and they also seem to prefer relatively clear waters. They are commonly abundant in reef environments. The sporadic distribution of fossil crinoid localities, with great numbers of individuals concentrated in local areas in bioherms and biostromes, indicates that ancient forms were also gregarious and that they flourished in reef environments.

The skeletal structures of the Crinoidea show a great range in complexity, and this makes them exceptionally suitable for study of evolutionary changes (Moore, 1952) and as stratigraphic indicators. Moore (1948) summarized this aspect of the class and in his work mentioned as one disadvantageous feature the prevailing lack of information concerning the ontogenetic development of fossil crinoids. Fortunately, however, something is known about the ontogeny of certain living crinoids, and a general knowledge of the essential aspects of the class may be obtained by a brief review of the life history of one of these, *Antedon* Fig. 14-16, as worked out by Seeliger (1892) and others and outlined by Swinnerton (1947).

The Life History of Antedon, a Modern Crinoid

The adult *Antedon* has a hemispheroidal skeleton composed of numerous calcareous plates arranged in three circlets with pentamerous symmetry. The flat ventral surface is uppermost and has the mouth in the center, hence is designated **oral;** the convex or dorsal side is down and is designated **aboral.** As a matter of fact, it is more accurate to use the terms oral and aboral, a practice followed in the present work, because the parts so designated actually represent, respectively, the left and right sides of the larva, not the ventral and dorsal. From the circular line along which the oral and aboral surfaces join, there arise five and then, by bifurcation of these, 10 long slender arms that consist of a single series of alternating plates, each of which typically bears a small arm, or pinnule. The oral side of the animal is covered by a soft tough integument, the **tegmen,** which has no calcareous plates (in many ancient crinoids the tegmen was a complex of plates). The mouth lies at the center of the oral surface, and from it radiate five food grooves which branch near the periphery of the peristomial area, with each branch leading onto one of the 10 arms. On the arm the groove lies in a channel on the upper side and gives off short branchlets to the alternating pinnules. From the margin of a **centrodorsal** plate at the center of the aboral surface are given off 20 to 30 unbranched armlike appendages, the **cirri,** which have a clawlike terminal plate and serve to attach the animal to some object. At times, as during the breeding season, the animal may release its hold and swim away by gently moving its pinnulate arms.

The eggs of *Antedon* are produced by generative organs residing in the arms and reach the exterior through the ambulacral grooves. They are fertilized externally while attached to the pinnules or the arms. Soon after fertilization the egg membrane bursts and an ovoidal, bilaterally symmetrical, gastrula-like larva is released. This larva, more or less like the larva of other echinoderms at this stage, is supposedly similar to the hypothetical ancestor, the *Dipleurula* (Fig. 14-55), of all the Echinoderma. Eventually, the larva attaches itself to some object by the anterior end and continues to

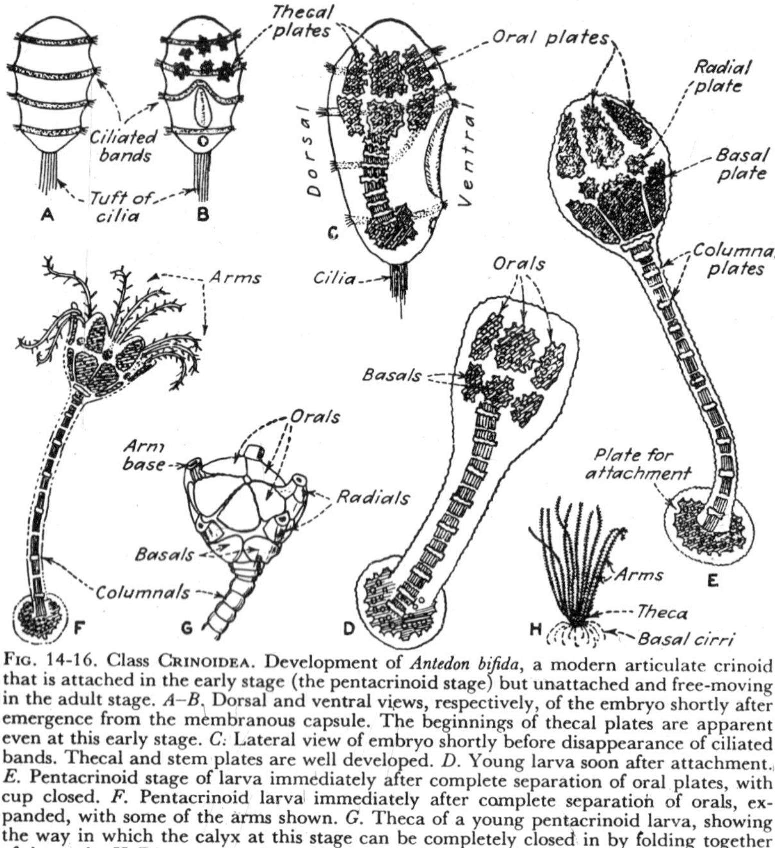

FIG. 14-16. Class CRINOIDEA. Development of *Antedon bifida*, a modern articulate crinoid that is attached in the early stage (the pentacrinoid stage) but unattached and free-moving in the adult stage. *A–B*. Dorsal and ventral views, respectively, of the embryo shortly after emergence from the membranous capsule. The beginnings of thecal plates are apparent even at this early stage. *C*. Lateral view of embryo shortly before disappearance of ciliated bands. Thecal and stem plates are well developed. *D*. Young larva soon after attachment. *E*. Pentacrinoid stage of larva immediately after complete separation of oral plates, with cup closed. *F*. Pentacrinoid larva immediately after complete separation of orals, expanded, with some of the arms shown. *G*. Theca of a young pentacrinoid larva, showing the way in which the calyx at this stage can be completely closed in by folding together of the orals. *H*. Diagram based on an adult specimen, which, it is to be noted, is stemless and provided with basal cirri. (*Adapted from A. H. Clark*, 1921, *chiefly after W. Thomson*, 1865, *and W. B. Carpenter*, 1866.)

undergo change as it develops into an adult. The anterior end becomes drawn out into a long stem, and the organs of the body shift in position until the mouth, anus, and hydropore lie in a vertical plane on the summit of the pear-shaped body. During this growth period the calcareous plates of the skeleton appear as two sets. One set, the columnals, makes a single series of 11 plates along the axis of the stem. The second set makes an ellipsoidal **theca,** which encloses the vital organs. The plates of this theca are arranged in three cycles of five each—the **orals** surrounding the mouth; the **basals** surrounding the main part of the vital organs; and the **infrabasals** lying above the upper end of the stem. Later, a centrodorsal plate develops in the center of the circlet of infrabasals, but not until all columnals have been formed. Meanwhile, a circlet of six plates appears between the orals and basals, and these grow until they are as large as the basals. Five of these new plates, the **radials,** alternate with the basals, and from each grows a single series of small plates, the **brachials,** that ultimately develop into the proximal parts of the five pairs of arms. The sixth, or **anal**, plate, which is smaller than the radials and is associated with the anus, marks the posterior interradius. The three circlets of plates—radials, basals, and infrabasals—constitute the **patina,** which, as new plates are added, becomes the dorsal cup of the adult crinoid.

At this stage the young *Antedon* has all the essential features of a typical crinoid—stem, theca (a patina, which later becomes the calyx—dorsal cup and tegmen), and a system of arms. With further growth the orals remain small and near the mouth, the anal plate rises out of the circlet of radials and ultimately degenerates and disappears as a terminally perforated **anal tube** grows above the tegmen, and the arms and stem lengthen. After two arm plates, the **primibrachs,** have been added, the arm bifurcates, and all additional plates, the **secundibrachs,** are added at the ends of the 10 arms. It is seen, therefore, that the 10 slender arms are actually five paired arms. As new plates are added to the distal extremities of the arms, pinnules arise alternately on opposite sides of the arm on successive plates. New columnals are added between the stem and the patina until there are about two dozen; thereafter plate formation ceases, and the last columnal to be added (the **proximale**) increases in size, becomes convex downward and concave upward, first covers the infrabasals and ultimately the basals also, and may even extend over the radials. Thus much of the adult calyx consists of this greatly enlarged plate. Meanwhile, the apical components fuse with i , and if cirri develop around its margin, it is designated the **centrodorsal** plate. Soon after the stem ceases to grow, the calyx, together with the arms and aboral cirri, breaks off below the centrodorsal plate, and the animal thereafter is free-swimming. The stem is left behind to disintegrate, and this may be one reason why many ancient crinoidal limestones are composed of columnals only.

During the development of the arms and stem, there is a simultaneous modification of the peristomial part of the theca, in addition to the transformation of the patina into the calyx. The food grooves are incorporated in an expanded tegmen, the enlarged interradial areas both orally and suborally are filled in with integument, and the adult stage is completed. Hereafter, the adult *Antedon* may be temporarily attached or free-swimming as it chooses.

Morphology

The complete crinoid consists of three distinct parts—the theca, or calyx, which encloses the main mass of the vital organs; the arms, which constitute the convective system over which food particles are carried to the peristomial part of the theca; and the stem or system of cirri by which the theca is attached to some object or to the bottom (Fig. 14-17*A*).

The **calyx** consists of the **dorsal cup,** which is composed of several cycles of calcareous plates, and its oral covering, the **tegmen,** which may be solely integumental or a combination of plates and integument. The calyx and surmounting arms constitute the **crown.** In attached crinoids, the stem, or

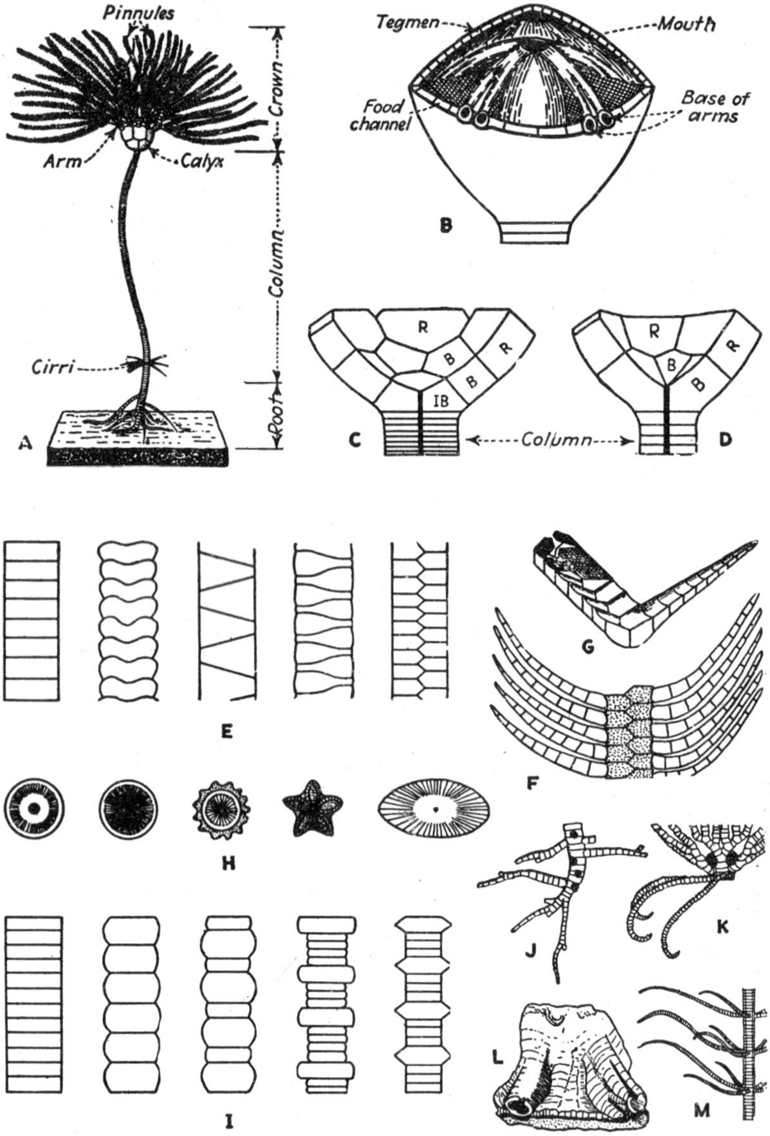

FIG. 14-17. Class CRINOIDEA. Diagrams showing variation in the structure of the different parts of a crinoid skeleton. (*See opposite page for further information.*)

column, supports the crown at its distal end and is anchored at the proximal end by a root system. Girdles of cirri on the stem may also aid in attachment. In forms that are intermittently attached, the stem is atrophied or abandoned, and cirri around the margin of the centrodorsal plate serve for temporary fixation.

The mass of vital organs lies in the dorsal cup and is protected from above by the tegmen. The mouth lies in the center of the oral surface and opens directly through the tegmen or is beneath it (subtegminal; Fig. 14-17*B*). Five ciliated food grooves radiate from the mouth to the bases of the arms on the periphery of the calyx, thence along the axial channel of each arm, its branches, and their pinnules to terminate at their distal extremities. Parts or all of the convective system of certain ancient crinoids may have tiny calcareous covering plates (Fig. 14-17*G*). Food particles, acquired throughout the feeding area, are urged along the food grooves by ciliary motion and finally reach the mouth, whence they pass downward into the U-shaped alimentary tract. The waste products are expelled from the anus or anal vent situated in the posterior interradius on the oral surface within the circle of arms. One or more **anal plates** may cover the opening. The anus of some species lies at the end of an anal tube that rises conspicuously above the oral surface.

The arms contain prolongations of the coelomic cavity in the form of generative organs, nerve fibers, vessels of the water-vascular system, and podia. These fleshy parts lie upon or within the calcareous pieces that make the arm. The interradial plates of some crinoids are perforated to admit osmotic contact of sea water with coelomic fluids, and the members of one group, the **Fistulata,** have a madreporite.

When the crinoid is alive, the entire skeleton is covered with integumental tissue, and it is believed that this was also true of ancient forms.

Crinoids, like most other echinoderms, display a marked radial symmetry (usually pentamerous, but rarely hexamerous) in the arrangement of certain soft parts and of most skeletal elements, but, again as in other echinoderms, they have a fundamental bilateral symmetry with respect to a plane that includes the anterior ray and bisects the posterior interradius. This plane passes through both mouth and anus as well as through the hydropore between (see Fig. 14-4).

The Crinoid Skeleton

Terminology (Figs. 14-17, 14-18, 14-19). If all soft tissue were stripped away from a crinoid, the remaining skeletal structure would consist of five groups of calcareous plates composing the anchoring system, the stem, the

FIG. 14-17. *A*. A complete crinoid skeleton in living position. *B*. Diagram of calyx with tegmen and subtegminal mouth. *C*. A dicyclic calyx (*IB* = infrabasal; *B* = basal; *R* = radial). *D*. A monocyclic calyx (cf. *C*). *E*. Uniserial and biserial arms. *F*. A biserial arm with pinnules. *G*. Diagram of arm and pinnule with a few covering plates shown. *H*. Columnals viewed along columnar axis. *I*. Columnals viewed from the side. *J*–*M*. Roots and cirri: *J*, root system of *Platycrinites; K*, cirri of *Neometra; L*, stumplike root of *?Barycrinus; M*, part of cirrate column of *Comastrocrinus*. (*A*–*J*, *L from several works of Wachsmuth and Springer, and of Springer alone; K, M after A. H. Clark,* 1921.)

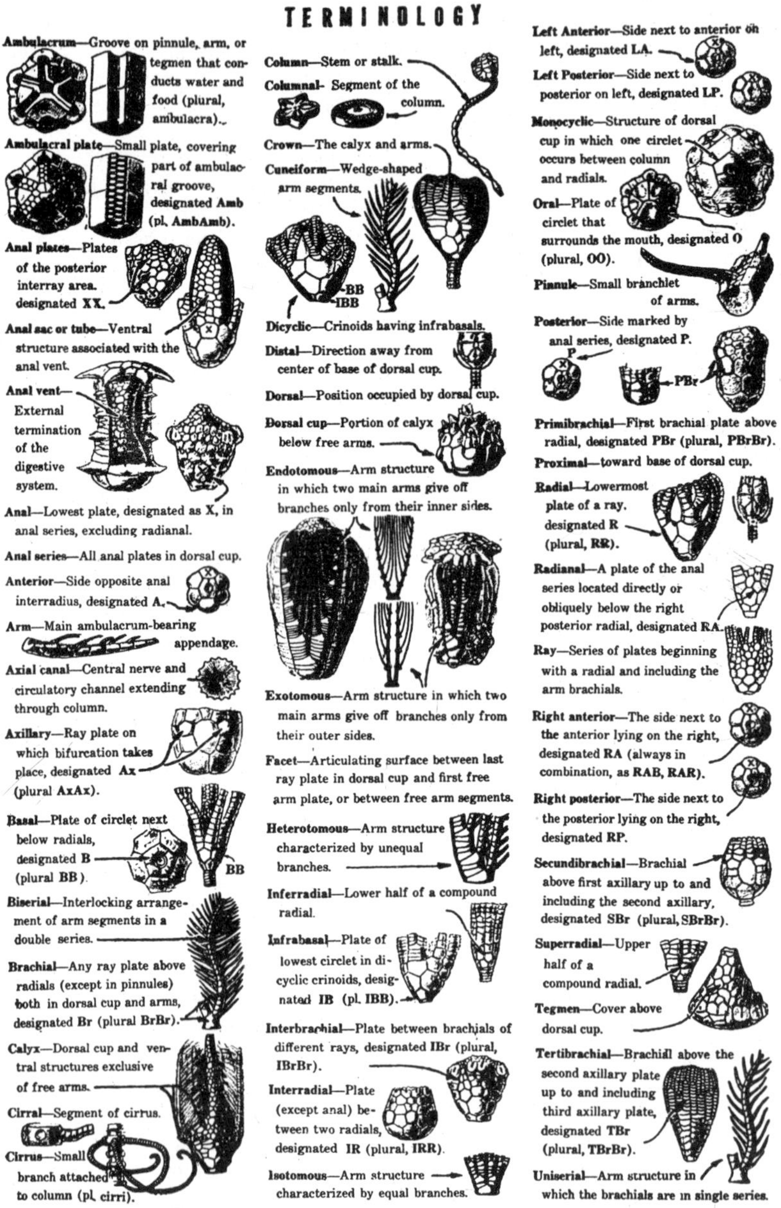
TERMINOLOGY
Ambulacrum—Groove on pinnule, arm, or tegmen that conducts water and food (plural, ambulacra).
Ambulacral plate—Small plate, covering part of ambulacral groove, designated Amb (pl. AmbAmb).
Anal plates—Plates of the posterior interray area. designated XX.
Anal sac or tube—Ventral structure associated with the anal vent.
Anal vent—External termination of the digestive system.
Anal—Lowest plate, designated as X, in anal series, excluding radianal.
Anal series—All anal plates in dorsal cup.
Anterior—Side opposite anal interradius, designated A.
Arm—Main ambulacrum-bearing appendage.
Axial canal—Central nerve and circulatory channel extending through column.
Axillary—Ray plate on which bifurcation takes place, designated Ax (plural AxAx).
Basal—Plate of circlet next below radials, designated B (plural BB).
BB
Biserial—Interlocking arrangement of arm segments in a double series.
Brachial—Any ray plate above radials (except in pinnules) both in dorsal cup and arms, designated Br (plural BrBr).
Calyx—Dorsal cup and ventral structures exclusive of free arms.
Cirral—Segment of cirrus.
Cirrus—Small branch attached to column (pl. cirri).
Column—Stem or stalk.
Columnal- Segment of the column.
Crown—The calyx and arms.
Cuneiform—Wedge-shaped arm segments.
BB
IBB
Dicyclic—Crinoids having infrabasals.
Distal—Direction away from center of base of dorsal cup.
Dorsal—Position occupied by dorsal cup.
Dorsal cup—Portion of calyx below free arms.
Endotomous—Arm structure in which two main arms give off branches only from their inner sides.
Exotomous—Arm structure in which two main arms give off branches only from their outer sides.
Facet—Articulating surface between last ray plate in dorsal cup and first free arm plate, or between free arm segments.
Heterotomous—Arm structure characterized by unequal branches.
Inferradial—Lower half of a compound radial.
Infrabasal—Plate of lowest circlet in dicyclic crinoids, designated IB (pl. IBB).
Interbrachial—Plate between brachials of different rays, designated IBr (plural, IBrBr).
Interradial—Plate (except anal) between two radials, designated IR (plural, IRR).
Isotomous—Arm structure characterized by equal branches.
Left Anterior—Side next to anterior on left, designated LA.
Left Posterior—Side next to posterior on left, designated LP.
Monocyclic—Structure of dorsal cup in which one circlet occurs between column and radials.
Oral—Plate of circlet that surrounds the mouth, designated O (plural, OO).
Pinnule—Small branchlet of arms.
Posterior—Side marked by anal series, designated P.
P
PBr
Primibrachial—First brachial plate above radial, designated PBr (plural, PBrBr).
Proximal—toward base of dorsal cup.
Radial—Lowermost plate of a ray, designated R (plural, RR).
Radianal—A plate of the anal series located directly or obliquely below the right posterior radial, designated RA.
Ray—Series of plates beginning with a radial and including the arm brachials.
Right anterior—The side next to the anterior lying on the right, designated RA (always in combination, as RAB, RAR).
Right posterior—The side next to the posterior lying on the right, designated RP.
Secundibrachial—Brachial above first axillary up to and including the second axillary, designated SBr (plural, SBrBr).
Superradial—Upper half of a compound radial.
Tegmen—Cover above dorsal cup.
Tertibrachial—Brachial above the second axillary plate up to and including third axillary plate, designated TBr (plural, TBrBr).
Uniserial—Arm structure in which the brachials are in single series.

FIG. 14-18. Class CRINOIDEA. Terminology of the plates, structural and architectural features, and morphology of crinoid skeletons. (*After Moore and Laudon*, 1944, *in "Index Fossils of North America."*)

calyx, the tegmen, and the arms (Figs. 14-17, 14-18). These skeletal parts range in number from a few tens in simple forms to many hundreds or even thousands in complex species, and although they can be grouped into the several categories just mentioned, they cannot be satisfactorily differentiated without the use of an elaborate terminology. In the present discussion technical terms are kept to a minimum, but anyone intending to study crinoids at all extensively will find it necessary to consult glossaries for definitions. For this purpose the reader is referred to recent discussions by Moore and Laudon (1941) and Moore (1948). The abbreviated terminology used in the present discussion is defined in Table 14-3 and Fig. 14-18 and illustrated in Figs. 14-18, 14-19, and 14-20.

The Root System (Figs. 14-17, 14-18). Most extinct crinoids seem to have been attached, at least throughout much of their life; in contrast, most modern forms are stemless and free-swimming in the adult stage, though, as illustrated by *Antedon* (Fig. 14-16), they can attach themselves if desired by means of the cirri encircling the centrodorsal plate.

In stemmed forms the anchoring device may be a spray of segmented **rootlets** at the bottom of the column (Fig. 14-17*J*), a stumplike spreading of the columnar base (Fig. 14-17*L*), a grapnel with several spurs, as in *Ancyrocrinus*, or some other modification of the columnar base (*Scyphocrinus* is supposed to have had a bulbous root or a float, which has been given the generic name *Camarocrinus*). Part or all of the column may also have girdles of flexible cirri (Fig. 14-17*M*), which can wrap themselves around objects. Root systems and other anchoring devices are commonly preserved[1] along with column and calyx, and in such cases leave no doubt as to how the crinoid was attached.

The Stem or Column. The typical crinoid stem is a more or less flexible column of variously shaped and sized calcareous plates generally arranged in a single series. Each plate originally has a small central perforation, and these perforations together form an axial canal. Most plates also have radial grooves for connective tissue. In the living crinoid, dermal tissue encloses the stem, a rod or tube of fleshy matter occupies the axial canal, and radial strands lie between the plates. This tissue holds the plates together, gives flexibility to the stem, and participates in the nutrition of the stem. Upon death of the animal, the tissue generally disappears rapidly, and the column is quite likely to fall apart to a greater or less extent. Some fossil crinoids have the entire stem and root system preserved, but fragmental stems and single columnals are much more typical.

The stem is short in all living stemmed crinoids, if it is retained at all

[1] Crinoid stems or rootlets have been found adherent to fragments of Devonian wood in such manner as to suggest that the living crinoid was attached to a piece of floating wood. See Wickwire, G. T., 1936, Crinoid stems on fossil wood, *Amer. Jour. Sci.*, 5th ser., vol. 32, pp. 145–146.

A Cincinnati physician, curious as to whether several species of Upper Ordovician crinoids were attached forms or floaters, cleaned off the shale down the stem. He discovered well-developed root systems and thus established that these forms were attached. The stems from calyx to roots ranged in length from ½ to 12½ in. (Dyche, D. T. D., 1892, Roots of crinoids from the Cincinnati group, *Amer. Geol.*, vol. 10, p. 130.)

after the larval stage, but in extinct forms it shows great range—from a few centimeters to more than 21 m. (70 ft.). In cross section the stem may be circular, elliptical, quadrangular, pentagonal, or polygonal. The columnal plates are usually thin as compared to their diameter and may all be of the same size in a stem, or smaller and larger plates may alternate (Fig. 14-17*I*).

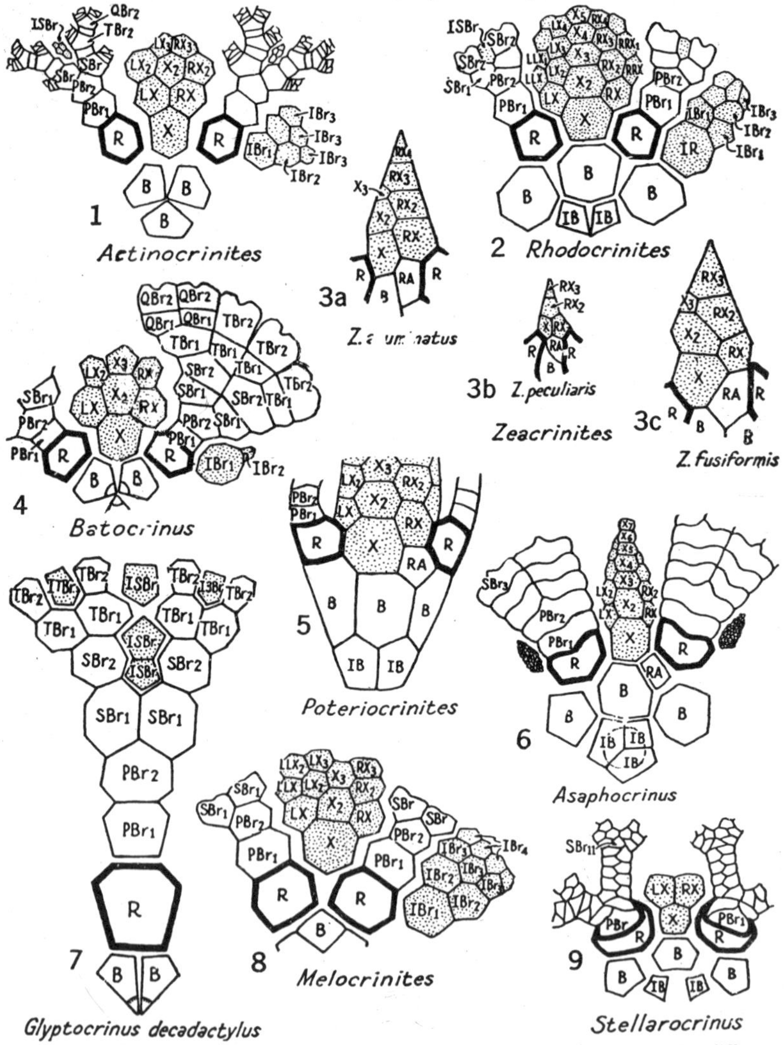

FIG. 14-19. Class CRINOIDEA. Diagrams showing arrangements of plates in the different types of crinoids and the symbols used for designating the plates. Radial plates are differentiated from others by a heavier border line, and plates belonging to interbrachial positions are stippled. (*See opposite page for further details.*)

The upper and lower flat surfaces along which adjacent columnals join are typically sculptured. Columnals are common in marine sedimentary rocks of all ages from Ordovician to Jurassic, after which they are relatively rare. In some Paleozoic limestones they constitute the major part of the rock.

The Calyx (Fig. 14-17, 14-18, 14-19, 14-20). The main part of the fleshy body of a crinoid is enclosed in a hemiellipsoidal multiplated theca, generally designated the calyx. The part of the calyx below the arm bases is the **dorsal cup;** the remaining part above is the **ventral disk,** or tegmen.

The dorsal cup, or simply **cup,** as it is commonly termed, consists of plates that have a constant arrangement in each genus (Fig. 14-19). In all crinoids

TABLE 14-3. RECOMMENDED SYMBOLS FOR THE PARTS OF CRINOIDS, TO ACCOMPANY FIG. 14-19

After Moore and Laudon, 1941

Infrabasal (*IB*), infrabasals (*IBB*); infrabasal circlet (*IB* circlet—not *IBB* circlet)

Basal (*B*), basals (*BB*); basal circlet (*B* circlet—not *BB* circlet)

Radial (*R*), radials (*RR*); radial circlet (*R* circlet—not *RR* circlet)

Anterior (*A*); posterior (*P*); left (*L*); right (*R*); (posterior basal), *PB*; left posterior basal, *LPB*; anterior radial, *AR*; right anterior radial, *RAR*; etc.

Anal (*X*), also called first anal or special anal, anals (*XX*), plates of the posterior interradius, excluding radianal; right second anal (tube plate) (*RX*), middle second anal (X_2), left second anal (*LX*)

Radianal (*RA*); regarded as synonymous with right posterior inferradial as currently defined in some inadunate crinoids

Brachial (*Br*), brachials (*BrBr*)

Primibrachial (*PBr*), primibrachials (*PBrBr*); first primibrachial (PBr_1), first primibrachials ($PBrBr_1$), second primibrachial (PBr_2), etc.; axillary primibrachial or primaxil (*PAx*), primaxils (*PAxAx*); axillary first primibrachial or first primibrachial axillary (PBr_1ax), axillary second primibrachial (PBr_2ax), etc.

Secundibrachial (*SBr*), secundibrachials (*SBrBr*); first secundibrachial (SBr_1), etc.

Tertibrachial (*TBr*), tertibrachials (*TBrBr*), etc.

Interradial (*IR*), interradials (*IRR*), plates between rays

Interbrachial (*IBr*), interbrachials (*IBrBr*), plates between primibrachials belonging to the same ray

Intersecundibrachial (*ISBr*), intersecundibrachials (*ISBrBr*), plates between secundibrachials belonging to the same half-ray

Oral (*O*), orals (*OO*)

Ambulacral (*Amb*), ambulacrals (*AmbAmb*)

Interambulacral (*IAmb*), interambulacrals (*IAmbAmb*)

FIG. 14-19. 1. *Actinocrinites*, showing basal circlet, posterior rays, anal interradius, and right posterior interradius. 2. *Rhodocrinites*, posterior part of dicyclic dorsal cup, showing interradial, interbrachial, and intersecundibrachial plates. 3. *Zeacrinites*, posterior interradius of three species, showing designation of anal plates. 4. *Batocrinus*, posterior part of dorsal cup showing three bifurcations in one ray. 5. *Poteriocrinites*, posterior view of dorsal cup showing typical arrangement of anal plates. 6. *Asaphocrinus*, infrabasal circlet and posterior plates of a flexible crinoid, showing very numerous small interbrachial plates of the left and right posterior interrays. 7. *Glyptocrinus decadactylus*, plates of a ray, showing presence of intersecundibrachials and intertertibrachials. 8. *Melocrinites*, showing plates of the posterior part of the cup. 9. *Stellarocrinus*, an inadunate crinoid having three anal plates in the dorsal cup and upward extensions of the radials between the arms. (*After Moore and Laudon*, 1941; 3 revised from Sutton Hagan.)

the lowermost cup plates are arranged in horizontal circlets, with the plates of one circlet alternating in position with those of adjoining circlets. The upper part of the cup may be composed in part of plates arranged in a vertical series. The lowermost cup plate in direct vertical series with one of the five (typically five, and never more than five) arm trunks or rays (Fig. 14-4) is termed a **radial** and, following almost universal practice, is desig-

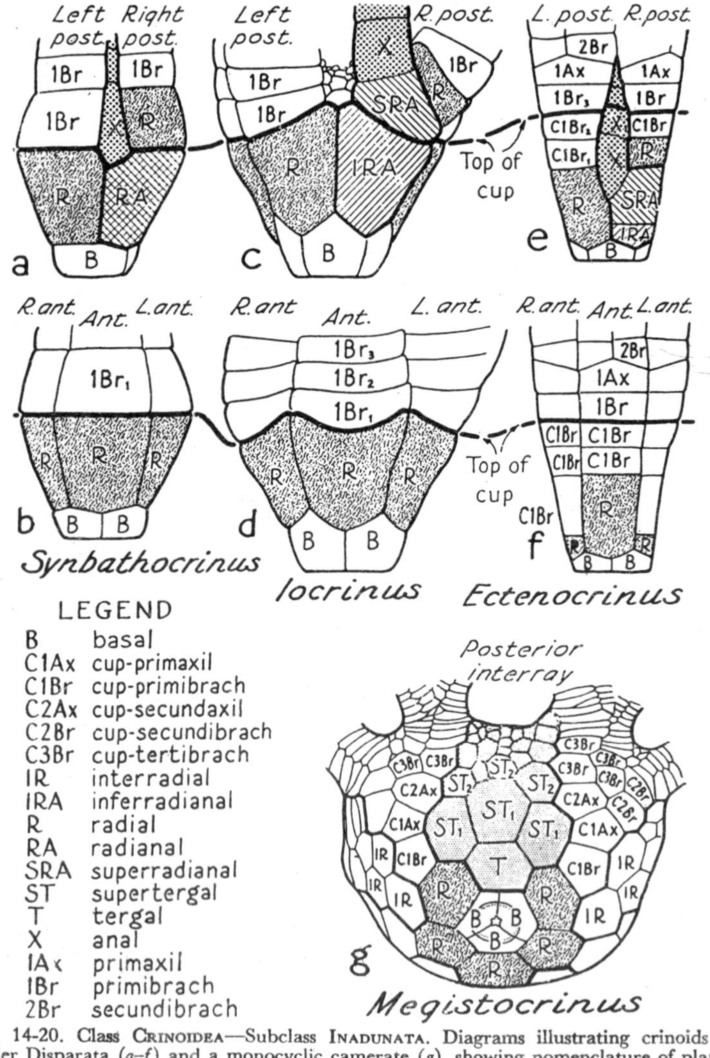

FIG. 14-20. Class CRINOIDEA—Subclass INADUNATA. Diagrams illustrating crinoids of Order Disparata (*a–f*) and a monocyclic camerate (*g*), showing nomenclature of plates. (*After Moore*, 1948.)

nated by the letter *R* (radials = *RR*).[1] In certain primitive types of Paleozoic crinoids, however, the two lowermost plates are collectively referred to as a **compound radial,** with the lower designated **inferradial** and the upper, **superradial.** Next below the radials, and alternating with them in position, is the circlet of **basals** (*B*; *BB*). Crinoids that have only *BB* and *RR* in the cup, with the *BB* attached to the stem, are described as **monocyclic** (Fig. 14-17*D*). All living crinoids and a great many fossil forms have a plate (**centrodorsal** = *CD*) or a circlet of plates, the **infrabasals** (*IB*; *IBB*), below the *BB*, and, in the case of the *IBB*, in alternate position. These are termed **dicyclic crinoids** (Fig. 14-17*C*).

Plates of the rays that lie above the *RR* are designated **brachials** (*Br*; *BrBr*); these form the arms, and if any are incorporated in the cup, they are **fixed brachials.** Some cups have no fixed brachials; others have as many as five series.

If plates intervene between the *RR* so as to separate them, these are termed **interradials** (*IR*; *IRR*). Numerous plates of taxonomic importance are developed in the posterior interradius where the anal opening is situated. There may be only a single plate, the **radianal** (*RA*), or numerous plates, all **anal;** or the anus may lie on the summit of a multiplated anal sac or anal tube. Collectively, anal plates are designated as *XX*; they may or may not be a part of the dorsal cup.

The simplest crinoid tegmen consists of five interradially disposed circumoral plates, the **orals** (*O*; *OO*). Their contiguous edges mark the position of food grooves (**ambulacra**). In crinoids that have the tegmen composed of many plates, the latter may be divisible into **ambulacrals** (*Amb*; *AmbAmb*) and **interambulacrals** (*IAmb*; *IAmbAmb*). In some forms the tegmen is distended into a multiplated hemiellipsoidal anal sac or anal tube with the anal vent near or on the summit.

All brachials (*BrBr*) up to the first forking of plates in a ray are designated **primibrachials** (*PBr*; *PBrBr*); those from this point to the next forking, **secundibrachials** (*SBr*; *SBrBr*); and successively from there on as **tertibrachials** (*TBr*; *TBrBr*); **quartibrachials** (*QBr*; *QBrBr*); **quintibrachials** (*VBr*; *VBrBr*), etc. The ray plates that support two branches are designated axillaries (*Ax*; *AxAx*). Plates occupying a position between *BrBr* of different rays are termed **interbrachials** (*IBr*; *IBrBr*); those between *SBrBr*, **intersecundibrachials** (*ISBr*; *ISBrBr*), etc. The individual plates in any category of the vertical series are indicated by subscript numbers, with numbering away from the cup (*e.g.*, PBr_1, PBr_2; Fig. 14-18).

Certain modifications of the terminology just discussed have been proposed by Moore (1948), but these apply to future usage rather than to past descriptions and need no further consideration here.

[1] It is now general practice to use a letter or a combination of letters for every plate in the dorsal cup and for those in the tegmen and arms (or combinations of capital and lower case letters or a single capital to stand for the singular of the term involved; duplicate letters indicate the pural; *e.g.*, *B* = singular and *BB* = plural, or *Br* = singular and *BrBr* = plural). The more important of these plates and their symbols are defined and illustrated in Table 14-3 and Figs. 14-17, 14-18, 14-19, and 14-20. For more detailed terminology see Moore and Laudon (1941) and Moore (1948).

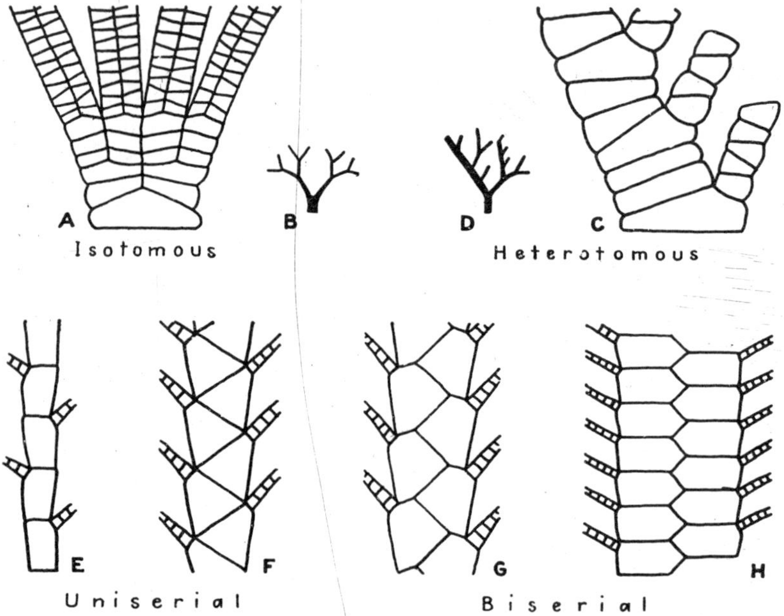

FIG. 14-21. Class CRINOIDEA. Arm structure. *A–B*. Isotomous structure, characterized by equal splitting and equal branches. *C–D*. Heteronomous structure, characterized by irregular splitting and unequal branches. Pinnules are omitted in *A–C*. *E–H*. Diagrams illustrating evolution of arm structure from a simple uniserial form (*E*), through zigzag (*F*), to biserial arrangements (*G–H*). (*Highly diagrammatic and adapted from Bather*, 1900, *and Moore and Laudon*, 1944.)

The Arms (Figs. 14-17, 14-18, 14-21). Arms are present on all crinoids, usually five, or by branching multiples of five, in number, and are to be regarded as appendages of the calyx that are given off around the marginal rim of the dorsal cup. They may be branched or unbranched and generally are freely movable or at least flexible. Branching is **isotomous** if the two divisions are essentially equal and **heterotomous** if strongly unequal (Fig. 14-21). In some forms the arms branch only once and are pinnulate, whereas in others they branch several times.

The upper (*i.e.*, inner or ventral) side of each arm has a food groove (**ambulacrum**) that is roofed over by tiny covering plates (**ambulacrals** and **adambulacrals,** or side plates) (Fig. 14-17*G*, 14-18). Food particles are driven down the grooves to the mouth by ciliary motion. If the main arm plates, or brachials (*BrBr*), are arranged in a single series, the arm is described as **uniserial;** if in two series, as **biserial** (Figs. 14-17*E*, 14-21).

In all living and fossil crinoids the arm plates are perforated by a single or double canal that contains nerve cords in living forms and is thought to have been occupied by similar cords in fossil forms. This canal extends from the arms through the radials, or lies in a furrow upon the inner surface of these plates, and ends in the basals.

Classification

The several more widely accepted classifications of the Crinoidea are based on some or all of the following features, which when considered together give a basis for differentiating supposed natural major groups:

1. Structure of the calyx as a whole.
2. Structure of the dorsal cup—*i.e.*, number of cycles of plates (monocyclic and dicyclic).
3. Nature and structural features of the food-gathering or convective system—*i.e.*, arms, mouth, gut, and anal vent.
4. Nature of tegmen.
5. Nature of respiratory and nervous systems.
6. Nature of musculature of arms and body.

The classification used in the present work is a combination of those proposed by Wachsmuth and Springer (1897) and Moore and Laudon (1944) and is based on:

1. Structure of base of dorsal cup—*i.e.*, whether monocyclic or dicyclic.
2. Structural features of the rays—*i.e.*, organization of ray plates and nature of the articulation of free arm segments.
3. Number and arrangement of interradial and interbrachial plates.
4. Structure of posterior side of dorsal cup.
5. Nature of tegminal structures.
6. Peculiarities of the column and holdfasts.

On the basis of these, the Crinoidea are divided into the following subclasses and orders:

Subclass Inadunata (Middle Ordovician—Triassic).
  Order 1. Disparata.
  Order 2. Cladoidea.
Subclass Flexibilia (Middle Ordovician—Middle Permian).
  Order 1. Taxocrinoidea.
  Order 2. Sagenocrinoidea.
Subclass Camerata (Lower Ordovician—Middle Permian).
  Order 1. Diplobathra.
  Order 2. Monobathra.
Subclass Articulata (Triassic—Recent).

Subclass Inadunata

Definition. This small subclass includes relatively primitive crinoids that have the plates of the calyx joined firmly together, the mouth subtegminal, and the arms typically free above the radials. The subclass had its great development during the Paleozoic, and only a single family continued into the Triassic, when it too became extinct. Two orders are recognized: Disparata (monocyclic) and Cladoidea (dicyclic).

Order 1. Disparata. This order includes monocyclic inadunates having a dorsal cup composed of basals, radials, and generally an anal plate and a radianal, or in place of the latter (RA), an inferradianal or superradianal.

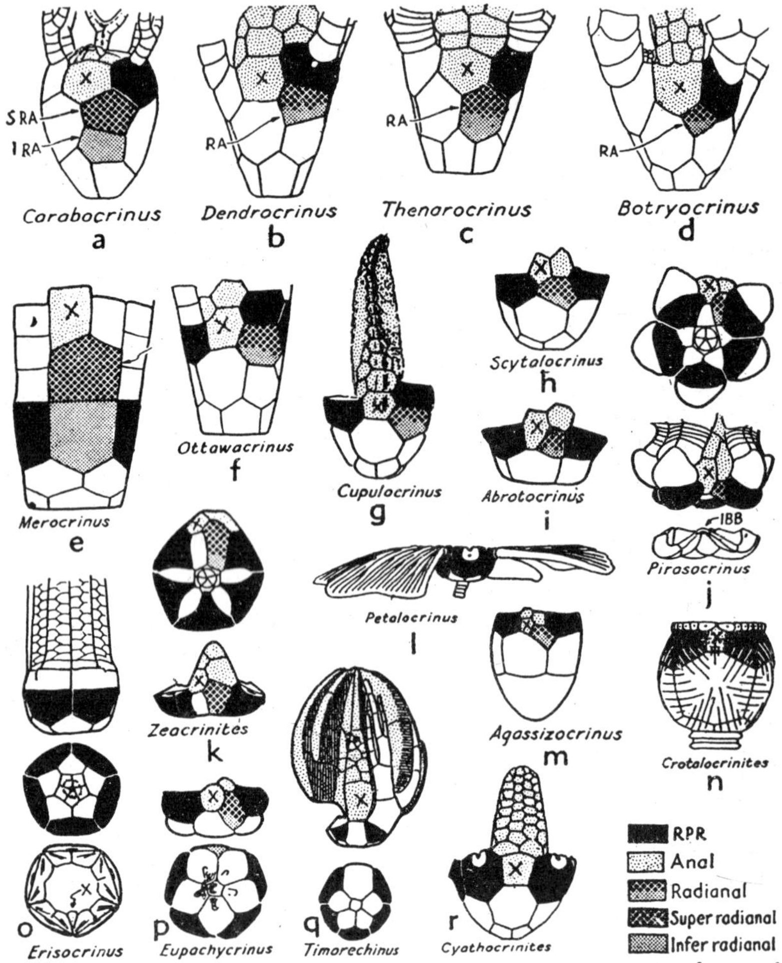

FIG. 14-22. Class CRINOIDEA—Subclass INADUNATA. Diagrams illustrating form and structure of representative crinoids of Order Cladoidea. (*After Moore*, 1948.)

The arms are free above the radials, and dissimilarity of structure in different rays is characteristic. *Middle Ordovician to Permian.*

Typical genera are illustrated in Fig. 14-20*a–f*.

Order 2. Cladoidea. This order includes dicyclic inadunates and constitutes one of the most important groups of Paleozoic crinoids. The earliest representatives are simple; later forms are much more complex with a high degree of specialization in some particular structure. A single family survived the Permian but became extinct during the Triassic. *Middle Ordovician to Triassic.*

Typical Paleozoic genera are illustrated in Fig. 14-22; *Agassizocrinus* (Fig. 14-25*C–D*) is an index genus of the Upper Mississippian of North America.

Subclass Flexibilia

Definition. The Subclass Flexibilia consists of dicyclic crinoids in which the lower brachials are incorporated in the dorsal cup, but not rigidly. The lowermost circlet consists of three infrabasals, two large and one small, and the latter is invariably situated in the right posterior ray. The tegmen is flexible and bears the exposed food grooves and the mouth; there is no anal sac. The calyx as a whole is flexible, because of the way in which the plates are joined together, and the subclass takes its name from this characteristic. The arms are uniserial and nonpinnulate, and the stem has a circular cross section. The Flexibilia are thought to have evolved from a cladoid ancestor early in the Paleozoic, or even before; they appear first in Middle Ordovician rocks and became extinct sometime during the later Permian. *Middle Ordovician to Middle Permian.*

Two orders are recognized—Taxocrinoidea and Sagenocrinoidea.

Order 1. Taxocrinoidea. This order includes flexible crinoids with an elongate crown and relatively weak calyx. The anal plates are not joined to adjacent rays. The oldest flexible crinoid, *Protaxocrinus*, from the Middle Ordovician, belongs in this order; *Taxocrinus* (Dev.–Miss.) is another representative genus (Fig. 14-23). *Middle Ordovician to Middle Permian.*

Order 2. Sagenocrinoidea. This order includes most flexible crinoids, and in these the calyx plates, including those of the posterior interray, are rather firmly united. The anal *X* is joined by close sutures to the posterior basal and adjoining ray plates, or *X* may be absent, and there is no series of anals as in the Taxocrinoidea. The order is thought to have been derived from early taxocrinoids. *Silurian to Permian.*

The following representative genera are illustrated in Fig. 14-23: *Ichthyocrinus* and *Lecanocrinus* (Sil.–L. Dev.); *Forbesiocrinus* and *Wachsmuthicrinus* (Miss.); *Talanterocrinus* (U. Penn.); and *Cibolocrinus* (L. Penn.–M. Perm.)

Subclass Camerata

Definition. The camerates are characterized by a calyx of variable form in which all plates are united by rigid suture. The mouth and food grooves lie beneath a rigid tegmen and no radianal is present. The arms are pinnulate and uniserial or biserial. The fundamental characters of the most primitive genera of camerates are the same except that some have a monocyclic dorsal cup, whereas others are dicyclic. This important difference is the basis for subdividing the subclass into the two orders—Monobathra (monocyclic) and Diplobathra (dicyclic). The Camerata constitute a large and important group of crinoids and range from Lower Ordovician to Middle Permian.

Order 1. Diplobathra. These are dicyclic camerate crinoids with the dorsal cup having five infrabasals and five basals. *Middle Ordovician to Upper Mississippian.*

Representative genera showing the evolution of the order are illustrated in Fig. 14-24.

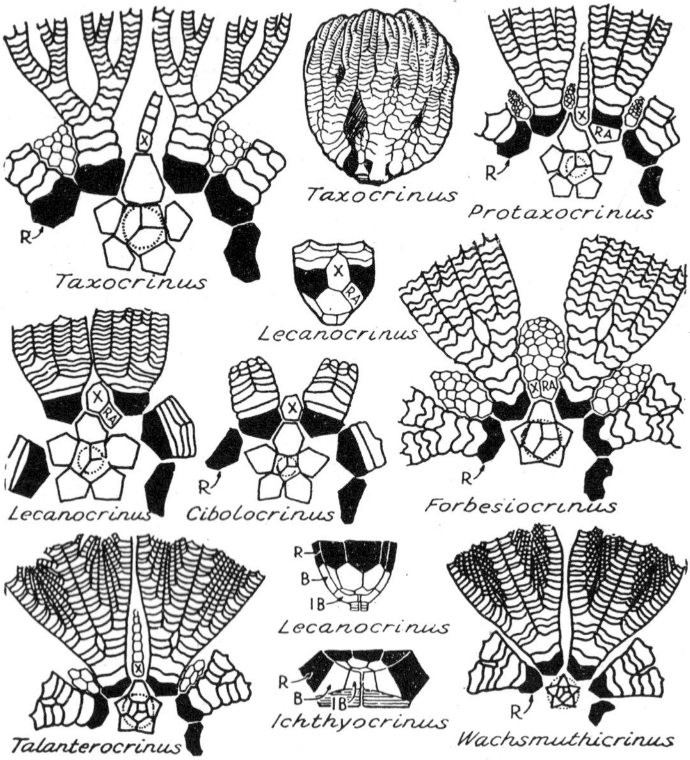

FIG. 14-23. Class CRINOIDEA—Subclass FLEXIBILIA. Diagrams illustrating range of structure in representative genera. (*After Moore*, 1948.)

Order 2. Monobathra. This order consists of monocyclic camerates and ranges from Lower Ordovician to Middle Permian. Representative genera illustrating the evolution of the order are illustrated in Fig. 14-24. Of particular interest is the long-ranging genus *Platycrinites*, (M. Sil.–M. Perm.), of which some Mississippian species have the cap-shaped shell of a small gastropod, *Orthonychia* (= *Igoceras*), perched over the anal opening (Fig. 14-25*A*).

Subclass Articulata

With the exception of the cladoid family Erisocrinoidae, all post-Paleozoic crinoids are referred to Subclass Articulata, which is a highly organized group of stemmed and stemless crinoids exhibiting wide diversity of form and structure. In a typical articulate the dorsal cup is reduced in size and consists of circlets of radials, basals, and (actually or potentially) infrabasals. The calyx plates are united by articulation, a characteristic indicated by the name of the subclass, and the calyx, as a consequence, is relatively flexible. The mouth and food grooves are exposed on the tegmen, and the tegmen

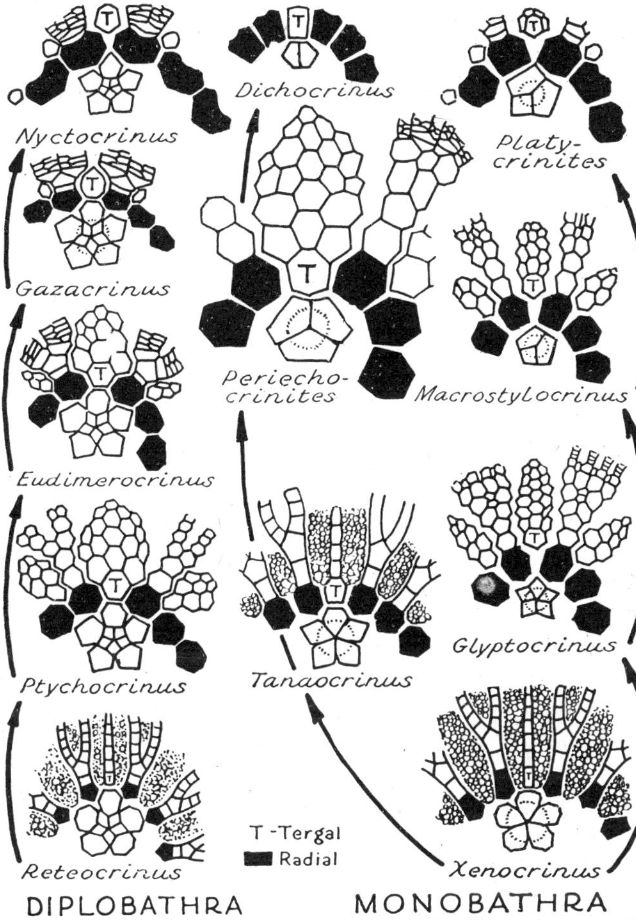

FIG. 14-24. Class CRINOIDEA—Subclass CAMERATA. Diagrams illustrating range of structure in representative genera (see also Fig. 14-20*g*). (*After Moore*, 1948.)

itself, which is invariably a flexible integument, may be studded with calcareous ossicles or carry well-defined plates. Although many genera bear a stem throughout life (*e.g.*, the Pentacrinidae and Apiocrinidae; Fig. 14-26), most (chiefly the Comatulida) are stemless in the adult stage. The post-embryonic life history of *Antedon* (Fig. 14-16), an example of the stemless type, is discussed at some length on preceding pages. A point of some interest is the fact that all Articulata have uniserial arms, suggesting that the Paleozoic inadunate from which they are thought to have been derived[1] must also have had such arms.

[1] Some investigators believe that the Articulata are polyphyletic (Moore, 1948).

The living representatives of the Articulata were once considered to be a small and unimportant group of crinoids that represented the waning vestiges of a dying race. It is now known, thanks to the great monographs of Carpenter (1884; 1888) and A. H. Clark (1915 to date), that they are probably as important a part of the existing echinoderm fauna as their ancient relatives were in the Paleozoic faunas.

The earliest of both stemmed and stemless articulates appear in Triassic rocks, and both types are still living; hence, the range of the subclass is Triassic to Recent. No orders have been recognized, but three suborders have been proposed as follows:

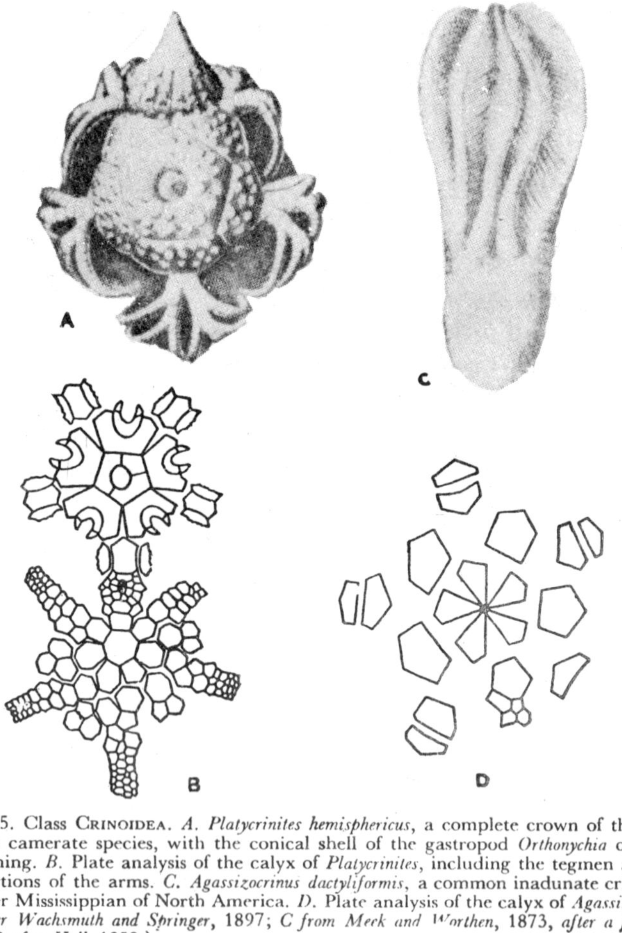

FIG. 14-25. Class CRINOIDEA. *A. Platycrinites hemisphericus*, a complete crown of this Mississippian camerate species, with the conical shell of the gastropod *Orthonychia* over the anal opening. *B.* Plate analysis of the calyx of *Platycrinites*, including the tegmen and the basal portions of the arms. *C. Agassizocrinus dactyliformis*, a common inadunate crinoid in the Upper Mississippian of North America. *D.* Plate analysis of the calyx of *Agassizocrinus*. (*A–B after Wachsmuth and Springer*, 1897; *C from Meek and Worthen*, 1873, *after a figure by Roemer; D after Hall*, 1858.)

Suborder 1. **Apiocrinida.** Characterized by a long, multisegmented, generally noncirrate stem that is retained throughout life. *Triassic to Recent. Apiocrinus* (Fig. 14-26*A–B*) is typical.

Suborder 2. **Pentacrinida.** Stem is present or wanting, but cirri usually numerous, either as whorls on stem or as a cluster about the centrodorsal plate; if cirri are lacking, the centrodorsal plate is smooth and pentagonal. *Triassic to Recent. Antedon* (Fig. 14-15) is a well-known modern form, and *Pentacrinus* (Figs. 14-15, 14-26*D*) has the same range as the suborder.

Suborder 3. **Holopocrinida.** Cirri are wanting; stem short with a few long segments or wanting, or replaced by a short, thick unsegmented support.

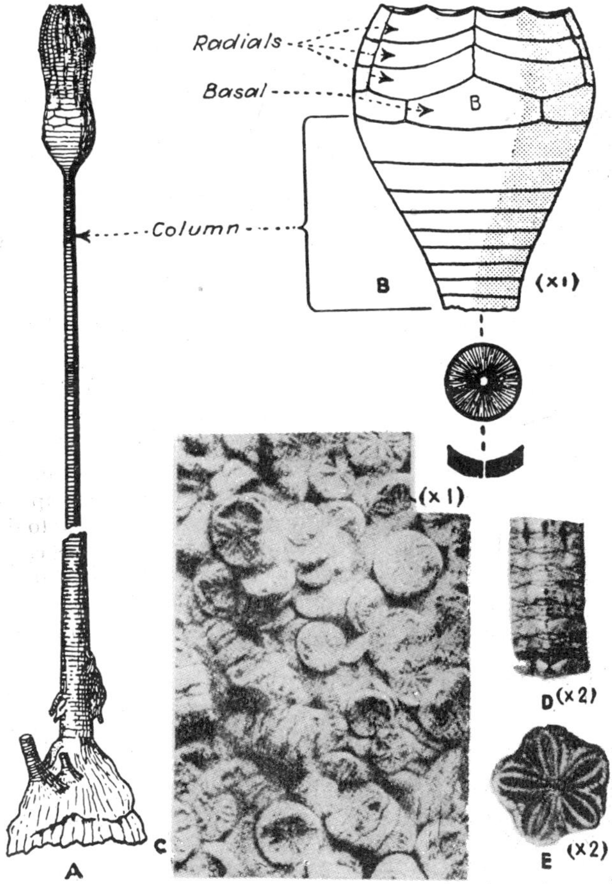

Fig. 14-26. Class Crinoidea—Subclass Articulata. *A. Apiocrinus roissyanus* from the Jurassic. *B. Apiocrinus parkinsoni* from the Jurassic of England, showing the calyx, part of the column, and one columnal as viewed along the columnar axis and as sectioned transversely. *C–E. Pentacrinus* from the Jurassic: *C*, stem fragments of *P. subangularis alaska; D*, a stem fragment of *P. rotiensis; E*, typical joint face showing characteristic pentamerous petaloid aspect in *P. subangularis.* (*A after Nicholson and Lydekker*, 1889; *C–E after Springer*, 1925.)

Ontogeny, Phylogeny, and Geologic History

Ontogeny, Phylogeny, and Evolution. The complete life history is known for only a few living crinoids. That of *Antedon* is outlined briefly on preceding pages, and from it certain deductions may be made concerning the ancestry and evolution of the Crinoidea as a whole. From this ontogeny it is suggested that one group of early crinoids had a dicyclic dorsal cup from the beginning, a circlet of orals on the ventral surface, and simple non-pinnulate arms that branched from the third primibrach and were free from the base up. It has been assumed that an unbranched arm preceded the branched type, and that pinnules appeared as a later development. Whether or not a monocyclic dorsal cup preceded a dicyclic is a moot question. Some specialists believe that the monocyclic condition was the earlier, and that dicyclic cups developed by the addition of another circlet of plates. Other investigators favor the idea that monocyclic cups were developed from primitively dicyclic forms by loss of the circlet of infrabasals. In this connection it is interesting to note that the oldest known crinoids, which come from Early Ordovician (Chazyan) rocks, include both monocyclic and dicyclic inadunates and dicyclic camerates. It is of further significance that typical representatives of the Flexibilia occur in Middle Ordovician rocks. Finally, one of the oldest, if not the most ancient, crinoid is a Lower Ordovician camerate, probably dicyclic, which has an advanced type of arm structure and a stem composed of well-differentiated nodal and internodal segments (Moore, 1948).

From the preceding statements it is seen that virtually at the beginning of the crinoid record (*i.e.*, Early Ordovician), all three Paleozoic subclasses were represented, and were by this time so well developed that divergence from an ancestral stock must have occurred much earlier, either in the Cambrian or possibly even in the later Pre-Cambrian. Although time of appearance in the stratigraphic column, therefore, gives no satisfactory clue as to the genetic interrelationships of the three Paleozoic subclasses, similarity of certain structural features can be used to determine origins. For example, the Flexibilia were probably derived from earlier inadunates, and the more recent Articulata are thought to have evolved from Late Paleozoic inadunates with uniserial arms (Fig. 14-56).

The origin of the Crinoidea as a whole is uncertain, partly because so little is known about the ontogeny of the extinct Paleozoic forms and partly because there are no satisfactory connecting links with the fossil pelmatozoans found in rocks contemporaneous with or older than those containing the most ancient crinoid remains. It was formerly thought that the Crinoidea were evolved from primitive cystoids, but doubt has been cast on this origin because of the presence among cystoids of structures that are foreign to crinoids, and also because the two groups were essentially contemporaneous in their early development (Crinoidea, Lower Ordovician to Recent; Cystoidea, Middle Ordovician to Lower Permian). Regnell (1945) has suggested that the Class Eocrinoidea (L. Camb.–M. Ord.) may be the ancestral or root stock of all pelmatozoans, and this view seems to have merit (Moore, 1948) (see discussion of Eocrinoidea on pages 666 and 667).

In their evolution, the Crinoidea display several prominent tendencies and series of changes. In general, the more primitive conical or bowl-shaped dorsal cup that is typically higher than wide tends to become a flatter discoidal structure that is wider than high. However, this tendency does not apply to many camerates that retain primitive characters of cup form. There is a general tendency toward fusion of plates throughout the calyx. Undivided or unbranched arms are considered primitive or secondarily simplified; the general tendency was toward more and more branching, with isotomous division being followed by heterotomous (hence the last type suggests evolutionary development). Pinnules are formed only on crinoids showing advanced evolution of arms; the arms of the most primitive crinoids do not have pinnules, and the same condition obtains in the earliest growth stages of *Antedon*. A universal trend is the upward displacement and ultimate elimination from the dorsal cup of all plates except the radials and one or two circlets of plates below them (*e.g.*, *Apiocrinus*, Fig. 13-26*B*). There seems to have been little tendency among Paleozoic crinoids to lose the stem, but this tendency is quite apparent in Mesozoic and Cenozoic species. Modern crinoids include both stemmed and partly or wholly stemless types; many of the latter have a stem during the early growth stages but abandon it and become free-swimming later. Finally, the evolutionary end form of many branches in all the subclasses is a simply constructed crown having perfect pentamerous symmetry.

Geologic History and Stratigraphic Range. The oldest known unquestioned crinoid is of Lower Ordovician age; isolated plates of supposed Cambrian crinoids are of unknown origin. Soon after their appearance, the crinoids became a diverse and flourishing group of organisms and by Middle Ordovician there were many representatives of each of the three Paleozoic subclasses—Inadunata, Flexibilia, and Camerata. Throughout the remainder of the Paleozoic, they were an important component of marine bottom faunas in every period. The class as a whole seems to have had one climax of development during the Mississippian, when the shallow seas of that period teemed with them, and another, though perhaps of less magnitude, during the first half of the Permian, when an unusually rich and varied fauna filled the seas over the present island of Timor. Inexplicably, Paleozoic crinoid groups vanished, almost to the last genus, before the end of the Permian, with only one family of inadunates surviving into the Triassic. These stragglers died out before the end of the period. Meanwhile, the first of the Articulata had appeared—stemmed forms in the Early Triassic and stemless forms in the late part of the period—and this subclass alone left the crinoid record that exists in post-Paleozoic rocks. All living crinoids, numbering more than 100 genera and perhaps 1,000 species, belong to the Articulata, and of these the stemless forms constitute by far the larger group and include more than 90 per cent of all known living species.

Without exception living crinoids are marine, and it is universally accepted that extinct species were likewise. Consequently, finding an abundance of crinoidal remains in a rock is taken as evidence that the enclosing sedimentary materials were deposited in a marine environment. Because of the gregarious habit of crinoids, both living and extinct, their remains are likely to be found in great profusion at one locality and only sporadically or not at all in surrounding areas. This local abundance is first displayed in Ordovician rocks, and localities of Silurian strata in Dudley, England, on the islands of Gotland and Anticosti, and in eastern United States (New York to Iowa) have yielded large faunas. Well-known localities for Devonian crinoids are in the Eifel district of Germany, in New York and Michigan, and near the falls of the Ohio at Louisville, Kentucky. In addition to Mississippian localities in Belgium, Great Britain, and Russia, there are the world-famous "crinoid gardens" at Crawfordsville, Indiana, and in the Burlington and Keokuk areas of Iowa. In the Mississippian beds of Indiana, crinoids have left large mounds and lenses of columnals and fragmental crowns, which are examples of **crinoidal bioherms** (Stockdale, 1931). In these accumulations columnals may dominate in one deposit and crowns in another; they rarely abound together. It is thought that this distribution may be accounted for by assuming that in later life the crown broke free from the stem and floated or swam away, much as *Antedon* and certain other living forms now do. Under such circumstances, root systems and columnals accumulated where gardens persisted, whereas the mobile crowns accumulated elsewhere (Kirk, 1911). Mid-continental United States has yielded many fine Pennsylvanian crinoids, and no area surpasses the island of Timor in its remarkable Permian fauna. There are few good Mesozoic localities; one of the best known crinoidal formations is the Cretaceous Niobrara chalk of Kansas which contains large but very local concentrations of the planktonic genus *Uintacrinus* (a dozen tangled specimens are commonly present on a square foot of bedding plane). As previously stated, existing crinoids are now known to constitute an important component of the echinoderm fauna, thanks to the dredging activities of the *Challenger*, *Albatross*, and other similar ships and to the monographic studies of Carpenter (1884; 1888) and A. H. Clark (1915 to date).

Crinoids are excellent index fossils for some formations, and they would be even more useful if found more abundantly and over wider areas. Species, and to a large extent genera also, have quite limited vertical ranges, thus adding to their use for stratigraphic purposes. However, the strongly gregarious habit and random geographic distribution of crinoids definitely lessen the actual usefulness of their remains.[1]

Class Edrioasteroidea[2]

The Edrioasteroidea constitute a small group of unique, exclusively Paleozoic pelmatozoans characterized by a flexible sessile test with five prominent food grooves radiating from the mouth (Fig. 14-27). The test is composed of an indefinite number of irregular plates, normally arranged without definite symmetry. In a few forms the plates imbricate to some extent, and

[1] The interested reader will find comprehensive discussions of the stratigraphic uses and occurrences of crinoids and their remains in several articles by Moore (1938; 1948).

[2] Edrioasteroidea—Gr. *edrion*, diminutive of *edra*, seat, + *aster*, star, + *-oid*, like; referring to the fact that the test has the appearance of a little cushion with a star on top. The star is formed by the five recumbent ambulacral grooves.

most tests seem to have been more or less flexible. The discoidal shape or flattened condition of many fossil specimens is no doubt due in part to crushing at the time of burial.

There are no subvective appendages rising from the test. Normally, five straight or curved unbranched ambulacra radiate through the test from the central mouth. These are floored with small plates, between which are pores believed to have permitted the passage of extensions (*i.e.*, tube feet) from the water vessels of a water-vascular system. They are also covered by a double series of alternating plates, the cover plates typical of many echinoderms.

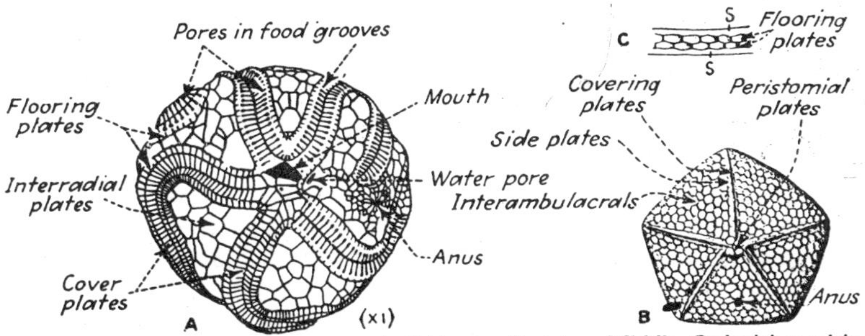

FIG. 14-27. Class EDRIOASTEROIDEA. *A. Edrioaster bigsbyi*, a Middle Ordovician edrioasteroid, showing aspect of oral side. Cover plates are shown on only two ambulacra. *B–C. Stromatocystites*, the earliest edrioasteroid: *B*, *S. pentangularis* from the Middle Cambrian of Bohemia; a restoration of the upper surface by Bather; *C*, small portion of an ambulacrum (*S* = ridges of side plates), showing flooring plates in *S. walcotti*, the oldest known edrioasteroid, from the Lower Cambrian of western Newfoundland. (*A modified after Bather*, 1914; *B–C after Schuchert*, 1919.)

The ambulacra show considerable variation in their orientation. Typically, four curve to the left and the fifth (the right posterior) to the right, as in *Edrioaster* (Fig. 14-27), so that the anal interradius and anal opening lie between the right and left posteriors. In some genera, however, the ambulacra are straight, whereas in others they all curve to the right or to the left, or, as in *Agelacrinites*, three to the left and two to the right. The number of ambulacra, although typically five, may vary in the same species, just as the number of arms varies in modern starfish.

A water pore and the anus, which has a valvular pyramid, lie in the posterior interambulacral area, between rays 1 and 5 that constitute the bivium of a pentamerous symmetry (Fig. 14-4).

Although the Edrioasteroidea are present in Lower Cambrian rocks, they are not regarded as primitive echinoderms. It is true that they present a primitive aspect in their saclike test attached by its aboral surface and with mouth, anus, and hydropore on the oral surface. However, in the structure and relations of the ambulacra they are far from primitive simplicity, and specialists now almost universally regard them as a distinct and unique class of pelmatozoan echinoderms that traveled a road different from the one followed by other Pelmatozoa. They seem closest to the cystoids in the irregularity of plate form and arrangement and in the mode of attachment, but they differ in the highly organized ambulacral system—a feature in which they closely resemble the asteroids—and in several other important respects.

The Class Edrioasteroidea ranges from the Lower Cambrian to Pennsylvanian and is well illustrated by the Ordovician *Edrioaster* (Fig. 14-27), which has a flexible, discoidal test of thin somewhat imbricating plates, and five prominent sickle-shaped ambulacra.

SUBPHYLUM HOMALOZOA[1]

General Considerations

Included in the Echinoderma are two extinct anomalous groups—*Carpoidea* (M. Camb.–L. Dev.) and *Machaeridia* (Ord.–Dev.)—of seemingly primitive Early Paleozoic animals that have not found satisfactory resting places in any of the generally recognized classes, because they differ from all others in lacking the characteristic radial symmetry of thecal plates. Whitehouse (1941, page 22) has proposed that they be kept apart from all other echinoderms and that they be recognized as separate classes, together constituting a new subphylum Homalozoa (a term emphasizing the flattened nature of the animal). The geologic range of the subphylum thus constituted would be Middle Cambrian to Devonian.

Both the subphylum and the two classes are of uncertain taxonomic validity, and it is not at all certain that the several genera included in the two classes are even echinoderms. However, they are interesting and controversial fossils that should be brought to the student's attention somewhere; hence they are included in the Echinoderma, because they seem most closely related to members of that phylum. The reader will understand, therefore, that including them in the present discussion does not automatically establish that they are echinoderms!

Class Carpoidea[2]

The Carpoidea are extinct primitive organisms having a well-developed stem, a bilaterally compressed body, and seemingly only one or two rays. They were formerly included in Class Cystoidea, but specialists now regard them as a separate class because they seem to lack a water pore, a genital pore, and most important of all, any trace of the radial symmetry that is so characteristic of all other classes of echinoderms. On the other hand, the composition, structure, and general relations of the skeletal elements seem to leave little doubt as to their echinodermal affinities.

Most typical carpoids are bilaterally compressed, with the plane of compression including the thecal openings, and are supposed to have been attached to the bottom with orientation such that the plane of compression was parallel to the sea floor. In some forms the entire thecal wall was flexible, whereas in others only one side remained so; in either case, it is assumed

[1] Homalozoa—Gr. *homalos*, flat, + *zoon*, animal; referring to the flattened nature of the bilaterally compressed organisms included in this subphylum.

[2] Carpoidea—Gr. *karpos*, fruit, + *-oid*, like; referring to the resemblance of carpoids to certain fruits.

that the theca expanded and contracted as the animal drew in and expelled water. The water contained food as well as the oxygen that the animal extracted in its respiration.

The mouth and anal vent varied in position with reference to the stem and to one another according to the particular habitat and mode of life of each genus. Originally they may have been at opposite ends of the body with the stem arising between them, as suggested by Bather (1929). In any case their arrangement differed from that of all other Pelmatozoa.

Description of typical representatives will suffice to indicate the more characteristic features of the class. *Dendrocystis* and *Mitrocystella* may be taken for this purpose; their chief features are compared in Fig. 14-28. In each there is a short stem composed of several vertical rows of plates and a flattened saclike theca made up of many irregular plates variously arranged. The thecal openings in *Dendrocystis* seem to be a lateral anal vent low down on the body and a probable intake opening on the summit of the theca at the base of the armlike appendage that rises above the supposed oral surface. In *Mitrocystella*, side pores are present on the lateral lobes that flank the deep concavity where the stem is attached; since no other thecal openings are known, it is assumed that somehow water was taken in and expelled through these pores.

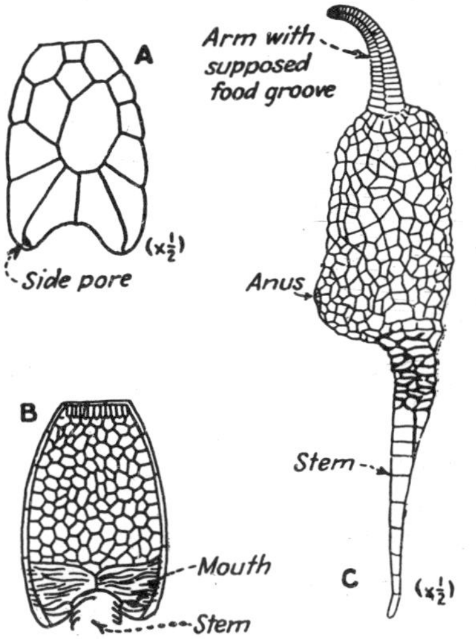

FIG. 14-28. Class CARPOIDEA. *A–B*. Inferior and superior faces, respectively, of *Mitrocystella incipiens* from the Ordovician. This species had no oral appendages. *C*. *Dendrocystis sedgwicki*, from the Ordovician of Bohemia, restored to show prominent stem and tubular oral appendage (arm) with supposed food groove. (*A–B redrawn from Chauvel*, 1941; *C modified after Bather*, 1900.)

Dendrocystis and *Mitrocystella*, together with half a dozen closely related genera, are believed to constitute a group of primitive echinoderms which retained and in some cases perfected the fundamental bilateral symmetry that is supposed to have been characteristic of the postulated archetypal or ancestral echinoderm, the *Dipleurula* (Fig. 13-55). Because they traveled an evolutionary road entirely different from that followed by typical echinoderms, they lack the familiar radial symmetry and other features that characterize the other classes (Bather, 1930). They are, to say the least, enigmatic organisms, and much more study will be required before their taxonomic affinities are entirely clear.

The Carpoidea first appeared in the Middle Cambrian (*Trochocystites*) and became extinct in the Early Devonian (*Anomalocystites* and *Placocystella*). So far as is known they left no descendants.

Class Machaeridia

Included in Class Machaeridia are a few peculiar multiplated, more or less bilaterally symmetrical structures that are now commonly referred to the Echinoderma largely because the individual plates of some forms are unit crystals of calcite and display the cleavage so characteristic of echinoderms. Certain of these fossils were customarily referred to the Cirripedia until Withers (1926) showed they did not possess the features characteristic of that order and suggested they were better classified as echinoderms—a view accepted by Bather (1929) and certain other later investigators. However, both cirriped and echinoderm affinities have been strongly denied by Wolburg (1938), who inclines to the view that they may be an unknown group of the Mollusca. Obviously, the group is a controversial one, as stated in a preceding paragraph, and this fact should be kept in mind in reading the following discussion.

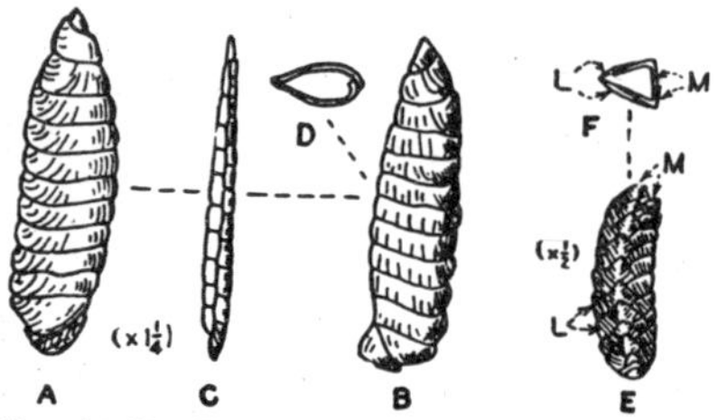

Fig. 14-29. Class Machaeridia. *A–D*. *Lepidocoleus sarlei* from the New York Silurian: *A–B*, opposite sides of specimen; *C*, ventral edge, showing alternation of plates in the two rows; *D*, diagrammatic transverse section. *E–F*. *Turrilepas wrighti* from Silurian of England: *E*, basal half of a specimen, showing the two median rows of intersecting keeled plates (*M*) and a row of lateral plates (*L*); *F*, diagrammatic transverse section. (*Adapted from Clarke*, 1896, *and Withers*, 1915.)

In the simpler machaeridians (*e.g.*, *Lepidocoleus;* Fig. 14-29) the skeleton is pod-shaped and consists of two rows of plates that overlap along the thicker rounded edge and open along the opposite thin edge. Scars on the inner faces of the plates have been interpreted as sites of attachment for muscles that pulled the two sides together. It has been inferred that the animal was elongate, flexible, and bilaterally symmetrical, with a mouth and some sort of sensory organ at the front end and an anal vent at or near the opposite end. Bather (1929) has suggested that forms like *Lepidocoleus* may have resembled the hypothetical *Dipleurula* in some respects. More advanced forms of the Machaeridia (*e.g.*, *Turrilepas;* Fig. 14-29) had four rows of plates, two on each side, and at one end of some forms a plate is modified in a way that suggests temporary fixation.

Lepidocoleus (Ord.–Dev.; Fig. 14-29*A–D*) and *Turrilepas* (M. Sil.; Fig. 14-29*E–F*) may be regarded as typical of the biserial and quadriserial skeletons, respectively. Other genera that have been provisionally referred to the Machaeridia are *Deltacoleus* (Ord.), *Lophocoleus* (Sil.), and *Strobilepas* (Dev.).

SUBPHYLUM HAPLOZOA[1]

General Considerations and Classification

The Subphylum Haplozoa was recently erected (Whitehouse, 1941) to embrace two small groups of early and extinct, unattached animals, supposedly primitive echinoderms, that had a simple skeleton composed of a

[1] Haplozoa—Gr. *haploos*, simple, + *zoon*, animal, referring to the simple nature of the skeleton.

few plates forming a theca about an unrestricted calyx. One group has bilaterally symmetrical skeletons (Class Cyamoidea); the other has radially symmetrical skeletons (Class Cycloidea). The microstructure of the two groups is also different—in one it is nonfibrous; in the other, fibrous. Since bilaterally and radially symmetrical stages are separate and distinct in the early ontogeny of living echinoderms, Whitehouse (1941) has suggested that the two groups be regarded as separate classes. He proposed the term Cyamoidea for the bilateral forms and Cycloidea for the radial forms. All members of both classes are, so far as known, restricted to rocks of Middle Cambrian age.

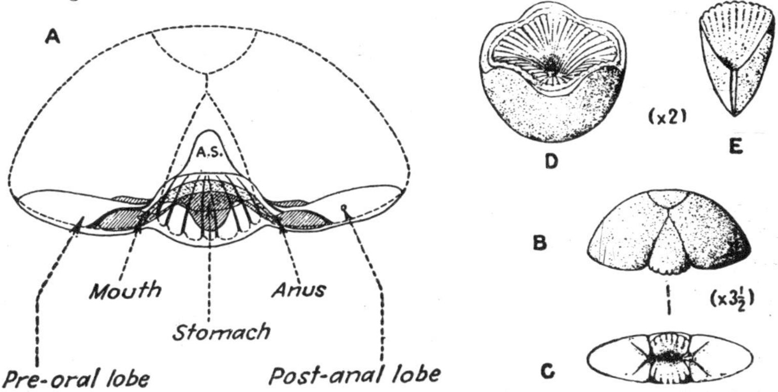

FIG. 14-30. Subphylum HAPLOZOA. *A–C. Peridionites navicula* (Class Cyamoidea): *A*, hypothetical reconstruction of the soft structures; the stippled area indicates the digestive tract, the lightly ruled areas are supposed coelomic sacs, and the heavily ruled areas represent muscles [*As* = aboral (?genital) sinus]; the presence of median coelomic sacs is questionable; Whitehouse (1941) considers *Peridionites* to have been morphologically similar to the hypothetical *Dipleurula* of all Echinoderma; *B–C.* Lateral and ventral aspects of the bilaterally symmetrical skeleton, which is composed of five plates—an apical plate (dorsocentrale), two end plates, and two mediolateral plates, the latter bounded by the converging end plates. *D–E. Cymbionites craticula* (Class Cycloidea): *D*, the complete skeleton, which is composed of five similar plates disposed with radial symmetry; *E*, one of the radial plates. Both *Peridionites* and *Cymbionites* are thought to have been unattached echinoderms, and Whitehouse (1941) has suggested that the Cyamoidea were ancestral to the Cycloidea (Fig. 14-57). Both fossils are found in the Middle Cambrian of Queensland. (*After Whitehouse*, 1941.)

CLASS CYAMOIDEA[1]

The Class Cyamoidea embraces bilaterally symmetrical haplozoans that have a small pouchlike or bean-shaped theca composed of five plates. The **stereom,** or skeletal material, is not formed of prismatic fibers as in the Cycloidea. *Peridionites* (Fig. 13-30*A–C*), the only genus thus far known, has a bean-shaped calcareous theca that is bilaterally symmetrical about two planes, is closed orally, and consists of five plates—an apical or dorsocentral plate, two end plates, and two mediolateral plates bounded by the converging end plates. The animal that made the theca is assumed to have had the features shown in Fig. 13-30*A* and to have been unattached. *Middle Cambrian.*

[1] Cyamoidea—Gr. *kuamos*, bean, + *-oid*, like; referring to the beanlike appearance of the theca.

Class Cycloidea[1]

The Class Cycloidea embraces those members of the Haplozoa that had the five thecal plates arranged with radial symmetry and composed of prismatic fibers of calcite. The theca has the shape of a small cup and is composed of five equal, thick, curved, wedgelike plates united laterally and apically to form the structure shown in Fig. 14-30 *D–E* and named *Cymbionites*. Whitehouse (1941, page 10) concludes that the body of the animal was seated in the fluted, radially symmetrical calyx that was probably closed aborally by a membranous tegmen, and that all essential openings of the body were ventral (oral) in position. The grooves of the calyx may represent sites where muscles were seated.

Cymbionites (Fig. 14-30), the only known genus thus far described, is limited to the Middle Cambrian.

Phylogeny of the Haplozoa

Great importance is to be attached to the members of the Haplozoa for the light they may throw on the much-discussed problems of echinoderm evolution. Whitehouse (1941) has reached some revolutionary conclusions of far-reaching significance as a result of his study of the nature and structure of the haplozoans. Since these conclusions concern all major divisions of Echinoderma, discussion of them is postponed to a later section on the phylogeny of the phylum as a whole.

Here it suffices to emphasize that, as early as the Middle Cambrian, simple unattached organisms of two types were building quite different calcitic skeletons. One (*Peridionites*), presumably a bilaterally symmetrical animal, built a theca of five symmetrically arranged plates, each of which is a nonfibrous single calcite crystal. The other (*Cymbionites*) built a radially symmetrical theca composed typically of five variously arranged plates (also additional accessory plates—centrals and radials—in some specimens), which have conspicuous fibrous structure.

SUBPHYLUM ELEUTHEROZOA[2]

General Considerations

The Eleutherozoa are free-moving, bottom-dwelling echinoderms that are customarily divided into the following three classes on the basis of bodily structure, nature of food-gathering system, and morphology of the theca (or test): *Stelleroidea* and *Holothuroidea*, both with the body encased in a tough flexible integument; and *Echinoidea*, in which the theca is a more or less rigid semiglobular structure composed of many calcareous plates with radially symmetrical arrangement.

The subphylum includes the familiar starfishes, brittle stars, sea urchins, and sea cucumbers and has a long and extensive fossil record that reaches

[1] Cycloidea—Gr. *kuklos*, a circle, + *-oid*, like; referring to the circular outline and radial symmetry of the theca.

[2] Eleutherozoa—Gr. *eleutheros*, free, + *zoon*, animal; referring to the fact that members of this subphylum are free-moving rather than attached.

back to the Cambrian and includes thousands of fossil species. Living eleutherozoans move about over the bottom with the mouth on the under side and the anal vent on the top side, or in a posterior position in elongated forms. Their living position, therefore, is exactly opposite that of the attached pelmatozoans, as illustrated in Fig. 14-1*H–O*.

Class Stelleroidea[1]

The Class Stelleroidea comprises four subclasses of star-shaped, free-living, stemless echinoderms—*Asteroidea* (starfishes, or sea stars), *Ophiuroidea* (brittle stars), *Auluroidea* (extinct forms resembling both the preceding in some respects), and *Somasteroidea* (extinct and including most ancient and primitive forms)—all characterized by a flexible, star-shaped body consisting of a **central disk** and five or more radiating **arms** (or **rays**) (Figs. 14-3, 14-4, 14-32, 14-34). In the ophiuroids and auluroids the arms are sharply set off from the disk and are composed of small calcareous plates arranged in series; in the asteroids the arms merge into the disk and in some forms are so short as to be practically indistinguishable from the disk, in which cases the entire body is perfectly pentagonal or hexagonal; in the somasteroids the arms are merely differentiated portions of the oral surface. The arms of asteroids are hollow but are studded with numerous marginal and dorsal plates. Arm structure in three subclasses of stelleroids is illustrated in Fig. 14-31.

All stelleroids have a water-vascular system, and the arms are equipped with radial water vessels and podia, as shown in Fig. 14-31.

The mouth is centrally situated on the oral side and the anal vent directly opposite on the summit of the aboral surface. Typically, five food or ambulacral grooves radiate from the mouth with pentamerous symmetry (Figs. 14-2, 14-4). They are confined to the central disk in ophiuroids but extend to the tips of the arms in asteroids and auluroids; they were undeveloped in

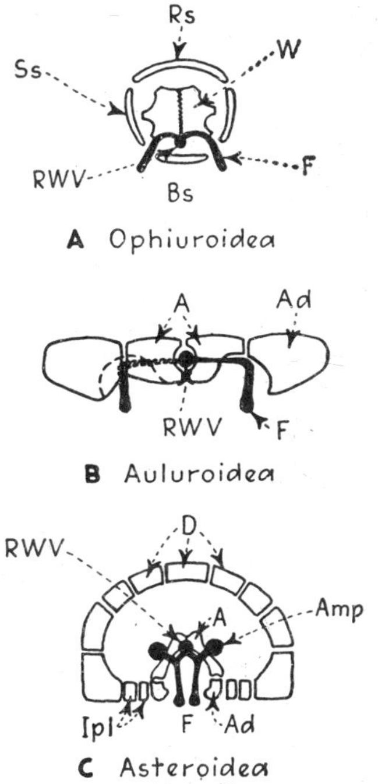

FIG. 14-31. Class STELLEROIDEA. Arm structure. *A*. Cross section of an ophiuroid arm, showing the ventral (*Bs*), lateral (*Ss*), and dorsal (*Rs*) shields, the vertebral ossicles (*W*), the radial water vessel (*RWV*), and the podia (*F*). *B*. Cross section of an auluroid arm, showing the alternating ambulacralia (*A*), the adambulacralia (*Ad*), radial water vessel (*RWV*), and podia (*F*). *C*. Cross section of an asteroid arm, showing dorsal (*D*) and marginal skeletal elements, accessory ossicles (*Ipl*), adambulacralia (*Ad*), ambulacralia (*A*), ampullae (*Amp*), radial water vessel (*RWV*), and podia (*F*). (*Simplified after Lyman*, 1882, *and Schoendorf*, 1910.)

[1] Stelleroidea—N. L., from Fr. *stellérides*, from L. *stella*, star, + Gr. *-oid*, like; referring to the starlike appearance of members of the class.

the somasteroids. Prolongations of the viscera extend into the hollow arms of the asteroids, but not into the slender segmented arms of the ophiuroids and auluroids.

The body of the stelleroid is generally not enclosed within a rigid or flexible calcareous box, as is the case with most pelmatozoans and with the eleutherozic echinoids; instead, it has a tough integument that is given strength by small, irregular calcareous plates (**ossicles**), which may or may not be united, and by calcareous spines (Fig. 14-33).

The oldest known fossil stelleroids come from Lower Ordovician rocks. Fossil specimens are relatively rare but have been found in marine formations in almost every period since the Cambrian. In existing seas starfish and brittle stars are extremely abundant on many bottoms where they carry on predatory activities.

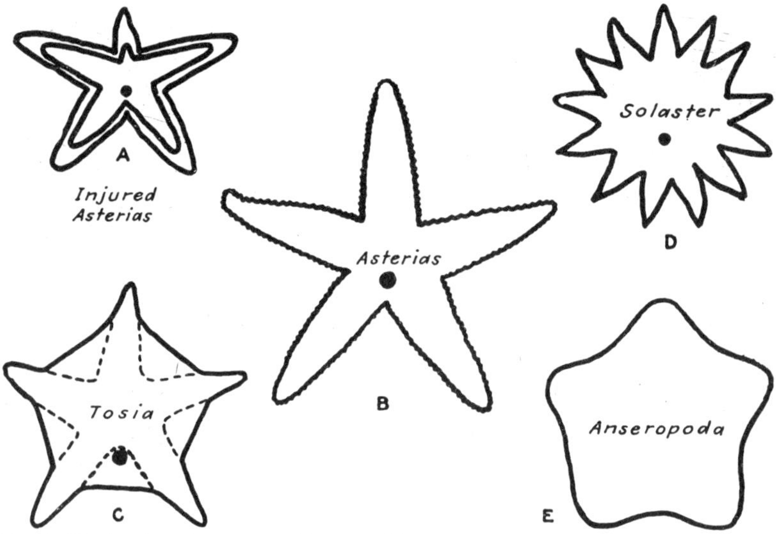

FIG. 14-32. Subclass ASTEROIDEA. Diagrams illustrating variation in shape of test among asteroids. In *A* the smaller inner outline shows the specimen soon after a lost arm had started to regenerate. The outer outline shows the same starfish after a few weeks of growth. All figures are about one-third to one-fourth the size of a fully grown adult. The madreporite is shown by a circular black spot.

Subclass Asteroidea[1]

The Asteroidea have star-shaped or polygonal (typically pentagonal) bodies consisting of 5 to 40 or more simple, more or less flattened, unbranched arms that radiate from an indefinitely outlined central disk (Fig. 14-2). Prolongations of the viscera extend into the hollow arms, and if one of the arms is torn off another may be generated to replace it (Fig. 14-32). If the whole animal is torn apart in such manner that each of the two portions includes some of the central disk, two individuals may ultimately develop, although

[1] Asteroidea—Gr *aster*, star, + *-oid*, like; referring to the starlike appearance of the typical sea star, or starfish.

there is a good chance that both mutilated specimens will perish. Contrary to the oft-repeated statement that an arm alone will regenerate a complete organism, such a phenomenon is not known to have been produced under controlled experimental conditions (Mead, 1900).

The morphology of the familiar starfish, *Asterias forbesi* (Fig. 14-32), is typical of the subclass. On the dorsal or aboral surface in central position is the tiny anal vent; near it lies the prominent madreporite at the junction of the two posterior rays (Fig. 14-4). On the oral side the ambulacral grooves extend from the centrally located mouth to the tips of the arms, where they terminate in a tiny calcareous **ocular plate,** which carries a light-sensitive organ. The ambulacra are plentifully provided with tube feet equipped with suction disks (Fig. 14-3). The mouth, situated in the center of the oral surface, has a pentagonal outline made by the inward projection of five pairs of oral plates (**orals**).

The **skeleton** of an asteroid consists of numerous detached or united ossi-

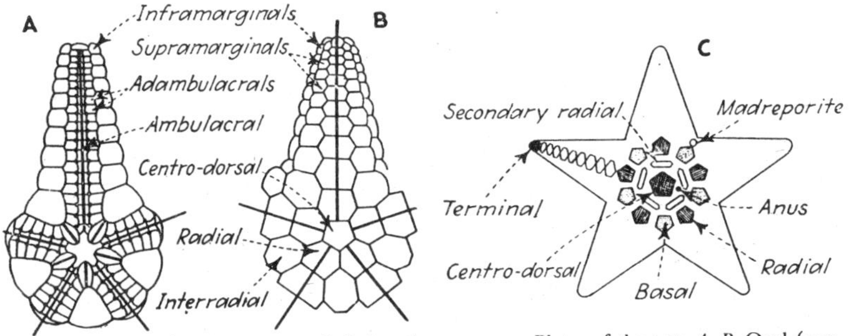

FIG. 14-33. Class STELLEROIDEA—Subclass ASTEROIDEA. Plates of the test. *A–B*. Oral (ventral) and aboral (dorsal) views of the hypothetical phylembryo of the Stelleroidea, based on *Hudsonaster* (cf. Fig. 14-34*A*). *C*. Diagram showing relations of the chief plates of the apical system in a young starfish. (*A–B after Schuchert*, 1915; *C after Parker and Haswell*, 1928.)

cles that touch along the edges, overlap, or join in reticulate fashion. These ossicles lie in the fleshy mesoderm, which produces them, and are covered by ciliated epidermis. They are of various but definite shapes and have a regular arrangement (Fig. 14-33*C*). The aboral surface carries many blunt calcareous spines, which are parts of the skeleton, and in many species minute pincerlike **pedicellariae** between the spines. Each pedicellaria has two tiny jaws that open and snap shut when disturbed; these unusual structures, found also on the echinoids (Fig. 14-44), keep the body surface free of debris and of small organisms that may be looking for a place of attachment.

The skeletal elements of recent and fossil asteroids are compared in Fig. 14-33. In recent species the ends of the ambulacral plates are opposed along the median line of the ambulacrum like the rafters of a roof, but in Paleozoic species they are arranged in alternate rows and are inclined toward each other at a small angle. Pores through which the tube feet protrude are present along the junctions of adjacent plates. The ends of the flooring and roofing plates fit against a series of **adambulacral plates** (**adambulacralia**) and these in turn may be bounded by several kinds of marginal plates (**marginalia**) (Fig. 14-33*A–B*).

It is customary to base the classification of the Asteroidea on the size of the marginal plates in the arms. Two orders are recognized—*Phanerozonia*, with large and conspicuous plates; and *Cryptozonia*, with small plates. Most fossil asteroids are referred to the first order; existing species include representatives of both orders.

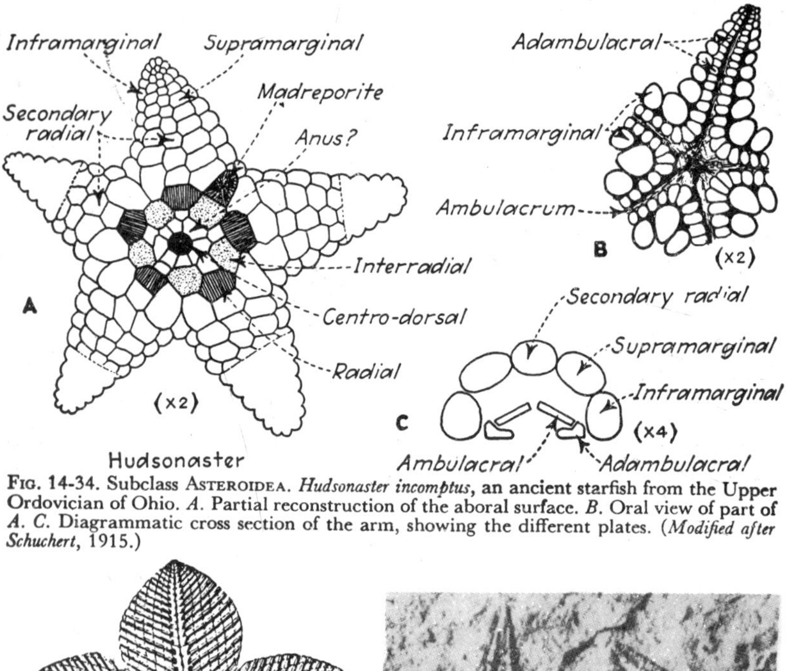

FIG. 14-34. Subclass ASTEROIDEA. *Hudsonaster incomptus*, an ancient starfish from the Upper Ordovician of Ohio. *A.* Partial reconstruction of the aboral surface. *B.* Oral view of part of *A. C.* Diagrammatic cross section of the arm, showing the different plates. (*Modified after Schuchert*, 1915.)

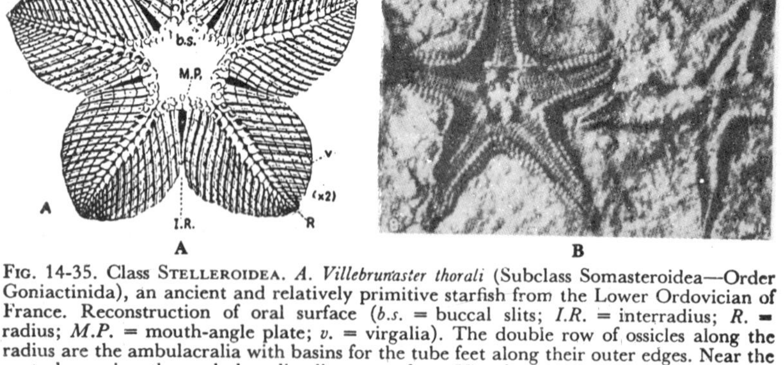

FIG. 14-35. Class STELLEROIDEA. *A. Villebrunaster thorali* (Subclass Somasteroidea—Order Goniactinida), an ancient and relatively primitive starfish from the Lower Ordovician of France. Reconstruction of oral surface (*b.s.* = buccal slits; *I.R.* = interradius; *R.* = radius; *M.P.* = mouth-angle plate; *v.* = virgalia). The double row of ossicles along the radius are the ambulacralia with basins for the tube feet along their outer edges. Near the central opening the ambulacralia diverge to form V's of the mouth frame. The spaces between the ambulacralia are filled in by rod-shaped interambulacralia (virgalia). The mouth-angle plates are in the interradii and adjoin the central opening. *B.* Two specimens of *Astropecten* (Subclass Asteroidea), from the Cretaceous of California. (*A after Spencer*, 1951; *B after Asaeda in Durham*, 1948.)

The oldest and simplest representatives of the Asteroidea seem to have appeared in the Middle Ordovician. *Hudsonaster* (M. Ord.–Sil.), whose detailed structure is illustrated in Figs. 14-33*A*–*B* and 14-34, is typical of the early forms. It is to be noted that the plates of its aboral surface (Figs. 14-33*B*, 14-34*A*) closely resemble those that appear in young individuals of recent starfish (Fig. 14-33*C*), of which *Asterias* (Fig. 14-2) is a familiar example. Figure 14-35*B* shows an exceptionally well preserved fossil specimen of *Astropecten*, a Cretaceous starfish. Inasmuch as the remains of 35 individuals of this species were found in about a square foot on a single bedding plane,[1] and no others in the immediate area, it would seem that ancient starfish, like their relatives in modern seas, also had the gregarious habit so characteristic of the echinoderms as a whole.

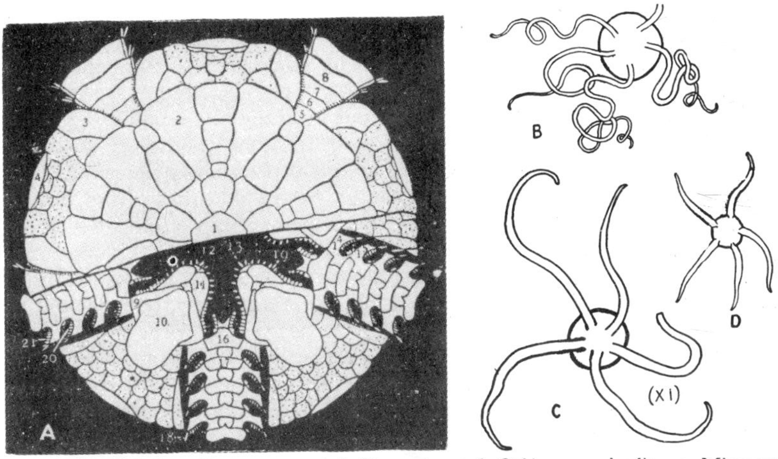

FIG. 14-36. Subclass OPHIUROIDEA. *A*. Reconstructed *Ophiura marylandica*, a Miocene ophiuroid. The upper half shows the aboral side; the lower, the oral side. (1 = central disk plate; 2 = radial shield; 3 = comb plate; 4 = interradial marginal plate; 5–8 = first to fourth dorsal arm plates; 9 = side mouth shield; 10 = oral shield; 11 = jaw; 12 = torus angularis; 13 = teeth; 14–15 = first and second lateral arm plates; 16–17 = first and second ventral arm plates; 18 = spine; 19 = mouth papillae; 20 = tentacle pore; 21 = tentacle scales; 22 = genital slit.) The stippled plates are hypothetical. *B*–*D*. Modern forms: *B*, a typical "serpent star" with characteristic long and serpentlike arms; *C*, *Amphipholis; D*, *Ophioglypha*. (*A after Berry*, 1934; *B*, *C traced from H. L. Clark*, 1915; *D traced from Nicholson and Lydekker*, 1889.)

Subclass Ophiuroidea[2]

The ophiuroids have a sharply defined central disk, which contains the digestive, genital, and other organs of the body, and five or more snakelike flexible arms that radiate from the disk with radial symmetry (Fig. 14-36)

[1] DURHAM, J. W., Fossil starfish, *Pacific Discovery*, vol. 1, pp. 28–29. DURHAM, J. W., and ROBERTS, W. A., 1948, Cretaceous asteroids from California, *Jour. Paleontology*, vol. 22, pp. 432–439.

[2] Ophiuroidea—Gr. *ophis*, a serpent, + *oura*, tail, + *-oid*, like; referring to the slender, flexible arms that have a snakelike appearance and are responsible for the popular designation, "serpent star."

and lack any extensions of the viscera. There is no true anal vent, the madreporite lies near the mouth on the oral surface, there are no ambulacral grooves, and the ambulacra lack ampullae and podia.

The skeletal ossicles lie in the mesoderm and are held together by connective tissue and muscles; a tough spiny epidermis covers the entire animal. The central disk is composed of many different plates, some with regular shape and arrangement and others irregular in both respects. In some forms there are five pairs of **radial shields** at the places where the arms originate. Five **mouth shields** form a circlet around the mouth, and one of these is modified to function as a madreporite. An inner circlet of **side mouth shields** bear dentate jaws. These and other disk plates are illustrated in Fig. 14-36*A*.

The arms, which may reach a length of 60 cm. (2 ft.) in the largest specimens, are flexible, in many cases branched, and are capable of a snakelike wriggling motion that has given the popular designation "serpent stars" to the group. They are composed of a single axial series of tiny calcareous **vertebral ossicles** surrounded by four series of marginal plates as shown in Figs. 14-31*A* and 14-36. Each ossicle results from the fusion of four adjacent plates, and the ossicles themselves articulate with one another by means of domelike elevations on each side. The marginal plates are arranged in four series—a ventral, a dorsal, and two lateral. No ambulacral furrows are developed on the arms, but the undersides of the ossicles are indented to receive the radial water vessel. Small podia pass to the exterior through tubes in the lateral margins of the vertebral ossicles; these podia lack terminal suction disks, hence cannot perform a locomotory function, but they aid in respiration. Blood vessels and nerve cords extend into the arms, but no caeca of the intestines or prolongations of the genital organs. The arms are quite brittle and easily broken or shed, so that the designation "brittle star" is widely used for ophiuroids. Complete fossil specimens are rare and were preserved only under unusual depositional conditions.

Some 1,500 species of living ophiuroids (200 genera) have been described; these are largely confined to deeper waters (*i.e.*, more than 500 m., or 1,650 ft.) where they have been encountered in prodigious numbers. The *Challenger* expedition (Lyman, 1882) dredged specimens by the hundred-

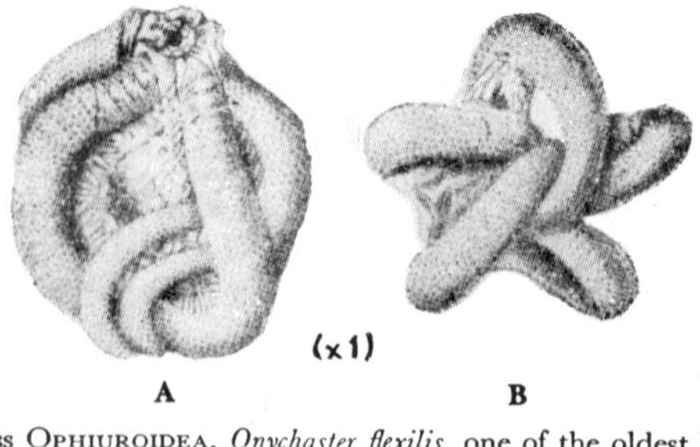

FIG. 14-37. Subclass OPHIUROIDEA. *Onychaster flexilis*, one of the oldest known ophiuroids, from the Mississippian of Indiana. *A–B*. Two views of a specimen with arms intertwined. (*After Meek and Worthen*, 1868.)

weight from the Atlantic and more recently expeditions to the Mid-Atlantic Ridge have photographed specimens on bottoms several miles deep.

The earliest known fossil ophiuroids (*Onychaster;* Fig. 14-37) come from Mississippian rocks, but the subclass is believed to have evolved earlier, probably during the Devonian, from some auluroid ancestor. They do not seem to have become abundant until the Mesozoic, and complete fossil specimens are rare in any rocks. More common as fossils are the detached plates, but these are generally of little paleontological interest (Berry, 1934; Howe, 1942).

Because of the sparse fossil record, ordinal and lower taxonomic divisions have been based upon the outward appearance and arrangement of plates in existing species (Lyman, 1882; Clark, H. L., 1915). It is possible, as suggested by Berry (1934), that critical study of the plates themselves might result in a more natural classification.

Typical of the subclass, which ranges from Middle Mississippian to Recent, are *Onychaster* (Miss.; Fig. 14-37), *Ophiura* (L. Cret.–Recent; Fig. 14-36), and *Ophioglypha* (Recent).

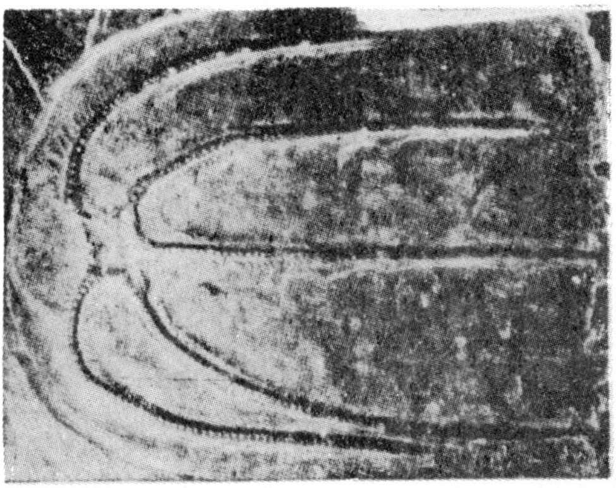

FIG. 14-38. Class STELLEROIDEA—Subclass AULUROIDEA. "*Ophiura*" *decheni*, a well-preserved auluroid (×1½) from the Lower Devonian black shale of the Bundenbach region in the German Rhineland. The pentagonal central disk exhibits the interradial concavities characteristic of representatives of the subclass.

Subclass Auluroidea[1]

The Auluroidea constitute a group of extinct starfish that have some characteristics of both asteroids and ophiuroids. They are believed to have evolved from an ancestral asteroid in the Early Paleozoic and to have given rise to the Ophiuroidea in the Mid-Paleozoic (Fig. 14-39). Their own geologic range is from Early Ordovician to Late Mississippian.

The auluroids resemble the ophiuroids in having a distinct disk, typically with concave margins with or without marginal ossicles, and slender ophiuran-like arms of such structure that intestinal caeca and prolongations of

[1] Auluroidea—Gr. *aulos*, flute or oboe, + *-oid*, like; referring to the tubular nature of the auluroid arms.

the genital organs do not seem to have extended into them. As in the Ophiuroidea, also, the radial water vessel lay within the ossicles (Fig. 14-31*B*). The ambulacral furrows were open and a true madreporite was present (although it lies on the oral surface rather than on the aboral as in the asteroids); these features indicate relationships with the Asteroidea. Because of the peculiar combination of stelleroid features, the Auluroidea are generally regarded as an intermediate link between the Asteroidea and Ophiuroidea (Fig. 14-39).

Five simple unbranched arms radiate from a fairly definite pentagonal central disk, which typically has concave interradial margins (Fig. 14-38). The arm itself consists of four rows of plates, with the plates of adjacent rows in alternating position (Fig. 14-31). The two axial rows are ambulacrals, and the two marginal are adambulacrals. The arm plates are in no case united to form vertebral ossicles, as in the Ophiuroidea, and the arms were probably much less flexible than in that group.

The Subclass Auluroidea has been divided into two orders based on the arrangement of the ambulacral plates—Lysophiuroida (= Lysophiurae) with the plates arranged alternately, and Streptophiuroidae (= Streptophiurae) with the plates lying opposite one another.

The auluroids seem to have appeared fairly early in the Ordovician and to have become extinct by the end of the Mississippian. Many excellently preserved specimens have been taken from the famous Lower Devonian black shales of Germany (Fig. 14-38).

Subclass Somasteroidea

Representatives of this extinct Lower Ordovician subclass have the central part of the body large and the arms merely differentiated portions of the oral surface. In the earliest genera the oral side has only two kinds of ossicles: (1) **ambulacralia,** which are arranged in a double alternating row stretching from the central mouth to an arm extremity; and (2) **interambul cralia,** which consist of rod-shaped ossicles (**virgalia**) placed in a linear series at an angle to the ambulacralia (Fig. 14-35A). Later forms show differentiation of the interambulacral rows into **marginalia** and **adambulacralia,** and still later species lack an interambulacral skeleton altogether.

If an aboral skeleton is present, it is of primitive structure and consists of ossicles with a small center from which radiate slender branches, so that the whole skeleton forms a net with a wide mesh. This is the most primitive skeletal structure known in the Echinoderma, and it forms the basic framework for the aboral skeletons of all later stelleroids.

The Subclass Somasteroidea[1] includes the earliest known starfish (*Villebrunaster* from the Lower Ordovician) and is of unusual interest because its members display peculiar and primitive features of skeletal structure not found in later stelleroids. Somasteroids are believed to show the first stages in the differentiation of a starfish, which possibly evolved from a ciliary feeding pelmatozoan (Spencer, 1951). The earliest representatives show no evidence of an ambulacral groove—a feature previously regarded as primitive

[1] With a single order, Goniactinida

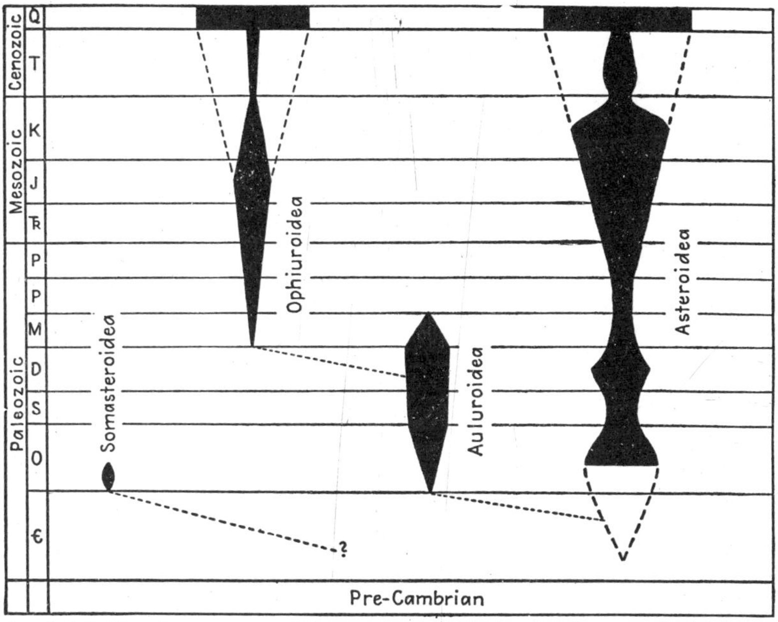

FIG. 14-39. Class STELLEROIDEA. Diagram showing relative abundance, geologic history, and supposed relationships of the four orders of stelleroids. The heavy dashed lines indicate postulated extensions of the known fossil record, which is indicated in solid black. It has been postulated that the Auluroidea arose from a Cambrian asteroid and gave rise to the ophiuroids sometime during the Devonian. The relationship of the somasteroids to other stelleroids is uncertain. (*Adapted in part from Berry*, 1934.)

and used as a basis for linking stelleroids with edrioasteroids. The tube feet, each housed on two ambulacralia as in all asteroids and auluroids, were placed in position by changes in body shape. Somasteroids seem to have lived on small food either from the planktonic shower or from that in the upper layers near the bottom, and some forms probably burrowed in soft mud.

Characteristic of the subclass, whose members have many primitive skeletal features in addition to the essential characters of all stelleroids, is *Villebrunaster* (Fig. 14-35*A*) from the Lower Ordovician, the earliest known starfish.

CLASS ECHINOIDEA[1]

General Considerations

The Echinoidea, variously known as sea urchins, heart urchins, sand dollars, sand shillings, shield urchins, and echini, are echinoderms having the body encased in a firm, usually rigid and spinose, flattened globular test composed of many symmetrically arranged plates and without free arms or

[1] Echinoidea—Gr. *echinos*, used as a designation for both hedgehog and a sea urchin, + *-oid*, like; referring to the resemblance to the latter.

stem. The test is commonly more or less hemispheroidal with the lower oral side flattened, but the shape ranges from ellipsoidal (vertical axis longer than horizontal) to discoidal (horizontal axis several times as long as vertical) (Fig. 14-40). In life the test encloses the organs of the body and the complicated water-vascular system and is itself enclosed in a tough dermal integument. The spines are much diversified in size and structure and have been studied extensively because of their common occurrence as fossils.

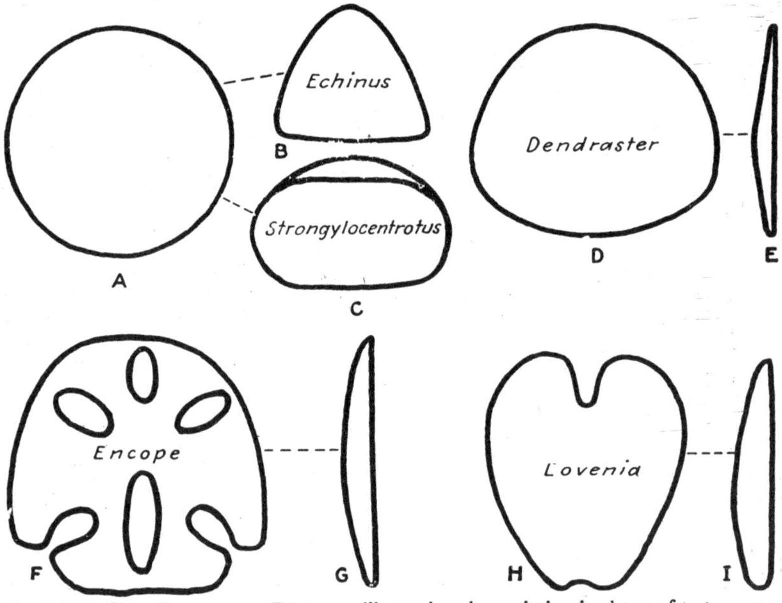

Fig. 14-40. Class Echinoidea. Diagrams illustrating the variation in shape of test among the echinoids. The first figure is the outline of the test when viewed from above; the second is a profile. *A–C.* Regular echinoids illustrated by *Echinus* and *Stongylocentrotus*, either of which may vary somewhat from the perfect circular outline. *D–I.* Irregular echinoids: *D–E,* "sand dollar"; *F–G,* "keyhole urchin"; *H–I,* "heart urchin."

Echinoids exhibit considerable variation in symmetry; **regular** forms typically have radial pentamerous symmetry superimposed on an original bilateral symmetry, the plane of which passes through the mouth and anus so oriented that the madreporite lies in the right anterior interradius or interambulacrum (Fig. 14-45). Later **irregular** forms have a secondary bilateral symmetry that is superimposed on the first and distorts the radial pentamerous symmetry. In these the anal opening has migrated to a posterior position and other important modifications have altered the symmetry.

Echinoids differ from pelmatozoans in being stemless and free-moving; from stelleroids in lacking free arms and having interradial instead of radial reproductive organs; and from holothuroids in having a relatively rigid test and a different living position with respect to the substratum (Fig. 14-1).

Morphology of Soft Parts

The individual echinoid has the organs typical of all echinoderms—alimentary, water-vascular, nervous, and circulatory systems. The mouth lies at or anterior to the center of the oral side of the globular test and opens into a coiled digestive tract consisting of esophagus, stomach, and ciliated intestine. The tract discharges through an anal opening that may lie in a central position on the summit of the test or in a posterior position

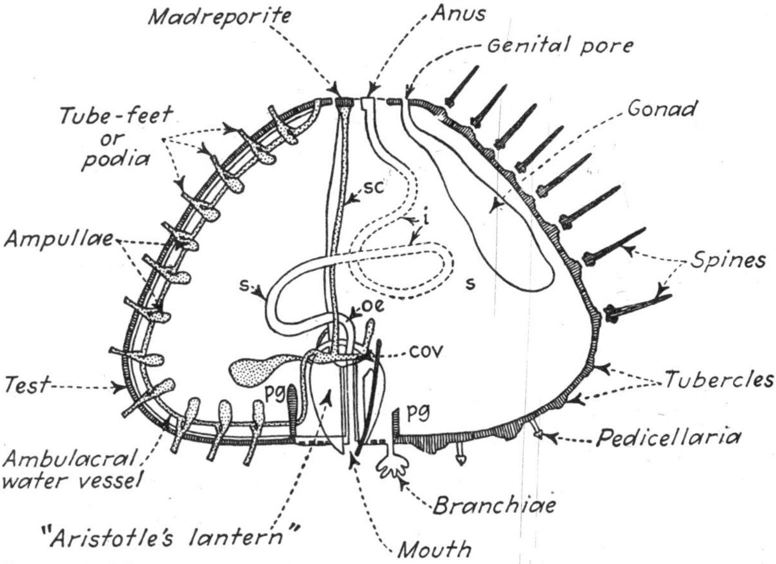

FIG. 14-41. Class ECHINOIDEA. Diagrammatic cross section of *Echinus*, a modern endocyclic echinoid. (*cov* = circumesophageal vessel of the water-vascular system; *i* = intestine; *oe* = esophagus; *pg* = perignathic girdle; *s* = stomach; *sc* = stone canal.) (*Modified after Bather*, 1900.)

elsewhere on the test—on the aboral surface, on the posterior margin, or even on the oral surface in some forms (Fig. 14-42*A–B*).

The mouth is surrounded by a tough plate-studded skin, the **peristome**, and the anal opening by a similar structure, the **periproct** (Fig. 14-47). One plate near the periproct in the anterior interambulacrum serves as a madreporite through which water enters the water-vascular system. The water passes down the stone canal to a ring canal surrounding the esophagus. From this latter canal a radial vessel extends along and under the plates of each ambulacral area, giving off small saclike tubes that extend through pores in the test wall to the surface, where they become **tube feet** or **branchiae** (Fig. 14-41). The tube feet are provided with suction disks and perform a locomotory function, as in the asteroids; the branchiae aid in respiration.

A nerve ring surrounds the esophagus and sends a cord along each ambulacrum beneath the circulatory tube. The genital organs have interradial position in the aboral part of the test, and their products leave the animal through perforate genital plates on the summit of the test (Figs. 14-41, 14-47).

The Test

The echinoid test is a semiglobular or discoidal body composed of hundreds of calcareous plates firmly united by their edges and systematically arranged to form a more or less rigid hollow structure—a few echinoids have a flexible test in which the plates imbricate. The complete test may be divided into four systems of plates: (1) the corona; (2) the plates of the peristome; (3) those of the oculogenital system; and (4) those of the periproct. The periproct and oculogenital system together constitute the apical system in endocyclic echinoids.

The Corona. The **corona** constitutes the main part of the test and is composed of 10 meridional bands of plates—5 of these bands, the radial, or ambulacral, areas, alternate with 5 interradial, or interambulacral, areas. Ambulacral areas have from 2 to 20 columns of plates; interambulacral, from 1 to 14 columns. The plates of the former are perforate and lie above the internal radial vessels of the water-vascular system; those of the interambulacral areas are imperforate (except in a few early and supposedly primitive forms; *e.g.*, *Bothriocidaris* from the Ordovician—Fig. 14-50).

In all modern and almost all Tertiary and Mesozoic echinoids (*Tetracidaris* excepted) there are 2, and only 2, rows of plates in each of the coronal series, although some of the plates may be compound. In Paleozoic forms, by contrast, the number of rows in the interambulacral series ranges from 1 in *Bothriocidaris* (Fig. 14-50), an Ordovician genus, to 14 in some species of the Mississippian *Hyattechinus* (Fig. 14-52*B*) and in the ambulacral series from 2 to 16 in *Lepidesthes* (Fig. 14-52*A*) from the Carboniferous. The zone of greatest horizontal circumference of the corona as seen from above is designated the **ambitus.**

The corona has two primary openings—the oral, with the mouth and its peristomial integument; and the aboral, with the anus (except in irregular forms) and the surrounding periproct and oculogenital systems of plates. The cover for both openings is tough epidermis more or less studded with calcareous plates, which tend to be arranged symmetrically with reference to the plates of the coronal series.

If the mouth lies at the lower and the anus at the upper end of a vertical oral-aboral axis, the corona and the test are described as regular, or **endocyclic,** and are usually radially symmetrical (Figs. 14-42*C–D*, 14-45). In those forms in which the anal opening has migrated from the summit of the test to a "posterior" position (and the mouth in most cases to a somewhat "anterior" position), the corona and test are described as irregular, or **exocyclic** (Fig. 14-42*A–B*), and the symmetry is likely to be bilateral.

The plates of the corona commonly bear tubercles and granules to which are attached various types of movable spines and small pedicellariae. The spines are greatly diversified in character and range from almost microscopic size to heavy clublike forms as much as 30 cm. (1 ft.) in length and 25+ mm. (1 in. or more) in diameter[1] (Fig. 14-43). The larger tubercles and surmounting spines are **primary;** the smaller, **secondary.** Some spines are used in walking; others probably serve for defense or protection. The different parts of a spine and the tubercle to which it is attached are shown in Fig. 14-43.

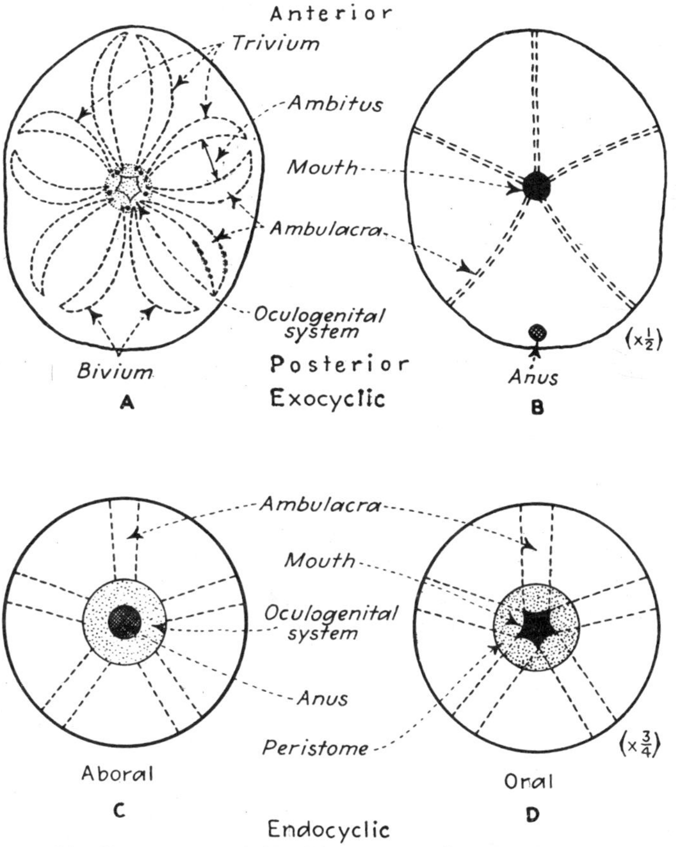

FIG. 14-42. Class ECHINOIDEA. *A–B*. Simplified diagram of an exocyclic echinoid, based on *Clypeaster: A*, aboral aspect; *B*, oral aspect. *C–D*. Simplified diagrams of an endocyclic echinoid, based on *Cidaris: C*, aboral aspect; *D*, oral aspect.

Fascioles are narrow bands of close granular ornamentation, and such an area, the **anal fasciole,** surrounds the anus in some forms.

Pedicellariae are mentioned on preceding pages as being present on asteroids. On echinoids they consist of a short stem surmounted by a head that is composed of two or more pincerlike valves (Fig. 14-44). They may

[1] A large urchin living in Florida waters is 15 cm. (6 in.) in diameter and has spines as much as 30 cm. (12 in.) long.

be present on any part of the test and are known from experimentation to perform at least two functions: (1) defending the echinoid against large predatory organisms and small parasites by snapping the valves open and shut; and (2) ridding the test of exuviae and foreign debris by passing the particles along from one head to the next. Although tiny and delicate structures, pedicellariae have been preserved, and Geis (1936) mentions fossil forms from rocks as old as Mississippian. The pedicellariae of modern

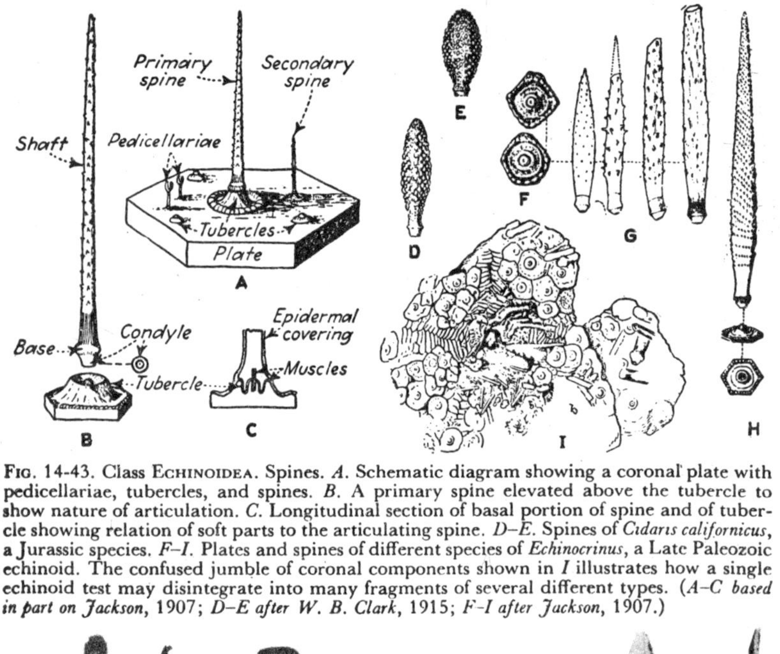

FIG. 14-43. Class ECHINOIDEA. Spines. *A*. Schematic diagram showing a coronal plate with pedicellariae, tubercles, and spines. *B*. A primary spine elevated above the tubercle to show nature of articulation. *C*. Longitudinal section of basal portion of spine and of tubercle showing relation of soft parts to the articulating spine. *D–E*. Spines of *Cidaris californicus*, a Jurassic species. *F–I*. Plates and spines of different species of *Echinocrinus*, a Late Paleozoic echinoid. The confused jumble of coronal components shown in *I* illustrates how a single echinoid test may disintegrate into many fragments of several different types. (*A–C based in part on Jackson*, 1907; *D–E after W. B. Clark*, 1915; *F–I after Jackson*, 1907.)

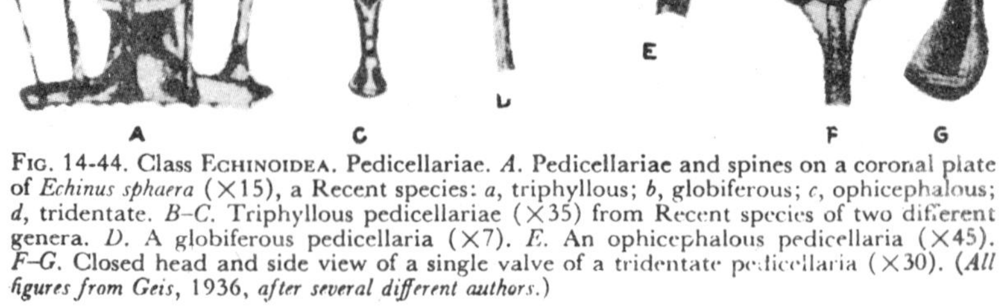

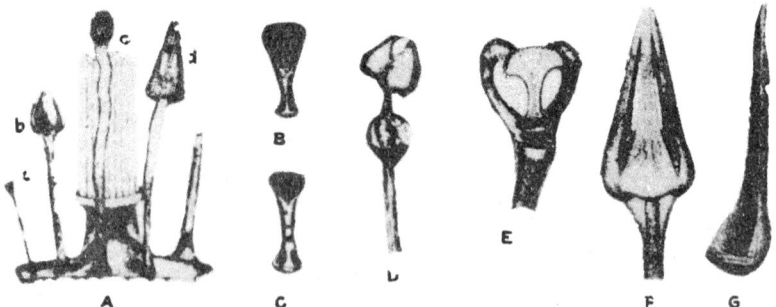

FIG. 14-44. Class ECHINOIDEA. Pedicellariae. *A*. Pedicellariae and spines on a coronal plate of *Echinus sphaera* (×15), a Recent species: *a*, triphyllous; *b*, globiferous; *c*, ophicephalous; *d*, tridentate. *B–C*. Triphyllous pedicellariae (×35) from Recent species of two different genera. *D*. A globiferous pedicellaria (×7). *E*. An ophicephalous pedicellaria (×45). *F–G*. Closed head and side view of a single valve of a tridentate pedicellaria (×30). (*All figures from Geis*, 1936, *after several different authors*.)

echinoids are generally placed in one of four morphological categories—**globiferous, tridentate, ophiocephalus,** and **triphyllous** (Fig. 14-44). These are intergradational to some extent, and fossil pedicellariae usually cannot be assigned to a definite category because diagnostic soft parts are missing.

The echinoid test increases in size through addition of plates at the aboral ends of the ambulacral and interambulacral bands; these plates are inserted outside those of the oculogenital system. The test may also be increased in size by enlargement of plates already present.

Ambulacra. Ambulacra are continuous from peristome to periproct in regular echinoids and constitute a band stretching from one pole to the other; in most irregular echinoids they are confined to the aboral part of

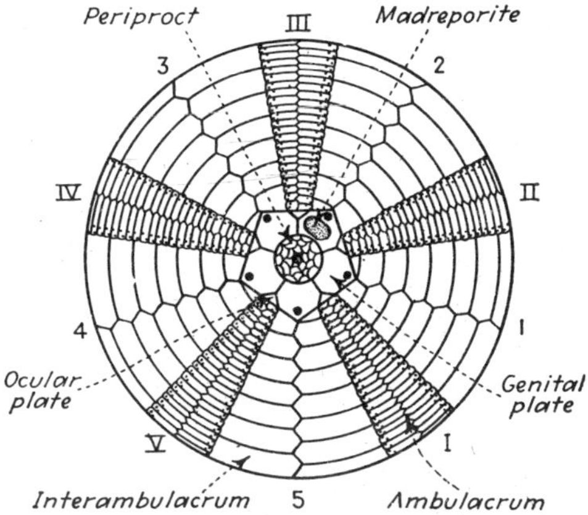

FIG. 14-45. Class ECHINOIDEA. Diagram of the aboral part of a regular echinoid, showing the numbering system used in designating the ambulacra and interambulacra. The posterior interambulacrum is 5. The madreporite lies in interambulacrum 2.

the test and radiate from the periproct like petals of an open flower (Fig. 14-42). They are **simple, petaloid,** or **subpetaloid.** Simple ambulacra are essentially parallel-sided except near the summit. Petaloid types are considerably wider in the middle part than at the two extremities. The widest part of the ambulacrum is designated the **ambitus;** this use of the term must not be confused with the other definition in which it refers to the greatest latitudinal diameter of the corona.

It is now customary to number the ambulacral and interambulacral areas of echinoids according to a system devised by Loven and illustrated in Fig. 14-45. Roman numerals are used for the ambulacra and Arabic for the interambulacra. The test is placed in living position and oriented so that the madreporite lies in the upper right-hand interradius. Ambulacrum III is then anterior, and interambulacrum 5 is posterior. According to Loven, the plates of the interambulacral areas are symmetrical with respect to a plane including ray B and interray E–D; this plane has been designated the **echinoid plane** (Bather, 1929). (See Fig. 14-4*B*.)

In regular echinoids the madreporite and vertical axis determine the so-called **M plane,** and the anal vent is shifted slightly from the apical pole along radius B (Fig. 14-4). In most exocyclic echinoids the anal vent passes along the interradius between radii B and A (Fig. 14-4), whereas the mouth position moves along radius D. The plane determined by these two radii has been designated **Loven's plane.** In exocyclic echinoids the ray D is termed anterior (Fig. 14-42*A*), hence rays C, D, and E form the **trivium** and rays B and A the **bivium**—it is to be noted, however, that these are not the same rays that form the bivium and trivium in other classes of echinoderms. (See Fig. 14-4*A*.)

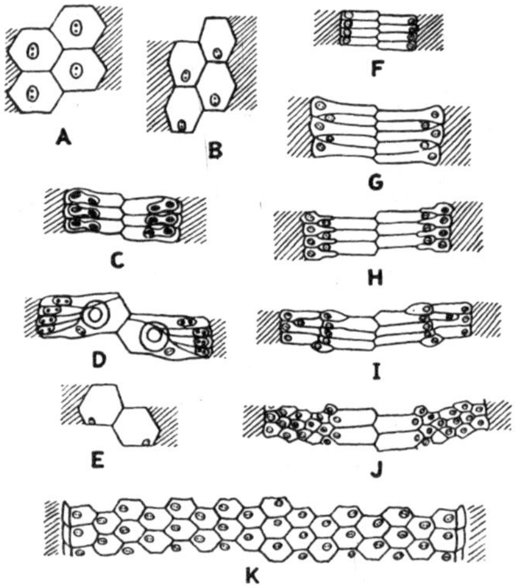

FIG. 14-46. Class ECHINOIDEA. Structure of the ambulacrum in representative echinoids. The lateral portions of the interambulacra are indicated by lining. No plates of the interambulacra are shown. *A. Bothriocidaris. B. Goniocidaris. C. Eucidaris. D. Strongylocentrotus. E. Micraster. F. Palaeechinus. G. Maccoya. H. Lovenechinus. I. Oligoporus. J. Melonechinus. K. Lepidesthes.* (*A*, Ord.; *B–D*, Recent; *E*, Cret.; *F–K*, Miss.) (*Adapted after Jackson*, 1907.)

The pores in the plates of the ambulacral areas are typically in pairs, though in a few cases they are unpaired (Fig. 14-46). They tend to lie along the plate nearest the adjacent interambulacrum, so that the two columns of ambulacral plates have a broad imperforate band down the median line of the series. If the pairs of pores are situated one above the other in a single continuous series stretching from peristome to periproct, the arrangement is described as **uniserial.** If there are two lines of pores on each side of the ambulacrum, the arrangement is **biserial;** and if more than two, **polyserial** (= **pleuriserial**) (Fig. 14-46). A pore pair may be surrounded by an elevated rim, the **peripodium,** and if the two pores are united by a transverse furrow, they are **conjugate.**

The Oculogenital System (Fig. 14-47). Each ambulacrum terminates aborally in a porous plate, the **ocular,** which contains a light-sensitive organ.

This plate has a single pore except in some Paleozoic forms in which there seem to be either two or none. Each interambulacrum ends aborally in a **genital plate,** which has one pore. The genital plate in the right anterior interambulacrum of endocyclic echinoids is greatly perforated and serves as a madreporite. The openings in the genital plates provide a means of escape for the products of the reproductive glands (Fig. 14-41).

The five oculars and five genitals together constitute the **oculogenital system,** which forms a ring around the periproct. If the 10 plates make a single continuous ring—an arrangement characteristic of Paleozoic echinoids—the oculars are described as **insert** (Fig. 14-47*A–B*). If, however, the oculars lie outside the genital ring, so that the plates of the latter are in

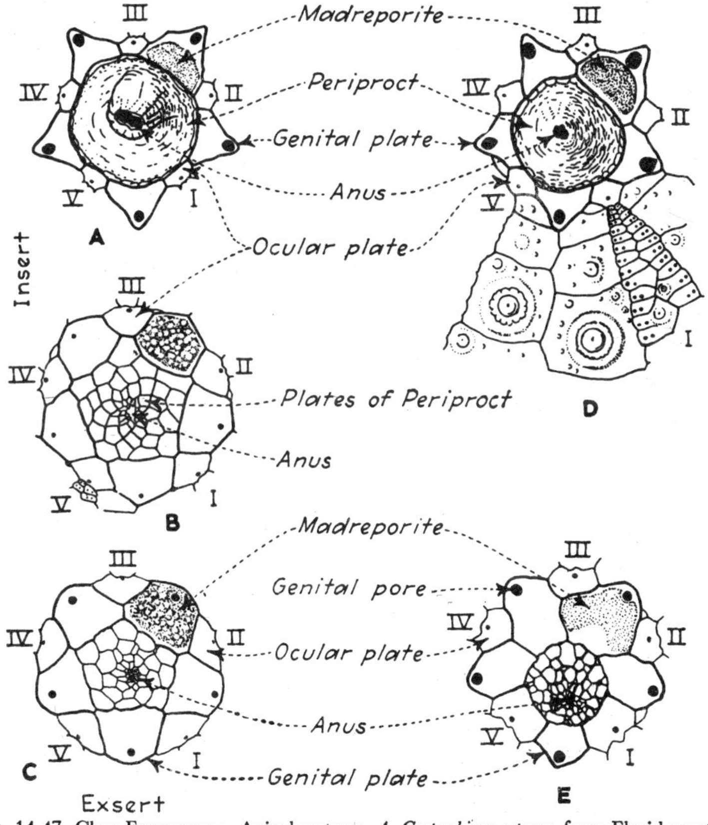

FIG. 14-47. Class ECHINOIDEA. Apical systems. *A. Centrechinus setosus*, from Florida waters, with oculars insert; *D*, same species, from West Indian waters, with oculars I, IV, and V insert. *B. Cidaris affinis*, a modern species, with all oculars insert and with periproctal plates. *C. Eucidaris tribuloides*, from Jamaican waters, with all oculars exsert. *E. Strongylocentrotus dröbachiensis*, from Massachusetts waters, with oculars I and IV insert, the typical condition in the species. (*After Jackson*, 1907.)

contact with themselves and the periphery of the periproct, the oculars are **exsert** (Fig. 14-47*C*). This arrangement is characteristic of many Mesozoic forms. In some species with exsert oculars, two or more of the genital plates may unite, and in one group all five plates are fused into a single pentagonal plate.

It is important to note that both ocular and genital plates retain an apical position independent of the migration of the periproct and anal vent.

The Periproct. The periproct, which always bears the anal vent, lies inside the oculogenital ring in endocyclic echinoids and outside in exocyclic forms (Fig. 14-42). In the latter it is situated some distance posteriorly from the aboral pole in the median line of the posterior interambulacrum, and in some forms it lies on the oral surface (Fig. 14-42*B*). It may be a simple tough integument, or such studded with many calcareous plates. It is rarely preserved, but its position in the test is generally obvious and is useful in systematic classification.

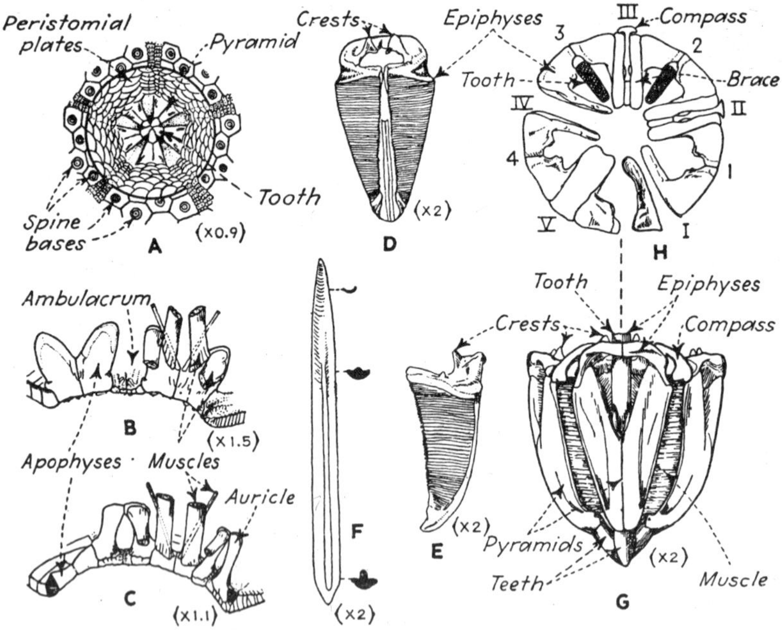

FIG. 14-48. Class ECHINOIDEA. The peristome and associated structures. *A*. *Echinocrinus wortheni* (Mississippian). The complete peristome and basal part of corona restored. *B*. *Eucidaris tribuloides* from Bahaman waters. Several elements of the perignathic girdle attached to basal part of corona. The muscles are shown attached to one set of elements. *C–H*. *Strongylocentrotus dröbachiensis* from Maine waters: *C*, same as *B* but somewhat more complex; *D*, pyramid of interradius 4 without the tooth; E, lateral view of *D*; *F*, tooth, flattened out, with sections at three points to show nature of cross section; *G*, complete lantern, showing teeth in position in the pyramids; *H*, top of lantern to show structure. Teeth are in position in areas 2 and 3 only. Certain other structures are removed. (*After Jackson*, 1907.)

The Peristome (Fig. 14-48*A*). The peristome, as the name implies, is the structure around the mouth. It varies in outline from circular to polygonal (pentagonal or decagonal) and always lies on the oral surface—centrally located in endocyclic species and somewhat anteriorly situated in exocyclic forms. It may be a simple naked integument with the mouth at the center, but quite commonly it is partly or almost completely clad with small imbricating plates, the arrangement of which has taxonomic importance.

Many echinoids have a unique masticatory apparatus known as **Aristotle's lantern** or simply **lantern** (Fig. 14-48). This structure (except in the Clypeastroids) is composed of 40 calcareous pieces, of which 5 are powerful **teeth** arranged in a circlet around the mouth (Fig. 14-48*A*). The other 35 pieces, which vary from species to species, consist of 10 pieces making 5 **compasses,** 10 **epiphyses,** 10 pieces making 5 **pyramids,** and 5 **braces.** These are illustrated in Fig. 14-48*D–G*. Certain of the pieces articulate with one another, and the entire lantern articulates with an internal skeletal process, the **perignathic girdle,** that surrounds the mouth. The teeth are independently moved by means of powerful muscles and are capable of cutting many kinds of materials, including hard rocks like quartzite and granite.

Ecology

Modern echinoids are exclusively marine, and the group seems always to have been confined to that habitat. They live on all sorts of bottoms and at all depths from the shallowest waters to the abyss. Typically gregarious, they commonly assemble in immense numbers and are active predators and scavengers. Many species eat bottom materials for contained organic matter; some burrow in sand and mud; and still others prefer rocky bottoms. On wave-beaten, rocky coasts certain species cut holes in the hardest rocks (*e.g.*, quartzite and granite) and nestle in the depressions more or less protected from the waves (*e.g.*, *Strongylocentrotus dröbachiensis* along the Maine coast). Since the same species do not cut holes on rocky bottoms in quiet waters, it is assumed that the tendency to cut is an adaptive one.[1]

Stratigraphic Range and Geologic History

The earliest known echinoids come from Middle and Upper Ordovician rocks; *e.g.*, *Bothriocidaris* and *Myriastiches* (M. Ord.), *Aulechinus* and *Ectinechinus* (U. Ord.). In general, Paleozoic echinoids, at least in recognizably large fragments, are rare fossils; most genera and species thus far described have come from Devonian and Mississippian[2] rocks. Echinoid spines and isolated

[1] Those who have made a special study of urchins and their excavations conclude that the holes are made in part by action of the teeth and coronal spines and in part by the motions of the whole animal as it is rotated in the excavation by turbulent water. In the first case the animal intentionally uses its teeth and spines; in the second it is a passive object in the grasp of the water. Burrowing and grinding are believed to result from a combination of the two activities (Grew, 1890).

[2] Specimens from the prolific assemblage of *Melonechinus* in Mississippian limestone at St. Louis, Mo., are familiar objects in larger museums all over the world.

plates or small clusters of plates are abundant in Pennsylvanian strata in many areas, but entire or even partial coronas are rare. During the Mesozoic and Cenozoic the class flourished, but fossil specimens are more or less restricted to local assemblages in certain formations. As an example, individuals of the genus *Enallaster* can be collected by the thousands from certain layers of Lower Cretaceous limestone in Texas.

Most fossil echinoids have limited vertical distribution and would, therefore, be excellent index fossils were it not for the facts that they are not particularly abundant and that when they are found in large numbers they are almost certain to have limited areal distribution. From a practical point of view, therefore, fossil echinoids are of less stratigraphic importance in most formations than pelmatozoan echinoderms. In recent years, however, as the attention of paleontologists has turned to microfossils, echinoid fragments have become of considerable interest and importance, and their possible future use, particularly in subsurface correlations, has been pointed out by several investigators (Smiser, 1933; Geis, 1936; and Howe, 1942).

The earliest fossil echinoids had more or less than 20 meridional rows of plates in the corona, but the number 20 seems to have become established in the Early Mesozoic and has been maintained to the present time. The earliest echinoids were exclusively endocyclic; exocyclic forms did not appear until Mid-Mesozoic time, but they are now an important group and are quite numerous and varied.

Phylogeny and Classification

The early larval stages of the Echinoidea are similar to those of the Stelleroidea and Holothuroidea but quite unlike those of the Crinoidea, and it has been customary to accept the view that the echinoids, along with the stelleroids and holothuroids, arose quite early in the Paleozoic from an ancestral echinoderm of simple structure.[1] Further comments on phylogeny are made at the end of this chapter.

Bothriocidaris (Fig. 14-50), from the Middle Ordovician of Estonia, was long considered to be either the actual ancestor of the Echinoidea or at least quite near the ancestral stock. This view has been seriously challenged in recent years with the result that *Bothriocidaris* is now a controversial genus (the controversy is briefly reviewed by H. L. Clark, 1932). With the discovery of other Ordovician echinoids of the same age (Fig. 14-49), or only slightly younger, that have characteristics quite different from those of *Bothriocidaris*, it now seems much more likely that the Echinoidea were by Middle Ordovician time a well-differentiated group and that the ancestral stock will have to be sought among Early Ordovician or Cambrian echinoderms of simple nature. (See Jackson, 1907; Mortensen, 1928; H. L. Clark, 1932; MacBride and Spencer, 1938.)

Most students of the Echinoidea have sharply distinguished Paleozoic genera from later forms, using as a basis of distinction the fact that post-

[1] Mortensen (1938, p. 7), for example, believes that the ancestor of the echinoids must be sought for among the oldest edrioasteroids, like *Stromatocystites*, which have a great number of indefinitely arranged interambulacrai plates.

Paleozoic genera have two and only two meridional rows of coronal plates in each ambulacral and interambulacral area, whereas Paleozoic forms have either one row or more than two rows of interambulacral plates. On this basis Zittel and others have recognized two subclasses—*Paleechinoidea*, limited to the Paleozoic (M. Ord.–Perm.) and including all Paleozoic genera except *Miocidaris* (Miss.–Perm.); and *Euechinoidea*, including all post-Paleozoic forms and the single afore-mentioned Paleozoic genus *Miocidaris*. This division of the Echinoidea, which is not used in the present work, has been attacked vigorously by some investigators, particularly Mortensen (1938, etc.), and just as vigorously defended by others, particularly those accepting the views upon which Jackson (1912) in his classic "Phylogeny of the Echini" derived

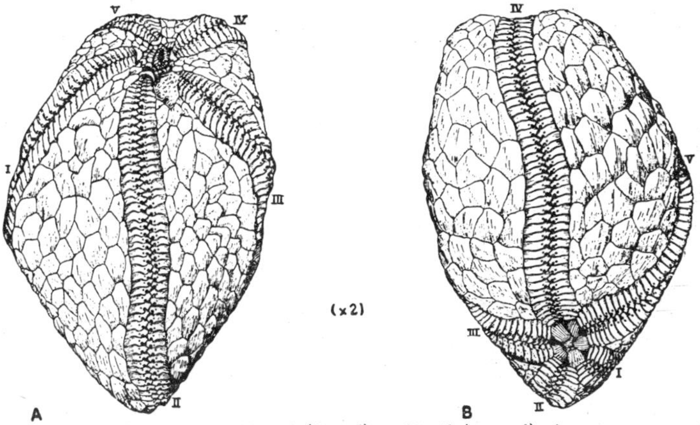

FIG. 14-49. Class ECHINOIDEA. Aboral (dorsal) and oral (ventral) views reconstructed of *Ectinechinus lamonti* from the Upper Ordovician of Scotland. Numbers I–V indicate the ambulacra. The madreporite-genital plate lies between II and III. The plates of the ambulacra are regularly arranged, but those of the interambulacral areas are irregularly disposed. The five teeth of the simple lantern are conspicuous. (*After MacBride and Spencer*, 1939.)

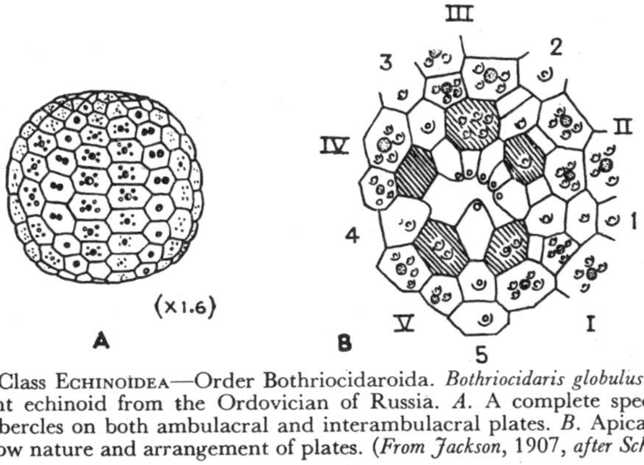

FIG. 14-50. Class ECHINOIDEA—Order Bothriocidaroida. *Bothriocidaris globulus*, an ancient and aberrant echinoid from the Ordovician of Russia. *A*. A complete specimen, with perforate tubercles on both ambulacral and interambulacral plates. *B*. Apical region enlarged to show nature and arrangement of plates. (*From Jackson*, 1907, *after Schmidt*, 1874.)

both subclasses from *Bothriocidaris.* Inasmuch as no complete ordinal classification has appeared since Jackson's (1912), although a new one is taking shape as successive volumes of Mortensen's (1928–1948) great "A Monograph of the Echinoidea" appear, no course is left other than to continue using the former. We are fully aware, however, that when Mortensen's classification is complete it may well be preferable to that of Jackson. The seven orders recognized by Jackson are based on the nature and structure of the test—corona, peristome and associated plates, periproct, and oculogenital system—and its external spines and pedicellariae. To them has been added Mortensen's Order Lepidocentroida for the reception of *Myriastiches* (M. Ord.), *Ectinechinus* (U. Ord.; Fig. 14-49) and *Aulechinus* (U. Ord.).

The eight orders are synoptically compared in Table 14-4.

TABLE 14-4. CHIEF CHARACTERISTICS OF THE DIFFERENT ORDERS OF ECHINOIDEA
Data from Jackson, 1912, and MacBride and Spencer, 1938

Order	Test	Rows of plates		Kinds of plates	Range
		Ambulacrum	Interambulacrum		
1. Lepidocentroida	Endocyclic or exocyclic	2 to many	2 to many	Simple	Ord.–Perm.
2. Bothriocidaroida	Regular and endocylic	2	1	Simple	Ord.
3. Cidaroida	Regular and endocyclic	2	2 to 4	Simple	Miss.–Recent
4. Centrechinoida	Regular and endocyclic	2	2	Compound	Jura.–Recent
5. Exocycloida	Irregular and exocyclic	2	2	Simple or compound	Jura.–Recent
6. Plesiocidaroida	Regular and endocyclic	2	3	Simple	Trias.
7. Echinocystoida	Irregular and exocyclic	2 to 4	8 to 9	Simple	Sil.
8. Perischoechinoida	Regular and endocyclic	2 to 20	3 to 14	Simple	Sil.–Perm.

Order 1. Lepidocentroida. Tests are hemiellipsoidal, spheroidal, or flattened globular; ambulacral plates are in two to many series and pore pairs are uniserial or pleuriserial; interambulacral plates are in two to many columns; the peristome is covered with series of ambulacrals, but not of

interradial plates; teeth are grooved; spines are simple and tubercles perforate. *Ordovician to Permian.*

This order, as defined by Mortensen (1935), includes most or all of the genera that Jackson assigned to his orders Echinocystoida and Perischoechinoida, both of which are discussed in following paragraphs. In the present work only the three following genera are assigned to Mortensen's order—*Myriastiches*, *Ectinechinus* (Fig. 14-49), and *Aulechinus*.

Order 2. Bothriocidaroida. These are regular echinoids in which the periproct is enclosed by the oculogenital ring. The globular test has two columns of plates in each ambulacral area and one column in each interambulacrum. The order is based on the single genus *Bothriocidaris* (Fig. 14-50) from the Middle Ordovician of Estonia. Mortensen (1938) believes that *Bothriocidaris* is not a true echinoid but rather a representative of the cystoidean Diploporita that left no descendants; he would place the genus in a new subclass, the Pseudechinoidea. *Middle Ordovician.*

Order 3. Cidaroida. In this order the regular endocyclic test has two columns of low, narrow ambulacral plates and two[1] columns of pentagonal interambulacral plates. Each interambulacral plate has a large, primary central tubercle and spine and secondary marginal spines. The periproct is enclosed within the oculogenital ring. The earliest cidaroids, by some placed in a separate family, the Archeocidaridae, appeared in the Mississippian; the order seemingly reached its maximum development in the Jurassic and Cretaceous, though it is still represented by several hundred species assigned to more than fifty living and extinct genera. *Dorocidaris* (Cret.–Recent; Fig. 14-51*H–K*) and *Cidaris* (U. Trias.–Recent) are typical living genera with fossil representatives.

Order 4. Centrechinoida. In this order the regular endocyclic test has two columns of compound plates (2 to 10 elements in each) in each ambulacrum, a like number in each interambulacrum, and a thick cover of primary spines. Each part of a compound plate is provided with pore pairs. The peristome is complex, with a lantern and associated structures. The order embraces many species included in a dozen families. The familiar green Atlantic sea urchin, *Strongylocentrotus* (Tert.–Recent), is a representative living genus with fossil species; *Phymosoma* (Cret.–Tert.; Fig. 14-51*L–M*) is an extinct genus.

Order 5. Exocycloida. In this order the test is irregular and exocyclic, and the periproct lies outside the oculogenital ring in interambulacrum 5. The ambulacra are typically petaloid, with a bivium and trivium. Most species have Loven's plane conspicuous. The order includes the so-called heart urchins and sand dollars, which are divided among a dozen families, and range from Jurassic to Recent.

In view of the usefulness and importance of the many Mesozoic and Cenozoic fossil representatives of this order, it seems desirable to characterize briefly the three major groups generally considered as suborders.

Suborder 1. *Holectypina.* Peristome central; lantern and perignathic girdle present; ambulacral pores in an unbroken series from summit to peristome, hence ambulacra non-

[1] Except for *Tetracidaris*, which has four in the wider part.

petaloid dorsally; in a progressive series the periproct migrates out of the apical system and in the final stage is on the oral surface. *Jurassic to Tertiary.*

Holectypus (Jura.–Cret.; Fig. 14-51*D–E*) is representative.

Suborder 2. *Clypeastrina.* Peristome central or slightly eccentric; lantern highly modified; anus on oral surface; ambulacral areas petaloid; ocular and genital plates fused in a mass with genital 5 generally lacking; test commonly flattened and strengthened by internal ribs and pillars. *Cretaceous to Recent.*

Dendraster (Plioc.–Recent; Fig. 14-51*F–G*) is representative.

Suborder 3. *Spatangina.* Peristome shifted forward; lantern and perignathic girdle lacking; anus on posterior of oral surface; ambulacral areas continuous, subpetaloid; Ambulacrum III commonly different from others and V variously modified according to position of anus. *Jurassic to Recent.*

Enallaster (Cret.; Fig. 14-51*A–C*) is a representative genus.

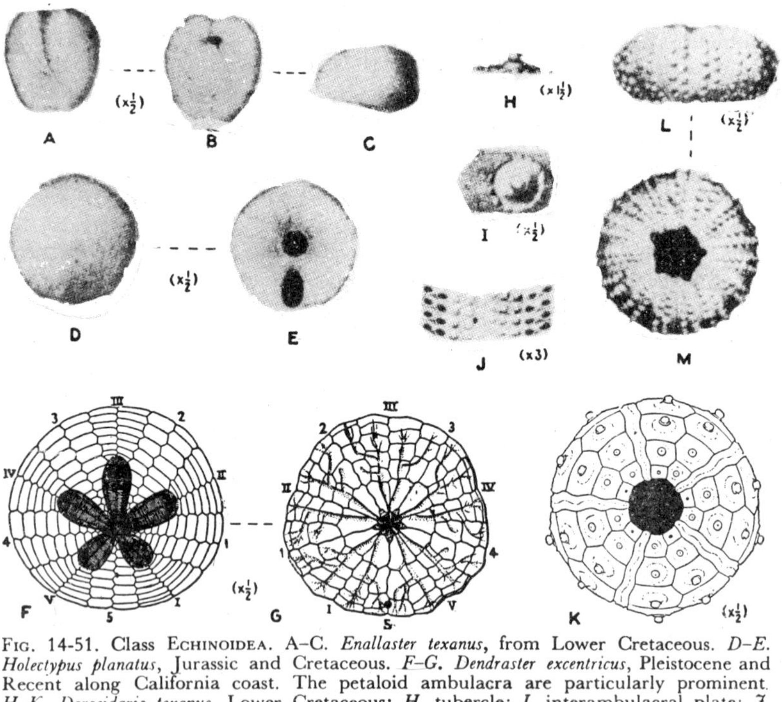

FIG. 14-51. Class ECHINOIDEA. A–C. *Enallaster texanus*, from Lower Cretaceous. *D–E*. *Holectypus planatus*, Jurassic and Cretaceous. *F–G*. *Dendraster excentricus*, Pleistocene and Recent along California coast. The petaloid ambulacra are particularly prominent. *H–K*. *Dorocidaris texanus*, Lower Cretaceous: *H*, tubercle; *I*, interambulacral plate; *J*, portion of ambulacrum; *K*, oral view of corona. *L–M*. Lateral and aboral views of *Phymosoma texanum*, Cretaceous and Tertiary. (*A–E, H–M after Clark*, 1915, *and Clark and Twitchell*, 1915; *F–G after Grant and Hertlein*, 1938.)

Order 6. Plesiocidaroida. Members of this extinct order have a regular endocyclic test with two columns of plates in each ambulacrum and three columns in each interambulacrum. The genital plates largely cover the aboral surface, and the oculars are exsert. Periproct and peristome are centrally located, but their structures are unknown. The order is based on a tiny specimen (*Tiarechinus*) from the Triassic of Tyrol.

Order 7. Echinocystoida. The members of this extinct order have small, seemingly exocyclic, spheroidal or flattened tests, with the periproct situated in an interambulacral area. There are two to four columns of low plates in each narrow ambulacral area and eight to nine columns of rather irregular, polygonal plates in each broad interambulacral area. There is a well-developed typical echinoid lantern, but nothing is known about the structure of either periproct or peristome. The order is based on two imperfectly known genera, *Echinocystites* and *Palaeodiscus*, from the Silurian of England.

Order 8. Perischoechinoida. This exclusively Paleozoic order includes rather large regular echinoids in which the periproct lies inside the oculogenital ring. The ambulacral areas are narrow or wide, with 2 to 20 columns of simple plates all bearing a single pore pair. Interambulacral areas have 3 to 14 columns of plates. Coronal plates may imbricate, and the madreporite usually is not recognizable. Genera typical of the order, which ranges

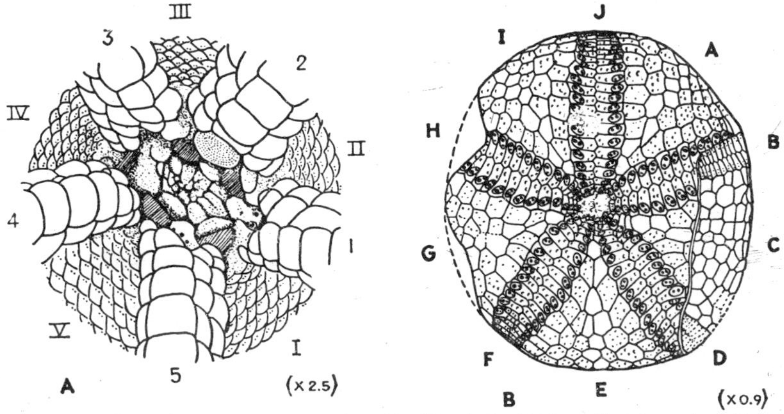

FIG. 14-52. Class ECHINOIDEA. Paleozoic echinoids. *A. Lepidesthes colletti* from the Mississippian of Indiana. This aboral view shows the simple ambulacra (I–IV), the interambulacra (1–5), and the broad, low, rounded genital and ocular plates. The genital plate in interambulacrum 2 is perforated with madreporic pores. *B. Hyattechinus rarispinus* from the Mississippian of Pennsylvania. An external mold of the ventral (oral) surface, and in areas *B*, *C*, and *D* an internal mold of the dorsal (aboral) surface. The ambulacra and interambulacra are lettered instead of numbered because the orientation is uncertain. (*After Jackson*, 1907.)

from Silurian to Permian, are *Echinocrinus* (Miss.–Penn.; Fig. 14-48*A*), *Hyattechinus* (Miss.; Fig. 14-52*B*), *Lepidesthes* (Dev.–Perm.; Fig. 14-52*A*), and *Melonechinus* (Miss.).

CLASS HOLOTHUROIDEA[1]

General Considerations

The Holothuroidea have a more or less elongate form and a tough flexible body wall in which the skeleton typically consists of scattered ossicles or

[1] Holothuroidea (also spelled Holothurioidea)—L. *holothuria*, plural, a sort of water polyp, from Gr. *holothurion*, + *-oid*, like; referring to the polyp-like appearance of many members of this group of animals.

spicules. In a few rare living and fossil forms the surface of the body wall is covered with tiny plates. Bilateral symmetry and a ventral and dorsal surface are commonly well marked. Members of several orders have radial

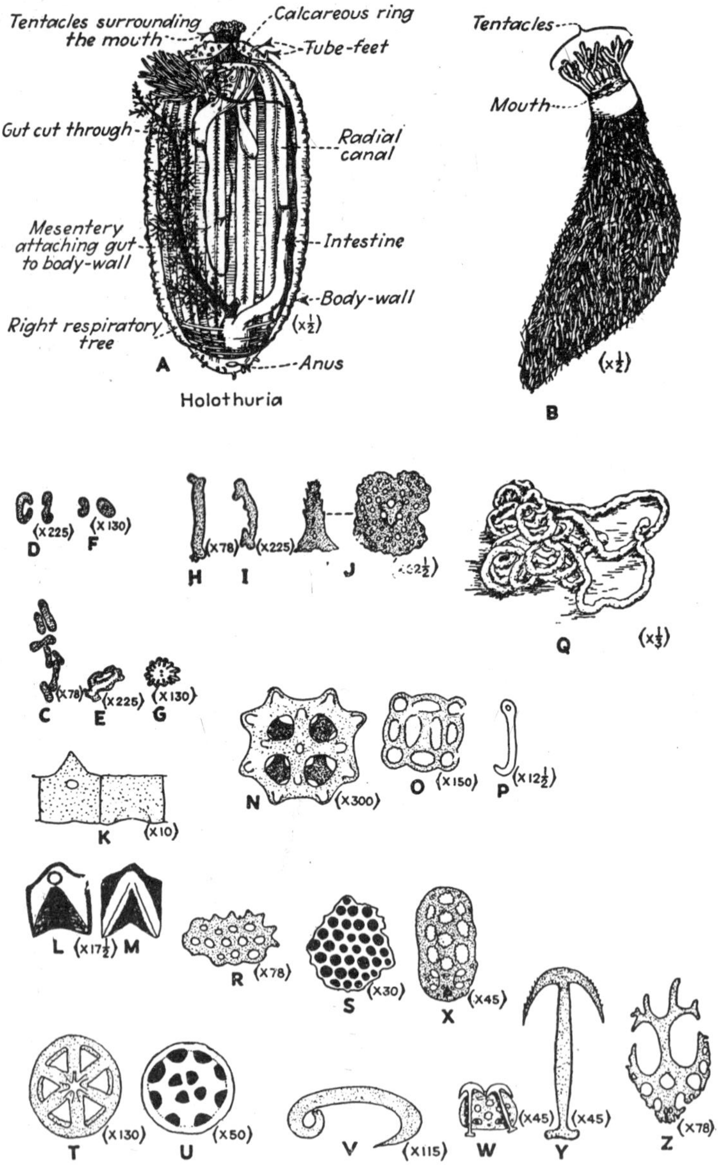

FIG. 14-53. Class HOLOTHUROIDEA. (*See opposite page for detailed description.*)

ambulacral vessels, tube feet (**pedicels**) used in creeping, respiratory trees, and special tentacles, the **Cuvierian organs,** for entangling food. In Order Apoda these structures are lacking and the body is slender and worm-like. A third group, Order Megalopoda, includes extinct plated members. Holothuroids range in length from 25 mm. (1 in.) to almost 1 m.; those superficially resembling cucumbers in body form are called sea cucumbers; and certain Pacific forms, known as bêche-de-mer or trepang, are highly esteemed as food.

The holothuroid body is elongated on the oral-aboral axis, with the mouth anterior and anus posterior. Surrounding the mouth are 10 to 30 retractile **tentacles,** which are comparable to the oral tube feet of other echinoderms. The body wall contains both longitudinal and circular muscles, and by action of these the animal can extend or contract its body and perform worm-like movements.

Modern holothuroids constitute a sharply defined group of marine organisms numbering about 800 species. They belong to the vagrant benthos and are particularly abundant in tropical waters, though also present in both temperate and polar regions. Their depth range is from tide level to more than 10,360 m. (34,000 ft.). They move sluggishly over the bottom by means of tube feet and muscular movements of the body, or burrow in the soft bottom sediment with only the ends of the body exposed. Their food is organic matter in the sediments and is obtained by swallowing the mud and sand. Voiding of the latter materials after the contained organic substances have been extracted is effected by violent contraction of the body in which it may be broken into several pieces or turned inside out. The excrement emerges as a solid thread several millimeters in diameter, and if the animal is irritated, excretion may be so violent that the excrement is thrown away from the animal as much as 7 cm. The first discharge is in the form of an irregular pile, but in the closing stages loosely looped coils are formed (Fig. 14-53*Q*). Some of the latter resemble the fossil *Lumbricaria* commonly interpreted as a worm casting, and it seems quite probable that many fossil worm-like markings may have been made by holothuroids.[1]

[1] The worm-casting interpretation of *Lumbricaria* has been questioned by C. L. and M. A. Fenton (1934, *Lumbricaria;* a holothuroid casting? *Pan-Amer. Geol.*, vol. 61, pp. 291–292), who observed a burrowing holothuroid, *Leptosynapta inhaerens*, make castings of similar nature. In a typical specimen of *Lumbricaria* (Fig. 11-13), the animal which formed it must have moved in and out through the ropy excrement, whereas the castings illustrated by the Fentons seem to have been formed during a single rapid discharge (Fig. 14-53*Q*). For this reason the holothuroid origin of *Lumbricaria* seems questionable.

FIG. 14-53. Class HOLOTHUROIDEA. Modern and fossil holothuroids; living unless otherwise indicated. *A*. Ventral view of a dissected specimen of *Holothuria tubulosa*. *B*. Exterior view of *Thyone briarcus*, showing spinose surface. *C–F*. Miliary granules of several genera. *G*. Rosette. *H–I*. Supporting rods from the tentacles. *J*. Table viewed from above and spine viewed from the side. *K–M*. Radial and interradial pieces of the calcareous ring. The former are perforated. *N*. Normal closed cup. *O*. Accessory perforated plate. *P*. Fossil hook from the Pennsylvanian of Texas. *Q*. Casting made by *Leptosynapta*, a living holothuroid (cf. *Lumbricaria*, Fig. 11-13). *R*. Perforated plate from the skin. *S*. Perforated plate from the Pennsylvanian of Texas. *T*. Wheel. *U*. Fossil wheel from the Mississippian of Illinois. *V*. Sigmoid body. *W–Z*. Anchors and anchor plates. (*A after Thomson*, 1899; *C–O, R, T, V–Z after H. L. Clark, et al.; P, S, U after Croneis and McCormack*, 1932; *Q after C. L. Fenton*, 1934.)

The body wall is thick and tough; typically it contains numerous tiny calcareous spicular elements (Fig. 14-53*C*–Z), which in rare cases are united to form a delicate skeletal structure, and it is commonly prickly because of surface spines. In a few rare living and fossil forms there is a true and typical echinodermal test of many calcareous scales or plates; *e.g.*, *Psolus* (Recent) and *Eothuria* (U. Ord.; Fig. 14-54).

The skeletal elements are largely of microscopic size, exhibit considerable diversity of form and structure, and are designated by an extensive terminology. The simplest are tiny irregular grains, designated **miliary granules** (Fig. 14-53*C*–*F*). Small straight or curved spicular bodies, or **supporting rods** (Fig. 14-53*H*–*I*), lie in the tentacles and other parts of the body. Irregularly branched elements in the body wall are **rosettes** (Fig. 14-53*G*), and if these have the form of a perforated disk, they are termed **plates.** A plate having a definite projection rising from its mid-point is described as a **table,** in which case the latter is said to consist of a **basal disk** and a **spire** (Fig. 14-53*J*). Other skeletal elements have the shapes of hooks, wheels, anchors, cups, buttons, etc., and are named after the objects they resemble (Fig. 14-53*N*–*P*,*R*–Z). All the elements thus far described lie in the body wall, in the tentacles, or in the tube feet.

In some holothuroids additional plates form a ring around the anus and esophagus. The five calcareous plates encircling the anal vent are designated **anal teeth.** In most holothuroids the esophagus is encircled by a **calcareous** or **dental ring** consisting of 10 calcareous plates—5 of these are termed **radials,** because radial muscles are attached to them, and the other 5, alternating with the radials, **interradials** (Fig. 14-53*K*–*M*).

Plated holothuroids have a circle of 10 circumoral **valves** resembling the five teeth of the echinoid lantern and many plates (which, in extinct *Eothuria*, Fig. 14-54, are arranged in distinct echinoid fashion) in ambulacral, interambulacral, and apical areas, strongly suggesting a close relationship between the early Echinoidea and Holothuroidea.

Superficially the typical holothuroid shows little or no indication of the radial symmetry so obvious in other classes of echinoderms, but a transverse section through the animal, midway between the mouth and anus (*i.e.*, between oral and aboral poles), reveals a pentameral radial symmetry essentially like that of a regular echinoid or of a pentagonal asteroid cut in a similar plane.

Classification

Special students of the Holothuroidea almost universally recognize five orders of living forms, but they are not yet agreed upon how these should be subdivided (a problem of no concern here).[1] To these five must be added a sixth, the extinct order *Megalopoda*, based entirely on a plated Upper Ordovician genus.

Order 1. Megalopoda (extinct).
Order 2. Dendrochirota.
Order 3. Elasipoda.
Order 4. Aspidochirota.
Order 5. Molpadonia.
Order 6. Apoda.

[1] Matsumoto (1915) proposed an elaborate classification of 2 subclasses, 4 orders, and 13 families, but this arrangement has not been accepted.

The five orders with living representatives are based largely on the nature of the tentacles, the number and arrangement of the appendages, and the general body form.[1] Families are based on body form and various internal organs; genera, on a combination of characteristics of both soft parts and hard skeletal elements. It is obvious, therefore, that only in rare cases can the fossil remains of Holothuroidea be given a taxonomic rank higher than genus, because of the fact that they are almost invariably in the form of scattered skeletal elements or vague impressions and outlines of no taxonomic importance. The notable exception is the beautifully preserved plated form, *Eothuria* (Fig. 14-54).

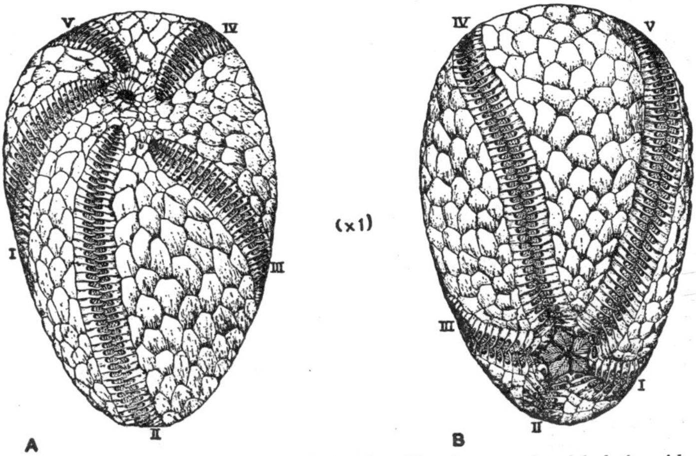

FIG. 14-54. Class HOLOTHUROIDEA. *Eothuria*, the oldest known plated holothuroid, from the Upper Ordovician of Scotland. *A–B*. Aboral (dorsal) and oral (ventral) views reconstructed. This ancient form resembles an equally old echinoid (*Ectinechinus;* Fig. 14-49), in many respects, but differs in having 10 valvelike plates instead of five teeth. The ambulacral pores are unlike those of any known eleutherozoan. The five ambulacra are indicated by numerals I to V. (*After MacBride and Spencer*, 1938.)

Order 1. Megalopoda. This extinct order based on a single Upper Ordovician genus and species, *Eothuria beggi* (Fig. 14-54), includes plated holothuroids that had 10 circumoral valves and several hundred skeletal plates arranged more or less systematically in five ambulacral and five interambulacral areas and in a simple apical system lacking oculars. The pores of the

[1] It is interesting to note that there is a close relationship between feeding habits and nature of the soft parts just mentioned. Holothuroids are typically mud eaters, and their classification reflects the environments of life, the modes of eating, and the modifications of the tentacles used in eating. The Elasipoda are deep-sea forms with tiny shield-shaped tentacles; the Aspidochirota live in both deep and shallow water and use their pedicels to stuff mud into the mouth; the Dendrochirota are plankton feeders and the microscopic organisms used as food are swept into the mouth from a cone made by the tentacles; the Molpadonia are simplified forms with small tentacles, no sucker feet, and a reduced water-vascular system; and the Apoda have leaf-shaped tentacles, a much-reduced water-vascular system, and a long worm-like body.

ambulacra are large and complex, and it is assumed that there was no differentiation of the podia into tentacles around the mouth and walking feet.

Eothuria closely resembles two echinoid contemporaries, *Aulechinus* and *Ectinechinus* (Fig. 14-49), but differs from them and from all other echinoids in having the five teeth of the echinoid lantern replaced by 10 circumoral valves. MacBride and Spencer (1938), who first described *Eothuria*, consider that its features help to confirm a long-suspected close relationship of the Echinoidea and the Holothuroidea.

Order 2. Dendrochirota. Tube feet (pedicels) and papillae are present; tentacles are dendritic; retractor muscles are present; and the body is cylindrical, or broad and flattened ventrally into a well-defined creeping sole. No fossils are known. *Thyone* (Fig. 14-53*B*) is a living representative.

Order 3. Elasipoda. Body with well-marked bilateral symmetry; tube feet on the flattened ventral surface and papillae on the dorsal surface. Members of this order are confined to deep water. No fossils are known. *Recent.*

Order 4. Aspidochirota. Tentacles shaped like a shield; respiratory trees well developed; and tube feet or papillae conspicuous. All commercially important forms belong in this order, which may go back as far as the Jurassic (see following discussion under The Fossil Record). *Holothuria* (Fig. 14-53*A*) is a typical living genus; *Protholoturia* (Jura.) may be an extinct representative.

Order 5. Molpadonia. These are small worm-like burrowing holothuroids, with digitate tentacles and without tube feet, that live buried in mud or soft sand with mouth and anus flush with the bottom and elongate body in a broad concave arc several centimeters below the surface. No fossils are known, although some of the smaller U-shaped burrows in ancient rocks may have been made by representatives of this order. The familiar living *Caudina* is a representative genus.

Order 6. Apoda. This order includes long snake-like or worm-like holothurians that have pinnate or lobate leaflike tentacles and that lack tube feet and radial ambulacral vessels. Of particular interest to paleontologists are the perforated plates, anchors, and sigmoid skeletal elements found in the body. Fossil spicules assigned to Family Synaptidae have been found in Permian rocks and other fossil spicules have been reported from Jurassic (Frentzen, 1944) and Tertiary rocks. The order, therefore, ranges at least as far back as the Permian. *Leptosynapta* (Fig. 14-53*Q*), *Synapta*, and *Chirodota* are familiar living representatives.

The Fossil Record

Holothuroids may leave four different types of fossil record: (1) mud fillings or impressions and outlines of a part or all of the body; (2) calcareous skeletal elements of diverse shape and structure; (3) complete skeletons of many polygonal plates; and (4) excremental castings imilar to those of some worms. Examples of all four types have been found as fossils.

Mud fillings and external impressions by their very nature and mode of formation should be extremely rare. Some such objects have been referred to the Holothuroidea, but there seems to be no present consensus as to the accuracy of identification. Chief among the older specimens are those of *Eldonia* from the Middle Cambrian Burgess shale (Walcott, 1911). A. H. Clark (1913) concluded that *Eldonia* was a free-swimming holothuroid structurally related to the Elpidiidae, a family belonging to Order Elasipoda, although H. L. Clark (1912) had previously rejected the possibility that the fossil was an echinoderm at all. No similar doubt seems to exist about *Protholoturia* from the Jurassic lithographic limestone of Solenhofen. The impressions of this form have characteristic calcareous skeletal elements on their surface, and though they cannot be assigned with certainty to any modern order, it is believed that they were most probably made by extinct members of Order Aspidochirota.

Scattered skeletal elements in the form of wheels, anchors, and sievelike fragments (Fig. 14-53) have been reported from the Middle Devonian of Bohemia (Prantl, 1947), from later Paleozoic rocks, in widely separated areas, and from a few Mesozoic (Frentzen, 1944) and Cenozoic formations. The characteristics of these interesting microfossils, assigned to both living and extinct genera, are discussed at considerable length by Croneis and McCormack (1932) in an excellent summary and evaluation of all supposed holothuroid fossils reported before 1932 and more recently by Frentzen (1944), who states that the Jurassic of Baden contains remains of *Chirodota* (Lias), *Stichopus* (Dogger), and *Myriotrochus* (Malm).

Reference has been made in preceding paragraphs to the Middle Ordovician plated holothuroid *Eothuria* (Fig. 14-54). Inasmuch as modern *Psolus* has a plated test, it seems possible that fossil forms like it may also be found.

Fossil excrements of holothuroids will always be questionable because of the difficulty of differentiating this type of material from the castings of certain worms.

The Holothuroidea probably evolved early in the Paleozoic, possibly even in the Pre-Cambrian, and are thought to have arisen, along with the Echinoidea which they closely resemble in several important particulars, from a primitive and simple edrioasteroid type of echinoderm (Fig. 14-56). It seems reasonable to expect spicules from rocks older than the Devonian, but it seems rather unlikely that well-preserved entire bodies will often be found.

PHYLOGENY OF THE ECHINODERMA

The Echinoderma as a group have an amazing embryology that has long attracted the attention of both neontologists and paleontologists, because of the evolutionary problems involved and the possibility that the earliest beginnings of the chordates may be suggested by certain larval stages. Another fact of interest is that the young echinoderm has actually been two quite different animals by the time it has developed into an adult. (As a matter of fact, the earliest larval stages of the different classes were once thought to be definite animals and were given different names; see Fig. 14-55.) It begins life as though it would become some kind of worm; then suddenly it develops many tactile lobes and undergoes an amazing metamorphosis in which it sheds parts of its body and reconstitutes its skeletal elements; only then does it emerge as a recognizable holothuroid, stelleroid, or echinoid.

After fertilization of the egg, development of the larval form follows much the same course in all known Echinoderma. Internal changes result in a blastula and this in turn passes into the gastrula stage as the blastopore becomes the anal vent and a mouth opening breaks through the side wall (Fig. 14-55). There next develops a ciliated, free-swimming barrel-shaped

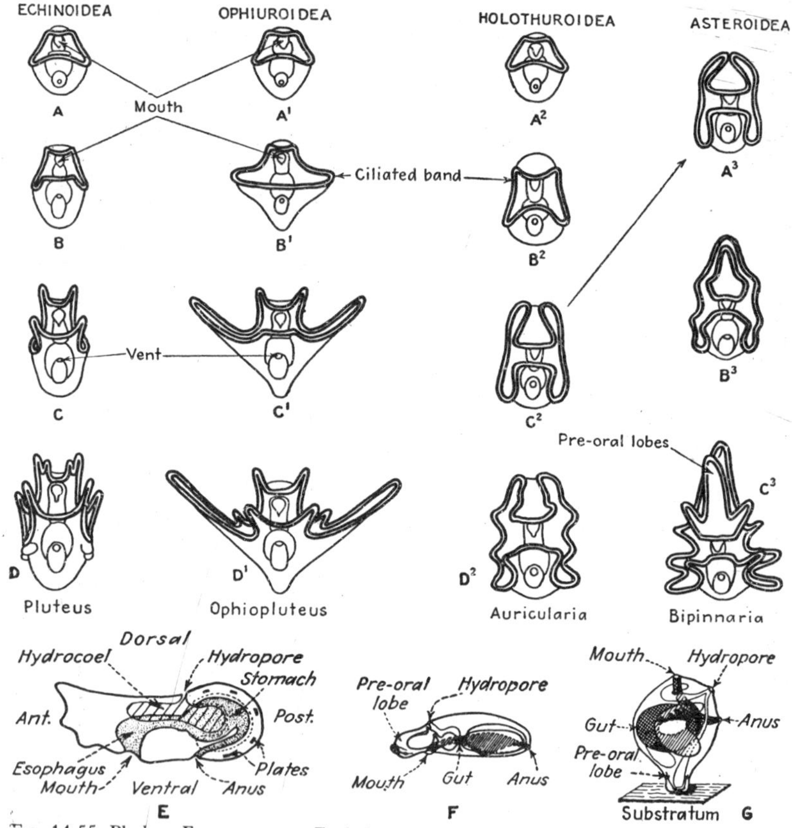

FIG. 14-55. Phylum ECHINODERMA. Early larval stages of different echinoderms. *A–D*. The *Pluteus*, or *Echinopluteus*, of echinoids: A^1–D^1, the *Ophiopluteus* characteristic of ophiuroids; A^2–D^2, the *Auricularia* of holothuroids; A^3–D^3, the *Bipinnaria* of asteroids. It is assumed that A^3 of the bipinnaria form arose from the auricularia series as indicated by the arrow. It is to be noted that the *D* forms are quite dissimilar as compared to the essentially similar *A* forms. *E*. Diagrammatic view of the left side of an early larval stage of *Echinus multituberculatus*, with a few skeletal plates present. *F*. Diagrammatic reconstruction of the hypothetical ancestral *Dipleurula*, with a straight simple alimentary tract, a well-developed water-vascular system, and a ciliated anterior end (to the left). *F* should be compared with *E*. *G*. Diagrammatic reconstruction of the supposed primitive ancestor of the Pelmatozoa. With attachment the alimentary tract has become looped, with the mouth on the upper surface and the anal opening on the posterior side. The hydropore occupies a position intermediate between the mouth and anus. The preoral lobe has been modified for attachment. *Aristocystites* is thought to have represented about this stage of development. (*A–D adapted from Mortensen*, 1901, *after Müller*, 1852; *E adapted from Bury*, 1889; *F–G adapted from Bather*, 1900.)

larva, which has a mouth, a large stomach, a short intestine, and a small anus and which possesses marked bilateral symmetry. This type of larva, which appears in the embryological development of each of the classes of existing Echinoderma, has been designated the *Dipleurulalarve* by Semon (1888, page 112). It has now become customary to use the shortened term *Dipleurula*,[1] which emphasizes the bilateral symmetry of the larva.

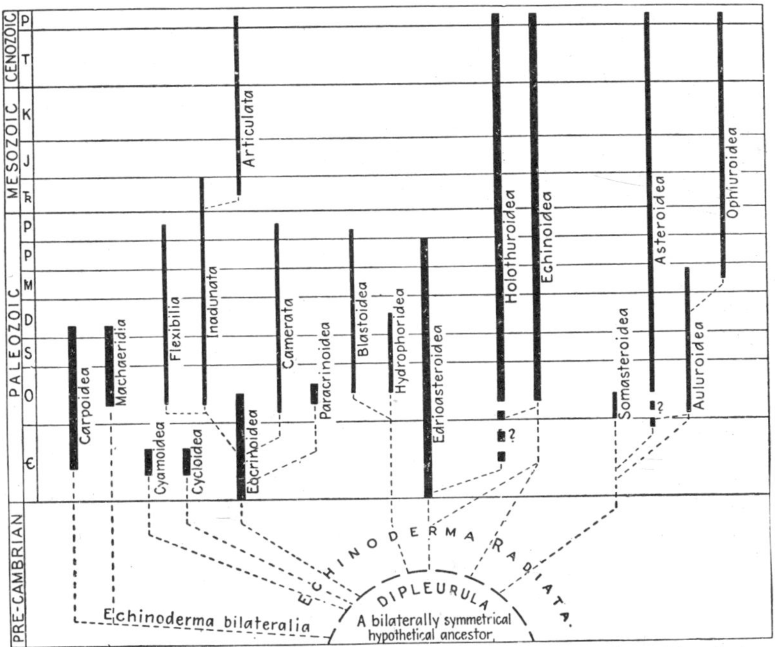

FIG. 14-56. Phylum ECHINODERMA. Stratigraphic range and supposed possible relationships of the classes (broader vertical bands) and subclasses (narrower vertical bands) of the phylum. Uncertain fossil record is indicated by broken heavy bands; possible genetic relationships are indicated by dashed lines. All major groups are shown as arising from the hypothetical bilaterally symmetrical ancestor—the *Dipleurula;* in some cases they may be interrelated. The times when different groups differentiated are not known exactly; some may have evolved far back in the Pre-Cambrian, whereas others may have arisen in later Pre-Cambrian time. (*Based on suggestions and data from many writers.*)

Inasmuch as a *Dipleurula* stage appears in the development of each of the existing classes of echinoderms, zoologists and paleontologists have postulated a phylogenetic stage of the same nature and have reconstructed the imagined *Dipleurula* ancestor. A diagrammatic restoration is shown in Fig. 14-55. It is supposed that some such hypothetical ancestor gave rise to the main classes of the Echinoderma in the Pre-Cambrian, for already in the Early Cambrian the edrioasteroids were a well-defined group and the other classes quickly followed (Fig. 14-56).

[1] *Dipleurula*—diminutive of *dipleura*, two-sided (*i.e.*, bilaterally symmetrical in at least some respects); from Gr. *dis*, twice or two, + *pleura*, sides.

The *Dipleurula* at first resembles the trochophore of annelid worms, and this resemblance has led to the suggestion that echinoderms descended from a worm-like ancestor, which itself had evolved from an earlier and simpler bilaterally symmetrical animal. It also bears striking resemblance to the larval form (*Tornaria*) of the primitive hemichordate *Balanoglossus* (Class Enteropneusta) and its allies, and this resemblance between larvae has led many investigators to the conclusion that the earliest chordates were closely related genetically to the earliest echinoderms.

If only adult echinoderms be considered, it is obvious at once that they are strikingly different from all other groups of animals. Once metamorphosis of the larva has been completed, the bilaterally symmetrical *Dipleurula* has become a radially symmetrical adult. It has been suggested that this metamorphosis represents the changes that took place in the early history of the different classes, and that its complexity is due to the enormous compression of the different stages in that history. The interested reader will find an extended discussion of this history in several important works by Bather (1900; 1929).

It may be postulated that several quite distinct lines arose from the hypothetical *Dipleurula*. Certain lines quickly vanished without leaving descendants (*e.g.*, the Homalozoa). Others persisted for long periods; in some cases

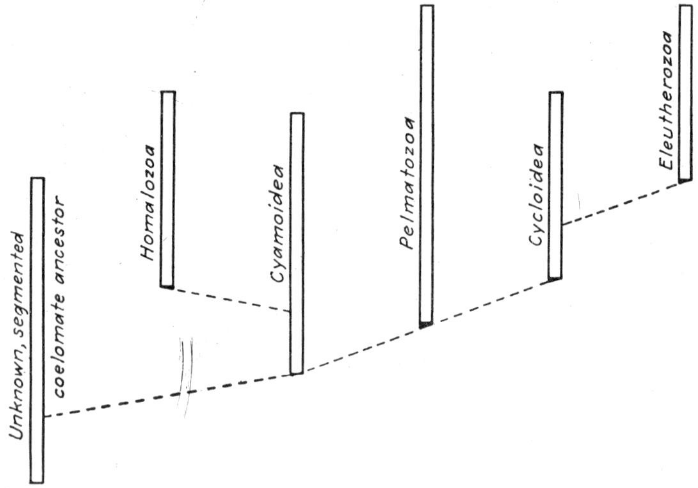

FIG. 14-57. Phylum ECHINODERMA. A conjectural ancestrai tree for the phylum, according to Whitehouse (1941). Some annelid-like ancestor supposedly gave rise to the bilaterally symmetrical Cyamoidea via the hypothetical *Dipleurula*. The Cyamoidea in turn gave off two branches. One branch, the Homalozoa, retained the bilateral symmetry of the ancestral cyamoids and soon became extinct without leaving any descendants. The other branch developed a secondary radial symmetry that was superimposed on the original symmetry of the Cyamoidea. Some members adopted the habit of attaching the calyx to the substratum and developed into the Pelmatozoa. Others continued to be free-living, and it is supposed that the Paleozoic and subsequent eleutherozoans had their beginning in the radially symmetrical Cycloidea, which already in the Middle Cambrian had diverged somewhat from the hypothetical *Dipleurula*.

they gave rise to descendants that have lived on to the present time, whereas in other cases they became extinct without leaving known descendants.

Once the phylogenetic history of Echinoderma seemed relatively simple and straightforward, and clean evolutionary trees could be drawn. In recent years, however, discoveries of great phylogenetic importance have made it obvious that the history of the Echinoderma is most complex and that any tree showing postulated genetic relationships must have certain extinct branches "grafted" on somewhere because we simply do not know for certain how these branches are connected with other parts of the echinoderm tree.

The stratigraphic ranges of the twelve classes of Echinoderma are shown in Fig. 14-56, and a highly conjectural evolutionary tree is suggested by connecting dotted lines. It is to be emphasized that tomorrow new discoveries may extend the range of an order, a class, or a subphylum and make necessary readjustments in the postulated relationships.

The different views as to genetic relationships between the major taxonomic divisions are considered in some detail in preceding paragraphs and need not be repeated here. Suffice it to say that one main line of echinoderm evolution led to a fixed habit and development of a great race of radially symmetrical forms, the Pelmatozoa, that were dominant during the Paleozoic, whereas the other led to an equally great and varied group of free-moving forms, the Eleutherozoa, in which radial symmetry is modified by a second bilateral symmetry.

The evolutionary trends in the artificial subphylum Homalozoa are uncertain. As to those in the subphylum Haplozoa, Whitehouse (1941) suggests the evolutionary relationships diagrammed in Fig. 14-57.

ECHINODERMA AS GEOLOGIC AGENTS

The ecology of living echinoderms and the paleoecology of extinct forms are considered at some length in a preceding part of the present chapter. Here attention is given to the action of echinoderms as geologic agents.

Both crinoids and echinoids have been important rock builders at certain times in the geologic past. The accumulations (biostromes and bioherms) made by them are almost certain to be local because of the gregarious habit of the builders. The Mississippian bioherms of southern Indiana, composed to a large extent of coarsely crystalline, crinoidal limestone, illustrate this type of deposit (Stockdale, 1931).

Crinoidal remains are almost invariably associated with Paleozoic reef deposits, especially in the stratified interreef rock. Such crinoidal limestones have been reported from the Niagaran of New York, Indiana, and Wisconsin and from the Silurian of the islands of Anticosti and Gotland. Some investigators have suggested that these extensive crinoidal deposits were formed from sessile species torn loose from the bottom during severe storms or from free-moving species which drifted into quiet, shallow water to die because of the lack of oxygen. It seems possible, also, that while the main organic reef was being constructed many species during early growth stages were attached to the bottom, either on the main reef mound or closely adjacent to it in shallow water, and that, upon attaining maturity, some of these may have abandoned the stems, which are the remains most often found in the vicinity of ancient reefs, and floated away as free-moving organisms to meet death and to undergo burial far from the places of origin. It is also possible that certain fish browsed over the "gardens" like cows in a pasture and bit off the crinoid heads because they contained more food than the stems.

In the Lower Pennsylvanian (Upper Rundle) of Alberta a 60-m. (200-ft.) unit of limestone is composed almost entirely of separate crinoid columnals. Similar rock, of much less thickness, is found in southern Indiana in the Harrodsburg (Mississippian) limestone.

Cystoids are rarely abundant enough to be of importance as rock builders. The same is true of stelleroids and holothuroids.

Certain echinoderms are of great geologic importance because of predatory and scavenging activities. Predatory habits determine to a large extent where certain forms exist, and scavenging is an important contribution to the sanitation of the ocean bottom. In addition to the activities just mentioned, some echinoderms pass great quantities of bottom materials through their digestive tracts for the purpose of extracting the contained organic matter. Repeated passages through a digestive tract must have important chemical and physical effects on the sedimentary materials.

REFERENCES

General

BATHER, A. F. 1900. The Echinoderma, Pt. III, *in* Lankester's "A Treatise of Zoology." London, A. & C. Black, Ltd.

——— 1929. Echinoderma *in* Encyclopaedia Britannica, 14th ed., vol. 7, pp. 894–904.

BURY, H. 1889. Studies in the embryology of the Echinoderms. *Quart. Jour. Micros. Sci.*, vol. 29 (n.s.), pp. 409–445.

CLARK, W. B., and TWITCHELL, M. W. 1915. The Mesozoic and Cenozoic Echinodermata of the United States. *U.S. Geol. Surv., Mon.* 54, 341 pp.

HOWE, H. V. 1942. Neglected Gulf Coast Tertiary microfossils. *Bull. Amer. Assoc. Petrol. Geol.*, vol. 26, pp. 1188–1199.

RAYMOND, P. E. 1941. Review of Echinodermata *in* Fiftieth Anniversary Volume, Geological Society of America, pp. 80–81.

SEMON, R. 1888. Die Entwickelung der *Synapta digitata* und die Stammesgeschichte der Echinodermen. *Zeitschr. Naturwiss.*, vol. 22, pp. 1–135.

STOCKDALE, P. B. 1931. Bioherms in the Borden group of Indiana. *Bull. Geol. Soc. Amer.*, vol. 42, pp. 707–718.

TERMIER, H., and TERMIER, G 1949. Les Échinodermes du Paléozoïque inférieur. *Rev. scientifique*, vol. 86, pp. 613–626.

WANNER, J. 1924. Palaeontologie von Timor, Lief. XIV, No. XXIII. Die Permischen Echinodermen von Timor, Teil. II. 81 pp.

WEST, C. D. 1937. Note on the crystallography of the echinoderm skeleton. *Jour. Paleontology*, vol. 11, pp. 458–459.

WITHERS, T. H. 1926. "Catalogue of the Machaeridia." London, British Museum (Natural History), 99 pp.

Pelmatozoa

BASSLER, R. S. 1935. The classification of the Edrioasteroidea. *Smith. Misc. Coll.*, vol. 93, pub. 3301, 11 pp.

———. 1936. New species of American Edrioasteroidea. *Ibid.*, vol. 95, pub. 3385, 33 pp.

——— and MOODEY, M. W. 1943. Bibliographic and faunal index of Paleozoic pelmatozoan echinoderms. *Geol. Soc. Amer., Spec. Paper* 45, 734 pp.

BATHER, F. A. 1898–1914. Studies in Edrioasteroidea I–IX. Wimbledon, England, published by the author. (Reprinted from *Geol. Mag.* for 1898, 1899, 1900, 1908, 1914, and 1915.)

BRANSON, E. B., and PECK, R. E. 1904. A new cystoid from the Ordovician of Oklahoma. *Jour. Paleontology*, vol. 13, pp. 89–92.

CARPENTER, P. H. 1884. Report on the Crinoidea—the stalked crinoids collected by *H.M.S. Challenger* during the years 1873–1876. *Zool.*, vol. 11, pt. 32, 440 pp.

———. 1888. Report on the Crinoidea—the Comatulae. *Ibid.*, vol. 26, pt. 60, 399 pp.

CHAUVEL, J. 1941. Recherches sur les Cystoïdes et les Carpoïdes armoricains. (Thèses présentées à la Faculté des Sciences de l'Université de Rennes, No. 1:3; Ser. C.) Rennes. 286 pp.

CLARK, A. H. 1915 to date. A monograph of existing crinoids: Vol. 1, The comatulids. *U.S. Nat. Mus., Bull.* 82, pt. 1, pp. 1–406 (1915); pt. 2, pp. 1–795 (1921); pt. 3, pp. 1–816 (1931); pt. 4*a*, pp. 1–603 (1941); pt. 4*b*, pp. 1–473 (1947); pt. 4*c*, pp. 1–383 (1950); pt. 5 (in preparation).

CRONEIS, C., and GEIS, H. L. 1940. Microscopic Pelmatozoa: Part I, Ontogeny of the Blastoidea. *Jour. Paleontology*, vol. 14, pp. 345–355.

HUDSON, G. H. 1907. On some Pelmatozoa from the Chazy limestone of New York. *N.Y. State Museum., Bull.* 107, pp. 97–152.

JAEKEL, O. 1918. Phylogenie und System der Pelmatozoen. *Pal. Zeitschr.*, vol. 3, pp. 1–128.

KIRK, E. 1911. The structure and relationships of certain eleutherozoic Pelmatozoa. *Proc. U.S. Nat. Mus.*, vol. 41, pp. 1–137.

MOORE, R. C. 1938. The use of fragmentary crinoidal remains in stratigraphic paleontology. *Denison Univ. Bull., Jour. Sci. Lab.*, vol. 33, pp. 165–250.

———. 1940. Early growth stages of Carboniferous microcrinoids and blastoids. *Jour. Paleontology*, vol. 14, pp. 572–583.

———. 1948. Evolution of the Crinoidea in relation to major paleogeographic changes in earth history. *Int. Geol. Cong., Rept. 18th Sess., Great Britain*, pt. 12, pp. 27–53.

———. 1952. Evolution rates among crinoids. *Jour. Paleontology*, vol. 26, pp. 338–352.

——— and Laudon, L. R. 1941. Symbols for crinoid parts. *Jour. Paleontology*, vol. 15, pp. 412–423.

——— and ———. 1943. Evolution and classification of Paleozoic crinoids. *Geol. Soc. Amer., Spec. Paper* 46, 153 pp.

——— and ———. 1943*a*. *Trichinocrinus*, a new camerate crinoid from Lower Ordovician (Canadian?) rocks of Newfoundland. *Amer. Jour. Sci.*, vol. 241, pp. 262–268.

——— and ———. 1944. Class Crinoidea (pp. 137–209) *in* Shimer and Shrock's "Index Fossils of North America." New York, John Wiley & Sons, Inc.; Cambridge, Mass., The Technology Press. This work contains a comprehensive list of references to North American works on crinoids.

REGNELL, G. 1945. Non-crinoid Pelmatozoa from the Paleozoic of Sweden—a taxonomic study. *Medd. Lunds Geol.-Miner. Inst.* 108, 255 pp.

SCHUCHERT, C. 1919. A Lower Cambrian edrioasteroid—*Stromatocystites walcotti*. *Smith. Misc. Coll.*, vol. 70, no. 1, 9 pp.

SINCLAIR, G. W. 1948. Three notes on Ordovician cystids. *Jour. Paleontology*, vol. 22, pp. 301–314.

STAINBROOK, M. A. 1941. Last of great phylum of the cystids. *Pan-Amer. Geol.*, vol. 76, pp. 83–98.

TERMIER, H., and TERMIER, G. 1949. Chez les Crinoïdes fossiles. *Bull. serv. cart. géol. de l'Algérie*, 1er ser., Paleontologie, no. 10, pp. 1–91.

Homalozoa

BATHER, F. A. 1930. A class of Echinoderma without trace of radiate symmetry. *Arch. zool. ital.*, vol. 14, pp. 431–439.

RUEDEMANN, R. 1942. Notes on Ordovician Machaeridia of New York. *N.Y. St. Mus., Bull.* 327, pp. 33–44.

WITHERS, T. H. 1926. "Catalogue of the Machaeridia (*Turrilepas* and its Allies) in the Department of Geology." London, British Museum (Natural History). 99 pp.

WOLBURG, J. 1938. Beitrag zum Problem der Machaeridia. *Pal. Zeitschr.*, vol. 20, pp. 289–298.

Haplozoa

SCHMIDT, H. 1951. Whitehouse's Ur-Echinodermen aus dem Cambrium Australiens. *Pal. Zeitschr.*, vol. 24, pp. 142–145.

WHITEHOUSE, F. W., 1941. Early Cambrian echinoderms similar to the larval stages of Recent forms. (Pt. 4 of Cambrian faunas of N.E. Australia.) *Mem. Queensland Mus.*, vol. 12, pt. 1, 64 pp.

Eleutherozoa

BATHER, F. A., and SPENCER, W. K. 1934. An Ordovician echinoid from Girvan, Ayrshire. *Ann. Mag. Nat. Hist.*, vol. 13, pp. 557–558.

BERRY, C. T. 1934. Miocene and Recent *Ophiura* skeletons. *Johns Hopkins Univ., Stud. Geol.*, no. 11, pp. 9–136.

———. 1939. More complete remains of *Ophiura marylandica*. *Proc. Amer. Phil. Soc.*, vol. 80, pp. 87–94.

CLARK, A. H. 1913. Cambrian holothurians. *Amer. Nat.*, vol. 47, pp. 488–507.

CLARK, H. L. 1912. Fossil holothurians. *Science* (n.s.), vol. 35, pp. 274–278.

———. 1913. Class 2. Holothuroidea *in* Zittel-Eastman's "Textbook of Paleontology," pp. 312–313. New York, The Macmillan Company; London, Macmillan & Co., Ltd.

———. 1915. Catalogue of Recent Ophiurans: based on the collection of the Museum of Comparative Zoology. *Mem. Mus. Comp. Zool.* (Harvard), vol. 25, pp. 163–376.

———. 1932. The ancestry of echinii. *Science* (n.s.), vol. 76, pp. 591–593.

CRONEIS, C., and MCCORMACK, J. 1932. Fossil Holothuroidea. *Jour. Paleontology*, vol. 6, pp. 11–148.

FEWKES, J. W. 1890. On excavations made in rocks by sea urchins. *Amer. Nat.*, vol. 24, pp. 1–21.

FRENTZEN, K. 1944. Ueber Massenvorkommen von Holothurien-Resten im Jura Badens. *Neues Jahrb. Min. Geol.*, Abt. B, Heft 4, pp. 99–104.

GEIS, H. L. 1936. Recent and fossil Pedicellariae, *Jour. Paleontology*, vol. 10, pp. 427–448.

JACKSON, R. T. 1912. Phylogeny of the echini, with a revision of Paleozoic species. *Boston Soc. Nat. Hist., Mem.* 7, 491 pp. (Also see Schuchert's review in *Amer. Jour. Sci.*, 5th ser., vol. 34, pp. 251–263.)

LYMAN, T. 1882. Report on the Ophiuroidea dredged by *H.M.S. Challenger* during the years 1873–76. *Zool.*, vol. 5, 386 pp.

MACBRIDE, E. W., and SPENCER, W. K. 1938. Two new Echinoidea, *Aulechinus* and *Ectinechinus*, and an adult plated holothurian, *Eothuria*, from the Upper Ordovician of Girvan, Scotland. *Phil. Trans. Roy. Soc. London*, vol. 229, pp. 91–136.

MATSUMOTO, H. 1915. A new classification of the Ophiuroidea: with descriptions of new genera and species. *Proc. Acad. Nat. Sci. Philadelphia*, vol. 67, pp. 43–92.

MEAD, A. D. 1900. The natural history of the star-fish. *U.S. Fish Comm. Bull.* 1899, pp. 203–224.

MORTENSEN, T. 1928–1948. "A Monograph of the Echinoidea." I, Cidaroida (1928); II, Bothriocidaroida, Melonechinoida, Lepidocentroida, and Stirodonta (1935); III,

Aulodonta (1940) and Camarodonta (1943); IV, Holectypoida and Cassiduloida (1948), and Clypeastroida (1948). Copenhagen, C. A. Reitzel; London, Oxford University Press.

PRANTL, F. 1947. Some holothurian remains from the Devonian of Bohemia. *Časopis Národního musea* (Praha), vol. 116, pp. 26–35. (English summary, pp. 33–35.)

SCHOENDORF, F. 1910. Palaeozoische Seesterne Deutschlands II. Aspidosomatiden des deutschen Unterdevon. *Palaeontographica*, vol. 57, pp. 1–66.

SCHUCHERT, C. 1915. Revision of the Paleozoic Stelleroidea with special reference to North American Asteroidea. *U.S. Nat. Mus., Bull.* 88, 301 pp.

SMISER, J. S. 1933. A study of the echinoid fragments in the Cretaceous rocks of Texas. *Jour. Paleontology*, vol. 7, pp. 123–163.

SPENCER, W. K. 1951. Early Palaeozoic starfish. *Phil. Trans. Roy. Soc. London*, vol. 235, pp. 87–129.

WALCOTT, C. D. 1911. Middle Cambrian holothurians and medusae. *Smith. Misc. Coll.*, vol. 57, no. 3.

CHAPTER 15

SUBPHYLUM HEMICHORDATA

INTRODUCTION

The organisms constituting this branch of the animal kingdom fall between the invertebrates on the one hand and the true chordates on the other, thus having characteristics of both divisions. A chapter is devoted to them because recent investigations have shown that the ancient and extinct graptolites, long considered to be closely allied to hydrozoan coelenterates, seemingly are much more closely related to certain living hemichordates (the pterobranchs) and should be placed in this branch or division, perhaps as an extinct and independent class.[1]

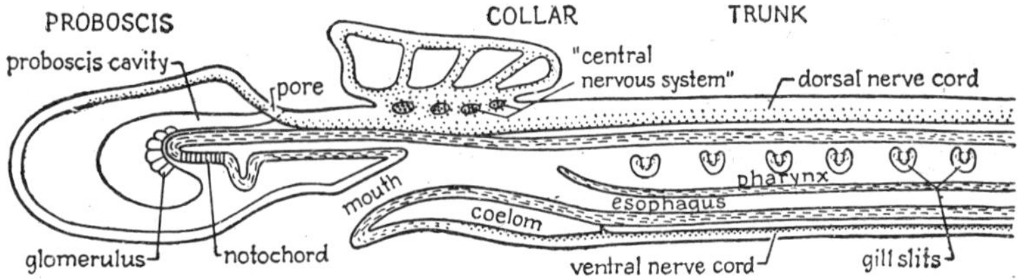

FIG. 15-1. Subphylum HEMICHORDATA. Dorsal view and median section of anterior portion of the tongue worm, *Dolichoglossus*, showing the essential features of a hemichordate. Note especially the gills, notochord, and dorsal nerve cord, a combination unknown among the invertebrates. (*After Storer*, 1943.)

The Hemichordata,[2] commonly called tongue worms, are small, soft-bodied worm-like animals constituting the simplest of the chordates.[3] They are called Hemichordata because they have partly chordate and partly invertebrate (*i.e.*, nonchordate) characteristics. They have long been considered the most primitive members of the Phylum Chordata and until recently were unknown in the fossil state.

Hemichordates have paired gill slits, a dorsal and ventral nerve trunk, and a peculiar preoral structure that has been called a "notochord" though it is not like the true notochord of all other chordates. Other internal and external structures are shown in Fig. 15-1.

[1] In this revolutionary change of taxonomic assignment of the graptolites, we follow the recent suggestions of Kozlowski (1938, 1947, 1948), Dawydoff (1948), Bulman (1949), and Thomas and Davis (1949*a*).

[2] It has been suggested recently (Dawydoff, 1948) that this subphylum be renamed Stomochordata, after the term **stomocord,** applied to a small anterior extension of the pharynx near the mouth (Fig. 15-1), because representatives of the subphylum do not have a true dorsal cord like that present in all other chordates. We prefer, however, to retain the older and more familiar term for the present discussion.

[3] Chordates are characterized by three distinct features: (1) a single **dorsal tubular nerve cord;** (2) the **notochord;** and (3) **gill slits** in the pharynx.

Three classes of Hemichordata are now recognized—two represented by living members, Class Enteropneusta and Class Pterobranchia; and one extinct, the ancient Class Graptozoa. These are discussed on following pages, with greatest emphasis on the Graptozoa, because of their extensive and important fossil record.

Class Enteropneusta

Typical representatives of this small class, which includes about a dozen genera and 60 or more species, are worm-like organisms a few inches long that are generally found in shallow marine waters. The more important features of a typical genus (*Dolichoglossus*) are shown in Fig. 15-1.

Of special interest to paleontologists and biostratigraphers is the well-known and world-widely distributed *Balanoglossus* which occupies a complex

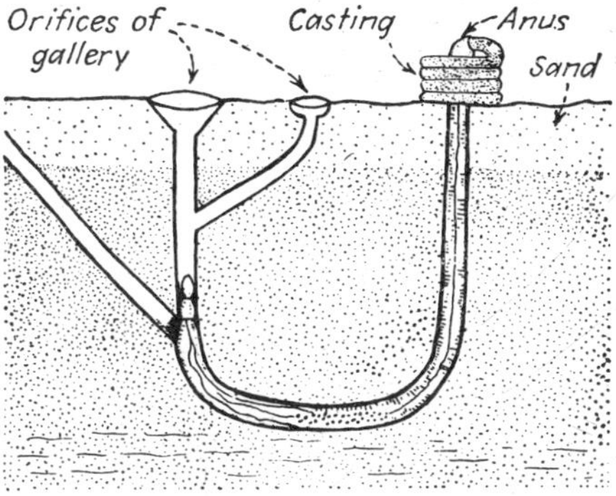

Fig. 15-2. Subphylum Hemichordata. Diagram showing *Balanoglossus*, Class Enteropneusta, in its complicated U-shaped burrow that it has excavated in sand. The castings make a ropelike pile around the posterior opening of the burrow. (*After Stiasny*, 1914, *with modifications*.)

U-shaped burrow that is excavated in sand and usually has ropelike excreta around one opening (Fig. 15-2). The organism lines the galleries with a viscous mucoid substance that rather firmly cements the sand grains together and thus strengthens the wall. This characteristic is similar to the burrowing habit of the marine lugworm, *Arenicola marina* (Fig. 11-12), and it seems possible that the burrows of the two greatly different organisms might well be confused if found in the fossil state.

So far as is known to us no fossil record of the Class Enteropneusta has been reported.

Class Pterobranchia

The pterobranchs are tiny colonial organisms that live in both shallow and deep marine waters, where they build dendritic colonies of transparent, chitinous tubes that constitute the **coenoecium.** They resemble the typical enteropneustan in general structure, but have only one pair of gill slits or none at all and reproduce both asexually (budding) and sexually.

Two orders of Pterobranchia are recognized, each with a single family.

Order 1. Rhabdopleuridea. Characterized by a hollow internal black stolon, absence of gill slits, and lophophore with a single pair of arms, and with the zooidal peduncle attached to the internal black stolon. The order has only one genus, *Rhabdopleura* (Figs. 15-3–15-7), which ranges from Cretaceous to Recent.

Order 2. Cephalodiscidea. Characterized by free-living individuals having a pair of gill slits, a lophophore of many pairs of tentacle-bearing arms, paired gonads, and a free peduncle. *Cephalodiscus* (Fig. 15-8), with four subgenera and 15 species, and *Atubaria*, with a single species, are the only living genera of the order. *Eocephalodiscus* (Fig. 15-9), from the Ordovician, is the only fossil form yet reported.

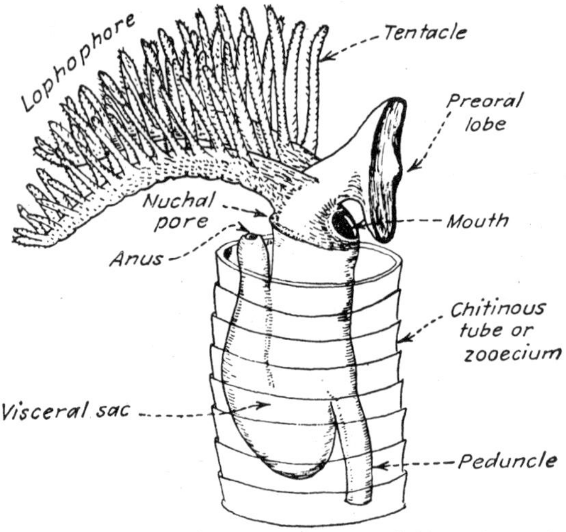

FIG. 15-3. Class PTEROBRANCHIA. Schematic diagram of *Rhabdopleura*, viewed from the right side. The individual zooid is not attached to the wall of the chitinous tube but is connected to the main black stolon of the colony by means of the peduncle (cf. Fig. 15-5). (*After Delage and Hérouard*, 1897, *with modifications*.)

Order 1. Rhabdopleuridea. The chief features of *Rhabdopleura*, one of the best known living pterobranchs, are shown in Figs. 15-3, 15-4, and 15-5. It is to be noted that this animal has some features also present in certain bryozoans.

The chitinous multitubular coenoecium, a system of branching tubes built by the individuals of a colony, is secreted by special glands in the disk of the preoral lobe (Fig. 15-3). The actual substance is a cuticular tissue, which is transparent in *Rhabdopleura* but more or less opaque and full of foreign particles in *Cephalodiscus* (Fig. 15-8*A*). The third and last genus, *Atubaria*, lacks a coenoecium altogether.

A complete coenoecium (Figs. 15-5, 15-6*A*) consists of several principal tubes and lateral branches from each of which are given off numerous short vertical tubes that are occupied by interconnected living individuals. Each of the short tubes, wide open at its free end and annulated posteriorly from this opening, houses an individual of the colony and is secreted by the preoral lobe of that individual. These vertical tubes are called **zooecia** and the individuals inhabiting them **zooids** (by analogy with the Bryozoa).

The coenoecium of *Rhabdopleura* has two quite different types of wall structure. The wall of the principal tube or tubes and of their lateral branches, and of the earlier part of the much shorter zooecial tubes, consists of semicircular bands of chitin that lie alternately one on top of the other and join at their extremities with interfingering relationships. The two zones of junction, one dorsal and the other ventral, appear as prominent zigzag lines on

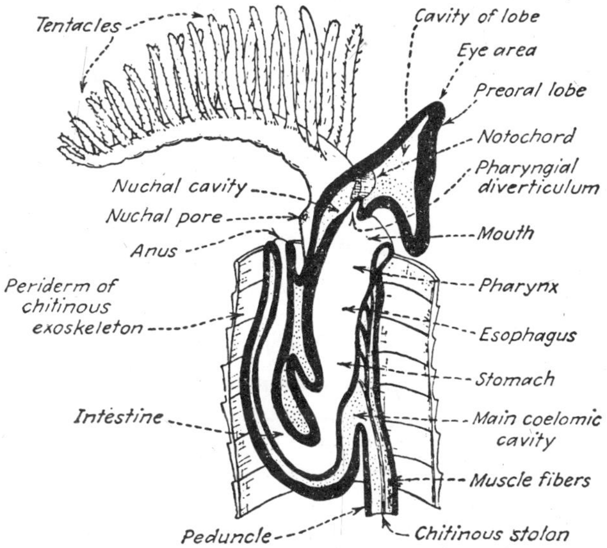

FIG. 15-4. Class PTEROBRANCHIA. Sagittal section showing internal structure of *Rhabdopleura*. The coelom is divided into three parts: the cavity of the preoral lobe, the nuchal cavity which opens to the exterior through the nuchal pore, and the main cavity in which the U-shaped alimentary tube is suspended. It is to be noted that the individual zooid is not attached to the wall of the chitinous tube that forms the exoskeleton. (*After Delage and Hérouard*, 1897, *with modifications*.)

the inner and outer surfaces of the tube (Fig. 15-5). The younger or distal part of the zooecial tubes consists of successive parallel annular bands joined in such manner that the line of suture is not visible (Fig. 15-5). These two types of structure have been discovered recently in the chitinous periderm of the ancient and long extinct graptolites (Fig. 15-7) and are one reason why the latter are now considered to be closely allied to living hemichordates.

Internally the principal tubes and their secondary branches have **transverse partitions,** or **septa,** at more or less regular intervals (Figs. 15-6*B*). These appear when buds develop into mature zooids and separate the last-formed zooid from earlier ones, so that only a single zooid ultimately occupies the chamber bounded by successive septa. The zooecial tubes, however, lack such partitions.

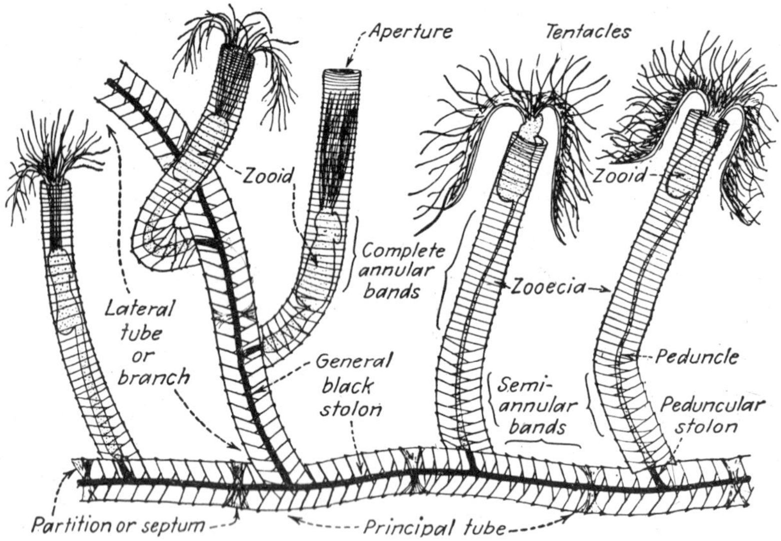

FIG. 15-5. Order Rhabdopleuridea. Small part of a colony of *Rhabdopleura*, much magnified and diagrammatic, showing a section of the principal tube with three zooecia and a section of a lateral branch with two zooecia. In the several zooecia the zooids, extended or retracted, are shown as if viewed through a transparent wall. It is to be noted that the principal and lateral tubes are divided into compartments by partitions, whereas the zooecia lack such transverse septa. These same tubes show only semiannular bands, whereas the zooecia have this kind of band in only the earliest part, being composed largely of complete annular bands in the latest part. (*After Schepotieff*, 1905, *highly generalized and diagrammatic.*)

The coenoecium of *Rhabdopleura* has a internal **black stolon** (**pectocaulus**) which occupies a kind of tunnel in the wall that lies in contact with the substratum. This stolon extends the length of the principal tubes and their secondary branches and sends a short extension into the base of each zooecium. This extension, to which the peduncle of the zooid is attached, is designated the **peduncular stolon** (Fig. 15-5). The stolon system is considered to be the prolongation of the tube made by the first zooid of the colony and to be the result of colonial growth by budding.

The colony begins in a fertilized egg that develops directly into a mature zooid. This zooid then gives rise to buds that ultimately become the individuals of a colony. (In *Cephalodiscus* [Fig. 15-8*B*] budding is localized on the ventral face of the distal extremity of the peduncle, where from two to as many as 14 separate buds may form.) The buds develop directly from the **gymnocaulus** inside the chitinous internal black stolon, and as they grow they ultimately form the short zooecial tubes that rise from the principal tube, or **creeping stem.** The buds are localized exclusively at the distal ends of the lateral, or secondary, branches of the principal tube. At the end of each of these branches is always an immature zooid, the **blastozooid,** which has been arrested in its development (Fig. 15-6*B*). After a bud has

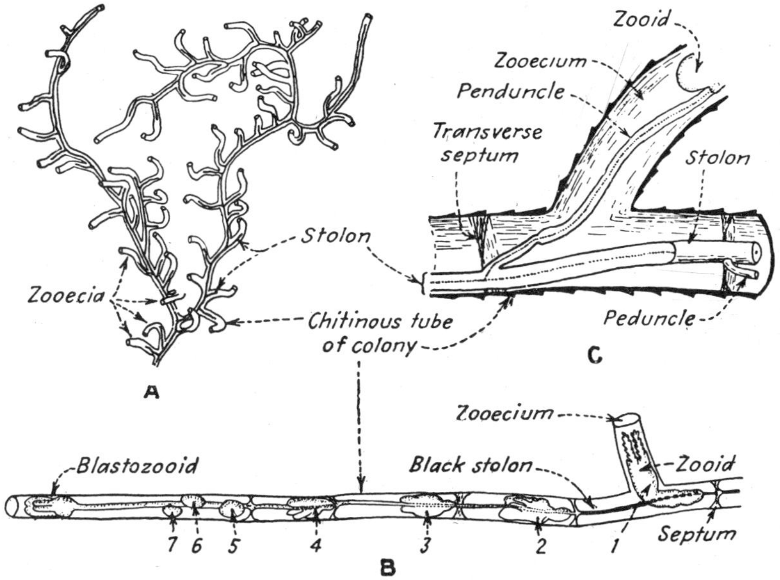

FIG. 15-6. Subphylum HEMICHORDATA—Genus *Rhaudopleura*. *A*. Part of a colony; such colonies characteristically are spread over the surfaces of shell fragments. *B*. Diagram showing nature of bud formation. The individual buds are numbered in the order of their appearance, 7 being the last to be formed. The oldest bud shown, 1, has started to build its zooecium. It is to be noted that each zooid ultimately is separated from the next older and younger buds by transverse septa. *C*. Diagram showing a small part of the stolonal system, much enlarged. The section follows the axis of the stolon except at the lower right, where the complete stolon and beginning of the peduncular stolon are shown. (*A and B after Lankester, considerably modified; C after Delage and Hérouard*, 1897, *modified.*)

reached a certain stage of development, it secretes behind itself a partition across the principal tube and then pierces the upper wall of the tube. Through this newly made opening it protrudes itself and immediately begins to secrete a new chitinous tube which ultimately will house it as a mature zooid. This newly developed zooid never leaves the colony to start a new one; it remains attached to the chief branch, having its peduncle, or **contractile stalk,** directly attached to the peduncular stolon. In this respect the rhabdopleurid differs from the individual cephalodiscid, which can leave the zooecium at will and move about freely over the coenoecium.

The recent discovery in several graptolites of a stolon system similar to that of *Rhabdopleura* provides further evidence for assigning the Graptozoa to the Hemichordata (see later discussion).

No fossil rhabdopleurids were known until 1930 when Thomas and Davis (1949, 1949*a*, 1950) discovered a fossil species in the Eocene London clay. This discovery was augmented by a second some years later (1938) when Kozlowski (1949) identified an Upper Cretaceous species from Poland. These recently discovered species, shown in Fig. 15-7, are particularly interesting

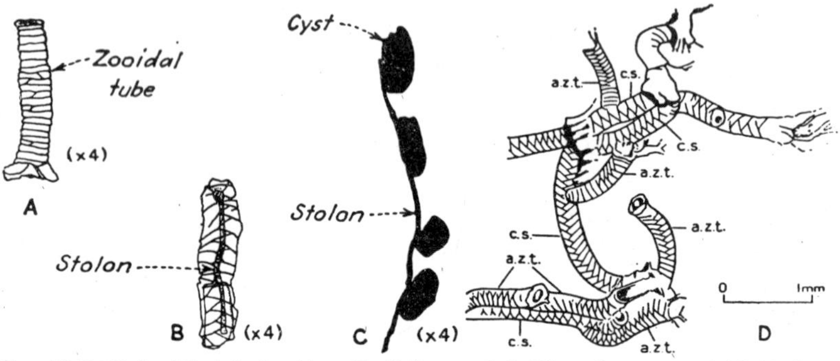

FIG. 15-7. Order Rhabdopleuridea. Fossil forms. *A–C.* Three fragments of *Rhabdopleura* from the Upper Cretaceous of Poland: *A*, zooidal tube; *B*, fragment of a main branch with stolon, consisting of two stolonal compartments of which the lower has a cyst of a sterile bud; *C*, fragment of stolon with four cysts of sterile buds. *D. R. eocenica* from the Eocene of England; a group of intersecting coenoecia (*a.z.t.* = adherent part of zooidal tube; *c.s.* = creeping stem). (*A–C after Kozlowski*, 1949, *with slight modifications; D after Thomas and Davis*, 1949, *by permission of the trustees of the British Museum.*)

because of the light they throw on the history and taxonomic relationships of the ancient graptolites.

Order 2. Cephalodiscidea. The chief external features of *Cephalodiscus*, from which the order takes its name, are shown in Fig. 15-8*B*. Although it bears considerable resemblance to *Rhabdopleura* (Figs. 15-3, 15-5), it should be noted that the lophophore is different, the peduncle is free, and the coenoecium (Fig. 15-8*A*) differs strikingly from the strongly tubular one of *Rhabdopleura* (Fig. 15-6*A*). The individual cephalodiscid zooids live in tiny cells, spaced along well-defined branches in some species but crowded together in irregular fashion in others (Fig. 15-9). The coenoecium is chitinous, com-

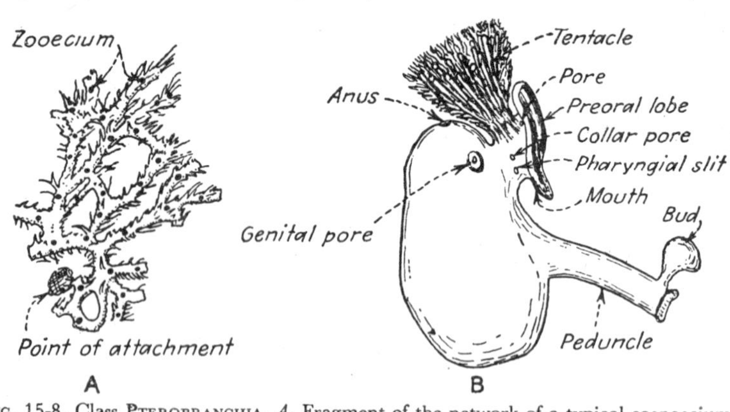

FIG. 15-8. Class PTEROBRANCHIA. *A.* Fragment of the network of a typical coenoecium of *Cephalodiscus*, showing one point of attachment and numerous zooecial apertures. The individual zooid lives in a zooecium but can leave it and move about freely over the surface of the coenoecium. *B.* The zooid of *Cephalodiscus* with a bud present at the extremity of the peduncle. (*After Delage and Hérouard*, 1897, *with modifications.*)

monly has extraneous sedimentary particles included in the chitin, and is characteristically spiny. It develops by budding from an initial zooid.

Until about a decade ago nothing was known of the geologic history of the Cephalodiscidea, as no fossil species had been reported. However, in 1938, and again in 1948, Kozlowski reported the discovery, in a graptolite-rich Ordovician shale of Poland, of a peculiar colonial fossil that he considered to be closely related to existing cephalodiscids. This fossil, which he named *Eocephalodiscus polonicus* (Fig. 15-9), has the shape of an irregular and asymmetrical ovoid with a basal expansion or disk of attachment and with eleven apertures arranged irregularly over the specimen. Morphologically the genus seems closest to living *Cephalodiscus*, and if the fossil is a true cephalodiscid, the Order Cephalodiscidea arose far back in geologic time (Fig. 15-27).

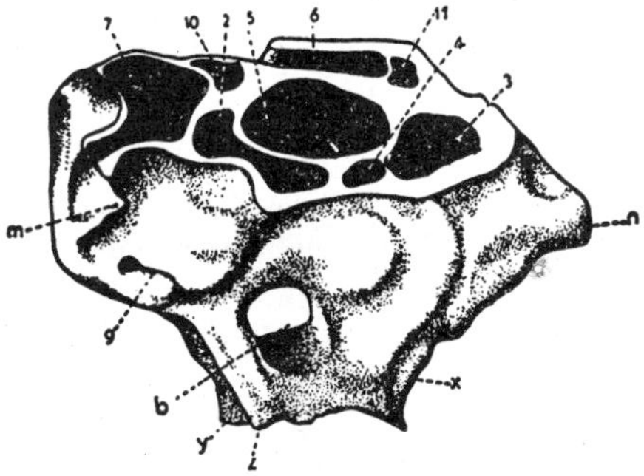

FIG. 15-9 Order Cephalodiscidea. Fossil form. Drawing of the holotype of *Eocephalodiscus polonicus* (×40), showing external appearance and internal structure of chitinous coenoecium which has been transected in the upper part to show the relation of the successive cells or chambers (numbered 1–11). The coenoecium started from a tiny cell (1, not shown on drawing) at the base (point *y*) and increased by budding, which resulted in the eleven irregular chambers. Internally, these chambers are interconnected by tiny tubes, the whole system ramifying the coenoecium in an irregular manner. (*After Kozlowski*, 1948.)

CLASS GRAPTOZOA[1] (GRAPTOLITHINA)[2]

The Graptozoa, or graptolites as they are almost universally called, are extinct marine Paleozoic organisms that built small dendritic or saw-blade-like colonial exoskeletons of chitinous material (Figs. 15-14, 15-25). The exoskeletons consist of a series of cups (or **thecae**) variously arranged along one or more branches (or **stipes**); the colony (or **rhabdosome**) originates in a conical **sicula** and increases by systematic budding. Compound rhabdosomes have been designated **synrhabdosomes.** More than 1,000 species have been described; these range from Middle Cambrian to Lower Mississippian in age.

Graptozoans are most typically preserved as black carbonaceous films resembling scroll-saw blades or hieroglyphics written on the rock, whence

[1] Graptozoa—Gr. *graptos*, written, + *zoon*, animal; referring to the fact that the carbonized remains of the originally chitinous exoskeletons resemble writing.

[2] The term Graptolithina is the usual designation for this group of organisms outside the United States of America.

the common name **graptolite** (Gr. *graptos*, written, + *lithos*, rock). Graptolites are most common and abundant in black shales, but they are also present in shales of other colors, as well as in sandstones and limestones, and in chert nodules. So far as is known, they seem to have been exclusively colonial and marine and to have been adapted to both benthonic and planktonic environments. As plankton they attained world-wide distribution. This, together with the fact that they evolved rapidly, make them among the most reliable of index fossils for intercontinental correlation (Ruedemann 1911, 1934; Grabau, 1929).

Although only the chitinous exoskeleton of the graptolite is found fossil, preservation of this structure is so perfect in certain limestones and cherts that investigators have been able to make faithful reconstructions of some of the soft parts and to determine the relationships of these to the **periderm,** as the material of the exoskeleton is called. Two of the best known of the earlier investigators, Holm and Wiman in Sweden, recovered well-preserved specimens from limestone by dissolving the rock in dilute hydrochloric acid. Bulman in England has used serial sectioning to excellent advantage on such materials. A recent investigator, Kozlowski of Poland, has been able to recover exquisitely preserved though highly fragmental specimens from chalcedonic lenses by dissolving away the enclosing silica with hydrofluoric acid.

The biological affinities of the graptolites have always been a somewhat debatable question. It is true that they have long been considered an extinct order of the Hydrozoa, but before they came to be so widely accepted as hydrozoans they had been assigned at one time or another to such widely diverse groups as plants, horny sponges, alcyonarian corals, cephalopods, and pterobranchs and had even been interpreted as inorganic markings. One small group of paleontologists (Neumayr, Wiman, and Elles) have even gone so far as to contend that the graptolites represent a wholly extinct group of organisms that cannot be assigned to any existing class of animals.

In 1931, Ulrich, a bryozoan specialist, and Ruedemann, a long-time investigator of the graptolites, found on comparing ideas that there was seemingly convincing evidence for considering graptolites as an extinct order of the Bryozoa, but this view has not gained wide acceptance among graptolite specialists. The most recent work bearing on the affinities of the group is that of Kozlowski (1938, 1948) mentioned in a preceding paragraph. On the basis of well-preserved fragments recovered from chert bands in certain Lower Ordovician beds, Kozlowski erected three new orders and concluded that the graptolites as a group show closest affinities with the pterobranch division of the Hemichordata. This conclusion is now being widely accepted by graptolite specialists, and we follow Kozlowski in assigning the Graptozoa to the Subphylum Hemichordata. (See following pages for a fuller discussion of graptolite affinities.)

Nature of Graptozoan Skeleton

So far as is known, the wall material (or periderm) of the typical graptozoan rhabdosome was a chitinoid substance of some sort (Ruedemann, 1947).

It [illegible]ely preserved in what is believed to be its original state; generally it is a somewhat carbonized transparent or opaque brown material with an opaque black "skin" on the outer surface. When in a perfect state of preservation, as Kozlowski found it in chert, it consists of two structurally different tissues: (1) the main substance, the **fusellus** (or **tissu fusellaire**), that is laid down in narrow fusiform bands (Fig. 15-10*A–B*) and produces the growth lines of the periderm; and (2) a **cortex** (or **tissu cortical**) consisting of thin, parallel laminae that were laid down successively on the outside of each theca, producing secondary thickening of the thecae and branches (or stipes). A somewhat modified form of this fundamental structure is present in the graptoloids (Fig. 15-10*C–D*). Kozlowski pointed out that the fusellar structure is such as to exclude the graptolites from either the Coelenterata or the Bryozoa but to relate them to the Pterobranchia of the hemichordates.

The graptozoan rhabdosome consists of one or more stipes composed of one, two, or four series of thecae. The thecae are of three different types: **autothecae** (or simply **thecae**), **bithecae**, and **stolothecae.** This polymorphism is thought to be the result of the modes of reproduction developed by the graptolites.

One of the common causes of dimorphism (and polymorphism) in animals (*e.g.*, Foraminifera; Ostracoda) is sexual, and although colonial animals are commonly hermaphroditic, both male and female zooids are developed in the Pterobranchia (*Rhabdopleura* and *Cephalodiscus*) and are believed by Kozlowski (1948) to have been developed in the dendroid graptolites. He interprets the autothecae as having housed female zooids and the bithecae as having been occupied by reduced males. In those orders lacking bithecae (Stolonoidea; Graptoloidea) it is assumed that the autothecal females became hermaphroditic, and that the bithecal males, and hence the bithecae themselves, were eliminated in the process. The stolothecae do not seem to have housed separate individuals ("stolozooids"); rather, they are thought to have been secreted by immature autozooids at an early stage of development, each autotheca being secreted by the same individual as was the stolotheca which preceded it (Bulman, 1949*a*). From the suggestions just made, it would follow that the dendroid colony consisted of two distinct types of individuals—autozooids and bizooids—whereas the typical graptoloid rhabdosome was made by autozooids alone, associated closely with an unchitinized stolon.

The three types of thecae are best developed in the Dendroidea and Tuboidea. Bithecae seem to be lacking in the Camaroidea, Stolonoidea, and Graptoloidea. Stolothecae are present in all graptolites, but the stolon system is excessively developed in the Stolonoidea.

Classification

It has been known for a long time that the dendroid graptolites differ greatly from the so-called true graptolites—those with rhabdosomes of a few stipes resembling saw blades. Kozlowski's (1938, 1948) important discovery indicates that the graptolites are a much more diverse group of organisms

TABLE 15-1. RANGE OF THE DIFFERENT GRAPTOLITE ORDERS AND FAMILIES FROM CAMBRIAN TO MISSISSIPPIAN.

ORDER FAMILY	PALEOZOIC											
	Camb.		Ord.			Sil.			Dev.			Miss.
	M	U	L	M	U	L	M	U	L	M	U	L
Dendroidea	—	—	—	—	—	—	—	—	—	—	—	—
Dendrograptidae	—	—	—	—	—	—	—	—	—	—	—	—
Acanthograptidae		—	—	—	—	—	—	—				
Ptilograptidae			—	—	—	—	—	—				
Tuboidea			—	- -	- -	—	—	—				
Idiotubidae			—									
Tubidendridae			—									
Cyclograptidae						—	—	—				
Camaroidea			—									
Stolonoidea			—									
Graptoloidea		—	—	—	—	—	—	—				
Dichograptidae		—	—									
Corynograptidae				—								
Leptograptidae			—	—	—							
Dicranograptidae			—	—	—							
Diplograptidae			—	—	—	—						
Glossograptidae			—	—	—							
Retiolitidae					—	—	—	—				
Dimorphograptidae						—						
Monograptidae						—	—	—				

than was formerly supposed. His new classification of the group is given below. The orders and families with their geologic range are shown in Table 15-1.

Order 1. Dendroidea. Dendroidea have dendritic rhabdosomes of uniserial stipes which anastomose or are connected by thin transverse processes; three types of thecae are present—autothecae, bithecae, and stolothecae, the last carrying a chitinized stolon from which successive triads of thecae spring. The colony originates from a sicula like that in the Graptoloidea, with the first bud forming well down on the lateral surface of the sicular cone (Fig. 15-13). *Middle Cambrian to Lower Mississippian.*

Order 2. Tuboidea. Tuboidea have rhabdosomes with three types of thecae that spring from a stolon as in the Dendroidea; bithecae and autothecae are essentially as in Dendroidea, but stolothecae are much less individualized. The stolothecae form a system of irregular tubes of variable length, with stolons equally variable as to size and structure. The regular triad budding of the Dendroidea is replaced by capricious budding in which the nodes and frequency of the thecae are both variable. The sicula and ontogeny of the order are unknown. *Lower Ordovician (Upper Tremadocian) to Upper Silurian.*

Order 3. Camaroidea. The rhabdosomes, which are incrusting and somewhat similar to certain bryozoans, are characterized by a complex autotheca composed of two distinct parts—an erect distal tubular part, the **collum,** and an enlarged creeping part, the **camara.** Some forms have bithecae; others do not. The colony increases by budding from stolons, but the development of the colony and its mature form are unknown. *Lower Ordovician (Upper Tremadocian).*

Order 4. Stolonoidea. The incrusting and branching rhabdosomes of this order differ from all other graptolites in the excessive development and extreme irregularity of the stolons, which are more developed than the autothecae. Thecae produced from stolons, besides the stolothecae themselves, seem to be autothecae only; bithecae seem to be lacking altogether. The order is based on exceedingly fragmentary material. *Lower Ordovician (Upper Tremadocian).*

Order 5. Graptoloidea. The rhabdosomes are composed of relatively few uniserial, biserial, or quadriserial stipes that hang free from a common float or disk to which they are generally attached by a **nema,** or **virgula.** The stipes contain only autothecae except in a few rare cases; bithecae do not seem to have been developed. The sicula hangs from a long nema. No internal system of stolons is known, but this condition is believed to be due to the fact that the stolon was not chitinized. The colony begins in a conical sicula, with the first bud appearing in the wall of the later part (**metasicula**) of the sicula; the way in which subsequent buds are formed and thecae added determines the ultimate architecture of rhabdosome and synrhabdosome. *Upper Cambrian to Upper Silurian.*

Order 1. Dendroidea.[1] *Morphology.* The typical dendroid graptolite is a small, grasslike or bushlike, black carbonaceous tracery which on careful examination proves to consist of serrated branches, or **stipes,** on which thecae are arranged uniserially. Adjacent stipes anastomose or are joined together at intervals by tiny transverse bars, termed **dissepiments,** and commonly bifurcate several times in a large rhabdosome. The thecae are **polymorphic** and take three different forms—autothecae, bithecae, and stolothecae (Fig 15-11)—all of which spring as a triad from the stolons that form a continuous

[1] Dendroidea—Gr. *dendron,* tree, + *-oid,* like; referring to the treelike appearance of the rhabdosome.

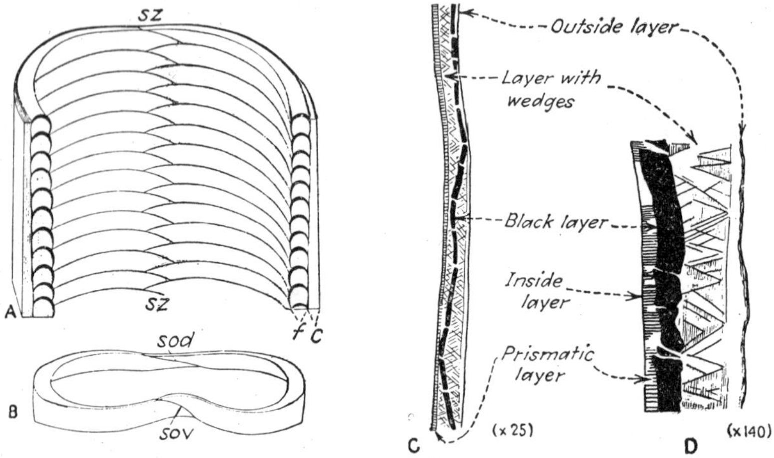

FIG. 15-10. Class GRAPTOZOA. Diagrams showing microscopic structure of periderm. *A*. Longitudinal section of earlier part of a dendroid autotheca, showing the inner fusellar layer (*f*), outer cortical layer (*c*), and zigzag suture (*Sz*) along which the terminal parts of successively overlapping fusellar collars join. *B* shows a complete ring composed of two collars which join along an inclined contact on the ventral (*sov*) and dorsal (*sod*) sides. *C–D*. Longitudinal section of the periderm of a graptoloid (*Monograptus*): *C* shows the cortical or black layer first in contact with the inner prismatic layer, then upward along the section in contact with the outside layer; *D* shows a small part of the wall greatly magnified. The prismatic layer is composed of many tiny pillars or columns and when well preserved is a dark yellowish-brown color. The layer with the prominent fibrous wedges, although considered to be homologous with the fusellar layer of the dendroids by at least one recent author (Waterlot, 1948), was, according to Ruedemann (1947), long ago found to be composed of calcite crystals formed secondarily during fossilization. Whether *Monograptus* had four distinct layers (*i.e.*, internal epidermis, prismatic layer, black layer, and external epidermis) or only three (*i.e.*, the black layer between an inner and an outer epidermis) is an unanswered question. (*A–B after Kozlowski*, 1938; *C–D adapted from Perner*, 1894.)

system of tubes reaching back to the embryonic sicula (Figs. 15-11, 15-12, 15-13). The colony develops from a conical sicula that is typically erect (Fig. 15-13) and generally more or less embedded in secondary chitinoid material forming a rootlike base. In life the sicula seems to have been attached by a long or short nema to some object on the bottom, to a disk of some sort, or to a seaweed (Figs. 15-12, 15-14).

Colonial Development (*Astogeny*). The dendroid graptolites, together with the graptoloids, have been much studied because of the great interest in their ontogeny, or more correctly **astogeny** (*i.e.*, the development of the colony from embryonic sicula to mature rhabdosome), and in the light that this remarkable development throws on some of the perplexing problems of organic evolution. They are the oldest fossil animals whose life history has been determined in any detail; hence their great age gives them added importance in this respect. A brief summary of the astogeny of two typical genera (*Dictyonema* and *Dendrograptus*), therefore, seems in order.

The development of a typical dendroid colony is well shown by *Dictyonema flabelliforme* (Figs. 15-11, 15-12), from the Upper Cambrian, the dendroid

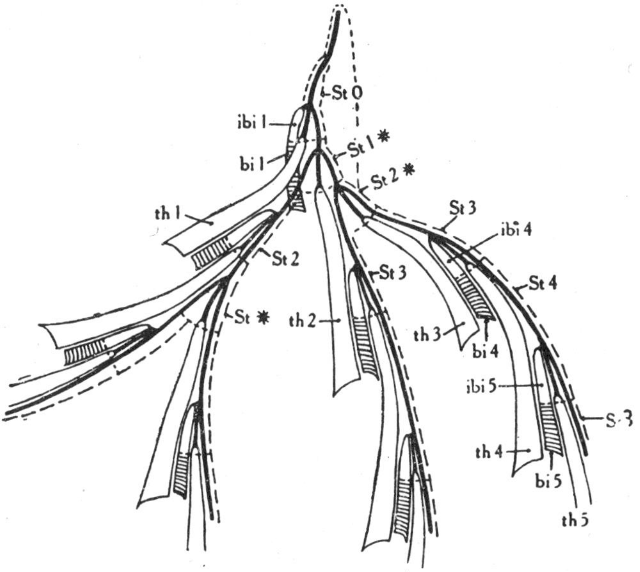

FIG. 15-11. Class GRAPTOZOA. Diagrammatic scheme of a right-handed proximal end of *Dictyonema flabelliforme*, with the stolon system shown in solid black. (*St*,0–5 = stolothecae; branching divisions marked with an asterisk; *th*,1–5 = autothecae; *bi*,1–5 = bithecae; proximal portions enclosed within the parent stolotheca lettered *ibi*, distal external portions shaded.) (*After Bulman*, 1949.)

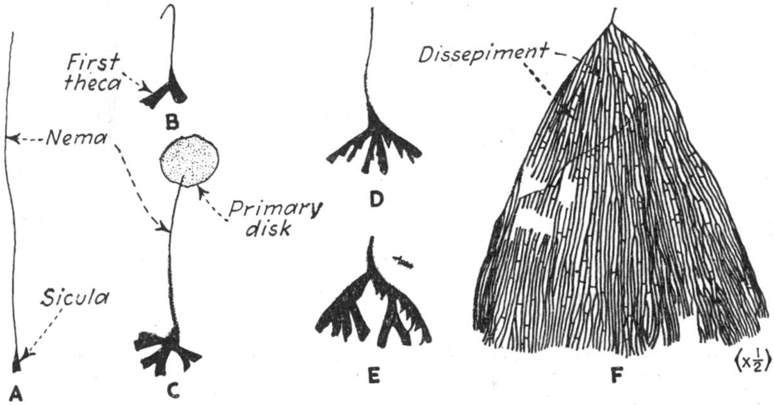

FIG. 15-12. Class GRAPTOZOA. Growth stages of a dendroid graptolite (*Dictyonema flabelliforme*) from the Upper Cambrian of New York. *A*. Sicula with long nema. *B*. Sicula with first mature theca. *C*. Young rhabdosome attached to primary disk by nema. *D–E*. Somewhat more advanced rhabdosomes, showing bifurcation of branches, and the first dissepiment in *E*. *F*. A normal and approximately mature rhabdosome. (*A–E*, ×2.5.) (*After Ruedemann*, 1904, *slightly modified*.)

which Bulman (1949) now considers ancestral to the true graptolites, or Order Graptoloidea, and by *Dendrograptus* (Fig. 15-13), a closely related genus. The first individual of the colony, supposedly the embryo, secreted a tiny chitinous cone, the **sicula**, which was attached by a hollow thread, the **nema** or **nemacaulus,** to some object on the bottom, or possibly to some floating organism such as a seaweed, depending on the habit of the particular species.

The sicula itself, which is none too well known in most Dendroidea (Figs. 15-13, 15-20), seems to be comparable with a similar structure of somewhat greater complexity in the Graptoloidea. It is a chitinous tube consisting of an earlier conical part, the **prosicula,** and a later tubular part, the **metasicula** (Figs. 15-20, 15-21, 15-22). The prosicula is a thin-walled conical tube tapering distally and passing into a tubular nema that served for attachment. The prosicular wall is strengthened by several longitudinal fibers and also has embedded in it a peculiar helical line which may be coiled either right-

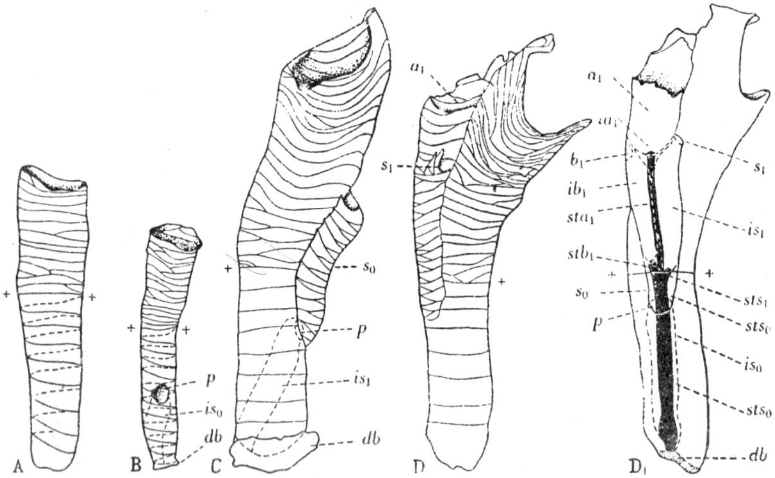

FIG. 15-13. Class GRAPTOZOA. Four specimens showing several different stages in the early development of the rhabdosome of *Dendrograptus* (*A*, *C–D*, are *D. communis; B, Dendrograptus* sp.). *A*. Incomplete sicula, showing most of the prosicula with its typical spiral line and the lower part of the metasicula with several semiannular rings. *B*. Sicula with metasicula somewhat more advanced than *A* (p = pore in wall of prosicula; is_0 = part of the sicular stolotheca—inside vertical dashed lines—formed inside the prosicula; db = outline of basal disk). *C*. Stage more advanced than *B*, showing part of sicular stolotheca (S_0) formed outside the sicula and arising from the pore (p). *D*. Sicula with the first triad of thecae: D_1, same specimen as though transparent, with the fusellar sutures omitted. There is shown inside the sicular stolotheca (is_0) a black stolon (sts_0) which at its upper external part divides into three stolons from which arise the three primary thecae (the first triad) of the colony. (a_1 = external part of autotheca, ia_1 = basal part of autotheca contained inside the stolotheca s_0; sta_1 = stolon of autotheca; b_1 = aperture of bitheca; ib_1 = basal part of bitheca contained inside the stolotheca s_0; stb_1 = stolon of bitheca; s_1 = aperture of the first stolotheca; is_1 = basal part of first stolotheca contained inside the stolotheca s_0; sts_1 = stolon of the first stolotheca. The tiny plus signs (+) mark the boundary between the prosicula below and the metasicula above.) All the figures are ×45. (*After Kozlowski*, 1947.)

or left-handedly. The metasicula is tubular and is set off sharply from the prosicula by its closely spaced transverse growth lines. As the metasicula nears completion a pore appears in its wall not far from the aperture, and it is from this pore or near it that the first bud or buds of the colony arise. Other features of the sicula are shown in Fig. 15-20.

The marked difference between the prosicula and metasicula has led some investigators to suggest that the embryo, after it had produced the prosicula, went through a metamorphosis or degeneration, giving rise to a new organism, which then proceeded to construct the metasicula. This suggestion recalls the well-known degeneration-regeneration phenomenon in Bryozoa, in which the protoecium-ancestroecium combination is formed before the first bud of the zoarium appears, and it has been used as a claim that the Graptozoa and Bryozoa are closely related (Ulrich and Ruedemann, 1931).

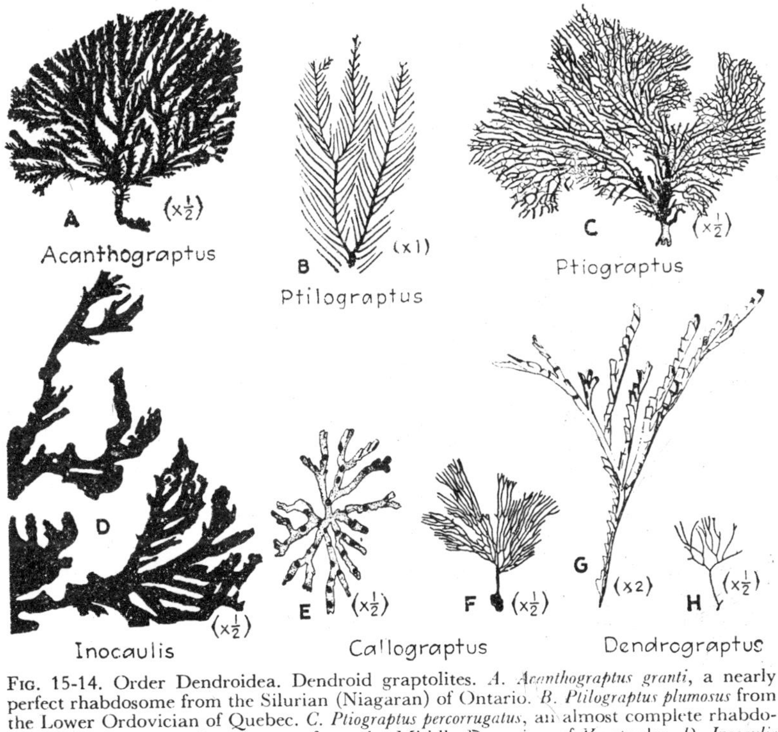

FIG. 15-14. Order Dendroidea. Dendroid graptolites. *A. Acanthograptus granti*, a nearly perfect rhabdosome from the Silurian (Niagaran) of Ontario. *B. Ptilograptus plumosus* from the Lower Ordovician of Quebec. *C. Ptiograptus percorrugatus*, an almost complete rhabdosome showing many dissepiments, from the Middle Devonian of Kentucky. *D. Inocaulis granti*, part of a large rhabdosome, from the Silurian (Niagaran) of Ontario. *E. Callograptus staufferi*, a compressed rhabdosome showing pits on the branches, from the Upper Cambrian (Trempealeau) of Minnesota. *F. Callograptus compactus*, an incomplete rhabdosome, with adhesion disk, from the Ordovician (Utica) of New York. *G. Dendrograptus hallianus*, an incomplete rhabdosome showing the thecae, from the Upper Cambrian (Trempealeau) of Wisconsin. *H. Dendrograptus ontarioensis* from the Silurian (Niagaran) of Ontario. (*B after Hall*, 1865; *D, H after Bassler*, 1909; *all others after Ruedemann*, 1908.)

From the side of the metasicula was developed the **initial bud,** which contained an extension of the stolon present in the sicular embryo (Figs. 15-11, 15-13, 15-20). This budding individual, inhabiting the first theca of the colony, stolotheca 0, gave rise to a triad of individuals that built an autotheca, a bitheca, and a stolotheca (Figs. 15-11, 15-13*D*). Other triads followed in orderly succession, each time springing from the stolon of an existing stolotheca. Hence, the stolon in stolotheca 0, which sprang from the initial bud that appeared on the sicular wall, gave rise to the first triad of the rhabdosome, which then built autotheca 1, bitheca 1 (the sicular bitheca), and stolotheca 1. Next, the stolon in stolotheca 1 underwent a branching division to produce the second triad, and from the stolotheca of this group (stolotheca 2) sprang the third triad. Stolotheca 2 passed across together with autotheca 1 to produce the first branch, or stipe, which shortly divided again. Budding continued, as shown in Fig. 15-11, until a complete rhabdosome had been formed. It should be noted that there is a system of stolons shown in solid black on Fig. 15-11 which originates in the sicula and sends an extension into each stipe. Where branching division takes place, the stolon trifurcates, with one branch leading to an autotheca, a second to a bitheca, and the third, its own extension, which continues to the next point of trifurcation. The stolon system is present in all dendroid graptolites and seems to have been developed in the embryo before the initial bud came into existence. Its chief function was for reproduction.

Habit. It is thought that the dendroid graptolites were dominantly sessile benthonic in habit, being rooted and growing erect from the sea bottom. It has been suggested, but not convincingly demonstrated for all supposed cases, that certain dendroids were attached to floating seaweeds or to floats of their own and hence were pseudoplanktonic or planktonic.[1] Whatever their habit, however, they did attain world-wide distribution, but they are not, in general, as reliable index fossils as the much shorter ranging graptoloid species have proved to be.

Geologic History. The earliest known dendroid graptolities have been reported from the European Middle Cambrian[2] (see Bulman, 1938, page D15). The youngest ones thus far described, which are also the last of all graptolites, are species of *Dictyonema* from the Lower Mississippian Chouteau limestone of Missouri and Englewood shale of South Dakota of North America (Ruedemann and Lochman, 1942) and species of *Desmograptus* and of *Callograptus* from the British Carboniferous Pendleside series.

The Order Dendroidea, to which some 25 genera and several hundred species have been referred, has the following three families: Dendrograptidae: *Dictyonema* (M. Camb.–L. Miss.; Fig. 15-12), *Dendrograptus* (U. Camb.–U. Sil.; Fig. 15-14*G–H*), *Callograptus* (U. Camb.–L. Miss.; Fig. 15-14*E–F*), and *Ptiograptus* (Dev.; Fig. 15-14*C*). Acanthograptidae: *Acantho-*

[1] STÖRMER, L., 1933, A floating organ in *Dictyonema*, *Norsk Geol. Tidsskrift*, vol. 13, pp. 102–112. BULMAN, O. M. B., 1938, Graptolithina, *in* Schindewolf's "Handbuch der Paläozoologie," vol. 2D, p. D9. Berlin, Gebrüder Borntraeger.

[2] *Dictyonema schucherti*, originally reported from the Lower Cambrian of Vermont, actually came from an Upper Cambrian slate (Ruedemann, 1947, p. 164).

graptus (U. Camb.–Sil.; Fig. 15-14*A*) and *Inocaulis* (Ord.–Sil.; Fig. 15-14*D*). Ptilograptidae: *Ptilograptus* (Ord.–Sil.; Fig. 15-14*B*).

Order 2. Tuboidea. This order was recently erected by Kozlowski (1938) to include two new families, Idiotubidae and Tubidendridae, and one already established family, Cyclograptidae (*Cyclograptus*, U. Sil.; Fig. 15-16), which was formerly included in the Dendroidea.

The rhabdosomes have the three types of thecae characteristic of the Dendroidea (although the stolothecae are much less individualized). The method of colonial increase from the stolon is similar (although the stolon varies in size and structure), but the thecae in different combinations appear in groups of two only and, being randomly distributed over the rhabdosome and commonly confined to only one side of the stipes, do not make regular and uniform rhabdosomes as in the Dendroidea.

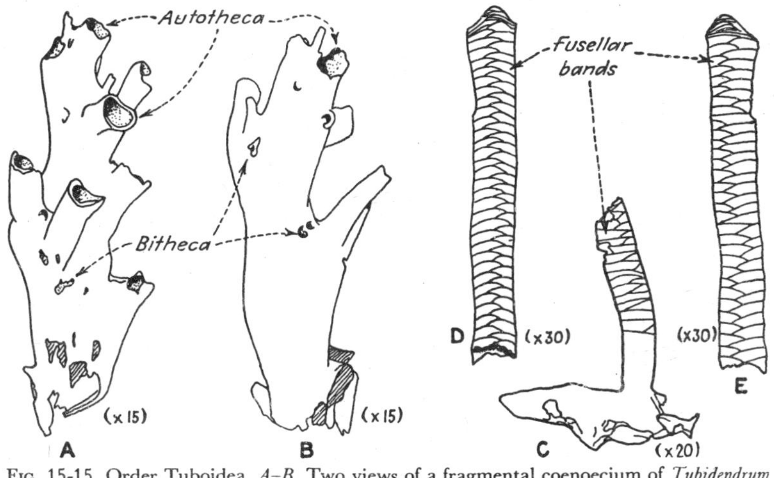

FIG. 15-15. Order Tuboidea. *A–B.* Two views of a fragmental coenoecium of *Tubidendrum bulmani*. It is to be noted that the apertures of the autothecae (larger tubes) are largely limited to one side only, whereas the bithecae (smaller openings) are present on both sides. *C–E. Idiotubus: C*, an autotheca with a fragment of the basal expansion; *D–E*, ventral and dorsal views, respectively, of the free part of an autotheca of *I. rectus*. (*After Kozlowski*, 1948, *slightly modified.*)

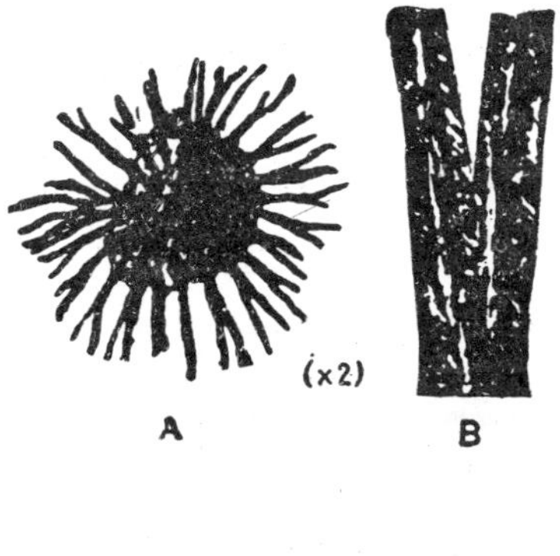

FIG. 15-16. Class GRAPTOZOA. *A.* Complete rhabdosome of *Cyclograptus rotadentatus*, showing the numerous forked stipes originating in the center of the disk. Originally, the stipes are thought to have formed a cup-shaped frond attached by its base to the prominent disk. In the rock the stipes are flattened so as to appear as the spokes in a wheel. *B.* A small portion of a stipe, showing the bifurcation, considerably enlarged. (*From Bassler*, 1909, *after Spencer*.)

The order seemingly ranges from Lower Ordovician to Upper Silurian (Fig. 15-27). *Idiotubus* (Fig. 15-15*C–E*) is characteristic of the Idiotubidae; *Tubidendrum* (Fig. 15-15*A–B*) of the Tubidendridae; and *Cyclograptus* (Fig. 15-16) of the Cyclograptidae.

Order 3. Camaroidea. This new order, based on a group of five genera, is reported by Kozlowski (1938, page 87) to have certain aspects of a bryozoan nature. The chief characteristic of this order lies in the form of the

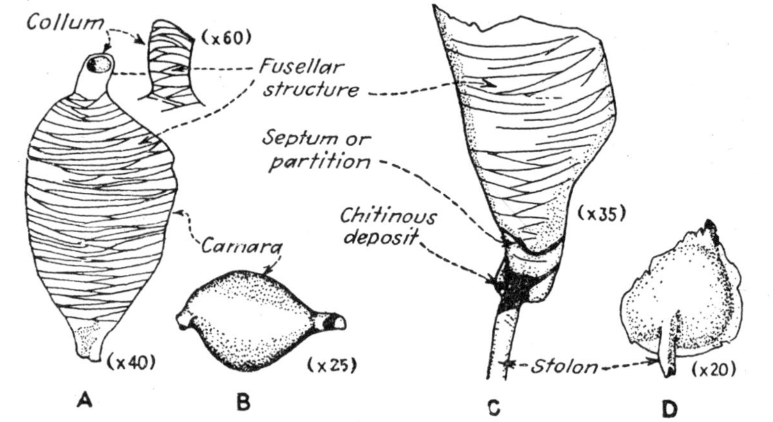

FIG. 15-17. Order Camaroidea. *A–B. Cysticamara? bicollis. A.* A single camara in transparent view. The fusellar structure of the ventral wall and of the ventral face of the collum is conspicuous. *B.* A typical camara. *C–D.* Separate camarae of *Bithecocamara gladiator: C,* posterior part of a camara with part of the stolon, seen as though transparent and showing the fusellar structure of the ventral wall as well as two partitions and a chitinous deposit in the early part; *D,* a single camara with a part of the stolon still intact. (*After Kozlowski,* 1948, *slightly modified.*)

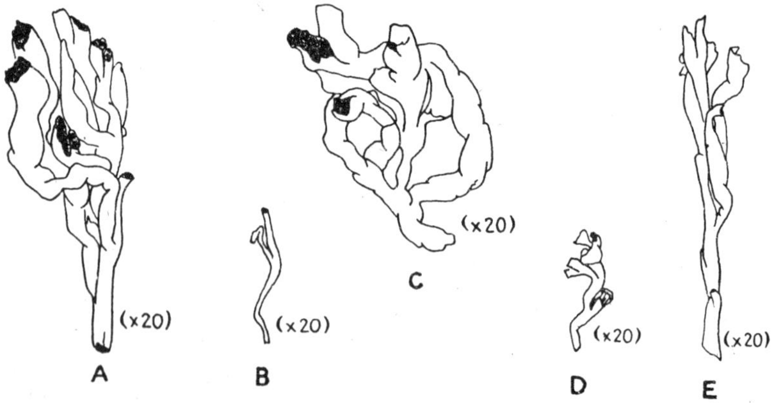

FIG. 15-18. Order Stolonoidea. Fragmental coenoecia of *Stolonodendrum.* The coenoecia consist largely of stolons and exhibit considerable variation in size and architecture. The Stolonoidea are characterized by the excessive development and irregular structure of the stolons. (*After Kozlowski,* 1948, *slightly modified.*)

autothecae, which have two different parts—a distal tuboidal collum, and a proximal expanded camara (Fig. 15-17*A*). Bithecae are present in some forms but lacking in others. The colonies, which are incrusting, increase by budding from a system of stolons as in other graptozoan orders.

All of the genera included in the order are confined to the Upper Tremadoc (Lower Ordovician) of Poland. *Cysticamara* (Fig. 15-17*A–B*) and *Bithecocamara* (Fig. 15-17*C–D*) are typical of the order.

Order 4. Stolonoidea. The Stolonoidea have as a distinctive feature the excessive development and marked irregularity of the stolon (Fig. 15-18). Bundles of stolons or single stolons are commonly present in stolothecal tubes composed of a thin transparent substance having fusellar structure, which is in complete rings, however, rather than in the half rings characteristic of the other graptolite orders. The thecae, excepting the stolothecae, seem to be autothecae exclusively. *Stolonodendrum* (Fig. 15-18), the single genus thus far described, is found only in the Upper Tremadoc (Lower Ordovician) of Poland.

Order 5. Graptoloidea. *Morphology.* The graptoloid rhabdosome generally consists of only a few stipes which are almost invariably composed of autothecae alone. The autothecae are arranged in a uniserial (*Monograptus*), biserial (*Glossograptus*), or rarely quadriserial (*Phyllograptus*) series. The colony arises from a sicula which is pendent from a nema (or virgula). The stipes are variously arranged relative to the sicula, as shown in Fig. 15-24. In a few forms several colonies are bound together to form a compound colony, or synrhabdosome (*e.g.*, *Glossograptus*, Fig. 15-25*O*).

The general structure of the graptoloid stipe is shown diagrammatically in Figs. 15-19 and 15-25*K*. The autothecae (or thecae) are essentially overlapping tubes open to the exterior distally at their apertures and merging at their inner extremity with what can be thought of as a common canal. The apertural part of the autothecae underwent elaborate modification as

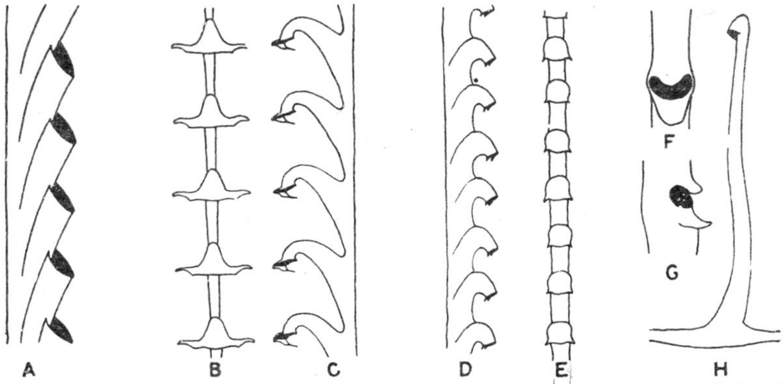

Fig. 15-19. Class Graptozoa. Schematic restorations showing different types of thecal elaboration in the monograptids. *A*. Simple thecae. *B–C*. Triangular thecae with apertural spines. *D–E*. Hooked thecae. *F–G*. Thecae with shelves. *H*. Isolate theca. (*Adapted from Bulman*, 1932 *and* 1933.)

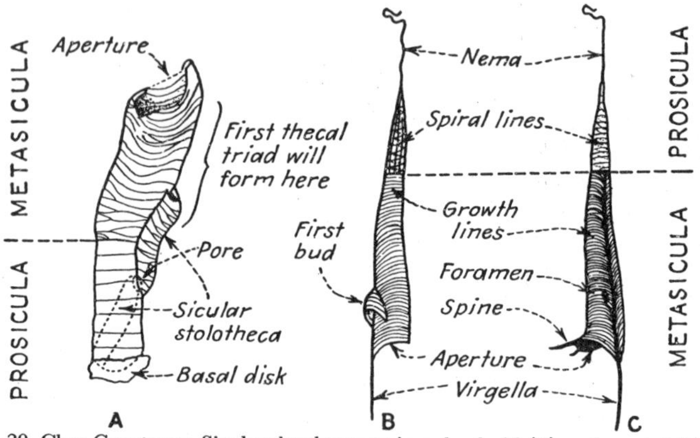

FIG. 15-20. Class Graptozoa. Sicular development in a dendroid (*A*) and graptoloid (*B–C*) graptolite. *A*. Well-advanced stage of the sicula of *Dendrograptus* (×30), just before bitheca 1 and autotheca 1 begin to develop from the sicular stolotheca (cf. Fig. 15-13). The sicula, which is oriented in supposed living position, is reversed with respect to *B* and *C*. *B*. Sicula of *Diplograptus* (*Orthograptus*) (×20), showing first bud just beginning. *C*. Fully developed sicula of *Diplograptus* (*Orthograptus*) (×30), showing foramen from which first bud will spring, virgella, and apertural spines. (*A after Kozlowski*, 1947; *B–C after Kraft*, 1926; *all with modifications and somewhat diagrammatic.*)

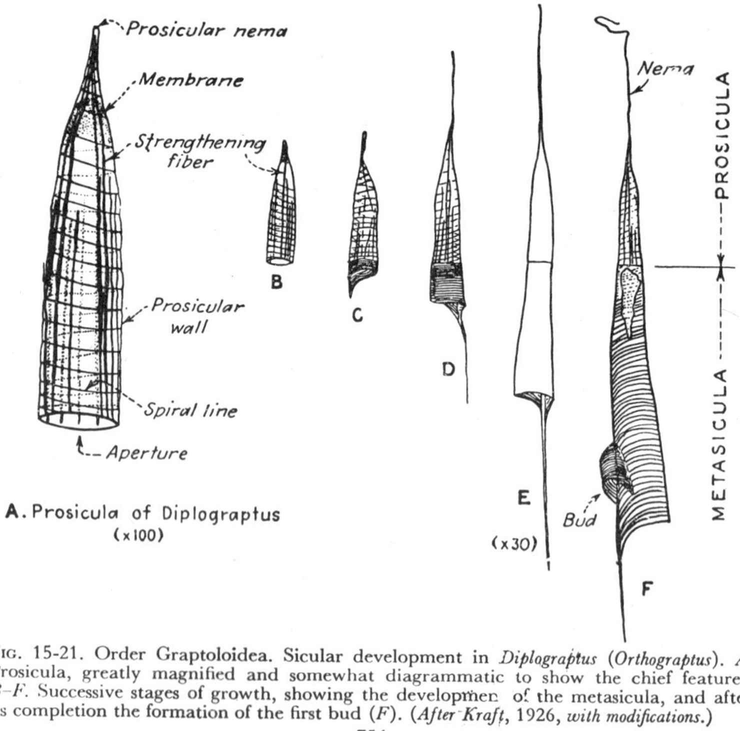

FIG. 15-21. Order Graptoloidea. Sicular development in *Diplograptus* (*Orthograptus*). *A*. Prosicula, greatly magnified and somewhat diagrammatic to show the chief features. *B–F*. Successive stages of growth, showing the development of the metasicula, and after its completion the formation of the first bud (*F*). (*After Kraft*, 1926, *with modifications.*)

the graptoloids evolved, and these variations are important for specific and even generic differentiation (Fig. 15-19).

Development of Sicula. In the development of a colony the embryo[1] first built a chitinous sicula consisting of a transparent cone, the prosicula, to which was added later a much larger tubular metasicula (Fig. 15-20). The

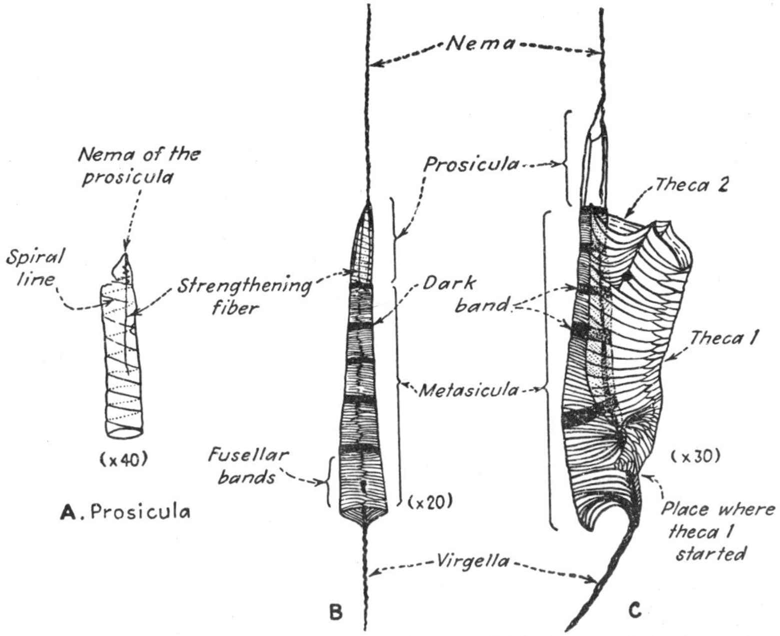

FIG. 15-22. Order Graptoloidea. Early growth stages of *Monograptus*. *A*. An incomplete prosicula showing the well-developed spiral line, or *Schraubenlinie*, and one longitudinal strengthening fiber (*Längsverstärkungsleiste*) extending downward from the **nema prosicula.** *B*. A complete sicula viewed from the side opposite the virgula. Four dark bands, the *Stillstandsgürtel*, are present in this species but they seem to be an exceptional feature. Increments to the proximal end of the metasicula are semicircular bands of chitin (the fusellus of Kozlowski) that wedge out along a median zigzag line on either side. Growth seems to have been by alternate deposition, first on one side then on another. *C*. Sicula with first and second thecae developed. The first theca sprang from a pore in the metasicula; the second from the lateral part of the first. (*Adapted from Kraft*, 1926.)

distal or pointed end of the sicula is prolonged as a slender chitinous nema, or virgula, which usually is attached to a disk or float (Fig. 15-25*M–N*). After the prosicula was completed, successive semicircular increments of chitinous material were added in alternating succession to make the tubular metasicula. The latter, therefore, has a strongly banded appearance with a

[1] It is still uncertain as to how the embryo of the graptolite was produced, whether sexually or asexually, and also whether or not there was alternation of generations. It seems likely, however, that whatever the origin of the embryo, it did not secrete the chitinous periderm of the sicula until it had been liberated from its parent. (Bulman, 1938.)

prominent zigzag line of suture along opposite (*i.e.*, the ventral and dorsal) sides. This is the **fusellus** of Kozlowski and is illustrated in Figs. 15-20, 15-21, and 15-22. Soon after the building of the metasicula started, the growth bands began to bend forward along one of the zigzag lines, and thereafter the rodlike **virgella** (Fig. 15-20) began to be laid down along this line, and the growth lines continued to run forward against it on either side. Upon completion of the metasicula, certain spinelike apertural projections were formed in some cases on the side opposite the projecting virgella.

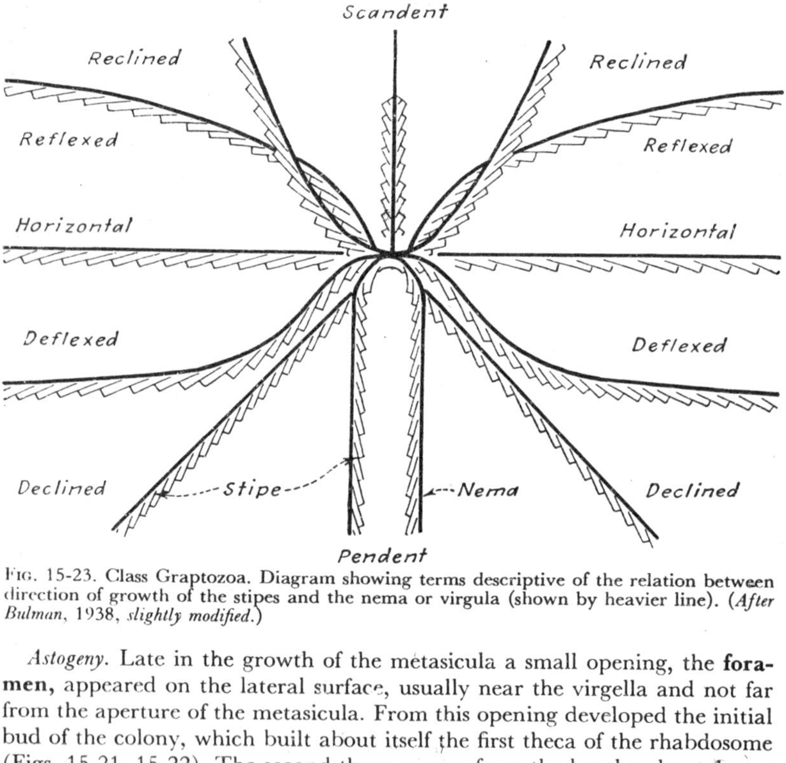

FIG. 15-23. Class Graptozoa. Diagram showing terms descriptive of the relation between direction of growth of the stipes and the nema or virgula (shown by heavier line). (*After Bulman*, 1938, *slightly modified.*)

Astogeny. Late in the growth of the metasicula a small opening, the **foramen,** appeared on the lateral surface, usually near the virgella and not far from the aperture of the metasicula. From this opening developed the initial bud of the colony, which built about itself the first theca of the rhabdosome (Figs. 15-21, 15-22). The second theca sprang from the basal or lateral part of the first, the third from the second, and so on until the complete rhabdosome had been formed (Figs. 15-24, 15-25). This simple early sequence was soon modified so that different kinds of rhabdosomes were evolved as series of thecae were added. The relation between the direction of growth of the stipes and the nema constitutes an important basis for subdividing the Graptoloidea (Fig. 15-24).

Two strong tendencies prevailed in this development. One tendency was that which led to a scandent direction of growth, from an original pendent direction, as indicated in Fig. 15-23. The second tendency was toward a reduction in the number of branches. The earliest graptoloids (*e.g.*, *Clono-*

graptus, Fig. 15-25*Y*) had many stipes, whereas the later forms had only two or even a single stipe (Diplograptidae; Monograptidae). In this evolution there was progressive change in the proximal end of the rhabdosome, with the production of a series of five distinct types as shown in Fig. 15-24. These types, entirely apart from the genera illustrating them, can be used to date beds because of their evolutionary significance (Bulman, 1933).

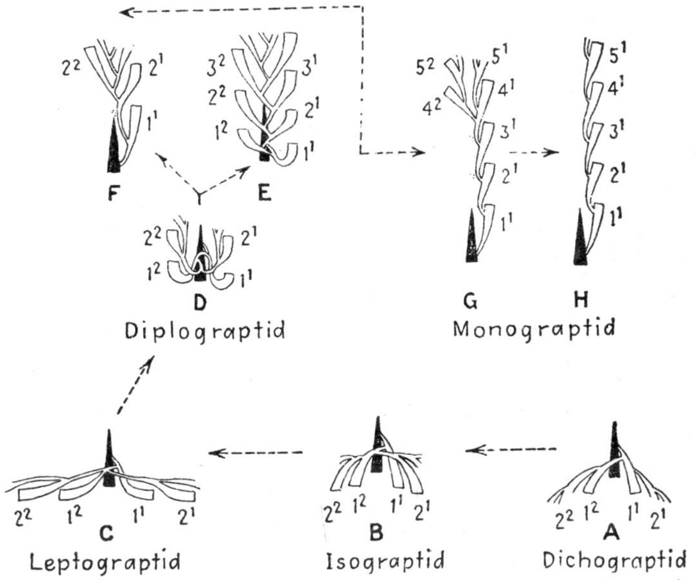

FIG. 15-24. Order Graptoloidea. Thecal diagrams showing progressive changes in the proximal end of the graptolite rhabdosome in the five different developmental types. The dashed lines with arrow points indicate the direction of change in different groups with the passage of time. In each diagram the sicula is shown as a black cone. *A*. The dichograptid type, illustrated by *Didymograptus bifidus*, is characterized by a single crossing canal (*th* 1^2 from *th* 1^1); *B*. The isograptid type, typified by *Isograptus gibberulus*, has two crossing canals (*th* 1^2 from *th* 1^1 and *th* 2^1 from *th* 1^2, and both primary stipes are derived from *th* 1^2). *C*. The leptograptid type, illustrated by *Leptograptus flaccidus*, with two or three crossing canals and a horizontal direction of growth of the distal parts of the initial thecae. *D–F*. The diplograptid type, illustrated by *Glyptograptus dentatus* (*D*), *Climacograptus typicalis* (*E*), and *Rhäphidograptus* [*Climacograptus*] *törnquisti*, all characterized by three or more crossing canals and a scandent direction of growth and with *F* showing the loss of *th* 1^2, a tendency more fully developed in *G* and *H*. *G–H*. The monograptid type, illustrated by *Dimorphograptus* (*G*) and *Monograptus* (*H*), and characterized by the upper growth of *th* 1^1, loss of *th* 1^2, etc., and final development of a completely uniserial rhabdosome. (*Adapted from Bulman*, 1936.)

The earliest representatives of the Graptoloidea have been found in rocks of Upper Cambrian (Lower Tremadocian) age and the last in Upper Silurian strata. Owing to their planktonic and pseudoplanktonic mode of life, they attained world-wide distribution in Ordovician and Silurian seas and are unexcelled as index fossils for intercontinental correlation. Locally, because of their obvious evolutionary changes—reduction in number of stipes (Fig. 15-24), change in direction of growth of the stipes (Fig. 15-23), and elabo-

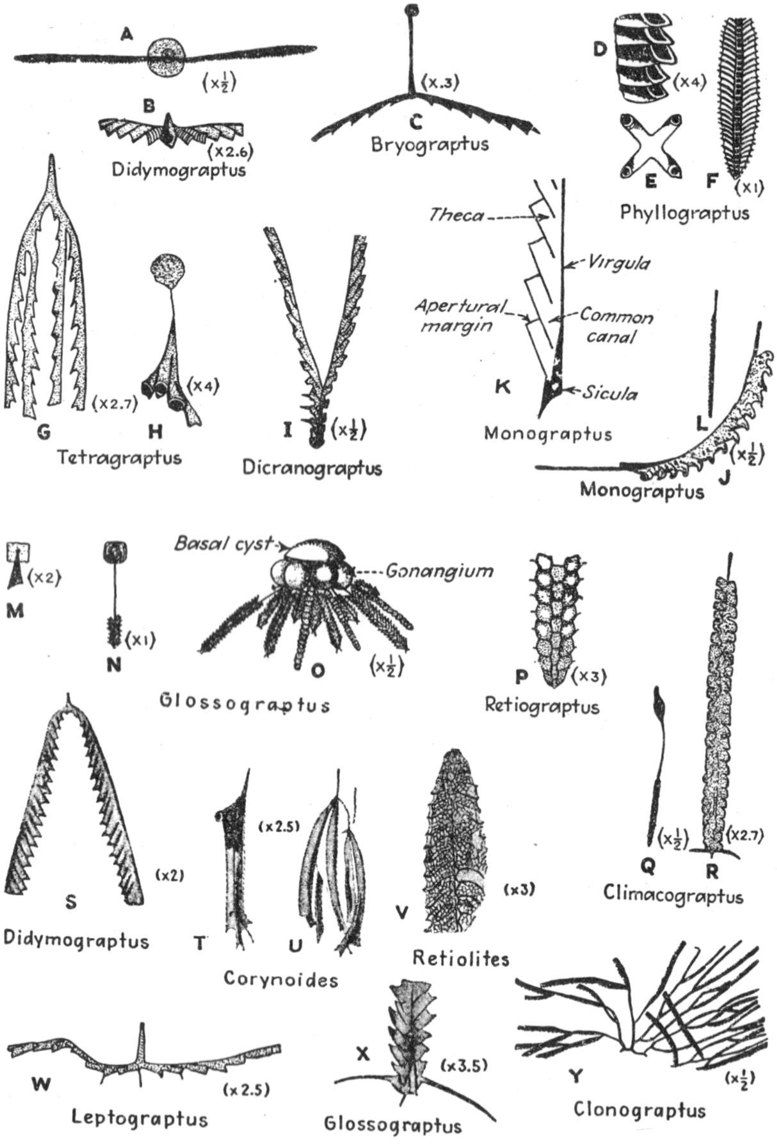

FIG. 15-25. Order Graptoloidea. (*See opposite page for detailed descriptions.*)

ration of the thecal aperture (Fig. 15-19)—they serve superbly for zoning the stratigraphic sequence.[1]

British paleontologists now recognize some 36 graptolite zones in the Upper Cambrian, Ordovician, and Silurian, of which 21 are in Silurian strata. Assuming that the graptolite-bearing Silurian sequence was deposited steadily during the 30,000,000 to 40,000,000 years usually assigned to the period, and postulating further that the 21 graptolite zones represent comparable periods of time, it follows that the average length of life of the short-ranging key species of a zone was of the order of 1,500,000 to 2,000,000 years. Although such an estimate is admittedly of uncertain accuracy, its order of magnitude is probably not too far from the truth.

The Order Graptoloidea, to which some 70 genera and several hundred species have been assigned, is composed of the following nine families[2] (see Table 15-1): Dichograptidae: *Clonograptus* (U. Camb.–L. Ord.; Fig. 15-25*Y*), *Bryograptus* (U. Camb.–L. Ord.; Fig. 15-25*C*), *Tetragraptus* (U. Camb.–L. Ord.; Fig. 15-25*G–H*), *Phyllograptus* (L. Ord.; Fig. 15-25*D–F*), and *Didymograptus*, (U. Camb.–L. Ord.; Fig. 15-25*A–B,S*). Corynograptidae: *Corynoides* (M. Ord.; Fig. 15-25*T–U*). Leptograptidae: *Leptograptus* (Ord.; Fig. 15-25*W*). Dicranograptidae: *Dicranograptus* (Ord.; Fig. 15-25*I*). Diplograptidae: *Climacograptus* (Ord.–L. Sil.; Fig. 15-25*Q–R*). Glossograptidae: *Glossograptus* (Ord.; Fig. 15-25*M–O,X*) and *Retiograptus* (Ord.; Fig. 15-25*P*).

[1] The evolutionary history of the graptolites as we know it today, thanks largely to the brilliant investigations of British and European paleontologists, is one of the great triumphs of paleontology. Better than any other known group of extinct invertebrates, the graptolites, as Bulman (1933) and others have so ably demonstrated, show "programme evolution"—*the tendency for a character to run through a similar developmental history independently in several lineages.*

[2] Most former classifications include two suborders *Axonolipa*, without an "axis" or virgula, and *Axonophora*, with a virgula. These divisions are not used here since no hard and fast line can be drawn between them and because graptolite specialists are not in agreement as to which families should be included in each suborder (Bulman, 1938).

FIG. 15-25. *A. Didymograptus patulus*, a nearly mature rhabdosome with central disk. *B.* A young rhabdosome of the same species, showing sicula and first few thecae. *C. Bryograptus lapworthi*, a young rhabdosome with fragment of primary disk. *D. Phyllograptus ilicifolius*, a few thecae much enlarged to show character of aperture. *E.* Diagram showing the quadriserial arrangement of the thecae in this genus. *F. Phyllograptus angustifolius*, a nearly complete rhabdosome. *G. Tetragraptus pendens*, proximal portion showing sicula and thecae. *H. T. fruticosus*, a young rhabdosome showing primary disk, nema, sicula, and first few thecae. *I. Dicranograptus nicholsoni*, part of a rhabdosome showing sicula (in black) and thecae. *J. Monograptus flexilis*, a rhabdosome showing nema, sicula, virgula, and thecae. *K.* Diagram of a rhabdosome of *Monograptus*, with structures indicated. *L.* Diagram to illustrate the usual appearance of fossils of this genus. *M. Glossograptus pristis*, the sicula. *N.* A young rhabdosome of *G. pristis* attached to the basal cyst. *O.* A complete synrhabdosome of *G. pristis* with basal cyst, gonangia or reproductive organs, and numerous complete or incomplete rhabdosomes. *P. Retiograptus geinitzianus*, a portion of the reticulated rhabdosome. *Q. Climacograptus parvus*, rhabdosome with inflation of nemacaulus. *R. C. modestus*, an almost complete rhabdosome. *S. Didymograptus protobifidus*, a widespread Ordovician species. *T–U. Corynoides. V. Retiolites*, a fragmental rhabdosome showing network. *W. Leptograptus flacidus*, sicular part only. *X. Glossograptus quadrimucronatus cornutus*, sicular part greatly enlarged. *Y. Clonograptus flexilis*, a rhabdosome with many stipes. (*E*, *K*, and *L* are *diagrammatic; J from the Wenlock shales of England after Elles and Wood*, 1901–1918; *S from American Ordovician after Decker*, 1941; *all other figures are of specimens from the Lower and Middle Ordovician of New York and are from Ruedemann's numerous papers.*)

Retiolitidae: *Retiolites* (L. and M. Sil.; Fig. 15-25*V*). Dimorphograptidae: *Dimorphograptus* (L. Sil.; Fig. 15-24*G*). Monograptidae: *Monograptus* (Sil.; Fig. 15-24*H*, 15-25*J–L*).

Habitat

Most dendroid colonies were probably fixed to the sea bottom and hence were a part of the benthos; some dendroids and many graptoloids were attached to seaweeds or other floating objects; and many graptoloids were attached to floats of their own making (Fig. 15-25), in which respect they are comparable to the present-day *Siphonophora*. The planktonic and pseudoplanktonic forms, especially, attained world-wide distribution and are important for intercontinental correlation and for delineation of ancient seaways.[1]

The fact that graptolites are found most abundantly in black shales does not indicate that they lived in any greater abundance in the waters beneath which such sediments were accumulating. The waters over such bottoms in some cases may not have been normally marine, hence destructive to most salt-water organisms.[2] When graptolites drifted into these waters from normal marine waters, they probably died. Bottoms covered with black muds indicate reducing conditions, low content of oxygen, high content of hydrogen sulphide, poor circulation, and, correlative with these conditions, general absence of scavenging organisms. Dead graptolites, as well as other dead organisms, on reaching the waters directly above black-mud bottoms, readily became entombed in the muds and were thus excellently preserved. On bottoms with circulation good enough to produce conditions favorable for scavenging benthonic animals, graptolite remains doubtlessly were generally eaten, and only in rare cases did they escape destruction. However, some excellently preserved rhabdosomes have been found in limestones and cherts.

Biological Affinities

Graptolites have been referred at one time or another to such widely separated groups as plants, horny sponges, alcyonarian corals, hydrozoans,

[1] DECKER, C. E., 1938, Transcendent value of graptolites for correlation demonstrated, *Bull. Amer. Assoc. Petrol. Geol.*, vol. 22, p. 221.

[2] The presence of marcasite and pyrite (FeS_2) in graptolite-bearing shales indicates that the muds and immediately overlying waters probably contained hydrogen sulphide and that the waters as a consequence were really poisonous. The blackness of the shales proves that reducing conditions were also present on the bottom. In this connection see the following references: RUEDEMANN, R., 1911, Stratigraphic significance of the wide distribution of graptolites, *Bull. Geol. Soc. Amer.*, vol. 22, pp. 231–237. GRABAU, A. W., and O'CONNELL, M., 1917, Were the graptolite shales as a rule deep or shallow water deposits? *Bull. Geol. Soc. Amer.*, vol. 28, pp. 959–964. MARR, J. E., 1925, The Stockdale shales of the Lake District, *Quart. Jour. Geol. Soc. London*, vol. 81, pp. 113–133. GRABAU, A. W., 1929, Origin, distribution, and mode of preservation of the graptolites, *Mem. Inst. Geol., Nat. Res. Council China*, pp. 1–52. RUEDEMANN, R., 1934, Paleozoic plankton of North America, *Geol. Soc. Amer., Mem.* 2, pp. 43–52. TWENHOFEL, W. H., 1939, Environments of origin of black shales, *Bull. Amer. Assoc. Petrol. Geol.*, vol. 23, pp. 1178–1198. RUEDEMANN, R., 1947, Graptolites of North America, *Geol. Soc. Amer., Mem.* 19, pp. 15–23 (an excellent summary of the paleoecology of the graptolites).

bryozoans, cephalopods, and pterobranchs.[1] Only the following groups merit serious consideration as possible relatives: hydrozoans, bryozoans, and pterobranchs.

Graptolites were first associated with hydrozoans more than a hundred years ago, and since that time most paleontologists have accepted this suggested relationship because the graptolites were thought to have been chitinous like some living hydrozoans and because the rhabdosomes, especially those of dendroid form, bore strong resemblance to some hydrozoan colonies. Finally, the thecae of the graptolites were considered to have housed individuals of the colony in the same way as the hydrothecae house certain polyps in living hydrozoans. Superficially, then, there were good grounds for considering the graptolites as ancient and extinct relatives of the Hydrozoa.

It has been pointed out (Kozlowski, 1948), however, that (1) the seemingly close resemblance of graptolites to hydrozoans breaks down when detailed structures are compared (*e.g.*, black stolon, and method of increase); (2) no known hydrozoan has structure comparable to that in the periderm of a graptolite; (3) the thecae of graptolites are bilaterally symmetrical, whereas those of a hydrozoan have a higher symmetry (radially symmetrical); and (4) certain ancient fossils (*e.g.*, *Archaeocryptolaria*, Fig. 4-3), originally included in the graptolites and cited as evidence in favor of hydrozoan affinities for them, are now known to be true hydrozoans and not graptolites at all. The recent discovery of the well-developed stolon system and of the peculiar structure of the wall of the stipe not only removes the graptolites from the Coelenterata altogether, but also indicates that in these two respects they are closely similar to living *Rhabdopleura* of the Pterobranchia. The preponderance of evidence, therefore, now seems to make a hydrozoan assignment of the graptolites untenable.

Ulrich and Ruedemann (1931) and Ruedemann (1947) alone have raised the question, *are graptolites bryozoans?* A strong similarity is seen in the nature of the early part of the exoskeleton in the two groups, the prosicula and succeeding metasicula of the graptolites being considered exactly similar to the protoecium and ancestroecium of the Bryozoa. Mode of budding and of increase of the colony are considered similar. The supposed muscle scars of the graptolites agree strikingly with those prevailing in certain Bryozoa. There is also some similarity between graptolites and bryozoans in the general nature of the colonies, but it is to be noted that, whereas the former are exclusively chitinous, the latter, although partly chitinous in a few of the more primitive forms, are dominantly calcareous.

In a critical examination of these supposed close relationships between graptolites and bryozoans, Bulman (1938) could not accept the suggestions of Ulrich and Ruedemann, and Kozlowski (1948, pp. 80–83) reached the conclusion that there is no compelling evidence to support the contention that "the graptolites represent a more primitive branch of the phylum [*i.e.*, Bryozoa] than the bryozoans themselves" (Ulrich and Ruedemann, 1931), and hence that the graptolites should not be included in the Bryozoa. Koz-

[1] An excellent history of these assignments is given in recent discussions by Ruedemann (1947, pp. 46–52) and Kozlowski (1947, pp. 93–107; 1948, pp. 59–83).

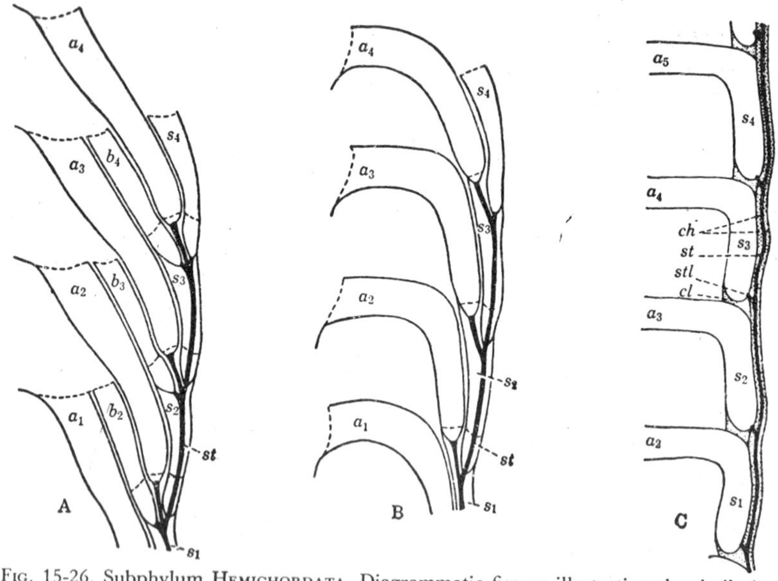

FIG. 15-26. Subphylum HEMICHORDATA. Diagrammatic figures illustrating the similarity of structure between the chitinized parts of *Dendrograptus* (*A–B*) and of *Rhabdopleura* (*C*). *A* shows the terminal part of a branch, or stipe, of *Dendrograptus*. *B* is similar to *A*, but has been simplified by omitting the bithecae. *C* is a diagram showing the structure of part of a branch of *Rhabdopleura* (cf. Fig. 15-5). (*a* = autothecae; *b* = bithecae; *ch* = chitinous deposits around the stolon; *cl* = partitions between the compartments of the stolon; *s* = stolothecae; *st* = stolon; *stl* = lateral branches of the stolon.) (*After Kozlowski*, 1947.)

lowski did, however, call attention to several significant characteristics common to both Graptozoa and Bryozoa.

This brings us finally to a consideration of the features of graptolites that make their reference to the Pterobranchia of the Hemichordata reasonable.

More than 45 years ago Schepotieff (1905) first called attention to the fact that the solid rod in the nemacaulus of *Monograptus* was similar to the prominent black stolon of *Rhabdopleura*, then assigned to the Bryozoa but now considered to be a pterobranch. This observation was generally overlooked or ignored, however, until Kozlowski made his discovery in Poland. The Polish material, fragmental though it be, is so well preserved that it shows clearly the existence of a characteristic stolon system (Figs. 15-11, 15-12, 15-13), which is lacking in both Hydrozoa and Bryozoa, a peculiar fusellar structure of the periderm (Fig. 15-10) like that in *Rhabdopleura* (Fig. 15-5) but unknown in any important invertebrate groups, and a mode of budding or colonial increase that is similar to the hemichordate just mentioned (Fig. 15-26). With the long-awaited appearance (1948) of Kozlowski's comprehensive work, the conclusions of which have been accepted by leading specialists (Dawydoff, 1948; Bulman, 1949*a*), there should be little doubt as to the close relation of the graptolites and pterobranchs of the rhabdopleurid type.

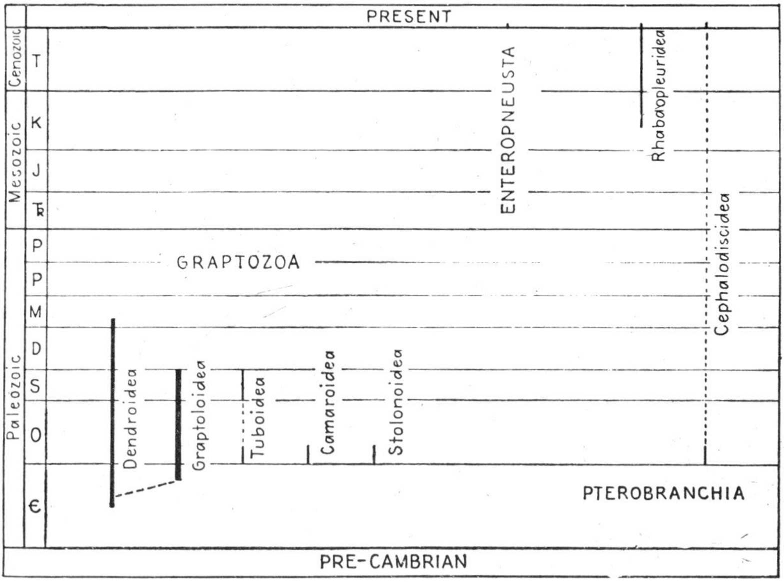

FIG. 15-27. Subphylum HEMICHORDATA. Diagram showing known geological range of the three classes of hemichordates. Vertical bars show range of the orders that have a fossil record.

Geologic History

Graptolites began as simple dendroid forms in the Middle Cambrian; attained their main acme of development in the Ordovician; arose during the Silurian to a second minor acme, with the dominant divisions different from those in the Ordovician; and then almost completely disappeared at the close of the Silurian, with the exception of a few dendroid stragglers that persisted through the Devonian into the Early Mississippian (Fig. 15-27). The disappearance of the planktonic forms at the close of the Silurian and the beginning of the Devonian possibly may be correlated with the advent of the fishes, which appeared in great numbers at that time. The dendroid or fan-shaped forms were the earliest to develop and also the last to survive. They are considered to have given rise to the more specialized types that are mainly confined to the Ordovician and Silurian.

REFERENCES

Hemichordata (General)

DAWYDOFF, C. 1948. Embranchement des Stomocordés *in* Grassé's "Traité de zoologie," vol. 11, pp. 365–532. Paris, Masson et Cie.

DELAGE, Y., and HÉROUARD, E. 1897. "Traité de zoologie concrète." Paris, Schleicher Frères, pp. 164–181.

KOZLOWSKI, R. 1949. Découverte du Ptérobranche *Rhabdopleura* à l'état fossile dans le Crétacé supérieur en Pologne. *Compt. rend. acad. sci.*, vol. 228, no. 19, pp. 1505–1507.

SCHEPOTIEFF, A. 1906–1908. Die Pterobranchier Anatomische und histologische Unterschungen ueber *Rhabdopleura normanii* Allman und *Cephalodiscus dodecalophus* M'Int. *Zool. Jahrb.*, vol. 23, pp. 463–534; vol. 24, pp. 553–608; vol. 25, pp. 405–494. (This is an excellent series of papers with an extensive bibliography.)

STIASNY, G. 1910. Zur Kenntnis der Lebensweise von *Balanoglossus clavigerus* Delle Chiaje. *Zool. Anz.*, vol. 35, pp. 561–565.

THOMAS, H. D., and DAVIS, A. G. 1949. *Abstracts of Proc. Geol. Soc. London*, No. 1450 (Session 1948–1949), p. 79.

——— and ———. 1949*a*. The Pterobranch *Rhabdopleura* in the English Eocene. *Bull. British Mus.* (*Nat. Hist.*), vol. 1, no. 1, pp. 1–19.

——— and ———. 1950. A fossil species of the pterobranch *Rhabdopleura*. *Quart. Jour. Geol. Soc. London*, vol. 106, p 133.

Graptozoa (Graptolithina)

BULMAN, O. M. B. 1927–1934. "Monograph of British Dendroid Graptolites." London, *Palaeontographical Society*, pts. 1–3, pp. 1–92.

———. 1929. The genotypes of the genera of graptolites. *Ann. Mag. Nat. Hist.*, vol. 4, pp. 169–185.

———. 1932, 1936. On the graptolites prepared by Holm, I–VII. *Arkiv. Zool.*, vol. 24A, (1932); vol. 26A (1934); vol. 28A (1936).

———. 1933. Programme-evolution in the graptolites. *Biol. Revs.*, vol. 8, pp. 311–334.

———. 1938. Graptolithina *in* Schindewolf's "Handbuch der Paläozoologie," vol. 2D, pp. D1–D92. Berlin, Gebrüder Borntraeger.

———. 1942. The structure of the dendroid graptolites. *Geol. Mag.*, vol. 79, pp. 284–290.

———. 1949. A re-interpretation of the structure of *Dictyonema flabelliforme* Eichwald. *Geol. Foren. Forhandl.*, vol. 71, pp. 33–40.

———. 1949*a*. Review (Kozlowski's Les Graptolithes, etc.). *Geol. Mag.*, vol. 86, pp. 329–332.

———. 1950. The structure and relations of *Cyclograptus* Spencer. *Jour. Paleontology*, vol. 24, pp. 566–570.

DECKER, C. E. 1939. Pneumatocysts on *Monograptus* (*Linograptus*) *phillipsi multiramosus*. *Jour. Paleontology*, vol. 13, pp. 49–51.

———. 1947. Additional graptolites and hydrozoan-like fossils from Big Canyon, Okla. *Ibid.*, vol. 21, pp. 124–130.

ELLES, G. L., and WOOD, E. M. R. 1901–1918. "Monograph of British Graptolites." *Palaeontographical Society*, 3 vols., pp. 1–700.

GRABAU, A. W. 1929. Origin, distribution, and mode of preservation of the graptolites. *Inst. Geol., Nat. Res. Council*, China, *Mem.* VII, 52 pp.

KOZLOWSKI, R. 1938. Informations préliminaires sur les Graptolithes du Tremadoc de la Pologne et sur leur portée théorique. *Ann. Mus. Zool. Polonici*, vol. 13, no. 16, pp. 183–196. (See comments on this article by Bulman in his Graptolithina *in* Schindewolf's "Handbuch der Paläozoologie," vol. 2D, pp. D91–D92, Berlin, Grebrüder Borntraeger.)

———. 1947. Les Affinités des Graptolithes. *Biol. Revs.*, vol. 22, pp. 93–108.

———. 1948. Les Graptolithes et quelques nouveaux groupes d'animaux du Tremadoc de la Pologne. *Pal. Polonica*, vol. 3, 235 pp. (See comments by Bulman in *Geol. Mag.*, vol. 86, pp. 329–332, 1949.)

KRAFT, P. 1926. Ontogentische Entwicklung und Biologie von *Diplograptus* und *Monograptus*. *Pal. Zeitschr.*, vol. 7, pp. 207–249.

PERNER, J. 1894–1899. Études sur les graptolites de Bohême. Pts. 1–3. Prague, Raimund Gerhard.

RUEDEMANN, R. 1913, 1927. Graptolitoidea, *in* Zittel-Eastman's "Textbook of Palaeontology," vol. 1, pp. 125–133. New York, The Macmillan Company: London, Macmillan & Co., Ltd. (This discussion has an extensive list of foreign references to graptolite literature.)

———. 1933. The Cambrian of the Upper Mississippi Valley. Pt. 3. Graptolitoidea. *Bull. Pub. Mus. City of Milwaukee*, vol. 12, pp. 307–348. (1934. *Science*, n.s., vol. 80, p. 15.)

———. 1934. Paleozoic plankton of North America. *Geol. Soc. Amer., Mem.* 2, 141 pp. (This work contains a good bibliography.)

———. 1947. Graptolites of North America. *Ibid., Mem.* 19, 652 pp. (This work contains an all-inclusive bibliography of North American graptolite literature.)

——— and LOCHMAN, C. 1942. Graptolites from the Englewood formation (Mississippian) of the Black Hills, S. Dak. *Jour. Paleontology*, vol. 16, pp. 657–659.

SCHEPOTIEFF, A. 1905. Ueber Stellung der Graptolithen im Zoologischen system. *Neues Jahrb. Min. Geol.*, vol. 2, pp. 79–90.

SHIMER, H. W., and SHROCK, R. R. 1944. "Index Fossils of North America." New York, John Wiley & Sons, Inc.; Cambridge, Mass., The Technology Press. Pp. 64–77. (This work includes an extensive bibliography of North American graptolite literature.)

SINCLAIR, G. W. 1948. The affinities of the Graptolites. *Amer. Jour. Sci.*, vol. 246, pp. 526–528.

ULRICH, E. O., and RUEDEMANN, R. 1931. Are the graptolites Bryozoans? *Bull. Geol. Soc. Amer.*, vol. 42, pp. 589–604.

WATERLOT, G. 1948. Classe des Graptolites *in* Grassé's "Traité de Zoologie," vol. 11, pp. 500–510. Paris, Masson et Cie.

WIMAN, C. 1893–1897. Ueber die Graptolithen. *Bull. Geol. Inst. Uppsala*, vols. I, III.

CHAPTER 16

CONODONTOPHORIDIA (CONODONTS)

INTRODUCTION

Conodonts are tiny, paired, toothlike fossils of uncertain origin that are common in Paleozoic rocks of Ordovician to Permian age, and less common as derived or admixed fossils in Mesozoic rocks. They are similar in general morphology to scolecodonts—teeth of annelid worms—but differ from them in being phosphatic rather than silicochitinous. This compositional difference makes it possible to separate the two rather easily by simple chemical means. A second striking difference is in their color; conodonts are typically light brown, with a glassy appearance, whereas scolecodonts are jet black because of having been carbonized.

Conodonts were first made known in 1856 by Pander, who considered them to be the teeth of fish but admitted that he could not determine from what parts of the mouth the several different forms came. They were little noticed during the ensuing seventy years until Ulrich and Bassler published a comprehensive study of them in 1926. This classic monograph stimulated widespread interest in the group, and during the past 25 years many investigators have taken up the study of conodonts, with the result that there has been a pronounced increase in the number of genera and species and in general literature pertaining to these fossils. Recent articles by Hass (1941), Branson and Mehl (1944), and Ellison (1946) summarize the more important work to date and include much information that may prove helpful to those wishing to investigate conodonts.

A separate chapter is here devoted to the conodonts for the following reasons: (1) they have assumed considerable importance among microfossils because of their common occurrence in well cuttings and because of their value in identifying and correlating beds; and (2) there is still some doubt as to whether the conodont-bearing animals were vertebrates or invertebrates.[1]

GENERAL NATURE OF CONODONTS

Conodonts are typically tusk-shaped or denticulate bodies of translucent or transparent nature composed of calcium phosphate.[2] They are commonly

[1] There is a certain humorous aspect to the fact that vertebrate paleontologists rather generally brush aside the idea that conodonts are vertebrate remains, whereas the invertebrate paleontologists, vigorously insisting that they are vertebrate remains, quickly become conservative and argumentive if the tables be turned by some colleague who suggests that they may be invertebrate remains.

[2] Ellison (1944*a*, p. 135) gives the following composition for conodonts:

CaO	48.05 per cent
P_2O_5	34.96 per cent
Insoluble (possibly adherent clay)	3.96 per cent
Remainder (possibly CO_2, H_2O, F_2, Fe_2O_3, organic matter, etc.)	13.03 per cent
Total	100.00 per cent

and states on the basis of spectrographic data that they have a composition similar to the minerals of the apatite group.

only a fraction of a millimeter long and seldom exceed 2 mm. Although they immediately suggest a toothlike structure, it is not universally agreed that they served this function, as is discussed more fully in later paragraphs. They can be distinguished readily from definitely known worm jaws (scolecodonts), which they resemble in some aspects, by their distinctive shapes, by their composition (calcium phosphate as compared to the silicochitinous nature of scolecodonts), and by their amber to light brown color, which stands in strong contrast to the jet-black appearance of the typical fossil worm jaw. They are quickly destroyed by nitric, sulphuric, and weak hydrochloric acids (scolecodonts are unaffected), but they are not affected by acetic or citric acid, by means of which they can be released from limestones and other calcareous rocks. Having a specific gravity (2.84 to 3.10) somewhat greater than that of calcite, dolomite, and quartz, conodonts can generally be separated easily from sediments composed of these minerals by use of heavy liquids.

Small spheroidal bodies, 0.01 to 0.02 mm. in diameter and with a tiny circular or elliptical opening, are commonly directly associated with conodonts and are believed to have belonged to the same animals that carried the conodonts. They seem to be composed of the same material as the toothlike structures and where abundant to be directly proportional to the number of conodonts, but far less numerous. Their exact nature is unknown, but it has been suggested that they may be **egg cases** (Stauffer, 1935, page 620) or **otoliths** (Youngquist and Miller, 1948, page 440). They have been reported from Ordovician and Devonian rocks.

CLASSIFICATION OF CONODONTS

Conodonts are divisible into **fibrous** and **nonfibrous** (**laminated**) types on the basis of internal structure. The fibrous group, constituting the Suborder Neurodontiformes, is composed of those conodonts that are made up of bundles of fibers. The nonfibrous group, Suborder Conodontiformes, is composed of those conodonts having laminated structure. Both groups include simple cones, denticulated blades and bars, and highly specialized platform types (Fig. 16-1).

More than 130 generic names have been proposed in the conodont literature for some 1,500 species. At least 50 of these genera are not considered valid by present-day specialists, leaving about 80 that are accepted. Accepted genera are sketched on Figs. 16-7 and 16-8, which were kindly furnished us by Dr. Samuel P. Ellison, Jr. In an attempt to indicate provisional genetic relationships, Branson and Mehl (1944), the two leading American students of conodonts, have proposed the following classification of the known valid genera (Fig. 16-1).

(Order) Conodontophoridia. Minute toothlike objects ranging in shape from simple recurved cones, through denticulate bars and blades, to highly specialized platforms; having either fibrous or laminated internal structure; composed of calcium phosphate; and attached to fragments of similar composition assumed to have been jaws. (*L. Ord.—Perm; Mesozoic?*)

Suborder 1. *Neurodontiformes* (Fig. 16-1). Conodonts composed of bundles of fibers. The following three families have been proposed: Coleodontidae, forms clasping the

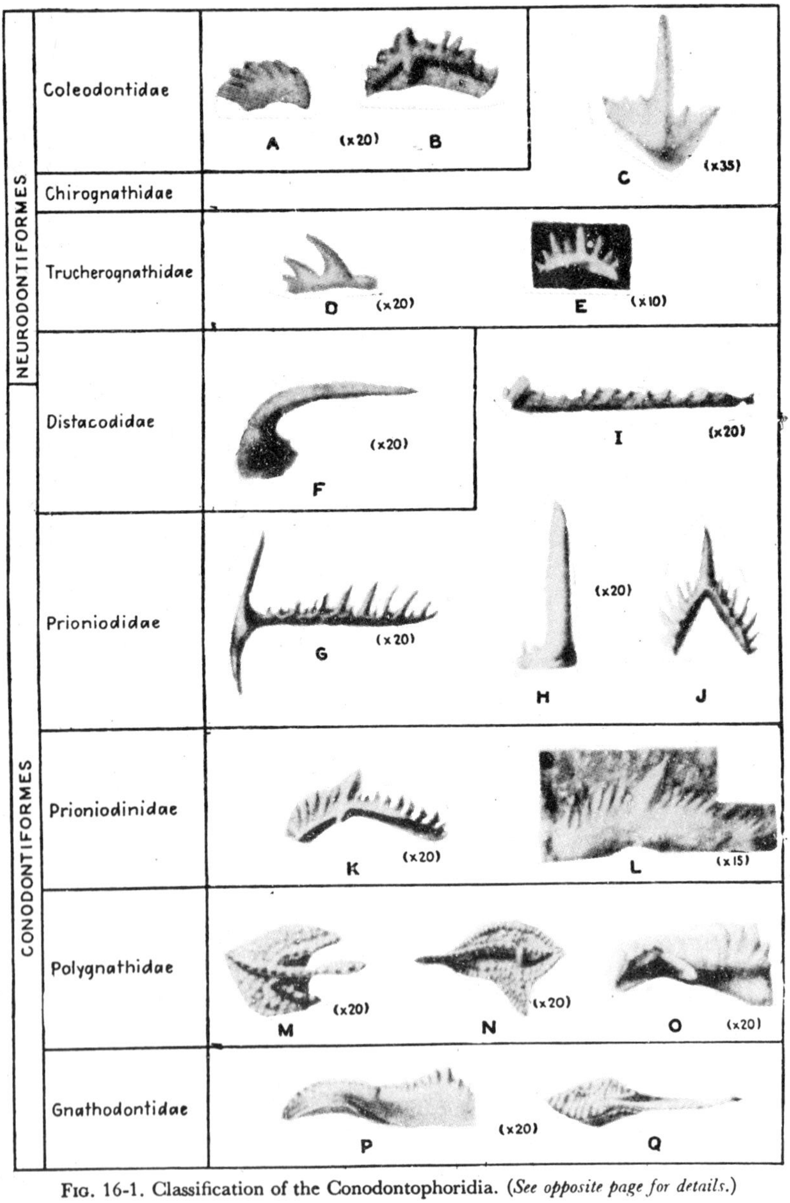

FIG. 16-1. Classification of the Conodontophoridia. (*See opposite page for details.*)

"jaw"; Chirognathidae, forms that clasp or tip the end of the "jaw ramus"; and Trucherognathidae, platelike forms resting on the "jaw ramus." *Ordovician.*

Suborder 2. *Conodontiformes* (Fig. 16-1). Conodonts with laminated structure. The following five families have been proposed: Distacodidae, simple recurved cones with deeply excavated base simulating a pulp cavity; Prioniodidae, denticulate bars with one major denticle anteriorly placed and an elongate posterior extension; Prioniodinidae, arched denticulate bars and blades, generally with a prominent apical denticle within middle third of length; Polygnathidae, bilaterally symmetrical leaflike plates with a median blade extending forward and a small attachment scar in the middle of the plate; and Gnathodontidae, elongate, platformlike or troughlike units with an anterior blade and broadly excavated aborally. *Ordovician to Permian.*

MICROSTRUCTURE AND MORPHOLOGY

Fibrous conodonts are composed of bundles of fibers of calcium phosphate in the form of minerals of the apatite group. Some have been found crushed and frayed rather than broken. The **nonfibrous conodonts** are composed of lamellae of apatite and in longitudinal section show a cone-in-cone structure owing to the fact that each denticle is composed of a series of slender nested cones (Fig. 16-2).

Distacodid Conodonts. The simplest nonfibrous conodonts are slightly recurved cones with a somewhat expanded base and with marked cone-in-cone structure internally (Figs. 16-1*E*, 16-2*A*). These belong to Family Distacodidae and are referred to as **distacodids.** The cone seems to have started as a minute cap, which then grew upward by adding successive conical lamellae along a single axis to form a cusp, and downward and outward to form the flangelike base that encloses the basal cavity.

Distacodid conodonts appear first in Lower Ordovician rocks, and the family seems to have died out sometime during the Silurian. Genera are based upon the shape and depth of the attachment scar and the shape of the transverse section. Typical genera, together with their range, are shown in Figs. 16-1, 16-7, and 16-8.

Compound and Platelike Types. Conodonts of these types differ from the distacodids only in that they show greater complexity of growth habit (Fig. 16-1*G–P*). Instead of uniaxial growth, as in the typical distacodid, growth took place simultaneously along several axes and in different directions. This polyaxial mode of growth resulted in the formation of denticulated blades and bars (Families Prioniodidae and Prioniodinidae) and in highly specialized platform types (Families Polygnathidae and Gnathodontidae). Regardless of their ultimate form, however, each started out as a laminated cone-in-cone structure that was built up about the apex of the "pulp cavity" (Fig. 16-2*B*), with successive conical increments being ar-

FIG. 16-1. The Branson and Mehl (1944) classification, with typical examples of each family. *A. Coleodus*, a typical denticulate blade. *B. Erismodus*. *C. Chirognathus*. *D. Polycaulodus* *E. Curtognathus*. *F. Multioistodus*, a typical distacodid. *G. Ligonodina*. *H. Hibbardella*. *I. Hindeodella*, a typical hindeodell. *J. Hibbardella*. *K. Bryantodus*, a typical bryantod. *L. Lewistownella*, a so-called cavusgnath. *M. Ancyrodella*. *N Palmatolepis*. *O. Polygnathus*, a typical polygnath. *P–Q. Idiognathodus*. (*From the volumes of Jour. Paleontology, as follows: A–B, vol.* 20; *A–b, vol.* 20; *C, vol.* 16; *D, vol.* 20; *E, vol.* 17; *F, vol.* 20; *G, vol.* 22; *H–I, vol.* 21; *J, vol.* 22; *K, vol.* 21; *L vol.* 16; *M–O, vol.* 21; *P–Q, vol.* 22.)

ranged along the growth axis of what was to become the main denticle or cup. This simple uniaxial mode of growth, however, was soon replaced by a polyaxial one that gave rise to the many compound and platelike forms.

Hass (1941), to whom along with Pander (1856) we owe much of our present knowledge about the microscopic structure of conodonts, concludes that in the laminated conodonts the lamellae can be thought of as representing distinct growth stages in the ontogeny of the organism that made

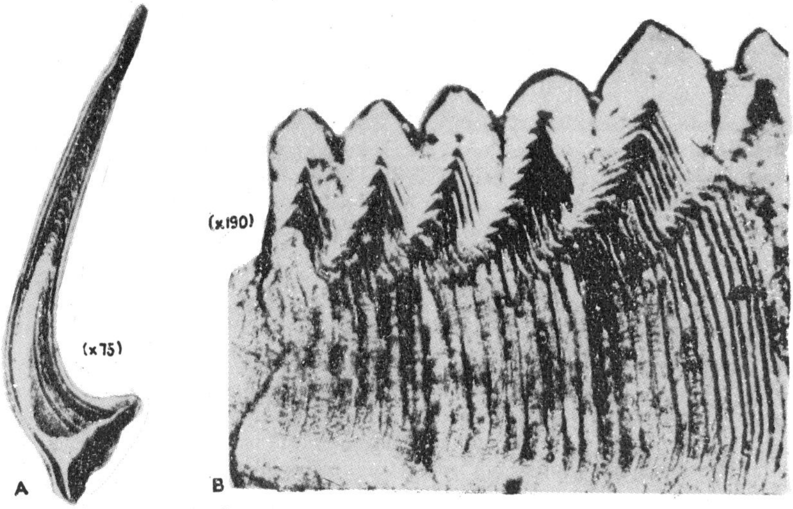

FIG. 16-2. Conodontophoridia. Microscopic internal structure. *A*. Longitudinal section of a distacodid conodont (*Oistodus*), showing typical cone-in-cone structure. *B*. Longitudinal section of part of a blade of *Gnathodus*, showing lamellar structure. (*A from Hass*, 1941, *after Pander*, 1856; *B after Hass*, 1941.)

the structure. Inasmuch as the first few growth stages of all laminated conodonts are quite similar, regardless of whether they later became distacodids, blades and bars, or platforms, the denticulated components are considered to form a morphologic unit. Hass further emphasizes that the pulp cavity is of prime taxonomic importance because it is a visible feature and also because it is the area about which the conodont was built. Using it as a starting point, it is possible to work out evolutionary series from distacodids, through bars and blades, to the more highly specialized platform types.

NATURE OF CONODONT-BEARING ORGANISM

Nothing definite is known about the zoological affinities of the organisms that left conodonts, nor is there a consensus as to what function or functions the conodonts served. No attempt will be made to resolve these perplexing questions, but it may be observed that whatever taxonomic affinities are ultimately accepted for the conodonts, the following characteristics will have to be taken into account:

1. Conodonts are composed of calcium phosphate, a substance not common among invertebrates but common among vertebrates.

2. They are minute in size and conical or multiconical in shape.

3. The internal structure is fibrous or laminated (cone-in-cone).

4. Denticles never show evidence of wear, though they are commonly broken off or frayed and some show subequent repair. From the latter fact, it has been postulated that the conodont was largely embedded in soft tissue, and that it grew by successive external increments.

5. Conodonts are commonly attached to basal "bony" fragments of composition identical to that of the conodont itself (Fig. 16-3), or are closely associated with such fragments. These relationships have led to the assumption that the detached and scattered conodonts were once firmly attached to some sort of basal structure.

6. Assemblages of little-disturbed conodonts of several distinctly different types are not uncommon and would seem to indicate that such a grouping represents a definite mechanism of some kind (Fig. 16-4). Undisturbed conodont assemblages consist of pairs or symmetrically arranged groups on slabs of fissile shale, indicating still waters. The actual spatial arrangement of the several different types of plates has suggested a complex dental or masticatory array, a straining or filtering apparatus, a cluster of scalelike elements, and an internal "skeleton" or loosely knit framework.

7. Conodonts are associated in the rocks with marine assemblages consisting of fish teeth, fish jaws, fish plates and scales, crustaceans, gastropods, cephalopods, pelecypods, inarticulate brachiopods, bryozoans, scolecodonts, worm castings, and worm burrows.

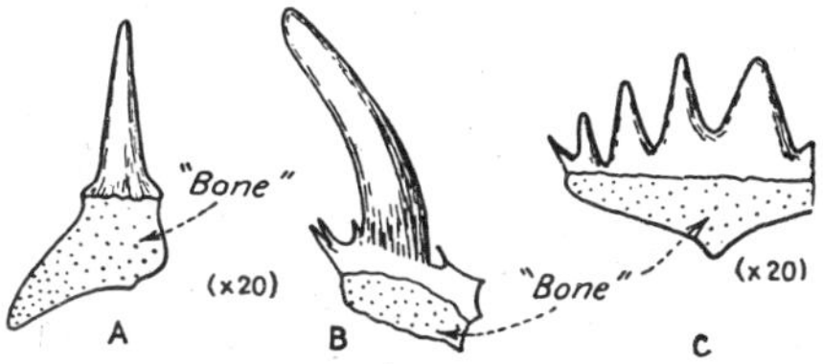

FIG. 16-3. Conodontophoridia. Conodonts from the Ordovician Harding sandstone of Colorado, showing the denticulate bodies attached to a "bony" plate which is similar in composition to that of the ostracoderm plates found in the same rock. This association has been used as evidence to support the contention that conodonts are parts of an early vertebrate. (*Diagrammatic representation after Kirk*, 1929.)

The following interpretations show how widely investigators have differed in their opinions of the zoological affinities and functions of the conodonts.[1]

Fish Origin. Pander (1856), who was the earliest student of conodonts, considered them to be the teeth of an ancient fish, though he was uncertain as to what parts of the mouth the denticles occupied. Ulrich and Bassler (1926), and in more recent years Branson and Mehl (1944), have followed Pander in considering them fish remains. Branson and Mehl have found hundreds of conodonts firmly attached to phosphatic fragments which not only are identical to the conodonts in composition but also are indistinguishable from much larger associated fragments and plates that show supposed bone structure (Kirk, 1929; Fig. 16-3).

On the basis of chemical and mineralogical composition, associated "bone"

[1] The reader will find an excellent summary of the different views of conodont affinities in an article by Scott (1934).

material, assemblage associations, nature of internal structure and growth additions, size, shape, morphology, and stratigraphic range, most students of conodonts agree with Ellison (1944) that the conodont-bearing animals were vertebrates of a low order, either primitive fish or some even lower vertebrate form. There is not, however, the same unanimity of opinion as to the **function** of conodonts.

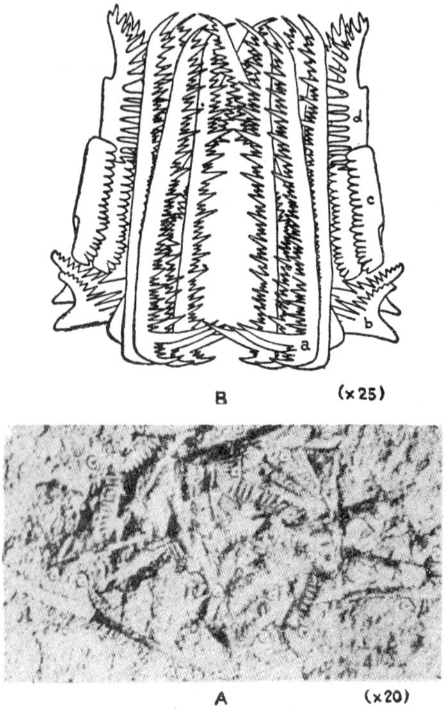

FIG. 16-4. Conodontophoridia. Conodont assemblages. *A*. Positive of an assemblage of *Lochriea* containing 10 hindeodells (*a*), 3 prioniods (*b*), 4 spathognaths (*c*), and 4 prioniodells (*d*). *B*. Schematic diagram, based on assemblages like *A*, showing probable arrangement of the different elements in a supposedly complete assemblage of the genus *Lochriea* (*a*, hindeodell; *b*, prioniod; *c*, spathognath, and *d*, prioniodell). Positions of *b*, *c*, and *d* are inferred. (*After Scott*, 1942.)

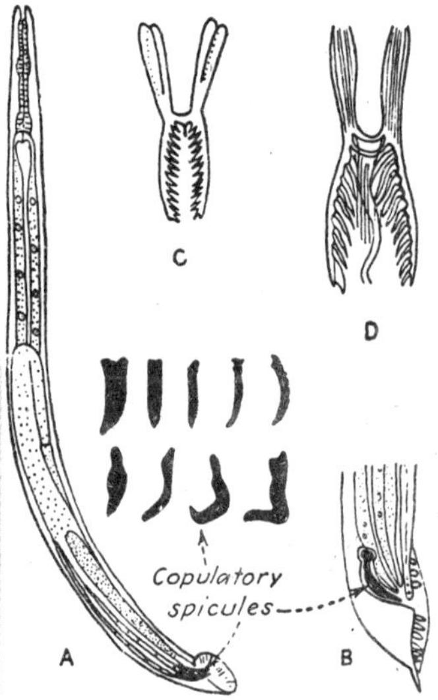

FIG. 16-5. Conodontophoridia vs. Nematoda and Turbellaria. *A–B*. Diagrams of a mature male nematode, *Rhabditis*, showing position and relations of the simple chitinous copulatory spicule, and a group of typical spicules. *C–D*. Elaborate chitinous copulatory process of the flatworm genus *Dalyellia*, greatly enlarged. (*After Denham*, 1944, *from several sources*.)

Conodonts have been considered to represent the dental or jaw armor of a primitive fish;[1] the lingual teeth of extinct cyclostomes; scale clusters of an

[1] Evidence in support of this conclusion is Kirk's (1929) report that conodonts and attached jaw substance in the Ordovician Harding sandstone of Colorado are identical in composition and intimately associated with ostracoderm plates. Miller, Cullison and Youngquist (1947) recently reported a denticulate jaw, supposedly a jaw of a primitive vertebrate, from the Lower Ordovician Dutchtown formation of Missouri.

extinct fish; and skeletal elements of the gill arches of certain fish (Demanet, 1939), but none of these suggestions has met with universal acceptance (DuBois, 1943).

Hass (1941), Huddle (1934), and many others doubt that conodonts functioned as teeth or other ingestive organs on the grounds that they never show evidence of wear, that they could be broken and then subsequently repaired, and that they seem to have been largely embedded in soft tissue. Hass (1941) is therefore led to the conclusion that conodonts probably functioned as internal supports for tissues that were located at a place exposed to stress, either upon the exterior of or within the bodies of genetically related marine animals. He did not, however, further identify the animals that bore the conodonts.

Thus the widespread acceptance of a lower vertebrate ("fish") origin for conodonts contrasts strikingly with the divergence of opinion as to the function of the conodonts.

Worm Origin. Conodonts were formerly assigned to annelid worms by some paleontologists, largely because of their striking resemblance to the jaws and teeth of many living species. In recent years, however, support for a worm origin has waned, and most paleontologists now reject such an origin because conodonts (1) are much smaller than most scolecodonts; (2) consist of calcium phosphate, whereas scolecodonts are silicochitinous;[1] (3) are light to dark brown and translucent, whereas scolecodonts are jet black and opaque; and (4) are commonly attached to basal fragments, whereas scolecodonts seem to have been bound together loosely by muscles.

Scott (1934, 1942) has most ably presented the case for the worm affinities of the conodonts. He has found dozens of "assemblages," which consist of several distinct types of denticulated plates (Fig. 16-4), each type of which had previously been given a generic name. He concluded that the assemblage was a natural one, in that the component parts were preserved with essentially *in situ* arrangement, and that all of the components belonged to a single individual. In 1934 he presented seemingly convincing evidence that the assemblages were of worm origin and could not be of fish origin. By 1942, however, on the basis of many more assemblages from other formations, he no longer emphasized the worm origin, stressing instead the fact that the components of an assemblage formed a mechanism that could operate with equal ease either as a jaw apparatus of an annelid or as gill rakes of a fish. It is interesting to note that Schmidt (1934), who reported similar assemblages in the same year that Scott first described his discoveries, concluded that the conodonts were parts of the straining device in the branchial organs of an ancient fish.

Denham (1944) has called attention to the tiny chitinous spicular structures associated with the reproductive organs of nematode and turbellarian worms, and has raised the question as to whether or not conodonts are copulatory structures of long extinct Paleozoic worms. These objects take the form of single spines, or spinose stalks, and occur in an individual worm singly,

[1] DuBois (1943), who supports a worm origin for condonts, points out that some annelids contain phosphorus and build it into their tubes, hence he suggests the possibility that certain extinct worms may have used phosphorus in their teeth.

paired, or in groups (Fig. 16-5). This suggestion, while ingenious and interesting, is not supported by the known facts. The spicules are chitinous rather than phosphatic, simple rather than complex in their morphology, and never have any hard base of attachment as do so many conodonts. Furthermore, modern roundworms and flatworms having the spicules are largely freshwater forms, whereas the conodont-bearing animals were marine.

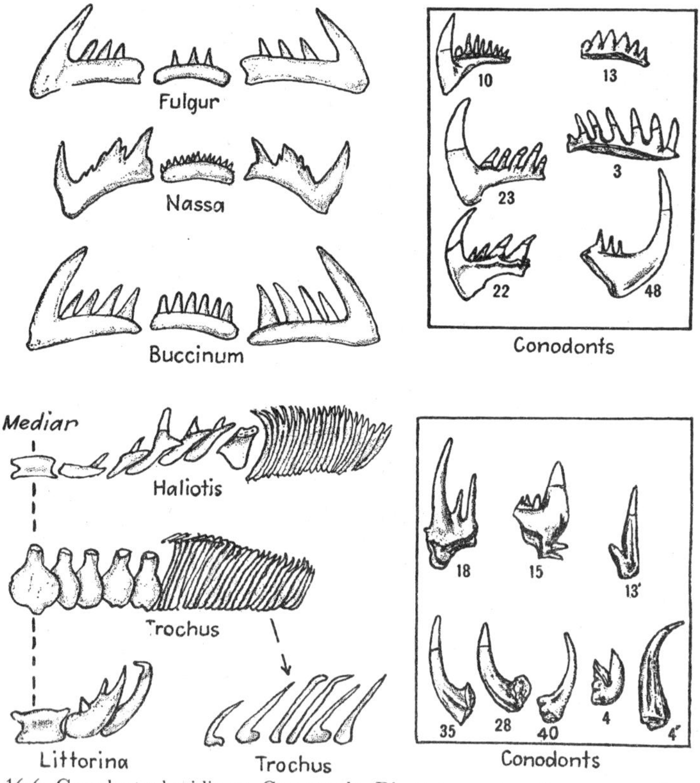

FIG. 16-6. Conodontophoridia vs. Gastropoda. Diagrams supposedly demonstrating that conodonts are but the radular denticles of extinct gastropods. The teeth in the left column are from the radulae of the modern gastropod genera indicated, whereas the conodonts within the two rectangles are from Ordovician rocks. All figures are approximately ×20. (*All drawings after Loomis*, 1936, *those of the conodonts being based on figures by Stauffer*, 1935.)

Molluscan Origin. It has been suggested that conodonts may be the radular teeth of ancient and extinct mollusks of some sort (James, 1884; Loomis, 1936). It is a fact that certain types of conodonts have a size and an external form quite similar to the radular teeth of several different genera of gastropods, as shown in Fig. 16-6, and that they, like radular hooks, can be replaced as lost. However, there is a striking difference in composition, conodonts being phosphatic whereas radular elements are chitinous; some radular components have no representation among the known conodonts; and many types of conodonts are not duplicated in form by any teeth of known radulae. Because of these and other objections, the suggestion that conodonts are molluscan radular elements has not met with much favor (DuBois 1943).

Summary. Whatever the nature of the conodont-bearing animal, the finding of natural assemblages of components by Schmidt (1934), Scott (1934, 1942), Jones (1938), Demanet (1939), and DuBois (1943) has thrown new light on the possible zoological significance of conodonts as animal hard parts. It has been general practice to assign a generic name to each distinct morphologic type of conodont, though most investigators have long recognized the possibility that more than one type might be present in a single individual. Scott's (1942) latest discovery is an assemblage that contains components referable to four previously recognized genera—*Hindeodella*, *Spathognathodus*, *Prioniodus*, and *Prioniodella*. Rather than continue the use of the generic names as such, Scott proposes to form a common noun from the name and apply this to the appropriate type. The assemblage would thus be described as consisting of paired hindeodells, spathognaths, prioniods, and prioniodells (Fig. 16-4).

Natural assemblages of conodonts of the sort just described obviously put into confusion the generic and specific nomenclature of this group of fossils, because it is quite probable that some ancient animal actually possessed hard parts that are now included in several genera. It is not within the purpose of the present work to suggest ways of overcoming the confusion, but probably as good a course as any is to continue the usual practice of creating new generic names for newly discovered types until these can be found in natural association with other types. If assemblages prove to be consistent in the kinds of conodonts they contain, then they will have to become the basis for a true genus. This question has been considered at length by several investigators whose references are listed at the end of this chapter (Croneis, 1941; Branson and Mehl, 1944; Ellison, 1944; etc.).

If conodonts do represent the remains of ancient fish-like animals, then the several items listed on page 773 should be reexamined to see if they are in harmony with a fish origin. Such an examination shows that there are no serious difficulties in reconciling the different aspects with such an origin. It may be suggested, incidentally, that if conodonts were carried by ancient fish, then they may have served several different functions within the body, as a result of which they should show considerable diversity in morphology and may well have been situated in several different parts of the animal.

GEOLOGIC HISTORY OF CONODONTS

In the absence of definite knowledge about the function of conodonts and the exact nature of the animals that bore them, it is not possible to determine much about the conditions under which the creatures lived and died. Inferences as to these environmental conditions have been based mainly on the lithology of the enclosing rock and the composition of the associated faunal elements (Ellison, 1944).

Conodonts seem to be limited to marine deposits, yet some investigators believe that the conodont-bearing organisms may have lived in fresh and brackish waters as well as in normal marine waters. Most evidence favors a marine habitat.

Conodonts have been recovered from conglomerates; sandstones; silt-

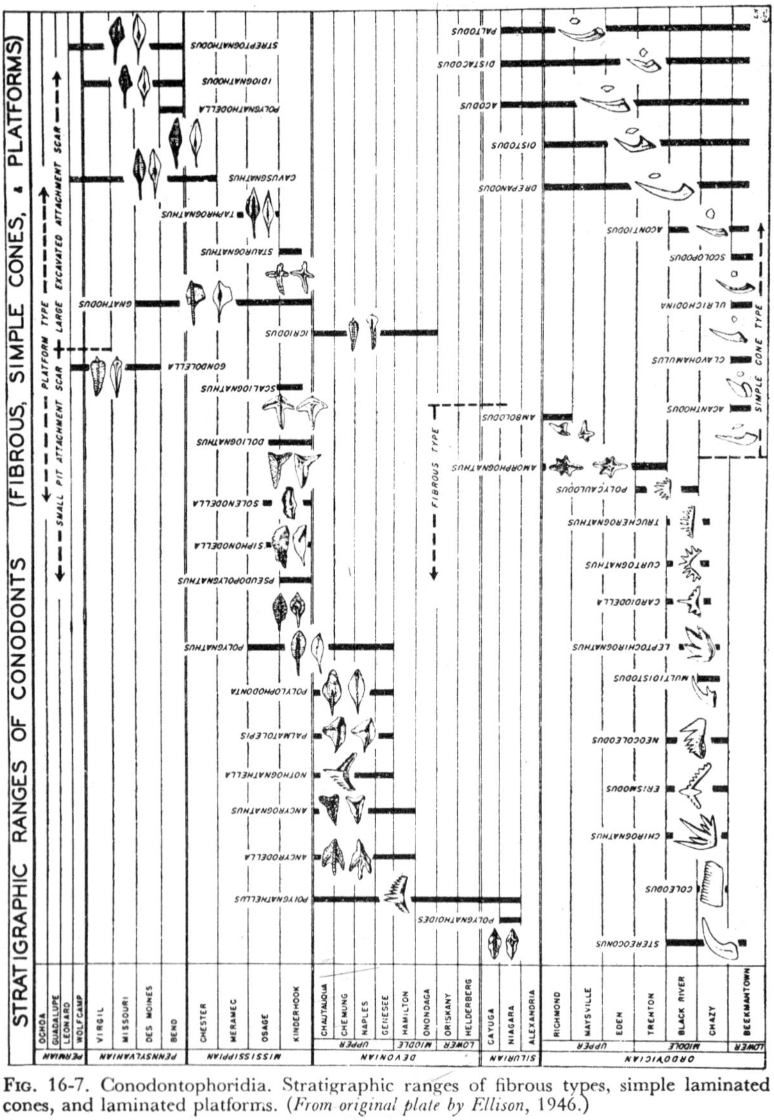

FIG. 16-7. Conodontophoridia. Stratigraphic ranges of fibrous types, simple laminated cones, and laminated platforms. (*From original plate by Ellison*, 1946.)

stones; red, gray, green, and black shales; sandy shales; and limestones. They are probably most abundant in black shales which generally do not contain many other types of fossils. Some investigators have suggested that the wide distribution of conodonts in so many different types of sedimentary rocks is evidence that the conodont-bearing organisms were nektonic, living

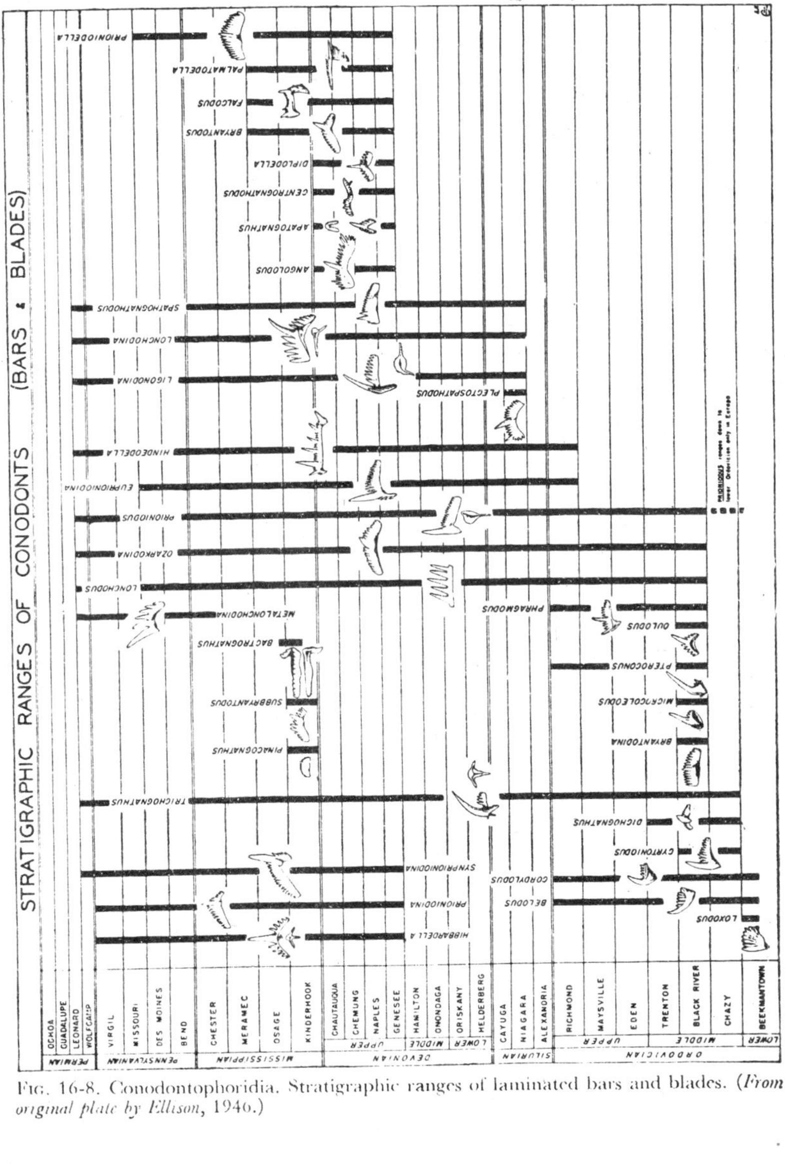

FIG. 16-8. Conodontophoridia. Stratigraphic ranges of laminated bars and blades. (*From original plate by Ellison*, 1946.)

in the open ocean (*i.e.*, pelagic) as well as in shallow-water areas near shore. It is possible that some were also planktonic or even benthonic, though the latter seems unlikely.

The organic remains with which conodonts are associated constitute a remarkably diverse group. In the Ordovician Decorah shale of Minnesota, for example, multitudes of conodonts, piled one on another and oriented in random fashion, are intimately associated with bryozoan fragments; brachiopod, gastropod, and ostracode shells; graptolite shreds; and amber-colored scalelike objects (Stauffer, 1930). In other rocks conodonts have been found closely associated with inarticulate brachiopods, worm trails and burrows, and vertebrate remains such as spines, scales, teeth, and bone fragments. Although conodonts most commonly are randomly distributed through a rock or concentrated to some extent on the surfaces of individual layers, little-disturbed assemblages are being discovered in increasing numbers. In these latter, pairs of conodonts of more than one type are so arranged as to suggest that they were buried without much disturbance and with essentially the interrelationships they had when they were hard parts in an ancient organism.

It should be obvious from the foregoing statements that it is not yet possible to draw any unqualified conclusions as to the environmental and faunal conditions under which the conodont-bearing organisms lived.

STRATIGRAPHIC DISTRIBUTION AND USE AS INDEX FOSSILS

The oldest known conodonts come from rocks of Lower Ordovician age in North America (Furnish, 1938) and Europe; the youngest from Permian[1] formations (Branson, C. C., 1932). Conodonts have been found in Triassic and Cretaceous formations, but in every case these seem to have been derived from Paleozoic rocks and to have been redeposited with the later sediments.

The youngest conodont-bearing formation that has come to our attention is a supposed Cretaceous clay lying between the Devonian Cedar Valley limestone and the Pleistocene glacial drift in Minnesota. Inasmuch as this clay contains both angiosperm and gymnosperm leaves, and conodonts of both Ordovician and Devonian age, Stauffer (1940) concluded that the mixed conodont assemblage was actually derived from Paleozoic rocks and redeposited in the Cretaceous clay. The general problem of mixed conodont assemblages has long troubled specialists and has to be taken into account in many cases if errors are to be avoided (Branson and Mehl, 1940).

The microconodonts reported by Wetzel[2] from Baltic Cretaceous chert raise an interesting question. These tiny objects, which strongly resemble certain types of conodonts, are preserved in chert and seem to have been buried originally in gelatinous silica. Whether or not they represent redeposited fossils or true Cretaceous specimens is an unanswered question. If

[1] Branson, C. C., 1932. A point of interest is Branson's statement that the Permian conodonts seem to be "senile specializations of a rapidly declining group."

[2] Wetzel, O., 1933, Die in organischer Substanz erhaltenen mikrofossilen des baltischen Kreidefeuersteins, *Paleontographica*, Bd. 78A, S1-110.

we accepted Glaessner's[1] identification of them as scolecodonts, the age problem would be resolved, but it must be admitted that the size, color (brown), and morphology of the fossils argue for a conodont identification.

At least one recognizable genus (*Gondolella*) has been reported from supposed Triassic rocks (Trochitenkalk zone of the Unter Haupmuschelkalk) of Germany, and authentic conodonts have also been found in Middle Triassic rocks in Egypt, but these occurrences seem to be cases of redeposition of later Paleozoic (Penn. or Perm.) specimens (Branson and Mehl, 1940).

Dr. P. E. Cloud, Jr., has informed us that he has recovered tiny coal-black denticulated objects from the Karnic (U. Trias.) of Nevada. It is not certain that these are conodonts, but it is worth noting that they are almost exactly similar to the widespread Devonian genus *Icriodus* and may well have been derived from neighboring rocks of that age.

Conodonts are relatively abundant in Ordovician, Devonian, and Mississippian rocks, and certain genera are reliable index fossils. The stratigraphic ranges of most of the known genera are indicated in Figs. 16-7 and 16-8. It will be noted that a few genera are quite restricted in vertical range, whereas many others are long ranging.

Because of small size and resistance to chemical destruction, conodonts can persist under conditions of weathering that would destroy many other fossils. Not uncommonly they remain in the residual soil left on a weathered limestone or shale terrane, and in the past there are some cases where such clay was carried downward into older formations to make a so-called **stratigraphic leak.** Under other conditions the clay was washed into adjacent basins of deposition where the conodonts from the previously weathered rock were mingled with more recent forms to make a **stratigraphic admixture.** Such leaks and admixtures are well known and have been interpreted accordingly (Branson and Mehl, 1935, 1940, 1944).

REFERENCES

Branson, C. C. 1932. Discovery of conodonts in the Phosphoria Permian of Wyoming. *Science*, vol. 75, pp. 337–338.

Branson, E. B., and Mehl, M. G. 1935. Value of conodonts in stratigraphic determinations (abstract). *Geol. Soc. Amer. Proc.*, 1934, p. 375.

——— and ———. 1935. Methods, problems, and results of conodont studies (abstract). *Ibid.*, p. 441.

——— and ———. 1936. Geological affinities and taxonomy of the conodonts (abstract). *Ibid.*, 1935, p. 436.

——— and ———. 1940. The recognition and interpretation of mixed conodont faunas. *Denison Univ. Bull., Jour. Sci. Lab.*, vol. 35, pp. 195–209.

——— and ———. 1944. Conodonts *in* Shimer and Shrock's "Index Fossils of North America," Chap. 7, pp. 235–246. New York, John Wiley & Sons, Inc.; Cambridge, Mass., The Technology Press.

Croneis, C. 1941. Micropaleontology—past and future. *Bull. Amer. Assoc. Petrol. Geol.*, vol. 25, pp. 1208–1255.

Demanet, F. 1939. Filtering appendages on the branchial arches of *Coelacanthus lepturus* Agassiz. *Geol. Mag.*, vol. 76, pp. 215–217.

[1] Glaessner, M. F. 1947, "Principles of Micropalaeontology," New York, John Wiley & Sons, Inc., p. 30.

DENHAM, R. L. 1944. Conodonts. *Jour. Paleontology*, vol. 18, pp. 216–218.

DUBOIS, E. P. 1943. Evidence on the nature of conodonts. *Jour. Paleontology*, vol. 17, pp. 155–159.

EICHER, D. B. 1946. Conodonts from Triassic of Sinai, Egypt. *Bull. Amer. Assoc. Petrol. Geol.*, vol. 30, pp. 613–616. (Also published in *Bull. inst. Egypte*, vol. 28, pp. 88–92.)

ELLISON, S. P., JR. 1944. Ecology of conodonts. *Nat. Res. Council, Rept. Comm. Marine Biol. as Related to Paleontology*, 1943–1944, pp. 1–4. (This paper contains a good bibliography.)

———. 1944*a*. The composition of conodonts. *Jour. Paleontology*, vol. 18, pp. 133–140.

———. 1946. Conodonts as Paleozoic guide fossils. *Bull. Amer. Assoc. Petrol. Geol.*, vol. 30, pp. 93–110.

HASS, W. H. 1941. Morphology of conodonts. *Jour. Paleontology*, vol. 15, pp. 71–81.

HUDDLE, J. W. 1934. Conodonts from the New Albany shale of Indiana. *Bull. Amer. Pal.*, vol. 21, no. 72.

JAMES, U. P. 1884. On conodonts and fossil annelid jaws. *Jour. Cin. Soc. Nat. Hist.*, vol. 7, pp. 143–149.

KIRK, S. R. 1929. Conodonts associated with the Ordovician fish fauna of Colorado—a preliminary note. *Amer. Jour. Sci.*, 5th ser., vol. 18, pp. 493–496.

LOOMIS, F. B. 1936. Are conodonts gastropods? *Jour. Paleontology*, vol. 10, pp. 663–664.

MILLER, A. K., CULLISON, J. S., and YOUNGQUIST, W. L. 1947. Lower Ordovician fish remains from Missouri. *Amer. Jour. Sci.*, vol. 245, pp. 31–34.

——— and YOUNGQUIST, W. L. 1947. Conodonts from the type section of the Sweetland Creek shale in Iowa. *Jour. Paleontology*, vol. 21, pp. 510–517.

PANDER, C. H. 1856. "Monographie der fossilen Fische des silurischen Systems der russisch-baltischen Gouvernements." St. Petersburg, 91 pp.

RAYMOND, P. E. 1941. Invertebrate Paleontology *in* Fiftieth Anniversary Volume, Geological Society of America. (Conodonts, pp. 77–78.)

SCOTT, H. W. 1934. The zoological relationships of the conodonts. *Jour. Paleontology*, vol. 8, pp. 448–455.

———. 1942. Conodont assemblages from the Heath formation, Montana. *Ibid.*, vol. 16, pp. 293–300.

SCHMIDT, H. 1934. Conodonten-Funde in ursprunglichen Zusammenhang. *Pal. Zeitschr.*, vol. 16, pp. 76–85.

STAUFFER, C. R. 1930. Conodonts from Decorah shale. *Jour. Paleontology*, vol. 4, pp. 121–128.

———. 1935. The conodont fauna of the Decorah Shale (Ordovocian). *Ibid.*, vol. 9, pp. 596–620.

———. 1940. Conodonts from the Devonian and associated clays of Minnesota. *Ibid.*, vol. 14, pp. 417–435.

ULRICH, E. O., and BASSLER, R. S. 1926. A classification of the toothlike fossils, conodonts, with descriptions of American Devonian and Mississippian species. *Proc. U.S. Nat. Mus.*, vol. 68, art. 12, 63 pp.

YOUNGQUIST, W., and MILLER, A. K. 1948. Additional conodonts from the Sweetland Creek shale of Iowa. *Jour. Paleontology*, vol. 22, pp. 440–450.

INDEX

Numbers in **boldface** type refer to pages where illustrations appear. Italics are used for generic names only. Names of taxonomic groups higher than genus and proper names are capitalized. The index omits incidental reference to common features and structures, and to the many technical terms that are grouped under a single topic, *e.g.*, the coiling of cephalopod shells, or the terminology of crinoid skeletons.

D

I

J

K

L

N